ELECTRONS IN MOTION

Attosecond Physics Explores Fastest Dynamics

World Scientific Series on Atomic, Molecular and Optical Physics

ISSN: 2010-2828

Series Editor: Phillip L Gould *(University of Connecticut, USA)*

Published:

Vol. 4 *Electrons in Motion: Attosecond Physics Explores Fastest Dynamics*
by Ferenc Krausz

Vol. 3 *Atoms in Electromagnetic Fields (Second Edition)*
by C Cohen-Tannoudji

Vol. 2 *Semiconductor Quantum Dots*
by L Bányai and S W Koch

Vol. 1 *Atoms in Electromagnetic Fields*
by C Cohen-Tannoudji

ELECTRONS IN MOTION

Attosecond Physics Explores Fastest Dynamics

FERENC KRAUSZ

Ludwig-Maximilians-Universität München, Germany
Max Planck Institute of Quantum Optics, Garching

World Scientific

NEW JERSEY • LONDON • SINGAPORE • BEIJING • SHANGHAI • HONG KONG • TAIPEI • CHENNAI • TOKYO

Published by

World Scientific Publishing Co. Pte. Ltd.
5 Toh Tuck Link, Singapore 596224
USA office: 27 Warren Street, Suite 401-402, Hackensack, NJ 07601
UK office: 57 Shelton Street, Covent Garden, London WC2H 9HE

Library of Congress Cataloging-in-Publication Data
Names: Krausz, Ferenc, author.
Title: Electrons in motion : attosecond physics explores fastest dynamics / author,
Ferenc Krausz (Ludwig-Maximilians-Universität München, Germany & Max Planck Institute of Quantum Optics, Germany).
Other titles: World Scientific series on atomic, molecular, and optical physics ; vol. 4.
Description: Singapore ; Hackensack, NJ : World Scientific Publishing Co. Pte. Ltd., [2019] |
Series: World Scientific series on atomic, molecular and optical physics ; vol. 4
Identifiers: LCCN 2018060709| ISBN 9789811201868 (hardcover ; alk. paper) |
ISBN 9811201862 (hardcover ; alk. paper)
Subjects: LCSH: Laser pulses, Ultrashort. | Electrons.
Classification: LCC QC689.5.L37 E44 2019 | DDC 535.8/4--dc23
LC record available at https://lccn.loc.gov/2018060709

British Library Cataloguing-in-Publication Data
A catalogue record for this book is available from the British Library.

For any available supplementary material, please visit
https://www.worldscientific.com/worldscibooks/10.1142/11322#t=suppl

Desk Editor: Nur Syarfeena Binte Mohd Fauzi

Typeset by Stallion Press
Email: enquiries@stallionpress.com

Printed in Singapore

To Arnold Schmidt, for his mentorship and guidance.

The author and publisher would like to thank the following publishers of the various journals for their assistance and permission to include the selected reprints found in this volume:

American Association for the Advancement of Science (Science); American Institute of Physics (Review of Scientific Instruments); American Physical Society (Physical Review Letters, Reviews of Modern Physics); Elsevier Science Publishers (Journal of Electron Spectroscopy and Related Phenomena); IEEE Intellectual Property Rights (IEEE Journal of Quantum Electronics, IEEE Journal of Selected Topics in Quantum Electronics); Institute of Physics (Physics World, Journal of Physics B); Optical Society of America (Optica, Optics Express, Optics Letters, Optics and Photonics News); The Royal Swedish Academy of Sciences (Physica Scripta); Springer Nature (Applied Physics B, Nature, Nature Communications, Nature Photonics).

Contents

Prologue

Curiosity—a simple word describing a sophisticated quality of living creatures. Its power is almost unlimited. It opens new frontiers in arts (see e.g. the genre opened by Nils Frahm through musical experiments)[1] and is the driving force behind almost all scientific progress. It has led to the discovery of electromagnetic fields by Michael Faraday and James Clerk Maxwell as well as of the electron by Joseph John Thomson. The propagation and current of electrons serve as a basis for transmitting and processing information, respectively. Electromagnetic waves, in turn, led to the discovery of the molecular structure of DNA by Francis Crick, James Watson, and Rosalind Franklin, which carries information fundamental to the development, functioning, and reproduction of all living organisms.

The field of attosecond science has been spawned by curiosity as well. It has urged a better understanding of the microscopic motions of electrons, which initiate the change of molecular structure and are responsible for the emission of light and the processing of information. It is the eagerness of enquiring minds that has enabled attosecond metrology to uncover fingerprint-like information of organic molecules. This very urge is inherent to the origin of all the efforts that resulted in the work collected in this volume.

Electrons and the products of their motions, electromagnetic waves, were already enchanting me in the early years of my life. I was fascinated by the fact that radio waves created thousands of miles away could be captured in a 15-m long wire that I spanned to form a long-wavelength antenna. The electrons brought into motion in that very wire made it into little circuits converting their current into signals at acoustic frequencies and carrying music and news from distant locations to my home. The "magic" happening in my self-made radio triggered my curiosity and focused this curiosity on the electrons, their motions and interactions with electromagnetic waves. This focus turned out to be a long-lasting one that persists even today.

I have had the privilege to be educated and fostered by teachers who spurred this curiosity and "infected" me with a never-ceasing urge to gain insight into the motion of electrons. They include Károly Simonyi (electron physics and electrodynamics) at the Technical University of Budapest and György Marx (quantum physics) at the Eötvös Loránd University in Budapest. Sometime later, my diploma thesis supervisors, József Bakos and Tibor Juhász attracted my attention to lasers. My PhD supervisor and mentor, Arnold Schmidt at the Technical University of Vienna eventually guided me to pursue developing tools and techniques that allow probing and controlling the atomic-scale motion of electrons as well as the force fields of light waves produced by these motions. I am deeply indebted to all of them. The story told in this volume about my voyage towards the birth of attosecond science could not have happened without their influence and impact.

[1] www.nilsfrahm.com

With the direction of my voyage being established, progress towards these goals was still not warranted. Successful achievement of goals in experimental physics relies on the dedication, single-mindedness, persistence, and perseverance – as well as curiosity – of a number of co-workers and collaborators. The names of these determined individuals appear in the introductions of the forthcoming chapters. In appreciation for their invaluable contributions, I append here a personal perspective solicited by the Royal Swedish Academy of Sciences in 2015, paying tribute to their feats.

All the editors who were contacted have granted permission to reprint the papers, which I gratefully acknowledge here. I am indebted to Mihaela Žigman for her help in preparing the prologue and for co-authoring the final chapter, Matthew Weidman for his support in preparing the introductory remarks and Corin Abert for managing the project from the beginning until its completion. I thank Dennis Luck and Thorsten Naeser for the graphical illustrations and their careful preparation of a camera-ready copy of the book, Jana Gerding and Nils Haag for their support in final proofreading and UltraFast Innovations GmbH[2] for their generous sponsorship of this production. Last but not least, I am grateful to David Schick from World Scientific Publishing Co. for his enthusiastic engagement in making this project happen.

Curiosity has spawned all the developments and discoveries described in this book. It drives the evolution of attosecond science up to the present day. Approaching its twentieth birthday, attosecond metrology is on the brink of opening entirely new prospects. One of them may possibly have a profound effect on the future of molecular science and medicine.

Just as insight into the structure of DNA has been driving a revolution in life sciences, the capability of sensing miniscule changes in the molecular composition of cells and systemic biofluids holds promise for revolutionizing the future of personalized medicine. Access to all kinds of biological molecules can only be ensured by access to the molecules' nuclear and electronic structure. The most direct probe of this structure known to physics is motion, namely the motion of nuclear and electronic wavepackets whose periodic acceleration radiates electromagnetic fields. Attosecond technology has meanwhile spawned the capability of measuring those fields as they vary in time – with a sensitivity approaching the ultimate detection limit dictated by the laws of nature.

Acquisition of such universal molecular fingerprints is a most ambitious goal that may require a long time to be achieved, possibly longer than my remaining lifetime. Yet, I cannot help but start this journey and infect as many young researchers as possible with my enthusiasm, so they may continue the voyage. That very voyage ahead of us, the inception of which is addressed in the final chapter of this book, is transforming the field of attosecond science from basic research to a field opening real-world applications, as it may similarly transform my life. Should there be a second volume of this book, it shall address those very transformations.

Munich, March 2019

[2] www.ultrafast-innovations.com

IOP Publishing | Royal Swedish Academy of Sciences Physica Scripta
Phys. Scr. 91 (2016) 063011 (12pp) doi:10.1088/0031-8949/91/6/063011

Invited Comment

The birth of attosecond physics and its coming of age

Ferenc Krausz

Max-Planck-Institut für Quantenoptik, Hans-Kopfermann-Strasse 1, D-85748 Garching, Germany
Fakultät für Physik, Ludwig-Maximilians-Universität, Am Coulombwall 1, D-85748 Garching, Germany

E-mail: Ferenc.Krausz@mpq.mpg.de

Received 27 October 2015, revised 17 December 2015
Accepted for publication 29 December 2015
Published 25 May 2016

Abstract

Classical electromagnetism allows the rapidity of light field oscillations to be inferred from measurement of the speed and wavelength of light. Quantum mechanics connects the rapidity of electronic motion with the energy spacing of the occupied quantum states, accessible by light absorption and emission. According to these indirect measurements, both dynamics, the oscillation of light waves as well as electron wavepackets, evolve within attoseconds. With the birth of attosecond metrology at the dawn of the new millennium, light waving and atomic-scale electronic motion, being mutually the cause of each other, became directly measurable.
These elementary motions constitute the primary steps of any change in the physical, chemical, and biological properties of materials and living organisms. The capability of observing them is therefore relevant for the development of new materials and technologies, as well as understanding biological function and malfunction. Here, I look back at milestones along the rocky path to the emergence of this capability, with some details about those my group had the chance to make some contributions to. This is an attempt to show—from a personal perspective—how revolution in science or technology now relies on progress at a multitude of fronts, which—in turn—depend on the collaboration of researchers from disparate fields just as on their perseverance.

Keywords: quantum optics, international year of light, attosecond physics

'Manfred Eigen, then director of the Max Planck Institute for Biophysical Chemistry in Göttingen, Germany, accepted the 1967 Nobel prize in chemistry with a lecture about his work on 'immeasurably fast reactions'. Ahmed Zewail, recipient of the 1999 Nobel prize in chemistry, who analyzed reactions over timescales a billion times shorter than Eigen, later joked about this title. If attosecond techniques do become reality, the immeasurable will have to be redefined once again'. I cited these sentences from a feature article of Yudhijit Bhattacharjee entitled 'Measuring the Immeasurable' and published in Nature [1] in August 2001. Several weeks later, the first observation of isolated attosecond pulses was reported [2].

Attosecond technology almost abruptly improved the temporal resolution of time-resolved spectroscopy by two orders of magnitude, following a one and a half decade long stagnation in resolving fastest microscopic phenomena. By doing this, it rendered a wealth of previously immeasurably fast microscopic phenomena driven by electronic motions measurable [3]. It is instructive to review the conceptual and experimental advances that laid the groundwork for the birth of attosecond science, revealing the need for progress on many frontiers to allow such a revolution in technology to happen. This article attempts to pay tribute to those who pushed the relevant frontiers with its author being aware that the work of many more may have deserved to be appreciated than possible in the framework of this essay. My apologies go to all those whose important work could not be explicitely acknowledged below.

0031-8949/16/063011+12$33.00

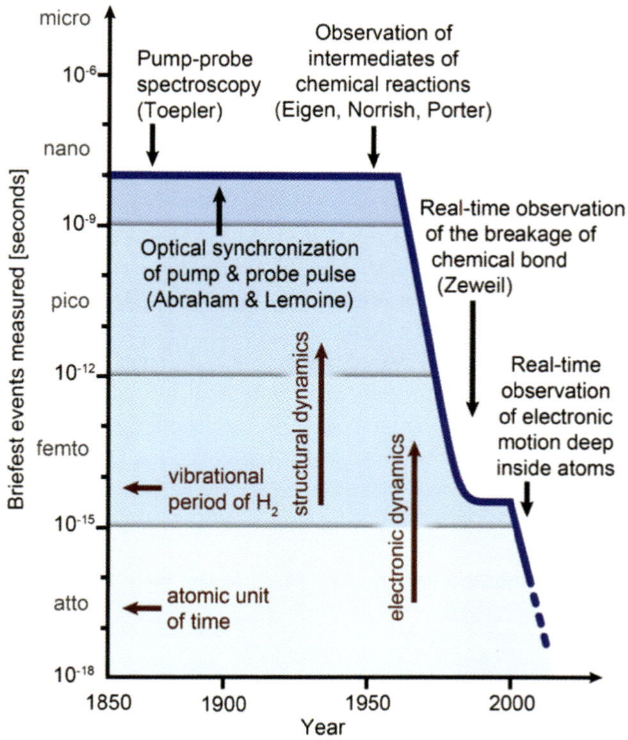

Figure 1. Evolution of techniques for real-time observation of microscopic processes. Discontinuities in the slope of briefest measured events versus years indicate revolutions in technology.

Capturing microscopic processes—early developments

In 1864, Toepler extended spark photography, already known as a method for recording rapid motion of macroscopic objects, to study microscopic dynamics [4]. He generated sound waves with a short light spark and subsequently photographed them with a second spark that was delayed electronically with respect to the first one initiating the motion. By taking pictures of the sound wave as a function of the delay time, he obtained a complete history of sound-wave phenomena. The pump–probe technique was born. In 1899, Abraham and Lemoine improved the technique by deriving the pump and probe flash from the same spark with a variable optical delay between them, improving the resolution of pump–probe spectroscopy to the limit dictated by the flash duration [5]. With these milestones, the conceptual framework for studying transient microscopic phenomena was complete. Subsequent progress was driven—for about hundred years, until the end oft he XXth century—by producing ever shorter flashes and techniques for their measurement.

The first revolution: laser pulses for pump–probe spectroscopy

The resolving power of pump–probe techniques was limited by the nanosecond duration of light pulses for more than half a century, before laser-based techniques improved it by six orders of magnitude within merely two and a half decades (figure 1). Femtosecond laser pulse generation and measurement techniques permitted—four decades after the observation of intermediates of chemical reactions by Norrish and Porter [6] and Eigen [7]—real-time observation of the breakage and formation of chemical bonds, a field triggered by the pioneering work of Zewail [8, 9].

Progress in temporal resolution was again halted in the mid-1980s, when the duration of the shortest laser pulses approached the several-femtosecond oscillation period of the lightwave carrying the pulse. Two approaches have been considered for breaking this barrier: (A) phase locking of coherent optical waves over several octaves to produce sub-cycle optical transients over the visible (VIS) and near-infrared (NIR) [10–13] and (B) high-order harmonic generation (HHG) [14–17] and their coherent combination to create several-cycle pulses in the extreme ultraviolet (XUV) [18–20]. In principle, both concepts permit pushing the frontiers of pump–probe spectroscopy into the sub-femtosecond and attosecond regime. In practice, none of them worked, at least initially, for entirely different reasons. (A) Failed because of the lack of precision multi-octave dispersion control which is required to synthesize a bandwidth-limited pulse over a multi-octave spectral range and (B) provided pulses too weak to be measured by standard nonlinear autocorrelation techniques and utilized for both triggering and probing ultrafast phenomena (i.e. being used both as a pump and as a probe pulse). These failures had the resolving power of ultrafast spectroscopy stagnated over one and a half decade long at a level of several femtoseconds (figure 1). Again, a radically new approach was required to end stagnation.

The second revolution: controlling electrons with the field of light

The second discontinuity in the rate of progress of capturing ultrafast processes occured around the turn of the millenium (figure 1) and was brought about by the use of the electric field of light for controlling electrons on sub-femtosecond time scales. This enabled the generation *and* measurement of isolated sub-femtosecond pulses of XUV light. With the full control of the electric field of VIS/NIR laser light changing its strength and direction within attoseconds, subsequently XUV flashes shorter than 100 as became routinely available. These attosecond pulses come in perfect synchronism to the controlled electric field waveform of the VIS/NIR laser pulse used for its generation. Hence, the attosecond electric force of light has been able to replace the weak XUV pulse *either* as a pump *or* as a probe in the scrutiny of microscopic dynamics, furnishing the century-old pump–probe spectroscopy with a resolution permitting direct observation of electronic motions for the first time. In what follows I shall briefly review the major milestones in laser physics that allowed this to happen. I will address in more detail those my group has been directly involved in along with the (often quite cumbersome) struggles that preceeded relevant progress.

Early milestones: femtosecond solid-state lasers and high-order harmonics

As already mentioned above, the invention of the lasers along with nonlinear optical techniques for generating and measuring short pulses from them entailed the improvement of the resolving power of pump–probe spectroscopy from nanoseconds to femtoseconds by the mid 80'es. This first revolution provided direct time-domain access to the chemical-bond-scale motion of atoms, allowing real-time observation of dynamic changes in molecular and crystalline structure (figure 1). Whilst this key performance indicator stagnated over the following one and a half decades, many important developments occured during this period that layed the groundwork for the second revolution, which opened the door for observing electronic phenomena.

Laser pulses as short as 6 fs [21] could be produced and measured by the mid 80s. They originated from first-generation femtosecond lasers employing organic dyes as the active medium [22]. Their nanojoule-scale energy was insufficient to extend ultrashort-pulse generation to much shorter wavelengths via nonlinear frequency conversion. To this end, femtosecond pulse generation needed to be scaled to millijoule-scale energies, by some 4–5 orders of magnitude. The prerequisities for this progress were created by three great inventions, (i) the broadband solid-state laser medium, titanium-doped sapphire (Ti:S) by Peter Moulton [23], (ii) 'self mode locking' (later referred to as Kerr-lens mode locking, KLM) by Wilson Sibbett and coworkers [24] and (iii) chirped-pulse amplification (CPA) by Gerard Mourou and coworkers [25, 26]. These inventions spawned the second generation of femtosecond technology based on solid-state (primarily Ti:S) laser media and capable of scaling the peak as well as the average power of femtosecond laser pulses by several orders of magnitude with respect to first-generation technology [27].

Parallel to these achievements, inspired by Charles Rhodes and coworkers [14], Anne L'Huillier and coworkers demonstrated, in a series of pioneering experiments, the feasibility of extending the generation of coherent radiation from the VIS/IR spectral range (covered by lasers) to several orders of magnitude shorter wavelengths, well into the XUV, via HHG by powerful ultrashort laser pulses from solid-state lasers [28–32]. With Ti:S lasers entering the scene, HHG became a standard approach to extending coherent radiation into the vacuum and XUV regime [16]. Although coherent superposition of the harmonics generated at the highest harmonic photon energies were predicted to yield attosecond bursts [20], the multi-cycle nature of the driver pulses implied that even if these burst existed, they must have come in the form of PHz-repetition-rate trains extending over a period comparable to that of the driver pulse. Opening the attosecond shutter of an 'attosecond camera' every femtosecond many times after each other apparently greatly limited its utility for capturing electronic phenomena which often unfold within attoseconds but extend over femtoseconds.

The importance of isolating a single attosecond burst for spectroscopic applications was recognized shortly after HHG driven by Ti:S lasers yielded broadband coherent XUV emission [16] the time structure of which was measured to exhibit bursts of attosecond pulses many years later [33]. Nevertheless, it took quite a few years to actually provide the experimental means for performing this feat. It required the control of the evolution of the electric field of light on a time scale of the field oscillation period. In a first step, one needed to make the cycles of a strong-field ultrashort laser pulse distinguishable. The most straightforward approach to this is to shorten the pulse duration to a value comparable to the carrier wave cycle. To complete control, one needed to gain full control over the waveform in a second step. Our group was involved in both steps and I comment in more detail on these steps to pay tribute to those colleagues involved in realizing them and to show that persistence—on the long run —pays off in science just as it does in other areas of life.

Making field cycles of powerful laser light distinguishable

The first Kerr-lens-mode-locked (KLM) Ti:S laser produced pulses of 60 fs duration [24], more than an order of magnitude longer than the Fourier-limit dictated by the bandwidth of the gain medium (stretching over 600–1100 nm) [23]. Recognizing that the actual pulse duration was limited by higher-order dispersion introduced by the prism sequence incorporated in the cavity for group-delay-dispersion (GDD) control [34, 35], intense research began to select the best possible prism material for minimizing this undesirable high-order dispersion. By the early 90s this research, work of Margaret Murnane [36], Frank Wise [37] and their coworkers, and that of our group [38] led to 10 fs-scale laser pulses (carried at a wavelength of about 800 nm) from KLM Ti:S lasers in which GDD control was implemented by low-loss Brewster-angled prisms selected for minimum high-order dispersion. Since the neighboring central field cycles differ only by a few percent in amplitude in such a pulse, significantly shorter pulses were needed to isolate a single laser cycle for extreme nonlinear optics and attosecond science.

Motivated by this goal, our collaborators Robert Szipöcs and Kárpát Ferencz from Budapest came up with the idea to develop and utilize multilayer mirrors for not only providing feedback by high reflectivity over the required bandwidth but also tailored dispersion for shaping the pulse emerging from the mode-locking process in the laser cavity. The project started back in 1991 with pertinent design efforts by Robert and was followed by several attempts to manufacture the designed prototypes. As a result of these struggles, by late 1992 we had the first mirrors with aperiodic multilayer structures (which later became known as 'chirped mirrors', figure 2) at hand, which appeared promising enough to be tested in a Kerr-lens-mode-locked Ti:S laser oscillator. This attempt proved unsuccessful, just as numerous other following ones. Whilst the mirrors offered a great return in form of more bandwidth with well-behaved dispersion and a much more compact and robust laser, they suffered from a severe shortcoming, too: unlike the prism pair routinely used in

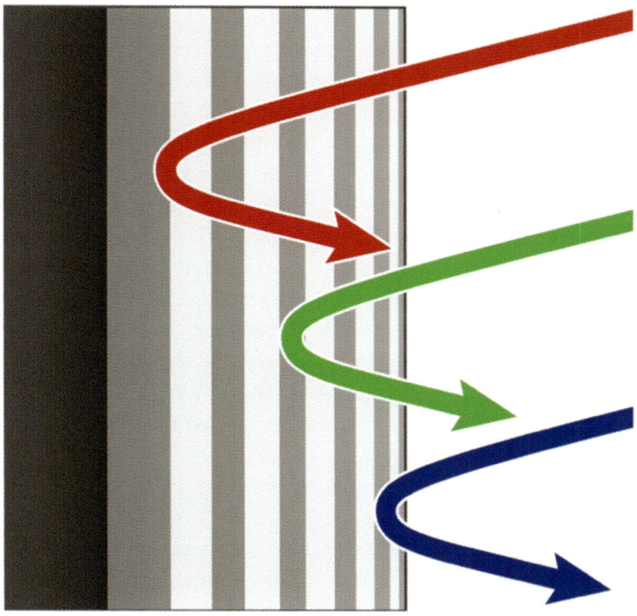

Figure 2. Schematic illustration of the origin of group-delay dispersion (GDD) in a chirped multilayer mirror. Narrow-band wave packets carried at different wavelengths penetrate to different depths before being reflected, as a consequence of a modulation of the multilayer period across the layer stack. In the illustrated example, the increasing multilayer period with increasing distance from the mirror surface implies that radiation with increasing wavelength has to penetrate deeper before being reflected. The result is a group delay that raises with increasing wavelength, i.e., decreasing frequency, giving rise to negative GDD, which is needed in laser oscillators and pulse compressors to compensate for the chirp imposed by self-phase modulation and material dispersion.

femtosecond oscillators for GDD control, they did not provide any degree of freedom for adjusting the GDD. This capability was crucial in prism-controlled systems, it allowed to set the negative GDD first to a higher value to facilitate soliton-like pulse build-up in the mode-locked laser and then minimize it for minimum steady-state pulse duration.

Andreas Stingl and Christian Spielmann, the first experimentalists under my supervision, spent numerous days and nights over many months in our laboratory, just as Robert and Kárpát spent on developing mirrors with improved characteristics before the first mirror-dispersion-controlled femtosecond laser (based on Ti:sapphire as the active medium and employing KLM as the mode locking mechanism) could eventually be successfully demonstrated in the summer of 1993 [39, 40]. As is often the case in science, serendipity played a role in the breakthrough: the unavoidable GDD fluctuation (versus wavelength) of chirped mirrors led, with a specific combination of mirrors, to an overall GDD curve that exhibited enhanced negative GDD over a narrow spectral range around the wavelength of the free-running laser (relevant at the beginning of the mode locking process). This dip in the GDD curve facilitated soliton-like pulse build-up, whilst the substantially lower average GDD over the mirror bandwidth ensured a very short steady-state pulse duration. Serendipity was helpful but did not come accidentally, it was compelled by perseverant search for the solution.

Thanks to the dramatic improvement of the efficiency of the KLM-based self-amplitude modulation owing to the pioneering theoretical work of Hermann Haus [41] and Thomas Brabec [42] as well as subsequent improvements of the bandwidth and flatness of GDD of chirped mirrors by Robert Szipöcs [43], Franz Kärtner [44], Gabriel Tempea [45], mirror-dispersion-controlled Ti:sapphire laser technology rapidly entered the sub-10 fs regime [46]. With semiconductor saturable absorbers pioneered by Ursula Keller they became self starting and outperformed any other femtosecond oscillator technology [47, 48]. The new technology rapidly proliferated all over the world and spawned, up to the present day, the only commercially available sub-10 fs laser oscillator, with the shortest pulses approaching the gain bandwidth limit of Ti:S, below 5 fs [49].

The power of these typically several-nanojoule-energy pulses have been boosted in Ti:sapphire chirped pulse amplifiers into the millijoule regime. However, gain narrowing in these amplifiers limited the duration of such pulses to about 20 fs from the high-power systems. The invention of gas-filled hollow-core fibers for spectral broadening of high-energy pulses via self-phase modulation by Orazio Svelto and Sandro de Silvestri from Milano [50] and subsequent compression of the self-phase-modulated pulses by chirped mirrors have spawn the first powerful few-cycle pulses of laser light [51].

With this milestone the above mentioned goal, the isolation of a single cycle of strong laser field, was achieved. Indeed, in a laser pulse confining about 1.5 field cycles within its full width of intensity half maximum, the amplitude of the central wave cycle exceeded that of neighboring cycles by some 20%. Whenever the laser-field driven process (e.g. strong-field ionization) was a highly nonlinear function of the strength of the laser electric field, a few-cycle pulse, made available by the above mentioned advances, permitted *confinement of the interaction to a single field cycle*, constituting an unprecedented amount of control of strong-field-matter interactions. However, this was not yet entirely satisfactory for sub-cycle (attosecond) control of laser–electron interactions. Uncontrolled shifts of the phase of the carrier-wave with respect to the amplitude envelope (dubbed carrier-envelope phase, CEP) of the pulse led to small variations in the sub-cycle evolution of a few-cycle laser field (figure 3) and might cause significant pulse-to-pulse variations in the evolution of the electron processes triggered and driven by these fields. Apparently, this deficiency needed to be fixed.

Controlling the field of light: laser pulses with reproducible waveform

The quest for controlled light waveforms started in 1996 with the first study of the pulse-to-pulse variation of the CEP in the train of ultrashort light pulses from a femtosecond laser oscillator [52] with the goal realized seven years later [53]. Whilst the control of the pulse-to-pulse CEP-shift could already be demonstrated in the early work, in order to achieve full control over the CEP and thereby the waveform of

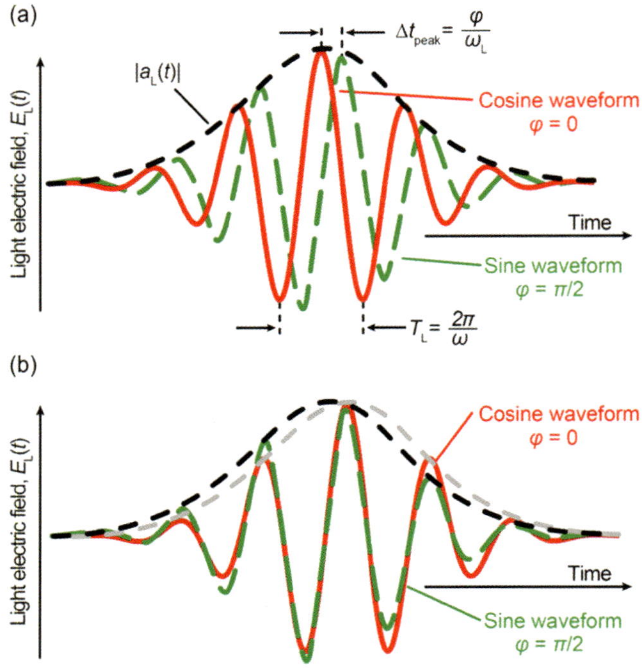

Figure 3. Few-cycle laser pulses containing two oscillation cycles within the full width at half maximum of their intensity profile, with sine- and cosine-shaped waveforms, with their (a) amplitude envelopes and (b) field oscillations coinciding.

individual pulses we utilized the frequency-comb technique of my colleague, Theodor Hänsch, who received the 2005 Nobel Prize for this invention [54].

The work that brought about this breakthrough had resulted from a collaborative effort between groups representing two communities (time-domain and frequency-domain metrology) that had little to do with each other before. As a matter of fact, the work of Baltuska *et al* [53] did not only report on controlled strong light fields for the first time and demonstrated their relevance for attosecond science. Likewise, it represents a culmination of synergies between frequency-domain and time-domain techniques, resulting from a fruitful collaboration of the two communities that continues up to the present day. The benefits are mutual: octave-spanning laser mode locking provides a 'ruler' for precision frequency metrology, whereas a frequency-domain self-referencing technique invented by Theodor Hänsch opened the door to controlling the electric and magnetic fields of ultrashort light pulses.

Isolated attosecond pulses and their first applications

First experimental hints at isolated sub-femtosecond pulses appeared in 1997, when Christian Spielmann and coworkers generated high-order harmonics in neon with few-cycle laser pulses [55]. The highest-order (cut-off) harmonics were observed to coalesce, suggesting the absence of periodicity, i.e. the temporal confinement of cut-off radiation to a fraction of the half cycle of the driving laser field. However, it took us several years to deliver the 'smoking gun' for the existence of isolated pulses shorter than 1 fs. To this end, we had to extend the relatively simple setup for generation (left-hand side of figure 4) with an instrument for temporal characterization (right-hand side of figure 4). The two-component mirror incorporating a nanometer-precision delay stage has been a key component enabling the few-cycle NIR pulse generating the XUV high-harmonic pulse to be delayed with respect to the XUV pulse with attosecond accuracy. Our collaborators from Bielefeld (Germany), Ulf Kleineberg, Markus Drescher and Ulrich Heinzmann have developed both the Mo/Si multilayer coating reflecting the 90–95 eV XUV and the 1.2–2.0 eV NIR radiation for the mirror and the time-of-flight (TOF) spectrometer.

The TOF electron energy analyzer was central for implementing the concept originally proposed by Paul Corkum (NRC Canada, Ottawa) for measuring a short XUV pulse: having photoelectrons from atoms released in the presence of a strong laser field serving as a chronoscope by decelerating or accelerating the electrons, depending on their moment of release (figure 5). These collaborations and several years of preparation allowed the first attosecond streaking experiment to be carried out in Vienna in 2001. Michael Hentschel and Reinhard Kienberger have succeeded in recording the first attosecond streaking trace (figure 6) during the night from 10 to 11 September 2001.

It was beyond description to see chronoscopy working on the sub-femtosecond scale and revealing an isolated sub-1 fs flash of XUV light in the morning of 11 September (unfortunately, our elation was dramatically damped by the news from New York several ours later). The manuscript [2] reporting 'the first attosecond measurement' [56] was accepted for publication within 18 days and featured in newspapers all over the world.

In their first application, these sub-femtosecond flashes measured the lifetime of an atomic inner-shell vacancy a year later [57] (figure 7) and these accomplishments have been highlighted among the most important breakthroughs of the year of 2002 by the editors of Nature [58] and Science [59].

Precision attosecond metrology

It took yet another year to get the waveform of intense few-cycle laser pulses under control. The resultant strong laser fields with a precisely controlled sub-cycle (e.g. sub-femtosecond) temporal evolution have been decisive for both the reproducible generation [53] and precision measurement of isolated attosecond pulses [60] as well as of the few-cycle waveforms driving them [61].

Again, these feats would not have been feasible without collaborations and full dedication and persistance of a number of colleagues. Thomas Udem, Theodor Hänsch and coworkers provided invaluable support in measuring and stabilizing the CEP for waveform control, whereas Paul Corkum's idea about an attosecond streak camera served as a basis for the temporal characterization of the tools. Equally

Phys. Scr. **91** (2016) 063011 Invited Comment

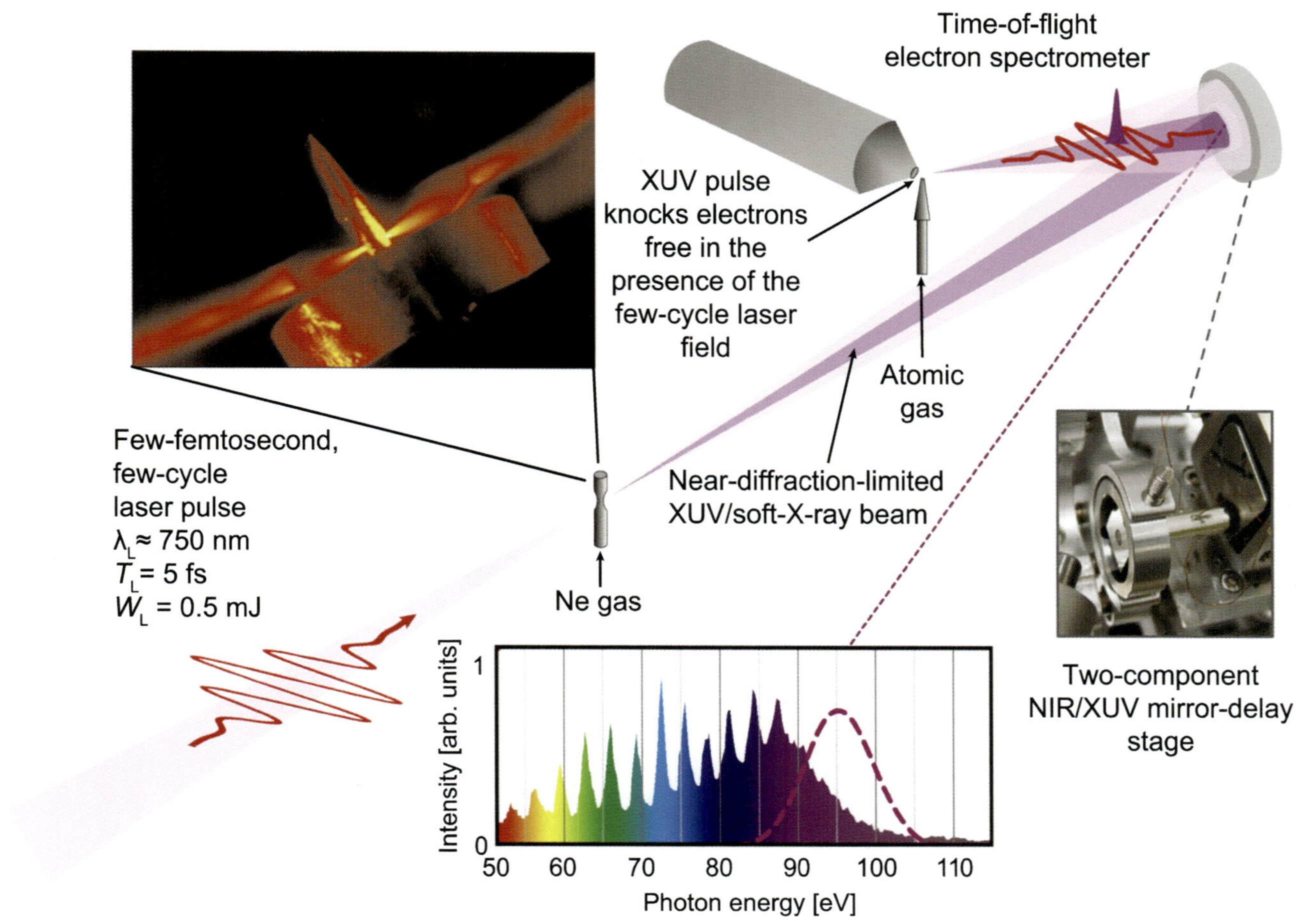

Figure 4. Schematic illustration of the experimental apparatus set up at the Vienna University of Technology and used for the generation, measurement and first applications of isolated sub-femtosecond light pulses. A few-cycle laser pulse generates high-order harmonics in a neon gas medium the cut-off spectral part of which is filtered out by the molybden–silicon multilayer mirror, which also focuses, via two independent concentric parts, both the few-cycle laser beam and the generated XUV burst into a second target. Nanometer-accuracy relative positioning of the two mirror parts permits delaying the XUV burst with respect to the few-cycle laser field previously used for its generation. The apparatus can be used both for temporally characterizing the laser field and the XUV pulse by attosecond streaking as depicted in figure 5 and, subsequently, for real-time observation and control of electronic motions.

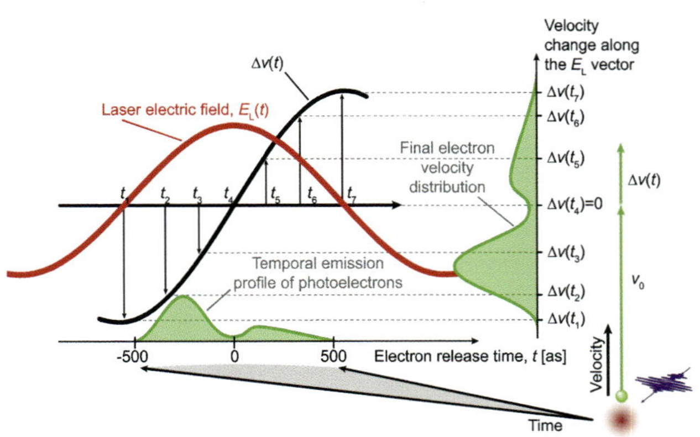

Figure 5. Concept of optical-field-driven attosecond streaking. Photoelectrons released in a laser field parallel to the direction of electric field suffer a change of their initial velocities that is proportional to the vector potential of the field at the instant of release. This function is monotonic within a half wave cycle of the field, mapping the intensity profile of a sub-fs XUV pulse into a corresponding final velocity (or energy) distribution of photoelectrons.

importantly, Alexander Apolonsky, Andrius Baltuska, Eleftherios Goulielmakis, Reinhard Kienberger and Michael Hentschel pushed laser technology to its limit to make these breakthroughs happen.

Attosecond technology and spectroscopy coming of age

Attosecond streaking quickly established itself as a 'gold standard' of attosecond metrology, not only for characterization of the key attosecond tools, waveform-controlled few-cycle laser fields and attosecond XUV pulses, but also for probing electron processes with these tools on attosecond-to-femtosecond time scales [62–64]. The temporal information about these phenomena triggered by an isolated attosecond XUV pulse and captured by the few-cycle laser field via attosecond streaking, or—analogously—triggered by a train of attosecond bursts and captured by multi-cycle laser fields via spectral interferometry [65–70], is encoded in the

Phys. Scr. **91** (2016) 063011 Invited Comment

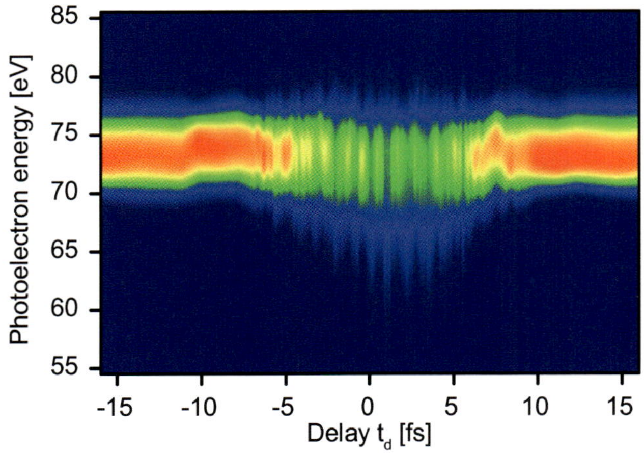

Figure 6. First attosecond streaking spectrogram recorded by the apparatus sketched in figure 4. The spectrogram consists of a series of energy distributions (shown in false color plots) of photoelectrons released by the XUV pulse and modified ('streaked') by the rapidly varying electric field of few-cycle NIR laser pulse previously used for the generation of the XUV pulse as a function of delay between the XUV and the NIR pulse. In this first experiment, the direction of the streaking electric field was set perpendicular to the initial velocity of the electrons (selected by the alignment of the time-of-flight detector) but otherwise the measurement relied on the same basic concept sketched in figure 5. Whilst the parallel detection geometry shown in figure 5 results primarily in an up- and down-shift of the photoelectron energy spectra, the dominant effect of the field in the orthogonal geometry is a broadening and narrowing of the spectrum depending on the delay. In both cases, the series of streaked spectra allows retrieval of the temporal evolution of the laser electric field and the intensity profile and chirp of the attosecond XUV pulse. For more details see [2, 60].

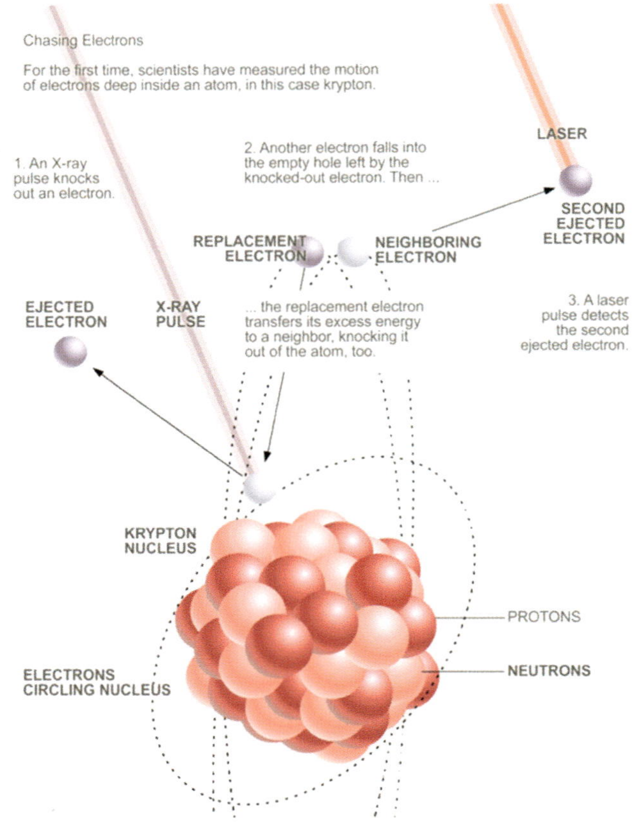

Figure 7. First attosecond time-resolved measurement of electronic motion deep inside an atom (using the apparatus and concept shown in figures 4 and 5). An attosecond XUV ('x-ray') pulse liberates an electron from an inner shell of the electronic system of a krypton atom. The created vacancy survives for an extremely short period of time: within several femtoseconds, i.e. several thousand attoseconds it is filled with an electron from an outer shell. This electron, upon doing this, transfers its excess energy to another outer electron that is thereby catapulted from the atom. The resultant secondary electron emission lasts as long as it takes the vacancy to be filled. Hence, measuring the temporal evolution of this secondary electron emission with a delayed few-cycle laser field via attosecond streaking reveals the lifetime of the inner-shell vacancy just as streaking of the emission of primary photoelectrons yields the duration of the attosecond XUV pulse. Figure reproduced with permission. Copyright PARS International Corp.

properties of photoelectrons released by the XUV pulse in the presence of the laser field. The more electrons of this kind are delivered per second the faster and the more reliably attosecond metrology can be performed.

Today, more than a decade after the first experiments, it is hard to imagine that the first streaking spectrograms [2] could be recorded at a photoelectron count rate of some 6 electrons/second. State-of-the-art Ti:sapphire laser technology [71] meanwhile offers count rates up to and beyond 10 000 electrons/second (from a gas of neon atoms, the standard target in an attosecond streak camera), permitting the recording of a streaking spectrogram within *less than a minute* and with a signal-to-noise level much higher than offered by the first measurements lasting *several hours*. The technological advances that have permitted this progress also increased the sensitivity of attosecond spectroscopy correspondingly, allowing, very recently, the first attosecond measurements to be performed in helium and providing a direct time-domain insight into the interaction of the two electrons during ionization [72].

The strong-field laser waveforms can also be used as a 'starter gun' to release electronic motions. Two different methods established themselves for probing these motions in real time.

(A) The electrons released via ionization in a linearly-polarized field return to their parent ion and their recollision results in high-order harmonic emission [73–76] and electron diffraction [77] that are sensitive to changes in the state of the electronic system during the excursion of the electron. The usage of the recolliding sub-femtosecond electron wavepacket for probing these changes was pioneered by Paul Corkum, Manfred Lein and Jon Marangos and the method has been successfully used for studying electron dynamics in a number of ionizing systems [73–86].

(B) The isolated attosecond XUV pulses offer even more flexibility for probing electron dynamics induced by strong laser fields. XUV probe pulses as short as 80 as [87] and meanwhile even less than 70 as [88] in second-generation beamlines (figure 8) have provided unprecedented real-time insight into fundamental phenomena such as electron

Phys. Scr. **91** (2016) 063011 Invited Comment

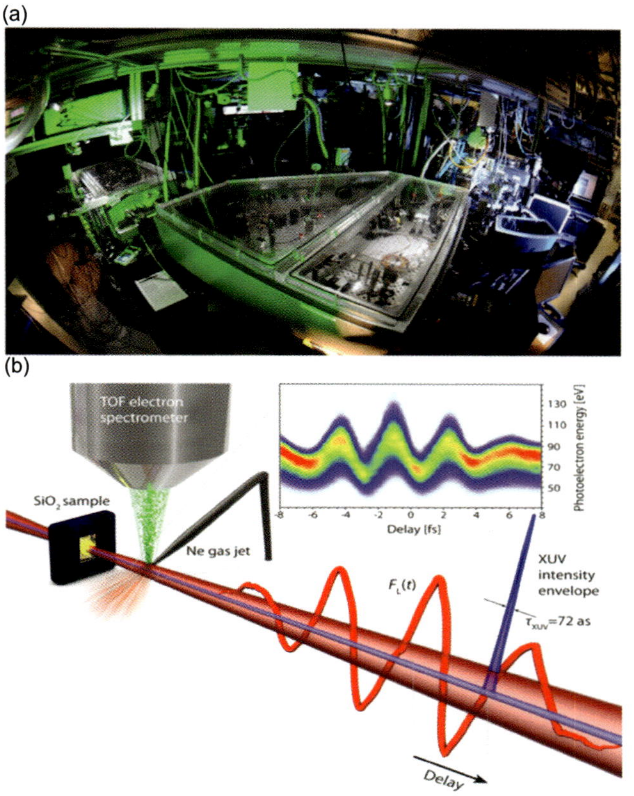

Figure 8. Second-generation attosecond beamline at the Max Planck Institute of Quantum Optics (a) uses sub-4 fs waveform-controlled laser pulses and near-70 as XUV pulses for the time-resolved investigation of ultrafast electron processes (b). The tools are temporary characterized by attosecond streaking (see recorded attosecond streaking spectrogram) and subsequently used for pump–probe spectroscopy, e.g. for the scrutiny of strong-field-induced electron processes in dielectrics, by means of attosecond XUV absorption spectroscopy. Courtesy of Thorsten Naeser panel (a) copyright IOP Publishing and Martin Schultze panel (b). Copyright Nature Publishing Group 2012.

tunneling [89], valence electron wavepacket dynamics [90, 91] and, more recently, strong-field-induced processes in solids [92–94]. Controlling the waveform of few-cycle pulses was also further advanced to the super-octave synthesis of light transients [95], which recently culminated in the demonstration of optical pulses with a duration of less than 1 fs [96] and provides a force engineerable on an attosecond time scale for steering atomic-scale electron motions. Such short optical pulses also open the door to sub-fs-pump/sub-fs-probe spectroscopy for the first time.

Presence-state of the art

In spite of all the progress reviewed above, the state of the art of attosecond physics is far from getting even close to exploiting the potential it offers for exploring microscopic phenomena. Synthesized optical waveforms of sub-cycle-to-few-cycle light [95, 96] and isolated attosecond XUV pulses have been providing unprecedented insight into a number of electron processes in atoms, molecules and condensed-matter systems, as well as an ever increasing control over their evolution and outcome. But there is a large number of phenomena and physical questions that cannot be addressed with the current tools. Dynamics involving transitions with small cross sections and/or bound states separated by more than several electronvolts in energy have remained largely inaccessible just as the temporal evolution of the electrons' probability distribution in atomic systems and the related picometer-scale (mean) electron trajectories. The effect of tailored light forced on the outcome of electron dynamics has remained modest, mainly due to the limited variety of waveforms being made available by the limited bandwidth (2 octaves) of current-generation optical waveform synthesizers.

These deficiencies unveil the frontiers where attosecond technology can and needs to be further advanced: the power and photon energy range of attosecond pulses and the bandwidth of optical field synthesizers. The attosecond technology living up to these expectations will be so radically different that it deserves the attribute 'next-generation'.

Future-next-generation attosecond technology

Attosecond technology is based on cutting-edge femtosecond technology. This has so far been provided by Ti:sapphire-based laser systems supplemented with nonlinear pulse compression techniques to overcome the gain-bandwidth limit in our strive for super-octave waveform generation. This few-cycle laser technology has now closely approached its limits in terms of average and peak power, in the range of several watts and a fraction of a terawatt, respectively. Revolutionizing attosecond technology hence calls for a new generation of femtosecond technology, which is capable of scaling both the average and the peak power of few-cycle laser pulses by several orders of magnitude. The only approach offering sufficient gain and scalability over the required bandwidth is optical parametric chirped-pulse amplification (OPCPA).

Although parametric amplification has been known for decades, powerful, efficient and cost-effective picosecond sources, required for pushing the process further ahead, have been unavailable. Thanks to major advances over the past several years triggered by Thomas Metzger at MPQ and continued by Thomas Nubbemeyer at the Ludwig Maximilians University with crucial support of Klaus Wallmeroth and Dirk Sutter from Trumpf Laser GmbH Laser, diode-pumped ytterbium-doped solid-state lasers based on thin-disk geometry (figure 9) are now able to offer near-1 ps pulses at kilowatt-scale average power levels, demonstrated by 10 mJ pulses at 100 kHz and 150 mJ pulses at a 5 kHz repetition rate [97]. The technology is scalable beyond the kilowatt average power level and up to multi-terawatt peak powers [98].

This unique performance comes in combination with the reliability of industrial lasers and renders this technology the best candidate for pumping high-power, ultra-broadband OPCPA systems. Yb:YAG-laser-driven multi-channel broadband OPCPA systems offer the potential for vastly outperforming previous-generation femtosecond technologies

Phys. Scr. **91** (2016) 063011 Invited Comment

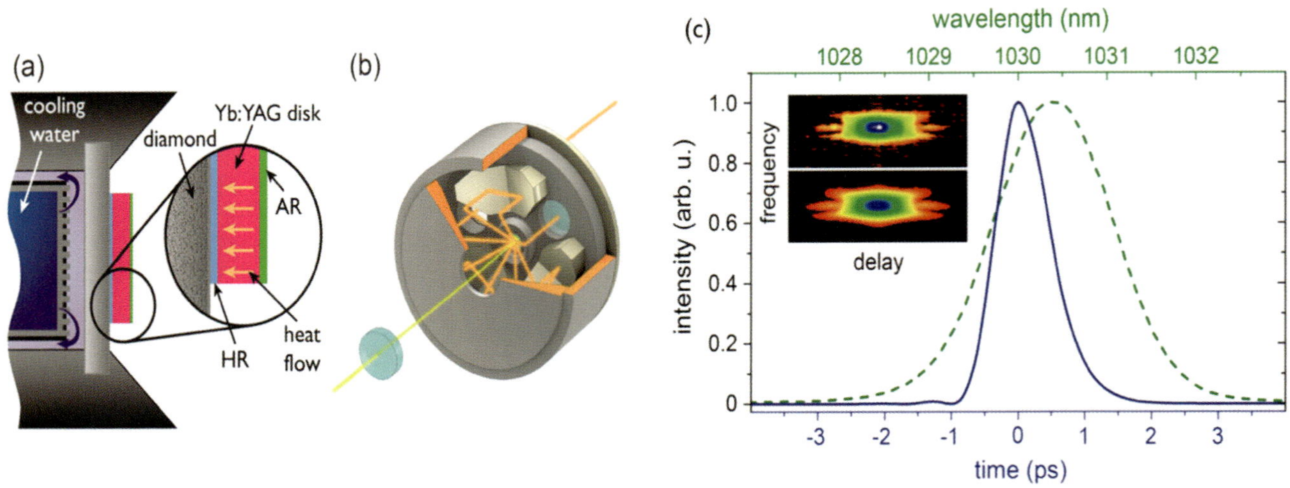

Figure 9. Schematics of thin-disk laser technology. (a) Cold finger with substrate showing the cooling mechanism in a disk laser head, (b) disk laser head showing the principle of the pump light reimaging technique onto the thin-disk active medium (courtesy of TRUMPF Laser GmbH). (c) Measured spectrum (green dashed line) and temporal intensity profile (blue line) from a system equipped with two disk modules (disk parameters: doping concentration >7%, thickness ~100 μm, beveled and roughened edge, TRUMPF Laser GmbH). These 0.97 ps pulses carry energy of 40 mJ at a repetition rate of 5 kHz, corresponding to an average power of 200 W. The corresponding measured (top) and reconstructed (bottom) FROG traces are shown as insets. Source: Fattahi *et al* 2014 *Optica* 1 45 . The system is currently being upgraded to 200 mJ pulse energy for driving the multi-channel OPCPA system outlined in figure 10.

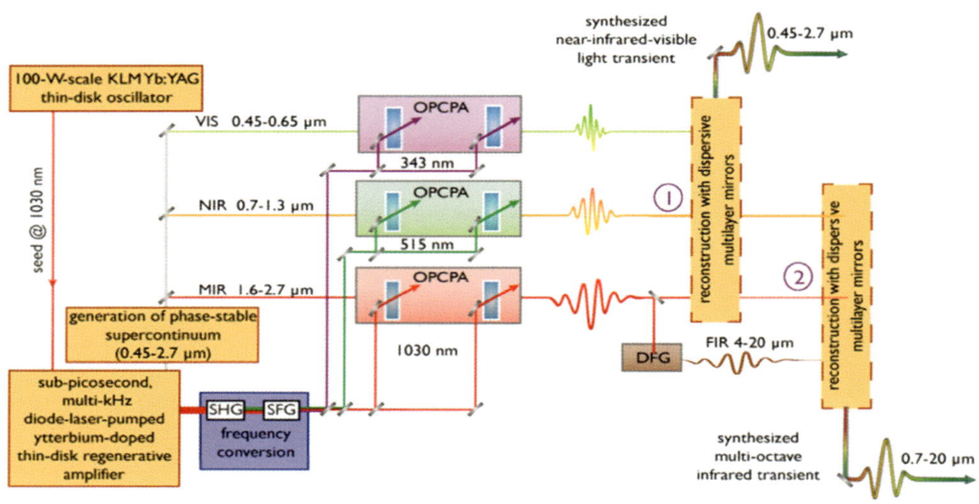

Figure 10. Schematic architecture of a three-channel OPCPA system seeded and pumped by a 1 ps ytterbium laser. A part of its output is used for generating a phase-stable multi-octave supercontinuum signal, which is split into three channels, centered at 550 nm, 1 μm, and 2 μm, respectively. The different channels are pumped by different low-order harmonics of the multi-mJ level, kHz, Yb:YAG regenerative amplifier output. Each channel supports few-cycle pulses. By coherently combining the three few-cycle channels in VIS, NIR, and MIR, amplified, non-sinusoidal, multi-octave light transients spanning from 0.45 to 2.7 μm can be generated (1). Extension to FIR can be achieved by intrapulse DFG of the MIR channel. Coherent synthesis of the NIR, MIR and FIR channels will result in unique multi-octave infrared transients with a spectral coverage of 0.7–20 μm (2). Courtesy of Hanieh Fattahi.

in terms of bandwidth, peak *and* average power and, after femtosecond pulses being generated and amplified in dye and Ti:sapphire lasers, open the era of *third-generation femtosecond technology* (3FST) [98]. The basic architechture of the prototypical 3FST source, which has been dubbed ACCORD, is currently in Garching under construction by a team led by Hanieh Fattahi and Nicholas Karpowicz in close cooperation with Vladimir Pervak and Michael Trubetskov in charge of the development of super-octave chirped mirrors with tailored reflectivity and dispersion [99, 100], as schematically illustrated in figure 10. Once completed (hopefully in 2017), it will deliver synthesized several-octave light transients [101] with multi-terawatt peak power in the visible–near-infrared (VIS–NIR) wavelength range of 0.45–2.7 μm and in the near-infrared-mid-infrared range of 0.7–20 μm.

Some of the possible VIS–NIR waveforms, simulated by Hanieh Fattahi, are shown in figure 11. They can—among others—be optimized for extending HHG into the keV regime of photon energies. The generation of isolated attosecond pulses at highest possible photon energies calls for a

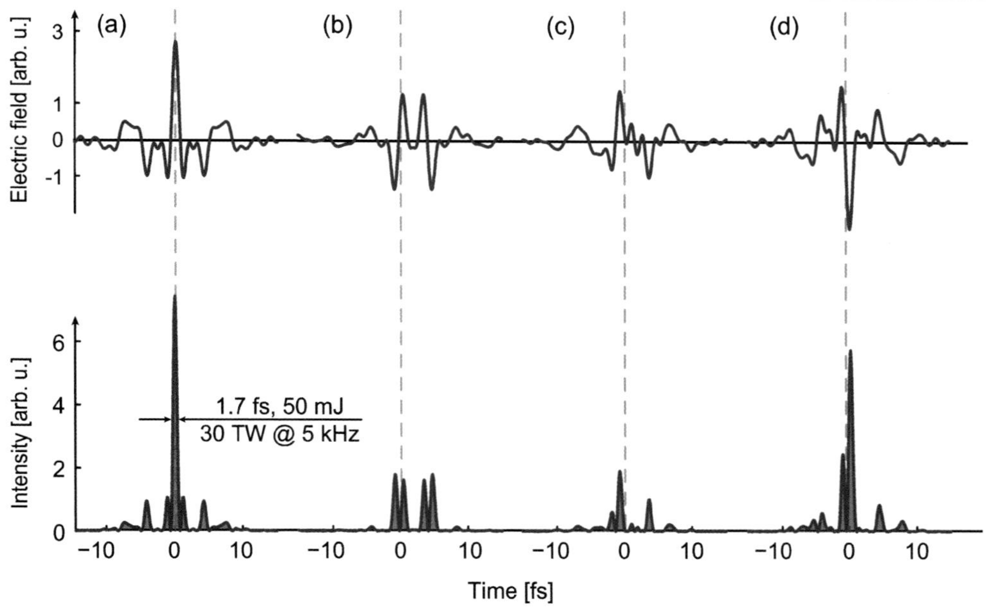

Figure 11. Multi-octave visible-infrared multi-terawatt laser waveforms anticipated from the prototypical three-channel optical parametric amplifier system currently under construction at the Ludwig-Maximilians Universität in Munich and outlined in figure 10. Panel (a) shows the transform-limited waveform that can be constructed from the three channels described in figure 10, whereas (b)–(d) feature a representative set of waveforms that can be created by merely varying the delay between the three OPA channels compressed to their Fourier limit. Courtesy of Hanieh Fattahi.

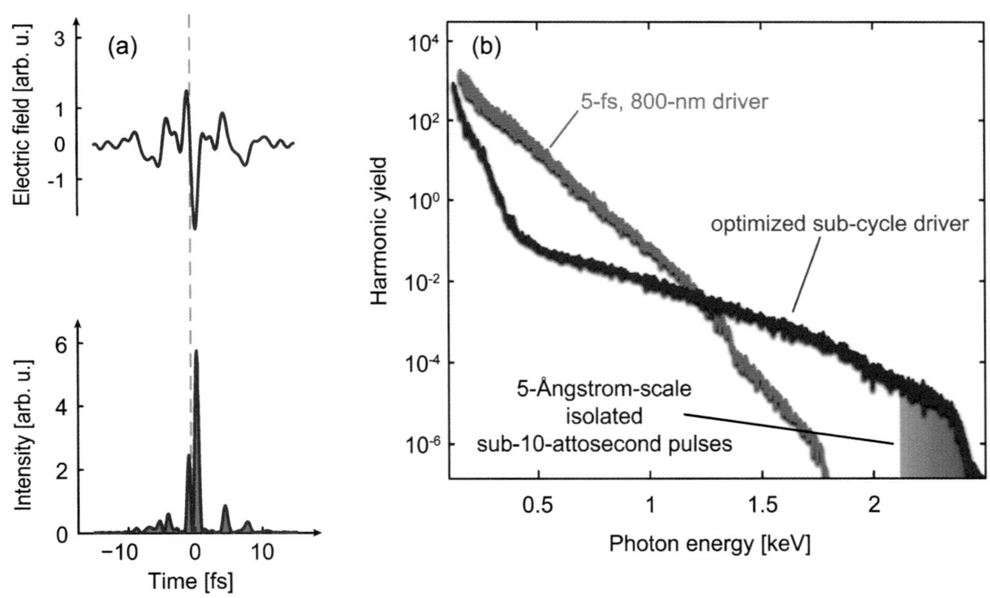

Figure 12. Multi-octave VIS–IR waveform from the multi-channel optical parametric amplifier-synthesizer system sketched in figure 10 optimized for generating high-order harmonics at the highest possible photon energy with the highest possible efficiency. The optimized near-single-cycle waveform dramatically extends the cut-off as compared to a 5 fs, 800 nm driver pulse of the same peak power (and hence larger pulse energy). This is due to the reduced ionization and better resultant phase matching, owing to the abrupt turn-on of the strong field in the single-cycle waveform. The simulation predicts the generation of 10 fs scale isolated attosecond pulses at wavelengths of 0.5 nm.

waveform consisting of a single field cycle, with its first half pushing electrons away from their parent ions and its second half pulling them back. Radiation before this time window would cause undesirable pre-ionization of the atoms in the interaction medium, reducing HHG efficiency due to impaired phasematching and/or depletion of the ground state. A multi-octave bandwidth is a necessary prerequisite for realizing the ideal, abrupt turn-on of the field amplitude within a fraction of the optical cycle. Figure 12 compares the HHG spectra yielded by the numerical studies of Hanieh Fattahi and Marcelo Ciappina for a 5 fs Gaussian driver pulse and a single-cycle VIS–NIR (0.45–2.7 μm) driving waveform of the same peak intensity. The optimum waveform that our algorithm has found so far, indeed, carried most of its energy in a single intense wave cycle, which is turned on as abruptly as possible with the available bandwidth.

The predicted isolated 2.5 keV attosecond pulses will provide access to both strongly bound electronic states and atomic-scale electron trajectories by means of keV attosecond electron spectroscopy and diffraction, respectively. This capability, combined with the ability to induce and control strong-field electron phenomena in dielectrics and semiconductors with the unique multi-octave VIS–NIR and NIR–MIR waveforms, offers extremely exciting research opportunities of fundamental and technological importance likewise.

Outlook

Next-generation femtosecond and attosecond technology promises to dramatically increase the flux and/or photon energy of attosecond light and the bandwidth available for engineering light forces. Orders of magnitude more powerful than nowadays available vacuum/XUV or soft-x-ray attosecond pulses will permit attosecond probing of processes with much smaller cross-sections and attosecond-pump/attosecond-probe spectroscopies of a wide range of bound electron excitation and relaxation dynamics, which are inaccessible today.

On the other hand, extending attosecond pulses (at useful flux levels) to x-ray (several keV) photon energies will allow capturing the picometer-attosecond-scale motions of electrons via attosecond x-ray diffraction.

Direct visualization of dynamic changes in electron density may thereby become feasible in arbitrarily complex systems, providing never-before-attained insight into the working of biological systems at the most fundamental level: at the level of (coupled) electronic and nuclear motions. Last but not least, multi-octave waveforms spanning from the visible (Petahertz) to infrared (Terahertz) spectral range will be instrumental in exploring electron phenomena in semiconductors and dielectrics that may be essential for advancing signal processing and metrology to their ultimate limit, to optical (Petahertz) frequencies. These advances would entail a new revolution in exploring microscopic dynamics. All motions outside the atomic core would, in principle, become accessible and subject to human control to a never-before-achieved extent.

Acknowledgment

The author gratefully acknowledges the support of Mandy Singh and Christian Hackenberger in the final editing of the text and the figures, respectively.

References

[1] Battacharjee Y 2001 *Nature* **412** 474
[2] Hentschel M *et al* 2001 *Nature* **414** 509
[3] Krausz F and Ivanov M 2009 *Rev. Mod. Phys.* **81** 163
[4] Krehl P and Engemann S 1995 *Shock Waves* **5** 1
[5] Abraham H and Lemoine T 1899 *Compt. Rend.* **129** 206
[6] Norrish R G W and Porter G 1949 *Nature* **164** 658
[7] Eigen M 1954 *Discuss. Faraday Soc.* **17** 194
[8] Zewail A H 1994 *Femtochemistry* vol I and II (Singapore: World Scientific)
[9] Zewail A 2000 *J. Phys. Chem.* A **104** 5660
[10] Hänsch T W 1990 *Opt. Commun.* **80** 71
[11] Farkas G and Tóth C 1992 *Phys. Lett.* A **168** 447
[12] Harris S E, Macklin J J and Hänsch T W 1993 *Opt. Commun.* **100** 487
[13] Harris S E and Sokolov A V 1998 *Phys. Rev. Lett.* **81** 2894
[14] McPherson A *et al* 1987 *J. Opt. Soc. Am.* B **4** 595
[15] L'Huillier A and Balcou P 1993 *Phys. Rev. Lett.* **70** 774
[16] Macklin J, Kmetec J D and Gordon C L III 1993 *Phys. Rev. Lett.* **70** 766
[17] Lewenstein M, Balcou P, Ivanov M Y, L'Huillier A and Corkum P B 1994 *Phys. Rev.* A **49** 2117
[18] Corkum P B, Burnett N H and Ivanov M Y 1994 *Opt. Lett.* **19** 1870
[19] Ivanov M Y, Corkum P B, Zuo T and Bandrauk A 1995 *Phys. Rev. Lett.* **74** 2933
[20] Antoine P, L'Huillier A and Lewenstein M 1996 *Phys. Rev. Lett.* **77** 1234
[21] Fork R L *et al* 1987 *Opt. Lett.* **12** 483
[22] Shank C V 1988 *Ulrashort Laser Pulses and Applications* (*Topics in Applied Physics* No. 60) ed W Kaiser (New York: Springer) p 5
[23] Moulton P F 1986 *J. Opt. Soc. Am.* B **3** 125
[24] Spence D E, Kean P N and Sibbett W 1991 *Opt. Lett.* **16** 42
[25] Strickland D and Mourou G 1985 *Opt. Commun.* **56** 219
[26] Maine P, Strickland D, Bado P, Pessot M and Mourou G 1988 *IEEE J. Quantum Electron.* **24** 398
[27] Brabec T and Krausz F 2000 *Rev. Mod. Phys.* **72** 545
[28] Li X F, L'Huillier A, Ferray M, Lompré L A and Mainfray G 1989 *Phys. Rev.* A **39** 5751
[29] L'Huillier A, Li X F and Lompré L A 1990 *J. Opt. Soc. Am.* B **7** 527
[30] L'Huillier A, Schafer K J and Kulander K C 1991 *Phys. Rev. Lett.* **66** 2200
[31] L'Huillier A, Balcou P, Candel S, Schafer K J and Kulander K C 1992 *Phys. Rev.* A **46** 2778
[32] L'Huillier A, Lompré L-A, Mainfray G and Manus C 1992 *Atoms in Intense Laser Fields* ed M Gavrila (New York: Academic) p 139
[33] Paul P M *et al* 2001 *Science* **292** 1689
[34] Fork R L, Martinez O E and Gordon J P 1984 *Opt. Lett.* **9** 150
[35] Martinez O E, Gordon J P and Fork R L 1984 *J. Opt. Soc. Am.* A **1** 1003
[36] Asaki M T *et al* 1993 *Opt. Lett.* **18** 977
[37] Proctor B and Wise F 1993 *Appl. Phys. Lett.* **62** 470
[38] Curley P F *et al* 1993 *Opt. Lett.* **18** 54
Spielmann C, Curley P F, Brabec T and Krausz F 1994 *IEEE J. Quantum Electron* **30** 1100
[39] Szipöcs R, Ferencz K, Spielmann C and Krausz F 1994 *Opt. Lett.* **19** 201
[40] Stingl A, Spielmann C, Krausz F and Szipöcs R 1994 *Opt. Lett.* **19** 204
[41] Haus H A, Fujimoto J G and Ippen E P 1992 *IEEE J. Quantum Electron.* **28** 2086
[42] Brabec T *et al* 1992 *Opt. Lett.* **17** 748
Brabec T *et al* 1992 *Opt. Lett.* **17** 1292
[43] Mayer E J, Möbius J, Euteneuer A, Rühle W W and Szipöcs R 1997 *Opt. Lett.* **22** 528
[44] Kärtner F X *et al* 1997 *Opt. Lett.* **22** 831
[45] Tempea G *et al* 1998 *J. Sel. Top. Quantum Electron.* **4** 193
[46] Stingl A *et al* 1995 *Opt. Lett.* **20** 602

[47] Sutter D H *et al* 1999 *Opt. Lett.* **24** 631
[48] Ell R *et al* 2001 *Opt. Lett.* **26** 373
[49] Binhammer T, Rittweger E, Morgner U, Ell R and Kärtner F X 2006 *Opt. Lett.* **31** 1552
[50] Nisoli M, De Silvestri S and Svelto O 1996 *Appl. Phys. Lett.* **68** 2793
[51] Nisoli M *et al* 1997 *Opt. Lett.* **22** 522
[52] Xu L *et al* 1996 *Opt. Lett.* **21** 2008
[53] Baltuška A *et al* 2003 *Nature* **421** 611
[54] Hänsch T W 2006 *Rev. Mod. Phys.* **78** 1297
[55] Ch S *et al* 1997 *Science* **278** 661
[56] Silbergerg Y 2001 *Nature* **414** 494
[57] Drescher M *et al* 2002 *Nature* **419** 803
[58] The editors of Nature *Nature* 2002 **420** 737
[59] The editors of Science *Science* 2002 **298** 2297
[60] Kienberger R *et al* 2004 *Nature* **427** 817
[61] Goulielmakis E *et al* 2004 *Science* **305** 1267
[62] Cavalieri A *et al* 2007 *Nature* **449** 1029
[63] Schultze M *et al* 2010 *Science* **328** 1658
[64] Neppl S *et al* 2015 *Nature* **517** 342
[65] Johnsson P *et al* 2005 *Phys. Rev. Lett.* **95** 013001
[66] Remetter T *et al* 2006 *Nat. Phys.* **2** 323
[67] Mauritsson J *et al* 2008 *Phys. Rev. Lett.* **100** 073003
[68] Sansone G *et al* 2010 *Nature* **465** 763
[69] Neidel C *et al* 2013 *Phys. Rev. Lett.* **111** 033001
[70] Mansson E P *et al* 2014 *Nat. Phys.* **10** 207
[71] Cavalieri A *et al* 2007 *New J. Phys.* **9** 242
[72] Schultze M *et al* 2016 private communications
[73] Niikura H *et al* 2002 *Nature* **417** 917
[74] Lein M, Hay N, Velotta R, Marangos J P and Knight P L 2002 *Phys. Rev. Lett.* **88** 183903
[75] Niikura H *et al* 2003 *Nature* **421** 826
[76] Itatani J *et al* 2004 *Nature* **432** 867
[77] Meckel M *et al* 2008 *Science* **320** 1478
[78] Niikura H, Villeneuve D M and Corkum P B 2005 *Phys. Rev. Lett.* **94** 083003
[79] Kanai T, Minemoto S and Sakai H 2005 *Nature* **435** 470
[80] Lein M 2005 *Phys. Rev. Lett.* **94** 053004
[81] Vozzi C *et al* 2005 *Phys. Rev. Lett.* **95** 153902
[82] Baker S *et al* 2006 *Science* **312** 424
[83] Smirnova O *et al* 2009 *Nature* **460** 972
[84] Woerner H J *et al* 2010 *Nature* **466** 604
[85] Woerner H J *et al* 2011 *Science* **334** 208
[86] Hammond T J *et al* 2015 *Nature* **522** 462
[87] Goulielmakis E *et al* 2008 *Science* **320** 1614
[88] Zhao K *et al* 2012 *Opt. Lett.* **37** 3891
[89] Uiberacker M *et al* 2007 *Nature* **446** 627
[90] Goulielmakis E *et al* 2010 *Nature* **466** 739
[91] Chini M *et al* 2013 *Sci. Rep.* **3** 1105
[92] Schiffrin A *et al* 2013 *Nature* **493** 70
Schiffrin A *et al* 2014 *Nature* **507** 386
[93] Schultze M *et al* 2013 *Nature* **493** 75
[94] Schultze M *et al* 2014 *Science* **346** 6
[95] Wirth A *et al* 2011 *Science* **334** 195
[96] Luu T *et al* submitted for publication
[97] Thomas Metzger and Nubbemeyer T 2016 private communications
[98] Fattahi H *et al* 2014 *Optica* **1** 45
[99] Pervak V *et al* 2007 *Appl. Phys.* B **87** 5
[100] Chia S-H *et al* 2014 *Optica* **1** 315
[101] Mücke O D *et al* 2015 *J. Sel. Top. Quantum Electron.* **21** 8700712

General Introduction

Throughout my career, basic research has brought numerous moments of satisfaction and childlike pleasure. Some of them are more special than others. One such moment happened in 1995, in a small laser laboratory at the Technical University of Vienna, with the appearance of a shift in the phase of the electric field oscillations of a light pulse. The phase shift occurred in an ultrashort laser pulse after one round trip in the laser cavity[3]. Its existence was not surprising, nor was it useful for anything. However, a great deal of satisfaction was brought by the very fact that the measurement of this phase (usually denoted by φ), or – more precisely – of its change, constituted the very first step towards gaining experimental access to the oscillation of the electric field of light.

Why was this prospect exciting? Simply because controlling this force field would be directly yielding a force $F_L(t)$ acting on electrons that is governed *within* the period of these oscillations, T, lasting about 1–2 fs = 1000–2000 as (1 as = 10^{-18} s) in the visible spectral range,

$$F_L(t) = eE_L(t) = \frac{1}{2}\, eE_{peak}\, a_L(t)\, e^{-i(2\pi t/T+\varphi)} + c.c.$$

As a consequence, the force $F_L(t)$ is evolving on an attosecond time scale. This happens to be the natural time scale of electronic motions on atomic length scales. The coincidence is not accidental. It is the atomic-scale oscillatory motion of electrons that forms the antennas responsible for the emission of light waves in the visible and nearby spectral ranges. Hence, the force fields of light naturally evolve on the electronic time scale.

Control of the phase of ultrashort pulses and thereby control of their waveform could provide an attosecond-varying light force, which, if strong enough to rival the (static) intra-atomic and intra-molecular electric forces acting between electrons and nuclei, would provide a means of controlling the motion of electrons at the atomic scale.

At the dawn of the new millennium, this strong attosecond light force indeed became available in a controlled fashion and permitted the establishment of experimental attosecond physics, the science of controlling and measuring light waves and electronic motions. Experimental attosecond physics or *attosecond metrology* provides real-time access to the dynamics of electrons and light waves as well as their interactions, on their natural attosecond-to-femtosecond time scale.

The femtosecond-scale control of the amplitude and frequency of laser light, both of which are encoded in the complex amplitude $a_L(t)$, gave birth to *femtochemistry* by permitting real-time control and interrogation of molecular dynamics[4]. At the same time, the attosecond-scale control of the oscillating laser fields, which was allowed by controlling φ in addition to $a_L(t)$, opened the door for observing and controlling electron processes in atomic systems, initiating the era of *attophysics*.

Access to the light phase paved the way from real-time control and interrogation of the dynamics of molecular structure in femtochemistry to those of atomic-scale electron dynamics in attophysics. Briefly, light-phase control paved the way from femtochemistry to attophysics. As a result, this volume, which is devoted to the birth and coming of age of attosecond science, begins with a chapter reviewing the first steps towards measuring and controlling the light phase.

This access would not have been possible without the generation of laser pulses with their energy substantially

[3] L. Xu *et al.* Route to phase control of ultrashort laser pulses. *Opt. Lett.* **21**, 2008 (1996).

[4] A. H. Zewail. Femtochemistry: atomic-scale dynamics of the chemical bond. *J. Phys. Chem. A* **104**, 5660 (2000).

confined to a single wave cycle. There have been many advances which led to the generation of near-single-cycle light pulses (Chapter 2). This extreme form of light has spawned extreme manifestations of optical nonlinearities. Examples include optical-field ionization with a rate approaching the frequency of light and the generation of extremely high-order harmonics of the laser field driving the ionization process. Thorough understanding of these nonlinearities (Chapter 3) has been instrumental in their exploitation for the first attosecond-scale measurements (Chapter 4).

The extreme temporal confinement of light fields together with their reproducible synthesis were able to eventually spawn force fields that oscillate in a controlled fashion on the attosecond time scale (Chapter 5). They have permitted the reproducible generation of isolated attosecond pulses of extreme ultraviolet (XUV) light and based on them the birth of attosecond metrology of the first generation, or AM 1.0 (Chapter 6).

Attosecond metrology constitutes a revolution in technology. Controlled microwave fields permit electronic currents to be switched on in a controlled fashion within a fraction of a nanosecond, forming the basis of modern analog and digital electronics. The controlled amplitude and chirp, both encoded in the complex amplitude $a_L(t)$, of ultrashort laser pulses have enabled the control and interrogation of the motion of nuclei in molecules and condensed matter. Control of the field oscillations of few-cycle laser pulses along with attosecond XUV pulses synchronized to them has provided access to the fastest motions outside the atomic core – the motion of electrons (Fig. 1).

AM 1.0 has been relying on (i) the controlled attosecond force exerted by the electromagnetic field of waveform-controlled light on electrons and (ii) isolated attosecond pulses synchronized to this field with attosecond precision. The emergence of AM 1.0 (Chapter 7) has enabled experimental physics to track and steer atomic-scale electronic motions in real-time (Chapters 8 and 9). The price for this progress was the need for a complex instrumentation embedded in a vacuum apparatus in addition to demanding laser technology (Chapter 10).

The controlled attosecond force of near-single cycle light and extreme nonlinearities exploited within solid matter has yielded yet another attosecond probe that consists of carriers being injected into an insulating material within a fraction

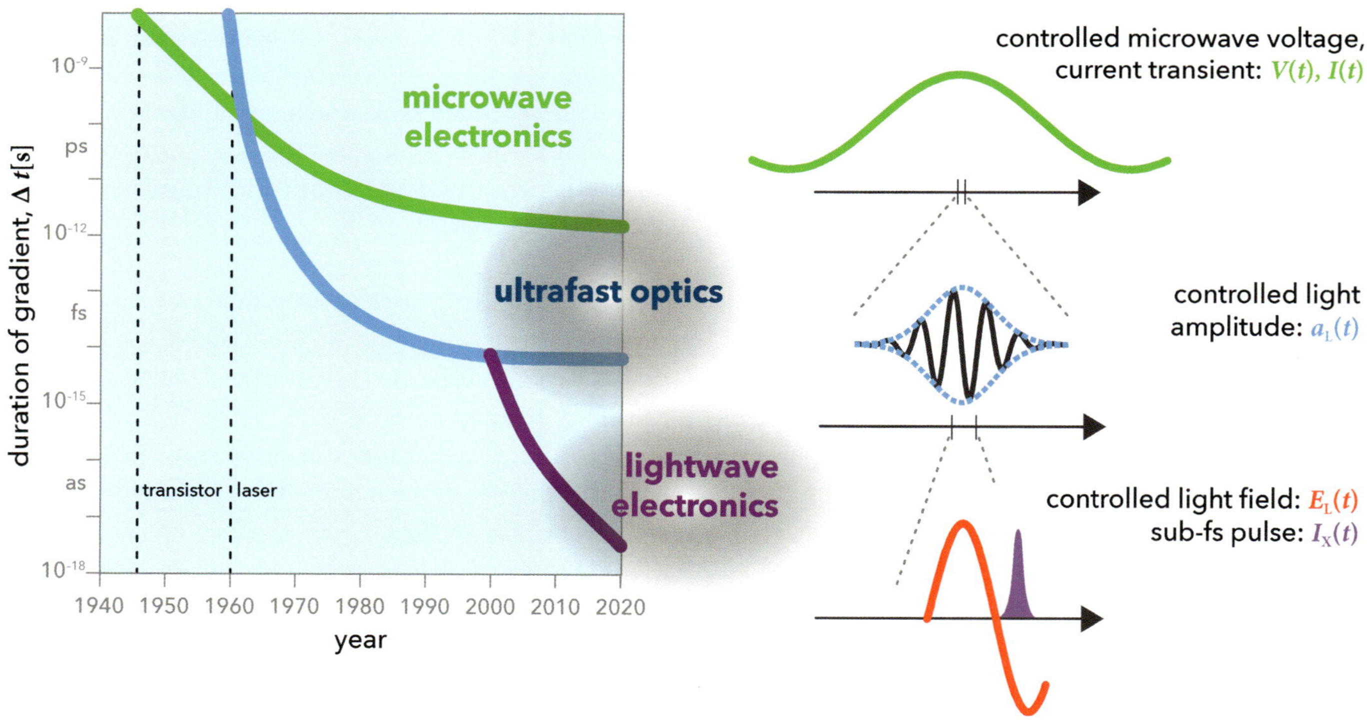

Figure 1 From microwave electronics, controlling and measuring electric current within picoseconds in nanometer-scale circuits, to attosecond technology, controlling and tracking electron motions in atomic dimensions.

of a femtosecond. Attosecond carrier injection can be used for sampling the electric field of light waves up to terahertz frequencies and doing so in solid-state circuitry, obviating the need for vacuum instrumentation. The resultant fully-characterized optical fields can be used, again without the need for vacuum, for exciting and probing the linear and nonlinear polarization response of matter with attosecond resolution. This capability opens the door for attosecond measurements with a simple measurement apparatus and is applicable to all forms of matter. These include, in particular, the interior of condensed matter (inaccessible to attosecond probes coming in the form of XUV light), where processes of the highest relevance for real-world applications such as digital electronics and molecular diagnostics happen. Hence, this new form of extreme chronoscopy deserves a new name: next-generation attosecond metrology, known as AM2.0 (Chapter 11).

The novel compact solid-state attosecond instrumentation paves the way for real-world applications, in particular methods for exploring viable routes towards THz-to-PHz-scale optical-field-driven electronics and sensing miniscule changes in the molecular composition of complex biological samples via femtosecond-attosecond sampling of infrared-to-ultraviolet fields emitted by excited molecules. In the concluding chapter, co-authored by Mihaela Žigman, we envision a scenario where attosecond physics may one day affect our everyday lives as well as improve the quality of those for future generations.

The new discipline of *attosecond science* could not have emerged without the contributions of numerous research groups from all over the world. In addition to Nobel Prize winning breakthroughs such as chirped-pulse amplification (Physics 2018), femtochemistry (Chemistry 1999), and frequency combs (Physics 2005), many other discoveries, inventions and technological innovations have been indispensable in spawning the techniques that eventually provided access to the dynamics of electrons in atomic dimensions. This volume is a collection of my work as well as that of my co-workers and collaborators that have contributed to the birth and coming of age of attosecond science. The extensive reference lists of the reprinted papers provide an overview of the vast number of important contributions of others which were instrumental in the emergence of the new discipline.

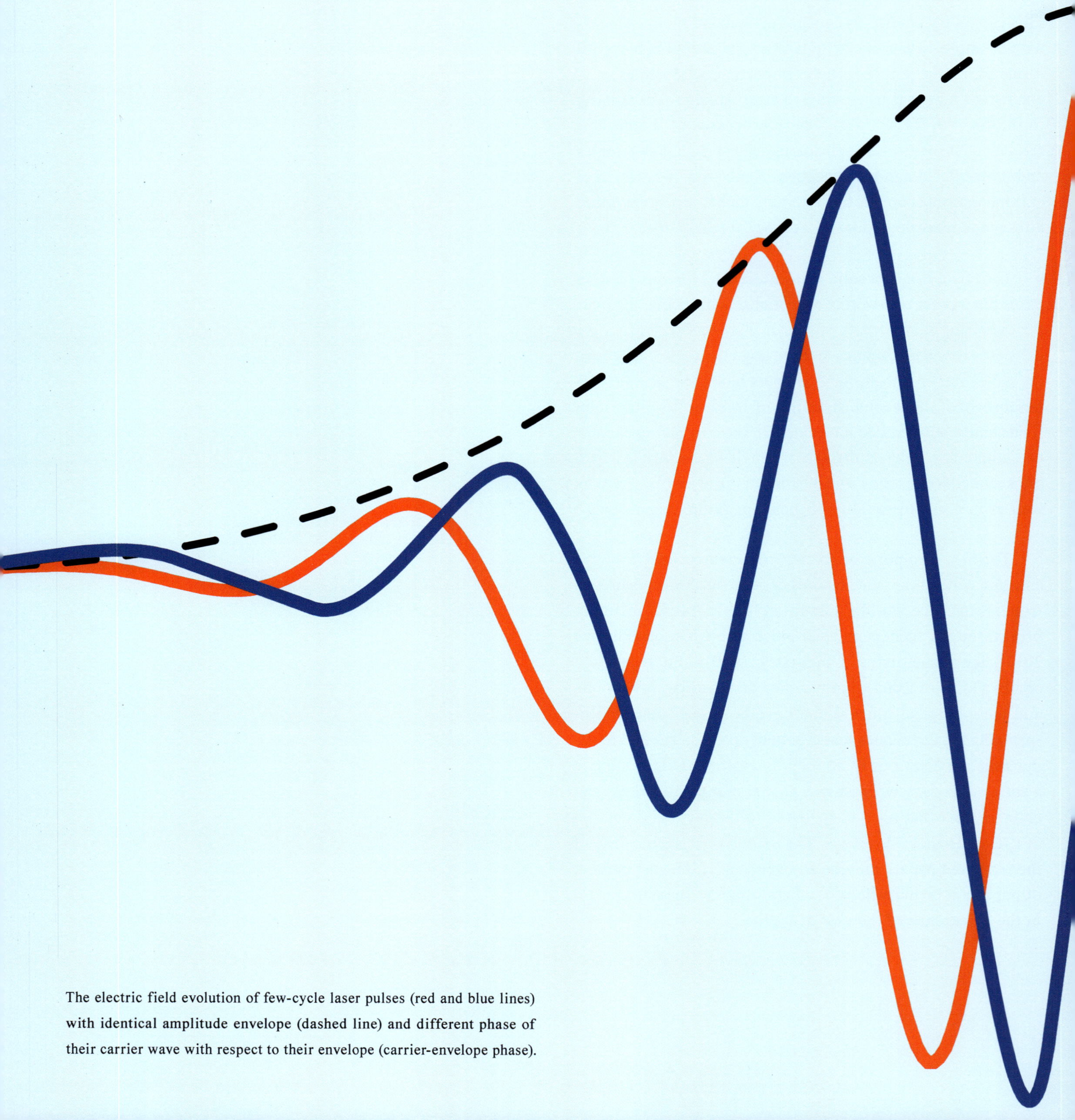

The electric field evolution of few-cycle laser pulses (red and blue lines) with identical amplitude envelope (dashed line) and different phase of their carrier wave with respect to their envelope (carrier-envelope phase).

Looking Inside the Wave Cycle of Light

1

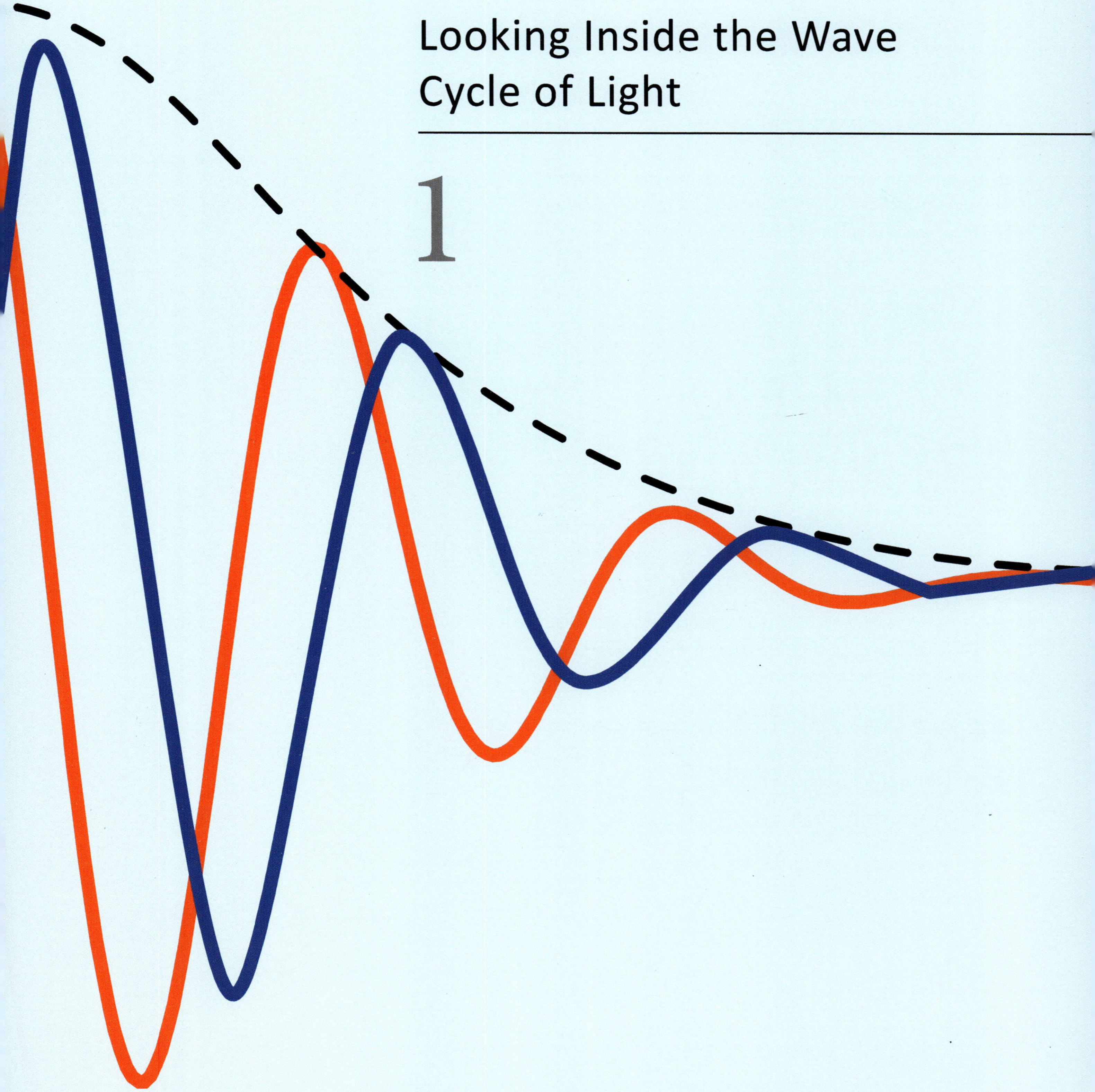

Ultrashort-pulse laser techniques have evolved over several decades to provide laser pulses

$$E_L(t) = \frac{1}{2} E_{peak} a_L(t) e^{-i(2\pi t/T+\varphi)} + c.c.$$

with ever better reproducibility in terms of their complex amplitude $a_L(t)$, including the amplitude envelope and frequency sweep of the laser pulse, both varying on a time scale longer than the cycle period, T. The phase φ, determining whether the oscillating electric field is maximum, minimum (zero) (Fig. 1-1), or something in between at the peak of the light pulse, later referred to as the carrier-envelope phase, has not been accessible. The paper in this section describes the first step towards experimentally accessing this phase by measuring its change upon circulation of the laser pulse inside the cavity of a femtosecond titanium-doped sapphire laser.

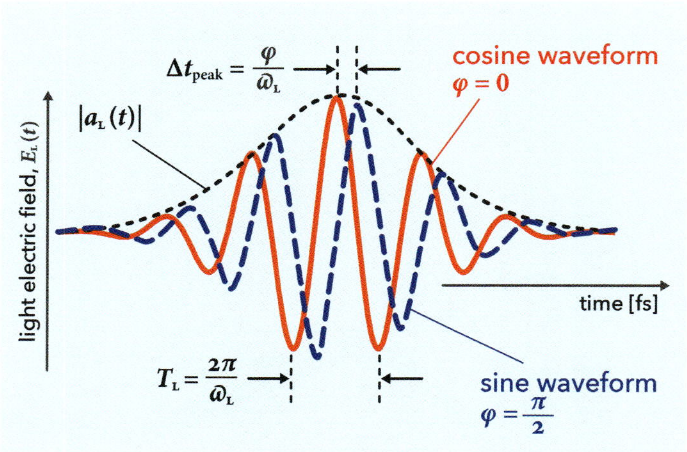

Figure 1-1 Laser pulses containing two oscillation cycles within the full width at half maximum of their intensity profile, with sine- and cosine-shaped waveforms.

[1] L. Xu, C. Spielmann, A. Poppe, T. Brabec, F. Krausz, T. Hänsch. Route to phase control of ultrashort light pulses. *Optics Letters* **21**, 2008 (1996).
(Page: 33)

OPTICS LETTERS / Vol. 21, No. 24 / December 15, 1996

Route to phase control of ultrashort light pulses

L. Xu, Ch. Spielmann, A. Poppe, T. Brabec, and F. Krausz

Abteilung Quantenelektronik und Lasertechnik, Technische Universität Wien, Gusshausstrasse 27, A-1040 Wien, Austria

T. W. Hänsch

Max-Planck-Institut für Quantenoptik, Hans-Kopfermann-Strasse 1, D-85748 Garching, Germany

Received July 12, 1996

A feasibility study of controlling the carrier phase in ultrashort light-wave packets emitted by a sub-10-fs laser is reported. An experimental apparatus capable of exploring the phase sensitivity of nonlinear-optical interactions is presented.

The electric field of pulsed radiation can generally be written as

$$E(t, z) = A(t, z)\exp[-i\omega_0 t + i\psi(z)] + \text{c.c.}, \qquad (1)$$

where $A(t, z)$ is the complex amplitude envelope of the pulse and the carrier frequency ω_0 is defined as the center of gravity of the frequency spectrum of $E(t, z = 0)$. The concept of the envelope can by physically meaningful for pulse durations that are comparable with a carrier oscillation cycle. If t is measured in a frame of reference retarded by z/v_g, where v_g is the group velocity, $\psi(z)$ defined by requiring $\text{Im}[A(t = 0, z)]$ has a simple physical meaning: It determines the position of the carrier wave with respect to the envelope. Whereas several techniques have been developed to obtain the complex envelope $A(t, z)$ of ultrashort light pulses, to our knowledge ψ has not been obtained. In light pulses of a duration comparable with the carrier oscillation cycle, a change in ψ significantly affects the evolution of $E(t)$. As a consequence, ψ is expected to become an essential parameter for nonperturbative (strong-field) interactions of few-cycle ultrashort light pulses with atomic systems. In this Letter we report on what is believed to be the first experiment that provides information on the evolution of the carrier phase in laser pulses and permits exploration of the influence of this parameter on atomic processes.

A prerequisite for exploiting the potential of phase control in nonlinear optics and high-field physics is the availability of a stationary train of ultrashort pulses with a fixed ψ. Mode-locked lasers generally fail to meet this requirement because the relative carrier phase ψ picks up a shift of

$$\Delta\psi = \phi_R - \omega_0 T_R \qquad (2)$$

on each round trip in the cavity. Here ϕ_R and T_R stand for the round-trip phase and the group retardation, respectively. We investigated the evolution of ψ in sub-10-fs wave packets emitted by a mode-locked Ti:sapphire ring oscillator.[1] To this end, we constructed a Michelson-type correlator (Fig. 1) to measure the interferometric cross correlation $G(\tau)$ of successive pulses from the laser.[2] $\Delta\psi - 2k\pi$ can be determined from $G(\tau)$ (Fig. 2), where k is a positive integer, which does not affect the physical properties of the wave packet.

First, we investigated the influence of dispersive material on $\Delta\psi$ by varying the path length l_g in a thin fused-silica glass wedge incorporated into the laser.[1] The filled circles in Fig. 3 show the measured variation of $\Delta\psi$ with l_g, and the line depicts $\Delta\psi(l_g) = \Delta\psi_0 + \phi(\omega_0) - \omega_0(\partial\phi/\partial\omega)_{\omega_0} = \Delta\psi_0 + 2\pi(\text{d}n/\text{d}\lambda)_{\lambda_0} l_g$. For fused silica $\text{d}n/\text{d}\lambda = 0.017\ \mu\text{m}^{-1}$ at 800 nm. The small change in l_g that changes $\Delta\psi$ by 2π permits $\Delta\psi$ to be controlled without notably affecting the pulse shape.

Stationary output pulses with a fixed relative carrier phase call for setting $\Delta\psi = 2k\pi$. However, this adjustment will freeze ψ for only a limited period because nonlinear contributions to ϕ_R and T_R [see Eq. (2)] translate fluctuations of the pulse parameters into small fluctuations of $\Delta\psi$, resulting in $\Delta\psi(T) = 2k\pi + \delta\psi(T)$. Over many round trips $\delta\psi$ can lead to a substantial jitter of the relative carrier phase $\psi(T) - \psi_0 = (1/T_R)\int_0^T \delta\psi(T')\text{d}T'$. The rms of this jitter, σ_ψ, can be expressed as

$$\sigma_\psi^2 = \langle[\psi(T) - \psi_0]^2\rangle = \frac{1}{\pi}\int_0^{\omega_R/2} \frac{S_{\delta\psi}(\omega')}{\omega'^2 T_R^2}\text{d}\omega', \qquad (3)$$

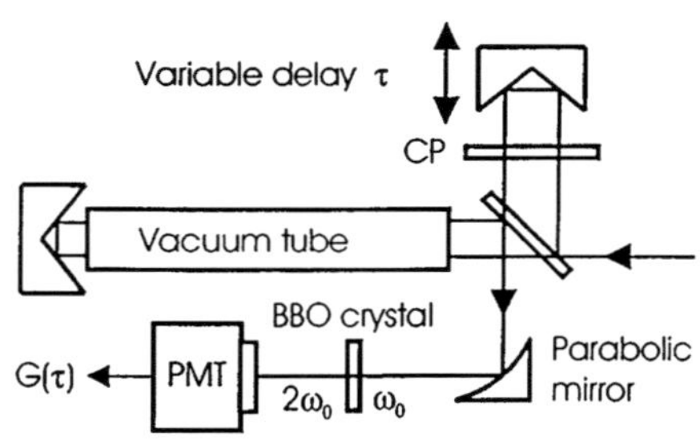

Fig. 1. Schematic diagram of the correlator used to the measure $\Delta\psi$. To balance dispersion in the correlator arms precisely, we evacuate a delay section equal to the resonator round-trip length in the long arm, a compensation plate (CP) is used to introduce the same amount of fused silica into the short arm as the tube windows introduce into the long arm, and pulse splitting and recombination are implemented by identical broadband dielectric coatings on opposite sides of the beam splitter. PMT, photomultiplier tube; BBO, β-barium borate.

0146-9592/96/242008-03$10.00/0

December 15, 1996 / Vol. 21, No. 24 / OPTICS LETTERS

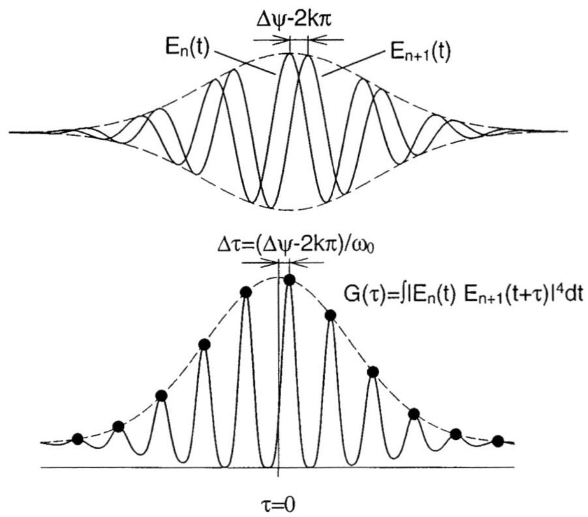

Fig 2. Principle of the measurement of $\Delta\psi$. $E_n(t)$ and $E_{n+1}(t)$ describe successive pulses from the laser. $\Delta\psi - 2k\pi$ can be determined from the position of the fringe peaks on the envelope of $G(\tau)$.

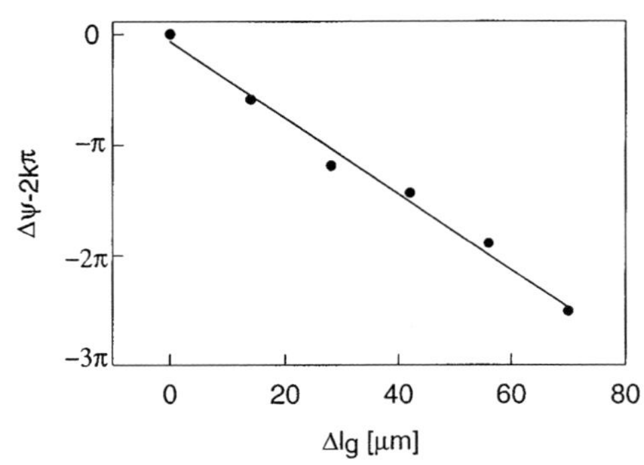

Fig. 3. Measured (filled circles) and calculated (line) changes in the round-trip carrier phase shift with increasing propagation length through the intracavity fused-silica wedge. The initial value of l_g was set to yield $\Delta\psi(\Delta l_g = 0) \approx 2k\pi$.

where the angle brackets denote time average, $\omega_R = 2\pi/T_R$, and $S_{\delta\psi}(\omega)$ is the power spectral density of $\delta\psi(T)$. Replacing the lower boundary of integration by $\omega(0 < \omega < \omega_R/2)$ yields $\sigma_\psi(\omega)$, the rms jitter of ψ that originates from noise components of frequencies higher than ω. $\sigma_\psi(\omega)$ is expected to decrease rapidly with increasing ω. The critical frequency ω_{crit} at which $\sigma_\psi(\omega)$ drops below a fraction of π determines the dephasing rate of the emitted wave packets and hence the speed of optical control required for phase stabilization.

Because $S_{\delta\psi}(\omega)$ cannot be directly measured, we relate $\delta\psi(T)$ to variations of the pulse parameters. This relationship will permit an evaluation of $S_{\delta\psi}(\omega)$ from the noise spectra of the respective pulse parameters. Owing to a short soliton period, the solitonlike pulse circulating in a sub-10-fs Ti:sapphire oscillator adiabatically follows changes in the laser parameters that occur on a long time scale compared with T_R. According to soliton perturbation theory,[3] under these circumstances two invariants, the changes in pulse energy $\delta W(T)$ and in the center frequency $\delta\omega_0(T)$, along with external noise sources, govern the variation of other pulse parameters. This also holds for the relative carrier phase, which can be written as

$$\delta\psi(T) = \phi_{\text{Kerr}}\left(\frac{\mathrm{d}P_{\text{peak}}}{\mathrm{d}W}\right)_{W_0} \delta W(T) - \omega_0 D\,\delta\omega_0(T)\,. \quad (4)$$

The first term is a perturbation to ϕ_R and accounts for the soliton phase shift that is due to the optical Kerr effect in the gain medium, where ϕ_{Kerr} gives the phase shift per unit power and P_{peak} is the peak power of the pulse inside the gain medium. The second term results from a variation of T_R owing to a shift of the laser spectrum and scales with D, the net intracavity group delay dispersion (GDD). The influence of other noise sources (e.g., cavity length fluctuations) on $\delta\psi(T)$ can be shown to be negligible.

The filled squares and triangles in Fig. 4 represent the measured variation of $\Delta\psi$ with intracavity pulse energy for two different values of the intracavity GDD. Our measurements yield a $\delta\psi$ that decreases with increasing pulse energy. This is in apparent contradiction with the positive sign of the soliton phase shift that is due to $\phi_{\text{Kerr}} > 0$. One can resolve the paradox by assuming a coupling between δW and $\delta\omega_0$. As a matter of fact, the evaluation of the center of gravity of the corresponding laser spectra reveals an appreciable variation of ω_0 with W, as shown by the filled circles in Fig. 4.[4] For increasing pulse energy, the negative cavity GDD translates the red shift of ω_0 into a contribution of negative sign that dominates over the positive soliton phase shift.

Integration of Eq. (4) yields

$$\Delta\psi(W) - \Delta\psi(W_0) \approx \phi_{\text{Kerr}}[P_{\text{peak}}(W) - P_{\text{peak}}(W_0)] - \omega_0(W_0)D[\omega_0(W) - \omega_0(W_0)]\,, \quad (5)$$

which permits a comparison of the experimental results with the prediction. Using the measured values of the pulse width $\tau_p(W, D)$, which vary between 8 and 14 fs, one can calculate $P_{\text{peak}}(W, D)$ for both settings of

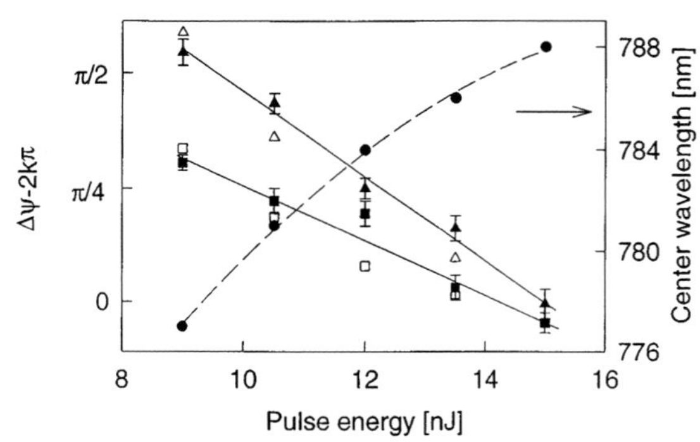

Fig. 4. Measured round-trip carrier phase shift for different values of the cavity GDD (filled squares and triangles) and center of the laser spectrum (filled circles) as a function of intracavity pulse energy. The open squares and triangles depict $\Delta\psi - 2k\pi$ obtained from Eq. (4) as described in the text. The solid lines and the dashed curve are guides to the eye.

OPTICS LETTERS / Vol. 21, No. 24 / December 15, 1996

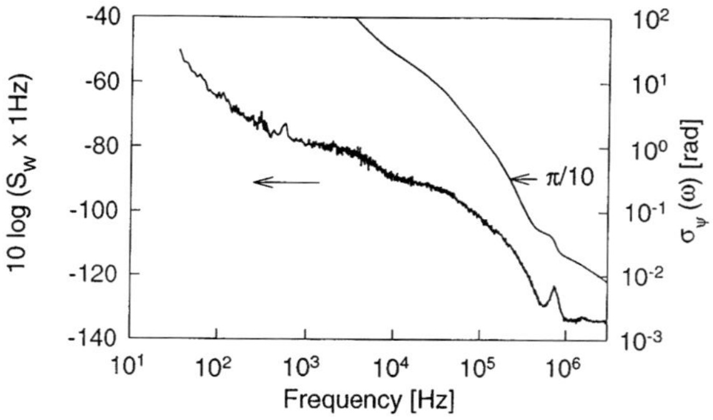

Fig. 5. Power spectral density $S_W(\omega)$ of $\delta W(T)/W_0$ (normalized to a 1-Hz bandwidth) on a logaritmic scale and the calculated $\sigma_\psi(\omega)$ as defined in the text.

the cavity GDD by taking into account the variation of τ_p inside the resonator. $\omega_0(W)$ is directly obtained from $\lambda_0(W)$, as shown in Fig. 4. Inserting $P_{\text{peak}}(W, D)$, $\omega_0(W)$, and $\phi_{\text{Kerr}} = 0.4 \times 10^{-6}\ \text{W}^{-1}$ (Ref. 2) into expression (5) leads to the values represented by the open squares and triangles in Fig. 4 for $D_1 = -17\ \text{fs}^2$ and $D_2 = -24\ \text{fs}^2$, respectively. These values of the cavity GDD were obtained from least-squares fits to the measured data (filled squares and triangles in Fig. 4). They are consistent with previously evaluated values of the cavity GDD,[1] and $D_1 - D_2$ is reasonably close to 5 fs^2, which we obtained from the path-length change $l_{g,1} - l_{g,2}$ in the fused-silica wedge.

Having identified pulse-energy fluctuations as the primary source of $\delta\psi$ by measurement of $S_W(\omega)$, the power spectrum of $\delta W(T)/W_0$, we can determine $S_{\delta\psi}(\omega)$ by using $\delta\psi = (\text{d}\Delta\psi/\text{dW})_{W_0}\delta W$ in the adiabatic approximation. This relation yields $S_{\delta\psi}(\omega) = \alpha S_W(\omega)$, where $\alpha = [W_0(\text{d}\Delta\psi/\text{dW})_{W_0}]^2$. We obtained $S_W(\omega)$ by measuring the sideband of the first harmonic of the laser power spectrum with a rf spectrum analyzer[5] and is shown in Fig. 5. For $D = -17\ \text{fs}^2$ and $W_0 = 15$ nJ we evaluate $\alpha = 7.5$ from the filled squares in Fig. 4. Substitution of $S_{\delta\psi}(\omega) = \alpha S_W(\omega)$ into Eq. (3) and replacing the upper boundary of integration by ω yield the rms carrier phase jitter $\sigma_\psi(\omega)$ depicted in Fig. 5. It drops to $\pi/10$ at $\omega_{\text{crit}}/2\pi \approx 2 \times 10^5$ Hz; hence we may conclude that a sub-10-fs Ti:sapphire laser is capable of generating trains consisting of $>10^3$ stationary light pulses with a fixed carrier phase, provided that the nominal value of $\Delta\psi(W_0, l_g)$ is set equal to $2k\pi$. The fact that notable noise-induced dephasing occurs on a time scale longer than 5 μs implies that the 3-μs gain relaxation time of Ti:sapphire allows one to control the intracavity pulse energy at a sufficiently high speed to prevent dephasing of the few-cycle wave packets. As a consequence, our experiments lead us to the conclusion that the physical characteristics of a sub-10-fs Ti:sapphire oscillator potentially permit the generation of phase-stabilized light pulses.

The most important consequence of the above investigations is that the experimental system that is presented offers the possibility of exploring the influence of ψ on light–matter interactions. As noise-induced dephasing occurs on a microsecond time scale, setting $\Delta\psi - 2k\pi$ equal to π/N, i.e., a submultiple of π (with N being currently limited to $\approx$20 by the measurement accuracy of $\Delta\psi$) will result in a periodic modulation of ψ of the output pulses at the corresponding submultiple $1/2NT_R$ of the pulse-repetition rate. Identifying and quantifying a corresponding modulation in any of the physical measurables of the irradiated atomic system will provide direct information on the role of the phase of the light field in the respective interaction. One intriguing example is photoemission in the optical tunneling regime.[6] The observation of a modulation of the photocurrent at $1/2NT_R$ may provide not only the first direct evidence for optical tunneling but also the ψ-dependent signal needed to control the carrier phase in laser pulses. These prospects together with recent progress in approaching the single-cycle limit in optical pulse generation[7] hold out the promise of exploring and controlling atomic processes to an unprecedented depth by use of precisely controlled electromagnetic transients.

We thank R. Szipöcs and K. Ferencz (Research Institute for Solid State Physics, Budapest, Hungary) for providing the special dispersive mirrors used in the Ti:sapphire laser. This research was supported by Fonds zur Förderung der wissenschaftlichen Forschung grants P-09710 and P-10409, and by Österreichische Nationalbank grants 5335 and 5124.

References

1. L. Xu, Ch. Spielmann, F. Krausz, and R. Szipöcs, Opt. Lett. **21,** 1259 (1996).
2. K. L. Sala, G. A. Kenny-Wallace, and G. E. Hall, IEEE J. Quantum Electron. **16,** 990 (1980).
3. H. A. Haus and A. Mecozzi, IEEE J. Quantum Electron. **29,** 983 (1993).
4. Our computer simulations revealed that the coupling between δW and $\delta\omega_0$ relates to the rapidly decreasing magnitude of negative cavity GDD below 0.7 μm (Ref. 1) and to the finite response time of the Kerr nonlinearity.
5. D. von der Linde, Appl. Phys. B **39,** 201 (1986).
6. L. V. Keldysh, Sov. Phys. JETP **47,** 1307 (1964).
7. M. Nisoli, S. De Silvestri, O. Svelto, R. Szipöcs, K. Ferencz, S. Sartania, Ch. Spielmann, and F. Krausz, "Compression of high energy pulses below 5 fs," submitted to Opt. Lett.

Newport U100
U L T I M

Confining Strong Light Fields to the Wave Cycle

2

Opposite: Chirped multilayer mirrors reflecting light over the wavelength range of 550 - 1100 nm and imposing a wavelength-dependent phase shift on the reflected light. The resultant wavelength-dependent group delay compresses octave-spanning visible-near-infrared laser pulses to their Fourier limit, to a duration approaching their wave cycle.

Ultrashort-pulse laser techniques have revolutionized as well as spawned several fields of laser science, such as nonlinear optics, femtochemistry and precision spectroscopy, each of which has been recognized with a Nobel Prize. In all these cases, it has been the amplitude envelope and frequency sweep of the laser pulse, both encoded in the complex amplitude, $a_L(t)$, that have fully determined and governed the underlying light-matter interaction driven by the electric field of the laser pulse

$$E_L(t) = \frac{1}{2} E_{peak} a_L(t) e^{-i(2\pi t/T+\varphi)} + c.c.$$

As a matter of fact, processes scrutinized or utilized in these disciplines have typically occurred in the perturbative regime of light-matter interactions, where the electric field "shakes" electrons with sub-atomic-scale amplitude over multiple field cycles. Due to the interaction extending over several field cycles, the phase φ is neither relevant nor measurable.

Strong light fields substantially confined in time to a single wave period dramatically change the nature of the interaction with matter. A force field rivaling the strength of the intra-atomic Coulomb fields, which is "switched on" abruptly within a wave cycle or two, is able to induce large-scale (multi-atomic-unit) electron excursion within an individual wave cycle. This is exemplified in above-threshold ionization and high-order harmonic generation. In contrast to perturbative interactions, the sub-cycle temporal evolution of the force field becomes decisive under these circumstances. This is increasingly affected by the value of φ for ever shorter (few-cycle to single-cycle) pulses.

By the mid 1990's, "strong" and "temporally confined" conditions, to the extent described above, were met through several advances in ultrafast laser technology. In this chapter a review of the factors that contributed to those advances is presented.

The invention of titanium-doped sapphire (Ti:Sa) as an efficient broadband gain medium by Peter Moulton in the 1980's opened the door for the generation of shorter pulses with higher peak and average power. This new pulse generation ability in a solid-state gain medium, in contrast to the dye lasers that constituted the primary sources of femtosecond pulses for two decades (1970–1990), was based on a near-instantaneous nonlinear interaction. Unlike dye lasers, solid-state gain media, due to their relaxation time being much longer than the cavity round-trip time, are unable to actively participate in

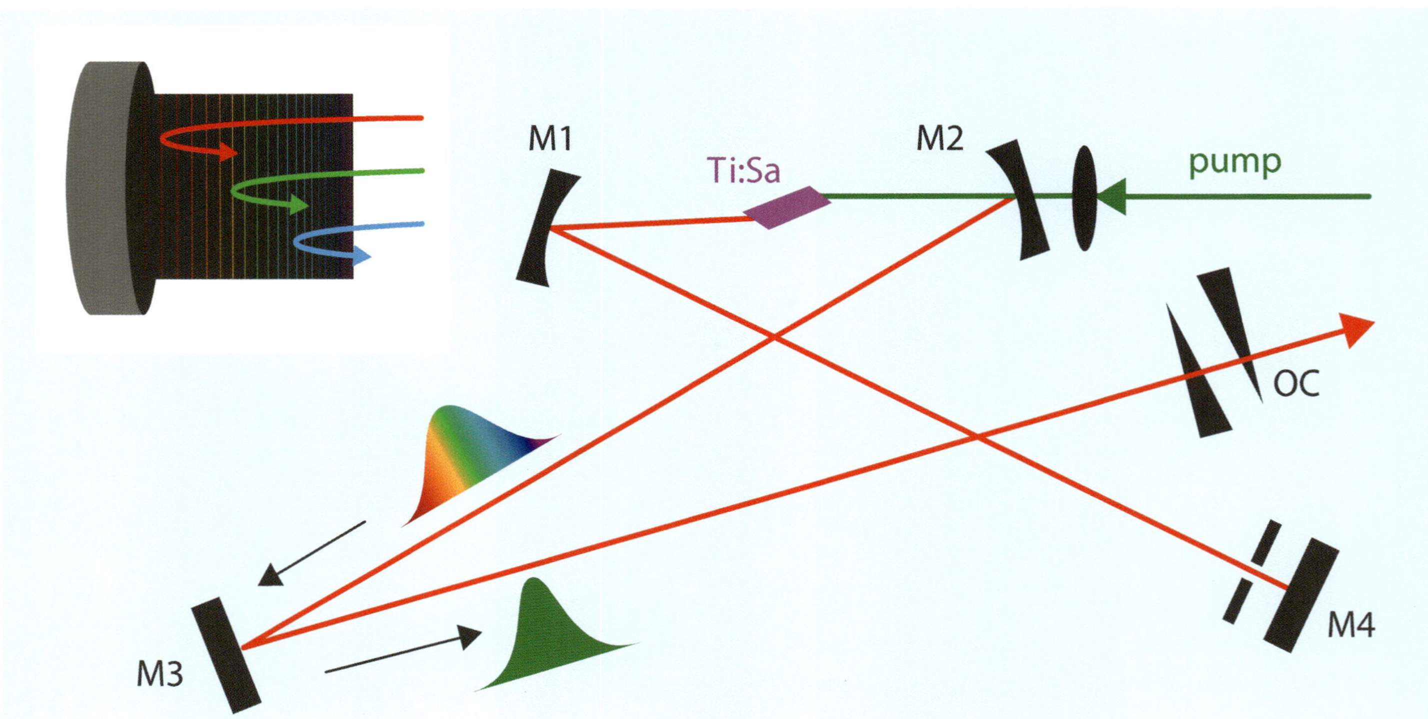

Figure 2-1 Kerr-lens mode-locked mirror-dispersion-controlled titanium-doped sapphire laser generating laser pulses shorter than 10 femtoseconds [10]. Mirrors M1–M4 not only provide high-reflectivity but also controlled group-delay dispersion, thanks to their chirped multilayer structures [8-9]. OC, output coupler; Ti:Sa, titanium-doped sapphire.

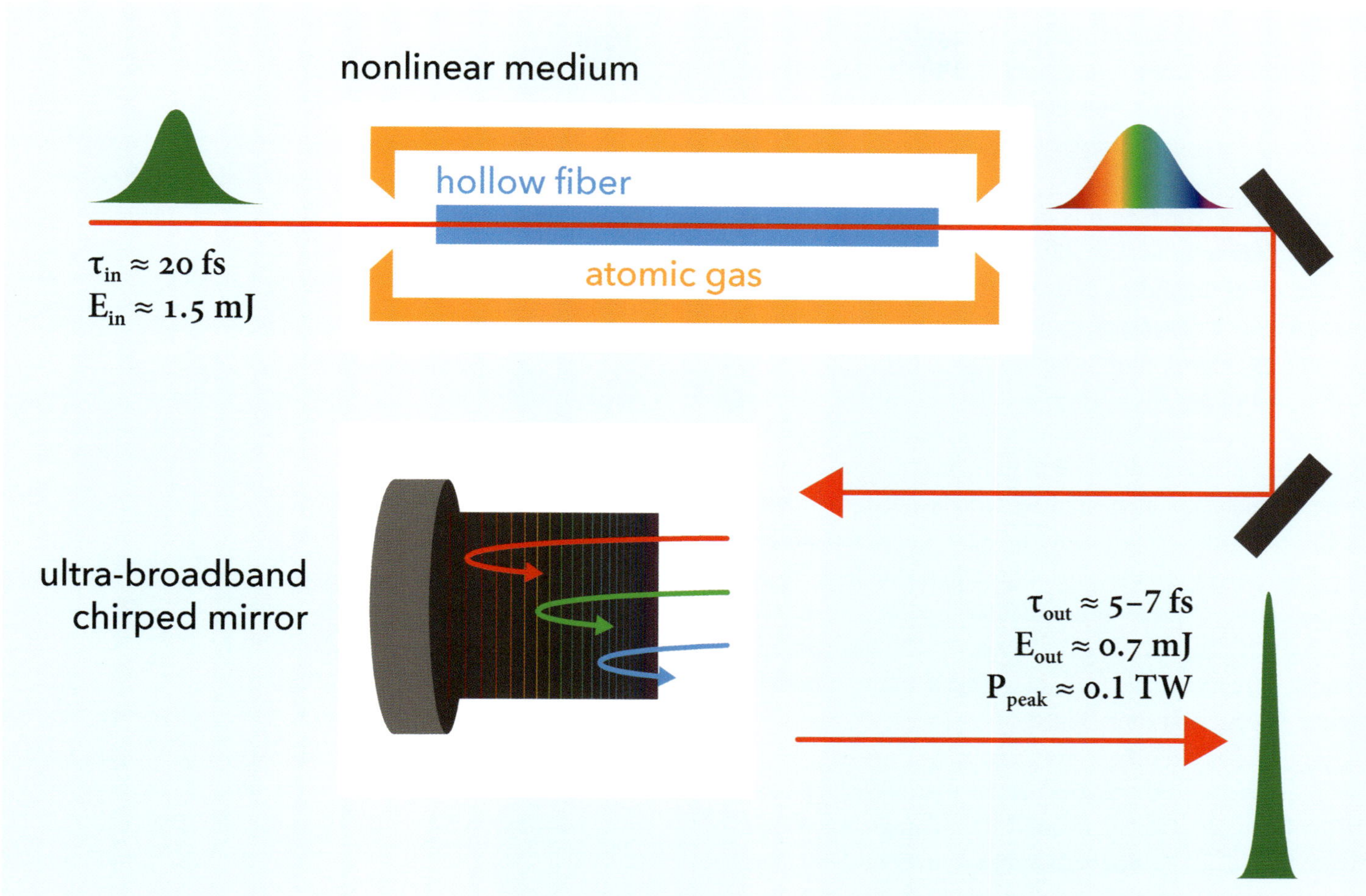

Figure 2-2 Nonlinear compression of millijoule-scale powerful laser pulses delivered by kHz-rate Ti:sapphire laser amplifiers. In a first step, self-phase modulation induced by the optical Kerr effect in a gas-filled hollow fiber broadens the spectrum of the pulses passing through the nonlinear waveguide. The spectrally-broadened, chirped output pulses are temporally compressed to a pulse duration of several femtoseconds in chirped multilayer mirrors [10].

shortening the pulse circulating in a mode-locked laser to the femtosecond regime. Passively mode-locked solid-state lasers rely on a near-instantaneous nonlinear effect for modulating the circulating radiation, as its amplitude is a function of its instantaneous intensity.

The most successful implementation of such amplitude-modulating nonlinearity in bulk solid-state lasers has been Kerr-lens mode locking (KLM), invented by Wilson Sibbett and his co-workers. KLM is inextricably linked to, and hence comes in combination with, self-phase modulation (SPM). In addition, unavoidable frequency-dependent group delay (group-delay dispersion or GDD) within the laser resonator requires precise compensation over ever-increasing bandwidth for ever-shorter pulse generation. These effects, KLM, SPM and GDD, need to be optimized and mutually balanced for stable, reliable femtosecond pulse generation.

With reprints [2-7], co-authored with my first PhD students, Thomas Brabec and Christian Spielmann, and my first post-doctoral co-worker, Peter Curley, we review the emergence of this technology, provide the relevant theoretical insight, and document how the ever-improving exploitation of these nonlinearities along with GDD control over ever increasing bandwidth led to uncompensated high-order dispersion becoming a limiting factor. In collaboration with Robert Szipöcs and Kárpát Ferencz from Budapest, this limitation was overcome with the invention of aperiodic multilayer mirrors, or chirped mirrors (CM), in 1993. Modulating the period of thin-film multilayer structures enabled the production of laser mirrors that provide not only high reflectivity but also tailored group-delay dispersion (GDD) over a wide frequency band. The first mirrors of this kind are described in [8-9]. Shortly after their first demonstration, CMs permitted the generation of pulses from Ti:sapphire laser oscillators well below 10 fs, as reported by Andreas Stingl and his co-workers in [10] (Fig. 2-1).

The subsequent systematic improvement of chirped mirrors and KLM oscillators using them, led to the generation of sub-5-fs pulses directly from laser oscillators. These constitute the most cutting edge ultrashort-pulse laser oscillators to date.

In order to combine this extreme temporal confinement with field strengths comparable to those in intra-atomic fields, the energy of the pulses delivered by the oscillators can be increased via chirped-pulse amplification (CPA), which was invented by Gérard Mourou and Donna Strickland, for which they were awarded the 2018 Nobel Prize in Physics. Although this process induces some temporal broadening, as revealed in [11], self-phase modulation in gas-filled hollow-core waveguides, which was pioneered by Orazio Svelto and Sandro De Silvestri, can subsequently broaden the spectrum of the amplified pulses. Application of the CM technology for chirp compensation of these ultra-broadband pulses, as described by Mauro Nisoli and his co-workers in [12], has pushed the frontier of femtosecond technology close to the single-cycle regime for the first time (Fig. 2-2).

These are the technologies that have permitted the substantial confinement of strong force fields to a single wave cycle in the 1990's and have since been routinely used for (i) exploiting extreme optical nonlinearities that are otherwise inaccessible during multi-cycle interactions and (ii) controlling the sub-cycle evolution of strong optical fields. Attosecond science and technology rely on both of these techniques, which will be reviewed in the next two chapters.

Femtosecond pulse generation from solid-state lasers

[2] F. Krausz, T. Brabec, C. Spielmann. Self-starting passive mode locking. *Optics Letters* **16**, 235 (1991).
(Page: 42)

[3] T. Brabec, C. Spielmann, F. Krausz. Mode locking in solitary lasers. *Optics Letters* **16**, 1961 (1991).
(Page: 45)

[4] T. Brabec, C. Spielmann, F. Krausz. Limits of pulse shortening in solitary lasers. *Optics Letters* **17**, 748 (1992).
(Page: 48)

[5] T. Brabec, C. Spielmann, P. F. Curley, F. Krausz. Kerr lens mode locking. *Optics Letters* **17**, 1292 (1992).
(Page: 51)

[6] P. F. Curley, C. Spielmann, T. Brabec, F. Krausz, E. Wintner, A. J. Schmidt. Operation of a femtosecond Ti:sapphire solitary laser in the vicinity of zero group-delay dispersion. *Optics Letters* **18**, 54 (1993).
(Page: 54)

[7] C. Spielmann, P. F. Curley, T. Brabec, F. Krausz. Ultrabroadband femtosecond lasers. *Journal of Quantum Electronics* **QE-30**, 1100 (1994).
(Page: 57)

Approaching the wave cycle: chirped multilayer mirrors control dispersion

[8] R. Szipöcs, K. Ferencz, C. Spielmann, F. Krausz. Chirped multilayer coatings for broadband dispersion control in femtosecond lasers. *Optics Letters* **19**, 201 (1994).
(Page: 72)

[9] R. Szipöcs, A. Stingl, C. Spielmann, F. Krausz. Pushing the limits of femtosecond technology: chirped dielectric mirrors. *Optics and Photonics News* **6**, 16 (1995).
(Page: 75)

[10] A. Stingl, M. Lenzner, C. Spielmann, F. Krausz, R. Szipöcs. Sub-10-fs mirror-dispersion-controlled Ti:sapphire laser. *Optics Letters* **20**, 602 (1995).
(Page: 81)

[11] M. Lenzner, C. Spielmann, E. Wintner, F. Krausz, A. J. Schmidt. Sub-20-fs, kilohertz-repetition-rate Ti:sapphire amplifier. *Optics Letters* **20**, 1397 (1995).
(Page: 84)

[12] M. Nisoli, S. De Silvestri, O. Svelto, R. Szipöcs, K. Ferencz, C. Spielmann, S. Sartania, F. Krausz. Compression of high-energy laser pulses below 5 fs. *Optics Letters* **22**, 522 (1997).
(Page: 87)

February 15, 1991 / Vol. 16, No. 4 / OPTICS LETTERS

Self-starting passive mode locking

F. Krausz, T. Brabec, and Ch. Spielmann

Abteilung Quantenelektronik und Lasertechnik, Technische Universität Wien, Gusshausstrasse 27, A-1040 Vienna, Austria

Received July 11, 1990; accepted November 19, 1990

We investigate the evolution of continuous-wave laser oscillation from free-running to mode-locked operation assuming a nonlinear device with an intensity-dependent transmittivity or reflectivity to be the mode-locking element. An intensity threshold for self-starting passive mode locking is predicted and related to the linewidth of the first beat note of the power spectrum of the free-running laser output. Experimental results confirm the predictions of the theory.

Passive mode locking of cw lasers has been a reliable technique for subpicosecond and femtosecond optical pulse generation. A common feature of these lasers is the existence of a threshold intracavity power required for self-starting mode locking. Laser media with short relaxation times (dyes, color centers) can efficiently participate in the pulse-shaping process themselves and therefore have low thresholds when mode locked with slow saturable absorbers. By contrast, solid-state laser materials with long upper-state lifetimes are incapable of taking part in pulse formation and necessitate fast-response nonlinear devices for mode locking. Although several nonlinear elements other than the conventional fast saturable absorber have been devised,[1–4] until recently passive mode locking could be demonstrated only in pulsed systems because of the high intensities needed to produce sufficient nonlinearity for coupling the axial modes. The exploitation of the Kerr nonlinearity of a single-mode optical fiber inserted in a coupled cavity has resulted in the first successful cw passive mode locking of a solid-state laser.[5] The technique has been termed additive-pulse mode locking[6] (APM). Owing to the tremendous enhancement in the efficiency of the nonlinear processes in optical fibers, self-starting APM has been easily achieved at intracavity powers as low as a few hundred milliwatts.[7,8] More recently, self-starting cw passive mode locking was also achieved with a bulk nonlinearity in a high-power laser.[9]

One of the most important questions in the development of new passive mode-locking techniques and in the design of cw passively mode-locked solid-state lasers is the cw intracavity power needed for the laser to become mode locked with a given nonlinear device or, conversely, the necessary nonlinearity that has to be introduced in a cw laser for self-starting mode locking with the available intracavity power. The purpose of this Letter is to derive a condition for self-starting cw passive mode locking of solid-state lasers by using a nonlinear element with an intensity-dependent transmittivity or reflectivity. We investigate the evolution of laser oscillation from the free-running state toward the steady-state mode-locked operation. Since the intensity-induced change in the round-trip loss is low during most of the transient pulse evolution, it may be expressed as a linear function of the intracavity intensity as long as its response time is short compared with the mode-locked pulse width. This approximation permits a unified characterization of different nonlinear devices and thus a unified treatment of the transient mode-locking process.

The evolution of the circulating radiation in a homogeneously broadened laser is governed by the equation[10]

$$\frac{\partial}{\partial T}v(t, T) = \frac{1}{2T_R}\left(g - l - q + D\frac{\partial^2}{\partial t^2}\right)v(t, T), \qquad (1)$$

where $v(t, T)$ is the electric-field amplitude of the circulating electromagnetic wave at a fixed position on the cavity axis as a function of the fast time parameter $t(0 \le t < T_R)$ and the slow time parameter $T(T \gg T_R)$. T_R is the round-trip time, l and g are the round-trip intensity loss and gain, respectively, q describes the effect of nonlinearity employed for pulse forming, and the differential operator accounts for dispersion. D represents the bandwidth limitation due to the finite gain bandwidth and other possible bandwidth-limiting elements.

The mode-locked laser pulse will evolve from the most intensive mode-beating fluctuation present in the multimode free-running laser. Neglecting the smaller fluctuations, we can write the circulating photon flux as $|v|^2 = S(T) + s(t, T)$, where $S(T)$ is the slowly varying average background intensity and $s(t, T)$ is the most intensive mode-beating fluctuation in the cavity. The intensity is normalized to the photon energy $\hbar\omega$ and is measured in the gain medium. The nonlinearity used for mode locking[1–9] causes an intensity-dependent change in the cavity round-trip loss, which is, to first order, given by

$$q(t, T) = -\kappa s(t, T). \qquad (2)$$

The S-dependent part of q is incorporated in l. Equation (2) applied to all types of passive absorberlike mode locker until the intensity becomes high and/or the pulse duration becomes comparable with the response time of the nonlinearity. κ is proportional to A_g/A_q, where A_g and A_q represent the effective beam cross sections in the gain and nonlinear medium, re-

0146-9592/91/040235-03$5.00/0

OPTICS LETTERS / Vol. 16, No. 4 / February 15, 1991

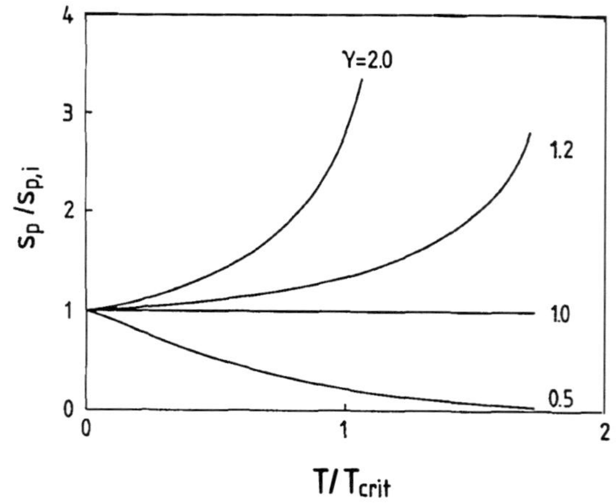

Fig. 1. Evolution of the normalized peak intensity $s_p/s_{p,i}$ as a function of the slow time parameter T for different values of $\gamma = \tau_c/T_{crit}$ at a fixed T_{crit}.

spectively. Sometimes it is useful to introduce $\kappa' = \hbar\omega A_g\kappa$, which is characteristic of the nonlinearity alone and relates q to the circulating intracavity power. In typical APM lasers using an $\approx$1-m length of single-mode fiber, $\kappa' \approx 10^{-4}\ \mathrm{W}^{-1}$. As a rule, κ' is orders of magnitude smaller for bulk nonlinearities.

Instead of a complete characterization of the evolution of passive mode locking, we restrict ourselves to following the development of the peak intensity $s_p(T) = s(0, T)$ of the fluctuation. This parameter is a good measure of the evolution of mode locking even in the last stage when the pulse energy $E_p(T)$ is approximately constant. From Eqs. (1) and (2) we can obtain after some manipulations the rate of the peak intensity growth of the perturbation,

$$\frac{\mathrm{d}s_p}{\mathrm{d}T} = \frac{1}{T_R}\left[\left(\kappa - \frac{1}{2}\alpha\sigma g_0\tau_p\right)s_p - \beta\frac{4D}{\tau_p^{\,2}} - l_0\right]s_p, \qquad (3)$$

where $\tau_p(T)$ is the pulse duration and $\alpha = s_p\tau_p/E_p$ and β are pulse-shape-dependent numerical factors of order unity. For a Gaussian pulse shape, $\alpha = 1.06$ and $\beta = 0.70$. The term containing the stimulated-emission cross section σ accounts for dynamic gain saturation.[11] For brevity, we introduced $l_0(T) = l - g_0$, where g_0 is the quasi-stationary gain. If we assume an absence of relaxation oscillations, the average intensity is approximately constant during the pulse evolution, i.e., $T_R(\mathrm{d}S/\mathrm{d}T) + \mathrm{d}E_p/\mathrm{d}T = 0$. In the early stage of the transient mode-locking process $E_p \ll T_RS$, and therefore $l_0 \approx 0$. The incipient growth of s_p is not affected by gain dispersion, because the initially oscillating modes experience the same round-trip gain. Thus $\mathrm{d}s_p/\mathrm{d}T > 0$ if $\kappa > \alpha\sigma g_0\tau_{p,i}/2$ $[\tau_{p,i} = \tau_p(0)]$, a condition similar to that obtained for $\mathrm{d}E_p/\mathrm{d}T > 0$ by other authors recently.[11] When dynamic gain saturation is negligible, Eq. (3) can be solved, and we find that $s_p \to \infty$ as $T \to T_{crit}$, where $T_{crit} = T_R/\kappa s_{p,i}$ and $s_{p,i} = s_p(0)$. The reaching of this point is prevented by the onset of dispersion owing to the shortened pulse duration τ_p at high peak intensities.

According to these results, mode locking is, regardless of the laser intensity, self-starting if the dynamic gain saturation is small enough compared with the nonlinearity. This is in contradiction with experimental observations. The insufficiency of Eq. (3) becomes obvious in the absence of nonlinearity and dynamic gain saturation (κ, $\sigma = 0$). For this case Eq. (3) predicts that s_p = const., which implies that the ratios of the complex mode amplitudes of the oscillating modes $V_k(t)/V_l(t)$ should be constant. In reality, however, the axial modes oscillating in the free-running laser have a finite correlation time τ_c. During this characteristic time the cavity modes tend to lose mutual coherence by random perturbations of the phase differences $(\phi_k - \phi_l)$ and/or the ratios of amplitudes $|V_k/V_l|$. Recalling that s_p is the most intensive fluctuation in the cavity, i.e., it originates from dominantly constructive interference of cavity modes, we conclude that s_p will decay with a characteristic lifetime τ_c. Simultaneously, an intensive fluctuation arises at another instant within the time interval $0 \le t < T_R$. We assume $s_p(T)$ to decrease exponentially with a time constant τ_c in the absence of nonlinearity. This corresponds to an additional term having the form $-s_p/\tau_c$ on the right-hand side of Eq. (3). This approximation may be justified by the ability of the model to predict experimental results. The fundamental equation governing the development of s_p at the early stage of the transient mode-locking process can now be written as

$$\frac{\mathrm{d}s_p}{\mathrm{d}T} = \frac{\kappa}{T_R}s_p^{\,2} - \frac{s_p}{\tau_c}, \qquad (3')$$

where dynamic gain saturation is neglected. The solution of Eq. (3′) is given by

$$s_p(T) = \frac{s_{p,i}}{\gamma + (1-\gamma)\exp(T/\tau_c)}, \qquad (4)$$

where $\gamma = \tau_c/T_{crit}$. Figure 1 shows $s_p(T)$ for different values of the parameter γ. $s_p(T)$ goes to zero for $\gamma < 1$, and $s_p(T) \to \infty$ for $\gamma > 1$. Thus self-starting mode locking requires that $\gamma > 1$. The development of $s_p(T)$ in this case is shown on a larger scale in Fig. 2. The dashed curves are hypothetical representations of the

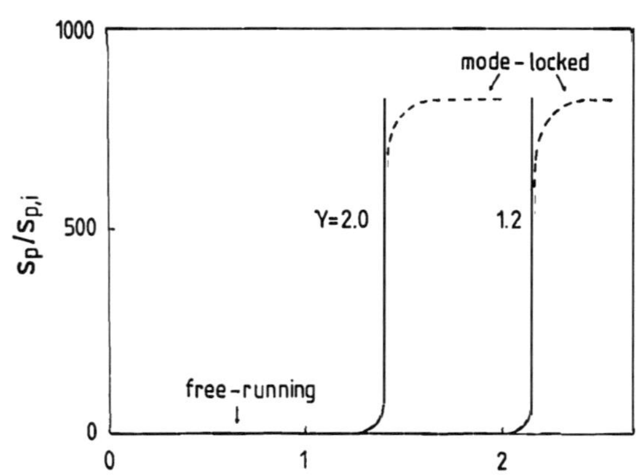

Fig. 2. Development of $s_p/s_{p,i}$ for different values of γ (>1) at a fixed T_{crit}. The dashed curves represent the effect of dispersion due to bandwidth limitation.

February 15, 1991 / Vol. 16, No. 4 / OPTICS LETTERS

effect of dispersion at the end of the pulse evolution. With an initial number of modes m_i, the peak intensity of the most intensive fluctuation in the cavity is given by $s_{p,i} \approx S_i \ln(m_i)$,[12] where $S_i = S(0)$ is the average photon flux in the free-running laser. Making use of this relationship, we can write the condition for self-starting as

$$\kappa' P_i > \frac{1}{\ln(m_i)} \frac{T_R}{\tau_c}, \tag{5}$$

where P_i is the circulating intracavity power in the free-running laser. Note that increasing m_i does not necessarily reduce the self-starting threshold because it generally gives rise to a reduction of τ_c.

Quantitative prediction of the threshold for $\kappa' P_i$ calls for the knowledge of τ_c. In order to obtain a rigorous definition of τ_c we introduce the correlation function

$$G_{kl}(\tau) = \langle V_k(t) V_l^*(t) V_k^*(t+\tau) V_l(t+\tau) \rangle \cos \omega_{kl}\tau, \tag{6}$$

where the angle brackets denote time average. Furthermore $\omega_{kl} = |k - l|\Delta\omega_0 + \delta\omega_{kl}$, where $\Delta\omega_0$ is the mode spacing at the line center and $\delta\omega_{kl}$ is some nonlinear frequency shift that depends on k and l rather than only on their difference due to nonlinear (quadratic or higher order) dispersion inside the cavity. The power spectrum of the laser output at the nth beat note is given by

$$P_n(\omega) = \frac{1}{\pi} \sum_{k-l=n} \int_{-\infty}^{+\infty} G_{kl}(\tau) \exp(i\omega\tau) d\tau. \tag{7}$$

Cavity length fluctuations do not perturb the initial fluctuation, i.e., they do not affect τ_c, but they do cause random shifts of $P_n(\omega)$, the magnitude of which is proportional to n. It is the smallest and usually negligible at $n = 1$, the first beat note. Therefore we use $P_1(\omega)$ for the definition of τ_c. If we assume $P_1(\omega)$ to have a Lorentzian line shape, the sum of the cross-correlation functions of the adjacent axial modes decays exponentially with a time constant, which we identify with τ_c and is given by

$$\tau_c = \frac{1}{\pi} \frac{1}{\Delta\nu_{3\,\mathrm{dB}}}, \tag{8}$$

where $\Delta\nu_{3\,\mathrm{dB}}$ is the 3-dB full width of the first beat note. Beat note line shapes other than Lorentzian result only in a slight change in the proportionality factor $1/\pi$.

Relations (5) and (8) now offer the possibility of testing our simple model by measurements. First, the general trends predicted by relation (5) have been investigated by using an APM Nd:glass laser. We found, in agreement with the results of others,[7] that a reduced κ' requires a higher P_i for self-starting and vice versa. A decrease in T_R definitely reduced the intensity needed for self-starting at a fixed κ'. In the Nd:phosphate glass laser with $T_R = 10$ nsec and $m_i \approx 1.5 \times 10^2$ we obtained 2.1 kHz $< \Delta\nu_{3\,\mathrm{dB}} <$ 3.5 kHz for different cavity parameters.[13] The self-starting threshold is predicted to take values of $1.3 \times 10^{-5} < (\kappa' P_i)_{\mathrm{th}} < 2.2 \times 10^{-5}$. Direct measurements of the threshold power yielded $2.2 \times 10^{-5} < (\kappa' P_i)_{\mathrm{th}} < 3.3 \times 10^{-5}$. In considering experimental uncertainties and the simplicity of the model, the predictions of the above analysis appear to be in good qualitative and quantitative agreement with experiments. Note that any kind of phase modulation has been left out in the approach presented. Actually, if the nonlinearity causes some chirp to accumulate on the circulating pulse, significant group-velocity dispersion in the cavity may considerably influence the transient pulse evolution. By careful design, one can take advantage of these additional pulse-shaping effects.

The presented condition for self-starting passive mode locking is not only an interesting general relationship between various laser parameters but also a useful tool in the design and development of cw passively mode-locked solid-state lasers. Simple spectral measurements of the free-running laser output permit predictions of the necessary amount of nonlinearity for a particular laser or the power requirements when a given nonlinearity is used. Profound understanding of the origins of the finite beat note linewidth necessitate further experimental and theoretical investigations. The results of this future research might be especially useful for cw passive mode locking of low- and moderate-power (e.g., diode-pumped) solid-state lasers.

We are indebted to H. A. Haus, D. C. Hanna, A. Dienes, and A. J. Schmidt for fruitful discussions and critical reading of the manuscript. Helpful conversations with M. E. Fermann and E. Wintner are gratefully acknowledged. This research was supported by the Fonds zur Förderung der wissenschaftlichen Forschung in Österreich grant P7282.

References

1. L. Dahlström, Opt. Commun. **5**, 157 (1972).
2. K. Sala, M. C. Richardson, and N. R. Isenor, IEEE J. Quantum Electron. **QE-13**, 915 (1977).
3. F. Ouellette and M. Piche, Opt. Commun. **60**, 99 (1986).
4. K. A. Stankov and J. Jethwa, Opt. Commun. **66**, 41 (1988).
5. J. Goodberlet, J. Wang, J. G. Fujimoto, and P. A. Schulz, Opt. Lett. **14**, 1125 (1989).
6. E. P. Ippen, H. A. Haus, and L. Y. Liu, J. Opt. Soc. Am. B **6**, 1736 (1989).
7. J. Goodberlet, J. Jacobsen, J. G. Fujimoto, P. A. Schulz, and T. Y. Fan, Opt. Lett. **15**, 504 (1990).
8. F. Krausz, Ch. Spielmann, T. Brabec, E. Wintner, and A. J. Schmidt, Opt. Lett. **15**, 1082 (1990).
9. T. F. Carruthers and I. N. Duling III, Opt. Lett. **15**, 804 (1990).
10. H. A. Haus, IEEE J. Quantum Electron. **QE-11**, 736 (1975).
11. E. P. Ippen, L. Y. Liu, and H. A. Haus, Opt. Lett. **15**, 183 (1990).
12. P. G. Kryukov and V. S. Letokhov, IEEE J. Quantum Electron. **QE-8**, 776 (1972).
13. Ch. Spielmann, F. Krausz, T. Brabec, E. Wintner, and A. J. Schmidt, "Experimental study of self-starting additive-pulse mode locking," IEEE J. Quantum Electron. (to be published).

December 15, 1991 / Vol. 16, No. 24 / OPTICS LETTERS

Mode locking in solitary lasers

T. Brabec, Ch. Spielmann, and F. Krausz

Technische Universität Wien, Abteilung für Quantenelektronik und Lasertechnik, Gusshausstrasse 27, A-1040 Wien, Austria

Received July 16, 1991

We present an analysis of passively mode-locked lasers in which pulse formation is dominated by the interplay between self-phase modulation and negative dispersion in separate cavity elements. Steady-state pulse parameters and stability issues are discussed. Stability in these solitary systems relies on some passive amplitude modulation, and the ultimate system performance is found to depend sensitively on the magnitude of amplitude modulation relative to that of phase modulation.

Self-phase modulation (SPM) that arises from a fast Kerr nonlinearity, in conjunction with appropriate dispersive compensation, may result in shortening of intensive optical pulses. This solitonlike pulse shaping, which is purely a phase effect, can efficiently support ultrashort-pulse formation in lasers passively mode locked by slow or fast saturable absorbers. The comprehensive theory of Martinez *et al.*[1] provides an adequate description of a wide range of cw passively mode-locked (PML) lasers incorporating either of the two types of passive amplitude modulation (PAM) and solitonlike shaping. The analytical solution found by Martinez *et al.* has been used recently for the study of fast absorber systems.[2,3] These analytical approaches[1–3] are all based on the assumption that the fractional change in pulse shape during the passage through the cavity elements is small compared with unity. This assumption is almost always fullfilled by PAM but may not be met by phase-affecting cavity elements under certain conditions. In this Letter we analyze cw passively mode-locked lasers in which femtosecond pulse formation is dominated by strong solitonlike shaping rather than by PAM.

This research was stimulated by the recent appearance of a new class of PML femtosecond laser based on solid-state gain media. The high intracavity power and the long optical path within the amplifier medium (compared with those of dye lasers) lead to strong SPM in these systems. A breakthrough in solid-state ultrafast technology was achieved by exploiting solitonlike shaping mechanisms with the addition of negative group-velocity dispersion[4–9] (GVD). A common feature of these systems is that the phase of the mode-locked pulse is substantially changed during one transit of the resonator, therefore the approximation used in the establishment of the theories in Refs. 1–3 is not justified. A correct theoretical treatment of PML lasers in which mode locking is dominated by solitonlike shaping requires the pulse-shaping elements to be described by operators that do not commute. As a consequence, the equation of motion for the mode-locked pulse is an operator equation that cannot be transformed to a common differential equation as in Refs. 1–3. A schematic diagram of this novel generation of femtosecond lasers, which we call solitary lasers, is shown in Fig. 1. In addition to SPM and negative dispersion, PAM has been included in our model. PAM is always needed in PML lasers to build up the mode-locked pulses and keep the steady state stable against noise perturbations. With the assumption of a broadband laser medium (e.g., Ti:sapphire), gain dispersion has been left out in the present analysis. This approximation allows for an instructive comparison between the steady state of a discrete solitary system and the fundamental soliton in an optical fiber, where SPM and negative GVD are continuously distributed along the optical path.

A transfer operator $\hat{T}$ can be assigned to each position (z_1, z_2, z_3, z_4) of the solitary system. $\hat{T}$ governs the pulse evolution,

$$v_{n+1}(t) = \hat{T}v_n(t), \tag{1}$$

where $v_n(t)$ is the complex electric-field amplitude after the nth round trip and t is the local time retarded by $n\tau_D$ with respect to the real time, with τ_D the group delay on one round trip. For the position z_2, $\hat{T}$ can be written as

$$\hat{T} = e^{\hat{D}}e^{\hat{A}/2}e^{\hat{N}}e^{\hat{A}/2}, \tag{2}$$

where $\hat{D} = iD/2\mathrm{d}^2/\mathrm{d}t^2$, $\hat{N} = i\phi|v|^2$, and $\hat{A}$ stands for PAM. D denotes the group delay dispersion of the dispersive element, $\phi|v|^2$ is the round-trip nonlinear phase shift in the Kerr medium, and $|v|^2$ is normalized to give the power. A solitary system can now be defined by $D/\phi < 0$ and $|\hat{A}v| \ll |\hat{N}v|$. Both conditions are satisfied by femtosecond solid-state lasers.[4–9] The amplitude modulation is assumed to be introduced by an ideal fast saturable absorber, i.e., $\hat{A} = g - l + \kappa|v|^2$, where $g - l$ is the resultant time-independent round-trip gain, and $\kappa|v|^2$ accounts for the amplitude nonlinearity. It is important to notice that the steady-state pulse parameters in the solitary system are not sensitive to the specific choice of $\hat{A}$. The gain g is adjusted to conserve pulse energy, which is an input parameter in our model. The steady-state solution of Eq. (1) can be found iteratively by computing $\hat{T}^n v_0(t)$, where the iteration has to be repeated until $v_{n+1}(t) = \exp(i\varphi)v_n(t)$ and φ is a constant phase shift. The

0146-9592/91/241961-03$5.00/0

OPTICS LETTERS / Vol. 16, No. 24 / December 15, 1991

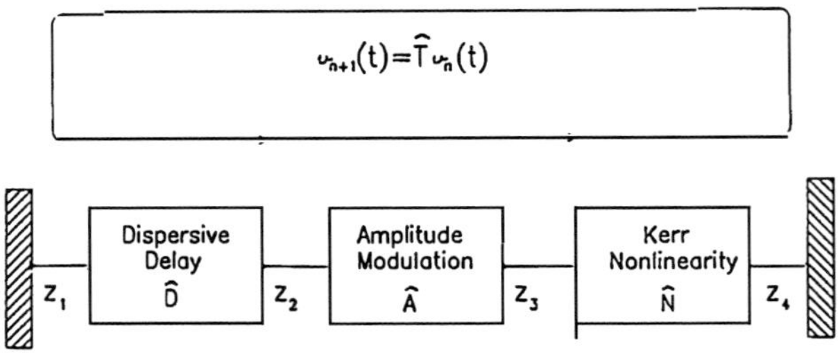

Fig. 1. Schematic of the solitary laser.

steady-state pulse at positions other than z_2 can be inferred from that at z_2 by applying the corresponding exponential operators.

Before evaluating our computer simulations it is expedient to quantify the difference between a soliton and solitary system. For this purpose we introduce the distributed parameters $\beta_2 = D/L_r$ and $\gamma = \phi/L_r$, where L_r is the round-trip length of the resonator. These parameters define a soliton-supporting system[10] that the solitary system goes into as the pulse energy approaches zero. A good measure for the discreteness of the solitary system is the ratio of L_r to the soliton period[10] $L_s = (\pi/2)(T^2/|\beta_2|)$, where $T = 2|\beta_2|/\gamma W$ and W is the pulse energy. Making use of these expressions and the definitions of β_2 and γ, we arrive at

$$r = \frac{L_r}{L_s} = \frac{1}{2\pi}\frac{(\phi W)^2}{|D|}. \tag{3}$$

This parameter plays a fundamental role in the behavior of a solitary laser. Our investigations have revealed that unlike a soliton-supporting system ($r = 0$), a solitary system ($r > 0$) cannot support stable pulses without PAM even in the absence of noise perturbations. This is because the pulse is subjected to SPM without simultaneous GVD compensation, which results in a nonlinearly chirped pulse. A negative dispersion external to the nonlinearity compresses only the central portion of the pulse while giving rise to splitting at the pulse tails, where the chirp is nonlinear. PAM is able to prevent instabilities by pushing the portions that tend to split off toward the middle of the pulse. An increased r parameter requires stronger PAM for the pulse to remain stable. Conversely, ideal soliton formation is achieved in the limits $r \to 0$ and $\hat{A} \to 0$.

The most important characteristic of a mode-locked system, the pulse width, is determined by D, ϕ, and W in a solitary laser. As long as $|\hat{A}v| \ll |\hat{N}v|$, PAM has a negligible influence on the pulse duration. The discrete pulse-shaping elements result in different pulse widths at different positions in the cavity. From a practical point of view, the pulse widths at z_1 and z_4 are the most interesting, because output coupling is usually achieved at these positions. $\tau(z_1)$ and $\tau(z_4)$ are plotted as a function of D for different pulse energies in Fig. 2. For reference, the durations $\tau_s = 3.53|D|/\phi W$ of the corresponding $N = 1$ solitons are also depicted. The numbers chosen for the calculations are typical for state-of-the-art solid-state femtosecond lasers, except for the large value of κ, which has been taken to achieve stable mode locking for $\tau < 50$ fs. Because the discreteness of solitonlike shaping is measured by the r parameter, the deviation of solitary pulse parameters from solitons can be interpreted in terms of r. To be more precise, we assume the ratio of the solitary pulse width τ to τ_s to be a function of r, i.e., $\tau = \tau_s f(r)$. For the reasons discussed above, we expect $f(r)$ to approach unity as $r \to 0$. Thus a Taylor expansion to first order in r yields for the duration of the solitary pulse $\tau = \tau_s(1 + \alpha' r)$, where α' is an expansion coefficient. By using the expressions for τ_s and r, τ can be cast in the form

$$\tau = \frac{3.53|D|}{\phi W} + \alpha\phi W, \tag{4}$$

where $\alpha = 0.56\alpha'$, i.e., the soliton pulse width is just increased by an amount proportional to ϕW independently of D. In fact, $\tau - \tau_s$ is linearly proportional to the pulse energy as revealed by Fig. 2. Equation (4) is accurate to within 10% for $r < 10$ and $\kappa/\phi < 0.2$. α depends only on the position, and we obtained $\alpha(z_1) = 0.1$ and $\alpha(z_4) = 0.25$. The connection between $\tau(z_1)$ and $\tau(z_4)$ can be written as $\tau(z_4) - \tau(z_1) = \beta(\phi W)$, where $\beta = 0.15$. This difference may become significant in powerful solitary lasers and should be taken into account in the design of such systems.

In addition to the pulse duration, the pulse shape and chirp are important characteristics of mode-locked lasers. To obtain some quantitative information of these parameters, we have computed the time–bandwidth products $\tau\Delta\nu$ of the steady-state pulses for different values of κ at z_1 and those $(\tau\Delta\nu)_{\mathrm{bl}}$ of the corresponding bandwidth-limited pulses, i.e.,

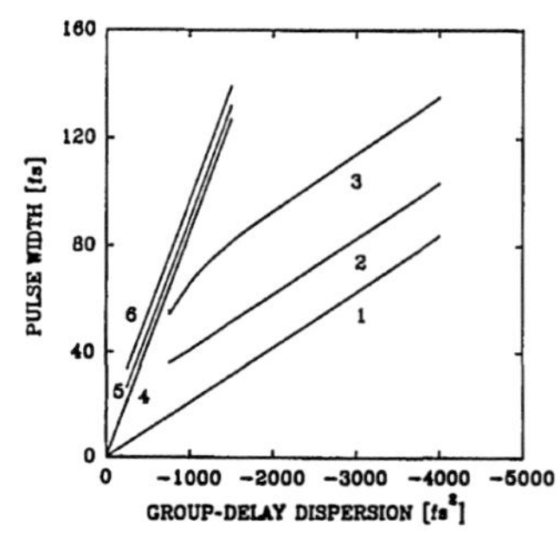

Fig. 2. Soliton and solitary pulse widths versus group-delay dispersion D for $\phi = 12.5 \times 10^{-7}\ W^{-1}$, $\kappa = 2 \times 10^{-7}\ W^{-1}$, and different pulse energies W. Curves 1, 2, and 3 illustrate τ_s, $\tau(z_1)$, and $\tau(z_4)$, respectively, for $W = 138$ nJ. Curves 4, 5, and 6 illustrate τ_s, $\tau(z_1)$, and $\tau(z_4)$, respectively, for $W = 34.5$ nJ.

December 15, 1991 / Vol. 16, No. 24 / OPTICS LETTERS

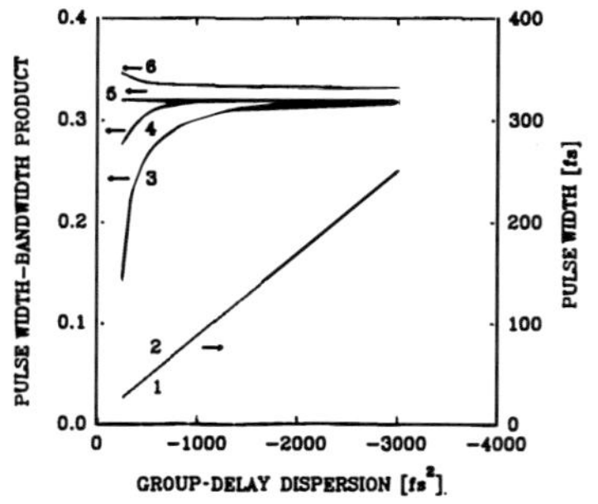

Fig. 3. Pulse width–bandwidth products $\tau\Delta\nu$ of solitary pulses and with the chirp removed $(\tau\Delta\nu)_{bl}$ for $\phi = 12.5 \times 10^{-7}\ W^{-1}$, $W = 34.5$ nJ, and different amplitude nonlinearities κ. Curves 1 and 2 denote τ; curves 3 and 4 denote $(\tau\Delta\nu)_{bl}$; curves 5 and 6 denote $\tau\Delta\nu$ for $\kappa = 2 \times 10^{-7}\ W^{-1}$ and $\kappa = 4 \times 10^{-7}\ W^{-1}$, respectively.

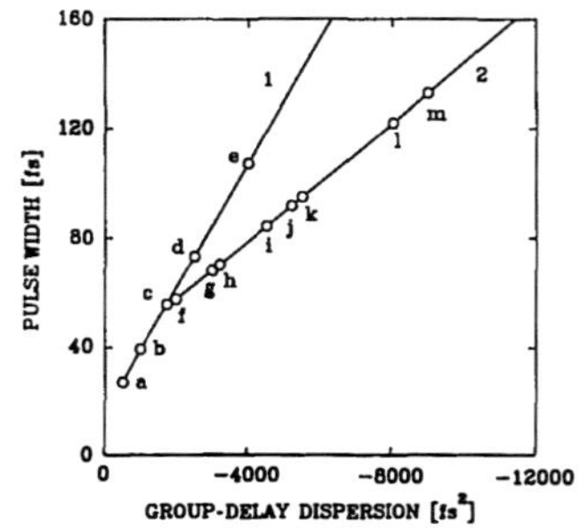

Fig. 4. Boundaries of regimes of stable mode locking on the $\tau(D, \phi W = \text{const.})$ curves for different amplitude and phase nonlinearities κ and ϕ. Curve 1, $\phi W = 172.5$ fs; curve 2, $\phi W = 345$ fs. The values of κ in units of $10^{-7}\ W^{-1}$ and the corresponding ratios κ/ϕ are as follows: a, $\kappa = 2$, $\kappa/\phi = 0.16$; b, 1, 0.08; c, 0.5, 0.04; d, 0.25, 0.02; e, 0.125, 0.01; f, 2, 0.16; g, 1, 0.08; h, 2, 0.08; i, 0.5, 0.04; j, 1, 0.04; k, 2, 0.04; l, 0.5, 0.02; m, 1, 0.02.

of pulses having the same shape and being free from phase modulation. $(\tau\Delta\nu)_{bl}$ provides information about the pulse shape, whereas $\tau\Delta\nu - (\tau\Delta\nu)_{bl}$ is a good measure of the chirp carried by the pulse. $(\tau\Delta\nu)_{bl}$ appears to decrease and $\tau\Delta\nu - (\tau\Delta\nu)_{bl}$ appears to increase with decreasing $|D|$ (see curves 3 and 5 in Fig. 3), which indicates that the pulse tails become longer and the chirp becomes greater for reduced negative dispersion. These results along with further investigations have shown that the pulse quality gets impaired in a solitary system as r increases. For $\kappa/\phi < 0.1$, solitary pulses significantly deviate both in shape and in phase from a bandwidth-limited hyperbolic-secant pulse once r becomes greater than 2. In this regime, increased PAM considerably improves the pulse quality by reducing both the wings and the chirp of the solitary pulses (curves 4 and 6 in Fig. 3). These results indicate that femtosecond solitary pulses must be subjected to a careful temporal and spectral analysis before making any statement about their shape and phase. Recently, femtosecond pulses with $(\tau\Delta\nu)_{bl} < 0.2$ have been generated in a high-power solitary laser.[11]

As expected, the gradual degradation of pulse quality with increasing r ends up in unstable operation when r reaches a critical value r_{crit}. This sets a limit to pulse shortening by reducing $|D|$. Figure 4 shows the minimum values of $|D|$ that allow stable mode-locked operation and the corresponding shortest pulse durations τ_{min} for different values of κ, ϕ, and W. Our simulations have revealed that r_{crit} is a sensitive function of κ/ϕ and only slightly dependent on ϕW. This implies that increased κ/ϕ not only improves pulse quality but also allows the generation of shorter pulses. Alternatively, shorter pulses can be produced by decreasing ϕW. As a consequence one may anticipate better femtosecond performance of a solitary system (free from bandwidth limitation and higher-order dispersion) at reduced pulse energies, but, clearly, W cannot be decreased without restriction because of the requirement of self-starting mode locking.[12] According to our results, stable sub-100-fs pulse generation in practical solitary systems with $\phi W \approx 100$ fs calls for $\kappa/\phi \approx 0.02$. This corresponds to a PAM of a few percent, in good agreement with experimental observations.[8]

In summary, we analyzed PML lasers in which pulse shortening is dominated by an interplay between a nonlinear phase shift in a Kerr medium and negative dispersion in a dispersive delay line. We introduced a parameter measuring the difference between a solitary and soliton system, which primarily determines the deviation of solitary pulses from solitons. Although pulse duration is not directly affected by amplitude modulation, the ratio of amplitude nonlinearity to phase nonlinearity has a strong influence on pulse quality and stability and sets an ultimate limit to pulse durations achievable in solitary systems.

This research was supported by the Jubiläumfonds der Österreichischen Nationalbank grant 3828.

References

1. O. E. Martinez, R. L. Fork, and J. P. Gordon, J. Opt. Soc. Am. B **2,** 753 (1985).
2. P. A. Bélanger, "Coupled-cavity mode-locking: a nonlinear model," J. Opt. Soc. Am. B (to be published).
3. H. A. Haus, J. G. Fujimoto, and E. P. Ippen, "Structures for additive-pulse mode locking," J. Opt. Soc. Am. B (to be published).
4. D. E. Spence, P. N. Kean, and W. Sibbett, Opt Lett. **16,** 42 (1991).
5. N. Sarukura, Y. Ishida, and H. Nakano, Opt. Lett. **16,** 153 (1991).
6. M. Hofer, M. E. Fermann, F. Haberl, M. H. Ober, and A. J. Schmidt, Opt. Lett. **16,** 502 (1991).
7. Ch. Spielmann, F. Krausz, T. Brabec, E. Wintner, and A. J. Schmidt, Appl. Phys. Lett. **58,** 2470 (1991).
8. L. Spinelli, B. Couillaud, N. Goldblatt, and D. K. Negus, in *Digest of Conference on Lasers and Electro-Optics* (Optical Society of America, Washington, D.C., 1990), paper CPDP-7.
9. Ch. Spielmann, F. Krausz, T. Brabec, E. Wintner, and A. J. Schmidt, Opt. Lett. **16,** 1180 (1991).
10. G. P. Agrawal, *Nonlinear Fiber Optics* (Academic, New York, 1989).
11. J. P. Likforman, G. Grillon, M. Joffre, C. Le Blanc, A. Migus, and A. Antonetti, Appl. Phys. Lett. **58,** 2061 (1991).
12. F. Krausz, T. Brabec, and Ch. Spielmann, Opt. Lett. **16,** 235 (1991).

OPTICS LETTERS / Vol. 17, No. 10 / May 15, 1992

Limits of pulse shortening in solitary lasers

T. Brabec, Ch. Spielmann, and F. Krausz

Technische Universität Wien, Abteilung für Quantenelektronik und Lasertechnik, Gusshausstrasse 27, A-1040 Wien, Austria

Received December 10, 1991

The influence of dispersive pulse-broadening effects on femtosecond pulse formation in solid-state lasers has been investigated. Empirical formulas are derived from computer simulations, which permits the estimation of the magnitude of performance-limiting effects in practical solid-state systems.

Recent developments in passively mode-locked solid-state lasers have revolutionized ultrafast laser technology. The strong self-phase modulation (SPM) inherent in mode-locked solid-state systems can be efficiently used for pulse shortening with the introduction of negative group-delay dispersion (GDD) in the resonator. In the presence of a weak passive amplitude modulation (PAM) the interplay between SPM and GDD leads to femtosecond pulse generation in broadband solid-state lasers.[1-10] Owing to the dominant role of solitonlike shaping in pulse formation, these systems have been referred to as solitary lasers.[11]

A distinctive feature of these solid-state lasers is that the femtosecond pulse is subjected to substantial changes during one round trip in the cavity. As a consequence, analytical approaches based on the assumption of a small change during one transit of the resonator cannot provide a quantitatively correct description of these systems. Recently we presented an operator formalism that permits a unifying characterization of femtosecond solid-state oscillators.[11] An operator equation governing pulse evolution in a solitary laser has been derived and solved numerically. A semiempirical relationship has been found between the steady-state pulse duration and laser parameters. In this Letter the original model that includes SPM, GDD, and PAM is supplemented with dispersive effects such as gain dispersion, third-order dispersion (TOD), and positive dispersion in the Kerr medium, and their effects on steady-state pulse duration are investigated. The numerical procedure is the same as in Ref. 11. The results of a large number of computer simulations carried out in parameter ranges relevant for practical systems have been summarized in empirical formulas that allow one to find the principal mechanisms limiting pulse duration in solid-state femtosecond sources.

In the absence of dispersion other than the negative GDD introduced on purpose, the steady-state pulse duration in the solitary laser is given by[11]

$$\tau = \frac{3.53|D|}{\phi W} + \alpha\phi W, \tag{1}$$

where $D < 0$ is the net round-trip GDD, $W = \int P(t)\mathrm{d}t$ denotes the pulse energy, $P(t)$ represents the instantaneous power of the pulse, $\phi P(t)$ is the round-trip nonlinear phase shift in the Kerr medium, and $\alpha = \alpha(z)$ is a numerical factor depending on the position where τ is measured in the cavity. α and thus τ are the smallest at the dispersive end of the cavity with $\alpha = 0.1$. The pulse duration has been calculated for this position throughout this Letter. Limitations to the pulse duration in an ideal solitary laser originate from the fact that neither ϕW nor $|D|$ can be reduced without restriction because of the requirements of self-starting[12,13] and stability,[11] respectively. In general, dispersion other than negative GDD affects femtosecond performance in two different ways: (i) the tendency to instabilities becomes greater, requiring a $|D|$ increased for stable operation, and (ii) the pulse width given by Eq. (1) is directly increased by dispersion. All the following results have been verified for $|D| < 5000$ fs^2 and $\phi W < 200$ fs.

First, consider a solitary laser incorporating a gain medium with a limited bandwidth. Mathematically the finite bandwidth can be taken into account by replacing D with $D' = D - iD_g$ in the relation for the dispersion operator $\hat{D} = i(D/2)\mathrm{d}^2/\mathrm{d}t^2$,[11] if the transmission function of the bandwidth-limiting element is expanded to second order in the optical frequency. In case the bandwidth is determined by gain dispersion alone, $D_g = 2g/(\pi\Delta\nu)^2$, where g is the saturated round-trip gain and $\Delta\nu$ is the width of the fluorescence line, if we assume a Lorentzian line shape. D_g can be increased by introducing further bandwidth-limiting elements in the cavity. Figure 1 shows the steady-state pulse duration versus GDD for different values of D_g and ϕW. The lower ends of the graphs represent the limit of stable mode locking under the conditions given in the figure captions. As expected, the effect of limited bandwidth becomes stronger as $|D|$ is reduced and the pulse gets shorter. We have deduced from our simulations that the steady-state pulse width deviates less than 10% from τ [given by Eq. (1)] as long as the requirement

$$D_g < 0.35|D| \tag{2}$$

is fullfilled. For instance, in a Ti:sapphire laser $D_g < 10$ fs^2, which implies that gain dispersion does not affect pulse formation even for intracavity GDD

0146-9592/92/100748-03$5.00/0

May 15, 1992 / Vol. 17, No. 10 / OPTICS LETTERS

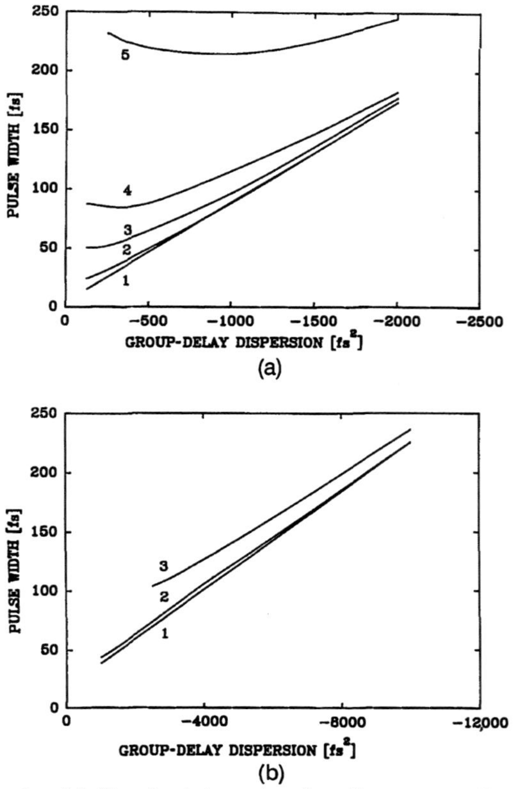

Fig. 1. (a) Steady-state pulse duration versus the net round-trip GDD in a solitary laser with limited bandwidth for different values of D_g and $\phi W = 43$ fs. Curve 1, $D_g = 0$; curve 2, $D_g = 112$ fs^2; curve 3, $D_g = 335$ fs^2; curve 4, $D_g = 632$ fs^2; curve 5, $D_g = 1760$ fs^2. (b) Steady-state pulse durations versus the net round-trip GDD in a solitary laser with limited bandwidth for different values of D_g and $\phi W = 172.5$ fs. Curve 1, $D_g = 0$; curve 2, $D_g = 335$ fs^2; curve 3, $D_g = 1760$ fs^2.

as low as $|D| \approx 100$ fs^2. Such low values of $|D|$ would, in principle, allow the generation of pulses shorter than 20 fs in the presence of sufficiently strong PAM,[11] if limiting effects discussed below could be eliminated.

With the bandwidth reduced the pulse duration appears to depend sensitively on ϕW [compare curves 5 and 3 in Figs. 1(a) and 1(b), respectively]. The pulses get shorter with increasing ϕW until instabilities occur as a result of the interaction of a nonlinearly chirped pulse with a bandwidth-limiting element, which tends to split off the nonlinearly chirped leading and trailing portions of the pulse.[14] Once the gain bandwidth is sufficiently narrow that pulse duration is limited by the onset of these instabilities, the minimum pulse width achievable by optimizing D and ϕW is found to be

$$\tau_{\min} \approx \beta/\Delta\nu. \tag{3}$$

The numerical factor β depends on g and κ/ϕ, where $\kappa P(t)$ represents the change in round-trip gain that is due to some PAM.[11] With $g = 0.05$ and $\kappa/\phi = 0.16$, where κ/ϕ is somewhat higher than that achieved in practical systems so far, we obtained $\beta \approx 0.3$, i.e., with efficient amplitude modulation the full gain bandwidth can be locked in solitary lasers. In fact, nearly system-bandwidth-limited pulse generation has recently been demonstrated in passively mode-locked Nd:glass lasers.[6,10]

An interesting general feature of bandwidth-limited solitary systems is that the influences of the real and imaginary parts of the dispersion parameter (D and D_g) on mode-locking stability are completely different although the pulse width increases with both $|D|$ and D_g. Whereas increasing D_g drives the laser unstably [compare curves 2 and 3 in Fig. 1(b)], enhanced negative GDD is capable of restoring stability. The reason for this behavior is that the negative GDD shifts the portions splitting off back toward the center of the pulse, because their carrier frequencies are different (owing to SPM) from that of the central pulse.

The influence of TOD on mode-locking performance can be investigated by replacing the dispersion operator $\hat{D}$ with $\hat{D}' = \hat{D} - \hat{D}_3$, where $\hat{D}_3 = (D_3/6)\mathrm{d}^3/\mathrm{d}t^3$ and D_3 is the net round-trip TOD. With $D_3 \neq 0$ the pulse duration stops shortening with decreasing $|D|$ when a certain value is reached depending on the ratio $|D_3/D|$, as shown in Fig. 2, curves 1–4. Clearly this behavior had to be expected by virtue of the fact that TOD results in GDD that varies linearly with the optical frequency. Once the change in D within the mode-locked bandwidth becomes comparable with or greater than the nominal value of D, the net GDD loses its role in determining the steady-state pulse width. Interestingly, TOD may lead to pulse shortening in a limited range of parameters. This has also been reported in Ref. 15, in which the effect of TOD on a colliding-pulse mode-locked dye laser has been investigated. However, it is illusory to expect an improved performance from TOD, since the pulse duration can be further reduced by simply reducing $|D|$ in the absence of TOD. In addition, TOD leads to an asymmetric pulse shape and imposes a highly nonlinear chirp on the pulse. For increased values of ϕW this anomalous pulse shortening is prevented from oc-

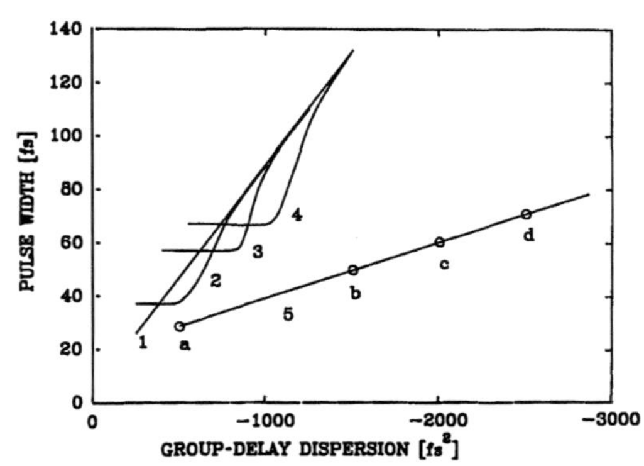

Fig. 2. Solitary pulse width versus the net GDD in the presence of third-order dispersion with $\phi W = 43$ fs: curve 1, $|D_3| = 0$ fs^3; curve 2, $|D_3| = 10{,}000$ fs^3; curve 3, $|D_3| = 20{,}000$ fs^3; curve 4, $|D_3| = 30{,}000$ fs^3. The circles on curve 5 represent the boundaries of stable mode-locking regimes for the same values of D_3 with $\phi W = 172.5$ fs: a, $|D_3| = 0$ fs^3; b, $|D_3| = 10{,}000$ fs^3; c, $|D_3| = 20{,}000$ fs^3; d, $|D_3| = 30{,}000$ fs^3.

OPTICS LETTERS / Vol. 17, No. 10 / May 15, 1992

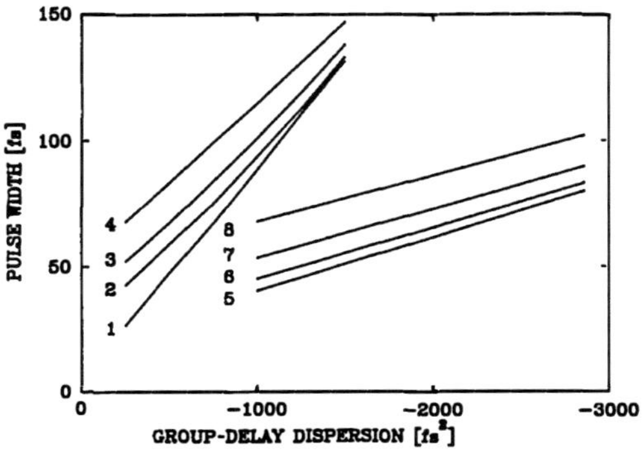

Fig. 3. Steady-state pulse duration versus the net GDD in a solitary laser with a dispersive Kerr medium. Curve 1, $D_n = 0$ fs^2; curve 2, $D_n = 800$ fs^2; curve 3, $D_n = 1600$ fs^2; curve 4, $D_n = 3200$ fs^2; curve 5, $D_n = 0$; curve 6, $D_n = 800$ fs^2; curve 7, $D_n = 1600$ fs^2; curve 8, $D_n = 3200$ fs^2. $\phi W = 43$ fs for curves 1–4, and $\phi W = 172.5$ fs for curves 5–8.

curring by the onset of instabilities (Fig. 2, curve 5). According to our simulations the condition for TOD to be negligible in femtosecond pulse formation can be cast in the form

$$|D_3| < 0.2|D|\tau. \tag{4}$$

In solid-state sytems incorporating a pair of heavy flint glass prisms, D_3 is negative with an absolute value of typically well beyond 10^4 fs^3. Such a high value of TOD becomes a limiting factor when pulses shorter than 50 fs should be generated.

The positive dispersion of the Kerr medium is the third effect that has to be considered when characterizing real solitary lasers. For this purpose the operator describing the Kerr nonlinearity $\hat{N} = \phi P(t)$ has to be supplemented with the dispersion operator $\hat{D}_n = D_n \mathrm{d}^2/\mathrm{d}t^2$, where D_n is the round-trip GDD of the nonlinear medium. Since dispersion and SPM occur simultaneously, the split-step technique has been used to model pulse propagation in the nonlinear medium, i.e., $\hat{N}$ and $\hat{D}_n$ have been divided into n equal parts and alternately applied to the propagating pulse. It is just this simultaneous action of $\hat{N}$ and $\hat{D}_n$ that does not allow $\hat{D}_n$ to be simply added to $\hat{D}$ but makes a separate treatment necessary. The results of our simulations for different values of D_n and ϕW are depicted in Fig. 3. In all cases $D_n > 0$ results in a lengthening of the steady-state pulse duration without significantly affecting the time–bandwidth product of the laser output. As a consequence, positive dispersion in the Kerr medium reduces the mode-locked bandwidth of solitary lasers at a fixed net round-trip GDD in the cavity, which can be explained as follows. The investigation of an ideal solitary system[11] has shown that the pulse duration of the $N = 1$ soliton of the corresponding soliton system is increased by the pulse formation's occurring in discrete steps. The discrete steps are increased at a constant net cavity dispersion D by the introduction of positive dispersion in the Kerr nonlinearity, thereby leading to further pulse broadening. For the effect to cause a change not greater than 10% in pulse duration, the condition

$$D_n < 0.25\tau^2 \tag{5}$$

has to be fulfilled. This is indeed a severe constraint on D_n and thus on the length of the nonlinear medium if short pulses have to be generated. For example, in a Ti:sapphire laser with a gain medium 15 mm in length the effect of positive dispersion becomes noticeable at pulse durations as long as ≈100 fs. From these results we conclude that the length of the gain medium is an important parameter in sub-100-fs pulse generation.

In summary, we have analyzed the effects of bandwidth limitation, third-order dispersion, and positive dispersion in the Kerr medium on femtosecond pulse formation in solid-state lasers. The empirical formulas presented are helpful in the design of femtosecond solid-state oscillators as well as in finding the limitations in existing systems and providing tips for their improvement. By using the inequalities for optimization of our Ti:sapphire laser system,[16] 32-fs pulses could be obtained. The experimental data of Ref. 16 fit well to the numerical results of this Letter.

This research was supported by the Jubiläumsfonds der Österreichischen Nationalbank grant 3828.

References

1. D. E. Spence, P. N. Kean, and W. Sibbett, Opt. Lett. **16,** 42 (1991).
2. N. Sarukura, Y. Ishida, and H. Nakano, Opt. Lett. **16,** 153 (1991).
3. M. Hofer, M. E. Fermann, F. Haberl, M. H. Ober, and A. J. Schmidt, Opt. Lett. **16,** 502 (1991).
4. K. Naganuma and K. Mogi, Opt. Lett. **16,** 738 (1991).
5. L. Spinelli, B. Conillaud, N. Goldblatt, and D. K. Negus, in *Digest of Conference on Lasers and Electro-Optics* (Optical Society of America, Washington, D.C., 1991), paper CPDP7.
6. Ch. Spielmann, F. Krausz, T. Brabec, E. Wintner, and A. J. Schmidt, Appl. Phys. Lett. **58,** 2470 (1991).
7. P. F. Curley and A. I. Ferguson, Opt. Lett. **16,** 1016 (1991).
8. U. Keller, G. W. t'Hooft, W. H. Knox, and J. E. Cunningham, Opt. Lett. **16,** 1022 (1991).
9. Ch. Spielmann, F. Krausz, T. Brabec, E. Wintner, and A. J. Schmidt, Opt. Lett. **16,** 1180 (1991).
10. M. Hofer, M. H. Ober, F. Haberl, and M. E. Fermann, "Characterization of ultrashort pulse formation in passively mode-locked fiber lasers," IEEE J. Quantum Electron. (to be published).
11. T. Brabec, Ch. Spielmann, and F. Krausz, Opt. Lett. **16,** 1961 (1991).
12. F. Krausz, T. Brabec, and Ch. Spielmann, Opt. Lett. **16,** 235 (1991).
13. H. A. Haus and E. P. Ippen, Opt. Lett. **16,** 1331 (1991).
14. H. Avramopoulos, P. M. W. French, J. A. R. Williams, G. H. C. New, and J. R. Taylor, IEEE J. Quantum Electron. **24,** 1884 (1988).
15. M. R. X. de Barros, R. S. Miranda, and C. H. Brito Cruz, Opt. Lett. **15,** 127 (1990).
16. F. Krausz, Ch. Spielmann, T. Brabec, E. Wintner, and A. J. Schmidt, Opt. Lett. **17,** 204 (1992).

OPTICS LETTERS / Vol. 17, No. 18 / September 15, 1992

Kerr lens mode locking

T. Brabec, Ch. Spielmann, P. F. Curley, and F. Krausz

Technische Universität Wien, Abteilung für Quantenelektronik und Lasertechnik, Gusshausstrasse 27, A-1040 Wien, Austria

Received May 28, 1992

Self-focusing in conjunction with an intracavity aperture creates a power-dependent amplitude modulation in laser oscillators, which allows passive mode locking. A simple analytical formalism yields closed-form expressions for the depth of passive amplitude modulation introduced by either the spatial gain profile or a hard aperture inserted in the resonator. Design issues for this mode-locking technique are discussed.

It has been known for a long time that intense laser beams are subject to self-focusing when they propagate through optical media that have a nonlinear index of refraction. This Kerr self-focusing effect leads to slight changes in the spatial intensity profile of the resonator mode in laser oscillators. As a consequence, by introducing an intracavity aperture, a power-dependent loss can be created. When carefully optimized, this passive amplitude-modulation mechanism favors a high intracavity power, i.e., mode-locked operation over free-running laser oscillation. Owing to the quasi-instantaneous response of nonresonant Kerr nonlinearities, the amplitude modulation induced by self-focusing is able to simulate ultrafast saturable-absorber action and support pulse formation down to the femtosecond regime in solid-state lasers that have long gain-relaxation times.[1-8] The technique has been termed self-mode-locking[1] or Kerr lens mode locking[2] (KLM). Recently computer simulations have verified the mechanisms responsible for KLM.[9-11] In this Letter a simple analytical description based on the self-similar solution of the nonlinear wave equation describing self-focusing[12,13] is presented. The derived results give insight into the basic physical processes in KLM and provide guidelines for the design and optimization of KLM lasers.

The analysis makes the following assumptions and approximations: (1) The resonator consists of two focusing mirrors (or lenses) and two flat end mirrors (Fig. 1). The Kerr medium is inserted into the tightly focused section of the cavity. The nonlinear medium is assumed to be plane cut, so that astigmatism can be neglected. (2) Only the fundamental Gaussian mode is oscillating in the resonator. (3) A self-similar solution of the nonlinear wave equation in the aberrationless approximation (quadratic index profile) is used for the description of propagation in the nonlinear medium. (4) Perturbations to the mode profile by the gain aperture or some additional hard apertures are negligible.

Figure 1 shows a schematic diagram of the resonator commonly used for KLM. The focal length of the lenses are f_1 and f_2, and d_1 and d_2 ($d_1 \le d_2$) are the distances between the lenses and the output couplers. The spacing between the two lenses is $d_f = f_1 + f_2 + \delta$, where δ is the parameter defining the range of stable operation. Note that d_f is somewhat different from the real physical separation of the lenses in the presence of some dense optical medium in between. The position of the beam waist in the linear resonator can be obtained by determining the solutions of the equivalent two-mirror resonator[14] for which standard solutions exist. The radii of curvature of the mirrors and the length of the equivalent resonator are $R_1 = -f_1^2/(d_1 - f_1)$, $R_2 = -f_2^2/(d_2 - f_2)$, and $t = R_1 + R_2 + \delta$, respectively. If we assume that $|R_2| < |R_1|$, $d_1 > f_1$, and $d_2 > f_2$, stable operation for this configuration is possible for two separate ranges, namely for $0 < \delta < -R_2$ and for $-R_1 < \delta < -R_1 - R_2$. The different stability ranges can be quickly identified by comparison of the beam parameters at the resonator ends. The beam radius at the output coupler of the short arm is notably larger for the range $0 < \delta < -R_2$ than for $-R_1 < \delta < -R_1 - R_2$, whereas the radius at the output coupler of the long arm is comparable for both stability ranges.

In order to obtain high nonlinearity the Kerr medium has to be situated near the focal plane of the linear resonator. The optical length is l/n, where n is the linear refractive index, and lx_1/n and lx_2/n denote the distances between the focal point and the two surfaces of the Kerr medium ($x_1 + x_2 = 1$). The center of the nonlinear medium and the focal point coincide for $x_1 = x_2 = 1/2$. The self-similar solution of the propagation equation[12,13] implies that the nonlinearity modifies the confocal parameter of a Gaussian beam by a factor $1/\eta$, where $\eta = \sqrt{1 - P/P_{cr}}$. Consequently the position of the beam waist x_{10} and confocal parameter $z_{r0} = \pi n w_{r0}^2/\lambda$ of the linear resonator mode are modified to $x_1(\eta)$ and $z_r(\eta)$. The vacuum laser wavelength is λ, n_2 is the nonlinear refractive index, and w_r is the radius (half-width at $1/e^2$ of the maximum intensity) of the beam waist. Here and throughout this Letter the subscript 0 refers to the solution of the linear resonator for $\eta = 1$. The critical power for self-trapping is $P_{cr} = a\lambda^2/(8\pi n n_2)$, where a is an experimentally obtained correction factor[15] that extends the validity of the aberrationless model in the presence of a strong nonlinearity, and n_2 is the nonlinear refractive index. This expression yields for Ti:sapphire a power of 2.6 MW. Numerical investi-

0146-9592/92/181292-03$5.00/0

September 15, 1992 / Vol. 17, No. 18 / OPTICS LETTERS

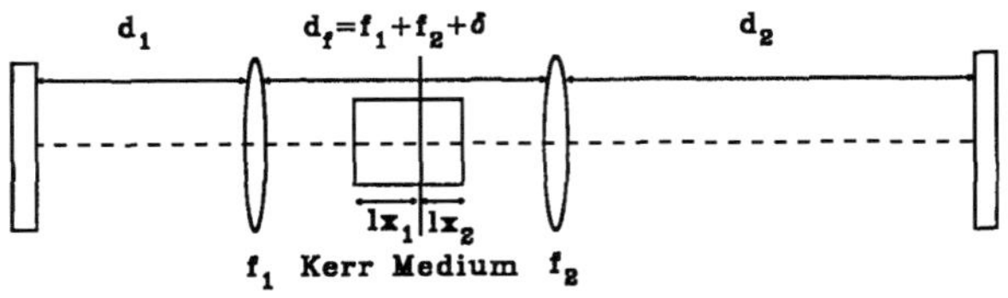

Fig. 1. Schematic of the cavity used in our calculations. The distances d_1 and d_2 and the length of the Kerr medium l are taken to be 80, 120, and 1.6 cm, respectively, throughout this Letter. The refractive index of the Kerr medium is assumed to have the value $n = 1.76$.

gations[16] have shown that owing to the inclusion of the correction factor the Gaussian self-similar description is valid for $\bar{P} = P/P_{cr} < 0.25$. From the discussion above, the radii of curvature of the phase fronts $\bar{R}_1$ and $\bar{R}_2$ at both surfaces of the Kerr medium can then be written as

$$\bar{R}_1(\eta) = \frac{lx_1(\eta)}{n}\left\{1 + \left[\frac{z_r(\eta)}{lx_1(\eta)\eta}\right]^2\right\}, \tag{1}$$

where $\bar{R}_2$ is obtained by replacing the subscript 1 with 2 in Eq. (1). Outside the nonlinear index medium the Gaussian beam can be traced by the *ABCD* matrix formalism. Because of the reciprocity of the resonator mode the radii of curvatures of the mode are required to be infinite at the flat end mirrors. Reciprocity can be assumed as long as the nonlinearity only introduces a rescaling of the linear Gaussian propagation. From this requirement one obtains the following equations for the beam parameters in the nonlinear resonator:

$$\bar{R}_1^2(\eta) - \bar{R}_1(\eta)\left\{\bar{R}_{10} + \left[\frac{1}{x_{10}} - \frac{1}{x_1(\eta)\eta^2}\right]\frac{n}{l}\frac{B_1D_1}{A_1C_1}\right\} - \frac{1-\eta^2}{\eta^2}\frac{B_1D_1}{A_1C_1} = 0, \tag{2}$$

$$\hat{T}_1 = \begin{bmatrix} A_1 & B_1 \\ C_1 & D_1 \end{bmatrix} = \begin{bmatrix} \dfrac{f_1}{R_1} & \dfrac{f_1}{R_1}\left\{R_1 - \left[\dfrac{lx_{10}}{n} - \dfrac{\delta(\delta + R_2)}{2\delta + (R_1 + R_2)}\right]\right\} \\ -\dfrac{1}{f_1} & \dfrac{1}{f_1}\left[\dfrac{lx_{10}}{n} - \dfrac{\delta(\delta + R_2)}{2\delta + (R_1 + R_2)}\right] \end{bmatrix}. \tag{3}$$

The equations for $\bar{R}_2$ and $\hat{T}_2$ can again be obtained by replacing 1 with 2. The matrices $\hat{T}_1$ and $\hat{T}_2$ relate the Gaussian beam parameters at the surfaces of the Kerr medium to the beam parameters at the output couplers. Combination of Eqs. (1) and (2) and the equivalent equations for side 2 of the resonator yields two equations for $w_r(\eta)$ and $x_1(\eta)$, which provide a complete description of the mode of the nonlinear resonator. Solution of the equations to first order in $\bar{P}$ yields the power-dependent change in position (Δx_1) and size (Δw_r) of the beam waist, which can be transformed into a power-dependent round-trip gain by employing some kind of aperturing in the cavity.

First, consider the case of aperturing the beam within the Kerr medium. In longitudinally pumped lasers the spatial gain profile can play the role of a soft aperture in the Kerr medium. For the evaluation of gain-based KLM the gain integral[17] has to be solved, where the beam waists and beam axes of resonator and pump beams are assumed to be collinear. For $z_p < z_{r0} < l/2$, $\delta = -R_2/2$, and $x_{10} = x_{20} = 1/2$, where z_p is the confocal parameter of the pump beam, a simple relation for the absolute change in power round-trip gain Δg can be derived:

$$\Delta g \approx \frac{g_s}{2}\frac{\bar{P}}{1 + S}, \tag{4}$$

where g_s is the saturated power round-trip gain of the linear resonator and $S = 2P_{av}/(\pi w_{r0}^2 I_s)$ is the gain saturation parameter, with P_{av} and I_s the average power and saturation intensity, respectively. As $\Delta g \to 0$ for large intracavity powers, the gain saturation limits the maximum obtainable amplitude modulation. This dependence is depicted in Fig. 2, where analytical (dashed curves) and numerical (solid curves) results of Δg versus $\bar{P}$ are plotted for different confocal parameters of pump and resonator beams for $P_{av} = 2$ W and $I_s = 300$ kW/cm^2. From Fig. 2, it can be seen that gain saturation reduces the efficiency of the amplitude modulation depending on the radius of the resonator mode. This reveals that for the design of a KLM system z_{r0} is an important parameter, and calculations have shown that a reasonable choice, combining a high gain and high modulation efficiency, is $z_{r0} < \approx l/2$. In the case of $z_p > l/2$, Δg becomes dependent on z_p, and the efficiency of the gain modulation is reduced. Furthermore KLM in combination with gain aperture yields amplitude modulation over most parts of the stability range, and optimum operation is achieved around the centers of both stability ranges.

Alternatively, a hard aperture can be introduced into one of the collimated resonator arms. For a low linear loss L introduced by the aperture inserted near one of the end mirrors the perturbations to the

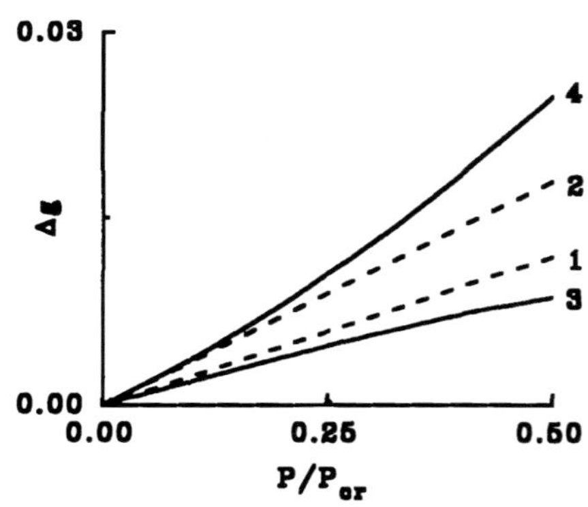

Fig. 2. Absolute change in gain Δg versus normalized intracavity power for a gain aperture within the Kerr medium. $g_s = 0.1$, $I_s = 300$ kW/cm^2, $P_{av} = 2$ W, and $\delta = R_2/2$. Curves 1 and 3 are the analytical and numerical solutions for $f_1 = f_2 = 6$ cm, $z_{r0} = 0.26$ cm, and $z_p = 0.22$ cm; curves 2 and 4 are the analytical and numerical solutions for $f_1 = f_2 = 10$ cm, $z_{r0} = 0.73$ cm, and $z_p = 0.22$ cm.

OPTICS LETTERS / Vol. 17, No. 18 / September 15, 1992

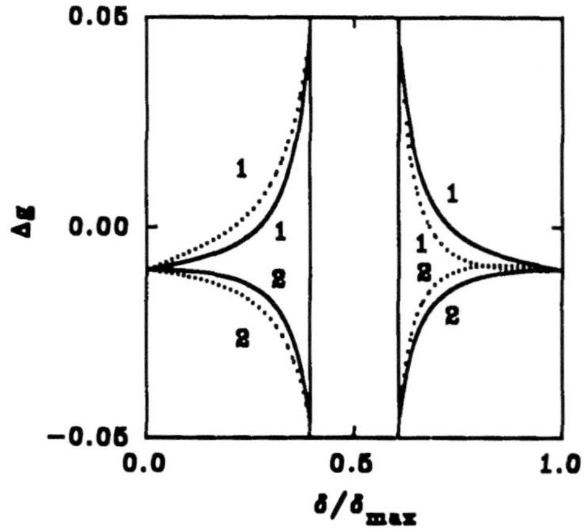

Fig. 3. Absolute change in gain versus the normalized stability parameter. $\delta_{max} = R_1 + R_2$ and $L = 0.1$, $P = 0.25$, and $f_1 = f_2 = 6$ cm. The vertical lines denote the limits of the stability range. Curves 1 and 2 show the results for the aperture inserted at the output coupler of the short arm and of the long arm, respectively. The solid curves are for $\epsilon = 0$, and the dotted curves are for $\epsilon = 0.25$.

linear resonator mode may be assumed negligible and the nonlinear change in gain can be shown to be equal to $1 - [w_{1,2}^2(\eta)/w_{1,2}^2(\eta = 1)]$, where $w_{1,2}$ are the beam radii at the output couplers. By using the matrices $\hat{T}_{1,2}$ one can determine the variation of the radii w_1 and w_2 at the end mirrors from the change in the size and the position of the beam waist within the gain material. Hence the gain variation arising from an aperture in the short arm (index 1) and in the long arm (index 2) is given by

$$\Delta g_{1,2} \approx P \frac{nLR_{1,2}}{lP_{cr}\left[1 + \left(\frac{l}{2z_{r0}}\right)^2\right]} \left\{ 1 \mp \frac{R_1^2 - R_2^2}{(R_1 + R_2 + 2\delta)^2} \pm \frac{8l\epsilon}{n\left[1 + \left(\frac{l}{2z_{r0}}\right)^2\right]} \frac{1}{R_1 + R_2 + 2\delta} \right\}, \quad (5)$$

where ϵ is defined by $x_{10} = 1/2 + \epsilon$, which denotes a possible shift of the nonlinear medium. Positive ϵ means a shift of the nonlinear medium toward the shorter resonator arm. From relation (5), which is plotted in Fig. 3 for a given set of resonator parameters, the most important features of hard-aperture KLM can be obtained immediately. Mode-locked operation depends mainly on three parameters, namely, the position of the Kerr medium (ϵ), the confocal parameter (z_{r0}), and the position in the stability range. Relation (5) reveals that for a large confocal parameter high amplitude modulation can be achieved. However, if z_{r0} becomes greater than $l/2$ this results in a reduction of the gain, which decreases the maximum loss bias that can be produced by the hard aperture. A reasonable trade-off can be realized by choosing $z_{r0} \approx l/2$. As can be seen from relation (5) and the curves labeled 1 in Fig. 3, mode-locked operation ($\Delta g > 0$) for $\epsilon = 0$ can only be obtained with the aperture in the short resonator arm.

Figure 3 demonstrates that enhanced gain modulation can be obtained by shifting the laser medium toward the shorter arm, if the laser is operated at $\delta/\delta_{max} < 0.5$. For $\delta/\delta_{max} > 0.5$ the laser medium has to be shifted toward the longer arm in order to gain higher modulation efficiency. The stability parameters for optimum amplitude modulation are the limits of the stability ranges at $\delta = -R_2$ and $\delta = -R_1$. Clearly, the hard-aperture KLM laser operates most efficiently near the limits of the stability ranges, and modulation efficiency has to be traded off against laser stability.

In conclusion, we have developed a self-consistent formalism for treating nonlinear resonators. This has allowed the derivation of analytical expressions for the passive amplitude modulation introduced by the combined effect of self-focusing and hard or soft aperture. For the gain aperture the negative influence of gain saturation on the working efficiency has been demonstrated, and design considerations have been given. The most important parameters for determining hard-aperture KLM have also been identified, and optimum design considerations have been discussed.

This research was supported by the Fond zur Förderung der wissenschaftlichen Forschung in Österreich grant P8391 and by the Österreichische Nationalbank Jubiläumsfondprojekt 4418. P. F. Curley acknowledges the support of the Royal Society.

References

1. D. E. Spence, P. N. Kean, and W. Sibbett, Opt. Lett. **16,** 42 (1991).
2. L. Spinelli, B. Couilland, N. Goldblatt, and D. K. Negus, in *Conference on Lasers and Electro-Optics,* Vol. 10 of 1991 OSA Technical Digest Series (Optical Society of America, Washington D.C., 1991), paper CPDP7.
3. G. P. A. Malcolm and A. I. Ferguson, Opt. Lett. **16,** 1967 (1991).
4. P. F. Curley and A. I. Ferguson, Opt. Lett. **16,** 1016 (1991).
5. Ch. Spielmann, F. Krausz, T. Brabec, E. Wintner, and A. J. Schmidt, Opt. Lett. **16,** 1180 (1991).
6. A. Miller, P. Li Kam Wa, B. H. T. Chai, and E. W. Van Stryland, Opt. Lett. **17,** 195 (1992).
7. G. Gabetta, D. Huang, J. Jacobson, M. Ramaswamy, E. P. Ippen, and J. G. Fujimoto, Opt. Lett. **16,** 1756 (1991).
8. U. Keller, G. W. 'tHooft, W. H. Knox, and J. E. Cunningham, Opt. Lett. **16,** 1022 (1991).
9. S. Chen and J. Wang, Opt. Lett. **16,** 1689 (1991).
10. F. Salin, J. Squier, and M. Piché, Opt. Lett. **16,** 1674 (1991).
11. M. Piché, Opt. Commun. **86,** 156 (1991).
12. J. H. Marburger, Prog. Quantum Electron. **4,** 35 (1975).
13. P. A. Belanger and C. Paré, Appl. Opt. **22,** 1293 (1983).
14. H. Kogelnik, Bell Syst. Tech. J. **44,** 455 (1965).
15. D. Huang, M. Ulman, L. H. Acioli, H. A. Haus, and J. G. Fujimoto, Opt. Lett. **17,** 511 (1992).
16. E. L. Dawson and J. H. Marburger, Phys. Rev. **179,** 862 (1969).
17. F. Krausz, E. Wintner, A. J. Schmidt, and A. Dienes, IEEE J. Quantum Electron. **26,** 158 (1990).

OPTICS LETTERS / Vol. 18, No. 1 / January 1, 1993

Operation of a femtosecond Ti:sapphire solitary laser in the vicinity of zero group-delay dispersion

P. F. Curley, Ch. Spielmann, T. Brabec, F. Krausz, E. Wintner, and A. J. Schmidt

Technische Universität Wien, Abteilung Quantenelektronik und Lasertechnik, Gusshausstrasse 27, A-1040 Wien, Austria

Received August 31, 1992

We report the operating characteristics of a self-mode-locked Ti:sapphire solitary laser at reduced group-delay dispersion. The generation of ≈12.3 fs near-sech^2 optical pulses at 775 nm is reported, together with experimental evidence for the dominant role of third-order dispersion (TOD) as a limiting factor to further pulse shortening in the oscillator. At reduced second-order dispersion excessive residual TOD is shown to lead to dispersive wave generation, and the position of the dispersive resonance is used to determine the ratio of the net second- and third-order intracavity dispersions. Since the magnitude of TOD rapidly decreases with increasing wavelength in prism-pair dispersion-compensated resonators, the oscillator presented has the potential for producing sub-10-fs pulses in the 800-nm wavelength region.

The operation of self-sustaining mode-locked Ti:sapphire lasers[1] at reduced net second-order group-delay dispersion (GDD) previously permitted the generation of sub-20-fs pulses directly from a laser.[2,3] The pulse shaping in these lasers is a result of the discrete solitonlike interplay between the GDD and self-phase modulation in the gain medium, with the shortest pulses generated for small net round-trip negative GDD.[4–6] Since a reduction of the net negative GDD in solitary lasers calls for low round-trip intracavity third-order dispersion[7] (TOD), the resonator has to be optimized for minimum TOD if short pulses are to be generated.[8–10] In this Letter we report the generation of 12-fs pulses from a Kerr lens mode-locked Ti:sapphire laser optimized for maximum self-amplitude modulation and minimum intracavity TOD. We discuss the operating characteristics of the laser at reduced GDD that provide evidence for the limiting role of TOD and report what is to our knowledge the first observation of dispersive waves as a result of TOD in a solitary laser.

The laser is constructed by using an X cavity configuration and a 4-mm-long Brewster-cut Ti:sapphire rod (Crystal Systems). The cavity design follows the layout of a previous design[9] and uses single-stack reflective optics comprising two 100-mm radius-of-curvature high reflectors, a flat high-reflector fold mirror and a flat high-reflector end mirror. Two Brewster-angled fused-silica prisms are positioned in the longer arm of the resonator, with an apex-to-apex separation of 59.5 cm, and a flat 5% wedged output coupler is positioned in the dispersive cavity end to couple out the solitary pulse with the shortest duration.[6] The laser is pumped with 5 W of 527.5-nm pump light from a cw actively mode-locked (76 MHz), frequency-doubled Nd:YLF laser (Quantronix 4216D). Mode locking of the Ti:sapphire laser is achieved by using synchronous pumping and a vertical slit in the short nondispersive cavity arm, with the cavity focusing mirrors positioned to maximize the passive amplitude modulation.[11] A second external prism pair is used to remove residual spatial chirp and permits compensation for chirp introduced by the output mirror substrate and precompensation for the finite thickness of the autocorrelator beam splitter. The pulse duration is measured by collinear interferometric autocorrelation by use of a 30-μm-thick barium borate doubling crystal. In Fig. 1 the variation of the mode-locked pulse spectrum is displayed as a function of increasing intracavity glass, with optimum mode locking corresponding to the central region of the diagram. It can be seen that, as the net negative GDD is reduced, the spectrum

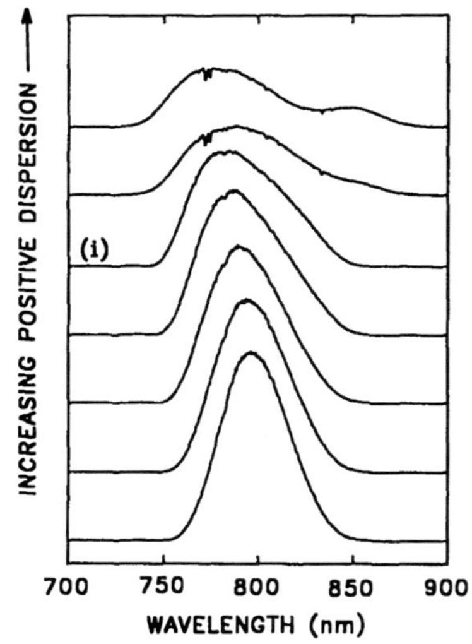

Fig. 1. Mode-locked spectra recorded by an optical multichannel analyzer at the dispersive cavity end as a function of increasing prism glass path for fixed values of intracavity pulse energy and aperture width. The relative change in net intracavity GDD between the adjacent spectra is $\Delta D \approx 10$ fs^2 for the initial reference $\lambda = 0.8$ μm. The discontinuities in some of the spectra at 775 nm are due to damaged detector array elements.

0146-9592/93/010054-03$5.00/0

January 1, 1993 / Vol. 18, No. 1 / OPTICS LETTERS

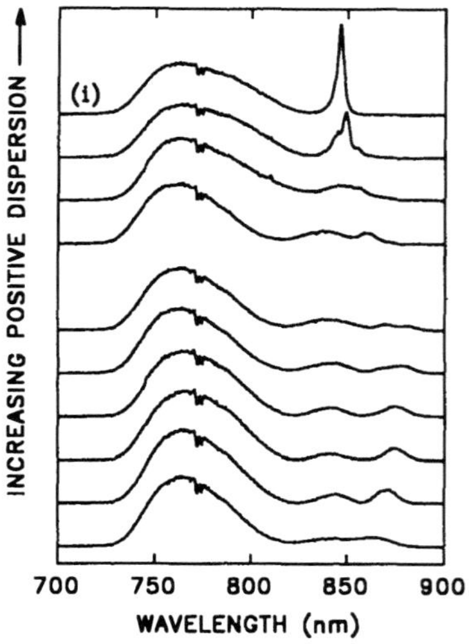

Fig. 2. Mode-locked spectra as a function of increasing prism glass path as the minimum operating GDD for stable operation is approached. The relative change in GDD from the top scan of Fig. 1 to the first scan in Fig. 2 is $\approx 5\ \mathrm{fs}^2$. Here $\Delta D \approx 1\ \mathrm{fs}^2$ at $\lambda = 0.8\ \mu\mathrm{m}$ and was controlled by translating the second prism on a precision micrometer stage. In curve (i) the shift of energy is into a dispersive resonance at a higher wavelength.

broadens and the peak wavelength shifts to shorter wavelengths; this change is understood to be due to the blue shift of the negative dispersion regime with increasing prism insertion. The lack of any significant spectral broadening with subsequent reduction in GDD, from curve (i) of Fig. 1 onward, is a strong indication of the dominance of TOD in limiting further pulse shortening. The spectra in Fig. 2 continue on from Fig. 1 and display the rapid spectral variation observed as a function of glass insertion as the zero-dispersion wavelength approaches the solitary pulse spectrum. The secondary spectral features are in the positive dispersive regime and correspond to a longer pulse copropagating with the short solitary pulse, a behavior similar to that observed in colliding-pulse mode-locked dye lasers.[12,13] In proximity to the minimum operating GDD, energy extraction from the main pulse by means of dispersive waves becomes dominant [curve (i) in Fig. 2]. Further reduction of the GDD from this point causes the laser to stop mode locking. The appearance of a sharp dispersive resonance[14,15] (dispersive wave generation) has not been observed in femtosecond dye lasers, presumably because of the strong intracavity passive amplitude modulation present in the cavity.

The *in situ* measurement of the operating GDD by the method described by Knox[16] was attempted; however, reproducible wavelength tuning with a slit in the dispersive arm of our laser has not been possible with bandwidths in excess of 50 nm. Nevertheless, dispersion characteristics at the pulse center frequency ω_0 can be obtained by considering the relative positions of the spectral features in curve (i) of Fig. 2. A rigorous relation in the presence of dispersive waves can be obtained by using the knowledge that the dispersive wave can efficiently extract energy from the main solitary pulse only if the two waves are phase matched. A phase-matching condition can thus be derived by using an extension to the treatment given by Gordon[17] and is depicted graphically in Fig. 3. Phase matching occurs where the dispersive wave number $k_r = (1/L_r)[(D\Delta\omega^2/2) + (D_3\Delta\omega^3/6)]$ and the solitary-pulse wave number $k_s = -(1/L_r)(D/2T^2)$ coincide. Here L_r is the round-trip resonator length, D and D_3 represent the net round-trip GDD and TOD at the solitary pulse center frequency ω_0, respectively, and $\Delta\omega$ is the frequency shift from ω_0. The constant $T = \tau_p/1.76$, where τ_p is the solitary pulse duration. The phase-matching condition can now be written as

$$\Delta\omega_r^2 + \beta\Delta\omega_r^3 = -(1/T^2), \qquad (1)$$

where $\Delta\omega_r$ is the frequency spacing between the dispersive resonance ω_r and the center frequency ω_0 and $\beta = D_3/3D$. When the measured solitary pulse width $\tau_p \approx 18$ fs and $\Delta\omega_r \approx -0.24\ \mathrm{fs}^{-1}$ corresponding to curve (i) of Fig. 2 are used, substitution into Eq. (1) yields $\beta \approx 4.9$ fs. Now we can exploit the relative insensitivity of the calculated net TOD to errors in the measurement of the prism insertion to determine the net intracavity GDD. Substituting the calculated TOD of $D_3 \approx -800\ \mathrm{fs}^3$ into the expression for β gives $D \approx -50\ \mathrm{fs}^2$ for curve (i) of Fig. 2.

The shortest mode-locked pulses are produced under conditions corresponding to those shown by curve (i) of Fig. 1, where the broadest structureless spectra are generated. For this operating regime, typical single-scan interferometric autocorrelations of pulses from the dispersive arm of the laser

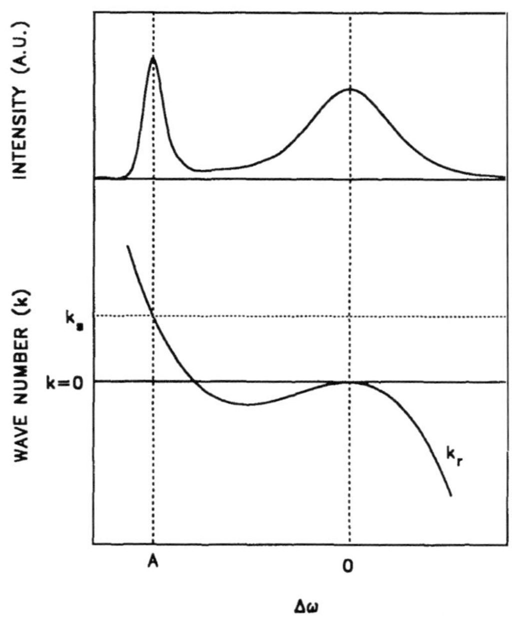

Fig. 3. (Top) Plot of an ideal solitary pulse mode-locked spectrum in the presence of dispersive waves. (Bottom) Dispersion diagram for a solitary pulse (k_s) and a dispersive wave (k_r) in the presence of GDD and TOD in the cavity. Dashed line A denotes the dispersive resonance position, $\Delta\omega_r$.

OPTICS LETTERS / Vol. 18, No. 1 / January 1, 1993

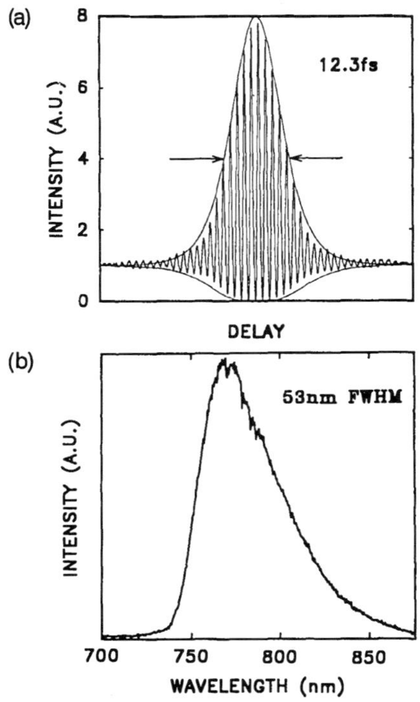

Fig. 4. (a) Interferometric autocorrelation of the shortest mode-locked pulses from the dispersive cavity arm of the laser, together with an ideal sech^2 pulse envelope. (b) The corresponding pulse spectrum.

and the pulse spectrum are shown in Figs. 4(a) and 4(b), respectively. A sech^2 fit to the envelopes of the fringes is displayed with the measured autocorrelation trace, yielding a pulse duration of 12.3 fs and a time–bandwidth product of ≈ 0.33.

For an absorbed pump power of ≈ 2.5 W the average mode-locked output power is 160 mW, yielding an intracavity pulse energy of $W \approx 40$ nJ and an output pulse peak power of 0.17 MW. Using the value of GDD determined from Eq. (1) and curve (i) of Fig. 2, and knowledge of the relative prism insertion, we estimate the operating GDD for optimum mode locking at 775 nm to be $D \approx -80\ \mathrm{fs}^2$. From the calculated value for the round-trip nonlinear phase shift $\phi = 1.5 \times 10^{-6}\ \mathrm{W}^{-1}$, the predicted solitary-pulse duration[6] for the described system is ≈ 10.5 fs, in reasonable agreement with the experimental result.

In summary, we have reported the reproducible operation of a sub-15-fs Ti:sapphire laser in the vicinity of the zero-dispersion point and confirmed the observation of spectral splitting as a result of third-order dispersion. The results provide evidence that further pulse shortening is limited by residual intracavity TOD in our Ti:sapphire solitary laser. Since the magnitude of the net intracavity TOD rapidly decreases with increasing wavelength, it is anticipated that shorter pulses may still be obtained by using suitable reflective optics to force the laser to oscillate at higher wavelengths (≥ 800 nm).

This research was supported by the Fonds zur Förderung der Wissenschaftlichen Forschung in Österreich grant 8391. P. F. Curley acknowledges the support of the Royal Society, London, in the form of a European Exchange Fellowship. The authors are indebted to Karpat Ferencz at the Research Institute for Solid State Physics, Budapest, Hungary, for providing the low-dispersion dielectric coatings.

References

1. D. E. Spence, P. N. Kean, and W. Sibbett, Opt. Lett. **16,** 42 (1991).
2. C. P. Huang, M. T. Asaki, S. Backus, H. Nathel, M. M. Murnane, and H. C. Kapteyn, presented at Ultrafast Phenomena Conference, Antibes, France, June 8–12, 1992.
3. Ch. Spielmann, P. Curley, T. Brabec, E. Wintner, A. J. Schmidt, and F. Krausz, presented at Ultrafast Phenomena Conference, Antibes, France, June 8–12, 1992.
4. O. E. Martinez, R. L. Fork, and J. P. Gordon, Opt. Lett. **9,** 156 (1984).
5. H. A. Haus, J. G. Fujimoto, and E. P. Ippen, J. Opt. Soc. Am. B **8,** 2068 (1991).
6. T. Brabec, Ch. Spielmann, and F. Krausz, Opt. Lett. **16,** 1961 (1991).
7. T. Brabec, Ch. Spielmann, and F. Krausz, Opt. Lett. **17,** 748 (1992).
8. C. P. Huang, H. C. Kapteyn, J. W. McIntosh, and M. M. Murnane, Opt. Lett. **17,** 139 (1992).
9. F. Krausz, Ch. Spielmann, T. Brabec, E. Wintner, and A. J. Schmidt, Opt. Lett. **17,** 204 (1992).
10. J. M. Jacobson, A. G. Jacobson, K. Naganuma, H. A. Haus, and J. G. Fujimoto, in *Conference on Lasers and Electro-Optics,* Vol. 12 of 1992 OSA Technical Digest Series (Optical Society of America, Washington, D.C., 1992), paper CTuU2.
11. T. Brabec, Ch. Spielmann, P. Curley, and F. Krausz, Opt. Lett. **17,** 1292 (1992).
12. F. W. Wise, I. A. Walmsley, and C. L. Tang, Opt. Lett. **13,** 129 (1988).
13. F. Salin, P. Grangier, P. Georges, and A. Brun, Opt. Lett. **15,** 1374 (1990).
14. P. K. A. Wai, C. R. Menyuk, Y. C. Lee, and H. H. Chen, Opt. Lett. **11,** 464 (1986).
15. S. Wen and S. Chi, Opt. Quantum Electron. **21,** 335 (1989).
16. W. H. Knox, Opt. Lett. **17,** 514 (1992).
17. J. P. Gordon, J. Opt. Soc. Am. B **9,** 91 (1992).

IEEE JOURNAL OF QUANTUM ELECTRONICS, VOL. 30, NO. 4, APRIL 1994

Ultrabroadband Femtosecond Lasers

Christian Spielmann, Peter F. Curley, Thomas Brabec, and Ferenc Krausz, *Member, IEEE*

***Abstract*—The exploitation of soliton-like pulse shaping mechanisms and the optimization of phase dispersion in broadband self-modelocked solid state laser oscillators have led to unprecedented advances in ultrafast laser technology. This paper reviews the basic physical mechanisms governing pulse formation and addresses the requirements for optimum performance of these novel ultrashort pulse laser sources. These considerations together with the demonstration of ≈10 fs pulse generation from a Ti:sapphire laser suggest that a family of solid-state laser oscillators with comparable performance can be developed.**

I. Introduction

The appearance of broadband solid-state laser media has recently provoked much interest in the generation of ultrashort optical pulses. The broad fluorescence emission [1] of novel transition-metal-doped laser crystals (Fig. 1 and Table I) can potentially support light pulses of a few femtoseconds in duration. Efficient exploitation of these bandwidths for modelocking calls for the presence of a sufficiently fast self-amplitude modulation (SAM) mechanism in the cavity [2], [3]. There are two distinctively different SAM techniques which can provide intensity discrimination with a suitably fast response: a) the combined effect of saturable absorption and saturable gain termed "slow saturable absorber" modelocking [4], and b) the combination of a suitable linear optical element with a medium exhibiting a broadband nonresonant optical nonlinearity, which can simulate a fast saturable absorber in the cavity. The first scheme imposes a severe constraint on the resonator design since the cavity round-trip time is required to be comparable to the upper state lifetime of the laser medium [4]. The μs relaxation time of current solid-state gain media does not permit practical systems to meet this requirement and so nonresonant all-optical modulators need to be employed. The great advantage of nonresonant passive modulation techniques is their universal applicability over a broad wavelength range irrespective of the specific properties of different laser materials [5]–[22].

In addition to SAM the modelocked pulses always experience self-phase modulation (SPM) arising from the nonlinear index of the solid-state gain medium, in which the cavity mode is tightly focused. Generally this Kerr-effect-induced SPM introduces a much stronger modification of the complex pulse envelope than any sort of nonresonant SAM mechanism. SPM can be converted into efficient pulse shortening in the presence of a net negative group delay dispersion (GDD) in the cavity. The steady-state pulsewidth in these systems is primarily determined by the phase-modifying effects of SPM and GDD as a result of solitary pulse formation [23]. The achievable pulse duration is ultimately limited either by the onset of instabilities [23] or by the presence of gain dispersion and/or higher-order phase dispersion [24] in the cavity. Intensive experimental research has resulted in the efficient exploitation of the solitonlike shaping mechanisms and a suppression of the undesirable effects that adversely affect pulse shortening [25]–[33]. This work culminated in the development of quartz-prism-controlled Ti:sapphire lasers [27], which are now capable of generating optical pulses close to 10 fs in duration [27], [30], [32], [33]. This paper gives a comprehensive description of this novel class of femtosecond lasers with the aim of providing a guide for the design and optimization of modelocked lasers operating in the 10-fs regime.

Manuscript received February 1, 1993; revised June 15, 1993. This research was sponsored by the Jubiläumsfonds der Österreichischen Nationalbank, grant 3828, and the Fonds zur Förderung der Wissenschaftlichen Forschung in Österreich, P8566. P. F. Curley acknowledges the support of the Royal Society, in the form of a European Exchange Fellowship.

The authors are with the Technische Universität Wien, Abteilung für Quantenelektronik und Lasertechnik, Gusshausstr. 27, A-1040 Wien, Austria.

IEEE Log Number 9215815.

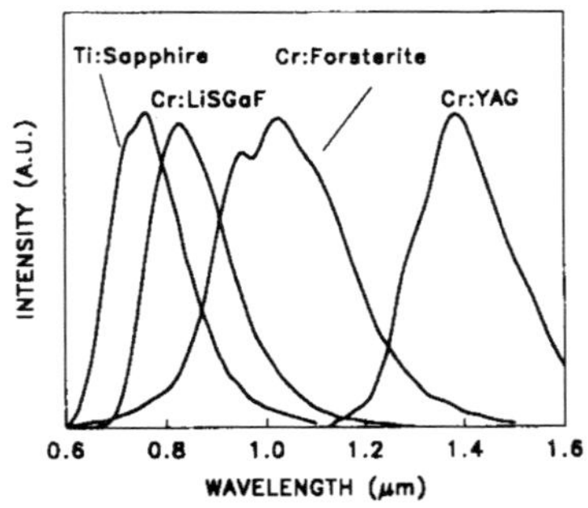

Fig. 1. Fluorescence emission spectra for some transition metal-doped broadly tunable near-infrared solid-state laser materials.

TABLE I
Spectroscopic Data for Transition Metal-Doped Broadly Tunable Solid-State Laser Materials

Material	Pump band	Emission cross section	Upper-state lifetime
	λ (μm)	σ (10−19 cm^2)	τ (μs)
Ti:sapphire	0.45-0.6	3.8	3.2
Cr:LiSGaF	0.60-0.7	0.33	88
Cr:forsterite	0.85-1.2	1.1	15
Cr:YAG	0-.88-1.1	8	4

II. Fundamentals

In this section we briefly outline the theoretical framework of femtosecond oscillators and summarize the implications of the theory for the design of practical systems.

0018-9197/94$04.00 © 1994 IEEE

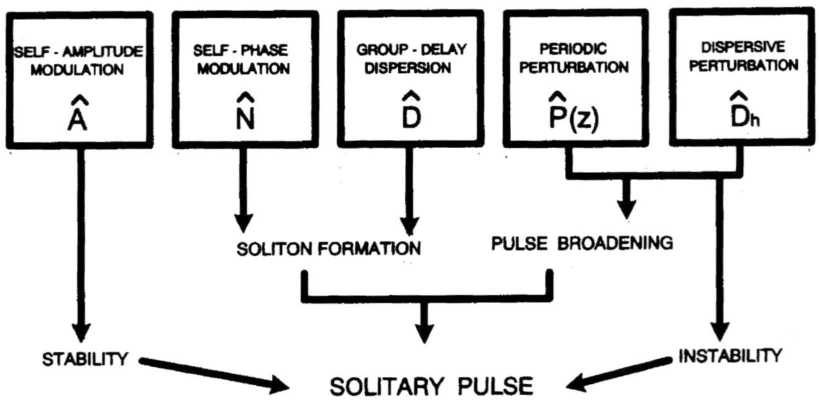

Fig. 2. Summary of the various processes that influence the generation and evolution of femtosecond pulses in solid-state systems. Since the influence of SPM is usually much stronger than SAM, pulse formation is dominated by a soliton-like interplay of SPM and negative GDD. The primary role of SAM as the steady state is approached is to provide stabilization for the modelocked pulse against perturbations arising from higher-order dispersion and the separated action of SPM and GDD.

The evolution of the modelocked pulse circulating in the cavity (see equations 22 and 23 in [3]) is governed by the transfer operator [3]:

$$\hat{T}(z) = \exp(\hat{A} + \hat{N} + \hat{D} + \hat{P}(z, \hat{N}, \hat{D}) + \hat{D}_h), \quad (1)$$

where the operators in the exponent represent the principal pulse shaping effects summarized in Fig. 2. The operator $\hat{A}$ stands for SAM, which is indispensable for establishing and sustaining a femtosecond pulse in the resonator [2], [3]. For moderate intracavity powers, the self-induced dynamic change in the round-trip gain is described by the operator $\hat{A} = \kappa|v|^2$, where $|v|^2$ is the (instantaneous) pulse power and κ (unit: W^{-1}) is a proportionality factor, which plays a determining role in the transient buildup of passive modelocking [34], [35], and stabilization of the steady state [23], [36]. In optimized Kerr-lens modelocked systems κ takes values of the order of $10^{-7}W^{-1}$ [21]. The operator $\hat{N} = -i\phi|v|^2$ accounts for SPM and describes the self-induced phase shift that is imposed on the pulse by the optical Kerr effect. Here ϕ gives the round-trip phase shift per unit power in the Kerr-medium (unit: W^{-1}). In solid-state lasers the gain medium typically gives rise to ϕ-values of the order of 10^{-6} W^{-1} (see Section V). The intracavity GDD is represented by the operator $\hat{D} = i(D/2)\partial^2/\partial t^2$, where $D = \varphi''(\omega_l)$ (unit: fs^2). Here $\varphi''(\omega)$ is the second derivative of the frequency-dependent cavity round-trip phase delay $\varphi(\omega)$, and ω_l is the center frequency of laser oscillation. The cavity GDD can also be expressed as $D = T_r'(\omega_l)$, the first derivative of the cavity round-trip time $T_r(\omega_l)$, where $T_r(\omega_l) = \varphi'(\omega_l)$ [37].

If $D > 0$, the interplay between SPM and GDD results in pulse broadening, which stops the pulse shortening introduced by SAM at the picosecond level [17]. By contrast, the combination of an overall *negative* GDD ($D < 0$) and SPM in the cavity can efficiently support pulse shortening into the fs regime. In practical solid-state systems $\phi \gg \kappa$ and the interplay between SPM and negative GDD, which is often referred to as soliton-like shaping, dominates the pulse shortening process [3]. If the influence of the perturbation operator $\hat{P}$, and the high-order dispersion operator $\hat{D}_h$ are negligible the stationary ed pulse corresponds to the fundamental soliton [38] and the steady-state pulse duration (full width at half maximum, FWHM) is given by

$$\tau_s = \frac{3.53|D|}{\phi W} \quad (2)$$

where W is the pulse energy in the cavity. With $\phi \approx 10^{-6}$ W^{-1} and $W \approx 50$ nJ, which are typical of Kerr-lens ed lasers, an operating $D \approx -10^3$ fs^2 yields a predicted ed pulse duration of $\tau_s \approx 70$ fs. In practice, the intracavity GDD is the only parameter that allows control of the ed pulse duration over a wide range. However for $|D| \to 0$, (2) predicts $\tau_s \to 0$, which represents an unphysical behavior. Reaching this "ideal" operating condition is inhibited by the onset of limiting effects described by the operators $\hat{P}$ and $\hat{D}_h$, which will be discussed further in the subsequent paragraphs.

The influence of SPM and negative GDD on the pulse shape can be so strong in the sub-100 fs region that the effects of the commutators of $\hat{N}$ and $\hat{D}$ become comparable to those of the primary pulse shaping operators $\hat{N}$ and $\hat{D}$, i.e., the order of their action becomes relevant. This fact is accounted for by the position-dependent perturbation operator $\hat{P}(z, \hat{N}, \hat{D})$, which can be expanded in commutators of increasing order in $\hat{N}$ and $\hat{D}$ [3]. The fractional change of the complex pulse envelope introduced by either SPM or GDD scales with the ratio of the resonator round-trip length L_r to that of the soliton period L_s, which can be written as [23]

$$r = \frac{L_r}{L_s} = \frac{1}{2\pi}\frac{(\phi W)^2}{|D|}. \quad (3)$$

The influence of the commutator of $\hat{N}$ and $\hat{D}$ on the pulse scales with r^2, that of the double commutators with r^3, and so on. Since r varies in inverse proportion to $|D|$, a reduction of the pulse duration by reducing $|D|$ results in an increased value of r. As the r parameter approaches unity, the effects of the first few terms in the expansion of $\hat{P}$ [3] become comparable to those of $\hat{D}$ and $\hat{N}$ and the higher-order operator $\hat{P}$ then

IEEE JOURNAL OF QUANTUM ELECTRONICS, VOL. 30, NO. 4, APRIL 1994

has to be taken into consideration. The primary implication of a nonnegligible $\hat{P}$ is that the transfer operator of the system (1) becomes position-dependent, implying stationary pulse parameters that are different at different positions in the resonator. An inspection of $\hat{P}$ up to third-order in r reveals that the perturbed soliton (solitary pulse) is longer but remains approximately bandwidth-limited at the ends of a linear (standing-wave) cavity containing a dispersive delay line (with negative GDD) and a Kerr medium. The theory predicts that the circulating pulse is at its shortest at the dispersive end of the cavity [3], [23] (see Section V).

The last term $\hat{D}_h$ in the expression of $\hat{T}$ represents undesirable dispersive effects originating from the finite bandwidth of the oscillator (gain dispersion) and frequency dependence of the cavity GDD (higher-order phase dispersion) [39]. Similarly to $\hat{P}$, the operator $\hat{D}_h$ increasingly affects steady-state modelocking performance as the negative GDD is reduced. The lowest-order dispersive perturbation [39] to solitary pulse formation originates from the finite bandwidth of the laser oscillator. The influence of gain dispersion, D_g [39], scales with the parameter

$$\beta_g = \frac{D_g}{|D|}. \tag{4}$$

For the oscillator bandwidth not to affect pulse formation, the condition $\beta_g \ll 1$ must be satisfied. Unless tuning or other bandwidth-limiting elements are introduced, gain dispersion primarily comes from the finite fluorescence linewidth of the amplifier medium. Thus we have $D_g \approx 2g/(\pi\Delta\nu_f)^2$, where g is the saturated round-trip gain and $\Delta\nu_f$ is the full width at half maximum (FWHM) of the fluorescence line (assuming a Lorentzian line shape). In solid-state systems with fluorescence bandwidths on the order of 100 THz (see Fig. 1), D_g typically takes values of < 5 fs^2 in the absence of bandwidth-limiting elements other than the amplifier and dielectric mirrors in the cavity. This low value of D_g implies that modelocking is unaffected by bandwidth limitation for $|D| > 50$ fs^2, i.e., for pulse durations down to 10 fs (see Section V).

The higher-order contributions to $\hat{D}_h$ describe the frequency-dependence of the intracavity GDD [39]. Third-order dispersion (TOD) gives a linear, and usually dominant contribution. The fractional change of the intracavity GDD over the ed spectrum due to TOD is given by:

$$\beta_3 = \frac{1}{3}\frac{|D_3|\Delta\omega_l}{|D|} \tag{5}$$

where $D_3 = \varphi'''(\omega_l)$ (unit: fs^3) is the net intracavity TOD, $\Delta\omega_l = 2\pi\Delta\nu_l$, and $\Delta\nu_l$ is the FWHM of the ed laser spectrum. The influence of TOD on modelocking rapidly increases with decreasing negative GDD and becomes a major limitation to pulse shortening as β_3 becomes of the order of unity [24]. In an analogous way, we can quantify the effect of fourth-order dispersion by defining the scaling parameter

$$\beta_4 = \frac{1}{48}\frac{|D_4|\Delta\omega_l^2}{|D|} \tag{6}$$

where $D_4 = \varphi''''(\omega_l)$ (unit: fs^4) denotes the overall fourth-order dispersion in the cavity. The scaling parameters β_3 and β_4 are defined such that they equal unity if the variation of the cavity GDD caused by D_3 or D_4 within the FWHM of the ed spectrum (i.e., the difference between the extremes of D within $\Delta\omega_l$) is equal to the magnitude of the nominal GDD at the center frequency ω_l.

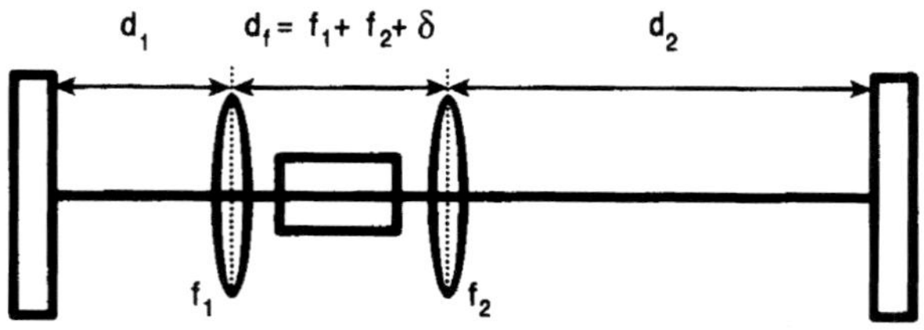

Fig. 3. Schematic of a tightly focused four-mirror laser cavity.

In summary, for sufficiently large values of the negative intracavity GDD the perturbation scaling parameters defined by (3)–(6) are much less than unity. As a consequence, $\hat{P}$ and $\hat{D}_h$ may be neglected in (1). This leads to an approximate analytic steady-state solution that is equivalent to the fundamental soliton propagating in optical fibers with anomalous dispersion. As the net negative GDD is reduced in the cavity and the steady-state pulse duration decreases, the influence of dispersive ($\hat{D}_h$) and discrete pulse shaping ($\hat{P}$) effects increase. If one of the scaling parameters r, β_g, β_3, and β_4 becomes of the order of unity, the soliton-like ed pulse is subject to modifications, and further pulse shortening is limited by the corresponding perturbation. The undesirable effects represented by $\hat{P}$ and $\hat{D}_h$ can adversely affect the pulse parameters and tend to break up the ed state, leading to unstable operation. The only mechanism opposing the emergence of instabilities is SAM in the cavity, and so an efficient SAM mechanism is crucial to the generation of stable pulses in the 10 fs regime.

In Section III we shall discuss the design and optimization of femtosecond laser resonators for minimized perturbations ($\hat{P}$, $\hat{D}_h$) and maximized self-amplitude modulation ($\hat{A}$). Special emphasis will be placed on Kerr-lens modelocking, the simplest SAM technique that is applicable to a wide range of lasers. Section IV is devoted to stability considerations and Section V addresses the effects limiting pulse shortening and discusses how these effects can be minimized.

III. Resonator Design*

A schematic diagram of the most widely used cavity configuration for femtosecond laser oscillators is shown in Fig. 3. The focusing elements, which are curved mirrors in practice, provide a tightly focused resonator mode in the gain medium to ensure a low pump threshold for laser oscillation. The collimated cavity arms can be used for the incorporation of optical elements that make the laser favor ed over free-running CW oscillation, e.g., saturable absorber, aperture, dispersive delay, etc.

For femtosecond modelocking the simplest embodiment of an artificial fast saturable absorber is self-modelocking [17] or Kerr-lens modelocking [18] (KLM). The technique makes use

*Further details are available on request.

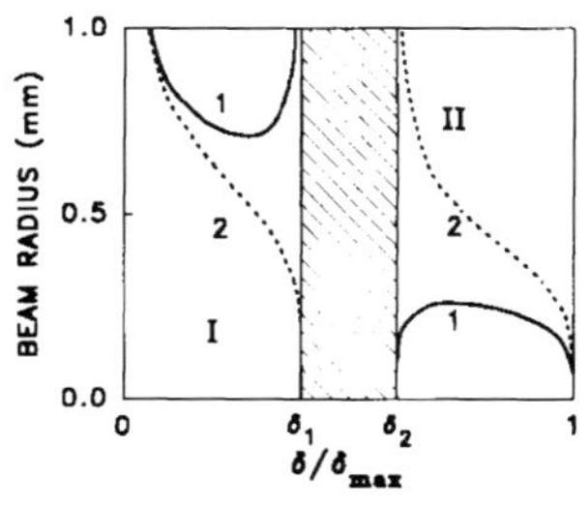

Fig. 4. Variation of the laser mode-spot sizes at the cavity ends versus the stability parameter δ for the representative case of $f_1 = f_2 = 6$ cm, $d_1 = 65$ cm,$d_2 = 120$ cm, and $\gamma = 1.54$. Plots 1 and 2 depict the spot sizes at the cavity ends corresponding to d_1 and d_2, respectively. The results have been obtained neglecting the effect of gain guiding.

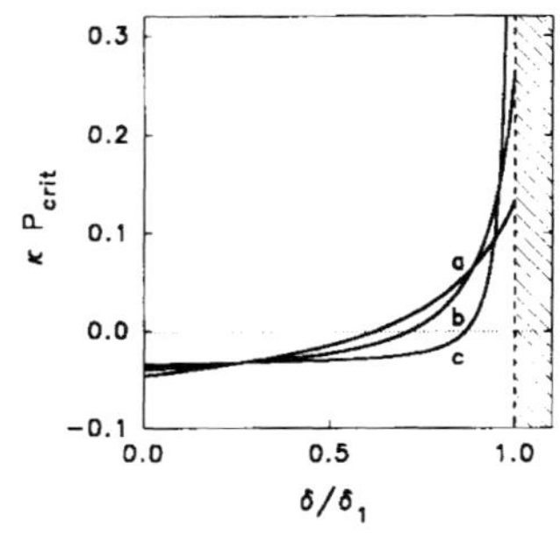

Fig. 5. Variation of κ normalized to the inverse of the critical power for self-trapping P_{cr}, versus the stability parameter δ in region I of the stability range for a linear loss of 10% introduced by the intracavity aperture. Plots are given for various values of the cavity asymmetry parameter γ as defined by (7): Plot (a), $\gamma = 2.11$; (b) $\gamma = 1.38$;and (c), $\gamma = 1.05$. Here $f_1 = f_2 = 6$ cm, and $d_2 = 120$ cm was kept constant and the length of the short collimated cavity arm was varied. The results have been obtained neglecting the effect of gain guiding.

of the spatial variation of the beam profile due to self-focusing caused by the optical Kerr-effect in the gain medium [20]–[22]. In combination with some aperturing of the beam (caused either by the spatial profile of the gain or by an additional "hard" aperture) the self-focusing can be converted into an efficient self-amplitude modulation. Since KLM relies on the nonresonant Kerr nonlinearity, it is applicable to a broad range of laser materials regardless of wavelength and spectroscopic parameters [17], [40]–[45]. In what follows, we summarize the main aspects of KLM laser cavity design and present guidelines for the optimization of the SAM mechanism.

Before describing the nonlinear behavior of the resonator we shall briefly discuss the linear properties of the system. In the four-mirror cavity outlined in Fig. 3, the effective optical distance d_f [46] between the two focusing mirrors (focal lengths: f_1 and f_2) must be close to $f_1 + f_2$ for the cavity to support a stable TEM_{00} Gaussian mode [47]. We therefore define a stability parameter δ using the relation $d_f = f_1 + f_2 + \delta$. The range of values of δ for which the resonator is stable is divided into two separate regions (I and II) of equal length as indicated in Fig. 4. Besides the stability parameter δ, the parameter

$$\gamma = \frac{f_1^2}{f_2^2}\frac{d_2 - f_2}{d_1 - f_1} \quad (7)$$

which we shall refer to as the cavity asymmetry factor, can sensitively affect the cavity mode. Without loss of generality we may assume $\gamma \geq 1$. The larger γ, the broader is the gap between regions I and II (Fig. 4) of the stability range.

The Kerr nonlinearity associated with the gain medium introduces a power-dependent change in both the position and diameter of the resonator mode waist between the focusing mirrors. As a consequence, the beam diameters plotted in Fig. 4 are subject to power-induced variations, which can be converted into a power-dependent loss by inserting an aperture at one of the end mirrors in the cavity. Recently, analytic theories have yielded closed-form algebraic expressions for the effective modulation index κ of the SAM action introduced by KLM [21], [22], [48]. Irrespective of the specific cavity configuration, κ scales with the small-signal loss introduced by the aperture and is inversely proportional to the critical power for self-trapping P_{crit} [21]. For Ti:sapphire P_{crit} was estimated to be ≈2.6 MW [49]. In order to achieve modelocking, κ must be positive, and it has been shown that aperturing the resonator mode at the cavity end corresponding to d_1 can lead to $\kappa > 0$, provided that δ is adjusted towards the edge of the stability range indicated by δ_1 and δ_2 in Fig. 4 [21].

Figure 5 plots κ normalized to P_{crit}^{-1} versus δ for different values of γ for region I of the stability range. It is clear from Fig. 5 that efficient saturable absorber action can only be obtained for δ close to δ_1. A resonator with an asymmetry factor close to unity allows a higher κ to be achieved, but the condition $\kappa > 0$ requires operation closer to the stability limit δ_1. Thus a tradeoff between κ (i.e., the strength of SAM) and laser stability is required and this can be achieved by a suitable choice of γ and δ. The dependence of κ on δ and γ is similar in region II (Fig. 4), except that the maximum SAM effect occurs at the lower boundary δ_2.

With these considerations in mind one can design and optimize a laser cavity for KLM operation. Using a Brewster-angled Ti:sapphire slab as the gain medium, we constructed a five-mirror cavity as shown in Fig. 6. For our system the asymmetry factor is $\gamma = 1.75$, which has been found to represent a good tradeoff between high SAM efficiency and robustness against environmental perturbations. The Brewster faces of the gain medium break the cylindrical symmetry of the resonator, a feature that has important implications for KLM. If the gain medium is longer than the confocal parameter of the focused cavity mode, which applies to almost all practical KLM systems operated close to the boundary of the stability range, the Kerr-induced change in the beam diameter at the resonator ends is larger *in* the plane of the cavity (x direction) than *perpendicular* to it (y direction) [50]. This Kerr-induced astigmatism is a consequence of the different self-focusing actions in the two transverse dimensions, which results from a difference in the corresponding confocal parameters of the cavity mode in the gain medium. Further aspects of KLM in four-mirror systems using Brewster-angled gain media are discussed in [50]. Fig. 7 schematically illustrates the low-power and high-power beam profiles at $M1$ for a δ-parameter ensuring $\kappa > 0$. As a consequence of the more

IEEE JOURNAL OF QUANTUM ELECTRONICS, VOL. 30, NO. 4, APRIL 1994

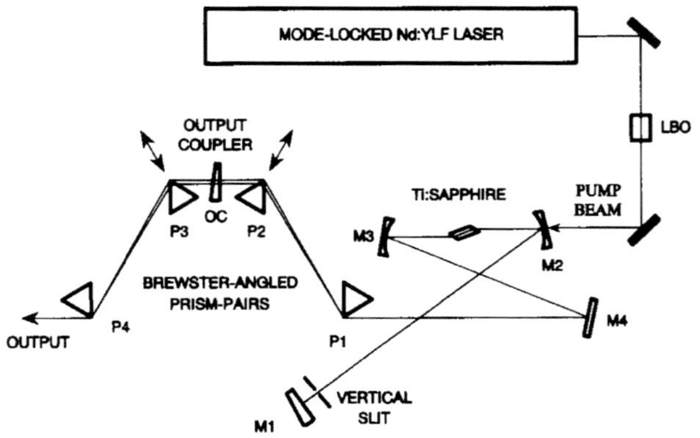

Fig. 6. Schematic of the experimental arrangement for a KLM Ti:sapphire laser. $M1, M4$ HR plane mirrors; $M2, M3$, HR curved mirrors with radii of curvature of 10 cm; OC, output coupler with 5% transmission; $P1$–$P4$, fused silica prisms. The distance between $M1$ and $M2$ is $d_1 = 65$ cm and the optical path length from $M3$ to the output coupler OC is $d_2 = 110$ cm.

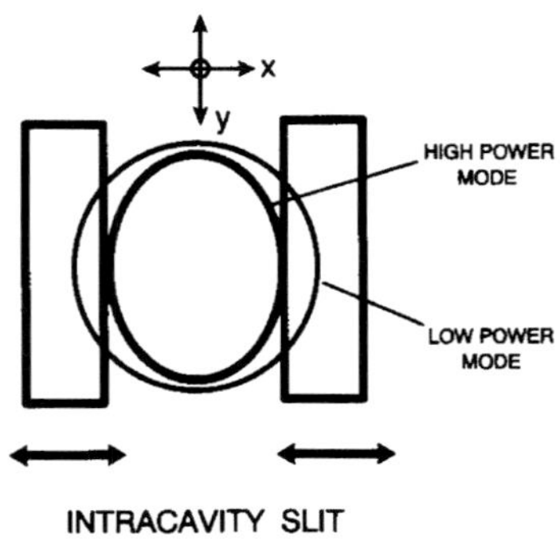

Fig. 7. Idealized schematic indicating the action of the hard aperture in the selection of the laser operation mode. In ed operation the action of self-focusing in the gain medium results in a reduced mode-size at the aperture and thus a reduced overall intracavity loss. The change in mode diameter at the cavity end is generally different for the different transverse directions x and y, hence the proper choice of the orientation of the slit is essential for optimum self-amplitude modulation.

pronounced KLM action along the x direction, a stronger SAM effect relative to the small signal loss can be achieved with a vertical slit than with a circular aperture. Following these considerations, we employ a vertical slit close to $M1$ for maximum SAM in the Ti:sapphire laser, and adjust the separation of the focusing mirrors so that δ is close to δ_1 in Fig. 4. The latter can be achieved by increasing the distance $M2$–$M3$ until the output power starts to notably decrease, whilst observing the spots on the cavity end mirrors (see Fig. 4). The SAM can be further increased by reducing the slit width, however this is ultimately limited by the increase in linear loss.

After discussing the optimization of the SAM mechanism we now address the perturbative effects, which need to be minimized for efficient exploitation of the soliton-like pulse shaping. To keep the adverse effects of higher-order phase dispersion in the cavity as low as possible, all cavity mirrors consist of a single stack of standard quarterwave layers [51]. Further, for intracavity GDD control a pair of fused silica Brewster-angled prisms [52] is employed, which exhibits a particularly low TOD relative to GDD at ≈ 0.8 μm as compared to other optical glasses (see Section V). In order to avoid excessive prism separation when using low-dispersion fused silica prisms, the positive intracavity material GDD needs to be minimized in the cavity. With a 4-mm-long Ti:sapphire crystal being the only other intracavity element, a net negative intracavity GDD can be attained using a prism apex-to-apex separation of ≈ 60 cm.

As discussed in the preceding section, another perturbation to ideal soliton formation arises from the fact that the actions of SPM and negative GDD are separated in space and time. The operator $\hat{P}(z)$, which accounts for this effect, introduces the smallest perturbation at the dispersive cavity end (at OC in Fig. 6) [3], leading to the shortest steady state pulse duration at this position inside the cavity (see Section V). Hence, output coupling should be realized at the dispersive resonator end for optimized performance. In this case a second pair of identical prisms with the same separation as the internal pair is needed outside the cavity ($P3, P4$ in Fig. 6) to eliminate the spatial frequency sweep introduced by $P1$–$P2$. Using a suitably thin output coupler substrate (BK7, 2 mm) that introduces significantly less positive GDD than the gain medium, the overall GDD outside the cavity can be controlled by varying the insertions of $P3$ and $P4$.

Unlike slow-saturable absorber modelocking, KLM does not start on its own from free-running operation even in an optimized cavity. This arises from the low value of the product of κ and the average intracavity power in practical KLM lasers [34], [35]. This problem can be circumvented by using some kind of active initiation of the passive modelocking in these systems. In the first experiments KLM was started by inducing a small but rapid change in the resonator length [17], [18]. Initiation of passive modelocking without the need for any additional intracavity elements can also be provided by synchronous pumping [53]. Further alternatives are: active modelocking [54], regenerative modelocking [55]–[57], moving mirror [58]–[61], semiconductor [19], or dye-saturable absorber [41], [62], [63] modelocking.

In the experiments presented in the following sections we have used synchronous pumping as the starting mechanism for self-sustaining KLM of a Ti:sapphire laser [53]. Unlike synchronously-pumped dye lasers such a system does not require a strict cavity length control because the ed pulse train loses synchronization with the pump train as soon as passive modelocking takes over. The pump source is a CW ed (80 MHz) Nd:YLF laser (Quantronix 4216D). The pump laser output is frequency-doubled using a temperature-phase-matched 20–mm-long LBO crystal, which yields up to 8 W of average power at 527 nm. At typical pump levels of 5–6 W the CW Ti:sapphire laser delivers 0.6–0.7 W output power at ≈ 0.8 μm with a pump absorption efficiency of $\approx 55\%$ (single pass) and output coupler transmission of 5%. The ed average output power ranges from 150 mW to 400 mW depending on the linear loss introduced by the slit. The low-dispersion cavity mirrors used in the experiments reported below are centered at 780 nm unless otherwise stated.

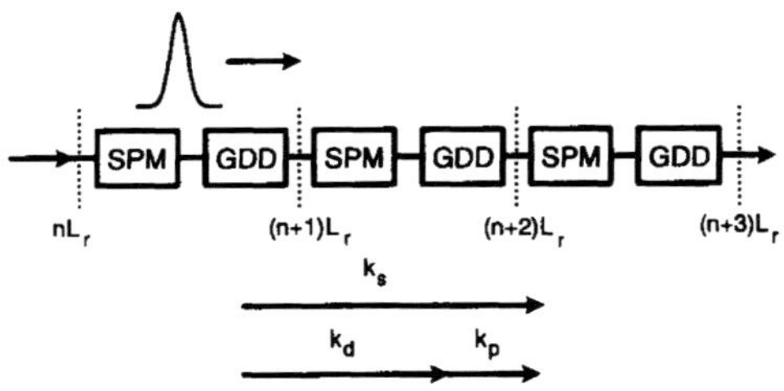

Fig. 8. Qualitative representation of the periodic action of self-phase modulation (SPM) and group delay dispersion (GDD) over repeated cavity round trips. Propagation through a periodic structure leads to light scattering at discrete frequencies where the phase matching condition between the incident and scattered light is fulfilled.

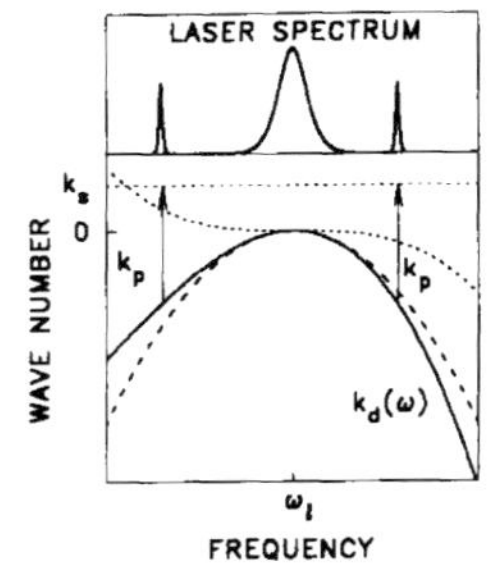

Fig. 9. Wave numbers of the solitary pulse and the dispersive wave versus angular frequency in the vicinity of the center laser oscillation frequency ω_l. The full line plots the overall wavenumber $k_d(\omega)$, whereas the dashed and dotted lines represent the GDD and TOD contributions ($D_3 < 0$)to $k_d(\omega)$, respectively. The upper trace sketches qualitatively the expected laser spectrum.

IV. Stable Mode Locking

In this section we will explore the physical mechanisms giving rise to the instabilities that ultimately limit the achievable pulse duration in femtosecond oscillators. The perturbations described by the leading terms in the expansion of the perturbation operators $\hat{P}$ and $\hat{D}_h$ can be shown to give rise to an energy growth in the pulse wings at the expense of an energy loss from the center of the pulse. These undesirable pulse shaping effects tend to increasingly impair modelocking as the negative GDD, and thus the pulsewidth is reduced. In lasers having gain bandwidths of the order of 100 THz the influence of gain dispersion on pulse formation is negligible, therefore we shall focus on the implications of the periodic perturbation and high-order phase dispersion in the cavity. In what follows we demonstrate that the ed spectrum sensitively reacts to these adverse perturbative effects and present a frequency-domain characterization of the perturbed ed system.

A. Effects of Periodic Perturbation

modelocking in the presence of separated SPM and negative GDD can be thought of as a solitary pulse propagating in a hypothetical optical fiber that consists of alternating segments of SPM and GDD, as shown in Fig. 8. It is well known from classical optics that coherent light traversing a periodic structure (which we shall simply refer to as a grating) is subject to coherent scattering processes, which extract energy from the propagating radiation. In linear scattering, with the grating parameters independent of the electromagnetic field, the time-dependence of the scattered light follows that of the incident radiation. The presence of a nonlinearity may lead to a rather different behavior. In the specific case depicted in Fig. 8 solitary pulse propagation relies on SPM, which cannot be exploited by the low-intensity scattered wave. Hence it necessarily will spread out in time due to dispersion and give rise to a quasi-CW background accompanying the source wave [64], [65]. In soliton physics this is often referred to as a dispersive wave.

In CW-pumped lasers pulse propagation is effectively indefinite, hence an arbitrarily small energy loss rate caused by scattering to the CW background is able to destroy the pulse. A long-relaxation-time gain medium cannot make up for any continuous energy loss from the pulse, since it equally amplifies the undesirable CW background. Therefore SAM is essential for the creation of a gain window that discriminates between the solitary pulse and the unwanted scattered radiation. This leads to a net gain for the solitary pulse and a net loss for the CW background. Stability prevails as long as SAM is capable of compensating for the energy loss from the solitary pulse caused by scattering.

In the steady state the scattered wave is made up of quasi-monochromatic components, whose frequencies are determined by the requirement of phase matching between the fundamental and the scattered waves [64], [65]: $k_d = k_s \pm k_p$, where $k_d = (1/L_r)\{(D/2)(\omega - \omega_l)^2 + (D_3/6)(\omega - \omega_l)^3$, $k_s = -(1/L_r)(D/2T^2)$, and $k_p = 2\pi/L_r$ are the wave numbers of the dispersive wave, solitary wave, and periodic perturbation, respectively. Here $T \approx \tau/1.76$, where τ is the FWHM pulsewidth. Terms linearly varying with ω have been ignored in the expressions of k_d and k_s because they cancel out of the phase matching condition. Fig. 9 illustrates this condition graphically together with the corresponding optical spectrum in the presence of TOD.

Fig. 10 shows the output spectrum of our KLM Ti:sapphire laser for reduced small signal aperture loss. This resulted in a weaker SAM and a higher intracavity pulse energy. Fig. 10 reveals that a reduction in the negative intracavity GDD increases the frequency separation between the sidebands and the principal spectrum, which is clear by inspection of the phase matching condition $k_d = k_s - k_p$ and Fig. 9. Simultaneously, the increase in the r-parameter of the discrete system makes the scattered radiation grow, eventually resulting in a splitting of the solitary pulse into two more or less identical pulses as shown in Fig. 11(a). Two pulses sharing the energy of one lead to an r value reduced by a factor of 4 [see (3)], and thus to a substantial reduction in the intensity of the CW background. This manifests itself in the disappearance of the sidebands in Fig. 11(b). For pulse durations close to 10 fs (i.e., at low values of $|D|$) pulse splitting has also been observed without preceding sideband generation. We surmise that in this operating regime the sidebands can be so far off the center laser frequency that they are subject to strong attenuation by the finite gain bandwidth of the oscillator.

Comprehensive experimental investigations have shown that an optimized SAM mechanism (see Section III) and operation

IEEE JOURNAL OF QUANTUM ELECTRONICS, VOL. 30, NO. 4, APRIL 1994

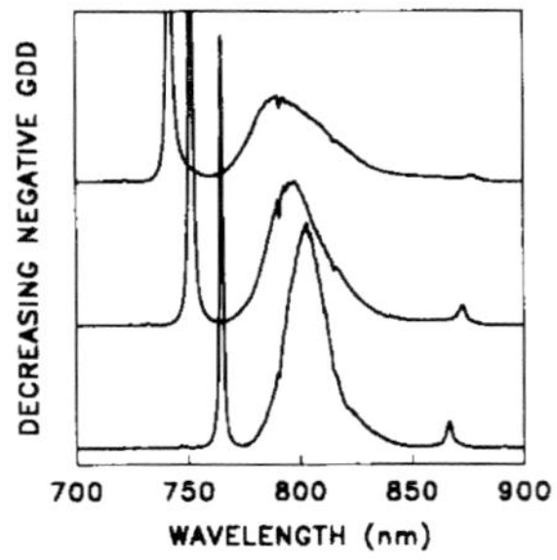

Fig. 10. ed spectra of the Ti:sapphire laser with reduced self-amplitude modulation (as compared to optimized conditions) for three different operating round-trip GDDs. The position of the sidebands observed depends on the magnitude of the GDD, with the sideband separation increasing with decreasing GDD.

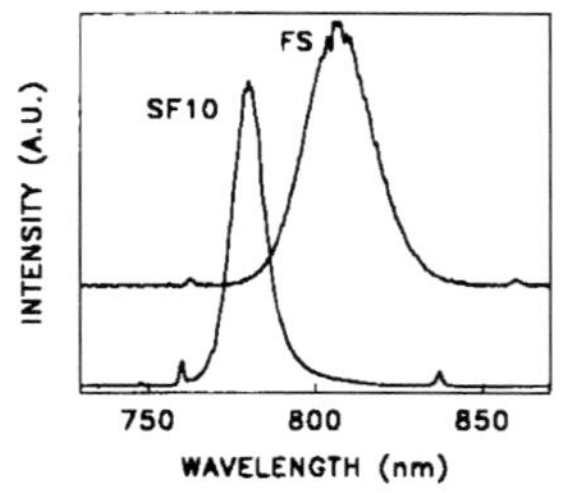

Fig. 12. ed spectra of the Ti:sapphire laser employing fused silica (FS) and SF10 prisms for non-optimized self-amplitude modulation. The strongly asymmetric location of the sidebands obtained with the SF10prisms indicates a substantially higher intracavity TOD than that introduced by the fused silica prisms.

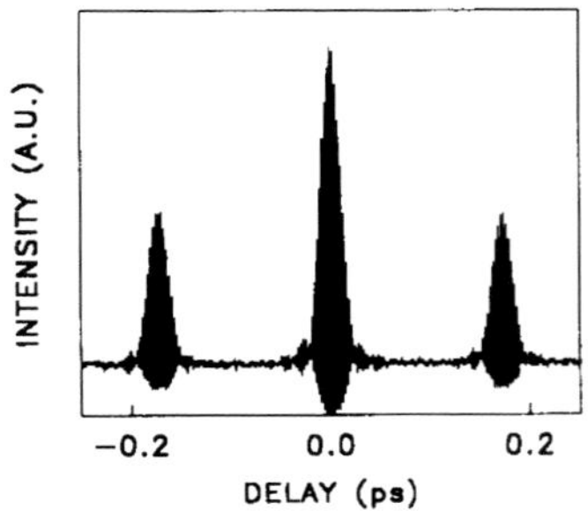

(a)

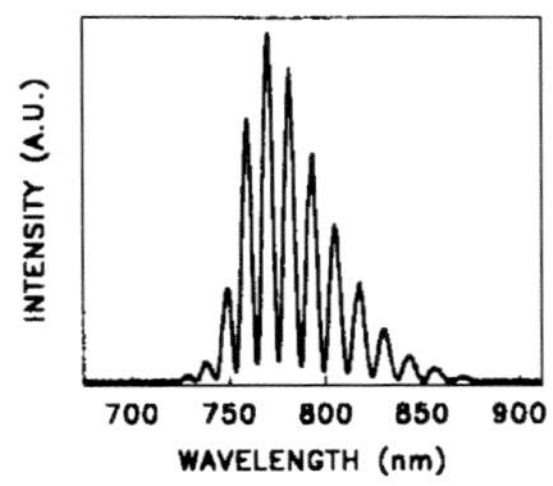

(b)

Fig. 11. (a) Collinear interferometric autocorrelation and (b) the corresponding spectrum of the laser output in double-pulse operation mode. The pulse separation can range from less than 100fs up to the picosecond regime and remain stable for indefinite periods. Neither the origin of coupling which keeps the separation of the two pulses constant nor the effects determining the pulse separation have been explored so far.

at moderate intracavity pulse energies can efficiently suppress dispersive waves that arise from discrete pulse shaping and thus prevent pulse splitting from occurring. Generally, the larger the value of r, the stronger SAM required to suppress the CW background and maintain a single stable pulse in the cavity. It has been found (see Section V) that the optimized self-modelocking action is capable of maintaining a stable solitary pulse down to the 10 fs regime in the Ti:sapphire laser as long as the intracavity pulse energy does not significantly exceed 50 nJ. Estimations of D and ϕ have yielded r-values up to 3 for these operating conditions.

Before leaving this discussion we shall briefly address the observed asymmetry in the location and height of the sidebands. In Fig. 10 the blue-shifted sideband is located closer to the center frequency than the red-shifted component. A comparison of the measured spectrum with Fig. 9 implies an overall negative TOD in the resonator, which is characteristic of prism-pair dispersion-controlled lasers in the wavelength range $\lambda < 0.9$ μm (see Section V). Fig. 10 reveals a sideband asymmetry becoming more and more pronounced as $|D|$ decreases, which points to an increasing role of TOD (due to increasing β_3) with decreasing GDD in the cavity. The unequal height of the sidebands is a direct consequence of the different spectral intensities of the main pulse spectrum they are fed by. Fig. 12 compares the sideband structure of ed Ti:sapphire lasers employing fused silica and SF-10 prisms. The striking difference in sideband asymmetry is the consequence of a far higher TOD of the SF-10 prism pair than that produced by the pair of fused silica prisms (see Section V).

B. Effects of Higher-Order Cavity Dispersion

Whilst instabilities originating from $\hat{P}(z)$ can be suppressed by optimizing SAM and keeping the intracavity pulse energies at moderate levels (by suitable choice of the output coupling) higher-order cavity dispersion can adversely affect femtosecond modelocking even for low W and high κ values. In this section we concentrate on the effect of TOD, which is the leading term in $\hat{D}_h$ [39], assuming that the gain dispersion can be neglected.

In the following analysis we retain terms up to the third order of the Taylor-expansion of the round-trip phase delay and neglect the higher-order contributions. Although fourth-order dispersion usually does have a notable effect if $\Delta\nu_l >$ ~30 THz, (see Section V), the following considerations are qualitatively valid, as long as $\beta_4 \ll \beta_3$. Neglecting fourth-order dispersion, the cavity GDD is a linear function of the frequency (see $D(\omega)$ in Fig. 13), and the separation of the zero-dispersion frequency ω_0 from the center laser frequency ω_l is given by $\Delta\omega_0 = \omega_0 - \omega_l = -3D/D_3$. For large values of $|D|$, i.e., relatively long pulses, $|\Delta\omega_0|$ is much larger than $\Delta\omega_l = 2\pi\Delta\nu_l$, the bandwidth of the ed laser and hence the influence of TOD is negligible. However, as $|D|$ is reduced, $|\Delta\omega_0|$ becomes comparable to $\Delta\omega_l$, and a part of the laser spectrum will experience an overall positive GDD in the cavity. In Fig. 13 this occurs on the low-frequency (long-wavelength) side of the spectrum as a consequence of the

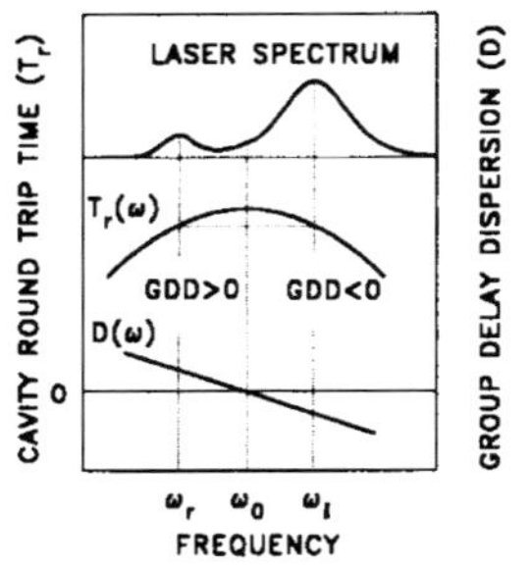

Fig. 13. Variation of the cavity round-trip time (group delay) and its derivative (GDD) versus frequency in the presence of TOD in the cavity. The upper trace is a qualitative representation of the corresponding laser spectrum.

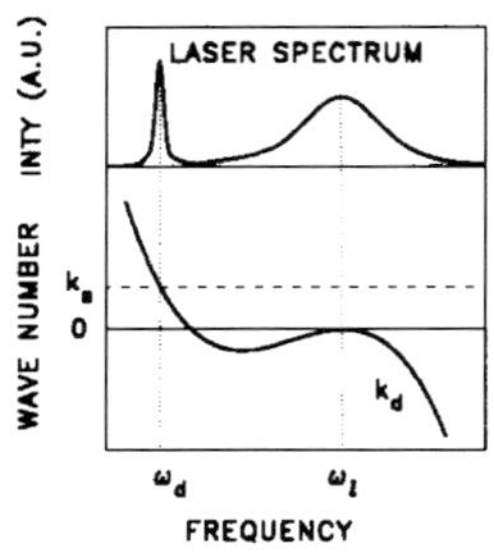

Fig. 14. Dispersion diagram for a solitary pulse and dispersive waves in the presence of strong TOD. The dashed line denotes the dispersive resonance position, ω_d. The top diagram shows qualitatively the solitary pulse spectrum in the presence of a dispersive wave originating from TOD.

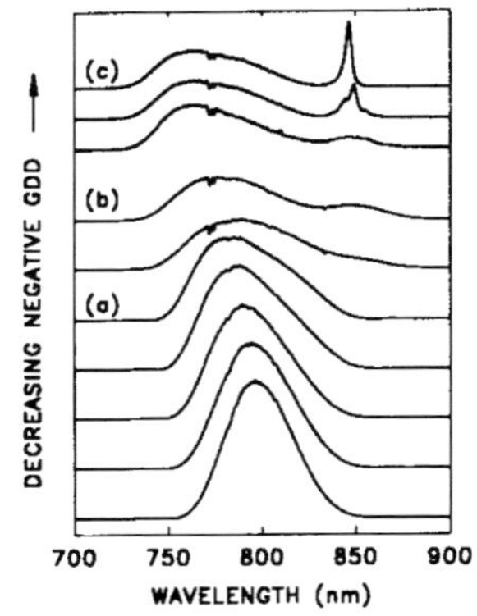

Fig. 15. ed spectra as a function of the intracavity GDD for a prism separation of ≈60 cm and an intracavity pulse energy of $W \approx 40$ nJ. The relative change in the net intracavity GDD between the adjacent spectra up to plot (b) is $\Delta D \approx 10$ fs^2. The appreciable blue-shift of the ed spectrum with decreasing negative GDD is a consequence of the corresponding blue shift in the zero dispersion frequency ω_0 (Fig. 13) and the fact that the ed laser favors oscillation in the negative GDD regime. This is because $D < 0$ supports a shorter pulse duration and hence a higher peak power, allowing reduction of the round-trip cavity loss (in the presence of SAM) and a maximization of the total field energy in the oscillator. The continuing blue-shift of the ed laser spectrum is eventually stopped, presumably by the sharply falling edge of the Ti:sapphire fluorescence line. The role of the gain dispersion here is not to limit the laser bandwidth but to prevent the ed spectrum from moving away from the approaching positive-dispersion wavelength regime.

assumed negative TOD ($D_3 < 0$). Since the pulse component experiencing positive GDD must copropagate with the main solitary pulse under steady-state conditions, one expects the appearance of a spectral resonance at $\omega_r = \omega_l - 6D/D_3$, where the cavity round-trip time equals that of the ed pulse, i.e., $T_r(\omega_r) = T_r(\omega_l)$ [66]. Broadening of the non-soliton wave due to (positive) dispersion is prevented by the gain window created by SAM.

In the presence of TOD, phase matching between linear dispersive waves and the solitary pulse can be achieved even in the absence of periodic perturbations [67]–[70]. For high values of the net intracavity TOD relative to GDD, the dispersive resonance is close to the center of the ed pulse spectrum, and efficient dispersive wave generation may occur (Fig. 14). The phase matching criterion [33], [69] can be written as

$$\Delta\omega_d^2 + \frac{D_3}{3D}\Delta\omega_d^3 = -\frac{1}{T^2} \tag{8}$$

where $\Delta\omega_d = \omega_d - \omega_l, T = \tau/1.76$, and τ is the solitary pulse duration. The measured value of $\Delta\omega_d$ can be used to determine the ratio of TOD to GDD in the cavity.

The evolution of the ed laser spectrum [33] with decreasing $|D|$ is shown in Fig. 15 starting with a spectrum corresponding to a pulse duration of ≈24 fs. With increasing prism insertion the spectrum broadens and the corresponding pulsewidth gets shorter until the broadest structureless spectrum is achieved (plot (a) in Fig. 15). The rapid broadening of the spectrum is followed by the appearance of a broad spectral resonance (plot (b) in Fig. 15), indicating that the GDD changes sign within the ed spectrum. As $\omega_r < \omega_l$, we have $D_3 < 0$, in agreement with the implication of the observed sideband asymmetry in Figs. 10 and 12. As a result of further reduction in $|D|$, the long wavelength tail tends to break away and causes a severe degradation of pulse quality.

Eventually a sharp resonance appears at $\Delta\omega_d = -0.24$ fs^{-1} (plot (c) in Fig. 15), which indicates the onset of dispersive wave generation. Substituting $\Delta\omega_d$ into (8), we obtain $D_3/D \approx 15$ fs for plot (c) in Fig. 15. With the calculated TOD of $D_3 \approx -8 \times 10^2$ fs^3 this gives $D \approx -50$ fs^2 for curve (c) of Fig. 15. Here we took advantage of the relative insensitivity of the net TOD to errors in the measurement of the prism insertion, which can severely affect the accuracy of the calculated net intracavity GDD. From these data we deduce $D \approx -80$ fs^2 and $\beta_3 \approx 0.55$ for the optimum operating point represented by plot (a) in Fig. 15. An increase in β_3 beyond this value (due to a reduction of $|D|$) leads to a rapid degradation of pulse quality, the emergence of a CW background, and finally to a dropout of ed operation.

In summary, this section has shown that the periodic perturbation resulting from discrete soliton-like pulse shaping and higher-order cavity dispersion are potential sources of instabilities in broadband solid-state femtosecond lasers. The presented experimental results provide evidence that Kerr-lens modelocking has the potential for suppressing instabilities introduced by the spatially separated actions of negative GDD and SPM, at least for $r < 3$. In the wavelength range around 0.8 μm the ultimate stability limit appears to be set by residual TOD in the prism-pair controlled cavity. As we shall see in the next section this is not necessarily the case for lasers operating at longer wavelengths in the near-infrared spectrum.

IEEE JOURNAL OF QUANTUM ELECTRONICS, VOL. 30, NO. 4, APRIL 1994

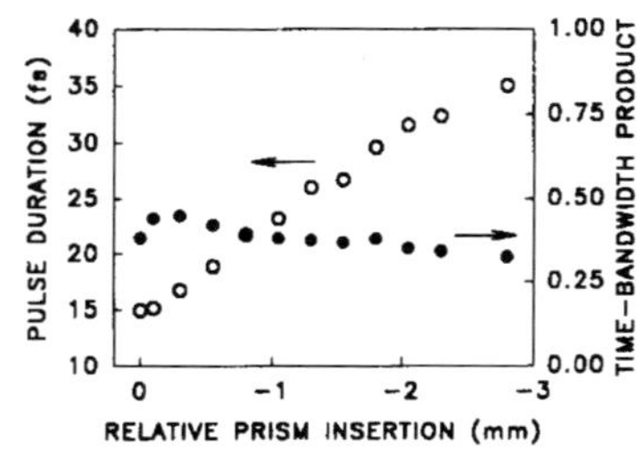

Fig. 16. Pulse duration (hollow circles) and time-bandwidth product (filled circles) as a function of glass insertion for a prism separation of 74 cm and an intracavity pulse energy of $W \approx$ 40nJ. The data have been evaluated by assuming a sech^2 pulse shape. A 1 mm decrease in prism insertion along the prism axis increases the net negative GDD by ≈ -95 fs^2 and the net negative TOD by ≈ -60 fs^3. At the zero reference point the calculated net negative GDD and TOD are $D = -110 \pm 40$ fs^2 and $D_3 = -760 \pm 30$ fs^3, respectively. The errors in the calculation of the net GDD and TOD originate from inaccuracies in the measurement of the optical path length in the prisms and its margin corresponds to an uncertainty of ± 1 mm over a full round trip. The change in the slope of the $\tau(D)$ curve near zero relative insertion is caused by TOD (due to the increased value of β_3).

V. Ultimate Limitations to Pulse Shortening

A. Pulse Duration in the Absence of High-Order Dispersion

If gain dispersion and higher-order cavity dispersion are negligible, the only perturbation to ideal soliton formation comes from the spatially separated actions of GDD and SPM. Retaining terms up to order r^3 in the expansion of the perturbation operator $\hat{P}(z)$, the solitary pulse can be generally shown to be bandwidth-limited at the ends of the cavity (mirrors $M1$ and OC in Fig. 6) with time-bandwidth products close to that of the unperturbed soliton [3]. The corresponding pulse duration can be written as:

$$\tau(z) = \tau_s + \alpha(z)\phi W \tag{9}$$

where τ_s is the soliton pulse duration given by (2) and the coefficient $\alpha(z)$ has been computed [23] to be $\alpha(z_1) = 0.10$ at the dispersive resonator end and $\alpha(z_4) = 0.25$ at the nondispersive end of the cavity (OC and $M1$ in Fig. 6, respectively) [71]. Although (9) is a reasonable approximation for $r < 10$, Fourier-limited pulses with time-bandwidth products of $\approx$0.3 at the resonator ends can be expected only up to $r \approx 1 - 3$ in practical systems, depending on the efficiency of SAM [23]. The first term on the RHS of (9) implies a soliton-like dependence of τ on D, whereas the position-dependent perturbation term $\alpha(z)\phi W$ resulting from $\hat{P}(z)$ is responsible for the different pulse parameters at the different cavity ends observed in recent experimental investigations [3], [29], [30], [60].

The influence of the intracavity GDD appears only in the soliton contribution to τ of (9), predicting a linear variation of the ed pulsewidth with D. For the measurement of τ as a function of D in a reasonable dynamic range, the separation of the fused silica prisms was increased to $\approx$74 cm. Fig. 16 shows the pulse duration and the time-bandwidth product, assuming a sech^2 pulse shape, versus prism insertion for an intracavity pulse energy of $W \approx 40$ nJ in the Ti:sapphire laser. We can see that the pulsewidth increases approximately linearly with increasing negative GDD, in agreement with the theoretical prediction. This linear dependence provides important evidence for the soliton-like nature of the pulse (i.e., the dominance of soliton-like shaping effects) even in the 10-fs regime. From the slope of $\tau(D)$ we evaluate $\phi W \approx 42$ fs. Using the above value of the intracavity pulse energy these measurements give $\phi = 1.05 \times 10^{-6}$ W^{-1} [72]. The zero reference position in Fig. 16 corresponds to a calculated net GDD of $D = -110 \pm 40$ fs^2, which yields $\tau \approx 14 \pm 3$ fs upon substitution in (9). This compares reasonably to the minimum measured pulse duration depicted in Fig. 16.

An important general conclusion from these considerations is that the parameter ϕ can be determined by measuring the change in pulse duration with respect to a change of the GDD by simply varying the prism insertion in a parameter range where τ is a linear function of D, i.e., where high-order dispersion can be neglected. Calculation or measurement of the intracavity GDD for a fixed prism position will then allow a prediction of the pulse duration for the whole range of $|D|$ values greater than $|D_{\text{crit}}|$, the critical operating value of negative intracavity GDD where one of the dispersion scaling parameters β_g, β_3 or β_4 approaches the order of unity.

Equation (9) predicts that pulse shortening by reducing the negative GDD is limited by the finite value of the perturbation term $\alpha\phi W$. However, because of increasing computational problems for $r \to \infty$ there is no strict evidence that (9) keeps validity for $D \to 0$. Experimentally a reduction in the value of $|D_{\text{crit}}|$ can be achieved by reducing high-order phase dispersion contributions in the laser cavity. At reduced values of $|D_{\text{crit}}|$ more reliable information concerning the ultimate limitations originating from the discrete nature of soliton-like pulse shaping can be obtained (see next section).

B. Implications of Higher-Order Dispersion

In practical systems, high-order dispersive effects become significant as $D \to 0$ and the pulse duration stops decreasing when D reaches the critical value D_{crit}, when one of the scaling parameters defined by (4)–(6) becomes the order of unity. With the prism separation of 74 cm the minimum achievable pulsewidth was $\approx$ 15 fs at $D_{\text{crit}} \approx -110$ fs^2. This value of D_{crit} along with the calculated values of D_g, D_3 and D_4, yields upon substitution into (4)–(6), $\beta_3 \approx 0.52$ and $\beta_g, \beta_4 \ll \beta_3$. This result demonstrates, in accordance with the spectral evolution shown in Fig. 15, that in the Ti:sapphire laser (at $\lambda \approx 0.8$ μm) TOD determines D_{crit} and thus limits the minimum achievable pulse duration.

It has been pointed out previously [63] that minimized prism insertion and separation ensures minimum TOD in the Ti:sapphire cavity. However, this comes at the expense of a reduction in the range over which the pulsewidth and bandwidth can be varied. With a prism separation reduced to $\approx$60 cm the Ti:sapphire laser routinely produces nearly bandwidth-limited pulses of $\approx$12–13 fs [33], but the maximum achievable pulse duration also reduces below 30 fs. The minimum pulse duration corresponds to plot (a) in Fig. 15, for which we have evaluated $D_{\text{crit}} \approx -80$ fs^2 and $\beta_{3,\text{crit}} \approx 0.55$ (see Section IV.B). This latter value compares favorably to

$\beta_{3,\mathrm{crit}} \approx 0.52$ for a 74-cm prism separation. Thus, we may conclude that the critical value of β_3 is approximately constant and the allowed reduction of the product of $|D_{\mathrm{crit}}|$ and the minimum achievable pulse duration $\tau_{\min}$ is [according to (5)] proportional to that of $|D_3|$ the net TOD in the cavity. This, in combination with the approximate expression of the pulse duration given by (5) [i.e., neglecting the perturbation term in (9)] yields $\tau_{\min} \sim \sqrt{D_3}$ [29]. It is noteworthy that the $\beta_{3,\mathrm{crit}}$-values evaluated from our experiments are notably higher than those obtained previously in longer-pulse-duration Ti:sapphire systems [26], [57], [63], where typically $0.2 < \beta_{3,\mathrm{crit}} < 0.3$ for the shortest reported pulses. The larger β_3-values for sub-20-fs systems is presumably due to the fact that $|D_3|$ decreases with increasing wavelength leading to a reduced effective value of TOD experienced by the broad long-wavelength edge of the ed spectrum.

Generally it is expected that the periodic variation of the pulse parameters predicted by the third-order theory of (3) (expansion of $\hat{P}(z, \hat{N}, \hat{D})$ up to the third order in r) tends to be masked to some extent by high-order dispersive effects (in Ti:sapphire by TOD) as the intracavity GDD approaches D_{crit}. In addition, the positive GDD of the Kerr-medium also influences pulse propagation in the sub-20-fs regime and can modify the predictions of the theory. Experimentally we have found that in the sub-20-fs domain the laser can operate in two distinct regimes that exhibit different chirp characteristics, depending on the specific operating conditions. By monitoring the fringe-resolved autocorrelation the residual linear chirp can be eliminated by proper adjustment of the prism insertions in the external prism pair. Despite the number of various effects influencing modelocking, our experiments with the Ti:sapphire laser have led to the conclusion that (9) with $\alpha = 0.1$ provides a reasonable prediction of the minimum achievable solitary pulse duration (after external compensation) for pulsewidths down to ≈ 10 fs.

In the rest of this section we explore the main contributions to high-order phase dispersion in visible and near-infrared femtosecond lasers. Generally, three different sources of high-order cavity dispersion need to be considered, namely the dielectric resonator mirrors, intracavity optical materials, and the prism pair.

The reflectivity and phase characteristics of dielectric coatings can be expressed in terms of the normalized frequency parameter $(\omega - \omega_0)/\omega_0$. Here ω_0 is an arbitrary reference frequency, which is chosen to coincide with the center frequency of the high-reflectivity range. This implies that the "usable" bandwidth of dielectric mirrors $\Delta\omega_m$, regardless of its specific definition, scales with ω_0. For the following discussion we write $\Delta\omega_m$ in the form

$$\Delta\omega_m = \frac{C_m}{\lambda_m} \tag{10}$$

where λ_m is the mirror center wavelength, and the constant C_m is determined by the specific design of the coating and the definition of $\Delta\omega_m$. As an example, we consider a standard single-stack HR mirror consisting of 21 alternating $\lambda/4$-layers of TiO_2 ($n_H = 2.30$) and SiO_2 ($n_L = 1.45$). The GDD and TOD of the mirror as a function of the normalized frequency

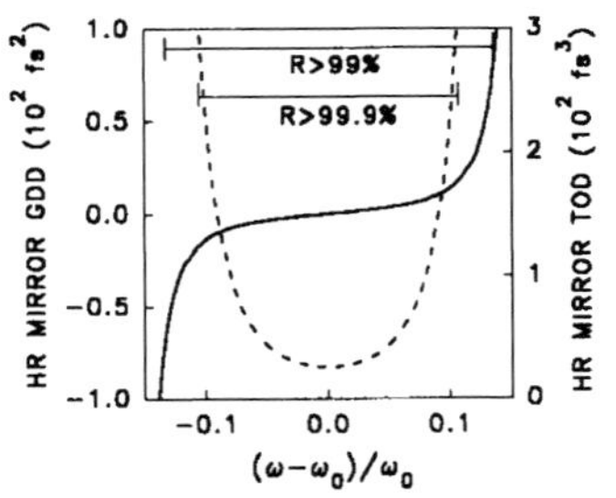

Fig. 17. The calculated GDD (full line) and TOD (dashed line) of a high-reflecting mirror consisting of 21 alternating $\lambda/4$ layers of TiO_2 and SiO_2 versus normalized frequency. The frequency ranges for a reflectivity $R > 99.9\%$ and $R > 99\%$ are indicated.

$(\omega - \omega_0)/\omega_0$ are plotted in Fig. 17. Since the low-dispersion region approximately coincides with that characterized by $R > 99.9\%$, we identify its width with the effective mirror bandwidth for femtosecond lasers. Using this definition of $\Delta\omega_m$ we obtain $C_m = 0.3$ μm/fs for the 21-layer coating of Fig. 17. This yields $\Delta\omega_m/2\pi = 50$–70 THz for $\lambda = 0.7$–0.9 μm (Ti:sapphire, Cr:LiSAF), which reduces to 25–35 THz for the 1.5 μm wavelength range. Similar values can be obtained for single-stack output couplers with transmissions of 5–10%.

For the consideration of intracavity material dispersion, Fig. 18 plots the second- and third-order dispersion of fused silica and the heavy flint glass SF10 for the near infrared using the data published in [73] and [74]. Whereas the material GDD may become negative for increasing wavelength, material TOD is positive throughout the near infrared exhibiting little variation over the bandwidth of any broadband laser oscillator. The positive GDD of the gain host and other intracavity materials is compensated by inserting a Brewster-angled prism pair [52] in the resonator. The round-trip GDD and TOD of fused silica and SF10 prism pairs are depicted in Fig. 18 for the ideal case of zero prism insertion. Unlike material dispersion, prism-pair delay lines have negative GDD over the near-infrared spectral range with a maximum value (minimum negative GDD) reached at a wavelength λ_{pr}, which is characteristic of the prism material [75]. The TOD of the prism pair equals to zero at this wavelength.

The approximate linear dependence of the material GDD on wavelength leads to a net intracavity GDD that exhibits a wavelength-dependence similar to that of the corresponding intracavity prism pair in Fig. 18. However, the maximum value of the net intracavity GDD is reached at a wavelength λ_{cav} that is blue-shifted with respect to λ_{pr}. Because of the relatively low value of the material TOD this blue-shift is typically less than 100 nm in the wavelength range $\lambda < 1$ μm, for operating net GDD's of the order of -100 fs^2. As a consequence, prism-pair dispersion-controlled lasers operating at $\lambda < 0.9$ μm suffer from a residual negative TOD in the cavity, even when using a fused silica prism pair, which has the shortest λ_{pr} among the current commercially available optical glasses.

Evaluating the net TOD in the cavity, we find that owing to its larger λ_{pr}, an SF-10 prism pair introduces over an order-of-magnitude higher TOD at 0.8 μm than a prism pair made of fused silica. This finding is supported qualitatively by the sideband measurements shown in Fig. 12. Nevertheless, the

IEEE JOURNAL OF QUANTUM ELECTRONICS, VOL. 30, NO. 4, APRIL 1994

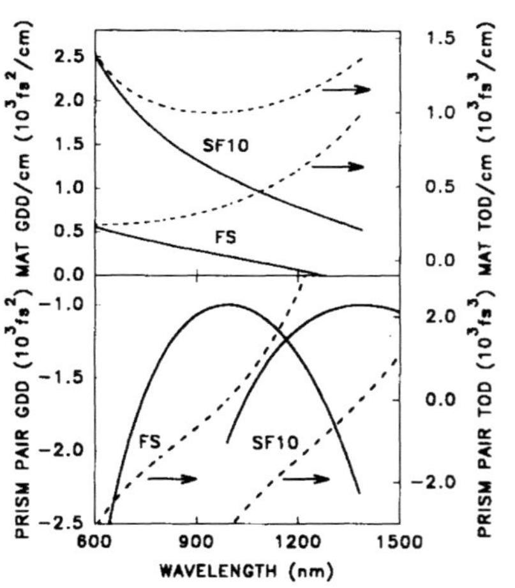

Fig. 18. Upper diagram: Calculated material GDD and TOD for unit lengths of fused silica and SF10. The Sellmeier coefficients of fused silica (FS) and SF10 were taken from [73] and [74], respectively. Lower diagram: Calculated double-pass GDD and TOD for fused silica and SF10 prism pairs for the hypothetical case of zero prism insertion. The prisms are Brewster-angled at λ_{pr}, and the prism separations of 47 cm (FS) and 11 cm (SF10) are chosen to produce a GDD of -1000 fs^2 at λ_{pr}, where the prism negative GDD minimum.

reduction of TOD by using fused silica prisms comes at the expense of a reduced negative GDD of the prism pair (see caption of Fig. 18), which makes larger prism separations necessary. Hence a prerequisite for the use of fused silica prism pairs is a minimization of positive material GDD in near-infrared lasers. For the Ti:sapphire laser to operate in the negative GDD regime in a manageable configuration with fused silica prisms, the gain medium had to be shortened to ≤ 5 mm to minimize the intracavity material dispersion.

The rapid decrease of the magnitude of prism TOD with wavelength around $\lambda \approx 0.8$ μm suggests that the mode-locking performance of the Ti:sapphire laser could be further improved by tuning the oscillator to longer wavelengths. Since reproducible wavelength tuning has not been possible with bandwidths >20 THz in the Ti:sapphire laser, the 780 nm center-wavelength reflective optics used in our laser were replaced by optics centered at 810 nm to force the laser to oscillate at longer wavelengths. Although the duration of the shortest sech^2-like pulse shown in Fig. 19(a) represents little improvement as compared to the 12–13 fs pulses obtained with the 780–nm optics [33], the reasonable fit to the sech^2-envelope even in the wings and a broader spectrum with a less pronounced asymmetry [Fig. 19(b)] provide support for the predicted trend of decreasing cavity TOD with wavelength.

Optimum performance of prism-pair dispersion-controlled solitary systems can be expected in the vicinity of λ_{cav}, where the cavity TOD is approximately zero. Consequently, in this wavelength range high-order phase dispersion is dominated by the fourth-order contribution [39]. Since $D_4 < 0$, the net negative GDD at λ_{cav} can be decreased without creating an overall positive cavity GDD for any part of the ed spectrum. If the center wavelength of the laser spectrum approximately coincides with λ_{cav}, the increase of the scaling parameter β_4 (6) rather than β_3 (5) will limit pulse shortening as the negative

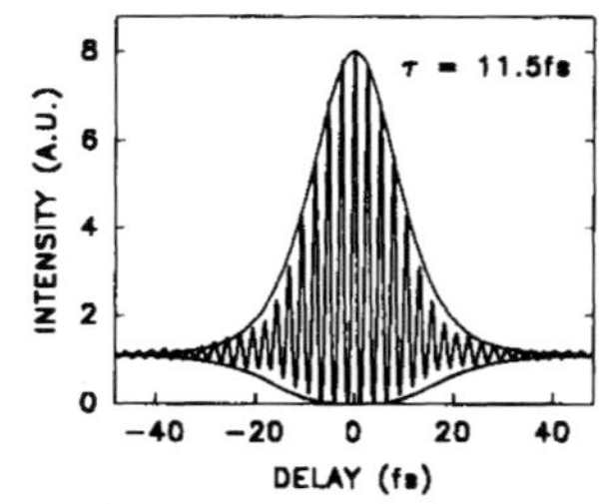

(a)

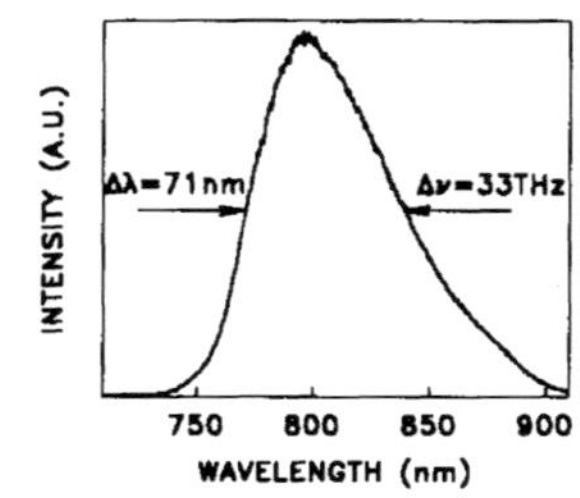

(b)

Fig. 19. (a) Fringe-resolved autocorrelation with a sech^2 fit to the envelopes of the autocorrelation of the shortest recorded near sech^2 pulse. (b) The corresponding spectrum. The time-bandwidth product is 0.38. Note the less pronounced spectral asymmetry and the greater spectral width as compared to the results of [33] as a consequence of the center wavelength shifted from ≈ 0.77 μm to ≈ 0.8 μm.

GDD is reduced and the laser spectrum broadens. Assuming a critical value of $\beta_{4,crit}$ the width of the ed spectrum cannot broaden beyond the "usable" bandwidth $\Delta\omega_{cav}$ as given by

$$\Delta\omega_{cav} = \sqrt{48\beta_{4,crit}}\sqrt{\frac{D}{D_{pr}}}\Delta\omega_{pr}, \tag{11}$$

where the prism-pair bandwidth

$$\Delta\omega_{pr} = \sqrt{\frac{D_{pr}}{D_{4,pr}}} \tag{12}$$

is a characteristic of the prism material alone. Here D_{pr} and $D_{4,pr}$ denote the round-trip second- and fourth-order dispersion of the prism pair (with zero prism insertion), respectively. The small material contribution to D_4 has been neglected in the derivation of (11). The ratio of the overall negative cavity GDD (D) to that introduced by the prism pair (D_{pr}) decreases with increasing positive material GDD in the cavity. Hence, as in the case with TOD as the dominant performance-limiting effect, positive material GDD is undesirable even if fourth-order dispersion limits the minimum achievable pulse duration.

The dots in Fig. 20 indicate $\Delta\omega_{pr}/2\pi$ at the corresponding λ_{pr} for various prism materials. This prism-pair bandwidth $\Delta\omega_{pr}$ monotonically decreases with increasing zero-TOD wavelength, λ_{pr}. The significance of $\Delta\omega_{pr}$ becomes clear when evaluating the usable cavity bandwidth $\Delta\omega_{cav}$ for practical systems. To produce sub-10–fs pulses, D should be reduced to ≈ -50 fs^2 in the Ti:sapphire laser, where $D_{pr} \approx -1300$ fs^2 for a 60–cm separation of the fused silica prisms. Substituting these values into (11) we obtain

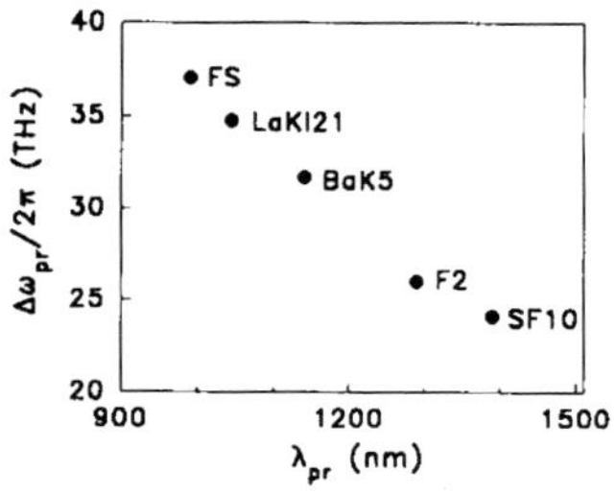

Fig. 20. The prism-pair bandwidth $\Delta\omega_{\rm pr}/2\pi$ as defined by (12) depicted at the corresponding values of $\lambda_{\rm pr}$ for different prism pairs with zero prism insertion.

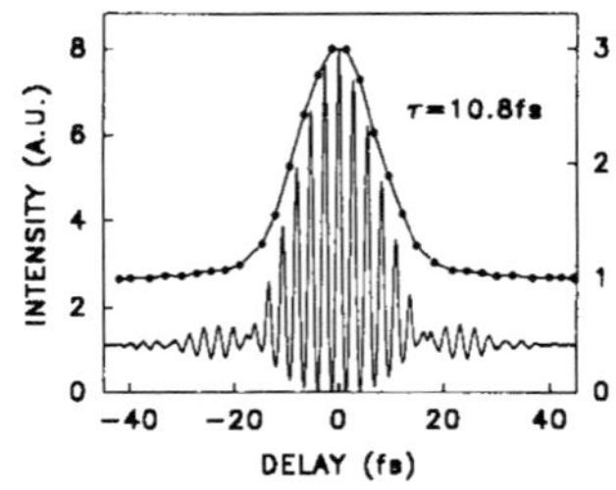

(a)

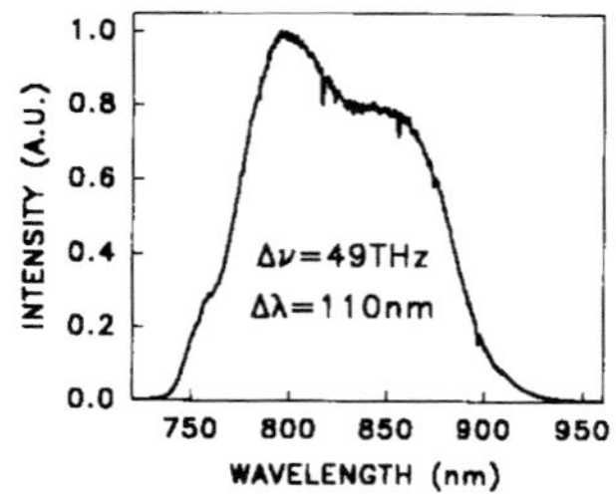

(b)

Fig. 21. (a) Measured interferometric and calculated intensity autocorrelations. (b) The corresponding spectrum. The best fit obtained from a Fourier-analysis yields $\tau = 10.8$ fs and a time-bandwidth product $\tau\Delta\nu = 0.53$. The dots in the intensity autocorrelation represent the mean values calculated by averaging between the peaks of adjacent fringes. Note the disappearance of the structure in the wings of the intensity autocorrelation.

$\Delta\omega_{\rm cav}/2\pi \approx \sqrt{\beta_{4,\rm crit}} \times 37$ THz. For the 1.5 μm range this value drops by almost a factor of 2 as a consequence of a decrease in $\Delta\omega_{\rm pr}$ for longer wavelengths (Fig. 20). Thus, we can conclude that both the prism pair and the dielectric cavity mirrors introduce increasing limitations for increasing wavelengths in the near infrared, and optimum performance (with the currently known prism materials) can be expected in the wavelength range of $0.8 - 1$ μm (e.g., Ti:sapphire, Cr: LISAF).

In order to back up these considerations with some experimental results, we removed the intracavity slit and modelocked the Ti:sapphire laser by misaligning the resonator. This modification removes a weak bandwidth limitation introduced by the intracavity aperturing of the beam, which becomes apparent for lasing bandwidths of ≈30 THz. This bandwidth limitation is a direct consequence of some transverse chirp carried by the pulse at mirror $M1$ in Fig. 6, which can be identified by a slight shift in the ed center frequency once the slit is translated perpendicularly to the resonator axis. The origin of this limitation at the nondispersive cavity end is not entirely clear. One possible explanation is wavelength-dependent refraction in the Brewster-angled gain medium. As is usual with this type of modelocking [17], [29], some 5% of the output is delivered in higher-order transverse modes.

Figures 21 and 22 show experimental results obtained with the Ti:sapphire laser ed by slight misalignment of the end mirrors of the resonator, which typically reduces the output power by ≈20%. The average output power and intracavity pulse energy in these experiments were ≈300 mW and ≈4 nJ, respectively. The spectra shown in Figs. 21(b) and 22(b) were recorded by an optical multichannel analyzer and have been corrected for grating and detector response. In spite of their broad bandwidth, the spectra exhibit no pronounced asymmetry, unlike plot (b) in Fig. 15. This supports the theoretical conclusion that the intracavity TOD becomes negligibly small beyond 850 nm. A comparison of the ed bandwidth of 65 THz with the predicted value of $\Delta\omega_{\rm cav}/2\pi \approx \sqrt{\beta_{4,\rm crit}} \times 37$ THz indicates that $\beta_{4,\rm crit}$ is substantially larger than $\beta_{3,\rm crit} \approx 0.5$. This is presumably due to the fact that for $D_3 = 0$ and $D_4 < 0$ GDD remains negative over the full ed spectrum as long as $D < 0$ at the center laser wavelength, in contrast with the case of $D_3 \neq 0$.

The dots in Figs. 21(a) and 22(a) represent mean values calculated by averaging the interferometric autocorrelation signal over full oscillation periods between the peaks of the adjacent fringes. Since the slowly varying envelope approximation does not strictly apply to optical pulses of ≈10 fs, the averaged autocorrelation function cannot be defined unambiguously, nor can the averaged intensity (i.e., the envelope of the pulse). Thus, the concept of intensity autocorrelation is not really meaningful in this time domain. In addition, valuable information about phase and/or amplitude modulation of the complex pulse envelope is lost during the averaging, as demonstrated by the disappearance of the wing-structure. Hence, a reliable determination of the parameters of optical pulses comprising a few cycles calls for the measurement of the fringe-resolved (electric-field) autocorrelation.

The evaluation of ed pulse parameters from the measured data in Figs. 21 and 22 requires careful analysis, since neither the autocorrelations nor the spectra fit to familiar pulse shapes (e.g., sech^2, Gaussian). Hence we have performed a Fourier-analysis taking into consideration both linear and quadratic chirp carried by the pulses. Using the measured intensity spectra the chirp parameters were varied in the analysis until good agreement between the calculated and measured autocorrelations were achieved. This approach yielded $\tau = 10.8$ fs and $\tau = 10.3$ fs for Fig. 21 and 22, respectively. Considering the fact that the pulses were found to be within 15% of their transform limit, the evaluated time–bandwidth product of 0.66 for Fig. 22 indicates a significant deviation from a sech^2 pulse shape predicted by current theories of passive modelocking. Hence, the assumption of a sech^2 pulse shape in evaluating the pulse width from the measured autocorrelation is not justified under these operating conditions and can lead to a significant underestimation of the pulse duration. (For the measured data

IEEE JOURNAL OF QUANTUM ELECTRONICS, VOL. 30, NO. 4, APRIL 1994

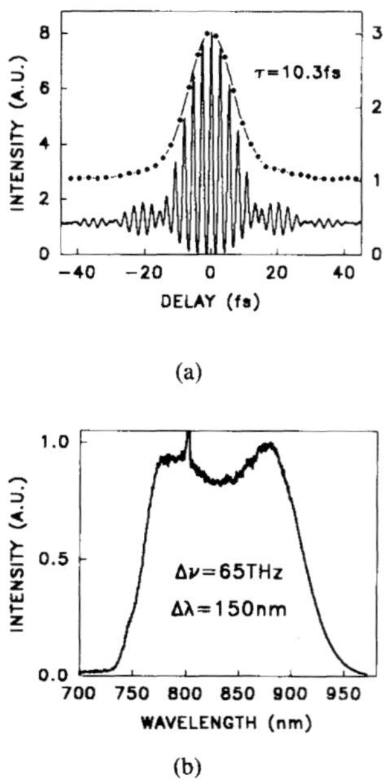

Fig. 22. (a) Measured interferometric and calculated intensity collinear autocorrelation. The best fit obtained from a Fourier-analysis yields $\tau = 10.3$ fs and a time-bandwidth product $\tau\Delta\nu = 0.66$ The dots in the intensity autocorrelation represents the mean value calculated by averaging between the peaks of adjacent fringes. Note the disappearance of the structure in the wings of the intensity autocorrelation. (b) The corresponding spectrum shows also an additional narrow spike, which originates from a weak pedestal extending over the synchronously pumped ed pulsewidth of $\tau_{\text{synch}} \approx 50$–$100$ ps with an intensity a factor of $\approx(\tau_{\text{synch}}/\tau)^2 > 10^7$ weaker than the peak intensity of the ed pulse. Note that the ed spectrum covers essentially the whole reflectivity range of the dielectric cavity mirrors.

shown in Fig. 22, this assumption would have yielded a pulse width of $\approx$ 9 fs.)

Recent theoretical investigations have revealed that the large time–bandwith product relates to coherent ringing on the trainling edge of the pulse, which is the result of the delayed response of the resonant atomic polarization in the gain medium and the broad ed spectrum detuned from gain line center [76]. As a consequence, if the pulse duration is to be minimized in a femtosecond solid-state laser without compromising the time–bandwidth product (i.e., pulse quality), the ed spectrum should be tuned close to gain line center. Unfortunately, no prism materials that allow the construction of cubic-dispersion-free cavities for the wavelength range $\lambda \leq 0.8\,\mu$m (e.g., Ti:sapphire) have been found yet.

VI. Summary

In this paper we have summarized the principal physical mechanisms responsible for the formation of ultrashort pulses in broadband solid-state lasers. Particular emphasis has been placed on stability issues and the effects limiting pulse shortening when the net negative group-delay dispersion is reduced in the cavity. The spatially separated actions of the soliton-like pulse shaping mechanisms and the influence of residual higher-order phase dispersion in the resonator have been identified as the major limitations to the femtosecond performance of broadband solid-state systems.

We have shown that prism-pair dispersion-controlled resonators exhibit a characteristic wavelength λ_{cav}, where the cavity group-delay dispersion is constant to a first-order approximation, i.e., cubic phase dispersion vanishes. Our experiments demonstrate that a Ti:sapphire oscillator is capable of producing 10 fs pulses in the vicinity of λ_{cav}. This performance should be more easily achievable in lasers operating beyond 0.85 μm, where λ_{cav} can be chosen to exactly coincide with the center laser wavelength by appropriate selection of the prism material. Our investigations have revealed that for wavelengths below 0.8 μm the residual intracavity third-order dispersion can impair 10-fs performance, whereas for higher wavelengths (>1 μm) residual fourth order dispersion from the prism pair and the finite bandwidth of the dielectric mirrors may provide limitations.

Concurrent advances in short-wavelength gain media, prism materials exhibiting λ_{pr} shorter than that of fused silica, and/or broadband cubic dispersion compensation techniques may extend 10 fs performance into the visible range. Owing to their long upper-state lifetime, the same gain media can also effectively be used in subsequent amplifier stages to boost the pulse energies up to microjoule or millijoule levels. The availability of 10-fs optical pulses at either high repetition rates or high pulse energies throughout the near infrared will open new avenues of investigation in ultrafast spectroscopy and in the study of interactions of intense electromagnetic fields with atomic systems.

Note Added in Proof

Recently, a novel approach to broadband intracavity dispersion control in femtosecond lasers using specially designed "chirped" multilayer dielectric mirrors was proposed and demonstrated [77]. Dispersion engineering of dielectric laser mirrors offers the potential for extending the "usable" cavity bandwidth $\Delta\omega_{cav}$ (over which the GDD is nearly constant) over the entire high-reflectivity bandwidth of the cavity mirrors at any laser wavelength in the visible and near-infrared spectral range. Beyond the elimination of high-order phase dispersion from the cavity, this mirror-dispersion-controlled (MDC) femtosecond oscillator technology also removes the need for intracavity prisms, allowing the construction of compact and reliable ultrashort-pulse sources. Using this technique, recently highly stable bandwidth-limited sub-10–fs pulses have been generated from a MDC Ti:sapphire laser [78].

Acknowledgment

The authors are deeply indebted to Professors A. J. Schmidt and E. Wintner for many helpful discussions and for their support. They also wish to thank S. M. J. Kelly for his valuable comments and critical reading of the manuscript, and acknowledge R. Szipöcs for computing the curves shown in Fig. 17.

References

[1] P. F. Moulton, "Spectroscopic and laser characteristics of Ti:Al_2O_3," *J. Opt. Soc. Am. B*, vol. 3, p. 125, 1986; L. K. Smith, S. A. Payne, W. L. Kway, L. L. Chase, and B. H. T. Chai, "Investigation of the

laser properties of Cr^{3+}:LiSrGaF$_6$," *IEEE J. Quantum Electron.*, vol. 28, p. 2612, 1992; V. Petricevic, S. K. Gayen, and R. R. Alfano, "Continuous-wave laser operation of chromium-doped forsterite," *Opt. Lett.*, vol. 14, p. 612, 1989; N. L. Borodin, A. G. Okrimchuk, and A. V. Shestakov, "Polarizing spectroscopy of $Y_3Al_5O_{12}$, $SrAl_2O_4$, and $CaAl_2O_4$ crystals containing Cr^{4+}," in *OSA Proceedings on Advanced Solid State Lasers*, vol. 13 (Optical Society of America, Washington, DC), p. 42, 1992.

[2] E. P. Ippen, *Ultrafast Phenomena VIII*, Springer Series in Chemical Physics, Springer, Heidelberg, 1992.

[3] F. Krausz, M. E. Fermann, T. Brabec, P. F. Curley, M. Hofer, M. H. Ober, Ch. Spielmann, E. Wintner, and A. J. Schmidt, "Femtosecond solid state lasers," *IEEE J. Quantum Electron.* vol. 28, p. 2097, 1992.

[4] H. A. Haus, "Theory of modelocking with a slow saturable absorber," *IEEE J. Quantum Electron.* vol. QE-11, p. 736, 1975.

[5] L. Dahlström, "Passive modelocking and *Q*-switching of high-power lasers by means of the optical Kerr effect," *Opt. Commun.*, vol. 5, p. 157, 1972.

[6] K. Sala, M. C. Richardson, and N. R. Isenor, "Passive modelocking of lasers with the optical Kerr effect modulator," *IEEE J. Quantum Electron.*, vol. QE-13, p. 915, 1977.

[7] F. Ouellette and M. Piché, "Pulse shaping and passive modelocking with a nonlinear Michelson interferometer," *Opt. Commun.*, vol. 60, p. 99, 1986.

[8] K. A. Stankov and J. Jethwa, "A new ing technique using a nonlinear mirror," *Opt. Commun.*, vol. 66, p. 41, 1988.

[9] L. F. Mollenauer and R. H. Stolen, "The soliton laser," *Opt Lett.*, vol. 9, p. 13, 1984.

[10] F. M. Mitschke and L. F. Mollenauer, "Ultrashort pulses from the soliton laser," *Opt. Lett.*, vol. 12, p. 407, 1987.

[11] K. J. Blow and D. Wood, "ed lasers with nonlinear external cavities," *J. Opt. Soc. Am. B*, vol. 5, p. 629, 1988.

[12] K. J. Blow and B. P. Nelson, "Improved modelocking of an F-center laser with a nonlinear nonsoliton external cavity," *Opt. Lett.,*, vol. 13, p. 1026, 1988; P. N. Kean, X. Zhu, D. W. Crust, R. S. Grant, N. Langford, and W. Sibbett, "Enhanced modelocking of color-center lasers," *Opt. Lett.*, vol. 14, p. 39, 1989; J. Mark, L. Y. Liu, K. L. Hall, H. A. Haus, and E. P. Ippen, "Femtosecond pulse generation in a laser with a nonlinear external resonator," *Opt. Lett.*, vol. 14, p. 48, 1989; P. M. W. French, J. A. R. Williams, and J. R. Taylor, "Femtosecond pulse generation from a titanium-doped sapphire laser using nonlinear external cavity feedback," *Opt. Lett.*, vol. 14, p. 668, 1989.

[13] T. F. Carruthers and I. N. Duling III, "Passive modelocking with an antiresonant nonlinear mirror," *Opt. Lett.*, vol. 15, p. 804, 1990; M. E. Fermann, M. Hofer, F. Haberl, A. J. Schmidt, and L. Turi, "Additive-pulse-compression modelocking of a neodymium fiber laser," *Opt. Lett.* , vol. 16, p. 244, 1991.

[14] M. Hofer, M. E. Fermann, F. Haber, M. H. Ober, and A. J. Schmidt, "modelocking with cross-phase and self-phase modulation," *Opt. Lett.*, vol. 16, p. 502, 1991; M. Hofer, M. H. Ober, F. Haberl, and M. E. Fermann, "Characterization of ultrashort pulse formation in passively ed fiber lasers," *IEEE J. Quantum Electron.* vol. 28, p. 720, 1992.

[15] E. P. Ippen, H. A. Haus, and L. Y. Liu, "Additive pulse modelocking," *J. Opt. Soc. Am. B* , vol. 6, p. 1736, 1989.

[16] M. Morin and M. Piché, "Interferential modelocking: Gaussian pule analysis," *Opt. Lett.*,vol. 14, p. 1119, 1990.

[17] D. E. Spence, P. N. Kean and W. Sibbett, "60-fsec pulse generation from a self-ed Ti:sapphire laser," *Opt. Lett.*vol. 16, p. 42, 1991.

[18] L. Spinelli, B. Couillaud, N. Goldblatt, and D. K. Negus, "Starting and generation of sub-100 fs pulses in Ti:Al_2O_3 by self-focusing," in *Digest of Conference on Lasers and Electro-Optics* (Optical Society of America, Washington, D.C., 1991), paper CPDP7.

[19] U. Keller, G. W. tHooft, W. H. Knox, and J. E. Cunningham, "Femtosecond pulses from a continuously self-starting passively ed Ti:sapphire laser," *Opt. Lett.*,vol. 16, p. 1022, 1991.

[20] M. Piché, "Beam reshaping and self-ing in nonlinear laser resonators," *Opt. Commun.* , vol. 86, p. 156, 1991.

[21] T. Brabec, C.. Spielmann, P. F. Curley, and F. Krausz, "Kerr-lens modelocking," *Opt. Lett.*,vol. 17, p. 1292, 1992.

[22] H. A. Haus, J. G. Fujimoto, and E. P. Ippen, "Analytic theory of additive pulse and Kerr lens modelocking," *IEEE J. Quantum Electron.* , vol. 28, p. 2086, 1992.

[23] T. Brabec, Ch. Spielmann, and F. Krausz, "modelocking in solitary lasers," *Opt. Lett.* vol. 16, p. 1961, 1991.

[24] T. Brabec, Ch. Spielmann, and F. Krausz, "Limits of pulse shortening in solitary lasers," *Opt. Lett.* ,vol. ;17, p. 748, 1992.

[25] C.-P. Huang, H. C. Kapteyn, J. W. McIntosh, and M. M. Murnane, "Generation of transform-limited 32-fs pulses from a self-ed Ti:sapphire laser," *Opt. Lett.* , vol. 17, p. 139, 1992.

[26] F. Krausz, Ch. Spielmann, T. Brabec, E. Wintner, and A. J. Schmidt, "Generation of 33-fs optical pulses from a solid-state laser," *Opt. Lett.* , vol. 17, p. 200, 1992.

[27] Ch. Spielmann, P. F. Curley, T. Brabec, E. Wintner, and F. Krausz, "Generation of sub-20 fs ed pulses from a Ti:sapphire laser," *Electron. Lett.* , vol. 28, p. 1532,1992.

[28] J. M. Jacobson, K. Naganuma, H. A. Haus, J. G. Fujimoto, and A. G. Jacobson, "Femtosecond pulse generation in a Ti: Al_2O_3 laser by using second- and third-order intracavity dispersion," *Opt. Lett.* , vol. 17, p. 1608, 1992.

[29] C.-P. Huang, M. T. Asaki, S. Backus, M. M. Murunane, H. C. Kapteyn, and H. Nathel, "17-fs pulses from a self-ed Ti:sapphire laser," *Opt. Lett.*, vol. 17, p. 1289, 1992.

[30] B. Proctor and F. Wise, "Quartz prism sequence for reduction of cubic phase in a ed Ti:Al_2O_2 laser," *Opt. Lett.* , vol. 17, p. 1295, 1992; B. Proctor and F. Wise, "Generation of 13-fs pulses from a ed Ti:Al_2O_3 laser with reduced third-order dispersion," *Appl. Phys. Lett.*, vol. 62, p. 470, 1993.

[31] B. E. Lemoff and C. P. J. Barty, "Generation of high-peak-power 20-fs pulses from a regeneratively initiated, self-ed Ti:sapphire laser," *Opt. Lett.* , vol. 17, p. 1367, 1992.

[32] M. T. Asaki, C. P. Huang, D. Harvey, J. Zhou, H. Nathel, H. C. Kapteyn, and M. M. Murnane, "11 femtosecond pulses from a modelocked Ti:sapphire laser," *Opt. & Photonics News* , vol. 12, p. 65, 1992.

[33] P. F. Curley, Ch. Spielmann, T. Brabec, F. Krausz, E. Wintner, and A. J. Schmidt, "Operation of a femtosecond Ti: sapphire solitary laser in the vicinity of zero group-delay dispersion," *Opt. Lett.* , vol. 18, p. 54, 1993.

[34] F. Krausz, T. Brabec, and Ch. Spielmann, "Self-starting passive modelocking," *Opt. Lett.*, vol. 16, p. 235, 1991.

[35] H. A. Haus and E. P. Ippen, "Self-starting of passively ed lasers," *Opt. Lett.*, vol. 16, p. 1331, 1991.

[36] H. A. Haus, J. G. Fujimoto, and E. P. Ippen, "Structures for additive pulse modelocking," *J. Opt. Soc. Am. B*, vol. 8, p. 2068, 1991,.

[37] This relationship offers a convenient means for determining the cavity GDD by measuring T_r as a function of laser frequency in broadband tunable lasers, as it was pointed out recently by W. H. Knox, "*In situ* measurement of complete intracavity dispersion in an operating Ti:sapphire femtosecond laser," *Opt. Lett.*, vol. 17, p. 514, 1992.

[38] G. P. Agrawal, *Nonlinear Fiber Optics*, Academic, New York, 1989.

[39] Since $\hat{D}_h$ has been introduced neither in [3] nor elsewhere previously, we give its expression here for the sake of completeness, $\hat{D}_h = (D_g/2)\partial^2/\partial t^2 - (D_3/6)\partial^3/\partial t^3 - i(D_4/24)\partial^4/\partial t^4 + \ldots$ Here D_g is the second derivative of the imaginary part of the round-trip phase delay with respect to frequency (gain dispersion). The coefficients of the higher-order terms $D_3 = \varphi'''(\omega_l)$ and $D_4 = \varphi''''(\omega_l)$, etc. account for the frequency-dependence of the cavity GDD (higher-order phase dispersion). The above form of $\hat{D}_h$ and the definitions of the dispersion parameters D_g, D_3, and D_4 imply that the latter are real quantities. Higher-order gain dispersion (i.e., the imaginary part of D_3 and D_4) have been neglected.

[40] G. P. A. Malcolm and A. I. Ferguson, "Self-modelocking of a diode-pumped Nd:YLF laser," *Opt. Lett.* , vol. 16, p. 1967, 1991.

[41] N. H. Rizvi, P. M. W. French, and J. R. Taylor, "Generation of 33-fs pulses from a passively ed Cr^{3+}:LiSrAlF$_6$ laser," *Opt. Lett.* , vol. 17, p. 1605, 1992.

[42] P. LiKamWa, B. H. T. Chai, and A. Miller, "Self-ed Cr^{3+}:LiCaAlF$_6$ laser," *Opt. Lett.* , vol. 17, p. 1438, 1992.

[43] J. M. Evans, D. E. Spence, W. Sibbett, B. H. T. Chai, and A. Miller, "50-fs pulse generation from a self-ed Cr^{3+}:LiSrAlF$_6$ laser," *Opt. Lett.* , vol. 17, p. 1447, 1992.

[44] A. Seas, V. Petricevic, and R. R. Alfano, "Generation of sub-100-fs pulses from a CW ed chromium-doped forsterite laser," *Opt. Lett.* , vol. 17, p. 937, 1992.

[45] K. X. Liu, C. J. Flood, D. R. Walker, and H. M. van Driel, "Kerr lens modelocking of a diode-pumped Nd:YAG laser," *Opt. Lett.*, vol. 17, p. 1361, 1992.

[46] The effective optical length is defined by $\sum d_i/n_i$, where d_i and n_i are the physical lengths and indices of refraction of the different optical elements, respectively.

[47] H. Kogelnik, "Imaging of optical modes—resonators with internal lenses," *Bell. Syst. Tech. J.*, vol. 44, p. 455, 1965.

[48] D. Georgiev, J. Herrmann, and U. Stamm, "Cavity design for optimum nonlinear absorption in Kerr-lens ed solid-state lasers," *Opt. Commun* , vol. 92, p. 368, 1992.

[49] D. Huang, M. Ulman, L. H. Acioli, H. A. Haus, and J. G. Fujimoto, "Self-focussing-induced saturable loss for laser modelocking," *Opt. Lett.*, vol. 17, p. 511, 1992.

IEEE JOURNAL OF QUANTUM ELECTRONICS, VOL. 30, NO. 4, APRIL 1994

[50] T. Brabec, P. F. Curley, Ch. Spielmann, E. Wintner, and A. J. Schmidt, "Hard aperture Kerr lens modelocking," *J. Opt. Soc. Am. B.*, vol. 10, June 1993.

[51] S. De Silvestri, P. Laporta, and O. Svelto, "The role of cavity dispersion in CW ed lasers," *IEEE J. Quantum Electron.* , vol. QE-20, p. 533, 1984; P. Laporta and V. Magni, "Dispersive effects in the reflection of femtosecond optical pulses from broadband dielectric mirrors," *Appl. Opt.*, vol. 24, p. 2014, 1985; D. N. Christodoulides, E. Bourkoff, R. I. Joseph, and T. Simos, "Reflection of femtosecond optical pulses from multiple-layer dielectric mirrors-analysis," *IEEE Transact. Quantum Electron.*, vol. 22, p. 186, 1986; R. Szipöcs and K. Ferencz, "Advanced optical coatings," *Opt. Engineering*, vol. 32, p. 2525, 1993.

[52] R. L. Fork, O. E. Martinez, and J. P. Gordon, "Negative dispersion using pairs of prisms," *Opt. Lett.* vol. 9, p. 150, 1984; O. E. Martinez, J. P. Gordon, and R. L. Fork, "Negative group-velocity dispersion using refraction," *J. Opt. Soc. Am. A.*, vol. 1, p. 1003, 1984.

[53] Ch. Spielmann, F. Krausz, T. Brabec, E. Wintner, and A. J. Schmidt, "Femtosecond pulse generation from a synchronously pumped Ti:sapphire laser," *Opt. Lett.*, vol. 16, p. 1180, 1991.

[54] P. F. Curley and A. I. Ferguson, "Actively ed Ti:sapphire laser producing transform-limited pulses of 150-fs duration," *Opt. Lett.*, vol. 16, p. 1016, 1991.

[55] L. Turi and F. Krausz, "Amplitude modulation modelocking of lasers by regenerative feedback," *Appl. Phys. Lett.*, vol. 58, p. 810, 1991.

[56] D. E. Spence, J. M. Evans, W. E. Sleat, and W. Sibbett, "Regeneratively initiated self-ed Ti:sapphire laser," *Opt. Lett.*, vol. 16, p. 1762, 1991.

[57] J. D. Kafka, M. L. Watts, and J.-W. J. Pieterse, "Picosecond and femtosecond pulse generation in a regeneratively ed Ti:sapphire laser," *IEEE J. Quantum Electron.*, vol. 28, p. 2151, 1992.

[58] N. H. Rizvi, P. M. W. French, and J. R. Taylor, "Continuously self-ed Ti:sapphire laser that produces sub-50-fs pulses," *Opt. Lett.*, vol. 17, p. 279, 1992.

[59] Y. M. Liu, K. W. Sun, P. R. Pruncal, and S. A. Lyon, "Simple method to start and maintain self-ing of a Ti:sapphire laser," *Opt. Lett.*, vol. 17, p. 1219, 1992.

[60] W. S. Pelouch, P. E. Powers, and C. L. Tang, "Self-starting ed ring-cavity Ti:sapphire laser," *Opt. Lett.*, vol. 17, p. 1581, 1992.

[61] M. H. Ober, M. Hofer, and M. E. Fermann, "42-fs pulse generation from a ed fiber laser started with a moving mirror," *Opt. Lett.*, vol. 18, p. 367, 1993.

[62] N. Sarukura, Y. Ishida, and H. Nakano, "Generation of 50-fsec pulses from a pulse-compressed, CW, passively ed Ti:sapphire laser," *Opt. Lett.*, vol. 16, p. 153, 1991.

[63] K. Naganuma and K. Mogi, "50-fs pulse generation directly from a colliding-pulse ed Ti:sapphire laser using an antiresonant ring morror," *Opt. Lett.*, vol. 16, p. 738, 1991.

[64] J. P. Gordon, "Dispersive perturbations of solitons of the nonlinear Schrödinger equation," *J. Opt. Soc. Am. B*, vol. 9, p. 91, 1992.

[65] S. M. J. Kelly, "Characteristic sideband instability of periodically amplified average soliton," *Electron. Lett.*, vol. 28, p. 806, 1992.

[66] F. Salin, P. Grangier, P. Georges, and A. Brun, "Pulse propagation near zero group-velocity dispersion in femtosecond dye laser," *Opt. Lett.*, vol. 15, p. 1374, 1990.

[67] P. K. A. Wai, C. R. Menyuk, Y. C. Lee and H. H. Chen, "Nonlinear pulse propagation in the neighborhood of the zero-dispersion wavelength of monomode optical fibers," *Opt. Lett.*, vol. 11, p. 464, 1986.

[68] S. Wen and S. Chi, "Approximate solution of optical solton in lossless fibers with third-order dispersion," *Opt. Quantum Electron.*, vol. 21, p. 335, 1989.

[69] J. N. Elgin, "Soliton propagation in an optical fiber with third-order dispersion," *Opt. Lett.*, vol. 17, p. 1409, 1992.

[70] H. A. Haus, J. D. Moores, and L. E. Nelson, "Effect of third-order dispersion on passive modelocking," *Opt. Lett.*, vol. 18, p. 51, 1993.

[71] The fact that the solitary pulse is the shortest at the dispersive cavity end seems to be in contradiction with the principle of maximum field energy in the oscillator, since minimum pulse duration at the nondispersive end (i.e., in the Kerr-medium) would ensure minimum cavity losses due to SAM. Nevertheless, the generation of the shortest pulse at the dispersive resonator end is an inherent feature of an ideal solitary laser, which incorporates only negative GDD, and SPM and some relatively weak SAM for stabilization. Simple symmetry considerations [3] imposes the constraint on the ed steady state that the pulse must be bandwidth-limited at the resonator ends, which in turn implies that the shortest pulse duration is achieved at the dispersive end in the system. In the passively ed steady state the laser will then produce maximum intracavity field energy *under this constraint*. Experimental evidence for these considerations is given in P. F. Curley, Ch. Spielmann, T. Brabec, E. Wintner, and F. Krausz, "Periodic evolution of femtosecond pulses in solitary lasers," *J. Opt. Soc. Am. B*, vol. 10, p. 1025, 1993.

[72] This value is $\approx 30\%$ lower than that of ϕ previously estimated [3] using $n_{2,\mathrm{sapphire}} = 3.45 \times 10^{-16}$ cm^2/W (R. W. Boyd, *Nonlinear Optics*, Academic Press, 1992) and assuming resonator beam parameters that corresponded to the middle of one of the stability regions in Fig. 4. Hence, the r parameters deduced by using the calculated value of ϕ are subject to corresponding corrections in [3] and [26]. It is not known whether the discrepancy in the measured and calculated ϕ is due to the accuracy of $n_{2,\mathrm{sapphire}}$ or it originates from an increase in the average beam diameter in the gain medium when the resonator is adjusted towards the edge of the stability range.

[73] I. H. Malitson, "Interspecimen comparison of the refractive index of fused silica," *J. Opt. Soc. Am.*, vol. 55, p. 1205, 1965.

[74] *Optical Glass Catalogue*, Schott Inc. Mainz, Germany.

[75] This general feature of prism pairs has also recently been pointed out by B. E. Lemoff and C. P. J. Barty, "Cubic-phase-free dispersion compensation is solid-state ultrashort-pulse lasers," *Opt. Lett.*, vol. 18, p. 57, 1993.

[76] J. D. Harvey, J. M. Dudley, P F. Curley, Ch. Spielmann, and F. Krausz, "Coherent effects in a self-modelocked Ti:sapphire laser," *Opt. Lett.*, *vol. 19, 1994.*

[77] R. Szipöcs, K. Ferencz, Ch.; Spielmann, and F. Krausz, "Chirped multilayer coatings for broadband dispersion control in femtosecond lasers," *Opt. Lett.*, *vol. 19, p. 201, 1994; A. Stingl, Ch. Spielmann, F. Krausz, and R. Szipöcs, "Generation of 11-fs pulses from a Ti:sapphire laser without the use of prisms,"* Opt. Lett., *vol. 19, po. 204, 1994.*

[78] F. Krausz, Ch. Spielmann, P. F. Curley, T. Brabec, S. M. J. Kelly, A. Stingl, R. Szipöcs, E. Wintner, and A. J. Schmidt, "Ultrafast solid state lasers: status and prospects," in *Ultrafast Phenomena IX,eds. W. H. Knox, P. F. Barbara, G. A. Moutrou, and A. H. Zewail. Heidelberg: Springer, 1994.*

Christian Spielmann, photograph and biography not available at the time of publication.

Peter F. Curley, photograph and biography not available at the time of publication.

Thomas Brabec, photograph and biography not available at the time of publication.

Ferenc Krausz (M'92), photograph and biography not available at the time of publication.

February 1, 1994 / Vol. 19, No. 3 / OPTICS LETTERS

Chirped multilayer coatings for broadband dispersion control in femtosecond lasers

Robert Szipöcs and Kárpát Ferencz

Optical Coating Laboratory, Research Institute for Solid State Physics, P.O. Box 49, H-1525 Budapest, Hungary

Christian Spielmann and Ferenc Krausz

Abteilung Quantenelektronik and Lasertechnik, Technische Universität Wien, Gusshausstrasse 27, A-1040 Wien, Austria

Received August 23, 1993

Optical thin-film structures exhibiting high reflectivity and a nearly constant negative group-delay dispersion over frequency ranges as broad as 80 THz are presented. This attractive combination makes these coatings well suited for intracavity dispersion control in broadband femtosecond solid-state lasers. We address design issues and the principle of operation of these novel devices.

The relevance of intracavity dispersion control in passively mode-locked ultrashort-pulse laser was recognized soon after the appearance of the first systems operating in the femtosecond domain.[1] Negative dispersion that is due to wavelength-dependent refraction in a pair of Brewster-angled prisms combined with positive material dispersion proved to be an efficient and convenient means of controlling the net group-delay dispersion (GDD) inside the laser cavity.[2]

In solid-state lasers femtosecond pulse generation invariably relies on a net negative intracavity GDD because of an ultrafast self-phase modulation caused by the optical Kerr effect in the laser medium. Hence prism pairs have become standard components in these systems. The interplay between negative GDD and Kerr-induced self-phase modulation, often referred to as solitonlike shaping, appears to be the dominant pulse-forming mechanism that determines the steady-state pulse duration in femtosecond solid-state lasers.[3] In practical prism-pair-controlled broadband laser systems a major limitation to ultrashort pulse generation originates from the variation of the intracavity GDD with wavelength. The principal source of this high-order dispersion was found to be the prism pair.[3,4] In this Letter we report the novel development of chirped multilayer mirror coatings that can exhibit essentially constant negative GDD over a frequency range as broad as 80 THz. Careful design permits higher-order contributions to the mirror phase dispersion to be kept at low values or to be chosen such that high-order phase errors introduced by other cavity components (e.g., the gain medium) are canceled. Replacing the prism pair with these novel devices offers the potential of generating pulses that are shorter than previously achievable directly from the laser. In addition, this simplifies the cavity design and may permit the construction of more compact and reliable femtosecond sources.

The first thorough investigations of the frequency-dependent phase retardation (phase dispersion) of multilayer dielectric coatings date back to the early 1960's.[5,6] The emergence of femtosecond lasers in the 1980's has led to a revival of interest in this field.[7–13] Whereas standard quarter-wave dielectric mirrors were shown to introduce negligible dispersion at the center of their reflectivity bands,[7,8,11] various specific high-reflectivity coatings (Gires–Tournois interferometers, double-stack mirrors, etc.) with adjustable GDD (through angle tuning) were devised and used for the precise control of intracavity dispersion in femtosecond dye lasers.[14,15] However, the GDD introduced by these mirror coatings is accompanied by high cubic and higher-order dispersive contributions. As a consequence, a constant GDD could be obtained only over a limited wavelength range (<10 THz). The difficulty in realizing broadband GDD control relates to the physical origin of dispersion in these devices; different frequency components are trapped for different periods of time in Fabry–Perot-like resonant structures.

A distinctly different type of GDD arises from the wavelength dependence of the penetration depth of the incident optical field in multilayer coatings. This effect does not rely on the presence of resonant structures and offers the possibility of realizing a GDD that is a slowly varying function of wavelength over a broad bandwidth. A constant GDD requires a group delay that varies approximately linearly with the wavelength. A wave packet of a given center wavelength is most efficiently reflected by a corresponding quarter-wave stack. Therefore a monotonic variation of the multilayer period throughout the deposition process (chirped coating) should result in a penetration depth[16] (and thus group delay) that varies monotonically with the wavelength. However, a previous study of chirped multilayer coatings with layer thicknesses following monotonic variations revealed that the GDD is strongly perturbed by some Fabry–Perot-like resonances in these simple structures.[17] Our studies have indicated that the undesirable resonant features can be almost completely eliminated by

0146-9592/94/030201-03$6.00/0

OPTICS LETTERS / Vol. 19, No. 3 / February 1, 1994

slight adjustment of the layer thicknesses. This finding has been a most important step toward the practical realization of dispersive mirrors having a group delay that is a linear function of optical frequency. In addition, chirped multilayer coatings have the potential for extending the bandwidth of standard low-dispersion quarter-wave mirrors.

A preliminary design utilized a structure consisting of 42 alternating layers of SiO_2 ($n = 1.45$) and TiO_2 ($n = 2.3$) with optical thicknesses close to a quarter of 0.8 μm, our selected center wavelength. The multilayer period was slightly increased near the substrate and decreased at the air–coating interface to produce a group delay that increased with wavelength, i.e., negative GDD. The mirrors were required to be highly reflecting and have a constant negative GDD over the wavelength range 710–900 nm. Alternatively the GDD could be required to exhibit a slight linear variation with a slope suitable for compensating the cubic phase dispersion of the gain medium. This is in strong contrast to prism pairs, in which the ratio of cubic to quadratic dispersion is determined by material parameters and the operation wavelength.[18] A simple computer refinement algorithm[19] was used to minimize the quadratic deviation of the complex reflectivity-versus-frequency function of the actual mirror design from the required specification. The quadratic error was made up of two parts, for the amplitude and the phase characteristics, whose relative weights could be independently adjusted. The prescribed value of negative GDD was increased stepwise until the quadratic error started to increase rapidly. Accordingly the operating negative GDD was found to trade off against a high reflectivity and a low variation of the prescribed GDD-versus-wavelength function.

This optimization procedure enabled us to realize the dispersive mirror design shown in Fig. 1 (Ref. 20) by use of a standard electron-beam evaporation technique.[13] Our obtained structure preserved the feature of increasing layer period toward the substrate, even though the variation is far from what we consider as linear. The calculated reflectivity of this design is larger than 99.9% around the center wavelength and drops to ≈99.5% at 710 and 900 nm. The nominal GDD of the mirror is ≈ -45 fs^2 at $\lambda = 800$ nm, with a slight linear variation yielding a cubic dispersion of ≈ -33 fs^3, which permits a simultaneous compensation of GDD and cubic dispersion in a Ti:sapphire laser.

The wavelength dependence of the GDD exhibits a weak oscillatory behavior, with deviations less than ± 5 fs^2 from the required (linear) GDD function over the wavelength range of 720–890 nm (≈80 THz). The GDD of -45 fs^2 corresponds to a group-delay difference of ≈22 fs between the extremes of this spectral range. As we show below, for a 42-layer design this is close to the maximum delay difference that can be attained without making use of resonances. Any further increase in GDD could be achieved only at the expense of resonant structures (e.g., Gires–Tournois interferometers) appearing in the multilayer design, which implies large variations in the GDD over the mirror reflectivity range.

Figure 2 plots the computed electric-field distribution inside the dielectric mirror as a function of wavelength. As required, the penetration depth (and thus the group delay) increases approximately linearly with the wavelength. The figure also gives clear evidence of the high reflectivity of the mirror between 700 and 900 nm, as indicated by the disappearance of the optical field at the substrate–coating interface (optical distance 0).

The computed group delay (first derivative of the reflected phase with respect to frequency) as a function of wavelength is depicted in Fig. 3 along with measured data. The experimental data were obtained by use of the white-light interferometer technique of Knox *et al.*[11] To reduce the experimental errors, we used 16 reflections at a ≈10° angle of incidence in the interferometer, resulting in a measurement accuracy of ≈1 fs. These results demonstrate that dielectric mirrors with approximately constant GDD over a broad spectral range can be designed and fabricated by use of standard computation and evaporation techniques, respectively.

The maximum achievable negative GDD is limited by the maximum group-delay difference that can be obtained between the extremes of the reflectivity range. This in turn relates to the optical thickness

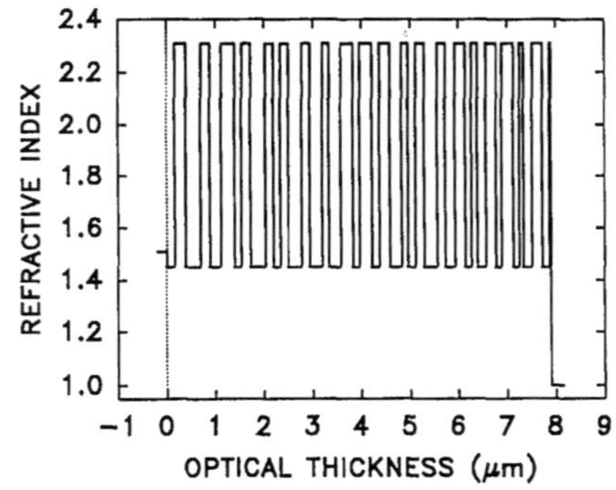

Fig. 1. Theoretical refractive-index profile of a high-reflectivity TiO_2–SiO_2 multilayer coating designed specifically for broadband GDD control in femtosecond lasers.

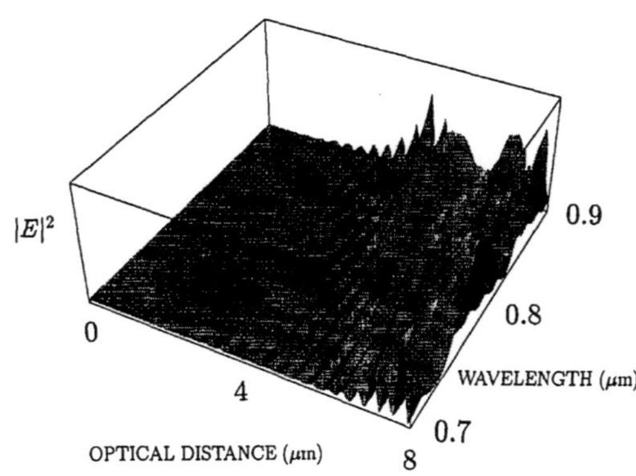

Fig. 2. Computed electric-field distribution as a function of wavelength in the chirped dielectric structure shown in Fig. 1.

February 1, 1994 / Vol. 19, No. 3 / OPTICS LETTERS

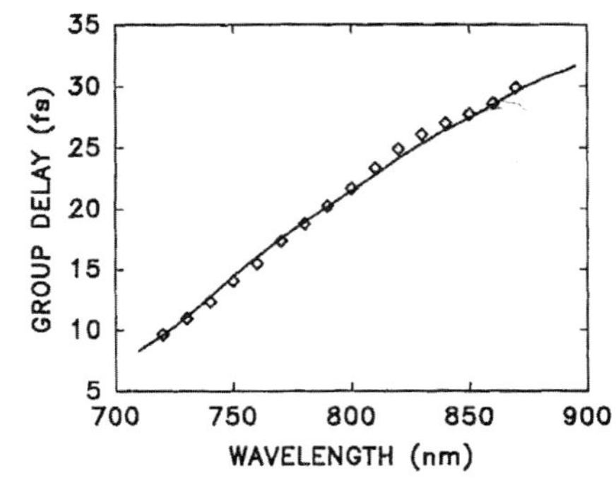

Fig. 3. Computed group delay as a function of wavelength (solid curve) together with experimental data (squares) for the multilayer design of Fig. 1. Note that the absolute delay could not be measured; therefore a wavelength-independent constant delay was added to the measured relative data.

of the coating. We have found that a simple approximate expression for the maximum achievable group-delay difference can be written as

$$\Delta\tau_{max} = \frac{2(t_{chirped} - t_{qw})}{c}, \tag{1}$$

where $t_{chirped}$ is the optical thickness of the chirped multilayer coating and t_{qw} is that of a standard quarter-wave high reflector ($R > 99.9\%$) consisting of the same pair of alternating layer materials. In physical terms, the required high reflectivity of the dispersive mirror calls for a minimum optical thickness of $t \approx t_{qw}$, and only excess layers can introduce an appreciable frequency-dependent group delay around the center of the high reflectivity band. Assuming that the group delay varies in a linear manner with frequency, we see that the corresponding upper estimate for the GDD is given simply by the ratio of $\Delta\tau_{max}$ to the mirror bandwidth $\Delta\omega$.

For the specific case of TiO_2-SiO_2 mirrors centered around $\lambda \approx 0.8$ μm we have $t_{qw} \approx 4$ μm, yielding $\Delta\tau_{max} \approx 27$ fs for our 8-μm-thick structure, in reasonable agreement with the results presented in Fig. 3. With the number of layers fixed, $\Delta\tau_{max}$ scales linearly with the chosen center wavelength of the dispersive mirror. For a selected operating wavelength, we can increase $\Delta\tau_{max}$ and thus the magnitude of broadband negative GDD only by increasing the number of layers, which is limited by scattering and absorption losses that are due to structural defects and impurities in the deposited layers, respectively.[13] It is expected that more sophisticated coating techniques will permit the production of higher-quality layers and thereby open the way toward the realization of more complex structures with higher values of the negative GDD over bandwidths approaching 100 THz.

In summary, we have reported a novel dielectric mirror providing approximately constant negative dispersion over a bandwidth as broad as 80 THz. The presented mirror design can be adopted for any broadband mode-locked solid-state laser operating in the wavelength range of 0.5–2 μm by a simple rescaling of the layer thicknesses. With suitable layer materials this spectral range can be extended well into the ultraviolet and infrared spectra. Furthermore, engineering the wavelength dependence of the penetration depth might lead to interesting applications in other areas of physics in which scattering and interference in quasi-periodic structures take place.

We thank S. M. J. Kelly, P. Curley, P. Apai, and Zs. Bor for helpful discussions and A. J. Schmidt, E. Wintner, and N. Kroo for their support. This research was sponsored by the Austrian and Hungarian Science Foundations under grants P-09710, T-007376, and WTZ-Ö-U Project A17.

References

1. R. L. Fork, C. V. Shank, R. Yen, and C. A. Hirlimann, IEEE J. Quantum Electron. **QE-19**, 500 (1983).
2. R. L. Fork, O. E. Martinez, and J. P. Gordon, Opt. Lett. **9**, 150 (1984).
3. F. Krausz, M. R. Fermann, T. Brabec, P. F. Curley, M. Hofer, M. H. Ober, Ch. Spielmann, E. Wintner, and A. J. Schmidt, IEEE J. Quantum Electron. **28**, 2097 (1992).
4. C. P. Huang, H. C. Kapteyn, J. W. McIntosh, and M. M. Murnane, Opt. Lett. **17**, 139 (1992); F. Krausz, Ch. Spielmann, T. Brabec, E. Wintner, and A. J. Schmidt, Opt. Lett. **17**, 200 (1992); C. P. Huang, M. T. Asaki, S. Backus, M. M. Murnane, H. C. Kapteyn, and H. Nathel, Opt. Lett. **17**, 1289 (1992); B. Proctor and F. Wise, Opt. Lett. **17**, 1295 (1992); B. E. Lemoff and C. P. J. Barty, Opt. Lett. **17**, 1367 (1992); J. M. Jacobson, K. Naganuma, H. A. Haus, J. G. Fujimoto, and A. G. Jacobson, Opt. Lett. **17**, 1608 (1992); P. F. Curley, Ch. Spielmann, T. Brabec, F. Krausz, E. Wintner, and A. J. Schmidt, Opt. Lett. **18**, 54 (1993).
5. C. F. Bruce and P. E. Ciddor, J. Opt. Soc. Am. **50**, 295 (1960).
6. J. M. Bennett, J. Opt. Soc. Am. **54**, 612 (1964).
7. W. Dietel, E. Döpel, K. Hehl, W. Rudolph, and E. Schmidt, Opt. Commun. **50**, 179 (1984).
8. S. De Silvestri, P. Laporta, and O. Svelto, Opt. Lett. **9**, 335 (1984).
9. D. N. Christodoulides, E. Bourkoff, R. I. Joseph, and T. Simos, IEEE J. Quantum Electron. **QE-22**, 186 (1986).
10. A. M. Weiner, J. G. Fujimoto, and E. P. Ippen, Opt. Lett. **10**, 71 (1985).
11. W. H. Knox, N. M. Pearson, K. D. Li, and Ch. A. Hirlimann, Opt. Lett. **13**, 574 (1988).
12. M. Beck and I. A. Walmsley, Opt. Lett. **15**, 492 (1990).
13. K. Ferencz and R. Szipöcs, Opt. Eng. **32**, 2525 (1993).
14. J. Heppner and J. Kuhl, Appl. Phys. Lett. **47**, 453 (1985).
15. M. Yamashita, K. Torizuka, and T. Sato, IEEE J. Quantum Electron. **QE-23**, 2005 (1987).
16. D. I. Babic and S. W. Corzine, IEEE J. Quantum Electron. **28**, 514 (1992).
17. P. Laporta and V. Magni, Appl. Opt. **24**, 2014 (1985).
18. B. E. Lemoff and C. P. J. Barty, Opt. Lett. **18**, 54 (1993).
19. See, e.g., J. A. Dobrowolski and R. A. Kemp, Appl. Opt. **29**, 2879 (1990).
20. Further details about the mirror design and fabrication are available on request.

PUSHING THE LIMITS of Femtosecond Technology: Chirped Dielectric Mirrors

By Robert Szipöcs, Andreas Stingl, Christian Spielmann, and Ferenc Krausz

1047-6938/95/6/0016/05-$06.00

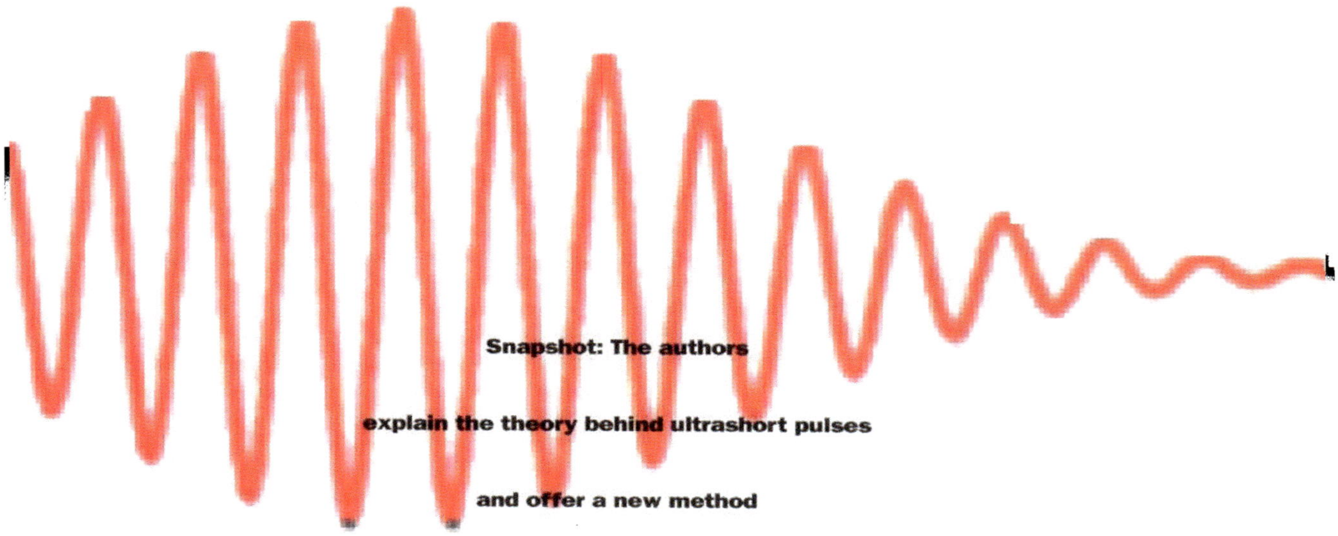

Snapshot: The authors explain the theory behind ultrashort pulses and offer a new method — chirped dielectric mirrors — for achieving sub-10 fsec pulses.

One of the major trends in laser physics today is the research and development of femtosecond laser sources. The motivation for generating short electromagnetic waveforms comes from many areas of science and technology. Ultrashort optical pulses are capable of taking "snapshots" of the state of matter and hence following the evolution of ultrafast processes at the microscopic level. Probing charge-carrier dynamics in semiconductors, the formation and breaking of chemical bonds, or time-resolved studies of photoisomerisation in biology are just a few examples of a number of intriguing applications. The generation of pulses shorter than previously possible would benefit many application fields and calls for a precise dispersion control over increasingly broad bandwidths in femtosecond laser oscillators as well as in subsequent optical systems. With the advent of chirped multilayer dielectric mirrors, feedback and phase dispersion control over unprecedented bandwidths have become feasible, thus opening the way for further advances in femtosecond technology.

Optical pulse propagation in dispersive media

Short electromagnetic waveforms (henceforth pulses) can be thought of as a superposition of long, quasi-monochromatic wavepackets of different carrier frequencies. For a particular spectral intensity distribution, minimum pulse duration is achieved when the centers of the wavepackets coincide in space. Under these conditions the pulse is referred to as Fourier-limited or transform-limited because the shortest pulse attainable with the spectral intensity distribution given has been realized.

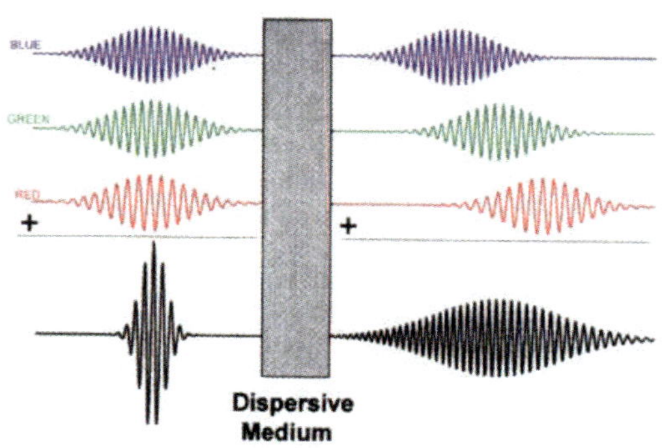

Figure 1. Optical pulse propagation through a dispersive medium: a frequency-dependent group delay leads to a pulse broadening and to a carrier frequency sweep.

If such a transform-limited pulse is sent through a dispersive medium, in which wavepackets of different carrier frequency propagate at different velocities, the pulse envelope will be broadened and carried by a time-varying instantaneous frequency at the output, as illustrated in Figure 1. Dispersion can be quantified by expanding the phase retardation $\phi(\omega)$ of a dispersive system about the center of the pulse spectrum ω_0 in the form

$$\phi(\omega) = \phi_0 + \phi'(\omega-\omega_0) + \tfrac{1}{2}\phi''(\omega-\omega_0)^2 + \tfrac{1}{6}\phi'''(\omega-\omega_0)^3 + \ldots \quad (1)$$

where $\phi_0 = \phi(\omega_0)$, and the derivatives ϕ', ϕ'', etc. are also evaluated at ω_0. The first derivative ϕ' is called group delay because it gives the time taken by the center of the pulse to reach the output of the wave-propagating medium. The higher-order terms in the expansion describe a frequency dependence of the group delay,

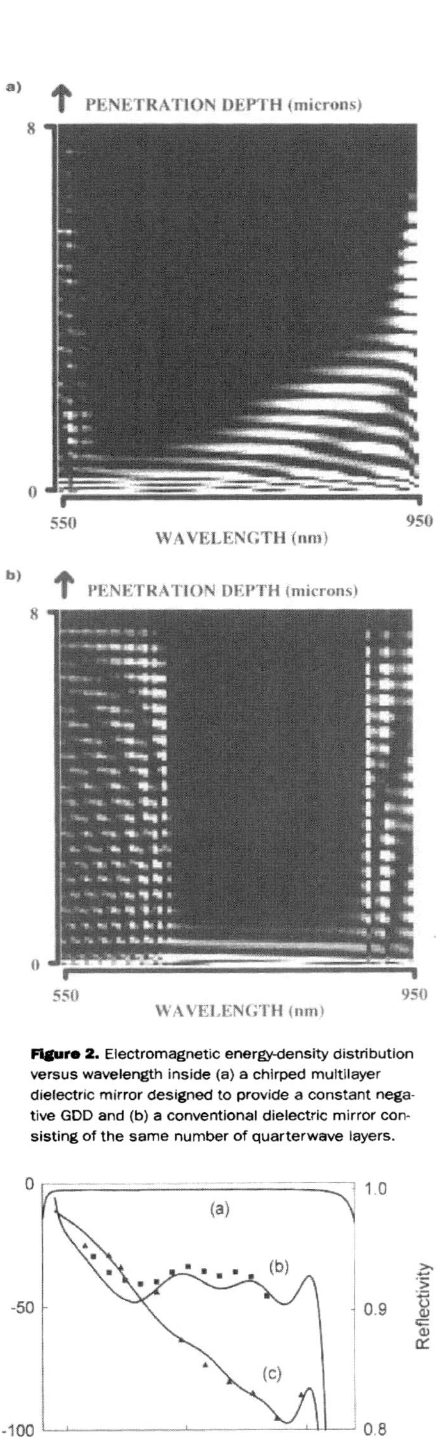

Figure 2. Electromagnetic energy-density distribution versus wavelength inside (a) a chirped multilayer dielectric mirror designed to provide a constant negative GDD and (b) a conventional dielectric mirror consisting of the same number of quarterwave layers.

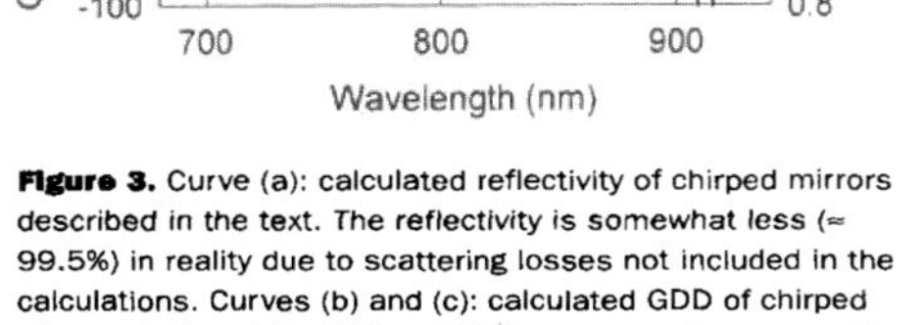

Figure 3. Curve (a): calculated reflectivity of chirped mirrors described in the text. The reflectivity is somewhat less (≈ 99.5%) in reality due to scattering losses not included in the calculations. Curves (b) and (c): calculated GDD of chirped mirrors designed for GDD and TOD compensation, respectively. The squares and triangles represent measured data points obtained with spectrally-resolved[15] and conventional[16] white-light interferometry, respectively.

and hence are responsible for dispersive effects. The lowest-order dispersion parameter ϕ'', which is termed group-delay dispersion (GDD), causes a broadening of the pulse without significant changes in the pulse shape and a carrier frequency varying linearly with time, which is often referred to as a linear chirp. The higher-order coefficients ϕ''' (third-order dispersion, TOD), ϕ'''' (fourth-order dispersion, FOD), etc. become only important for broad-bandwidth signals, and can give rise to a nonlinear chirp as well as to a distortion of the pulse. Critical values of the dispersion coefficients, above which dispersion causes a substantial change of the pulse, obey the simple scaling $\phi^{(n)} = \tau^n$, where τ is the pulse duration. For instance, a GDD of $\phi'' = \tau^2$ results in a pulse broadening by more than a factor of two. This scaling reveals a dramatic increase in the susceptibility to dispersion-induced broadening and distortion for decreasing pulse durations.

Dispersion in femtosecond laser oscillators

Ultrashort pulse generation from laser oscillators is ultimately limited by the finite oscillator bandwidth. However, both the gain bandwidth in organic dyes or vibronic solid-state gain media and the bandwidth of resonators formed by quarterwave multilayer dielectric mirrors support pulses much shorter than those produced previously. This is due to the fact that dispersive cavity components lead to a frequency-dependent cavity round-trip time (or group delay) $T_r(\omega)$, which restricts the "usable" bandwidth for ultrashort-pulse formation to a fraction of the full oscillator bandwidth. $T_r(\omega)$ can be expressed in terms of the dispersion parameters introduced above by deriving the Taylor series of the cavity-round-trip phase retardation $\phi_{cav}(\omega)$ as given by Eq. (1) with respect to ω.

The ideal $T_r(\omega)$ curve for optimum performance may be different in different femtosecond lasers due to differencies in the nonlinear optical processes responsible for pulse formation. In femtosecond dye lasers[1] oscillator-bandwidth-limited pulse generation requires a round-trip time nearly *constant* over the full oscillator bandwidth with a small amount of positive or negative cavity GDD (ϕ''_{cav}) depending on the operating conditions. The situation is quite different in the new generation of self-mode-locked femtosecond solid-state lasers.[2,3] In these systems, the nonlinear index of the laser crystal (Kerr-effect) gives rise to a strong self-phase modulation of the circulating pulse. If a bandwidth-limited pulse interacts with the nonlinear index of a transparent optical medium, the spectrum of the pulse gets broadened. Frequency components lower than the carrier arise on the leading edge of the pulse whilst the higher-frequency components lag behind on the trailing edge. If such a *positively* chirped pulse experiences even a very small amount of positive cavity GDD ($\phi''_{cav}(\omega) > 0$), the leading edge completes every cavity round trip faster than the trailing edge of the pulse, resulting in a rapidly accumulating pulse broadening. As a consequence, it is imperative that the net cavity GDD be negative for femtosecond pulse generation in solid-state systems. A soliton-like interplay between self-phase modulation and *negative* GDD can shorten the circulating pulse to the limit set by the oscillator bandwidth, provided the GDD is constant over the full oscillator bandwidth. Despite the differences in the pulse forming mechanisms, what femtosecond lasers have in common, is that *higher-order cavity dispersion (TOD, FOD) must be negligible if oscillator-bandwidth-limited performance is to be achieved.*

The invention of prism pairs as a source of negative GDD[4] marked a milestone in the evolution of femtosecond laser technology. In dye lasers, intracavity prism pairs have provided a compensation of the positive material dispersion of intracavity components (*e.g.*, gain medium) and allowed continuous adjustment of the cavity GDD in a region of small positive and negative values. In a cavity formed by low-dispersion dielectric mirrors consisting of quarterwave layers[5], they yielded pulses as short as 27 fsec from a colliding-pulse mode-locked Rhodamine 6G (Rh6G) dye laser.[6] Prism-pair dispersive delay lines have also been key components of femtosecond solid-state lasers.[3] In these systems, the role of the prisms has been to overcompensate the positive dispersion of the laser crystal for soliton-like shaping.

The recent evolution of prism-controlled femtosecond solid-state lasers culminated in the development of a self-mode-locked Ti:sapphire laser using fused silica prisms,[7] which is capable of generating optical pulses of 10-15 fsec.[8-10] Both in the optimized Rh6G and Ti:sapphire laser, TOD of the prism pair[11, 12] has been identified as the major effect that prevents further pulse shortening as well as impairs the quality of the shortest pulses achieved. Approaching 10 fsec, an additional limitation tends to arise due to the finite bandwidth of the quarterwave mirrors. Hence, the evolution of ultrashort-pulse lasers has arrived at the frontiers of conventional femtosecond technology. Nevertheless, the gain linewidth of Ti:sapphire and related vibronic systems supports optical pulses as short as 3 fsec. These enormous bandwidths issue a great challenge to the community interested in the generation and application of ultrashort pulses. To rise to this challenge, novel techniques for providing feedback and dispersion control over ultrabroad bandwidths must be developed.

Chirped multilayer dielectric mirrors

The idea of chirped (aperiodic) refractive-index-modulated distributed-feedback (Bragg) structures dates back to the early 1970s, when Hill proposed the use of chirped Bragg reflectors for dispersion compensation in optical fiber systems.[13] The operation principle of these devices is fairly straightforward: different spectral components penetrate to different depths before reflected back from appropriate resonant sections of the chirped distributed-feedback structure. This introduces a frequency-dependent group delay, *i.e.* dispersion, which can be engineered. As an additional benefit, chirping the modulation period of a Bragg reflector also increases the bandwidth of its high-reflectivity (HR) range, just as a chirped pulse exhibit a broader frequency spectrum than its chirp-free counterpart.

These findings are directly applicable to an important special class of Bragg reflectors, namely multilayer dielectric mirrors. The widely-used standard quarterwave dielectric mirrors consisting of alternating quarterwave-layers of low- and high-index materials exhibit a low dispersion over a substantial part of their HR range.[5] Varying the layer thicknesses during the evaporation process provides a simple means of modulating the period of the multilayer structure.[14] These chirped multilayer dielectric mirrors (henceforth chirped mirrors) offer the benefits of a broader HR range (as compared to quarterwave mirrors) as well as an engineerable dispersion over the HR band. Consequently, the chirped mirror technology has the potential for overcoming the limitations researchers previously encountered in their attempts to produce shorter pulses from femtosecond oscillators.

To put these ideas into practice, chirped mirrors for use in Ti:sapphire oscillators and amplifiers have been designed and manufactured at the Research Institute for Solid State Physics Budapest (Hungary). The mirrors realized so far consist of 42 alternating quasi-quarterwave layers of SiO_2 (n_L=1.45) and TiO_2 (n_H=2.3) and have been produced by using standard electron-beam evaporation. For improving the performance of self-modelocked Ti:sapphire oscillators, chirped mirrors with a nearly constant negative GDD have been developed. The energy-density distribution as a function of wavelength inside the mirror is shown in Figure 2a. The most striking features are an increased HR bandwidth as compared to a quarterwave mirror shown in Figure 2b and a penetration depth that increases approximately linearly with wavelength between 700 and 800 nm. Curves (a) and (b) in Figure 3 show the calculated reflectivity and GDD of these mirrors, respectively. The HR range is currently limited by narrow resonances, which are not resolved in Figure 2a. Work is under way to eliminate these resonances. Nevertheless, even this "restricted" HR bandwidth is ≈30% broader than that of standard quarterwave mirrors. The negative GDD is approximately constant over

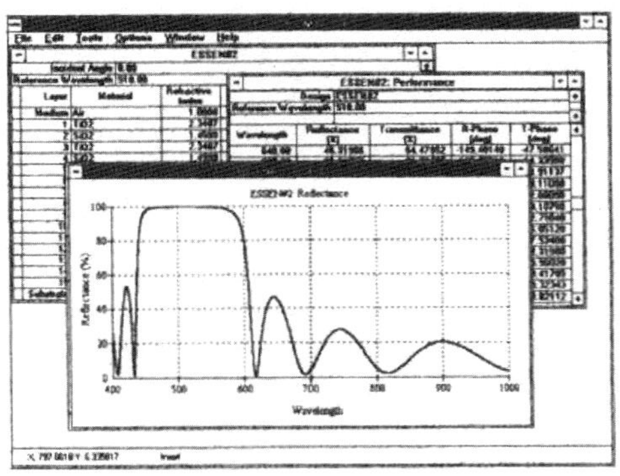

a bandwidth of ≈ 80 THz, opening the way for bandwidth-limited sub-10 fsec generation from Ti:sapphire lasers. Previous attempts to compensate dispersion with thin-film structures used resonant effects with much narrower bandwidths.[17, 18]

The broad constant-GDD bandwidth comes at the expense of a low value of the negative GDD. This relates to the fact that the overall group delay difference (= the product of bandwidth and GDD) is limited by the physical thickness of the mirrors, which must be traded off against low scattering losses.[14] One reflection off the broadband-GDD mirror (≈ -40 fsec^2) compensates the material GDD of 1 mm fused silica or 0.7 mm sapphire.[19] Hence, material dispersion must be minimized before these devices can be used for dispersion control.

Large amounts of material in the beam path are often unavoidable inside or outside the oscillator. In this case one has to resort to prism pairs limiting system performance by some residual TOD as discussed in the preceding section. Chirped mirrors can also be designed and fabricated to compensate for TOD, as shown in Figure 3. These mirrors exhibit a nominal GDD of ≈ -50 fsec^2 at 790 nm and a TOD of ≈ $+100$ fsec^3 and in combination with low-dispersion prisms can provide FOD-limited dispersion control in Ti:sapphire systems containing sizable amounts of optical material.

Chirped mirrors with similar or other dispersion characteristics can be produced for any wavelength range, where suitable (transparent) layer materials are available.[20] It is hoped that more sophisticated coating techniques will allow the fabrication of thicker stacks with tolerable scattering losses. These will provide higher dispersion parameters and/or broader bandwidths in the future.

Mirror-dispersion-controlled femtosecond systems

One of the most interesting application of the chirped mirrors is dispersion control in broadband femtosecond solid-state laser oscillators. To this end, the GDD-compensating mirrors characterized by the dispersion curve (b) in Figure 3 have been used to construct a prismless Ti:sapphire oscillator with a net negative GDD. Seven bounces on the dispersive mirrors (in one round trip) are necessary to overcompensate the positive GDD of a highly-doped 2-mm-thick Ti:sapphire crystal. Owing to the short gain length, a pump power of 3 to 4 W in the blue-green spectral region is sufficient (*e.g.*, from a small-frame argon or a frequency-doubled neodymium laser) for stable femtosecond pulse generation. With a 3%-transmitting output coupler, this mirror-dispersion-controlled (MDC) self-mode-locked Ti:sapphire oscillator delivers nearly bandwidth-limited 8 fsec optical pulses with a pulse energy of ≈ 1 nJ.[21] Figure 4 shows the fringe-resolved autocorrelation trace of the laser output. To our knowledge, these are the shortest pulses ever produced from a laser oscillator. They contain 3 cycles within their full width at half maximum just as the 6 fsec pulses obtained by extracavity pulse compression at 620 nm.[22]

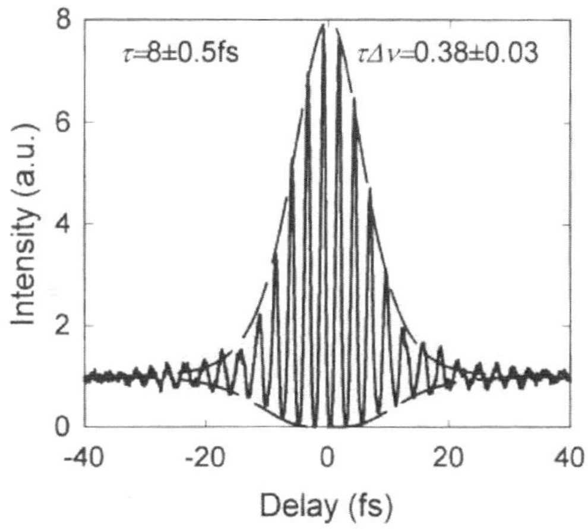

Figure 4. Interferometric second-order autocorrelation trace of the 8-fsec optical pulses generated by the mirror-dispersion-controlled Ti:sapphire oscillator described in Reference 21. The dashed-line around the pulse shows the calculated envelopes of an 8.2-fsec sech^2-shaped pulse.

The excellent stability of these sub-10-fsec pulses ≈1 % second harmonic fluctuations) offers an unprecedented sub-femtosecond resolution for ultrafast spectroscopy.[23] In strong contrast to prism-controlled Ti:sapphire systems producing pulses in this time domain,[12, 24] the 8 fsec pulses are generated at 800 nm, the center of the Ti:sapphire gain line, with a time-bandwidth product <0.4. These features make MDC Ti:sapphire oscillators ideal sources for both high-resolution ultrafast spectroscopy and seeding ultrashort-pulse Ti:sapphire amplifiers. The TOD-compensating chirped mirrors curve (c) in Figure 3 have recently been applied in combination with a pair of low-TOD prisms for recompressing the output of a kHz Ti:sapphire amplifier seeded with an MDC oscillator. This system currently produces 17-18 fsec pulses of an energy of 0.1mJ at a repetition rate of 1kHz.[25] These are the shortest optical pulses produced at multigigawatt power levels so far.

Conclusions

The availability of chirped dielectric mirrors permits the construction of more compact, stable, and user-friendly femtosecond oscillators and amplifiers than previously possible. In addition, they offer the potential for obtaining gain-bandwidth-limited pulses from Ti:sapphire and related broadband femtosecond solid-state oscillators. This means that the way is open toward the generation of optical pulses comprising only 1 cycle of light. We are entering exciting times in ultrafast science.

Acknowledgements

We wish to thank A. J. Schmidt and N. Kroo for their stimulating support and K. Ferencz for manufacturing the chirped mirrors. This research was sponsored by the Austrian and Hungarian *continued on page 59*

Pushing the Limits of Femtosecond Technology

Continued from page 20

Science Foundations, under grants P-9710, P-10409, and T-7376, respectively.

References

1. R. L. Fork *et al.*, "Generation of pulses shorter than 0.1 ps by colliding pulse modelocking," Appl. Phys. Lett. **38,** 671-673, (1981).
2. D. E. Spence *et al.*, "60-fsec pulse generation from a self-mode-locked Ti:sapphire laser," Opt. Lett. **16,** 41-43, (1991).
3. F. Krausz *et al.*, "Femtosecond solid-state lasers," IEEE Quantum Electron. **28,** 2097-2122, (1992).
4. R. L. Fork *et al.*, "Negative dispersion using pairs of prisms," Opt. Lett. **9,** 150-153, (1984).
5. S. De Silvestri *et al.*, "The role of cavity dispersion in cw mode-locked lasers," J. Quantum Electron. **20,** 533-539, (1984).
6. J. A. Valdmanis and R. L. Fork, "Design considerations for a femtosecond pulse laser balancing self phase modulation, group velocity dispersion, saturable absorption, and saturable gain," IEEE Quantum Electron. **22,** 112-118, (1986).
7. Ch. Spielmann *et al.*, "Generation of sub-20 fs mode-locked pulses from a Ti:sapphire laser," Electron. Lett. **28,** 1532-1533, (1992).
8. P. F. Curley *et al.*, "Operation of a femtosecond Ti:sapphire solitary laser in the vicinity of zero group-delay dispersion," Opt. Lett. **18,** 54-56, (1993).
9. B. Proctor and F. Wise, "Generation of 13-fs pulses from a mode-locked Ti:Al2O3 laser with reduced third-order dispersion," Appl. Phys. Lett. **62,** (1993) 470-473,.
10. M. T. Asaki *et al.*, "Generation of 11-fs pulses from a modelocked Ti:sapphire laser," Opt. Lett. **18,** 977-979 (1993).
11. B. E. Lemoff and C. P. J. Barty, "Cubic-phase-free dispersion compensation in solid-state ultrashort-pulse lasers," Opt. Lett. **18,** 57-59, (1993).
12. Ch. Spielmann *et al.*, "Ultrabroad-band femtosecond lasers," IEEE Quantum Electron. **30,** 1100-1114, (1994).
13. K. O. Hill, "Aperiodic distributed-parameter waveguides for integrated optics," Appl. Opt. **13,** 1853-1858, (1974).
14. R. Szipöcs *et al.*, "Chirped multilayer coatings for broadband dispersion control in femtosecond lasers," Opt. Lett. **19,** 201-203, (1994).
15. A. Kovacs *et al.*, "Group-delay measurement on laser mirrors by spectrally resolved white-light interferometry," Opt. Lett. **20,** 788-789 (1995).
16. W. H. Knox *et al.*, "Interferometric measurements of femtosecond group delay in optical components," Opt. Lett. **13,** 574-576, (1988).
17. J. Heppner and J. Kuhl, "Intracavity chirp compensation in a CPM laser using thin film interferometers," Appl. Phys. Lett. **47,** 453-455, (1985).
18. J. M. Jacobson *et al.*, "Femtosecond pulse generation in a Ti:Al_2O_3 laser using second- and third-order intracavity dispersion," Opt. Lett. **17,** 1608-1610, (1992).
19. This should be contrasted with the orders of magnitudes higher GDD achievable with in-fiber Bragg reflectors. The bandwidth of these devices is, however, by orders of magnitudes smaller because of the much weaker index modulation. For details see e.g. K. O. Hill *et al.*, " Chirped in-fiber Bragg gratings for compensation of optical-fiber dispersion," Opt. Lett. **19,** 1314-1316, (1994).
20. Recently, chirped mirrors have been used for dispersion control in a femtosecond infrared OPO: J. Hebling *et al.*, "Chirped-mirror dispersion compensated optical parametric oscillator," accepted by Opt. Lett.
21. A. Stingl *et al.*, "Sub-10-femtosecond, mirror-dispersion-controlled Ti:sapphire laser," Opt. Lett. **20,** 602-604 (1995).
22. R. L. Fork *et al.*, "Compression of optical pulses to six femtoseconds by using cubic phase compensation," Opt. Lett. **12,** 483-485, (1987).
23. Ch. Spielmann *et al.*, "Tunneling of optical pulses through photonic band gaps," Phys. Rev. Lett. **73,** 2308-2311, (1994).
24. P. Zhou *et al.*, "Pulse evolution in a broad-bandwidth Ti:sapphire laser," Opt. Lett. **19,** 1149-1151, (1994).
25. Lenzner *et al.*, "Sub-20-femtosecond, kHz-repetition-rate Ti:sapphire amplifier," accepted by Opt. Lett.

R. Szipöcs is a postdoctoral researcher with the Research Institute for Solid State Physics Budapest, Hungary. A. Stingl and Ch. Spielmann are postdoctoral researchers, and F. Krausz is assistant professor, in the Department of Quantum Electronics & Laser Technology, Technical University Vienna, Austria.

OPTICS LETTERS / Vol. 20, No. 6 / March 15, 1995

Sub-10-fs mirror-dispersion-controlled Ti:sapphire laser

A. Stingl, M. Lenzner, Ch. Spielmann, and F. Krausz

Abteilung Quantenelektronik und Lasertechnik, Technische Universität Wien, Gusshausstrasse 27-29, A-1040 Wien, Austria

R. Szipöcs

Research Institute for Solid State Physics, P.O. Box 49, H-1525 Budapest, Hungary

Received October 25, 1994

We demonstrate the generation of nearly bandwidth-limited 8-fs optical pulses near 0.8 μm from a self-mode-locked Ti:sapphire laser oscillator, using chirped dielectric mirrors for dispersion control. The mode-locking performance is described, and limitations are discussed.

The motivation for generating shorter, more intense, and higher-quality optical pulses comes from many fields in physics and, indirectly, from other areas of science and technology. Optical excitation (and probing) on a time scale of 10^{-14} s provides the only means of creating a coherent macroscopic polarization and investigating important related phenomena in semiconductors[1] as well as studying the dynamics of a number of chemical reactions or biological processes directly in the time domain. On the other hand, the availability of intense optical pulses approximately 10^{-14} s in duration should open the way to reversible nonlinear-optical experiments at intensity levels many orders of magnitude higher than previously feasible in the picosecond and subpicosecond time domains.[2,3]

In this Letter we report on what is to our knowledge the first ultrashort-pulse laser oscillator producing nearly bandwidth-limited optical pulses less than 10 fs in duration. The extremely broad bandwidth necessary to generate electromagnetic energy in such short intervals is provided by Ti-doped sapphire,[4] one of the most excellent broadband laser materials available these days. Exploitation of a substantial part of the enormous bandwidth of Ti:sapphire ($\approx$100 THz) has become possible with the discovery of self-mode locking (also termed Kerr-lens mode locking),[5] a unique method in that a passive amplitude modulation with virtually instantaneous response is introduced without the need for inserting any additional dispersive element in the cavity. Last but not least, femtosecond pulse formation in a broadband solid-state laser relies on a cavity round-trip time decreasing with the optical frequency near the gain line center [negative group-delay dispersion (GDD)] because of the strong self-phase modulation inherent in theses systems. In our self-mode-locked Ti:sapphire laser negative GDD is introduced by the same high-reflectivity dielectric mirrors that provide the feedback necessary for oscillation. These specifically designed chirped multilayer mirrors[6] can produce approximately constant negative GDD over a broader-bandwidth $\Delta\nu_{\text{GDD}}$ than any other low-loss dispersive optical system demonstrated to date.

The finite value of $\Delta\nu_{\text{GDD}}$ has been recognized to be a major effect limiting the performance of practical self-mode-locked oscillators. In systems employing prism pairs for controlling cavity dispersion,[7] third-order dispersion introduced by the prism pair itself was identified as the dominant limitation. Improved performance obtained by use of selected prism materials provided evidence for this finding.[8] The evolution of prism-controlled self-mode-locked Ti:sapphire lasers culminated with the development of a Ti:sapphire oscillator using fused-silica prisms,[9] which were shown to introduce the lowest cubic phase distortion near 0.8 μm among the commercially available optical materials. These systems can now produce pulses in the 10–15-fs region[10–12] but are still limited by the residual cavity third-order dispersion, as evidenced by their pronounced spectral asymmetry.

More recent investigations revealed that the overall cavity third-order dispersion vanishes at $\approx$0.85 μm in a Ti:sapphire laser using fused-silica prisms.[13,14] In fact, extremely broad double-peaked spectra centered at 0.85 μm could be obtained.[14,15] However, the increase in mode-locked bandwidth (beyond 150 nm) was not accompanied by a corresponding reduction in pulse duration because of fourth-order dispersion of the prisms[16] and possible coherent ringing owing to detuning.[17] A careful spectral analysis and fit to the interferometric autocorrelation yielded pulse durations of $\approx$10 fs,[14] whereas 8.5–9-fs pulse durations were evaluated by the simple assumption of a sech^2 pulse shape.[14,15]

In summary, prism-controlled Ti:sapphire lasers are capable of generating high-quality nearly sech^2-shaped pulses with durations as low as $\approx$15 fs. Below 15 fs the spectrum tends to become increasingly asymmetric, and pulse durations near 10 fs can be achieved only at the expense of detuning and a poor pulse quality characterized by time–bandwidth products of $\approx$0.6. In contrast, the mirror-dispersion-controlled (MDC) Ti:sapphire laser presented here delivers 8-fs nearly sech^2-shaped pulses with approximately symmetric spectra and time–bandwidth products of $\approx$0.38.

0146-9592/95/060602-03$6.00/0

March 15, 1995 / Vol. 20, No. 6 / OPTICS LETTERS

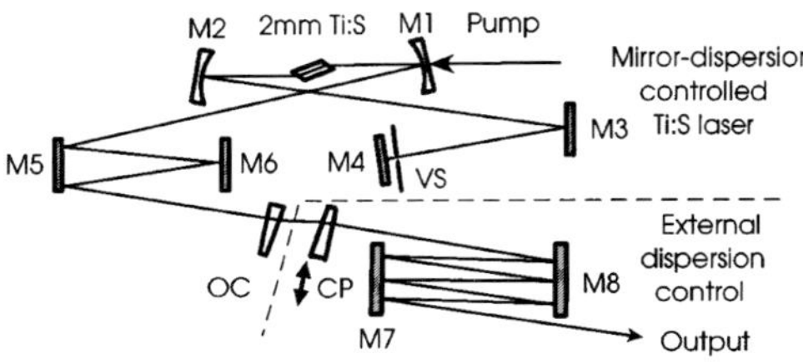

Fig. 1. Schematic of the MDC femtosecond Ti:sapphire (Ti:S) laser. The pump beam is focused with a 40-mm lens onto a 2-mm-thick highly doped Ti:S crystal (absorption coefficient, 5 cm^{-1} at 514 nm). M1, M2, single-stack quarter-wave dichroic mirrors highly transmitting at the pump wavelengths with radii of curvature of 5 cm; M3–M8, chirped dispersive mirrors; VS, vertical slit; OC, 3.3% broadband output coupler; CP, wedged glass plate.

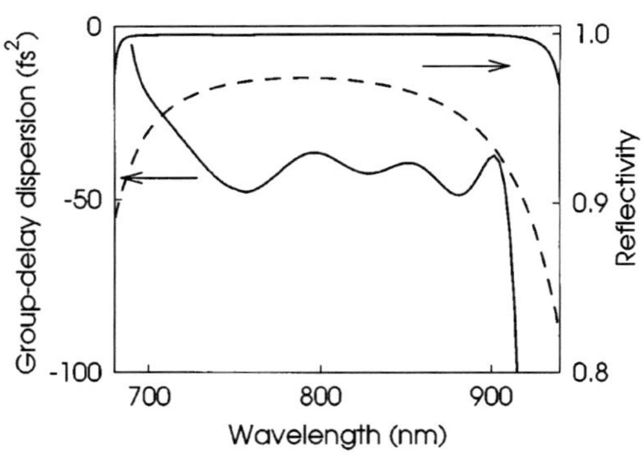

Fig. 2. Reflectivity and GDD of the chirped dielectric mirrors (solid curves) and the reflectivity of the output coupler (dashed curve).

A schematic of the MDC laser is shown in Fig. 1. To increase $\Delta\nu_{\mathrm{GDD}}$, we have redesigned the chirped multilayer mirrors used in a forerunner of the present system.[18] As revealed by Fig. 2, the new mirrors exhibit a $\Delta\nu_{\mathrm{GDD}}$ extending over ≈200 nm. The broader negative-GDD bandwidth, however, comes at the expense of a somewhat reduced value of average negative GDD. Because of the reduced mirror dispersion the number of reflections off the dispersive mirrors per round trip has been increased from 7 to 9 by the insertion of an additional folding mirror in the short collimated cavity arm. This has also resulted in a more balanced distribution of the negative GDD in the cavity, reducing the discreteness of solitonlike shaping and the related perturbations to the steady-state pulse.[14] However, this MDC cavity still has a drawback in that we are unable to control the net cavity GDD continuously. To improve this situation, we have adopted two different types of mirror with comparable bandwidths but with slightly different GDD that permit us to adjust the net cavity dispersion in small steps ($\Delta D \approx 10$ fs^2).

Because of the broad bandwidth of the chirped mirrors a conventional low-transmittivity quarter-wave output coupler would introduce a severe limitation to the oscillator bandwidth. Hence a specific broadband output coupler has been developed that not only exceeds the bandwidth of a quarter-wave output coupler having comparable transmittivity by some 30 nm (see the dashed curve in Fig. 2) but also introduces a broadband negative cubic dispersion for partial compensation of the positive cubic dispersion introduced by the laser crystal.

The laser is pumped by the blue-green lines (488 and 514 nm) of a small-frame Ar laser. At 3 W of pump power the cw and mode-locked output powers with the 3.3% output coupler are 150–200 and 60–100 mW, respectively. This mode-locked average output power corresponds to an output pulse energy of ≈1 nJ at a repetition rate of ≈80 MHz. Mode locking is accomplished in the usual manner[14]; i.e., the distance between the two curved mirrors is adjusted close to the upper boundary of the lower-stability region (corresponding to shorter mirror spacing), and a vertical slit is placed in the shorter collimated arm close to the end mirror. An optimum trade-off between stability and maximum self-amplitude modulation was achieved with a ratio of the lengths of collimated cavity arms of ≈2:1, yielding 0.45-mm-long stability regions 0.25 mm apart.[14] With the resonator optimized for Kerr-lens-induced amplitude modulation we can start pulse formation by tapping one of the cavity mirrors, and the laser stays mode locked without any notable degradation of performance for many hours.

We must take special care in steering and characterizing the broadband optical pulses leaving the mode-locked MDC laser. The spatial chirp and associated pulse front tilt induced by the wedged output coupler is eliminated by a second wedged glass plate identical to the output coupler substrate. The same glass plate (of varying thickness) also serves, in combination with chirped dispersive mirrors, to tune continuously the extracavity dispersion in the vicinity of a net zero GDD. The widely used quarter-wave dielectric beam splitter for p-polarized light is inadequate for distortion-free splitting of optical pulses with spectra extending over 200 nm. Hence we rotate the originally horizontal polarization of the laser output by 90° with a pair of turning mirrors and use a broadband quarter-wave beam splitter for s polarization in the autocorrelator. This beam splitter exhibits less than 5% variation in the reflectivity and less than 0.5-fs variation in the reflected and transmitted group delay over the wavelength range of 650–950 nm. We balance the dispersion of the 0.5-mm-thick fused-silica beam-splitter substrate by placing an identical antireflection-coated plate in the opposite arm of the autocorrelator. With the exception of the chirped mirrors and the beam splitter, the extracavity beam-steering and focusing optics are composed exclusively of Au-coated unprotected mirrors. The pulses are focused by a mirror of 25-mm focal length onto a 25-μm-thick beta-barium borate frequency-doubling crystal, in which the second-harmonic correlation signal is generated. Care has been taken to minimize the folding angle at the focusing mirror to keep astigmatism at a minimum.

The fringe-resolved autocorrelation (FRAC) trace and the spectrum of the mode-locked laser output are shown in Fig. 3. A sech^2 fit to the measured

OPTICS LETTERS / Vol. 20, No. 6 / March 15, 1995

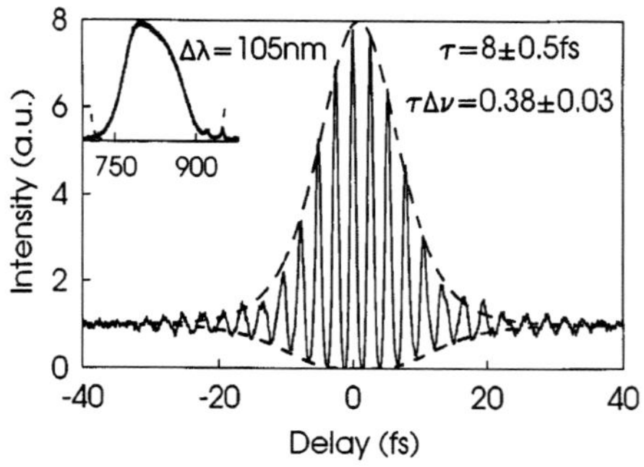

Fig. 3. Single-scan FRAC trace of the output of the MDC Ti:sapphire laser. The dashed curves represent the calculated envelopes of an 8.2-fs $sech^2$-shaped pulse. The collinear autocorrelator has been calibrated by use of a He–Ne laser. The fringe asymmetry is presumably caused by a slight bend of the thin beam splitter or the astigmatism introduced by the focusing mirror. Inset: spectrum of the mode-locked laser (solid curve) and transmittivity of the dichroic curved mirrors (dashed curve). The ends of the dashed curve correspond to a transmission of 10%.

FRAC trace yields a pulse duration of $\tau = 8 \pm 0.5$ fs (FWHM) and a time–bandwidth product of $\tau\Delta\nu = 0.38 \pm 0.03$.[19] The high visibility of the fringes in the wings of the FRAC trace and the absence of substructure in the envelopes provide clear evidence for the high quality and the nearly transform-limited nature of the generated pulses.

Although these optical pulses are still somewhat longer than the shortest pulses obtained by external pulse compression (6 fs),[20] they offer what we believe is an unprecedented time resolution for spectroscopy owing to their high stability and repetition rate. In a recent experiment subfemtosecond time resolution was demonstrated by use of 12-fs pulses generated by a forerunner of the present system.[21]

A comparison of the mode-locked spectrum with the transmittivity of the quasi-quarter-wave dichroic curved mirrors (dashed curve inset of Fig. 3) indicates that the major limitation preventing further pulse shortening in the current system is the finite bandwidth of the two standard quarter-wave cavity mirrors. Because dispersion-engineered mirrors cannot currently be produced with high transmittivity at the pump wavelengths, a replacement of the dichroic focusing mirrors with broadband chirped mirrors calls for a new pump configuration.

In conclusion, we have demonstrated what to our knowledge is the first laser oscillator that generates nearly bandwidth-limited high-quality optical pulses in the sub-10-fs regime. The mode-locked spectrum is centered close to the peak of the gain line in Ti:sapphire.[22] These features, along with the high stability and reproducibility of performance, make this system a unique tool both for high-resolution time-resolved spectroscopy and for seeding ultrashort-pulse Ti:sapphire amplifiers.

We are deeply indebted to A. J. Schmidt for his stimulating support and to K. Ferencz for manufacturing the dielectric coatings. We also thank E. Wintner for his support, S. M. J. Kelly for careful reading of the manuscript, and L. Xu for helpful comments and assistance in the measurements. This study was supported by the Austrian and Hungarian Science Foundations under grants P09710 and P10409, and T-007376, respectively.

References

1. See, for example, B. B. Ju, E. A. De Souza, W. H. Knox, M. C. Nuss, and J. E. Cunningham, in *Ultrafast Phenomena*, Vol. 7 of 1994 OSA Technical Digest Series (Optical Society of America, Washington, D.C., 1994), paper PDP12.
2. P. B. Corkum, F. Brunel, and N. K. Sherman, Phys. Rev. Lett. **61,** 2886 (1988).
3. D. Du, X. Liu, G. Korn, J. Squier, and G. Mourou, Appl. Phys. Lett. **64,** 3071 (1994).
4. P. F. Moulton, J. Opt. Soc. Am. B **3,** 125 (1986).
5. D. E. Spence, P. N. Kean, and W. Sibbett, Opt. Lett. **16,** 42 (1991).
6. R. Szipöcs, K. Ferencz, Ch. Spielmann, and F. Krausz, Opt. Lett. **19,** 201 (1994).
7. R. L. Fork, O. E. Martinez, and J. P. Gordon, Opt. Lett. **9,** 150 (1984).
8. C. P. Huang, H. C. Kapteyn, J. W. McIntosh, and M. M. Murnane, Opt. Lett. **17,** 139 (1992); F. Krausz, C. Spielmann, T. Brabec, E. Wintner, and A. J. Schmidt, Opt. Lett. **17,** 204 (1992); C. P. Huang, M. T. Asaki, S. Backus, M. M. Murnane, H. C. Kapteyn, and H. Nathel, Opt. Lett. **17,** 1289 (1992); B. Proctor and F. Wise, Opt. Lett. **17,** 1295 (1992); B. E. Lemoff and C. P. J. Barty, Opt. Lett. **17,** 1367 (1992).
9. Ch. Spielmann, P. F. Curley, T. Brabec, E. Wintner, and F. Krausz, Electron. Lett. **28,** 1532 (1992).
10. P. F. Curley, Ch. Spielmann, T. Brabec, F. Krausz, E. Wintner, and A. J. Schmidt, Opt. Lett. **18,** 54 (1993).
11. B. Proctor and F. Wise, Appl. Phys. Lett. **62,** 470 (1993).
12. M. Asaki, C. Huang, D. Garvey, J. Zhou, H. C. Kapteyn, and M. M. Murnane, Opt. Lett. **18,** 977 (1993).
13. B. E. Lemoff and C. P. J. Barty, Opt. Lett. **18,** 57 (1993).
14. Ch. Spielmann, P. F. Curley, T. Brabec, and F. Krausz, IEEE J. Quantum Electron. **30,** 1100 (1994).
15. J. Zhou, G. Taft, C. P. Huang, M. M. Murnane, H. C. Kapteyn, and I. Christov, Opt. Lett. **19,** 1149 (1994).
16. I. P. Christov, M. M. Murname, H. C. Kapteyn, J. Zhou, and C.-P. Huang, Opt. Lett. **19,** 1465 (1994).
17. J. D. Harvey, J. M. Dudley, P. F. Curley, Ch. Spielmann, and F. Krausz, Opt. Lett. **19,** 972 (1994).
18. A. Stingl, Ch. Spielmann, F. Krausz, and R. Szipöcs, Opt. Lett. **19,** 204 (1994).
19. Although the Fourier transform of the measured spectrum would give a pulse width of 8.6 fs in the absence of spectral phase modulation, this value is likely to be modified by the spectral response of the monochromator, which is not precisely known. Even with a constant spectral FWHM, the calculated minimum pulse width is utterly sensitive to slight changes in the wings of the spectrum.
20. R. L. Fork, C. H. Brito Cruz, P. C. Becker, and C. V. Shank, Opt. Lett. **12,** 483 (1987).
21. Ch. Spielmann, R. Szipöcs, A. Stingle, and F. Krausz, Phys. Rev. Lett. **73,** 2308 (1994).
22. This should be contrasted with the center wavelength of ≈850 nm in prism-controlled Ti:sapphire lasers offering comparable performance (Refs. 14 and 15).

June 15, 1995 / Vol. 20, No. 12 / OPTICS LETTERS

Sub-20-fs, kilohertz-repetition-rate Ti:sapphire amplifier

M. Lenzner, Ch. Spielmann, E. Wintner, F. Krausz, and A. J. Schmidt

Abteilung Quantenelektronik und Lasertechnik, Technische Universität Wien, Gusshausstrasse 27-29, A-1040 Wien, Austria

Received February 15, 1995

A simple four-pass Ti:sapphire amplifier is seeded by sub-10-fs pulses generated from a mirror-dispersion-controlled Ti:sapphire laser. Pulses of 17–18-fs duration with energies up to 50 and 100 μJ have been produced at repetition rates of 2 and 1 kHz, respectively. Because of the absence of a pulse stretcher, this performance is achieved from an extremely compact system.

Ultrashort-pulse kilohertz-repetition-rate amplifiers have been powerful tools for ultrafast spectroscopy because they simultaneously offer high pulse energies and the possibility of use of sensitive signal-averaging techniques.[1] Until recently, kilohertz-rate amplification of femtosecond pulses could be performed only in complex dye laser systems. The evolution of kilohertz femtosecond dye amplifiers culminated in the generation of submicrojoule-energy 16-fs pulses from a sophisticated multistage amplifier system.[2] In this Letter we report amplification of pulses of comparable duration to the 100-μJ-level in a much more compact, reliable, and user-friendly solid-state laser, which uses Ti:sapphire as the gain medium.

The discovery of self-mode locking in a Ti:sapphire oscillator[3] was soon followed by a demonstration of kilohertz amplification of 150-fs pulses up to 1 mJ in a Ti:sapphire laser system.[4] Subsequent advances in femtosecond pulse generation from Ti:sapphire oscillators led to the generation of pulses in the 10–20-fs range,[5] thus provoking much interest in amplifying pulses well below 100 fs. Recently, regenerative amplification of 55- and 30-fs pulses was demonstrated at kilohertz repetition rates.[6,7] Regenerative amplification is a well-established technique for the efficient generation of microjoule- and millijoule-energy ultrashort pulses from solid-state lasers with excellent beam quality. To preclude excessive peak intensities in the amplifier, the seed pulse is temporally expanded by a dispersive delay line and recompressed after amplification (chirped-pulse amplification).[8] After being seeded into a resonator containing the amplifying medium, the pulse experiences gain over many round trips. It enters and leaves the cavity via a Pockels cell/dielectric polarizer system, which is passed in every round trip. The extinction ratio of the polarizer must be severely compromised if cumulative higher-order dispersion effects are to be kept at a low level for broad-bandwidth pulses, constituting a major limitation to shorter pulse generation from regenerative amplifiers.

This limitation is absent in multipass amplifiers, in which the pulse slicer is passed only once and the selected pulse then propagates a number of times along slightly different paths through the gain region of the amplifier medium. The necessary overlap between the seed beam and the pumped region limits the number of passes; therefore high single-pass gain is required. Recently, the amplification of 26-fs pulses to the terawatt level at a repetition rate of 10 Hz was demonstrated in a two-stage multipass amplifier.[9] Another limitation originates from previous implementations of the chirped-pulse-amplification concept, which employed sophisticated pulse expansion–compression systems.[6,7,10] These setups not only add considerable complexity but require a higher overall gain to compensate for the losses. The higher gain, in turn, implies a more severe gain-narrowing effect, which ultimately limits the shortest pulse duration achievable from an amplifier at a given energy level. Whereas these systems are indispensable in terawatt-scale systems,[9,11] they can be replaced by a more compact and efficient scheme, as we demonstrate below.

In this Letter we report a novel approach to amplifying extremely short optical pulses to moderate (submillijoule) energy levels. Seeding a multipass Ti:sapphire amplifier with ultrabroad-bandwidth (90–100-nm, 40–45-THz) pulses leads to a substantial natural pulse broadening because of the material dispersion of the system components and hence to an automatic implementation of the chirped-pulse-amplification concept. Pulse stretching by the dispersion of additional optical materials was demonstrated recently.[12] The moderately stretched amplified pulses can then be recompressed with a low-dispersion prism pair in combination with standard and specially designed high-reflecting mirrors for cubic dispersion control. This approach combines an extremely high overall system throughput and bandwidth with the benefits of an unprecedented simplicity and compactness.

The seed pulses for our amplifier are generated from a mirror-dispersion-controlled Kerr-lens mode-locked Ti:sapphire laser.[13] The oscillator reported in Ref. 13 has been improved and is now capable of generating nearly bandwidth-limited 8-fs pulses if operated with a 3%-transmitting output coupler.[14] For seeding the amplifier we use an 8% output coupler, which yields slightly longer 9–10-fs pulses at a 80-MHz repetition rate with an average power of 100–150 mW when pumped with 3–3.5 W of power from a small-frame argon laser (all lines). As a unique feature of the mirror-dispersion-controlled Ti:sapphire laser, the mode-locked spectrum is as broad as 90–100 nm and centered between 780 and 790 nm. The same spectral

0146-9592/95/121397-03$6.00/0

OPTICS LETTERS / Vol. 20, No. 12 / June 15, 1995

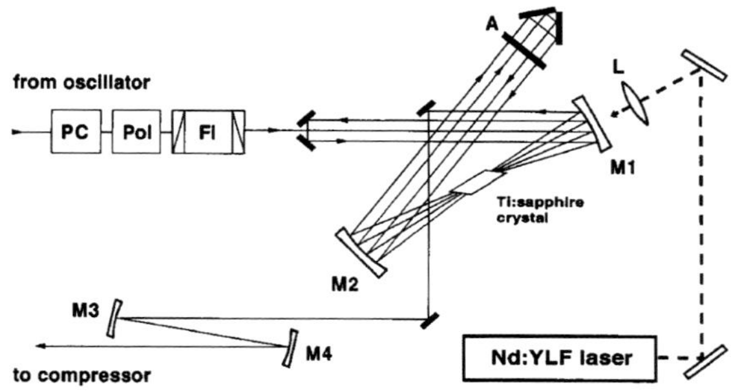

Fig. 1. Schematic of the confocal multipass amplifier. The radius of curvature of focusing mirrors M1 and M2 is 30 cm. PC, Pockels cell; Pol, Berek's polarization compensator; FI, Faraday isolator including two polarizers; A, multiple aperture; M3, M4, collimating telescope; L, lens ($f = 15$ cm).

width can be obtained from prism-dispersion-controlled Ti:sapphire lasers only in the wavelength range centered around 850 nm,[15,16] which is poorly matched with the gain profile of the amplifier.

A schematic diagram of the amplifier is shown in Fig. 1. The Pockels cell selects single pulses from the 80-MHz pulse train at a repetition rate between 1 and 5 kHz, depending on that of the pump laser of the amplifier, with an extinction ratio of $>10^3$. A Berek compensator allows for any polarization transformation between the Pockels cell and the Faraday rotator, providing a convenient means of changing the polarization for transmitting the full train for alignment purposes or switching single pulses for amplification. A special broadband Faraday rotator (Gsaenger FR 820 BB) surrounded by two MacNeille polarizers[17] provides an isolation of >50 dB over a wavelength range of 680–1200 nm, with a transmittivity peaking near 820 nm and having a 50% bandwidth of $\approx$350 nm. The finite bandwidth and some wavelength mismatch of the isolator and reflective optics in the amplifier reduce the spectral width of the pulse to 65–70 nm. This time-independent isolation was found to be necessary (in addition to the time-dependent isolation provided by the Pockels cell) to prevent amplified spontaneous emission (ASE) in the reverse direction from reaching the oscillator. This reverse ASE did not disturb the oscillator noticeably, but when it is reflected back by the output coupler it experiences the same high gain as the seed pulse and hence gives rise to considerable background.

The selected pulses passing through the pulse slicer and isolator are dispersively broadened by a factor of $\approx$300 to a duration of 3 ps. The amplifier uses a 6-mm-long, highly doped (0.20%) Brewster-cut Ti:sapphire slab (Crystal Systems) at the focus of an astigmatically compensated confocal four-pass amplifier formed by two dichroic mirrors ($f = 150$ mm, 40-mm diameter) and retroreflectors, as shown in Fig. 1. A multiple aperture in the beam path prevents the onset of laser action in the high-gain amplifier. The Ti:sapphire crystal is pumped by a cw-pumped, intracavity frequency-doubled, Q-switched Nd:YLF laser (BMI, Model 621-D), which deliver 300-ns pulses at a repetition rate variable between 1 and 5 kHz with a maximum average power of 12 W. The crystal absorbs 64% of the incident pump energy in the first pass, with the transmitted portion backreflected into the crystal to achieve an overall absorption efficiency of $>80\%$. Pumping the amplifier with an energy of 6 mJ at a 2-kHz repetition rate produces an effective single-pass gain of $\approx$18, yielding a total gain of $\approx 10^5$. From the single-pass gain we calculate the pump fluence on the first crystal face as $\approx$5 J/cm^2, at which safe operation without the risk of damage is possible owing to the comparatively long pump pulses.

After amplification the chirped 3-ps pulses are recompressed in a compressor that comprises dispersive mirrors and a low-dispersion FK5 (Schott) prism pair. An additional source of positive third-order dispersion (TOD) is required because the positive TOD of system components (materials and mirrors) is overcompensated by the negative TOD of the prism pair when the overall group-delay dispersion is zero. To this end chirped dielectric mirrors[18] with negative group-delay dispersion and positive TOD have been designed and manufactured at the Research Institute of Solid State Physics in Budapest, Hungary. The dispersive characteristics of these special mirrors along with those of the prisms and other system components will be described in detail elsewhere.[19] With 16 reflections off the chirped mirrors (group-delay dispersion approximately -50 fs^2, TOD ≈ 100 fs^2) the prisms are 5 m apart. This rather large separation was achieved with a folding mirror and leads to a beam size at the second prism of $\approx$4 cm. This combination of dispersive elements results in a virtually complete elimination of

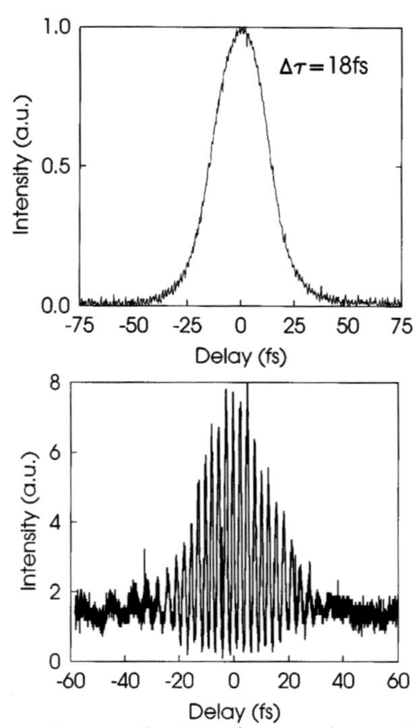

Fig. 2. Intensity and interferometric autocorrelation traces of the amplified pulses. A sech^2 shape was assumed for determination of the pulse width. The signal-to-noise ratio of this interferometric autocorrelation is at the limit of what can be measured. The difficulties arise from the fact that resolution of the fringes requires a slow scan speed ($<$1 Hz).

June 15, 1995 / Vol. 20, No. 12 / OPTICS LETTERS

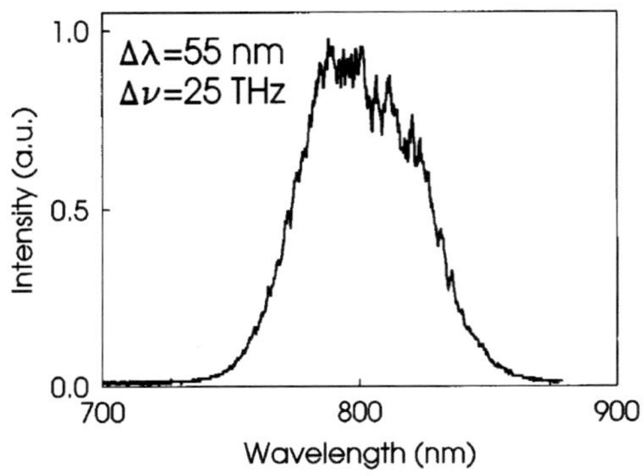

Fig. 3. Spectrum of the amplified pulses.

TOD in the system, with a residual fourth-order dispersion of $\approx -2 \times 10^4$ fs^4.

This simple, low-loss compressor is capable of recompressing the unamplified pulses after a stretching by more than a factor of 300 to $\approx$15 fs (spectral width 65–70 nm.[19] With the amplifier in operation, the pulse duration increases to 17–18 fs as evaluated from the fringe-resolved and background-free autocorrelation traces shown in Fig. 2. This is a consequence of gain narrowing in the amplifier, which reduces the spectrum to approximately 55 nm (Fig. 3). The corresponding time–bandwidth product of $\approx$0.45 indicates that the pulse carries some residual chirp, a conclusion that is consistent with the calculated dispersion parameters of the amplifier system.

These measurements were carried out at a repetition rate of 2 kHz and with a pump energy of $\approx$6 mJ, leading to a compressed pulse energy of $\approx$50 μJ owing to the high throughput (>80%) of the compressor. The ASE level is typically between 3% and 5%, depending on the alignment of the amplifier. This can be substantially reduced by use of either a Pockels cell or spatial filtering. If the repetition rate is lowered to 1 kHz the pump laser can deliver higher-energy pulses. However, increasing the pump energy gives rise only initially to an increase in the compressed pulse energy to approximately 100 μJ (at a pump energy of $\approx$7 mJ). Under these operating conditions the relative ASE energy content is 10%, which grows dramatically if the pump energy is further increased, inhibiting any notable increase of the amplified pulse energy. Therefore we may conclude that effective ASE suppression inside the multipass amplifier is required if higher output pulse energies are to be achieved. The pulse-to-pulse energy fluctuation is <10%, which is attributed to the 1–2% energy fluctuation of the pump source and to the fact that the amplifier is far from being saturated. No evidence of nonlinear effects was observed, in agreement with the estimated value of the B integral of 0.2.

In conclusion, we have demonstrated 18-fs optical pulses amplified to the 100-μJ energy level, yielding a peak power in excess of 5 GW. These represent the most powerful sub-20-fs pulses ever obtained to our knowledge. With some modifications and improvements this system should be capable of amplifying even shorter pulses to the millijoule level. Owing to its unique performance, compactness, and reliability, this kilohertz amplifier promises to become an important tool for high-resolution ultrafast spectroscopy.

We are grateful to R. Szipöcs for designing and K. Ferencz for manufacturing the chirped mirrors. This research was supported by the Fonds zur Förderung der Wissenschaftlichen Forschung in Österreich, grants P9710 and P8566, and by the Jubiläumsfond der Österreichischen Nationalbank, grants 4915 and 5335. M. Lenzner acknowledges a grant by the Deutsche Forschungsgemeinschaft.

References

1. W. H. Knox, M. C. Downer, R. L. Fork, and C. V. Shank, Opt. Lett. **9,** 552 (1984).
2. G. Boyer, M. Franco, J. P. Chambaret, A. Migus, A. Antonetti, P. Georges, F. Salin, and A. Brun, Appl. Phys. Lett. **53,** 823 (1988).
3. D. E. Spence, P. N. Kean, and W. Sibbett, Opt. Lett. **16,** 42 (1991).
4. F. Salin, J. Squier, G. Mourou, and G. Vaillancourt, Opt. Lett. **16,** 1964 (1991).
5. Ch. Spielmann, P. F. Curley, T. Brabec, E. Wintner, and F. Krausz, Electron. Lett. **28,** 1532 (1992); B. E. Lemoff and C. P. J. Barty, Opt. Lett. **17,** 1367 (1992); P. F. Curley, Ch. Spielmann, T. Brabec, F. Krausz, E. Wintner, and A. J. Schmidt, Opt. Lett. **18,** 54 (1993); B. Proctor and F. Wise, Appl. Phys. Lett. **62,** 470 (1993); M. Asaki, C. Huang, D. Garvey, J. Zhou, H. C. Kapteyn, and M. M. Murnane, Opt. Lett. **18,** 977 (1993).
6. J. V. Rudd, G. Korn, S. Kane, J. Squier, G. Mourou, and P. Bado, Opt. Lett. **18,** 2044 (1993).
7. K. Wynne, G. D. Reid, and R. M. Hochstrasser, Opt. Lett. **19,** 895 (1994).
8. D. Strickland and G. Mourou, Opt. Commun. **56,** 219 (1985).
9. J. Zhou, C. P. Huang, M. M. Murnane, and H. C. Kapteyn, Opt. Lett. **20,** 64 (1995).
10. B. E. Lemoff and C. P. J. Barty, Opt. Lett. **18,** 1651 (1993).
11. C. P. J. Barty, C. L. Gordon, and B. E. Lemoff, Opt. Lett. **19,** 1442 (1994).
12. T. Joo, Y. Jia, and G. R. Fleming, Opt. Lett. **20,** 389 (1995).
13. A. Stingl, Ch. Spielmann, F. Krausz, and R. Szipöcs, Opt. Lett. **19,** 204 (1994).
14. A. Stingl, M. Lenzner, Ch. Spielmann, F. Krausz, and R. Szipöcs, Opt. Lett. **20,** 602 (1995).
15. Ch. Spielmann, P. F. Curley, T. Brabec, and F. Krausz, IEEE J. Quantum Electron. **30,** 1100 (1994).
16. J. Zhou, G. Taft, C. P. Huang, M. M. Murnane, H. C. Kapteyn, and I. Christov, Opt. Lett. **19,** 1149 (1994).
17. R. Wynands, F. Diedrich, D. Meschede, and H. R. Telle, Rev. Sci. Instrum. **53,** 5586 (1992).
18. R. Szipöcs, K. Ferencz, Ch. Spielmann, and F. Krausz, Opt. Lett. **19,** 201 (1994).
19. Ch. Spielmann, M. Lenzner, F. Krausz, and R. Szipöcs, "Compact, high-throughput expansion–compression scheme for chirped-pulse amplification in the 10-fs range," submitted to Opt. Lett.

OPTICS LETTERS / Vol. 22, No. 8 / April 15, 1997

Compression of high-energy laser pulses below 5 fs

M. Nisoli, S. De Silvestri, and O. Svelto

Centro di Elettronica Quantistica e Strumentazione Elettronica—Consiglio Nazionale delle Ricerche, Dipartimento di Fisica, Politecnico, Piazza L. da Vinci 32, 20133 Milano, Italy

R. Szipöcs and K. Ferencz

Szilárdtestfizikai Kutatóintézet, Pf. 49, H-1525 Budapest, Hungary

Ch. Spielmann, S. Sartania, and F. Krausz

Abteilung Quantenelektronik und Lasertechnik, Technische Universität Wien, Gusshausstrasse 27, A-1040 Wien, Austria

Received October 25, 1996

High-energy 20-fs pulses generated by a Ti:sapphire laser system were spectrally broadened to more than 250 nm by self-phase modulation in a hollow fiber filled with noble gases and subsequently compressed in a broadband high-throughput dispersive system. Pulses as short as 4.5 fs with energy up to 20-μJ were obtained with krypton, while pulses as short as 5 fs with energy up to 70 μJ were obtained with argon. These pulses are, to our knowledge, the shortest generated to date at multigigawatt peak powers.

Spectral broadening of laser pulses by self-phase modulation (SPM) in a single-mode optical fiber followed by chirp compensation in suitable phase-dispersive elements is a well-established technique for pulse shortening. Pulses down to 6 fs were obtained in 1987 from 50-fs pulses of a mode-locked dye laser,[1] and more recently pulses of ~5 fs were generated from 13-fs pulses of a cavity-dumped Ti:sapphire laser.[2] However, the use of single-mode fibers limits the pulse energy in both cases to a few nanojoules. During the past few years great technological advances have occurred in the field of ultrafast-pulse generation, in particular the development of high-energy solid-state femtosecond lasers. Ti:sapphire amplifiers seeded by 10-fs laser oscillators can now generate pulses of ~20 fs with gigawatt[3,4] or terawatt[5,6] peak power at repetition rates in the kilohertz and 10-Hz regimes, respectively. Recently, a powerful pulse-compression technique based on spectral broadening in a hollow fiber filled with noble gases demonstrated the capability of handling high-energy pulses (submillijoule range).[7] This technique presents the advantages of a guiding element with a large-diameter single mode and of a fast nonlinear medium with a high threshold for multiphoton ionization.

In this Letter we show that combination of the hollow-fiber technique with a broadband, high-throughput dispersive system allows the compression of 20-fs pulses down to a duration as short as 4.5 fs in the pulse-energy range of tens of microjoules, corresponding to multigigawatt peak powers. This result, along with the potential scalability of the system to significantly higher pulse energies, is expected to open up new prospects in high-field light–matter interaction.[8]

We carried out the experiments using a Kerr-lens mode-locked mirror-dispersion-controlled Ti:sapphire oscillator, which provides nearly transform-limited 8-fs pulses.[9] These pulses were amplified at a repetition rate of 1 kHz in a multipass amplifier pumped by the second harmonic of a Q-switched Nd:YLF laser.[3] The output pulses have a duration of 20 fs, energy up to 300 μJ, and a spectrum centered at 780 nm. The pulses were almost transform limited. The amplified pulses were coupled into a 160-μm-diameter, 60-cm-long fused-silica hollow fiber. The fiber was kept straight in a V groove made in an aluminum bar that was placed in a high-pressure chamber with fused-silica windows (1 mm thick) coated for broadband antireflection. The hollow fiber was filled with argon or krypton at different pressures. By properly matching the input beam to the EH_{11} mode of the fiber, we measured an overall fiber transmission of 65%, which is close to the value (~73%) predicted by the theory.[10]

The frequency-broadened pulses emerging from the hollow fiber were compressed by a double pass through two pairs of fused-silica prisms of small apex angle (20°) and by two reflections on a broadband chirped mirror for compensation of quadratic as well as cubic phase distortion. The use of thin prisms with a broadband antireflection coating instead of Brewster-angle prisms allows for a smaller propagation length through glass. This results in a smaller amount of material group delay dispersion and, correspondingly, in a reduction of the negative group delay dispersion required by the prism pairs as well as higher-order dispersion terms. A chirped mirror was introduced to compensate for the cubic phase distortion arising from propagation through the hollow fiber and the prism sequence. This mirror provides at each reflection a negative group delay dispersion with a positive cubic and quartic dispersion. The prism-chirped mirror combination ensures control of second- and third-order dispersion over ~120 THz and provides a high transmission efficiency (>80%) in the range 630–1030 nm.[8]

A typical shape of the spectrum measured at the output of the fiber-compressor system for an argon pressure $p = 3.3$ bars and an input peak power $P_0 = 4$ GW is shown in Fig. 1(a). The shape of the

April 15, 1997 / Vol. 22, No. 8 / OPTICS LETTERS

spectrum is much more uniform than that previously obtained with longer pulses (140 fs), for which a pure SPM spectrum was observed.[7] This indicates that, at this shorter pulse duration, gas dispersion, besides SPM, also plays an important role. The relative weights of SPM and dispersion can be evaluated by use of characteristic parameters such as the nonlinear length L_{NL} and the dispersion length L_D, defined as[11] $L_{NL} = 1/\gamma\, P_0$ and $L_D = T_0{}^2/|\beta_2|$, where T_0 is the half-width (at the $1/e$ intensity point) of the pulse and $\beta_2 = \mathrm{d}^2\beta/\mathrm{d}\omega^2$ is the group-velocity dispersion of the fiber filled with gas. The nonlinear coefficient γ is given by $\gamma = n_2\omega_0/cA_{\text{eff}}$, where n_2 is the nonlinear index coefficient of the gas ($n = n_0 + n_2 I$, where I is the field intensity), ω_0 is the laser central frequency, c is the speed of light in vacuum, and A_{eff} is the mode effective area. For best pulse compression, an optimum fiber length L_{opt} exists, and its value is well approximated by $L_{\text{opt}} \approx (6L_{NL}L_D)^{1/2}$.[12] Assuming that, for argon, $n_2/p = 9.8 \times 10^{-24}\,\mathrm{m}^2/\mathrm{W\,bars}$ (Ref. 13), one obtains $L_{NL} \approx 0.92$ cm. On considering the contributions to second-order dispersion from both gas[14] and waveguide,[10] one obtains a value of $\beta_2 \approx 40\,\mathrm{fs}^2/\mathrm{m}$, which gives $L_D \approx 320$ cm. Then, the optimum length turns out to be $L_{\text{opt}} \approx 42$ cm, somewhat shorter than that used in the experiments. However, if one takes into account the peak-power reduction during propagation, which tends to increase L_{NL}, the length of the fiber is not too far from optimum.

We sent the self-phase-modulated output to the compressor and then measured the compressed pulses by an interferometric autocorrelator with silver mirrors and a very thin (15-μm) β-BaB_2O_4 crystal. Figure 1(b) shows the measured second-harmonic interferometric autocorrelation trace obtained when separation between the two couples of prisms was set at 2 m. The chirped mirror was used for folding the prismatic delay line. The time delay in the autocorrelator was calibrated with an He–Ne laser. To evaluate the pulse duration, we took the inverse Fourier transform of the spectrum of Fig. 1(a) and assumed, as a free parameter, some residual cubic phase distortion, ($\mathrm{d}^3\phi/\mathrm{d}\omega^3$). A good fit to the experimental data was then obtained with a pulse duration (FWHM) of 5.3 fs and $|\mathrm{d}^3\phi/\mathrm{d}\omega^3| = 20\,\mathrm{fs}^3$. The precision of this evaluation is mainly affected by possible errors in the measured spectral shape (the spectrograph was calibrated with a standard tungsten lamp) and in the assumed spectral phase; we expect to introduce errors of less than $\pm 10\%$. The minimum pulse duration, as calculated on assuming optimum phase-distortion compensation, was 5.2 fs. Therefore the pulses can be considered to be almost transform limited. The results shown in Fig. 1 refer to the case in which output pulse energy from the compressor was 40 μJ, corresponding to an input energy to the fiber of 80 μJ. By increasing the input pulse energy to 140 μJ, we generated pulses as short as 5 fs, with an energy of 70 μJ.

A typical broadened spectrum measured for a krypton pressure $p = 2.1$ bars and an input peak power $P_0 = 2$ GW is shown in Fig. 2(a). Under these conditions L_{NL} and L_D for krypton almost match those for argon, since the increased cubic nonlinearity[13] and dispersion[14] for krypton are balanced by the reduction in input peak power and pressure. The shape of the krypton spectrum is more uniform than that of argon and presents a somewhat greater extension. The different spectral-broadening features of krypton confirm previous observations performed with longer pulse duration (140 fs).[7] By best compression of the pulse whose spectrum is shown in Fig. 2(a), we measured the interferometric autocorrelation trace of Fig. 2(b). From this trace, a pulse duration of 4.5 fs (FWHM) was evaluated, assuming a residual cubic phase distortion $|\mathrm{d}^3\phi/\mathrm{d}\omega^3| = 10\,\mathrm{fs}^3$. Pulse energy after compression was 20 μJ. These pulses represent the shortest generated to date at the tens-of-microjoules energy level. The minimum pulse duration, estimated from the spectrum shown in Fig. 2(b) was 4.3 fs; therefore, the pulses are almost transform limited. On increasing the input peak power to 4 GW and decreasing the pressure to 1.1 bars to maintain constant L_{NL}, slightly longer pulses (5.3 fs) with twice as much energy (40 μJ) were obtained.

The output beam was found to be linearly polarized just like the input beam. We tested the spatial coherence of the beam emerging from the fiber by measuring the transverse profile at different distances from the tip of the fiber. The measured beam profiles were compared with the calculated intensity profiles, assuming free-space propagation of a beam with an initial shape equal to that of the EH_{11} mode of the fiber. A typical result obtained at a given distance from the

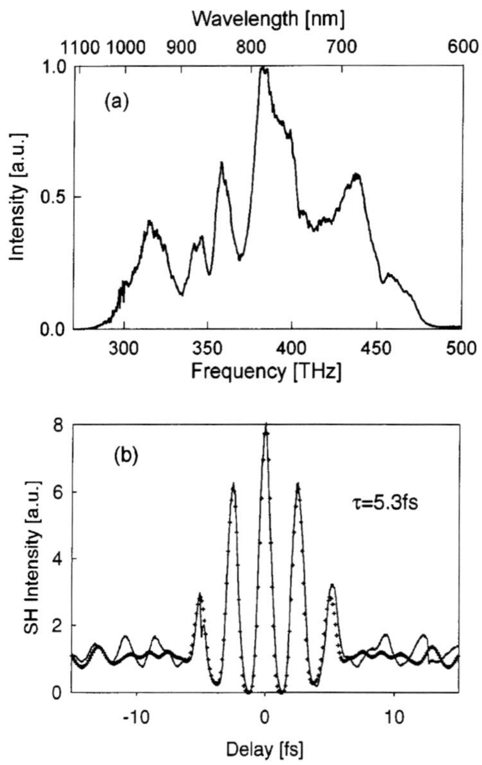

Fig. 1. (a) Spectral broadening in argon at $p = 3.3$ bars and $P_0 = 4$ GW. A low-intensity pedestal (~1% of the peak) extends below 600 nm. (b) Measured (solid curve) and calculated (crosses) autocorrelation trace; an evaluation of pulse duration (FWHM) is also given.

OPTICS LETTERS / Vol. 22, No. 8 / April 15, 1997

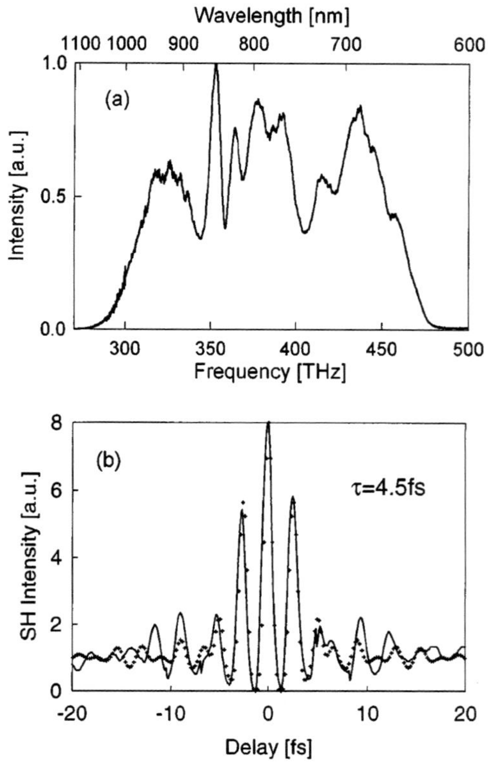

Fig. 2. (a) Spectral broadening in krypton at $p = 2.1$ bars and $P_0 = 2$ GW. A low-intensity pedestal ($\sim 1\%$ of the peak) extends below 600 nm. (b) Measured (solid curve) and calculated (crosses) autocorrelation trace; an evaluation of the pulse duration (FWHM) is also given.

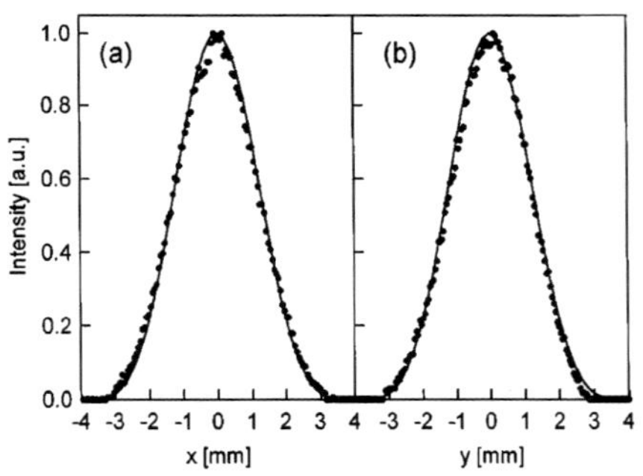

Fig. 3. Measured beam profiles (dots) along orthogonal directions (a) x and (b) y at a distance of 43 cm from the output of fiber. The calculated beam profiles (solid curves) are also shown.

fiber is reported in Fig. 3. Since the measured beam shapes along two orthogonal directions closely follow the calculated curves, the output beam is diffraction limited.

The scalability of the system toward higher pulse energy is an important issue considering the current availability of 20-fs laser pulses with peak powers up to the terawatt level or more. The following two considerations play the most important role: (1) The laser peak power must be smaller than the self-focusing value. This sets a constraint on the type of noble gas to be used and its pressure. (2) The laser peak intensity must be smaller than the multiphoton ionization threshold which applies for the given pulse duration. This represents a constraint on the hollow-fiber diameter and on the type of gas to be used. Since the threshold for multiphoton ionization increases when pulse duration is decreased,[15] we are somewhat confident that the hollow fiber technique can be scaled up to pulse energies of a few millijoules.

The authors thank A. J. Schmidt for his stimulating support, M. Lenzner and Z. Cheng for their important contributions to the development of the Ti:sapphire laser system, and L. Pallaro for valuable technical support. This research was supported by Consiglio Nazionale delle Ricerche and Istituto Nazionale per la Fisica della Materia in Italy, by the Austrian Science Foundation and the Ministry of Science and Arts in Austria under grants P9710 and P11109, and by the Orzsagos Tudmanyos Kutatasi Alap in Hungary.

References

1. R. L. Fork, C. H. Brito Cruz, P. C. Becker, and C. V. Shank, Opt. Lett. **12,** 483 (1987).
2. A. Baltuska, Z. Wei, M. S. Pshenichnikov, and D. A. Wiersma, Opt. Lett. **22,** 102 (1997).
3. M. Lenzner, Ch. Spielmann, E. Wintner, F. Krausz, and A. J. Schmidt, Opt. Lett. **20,** 1397 (1995).
4. S. Backus, J. Peatross, C. P. Huang, M. M. Murnane, and H. C. Kapteyn, Opt. Lett. **20,** 2000 (1995).
5. J. Zhou, C. P. Huang, M. M. Murnane, and H. C. Kapteyn, Opt. Lett. **20,** 64 (1995).
6. C. P. J. Barty, T. Guo, C. Le Blanc, F. Raksi, C. R.-P. Petruck, J. Squier, K. R. Wilson, V. V. Yakovlev, and K. Yamakawa, Opt. Lett. **21,** 668 (1996).
7. M. Nisoli, S. De Silvestri, and O. Svelto, Appl. Phys. Lett. **68,** 2793 (1996).
8. M. Nisoli, S. Stagira, S. De Silvestri, O. Svelto, S. Sartania, Z. Cheng, M. Lenzer, Ch. Spielmann, and F. Krausz, "A novel high energy pulse compression system: generation of multigigawatt sub-5-fs pulses," Appl. Phys. B (to be published).
9. L. Xu, Ch. Spielmann, F. Krausz, and R. Szipöcs, Opt. Lett. **21,** 1259 (1996).
10. E. A. J. Marcatili and R. A. Schmeltzer, Bell. Syst. Tech. J. **43,** 1783 (1964).
11. G. P. Agrawal, *Nonlinear Fiber Optics* (Academic, San Diego, Calif., 1995).
12. W. J. Tomlinson, R. H. Stolen, and C. V. Shank, J. Opt. Soc. Am. B **1,** 139 (1984).
13. H. J. Lehmeier, W. Leupacher, and A. Penzkofer, Opt. Commun. **56,** 67 (1985).
14. A. Dalgarno and A. E. Kingston, Proc. R. Soc. London Ser. A **259,** 424 (1966).
15. I. P. Christov, J. Zhou, J. Peatross, A. Rundquist, M. M. Murnane, and H. C. Kapteyn, Phys. Rev. Lett. **77,** 1743 (1996).

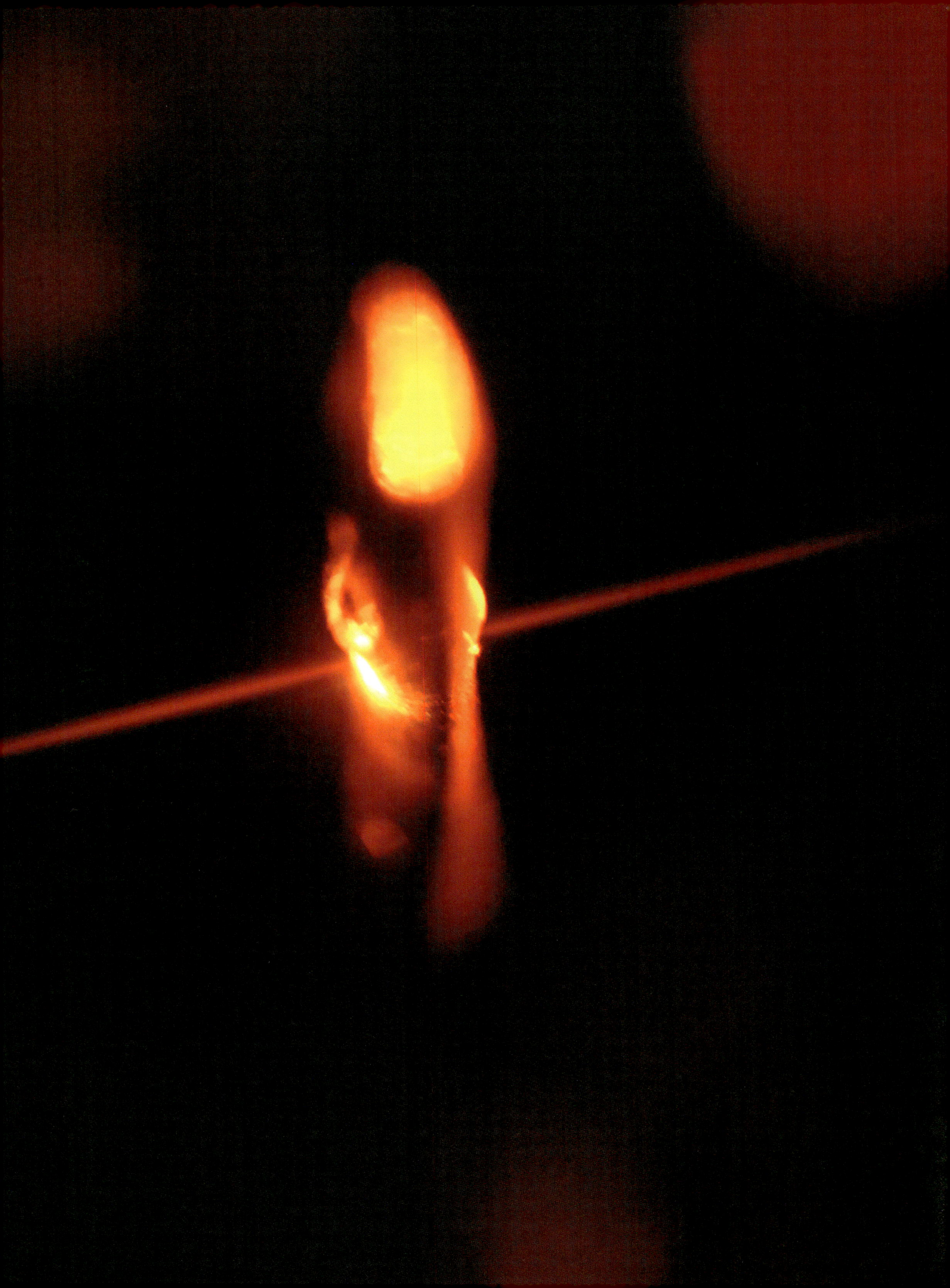

Extreme Nonlinear Optics

3

Few-cycle laser light propagating through a tube filled with neon gas ionizes the atoms, the fluorescence of which renders them visible as they stream out of the tube. © Thorsten Naeser

A multi-volt-per-ångström electric field, when "turned on" within a single optical cycle (briefly: extreme light) is capable of creating a polarization response that cannot be induced in any other way. The sudden "turning on" of a force field that rivals the static Coulomb forces of the nuclei permits extreme nonlinearities to be induced and utilized for creating the tools of attosecond metrology. Modeling the interaction of such fields with matter and their propagation is not feasible in the framework of the usual approximations relying on slowly-varying amplitude and frequency and/or moderate field strength.

In order for these extreme nonlinearities to be explored and exploited, it is usually necessary to propagate the strong, few-cycle laser pulses over extended interaction lengths. The nonlinear polarization induced at field strengths in the non-perturbative regime may significantly modify the electric-field waveform of the laser pulse as it traverses the medium. These propagation effects need to be accurately modeled to be able to properly calculate the response of matter to the extreme light, studied in a macroscopic interaction volume. Thomas Brabec, Michael Geissler, Gabriel Tempea, and Armin Scrinzi performed this modeling and reported their results in [13-14]. The equations derived by Thomas Brabec and his co-workers for the propagation of few-cycle light in the framework of what they dubbed the slowly-evolving wave approximation became a standard tool in the theoretical treatment of extreme light-matter interaction confined to several field cycles, see e.g. its treatment in Robert Boyd's classic textbook[5].

Extreme nonlinearities may result in a fs^{-1}-scale ionization rate (Fig. 3-1) and the generation of coherent light with photon energy exceeding that of the incident photons by a factor of a hundred or more (Fig. 3-2). First demonstrated with multi-cycle driving fields, the latter phenomenon has been dubbed high-order harmonic generation (HHG). Christian

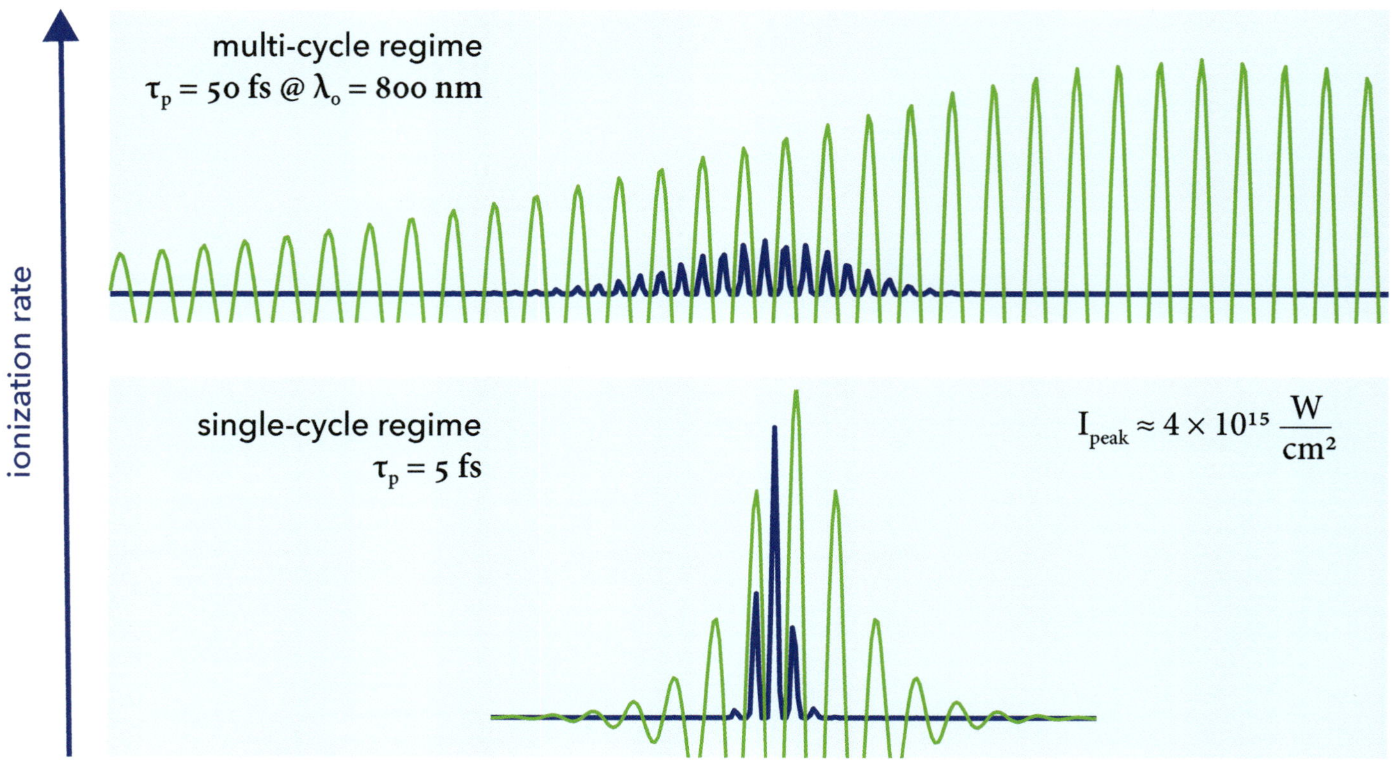

Figure 3-1 Rate of ionization induced in helium atoms by linearly-polarized near-infrared laser fields with duration of 50 fs and 5 fs. In sharp contrast with the response to multi-cycle exposure, ionization by strong few-cycle light is depleted within a single wave cycle (as apparent from its substantially reduced rate at the pulse peak with respect to that induced by the preceding half cycle), indicating ionization rates of the order fs^{-1}. The calculations have been performed by Armin Scrinzi and his co-workers[6].

[5] R. Boyd, Nonlinear Optics, Second Edition (*Academic Press*, 2003).

[6] A. Scrinzi *et al.* Ionization Above the Coulomb Barrier, *Phys. Rev. Lett.* **83**, 706 (1999).

Spielmann and his collaborators implemented HHG with near-single-cycle light and reported their results in [15-16]. Under these conditions, HHG yields a spectral continuum, indicative of the emission being substantially confined (just as the underlying ionization process) to less than a wave cycle of the driving field. These processes, unfolding on a sub-cycle time scale, paved the way towards attosecond technology and have been reviewed in [17].

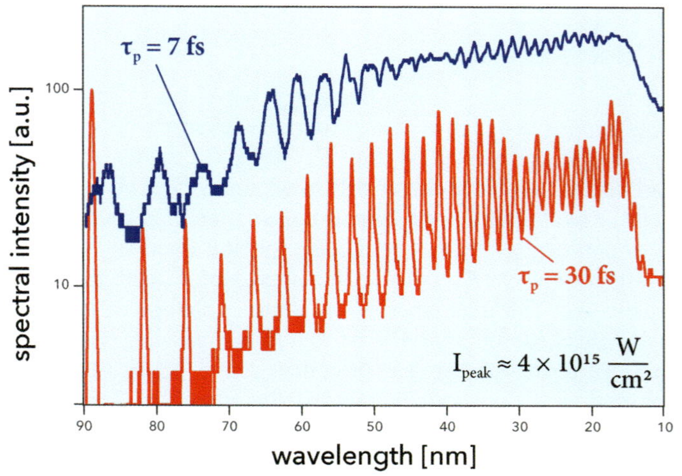

Figure 3-2 High-order harmonic generation by linearly-polarized near-infrared laser pulses carried at a wavelength of 780 nm, with a pulse duration of 30 fs and 7 fs. These unpublished results were obtained by Christian Spielmann and his co-workers in 1997.

[13] T. Brabec, F. Krausz. Nonlinear optical pulse propagation in the single-cycle regime. *Physical Review Letters* **78**, 3282 (1997).
(Page: 94)

[14] M. Geissler, G. Tempea, A. Scrinzi, F. Krausz, M. Schnürer, T. Brabec. Light propagation in field-ionizing media: extreme nonlinear optics. *Physical Review Letters* **83**, 2930 (1999).
(Page: 98)

[15] C. Spielmann, H. Burnett, S. Sartania, R. Koppitsch, M. Schnürer, C. Kan, M. Lenzner, P. Wobrauschek, F. Krausz. Generation of coherent X-rays in the water window using 5-femtosecond laser pulses. *Science* **278**, 661 (1997).
(Page: 102)

[16] C. Spielmann, C. Kan, H. Burnett, T. Brabec, M. Geissler, A. Scrinzi, M. Schnürer, F. Krausz. Near-keV coherent X-ray generation with sub-10-fs lasers. *Journal of Selected Topics of Quantum Electronics* **4**, 249 (1998).
(Page: 106)

[17] T. Brabec, F. Krausz. Intense few-cycle laser fields: frontiers of nonlinear optics. *Reviews of Modern Physics* **72**, 545 (2000).
(Page: 123)

VOLUME 78, NUMBER 17 PHYSICAL REVIEW LETTERS 28 APRIL 1997

Nonlinear Optical Pulse Propagation in the Single-Cycle Regime

Thomas Brabec* and Ferenc Krausz

Abteilung Quantenelektronik und Lasertechnik, Technische Universität Wien, Gusshausstrasse 27-29, A-1040 Wien, Austria

(Received 27 September 1996)

A general three-dimensional wave equation first order in the propagation coordinate is derived covering a broad range of phenomena in nonlinear optics. This equation provides an accurate description of the evolution of the wave packet envelope down to pulse durations as short as one carrier oscillation cycle. The concept of envelope equations is found to be applicable to the single-cycle regime of nonlinear optics. [S0031-9007(97)02995-5]

PACS numbers: 42.65.−k, 31.15.−p

The propagation equation governing the evolution of the complex envelope of optical pulses plays a key role in nonlinear optics. Owing to the assumption of a slowly varying envelope, this partial differential equation contains only the first derivative with respect to the spatial coordinate along the propagation direction. Hence, it can be solved with substantially smaller computational effort than Maxwell's wave equation, which is of second order in the propagation coordinate. This benefit has been exploited in the investigation of a vast number of nonlinear optical phenomena [1−10].

The evolution of ultrashort pulse optics has now arrived at a point where light pulses with durations comparable to the carrier oscillation cycle have become available [11]. This progress opens up new prospects in nonlinear optics as well as strong-field physics and prompts the question of whether the first-order propagation equation is valid in this new and important regime of optics. The purpose of this Letter is to show that (i) the powerful concept of the envelope can be extended to pulse durations equal to the carrier oscillation period T_0, and (ii) the validity of the first-order envelope equation, commonly used to model one-dimensional nonlinear pulse evolution, extends down to pulse durations as short as T_0. Furthermore, (iii) our approach leads, in a natural way, to a generalized, scalar, three-dimensional, first-order propagation equation, which obviates the need for the Maxwell equations in tackling a number of problems, where transverse effects cannot be neglected in the interaction of ultrashort light pulses with matter.

A prerequisite for introducing an envelope equation is the unambiguous definition of the envelope. To this end, the electric field $E(t) = \tilde{E}(t) + \text{c.c.}$ is represented by the complex electric field which is written as $\tilde{E}(t) = A(t)\exp(-i\omega_0 t + i\psi)$. Here, $\omega_0 = \int_0^\infty \omega |E(\omega)|^2 d\omega / \int_0^\infty |E(\omega)|^2 d\omega$ is the carrier frequency, $E(\omega)$ is the Fourier transform of $E(t)$, and ψ is defined such that the imaginary part of the complex envelope $A(t)$ is zero at $t = 0$. The above definition of $A(t)$ is only physically meaningful as long as the envelope remains *invariant* under a change of ψ. This condition is equivalent to the requirement that a phase shift of the electric field $\tilde{E}'(t) = \tilde{E}(t)\exp(i\Delta\psi)$ does not change the center frequency $\omega_0' = \omega_0$. That indeed applies to a high accuracy for pulse durations down to $\tau_p \approx T_0$, as revealed by Fig. 1. The pulse width τ_p is defined as the full width at half maximum (FWHM) of $|A(t)|^2$. For shorter wave packets, the variation of ψ gives rise to non-negligible changes of the spectral intensity at low frequencies and results in the increased sensitivity of ω_0 to ψ. The invariance of ω_0 under a change of ψ can be checked for arbitrary wave forms following the procedure presented above. In conclusion, a physically meaningful envelope can be assigned to ultrashort light wave packets that contain at least one carrier cycle within the FWHM of their intensity envelope. At the 780 nm center wavelength of the Ti:sapphire laser this implies $\tau_p = 2.6$ fs.

Our derivation of the envelope equation starts with Maxwell's equations. For small transverse inhomogeneities of the medium polarization, the three-

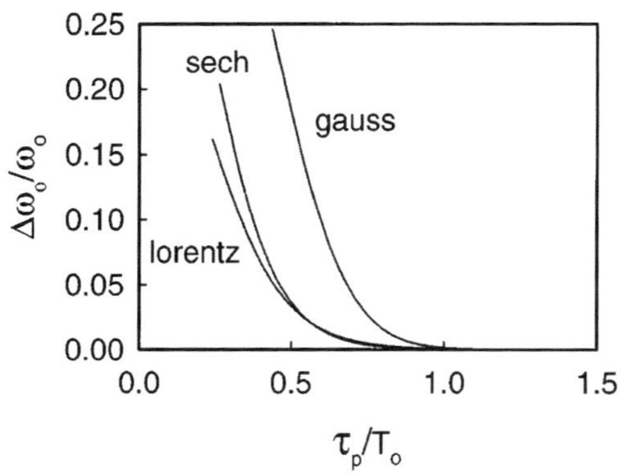

FIG. 1. Phase dependence of the carrier frequency $\Delta\omega_0/\omega_0 = |\omega_0(\psi = 0) - \omega_0(\psi = \pi/2)|/\omega_0$ versus the normalized pulse duration for various analytic pulse shapes, $E(t) = A(t)\sin(\omega_c t + \psi)$; the envelopes are described by a Gaussian, $A_g(t) = \exp[-(1.67t/\tau_p)^2]$, a hyperbolic secant, $A_s(t) = \text{sech}(1.76t/\tau_p)$, and a Lorentzian, $A_l(t) = 1/[1 + (1.29t/\tau_p)^2]$ pulse. The optical cycle $T_0 = 2\pi/\omega_c = 2.67$ fs at a wavelength of 800 nm. Here, ω_c is the carrier frequency of the analytical pulse shape and ω_0 is the center of gravity of the spectrum, as defined in the text. For $\tau_p \geq T_0$ the phase sensitivity is negligible and $\omega_c \approx \omega_0$.

0031-9007/97/78(17)/3282(4)$10.00

VOLUME 78, NUMBER 17 PHYSICAL REVIEW LETTERS 28 APRIL 1997

dimensional wave equation can be written as [1–7]

$$(\partial_z^2 + \nabla_\perp^2)E(\mathbf{r},t) - \frac{1}{c^2}\partial_t^2\int_{-\infty}^{t} dt'\varepsilon(t-t')E(\mathbf{r},t') = \frac{4\pi}{c^2}\partial_t^2 P_{nl}(\mathbf{r},t), \tag{1}$$

where $\nabla_\perp^2 = \partial_x^2 + \partial_y^2$ is the transversal Laplace operator, $\partial_{i=x,y,z,t}$ stands for the respective partial derivatives, $\varepsilon(t) = (2\pi)^{-1}\int_{-\infty}^{\infty} d\omega\,\varepsilon(\omega)e^{-i\omega t}$, $\varepsilon(\omega) = 1 + 4\pi\chi(\omega)$, and $\chi(\omega)$ is the linear electric susceptibility. The electric field E propagates along the z direction. Both E and the nonlinear polarization P_{nl} are polarized parallel to the x axis.

In order to introduce a first-order propagation equation for the wave packet envelope defined above, we use the ansatz $\mathbf{E}(\mathbf{r},t) = A(\mathbf{r}_\perp,z,t)e^{i(\beta_0 z-\omega_0 t+\psi_0)} + \text{c.c.}$, where ω_0 and ψ_0 are determined at the input reference plane $z=0$ on the beam axis $\mathbf{r}_\perp = 0$ as prescribed above. Further, $\beta_0 = \text{Re}[k(\omega_0)] = (\omega_0/c)n_0$, where $k(\omega) = (\omega/c)\sqrt{\varepsilon(\omega)}$ is the complex propagation constant and n_0 is the refractive index of the medium at ω_0. The induced nonlinear polarization is written as $P_{\text{nl}}(\mathbf{r},t) = B(\mathbf{r}_\perp,z,t,A)e^{i(\beta_0 z-\omega_0 t+\psi_0)} + \text{c.c.}$, where the complex amplitude B depends nonlinearly on the amplitude of the electric field. The neglect of backward propagating waves is consistent with the approximations that will be made in the following derivation of the envelope equation and will be commented on later.

The substitution of the above expressions of $\mathbf{E}(\mathbf{r},t)$ and $P_{\text{nl}}(\mathbf{r},t)$ in Eq. (1), Fourier transform of the integral term with respect to the time coordinate and the subsequent Taylor-expansion of $k(\omega)$ about ω_0 followed by an inverse Fourier transform yields

$$(-\beta_0^2 + 2i\beta_0\partial_z + \partial_z^2 + \nabla_\perp^2)A + \left(\beta_0 + i\frac{\alpha_0}{2} + i\beta_1\partial_t + \hat{D}\right)^2 A = \frac{4\pi\omega_0^2}{c^2}\left(1+\frac{i}{\omega_0}\partial_t\right)^2 B. \tag{2}$$

Here we assumed P_{nl} to be a small perturbation to the linear polarization, again, in conformance with later approximations, the dispersion operator $\hat{D}$ is given by

$$\hat{D} = -\frac{\alpha_1}{2}\partial_t + \sum_{m=2}^{\infty}\frac{\beta_m + i\alpha_m/2}{m!}(i\partial_t)^m, \tag{3}$$

and $\beta_m = \text{Re}[(\partial^m k/\partial\omega^m)_{\omega_0}]$ and $\alpha_m = \partial\,\text{Im}[(\partial^m k/\partial\omega^m)_{\omega_0}]$. In the moving reference frame $\tau = t - \beta_1 z$, $\xi = z$, Eq. (2) can be written as

$$\left(1+\frac{i}{\omega_0}\partial_\tau\right)\left[\left(\partial_\xi + \frac{\alpha_0}{2} - i\hat{D}\right)A + \frac{2\pi\beta_0}{in_0^2}\left(1+\frac{i}{\omega_0}\partial_\tau\right)B\right] + \frac{1}{2i\beta_0}\nabla_\perp^2 A$$
$$\simeq \left(\frac{\beta_0-\omega_0\beta_1}{\beta_0}\right)\frac{i}{\omega_0}\partial_\tau\left(\partial_\xi + \frac{\alpha_0}{2} - i\hat{D}\right)A - \frac{1}{2i\beta_0}\left(\partial_\xi^2 + \hat{D}^2 - \frac{\alpha_0^2}{4} - \alpha_0\beta_1\partial_\tau\right)A, \tag{4}$$

where $\hat{D}$ is obtained by replacing ∂_t with ∂_τ in (3). The terms on the right side are small as compared with the left side if

$$|\partial_\xi A| \ll \beta_0|A| \tag{5a}$$

and

$$|\partial_\tau A| \ll \omega_0|A| \tag{5b}$$

or

$$\left|\frac{\beta_0-\omega_0\beta_1}{\beta_0}\right| \ll 1. \tag{5c}$$

If either (5a) and (5b) or (5a) and (5c) are satisfied, (4) simplifies to

$$\partial_\xi A = -\frac{\alpha_0}{2}A + i\hat{D}A + \frac{i}{2\beta_0}\left(1+\frac{i}{\omega_0}\partial_\tau\right)^{-1}\nabla_\perp^2 A + i\frac{2\pi\beta_0}{n_0^2}\left(1+\frac{i}{\omega_0}\partial_\tau\right)B, \tag{6}$$

where $[1 + (i/\omega_0)\partial_\tau]^{-1}$ can be evaluated in the frequency domain. This generic nonlinear envelope equation (NEE) *first-order in the propagation coordinate* ξ provides a powerful means of describing light pulse propagation in dispersive nonlinear media. In the specific case of one-dimensional propagation (i.e., with the diffraction term discarded) with $\alpha_0 \approx 0$ and $\hat{D} \approx -(\beta_2/2)\partial_\tau^2$ Eq. (6) reduces to a *nonlinear Schrödinger equation*, which has been widely used to describe ultrashort pulse propagation in Kerr media ($B \propto |E|^2$) [1,3,5].

From the solution $A(\mathbf{r}_\perp,\xi,\tau)$ of Eq. (6) the electric field can be reconstructed as

$$E(\mathbf{r}_\perp,\xi,\tau) = Ae^{-i\omega_0\tau+i\psi(\xi)} + \text{c.c.},$$
$$\psi(\xi) = \psi_0 + (\beta_0 - \omega_0\beta_1)\xi. \tag{7}$$

These equations describe the evolution of light wave packets in terms of a *fixed* carrier frequency ω_0 defined at $\xi = 0$, the entrance of the propagation medium, and an *evolving* complex envelope $A(\mathbf{r}_\perp,\xi,\tau)$ and phase $\psi(\xi)$, which determines the "position" of the carrier wave relative to the envelope. The complex envelope $A(\mathbf{r}_\perp,\xi,\tau)$ evolves due to absorption (or gain), dispersion, diffraction, and nonlinearities, whereas $\psi(\xi)$ evolves due to a difference between group velocity (β_1^{-1}) and phase velocity in the propagation medium.

Traditionally, specific forms of the NEE such as the nonlinear Schrödinger equation have been derived by making use of the *slowly-varying-envelope approximation* (SVEA), $|\partial_z A| = |\partial_\xi A - \beta_1 \partial_\tau A| \ll \beta_0 |A|$. This condition can be decomposed into (5a) and (5b) by utilizing $\beta_0/\beta_1 \approx \omega_0$, which impose distinctly different requirements on the physical system. The former, "nonlocal" part of the SVEA requires that the complex amplitude does not excessively change *during propagation*, a requirement whose violation was previously shown to result in the emergence of a backward propagating wave [2]. The latter "local" contribution demands that the pulse duration must be much longer than the carrier oscillation period. The essential new finding from our derivation [12] of the NEE is that this latter requirement may be dropped if (5c) is satisfied. As a matter of fact, condition (5c) is met if the difference between group and phase velocity relative to the latter is small compared to one, which is fulfilled in a wide range of propagation phenomena. Drawing on (7), condition (5c) can be reexpressed as $|\partial\psi/\partial\xi| \ll \beta_0$. Consequently, (5a) and (5c) can be merged into a single mathematical requirement

$$|\partial_\xi E| \ll \beta_0 |E|, \tag{8}$$

which we refer to as the *slowly-evolving-wave approximation* (SEWA). The SEWA requires more from the propagation medium than the SVEA: not only the envelope A but also the relative carrier phase ψ must not significantly vary as the pulse covers a distance equal to the wavelength $\lambda_0 = 2\pi c/\omega_0$. In return, it does not explicitly impose a limitation on the pulse width. Therefore, *in the frame of the SEWA the nonlinear envelope equation accurately describes light pulse propagation down to the single cycle regime.* The region of validity of the SEWA can be easily assessed by introducing the characteristic propagation lengths $L_\psi = (2\pi)^{-1}|dn/d\lambda|_{\lambda_0}^{-1}$, over which ψ is changed by 1, further $L_{\alpha,m} = \tau_p^m/|\alpha_m| (m = 0, 1, \ldots)$, $L_{\beta,m} = \tau_p^m/|\beta_m| (m = 2, 3, \ldots)$, and $L_\perp = \beta_0 w_0^2$, and $L_{\mathrm{nl}} = |A/B|(n_0^2/2\pi\beta_0)$, over which the envelope is significantly modified due to absorption (α_0) dispersion ($\alpha_m, m \geq 1$; $\beta_m, m \geq 2$), diffraction, and nonlinearities, respectively. Here τ_p is the pulse duration, w_0 is the beam radius at the beam waist for a hypothetical linear propagation, whereas A and B are the respective field amplitudes at an arbitrary instant and position. The SEWA is applicable as long as each of these characteristic length scales meets the condition $\beta_0 L_{\mathrm{char}} \gg 1$. Careful inspection of the parameters of various nonlinear media yields the remarkable finding that these conditions are well satisfied for a wide range of phenomena in the parametric regime (i.e., ω_0 is far off resonances) below the ionization threshold. Strong resonant coupling may, however, in specific cases give rise to extremely short $L_{\psi,\alpha,\beta,\mathrm{nl}}$, violating the condition for the SEWA [13].

The validity of the NEE down to the single-cycle regime has been tested by solving (6) and the Maxwell equations numerically for a Kerr medium characterized by the constitutive law $B = (2\pi)^{-1} n_0 n_2 |A|^2 A$, where n_2 is the nonlinear index of refraction [1] and $|A|^2$ is normalized to give the intensity. For simplicity, we considered one-dimensional propagation, assumed the Kerr response to be instantaneous [1], and neglected the emergence of harmonic radiation [10]. A sech-shaped pulse with an FWHM duration of $\tau_p = 2.67$ fs, which corresponds to one optical cycle at the carrier wavelength of 0.8 μm, was propagated through a hypothetical dielectric medium with $\beta_2 = 0.0385$ fs$^2/\mu$m ($\beta_m = 0$ for $m > 2$) and $n_2 = 3 \times 10^{-16}$ W/cm^2 ($dn_2/d\omega = 0$), which correspond to the respective parameters of fused silica. In order to simulate an extremely strong parametric interaction the peak intensity of the incident pulse was chosen as $|A(\xi = 0, \tau = 0)|^2 = 4 \times 10^{13}$ W/cm^2, which is close to the expected critical intensity level for optical damage in fused silica and other dielectric materials. In advancing the coupled Maxwell's equations, from which (1) has been derived, in space the leap frog method [14] was used. The time derivatives were evaluated in the ω space using a fast Fourier transform. The NEE was integrated by applying the split-step Fourier method [1].

In Fig. 2 the solutions of the Maxwell's equations and Eq. (6) as given by (7) are depicted for the case of $\beta_2 = 0$. In the absence of dispersion, self steepening due to the time derivative of the nonlinearity creates an optical shock at the trailing edge of the pulse. The critical distance for self steepening [1] is given by $z_s = 0.43 L_{\mathrm{nl}} \omega_0 \tau = 23.36$ μm, which is close to the chosen propagation distance of $z = 20$ μm. Even along the shock front, where the change of the envelope is comparable to the change of the carrier frequency, the two solutions are virtually identical. Figure 3 shows the electric field

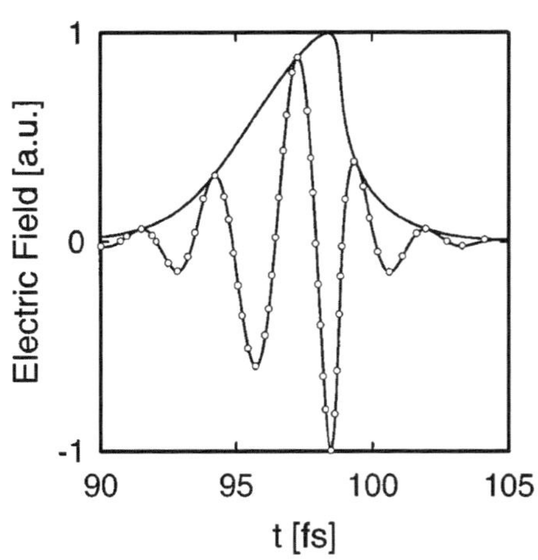

FIG. 2. Shows the electric field strength and the field envelope in arbitrary units versus propagation time. The electric fields obtained by the solution of the Maxwell equation and of Eq. (4) are depicted by the full line and by the open circles, respectively; the initial pulse has a sech-shape and the parameters are, $I = 4 \times 10^{13}$ W/cm^2, $n_2 = 3 \times 10^{-16}$ cm^2 W^{-1}, $\omega_0 = 2.35$ fs^{-1} (0.8 μm), $\tau_p = 2.67$ fs, $n_0 = 1.45$, $dn/d\omega\mid_{\omega_0} = 0$ fs (no dispersion), and a propagation distance $z, \xi = 20$ μm.

VOLUME 78, NUMBER 17 PHYSICAL REVIEW LETTERS 28 APRIL 1997

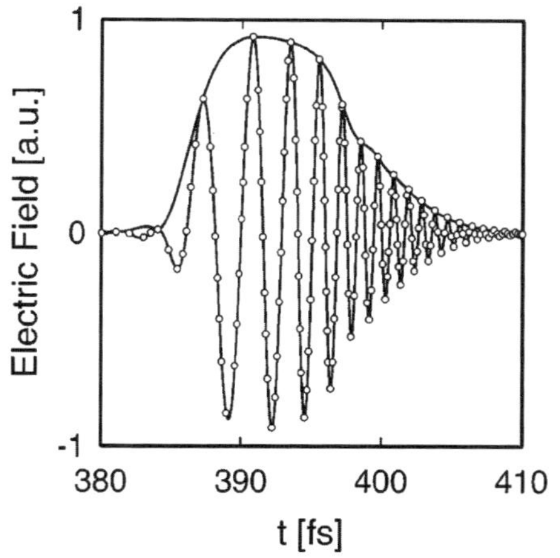

FIG. 3. The same as in Fig. 1, with a propagation distance $z = 80\ \mu m$ and $dn/d\omega \mid_{\omega_0} = 5.776 \times 10^{-3}$ fs, which gives $\beta_1 = 4.882$ fs/μm and $\beta_2 = 3.853 \times 10^{-2}$ fs$^2/\mu$m.

for the case of $\beta_2 > 0$ and $z = 80\ \mu$m. Dispersion suppresses the formation of a shock front in this case. Again, an excellent agreement between the solution of Maxwell's equations and that of the NEE is found, which has also been verified for longer propagation distances.

The NEE incorporates the "correction" operator $[1 + (i/\omega_0)\partial_\tau]$ in the terms accounting for diffraction and nonlinear effects when compared with the paraxial wave equation describing the propagation of stationary or long-pulse radiation [2–6]. Physically, the appearance of the correction operator in the nonlinear term is responsible for introducing a phase modulation if the nonlinearity modulates primarily the amplitude in the long-pulse limit (as is the case in a near-resonant interaction), and vice versa (e.g., self steepening versus self phase modulation in the case of a Kerr nonlinearity). Mathematically, the correction operator becomes zero at $\omega = 0$ in the frequency domain, frustrating the emergence of a dc component of the wave packet during propagation. The same operator in the denominator of the diffraction term in (6) accounts for an enhanced beam divergence at lower-frequency components. Since this generalized diffraction operator [15] has a singularity at $\omega = 0$ in the frequency domain, the use of Eq. (6) is restricted to initial pulses having a negligible dc spectral component, a requirement which is consistent with the definition of the envelope.

So far, our discussion was confined to studying the evolution of the incident wave packet carried at ω_0. The nonlinear response of the medium may arise to the emergence of waves at new (e.g., harmonic) frequencies. The concept of the SEWA can be extended to nonlinear frequency mixing processes, such as harmonic generation and parametric amplification, as long as the difference between phase and group velocities of the waves involved in the interaction is small compared to the phase (or group) velocity of the fundamental pulse [13]. Far from resonances, this requirement is fulfilled in the majority of parametric processes in which all waves propagate in the same direction. The condition is clearly violated in processes such as Brillouin scattering [2,3], where the pulses propagate in opposite directions. Therefore Brillouin scattering can be addressed using the NEE only in the frame of the SVEA, i.e., for pulse durations much longer than the optical cycle.

In conclusion, the concept of the envelope in the description of nonlinear light wave propagation has been extended to a regime, where the pulse duration is comparable to the carrier oscillation cycle. Drawing on this concept, a three-dimensional first-order envelope propagation equation has been derived, which, in the perturbative limit of small population transfer, provides an accurate description of the propagation of ultrashort light transients in nonlinear media.

The authors gratefully acknowledge illuminating discussions with G. A. Mourou and P. B. Corkum. This work was supported by the ONB grants No. 5124 and No. 5335 and by FWF Grant Y44-PHY. T. Brabec is supported by the Österreichische Akademie der Wissenschaften, APART Grant.

*Electronic address: brabec@rs6.iaee.tuwien.ac.at

[1] G. P. Agrawal, *Nonlinear Fiber Optics* (Academic Press, New York, 1994), 2nd ed.
[2] Y. R. Shen, *The Principles of Nonlinear Optics* (J. Wiley & Sons, New York, 1984).
[3] R. W. Boyd, *Nonlinear Optics* (Academic Press, New York, 1994).
[4] A. Yariv, *Quantum Electronics* (John Wiley & Sons, New York, 1989), 3rd ed.
[5] H. A. Haus, *Waves and Fields in Optoelectronics* (Prentice-Hall, Englewood Cliffs, New Jersey, 1984).
[6] A. E. Siegman, *Lasers* (University Science Books, Mill Valley, California, 1986).
[7] J.-C. Diels and W. Rudolph, *Ultrashort Laser Pulse Phenomena* (Academic Press, New York, 1996).
[8] K. J. Blow and D. Wood, IEEE J. Quant. Electron. **25**, 2665 (1989).
[9] R. M. Joseph *et al.*, Opt. Lett. **18**, 491 (1993); R. W. Ziolkowski and J. B. Judkins, J. Opt. Soc. Am. B **10**, 186 (1993); C. V. Hile and W. L. Kath, J. Opt. Soc. Am. B **13**, 1135 (1996).
[10] R. G. Flesch *et al.*, Phys. Rev. Lett. **76**, 2488 (1996).
[11] A. Baltuska *et al.*, Opt. Lett. **22**, 102 (1997); M. Nisoli *et al.*, Opt. Lett. (to be published).
[12] In contrast with previous approaches, we performed the coordinate transformation before making any approximation.
[13] T. Brabec, Ch. Spielmann, and F. Krausz, "Femtosecond Light Pulses," Rev. Mod. Phys. (to be published).
[14] William H. Press *et al.*, *Numerical Recipes* (Cambridge University Press, Cambridge, England, 1992), 2nd ed.
[15] Expanded to first order in $(i/\omega_0)\partial_\tau$ it is equivalent to Eq. (5) in J. E. Rothenberg, Opt. Lett. **17**, 1340 (1992).

VOLUME 83, NUMBER 15 PHYSICAL REVIEW LETTERS 11 OCTOBER 1999

Light Propagation in Field-Ionizing Media: Extreme Nonlinear Optics

M. Geissler, G. Tempea, A. Scrinzi, M. Schnürer, F. Krausz, and T. Brabec*

Institut für Angewandte Elektronik und Quantenelektronik, Technische Universität Wien, Gusshausstrasse 27-29, A-1040 Wien, Austria

(Received 10 June 1999)

A novel equation for light propagation in underdense field-ionizing media is derived by extending concepts of nonlinear optics into the strong-field domain. The equation is first order in the propagation coordinate and is valid for arbitrarily short pulse durations. Solutions of the first-order wave equation are found to be in excellent agreement with solutions of the scalar wave equation. Furthermore, the polarization response of a field-ionizing medium is derived from semiclassical considerations. The polarization reveals, in agreement with experiments, a previously unrecognized contribution that is shown to significantly affect the absorption loss in the presence of ionization.

PACS numbers: 42.65.–k, 32.80.–t, 52.35.Mw

Field ionization is the primary process in matter exposed to high-intensity ($I > 10^{14}$ W/cm^2) radiation triggering all strong-field processes. Whereas ionization plays a central role in strong-field atomic processes, such as high harmonic generation (HHG) [1], it did not appear to be important for high-field plasma physics so far. This is because ionization saturates at the leading edge of the previously used subpicosecond or longer pulses, so that the dominant part of the pulse interacts with a fully ionized plasma. The availability of ultrashort, few-cycle laser pulses has changed the situation, opening a novel parameter regime of plasma physics. Saturation of ionization is shifted to considerably higher intensities and ionization continues playing a role until instants close to the pulse peak. In spite of the central role of ionization in strong-field physics, there exists no simple and complete model for the evolution of laser pulses in field-ionizing media.

An exact description of strong-field atomic and plasma phenomena requires solution of Maxwells equations in combination with an analysis of the microscopic processes in the medium. This formidable task often exceeds current computer capacities and makes the use of simplified models indispensable. In this Letter we derive a novel propagation equation for the strong-field regime that simplifies and improves existing models in the following two respects.

(i) Our analysis extends concepts used in the perturbative regime of nonlinear optics into the strong-field regime. Drawing on the slowly evolving wave (SEW) approximation [2], a first-order propagation (FOP) equation for strong-field atomic and plasma physics is derived that is valid for underdense media and for arbitrarily short pulse durations. Numerical results obtained from a solution of the FOP equation and of the scalar wave equation reveal excellent agreement. *The FOP equation can be solved by more than 2 orders of magnitude faster than the scalar wave equation,* greatly facilitating the analysis of a wide range of strong-field atomic and plasma phenomena, such as high harmonic generation [3], x-ray lasing [4], and processes in relativistic nonlinear optics [5–7].

(ii) The polarization response of ionizing media [8–10] used in previous macroscopic propagation studies is incomplete [11] and is not capable of describing fundamental processes, such as the absorption loss introduced by field ionization, correctly. Our semiclassical analysis reveals an additional loss term. A comparison to absorption measurements in helium yields excellent agreement and corroborates the validity and importance of the additional term that increases the absorption loss by a factor of 3 in our experiments.

The propagation of high-intensity pulses in matter is modeled by Maxwells equations in combination with a model for the polarization response of the medium P. As long as $P(\mathbf{r}, t)$ in the coordinates ($r_\perp = x, y$) transversal to the propagation direction (z) exhibits little variation over a distance comparable to the center wavelength λ_0, Maxwells equations may be substituted by the scalar wave equation. In Gaussian units and for a linearly polarized electric field the scalar wave equation is given by

$$\partial_z^2 E + \nabla_\perp^2 E - \frac{1}{c^2}\,\partial_t^2 E = \frac{4\pi}{c^2}\,\partial_t^2 P\,, \tag{1}$$

where c is the vacuum velocity of light, $\partial_{z,t}$ refers to the partial derivatives with respect to z and t, and $\nabla_\perp^2$ is the transversal Laplace operator. The derivation of the FOP equation is performed in the spectral domain by inserting the ansatz $E(\mathbf{r}, \omega) = U(\mathbf{r}, \omega)\exp(i\omega z/c)$ into the Fourier transformed Eq. (1), where $E(\mathbf{r}, \omega) = \int_{-\infty}^{\infty} dt\, E(\mathbf{r}, t)\exp(-i\omega t)$. This yields

$$\left(\frac{2i\omega}{c}\,\partial_z + \nabla_\perp^2\right)U(\omega) = -\frac{4\pi\omega^2}{c^2}\exp\!\left(-\frac{i\omega z}{c}\right)\hat{F}[P], \tag{2}$$

where the operator $\hat{F}$ denotes the Fourier transform and the term $\partial_z^2 U$ was neglected. Neglect of the second space derivative has been termed SEW approximation [2] and is equivalent to elimination of backward propagating wave solutions. The SEW approximation is applicable as long as changes to the electric field induced by the polarization of the medium over a distance comparable to λ_0 are small, i.e., $|\partial_z E| \ll (\omega/c)E$. Substituting U by E followed by a

VOLUME 83, NUMBER 15 PHYSICAL REVIEW LETTERS 11 OCTOBER 1999

transformation of Eq. (2) back into the time domain yields

$$\partial_\xi E(r_\perp, \xi, \tau) - \frac{c}{2}\nabla_\perp^2 \int_{-\infty}^{\tau} d\tau' E(r_\perp, \xi, \tau') = -\frac{2\pi}{c}\partial_\tau P[E(r_\perp, \xi, \tau)]. \quad (3)$$

Here, we have introduced the moving coordinate frame $\xi = z$ and $\tau = t - z/c$ and have utilized the fact that the Fourier transform of $U(r_\perp, z, \omega)$ is identical to the electric field in the retarded coordinate system, $E(r_\perp, \xi, \tau)$. In contrast with the perturbative regime of nonlinear optics, where the FOP equation is derived for a complex envelope, the strong-field FOP equation (3) governs evolution of the electric field. In the strong-field regime of nonlinear optics, nonlinear effects dominate linear dispersion and a decomposition of the electric field into carrier wave and complex envelope is no longer beneficial. As our derivation contains no assumptions with respect to the pulse width, Eq. (3) is valid for arbitrarily short pulse durations. The FOP equation (3) is the first major result of this paper presenting a powerful tool for the analysis of complex propagation phenomena in strong-field atomic and plasma physics.

In the remainder of this paper the FOP equation is applied to the analysis of laser pulse evolution in the presence of field ionization. Ionization is of particular importance constituting the primary event for all strong-field phenomena. In order to model evolution of a laser pulse in a field-ionizing medium, the polarization response in (3) has to be determined. Our derivation of P is restricted to moderate gas densities and moderate laser intensities, at which collisions, collective plasma processes, and relativistic effects may be neglected. For inclusion of these phenomena, see, e.g., Refs. [5,11]. Under the above assumptions the medium polarization can be written as $P = n_0 e\langle x\rangle$, where n_0 is the atomic particle density, $e\langle x\rangle = e\langle\Psi|x|\Psi\rangle$ is the single-atom dipole moment, e is the electron charge, and x represents the microscopic coordinate of the electron along the polarization of the linearly polarized laser electric field. The electron wave function Ψ is obtained by solving the time-dependent Schrödinger equation in the frame of the dipole approximation and of the single-active-electron approximation.

Although solution of the FOP equation (3) in combination with the Schrödinger equation is possible with current computer capacities, it is still a demanding problem calling for a further simplification. To this end, we derive an approximate closed-form expression for $P(\mathbf{r}, t)$ in terms of $E(\mathbf{r}, t)$ and $n(\mathbf{r}, t)$, the density of electrons set free by field ionization. The derivation draws on a simple quasistatic model [8,12], which is valid as long as the Keldysh parameter $\gamma = \sqrt{I_p/(2U_p)} < 1$ [13,14]. Here, I_p is the ionization potential of the atom, $U_p = (eE)^2/4m\omega_0^2$ is the ponderomotive potential, m is the electron mass, and ω_0 is the center frequency of the laser pulse. The quasistatic model is depicted schematically in Fig. 1. In the quasistatic limit the electron tunnels through the suppressed Coulomb barrier and the tunneling rate adiabatically follows variations of the laser field. Coupling to other bound states is negligibly small [12,14]. In this picture, the electron is "born" on the outer side of the Coulomb barrier at a distance of $x_0(t) \approx I_p/eE(t)$ from the nucleus [15] and with zero initial velocity, $v_0 \approx 0$ [12].

The classical trajectory $x(t)$ of the freed electron in the nonrelativistic regime is obtained by solving Newton's equation of motion, $m\ddot{x}(t) = eE(t)$. By using the classical trajectory the current is determined by $J = \dot{P} = e\dot{n}x + en\dot{x}$. Here, the dot/double dot refers to the first/second partial time derivative. As $\dot{n} \neq 0$ only when an electron is born, $x(t)$ is substituted by $x_0(t)$ in the first term of the current, yielding $J = e\dot{n}x_0 + en\dot{x}$. The second time derivative of the polarization is obtained from the current by $\ddot{P} = \dot{J} = e\partial_t(\dot{n}x_0) + en\ddot{x} + e\dot{n}\dot{x}$. As the free electron density is changed only when an electron is born due to ionization and as the birth velocity $v_0 = 0$, the last term in $\ddot{P}$ vanishes. After expressing x_0 and $\ddot{x}$ in terms of E, and integrating $\ddot{P}$ with respect to t, we obtain the current

$$\partial_t P(\mathbf{r}, t) = I_p\left(\frac{\partial_t n(\mathbf{r}, t)}{E(\mathbf{r}, t)}\right) + \frac{e^2}{m}\int_{-\infty}^{t} dt'\, n(\mathbf{r}, t')E(\mathbf{r}, t'), \quad (4)$$

where

$$n(\mathbf{r}, t) = n_0\left(1 - \exp\left[-\int_{-\infty}^{t} dt'\, w\{E(\mathbf{r}, t')\}\right]\right), \quad (5)$$

and $w(E)$ is the quasistatic ionization rate. Equations (4) and (5) in combination with the closed-form expressions for $w(E)$ [16] given by the ADK (Ammosov-Delone-Krainov) theory or the tabulated values of $w(E)$ [14] provide an explicit constitutive law for field-ionizing media and represent the second major result of this paper.

In comparison to previous theoretical work [8–10], our derivation reveals an additional term proportional

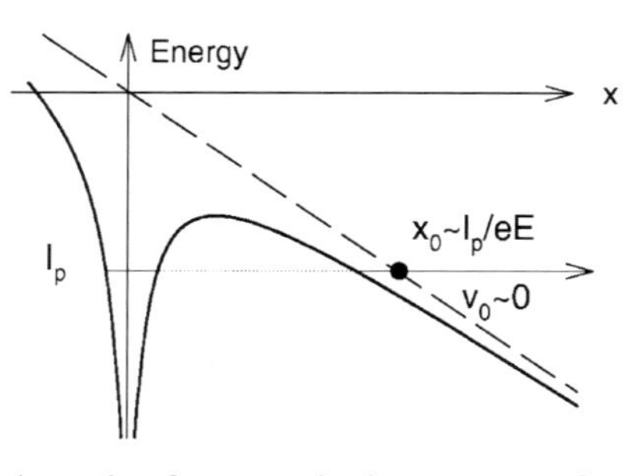

FIG. 1. Schematic of an atom in the presence of a laser pulse. The dashed and the full lines denote the electric field and the combined potential of the nucleus and of the electric field, respectively. The dot refers to the position of birth x_0.

to the ionization potential. Whereas the second term in (4) introduces predominantly phase effects, such as blueshifting and broadening of the spectrum, the first term is a pure loss term. To identify the physical effects underlying the two terms, the loss of pulse energy per volume in the presence of ionization $\Delta\rho_E = \int_{-\infty}^{\infty} dt\, EJ$ is determined. Inserting J from Eq. (4) in this expression yields

$$\Delta\rho_E = -\frac{e^2}{2mc^2}\int_{-\infty}^{\infty} A^2(t)\,[\partial_t n(t)]\,dt - n(t \to \infty)I_p\,. \tag{6}$$

Here, the first term on the right hand side has been obtained by performing partial integration twice, and the vector potential is defined by the relation $E = -(1/c)\dot{A}$ subject to the condition $A(t \to \pm\infty) = 0$. As the total energy of electrons and laser field has to be conserved, the energy gain of the electrons during ionization must be compensated by an energy loss of the laser field. During the interaction, the ionized electrons gain the potential energy I_p and a kinetic energy resulting from a drift velocity, the free electrons are left behind with after the laser field vanished. The drift velocity is given by $v_d = \lim_{t\to\infty}\dot{x}(t) = -\lim_{t\to\infty} e/(mc)\,[A(t) - A(t_0)] = e/(mc)A(t_0)$ [8], where t_0 is the instant of birth of the free electron and the velocity $\dot{x}$ is obtained from time integration of the Newton equation subject to the condition $\dot{x}(t_0) = v_0 = 0$. Indeed, we find that the energy gain of the ionized electrons corresponds to the energy loss of the laser pulse given by Eq. (6). The first term accounts for the energy loss due to the drift velocity and the second term on the right hand side introduces an energy loss I_p per ionized electron. The two terms are referred to as the drift loss and as the ionization loss from hereon. From Eq. (6) we find that the ratio of the ionization loss term and of the drift loss contribution scales with the square of the Keldysh parameter $\gamma^2 = I_p/2U_p$. Therefore, the influence of the ionization loss is strongest for short wavelength laser pulses and decreases with increasing field strength.

Substituting the constitutive law as given by (4) in (3) yields a novel propagation equation for the evolution of laser pulses in field-ionizing media,

$$\partial_\xi E(r_\perp,\xi,\tau) - \frac{c}{2}\nabla_\perp^2\int_{-\infty}^{\tau} d\tau'\,E(r_\perp,\xi,\tau') = -\frac{1}{2c}\int_{-\infty}^{\tau} d\tau'\,\omega_p^2(r_\perp,\xi,\tau')E(r_\perp,\xi,\tau') - \frac{2\pi I_p}{c}\,\frac{\partial_\tau n(r_\perp,\xi,\tau)}{E(r_\perp,\xi,\tau)}\,, \tag{7}$$

where $\omega_p^2(r_\perp,\xi,\tau) = 4\pi n(r_\perp,\xi,\tau)e^2/m$ is the plasma frequency squared [17]. Using the validity condition derived above we find that in the presence of ionization the FOP equation is valid for underdense plasmas, i.e., as long as $\omega_p \ll \omega_0$.

To corroborate the validity of this equation we have performed several tests. Figure 3 shows a numerical solution of the FOP equation (3), propagated by a fourth-order Runge-Kutta method, and of the scalar wave equation (1), solved by the commonly used leap-frog method, in one space dimension. The two solutions are nearly identical proving the applicability of the SEW approximation for underdense plasmas and the validity of the FOP equation. Small deviations come from a reflection at the vacuum-gas surface that cannot be accounted by the FOP equation. In the scalar wave equation, vacuum propagation of the laser pulse limits the maximum spatial step size to <0.005 μm. The major advantage of the FOP equation is that this limitation of the step size is eliminated by use of the moving coordinate frame. Consequently, for the integration of Eq. (3) a considerably larger step size (≈ 1 μm) can be used *speeding up solution of the FOP equation by more than 2 orders of magnitude.*

To demonstrate the validity of the ionization loss term revealed by our analysis we compared the experimentally measured energy loss in helium (500 Torr) with the loss obtained from a solution of the FOP equation (7) in two space dimensions assuming radial symmetry. The pulse parameters were $\lambda_0 = 800$ nm, $I_0 = 2 \times 10^{15}$ W/cm^2, and a FWHM pulse duration $\tau_p = 5$ fs; for a more detailed description of the laser system, see, e.g., Ref. [1].

The experimentally (full circles) and theoretically (full line) obtained energy loss is plotted versus gas interaction length in Fig. 2. The dashed line represents the numerical solution without the ionization loss term, which is the second term on the right hand side of (7). In order to eliminate possible errors originating from the limited validity of the ADK model we used the exact static

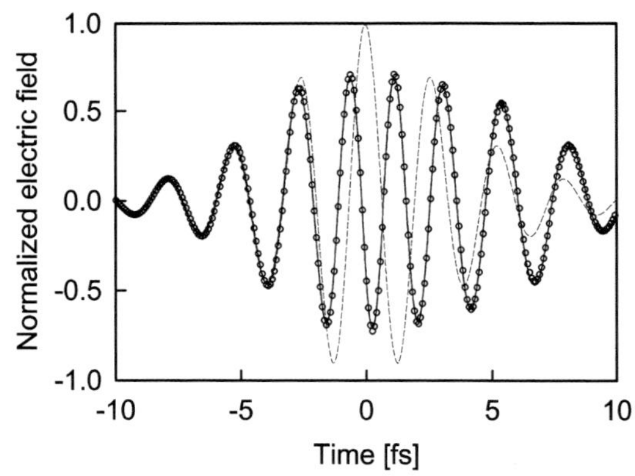

FIG. 2. Thin-dashed line: Normalized input electric field of a Ti:sapphire laser pulse. Empty circles and full line: Solution of the FOP equation (7) and of the scalar wave equation (1) with the polarization (4) in one space dimension, respectively. The pulse parameters are $I_0 = 2 \times 10^{15}$ W/cm^2, $\lambda_0 = 0.8$ μm, and full width at half maximum (FWHM) pulse duration $\tau_p = 5$ fs; the remaining parameters are propagation through a 1-mm-thick He gas target at a pressure of 500 Torr corresponding to $n_0 = 1.75 \times 10^{19}$ cm^{-3}. At full ionization the electron density corresponds to a ratio $\omega_p/\omega_0 = 0.1$. The excellent agreement between the two solutions proves the validity of the FOP equation (7) for underdense plasmas.

VOLUME 83, NUMBER 15 PHYSICAL REVIEW LETTERS 11 OCTOBER 1999

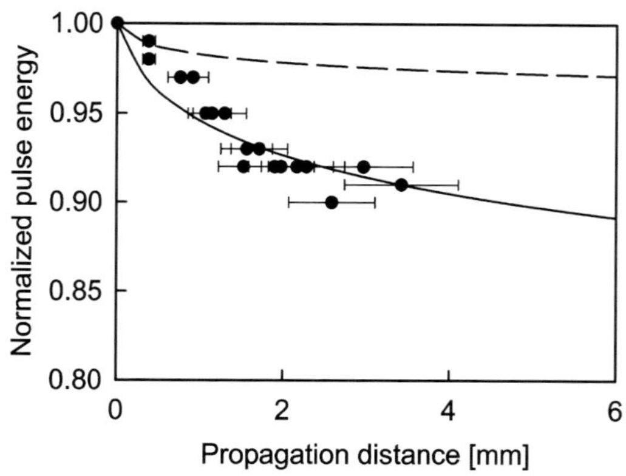

FIG. 3. Output pulse energy normalized to input pulse energy versus propagation distance for a 5 fs Ti:S laser pulse propagating in helium under conditions described in the caption of Fig. 3. The full circles denote the experimental values and the error bars refer to an estimated ±20% uncertainty in the measurement of the gas interaction length. The full/dashed lines denote the theoretical values calculated from a solution of the 2D wave equation (7) using cylindrical symmetry with/without the ionization loss term. Our analysis clearly reveals the necessity of the ionization loss term to account for the experimental data.

ionization rates w of helium in Eq. (5) that were derived by a full solution of the two-electron Schrödinger equation [14]. The agreement between our model's prediction and experimental data is excellent and clearly reveals the essential role of the ionization loss term in the constitutive law (4) for a proper description of the evolution of a laser pulse in a field-ionizing medium. *The ionization loss term increases the energy loss by a factor of 3 under our experimental conditions.* After 4–6 mm interaction length the pulse peak intensity has been reduced so much due to dispersion-induced pulse broadening that ionization stops, limiting the maximum loss to around 10%. We recall that the ratio of ionization loss to drift loss scales with $\gamma^2 \propto 1/\lambda_0^2$. Furthermore, the free electron induced group velocity dispersion decreases with λ_0^3, leading to longer interaction distances over which the pulse experiences loss. As a result, the ionization loss increases for shorter wavelengths. Our calculations show that a frequency doubled Ti:sapphire pulse ($\lambda_0 = 400$ nm, $\tau_p = 10$ fs, $I_0 = 3 \times 10^{15}$ W/cm^2) loses ≈45% of its pulse energy over a propagation distance of 12 mm mainly due to the first term in (4) missing in previous theoretical models.

As more electrons are ionized at the peak of the pulse than in the pulse wings, the ionization loss leads to a stronger reduction of the pulse peak and therewith, to a lengthening of the pulse. As a result, the nonlinear interaction saturates faster and the blueshifting and spectral broadening of the pulse is reduced. The observed behavior has important implications on strong-field phenomena, such as a modification of the maximum interaction length for x-ray lasing and high harmonic generation, which will have to be analyzed in future investigations.

In conclusion, we have derived a first-order propagation equation for strong-field atomic and plasma physics which is valid for ionizing media below the critical density and for arbitrarily short pulse durations. The simplified wave equation is accurate and expedites computing by more than 2 orders of magnitude. In combination with a closed-form constitutive law for optical field ionization this first-order wave equation constitutes a powerful tool for predicting the *macroscopic* response of atomic media driven by strong laser fields.

The authors gratefully acknowledge the support of Professor A. J. Schmidt. This work was supported by the Austrian Science Fund, Grant No. P12631-PHY, and by the Austrian Nationalbank, Jubiläumsfondsprojekt 7178.

*Email address: brabec@rs6.iaee.tuwien.ac.at

[1] C. Spielmann *et al.,* Science **278**, 661 (1997); M. Schnürer *et al.,* Phys. Rev. Lett. **80**, 3236 (1998).
[2] T. Brabec and F. Krausz, Phys. Rev. Lett. **78**, 3282 (1997).
[3] A. L'Huillier and P. Balcou, Phys. Rev. Lett. **70**, 774 (1993); J. J. Macklin *et al.,* Phys. Rev. Lett **70**, 766 (1993).
[4] B. E. Lemoff, C. P. J. Barty, and S. E. Harris, Opt. Lett. **19**, 569 (1994).
[5] A. Pukhov and J. Meyer-ter-Vehn, Phys. Rev. Lett. **76**, 3975 (1996).
[6] D. Umstadter, S.-Y. Chen, A. Maksimchuk, G. Mourou, and R. Wagner, Science **273**, 472 (1996).
[7] G. Malka, E. Lefebvre, and J. L. Miquel, Phys. Rev. Lett. **78**, 3314 (1997).
[8] F. Brunel, J. Opt. Soc. Am. B **7**, 521 (1990).
[9] Wm. M. Wood *et al.,* Phys. Rev. Lett. **67**, 3523 (1991).
[10] C. Kan *et al.,* Phys. Rev. Lett. **79**, 2971 (1997); V. B. Gil'denburg *et al.,* Phys. Lett. A **203**, 214 (1995); V. B. Gil'denburg *et al.,* JETP Lett. **51**, 105 (1990).
[11] S. C. Rae and K. Burnett, Phys. Rev. A **46**, 1084 (1992); **46**, 2077 (1992).
[12] P. B. Corkum *et al.,* Phys. Rev. Lett. **62**, 1259 (1989); P. B. Corkum, Phys. Rev. Lett. **71**, 1994 (1993); M. Lewenstein *et al.,* Phys. Rev. A **49**, 2117 (1994).
[13] L. V. Keldysh, Sov. Phys. JETP **20**, 1307 (1965).
[14] A. Scrinzi, M. Geissler, and T. Brabec, Phys. Rev. Lett. **83**, 706 (1999).
[15] B. Walker *et al.,* Phys. Rev. Lett. **73**, 1227 (1994); N. B. Delone and V. P. Krainov, J. Opt. Soc. Am. B **8**, 1207 (1991).
[16] M. V. Ammosov *et al.,* Sov. Phys. JETP **64**, 1191 (1986).

[17] Equation (7) can be converted to SI units by multiplying the right hand side with $1/4\pi\epsilon_0$.

REPORTS

Generation of Coherent X-rays in the Water Window Using 5-Femtosecond Laser Pulses

Ch. Spielmann, N. H. Burnett, S. Sartania, R. Koppitsch, M. Schnürer, C. Kan, M. Lenzner, P. Wobrauschek, F. Krausz*

Coherent extreme-ultraviolet radiation extending to wavelengths below the carbon K edge at 4.37 nanometers (nm) has been generated at a repetition rate of 1 kilohertz by focusing 5-femtosecond near-infrared (780 nm) laser pulses into a helium gas jet. The incident light field performs just a few oscillations, which results in the emission of an x-ray supercontinuum rather than discrete harmonics. Owing to the extremely short rise time of the driving pulses, neutral atoms can be exposed to high fields before they are depleted by ionization. As a result, the observed x-ray radiation extends well into the water window and is delivered in a well-collimated beam (divergence less than 1 milliradian). The high repetition rate and spatial coherence result in a brightness of about 5×10^8 photons per square millimeter per square milliradian per second in a 1-percent bandwidth at 4.37 nm, the carbon edge of the water window. The compact laboratory system holds promise as a source for biological holography and nonlinear optics in the x-ray regime.

Motivation for the development of coherent x-ray sources comes from many areas of science and technology. One of the greatest challenges is the three-dimensional observation of microscopic biological structures (for example, cells and macromolecules) in the living state by means of x-ray holography (*1*). Sources intended for this application must exhibit a high degree of spatial coherence and a reasonable temporal coherence and must provide a good contrast of the substance to be imaged with respect to its surrounding. The latter requirement can be met in the wavelength range between the K absorption edges of carbon (4.37 nm) and oxygen (2.33 nm), which is referred to as the "water window," where carbon-containing biological objects absorb radiation efficiently but water is comparatively transparent. This spectral window for high-contrast biological spectroscopy has set, in terms of wavelength, one of the most important objectives for the research and development of sources of coherent soft x-rays.

So far, coherent radiation in the water window could be demonstrated only in large-scale facilities, through the use of either magnetic undulators in electron storage rings (*2*) or a multikilojoule laser (*3*) to produce a plasma-column x-ray amplifier (*4*). Here we describe a high–repetition-rate laboratory source of coherent water-window radiation. The x-rays emerge from an ensemble of helium atoms coherently driven by quasi–single-cycle light pulses produced by a table-top laser. The underlying physical mechanism is closely related to that responsible for the emergence of high-order harmonics from a gas jet exposed to intense laser radiation. It is hoped that this compact source will make the power of water-window radiation available to many more biochemists and biophysicists than was previously possible. In addition, it offers the possibility of extending nonlinear optics into the x-ray regime and studying atomic processes with attosecond time resolution.

Atoms ionized by strong laser fields can emit coherent harmonic radiation at photon energies much higher than the atomic binding energy (*5*). For femtosecond pulses (duration $\tau_p \approx 100$ fs), harmonic orders exceeding 100 with photon energies up to 150 eV have been observed (*6*). This phenomenon has subsequently been extensively studied both experimentally and theoretically (*7–9*). The semiclassical model of a single active electron put forward independently by Corkum (*8*) and Kulander *et al.* (*10*) explains the emission of high-harmonic radiation in terms of three discrete steps. First, the electron tunnels through the barrier formed by the atomic Coulomb potential and the laser field. The quasi-free electron subsequently acquires kinetic energy from the laser field, and some half of an optical cycle later, it is returned (by the linearly polarized field) to its parent ion. With some (small) probability, the electron recombines to the ground state, emitting a photon with the energy of the ionization potential of the atom plus the kinetic energy gained in the laser field. The process is repeated periodically in the field of a driving pulse consisting of many cycles and gives rise to the emission of harmonics of the incident radiation. The photon energy of the highest harmonic has been predicted as (*7, 8*)

$$(\hbar\omega)_{max} \approx I_p + 3.17U_p \qquad (2)$$

here, I_p is the atomic ionization potential and

$$U_p = e^2E^2/4m\omega_0^2 \qquad (3)$$

is the mean kinetic (or ponderomotive) energy of the electron quivering in the laser field (E is the amplitude of the electric field, ω_0 is the laser frequency, and e and m are the charge and rest mass of the electron, respectively). The magnitude of the second term in Eq. 1 is roughly dictated by the maximum laser intensity that neutral atoms can be exposed to before they are depleted by ionization (*11*). This critical intensity is higher for driver pulses of shorter duration (*12*); hence, they hold the promise of producing harmonics at higher photon energies. In fact, the shortest-wavelength harmonics have been previously generated with the shortest driver pulses available (τ_p = 25 fs) (*13*), in helium down to wavelengths as short as 5.2 nm at a 10-Hz repetition rate (*14*), and longer-wavelength harmonics were also demonstrated at repetition rates of a few hundred hertz (*15*).

Dramatic advances in ultrashort-pulse laser technology have recently led to the generation of laser pulses with a duration as short as 5 fs and a peak power of 0.1 TW in a diffraction-limited beam at a repetition rate of 1 kHz (*16*). This performance has been achieved with a compact, table-top, solid-state laser system using a Ti: sapphire oscillator, an amplifier, and a hollow-fiber–chirped-mirror compressor (*16, 17*). The electric field performs just a few oscillations in these pulses (Fig. 1), which are carried at wavelength $\lambda \approx 780$ nm and have an oscillation period of 2.6 fs. Here we show experimentally that being focused in a helium gas jet, these "light transients" are capable of generating spatially coherent extreme ultraviolet (xuv) supercontinuum extending below the carbon K edge.

The helium gas target is formed by a thin nickel tube having a wall thickness of 0.05 mm and an inner diameter of 0.6 mm, which is squeezed to about 30 μm (Fig. 2) (*18*). The laser beam is focused with a silver-coated spherical mirror (focal length, 20 cm) through an 0.3-mm-thick fused silica window into a vacuum chamber containing the target at focus. The

Ch. Spielmann, S. Sartania, R. Koppitsch, M. Schnürer, M. Lenzner, F. Krausz, Abteilung Quantenelektronik und Lasertechnik, Technische Universität Wien, Gusshausstrasse 27, A-1040 Wien, Austria.
N. H. Burnett and C. Kan, Department of Electrical Engineering, University of Alberta, Edmonton T6G 2G7, Canada.
P. Wobrauschek, Atominstitut der Österreichischen Universitäten, Schüttelstrasse 115, A-1020 Wien, Austria.

*To whom correspondence should be addressed. E-mail: krausz@ps1.iaee.tuwien.ac.at

holes in the tube wall were bored by the laser beam itself, minimizing the gas load to the surrounding vacuum. Relay imaging revealed a somewhat elliptic, nearly Gaussian intensity profile on target with $1/e^2$ diameters of 50 and 70 μm. The ellipticity is caused by the off-axis spherical focusing mirror. In the experiments described below, the pulse energy was ~0.3 mJ on target, implying a peak intensity of $\approx 4 \times 10^{15}$ W cm^{-2} on the beam axis (*19*). The tube was continuously backed with pressures of 3 to 5 bar, which resulted in a pressure of ≈0.5 bar in the interaction region (*20*). The target chamber was evacuated to ≈1 mbar. The x-ray radiation produced collinearly with the laser beam was passed through a 1-mm aperture (placed ≈35 cm behind the target) into a 1-m grazing-incidence monochromator (248/310G, McPherson, Acton, MA) evacuated to $<10^{-4}$ mbar. The monochromator is equipped with a platinum-coated grating with 300 lines per millimeter and an uncoated channeltron (4715G, Galileo, Sturbridge, MA). The detector signal was collected by a high–dynamic-range lock-in amplifier triggered by the laser pulse train.

Figure 3 depicts the xuv spectrum emitted in the wavelength range from 110 to 10 nm. The monochromator was scanned with the entrance and exit slit widths kept constant, implying a wavelength-dependent bandwidth. The entrance slit located ≈50 cm downstream of the target blocks an increasing fraction of the incident xuv beam for increasing wavelengths (because of wavelength-dependent beam size). Correction for these effects is expected to increase the spectral intensity at long wavelengths by more than an order of magnitude relative to the short-wavelength components. At photon energies below ≈25 eV (corresponding to a wavelength of ≈50 nm), well-resolved odd harmonics are observed. Above 25 eV, the harmonics merge, forming a supercontinuum, as predicted by recent theoretical work (*21*, *22*). Near the transition from harmonics to continuum, what appears to be even harmonics are also present (the most prominent peaking at a wavelength corresponding to the 12th harmonic of the driving radiation), the origin of which needs further investigation. The high-energy cutoff of the continuum cannot be assessed from this measurement, in which the full broadband xuv output from the target is directed into the monochromator, because of the rapidly increasing amount of stray xuv rays leaking through as the zeroth order is approached with the leakage becoming dominant at $\lambda < 8$ nm in the spectrum shown in Fig. 3.

To suppress the zeroth-order throughput, we placed a 100-nm silver foil (fixed on a nickel mesh) in front of the entrance slit. This filter has a nearly constant transmittance (≈50%) between 3 to 10 nm and blocks radiation with wavelengths greater than 10 nm. The short-wavelength tail of the xuv continuum transmitted through the silver filter, obtained by averaging three scans, is shown along with the standard deviation at discrete wavelengths in Fig. 4. Between wavelengths of 4.4 and 4.2 nm, the spectral intensity in a 0.05-nm bandwidth yields a signal that exceeds the noise floor (signal observed with either the laser or the gas flow blocked) of our detection system by a factor of ≈500. In order to check the calibration of our monochromator and provide additional evidence for the emergence of water-window radiation from the He target, the silver filter was replaced by a pair of 900-nm-thick Mylar foils ($C_{10}H_8O_4$). This filter provides an absorption contrast of $>10^3$ at 4.37 nm, the K edge of carbon. The measured soft x-ray spectrum (Fig. 5) exhibits a sharp edge at 4.37 nm, providing a conclusive evidence for water-window radiation. The signal contrast at the carbon K edge was >300, limited by the noise floor of the detection system. From these measurements, the number of photons per pulse emitted within a 1% bandwidth at the carbon K edge is estimated as 100 (*23*).

In the 3- to 4-nm range, we also observe a signal that exceeds the detection limit by more than an order of magnitude (Fig. 4). However, at wavelengths of around 3 nm or shorter, the exit slit is close to the central image (zeroth order) of the monochromator, and a reliable assessment of some possible contribution of zeroth-order stray light to the observed sig-

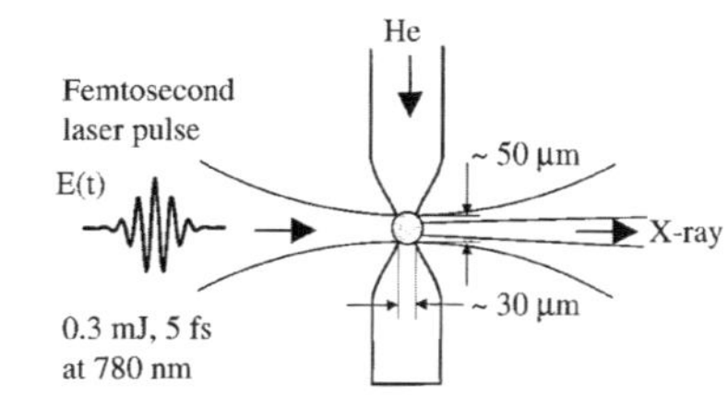

Fig. 2. Schematic of the target geometry for xuv continuum generation.

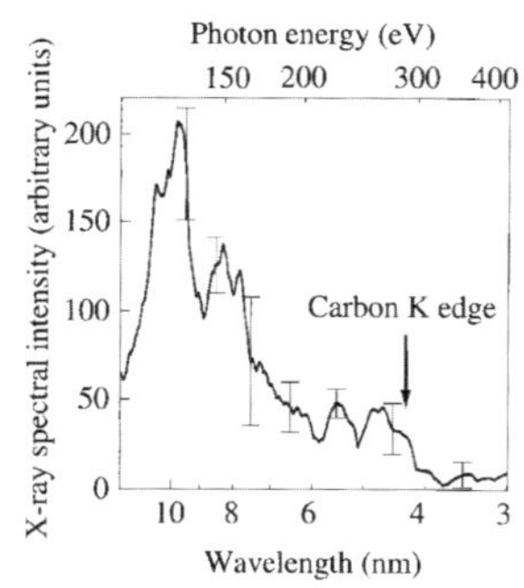

Fig. 4. Short-wavelength end of the xuv continuum as transmitted by a 100-nm-thick silver foil. The depicted spectrum is the average of three scans, and the bars give the respective standard deviation. The spectral components at $\lambda > 10$ nm are damped by the silver foil.

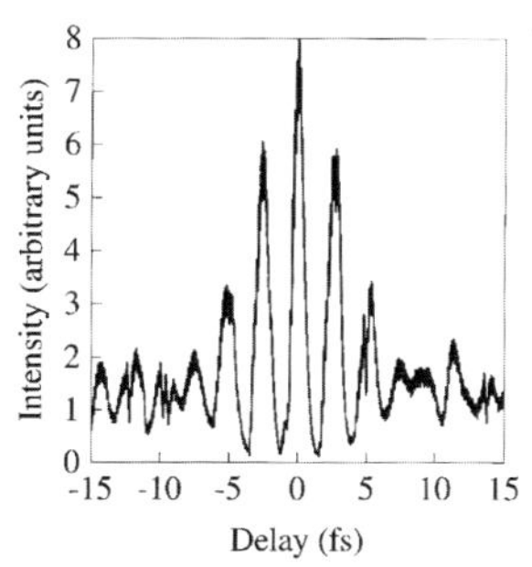

Fig. 1. Second-order interferometric autocorrelation trace of the 5-fs laser pulses used in the experiment. The fringe spacing is ≈2.6 fs, corresponding to the light oscillation cycle at wavelength $\lambda = 780$ nm.

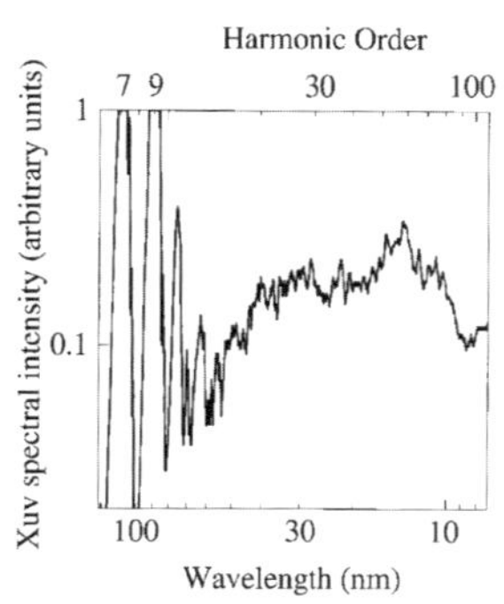

Fig. 3. Broadband (spatially) coherent xuv output from He on a logarithmic scale. The spectrum was recorded with an entrance and exit slit width of 100 μm, corresponding to a bandwidth of ~0.02 nm at $\lambda = 10$ nm and ~0.06 nm at $\lambda = 100$ nm.

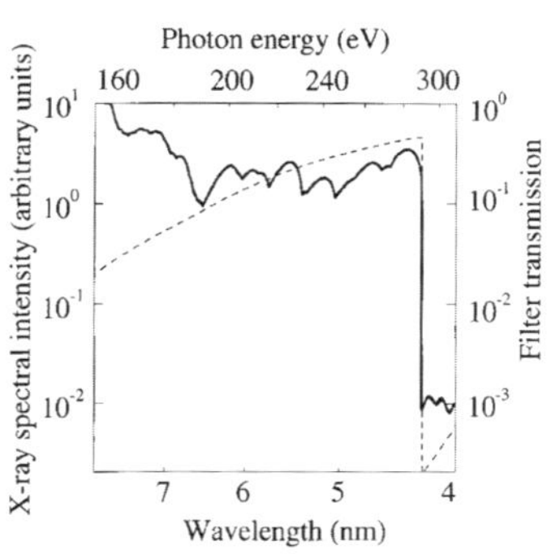

Fig. 5. Short-wavelength end of the xuv spectrum as transmitted by a 1.8-μm-thick Mylar foil (solid line) and the transmittance of the foil (dashed line).

nal is difficult. This short-wavelength edge of the spectrum has therefore also been studied with an energy-dispersive detection system (Z-MAX, NORAN Instruments, Middleton, WI) consisting of a cooled lithium-drifted silicon-crystal detector and a multichannel pulse height analyzer. After the number of counts per laser shot had been reduced to far below unity ($\approx$0.01) and the low-energy photons (E_{ph} < 100 eV) had been suppressed by suitable filtering, this system yielded reliable information on the photon energy distribution of the emitted x-ray radiation for photon energies higher than 0.2 keV (λ < 6 nm) (*24*). Under the experimental conditions described above, the coherent x-ray spectrum extended to photon energies exceeding 0.5 keV, which correspond to wavelengths shorter than 2.5 nm. Correction for filter transmittances resulted in x-ray intensities at around 2.5 nm, some three orders of magnitude lower than observed at 4.4 nm, which suggests that the spectrum shown in Fig. 4 is likely to contain a considerable stray-light component at around 3 nm.

The observed shape of the high-energy end of the emitted x-ray spectrum strongly deviates from that of the single-atom response as predicted by recent modeling (*22*). In fact, the radiation spectrum of a He atom driven by a 5-fs pulse exhibits a broad plateau of nearly constant intensity. The plateau extends to the photon energies given approximately by Eqs. 1 and 2, with E being the peak electric field of the laser pulse unless neutral atoms are substantially depleted by ionization before exposure to the pulse peak. Assuming the ADK model for tunnel ionization to be valid (*25*), the depletion is completed at a peak intensity of $\approx 4 \times 10^{15}$ W cm^{-2} with a 5-fs pulse. Drawing on the ADK model, our simulations yielded, in reasonable accordance with Eq. 1, a cutoff in the single-atom radiation spectrum above 0.5 keV (below 2.5 nm) for a peak intensity of 3×10^{15} W cm^{-2}. The striking discrepancy between the predicted plateau up to these photon energies and the experimentally observed rapid intensity roll-off for increasing photon energies points to the dramatically increasing role of dephasing between the fundamental- and high-frequency waves during propagation across the target. This dephasing is caused by an accumulating phase error due to plasma dispersion and wave front curvature, which rapidly increases for decreasing x-ray wavelengths.

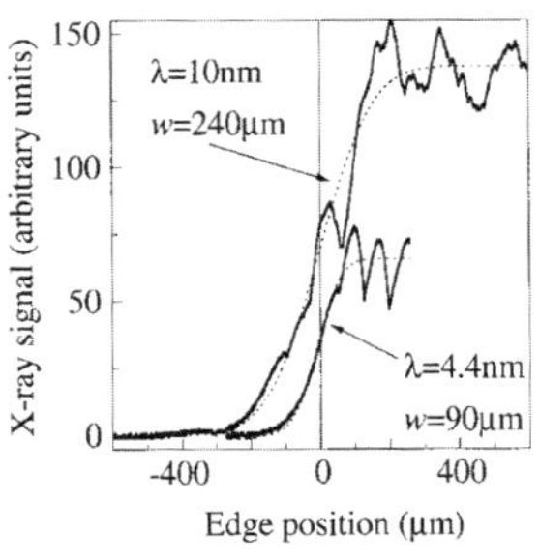

Fig. 6. Transmitted x-ray power as a function of the position of an opaque knife edge scanning transversely (in vertical direction) across the beam profile 29 cm from the He target at two different wavelengths (10 nm and 4.4 nm). The dotted lines depict Gaussian fits to the measured profiles.

Several experiments have been performed to check if incoherent plasma emission from the walls of the target tube or the high-pressure gas target, or both, contribute to the observed signal. First, deliberate misalignment of the laser beam to graze the tube wall in vacuum resulted in no measurable signal. Second, replacing helium with argon while keeping all other experimental parameters unchanged yielded no signal transmitted through the silver foil, although the electron density was significantly higher because of the lower ionization potential. Furthermore, increasing the pulse duration at constant pulse energy caused the short-wavelength edge of the spectrum to retreat to longer wavelengths. These tests, along with the well-collimated nature of the emission, provide conclusive evidence for the coherent harmonic emission process being responsible for the detected radiation.

The usefulness of this source for applications greatly depends on the spatial coherence of the emitted radiation. In order to estimate the degree of spatial coherence and, what is closely related with it, the brightness of the generated short-wavelength radiation, we have investigated the spatial profile of the x-ray radiation by scanning a knife edge transversely across the beam (*26*). The transmitted power measured as a function of the edge position with respect to the propagation axis at a position 29 cm from the He target is shown in Fig. 6 at two different wavelengths (10 and 4.4 nm). If we assume a Gaussian intensity profile $I(r) \propto \exp(-2r^2/w^2)$, the best fits to the measured data yield a $1/e^2$ diameter of $2w$ = 480 and 180 μm at λ = 10 and 4.4 nm, respectively. In the case of perfect spatial coherence, these data would imply a beam diameter of $2w_{0,\text{ideal}} \approx 10$ μm at the beam waist, a value that is merely some factor of 3 smaller than the diameter of the interaction region in which these high photon energies can be produced. The good spatial coherence results in a small divergence, which, together with the high repetition rate, gives rise to a high brightness. If we assume that the x-ray beam diameter is $2w_0 \approx 25$ μm at the beam waist (*27*), we estimate the brightness of the presented source at the carbon K edge within a 1% bandwidth as 5×10^8 photons mm^{-2} mrad^{-2} s^{-1}. Because the coherence length for generating water-window radiation is currently likely to be limited by geometric phase matching (*18*), higher pulse energies allowing more gentle focusing are expected to result in improved efficiency. Upgrade of the laser system to pulse energies of a few millijoules and further optimization of the xuv generation process should boost the brightness by at least 2 orders of magnitude, making this compact laboratory source a powerful tool for high-resolution biological microscopy and holography.

Our simulations including propagation (*28*) have revealed that a key parameter for optimizing the x-ray source for maximum brightness at high photon energies is the relative carrier phase ψ of the 5-fs pulses (*29*). The value of ψ appears to affect the x-ray fluence in the cutoff region by an order of magnitude. Because ψ is not controlled and hence cannot be kept constant in our laser pulses currently, this theoretically predicted phase sensitivity of the x-ray spectrum may explain the large fluctuations observed in the x-ray intensity at wavelengths < 10 nm in spite of the relatively small energy fluctuation of the 5-fs pulses (<5%). As a consequence, the carrier phase will have to be controlled (*30*) in the next generation of coherent soft x-ray sources, driven by sub–10-fs pulses.

Finally, we briefly address the temporal evolution of the radiation emitted at wavelengths less than 10 nm and its implications. Our simulations indicate, in accordance with previous work (*21, 22*), that the short-wavelength end of the continuum is emitted within a single burst lasting less than 1 fs. According to these simulations blocking radiation at wavelengths longer than 10 nm (silver filter) and 5 nm (titanium filter), the He ensemble irradiated by the 5-fs pulse emits a single attosecond x-ray pulse with a duration of <0.3 fs and <0.2 fs, respectively. Our analysis underlines the crucial role of propagation and phase matching in this single-pulse selection. For the wavelength range near the carbon K edge, this temporal evolution implies a peak intensity of $>10^6$ W cm^{-2} on the propagation axis behind the He target. Significantly higher x-ray intensities can be obtained at somewhat longer wavelengths (10 to 20 nm), owing to an increased coherence length related to weaker ionization. The predicted intense x-ray pulses delivered in a nearly diffraction-limited beam along with the sensitive detection techniques available at kilohertz repetition rates offer the potential for attosecond-resolution atomic spectroscopy and nonlinear optics in the x-ray regime.

REFERENCES AND NOTES

1. J. C. Solem and G. C. Baldwin, *Science* **218**, 229 (1982); M. Howells *et al.*, *ibid.* **238**, 514 (1987); J. E. Trebes *et al.*, *ibid.*, p. 517.
2. D. Attwood, K. Halbach, K.-J. Kim, *ibid.* **228**, 1265 (1985).
3. B. J. Macgowan *et al.*, *Phys. Rev. Lett.* **65**, 420 (1990).
4. Flash x-ray sources such as those demonstrated in (*3*) would be ideal for the avoidance of image blurring due to, for example, object motion during exposure [R. A. London *et al.*, *Appl. Opt.* **28**, 3397 (1989)]; however, the spatial coherence of the sources demonstrated in the water window so far are far from being sufficient for single-shot biological holography.
5. A. McPherson *et al.*, *J. Opt. Soc. Am. B* **4**, 595 (1987); X. F. Li *et al.*, *Phys. Rev. A* **39**, 5751 (1989).
6. A. L'Huillier and Ph. Balcou, *Phys. Rev. Lett.* **70**, 774 (1993); J. J. Macklin *et al.*, *ibid.*, p. 766.
7. J. L. Krause *et al.*, *ibid.* **68**, 3535 (1992).
8. P. B. Corkum, *ibid.* **71**, 1995 (1993).
9. M. Lewenstein *et al.*, *Phys. Rev. A* **49**, 2117 (1994).
10. K. C. Kulander *et al.*, in *Proceedings of the Workshop on Super-Intense Laser Atom Physics (SILAP) III*, P. Piraux, Ed. (Plenum, New York, 1993).
11. K. Miyazaki and H. Takada, *Phys. Rev. A* **52**, 3007 (1995).
12. I. P. Christov *et al.*, *Phys. Rev. Lett.* **77**, 1743 (1996); K. J. Schafer and K. C. Kulander, *ibid.* **78**, 638 (1997).
13. J. Zhou *et al.*, *ibid.* **76**, 752 (1996).
14. Z. Chang *et al.*, in *Applications of High Field and Short Wavelength Sources VII* (*OSA Tech. Digest Ser.*, vol. 7, Optical Society of America, Washington, DC, 1997), p. 187.
15. R. Haight and P. F. Seidler, *Appl. Phys. Lett.* **65**, 517 (1994).
16. S. Sartania *et al.*, *Opt. Lett.* **22**, 1562 (1997).
17. M. Nisoli *et al.*, *ibid.*, p. 522.
18. Given the finite tube wall thickness of 0.05 mm, the actual target thickness is estimated as 100 to 200 μm. The coherence length related to the phase error introduced by the tight focusing of the fundamental (*11*) is on the order of 10 μm for wavelengths shorter than 10 nm. The target thickness has been minimized in an attempt to keep the interaction length as close as possible to this "geometric" coherence length. It is this geometric coherence length limitation that dictates the necessity of the high pressures applied.
19. Note that this peak irradiance is reached only on the propagation axis in an infinitesimally small fraction of the cross section of the Gaussian beam. The majority of the helium atoms in the interaction volume are exposed to somewhat lower irradiances.
20. The pressure in the interaction region has been estimated as follows. The gas flow from the target region into the chamber is calculated from the known pumping speed and the measured background pressure in the target chamber. Gas flow and background pressure then determine uniquely the pressure in the target.
21. I. P. Christov *et al.*, *Phys. Rev. Lett.* **78**, 1251 (1997).
22. C. Kan *et al.*, *ibid.* **79**, 2971 (1997).
23. We used a detector quantum efficiency of 2%, an electron multiplication gain of 5×10^6, and a grating diffraction efficiency of 10% (data provided by the manufacturers) for this estimation.
24. M. Schnürer *et al.*, unpublished results.
25. M. V. Ammosov *et al.*, *Zh. Eksp. Teor. Fiz.* **91**, 2008 (1986) [*Sov. Phys. JETP* **64**, 1191 (1986)].
26. Siegman *et al.*, *IEEE J. Quantum Electron.* **27**, 1098 (1991).
27. Measurement of the x-ray beam profile at different positions or direct interferometric measurement of its spatial coherence will obviate the need for this assumption.
28. C. Kan and N. H. Burnett, unpublished results.
29. The electric field of a light pulse can be described in terms of a carrier of frequency ω_0 and a time-varying envelope of amplitude $A(t)$: $E(t) = A(t)\exp[-i(\omega_0 t + \psi)]$. This decomposition can be used to pulse durations down to the carrier oscillation cycle [T. Brabec and F. Krausz, *Phys. Rev. Lett.* **78**, 3282 (1997)]. The parameter ψ is the relative carrier phase. It determines the position of the carrier with respect to the envelope.
30. L. Xu *et al.*, *Opt. Lett.* **21**, 2008 (1996).
31. We are indebted to A. J. Schmidt for his encouragement and T. Brabec for useful discussions. K. Ferencz (Research Institute for Solid State Physics, Budapest, Hungary) is gratefully acknowledged for manufacturing the silver foils. This research was supported by the Austrian Science Foundation under grants P-11109 and Y44-PHY.

21 May 1997; accepted 9 September 1997

Superfluid Droplets on a Solid Surface

D. Ross, J. E. Rutledge, P. Taborek*

Photographs are presented of isolated superfluid helium-4 droplets prepared on a cesium surface, the only material known that is not wetted by superfluid helium. Although thermodynamic measurements show that the cesium surface is highly uniform, the contact angle of the droplets is extremely hysteretic and depends on whether the contact line is advancing or receding. Superfluid helium-4 droplets on an inclined surface do not flow downhill but rather are strongly pinned to the surface.

Superfluid He has unusual thermal and mechanical properties (*1*) and is well known for its ability to spread over surfaces and to flow without dissipation through even microscopic holes. Virtually all of the walls and surfaces used in earlier dissipationless flow experiments were observed to be wetted by superfluid He. This effect means that droplets on these substrates are unstable and immediately spread to form a smooth continuous film over the entire surface so that vapor and substrate are never in contact. Recent work (*2*) has shown that alkali metals are a special class of materials not completely wetted by superfluid He. In particular, Cs substrates can be used to prepare superfluid samples with a distinctly different topology consisting of a droplet with an edge where substrate, superfluid, and vapor meet at a three-phase contact line (*3*). We present here direct observations of isolated droplets of superfluid on a substrate (*4*). Both the static and the dynamic behaviors of the droplets were unusual. We found that the contact angle was an extremely hysteretic function of the volume of the drop. Perhaps most remarkable, superfluid droplets would not move across the surface until considerable force was applied to them. This result is surprising because solid surfaces are well known not to exert transverse forces on bulk superfluid or superfluid films without edges.

Our apparatus consisted of a substrate that can be rotated about a horizontal axis mounted in an optical cryostat with windows that provide an edge-on view of the substrate as well as a view from above at an angle of 60° from the normal. The substrate is a quartz microbalance with gold electrodes similar to those used in our earlier thermodynamic studies (*5*, *6*). Fifty atomic layers of Cs were vapor-deposited onto the quartz and gold surfaces of the microbalance at a rate of 0.01 layer per second. During the evaporation, the temperature of the substrate and the walls of the container were maintained below 6 K to maintain ultrahigh vacuum conditions. We used the microbalance to monitor the deposition and to perform thermodynamic characterizations of the surface; the wetting temperature was measured to be T_w = 2.04 K. A capillary tube (0.04 cm, outside diameter) attached to a source of room-temperature gas through a mass-flow controller provided a means of putting drops of superfluid on the surface. We maintained the system at liquid-vapor coexistence by filling the bottom of the container with bulk liquid ^4He. The drops were observed with a long-focal-distance microscope that provided a magnification of ~×30.

Figures 1 and 2 show a sequence of photos of superfluid drops on a Cs substrate at T = 1.16 K. Pictures taken with the microscope looking down on the substrate at an angle of 30° above the horizontal are shown in Fig. 1. The dark bar at the top of the pictures is the capillary tube, and the lower bar is its shadow. The tube was left in contact with the superfluid drop so that fluid could be added and withdrawn. This geometry is conventional for contact-angle measurements and typically yields the advancing and receding

Department of Physics and Astronomy, University of California, Irvine, CA 92697, USA.

*To whom correspondence should be addressed.

IEEE JOURNAL OF SELECTED TOPICS IN QUANTUM ELECTRONICS, VOL. 4, NO. 2, MARCH/APRIL 1998

Near-keV Coherent X-Ray Generation with Sub-10-fs Lasers

Christian Spielmann, Clarence Kan, Neal H. Burnett, Thomas Brabec, Michael Geissler, Armin Scrinzi, Matthias Schnürer, and Ferenc Krausz, *Member, IEEE*

(Invited Paper)

***Abstract*—Recent advances in solid-state laser technology and ultrafast optics have led to the generation of optical pulses as short as 5 fs with peak powers up to 0.1 TW from a compact kilohertz-repetition-rate all-solid-state laser. This source significantly pushes the frontiers of nonlinear optics. Intriguing possibilities include the development of a compact laser-driven coherent soft-X-ray source at photon energies near 1 keV, and the generation of attosecond XUV pulses. This paper analyzes strong-field ionization and high-harmonic generation in the few-cycle regime, and reviews the first experimental realization of high-harmonic generation with 5-fs pulses. These experiments have resulted in the emission of coherent X-rays at photon energies greater than 0.5 keV ($\lambda < 2.5$ nm), representing the highest photon energies achieved with a laser-driven coherent source so far. The generated XUV beam is well collimated and predicted to be delivered, after suitable spectral filtering, in a *single* burst of attosecond duration.**

***Index Terms*—Atomic physics, nonlinear optics, nonlinear wave propagation, tunneling, ultrafast optics.**

I. Introduction

THE UTILITY of ultrashort optical pulses originates mainly from their short duration and the fact that high-peak powers can be achieved with moderate pulse energies. Time-resolved spectroscopy is one of the "classic" application fields in which the short pulse duration is of prime importance. In nonlinear optics, the peak intensity has been appreciated as the key parameter with the shortness of the pulse regarded as necessary only to an extent which is needed to achieve the required peak powers. The possibility of "switching on" electric field strengths comparable to or higher than the atomic Coulomb field within a time comparable to the light oscillation cycle, which has now emerged with the development of a multigigawatt 5-fs laser source [1], promises to dramatically extend the frontiers of nonlinear optics and time-resolved spectroscopy. This extension relates to our ability to expose matter in its ground state to unprecedented field strengths. The implications are far reaching with respect to both fundamental aspects and applications. These include (but are not limited to) the feasibility of above-barrier ionization, the generation of coherent X-rays from a laboratory source in the wavelength range between the oxygen and carbon K edge (2.3–4.4 nm), the so-called water-window, and the generation of single attosecond XUV pulses.

A bright laboratory X-ray source in the water-window is expected to have great impacts on biophysics and biochemistry, allowing the study of biological samples in their natural environment with high contrast and resolution. This research previously relied on the use of synchrotron radiation [2] or an X-ray laser pumped by a multikilojoule laser [3], both available at large-scale facilities only. Recently laboratory sources of coherent X-rays in the water window have been demonstrated by Spielmann *et al.* [4] and Chang *et al.* [5] by means of high-harmonic generation in helium. The use of sub-10-fs near-infrared (NIR) pump pulses limits the effective interaction time relevant for the generation of the highest harmonics to essentially one optical cycle, which, in combination with propagation effects, can result in the emission of a *single* attosecond X-ray pulse near the cutoff of the emitted XUV spectrum, as recently predicted by Kan *et al.* [6]. These extraordinarily short pulses are likely to open the way to the "atomic time scale" characterization of the quantum-mechanical evolution of the electron wave function in bound atomic states as well as for electronic processes in chemical reactions. Furthermore, the concentration of the X-rays within an extremely short time interval and the high degree of spatial coherence of the sub-10-fs laser-driven XUV source give rise to an unprecedented peak brightness, which paves the way toward an extension of nonlinear optics into the X-ray regime.

In this paper, we shall review the new possibilities sub-10-fs pump pulses open up in the research and development of compact laser-driven ultrafast coherent X-ray sources. The remaining part of this section is devoted, after a brief overview of recent advances in the technology of short-pulse lasers, to the physical processes relevant to short-wavelength generation and to a discussion of various routes to coherent X-ray generation. In the subsequent sections we shall focus on one of the most promising routes, high-harmonic generation, and the underlying physical processes.

Ultrafast solid-state laser technology has been subject to dramatic advances over the last 10 years. The invention of titanium-doped sapphire (Ti:sapphire) [7] along with the development of Kerr-lens mode-locking [8], [83], [84] and chirped

Manuscript received November 13, 1997; revised March 12, 1998. This work was supported by the Austrian Science Foundation under Grant Y44-PHY. The work of C. Spielmann was supported by an APART Fellowship.

C. Spielmann, T. Brabec, M. Geissler, A. Scrinzi, M. Schnürer, and F. Krausz are with Abteilung Quantenelektronik und Lasertechnik, Technische Universität Wien, A-1040 Vienna, Austria.

C. Kan and N. H. Burnett are with the Department of Electrical Engineering, University of Alberta, Edmonton, AB, T6G2G7 Canada.

Publisher Item Identifier S 1077-260X(98)03840-4.

IEEE JOURNAL OF SELECTED TOPICS IN QUANTUM ELECTRONICS, VOL. 4, NO. 2, MARCH/APRIL 1998

dielectric mirrors for ultrabroad-band dispersion control [9] have resulted in a new generation of compact solid-state ultrafast oscillators, which can now routinely generate pulses in the sub-10-fs regime [10] with peak powers exceeding 1 MW [11], [85] and pulse durations as short as 6.5 fs [12]. Amplification of ultrashort pulses has also progressed rapidly. The combination of the technique of chirped pulse amplification [13] with Ti:sapphire allowed the construction of laboratory-scale high-power laser systems with unprecedented characteristics [1], [14]–[27]. Pulses as short as $\approx$20 fs have now become available with terawatt peak powers (pulse energy on the order of 100 mJ) at repetition rates of 10–50 Hz [22], [25]. The development of powerful kilohertz repetition rate pump lasers and the excellent thermal properties of Ti:sapphire opened the way to producing millijoule-energy 20-fs pulses at kilohertz frequencies [1], [27]. The output of these systems can now be spectrally broadened in a gas-filled capillary and subsequently compressed [28] at pulse energy levels many orders of magnitude higher than applicable in single-mode-fiber (SMF) compressors [29]. Using this technique (dubbed as the hollow-fiber technique), self-phase-modulated 20-fs pulses with energies of tens of microjoules have been compressed in a delay line consisting of chirped mirrors and fused-silica prisms to 4.5 fs [30], [86]. The more recent development of chirped mirrors providing proper dispersion control for self-phase modulated (SPM) pulses over a bandwidth of 150 THz (650–950 nm) [31] without relying on additional dispersive components has permitted upscaling the hollow-fiber-chirped-mirror compression technique to the submillijoule level while preserving the 5-fs duration [1], [32]. The electric field performs less than two oscillations within the full-width at half maximum (FWHM) of the intensity envelope of these pulses. They are delivered in a diffraction-limited beam and hence focusable to peak intensities in excess of 10^{17} W/cm^2.

The availability of light pulses comprising just a few oscillation cycles at intensities well above 10^{14} W/cm^2 opens up new prospects for nonperturbative nonlinear optics and coherent short-wavelength generation. Matter irradiated at these intensity levels undergoes tunnel ionization [33]–[35]. For pulses containing many oscillations, the number of ionized particles (e.g., atoms, molecules, or clusters) accumulates over many optical cycles, which may result in a depletion of the ground state well before the peak of the incident pulse impinges upon the target. Fig. 1(a) depicts the instantaneous ionization rate, as obtained from the theory of Ammosov, Delone, and Krainov (ADK) [34], [87][1] induced in helium by the linearly polarized electric field of a 50-fs near-infrared (800 nm) laser pulse with a peak intensity of 4×10^{15} W/cm^2. Fig. 1(b) shows the ionization rate under the same experimental conditions except that the pulse duration is 5 fs. Fig. 1(b) also depicts the ionization rate for circular polarization (dashed line). The consequences of the dramatically shortened exposure to the laser field are numerous and far reaching. First, the peak electron emission (or ionization) rate is much higher than in the multicycle regime and can considerably exceed the laser frequency ($\approx 3 \times 10^{14}$ s^{-1}). Secondly, the freed electrons are

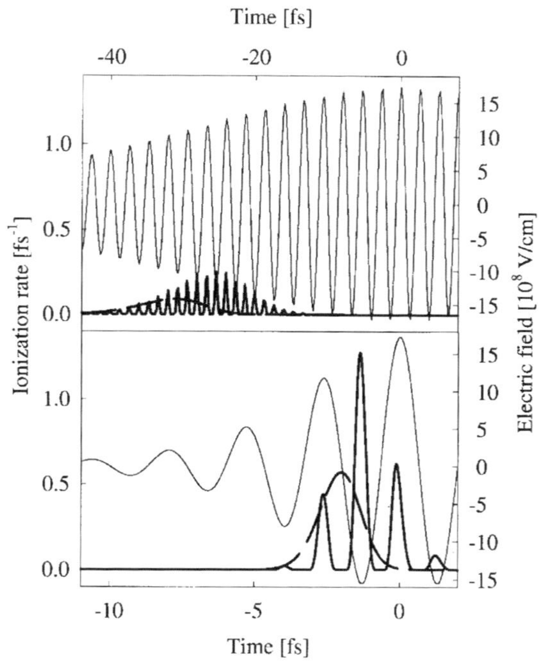

Fig. 1. (top) Instantaneous ionization rate induced in helium by a linearly (full line) or circularly (dashed line) polarized electric field (thin line) of a 50-fs NIR (800 nm) laser pulse with a peak intensity of 4×10^{15} W/cm^2. (bottom) Ionization rate under the same experimental conditions except that the pulse duration is 5 fs.

released into significantly stronger fields as compared to the multicycle regime. These implications suggest that coherent XUV generation resulting from a free-bound transition (high-order harmonic or continuum generation, henceforth, briefly HHG) [36], [37] will exhibit improved efficiency and generate higher XUV frequencies. In addition, electron-excited X-ray laser schemes (EEXL) [38], [39], [88] are likely to benefit from the improved energy coupling efficiency into an optically ionized plasma because the required high intensities can be achieved at much lower energy levels in the single cycle regime.

After ionization, the freed electron gains energy in the laser field. This is an essential process, because emission of high-energy photons is the result of the interaction of these electrons either with their parent ions (HHG) or with the surrounding ions in the ensemble (EEXL) in these schemes and the photon energy is limited by that of the recombining (HHG) or colliding (EEXL) electrons. Following Corkum *et al.* [40], [41], the electron motion in the laser field can be described by classical mechanics after tunneling. Fig. 2 shows electron trajectories for different polarization states of the laser field (linear and circular) and different release times t_r with respect to the phase of the electric field during the first laser period following ionization. The laser intensity was chosen as 4×10^{15} W/cm^2.

The electron reencounters its parent ion if and only if the laser field is linearly polarized and the electron is released at a suitable phase ωt_r of the laser field. The full line in Fig. 2 depicts the trajectory of the electron released at $\omega t_r = 17°$, which was previously shown to yield the electron passing the nucleus

[1] For hydrogen, the same expression was first obtained by [87].

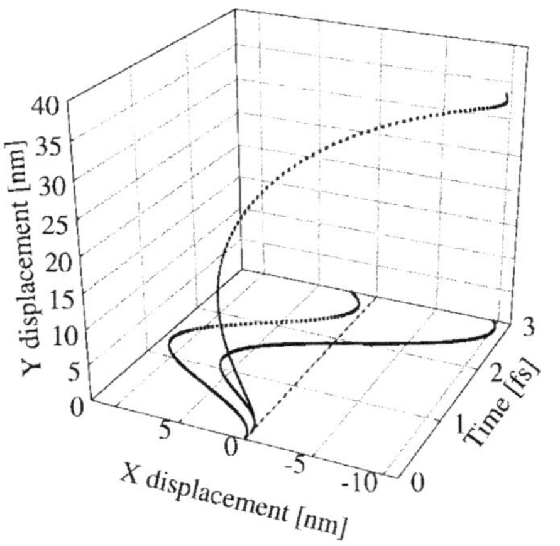

Fig. 2. Trajectories of the electron wavepacket in an electric field $E = E_0 \cos(\omega t)$ as a function of the time of ionization t_0 and polarization. The driving field has a cycle-averaged intensity of 4×10^{15} W/cm^2 and an oscillation period of 2.67 fs. An electron born at the field maximum (dotted line) will return to its parent ion with zero energy in a linearly polarized field. The electron born at $\omega t_r \approx 17°$ (full line) will reencounter the parent ion with the maximum energy and give rise to the emission of the highest XUV photon energies emitted by upon recombination into the ground state. An electron born at any time will never recombine with his parent ion in a circularly polarized field (dashed), but it can gain a much higher drift energy, which may be beneficial for collisional excitation of neighboring atoms.

with maximum kinetic energy, given by $(W_{\rm kin})_{\rm max} \approx 3.17\, U_p$ [41], [42], where U_p is the ponderomotive energy. In units of electronvolts it can be expressed as $U_p = 9.3 \times 10^{-14} I \lambda^2$, where I is the laser intensity and λ is the laser wavelength in units of W/cm^2 and μm, respectively. With some small probability the returning electron recombines in its original ground state upon emitting a photon with an energy equal to the sum of the ionization potential I_p and the electron kinetic energy $W_{\rm kin}$ gained in the laser field. Because the electron wavepacket evolution is driven by a spatially coherent laser field, the emission of the high-energy XUV photons from the particles in the ensemble also occurs in a spatially coherent manner, leading to a well-collimated XUV-beam collinear with the pump laser beam. In the multicycle regime, the process is repeated quasi-periodically over many laser periods, consequently the XUV emission spectrum is made up of discrete harmonics of the laser frequency. In the photon energy range close to the highest harmonics (cutoff region) the emission period has been predicted to be confined to a small fraction of the laser oscillation cycle. This prediction implies that the temporal evolution of the XUV output in the cutoff region can be characterized as a train of bursts of subfemtosecond duration [43]–[46]. It is obvious from the above considerations that the use of a quasi-single-cycle driver benefits HHG in several respects. First, higher photon energies (shorter wavelengths) can be achieved because the electrons are released into a stronger laser field (see Fig. 1). Second, The X-rays near cutoff can be is emitted in a *single* attosecond pulse, resulting in an unprecedented concentration of electromagnetic energy. Third, efficiency is increased due to the higher ionization rate and an increased coherence length [6]. These benefits will be discussed in more detail in Section III.

For circularly polarized light, the electron never returns to its parent ion [41] (see, e.g., the trajectory depicted by the dashed line in Fig. 2). However, the average kinetic energy of the electrons associated with the drift (nonoscillating) component of their motion, which the electrons are left behind with after the laser field is off, can be far higher than for linearly polarized laser field. The average electron energy in the plasma produced by circularly polarized light is $(W_{\rm kin})_{\rm av} \approx U_p$ and the energetic electrons can serve as an efficient pump source for X-ray lasers as proposed by Burnett and Corkum [47]. For instance, they can induce bound–bound transitions of the electron system of the ions via collisions [38], [39], [88] and thereby generate population inversion for X-ray lasing. After multiple ionization, the transition energies fall into the soft X-ray region. This EEXL scheme has been successfully demonstrated in Xe^{9+} produced by a circularly polarized 40-fs laser focused to intensities of $\approx 3 \times 10^{16}$ W/cm^2 (XUV emission wavelength $\approx$42 nm, photon energy $\approx$30 eV) [38], [88]. Recent theoretical studies suggest that electron-excited X-ray lasers could be scalable to photon energies approaching [48] or even exceeding [39] 1 keV.

Yet another route to coherent X-ray generation that opens up with the availability of few-cycle drivers is high-harmonic generation in crystalline solids. Recently Kalman and Brabec [49], [50] proposed a scheme which should allow the generation of coherent kiloelectronvolt X-rays by the use of very short pump pulses. Their approach draws on the idea that for sufficiently short pulse durations high intensity can be coupled into a crystal before the lattice structure is destroyed. The periodic potential of the crystal lattice is important for two reasons. First, the X-ray generation is enhanced by an additional momentum transfer from the lattice. Second, via Bragg coupling the X-ray modes with the minimum absorption losses are selected. These two effects will lead to substantial decrease of the threshold pump laser intensity required for kilohertz X-ray generation as compared to free electrons. For a successful realization of this scheme it is crucial that the pump pulses are much shorter than the electron–phonon relaxation time which is on the order of 100 fs. Numerical simulations have shown that focusing sub-30-fs pulses with a peak intensity of 10^{17} W/cm^2 onto a LiF crystal will result in the emission of coherent X-rays with photon energies of $\approx$3 keV having a brilliance comparable with synchrotron radiation [49].

The above examples demonstrate the tremendous potential for coherent X-ray generation that can be exploited with powerful ultrashort light pulses in the few-cycle regime. The new ultrafast drivers hold promise for both improving existing techniques significantly and for opening new ways to producing high-energy X-rays with laboratory sources. Furthermore, they also permit operation of laser-driven X-ray sources at laser pulse energies far smaller than in the multicycle regime. The relaxed pump energy requirements do not only allow the use of more compact laser drivers but also pave the way to significantly higher repetition rates, implications carrying much weight when it comes to applications of the novel X-ray sources.

The remaining part of this paper is organized as follows. Section II is devoted to strong-field ionization, the primary

IEEE JOURNAL OF SELECTED TOPICS IN QUANTUM ELECTRONICS, VOL. 4, NO. 2, MARCH/APRIL 1998

physical process in all high-field laser-atom interactions. For pulses consisting of many cycles and intensities significantly greater than 10^{14} W/cm^2, ionization occurs in the optical tunneling regime and is essentially completed in this same regime because the ground state is depleted before the laser field could become strong enough to suppress the Coulomb barrier below the ionization potential (in helium this occurs at an electric field of $E_{\text{bs}} \approx 10^9$ V/cm). As a consequence, ionization can be accurately described by the analytic theory of optical tunnel ionization [33]–[35]. Fig. 1(b) clearly reveals that the ground state remains populated up to electric fields well beyond E_{bs} in the single-cycle regime. Therefore, the validity of the widely used analytic expressions in this previously inaccessible regime is brought into question. In Section II predictions of the analytic ADK theory is compared with results obtained by numerically solving the time-dependent Schrödinger equation for hydrogenlike atoms in the presence of a strong laser field. Sections III and IV focus on the theory of HHG and its experimental realization in the quasi-single-cycle regime (XUV continuum generation). In Section III, we analyze the XUV emission of helium resulting from a (quasi-)free-bound transition induced by ultrashort laser pulses in the few-cycle regime. The analysis is not confined to an isolated atom, but also performed for an ensemble by using a quasi-classical description [41], [51] and the Maxwell equation for describing propagation of the pump and XUV wave. The XUV output is compared for different pump pulse durations. In Section IV, we review recent HHG experiments which were carried out with sub-10-fs pulses [4], [52]. The presented source emits an XUV continuum with excellent spatial coherence. Coherent XUV generation in the range of several hundred eV (extending beyond 0.5 keV) has been demonstrated for the first time at a kilohertz repetition rate in these experiments. Section V summarizes the challenges left to be addressed for the generation, control, and measurement of single attosecond XUV pulses and their use for spectroscopy and nonlinear optics.

II. Strong-Field Ionization in the Few-Cycle Regime

Ionization is of prime importance being the first step in high-intensity light-matter interaction processes. Ionization, acceleration by the electric field, and subsequent collision of the electron with its parent ion or with adjacent atoms is responsible for number of exciting phenomena in strong-field atomic, plasma, and solid-state physics, including beyond HHG and X-ray lasing, laser-induced damage of dielectrics [53], [54], molecular dissociation [55], [56], plasma heating [57], etc. The theoretical investigation of the above processes is simplified tremendously by the analytical description of ionization. This has been possible in two parameter ranges defined by $\gamma \gg 1$ and $\gamma \ll 1$, where γ is the Keldysh parameter [33]. It is defined as $\gamma = \omega_0/\sqrt{2m_e I_{\text{p}}}/eE$, where e is the electron charge, m_e is the electron mass, I_{p} is the ionization potential, and E is the amplitude of the oscillating electric field. If $\gamma \gg 1$, ionization can be described perturbatively in the regime of multiphoton ionization. The opposite limiting case is characterized by $\gamma \ll 1$, where ionization takes place

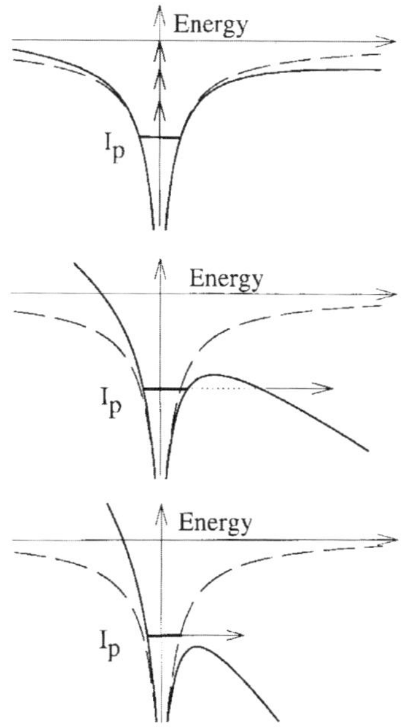

Fig. 3. Regimes of strong-field ionization. Exposing an atom to an intense laser field will result in a modified potential (solid line) composed of the Coulomb potential (dashed line) and the time dependent effective potential of the optical pulse. At moderate intensities the resulting potential is close to the unperturbed Coulomb potential and an electron can be liberated only upon simultaneous absorption of N photons, resulting in *(perturbative) multiphoton ionization*. The multiphoton ionization rate scales with the Nth power of the intensity of the optical pulse. At sufficiently high-field strengths the Coulomb barrier becomes narrow, allowing *optical tunnel ionization* to take over and resulting in a tunneling current that follows adiabatically the variation of the resultant potential. At very high fields, the electric field amplitude reaches values sufficient to suppress the Coulomb barrier below the energy level of the ground state, opening the way to *above-barrier ionization.*

as a result of tunneling [33], [34], [87]. The limiting cases of multiphoton ionization (at moderate intensities) and tunnel ionization (at high intensities) are illustrated schematically in Fig. 3(a) and (b). Tunneling takes over for $\gamma \leq 1/2$, which is fulfilled for neutral atoms at intensities ranging from 10^{14} W/cm^2 to 10^{15} W/cm^2 in the near infrared spectral range. In strong field experiments preferably noble gases are used because of their chemical inactivity. A typical representative is helium (having the highest ionization potential, I_p = 24.6 eV) for which $\gamma < 1/2$ at a wavelength of 800 nm calls for an intensity of $I > 8 \times 10^{14}$ W/cm^2.

As the duration of intense laser pulses approaches the light oscillation period, a new regime of ionization comes into play. Because the peak electric field in these pulse are "switched on" virtually abruptly, a significant fraction of the ground-state population survives until the electric field reaches the barrier-suppression limit, i.e., the Coulomb barrier in the direction of the electric field is suppressed below the ionization potential and the bound electron can escape freely, as illustrated in Fig. 3(c).

The critical field strength at which this phenomenon occurs in a hydrogen-like atom is given in atomic units (1 at.u. = 5.14×10^9 V/cm) by $E_{\text{bs}} = I_{\text{p}}^2/(4Z)$, where Z is the charge of the residual ion and I_{p} is the ionization potential in atomic units (1 at.u. = 27.21 eV) [58]. The amplitude of the optical field is related to the (cycle-averaged) intensity in SI units by $I[\text{W/cm}^2] = |E[\text{V/cm}]|^2/(2Z_0)$, where $Z_0 = 377\ \Omega$. Applying these simple formulas to He ($Z = 1$, $I_\text{p} = 24.6$ eV), the intensity needed for barrier suppression, also referred to as the appearance intensity, is $I_{ap} \approx 1.5 \times 10^{15}$ W/cm^2. Fig. 1(b) reveals that according to the optical tunneling rate [33]–[35], which was used for calculating the ionization dynamics depicted in Fig. 1, a dominant fraction of the helium atoms irradiated by a 5-fs pulse at $\lambda_0 = 800$ nm is ionized at field strengths greater than E_bs. This prompts the question if the optical tunnel rate [33]–[35] gives proper account for strong-field ionization in the quasi-single-cycle regime.

In what follows, we will investigate ionization numerically in the parameter range of above-barrier ionization and compare the results with the predictions of the analytic theory of optical tunnel ionization. For $\gamma \leq 1/2$, it was shown (see ADK-rate [34], [87] and the Keldysh-rate [33] corrected for the effect of the Coulomb potential by Krainov [35]) that optical tunneling can be described by DC tunneling [59], [60] with the dc electric field being replaced by the fast oscillating laser electric field. The ADK-theory [34], [87] as well as the Keldysh theory [33] after Coulomb correction [35] yield the following expression for the rate of optical tunnel ionization from the ground state

$$w(t) = \omega_\text{p}|C_{\text{n}^*}|^2\left(\frac{4\omega_\text{p}}{\omega_t}\right)^{2n^*-1}\exp\left(-\frac{4\omega_\text{p}}{3\omega_t}\right) \tag{1}$$

with

$$\begin{aligned}\omega_\text{p} &= \frac{I_\text{p}}{\hbar}\\ \omega_\text{t} &= \frac{eE_l(t)}{(2mI_\text{p})^{1/2}}\\ n^* &= Z\left(\frac{I_\text{ph}}{I_\text{p}}\right)^{1/2}\\ |C_{\text{n}^*}|^2 &= 2^{2n^*}[n^*\Gamma(n^*+1)\Gamma(n^*)]^{-1}.\end{aligned}$$

Here, I_p is the ionization potential of the species of interest, I_ph is the ionization potential of the hydrogen atom, and $E_l(t)$ is the electric field of the laser pulse. From the ionization rate $w(t)$, the density of freed electrons $n(t)$ can be obtained by solving a simple rate equation, yielding

$$\frac{n(t)}{n_0} = 1 - \exp\left[-\int_{-\infty}^{t} dt' w(t')\right] \tag{2}$$

where n_0 is the density of atoms.

The process of ionization can also be studied by solving numerically the time-dependent Schrödinger equation. The numerical solver is based on the complex scaling method. The wave equation is expanded in Laguerre polynomials, where the equations governing the time dependent amplitudes are solved by a Runge–Kutta scheme. The free electron density left behind by the pulse is then given by $n_f = n_0(1 - |\psi|_b^2)$, where $|\psi|_b^2$ denotes the projection of the wavefunction onto the bound states of the atom. Computer simulations have been carried out for hydrogen in the presence of a strong laser pulse for different pulse durations in the few-cycle regime. The results of this study are summarized in Fig. 4, which plots n_f/n_0 as a function of the peak electric field strength E_0 for different pulse durations.

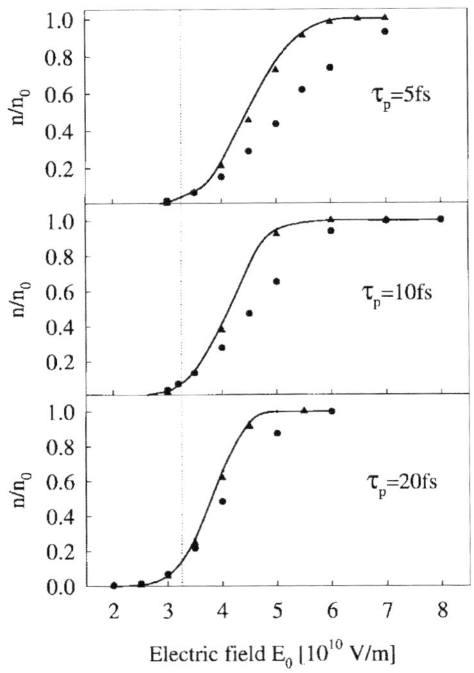

Fig. 4. Fraction of ionized hydrogen atoms in the regime of above-barrier ionization as function of the peak electric field amplitude for pulses of 5, 10, and 20 fs in duration. The full line shows the prediction of the ADK model, the triangles represent that of the Krainov model and the circles depict the exact solution of time-dependent Schrödinger equation. The dotted line depicts the barrier suppression field strength for hydrogen ($E_{bs} \approx 3.2 \times 10^8$ V/cm).

The full lines depict the results obtained by using the ADK-rate [34], [87] whereas the triangles represent the prediction of the theory of Krainov [35], who extended the tunneling theory into the barrier suppression regime. The dots show the results gained from the exact numerical solution of the Schrödinger equation. The appearance intensity for a hydrogen atom is given by $I_{ap} = 1.4 \times 10^{14}$ W/cm^2 and the corresponding electric field amplitude is indicated by the dotted line in Fig. 4. It is interesting that little difference is found between the barrier-suppression rate of Krainov and the tunneling rate of ADK even well above E_{bs}. However, we find that the analytic results increasingly deviate from the exact solutions for decreasing pulse durations for $E_0 > E_{bs}$. This is because, for shorter pulse durations a larger fraction of the electrons is ionized at intensities falling into the barrier suppression regime. Nevertheless, even at high intensities and close to the single-cycle regime ($\tau_p = 5$ fs) the predictions of the tunneling theory do not deviate from the exact calculations by more than a factor of 2. It is important to note that the analytic tunneling theory increasingly overestimates ionization as the pulse duration approaches the single-cycle regime. As a consequence, atoms can be exposed to intensities higher than predicted by the tunneling theory before the ground state is depleted. This finding is important for HHG, as will become apparent in the next section. For exact modeling of HHG from an atomic ensemble the coupled Schrödinger and Maxwell equations should be solved numerically. Because this

IEEE JOURNAL OF SELECTED TOPICS IN QUANTUM ELECTRONICS, VOL. 4, NO. 2, MARCH/APRIL 1998

is excessively time-consuming, the tunneling theory is used in the following section to determine the ionization rate [6]. It should be kept in mind that the tunneling theory overestimates the ionization rate for quasi-single-cycle drivers. However, the quantitative error introduced in the modeling in this manner does not significantly change the qualitative behavior of the HHG process.

III. Strong-Field-Driven Free-Bound Transition: High-Order Harmonic and XUV Continuum Generation

Coherent XUV emission through the process that has come to be understood as "ultrahigh-harmonic generation" has been studied for many years [61]. In this process, the XUV emission is thought to arise from the re-interaction of an essentially free electron wavepacket with its parent ion some fraction of an optical cycle after its appearance in the continuum [41], [42], [51]. The XUV emission is inextricably linked to the ionization event itself with the induced nonlinear polarization being proportional to the square root of the ionization rate. Coherent XUV emission can result from this interaction only if the coherence of the fundamental forces a large number of atoms to radiate in phase. The observed XUV emission is thus determined both by the individual atomic response and by the phase mismatch between the fundamental and harmonic fields. For very high-frequency emission such as we study here, the XUV field can be considered as propagating in vacuum (phase velocity c), the presence of free electrons however leads to a slight increase in the phase velocity of the fundamental. Time dependent ionization also leads to dispersion and self phase and amplitude modulation (SPAM) of the fundamental. These factors (the relative importance of which depends somewhat on pulselength) ultimately limit the coherent growth of the XUV emission. It is apparent that a phase advance of the fundamental of a fraction of a percent will interfere with the coherent growth of XUV whose frequency is more than one hundred times the fundamental frequency, i.e., very delicate distortions of the visible field will destroy the coherent growth of the high frequency XUV radiation.

Ultrashort (few-cycle) pulses offer several advantages in this process [45], [62]: First, the semiclassical picture predicts a cutoff in the harmonic spectrum at a harmonic energy corresponding to [41], [42]

$$(h\nu)_{\max} \approx I_p + 3.17U_p. \tag{3}$$

In a pulsed laser field, U_p is time-dependent and can be defined in terms of the instantaneous amplitude (envelope) of the electric field (which may not be strictly speaking correct in the few cycle regime). Note that for intense pump pulses U_p relevant to (3) may not be dictated by the peak intensity but the one at which the ground state is depleted on the leading edge of the pulse (see Fig. 1) [5]. For ultra-short pulses, ionization and the related XUV frequency conversion can persist to much higher intensities which allows coherent XUV generation at higher X-ray photon energies than is the case with longer pulses. Furthermore, the desired characteristics of a high nonlinear polarizability (high ionization rate) and a large coherence length (low level of ionization) appear on the surface to force one in the direction of shorter pulses for optimum conversion efficiency. The onset of SPAM at very short pulse durations will ultimately set a limit to this trend (see discussion below). Finally as we will show here, sub-10-fs driver pulses offer the prospect of concentrating the high frequency XUV in a single ≈100-as subpulse. Control of the phase of the fundamental will allow synchronization of this subpulse with external events and permit the achievement of previously impossible temporal precision in pump-probe experiments. We have examined numerically in the one-dimensional (1-D) limit these aspects of the high-frequency XUV conversion process. The main finding of our study is outlined in [6] and is the observation that for driver pulses as short as 5-fs SPAM does not interfere with the coherence growth of XUV emission for frequencies well beyond the carbon K-edge at 280 eV. In the present paper we present some additional results concerning the pulselength dependence of the coherent growth of high-frequency XUV emission. We observe that for a Ti:sapphire driver a pulselength in the vicinity of 5 fs is likely near optimum.

In what follows, we combine the model of the nonlinear polarization, developed by Lewenstein *et al.* [51] with ground state depletion based on the ADK tunnel ionization rate [34], [87] and a 1-D numerical solution of the Maxwell wave equation for the fundamental and harmonic fields. The 1-D aspect of this study limits its applicability, of course, to situations where the target is much thinner (a fraction of a percent) of the confocal parameter. In the experiments discussed in Section IV, this requirement is only marginally met, nevertheless we expect the 1-D model to be of some assistance in interpreting the results and planning future directions.

In the experiments, as described in the next section, the helium atoms were irradiated with intensities up to 4×10^{15}W/cm^2, which is almost three times the appearance intensity. In Section II, it was shown that the ionization rate is overestimated by the ADK approach at such high intensities. To estimate the error in the radiation spectrum of an isolated atom that is caused by this overestimation, the prediction of the model of Lewenstein *et al.* [51] has been compared with the square of the Fourier spectrum of the expectation of the dipole acceleration obtained from the exact numerical solution of the time-dependent Schrödinger equation for hydrogen. The chosen peak intensity of 5×10^{14} W/cm^2 of the 5-fs driver pulse ($\lambda_0 = 800$ nm) exceeds by approximately a factor of 3 the appearance intensity. The results are shown in Fig. 5. The prediction of the model of Lewenstein *et al.* [51] is found to be in good qualitative agreement with the exact numerical solution of the Schrödinger equation. However, due to the ionization rate significantly overestimated by the ADK-formula at these intensity levels, the model overestimates the intensity of high-harmonic radiation by approximately an order of magnitude at high frequencies. In order to reduce this deviation, the following simulations describing collective HHG in a helium ensemble are restricted to a peak pulse intensity of 2×10^{15} W/cm^2, which only moderately exceed the appearance intensity.

To calculate HHG in an atomic ensemble, we first solve numerically the 1-D Maxwell wave equation for the driver

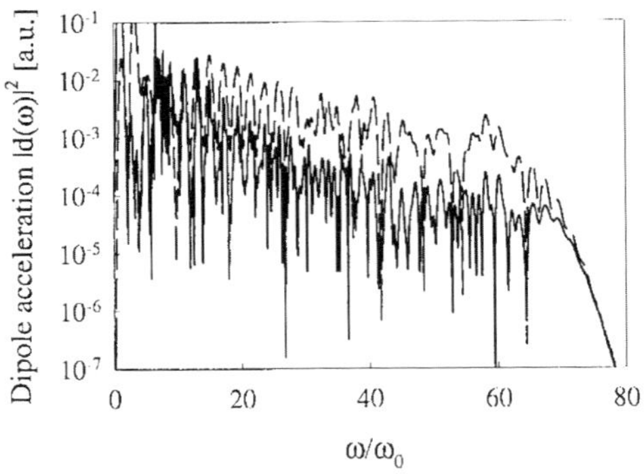

Fig. 5. Square of the dipole acceleration spectrum of a hydrogen atom exposed to a 5-fs pulse carried at $\lambda_0 = 800$ nm and having a peak intensity three times higher than the appearance intensity. The full line depicts the exact solution of the time dependent Schrödinger equation. The dashed line shows the prediction of the model of Lewenstein *et al.* combined with the ADK depletion of the ground state.

field in the presence of field dependent ionization as given by the ADK formula. This solution is used to compute the nonlinear polarization of the medium using the model of Lewenstein *et al.* and then finally the XUV emission is obtained by solving the wave equation with the induced nonlinear polarization as a source. In this final step, an approximate solution for the high frequency field is used where we neglect the influence of free electrons on propagation. This is justifiable since we are interested only in XUV frequencies for which $n/n_c \ll 1$, where n_c is the critical density.

First, we decompose the laser electric field in envelope and carrier

$$E_l(z, t) = A_l(z, t)e^{i(k_0 z - \omega_0 t + \psi_0)} + \text{c.c.} \tag{4}$$

This decomposition was shown to be physically meaningful down to pulse durations comparable to the carrier oscillation cycle $T_0 = 2\pi/\omega_0$ [63]. The Maxwell wave equation for the complex envelope $A_l(z, t)$ of the fundamental field in an ionizing medium can then be written as

$$\frac{\partial^2 A_l}{\partial z^2} - \frac{1}{c^2}\frac{\partial^2 A_l}{\partial t^2} + \frac{2i\omega_0}{c}\left(\frac{\partial A_l}{\partial z} + \frac{1}{c}\frac{\partial A_l}{\partial t}\right) - \frac{\omega_p^2(E_l, t)}{c^2}A_l = 0 \tag{5}$$

where $\omega_p^2(E_l, t) = 4\pi n(E_l, t)e^2/m_e$ is the square of the electron plasma frequency and $n(E_l, t)$ is given by (1) and (2). Solving the wave equation in envelope form besides affording some computational advantages provides direct access to phase and amplitude deviations of the fundamental from those corresponding to vacuum propagation. The solution for the fundamental field is then used to calculate the time and spatially dependent nonlinear polarization in the ionizing medium, which is subsequently inserted into the wave equation for the XUV electric field $E_h(z, t)$ neglecting dispersion, i.e.,

$$\frac{\partial^2 E_h}{\partial z^2} - \frac{1}{c^2}\frac{\partial^2 E_h}{\partial t^2} = \frac{4\pi}{c^2} n_0 d_a^h(E_l, t) \tag{6}$$

where d_a^h is the second time derivative of the atomic dipole moment, henceforth dipole acceleration, as calculated from the model of Lewenstein *et al.* [51]. This latter equation is

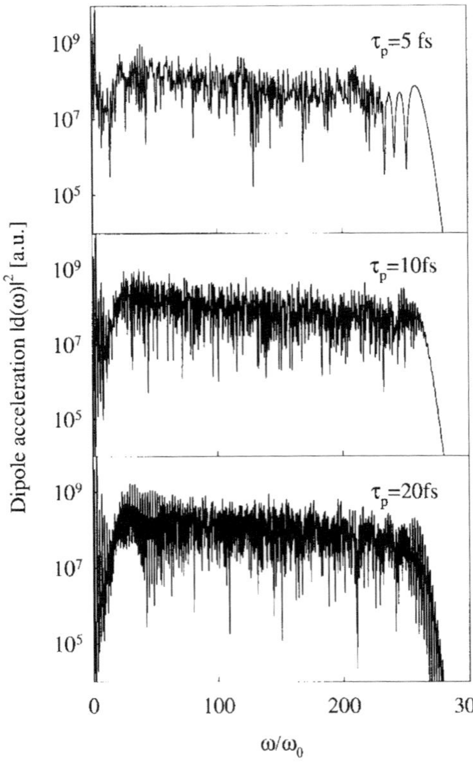

Fig. 6. Square of the dipole acceleration spectrum of helium exposed to laser pulses with a fixed peak intensity of 2×10^{15} W/cm^2 at $\lambda_0 = 0.8$ μm for different pulse durations.

solved using a Fourier transform method [64]. The phase and amplitude of the numerically propagated fundamental is defined at the entrance ($z = 0$) to the medium (500-torr He) as $E_l(0, t) = A_l(0, t)\exp(-i\omega_0 t + i\psi_0)$, where $A_l(0, t)$ was taken as Gaussian peaking at $t = 0$. The relative phase between the carrier and its envelope (ψ_0) turns out to be a significant parameter in our calculations. In what follows, unless otherwise noted, we have chosen $\psi_0 = 0$, i.e., a "cosine pulse" where the field peaks at the peak of the envelope.

In Fig. 6, we show the square of the Fourier transformed dipole acceleration (single atom response) for He atoms exposed to pulses with a peak intensity of 2×10^{15} W/cm^2 and carrier wavelength 800 nm of varying temporal duration. Note that the discrete harmonic structure is entirely absent near cutoff for $\tau_p = 5$ fs, indicating that emission of these high-energy photons is confined to one optical cycle in this case. Fig. 7 depicts the fundamental waveform at the entrance to the medium, time dependent ionization fraction from the ADK formula, and the square of the envelope of the inverse transformed dipole acceleration after applying a spectral filter to isolate components of frequency ranging from 240–280 times the carrier frequency. This latter curve corresponds to the radiation intensity envelope that would be emitted by an isolated atom at $z = 0$. In Fig. 8 we show these same parameters after propagating through 10 μm of the medium. In this case, we show the radiation intensity envelope of the forward propagating coherent XUV signal, again after a spectral filter applied to isolate components in the range h240–h280.

IEEE JOURNAL OF SELECTED TOPICS IN QUANTUM ELECTRONICS, VOL. 4, NO. 2, MARCH/APRIL 1998

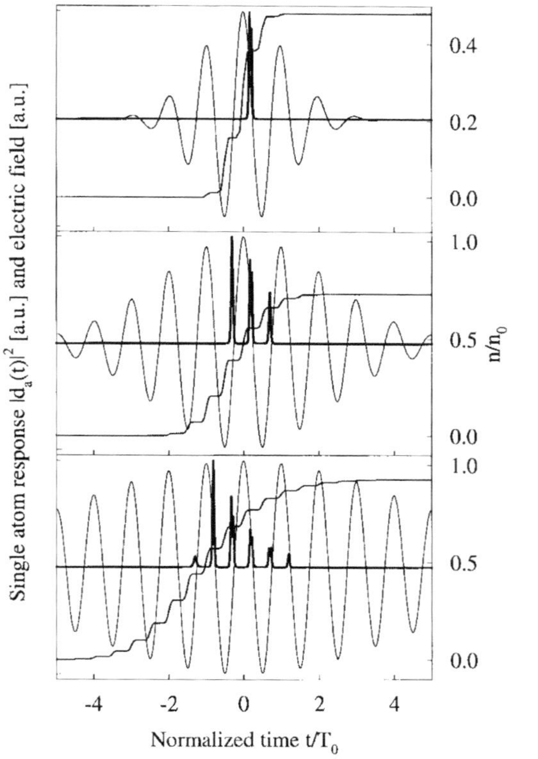

Fig. 7. Fundamental waveform, ionization profile and square of the dipole acceleration (single atom response) after applying a spectral filter to isolate components of frequency ranging from 240 to 280 times the carrier frequency for He atoms subjected to pulses of a peak intensity of 2×10^{15} W/cm^2 carried at $\lambda_0 = 800$ nm with different temporal duration. The scale for $|d_a(t)|^2$ is different for different pulse durations.

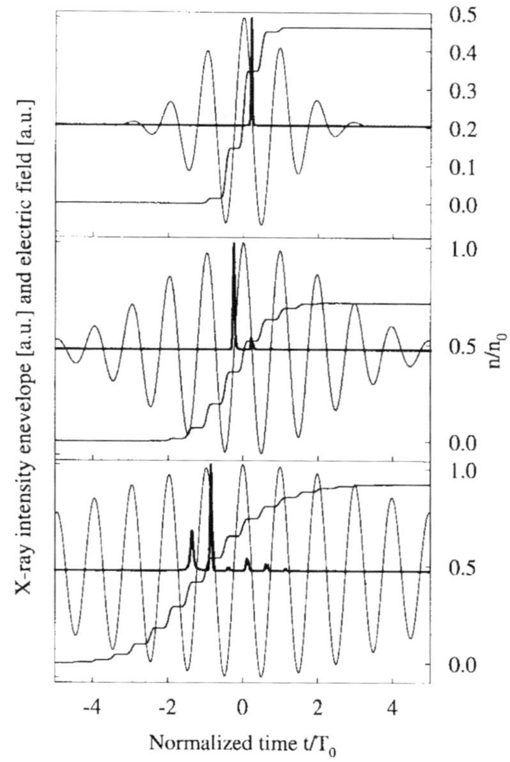

Fig. 8. Fundamental waveform, ionization profile and forward propagating radiation intensity after applying a spectral filter to isolate components of frequency ranging from 240 to 280 times the carrier frequency for He atoms subjected to pulses of peak intensity 2×10^{15} W/cm^2 and $\lambda_0 = 0.8$ μm of varying temporal duration. The results were obtained after a propagation length of 10 μm in 500-torr He. The scale for the XUV intensity is different for different pulse durations.

The number of 400-eV photons per unit area (full lines) and the peak intensity of the most intense micropulse (dotted lines) as a function of propagation distance are shown in Fig. 9 for pulse durations of 5, 10, and 20 fs. We have done a similar calculation for a 2.5 fs pulse, in this case the saturated peak intensity decreases to a value near that observed for the 10-fs case. We find that, in this latter case, self-phase and amplitude modulation (SPAM) and not simple free electron induced phase advance is the dominant limit to coherent growth. SPAM in a medium undergoing time dependent ionization leads to a frequency increase and amplitude reduction [65] of the fundamental which alters highest frequency XUV emission that is permitted by the 3.17 U_p cutoff rule, this also modifies the relative phase of the induced dipole moment at a given permitted XUV frequency. Fig. 9 indicates that 5-fs pulses are, owing to an enhanced coherence length, capable of generating by two orders of magnitude more 400-eV photons and three orders of magnitude higher peak intensities at these high-photon energies than 20-fs pulses. This is all the more remarkable when one considers that the comparison is made at constant peak powers, i.e., the energy of the 5-fs pulses is substantially lower as compared to the 20-fs pulses.

The micropulse structure seen in the XUV signal plotted in Figs. 7 and 8 is due to the nature of the classical trajectories of electrons which tunnel with small initial energy into an optical field. Electrons which re-encounter their parent ions with the highest energy are born in a narrow phase bin of

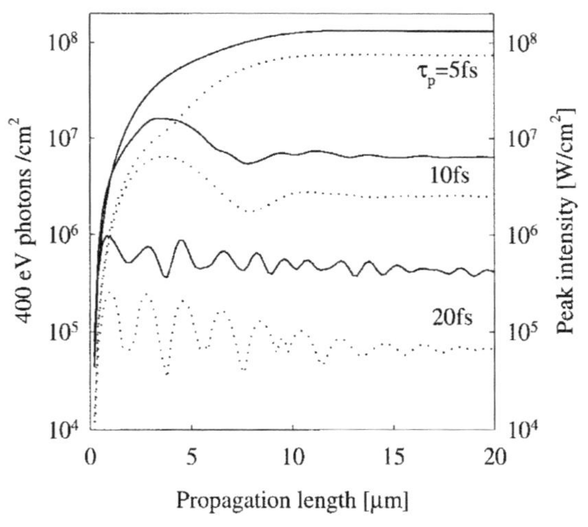

Fig. 9. Number of 400-eV photons per area within the filter band as given in Figs. 7 and 8 (solid lines) and peak intensity of the most intense micropulse (dotted lines) as a function of propagation distance for pulse durations of 5, 10, and 20 fs.

the fundamental just after ($\approx 17°$) a peak in the fundamental field (see full line in Fig. 2) and return about 2/3 of a cycle later in a narrow bin near a zero crossing of the electric field [41], [43]. Depending on the spectral window employed to isolate the emission, individual micropulses can have durations

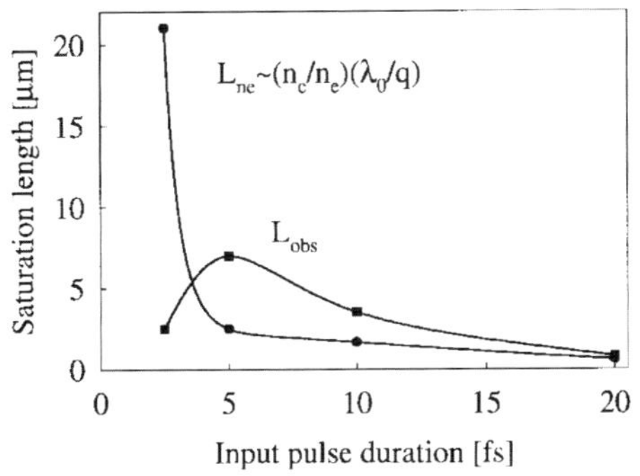

Fig. 10. Saturation length for the coherent growth of XUV radiation in the frequency range $240 < \omega/\omega_0 < 280$ as a function of the pulse duration for a fixed peak intensity of 2×10^{15} W/cm^2 in 500-torr He. The dots represent the "quasi-static" coherence length caused by ionization. The squares depict the computed coherence lengths as extracted from the results shown in Fig. 9.

considerably shorter than 100 as. In Fig. 9, we plot the peak intensity of the largest micropulse. The coherent growth of individual micropulses is roughly determined by the electron density that existed near the time when electrons responsible for that micropulse first appear in the continuum. So that generally the first micropulse that is allowed by the 3.17 U_p cutoff rule, seeing the lowest electron density, grows the most robustly.

Our calculations in regard to the pulselength dependence of the coherence length for growth of XUV emission in the frequency interval between h240 and h280 are summarized in Fig. 10. Here, we show the propagation length in 500-torr He over which the peak intensity of emission in this frequency interval grows to 90% of its saturated value as well as the expected coherence length L_{ne} for a plasma with electron density, n_e, corresponding to that existing when the electrons responsible for the most intense micropulse first appear in the continuum, i.e., $L_{ne} \approx n_c/n_e \lambda_0/q$ [66].[2] Here, n_c denotes the critical electron density at the fundamental laser carrier wavelength λ_0 and q denotes the "harmonic" order. Also plotted is the observed saturation length for the strongest micropulse from our propagation code. It is apparent that the limits to coherence for pulses longer than 20 fs can be thought of in terms of quasi-static plasma dispersion. For pulses in the range from 20–5 fs, the coherence length is improved, apparently as a consequence of the breakdown of the slowly varying envelope approximation [6]. Our results suggest that in the few-cycle regime the relevant velocity for phase matching is intermediate between the quasi-static phase velocity ($>c$) and the group velocity ($<c$). For pulselengths below 5 fs, the coherent growth of the XUV emission is limited by SPAM of the fundamental rather than by quasi-static plasma dispersion. It is apparent from Fig. 10 that the optimum pulse duration for efficient XUV emission is in the vicinity of 5 fs.

As is apparent from Fig. 8, this pulse duration has the coincidental benefit of allowing single-micropulse selection by means of simple spectral filtering. In practice both the XUV micropulse selectivity or contrast as well as the saturated peak intensity of the strongest micropulse depend somewhat on the relative carrier phase chosen for our calculation. We will not elaborate on this point here except to note that, somewhat surprisingly, for a 5-fs envelope, $\psi_0 = 0$ is not the optimum phase for efficient coherent XUV production. Likewise the micropulse contrast after propagation can be optimized by varying the carrier phase. In this case, the optimum phase is observed to vary somewhat depending on the chosen spectral window.

Although care must be taken in applying the results of a 1-D analysis such as used here to frequency conversion with a focused laser beam, the concept of a definable carrier phase (relative to its envelope) will still apply to a focused beam in the limit that the target thickness is much less than the confocal parameter. As few-cycle lasers will eventually have multimillijoule pulse energies, experiments will eventually closely approach this 1-D regime. In this case it is clear from Fig. 9 that, providing the fundamental has something approaching a "top hat" intensity profile and geometric phase advance is negligible in the nonlinear medium, stable, single attosecond pulse selection is possible through simple spectral filtering with a 5-fs driver. Present generation solid state lasers emit a stable (soliton-like) amplitude envelope and frequency spectrum but the relative phase between carrier and envelope is in practice random [67]. Stable single attosecond pulse selection will require access to and control over this phase. We will return to this point in Section V.

IV. Coherent XUV Continuum Generation with Sub-10-fs Laser Pulses

The ultrashort-pulse laser driver used for the XUV generation experiments reviewed below is an all-solid-state Ti:sapphire-based oscillator-amplifier system followed by a hollow-fiber-chirped-mirror pulse compressor. The system is described in detail elsewhere [1], [32]. Briefly, the front end of the laser system is a 10-fs Ti:sapphire oscillator (FemtoSource Pro, FemtoLasers GmbH) pumped by a frequency-doubled, diode-pumped Nd:YVO_4 laser (Millennia, Spectra Physics Inc.). The oscillator routinely delivers sub-10-fs pulses with an energy of about 5 nJ. The pulses are nearly transform limited and have a spectral width of about 100 nm centered at 780 nm. After stretching the pulses to several ps by a 10-cm-long slab of SF57, the energy is boosted by almost six orders of magnitude in a confocal eight-pass amplifier. The amplifier is pumped with an energy of 10 mJ at a 1-kHz repetition rate provided by a continuous-wave (CW) lamp-pumped intracavity frequency-doubled Q-switched Nd:YLF laser (Model 621D, BMI). Owing to a simple amplified spontaneous emission (ASE) suppression scheme the amplifier delivers pulses with energies up to 1.5 mJ and an ASE-to-amplified-energy ratio as low as $<2 \times 10^{-3}$. The pulse-to-pulse energy fluctuations are 1% rms [1]. The amplified pulses are compressed by a pair of double-fused silica prisms. The throughput is about ≈90%. The residual high-order dispersion is compensated by a pair of mirrors which are placed in front of the amplifier. This allows to keep the throughput of the compressor at maximum. The

[2]For a discussion of the coherence length dictated by a quasi-static free-electron density see, e.g., [66]. The situation for the short pulses we consider here is somewhat more complicated and is discussed by C. Kan *et al.* [6].

IEEE JOURNAL OF SELECTED TOPICS IN QUANTUM ELECTRONICS, VOL. 4, NO. 2, MARCH/APRIL 1998

compressed pulses have a duration of 20–25 fs and energy of typically 1.1–1.2 mJ. The performance of the stretcher and compressor is not alignment sensitive, therefore, these results are reproducible on a day-to-day basis without any time-consuming alignment.

Subsequently these pulses are coupled into a hollow waveguide filled with a noble gas. The gas acts as a nonlinear medium in which spectral broadening via self phase modulation takes place [28]. The best performance for the given input pulses was achieved with an argon-filled 85-cm-long fiber having an inner diameter of 260 μm. After passage through the fiber the 50-nm-wide input spectrum broadens to more than 200 nm and is slightly blue shifted. The blue shift may arise from self steepening and some plasma formation. The fiber serves not only as a nonlinear medium but also acts as a very efficient spatial filter. As a consequence the linearly polarized output beam is diffraction limited within 10% and has a nearly Gaussian far-field distribution. Finally, the pulses are compressed in a dispersive delay line consisting of chirped mirrors. They introduce a nearly constant group-delay dispersion (linear variation of the group delay) over a range of 650–950 nm. Five to seven bounces off these mirrors compensate the chirp induced by SPM and precompensate the material dispersion between compressor output and experimental target and result in pulse durations of typically 5–7 fs on target. The output pulse energy is 0.4–0.5 mJ, yielding peak powers up to $\approx$0.1 TW. Owing to the diffraction-limited nature of the beam [1], these intense kilohertz-rate 5-fs pulses can be focused to peak intensities well in excess of 10^{17} W/cm^2.

Special care has to be taken when employing 5-fs pulses in experiments, because the pulses broaden by almost a factor of three upon traveling a distance of 1 m in air. To ensure minimum pulse duration at the target, the dispersion of air (and of possible optical components) has to be precompensated. The right amount of compensation can be adjusted by setting up a reference path. In this reference path the pulses experience the same amount of dispersion (glass and air) before they are measured with an interferometric autocorrelator.

The experimental arrangement used for the HHG is schematically shown in Fig. 11. The laser beam was focused with a silver-coated spherical mirror (focal length 20 cm) through an 0.3-mm-thick fused-silica window into a vacuum chamber containing the target at focus. The focal region was imaged onto a CCD camera. The estimated 1/e^2 beam diameter were 50 and 70 μm in the vertical and horizontal direction, respectively. The slight ellipticity is caused by the off-axis focusing geometry using a spherical mirror. For a typical pulse energy of 0.3–0.4 mJ impinging the target, the on-axis peak intensity is estimated as 4–5 $\times$ 10^{15} W/cm^2 on target. The target itself is formed by a 1-cm-long thin nickel tube with an inner diameter of 0.6 mm and a wall thickness of 0.05 mm. To minimize the interaction length and the gas load, the tube was squeezed to a residual inner diameter of 30 μm and its end was sealed. The holes in the tube walls in the squeezed region were bored by the laser itself. The tube was continuously backed with helium or some other noble gas with pressures up to 5 bar. To avoid reabsorption of the generated radiation in the interaction chamber the pressure

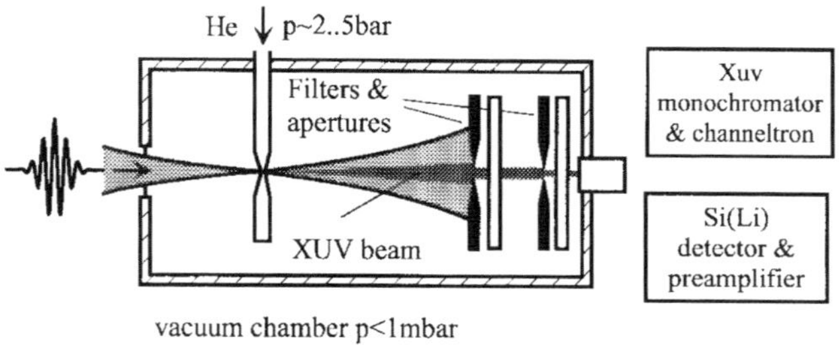

Fig. 11. Schematic of the experimental setup for measuring the high-harmonic XUV spectrum emitted by gas targets exposed to femtoseocnd pulses.

was kept below 1 mbar employing a root pump. From the pumping speed and the background pressure we were able to estimate the gas flow out of the target. Gas flow, target configuration, and background pressure [68] then determine uniquely the pressure in the interaction region. For a backing pressure of 5 bar, we estimated the average pressure in the interaction region as $\approx$500 mbar.

For wavelength-dispersive spectral analysis, the XUV radiation emerging from the target was passed into a 1-m grazing-incidence monochromator (Model 248/310G, McPherson Inc.) [4]. The monochromator was separated from the target chamber by a circular aperture of 1-mm diameter and evacuated by a turbo pump. This differential pumping scheme allowed the residual pressure to be reduced below 10^{-4} mbar in the monochromator, which was required by our single-channel electron multiplier detector (uncoated channeltron, Model 4175 G, Galileo). The monochromator was equipped with a 300-grooves-per-mm platinum-coated grating. The output of the detector was collected by a high-dynamic-range lock-in amplifier, which was synchronized by the laser pulse train. Alternatively, energy dispersive X-ray spectrometry (EDX) was employed for high-dynamic-range detection of X-rays with photon energies $>$170 eV [52]. To this end, the monochromator was replaced by a lithium-drifted silicon [Si(Li)] crystal detector cooled to 77 K. In this case, the X-ray beam must be strongly attenuated to make sure that no more than 1 X-ray photon hits the detector simultaneously. The impinging X-ray photon generates electron-hole pairs, whose number is proportional to the photon energy. The detector signal is preamplified and processed in a multichannel pulse height analyzer, yielding the photon energy spectrum of the emitted X-rays.

First, the HHG spectra for different noble gases were investigated using the wavelength-dispersive system. The entrance and exit slit width of the monochromator were fixed at 50 μm implying a resolution of 0.06 nm at 100 nm and 0.02 nm at 10 nm. The entrance slit and the aperture in front of the monochromator located 50 cm downstream of the target blocks an increasing part of the XUV signal for increasing wavelength because of the wavelength-dependent beam size. Correction for these two effects is expected to increase the spectral intensity of the long wavelength components by an order of magnitude in comparison to the short one. According to the information provided by the manufacturer the spectral response of the detector is fairly flat in the wavelength range

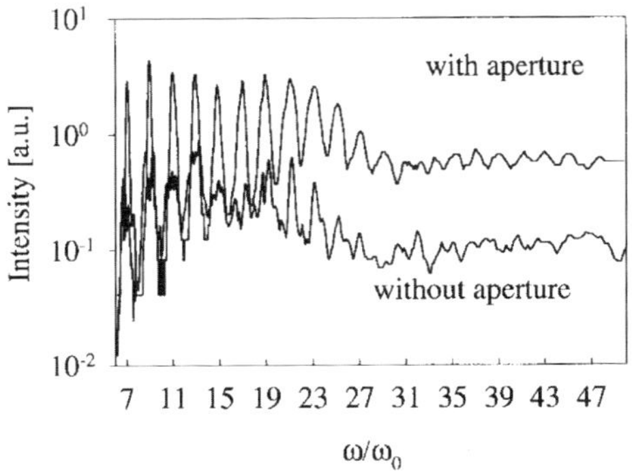

Fig. 12. Uncorrected spectrum of XUV radiation emerging from argon at a backing pressure of 0.5 bar. The upper curve was recorded after putting an aperture with a diameter of 5 mm into the laser beam prior to the focusing mirror. The beam diameter prior to the aperture was ≈7 mm.

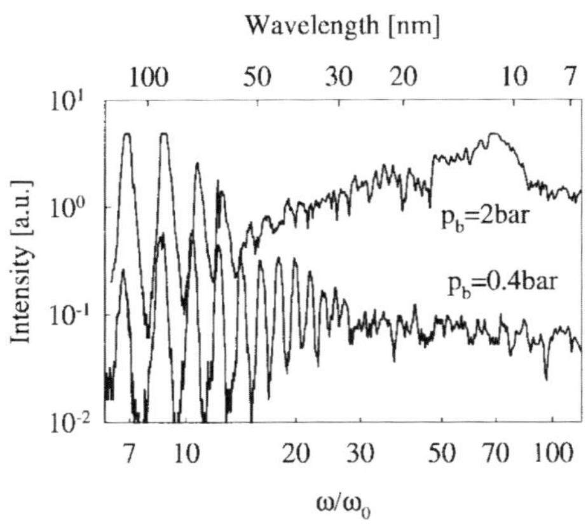

Fig. 13. Uncorrected spectrum of XUV radiation emitted by neon with a backing pressure of 0.5 and 2 bar.

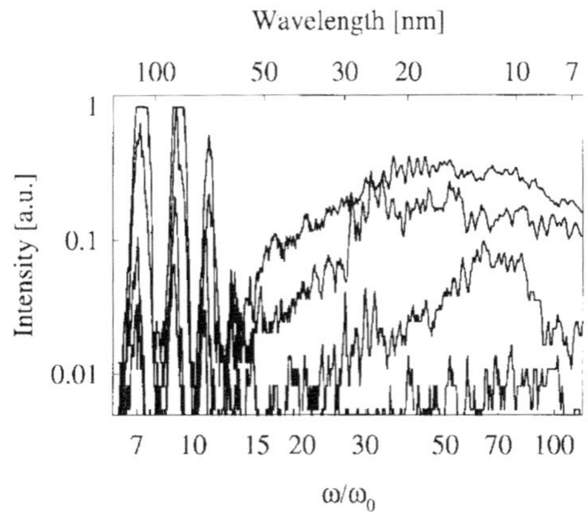

Fig. 14. Uncorrected spectra of XUV emission from helium for backing pressures of 1, 2, 4, and 5 bar. The spectral intensity increases with increasing pressure.

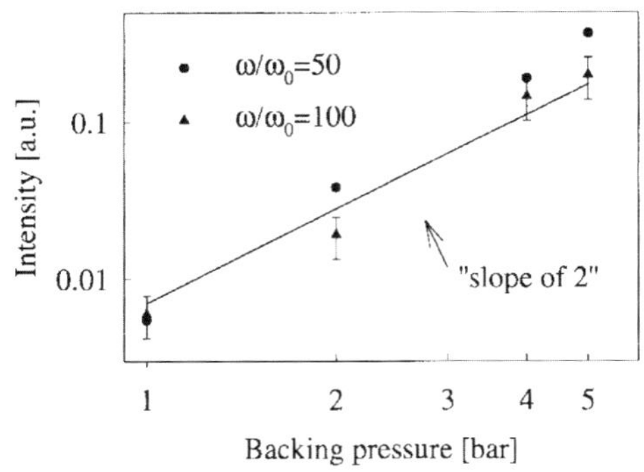

Fig. 15. Pressure dependence of the XUV intensity for different harmonic orders in helium.

of interest. In the absence of reliable information about the wavelength dependence of the grating efficiency the XUV spectra presented below have not been corrected for the spectrometer response.

The measured spectra for argon and neon are shown in Figs. 12 and 13. In these experiments, the pulse energy on target was about 300 μJ and f/30 focusing yielded an estimated on-axis peak intensity of $\approx 4 \times 10^{15}$ W/cm^2. The radiation spectrum of argon for a backing pressure of 0.5 bar shows well resolved discrete harmonics up to the 31st order. Above this frequency the harmonics merge to a continuum. Reducing the pulse energy by ≈50% by aperturing the laser beam before the focusing mirror resulted in a shift of the transition from discrete harmonics to continuum to the 43rd order. On the other hand, increasing the backing pressure up to 1 bar shifted the border of the continuum toward longer wavelengths. We repeated these measurements with neon at backing pressures of 0.4 and 2 bar, and observed a qualitatively similar behavior. These observations along with an increasing blue shift of the discrete harmonics with increasing backing pressure points to a nonnegligible role of plasma-induced SPM in these media [69]–[71].

Theory as well as previous experiments suggest that XUV radiation with the highest possible photon energy can be expected from helium. This is because helium has the highest ionization potential (24.4 eV) among all noble gases and hence can be exposed to the highest intensities before ionizing. The experiments reported below were performed using helium as the target material. Fig. 14 plots a wide-band scan extending from 100 to 7 nm. The spectral behavior differs from that observed with argon or neon. Discrete harmonics appear only up to the 13th order. In contrast with argon or neon, the number of discrete harmonics appears to be constant in helium up to backing pressures of 5 bar. To check whether the observed XUV radiation results from a coherent generation process (rather than, e.g., incoherent plasma emission), we measured the XUV signal as a function of the backing pressure. Fig. 15 depicts the variation of XUV spectral intensity with pressure at two different frequencies. The quadratic pressure dependence evaluated provides clear evidence for the coherent nature of the XUV emission from the ensemble [72].

The high-energy end of the continuum can not be accessed from the previous measurements because of the rapidly increasing amount of stray XUV light leaking through as the zeroth order is approached. The leakage becomes dominant for wavelengths below 7 nm. For a detailed investigation of the spectrum below 10 nm it is necessary to suppress the scatter from the central image (zeroth order) of the monochromator. To this end, a 100-nm silver foil, fixed on an electroformed nickel mesh (open area 80%) was placed in front of the entrance slit. The filter has a nearly constant transmission

IEEE JOURNAL OF SELECTED TOPICS IN QUANTUM ELECTRONICS, VOL. 4, NO. 2, MARCH/APRIL 1998

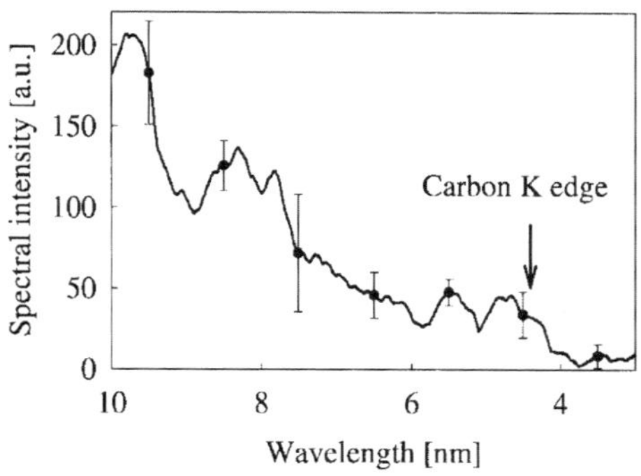

Fig. 16. Short-wavelength end of the XUV continuum emitted by the helium jet, as transmitted by a 100-nm-thick silver foil. The spectrum is the average of three scans. The error bars give the standard deviation of the signal at selected wavelengths averaged over several seconds. The spectral components at $\lambda > 10$ nm are damped by the silver foil.

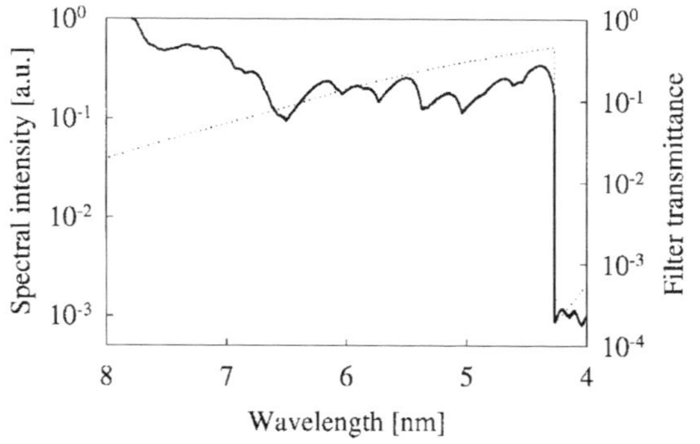

Fig. 17. Short-wavelength end of the XUV spectrum as transmitted by a 1.8-μm-thick mylar foil (solid line) and transmittance of the foil (dotted line).

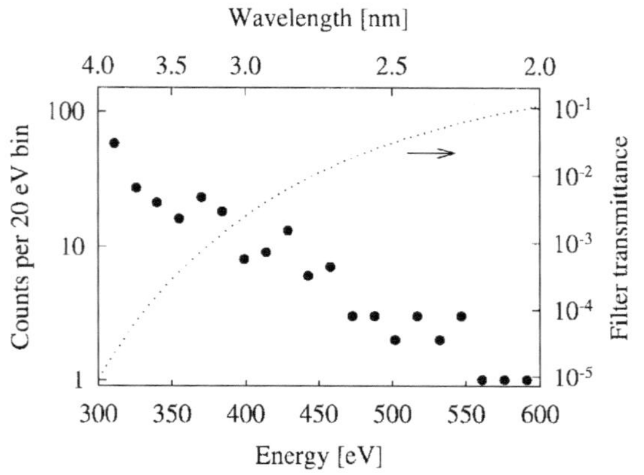

Fig. 18. High-energy end of the XUV spectrum for 7-fs pump pulses (9×10^5 shots) transmitted through 400-nm boron and 200-nm aluminum (dotted line). The photon energy spectrum has been obtained with the EDX system.

of about 50% in the wavelength range of 3–10 nm and blocks radiation with wavelengths greater than 10 nm. The short wavelength tail of the XUV continuum was obtained by averaging over three scans and is shown in Fig. 16. To clarify whether the observed structure of the spectrum is real or caused by fluctuations of the experimental parameters, we measured the signal for fixed wavelengths over several seconds. The rms deviations are displayed as error bars in Fig. 16. In the wavelength range of 4.2–4.4 nm, the detector signal exceeds the noise floor by a factor of $\approx$500. The noise floor is defined as the signal level with either the gas flow or the laser blocked. Further evidence for the signal originating from coherent XUV radiation rather than incoherent plasma emission can be gained from the fact that no measurable signal was observed after replacing helium with argon, although the ionization potential of argon is much lower than for helium, implying a much higher plasma density for argon. Systematic tests also ruled out the possibility that the tube wall contributes to the measured signal. Last but not least, increasing the pump pulse duration at constant pulse energy makes the short-wavelength tail of the spectrum to retreat to longer wavelengths, as expected from the theoretical considerations on HHG in the previous section.

In order to verify the calibration of our monochromator and provide additional evidence for the emergence of water-window radiation from the He target, we replaced the silver foil by a pair of 900-nm-thick Mylar foils ($C_{10}H_8O_4$). At 4.37 nm, the carbon K-edge, this filter provides an absorption contrast of $>10^3$. The measured spectrum shown in Fig. 17 exhibits a sharp cutoff at 4.37 nm, providing conclusive evidence for water window radiation. The signal contrast at the carbon K-edge of $>3 \times 10^3$ was limited by the noise floor of our detection apparatus. Drawing on data provided by the manufacturers of the spectrograph and the detector, a conservative estimate yields some 100 X-ray photons generated by a single laser pulse within a 1% bandwidth at the carbon K-edge.

The signal shown in Fig. 16 is found to exceed the detection limit by more than an order of magnitude down to 3 nm. At these short wavelengths, however, the exit slit is very close to the zeroth order of the diffraction grating. Therefore, some stray light contribution to the signal can not be ruled out. In order to get around this problem, the short wavelength end of the spectrum was also studied with an energy dispersive system (ZMAX, NORAN Instruments) [52]. The lithium-drifted silicon detector is able to reliably identify only single X-ray photons with an energy greater than 170 eV. Yet, the flux at low photon energies must be suppressed with appropriate filters to avoid pile-ups, i.e., the simultaneous detection of two or more photons, which would falsify the photon energy spectrum. For safe avoidance of pile-ups the number of counts per laser shot had been reduced far below unity ($\approx$0.01). After the appropriate filtering the system yielded reliable information of the photon energy spectrum of the emitted X-ray radiation. In order to gain access to the spectral region beyond 300 eV we used a 200-nm aluminum and a 400-nm boron foil to suppress low-energy radiation. With this filter combination and a count/shot probability as low as 0.5% pile-ups could be safely eliminated. The transmitted X-ray spectrum along with the filter transmittance (dashed line) is shown in Fig. 18. Fig. 18 reveals that the XUV spectrum extends beyond 500 eV, i.e., to the short-wavelength edge of the water window at 2.3 nm, the K-edge of oxygen.

The observed shape of the high energy end of the spectra strongly deviates from that of the single atom response, as shown in Fig. 6. The radiation spectrum of a He atom driven by 5-fs pulses exhibits a broad plateau of nearly constant intensity. For a peak laser intensity of 3×10^{15} W/cm^2, which finds a large fraction of the He atoms in their ground state for a 5-fs pulse as revealed by Fig. 1, (3) predicts this plateau to extend to a wavelength of around 2 nm. The striking discrepancy between the predicted plateau and the rapid roll off of the XUV spectral intensity for decreasing wavelengths (see Figs. 16 and 18) is attributed to a rapidly decreasing coherence length for increasing XUV frequency in our high-pressure target [6], [52]. The dephasing is caused by an accumulating phase error due to plasma dispersion and wave front curvature, both of which rapidly increase in importance with decreasing wavelengths. For our focusing geometry, the accumulating phase error introduced by the curvature of the fundamental wave front limits coherent accumulation of XUV radiation to an interaction length (geometric coherence length) on the order of 5–10 μm for wavelengths in the water window [61]. The high pressure used in our experiments is dictated by the short geometric coherence length associated with tight focusing of the pump laser, which was necessary to achieve the required peak intensities with our present source. Roughly, optimum signal can be expected at a pressure which makes the coherence length determined by plasma dispersion only approximately equal to the geometric coherence length. The role of plasma dispersion was extensively analyzed in the previous section. According to our computer simulations (see Fig. 9) the respective coherence length for radiation around the carbon K-edge is on the order of 5 μm for a target pressure of 0.5 bar. This compares well with the estimated geometric coherence length for the same XUV wavelength region and explains why we had to rely on the high pressures for maximum XUV emission in the water window in our experiments. Because the total interaction length, i.e., target thickness could not be confined to a few micrometers (rather it was estimated as 100–150 μm), the increasing coherence length for increasing wavelength resulted in a rapidly increasing spectral intensity. We expect that the availability of somewhat higher pulse energies (of 1–3 mJ) will allow a significant enhancement of the X-ray flux in the water window owing to the more gentle focusing and associated more relaxed conditions for optimizing coherence growth of the emitted radiation.

For practical applications of this X-ray source the spatial coherence is of great interest. To acquire some information about the spatial properties of the X-rays, a knife edge was transversely scanned across the beam. The transmitted power as a function of the knife edge position was recorded [73] for different wavelengths. The knife edge was positioned 29 cm downstream of the helium source. In Fig. 19, the transmitted power is shown for wavelengths at 4.4 and 10 nm. The diagram also depicts Gaussian fits to the measured data (dashed line), which yielded a 1/e^2-diameter of 480 and 180 μm at a wavelength of 10 and 4.4 nm, respectively. The corresponding beam divergence was estimated as <1 mrad and <0.5 mrad at 10 and 4.4 nm, respectively. Taking into account the transverse size of the XUV source, from these measurements we conclude that the emitted X-ray beam is within a factor of three diffraction limited (i.e., can be characterized with $M^2 < 3$ [74]). Assuming an X-ray beam

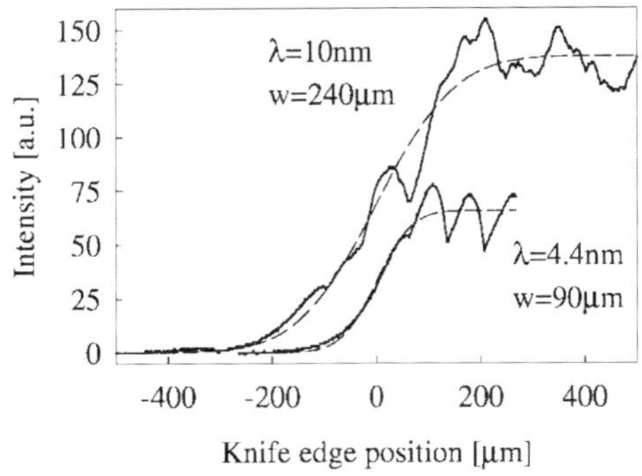

Fig. 19. Transmitted X-ray signal as a function of the position of an knife edge scanning transversely across the beam 29 cm downstream of the He target at two different wavelengths. The dashed lines depict Gaussian fits to the measured profiles.

diameter of 25 μm at the beam waist (in the source), we estimate the (time-averaged) brightness of the presented source at the carbon K-edge within a 1% bandwidth as 5×10^8 photons mm^{-2} mrad^{-2} s^{-1}. An upgrade of the laser driver to a pump pulse energy of several millijoules and further optimization of the XUV generation process is expected to increase both the peak and the average brightness of the present source by some two orders of magnitude. The former should then allow focusing the beam to peak intensities as high as 10^{15} W/cm^2 in the sub-10-nm wavelength range and open the way toward an extension of nonlinear optics into the X-ray regime [44], [75]. The latter should allow focusing the output to average intensities sufficiently high for permitting biological microscopy [76] and holography [77], [89] in the water window.

V. Towards the Generation of Single Attosecond XUV Pulses

In this section, we shall briefly address the temporal evolution of the coherent XUV radiation emitted at wavelengths below 10 nm and the challenges the predicted temporal behavior presents to the experimenter. The results of simulations presented in Section III suggest that the short-wavelength end of the X-ray spectrum emerging from helium driven by a 5-fs laser pulse is emitted in a single subfemtosecond burst. Transmitting the spectral components in the range of $240 < \omega/\omega_0 < 280$ (corresponding to wavelengths between 2.7 and 3.3 nm) through an appropriate filter, a single pulse with a duration of ≈100 as can be expected.

The theoretical analysis underlines the crucial role of propagation effects in selecting a clean single pulse for sub-10-fs laser drivers and suggests that the relative carrier phase ψ_0 of the laser pulse described by the electric field $E_l(t) = A_l(t)\exp(-i\omega_0 t + i\psi_0)$ plays an important role for the generation and control of single XUV attosecond pulses. In fact, numerical experiments clearly revealed that the evolution of the (filtered) instantaneous X-ray intensity on an attosecond time scale is a sensitive function of ψ_0. Both the peak intensity and the timing (with respect to the laser pulse) varies with ψ_0. As a consequence, the relative carrier phase (henceforth briefly: carrier phase) of the quasi-single-cycle laser

IEEE JOURNAL OF SELECTED TOPICS IN QUANTUM ELECTRONICS, VOL. 4, NO. 2, MARCH/APRIL 1998

pulse needs to be controlled for stable and fully controllable generation of single-attosecond XUV pulses. Unfortunately, CW mode-locked lasers, the primary sources of ultrashort pulses, do not produce pulses with a constant carrier phase. Recent experiments performed with a sub-10-fs Ti:sapphire laser revealed that even under optimum conditions the carrier phase is subject to rapid variations in the emitted pulse train [67]. Stabilization of the carrier phase calls for a process that yields a physical measurable which is sensitive to the phase. Our investigations indicate that phase sensitive effects can be explored in the strong-field regime of nonlinear optics, where the fast-oscillating electric field, rather than its slowly varying envelope governs light-matter interactions. Until recently, the intensity levels directly achievable with a CW mode-locked laser did not allow accessing the strong-field regime. Recent advances in sub-10-fs Ti:sapphire oscillator technology permitted focusing the mode-locked laser pulses to intensities higher than 5×10^{13} W/cm^2 [85]. These intensities should be sufficiently high for inducing photoemission in the tunneling regime from a gold photocathode. This process would yield a photocurrent (averaged over the pulse repetition time) which is dependent on the carrier phase in the sub-10-fs regime [78]. This φ_0-dependent signal could then be coupled into a feedback servo loop for stabilizing the carrier phase as proposed in [67]. This route represents a promising approach to realizing a phase-stabilized sub-10-fs pump source for coherent XUV generation.

Another issue that needs to be addressed is the spatial variation of the emitted XUV radiation. It is evident that the temporal evolution of the generated X-rays is subject to a strong spatial variation if the intensity of the driving laser field exhibits a radial variation on target. The adverse consequences include among others a spatially varying timing of the single attosecond pulse with a selected center frequency with respect to the peak of the laser pulse. As a consequence, focusing of the X-ray pulse would result in a pulse broadening up to several femtoseconds. This drawback can be eliminated by realizing a nearly flat-top intensity profile on target. This can, among others, be achieved by placing a hard aperture of suitable diameter in the collimated laser beam and imaging the aperture plane onto the target. The next question that arises is whether appropriate filters can be found which fulfill a couple of requirements: they 1) have to efficiently suppress longer-wavelength radiation with a sufficiently steep cut off and 2) may not cause significant dispersive broadening of the transmitted pulse. It can be shown, that a 200-nm aluminum filter in combination with a 200-nm titanium filter meet the above requirements.

Last but not least, once the attosecond pulses can be reliably produced they need to be characterized temporally before they can be used for applications. Recently, the ninth harmonic of a femtosecond Ti:sapphire laser was measured by autocorrelation [79]. The pulses were recombined in a helium jet and generated He^+ ions by two-photon ionization. In the absence of neutral atoms with correspondingly high ionization potential, this technique can not be extended to the wavelength range of around 10 nm or shorter. However, a vacancy could be produced in the K-shell of some light element (e.g., O, N, C, B, Be) by two-photon absorption, resulting in fluorescence at photon energies approximately equal to twice the absorbed photon energy. Unfortunately, the two-photon absorption cross section scales with the fourth power of the wavelength and hence, is much smaller than around 100 nm [80]. As a consequence, tight focusing and the use of a sensitive technique for detecting the two-photon-absorption-induced fluorescence will be inevitable. Estimations [80] show that the contemplated upgrade of the sub-10-fs laser driver along with the use of a state-of-the-art toroidal focusing mirror might result in a fluorescence yield that can be detected by a single-photon-counting EDX system. This would not only permit attosecond pulse measurement but also constitute, to the best of our knowledge, the first nonlinear optical experiment performed with X-rays. Finally, it should be noted that a corollary of single attosecond pulse generation will be the complete disappearance of harmonic structure in the XUV spectrum. This may be important for such spectroscopy applications as X-ray fine structure absorption spectroscopy (EXAFS) in the vicinity of the carbon K-edge.

VI. Summary

The above applications clearly indicate that the significance of intense few-cycle laser pulses is likely to go well beyond the frontiers of physical sciences. A laboratory-scale source of coherent X-rays in the water window will undoubtedly make a major impact on high-contrast live biological X-ray microscopy [76] and holography [77], [89]. Further optimization of the above demonstrated 5-fs-pulse-driven harmonic generator (e.g., by the implementation of quasi-phase matching) may allow the development of the first laboratory tool suitable for *in vivo* microscopy of living organizms with a resolution, which approaches the virus, DNA, and RNA scales. The predicted "concentration" of the high-energy XUV photons in a *single* burst of attosecond duration may not only extend nonlinear optics into the X-ray regime and allow studying the dynamics of bound electrons, but also provide a tool for studying chemical reactions and wave packet dynamics on a previously inaccessible time scale. Polarization control of the 5-fs infrared driver pulse allows the emission of a subfemtosecond electron burst with well-determined energy distribution [43]. The possibility of injecting an intense subfemtosecond electron burst into a plasma in a controlled manner is likely to benefit the development of laser-driven particle accelerators [81], [82] as well as electron-pumped X-ray lasers in the keV regime [39]. Nevertheless, at present we are just at the stage of exploring the new possibilities and phenomena, and a number of technical barriers are left to be overcome before these exciting prospects can materialize.

Acknowledgment

The contributions by S. Sartania, M. Lenzner, Z. Cheng, R. Koppitsch, P. Wobrauschek, and C. Streli to the above experiments are gratefully acknowledged. The authors would also like to thank A. J. Schmidt for his encouragement.

References

[1] S. Sartania, Z. Cheng, M. Lenzner, G. Tempea, C. Spielmann, F. Krausz, and K. Ferencz, "Generation of 0.1-TW 5-fs optical pulses at a 1-kHz repetition rate," *Opt. Lett.*, vol. 22, pp. 1562–1564, 1997.

[2] D. Attwood, K. Halbach, and K.-J. Kim, "Tunable coherent X-rays," *Science*, vol. 228, pp. 1265–1272, 1985.

[3] B. J. MacGowan, S. Maxon, L. B. Da Silva, D. J. Fields, C. J. Keane, D. L. Matthews, A. L. Osterheld, J. H. Scofield, A. Shimkaveg, and G. F. Stone, "Demonstration of X-ray amplifiers near the carbon K edge," *Phys. Rev. Lett.*, vol. 65, pp. 420–423, 1990.

[4] C. Spielmann, N. H. Burnett, S. Sartania, R. Koppitsch, M. Schnürer, C. Kan, M. Lenzner, P. Wobrauschek, and F. Krausz, "Generation of coherent X-rays in the water window using 5-fs laser pulses," *Science*, vol. 278, pp. 661–663, 1997.

[5] Z. Chang, A. Rundquist, H. Wang, M. Murnane, and H. C. Kapteyn, "Generation of coherent soft X-rays at 2.7 nm using high harmonics," *Phys. Rev. Lett.*, vol. 79, pp. 2967–2971, 1997.

[6] C. Kan, N. H. Burnett, C. Capjack, and R. Rankin, "Coherent XUV generation from gases ionized by several cycle optical pulses," *Phys. Rev. Lett.*, vol. 79, pp. 2971–2974, 1997.

[7] P. F. Moulton, "Spectroscopic and laser characteristics of Ti : Al_2O_3," *J. Opt. Soc. Amer. B*, vol. 3, pp. 125–133, 1986.

[8] D. E. Spence, P. N. Kean, and W. Sibbett, "60-fsec pulse generation from a self-modelocked Ti:sapphire laser," *Opt. Lett.*, vol. 16, pp. 42–44, 1991.

[9] R. Szipöcs, K. Ferencz, C. Spielmann, and F. Krausz, "Chirped multilayer coatings for broadband dispersion control in femtosecond lasers," *Opt. Lett.*, vol. 19, pp. 201–203, 1994.

[10] A. Stingl, M. Lenzner, C. Spielmann, F. Krausz, and R. Szipöcs, "Sub-10-femtosecond, mirror-dispersion-controlled Ti:sapphire laser," *Opt. Lett.*, vol. 20, pp. 602–604, 1995.

[11] L. Xu, G. Tempea, A. Poppe, M. Lenzner, C. Spielmann, F. Krausz, A. Stingl, and K. Ferencz, "High power sub-10fs Ti:sapphire oscillators," *Appl. Phys. B*, vol. 65, pp. 151–159, 1997.

[12] I. Jung, F. X. Kärntner, N. Matuschek, D. H. Sutter, F. Morier-Genoud, G. Zhang, U. Keller, V. Scheuer, M. Tilsch, and T. Tschudi, "Self-starting 6.5 fs pulses from a Ti:sapphire laser," *Opt. Lett.*, vol. 22, pp. 1009–1011, 1997.

[13] D. Strickland and G. Mourou, "Compression of amplified chirped optical pulses," *Opt. Commun.*, vol. 56, pp. 219–221, 1985.

[14] J. Squier, F. Salin, G. Mourou, and D. Harter, "100-fs pulse generation and amplification in Ti : Al_2O_3," *Opt. Lett.*, vol. 16, pp. 324–326, 1991.

[15] J. D. Kmetec, J. J. Macklin, and J. F. Young, "0.5TW, 125-fs Ti:sapphire laser," *Opt. Lett.*, vol. 16, pp. 1001–1003, 1991.

[16] A. Sullivan, H. Hamster, H. C. Kapteyn, S. Gordon, W. White, H. Nathel, R. J. Blair, and R. W. Falcone, "Multiterawatt, 100-fs laser," *Opt. Lett.*, vol. 16, pp. 1406–1408, 1991.

[17] K. Yamakawa, H. Siraga, Y. Kato, and C. P. J. Barty, "Prepuls-free 30-TW, 1 ps Nd: Glass laser," *Opt. Lett.*, vol. 16, pp. 1593–1595, 1991.

[18] T. B. Norris, "Femtosecond pulse amplification at 250 kHz with a Ti:sapphire regenerative amplifier and application to continuum generation," *Opt. Lett.*, vol. 17, pp. 1009–1011, 1992.

[19] J. V. Rudd, G. Korn, S. Kane, J. Squier, G. Mourou, and P. Bado, "Chirped-pulse amplification of 55-fs pulses at a 1-kHz repetition rate in a Ti:Al_2O_3 regenerative amplifier," *Opt. Lett.*, vol. 18, pp. 2044–2046, 1993.

[20] J. P. Zhou, C. P. Huang, C. Shi, H. C. Kapteyn, and M. M. Murnane, "Generation of 21-fs millijoule-energy pulses by use of Ti:sapphire," *Opt. Lett.*, vol. 19, pp. 126–128, 1994.

[21] C. P. J. Barty, C. L. Gordon, III, and B. E. Lemoff, "Multiterawatt 30-fs Ti:sapphire laser system," *Opt. Lett.*, vol. 19, pp. 1442–1444, 1994.

[22] J. P. Zhou, C. P. Huang, M. M. Murnane, and H. C. Kapteyn, "Amplification of 26-fs 2-TW pulses near the gain-narrowing limit in Ti:sapphire," *Opt. Lett.*, vol. 20, pp. 64–66, 1995.

[23] M. Lenzner, C. Spielmann, E. Wintner, F. Krausz, and A. J. Schmidt, "Sub-20 fs, kilohertz-repetition-rate Ti:sapphire amplifier," *Opt. Lett.*, vol. 20, pp. 1397–1399, 1995.

[24] S. Backus, J. Peatross, C. P. Huang, M. M. Murnane, and H. C. Kapteyn, "Ti:sapphire amplifier producing millijoule-level 21-fs pulses at 1 kHz," *Opt. Lett.*, vol. 20, pp. 2000–2002, 1995.

[25] C. P. J. Barty, T. Guo, C. Le Blanc, F. Raksi, C. Rose-Petruck, J. Squier, K. R. Wilson, V. V. Yakovlev, and K. Yamakawa, "Generation of 18-fs multiterawatt pulses by regenerative pulse shaping and chirped-pulse amplification," *Opt. Lett.*, vol. 21, pp. 668–670, 1996.

[26] J. P. Chambaret, C. Le Blanc, G. Cheriaux, P. Curley, G. Darpentigny, P. Rousseau, G. Hamoniaux, A. Antonetti, and F. Salin, "Generation of 25-TW, 32-fs pulses at 10 Hz," *Opt. Lett.*, vol. 21, pp. 1921–1923, 1996.

[27] S. Backus, C. G. Durfee, III, G. Mourou, H. C. Kapteyn, and M. M. Murnane, "0.2 TW laser system at 1 kHz," *Opt. Lett.*, vol. 22, pp. 1256–1258, 1997.

[28] M. Nisoli, S. De Silvestri, and O. Svelto, "Generation of high energy 10 fs pulses by a new pulse compression technique," *Appl. Phys. Lett.*, vol. 68, pp. 2793–2795, 1996.

[29] A. Baltuska, Z. Wei, M. S. Pshenichnikov, and D. A. Wiersma, "Optical pulse compression to 5fs at a 1-MHz repetition rate," *Opt. Lett.*, vol. 22, pp. 102–104, 1997.

[30] M. Nisoli, S. Stagira, S. De Silvestri, O. Svelto, S. Sartania, Z. Cheng, G. Tempea, C. Spielmann, and F. Krausz, "Toward a terawatt-scale sub-10fs laser technology," this issue, pp. 414–420.

[31] M. Nisoli, S. De Silvestri, O. Svelto, R. Szipöcs, K. Ferencz, C. Spielmann, S. Sartania, and F. Krausz, "Compression of high-energy laser pulses below 5 fs," *Opt. Lett.*, vol. 22, pp. 522–524, 1997.

[32] G. Tempea, F. Krausz, C. Spielmann, and K. Ferencz, "Dispersion control over 150 THz with chirped dielectric mirrors," this issue, pp. 193–196.

[33] L. V. Keldysh, "Ionization in the field of a strong electromagnetic wave," *Sov. Phys. JETP*, vol. 20, pp. 1307–1314, 1965.

[34] M. V. Ammosov, N. B. Delone, and V. P. Krainov, "Tunneling ionization of complex atoms and atomic ions by an alternating electromagnetic field," *Sov. Phys. JETP*, vol. 64, pp. 1191–1194, 1986.

[35] V. P. Krainov, "Ionization rates and energy and angular distribution at the barrier-suppression ionization of complex atoms and atomic units," *J. Opt. Soc. Amer. B*, vol. 14, pp. 425–431, 1997.

[36] A. L'Huillier and P. Balcou, "High-order harmonic generation in rare gases with a 1-ps 1053-nm laser," *Phys. Rev. Lett.*, vol. 70, pp. 774–777, 1993.

[37] J. J. Macklin, J. D. Kmetec, and C. L. Gordon, III, "High-order harmonic generation using intense femtosecond pulses," *Phys. Rev. Lett.*, vol. 70, pp. 766–769, 1993.

[38] B. E. Lemoff, C. P. J. Barty, and S. E. Harris, "Femtosecond-pulse-driven, electron excited XUV lasers in eight-times-ionized noble gases," *Opt. Lett.*, vol. 19, pp. 569–571, 1994.

[39] C. P. J. Barty, T. Guo, C. Le Blanc, F. Raksi, C. G. Rose-Petruck, J. A. Squier, B. C. Walker, K. R. Wilson, V. V. Yakovlev, and K. Yamakawa, "Sub-20-fs multiterawatt lasers and X-ray applications," in *Proc. 5th Int. Conf. X-ray Lasers*, S. Svaneberg and C. G. Wahlström, Eds. Bristol, U.K.: Inst. of Phys., 1996, pp. 282–288.

[40] P. B. Corkum, N. H. Burnett, and F. Brunel, "Above-threshold ionization in the long-wavelength limit," *Phys. Rev. Lett.*, vol. 62, pp. 1259–1262, 1989.

[41] P. B. Corkum, "Plasma perspective on strong-field multiphoton ionization," *Phys. Rev. Lett.*, vol. 71, pp. 1994–1997, 1993.

[42] K. C. Kulander, K. J. Schafer, and J. L. Krause, "Dynamics of short-pulse excitation, ionization and harmonic conversion," in *Proc. Workshop Super Intense Laser Atom Physics (SILAP III)*, B. Piraux, Ed. New York: Plenum, 1993, vol. 316, p. 95.

[43] P. B. Corkum, N. H. Burnett, and M. Yu. Ivanov, "Subfemtosecond pulses," *Opt. Lett.*, vol. 19, pp. 1870–1872, 1994.

[44] P. Antoine, A. L'Huillier, and M. Lewenstein, "Attosecond pulse trains using high-order harmonics," *Phys. Rev. Lett.*, vol. 77, pp. 1234–1237, 1996.

[45] K. J. Schafer and K. C. Kulander, "High harmonic generation from ultrafast pump lasers," *Phys. Rev. Lett.*, vol. 78, pp. 638–641, 1997.

[46] I. P. Christov, M. M. Murnane, and H. C. Kapteyn, "High harmonic generation of attosecond pulses in the 'single-cycle' regime," *Phys. Rev. Lett.*, vol. 78, pp. 1251–1254, 1997.

[47] N. H. Burnett and P. B. Corkum, "Cold-plasma production for recombination extreme-ultraviolet lasers by optical-field-induced ionization," *J. Opt. Soc. Amer. B*, vol. 6, pp. 1195–1199, 1989.

[48] S. M. Hooker and S. E. Harris, "Femtosecond-pulse-driven electron-excited extreme-ultraviolet lasers in Be-like ions," *Opt. Lett.*, vol. 20, pp. 1994–1996, 1995.

[49] P. Kalman and T. Brabec, "Generation of coherent hard X-ray radiation in crystalline solids by high-intensity femtosecond laser pulses," *Phys. Rev. A*, vol. 52, pp. R21–R24, 1995.

[50] ——, "Evolution of coherent hard X-ray radiation in crystalline solids by high-intensity femtosecond laser pulses," *Phys. Rev. A*, vol. 53, pp. 627–629, 1996.

[51] M. Lewenstein, P. Balcou, M. Yu. Ivanov, A. L'Huillier, and P. B. Corkum, "Theory of high-harmonic generation by low frequency laser fields," *Phys. Rev. A*, vol. 49, pp. 2117–2132, 1994.

[52] M. Schnürer, C. Spielmann, P. Wobrauschek, C. Streli, N. H. Burnett, C. Kan, K. Ferencz, R. Koppitsch, Z. Cheng, and F. Krausz, "Coherent 0.5-keV X-ray emission from helium driven by a sub-10 fs laser," *Phys. Rev. Lett.*, vol. 80, pp. 3236–3238, 1998.

IEEE JOURNAL OF SELECTED TOPICS IN QUANTUM ELECTRONICS, VOL. 4, NO. 2, MARCH/APRIL 1998

[53] D. Du, X. Liu, G. Korn, J. Squier, and G. Mourou, "Laser-induced breakdown by impact ionization in SiO_2 with pulse widths from 7ns to 150fs," *Appl.* Phys. Lett., vol. 64, pp. 3071–3073, 1994.

[54] B, C. Stuart, M. D. Feit, A. M, Rubenchik, B. W. Shore, and M. D, Perry. "Laser-induced damage in dielectrics with nanosecond to picosecond pulses," *Phys. Rev. Lett.*, vol. 74, pp. 2248–2251, 1995.

[55] T. Seidemann, M. Yu, Ivanov, and P. B. Corkum, "Role of electron localization in intense-field molecular ionization," *Phys. Rev. Lett.*, vol. 75, pp. 2819–2822, 1995.

[56] S. Chelkowski, A. Conjusteau, T. Zuo, and A. D. Bandrauk, "Dissociative ionization of H_2^+ in an intense laser field: Charge-resonance-enhanced ionization, Coulomb explosion, and harmonic generation at 600 nm," *Phys. Rev. A*, vol. 54, pp. 3235–3244, 1996.

[57] S. C. Rae and K. Burnett, "Possible production of cold plasma through optical-field-induced ionization," *Phys. Rev. A*, vol. 46, pp. 2077–2083, 1992.

[58] H. A. Bethe and E. E. Salpeter, *Quantum Mechanics of One- and Two-Electron Atoms.*, New York: Plenum, 1977.

[59] J. R. Oppenheimer, "Three notes on the quantum theory of aperiodic effects," *Phys. Rev.*, vol, 13, pp. 66–81, 1928.

[60] L. D. Landau and E. M. Lifshitz, *Quantum Mechanics,* 2nd cd, New York: Pergamon, 1965, p. 276.

[61] A. L'Huillier, L.-A. Lompre, G. Mainfray, and C. Manus, "High-order harmonic generation in rare gases," in *Atoms in Intense Laser Fields,* M. Gavrilla, Ed. New York: Academic, 1992, pp. 139–206.

[62] J. Zhou, J. Peatross, M. M. Murnane, H. C. Kapteyn, and I. P. Christov, "Enhanced high-harmonic generation using 25fs laser pulses," *Phys. Rev. Lett.*, vol. 76, pp. 752–755, 1996.

[63] T. Brabec and F. Krausz, "Nonlinear optical pulse propagation in the single-cycle regime," *Phys. Rev. Lett.*, vol. 78, pp. 3282–3285, 1997.

[64] A. L'Huillier, K. J. Schafer, and K. C. Kulander, "Theoretical aspects of intense field harmonic generation," *J. Phys. B: Atom. Mol. Phys.*, vol. 24, pp. 3315–3341, 1991.

[65] E. Esary, G. Joyce, and P. Sprangle, "Frequency up-shifting of laser pulses by copropagating ionization fronts," *Phys. Rev. A*, vol. 44, pp. 3908–3911, 1991.

[66] T. Ditmire, K. Kulander, J. K. Crane, H. Nguyen, and M. D. Perry, "Calculation and measurement of high-order harmonic energy yields in helium," *J. Opt. Soc. Amer. B*, vol. 13, pp. 406–411, 1996.

[67] L, Xu, C. Spielmann, A. Poppe, T. Brabec, F. Krausz, and T. Hänsch, "Route to phase control of ultrashort light pulses," *Opt. Lett.*, vol. 21, pp. 2008–2010, 1996.

[68] H. Zoebl and J. Kruschik, *Strömung durch Rohre und Ventile.* Vienna, Austria: Springer, 1982.

[69] S. C. Rae, K. Burnett, and J. Cooper, "Generation and propagation of high-order harmonics in a rapidly ionizing medium," *Phys. Rev. A*, vol. 50, pp. 3438–3446, 1994.

[70] K. Myazaki and H. Takada, "High-order harmonic generation in the tunneling regime," *Phys. Rev. A*, vol. 52, pp. 3007–3021, 1995.

[71] C. Altuucci, T. Starczewski, E. Mevel, C.-G. Wahlström, B. Carre, and A. L'Huillier, "Influence of atomic density in high-order harmonic generation," *J. Opt. Soc. Amur. B*, vol. 13, pp. 148–156, 1996.

[72] J. Peatross and D. D. Meyerhofer. "Intensity-dependent atomic-phase effects in high-order harmonic generation," *Phys, Rev. A*, vol. 52, pp. 3976–3987, 1995.

[73] A. E. Siegman, M. W. Sasnett, and T. F. Johnson, "Choice of clip levels for beam width measurements using knife-edge techniques," *IEEE J. Quantum Eleetroh.*, vol. 27, pp. 1098–1091, 1991.

[74] A. E. Siegman, "New development in laser resonators," in *Conference on Laser Resonators, SPIE,* 1990, vol. 1224, pp. 2–14.

[75] P. Corkum, "Harmonics to quicken the pulse," *Nature,* vol. 384, pp. 118–119, 1996.

[76] R, W. Eason, P. C. Cheng, R. Feder, A. G. Michette, R. J. Rosser, F, O' Neil, Y. Owanda, P. T. Rumsby, M. J. Shaw, and I. C. E, Turcu, "Laser X-ray microscopy," *Opticu Acta,* vol. 3, pp. 501–516, 1986.

[77] M. Howells, C. Jacobson, J. Kirz, R. Feder, K. McQuaid, and S. Rothman, "X-ray holograms at improved resolution: A study of zymogen granules," *Science,* vol. 238, pp. 514–517, 1987.

[78] F. Krausz and C. Spielmann, private communication.

[79] Y. Kobayashi, T. Seikikawa, Y. Nabekawa, and S. Watanabe, "27-fs XUV pulse generation by high-order harmonics," presented at the Conf. Lasers and Electro-Optics, 1997, postdeadline paper CPD4.

[80] P. Kalman, private communication.

[81] A. Modena, Z. Najmudin, A, E. Dangor, C, E. Clayton, K, A. Marsh, C. Joshi, V. Malka, C. B. Darrow, C. Danson, D. Neely, and F. N. Walsh, "Electron acceleration from the breaking of relativistic plasma waves," *Nature,* vol. 377, pp. 606–608, 1995.

[82] D. Umstadter, S.-Y. Chen, A. Maksimchuk, G. Mourou, and R. Wagner, "Nonlinear optics in relativistic plasmas and laser wake field acceleration of electrons," *Science,* vol. 273, pp. 472–475, 1996.

[83] L. Spinelli, B. Couillaud, N. Goldblatt, and D. K. Negus, "Starting and generation of sub-100fs pulses in Ti:Al_2O_3 by self focusing," in *Dig. Conf. Lasers and Electro-Optics,* Washington, DC: Opt. Soc. Amer., 1991, paper CPDP7.

[84] U. Keller, G. W. t'Hooft, W. H. Knox, and J. E. Cunningham, "Femtosecond pulses from a continuously self-starting passively modelocked Ti:sapphire laser," *Opt. Lett.*, vol. 16, pp. 1022–1024, 1991.

[85] L. Xu, G. Tempea, C. Spielmann, F. Krausz, A. Stingl, K. Ferencz, and S. Takano, "A CW mode-locked Ti:sapphire laser focusable to 5×10^{13} W/cm^2," *Opt. Lett.*, vol. 23, May 15, 1998.

[86] M. Nisoli, S. Stagira, S. De Silvestri, O. Svelto, S. Sartania, Z. Cheng, M. Lenzner, C. Spielmann, and F. Krausz, "A novel-high energy pulse compression system: Generation of multigigawatt sub-5 fs pulses," *Appl. Phys. B,* vol. 65, pp. 175–196, 1997.

[87] A. M. Perelomov, V. S. Popov, and M. V. Terent'ev, "Ionization of atoms in an alternating electric field," *Sov. Phys. JETP,* vol. 23, pp. 924–934, 1966.

[88] B. E. Lemoff, G. Y. Yin, C. L. Gordon, III, C. P. J. Barty, and S. E. Harris, "Demonstration of a 10-Hz femtosecond-pulse-driven XUV laser at 41.8 nm in Xe IX," *Phys. Rev. Lett.*, vol. 74, pp. 1574–1577, 1995.

[89] J. E. Trebes, S. B. Brown, E. M. Campbell, D. L. Matttews, D. G. Nilson, G. F. Stone, and D. A. Whelan, "Demonstration of X-ray holography with an X-ray laser," *Science,* vol. 238, pp. 517–520, 1987.

Christian Spielmann, for photograph and biography, see this issue, p. 184.

Clarence Kan, photograph and biography not available at the time of publication.

Neal H. Burnett, photograph and biography not available at the time of publication.

Thomas Brabec born in Linz, Austria, on April 10, 1964. He received the Ph.D. degree from Vienna University of Technology, Vienna, Austria, in 1992.

In 1994, he was awarded an APART Grant. He worked as a Post-Doctoral Fellow at the NRC Canada with Dr. Paul Corkum. Currently, he is a Research Associate at the Technical University of Vienna. His research interests are the theoretical modeling of ultrashort pulse generation and propagation, and the theory of high-intensity light-matter interactions.

Michael Geissler was born in Vienna, Austria, on July 18, 1972. He received the M.Sc. degree in electrical engineering at the Vienna University of Technology, Vienna, Austria, in 1998, and currently he is working towards the Ph.D. degree.

His research focuses on the theoretical investigation of high-field light-matter interactions.

Armin Scrinzi was born in Klagenfurt, Austria, in 1959. He received the M.Sc. degree in mathematical physics in 1985 and the Ph.D. degree in theoretical physics in 1989, at the University of Vienna, Austria.

He worked as a Post-Doctoral Fellow at the University of Florida, at the Kurchatov Institute in Moscow and at the Manne Siegbahn Institute, Stockholm, Sweden. Between 1994 and 1997, he was a Fellow of the APART Program of the Austrian Academy of Sciences. His current research interests center on mathematical and computational physics and atoms in strong external fields.

Matthias Schnürer was born in Magdeburg, Germany, in October 1956. He received the Diploma degree and the Ph.D. degree in physics on research of small gap semiconductor MOS structures at Humboldt-University in Berlin, Germany, in 1982 and 1984, respectively.

He joined the Central Institute of Optics in Berlin from 1984 to 1991 and the Max–Born-Institute in Berlin from 1992 to 1996, where he worked in the High Power Laser Departments. In 1997, he joined the Femtosecond Laser Group at the Vienna University of Technology, Austria. His research interests include laser–plasma interaction, especially the study of laser–plasma based coherent and incoherent X-ray sources.

Ferenc Krausz (M'92), for photograph and biography, see this issue, p. 184.

Intense few-cycle laser fields: Frontiers of nonlinear optics

Thomas Brabec and Ferenc Krausz*

Institut für Photonik, Technische Universität Wien, Gusshausstrasse 27/387, A-1040 Wien, Austria

The rise time of intense radiation determines the maximum field strength atoms can be exposed to before their polarizability dramatically drops due to the detachment of an outer electron. Recent progress in ultrafast optics has allowed the generation of ultraintense light pulses comprising merely a few field oscillation cycles. The arising intensity gradient allows electrons to survive in their bound atomic state up to external field strengths many times higher than the binding Coulomb field and gives rise to ionization rates comparable to the light frequency, resulting in a significant extension of the frontiers of nonlinear optics and (nonrelativistic) high-field physics. Implications include the generation of coherent harmonic radiation up to kiloelectronvolt photon energies and control of the atomic dipole moment on a subfemtosecond (1 fs$=10^{-15}$ s) time scale. This review presents the landmarks of the 30-odd-year evolution of ultrashort-pulse laser physics and technology culminating in the generation of intense few-cycle light pulses and discusses the impact of these pulses on high-field physics. Particular emphasis is placed on high-order harmonic emission and single subfemtosecond extreme ultraviolet/x-ray pulse generation. These as well as other strong-field processes are governed directly by the electric-field evolution, and hence their full control requires access to the (absolute) phase of the light carrier. We shall discuss routes to its determination and control, which will, for the first time, allow access to the electromagnetic fields in light waves and control of high-field interactions with never-before-achieved precision.

CONTENTS

I. INTRODUCTION

Lasers generating ultrashort light pulses have come of age. Their applications range from testing ultrahigh-speed semiconductor devices to precision processing of materials, from triggering and tracing chemical reactions to sophisticated surgical applications in opthalmology and neurosurgery. In the physical sciences, ultrashort light pulses enable researchers to follow ultrafast relaxation processes in the microcosm on never-before-accessed time scales and study light-matter interactions at unprecedented intensity levels.

Recent technological advances in ultrafast optics have permitted the generation of light wave packets comprising only a few oscillation cycles of the electric and magnetic fields. The spatial extension of these wave packets along the direction of their propagation is limited to a few times the wavelength of the radiation (≈ 0.5–$1\ \mu$m in the visible and near-infrared spectral range). The pulses are delivered in a diffraction-limited beam and hence focusable to a spot size comparable to the wavelength. As a consequence, radiation can be temporarily

*Electronic addresses: brabec@tuwien.ac.at ferenc.krausz@tuwien.ac.at

 0034-6861/2000/72(2)/545(47)/$24.40

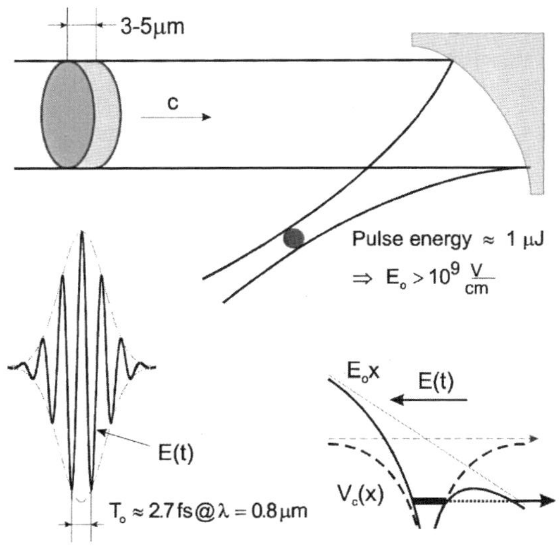

FIG. 1. Focusing of few-cycle ultrashort light pulses delivered in a collimated laser beam by a parabolic mirror, producing a "light bullet" with transverse and longitudinal dimensions of the order of a few microns. This extreme spatial and temporal confinement of light creates optical-field strengths sufficient to lower the Coulomb barrier of atoms and to tunnel-ionize an outer electron at moderate pulse energy levels.

confined to a few cubic micrometers at the focus of a parabolic mirror, forming a "light bullet," as shown in Fig. 1. Due to this extreme temporal and spatial confinement, moderate pulse energies of the order of one microjoule can result in peak intensities higher than 10^{15} W/cm^2. The amplitude of the electric field at these intensity levels approaches 10^9 V/cm. These field strengths exceed that of the static Coulomb field experienced by outer-shell electrons in atoms. As a consequence, the laser field is strong enough to suppress the binding Coulomb potential in atoms and triggers optical-field ionization (Keldysh, 1965).

This process temporarily enhances the nonlinear atomic polarization because the electron is far removed from the nucleus. Once the electron is set free, its response to the field becomes linear, terminating the nonlinear light-matter interaction (aside from a small residual nonlinearity originating from the ion polarizability). Significant new nonlinearities emerge only at substantially higher intensity levels by further ionization and/or when the free electrons become relativistic, both of which are beyond the scope of the present work. The nonlinear atomic polarization response culminates as the optical-field ionization rate peaks, shortly before the first weakly bound electron is detached. The intensity level at which this occurs strongly depends on the temporal evolution of the incident radiation.

Long laser pulses comprising many field oscillation cycles deplete the atomic ground state, i.e., detach an outer atomic electron with a probability approaching 100% at moderate intensity levels. In few-cycle laser fields high intensities can be "switched on" within a few optical periods lasting merely a few femtoseconds ($1\ \mathrm{fs}=10^{-15}\ \mathrm{s}$). Hence detachment of the first electron is completed at substantially higher field strengths and the optical-field ionization rate becomes comparable to the laser field oscillation frequency. As a result, the electrons gain unprecedented kinetic energies (up to and beyond the keV level) during the first field oscillation cycle following their detachment, and a substantial fraction of the atoms is ionized during one laser oscillation period T_o. The implications are numerous and far reaching. Generally speaking, the nonlinear response of the ionizing atomic medium is extended to unprecedented irradiation intensity levels and the induced instantaneous nonlinear current densities reach unprecedented values. Previously inaccessible regimes of nonlinear optics and high-field physics are being entered.

Somewhat more specifically, circularly polarized intense few-cycle fields can inject an ultraintense jet of high-energy electrons with a rise time comparable to T_o into a plasma. This may open the way to the development of electron-pumped inner-shell x-ray lasers in the keV range (Kim *et al.*, 1999). The high temporal gradient of the pulse front allows neutral atoms to survive to intensities of the order of 10^{16} W/cm^2. The oscillation spectrum of the dipole moment of these atoms may extend to frequencies some thousand times higher than that of the linearly polarized few-cycle driving laser field, giving rise to the emission of (spatially) coherent harmonic radiation up to photon energies of 0.5 keV (Schnürer *et al.*, 1998). The extremely high optical-field ionization peaks impose appreciable modulation on the driving few-cycle pulse within T_o during propagation. The resulting nonadiabatic self-modulation is predicted to permit phase-matched high-harmonic emission at kiloelectronvolt photon energies (Tempea *et al.*, 1999b). Few-cycle-driven harmonic x-ray radiation may be temporally confined to a single burst of subfemtosecond duration, which may result in unprecedented peak intensities in the x-ray regime and a never-before-achieved time resolution in physics.

Temporal confinement of irradiation of atomic systems to a few cycles allows atom-field interactions induced by visible or near-infrared radiation to take place for the first time in the *strong-field regime* (characterized by the onset of optical-field ionization) without any notable preionization in the multiphoton regime. This implies that the nonlinear atomic polarization P_{nl} at instant t can be analytically expressed as a function of the driving electric field $E(t')$ with $t' \leqslant t$, which greatly facilitates theoretical analysis of the interaction of intense radiation with an extended atomic medium (propagation effects). Even more importantly, the explicit dependence of P_{nl} on E indicates that the evolution of the atomic dipole moment (or the electron wave function) is governed directly by the electric field, $E(t) = E_a(t)\cos(\omega t+\varphi_0)$; hence it depends not only on the intensity envelope $E_a^2(t)$, but also on the *carrier phase* φ of a few-cycle laser pulse. Strong-field processes driven by few-cycle wave packets hold promise for gaining access to this parameter and thereby directly to the electric field in a light wave for the first time (Krausz *et al.*, 1998). Phase control will be one of the prerequisites for generating subfemtosecond extreme ultraviolet (xuv; 10–100 eV) /x-ray harmonic radiation in a reproducible manner. Generally, phase-controlled few-cycle light pulses will permit control of both the trajectories of freed electrons in strong-field interactions and bound-electron dynamics in atoms and molecules on a time scale of the light oscillation period.

This paper is devoted to the basic theoretical and experimental concepts underlying the generation of intense few-cycle laser pulses and high-field femtosecond light-atom interactions. Sections II and III review the underlying physics and evolution of ultrafast laser optics culminating in few-cycle pulse generation. Section IV addresses basic aspects of the nonlinear response of matter to intense radiation relevant to the generation of ultrashort light pulses (the perturbative regime) and their application to triggering optical-field ionization and concomitant phenomena (the strong-field regime). Section V introduces basic propagation equations for the above regimes. Section VI analyzes optical field ionization over a wide parameter range including the tunneling regime as well as the above-barrier regime. Section VII is devoted to few-cycle-driven high-order harmonic radiation, which can lead to the emission of single-attosecond xuv and x-ray pulses. Contrasting these phenomena with related processes driven by multicycle pulses beautifully shows the advantages of few-cycle drivers in the strong-field regime. In Sec. VIII it will be shown that the sensitivity of strong-field processes to the carrier phase of few-cycle light wave packets may open up access to the *absolute* phase of light for the first time. Finally, in Sec. IX we present some intriguing future prospects. In this paper we shall use units of the mksa system except for some cases, in which atomic units significantly simplify the treatment. Conversion between the two systems is given in Appendix A.

II. EVOLUTION OF ULTRASHORT-LIGHT-PULSE GENERATION

A. Basic concepts and early implementations

The development of ultrafast optics was triggered by the invention of *laser mode locking*, one of the most striking interference phenomena in nature (DiDomenico, 1964; Hargrove *et al.*, 1964; Yariv, 1965). Simultaneous oscillation of a vast number of highly coherent, phase-locked longitudinal modes in a laser yields a resultant field equal to zero most of the time except for very short intervals. The entire energy of the radiation field is concentrated within these short periods as a result of a short-lived constructive interference between the oscillating waves. Because the frequency spacing between adjacent phase-locked longitudinal modes is equal to $1/T_r$, where T_r is the resonator round-trip time, this temporary field enhancement is repeated periodically with a period T_r at a fixed position on the resonator axis. As a consequence, laser mode locking leads to the formation of a short light pulse circulating in the resonator. Each time the pulse hits a partially reflecting mirror, a small portion of its energy is coupled out of the oscillator, resulting in a train of ultrashort pulses at the output of the mode-locked laser, as illustrated in Fig. 2. Straightforward algebra relates the pulse duration and peak power to the number of phase-locked modes as summarized in Fig. 2.

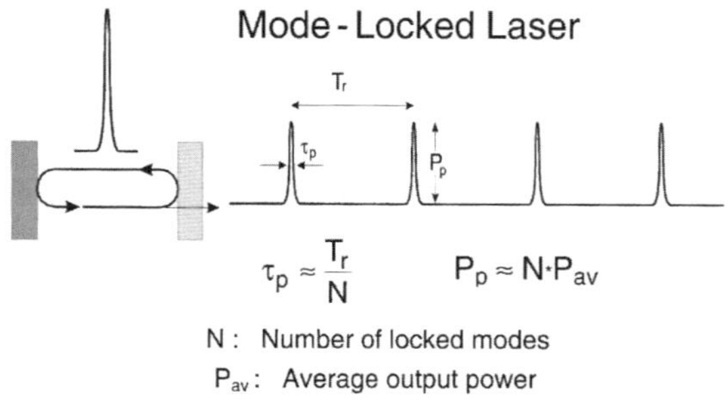

FIG. 2. Radiation power as a function of time at the output of a stationary mode-locked laser.

The large number of modes that can be locked using state-of-the-art mode-locking techniques give rise to light wave packets with unprecedented characteristics. In the rest of this section we shall survey the landmarks in the evolution of ultrashort-pulse laser physics and technology from the invention of laser mode locking up to the present day. When doing so we shall be able to cite only a tiny fraction of the relevant literature because of space limitations. Nevertheless, the referenced publications should provide a useful guide to the reader for finding further readings. Our apologies go to the numerous colleagues whose important contributions have not been credited. A comprehensive survey of the state of the art of primary (mode-locked laser) as well as secondary (frequency-converted) sources of ultrashort pulses is offered by conference proceedings, special issues of technical journals, and books.[1] The physics of the propagation, manipulation, and interaction of ultrashort laser pulses with matter (at low and moderate intensities) is thoroughly presented in two recent textbooks (Akhmanov *et al.*, 1992; Diels and Rudolph, 1996), whereas the physics and technology of the (most important) latest generation of ultrafast sources, namely, femtosecond solid-state lasers, have been reviewed by Krausz, Fermann *et al.* (1992), Keller (1994), Brabec *et al.* (1995), and French (1996).

The first generation of mode-locked lasers producing pulses of durations less than 100 ps used solid-state laser materials such as ruby, Nd:glass, or Nd:YAG as gain media. Mode locking was implemented either by active loss or frequency modulation driven by an external elec-

[1]These include Heritage and Nuss, 1992; Martin *et al.*, 1993; Barbara *et al.*, 1994; Krausz and Wintner, 1994; Barbara *et al.*, 1996; Zhang *et al.*, 1996; Keller 1997; Barty, White *et al.*, 1998; Elsässer *et al.*, 1998.

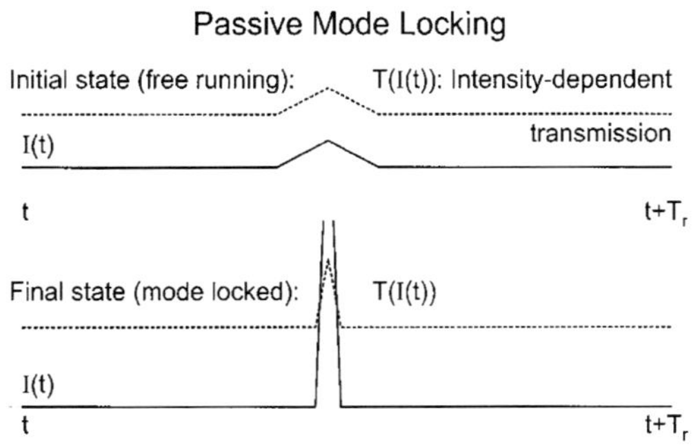

FIG. 3. Reliance of passive mode locking on a nonlinear element that exhibits an increased transmittivity for increased laser intensity. With such an element introduced in the oscillator, an initial amplitude fluctuation arising in the free-running laser (due to some mode beating, for instance) can grow upon repeated round trips in the cavity at the expense of the lower-intensity background. The resulting energy transfer continues until all the intracavity radiation energy is confined within a single pulse originating from the initial fluctuation. The steady-state pulse parameters are dictated by a balance between pulse-shortening and pulse-broadening effects. Saturable absorbers were the first (and for a long time the only) nonlinear optical elements employed for passive mode locking.

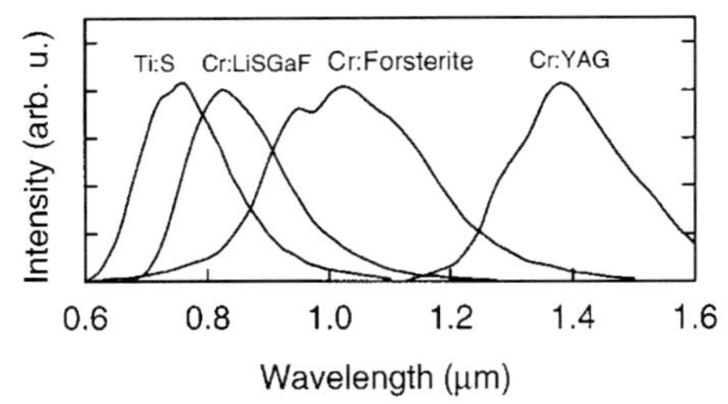

FIG. 4. Fluorescence emission spectra of some transition-metal-doped broadband solid-state laser crystals.

tronic oscillator (Deutsch, 1965; DiDomenico *et al.* 1966; Osterink and Foster, 1968; Kuizenga and Siegman, 1970a, 1970b), or by passive loss modulation with a fast-response saturable absorber (Mocker and Collins, 1965; DeMaria *et al.*, 1966). These techniques were termed active and passive mode locking, respectively. The principle of operation of the latter, which provides much more efficient pulse shaping and hence shorter pulses, is illustrated in Fig. 3. Nonlinear autocorrelation techniques needed to be developed for the temporal characterization of these pulses (Maier *et al.*, 1966; Armstrong, 1967; Weber, 1967). The passively mode-locked, flashlamp-pumped Nd:glass laser became the major "workhorse" for investigations in the field of nonlinear optics and opened up the entirely new field of time-resolved spectroscopy. The interested reader is referred to several excellent reviews devoted to these first-generation ultrafast sources and their characterization (DeMaria *et al.*, 1969; Bradley and New, 1974; Greenhow and Schmidt, 1974; Smith *et al.*, 1974).

The picosecond response time of organic saturable absorbers employed for passive mode locking has set a limit to the pulse duration. Continuous-wave operation (Peterson *et al.*, 1970) of an organic dye laser (Schäfer *et al.*, 1966; Sorokin and Lankard, 1966) and its mode locking by a saturable absorber (Ippen *et al.*, 1972) in a specifically designed cavity (Kogelnik *et al.*, 1972) triggered the development of the second generation of mode-locked lasers. In contrast with solid-state lasers, the response time of the absorber no longer constituted a limitation to the achievable pulse duration, owing to the active role of gain saturation in pulse formation (New, 1972, 1974; Haus, 1975a). As a result, optical pulses shorter than 1 ps could be produced for the first time (Ruddock and Bradley, 1976) and subsequent improvements in the cavity design allowed the breaking of the 100-fs barrier by utilizing a new concept, colliding-pulse mode locking (Fork *et al.*, 1981, 1983). Intracavity dispersion control by means of low-loss Brewster-angled prism pairs (Fork *et al.*, 1984; Martinez *et al.*, 1984) was the next major breakthrough, which together with improved insight into the relevant physical processes and limitations (Kühlke *et al.*, 1983; Stix and Ippen, 1983; De Silvestri *et al.*, 1984; Martinez, Fork, and Gordon, 1984; Diels *et al.*, 1985; Martinez *et al.*, 1985) allowed reproducible generation of sub-100-fs pulses and the demonstration of pulses as short as 27 fs directly from a laser oscillator (Valdmanis *et al.*, 1985; Valdmanis and Fork, 1986). This progress was achieved by using the organic dye Rhodamine 6G (Rh6G) as a gain medium emitting at around 620 nm. The dispersion-controlled colliding-pulse mode-locked Rh6G laser was the major workhorse for femtosecond spectroscopy until the late 1980s. Nevertheless, a number of other cw dye lasers have also been successfully mode locked to produce femtosecond pulses in the visible and near-infrared spectral range (French *et al.*, 1987, 1989; French and Taylor, 1988). For comprehensive reviews of ultrafast dye lasers see French (1985) and Shank (1988).

B. Continuous-wave passively mode-locked solid-state lasers

In the 1980s, continued work on the development of solid-state laser materials gave rise to the emergence of a number of new laser media. Various host crystals (YAG, sapphire, forsterite, LiSAF, etc.) doped with transition-metal (titanium, chromium) ions now provide laser transitions with enormous bandwidths on the order of 100 THz in the near-infrared wavelength range, as shown in Fig. 4 (Moulton, 1982, 1986; Petricevic *et al.*, 1989; Borodin *et al.*, 1992; Smith *et al.*, 1992). The development of novel techniques and devices suitable for passive mode locking of lasers with long gain relaxation and their application in these broadband solid-state systems led to the emergence of the third generation of ultrafast laser sources. Continuous-wave passive mode locking of broadband solid-state laser oscillators was accomplished by exploiting resonant (Keller, Knox, and 't Hooft 1992) and nonresonant (Krausz, Fermann, *et al.*, 1992) optical nonlinearities. The most successful devices for implementing the former technique have been semiconductor saturable absorber mirrors, referred to as SESAMs, and saturable Bragg reflectors (Keller *et al.*, 1990; Feldmann *et al.*, 1991; Keller, Miller, *et al.*, 1992; Rizvi *et al.*, 1993; Tsuda *et al.*, 1995; for reviews see Keller, Knox, and 't Hooft, 1992; Keller *et al.*, 1996; Tsuda *et al.*, 1996; Keller, 1999). The simultaneous exploitation of a solitonlike interplay between negative intracavity group delay dispersion (GDD) and self-phase modulation (SPM) induced by the nonresonant Kerr effect allowed the generation of optical pulses significantly shorter than the pico- or subpicosecond absorber recovery time (Kärtner and Keller, 1995; Kärtner *et al.*, 1996).

Because of the inability of long-relaxation-time solid-state gain media to participate in ultrashort pulse formation, fast-response "artificial" saturable absorbers based on nonresonant optical nonlinearities were devised by several researchers. The first attempt dates back to the early 1970s (Dahlström, 1972) and was followed by a number of proposals based on third-order $\chi^{(3)}$ type (Sala *et al.*, 1977; Ouellette and Piché, 1986) and second-order $\chi^{(2)}$ type (Stankov and Jethwa, 1988; Barr *et al.*, 1991) nonlinearities [for a definition of $\chi^{(2)}$ and $\chi^{(3)}$, see Eq. (19)]. The first breakthrough occurred when reseachers recognized that coupled cavities containing a Kerr nonlinearity [see Eq. (30)] introduced by a single-mode fused silica fiber that can be used for mode locking. This concept was first employed in the soliton laser (Mollenauer and Stolen, 1984; Mitschke and Mollenauer, 1986, 1987) and subsequently was shown to introduce a fast saturable-absorber effect over a wide wavelength range irrespective of fiber dispersion (Blow and Wood, 1988; Blow and Nelson, 1988; Kean *et al.*, 1989; Mark *et al.*, 1989). The technique was called coupled-cavity or additive-pulse mode locking (Ippen *et al.*, 1989; Morin and Piché, 1990; Haus *et al.*, 1991) because coherent superposition of the intracavity pulse with its self-phase-modulated replica introduced the saturable-absorber effect. Additive-pulse mode locking allowed cw self-starting passive mode locking (Ippen *et al.*, 1990; Krausz *et al.*, 1991) in a wide range of solid-state lasers. Implementing additive-pulse mode locking in all-fiber lasers (Duling 1991; Hofer *et al.*, 1991 1992; Richardson *et al.*, 1991; Tamura *et al.*, 1993; Fermann *et al.*, 1994) led to the development of compact turnkey femtosecond sources; for recent reviews see Duling and Dennis (1995) and Fermann (1995).

C. Kerr-lens and solitary mode locking: Routine generation of femtosecond pulses

The discovery of self-mode-locking (Spence *et al.*, 1991) in a titanium-doped sapphire (Ti:S) laser (Moulton, 1982, 1986) revolutionized ultrafast laser technology. Subsequent experimental (Keller *et al.*, 1991; Spinelli *et al.*, 1991) and theoretical (Piché, 1991; Salin *et al.*, 1991; Brabec, Spielmann, Curley *et al.*, 1992; Haus *et al.*, 1992) work revealed that an intracavity aperture translates self-focusing (Zakharov and Shabat, 1972; Marburger, 1975) introduced by the Kerr nonlinearity of the laser crystal into an ultrafast saturable-absorber-like

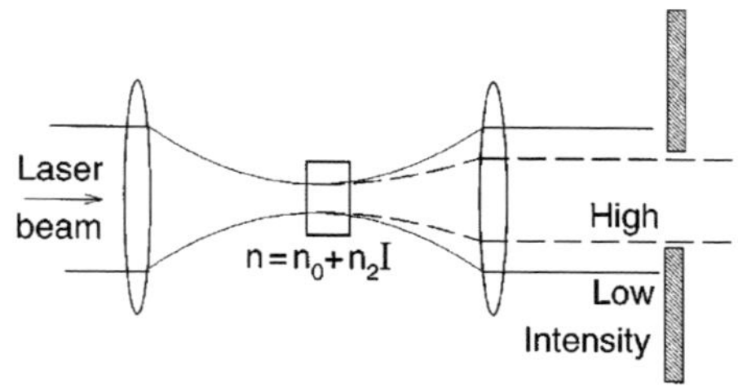

FIG. 5. Self-focusing-induced changes in the transmittivity of an aperture: the ultrafast-response saturable-absorber effect results in Kerr-lens mode locking.

self-amplitude-modulation (SAM), and the technique has been termed Kerr-lens mode locking (KLM). The operational principle of KLM is illustrated schematically in Fig. 5. The optical Kerr effect in the laser host crystal results in a fast-response intensity-induced change,

$$\Delta n(\mathbf{r},t)=n_2 I(\mathbf{r},t), \tag{1}$$

of the refractive index. Here n_2 [$\mathrm{cm^2/W}$] is the nonlinear index of refraction and I [$\mathrm{W/cm^2}$] is the cycle-averaged laser intensity. This effect transforms the radial intensity profile of a laser beam in a lensing effect, which tends to more strongly focus more intense temporal "slices" of the laser beam. An aperture of suitable size placed at a suitable position in the cavity can thus transmit a larger fraction of the laser beam at instants of higher intensity. The consequence is reduced loss for increased intensity, which constitutes a fast saturable-absorber effect (similar to additive-pulse mode locking). This ultrafast SAM effect can initiate and sustain the formation of an ultrashort pulse in Ti:S and other solid-state lasers. Simultaneously, $\Delta n(\mathbf{r},t)$ directly modulates the phase, which is termed self-phase-modulation (SPM).

The effect of SAM and SPM on the laser pulse circulating in the resonator can be simply described by the change they cause in the complex amplitude envelope $\tilde{E}_a(\tau)$, where the time τ is measured in a frame of reference retarded such that the pulse peaks at $\tau=0$ at any time [for definition see Eq. (14)] upon each round trip in the cavity:

$$\Delta\tilde{E}_a(\tau)=k_{SAM}p(\tau)\tilde{E}_a(\tau) \tag{2}$$

and

$$\Delta\tilde{E}_a(\tau)=ik_{SPM}p(\tau)\tilde{E}_a(\tau), \tag{3}$$

respectively. Here $p(\tau)\propto|\tilde{E}_a(\tau)|^2$ is the cycle-averaged time-dependent radiation power carried by the laser beam and k_{SAM},k_{SPM} [$\mathrm{W^{-1}}$] stand for the SAM and SPM coefficients, respectively. In KLM lasers, calculation of the intensity-dependent beam radius in the frame of the quadratic approximation[2] allows us to determine the coefficient k_{SAM} analytically. A more complete list of related literature is given in a recent paper of Kalosha

[2]See, for example, the calculations of (Piché, 1991; Salin *et al.*, 1991; Brabec, Spielmann, Curley *et al.*, 1992; Georgiev *et al.*, 1992; Haus *et al.*, 1992; Chilla and Martinez, 1993; Cerullo *et al.*, 1994; Herrmann, 1994; Agnesi, 1995; Lin *et al.*, 1995).

et al. (1998), which presents a comprehensive numerical analysis of Kerr-lens mode locking. For the SPM coefficient, straightforward algebra leads to

$$k_{SPM}=\gamma_{SPM}L, \quad \gamma_{SPM}=\frac{2n_2}{\lambda_0 w_0^2}. \tag{4}$$

Here λ_0 is the carrier wavelength of the laser pulse, L is the length of the Kerr medium, which is assumed to be shorter than the confocal parameter $2z_0=2\pi w_0^2/\lambda_0$, and w_0 is the $1/e^2$ beam radius.

In a typical KLM Ti:S laser, the SPM coefficient is on the order of $k_{SPM}\geqslant 10^{-6}\,\mathrm{W}^{-1}$, whereas SAM is weaker by more than an order of magnitude: $k_{SAM}\leqslant 10^{-7}\,\mathrm{W}^{-1}$ (Krausz, Fermann *et al.*, 1992). Typical intracavity pulse peak powers of the order of 10^5–10^6 W in the subpicosecond and femtosecond regimes hence introduce an amplitude modulation of merely a few percent by means of the KLM artificial saturable absorber. This comparatively weak modulation is incapable of overcoming the pulse broadening caused by dispersion of the laser medium in the femtosecond regime, which stops pulse shortening far from the limit set by the gain bandwidth in the Ti:S laser.

Accessing the regime well below 100 fs in the Ti:S laser becomes feasible when we introduce negative group delay dispersion (GDD) and exploit a highly efficient pulse-shortening mechanism resulting from an interplay between Kerr-induced SPM (e.g., in the laser crystal) and a net negative cavity GDD introduced, for example, by a pair of prisms (Fork *et al.*, 1984). This interplay is referred to as solitary mode locking (Brabec *et al.*, 1991) and will be discussed as the key technique for the generation of few-cycle light pulses in a broader context in the next section.

Because $k_{SPM}/k_{SAM}\gg 1$, solitary pulse shaping dominates pulse shortening in the presence of a net negative intracavity GDD and determines the steady-state pulse duration in the KLM Ti:S laser. In the frame of the weak pulse-shaping approximation, $|\Delta\tilde{E}_a(t)/\tilde{E}_a(t)|\ll 1$, which defines the range of validity of the powerful analytic master-equation approach of Haus *et al.* (1991, 1992), and in the limit $k_{SPM}/k_{SAM}\gg 1$, the master equation yields a steady-state pulse of the form $\tilde{E}_a(\tau)=E_0\mathrm{sech}(\tau/\tau_0)$ with

$$\tau_0=\frac{2|D|}{k_{SPM}W_p} \tag{5}$$

and a pulse duration of $\tau_p=1.76\tau_0$ [full width at half maximum of $p(\tau)\propto|\tilde{E}_a(\tau)|^2$]. Here W_p is the intracavity pulse energy and $D(<0)$ $[fs^2]$ denotes the net intracavity GDD [for a definition, see Eq. (8)]. Further, Eq. (5) is valid under the assumption that D is independent of frequency, i.e., that high-order dispersion is negligible. This latter approximation tends to fail increasingly for decreasing pulse durations, which rely on smaller and smaller values of $|D|$ according to Eq. (5), and this eventually sets a limit to pulse shortening. The lowest-order perturbation to a constant GDD is a small linear dependence on frequency, $D(\omega)=D+D_3(\omega-\omega_0)$, where $D=D(\omega_0)$, ω_0 is the center frequency of the mode-locked laser, and D_3 $[fs^3]$ is referred to as third-order dispersion, because it is equal to the third derivative of the phase with respect to frequency, as will be revealed by Eq. (8). Computer simulations yielded the explanation that mode locking becomes increasingly perturbed and eventually unstable as $\tau_p\rightarrow|D_3/D|$, setting a limit to the minimum pulse width achievable with solitary mode locking in the presence of third-order dispersion (Brabec, Spielmann, Curley, *et al.*, 1992). Overcoming this limitation would result in a mode-locked spectrum extending into a wavelength regime, where the cavity GDD $D(\omega)$ becomes positive, preventing stable solitary pulse formation (Spielmann *et al.*, 1994).

In the absence of high-order dispersion and limitations due to the finite high-reflectivity bandwidth of mirrors, the separate action of negative GDD and SPM is predicted to limit pulse shortening well before attainment of the gain-bandwidth limit of Ti:S, which would permit pulse durations approaching 1 fs at low cavity losses. Discrete solitary pulse shaping is a consequence of the Kerr-induced phase shift $k_{SPM}p(t)$ approaching (or even exceeding) unity at the pulse peak for decreasing $|D|$, leading to pulse shortening below 10 fs. As a result, the weak pulse-shaping approximation breaks down and a correction term proportional to $k_{SPM}W_p$, which is negligible for large values of $|D|$, appears on the right-hand side of Eq. (5) (Brabec *et al.*, 1991; Krausz *et al.*, 1992). This correction term is dependent on the position in the resonator, predicting a periodic pulse evolution in the (sub-)10-fs regime.

These findings resulted in a rapid evolution of prism-dispersion-controlled KLM Ti:S lasers, driven by the search for prism materials characterized by the smallest ratio of third-order dispersion to GDD. Fused silica was found to be the best choice and, as a result, fused-silica-prism-controlled KLM Ti:S lasers were the first laser oscillators producing pulses in the 10-fs regime (Asaki *et al.*, 1993; Curley *et al.*, 1993; Proctor and Wise, 1993; Spielmann *et al.*, 1994; Zhou *et al.*, 1994). Just as in femtosecond dye lasers (De Silvestri *et al.*, 1984; Salin *et al.*, 1990), high-order dispersion of the prisms limited the pulse duration (Lemoff and Barty, 1993a; Spielmann *et al.*, 1994). These technological limitations were to be overcome with the discovery of chirped multilayer dielectric mirrors (Szipöcs *et al.*, 1994, 1995), which will be treated below. With these devices, resonators providing high reflectivity and approximately constant negative GDD over the entire gain band of Ti:sapphire ($\approx$600–1000 nm) can be constructed.

Compact mirror-dispersion-controlled (MDC) KLM Ti:S laser oscillators now routinely generate high-quality sub-10-fs pulses (Stingl *et al.*, 1995; Kasper and Witte, 1996) with peak powers exceeding 1 MW (Xu *et al.*, 1997, 1998; Beddard *et al.*, 1999). Recently, hybrid dispersion control using prisms and chirped mirrors yielded pulses below 6 fs from Ti:S lasers (Gallmann *et al.*, 1999; Morgner *et al.*, 1999; Sutter *et al.*, 1999). Combining the concept of mirror dispersion control with extension of the laser cavity by a telescope (Cho *et al.*, 1999; Libertun

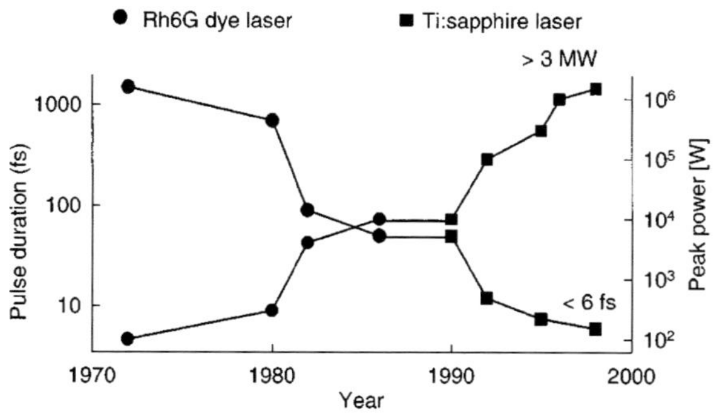

FIG. 6. Evolution of the passively mode-locked Rh6G dye laser and the passively mode-locked Ti:sapphire laser in terms of pulse duration and peak power from the mid 1960s to date. A comparison of the shortest pulse durations achieved with Kerr-lens mode locked, mirror-dispersion-controlled Ti:sapphire lasers with the round-trip time $T_r = 10–15$ ns reveals that more than one million phase-locked longitudinal resonator modes are oscillating in these systems (see Fig. 2).

et al., 1999) resulted in the generation of stable sub-10-fs pulses with peak powers exceeding 3 MW for the first time directly from a laser oscillator (Poppe, Lenzner, *et al.*, 1999). These technological advances led to a rapid evolution of femtosecond Ti:S lasers since the first demonstration of passive mode locking in a Ti:sapphire laser (French, Williams, *et al.*, 1989) and gave rise to an improvement of the pulse duration and peak power by more than two and four orders of magnitude, respectively, within the last 30 years, as depicted in Fig. 6. Due to the absence of intracavity components other than the gain medium in KLM/MDC Ti:S lasers, the noise performance of femtosecond sources could also be improved substantially (Poppe *et al.*, 1998). The KLM Ti:S laser is now commercially available with both prism and mirror dispersion control and has become the major workhorse for ultrafast time-resolved spectroscopy and nonlinear optics. More recently, it was also demonstrated as a promising tool for metrology applications in high-resolution spectroscopy (Udem *et al.*, 1999a; 1999b).

D. Chirped pulse amplification: Boosting the peak power to unprecedented levels

Many intriguing applications of ultrashort pulses call for intensities requiring peak powers far exceeding the power levels that can be directly obtained from cw mode-locked oscillators. To this end, the pulses delivered by mode-locked lasers need to be amplified. Femtosecond pulse amplification was first demonstrated by using dye cells or jets pumped by Q-switched Nd:YAG and copper-vapor lasers (Fork *et al.*, 1982; Sizer *et al.*, 1983; Koroshilov *et al.*, 1984; Knox *et al.*, 1985); for a review see Knox, 1988). The maximum pulse energy that can be practically achieved in dye amplifiers is restricted to less than ≈1 mJ. This limitation relates to the fact that the maximum possible energy that can be extracted per unit beam cross-sectional area without significant temporal distortion of the amplified pulse is given by the saturation fluence

$$F_{sat} = \hbar\omega_l/\sigma_e, \tag{6}$$

where ω_l and σ_e are the center frequency and the peak stimulated-emission cross section of the laser transition, respectively. The above limit is set by the low saturation fluence implied by the high emission cross section σ_e of dye lasers and the maximum dye volume that can be uniformly pumped in practice. The fundamental relationship

$$\tau_f \sigma_e \Delta\nu_l = k\frac{\lambda_l^2}{n^2} \tag{7}$$

connecting the characteristics of a laser transition ($\Delta\nu_l$ is the gain bandwidth, n is the refractive index of the gain medium, and k is a numerical factor of the order of unity that depends on the emission line shape) reveals that the high σ_e also implies a short fluorescence lifetime τ_f (Koechner, 1996). The related short energy-storage time and strong amplified spontaneous emission constitute further drawbacks of dye amplifiers. Excimer amplifiers have similar characteristics but offer significantly larger uniformly inverted apertures, allowing amplification up to hundreds of millijoules in the ultraviolet spectral range at pulse durations of the order of 100 fs (Glownia *et al.*, 1987; Szatmári *et al.*, 1987; Watanabe *et al.*, 1988; Taylor *et al.*, 1990; Mossavi *et al.*, 1993; for a review see Szatmári, 1994). Owing to their high peak power, of the order of 1 TW, excellent beam quality, and short wavelength, these pulses can be focused to a spot size below 1 μm, resulting in peak intensities in excess of 10^{18} W/cm^2 (Szatmári, 1994).

Novel solid-state gain media developed in the 1980s held promise for producing even higher peak powers due to their far higher energy fluences and broader bandwidths (see Fig. 4). However, amplification of femtosecond pulses in these media gives rise to catastrophic effects due to an accumulated intensity-dependent phase shift induced by the optical Kerr effect long before the saturation fluence can be reached (see, for example, Koechner, 1996). This limitation has been overcome by the ingenious concept of chirped pulse amplification (Strickland and Mourou, 1985; Maine *et al.*, 1988), the principle of which is illustrated in Fig. 7. By the early 1990s, chirped pulse amplification implemented with diffraction-grating-based pulse compressors (Treacy *et al.*, 1969) and stretchers (Martinez, 1987) had made it possible to generate pulses in the 100-fs range with peak powers of several terawatts from laboratory-scale systems based on Ti:S and Cr:LiSAF (Sullivan *et al.*, 1991; Beaud *et al.*, 1993; Ditmire and Perry, 1993), and the use of large-aperture Nd:glass power amplifiers resulted in ≈1-ps pulses having peak powers of tens of TW (Patterson and Perry, 1991; Sauteret *et al.*, 1991; Yamakawa *et al.*, 1991; Rouyer *et al.*, 1993; Blanchot *et al.*, 1995).

Advances in Ti:S seed oscillators (Asaki *et al.*, 1993; Curley *et al.*, 1993; Stringl *et al.*, 1995) and stretcher/compressor design (Lemoff and Barty, 1993b; Zhou

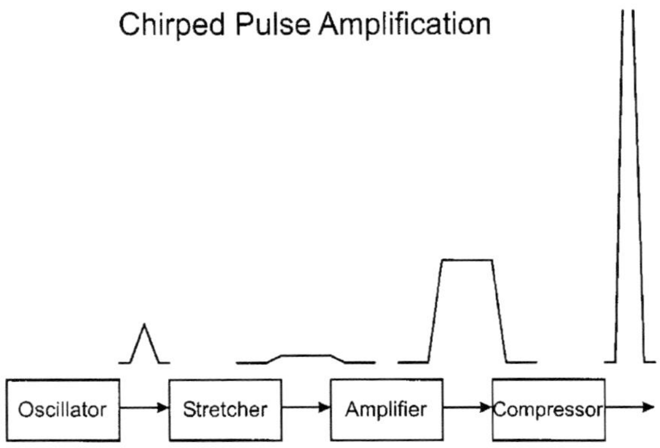

FIG. 7. Principle of chirped pulse amplification (Strickland and Mourou, 1985). A low-energy ultrashort seed pulse is temporally stretched before amplification and recompressed after amplification to avoid high peak powers in the amplifier system.

et al., 1995; Cheriaux *et al.*, 1996) led to the development of Ti:S chirped pulse amplification systems generating multiterawatt pulses in the 20-fs regime at 10 Hz (Barty *et al.*, 1994; Zhou *et al.*, 1995) and triggered a subsequent rapid evolution of these systems (Chambaret *et al.*, 1996; Yamakawa *et al.*, 1998a) to a peak power of 100 TW (Yamakawa *et al.*, 1998b) and (expected) peak intensities at focus in excess of 10^{20} W/cm^2. The evolution of laboratory chirped pulse amplification systems is depicted in Fig. 8 (for recent reviews see Perry and Mourou, 1994; Morou, 1997). This and similar laboratory-scale chirped pulse amplification architectures based on Ti:sapphire or Yb:glass (Nees *et al.*, 1998) are expected to be scalable to petawatt peak power levels, which could be demonstrated with a large laser system at Lawrence Livermore National Laboratory, Livermore (Perry *et al.*, 1999). Peak powers in excess of 10^{21} W/cm^2 can now be attained from both large-scale (Perry *et al.*, 1999 and lab-scale lasers (Patterson *et al.*, 1999). The excellent thermal conductivity and favorable thermo-optic properties of Ti:sapphire and the availability of kHz-rate Q-switched solid-state pump sources opened the way to implemeting chirped pulse amplification at kHz repetition rates. Related research recently culminated in the demonstration of kHz sources of sub-20-fs pulses with peak powers of several hundred gigawatts (Backus *et al.*, 1997; Nibbering *et al.*, 1997; Nabekawa *et al.*, 1998; for a review see Backus *et al.*, 1998) and suggest the feasibility of kHz-repetition-rate terawatt sources (Durfee *et al.*, 1998) in the near future.

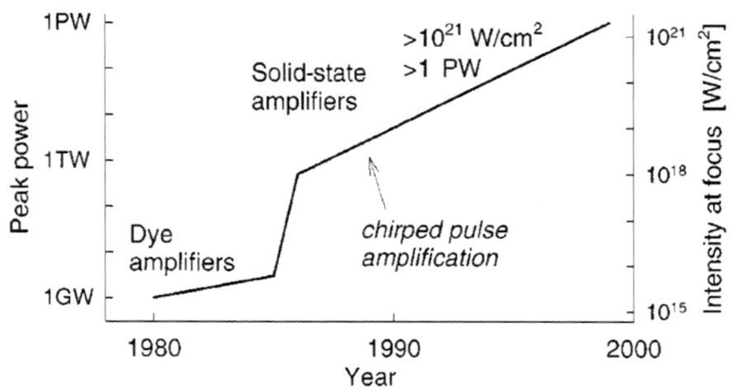

FIG. 8. Evolution of ultrashort-pulse amplification in terms of peak power and achievable peak intensity.

E. Chirped multilayer mirrors: Paving the way towards the single-cycle regime

In the preceding subsections we have addressed how optical nonlinearities can be efficiently exploited for the generation of femtosecond laser pulses in broadband solid-state oscillators and how they can be efficiently avoided when their energy is boosted in solid-state amplifiers. Apart from optical nonlinearities, dispersion becomes increasingly important as ever shorter pulses are sought. In fact, generating bandwidth-limited pulses of decreasing duration calls for controlling the frequency-dependent group delay over increasing bandwidth with increasing precision. Hence precision broadband dispersion control is a prerequisite for approaching the few-cycle regime in ultrashort pulse generation.

Dispersion can be quantified by expanding the group delay, which is equal to the first derivative of the phase retardation $\phi(\omega)$ of the optical system with respect to frequency, about the center of the pulse spectrum ω_0 in the form

$$T_g(\omega)=\phi'(\omega)=\phi'(\omega_0)+\phi''(\omega_0)(\omega-\omega_0) +\frac{1}{2}\phi'''(\omega_0)(\omega-\omega_0)^2+\frac{1}{6}\phi''''(\omega-\omega_0)^3 +\cdots . \quad (8)$$

Here $\phi'(\omega_0)$ gives the time it takes for the peak of the pulse to traverse the dispersive medium. The higher-order terms in the expansion describe the frequency dependence of the group delay and hence are responsible for dispersive effects. $\phi''(\omega_0)$ is the lowest-order (linear) group delay dispersion or second-order phase dispersion, most frequently referred to as group delay dispersion (GDD) in the literature and denoted by D, a convention that we also adopt in this work; $\phi'''(\omega_0)\equiv D_3$ and $\phi''''(\omega_0)\equiv D_4$ are termed third-order and fourth-order dispersion, respectively. Critical values of the dispersion coefficients, above which dispersion causes a substantial change of the pulse, obey the simple scaling $\phi^{(n)}=\tau_p^n$. For instance, a GDD of $\phi''=\tau_p^2$ results in a pulse broadening by more than a factor of 2. This scaling reveals a dramatic increase in susceptibility to dispersion-induced broadening and distortion for decreasing pulse durations.

Most laser and optical materials exhibit some *positive* GDD, implying a group delay that increases with frequency and hence imposing a positive frequency sweep or chirp on a pulse passing through the medium. Pulses tens of fs in duration tend to broaden significantly even upon passage through transparent optical media (such as quartz or sapphire) of merely a few mm in length. This broadening even gets accelerated at high intensities, leading to self-phase-modulation due to the optical Kerr

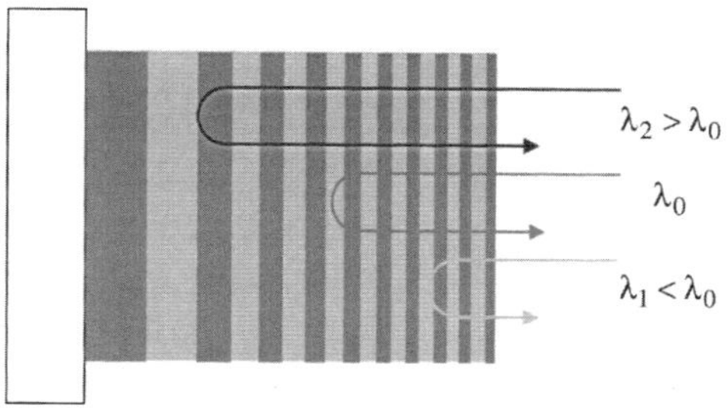

FIG. 9. Schematic illustration of the origin of dispersion in a chirped multilayer mirror. Quasimonochromatic wave packets carried at different wavelengths penetrate to different depths before being reflected, as a consequence of a modulation of the multilayer period across the layer stack. In the illustrated example, the increasing multilayer period with increasing distance from the mirror surface implies that radiation with increasing wavelength has to penetrate deeper before being reflected as a result of constructive interference of the partial waves reflected from the interfaces between the low- and high-index layers. The result is a group delay that increases with increasing wavelength, i.e., decreasing frequency, giving rise to negative group delay dispersion.

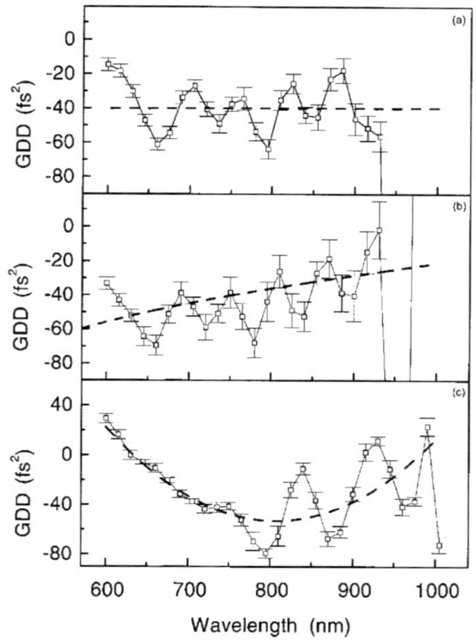

FIG. 10. Group delay dispersion (GDD) vs wavelength of ultrabroadband chirped mirrors aiming at introducing GDD as well as higher-order dispersion control in sub-10-fs Ti:sapphire-based systems: □, measured data (restricted to the range 600–1000 nm) obtained from white-light interferometry; dashed curves, nominal (averaged) or target GDD curves. The quasiperiodic fluctuation of the measured dispersion is inherent to chirped mirrors consisting of discrete layers and can be efficiently (by more than 50%) suppressed by complementary mirrors with opposite fluctuations. (a) Chirped mirror designed for nominally constant negative GDD of $-40\,\mathrm{fs}^2$. (b) Chirped mirror designed for compensating the dispersion introduced by fused silica up to (and including) fourth order over the wavelength range 580–930 nm; dashed line, dispersion introduced by fused silica (with reversed sign) over a 1-mm propagation length: $D=-36\,\mathrm{fs}^2$, $D_3=-27\,\mathrm{fs}^3$, and $D_4=-11\,\mathrm{fs}^4$ at $\lambda_0=800\,\mathrm{nm}$. (c) Chirped mirror designed for compensating high-order dispersion up to (and including) fourth-order dispersion over the wavelength range 560–1000 nm; dashed line, least-squares fit to the data yielding $D=-53\,\mathrm{fs}^2$, $D_3=9\,\mathrm{fs}^3$, and $D_4=400\,\mathrm{fs}^4$ at $\lambda_0=800\,\mathrm{nm}$. The high reflectivity (>99%) range of these mirrors extends over more than 200 THz, exceeding the bandwidth of standard quarterwave mirrors by more than a factor of two.

effect, which broadens the spectrum by imposing a positive temporal chirp. Clearly, restoring the duration of pulses passed through optical materials calls for negative GDD, just as does compressing pulses spectrally broadened by self-phase-modulation.

Treacy (1969) was the first to demonstrate negative GDD by passing light pulses through a pair of diffraction gratings, successfully achieving pulse compression with such a system. The Brewster-angled prism pairs of Fork *et al.* (1984) were the first low-loss sources of negative GDD, which have been extensively employed for dispersion control inside laser oscillators since their discovery. In both cases, negative GDD is accompanied by significant amounts of intrinsic high-order dispersion, which cannot be lowered or adjusted independently of the (useful) lowest-order dispersion (GDD), limiting the bandwidth over which adequate dispersion control can be provided. This drawback was overcome, to lowest order, by combining prism and grating pairs exhibiting third-order dispersion of opposite sign, thus allowing pulse compression to 6 fs in the mid 1980s (Fork *et al.*, 1987) and recently to less than 5 fs (Baltuska *et al.*, 1997a). Nevertheless, this approach cannot be used for few-cycle pulse generation either in oscillators (because of the high losses of gratings) or in external compressors at high power levels (because of unwanted nonlinearities in the prisms).

Recently, modulation of the multilayer period of dielectric laser mirrors was demonstrated to result in a wavelength-dependent penetration depth of the incident radiation (Szipöcs *et al.*, 1994). This implies a corresponding dependence of the group delay on frequency (see Fig. 9), which can be tailored to yield—within certain limits—required amounts of GDD as well as higher-order dispersion over almost the entire high-reflectivity band of the mirror, which can be substantially broader than that of a standard quarterwave stack. These devices have been referred to as chirped multilayer mirrors (Szipöcs *et al.*, 1994). Recent advances in the design and manufacturing of such structures have led to chirped mirrors exhibiting high reflectivity and approximately constant negative GDD over unprecedented bandwidths (Matuschek, 1998; Tempea, Krausz, *et al.*, 1998).

Figure 10 depicts the GDD versus wavelength for several chirped mirrors designed for second- and higher-order dispersion control in Ti:sapphire-based sub-10-fs systems. These results demonstrate the power of this technique for controlling dispersion over unprecedented bandwidths with unprecedented precision. The chirped mirrors' current capability of exhibiting high reflectance over a bandwidth >200 THz and controlling dispersion over a spectral range as broad as >150 THz has allowed the generation of sub-6-fs pulses directly from laser oscillators (Gallmann *et al.*, 1999; Morgner *et al.*, 1999; Sutter *et al.*, 1999) sub-5-fs pulses from optical parametric amplifiers (Shirakawa *et al.*, 1999), and pulse compression at high (subterawatt) power levels down to 4 fs (Cheng *et al.*, 1998). The latter pulses are carried at a wavelength of $\lambda_0 \approx 0.8\,\mu$m, implying a laser oscillation cycle of $T_o \approx 2.6$ fs. Hence chirped mirrors are able to provide adequate dispersion control for the generation of light pulses comprising merely one and a half field oscillation cycles within their intensity half maximum and hold promise for pushing the limits of ultrafast optics into the single-cycle regime in the near future.

III. GENERATION OF INTENSE LIGHT PULSES IN THE FEW-CYCLE REGIME

In this section we briefly review the key physical mechanisms and techniques that allow the generation of powerful light pulses with durations approaching the light oscillation period T_o and peak intensities penetrating far into the strong-field regime (see Fig. 24 below). After presenting the status of few-cycle optical pulse generation, we address a parameter that has not received attention until recently: the *absolute phase* of light wave packets becomes important in the interaction of intense few-cycle radiation with matter, and its control in ultrabroadband mode-locked laser oscillators will also benefit precision measurements of the frequency of light as well as atomic transitions.

A. Principles of optical pulse compression

The nonlinear refractive index n_2, as defined in Eq. (1), was recognized as a source of new frequency components during propagation of an intense short light pulse through a Kerr medium. The time-dependent nonlinearly induced phase shift $\Delta\varphi_{nl}(\tau) = (2\pi/\lambda_0)n_2 I(\tau)L$, where $I(\tau)$ is the intensity, τ the retarded time as defined in the previous section, and L the propagation length, manifests itself as a *self-phase-modulation* (SPM) and tends to broaden the spectrum of an initially bandwidth-limited pulse (Shimizu, 1967; Fisher *et al.*, 1969; Laubereau, 1969; Laubereau and von der Linde, 1970; Alfano and Shapiro, 1970). In high-power laser systems, this effect is undesirable because the dependence of I on the transverse space coordinates introduces phase aberrations leading to a degradation of the laser beam. However, under specific experimental conditions $\Delta\varphi_{nl}$ can be utilized for the generation of ultrashort pulses.

The feasibility of pulse compression arises from the sequential emergence of new redshifted and blueshifted spectral components at different positions of the pulse envelope. As a consequence, subsequent passage of the pulse through a delay line, introducing shorter group delay for the new spectral components riding on the trailing edge of the pulse, as compared to the delays suffered

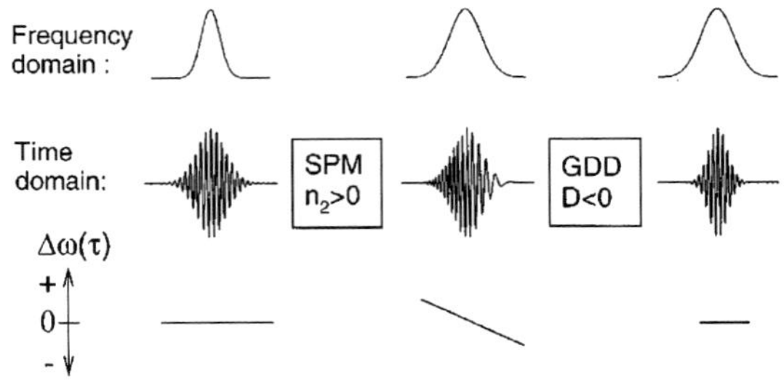

FIG. 11. Schematic representation of optical pulse compression. The graphs depict qualitatively the spectral intensity distribution (frequency domain), and evolution of the electric field $E(t)$ and change in instantaneous frequency with respect to the carrier frequency $\Delta\omega(\tau) = \omega_{\rm inst}(\tau) - \omega_0$ (time domain). The bandwidth-limited $[\Delta\omega(\tau)=0]$ input pulse propagates from the left to the right. The positive frequency sweep $\Delta\omega(\tau) = (d/d\tau)\Delta\varphi_{nl}(\tau)$ imposed on the intense light pulse by self-phase-modulation induced by the optical Kerr effect upon passage through a transparent optical medium can be removed by subsequently passing the pulse through a dispersive delay line with negative group delay dispersion, resulting in a temporal compression.

by the components emerging on the front edge, translates the pulse carrying $\Delta\varphi_{nl}$ into a temporally compressed pulse, as illustrated in Fig. 11. The frequency sweep (or chirp) $d\Delta\varphi_{nl}/d\tau$ is linear to good accuracy in the vicinity of the pulse center ($\tau=0$), where most of the energy is concentrated, particularly if $\Delta\varphi_{nl}$ emerges in the presence of GDD of the same sign as $\Delta\varphi_{nl}$ (Tomlinson *et al.*, 1984). As a consequence, optimum temporal compression calls for a group delay $T_g(\omega)$ exhibiting a near-linear dependence on frequency in the dispersive delay line. The nonlinear index n_2 is generally positive far from resonances, hence a negative group delay dispersion $(dT_g/d\omega)_{\omega_0}<0$ is required for pulse compression.

This interplay between Kerr-induced self-phase-modulation and negative GDD forms the basis of all pulse compression schemes that have been demonstrated to date. At low (nanojoule) pulse energy levels, compression can be most efficiently implemented by passing the pulse through a single-mode fiber and subsequently through a grating pair (Grischkowski and Balant, 1981; Johnson *et al.*, 1984), and somewhat later, through gratings and prisms for improved high-order dispersion control, resulting in nanojoule-energy 6-fs pulses at 620 nm (Fork *et al.*, 1987). This long-standing record has recently been improved by using a single-mode-fiber/chirped-mirror compressor seeded with 15-fs pulses generated by a cavity-dumped KLM Ti:sapphire laser (Ramaswamy *et al.*, 1993; Pshenichnikov *et al.*, 1994). This compact all-solid-state system is capable of generating nanojoule pulses down to 4.5 fs in duration at MHz repetition rates (Baltuska *et al.*, 1997a; 1997b; Pshenichnikov *et al.*, 1998), constituting a powerful tool for ultrafast spectroscopy with unprecedented temporal resolution.

Nanojoule-energy few-cycle pulses with durations between 5 and 6 fs have more recently been directly available from KLM Ti:S oscillators employing chirped multilayer resonator mirrors (Gallmann *et al.*, 1999; Morgner *et al.*, 1999; Sutter *et al.*, 1999). For restoring these pulse durations after amplification to millijoule energy levels, the implementation of self-phase-modulation in a gas-filled multimode hollow waveguide (Nisoli *et al.*, 1996) and subsequent temporal compression in chirped mirrors (Nisoli, De Silvestri, *et al.*, 1997; Nisoli, Stagira, *et al.*, 1997; Sartania *et al.*, 1997) have been proposed and successfully demonstrated. The formation of few-cycle pulses in Ti:S oscillators and hollow-fiber-based compressors is the consequence of the same physical mechanisms (SPM-GDD interplay) exploited in distinctly different parameter regimes. In the following subsections we provide a unified treatment of the formation of few-cycle pulses in solitary (bulk) laser oscillators and in multimode-waveguide/chirped-mirror-based compressors (at low and high energy levels, respectively) and present the state of the art of few-cycle pulse generation.

B. Self-phase-modulation in free space and guided-wave propagation

Self-phase-modulation (SPM) of ultrashort laser pulses and related spectral broadening can be most simply induced by focusing into a bulk nonlinear medium. If the length L of the medium is short compared to the confocal parameter of the beam, the wave front is approximately planar and the radial intensity distribution approximately constant in the nonlinear medium. Yet the radial intensity variation produces a spatially varying $\Delta\varphi_{nl}$, giving rise to self-focusing (for $n_2>0$) and small-scale instabilities (Campillo *et al.*, 1969). In addition, the extent of spectral broadening depends on the radial coordinate, frustrating a spatially uniform pulse compression. Creating a near-flat-top intensity profile in a thin nonlinear medium by beam truncation and suppressing small-scale instabilities by spatial filtering is one possible way of circumventing these problems (Rolland and Corkum, 1988). The penalty to be paid in this approach is very high losses.

Spatially uniform spectral broadening can be achieved much more efficiently by guiding the self-phase-modulated beam and distributing $\Delta\varphi_{nl}$ over an extended propagation length. To shed light on the crucial role of wave guiding and of a slow accumulation of $\Delta\varphi_{nl}$ we decompose the laser beam into transverse eigenmodes of the propagation medium and perform coupled-mode analysis to contrast nonlinear propagation in a multimode waveguide with that in free space. A unified theoretical treatment of these phenomena becomes feasible by replacing the transverse eigenmodes of the waveguide with the Hermite-Gaussian solutions of the paraxial wave equation (Haus, 1984) in the coupled-mode analysis of nonlinear wave propagation.

In most practical cases, the fundamental propagation mode is excited at the entrance of the nonlinear medium. The nonlinear index n_2 has two major implications in this scenario: (i) a spatially uniform phase shift $\Delta\varphi_{nl}(\tau)$ is imposed on the fundamental mode and (ii) a fraction of its energy is coupled to higher-order modes during propagation. Self-focusing and related beam deterioriation are consequences of the latter effect. It was recently shown that the coupled-mode equations can be solved analytically in the limit of small depletion of the fundamental mode for nonlinear light-pulse propagation in a multimode (hollow) waveguide (Tempea and Brabec, 1998b) as well as in free space (Milosevic *et al.*, 1999). In what follows we summarize the major results of these analyses and draw important conclusions for implementing few-cycle pulse generation.

First, we address free-space propagation. If the fundamental TEM_{00} mode is focused into a nonlinear medium, the nonlinear polarization response of the medium couples energy from the incident fundamental into higher-order TEM_{mn} modes. The upper diagram of Fig. 12 depicts qualitatively the fractional energy coupled from the TEM_{00} into the TEM_{01} mode during propagation in a bulk nonlinear medium of length L positioned between $z=-L/2$ and $z=L/2$, where $z=0$ is the position of the beam waist. A useful scale parameter for the nonlinear interaction is the nonlinear length, defined as

$$L_{nl}=\frac{1}{\gamma_{SPM}p_0}, \tag{9}$$

where γ_{SPM} is defined in Eq. (4) and p_0 stands for the peak power of the pulse. If $L_{nl}\gg z_0$, where $z_0=\pi w_0^2/\lambda_0$ denotes the Rayleigh range and w_0 is the $1/e^2$ radius at the beam waist, only a small fraction of the energy of the light pulse fed into the medium in the fundamental transverse mode is coupled to higher-order modes, and for a medium of length $L\gg z_0$ a perturbative approach allows us to solve the coupled-mode equations analytically.

From this solution we can obtain the fractional power $\Gamma(z)$, coupled from the fundamental TEM_{00} mode to the lowest-order higher transverse mode TEM_{01}, to which coupling due to n_2 is strongest, and the evolution of the peak nonlinear phase shift $\Delta\varphi_{nl}(z,r,\tau)$ imposed on the propagating beam. In the limit of $L_{nl}\gg z_0$, the fractional energy extracted temporarily from the TEM_{00} mode is small and the energy coupled to higher-order modes is fed back to the fundamental mode behind the beam waist in a lossless and dispersion-free medium. As a consequence, the output pulse is delivered in a Gaussian beam and carries a spatially uniform phase shift (i.e., $\Delta\varphi_{nl}$ is independent of the radial coordinate r). $\Gamma(z)$ (solid line) and $\Delta\varphi_{nl}(z,\tau=0)$ (dashed line) calculated under these conditions are plotted qualitatively in the upper diagram of Fig. 12. The energy flow between the TEM_{00} and TEM_{mn} modes decreases with increasing mode order and changes its sign an increasing number of times for increasing mode order between the fundamental and higher-order modes.

The maximum energy transferred to the mode TEM_{01} at the beam waist and the peak nonlinear phase shift carried by the pulse exiting the medium are given by (Tempea and Brabec, 1998b)

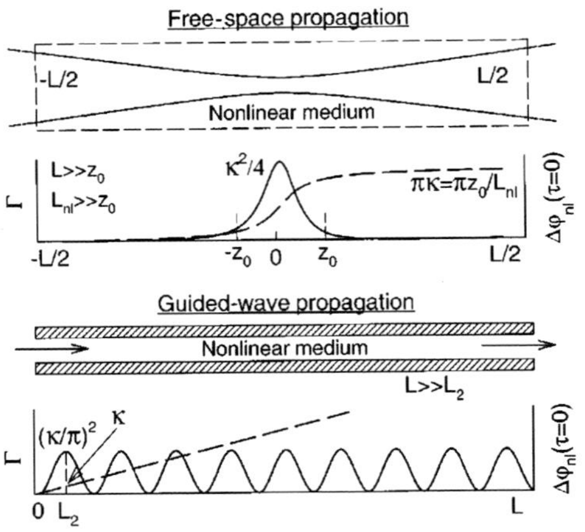

FIG. 12. Schematic of pulse evolution in a Kerr nonlinearity without and with guiding. Upper graphs, free space propagation in a Kerr nonlinear medium with a length L much longer than the nonlinear length L_{nl}, which in turn is assumed to be much longer than the Rayleigh range z_0 of the beam (for definitions see text) and the signal coupled from the fundamental (TEM_{00}) Gaussian mode into the next higher-order TEM_{01} mode vs propagation distance: solid curve, the fractional energy transfer $\Gamma(z)$ from the TEM_{00} into the TEM_{01} mode in the limit of $\Gamma(0)=\kappa^2/4\ll 1$. Under this condition coupling to higher-order modes may be neglected. Dashed curve, the nonlinear phase shift at the peak of the pulse, which is independent of the transverse coordinates as long as the condition of weak coupling, $\Gamma(0)\ll 1$, is met. The dominant contribution to the nonlinear phase shift $\Delta\varphi_{nl,out}(\tau=0)$, which determines Kerr-induced spectral broadening, originates from the region $-z_0<z<z_0$. Lower graphs, guided-wave propagation in a hollow fiber filled with some noble gas to introduce a Kerr nonlinearity. Here $\Gamma(z)$ represents the fraction of the energy of the pulse carried in the fundamental LP_{01} mode that is transferred into the next higher-order LP_{02} mode by means of n_2. In the limit of $\Gamma(L_2)=(\kappa/\pi)^2\ll 1$, only a small amount of energy is coupled from the fundamental into the next higher mode, which oscillates between the two modes with a periodicity equal to twice the coherence length L_2. Under this condition a spatially uniform nonlinear phase shift can grow to lengths much longer than the coherence length. This is in contrast with propagation in a bulk nonlinear medium and results from confinement of the beam, keeping the intensity high over propagation lengths far beyond z_0.

$$\Gamma(0)=\frac{\kappa^2}{4}=\left(\frac{p_0}{p_{c,fs}}\right)^2 \quad \text{and} \quad \Delta\varphi_{nl,out}(\tau=0)=\pi\kappa, \tag{10}$$

respectively, where

$$\kappa=\frac{z_0}{L_{nl}} \quad \text{and} \quad p_{c,fs}=\frac{\lambda_0^2}{\pi n_2}. \tag{11}$$

The requirement of small depletion of the fundamental mode can be formulated as $\kappa^2/4\ll 1$, which can be rewritten in terms of $(p_0/p_{c,fs})^2\ll 1$, where $p_{c,fs}$ determines the power at which self-focusing tends to introduce severe distortions to the laser beam propagating in free space. This power identified by coupled-mode analysis is in reasonable agreement with the well-known critical power for self-focusing in a bulk medium $p_c=\alpha_{fs}\lambda_0^2/(8\pi n_2)$, where $\alpha_{fs}\approx 3.8-6.4$ is a correction factor acquired from numerical investigations (Sheik-Bahae *et al.*, 1984).

The requirement of $\kappa^2/4\ll 1$, which is to be fulfilled for spatially uniform self-phase-modulation and weak beam deterioration, implies that $\Delta\varphi_{nl,out}(\tau=0)$ can at maximum be equal to $\approx\pi$. Such an SPM-induced peak nonlinear phase shift, when carried by a near-bandwidth-limited pulse, can induce relative spectral broadening by a factor of 2 (Agrawal, 1995), which determines the maximum compression factor in spatially uniform temporal compression that is achievable after a single pass through a bulk nonlinear medium. For generating a larger spatially uniform nonlinear phase shift, the process needs to be repeated several times by refocusing the beam with a suitable array of lenses or mirrors. For large compression factors requiring many passes, the resultant discrete "guiding" configuration tends to become impractical.

Guiding of high-power laser pulses can be implemented much more conveniently and efficiently in a hollow waveguide (Marcatili and Schmeltzer, 1964). Filling the waveguide with some gas can introduce a Kerr (or other) nonlinearity required for spectral broadening of high-power laser pulses and their subsequent temporal compression (Nisoli *et al.*, 1996). In the limit of small depletion, the energy transferred from the fundamental linearly polarized mode LP_{01}, which can be excited at the waveguide entrance by a Gaussian input beam with near-100% efficiency (Nisoli *et al.*, 1998), into the higher-order modes LP_{mn} ($m=0$ and $n\geq 2$, where for $m\geq 1$ the mode-coupling constant is zero) is fully returned to the fundamental mode over a propagation length comparable to z_0 of the input beam (Tempea and Brabec, 1998b).

The lower part of Fig. 12 depicts qualitatively the evolution of the energy coupled into the LP_{02} mode during propagation through the waveguide. The energy oscillation period is twice the mutual coherence length between LP_{01} and LP_{0n}, $L_n=\pi/|\beta_0^{(n)}-\beta_0^{(1)}|$. Under optimum input coupling conditions for exciting the fundamental mode, which is achieved for $w_0\approx(2/3)a$, where a is the bore radius of the hollow waveguide and w_0 is the beam waist of the incoming beam (Nisoli *et al.*, 1998), the coherence length between the two lowest-order modes L_2 can be approximately expressed as $L_2\approx 1.1z_0$. The propagation constant of the LP_{0n} mode, $\beta_0^{(n)}$, defined in Eq. (27), decreases with increasing mode order and so does the maximum transferred energy, which is proportional to L_n^2.

The bottom diagram of Fig. 12 depicts the periodic energy exchange between the fundamental mode and the LP_{02} mode, for which the oscillating energy is maximum, as a function of the propagation length in the waveguide. For low depletion of the fundamental mode, the maximum fractional energy coupled to the LP_{02} mode and the evolution of the peak nonlinear phase shift, which is uniform across the beam, can be written as

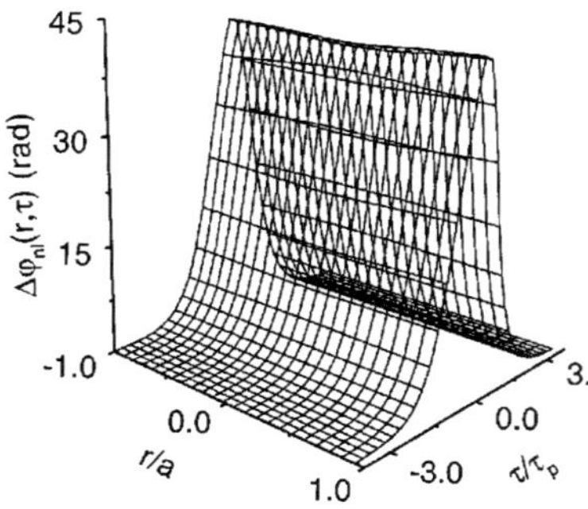

FIG. 13. Phase shift $\Delta\varphi_{nl}(z=L,r,\tau)$ imposed by the Kerr nonlinearity of argon gas at a pressure of 4 bars ($n_2=4\times10^{-19}$ cm/W) in a fused-silica hollow waveguide having a channel diameter of 140 μm and a length of $L_{nl}=29L_2\approx29$ cm upon a pulse of peak power $p_0=3.5$ GW injected into the fundamental LP_{01} mode at the entrance of the waveguide (Tempea and Brabec, 1998b). For these parameters, the pulse peak power is lower than the critical power for self-focusing, $p_{c,wg}\approx8$ GW. Under this condition, the power coupled into higher modes remains small ($\Gamma\approx0.25$), allowing the accumulation of spatially uniform self-phase-modulation on the pulse propagating primarily in the fundamental mode. This self-phase-modulation grows linearly with propagation distance until propagation losses become significant.

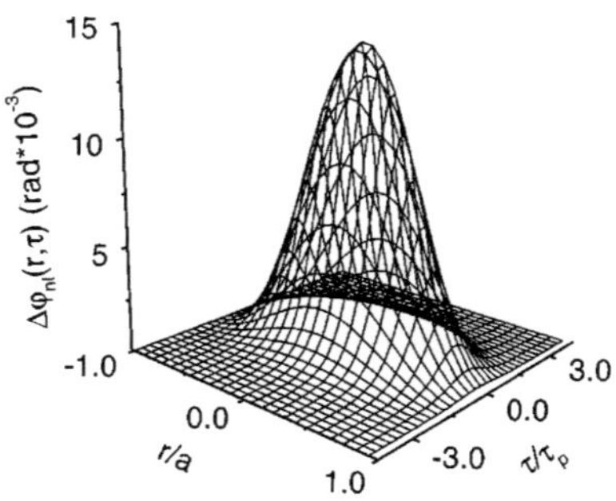

FIG. 14. Self-phase-modulation-induced phase $\Delta\varphi_{nl}(z,r,\tau)$ imposed by the Kerr nonlinearity on a pulse in a hollow waveguide as a function of the radial and the temporal coordinates under conditions described in the caption of Fig. 13 for a propagation length of $z=L_2/50$ (Tempea and Brabec, 1998b). For $z\ll L_2$, the signals coupled into different higher-order modes are comparable; therefore all coupling channels must be taken into account, resulting in a spatially varying phase shift that follows the transverse intensity profile of the fundamental LP_{01} mode. For pulse peak powers p_0 approaching $p_{c,wg}$ this phenomenon causes self-focusing of the beam, destroying propagation in the fundamental mode.

$$\Gamma(L_2)=\left(\frac{\kappa}{\pi}\right)^2=\left(\frac{p_0}{p_{c,wg}}\right)^2$$

and

$$\Delta\varphi_{nl}(z,\tau=0)=\kappa\frac{z}{L_2}\approx\kappa\frac{z}{z_0}, \tag{12}$$

respectively (see Fig. 12), where

$$p_{c,wg}=\frac{\lambda_0^2}{2n_2} \tag{13}$$

is the critical power for self-focusing in the multimode hollow waveguide and $p_{c,wg}\approx1.5p_{c,fs}$, i.e., self-focusing in waveguides and in free space sets in at comparable power levels (Milosevic *et al.*, 1999). For $w_0\approx\frac{2}{3}a$ the coupling coefficient κ is approximately defined by Eqs. (11), (9), and (4) with w_0 replaced by a (Tempea and Brabec, 1998b). Most importantly, meeting the requirement of weak depletion, $(\kappa/\pi)^2\ll1$, implies the feasibility of accumulating a spatially uniform SPM-induced phase shift over many times the Rayleigh range z_0 of the incident Gaussian beam. As a matter of fact, Fig. 13 reveals that a spatially uniform phase shift of $\Delta\varphi_{nl}\approx\pi$ can be imposed on the pulse over a propagation length of $2L_2$ without the onset of self-focusing, implying a spectral broadening of a factor of ≈10 over $L\approx20z_0$. Enhancing SPM-induced spectral broadening by increasing the propagation length is limited only by the propagation loss suffered by the fundamental mode in the leaky waveguide and/or by distortion of the temporal pulse shape.

It is instructive to look at $\Delta\varphi_{nl}(z,r,\tau)$ in the opposite limit, $L\ll L_2$, which is shown, under the same experimental conditions, in Fig. 14. The plot in Fig. 14 is the result of superimposing an infinite number of high-order modes, many of which with $L_n>L$ have comparable amplitudes, on the fundamental LP_{01} mode. For propagation lengths much shorter than the coherence length L_2, the nonlinear phase shift mimics not only the temporal shape of the pulse but also the transverse intensity distribution of the fundamental mode. It is this latter effect that gives rise to self-focusing if the peak power becomes so high that κ/π approaches unity and depletion of the fundamental mode becomes appreciable. Analogously, strong self-focusing emerges in a bulk nonlinear medium as $p_0\rightarrow p_{c,fs}$. In order to avoid catastrophic effects, L must be significantly shorter than z_0.

These are the experimental conditions relevant to passively mode-locked femtosecond Ti:sapphire and other solid-state laser oscillators. These systems are capable of delivering ultrashort pulses in the fundamental TEM_{00} mode, which is in apparent contradiction to coupling an appreciable amount of energy from the fundamental into higher-order transverse modes by the radially varying nonlinear phase shift $\Delta\varphi_{nl}(r,\tau)$ induced by the nonlinear index of the gain medium. This paradox can be resolved by considering the spatial filtering action of a laser resonator, which efficiently suppresses the emerging high-order modes and can lead to a TEM_{00} output beam even in the presence of strong nonlinearities in the cavity.

We may therefore conclude that the same nonlinear optical process in strongly different parameter regimes, $L<z_0$ and p_0 comparable to p_c on the one hand, and $L\gg z_0$ and $p_0\ll p_c$ on the other hand, is responsible, in combination with broadband GDD control, for the generation of low-energy few-cycle light pulses in laser oscillators and for the generation of few-cycle pulses at much higher power levels by guided-wave compression following amplification.

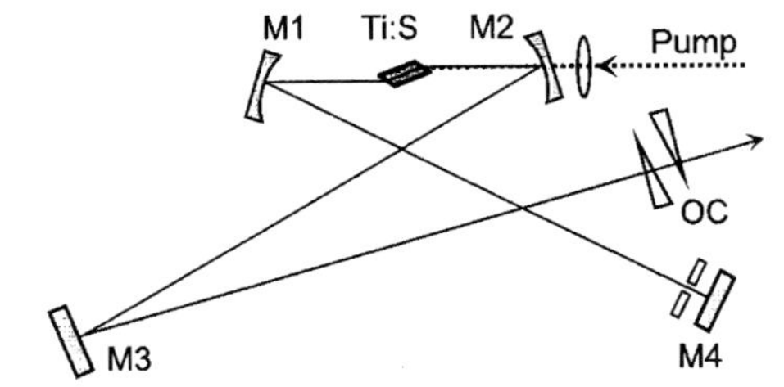

FIG. 15. Schematic of a mirror-dispersion-controlled Ti:sapphire laser made up of chirped mirrors (M1-M4), a broadband output coupler (OC), and a thin, highly doped Ti:sapphire crystal (Ti:S); for more details see, for example, Xu *et al.* (1997).

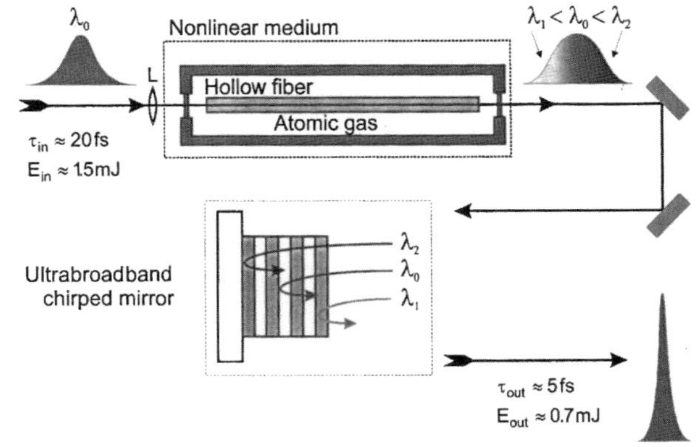

FIG. 16. Schematic experimental setup of a hollow-fiber chirped-mirror high-energy pulse compressor. The input pulse is coupled into a gas-filled hollow fiber, where its spectrum is broadened by self-phase-modulation imposed by the optical Kerr effect in the gas. Subsequently, the chirp of the broadened pulse is removed upon reflection off broadband chirped mirrors [see diagrams (a) and (b) in Fig. 10], which leads to a temporal compression of the laser pulse (courtesy of A. Poppe).

C. Few-cycle pulse generation: Current state of the art

Optical pulse compression based on the interplay between self-phase-modulation induced by the optical Kerr effect and negative group delay dispersion is the key concept for light-pulse generation in the few-cycle regime. Currently it is most efficiently implemented in mirror-dispersion-controlled (MDC) Kerr-lens mode-locked (KLM) Ti:sapphire laser oscillators and hollow-fiber chirped-mirror compressors.

In the former systems, schematically illustrated in Fig. 15, SPM in the laser crystal and negative GDD in the chirped mirrors M1–M4 act alternately many thousand times before the stationary pulse duration in the sub-10-fs regime is reached. The pulse is moderately chirped and spectrally broadened upon passage through the Ti:S crystal, which is converted in a temporal compression in the broadband chirped resonator mirrors. In the steady state this pulse, shortening process is stopped by the finite bandwidth over which nearly constant negative GDD can be introduced by the mirrors and/or balanced by the finite gain and resonator bandwidth.

Solitary pulse formation *inside* a laser oscillator needs to be assisted by a fast saturable-absorber-like self-amptitude modulation (SAM), which is introduced by KLM in the Ti:S oscillator, as described in the previous section. The role of this SAM action is (i) to initiate and support the formation of a short pulse in the cavity and (ii) to stabilize the SPM/GDD-dominated shortening process and eventually the stationary state against low-intensity noise. This noise arises inherently due to gain and positive feedback in the pulse wings and can be efficiently filtered by SAM outside the short interval comprising the mode-locked pulse (see Fig. 3).

When pulse compression external to a laser oscillator is implemented, both SPM and GDD usually act only once and in this sequence, as illustrated in Fig. 16. Hence spectral broadening in the Kerr medium is much stronger than in the case of intracavity solitary pulse formation. The absence of quasiperiodic pulse evolution and gain in extracavity pulse compression implies that instabilities and noise cannot grow. Hence filtering by SAM is not absolutely necessary. Summing up, both SPM and GDD tend to impose a much stronger modification on the light pulse in extracavity compressors than does compression inside the oscillator. In addition, intra- and extracavity spectral broadening in a KLM Ti:S laser and a multimode hollow waveguide take place in entirely different physical regimes, as described at the end of the previous subsection. Table I contrasts the parameter regimes of KLM/MDC Ti:S oscillators and multimode-hollow-waveguide/chirped-mirror compressors by listing typical values of the parameters relevant to few-cycle pulse formation.

KLM/MDC Ti:S laser oscillators routinely generate sub-10-fs pulses at typical pulse energy levels of a few nanojoules and multi-MHz repetition rates. The simplest embodiment of this system is depicted in Fig. 15. It is capable of producing pulses down to 7 fs in duration with peak powers up to 1 MW (Xu *et al.*, 1997; 1998; Beddard *et al.*, 1999) and exhibits unparalleled noise characteristics in the sub-50-fs regime while being pumped with a low-noise diode-pumped solid-state laser (Poppe *et al.*, 1998). Incorporating a pair of low-dispersion prisms as a source of adjustable GDD in addition to the chirped mirrors and employing specially designed output couplers in KLM/MDC Ti:S lasers recently pushed the record below 6 fs in laser oscillators (Morgner *et al.*, 1999; Sutter *et al.*, 1999). Figure 17 shows the intensity envelope and phase of sub-6-fs pulses (Gallmann *et al.*, 1999) obtained from a SPIDER measurement (spectral phase interferometry for direct electric-field reconstruction; Walmsley and Wong, 1996). The pulses are generated by a self-starting KLM/MDC Ti:S laser developed by U. Keller and co-workers at the Eidgenössische Technische Hochschule (ETH) Zurich with energies of 2–4 nJ at peak power levels of approximately 0.5 MW (Sutter *et al.*, 1999).

The pulse energy can be enhanced by extending the cavity with a telescope and reducing the repetition rate (Cho *et al.*, 1999; Libertun *et al.*, 1999). Implementing this technique in a KLM/MDC Ti:S laser has yielded

TABLE I. Parameter ranges for few-cycle pulse generation in Kerr-lens-mode-locked/mirror-dispersion-controlled Ti:S oscillators at low energy levels and in hollow-waveguide/chirped-mirror compressors at high energy levels.

Systems parameters	KLM/MDC Ti:sapphire oscillator	Multi-mode hollow-waveguide/ chirped-mirror compressor
Nonlinear interaction length	$<z_0$	$\gg z_0$
p_0/p_c	$\geqslant 1$	$\ll 1$
p_c	$\approx$2.5 MW	10 GW–1 TW
$\Delta\varphi_{nl}(\tau=0)$	0.1π–0.5π	5π–10π
Spectral broadening/pass	1.1–1.3	5–10
Net negative group delay dispersion $\|D\|$	10–30 fs^2	10–30 fs^2
Compression factor/pass	1.1–1.3	5–10
Self-amplitude modulation depth	1–3 %	—
Output pulse energy	3–30 nJ	0.01–1 mJ
Output pulse duration	5–8 fs	4–7 fs
Output peak power	0.3–3 MW	1 GW–0.2 TW
Peak intensity in focus	up to 10^{14} W/cm^2	up to 10^{18} W/cm^2

sub-10-fs pulses with energies in excess of 30 nJ, giving rise to peak powers of >3 MW at a repetition rate of 25 MHz (Poppe, Lenzner, *et al.*, 1999). This value approaches peak power levels which until recently could only be attained with cavity-dumped Ti:S lasers at high (>1 MHz) repetition rates (Ramaswamy *et al.*, 1993; Pshenichnikov *et al.*, 1994). The lower repetition rate (as compared to 80–100 MHz typically) of extended-cavity and cavity-dumped Ti:S oscillators not only enhances the peak power to levels allowing further pulse compression below 5 fs (Baltuska *et al.*, 1997a, 1997b; Pshenichnikov *et al.*, 1998), but also reduces thermal load to the sample in ultrafast spectroscopy.

The highest peak intensities that can be expected from oscillators in the few-cycle sub-10-fs regime approach 10^{14} W/cm^2 (Xu *et al.*, 1998; Jasapara and Rudolph, 1999; Poppe, Lenzner, *et al.*, 1999). Although strong-field effects may arise at these intensity levels in some systems, the exploration (and possibly exploitation) of a wide range of strong-field phenomena calls for substantially higher pulse energies, which can only be achieved by external amplification. This process is incapable of preserving the duration of few-cycle seed pulses even if laser media with the broadest amplification band available to date (see Fig. 4) are used. Gain narrowing is inherently coupled to a high gain, limiting the relative bandwidth $\Delta\omega/\omega_0$ of the amplified pulses typically to a few percent or less. Because wave packets comprising just a few field oscillation cycles are characterized by $\Delta\omega/\omega_0>0.1$, the generation of intense few-cycle light pulses requires pulse compression after amplification.

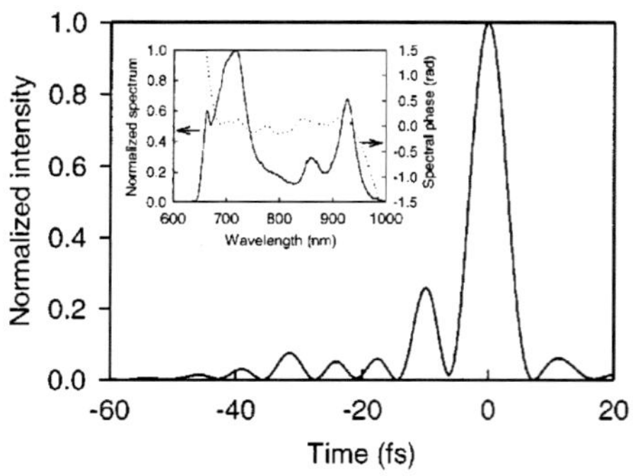

FIG. 17. Spectral intensity and phase (insert) and intensity envelope of sub-6-fs pulses produced by a semiconductor-saturable-absorber-initiated Kerr-lens mode-locking mirror-dispersion-controlled Ti:S laser (Sutter *et al.*, 1999). The data have been evaluated from a SPIDER (spectral phase interferometry for direct electric-field reconstruction) measurement (Gallmann *et al.*, 1999).

This can be implemented by spectral broadening in a multimode gas-filled hollow fiber (Nisoli *et al.*, 1996; Nisoli, De Silvestri, *et al.*, 1997; Nisoli, Stagira, *et al.*, 1997) followed by temporal compression upon reflection off chirped mirrors (Sartania *et al.*, 1997; Tempea and Brabec, 1998a), as illustrated schematically in Fig. 16. In strong contrast with pulse shortening in the oscillator, broadening of the laser-pulse spectrum can be effected dramatically upon one single passage through the nonlinear medium, owing to guided-wave propagation as explained above and shown in Fig. 18. Seeded with pulses of 20–25 fs in duration and 1.5–1.7 mJ in energy, the system described in the caption of Fig. 16 produces 5-fs pulses of energy of 0.7–0.8 mJ, implying a peak power of some 0.15 TW, at a repetition rate of 1 kHz. The technique is expected to be scalable to peak powers approaching the terawatt level. The pulses are delivered in a diffraction-limited beam and have recently been focused to intensities approaching 10^{18} W/cm^2. Figure 19 shows the temporal intensity envelope and the phase (Cheng *et al.*, 1999) as evaluated from a frequency-resolved optical gating measurement (Kane and Trebino, 1993; Trebino *et al.*, 1997) with φ_0 arbitrarily chosen to be zero. The clean front edge of the pulse is of prime importance for strong-field experiments. The time it takes for the intensity to rise from 10 to 90% of the peak is less than 5 fs and hence less than two field oscillation periods.

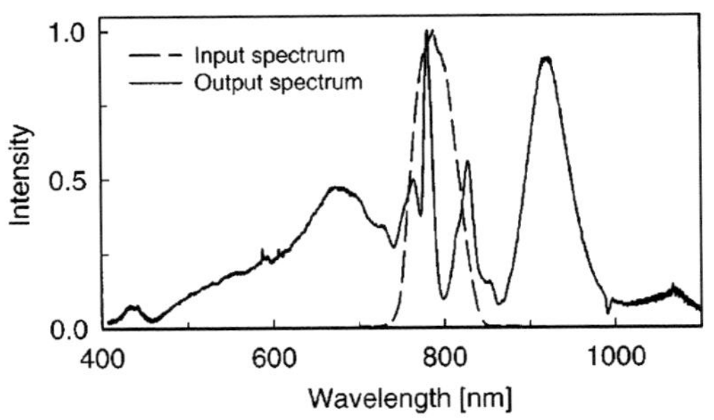

FIG. 18. Typical spectrum before (dashed line) and after (solid line) propagation through a hollow fiber having a channel diameter of 250 μm and a length of approximately 1 m. The waveguide is seeded with 25-fs, 1.5-mJ input pulses and filled with neon at a pressure of $\approx$1 bar.

To test the limits of currently available chirped mirror technology, sub-6-fs pulses from the compressor have been gently focused in atmospheric air at reduced energy levels and slightly broadened temporally (resulting in some positive chirp) such that the nonlinear index of air broadened the spectrum by some 30% uniformly across the laser beam, as predicted by the coupled-mode analysis reviewed in the previous subsection. Subsequent reflection off a few ultrabroadband chirped mirrors [see diagrams (a) and (b) in Fig. 10] compensated for the GDD and third-order dispersion of air and the SPM-induced chirp and resulted in the interferometric autocorrelation (Diels, Fontaine, *et al.*, 1985) depicted by the solid line in Fig. 20. The dotted line is calculated from the inverse Fourier transform of the measured spectrum (corrected for the spectrograph and detector response) under the assumption of no spectral phase modulation, yielding a pulse duration of 4 fs at a peak power of 30 GW (Cheng *et al.*, 1998). These are the shortest electromagnetic pulses demonstrated to date, comprising some one and a half field oscillation cycles within the full width at half maximum of their intensity envelope. These unique characteristics have far-reaching impacts on strong-field physics, some of which will be addressed in the remaining part of this paper.

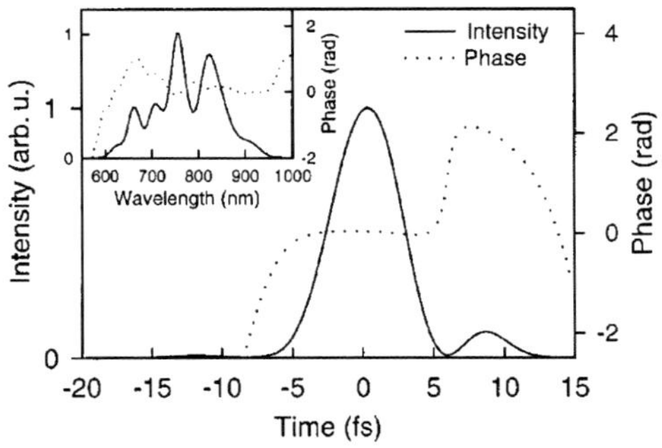

FIG. 19. Millijoule-scale pulses exiting the hollow-fiber chirped-mirror compressor: solid curve, intensity envelope; dotted curve, phase, with φ_0 chosen arbitrarily to be equal to zero; inset, spectrum. The data were retrieved from a frequency-resolved optical gating measurement (Cheng *et al.*, 1999). The pulse duration evaluated is $\tau_p = 5.3$ fs.

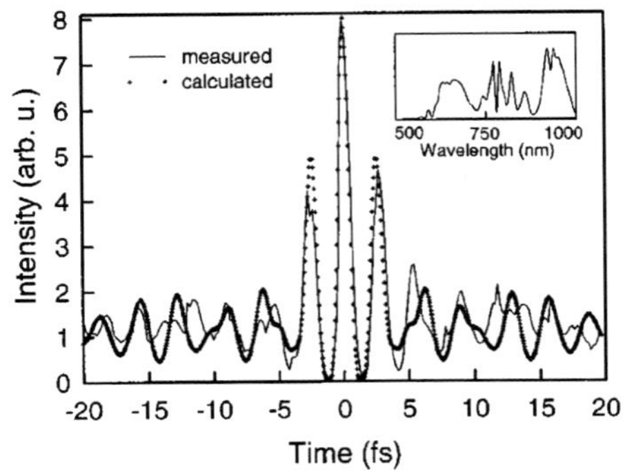

FIG. 20. Intense pulses originating from double-stage pulse compression, as described in the text and reported by Cheng *et al.* (1998): solid curve, measured interferometric autocorrelation; dotted curve, calculation from the inverse Fourier transform of the measured spectrum in the absence of a frequency-dependent phase, yielding a FWHM pulse duration of $\tau_p = 4$ fs; inset, spectrum.

D. Approaching the light oscillation period: Does the absolute phase of light matter?

Generating ultrashort pulses with durations approaching the light oscillation period $T_o = 2\pi/\omega_0 = \lambda_0/c$ brings up several questions for experimentalist and theorist alike: Are the techniques used for characterization in the multicycle regime still adequate and do they provide complete information about the characteristics of few-cycle wave packets? Does the description of ultrashort pulses in terms of carrier and envelope remain valid as the pulse duration τ_p approaches T_o? These questions are not quite independent of each other, given the fact that the determination of τ_p, defined as the full width at half maximum of the intensity envelope, relies on a physically meaningful definition of the intensity envelope.

First let us address the latter question, in order to provide a convenient mathematical framework for addressing the former. The complex amplitude (envelope) $\tilde{E}_a(t)$ of pulsed radiation permits the electric field $E(t)$ to be expressed as

$$E(t) = \tilde{E}_a(t) e^{-i\omega_0 t + i\varphi_0} + \text{c.c.} \tag{14}$$

For multicycle pulses, $\tilde{E}_a(t)$ can be derived from $E(t)$ by introducing the complex field $\tilde{E}(t) = (2\pi)^{-1/2} \int_0^\infty \tilde{E}(\omega) e^{-i\omega t} d\omega$, in which $\tilde{E}(\omega) = (2\pi)^{-1/2} \int_{-\infty}^\infty E(t) e^{i\omega t} dt$. This obeys $E(t) = \tilde{E}(t) + \text{c.c.}$, and hence $\tilde{E}_a(t)$ can be determined from $\tilde{E}(t) = \tilde{E}_a(t) e^{-i\omega_0 t + i\varphi_0}$, provided that ω_0 and φ_0 are known. The tilde represents a complex quantity throughout the paper. In order to avoid fast oscillations in $\tilde{E}_a(t)$, the reference frequency ω_0 must be roughly at the center of the spectrum of the wave packet. The precise choice of ω_0 is not critical for describing pulse propagation on the basis of Eq. (14). There are several possible definitions of ω_0, one of which,

$$\omega_0 = \bar{\omega} = \frac{\int_0^\infty \omega |\tilde{E}(\omega)|^2 d\omega}{\int_0^\infty |\tilde{E}(\omega)|^2 d\omega}, \tag{15}$$

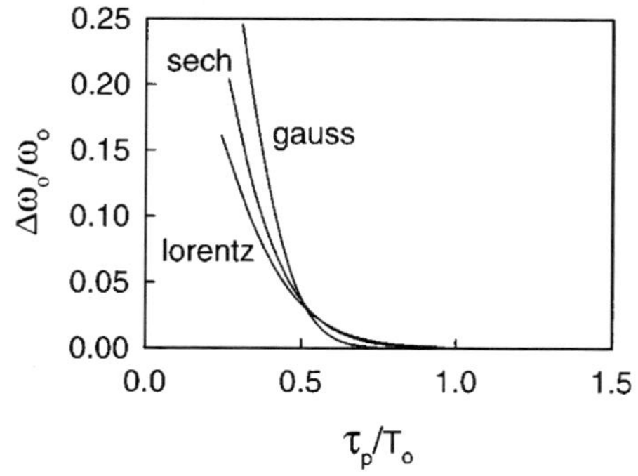

FIG. 21. Relative change $\Delta\omega_0/\omega_0$ in the carrier frequency, as defined in Eq. (15), caused by modifying the carrier phase by $\Delta\varphi=\pi/2$ as a function of the pulse duration for the following analytic pulse shapes: Gaussian, $A_g=\exp[-(1.17t/\tau_p)^2]$; Sechant hyperbolic, $A_s=\text{sech}(1.76t/\tau_p)$; and Lorentzian, $A_l=1/[1+(1.29t/\tau_p)^2]$.

stands out for its minimization of the intensity-weighted phase variation of $\tilde{E}_a(t)$ (Diels and Rudolph, 1996). In spite of this important characteristic, there is some arbitrariness in the above definition of ω_0. Therefore it is important to stress that the major conclusions from the following considerations are not affected by the specific choice of a definition for ω_0. Now with ω_0 defined and φ_0 chosen arbitrarily so that the imaginary part of $\tilde{E}_a(t)$ is zero at the reference instant $t=0$ (which is, for practical reasons, often adjusted to coincide with the center of gravity of $|E(t)|^2$), the route to determining $\tilde{E}_a(t)$ from $E(t)$ is unambiguously prescribed.

The above definition of the complex amplitude envelope of a transient wave form (wave packet) is apparently valid irrespective of the duration of the wave packet. However, to legitimize the concept of carrier and envelope we must require that ω_0 and $\tilde{E}_a(t)$ remain *invariant* under a change of φ_0, the physical significance of which is confined to determining the relative position of the carrier wave with respect to the envelope. A shift of φ_0 by some $\Delta\varphi$ yields the new wave form $\tilde{E}'(t)=\tilde{E}_a(t)e^{-i\omega_0 t+i(\varphi_0+\Delta\varphi)}$, i.e., translates the carrier with respect to the envelope. In order that the definition of $\tilde{E}_a(t)$ and ω_0 be self-consistent, the modified wave $\tilde{E}'(t)=\tilde{E}(t)e^{i\Delta\varphi}$ must yield $\omega_0'=\omega_0$, which also implies $\tilde{E}_a'(t)=\tilde{E}_a(t)$. Intuitively, one expects the concepts of carrier and envelope to fail as the temporal extension of the transient wave form becomes comparable to the oscillation cycle of the wave.

In order to develop some feel for the parameter range in which one can rely on the carrier-envelope concept, we have computed ω_0 for $E(t)=E_{\text{anal}}\sin(\Omega t)$ and ω_0' for its phase-translated counterpart $\tilde{E}'(t)=\tilde{E}(t)e^{i\Delta\varphi}$ for some of the most important analytic pulseforms. $\Delta\varphi=\pi/2$ was chosen to maximize the expected difference between ω_0' and ω_0 for short pulses. Figure 21 depicts $|\omega_0'-\omega_0|/\omega_0$ as a function of the pulse duration normalized to the oscillation period $T_{\text{osc}}=2\pi/\Omega$. As anticipated, the requirement of phase invariance of carrier frequency and hence that of the envelope is met with a high accuracy for pulse durations that are long compared to T_{osc}. It is, however, somewhat surprising that this requirement is still well satisfied for extremely short wave forms with durations comparable to the carrier oscillation period T_{osc} and tends to be increasingly violated only as the pulse duration τ_p [full width at half maximum (FWHM)] of the intensity envelope $|\tilde{E}_a(t)|^2$) becomes shorter than the carrier period. It can be seen by inspection that this finding applies irrespective of the precise choice of either ω_0 or the temporal profile of (bell-shaped) wave packets.

In the multicycle regime the complex envelope $\tilde{E}_a(t)$ can also be used to calculate the cycle-averaged radiation intensity $I(t)\propto|\tilde{E}_a(t)|^2$, which yields the energy fluence $F=\int I(t)dt$. To what extent this approach remains accurate for $\tau_p\rightarrow T_o$ can be assessed when we calculate the energy fluence by integrating the instantaneous energy density flow proportional to $E^2(t)$ over the pulse and investigate its phase dependence. For a Gaussian pulse, the maximum relative variation of F upon varying φ_0 can be expressed as $\Delta F/F=\exp[-\pi\tau_p/T_{\text{osc}}\ln 2)]$ (Reichert *et al.*, 1999), justifying the use of the cycle-averaged intensity from the envelope down to pulse durations approaching the carrier oscillation cycle. As a consequence, *the description of pulsed electromagnetic radiation in terms of carrier and envelope is self-consistent, unambiguous, and hence legitimate, for FWHM pulse durations down to the carrier oscillation cycle, $\tau_p\geqslant T_o$* (Brabec and Krausz, 1997). This finding allows the powerful mathematical and physical concepts based on the envelope and carrier [developed originally for pulses that obey $\partial|\tilde{E}_a(t)|/\partial t\ll|\tilde{E}_a(t)|/T_o$] to be extended into a new regime of nonlinear optics, in which the interaction time is limited to the order of the optical cycle.

The violation of the phase invariance of ω_0 and hence $\tilde{E}_a(t)$ for $\tau_p<T_o$ is found to correlate with a rapid increase in the variation of the spectral intensity $|\tilde{E}(\omega)|^2$ with φ_0 at low frequencies near $\omega=0$. This finding sheds light on the mathematical origin of the failure of the carrier-envelope concept: the amplitude of the low-frequency ($\omega\ll\omega_0$) spectral components of pulsed radiation are not invariant under the transformation $\tilde{E}'(t)=\tilde{E}(t)e^{i\Delta\varphi}$, but sensitively depend on φ_0. For sufficiently broadband signals, the fractional weight of the low-frequency components becomes significant, causing ω_0, and hence $\tilde{E}_a(t)$, to become dependent on φ_0. The low-frequency components also have important implications for laterally confined (three-dimensional) wave-packet propagation, because they suffer from a stronger diffraction when carried in a laserlike beam. As a result, the low-frequency components are subject to increased attenuation on the beam axis with increasing z, giving rise to a truncation of the low-frequency part of the pulse spectrum, i.e., to a deformation of the wave packet. This effect was recently observed in the propagation of single-cycle THz pulses (Hunsche *et al.*, 1999).

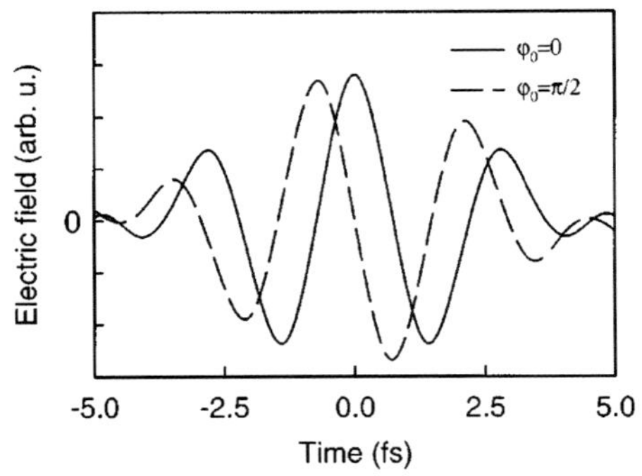

FIG. 22. Possible evolutions of the electric field in the 4-fs pulse obtained in the experiment summarized in Fig. 20 carried at a wavelength of 750 nm for two different pulse phases ($\phi_0 = 0, \pi/2$) with all other parameters $[\omega_0, \tilde{E}_a(t)]$ left unchanged. Variation of the absolute carrier phase ϕ_0 leaves the interferometric autocorrelation unchanged in Fig. 20.

We now return to the first question raised at the beginning of this subsection. Figure 22 depicts two (out of an infinite number of) possible evolutions of the electric field which yield the autocorrelation trace shown by the dotted line in Fig. 20. These evolutions are obtained by varying the absolute carrier phase φ_0 and leaving all other parameters of the pulse $[\omega_0, \tilde{E}_a(t)]$ unchanged. Clearly, the interferometric autocorrelation is invariant under changes of the absolute carrier phase and so are all other measurement techniques used for the characterization of ultrashort light pulses, including FROG (frequency-resolved optical gating) and SPIDER. This implies that none of the currently available techniques provides access to the absolute carrier phase φ_0, nor does a technique exist for controlling this parameter. In the multicycle regime of light-matter interactions, failure to gain access to φ_0 does not impair our ability to control nonlinear optical processes, because all known processes are insensitive to φ_0 for $\tau_p \gg T_o$. Hence *nonlinear optics may be fully controlled in terms of the amplitude envelope and frequency of intense light pulses in the regime of many cycle pulses.*

This situation changes dramatically as the interaction time between a light wave and atoms is confined to a period approaching the oscillation cycle of the light field, particularly for the strong-field regime (see Sec. IV), where the electric (and, at relativistic intensities, also the magnetic) field takes control over the evolution of nonlinear light-matter interactions. Under these conditions, interactions lasting merely a few optical cycles are greatly sensitive to the absolute phase φ_0 rather than only to the envelope $\tilde{E}_a(t)$ and frequency ω_0 of the incident radiation, as will be shown in Sec. VIII. As a consequence, gaining access to and controlling φ_0 in addition to $\tilde{E}_a(t)$ becomes indispensible for precise control of physical measurables originating from strong-field interactions in the few-cycle regime.

The issue of the absolute phase of few-cycle light pulses was first addressed by Xu *et al.* (1996). The experiments of Xu *et al.* revealed that the position of the carrier relative to the envelope is generally rapidly varying in the pulse train emitted from a mode-locked laser oscillator. This phenomenon arises because the carrier wave of the laser pulse gets shifted upon each round trip in the cavity, and the shift makes the carrier slide underneath the envelope in the output pulse train. To understand the origin of this carrier phase shift, we recall that initiating plane-wave propagation of the field given in Eq. (14) along the z axis from the plane $z=0$ in a linear medium results in a wave packet propagation that can be described by

$$E(z,t)=\tilde{E}_a(z,t-\beta_1 z)e^{i\beta_0 z - i\omega_0 t + i\varphi_0}+\text{c.c.}, \tag{16}$$

which can be obtained by solving the wave Eq. (29) in the absence of nonlinearities. Here $\beta_0=\omega_0 n_0/c$ and $n_0 = \text{Re}[n(\omega_0)]$ is the real part of the refractive index at $\omega = \omega_0$ and $\beta_1=(\omega_0/c)\text{Re}[(\partial n/\partial\omega)_{\omega_0}]$. The time retardation in the argument of the envelope indicates that the center of the wave packet propagates at a velocity of $v_g=\beta_1^{-1}$, which is referred to as the group velocity. It is therefore convenient to introduce a coordinate system moving with the pulse at the group velocity and following pulse evolution in this system. This can be simply done by performing the coordinate transformation $\tau=t - z/v_g$ and $\xi=z$. Equation (16) in this moving frame of reference can then be rewritten as

$$E(\xi,\tau)=\tilde{E}_a(\xi,\tau)e^{-i\omega_0 t + i\varphi(\xi)}+\text{c.c.},$$

where

$$\varphi(\xi)=\varphi_0+\omega_0\left(\frac{1}{v_{ph}}-\frac{1}{v_g}\right)\xi \tag{17}$$

and $v_{ph}=\omega_0/\beta_0$ represents the phase velocity.

By inspecting $\varphi(\xi)$, which determines the position of the carrier relative to the envelope, we can identify the difference between the phase delay ($\omega_0\xi/v_{ph}$) and the group delay ($\omega_0\xi/v_g$) as the reason that the carrier slides under the envelope as the pulse circulates in the mode-locked laser. The difference between phase and group delay in transparent optical materials originates from the wavelength dependence of the (real) refractive index, which becomes obvious when we express the change in the carrier phase shift upon passage through a medium of length L as $\Delta\varphi(\xi)=2\pi(\partial n/\partial\lambda)_{\lambda_0}\xi$, from which the dephasing length (Xu *et al.*, 1996)

$$L_{\text{deph}}=\frac{1}{2}\left|\frac{\partial n}{\partial\lambda}\right|_{\lambda_0}^{-1} \tag{18}$$

can be introduced. Here L_{deph} is the propagation length over which the carrier is offset by a phase of π with respect to the envelope. It can be as short as 29 and 19 μm in quartz and sapphire, respectively. As a consequence, $\varphi(\xi)$ is shifted by several dozen times 2π upon a full round trip of the laser pulse in a KLM/MDC Ti:S laser. The round-trip carrier phase shift (relative to the envelope) $\Delta\varphi_r=\varphi_{n+1}-\varphi_n$ causes a shift of the carrier with respect to the envelope by $\Delta\varphi_r - 2\pi m=\Delta\varphi_r'$ (where m is an integer chosen to yield $0<\Delta\varphi_r'<2\pi$)

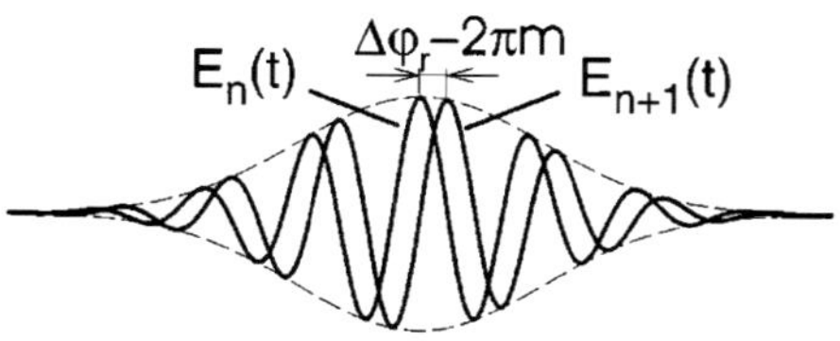

FIG. 23. Evolution of electric fields in two adjacent pulses in the pulse train emitted by a mode-locked laser. The phase shift $\Delta\varphi_r - 2\pi m$ can be measured by interferometric cross correlation (Xu *et al.*, 1996).

between any two laser pulses following one another in the output pulse train, as illustrated in Fig. 23. This shift can be measured by interferometric cross correlation (Xu *et al.*, 1996). It was recently pointed out that the physically relevant (relative) carrier shift $\Delta\varphi_r'$ offsets by $f_0 = (\Delta\varphi_r'/2\pi)T_r^{-1}$ the frequency comb emitted by the mode-locked laser, resulting in locked frequencies that can be written as $f_m = f_0 + mf_r$, where m is a (large) positive integer and $f_r = 1/T_r$ is the pulse repetition rate (Reichert *et al.*, 1999).

By introducing adjustable dispersion with a pair of thin quartz wedges in the cavity (Xu *et al.*, 1996), $\Delta\varphi_r$ can be set equal to an integer multiple of 2π. Unfortunately, the carrier cannot be "locked" to the envelope in the output pulse train by this measure for an extended period. This is because intensity-dependent contributions to both the round-trip phase delay and the round-trip group delay translate small energy fluctuations into a rapidly accumulating jitter of φ, destroying phase stabilization on a time scale of microseconds (a few hundred to a few thousand round trips) in the KLM/MDC Ti:S laser (Xu *et al.*, 1996).

Measuring and subsequently stabilizing the absolute phase of few-cycle light generated by a mode-locked laser might become feasible by exploiting the phase sensitivity of some strong-field processes, which will be discussed in detail in Sec. VIII. Unfortunately, well-known and well-understood candidates such as optical-field ionization (Sec. VI) and high-order harmonic generation (Sec. VII) in atomic gases come into play only at intensity levels ($I > 10^{14}\,\mathrm{W/cm^2}$) that are difficult to attain with pulses from an oscillator without subsequent amplification. Amplification can only be performed at strongly reduced repetition rates (every hundred-thousandth pulse or so), which would not allow carrier phase stabilization in the oscillator by a servo loop, if dephasing occurs during a few hundred or thousand round trips in the oscillator.

Replacing the atomic gas by a metal as the nonlinear medium and inducing photoemission from the surface of a metal in the tunneling regime (Sommerfeld, 1967) might provide low-energy oscillator pulses direct access to strong-field optics (Poppe, Fürbach, *et al.*, 1999). As a matter of fact, optical tunneling from a gold photocathode is predicted to occur (Toth *et al.*, 1991) and to be phase dependent at intensity levels (see Fig. 47 below) that can be reached by state-of-the-art sub-10-fs Ti:S oscillators (Xu *et al.*, 1998; Beddard *et al.*, 1999; Poppe, Lenzner, *et al.*, 1999). Once a phase-sensitive physical measurable (such as the photocurrent in the previous example) can be acquired, a servo loop for stabilizing φ_n can be constructed.

An entirely different approach to controlling the evolution of φ_n in the output pulse train was recently proposed (Telle *et al.*, 1999) and demonstrated (Reichert *et al.*, 1999) in the frequency domain. By locking different modes of the mode-locked lasers to different harmonics of a narrow-line reference source (Udem *et al.*, 1999a; Reichert *et al.*, 1999) or locking a mode to a frequency derived from another mode of the mode-locked laser by nonlinear conversion techniques (Telle *et al.*, 1999), the offset frequency $f_0 = (\Delta\varphi_r'/2\pi)T_r^{-1}$ could be measured (with a much higher accuracy than is feasible with the previously demonstrated interferometric cross-correlation technique) and could be controlled with suitable cavity adjustments (Reichert *et al.*, 1999). It could, in principle, also be set equal to zero, implying phase stabilization. However, the actual stabilized value of the absolute carrier phase cannot be acquired by these techniques. Exploiting some strong-field process that is field dependent seems to be imperative to achieve this objective. Because precise control of individual optical frequencies is highly developed in frequency-domain metrology, a combination of the above-described time-domain (strong-field) and frequency-domain techniques appears to be the most promising route to realizing phase-controlled few-cycle pulse generation. Our ability to measure the absolute phase at the output of a mode-locked laser will also allow us to test field-dependent nonlinearities *inside* the laser oscillator for achieving phase-stable (strong-field) mode locking.

The generation of phase-stabilized few-cycle light pulses from a mode-locked laser will have major impact on both precision frequency-domain spectroscopy and high-field physics. For the former, a precisely known comb of harmonics of the laser repetition rate (each being a frequency standard once one single laser mode is locked to a reference frequency) spanning an unprecedented spectral range will become available. In the latter, the evolution of atomic and molecular systems as well as plasmas will become controllable directly with the electric (and magnetic) fields. Gaining access to the absolute phase of light and developing viable techniques for its control is undoubtedly one of the major challenges facing researchers in modern optics in the coming years.

IV. NONLINEAR RESPONSE OF ATOMS TO STRONG LASER FIELDS

The nonlinear response of matter to intense radiation manifests itself in a nonlinear dependence of the induced polarization (atomic dipole moment times density of atoms) on the electric (and possibly magnetic) field of the incident radiation. The nonlinearity can originate from distinctly different processes, depending on the intensity. At low and moderate intensities, the external laser field is much weaker than the static atomic Coulomb field. As a consequence, the laser field only slightly perturbs the atomic quantum states under nonresonant excitation conditions. The energy levels suffer only a faint shift proportional to E_a^2, which is referred to as the ac Stark shift (Landau and Lifshitz, 1977). The atoms remain, with a high probability, in their ground state and the extension of the wave function of the ground state remains on the order of the Bohr radius a_B. Nonlinear interactions taking place under these conditions can be well described by a perturbative approach, and hence we refer to this parameter range as the *regime of perturbative nonlinear optics.*

If the electric-field strength becomes comparable to (or higher than) the binding atomic Coulomb field experienced by the outer-shell electrons, an electron can escape with a substantial probability from its bound state (via tunneling or above-barrier detachment) before the laser electric field reverses its sign. The electron wave packet liberated by optical-field ionization subsequently wiggles in the linearly polarized electric field. The amplitude of the wiggle exceeds the Bohr radius by several orders of magnitude and the cycle-averaged kinetic energy of the electron exceeds the binding energy W_b. The parameter range giving rise to these processes is referred to as the *strong-field regime* of nonlinear optics. The atomic polarization response is dominated by the ionization process and the contribution from the bound electrons is negligible.

Nonlinear polarization induced by optical-field ionization emerges only as long as the electron remains in close proximity to its parent ion. Once the electron is set free, its trajectory is governed by the Newtonian equations of motion, resulting in a linear response (with a small remaining nonlinearity originating from the parent ion's reduced polarizibility). A strong nonlinearity arises only at intensities order(s) of magnitude higher due to optical-field ionization stripping the next electron and/or to the wiggle energy of free electrons becoming comparable to their rest energy mc^2, indicating the onset of *relativistic nonlinear optics.*

Figure 24 assigns the relevant intensity regions to the above regimes of nonlinear optics for visible and near-infrared radiation. Here we focus on nonrelativistic light-matter interactions. We shall (i) review how exploitation of processes in the perturbative regime allows generation of intense light pulses in the few-cycle regime and (ii) demonstrate that these pulses are capable of significantly extending the frontiers of strong-field nonlinear optics. Whereas in the perturbative regime the intensity envelope governs the evolution of nonlinear optical processes, in the strong-field regime the electric (and at relativistic intensities also the magnetic) fields take control. Because the generation of ultrashort light pulses relies on perturbative processes, the phase φ_0 [see Eq. (14)] is unknown, and the electric field of these pulses is indefinite. The light fields are expected to become accessible when strong-field processes with few-cycle wave packets are induced (see Sec. VIII). In Secs. IV. A-IV.C, we analyze the polarization response of an atomic medium irradiated with strong laser fields and

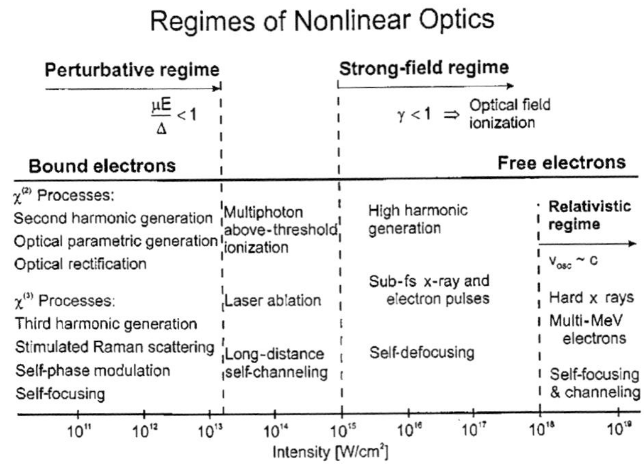

FIG. 24. Regimes of nonlinear optics. Only nonresonant interactions have been considered by assuming the detuning $\Delta\hbar$ to be larger than the photon energy ($\approx$1 eV). The boundaries between different regimes are not sharply defined. In the strong-field regime, the intensity scale applies to the visible and near-infrared spectral range.

introduce field-strength scale parameters that roughly define the borders of the above regimes. The relation of the induced medium polarization to the incident fields is indispensable for describing the propagation of intense radiation in atomic media, i.e., for the interaction of intense light with matter in a macroscopic volume.

A. Perturbative nonlinear optics

At low and moderate intensities the polarization, P [As/m^2] (where As represents Ampere seconds), of an atomic ensemble can be expanded into a Taylor series with respect to the electric field and can be given as the superposition of the linear and nonlinear responses $P = \varepsilon_0\chi^{(1)}E + P_{nl}$, where

$$P_{nl} = \varepsilon_0\chi^{(2)}E^2 + \varepsilon_0\chi^{(3)}E^3 + \varepsilon_0\chi^{(4)}E^4 + \cdots, \tag{19}$$

$\varepsilon_0 = 8.85\times10^{-12}$ As/V m is the vacuum permeability, and $\chi^{(k)}$ [(m/V)$^{k-1}$] is the kth-order susceptibility (Bloembergen, 1965). It is implicit in Eq. (19) that the atomic polarization instantly follows the change of the field, which is usually a good approximation even on a time scale of a few femtoseconds. This is because the induced atomic dipole moment is of purely electronic origin with a response time on the order of $1/\Delta$, where $\Delta = |\omega_{ik} - \omega_0|$, ω_{ik} represents the transition frequency from the initial (usually ground) quantum state i into some excited state k for which $|\omega_{ik} - \omega|$ is the minimum and the dipole transition matrix element μ_{ik} is nonzero, and ω_0 is the laser carrier frequency [see Eq. (14)]. Since the typical transition frequency from the atomic ground state to the lowest excited state significantly exceeds the laser frequency in the visible and near-infrared range, $1/\Delta$ is typically less than 1 fs. In molecules and condensed matter, nuclear motion may also provide a significant contribution to the induced dipole moment. This contribution has a response time of hundreds of femtoseconds to several picoseconds, leading to a more complicated expression for P_{nl} (Shen, 1984; Boyd, 1992). Moreover, the polarization response is generally anisotropic, with $\chi^{(k)}$ being a kth-rank tensor connecting the Cartesian components of E and P_{nl} (Shen, 1984; Boyd, 1992).

When we disregard bound-free transitions, the quantum theory of linear and nonlinear optical susceptibilities yields the simple approximate expression

$$\frac{\chi^{(k+1)}E^{k+1}}{\chi^{(k)}E^{k}} \approx \frac{\mu_{ik}E_a}{\hbar\Delta} \approx \frac{eE_a a_B}{\hbar\Delta} = \alpha_{bb} \tag{20}$$

for the ratio of the successive terms in Eq. (19), where E_a is the (time-dependent) amplitude of the (linearly polarized) radiation, which is related to the complex amplitude defined in Eq. (14) as $E_a=2|\tilde{E}_a|$, carried at the angular frequency ω_0, $\hbar=1.06\times10^{-34}$ Js is Planck's constant, and a_B is the Bohr radius. For $\alpha_{bb}\ll1$ bound-bound transitions are sufficiently weak to allow the expansion in Eq. (19) to converge.

For bound-free transitions the analysis of Keldysh (1965) yielded the scale parameter

$$\frac{1}{\gamma} = \frac{eE_a}{\omega_0\sqrt{2mW_b}} = \frac{eE_a a_B}{\hbar\omega_0} = \alpha_{bf}, \tag{21}$$

where $m=9.11\times10^{-31}$ kg is the electron rest mass, $e=-1.6\times10^{-19}$ C is the charge, and $W_b\gg\hbar\omega_0$ is the binding energy of the most weakly bound electron, i.e., the ionization potential of the atom. For the derivation of the second expression we utilized the general connection $a_B=\hbar/\sqrt{2mW_b}$, which can be used as a "generalized" Bohr radius for atomic number >1 in expressions (20) and (21). If $\alpha_{bf}\ll1$ bound-free transitions (i.e., ionization) can also be described perturbatively. In conclusion, *the regime of perturbative nonlinear optics is defined by* $\alpha_{bb},\alpha_{bf}\ll1$. A comparison of Eqs. (20) and (21) reveals that it is simply the ratio of detuning to laser frequency Δ/ω_0 that determines whether a bound-bound or bound-free transition violates the perturbative approximation for increasing field strengths.

In this paper we focus on the latter case, i.e., we assume $\Delta>\omega_0$, which is generally fulfilled in the interaction of visible and near-infrared radiation with atoms. In this spectral range the perturbative approach provides an adequate description of nonlinear processes up to intensities of around 10^{13} W/cm^2. Here we have also utilized the connection between intensity and electric field strength $I[\text{W/cm}^2]=(1/2Z_0)E_a^2[\text{V/cm}]$, where $Z_0=\sqrt{\mu_0\varepsilon_0}=377V/A$ is the vacuum impedance. In the above intensity range, Eq. (19) constitutes a good approximation, describing a wide range of nonlinear phenomena, a few of which are listed in Fig. 24. These include self-focusing and self-phase-modulation, which are the responsible nonlinear processes for femtosecond pulse generation in solid-state lasers (see Sec. II) and pulse compressors (see Sec. III).

For intensities of the order of 10^{13} W/cm^2, multiphoton ionization by simultaneous absorption of $N,N+1,N+2$, etc. photons comes into play (N being the smallest number of photons needed to reach the continuum), with the multiphoton channels of different order becoming comparable for increasing intensity. This is the phenomenon of above-threshold ionization (Agostini *et al.*, 1979; for a review, see Freeman *et al.*, 1992; Muller *et al.*, 1992). In this intensity regime the contributions of the electrons freed by multiphoton ionization and the induced atomic dipoles become comparable. One remarkable implication is that self-focusing due to the optical Kerr effect and self-defocusing due to free electrons tend to balance each other over long distances, resulting in self-channeling of intense femtosecond pulses over hundreds of meters (Braun *et al.*, 1995; Nibbering *et al.*, 1996; Wöste *et al.*, 1997). Unfortunately, in this intermediate regime between perturbative and strong-field nonlinear optics, the polarization response cannot be given analytically, and generally sophisticated numerical codes are needed for solving the time-dependent Schrödinger equation either directly (for a review see Kulander *et al.*, 1992) or indirectly by the Floquet ansatz (for a review, see Potvliege and Shakeshaft, 1992).

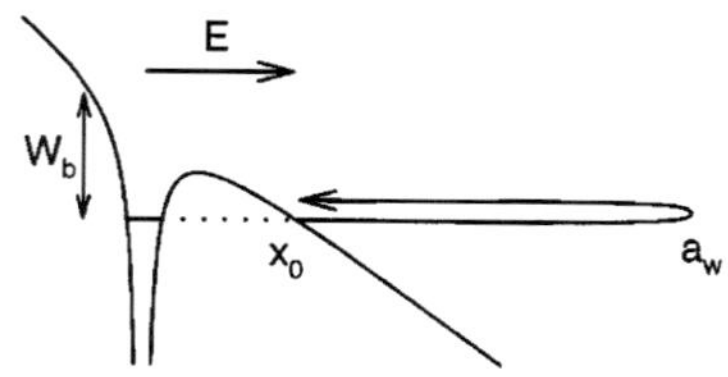

FIG. 25. Resultant (quasistatic) potential felt by the most weakly bound electron in the presence of a strong laser field. The electron can tunnel through the barrier. It is born at the position x_0 with some drift velocity (Corkum *et al.*, 1989) and follows the periodic motion of the laser field. The maximum amplitude during the first excursion is denoted by a_w.

B. The strong-field regime

For $\gamma^{-1}>1$, the laser field suppresses the Coulomb potential so strongly that the wave function of the (most weakly bound) electron of energy $-W_b$ penetrates the barrier and reaches its outer side at x_0, as illustrated in Fig. 25, within a fraction of the laser oscillation cycle T_o (Keldysh, 1965). As a consequence, the resultant escape rate adiabatically follows the variation of the optical field, giving rise to a "quasistatic" ionization rate $w(E)$ that is dependent only on the instantaneous electric field and the ground state from which the electron tunnels out. This process is generally termed optical-field ionization and will be addressed in more detail in Sec. VI.

Using classical mechanics to describe the evolution of the electron wave packet ejected into the continuum (Corkum *et al.*, 1989; Corkum, 1993) and neglecting the Coulomb field of the source ion, we obtain for the amplitude of the wiggling motion of an electron $a_w=eE_a/m\omega_0^2$ in the linearly polarized field and the release position of the electron $x_0\approx W_b/eE_0$. The cycle-averaged kinetic energy of the wiggling electron is given by $U_p=e^2E_a^2/4m\omega_0^2$, which is referred to as the *pon-*

deromotive potential ($U_p = 93$ eV and $a_w = 12.4$ nm at $I = 10^{15}$ W/cm^2 and $\lambda_0 = 1\ \mu$m). The Keldysh parameter can be reexpressed with these quantities as $1/\gamma^2 = a_w/2x_0 = 2U_p/W_b$. Consequently, for $1/\gamma > 1$, the freed electron is substantially removed from the position of its birth and acquires a large kinetic energy within a fraction of T_o. This indicates that the external field becomes dominant and the influence of the static Coulomb field becomes small immediately after ionization. These circumstances define the *strong-field regime of nonlinear optics*.

On the time scale of T_o, classical mechanics provides a quantum-mechanically meaningful and correct description of the evolution of the center of gravity of the liberated electron wave packet, because its extension grows at a rate of typically less than 1 nm/fs due to quantum diffusion (Delone and Krainov, 1991; Dietrich *et al.*, 1994). Thus it remains small as compared to the wiggle amplitude in the strong-field regime. The drift motion of the ejected wave packet is correctly accounted for by classical mechanics on essentially unlimited time scales. In fact, combining the quasistatic tunneling rate $w(E)$ obtained from quantum mechanics with a classical approach to tracing the motion of the center of mass of the electron wave packet, released from the atom with zero initial velocity provides an accurate description of strong-field phenomena including above-threshold-ionization electron spectra (Corkum *et al.*, 1989; Corkum, 1993) and high-order harmonic generation (Corkum, 1993; Lewenstein *et al.*, 1994; Spielmann *et al.*, 1998).

Here we use the same approach to derive the constitutive law for the *nonrelativistic* strong-field regime. The trajectory of the center of mass of the freed electron wave packet $x(t)$ fulfills $|x(t)| \gg a_B$; hence the first time derivative of the macroscopic polarization $P(t)$ of an atomic ensemble is dominated by the current density of the freed electrons, which for a linearly polarized strong electric field is given by $\dot{P} \approx \dot{P}_{nl} = J_{free} = e n_e \dot{x} + e \dot{n}_e x_0$. The dot represents the first partial time derivative and n_e is the density of free electrons. The second term originates from the appearance of electrons outside the Coulomb barrier, whereas the first term accounts for the contribution of free electrons moving away from the source atoms. This motion is governed by $m\ddot{x} = eE$. The relative weight of the second "localized" contribution to the current tends to decrease with respect to the first "plasma" contribution with increasing field strength due to the reversed scaling of x_0 and $\dot{x}$ with E. From $v_0 \approx 0$, it follows that the term $e\dot{n}\dot{x}$ vanishes in the expression of the second time derivative of the polarization, which therefore simplifies to

$$\ddot{P}_{nl}(t) = \frac{e^2}{m} n_e(t) E(t) + W_b \frac{\partial}{\partial t}\left(\frac{\dot{n}_e(t)}{E(t)}\right), \tag{22}$$

where

$$n_e(t) = n_a\left(1 - \exp\left[-\int_{-\infty}^{t} dt' w\{E(t')\}\right]\right), \tag{23}$$

n_a is the density of atoms, and $w(E)$ is the optical-field ionization rate (see Sec. VI). Equations (22) and (23) in combination with the optical-field ionization rate provide an explicit constitutive law for field-ionizing media. The first term in Eq. (22) predominantly modifies the phase of the propagating strong light wave, giving rise to a blueshift and broadening of its frequency spectrum. This is the only term showing up in previous work (Brunel, *et al.*, 1990; Gildenburg *et al.*, 1990; 1995; Wood *et al.*, 1991; Kan *et al.*, 1997). Our derivation yields an additional contribution proportional to W_b, giving proper account for the loss of light energy due to ionization. It will be shown in the next section that this term can be significant in regimes of practical importance. The same formalism can be used for deriving the constitutive law for arbitrary polarization. In deriving Eq. (22) we have neglected possible recombination of the electron into its original bound state approximately an optical cycle after its detachment, because this process occurs with a rather low probability and contributes negligibly to the medium polarization in the spectral range of an incident light field. Nevertheless, this recombination process plays a central role in high-order harmonic generation and hence will be discussed in detail in Sec. VII.

Whether or not Eq. (22) is energy conserving can be checked by calculating the change in field energy density due to interaction of the light pulse with the ionizing medium $\delta\rho_E = \int_{-\infty}^{\infty} dt\, E\dot{P}$. Using Eq. (22) we obtain

$$\delta\rho_E = -\frac{e^2}{2m}\int_{-\infty}^{\infty} dt_0 A^2(t_0)\dot{n}_e(t_0) - n_{ef}W_b, \tag{24}$$

where $A(t)$ is the vector potential defined by the relation $E = -\dot{A}$ and by the boundary condition $A(t \to \pm\infty) = 0$ and $n_{ef} = \lim_{t\to\infty} n_e(t)$. The second term in Eq. (24) is the energy density that needs to be transferred to the medium for producing free electrons with zero kinetic energy. Hence, for energy conservation, the first term must account for the kinetic energy of the free electrons. This results from a drift motion at a velocity $v_d = \lim_{t\to\infty} \dot{x}(t)$, which remains after the light pulse has left the (infinitesimal) interaction volume. Simple solution of $m\ddot{x} = eE$ by imposing the initial condition $v_0 = 0$ results in $v_d = (e/m)A(t_0)$ (Brunel, 1990). Therefore the first term in Eq. (24) totals the kinetic energy $mv_d^2/2$ of the free electrons ejected at all possible instants t_0, as required by energy conservation.

So far we have assumed that the polarization response of the atomic medium is *local*, i.e., at a fixed position in space $P(t)$ depends only on the electric-field evolution $E(t')$ for $t' \leqslant t$ at the same location. Therefore the spatial coordinate was dropped from the arguments of P and E for simplicity. This is the electric dipole approximation, which applies to good accuracy even after the onset of strong ionization as long as the interaction is nonrelativistic. For electric-field strengths sufficiently high to accelerate free electrons to velocities approaching c, this approximation breaks down, because the

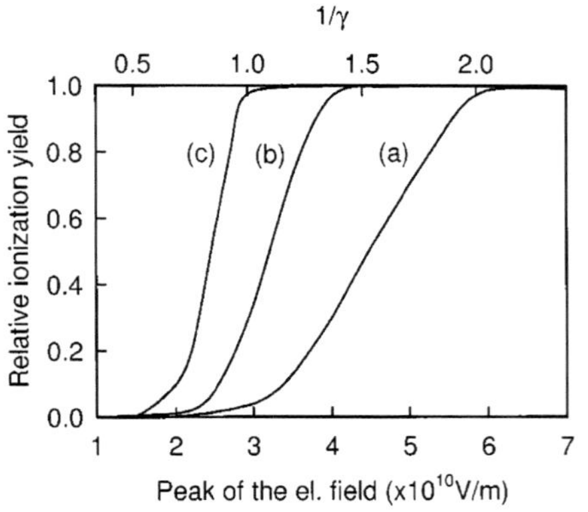

FIG. 26. Ionization yield in hydrogen after the laser pulse defined by Eq. (33) vs peak electric-field strength for laser pulses of durations of (a) $\tau_p=10\,\text{fs}$, (b) $\tau_p=100\,\text{fs}$, and (c) $\tau_p=1000\,\text{fs}$ carried at $\lambda_0=0.8\,\mu\text{m}$. Ionization was calculated by numerical solution of the time-dependent nonrelativistic Schrödinger equation.

magnetic component of the Lorentz force becomes comparable to the electric component and free electrons travel distances comparable to the wavelength during T_o. As a consequence, the polarization response becomes a *nonlocal* function of the fields. Light-matter interactions become relativistic as the parameter

$$\alpha_{rel}=\frac{eE_0}{mc\omega_0}=\sqrt{\frac{4U_p}{mc^2}} \tag{25}$$

approaches or exceeds unity (Sarachick and Schappert, 1970). In the visible and near-infrared spectral range, this condition is fulfilled for intensities on the order of $10^{18}\,\text{W/cm}^2$ and higher ($\alpha_{rel}\approx 0.9$ for $I=10^{18}\,\text{W/cm}^2$ at $\lambda=1\,\mu\text{m}$), giving rise to a number of striking phenomena including relativistic self-channeling (Pukhov and Meyer-ter-Vehn, 1996), the generation of collimated multi-MeV electron jets (Modena *et al.*, 1995; Umstadter *et al.*, 1996), and nonlinear Thomson scattering (Chen *et al.*, 1998).

C. The role of the pulse duration in strong field physics

Before we leave this section, a brief comment on the role of the pulse duration τ_p in strong-field interactions is in order. Because the intensity $I(t)$ starts from zero in each laser pulse, nonlinear interaction always comes into play in the perturbative regime first, which is possibly turned into a strong-field process at higher intensities on the leading edge of the pulse. Clearly, the extent to which the overall interaction occurs in the strong-field regime sensitively depends on τ_p. Figure 26 depicts the fraction of hydrogen atoms left behind ionized by laser pulses of different duration as a function of the electric field amplitude at the pulse peak E_0. The inverse Keldysh parameter γ^{-1} [defined by Eq. (21)], corresponding to this field strength, is also depicted. The results shown in Fig. 26 reveal that in the visible/near-infrared spectral range ionization is essentially completed before the strong-field regime ($\gamma^{-1}>1$) is entered for pulse durations of 1 ps or longer. Even for pulse durations as short as 100 fs, a significant fraction of the atoms is ionized in the intermediate region between the perturbative and strong-field regimes [Fig. (24)] via multiphoton channels of comparable magnitude (Agostini *et al.*, 1979). Only in the 10-fs regime does multiphoton preionization become negligible and optical-field ionization fully take over. As a consequence, *in the visible/near-infrared spectral range pure strong-field interactions can only be induced by few-cycle laser pulses.*

Figure 26 also reveals that the shorter the pulse, the stronger the laser field the electron experiences at the instant of its detachment. As a direct consequence, with a few-cycle laser pulse (i) the atom can be driven much more strongly before its dipole moment drops dramatically due to ionization and (ii) the detached electron can be ejected with a much higher drift velocity into the surrounding plasma. In addition, these processes can be confined temporally to a small fraction of T_o. Hence few-cycle light pulses open up previously inaccessible parameter regimes in high-field physics.

V. PROPAGATION OF INTENSE LIGHT PULSES

The constitutive laws derived in the previous section relate the induced nonlinear polarization P_{nl} to the driving electric field. The Fourier transform of $P_{nl}(t)$ generally contains frequency components within and around the spectral range of the incident light pulse. It may also contain frequency components far from ω_0, giving rise to high-order harmonics of the laser field, which will be addressed in Sec. VII. Here, we consider the components near ω_0, which modify the light pulse upon propagation through an extended interaction volume.

In the nonrelativistic regime, i.e., within the frame of the dipole approximation, a first-order propagation equation, which contains only the first-order derivative with respect to the propagation coordinate, can be derived. The significance of the first-order propagation equation lies in the fact that it can be solved with substantially less computational effort than can Maxwell's wave equation (Ziolkowski and Judkins, 1993), which is of second order in the propagation coordinate. In its most universal, although not directly usable, form the first-order propagation equation can be formally given in the complex Fourier domain as

$$[\partial_z - ik(\omega)]\tilde{E}(\mathbf{r},\omega)=\frac{i}{2k(\omega)}\nabla_\perp^2\tilde{E}(\mathbf{r},\omega) - \frac{i\omega}{2\varepsilon_0 n(\omega)c}\hat{F}[\tilde{P}_{nl}(\mathbf{r},t)]; \tag{26}$$

for a derivation see Appendix B. The first-order propagation equations applying specifically to the perturbative and strong-field regimes will be derived from Eq. (26) below. Here, $\tilde{E}(\mathbf{r},\omega)=\hat{F}[E(\mathbf{r},t)]$, the Fourier transform is defined by $\tilde{f}(\mathbf{r},\omega)=\hat{F}[f(\mathbf{r},t)]=\int dt\, f(\mathbf{r},t)\exp(i\omega t)$. Here the tilde again represents complex quantities. Further, the light wave is assumed to propagate along the z direction, $k(\omega)=\omega n(\omega)/c$ is the (generally complex) propagation constant, $n(\omega)$ is the linear refractive index, $\nabla_\perp=\partial_x^2+\partial_y^2$ is the transverse Laplace operator, and ∂_i stands for partial derivation with respect to coordinate i.

Two approximations lead to Eq. (26). First, the electric field is assumed to be slowly varying (i.e., exhibiting little variation over a distance of one wavelength) in the transverse dimensions, $|\partial_{x,y}\tilde{E}(\mathbf{r},\omega)|\ll\beta_0 E$, where $\beta_0=2\pi/\lambda_0$. This assumption allows wave propagation to be described with the *scalar wave equation* (Yariv and Yeh, 1984). Second, the wave packet is assumed to be subject to little change along a propagation distance equal to the wavelength, which has been referred to as the *slowly-evolving-wave approximation* (Brabec and Krausz, 1997).

It is generally rather inconvenient (in some cases impossible) to handle the Fourier transform of P_{nl}; therefore a transformation of Eq. (26) into the time domain is desirable. Fortunately, this is feasible in both the perturbative and the strong-field regimes of nonlinear optics, but for entirely different physical and mathematical reasons. Hence, starting from Eq. (26), different routes must be followed to reach the first-order propagation equation applicable to the respective parameter ranges.

A. Perturbative nonlinear optics

In the perturbative regime (assuming nonresonant interaction), the nonlinear response is usually much weaker than the linear one, $|P_{nl}|\ll|\varepsilon_0\chi^{(1)}E|$; hence dispersive effects resulting from the frequency dependence of the linear response tend to significantly affect femtosecond pulse propagation. Proper description of dispersive effects in the time domain relies on an expansion of $k(\omega)$ in a power series about the carrier frequency ω_0, which requires a physically meaningful definition of ω_0. This is feasible even for pulse durations approaching the carrier oscillation period T_o, as concluded in Sec. III, permitting a decomposition of wave packets into carrier and envelope,

$$E(\mathbf{r},t)=\tilde{E}_a(\mathbf{r},t)e^{i(\beta_0 z-\omega_0 t+\varphi_0)}+\text{c.c.} \tag{27}$$

Here $\tilde{E}_a$ is the complex envelope decribing both the amplitude envelope and possible phase modulation, $\beta_0=\text{Re}[k(\omega_0)]=\omega_0 n_0/c$ and $n_0=\text{Re}[n(\omega_0)]$. Owing to the simple power law connecting P_{nl} to E, as given by Eq. (19), the induced nonlinear polarization wave carried at ω_0 can be expressed as

$$P_{nl}(\mathbf{r},t)=\tilde{B}(\mathbf{r},t,\tilde{E}_a)e^{i(\beta_0 z-\omega_0 t+\varphi_0)}+\text{c.c.} \tag{28}$$

in the perturbative regime. These circumstances allow one to derive from Eq. (26) a time-domain first-order propagation equation for $\tilde{E}_a(\mathbf{r},t)$ that is free from the fast-oscillating component of E and P_{nl} (Appendix B):

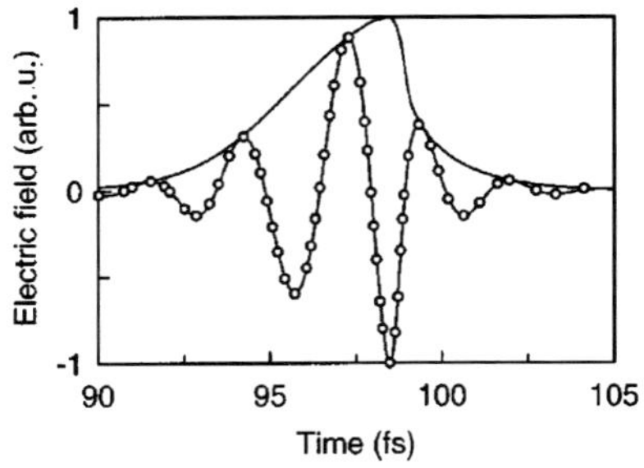

FIG. 27. Solution of the first-order propagation (FOP) equation [Eq. (29)] in a nonlinear Kerr medium, characterized by Eq. (30) in arbitrary units, vs propagation time: solid curve with circles, electric-field strength; solid curve without circles, the field envelope. The FOP equation was solved by a split-step Fourier method (Agrawal, 1995). ○, the electric field obtained from a solution of the Maxwell equation. In advancing the coupled Maxwell equations in space, a leapfrog method was used; the time derivatives were calculated by utilizing a fast Fourier transform. The initial pulse has a sech shape and the parameters are $I_0=4\times10^{13}\,\text{W/cm}^2$, $n_2=3\times10^{-16}\,\text{cm}^2/\text{W}$, $\omega_0=2.35\,\text{fs}^{-1}$ ($\lambda_0=0.8\,\mu\text{m}$, $T_0=2.67\,\text{fs}$), $\tau_p=T_0=2.67\,\text{fs}$ (one optical cycle within the FWHM width of the pulse), $n_0=1.45$, $dn/d\omega|_{\omega_0}=0$ (no dispersion), propagation distance $z=20\,\mu\text{m}$, and propagation time $t=96.7\,\text{fs}$. In the absence of dispersion, self-steepening due to the Kerr nonlinearity creates an optical shock wave at the trailing edge of the pulse. The critical distance for self-steepening is 23.26 μm, which is close to the chosen propagation distance. Even along the shock front, where the change of the envelope is comparable to the carrier frequency, the two solutions are virtually identical.

$$\begin{aligned}(\partial_z+\beta_1\partial_t)\tilde{E}_a(\mathbf{r},t) &= -\frac{\alpha_0}{2}\tilde{E}_a+i\hat{D}\tilde{E}_a+\frac{i}{2\beta_0}\hat{T}^{-1}\nabla_\perp^2\tilde{E}_a \\ &\quad+\frac{\beta_0}{2\varepsilon_0 n_0^2}i\hat{T}\tilde{B}(\mathbf{r},t,\tilde{E}_a),\end{aligned} \tag{29}$$

where $\hat{T}=1+(i/\omega_0)\partial_t$, $\hat{D}=-(\alpha_1/2)\partial_t+\Sigma_{m=2}^{\infty}(\beta_m+i\alpha_m/2)m!^{-1}(i\partial_t)^m$, $\beta_m=\text{Re}[(\partial^m k/\partial\omega^m)_{\omega_0}]$, and $\alpha_m=\text{Im}[(\partial^m k/\partial\omega^m)_{\omega_0}]$. The lowest-order and hence dominant contribution to $\tilde{B}$ is introduced by the optical Kerr effect:

$$\tilde{B}=2\varepsilon_0\Delta n\tilde{E}_a=\frac{4\varepsilon_0 n_0}{Z_0}n_2|\tilde{E}_a|^2\tilde{E}_a, \tag{30}$$

where Δn is given by Eq. (1) and the relationship $I=(2n_0/Z_0)|\tilde{E}_a|^2$ connecting the cycle-averaged intensity with the complex envelope of the electric field was used. Because the optical Kerr effect represents the dominant nonlinear effect in the perturbative regime, Eq. (29) in combination with Eq. (30) covers a wide range of phenomena, including those of central importance for the generation and manipulation of ultrashort pulses (see Sec. III). The results presented in Fig. 27 confirm that the first-order propagation Eq. (29) works well down to the single-cycle regime, and recent investi-

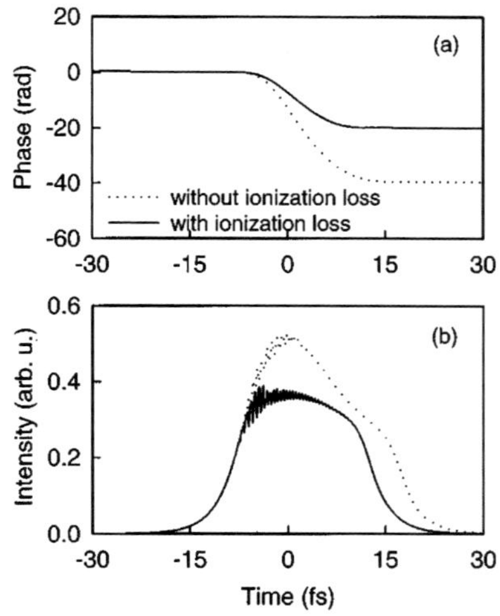

FIG. 28. A frequency-doubled Ti:sapphire laser pulse ($\lambda_0 = 0.4\,\mu$m) after a propagation distance of 8 mm in helium (500 Torr): (a) Phase; (b) intensity envelope for the following laser parameters: $I_0 = 2\times10^{15}$ W/cm^2, $\tau_p = 10$ fs. Dotted curve, the solution of the wave Eq. (31), in one space dimension without the ionization-loss term; solid curve, solution with the ionization-loss term. Ionization loss appears to have a strong influence on the evolution of a few-cycle laser pulse. Because more electrons are ionized at the peak of the pulse than in the pulse wings, the ionization loss leads to a stronger reduction of the pulse peak and hence, to lengthening of the pulse. As a result the nonlinear interaction saturates faster and the blue-shifting and spectral broadening of the pulse is reduced. The observed behavior has important implications for strong-field phenomena, such as a modification of the effective interaction length for x-ray lasing and high harmonic generation.

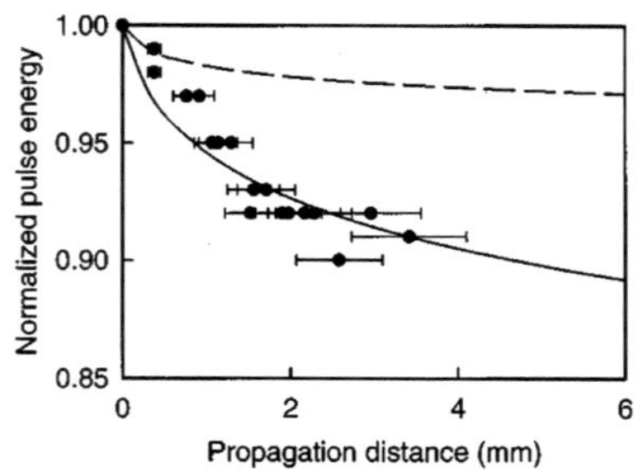

FIG. 29. Output pulse energy normalized to input pulse energy vs propagation distance for a laser pulse propagating in helium at a pressure of 500 Torr, showing the loss experienced by a laser pulse in a tunnel ionizing medium. ●, experimental values obtained with a laser pulse of $\tau_p \approx 5$ fs with $\lambda_0 = 0.8\,\mu$m and of $I_0 \approx 2\times10^{15}$ W/cm^2. The error bars relate to an estimated $\pm20\%$ uncertainty in the measurement of the gas-interaction length. Solid and dashed curves, the theoretical values calculated from a solution of the first-order propagation wave equation [Eq. (31)] with and without the ionization loss term for peak intensites 2 and 3×10^{15} W/cm^2. Our analysis clearly reveals the necessity of the ionization loss term to account for the experimental data.

gations reveal that it also accurately accounts for the spatiotemporal dynamics of tightly focused femtosecond pulses (Fibich and Papanicolaou, 1997; Ranka and Gaeta, 1998; Zozulya *et al.*. 1999). Its simplest one-dimensional form (with $\alpha_m = 0$ for $m \geq 0$ and $\beta_m = 0$ for $m \geq 3$) is known as the nonlinear Schrödinger equation and has been widely used for modeling ultrashort pulse propagation in optical fibers (Agrawal, 1995).

B. The strong-field regime

A strikingly different approach must be taken in the strong-field regime. A glance at Eqs. (22) and (23) reveals that P_{nl} cannot be cast in the form given in Eq. (28), hence the ansatz (27) is not beneficial. On the other hand, P_{nl} originating from the electrons set free in the atomic ensemble is much larger than the contribution of the remaining bound electrons to the overall polarization response. As a consequence, linear dispersion is negligible as compared to that introduced by P_{nl}. Using the approximation $n(\omega) \approx 1 \to k(\omega) \approx \omega/c$, the combination of Eqs. (26) and (22) yields

$$\left(\partial_z + \frac{1}{c}\partial_t\right)E(\mathbf{r},t) = \frac{c}{2}\nabla_\perp^2 \int_{-\infty}^{t} dt' E - \frac{e^2}{2\varepsilon_0 mc}\int_{-\infty}^{t} dt' n_e E - \frac{W_b}{2\varepsilon_0 c}\frac{\partial_t n_e}{E}, \tag{31}$$

where $n_e(\mathbf{r},t)$ is connected to E by Eq. (23) and by the quasistatic optical-field ionization rate introduced in Sec. VI. In contrast with Eq. (29), this first-order propagation equation applies directly to the real electric field (Esarey *et al.*, 1991). In combination with the optical-field ionization rate, this equation provides a powerful tool for describing ultrashort-light-pulse propagation in a field-ionizing gas medium (Geissler *et al.*, 1999a). Like Eq. (29), this equation is valid within the frame of the slowly-evolving-wave approximation (Brabec and Krausz, 1997), which applies to good accuracy as long as the condition $\omega_p^2/\omega_0^2 \ll 1$ is fulfilled, where

$$\omega_p = \left(\frac{e^2 n_e}{m\varepsilon_0}\right)^{1/2} \tag{32}$$

is the plasma frequency of the ionized propagation medium. If the plasma frequency is much lower than the laser carrier frequency, Eq. (31) provides an accurate description of the interaction of intense femtosecond light pulses with an extended atomic gas target in the intensity range confined by $\gamma < 1$ and $\alpha_{rel} \ll 1$. As Eq. (31) depends on the electric field and no complex envelope was defined, it is valid for arbitrarily short pulse durations. Figure 28 shows the intensity envelope and phase of initially 5-fs pulses after propagation through ionizing helium, calculated using Eq. (31). The obtained energy loss suffered by the laser pulse according to Eq. (24) is in good agreement with experimental observations (Fig. 29).

Equations (31) and (23) take proper account of nonlinear interactions in the strong-field regime where the use of intense few-cycle pulses pushes the nonlinear response of (ionizing) matter to unprecedented extremes. As a result, coherently generated harmonics of the driving laser radiation now exceed the 300th order. Modeling even predicts significant further improvement (in terms of both photon energy and conversion efficiency) in previously unaccessed parameter ranges and provides relevant guidelines for the experimenters (Sec. VII). This same theoretical framework also accounts for the fascinating temporal behavior of x-ray harmonic radiation (Sec. VII) as well as the dependence of strong-field processes on the carrier phase in few-cycle wave packets (Sec. VIII).

VI. OPTICAL-FIELD IONIZATION OF ATOMS

When matter is exposed to intense laser fields, a wealth of exciting phenomena can be observed, including high harmonic generation (Corkum, 1993), above-threshold ionization (Corkum *et al.*, 1989), atomic stabilization (Pont and Gavrila, 1990), x-ray lasing (Lemoff *et al.*, 1995; Hooker *et al.*, 1995), laser-induced damage of dielectrics (Du *et al.*, 1994; Stuart *et al.*, 1995; Lenzner *et al.*, 1998; Tien *et al.*, 1999) and molecular dissociation (Seidemann *et al.*, 1995; Chelkowski *et al.*, 1996). The key process triggering all of these strong-field phenomena is ionization. The theoretical investigation of these processes is significantly simplified by the analytic description of ionization that is available in the limiting cases characterized by $\gamma \gg 1$ and $\gamma \ll 1$ [see Eq. (21)]. The two limiting cases of multiphoton ionization and optical-field ionization are illustrated schematically in Figs. 30(a)–(c), respectively.

In what follows we shall restrict our discussion to the strong-field limit $\gamma \approx < 1$, for which significant ionization takes place. In this parameter range the quasistatic approximation is valid, which relies on the assumption that the perturbed electron wave function reaches a quasistatic state before the electric field changes significantly (Shakeshaft *et al.*, 1990). Then the fraction of electrons ionized in the laser field $E(t)$ as a function of time t may be calculated by Eq. (23). At the threshold for the detachment of the second (or higher) electrons, Eq. (23) must be generalized by calculating a sum over the individual ionization processes, $(1/n_a)\Sigma_i n_{ei}(t)$, determined by the ionization rates w_i for the ith electron. Furthermore, possible nonsequential ionization channels must be taken into account (Fittinghoff *et al.*, 1992; Corkum, 1993; Walker *et al.*, 1994; Augst *et al.*, 1995; Lablanquie *et al.*, 1995; Brabec *et al.*, 1996; Dorner *et al.*, 1998; Rosen *et al.*, 1999).

The power of the quasistatic approximation rests on the fact that ionization in time-varying laser fields may be calculated by using the static field ionization rate w. There exist two approaches for analytic calculation of static field ionization rates, namely, the Keldysh theory (Keldysh, 1965; Reiss, 1980) and the Ammosov-Delone-Krainov theory (Oppenheimer, 1928; Perelomov *et al.*, 1996; Ammosov *et al.*, 1986). In Fig. 31, the Ammosov-Delone-Krainov and Keldysh ionization rates for hydrogen and helium atoms exposed to a static electric field are compared with the rates obtained from an exact numerical solution of the time-independent Schrödinger

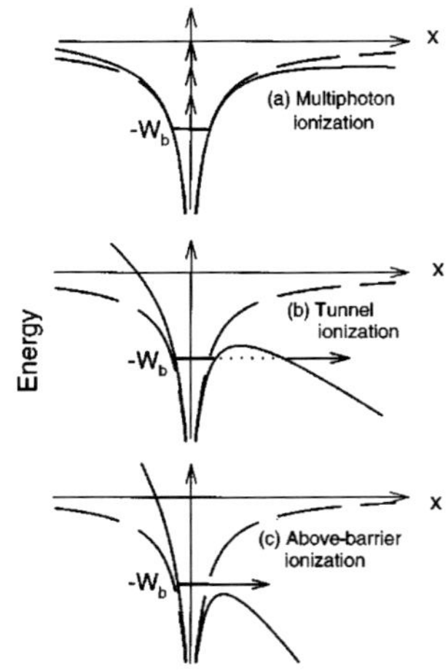

FIG. 30. Regimes of atomic ionization. Exposing an atom to an intense laser field will result in a modified potential (solid curve) composed of the Coulomb potential (dashed curve) and the time-dependent effective potential of the optical pulse. (a) At moderate intensities the resulting potential is close to the unperturbed Coulomb potential and an electron can be liberated only upon simultaneous absorption of N photons, resulting in multiphoton ionization. The multiphoton ionization rate scales with the Nth power of the intensity of the optical pulse. (b) At sufficiently high field strengths the Coulomb barrier becomes narrow, allowing optical tunneling ionization to take over and resulting in a tunneling current that follows adiabatically the variation of the resultant potential. (c) At very high field strengths, the electric field amplitude reaches values sufficient to suppress the Coulomb barrier below the energy level of the ground state, opening the way to above-barrier ionization.

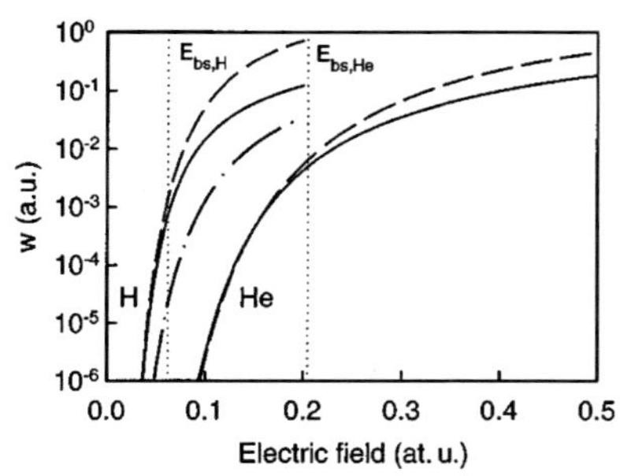

FIG. 31. Static field ionization rates in hydrogen and helium atoms vs the electric-field strength in atomic units: solid curve, numerical result; dashed curve, Ammosov-Delone-Krainov formula; dash-dotted curve, Keldysh theory for hydrogen. The dotted line denotes the barrier suppression field strength for H and He.

equation (Scrinzi *et al.*, 1999). The Keldysh theory predicts ionization rates that are smaller by 1–2 orders of magnitude. The reason for this discrepancy is the neglect of the Coulomb potential in the Keldysh theory.

By contrast, the Ammosov-Delone-Krainov rate is remarkably accurate in the tunneling regime characterized by $\gamma<1$ and $E<E_{bs}$. Here E_{bs} is the external field strength suppressing the peak of the binding potential to $-W_b$ along the direction of the electric field. For $E \geqslant E_{bs}$, the electron escapes "above the barrier" from its bound state (Augst *et al.*, 1989; Krainov, 1997) as illustrated in Fig. 30(c); therefore we refer to this process as above-barrier ionization. For $E \geqslant E_{bs}$, the Ammosov-Delone-Krainov theory increasingly overestimates the ionization rate and tends to lose its validity. In hydrogenlike atoms, the barrier supression field strength in atomic units is given by $E_{bs}=W_b^2/(4Z)$, where Z is the charge of the residual atom. Inserting the barrier supression field strength into the Keldysh parameter, we obtain $\gamma_{bs}=16\omega_0 Z/(2W_b)^{1.5}$. For a hydrogen atom and for a center wavelength $\lambda_0=0.8\,\mu$m, $\gamma_{bs}\approx 1$ so that tunneling does not take place. When the laser peak intensity is increased multiphoton ionization goes over directly into above-barrier ionization. Further, for the valence electrons of the noble gases and for the visible and near-infrared wavelength range, $\gamma_{bs}\approx 0.5$–2, showing the dominance of above-barrier ionization in this experimentally important parameter regime. Tunneling plays a dominant role only in the low-frequency limit that is realized for large λ_0 or W_b. For example, $\gamma_{bs}\approx 0.05$ for valence electrons of noble gases exposed to CO_2 laser radiation at $\lambda_0=10\,\mu$m.

It was shown recently that for $\gamma \approx <1$ above-barrier ionization may also be calculated by using the quasistatic approximation (Scrinzi *et al.*, 1999). Therefore, in order to describe above-barrier ionization correctly, numerically determined, static ionization rates for a broad range of atoms are required. The exact static ionization rate of helium plotted in Fig. 31 was obtained by a solution of the two-electron Schrödinger equation (Scrinzi *et al.*, 1999). Currently, exact static rates $w(E)$ exist only for hydrogen and helium. It is hoped that $w(E)$ can be computed for more complex atoms by using, e.g., R-matrix techniques (Burke and Burke, 1997) in the near future.

Finally, we use Eq. (23) in combination with the computed $w(E)$ as given in Fig. 31 to investigate strong-field ionization induced by light pulses containing only a few field oscillation cycles. The laser pulse used in our calculation is defined by the vector potential

$$A(t)=-(E_0 c/\omega_0)\mathrm{sech}(t/\tau_0)\sin(\omega_0 t+\varphi_0), \qquad (33)$$

from which the electric field is determined by $E=-dA/dt$. Here, the normalization parameter τ_0 is related to the FWHM pulse duration by $\tau_p=1.76\tau_0$. The peak intensity of the laser pulse in Fig. 32 is $I_0=4\times 10^{15}$ W/cm^2. Finally, the center wavelength is $\lambda_0=0.8\,\mu$m ($\omega_0=0.057$ a.u.) and the initial phase is chosen as $\varphi_0=0$.

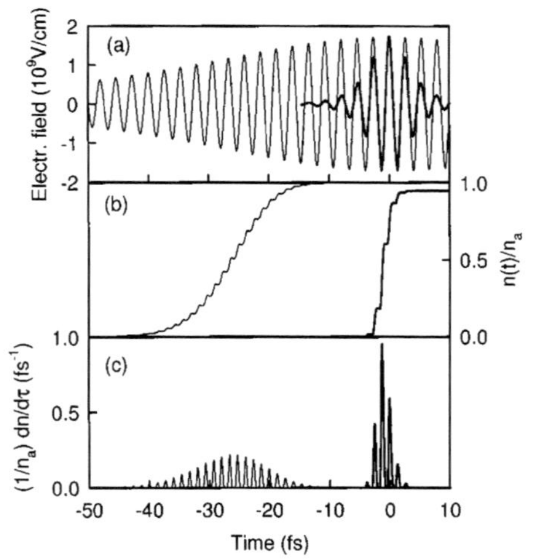

FIG. 32. Helium in the presence of a linearly polarized electric field of a laser pulse with $\lambda_0=0.8\,\mu$m and a peak intensity $I_0=4\times 10^{15}$ W/cm^2: (a) electric field; (b) fraction of ionized electrons; (c) instantaneous ionization rate. The thin and the heavy lines represent pulses of durations of $\tau_p=50$ fs and $\tau_p=5$ fs, respectively.

Figure 32 depicts ionization caused by a 50-fs and a 5-fs pulse in helium. In the multi-cycle regime, a number of ionized helium atoms can accumulate over many optical cycles. As a result, the ground state is depleted long before the peak of the laser pulse. For few-cycle pulses, saturation of ionization is shifted to considerably higher intensities. Consequently the ionization rate is enhanced and can even become comparable to the carrier frequency, $(dn_e/dt)/(n_a\omega_0)\approx 0.5$. The extreme ionization dynamics created by few-cycle pulses have important consequences for atomic and plasma physics, some of which which will be discussed in the following sections. In Sec. VII we shall analyze implications of few-cycle laser pulses for high harmonic generation, such as the role of nonadiabatic propagation effects, phase matching, and attosecond pulse generation. In Sec. VIII it will be shown that optical-field ionization, high harmonic generation, and other fundamental strong-field processes induced by few-cycle laser pulses become phase dependent, opening the way for control of atom-field interactions with phase-controlled laser fields.

VII. HIGH-ORDER HARMONIC GENERATION

The interaction of intense linearly polarized ultrashort laser pulses with atoms (McPherson *et al.*, 1987; Li *et al.*, 1989; Sarukura *et al.*, 1991; Crane *et al.*, 1992; Faldon *et al.*, 1992; Kondo *et al.*, 1993; L'Huillier and Balcou, 1993; Macklin *et al.*, 1993; Perry and Crane, 1993; Wahlström *et al.*, 1993; Tisch *et al.*, 1994; Myazaki and Takada, 1995), atom clusters (Donelly *et al.*, 1996; Hu and Xu, 1997), and molecules (Ivanov and Corkum, 1993; Liang *et al.*, 1994) results in the generation of high-order harmonic radiation in the extreme ultraviolet (xuv) and soft-x-ray spectral range. For recent reviews, see the articles of Joachain *et al.* (2000) and of Salières *et al.* (1999). The moderate pump-energy requirements and the excellent coherence make high harmonic generation a promising approach to developing a compact bright laboratory short-wavelength source. Rapid progress in high harmonic generation over the last few years has already led to the first applications in core-level (Haight and Seidler, 1994) and plasma (Theobald *et al.*, 1996) spectroscopy and in x-ray fluorescence analysis (Schnürer, Streli, *et al.*, 1999). Unfortunately, the efficiency of high harmonic generation is still too low for many other potential applications. A major limitation arises from the fact that generation of high-order harmonics is inextricably linked to the generation of free electrons, which leads to a phase mismatch between the fundamental and the harmonic beams, limiting the maximum coherence length over which harmonic radiation can grow. As shorter wavelengths are generated in the presence of a higher free-electron density, the phase mismatch increases rapidly with the harmonic order, thus also setting a limit to the highest achievable harmonics. Resolution of this fundamental problem would pave the way towards a number of revolutionary applications, including laboratory x-ray microscopy, holography, and femtosecond time-resolved x-ray diffraction and absorption experiments.

Recently it was demonstrated that the phase mismatch can be drastically reduced by the use of ultrashort driver pulses (Zhou *et al.*, 1996; Schafer and Kulander, 1997), which improved conversion efficiency and extended the range of high harmonic generation into the water window (2.3–4.4 nm) for the first time (Chang *et al.*, 1997; Spielmann *et al.*, 1997; Schnürer *et al.*, 1998). In addition, theoretical analysis has predicted the possibility of generating attosecond pulses via few-cycle-driven high harmonic generation (Christov *et al.*, 1997; de Bohan *et al.*, 1998; Tempea *et al.*, 1999a). In this section we shall review the fundamentals of high harmonic generation with particular emphasis on its implementation with few-cycle driver pulses. We begin with the (microscopic) analysis of the strongly driven atomic dipole moment, then address (macroscopic) propagation effects and phase-matching issues, investigate the temporal structure of the emitted short-wavelength radiation, and briefly discuss possible applications. Atomic units are used for the study of the microscopic response, whereas SI units are preferred in the macroscopic analysis.

The theoretical results are compared with available experimental data, obtained with the sub-10-fs-laser-driven kHz-rate high-harmonic source developed at the Vienna University of Technology (Spielmann *et al.*, 1997; Schnürer *et al.*, 1998 1999b). The laser-atom interaction volume in this source is formed by a thin-wall nickel tube, in which holes are made by the pump laser beam; this tube can be squeezed to yield an effective interaction length as short as <0.2 mm (see Fig. 33). Spectral characterization of high harmonic radiation has traditionally been implemented by means of wavelength dispersive spectrometry based on a grating spectrograph

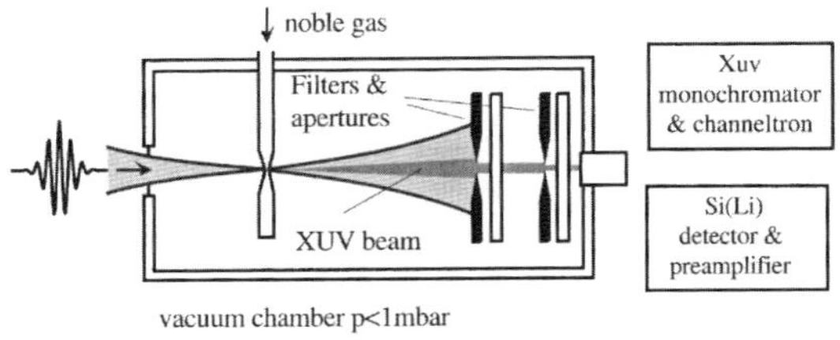

FIG. 33. Schematic of the experimental setup of a typical high harmonic generation experiment. The laser beam is focused into the target chamber. The target is formed by a tube, in whose walls are holes bored by the laser itself. The tube is continuously backed with some noble gas. The generated high harmonic radiation is measured either by wavelength-dispersive spectral analysis or by energy-dispersive x-ray spectrometry (courtesy of Ch. Spielmann).

all the way from vacuum ultraviolet to soft x-ray wavelengths. Alternatively, high pulse repetition rates ($\geqslant 1$ kHz) allow the use of energy dispersive spectrometry in the soft-x-ray regime ($\lambda < 10$ nm). Such a system consisting of a cooled semiconductor (e.g., lithium-drifted silicon) crystal detector and a multichannel pulse-height analyzer offers a significantly better signal-to-noise ratio and hence higher dynamic range in the x-ray regime (<10 nm) at the expense of lower resolution. In both cases, the driving laser beam and the low-order harmonics must be suppressed by apertures and filters in front of the spectrometer, as shown in Fig. 33.

A. Microscopic analysis: The single-atom dipole moment

Emission of harmonic radiation from a single atom or molecule is determined by the dipole acceleration, which in atomic units is $(d^2/dt^2)\langle\Psi|\mathbf{r}|\Psi\rangle$, where $\mathbf{r}$ is the space vector and Ψ is the electron wave function resulting from the solution of the nonrelativistic time-dependent Schrödinger equation.[3] The analysis of high harmonic generation can be considerably simplified by using the model of Lewenstein *et al.* (1994), which is based on two approximations first introduced by Keldysh for the calculation of strong-field ionization processes (Keldysh, 1965; Becker *et al.*, 1994); (i) In positive-energy (continuum) states the effect of the Coulomb potential is neglected so that the electron can be treated as a free particle. The main consequence of this approximation is that ionization is underestimated (see Fig. 31). We have corrected the Lewenstein model of high harmonic generation to remove this deficiency. (ii) The contribution of all other bound states except the ground state to the evolution of the system is neglected (Figueira *et al.*, 1998). This is fulfilled in the strong-field limit $\gamma<1$, where excited bound states are smeared out beyond redemption due to enormous Stark shifts, and intermediate resonances do not affect the transition

[3]See, for example, (Eberly *et al.*, 1989a 1989b; Kulander and shore, 1989; Potvliege and Shakeshaft, 1989; Krause *et al.*, 1992; Xu, 1992; Ben-Tal *et al.*, 1993; Sanpera *et al.*, 1995; Zuo *et al.*, 1995; Bandrauk *et al.*, 1997; Moisey and Weinhold, 1997; Bandrauk and Yu, 1999).

from the ground state to positive-energy (continuum) states. Under the above assumptions (i) and (ii), the quantum-mechanical expectation value of the high-frequency part of the atomic dipole moment responsible for high harmonic generation is obtained as a product of three probability amplitudes:

$$\tilde{a}_h(\tau)=\sum_{\tau_b}\frac{1}{\sqrt{i}}a_{ion}(\tau_b)\tilde{a}_{pr}(\tau_b,\tau)a_{rec}(\tau). \tag{34}$$

Here the tilde again stands for complex quantities and we use a retarded frame of reference $\tau=t-z/c$ to ensure translational invariance of the dipole moment for the macroscopic analysis in the next subsection. The three probability amplitudes in Eq. (34) are given by

$$a_{ion}(\tau_b)=\sqrt{\frac{dn(\tau_b)}{d\tau}}, \tag{35}$$

$$\tilde{a}_{pr}(\tau_b,\tau)=\left(\frac{2\pi}{\tau-\tau_b}\right)^{3/2}\frac{(2W_b)^{1/4}}{E_l(\tau_b)}\exp[-iS],$$

$$S=\int_{\tau_b}^{\tau}d\tau'\{[p(\tau_b,\tau)-A_l(\tau')]^2+W_b\}, \tag{36}$$

$$a_{rec}(\tau)=\frac{p(\tau_b,\tau)-A_l(\tau)}{[W_b+\{p(\tau_b,\tau)-A_l(\tau)\}^2]^3}, \tag{37}$$

where $n(\tau)$ is the free-electron density as given by Eq. (23),

$$p(\tau_b,\tau)=\frac{1}{\tau-\tau_b}\int_{\tau_b}^{\tau}d\tau' A_l(\tau'), \tag{38}$$

and A_l is the vector potential of the fundamental laser field. The instant of birth τ_b as a function of the time τ is determined by solution of the algebraic equation

$$p(\tau_b,\tau)-A_l(\tau_b)=0. \tag{39}$$

The sum in Eq. (34) accounts for the fact that there exist several possible times of birth for the generation of a particular harmonic. Note that for the calculation of the single-atom dipole moment the full electric field of the laser pulse is used, i.e., the original model (Lewenstein *et al.*, 1994) is generalized to account for nonadiabatic effects. Further, the original model is also improved by replacing the Keldysh ionization rate with accurate static ionization rates and allowing for ground-state depletion in Eq. (35) (Ivanov *et al.*, 1996). The model presented here is limited to linearly polarized light, for which the center of gravity of the electron wave packet returns directly to the nucleus as long as the magnetic component of the Lorentz force can be neglected (Appendix C). Elliptic polarization of the driving field makes the center of the electron wave packet miss the parent ion on its return. This reduces overlap with the atomic ground state and hence the recombination probability amplitude $a_{rec}(\tau)$ in Eq. (34), which decreases the harmonic efficiency (Budil *et al.*, 1993; Dietrich *et al.*, 1994; Burnett *et al.*, 1995; Weihe *et al.*, 1995). An extension of the model of Lewenstein *et al.* (1994) to account for elliptic polarization can be found in the literature (Long *et al.*, 1995; Ivanov *et al.*, 1996).

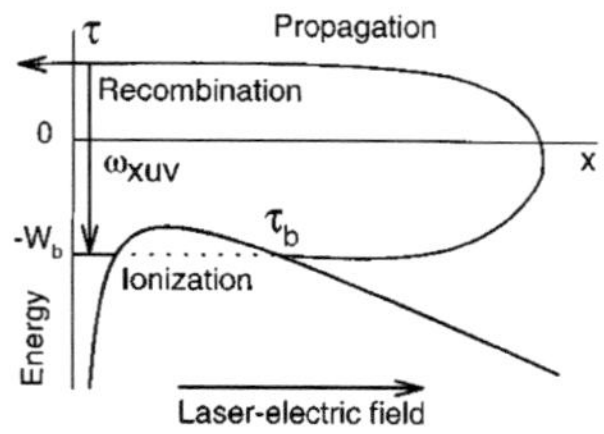

FIG. 34. Schematic illustration of the elementary processes responsible for high harmonic generation. A free electron is "born" at instant τ_b by tunnel ionization, is subsequently accelerated in the laser field, returns to the nucleus, and emits a high-energy extreme ultraviolet photon upon recombination to the ground state at the instant τ.

The probability amplitudes [Eqs. (35)–(37)] reflect the individual processes participating in high harmonic generation and can be interpreted in terms of a semiclassical model (Corkum, 1993; Kulander *et al.*, 1993; Lewenstein *et al.*, 1994) depicted graphically in Fig. 34. The electron wave packet is set free at an instant τ_b and a rate a_{ion} by tunnel ionization, then propagates in the strong laser field accounted for by $\tilde{a}_{pr}$, which brings it back to the nucleus approximately an oscillation period later at instant τ. It recombines to the ground state with probability a_{rec} upon releasing the energy it gained in the laser field plus W_b by emitting a high-energy photon. Due to quasiperiodic repetition of this process in a multicycle laser field, the resulting dipole emission spectrum is discrete, consisting of odd harmonics of the laser frequency ω_0. Figure 35 shows a typical high-order harmonic spectrum originating from Ne atoms irradiated by 30-fs pulses at $\lambda_0\approx 800$ nm, which exhibits well-resolved discrete harmonics. By contrast, the discrete structure is much less pronounced and even completely disappears in the cutoff region for a 7-fs driver. This is because for few-cycle pulses the highest harmonics are generated by a single-electron trajectory close to the peak, so that the periodicity of the high harmonic generation process is completely suppressed. In the time domain, the smooth

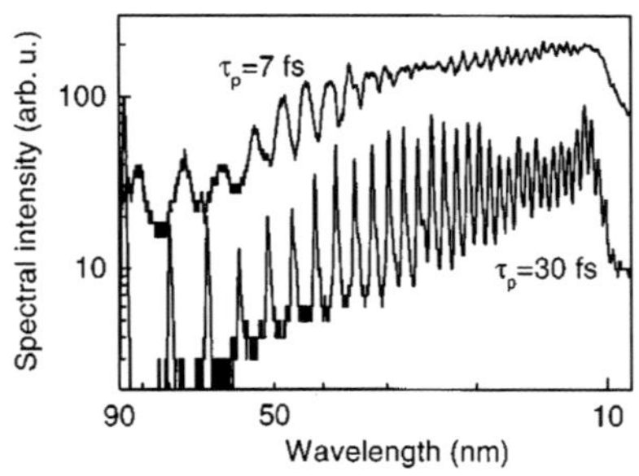

FIG. 35. Harmonic spectra generated in neon at a pressure of 250 Torr for Ti:sapphire laser pulses of 30 and 7 fs (FWHM) carried at $\lambda_0\approx 0.8$ μm. The peak intensity of both pulses was $\approx 3\times 10^{14}$ W/cm^2. At the peak of the 30-fs pulse roughly 10% of the neon atoms are ionized, while at the peak of the 7-fs pulse 1% are ionized (courtesy of M. Schnürer).

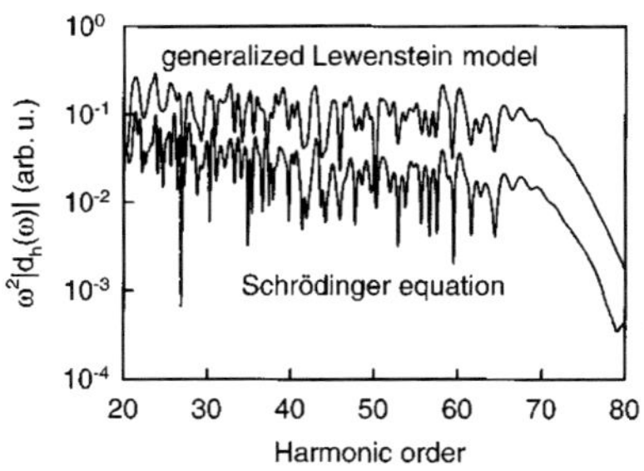

FIG. 36. Fourier transform of the single-atom dipole acceleration $\omega^2|\tilde{d}_h(\omega)|$ for high harmonic generation in hydrogen as determined by solving Eqs. (34)–(39) and the time-dependent nonrelativistic Schrödinger equation. The laser electric field is determined by Eq. (33) and the laser parameters are $I_0=5\times10^{14}$ W/cm^2, $\varphi_0=0$, $\lambda_0=0.8\,\mu$m, and $\tau_p=5$ fs.

spectrum at the cutoff was predicted to correspond to an isolated attosecond burst of radiation (Christov *et al.*, 1997), which will be discussed in more detail at the end of this section. Note that the smooth cutoff spectrum is currently the only (rather indirect) experimental evidence of the existence of subfemtosecond pulses.

The semiclassical picture predicts a cutoff in the harmonic spectrum at a harmonic photon energy (Krause *et al.*, 1992; Corkum, 1993; Kulander *et al.*, 1993; Lewenstein *et al.*, 1994; Milosevic and Starace, 1998, 1999) of

$$N_c\omega_0=W_b+3.17U_p(\tau), \tag{40}$$

where $U_p(\tau)$ is the ponderomotive potential, as defined in Sec. IV, to be inserted in atomic units in Eq. (40). If the ground state is not significantly depleted during the interaction, the fractional ionization is weak, which was the case in the experiments yielding the spectra in Fig. 35. In this limit, the cutoff harmonic N_c is determined by the ponderomotive potential at the pulse peak $U_p(0)$. As the pulses of different duration in Fig. 35 have the same peak intensity, the harmonic spectra are cut off at the same frequency, according to Eq. (40). In the limit of strong laser fields stripping the electron with a probability approaching one, well before the atoms are exposed to the pulse peak, N_c is no longer determined by the peak intensity, but by the laser intensity at which the ground state is depleted on the front edge of the pulse. As for shorter pulses, ground-state depletion is shifted to higher intensities (see Fig. 32), and the cutoff is shifted to higher harmonics (Christov *et al.*, 1996). In fact, the harmonic cutoff can be extended below 4.37 nm into the water window in neon by simply increasing the peak intensity of the 7-fs pulses. This could not be achieved with 30-fs pump pulses because of early depletion of the ground state.

In order to check the quality of the model drawing on Eqs. (34)–(39), we have compared the dipole moment for hydrogen as predicted by Eqs. (34)–(39) to that obtained from an exact solution of the Schrödinger equation for the electric field defined by Eq. (33) for various pulse durations and intensities. A typical result is depicted in Fig. 36. The model based on the theory of Lewenstein *et al.* (1994) overestimates the exact calculations by roughly one order of magnitude. The discrepancy is mainly due to the fact that for the chosen laser intensity, the electron is set free via above-barrier ionization and not via tunneling. Although the ionization rate in Eq. (35) is corrected to describe above-barrier ionization properly, the transverse distribution of the electron wave function at τ_b remains uncorrected and hence inaccurate. Nevertheless, aside from a quantitative difference, the structure of the exact harmonic spectrum is excellently reproduced by the model calculations for high photon energies near the cutoff. For low-order harmonics, the structural conincidence deteriorates due to the increasing influence of the Coulomb potential, which was neglected in the simplified model. Low-order harmonics are generated by electrons returning to the nucleus with low kinetic energies. The trajectory of these electrons is affected more strongly by the atomic Coulomb potential.

B. Macroscopic analysis: Propagation effects

Although many important features of high harmonic generation experiments can be understood from single-atom theory, for a detailed comparison with experiments a macroscopic theory including propagation effects, is required (L'Huillier *et al.*, 1990; 1991; L'Huillier, Lompré, *et al.*, 1992; Rae *et al.*, 1994; Kan *et al.*, 1997; Christov, Murnane, *et al.*, 1998; Tempea *et al.*, 1999a; 1999b). There are three important propagation effects, namely, (i) absorption, (ii) dephasing, and (iii) defocusing, setting a limit to the maximum achievable harmonic yield rate. These limitations are briefly discussed below.

(i) During propagation in the gas, the harmonic radiation can excite core electron states and be reabsorbed (Becker and Shirley, 1996). Numerical values of xuv/x-ray absorption coefficients can be found, for example, on the home page of the Center for X-ray Optics, Materials Sciences Division, Lawrence Berkeley National Laboratory (http://cindy.lbl.gov).

(ii) The difference between the phase velocities of the driving light wave and the high harmonic wave results in a phase mismatch between partial harmonic waves emitted at $z>0$ and $z=0$ in the gas medium. The propagation length at which the phase mismatch reaches π is referred to as the *coherence length*, indicating the propagation distance over which coherent growth of the macroscopic harmonic output tends to terminate. Three major effects contribute to the dephasing between fundamental and harmonic field:

(a) High harmonic generation is inextricably linked to ionization. The phase shift imposed by the free electrons on the laser pulse leads to dephasing because the effect of free electrons on the harmonic wave is negligible (Ditmire *et al.*, 1996; Shkolnikov *et al.*, 1996). The respective coherence length is given by

$$L_{fe}=\frac{2\pi c\omega_0}{\omega_p(\tau)^2 N}. \tag{41}$$

Here, $\omega_p(\tau)$ is the plasma frequency defined in Eq. (32), which is a function of the free-electron density at time τ where the Nth harmonic is generated (Kan *et al.*, 1997). For $n_a=1.75\times10^{19}\,\mathrm{cm}^{-3}$ (500 Torr), $\lambda_0=800$ nm, $N=100$, $n_e(\tau)/n_a=0.1$, the plasma frequency $\omega_p^2=0.005\,57\,\mathrm{fs}^{-2}$, and Eq. (41) gives a coherence length of $L_{fe}\approx8\,\mu$m.

(b) The curved wave front in the focused driving laser beam introduces a phase advance known as the Gouy phase shift, which is another "geometric" source of dephasing between fundamental and harmonic fields. A similar effect occurs in a hollow waveguide due to the wavelength dependence of the propagation constant. The related coherence length can be expressed as (L'Huillier *et al.*, 1991; L'Huillier, Lompré, *et al.*, 1992)

$$L_{g,fs}=\frac{\pi z_0}{N}=\frac{\pi^2 w_0^2}{N\lambda_0} \tag{42}$$

for propagation close to focus in free space, and as

$$L_{g,wg}=\frac{4\pi^2}{u_{11}^2}\frac{a^2}{N\lambda_0}, \tag{43}$$

where w_0 is the $1/e^2$ beam radius at focus, a is the bore radius, and $u_{11}=2.405$ (Marcatili and Schmeltzer, 1964). For optimum coupling, $w_0\approx(2/3)a$, which yields $L_{g,wg}\approx1.56L_{g,fs}$.

(c) The single-atom dipole moment for high harmonic generation depends on the pulse intensity, which is changed during propagation due to diffraction. As a result, the phase of the dipole moment decreases roughly linearly proportional to $-U_p/\omega_0$ with propagation, leading also to dephasing (Salières *et al.*, 1995). However, this contribution does not scale with the harmonic order and hence can usually be neglected for higher harmonics compared to the contributions (a) and (b).

(iii) During high harmonic generation, the laser pulse creates a free-electron density profile, in which the largest density is at the pulse peak and the density goes to zero at the pulse wings. Such a profile gives rise to defocusing, which reduces the laser pulse intensity, imposing a limitation on the effective interaction length over which a particular harmonic can be generated. For high harmonics, this effect is usually also weak compared to (iia) and (iib).

Limitations arising from geometric effects, such as dephasing by the Gouy phase shift (iib) and defocusing, depend on the confocal parameter and therefore can always be suppressed by increasing the pulse energy and the beam radius such that the peak intensity remains constant. These measures cannot remove free-electron-induced dephasing (iia) and absorption losses (i), which are more severe limitations. In what follows, we analyze the influence of these limiting effects on high harmonic generation by using a one-dimensional model (plane-wave propagation). The evolution of the laser field in an ionizing gas is described by

$$\partial_\xi E_l(\xi,\tau)=-\frac{1}{2c}\int_{-\infty}^{\tau}\omega_p^2(\xi,\tau')E_l(\xi,\tau')d\tau' - \frac{W_b}{2\varepsilon_0 c}\frac{\partial_\tau n_e(\xi,\tau)}{E_l(\xi,\tau)} \tag{44}$$

as it can be derived from Eq. (31). Here, we have introduced a coordinate frame moving at the speed of light in vacuum, $\tau=t-z/c$, $\xi=z$. The initial condition for solving Eq. (44) is assumed to be $E_l(0,\tau)=-dA_l/d\tau$, where $A_l(\tau)$ has the same form as given by Eq. (33). The wave equation for the generation and evolution of the harmonic field E_h,

$$\partial_\xi E_h(\xi,\tau)+\alpha_h E_h(\xi,\tau) = -\frac{1}{2\varepsilon_0 c}\partial_\tau\tilde{P}_h[E_l(\xi,\tau)]+\text{c.c.}, \tag{45}$$

is obtained from a derivation similar to that performed in Sec. V. Here, $\tilde{P}_h=kn_a\tilde{d}_h$ is the high-frequency part of the macroscopic nonlinear polarization, $\tilde{d}_h$ is given by Eq. (34), and $k=8.4773\times10^{-30}$ C m (Coulomb-meters) is the conversion factor between atomic and SI units. Further, α_h is the xuv absorption coefficient, which is frequency dependent and must be evaluated in the frequency domain.

C. Absorption-limited xuv and dephasing-limited soft-x-ray harmonic generation

Recently, it was demonstrated that for moderate harmonic orders ($\lambda\geqslant40$ nm) absorption limits the harmonic yield when using ultrashort, 40-fs driver pulses (Constant *et al.*, 1999). With a sub-10-fs driver absorption-limited high harmonic generation can be extended to the border of the soft-x-ray regime at ≈10 nm (Schnürer *et al.*, 1999b). In Fig. 37, the solid and dotted lines are obtained by solving Eqs. (44) and (45) including and neglecting xuv absorption, respectively. A comparison with experimental data (squares) provides conclusive evidence for the limiting role of absorption in few-cycle-driven high harmonic generation at the border of the xuv/soft-x-ray regime in neon. Owing to the steep front edge of the few-cycle driver, harmonics down to the 10–15-nm range can be produced at fractional ionization levels as low as $n_e(\tau_b)/n_a\approx0.5\%$, resulting in a coherence length L_{fe} according to Eq. (41) that exceeds the xuv absorption length. A further conclusive demonstration of absorption-limited high harmonic generation is presented in Fig. 38 showing the harmonic intensity generated by 7-fs pulses in argon as a function of wavelength. Strong harmonic emission is confined to an area of 30 nm as a result of rapidly increasing absorption for longer wavelengths and increasing ionization for shorter wavelengths.

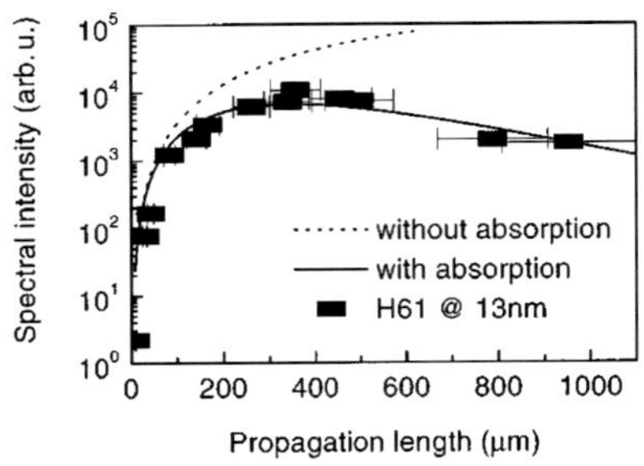

FIG. 37. Evolution of the 61st harmonic upon propagation of a few-cycle laser pulse in a neon target (300 Torr): ■, measured; solid curve, computed; dotted curve, result of the same calculation in the absence of absorption loss, $\alpha=0$. The laser parameters are $\lambda_0=0.8\,\mu$m, $I_0=5\times10^{14}$ W/cm^2, and $\tau_p=5$ fs. The error bars result from uncertainties in estimating effective interaction lengths and the pressure in the interaction region. The computed data are multiplied by a constant scaling parameter yielding best agreement between experiment and theory (Schnürer *et al.*, 1999b).

At sufficiently low ionization levels and for sufficiently large laser beam diameters, the phase advance of the driving laser radiation induced by free electrons and the Gouy phase shift (or waveguide phase shift), which lead to a finite L_{fe} and L_g, respectively, can be compensated for by the phase delay caused by the (positive) contribution of the neutral atoms to the refractive index. This compensation, which can be optimized by fine adjustment of the density n_a (i.e., pressure) of the atomic gas, results in a substantial (ideally infinite) enhancement of the coherence length over a limited fraction of the interaction time. This effect was first demonstrated at around 30 nm in an argon-filled hollow waveguide with 20-fs pump pulses carried at 800 nm (Rundquist *et al.*, 1998). More recently, it was also observed in the interaction of unguided (focused) 7-fs pulses with a neon gas jet, which led to absorption-limited harmonic emission down to 10–15 nm as reported above (Schnürer *et al.*, 1999). By extending the length of coherence growth to many times the xuv absorption length, this "pressure-tuned" phase-matching mechanism allows maximization of the yield (Constant *et al.*, 1999) in absorption-limited xuv harmonic generation with few-cycle laser pulses.

In Fig. 39, absolute measurements of the efficiency of few-cycle-driven high harmonic generation in Ar, Ne

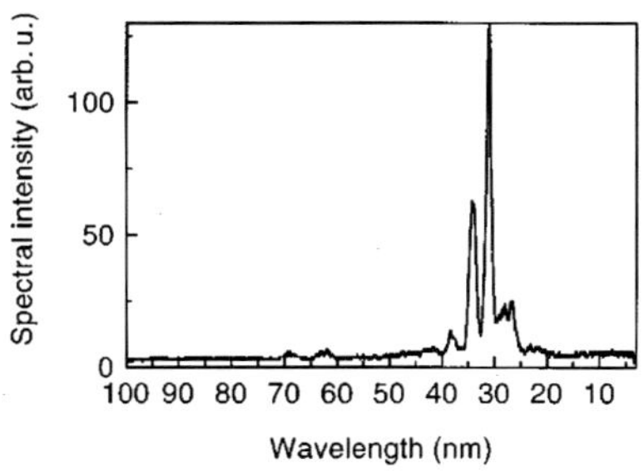

FIG. 38. Spectral harmonic intensity vs wavelength as generated in argon (300 Torr, interaction distance 2 mm). The laser parameters are $\lambda_0=0.8\,\mu$m, $I_0=5\times10^{14}$ W/cm^2, and $\tau_p=5$ fs.

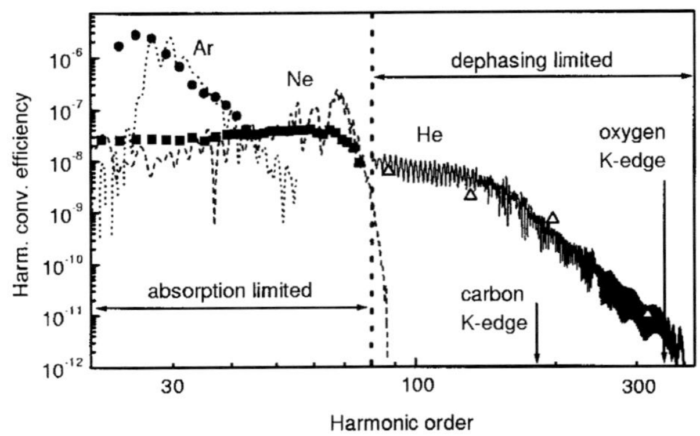

FIG. 39. Absolute measurement of the harmonic production efficiency in argon (225 Torr; effective gas length is 3 mm), in neon (300 Torr; effective gas length is 3 mm), and in helium (3000 Torr; effective gas length is 0.2 mm). The peak intensities in argon, neon, and helium are $I_0=5\times10^{14}$ W/cm^2, $I_0=2\times10^{15}$ W/cm^2, and $I_0=3\times10^{15}$ W/cm^2, respectively. The remaining pulse parameters are $\tau_p=7$ fs, $\lambda_0=0.8\,\mu$m (Ti:sapphire). Calculated spectra: dotted curve, argon; dashed curve, neon; solid curve, helium. For the calculations, a constant scaling factor was used to obtain optimum agreement between experiment and theory. The data obtained with Ar and Ne are accurate within a factor of 2, whereas those produced with He can be regarded as order-of-magnitude estimates (Schnürer *et al.*, 1999a).

and He, and a comparison with numerical simulations are presented. The efficiency of energy conversion into individual harmonics (within a bandwidth equal to twice the laser frequency) is plotted versus harmonic order (Schnürer *et al.*, 1999a; 1999b). Theoretical predictions from Eqs. (44) and (45) are found to be in reasonable agreement with experimental data. The numerical analysis also confirms that the maximum achievable harmonic yield in Ar and Ne is limited by xuv absorption. For few-cycle driver pulses absorption losses present the dominant limitation for wavelengths down to 10 nm. Within a 5% bandwidth, the few-cycle-driven neon harmonic source emits approximately 10^7 photons/pulse at 13.5 nm, a wavelength important for xuv lithography, and a repetition rate of 1 kHz within a time interval estimated as $\tau_x<3$ fs from the coalescence of discrete harmonics to a continuum in this wavelength range (Fig. 35). The enhanced photon yield together with its unprecedented temporal confinement results in peak powers on the order of 0.1 MW. These intense soft x-ray pulses are delivered in a near-diffraction-limited beam and hence should be focusable to peak intensities in excess of 10^{13} W/cm^2. These pulses are sufficiently intense to induce two-photon transitions in the x-ray regime, which may open the way to measuring the harmonic pulse duration. Such pulses will be discussed in Sec. VII E.

In the soft-x-ray regime ($\lambda\leqslant10$ nm) absorption rapidly increases with decreasing wavelength. As higher harmonics are generated in the presence of a larger free-electron density, according to Eqs. (23) and (40), dephasing increases with decreasing harmonic wavelength. Hence in the wavelength region $\lambda<10\,\text{nm}$ dephasing becomes the dominant limitation. Because the coherence length given by Eq. (41) decreases rapidly with increasing order, the high harmonic yield rapidly drops below the detection limit at wavelengths below 10 nm when generated with pulses in the multicycle regime. At these wavelengths, the short rise time of the driver pulse is even more crucial than in the xuv range addressed above. In fact, the extremely steep intensity gradient on the pulse front of few-cycle drivers has allowed the generation of high-order harmonics at significantly reduced free-electron densities and hence increased coherence lengths, resulting in an extension of high harmonic generation at signal levels several orders of magnitude above detection limit into the water window (Chang *et al.*, 1997; Spielmann *et al.*, 1997; Schnürer *et al.*, 1998). Figure 39 plots a measured harmonic spectrum (open triangles) emitted by a helium gas target irradiated by sub-10-fs pulses at $\lambda_0=780\,\text{nm}$ (Schnürer *et al.*, 1999a). Reasonable agreement with theoretical calculations is obtained. The propagated spectrum reveals a considerably stronger wavelength sensitivity than the single-atom dipole moment, which is introduced by the ionization-induced phase mismatch even for these extremely short driver pulses. The weak frequency dependence of the single-atom dipole radiation spectrum down to the water window offers the potential for substantially enhancing the harmonic conversion efficiencies in this wavelength range by devising techniques for making free-electron-induced dephasing less severe.

D. Phase matching of soft-x-ray harmonics

In the previous subsection we concluded that ionization-induced dephasing limits coherence growth of soft-x-ray laser harmonics even if the shortest (few-cycle) driver pulses are used for excitation. Due to this limitation, the efficiency of high harmonic generation is still too low for many important applications. During the last few years a number of phase-matching mechanisms have been proposed and demonstrated in the vuv-xuv range (Milchberg *et al.*, 1995; Kan *et al.*, 1996; Shkolnikov *et al.*, 1996; Peatross *et al.*, 1997; Christov, Kapteyn, *et al.*, 1998; Lange *et al.*, 1998; Rundquist *et al.*, 1998), but none of them could be extended into the x-ray regime. All the phase-matching schemes reported so far rely on the assumption that the phase of the harmonic wave emerging at a fixed position in the coordinate system moving at c grows linearly with propagation distance. In what follows, the phase-matching mechanisms relying on this linear growth are referred to as *adiabatic* (Geissler *et al.*, 1999b). Recent studies of propagation of few-cycle laser pulses in a field-ionizing atomic medium revealed distortions *within* an optical cycle, which can give rise to what we call *nonadiabatic phase matching* at wavelengths well below 10 nm (Tempea *et al.*, 1999b). On the other hand, nonadiabatic pulse shaping also sets a limit to adiabatic phase matching (Geissler *et al.*, 1999b). In what follows, we shall investigate how these effects can be exploited to substantially enhance the coherent growth of soft-x-ray laser harmonics.

All adiabatic phase-matching schemes rely on the approximation that the change in the free-electron density during one optical cycle may be neglected. Under this assumption the phase mismatch grows linearly with propagation distance and perfect phase matching may be achieved, as can be seen from an inspection of the spectral phase.

The spectral phase of a particular harmonic N at the frequency $N\omega_0$ is obtained by the Fourier transform of the second time derivative of the atomic dipole moment as given by Eq. (34), which yields

$$\phi_N=S(\tau_b,\tau_r)-N\omega_0\tau_r. \tag{46}$$

The primary change in ϕ_N during propagation results from the fact that the laser field experiences a phase shift in the presence of ionization that grows linearly with propagation distance, $\phi_l(\xi)=-\omega_p^2\xi/(2c\omega_0)$. In the frame propagating at the vacuum speed of light, which is comoving with the harmonic wave, this phase change shifts the local time τ_r, at which the harmonic is generated, as $\tau_r(\xi)=\tau_r(0)+\phi_l(\xi)/\omega_0$ with propagation distance. By virtue of Eq. (46), the shift of τ_r causes a change of the harmonic phase. In the adiabatic limit the change in the free-electron density during one laser cycle is negligible, hence $\Delta n_e=n_e(\tau_r)-n_e(\tau_b)\approx 0$. This implies that (i) τ_r and τ_b are shifted by the same amount and (ii) the laser electric field experienced by the electron between τ_b and τ_r remains unchanged, as illustrated in Fig. 40(a). As a consequence, the classical action integral S given by Eq. (36) remains constant. Inserting the resulting phase change into the condition $\phi_N(\xi)-\phi_N(0)=N\phi_l=\pi$ yields the coherence length L_{fe}, as given in Eq. (41). A similar procedure yields L_g if ϕ_l is dictated by the Gouy or waveguide phase shift.

The adiabatic solution is, strictly speaking, only valid as long as the laser pulse experiences a constant free-electron density. This is never the case, as high harmonic generation is inextricably linked to ionization. The *nonadiabatic* behavior originates from the variation of the free-electron density n_e [see Eq. (23)] during the interaction. The change in n_e within the optical cycle tends to slightly modify the laser electric-field evolution during the same time. As a result, the trajectory of the freed electrons is governed by slightly different electric-field evolutions at different positions in space during propagation, as illustrated in Fig. 40(b). As a result, the classical action S changes with ξ. This nonadiabatic contribution is opposite in sign to the above-discussed adiabatic contributions; therefore the (adiabatic) linear growth of the harmonic phase ϕ_N is reduced (Tempea *et al.*, 1999b). Nonadiabatic effects become increasingly significant as the pulse duration approaches and enters the few-cycle regime, which is revealed by the ionization rates depicted in Fig. 32. Subcycle modification of the laser electric field has two major impacts on high harmonic generation: (i) Adiabatic phase-matching mechanisms are adversely affected and ultimately limited by the nonadiabatically induced change in ϕ_N during

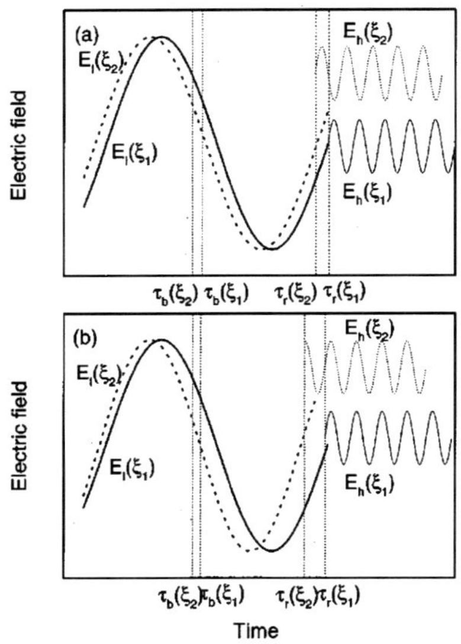

FIG. 40. Schematic of the laser field E_l and of the harmonic field E_h: solid curves, generated at ξ_1; dotted curves, generated at ξ_2, in the limit of (a) adiabatic evolution of a many-cycle laser pulse and (b) nonadiabatic evolution of a few-cycle laser pulse; the coordinate frame is moving with c, the approximate phase velocity of the harmonic wave. In this frame, the harmonic field remains unchanged with propagation and the laser field changes only due to the free-electron-induced phase mismatch. For comments see the text.

propagation. (ii) The nonadiabatic contribution to the change in ϕ_N with ξ can become strong enough that few-cycle drivers can compensate for adiabatic change of the harmonic phase, which is shown in Fig. 40(b), giving rise to nonadiabatic self-phase-matching (Tempea *et al.*, 1999b).

Let us consider a representative adiabatic phase-matching scheme, namely, quasi-phase-matching. A periodic gas-vacuum structure is realized by using an array of gas jets. In order to achieve quasi-phase-matching, one must choose the length of each gas cell to be equal to $L_{fe}L_g/(L_{fe}+L_g)$, so that the harmonic signals at the entrance and end of the gas cell are out of phase by π. For the sake of simplicity, we assume the geometric coherence length $L_g \gg L_{fe}$, for which the length of the cell is $L_{fe}L_g/(L_{fe}+L_g) \approx L_{fe}$. When the vacuum length L_v is equal to L_g, the Gouy phase shift brings the laser and harmonic pulses into phase again, resulting in a continuation of coherent growth during the next gas cell. The Gouy shift accumulated during vacuum propagation is incorporated into our one-dimensional model by multiplying the laser pulse by a factor $\exp[i\tan^{-1}(L_v/z_0)] \approx \exp[iL_v/z_0]$.

The limitations of quasi-phase-matching become apparent in Fig. 41, where the harmonic spectrum integrated over a band confined by $N=430$ and $N=450$ is plotted versus propagation distance for a 5-fs and for a 20-fs driver pulse. As anticipated, noadiabatic limitations become more severe for decreasing pulse duration. Whereas quasi-phase-matching with 20-fs pulses works properly over 35 gas-vacuum periods and enhances the

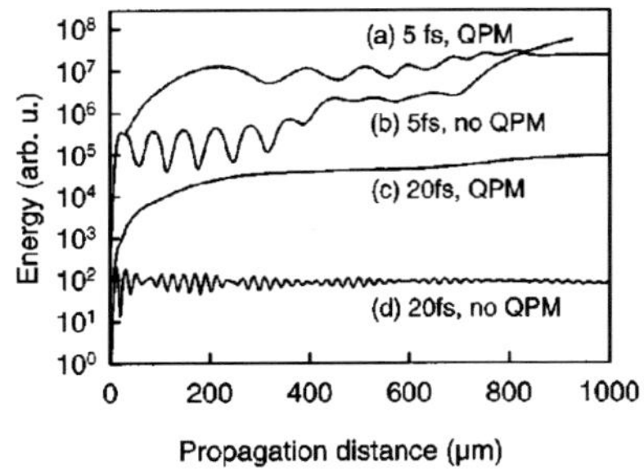

FIG. 41. Growth of the harmonic signal integrated between $N=430$ and $N=450$ vs propagation distance for the following parameters: $\lambda_0=0.8\,\mu$m for all plots; curves (c) and (d) the harmonic signal for $\tau_p=20$ fs, $I_0=3.5\times10^{15}$ W/cm^2 without and with quasiphase matching, respectively. For these parameters, the single-atom cutoff is at $N\approx445$. Curves (a) and (b) high harmonic generation for $\tau_p=5$ fs, $I_0=6.5\times10^{15}$ W/cm^2, without and with quasi-phase-matching (QPM), respectively. Both peak intensities were chosen to saturate ionization at the pulse peak. In order to make possible a comparison of calculations with and without QPM, the vacuum interaction distances were not plotted. Consequently the distance refers to the length propagated in the gas medium. The radius at the beam waist was assumed to be $w_0=0.5$ mm, corresponding to a confocal parameter of $z_0=98$ cm. The geometric coherence length of the harmonic $N=441$ is $L_g=7$ mm. To relax the pulse energy requirements, one could also use smaller confocal parameters for QPM. The helium gas density is 10 Torr, which corresponds to coherence lengths ($N=441$) of $L_{fe}=9.5\,\mu$m for $\tau_p=5$ fs and $L_{fe}=31.25\,\mu$m for $\tau_p=20$ fs, respectively.

harmonic yield by approximately three orders of magnitude, when 5-fs driver pulses are used quasi-phase-matched growth saturates after approximately seven periodic structures and the harmonic yield increases by a factor of ≈50. A comparison of plots (b) and (c) in Fig. 41 reveals that few-cycle pulses are able to produce soft-x-ray high harmonic radiation more efficiently even *without* phase matching than longer multicycle pulses *with* phase matching, owing to the significantly longer intrinsic coherence length L_{fe}, which results from the x-ray harmonic radiation emerging at a lower free-electron density. Adiabatic phase-matching schemes cannot compensate for the shortening of L_{fe} for longer pulses, due to the presence of nonadiabatic limitations (Geissler *et al.*, 1999b). From our investigations we may conclude that *x-ray laser harmonics can be most efficiently generated by few-cycle laser pulses even without having to rely on phase matching. This yield cannot be surpassed by applying an adiabatic phase-matching scheme based on a linear change of the harmonic phase.*

The reason for the sensitivity of nonadiabatic effects to the pulse duration becomes clear in Fig. 42, where the change in the free-electron density is plotted for the two pulse parameter sets of Fig. 41. The change between τ_b and τ_r for the harmonic $N=441$ for $\tau_p=5$ fs ($\Delta n_e \approx 0.07$) is found to be larger by a factor of 5 than that for a 20-fs pulse ($\Delta n_e \approx 0.015$). This explains the fact that in Fig. 41, quasi-phase-matching works for $\tau_p=20$ fs over a five times larger number of gas-vacuum periods than for a 5-fs pulse. Finally, Fig. 42 shows that quasi-phase-

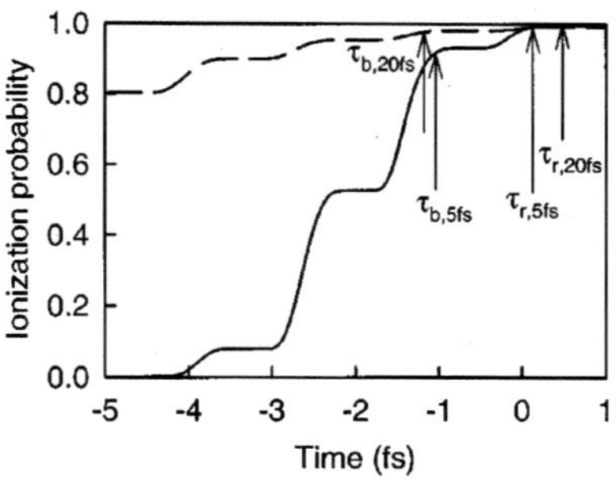

FIG. 42. Ionization profile for a 20-fs and 5-fs laser pulse and the parameters of Fig. 41. The arrows indicate the birth (τ_b) and recombination time (τ_r) of the electron, which generates the harmonic order $N=441$ close to the cutoff. Note that the harmonic $N=441$ is generated in several optical cycles. However, only the electron densities at the times τ_b and τ_r fulfill the quasi-phase-matching conditions chosen for the calculations in Fig. 41.

matching works best for cutoff harmonics produced in the vicinity of the laser pulse peak, where ionization saturates and Δn is small. For plateau harmonics Δn increases rapidly, which reduces the maximum gain achievable. In particular, for few-cycle laser pulses nonadiabatic limitations become so severe that adiabatic phase matching of plateau harmonics does not work at all.

As mentioned above, in the limit of high-intensity, few-cycle laser pulses, nonadiabatic effects can become strong enough to compensate for adiabatic dephasing, which makes nonadiabatic self-phase-matching possible. Nonadiabatic self-phase-matching already comes into play at the intensity level used in Fig. 41, as indicated by the growth of the harmonic signal [plot (b)] for $\xi > 300\,\mu$m. As a result, nonadiabatic self-phase-matching yields an output signal comparable to, or somewhat higher than, the output from the quasi-phase-matching scheme [plot (a)], because the former does not work in the presence of the latter. The performance of nonadiabatic self-phase matching becomes even more evident at somewhat higher intensities, $I_0 \approx 10^{16}\,\mathrm{W/cm^2}$. In Fig. 43, we have plotted the computed harmonic intensity spec-

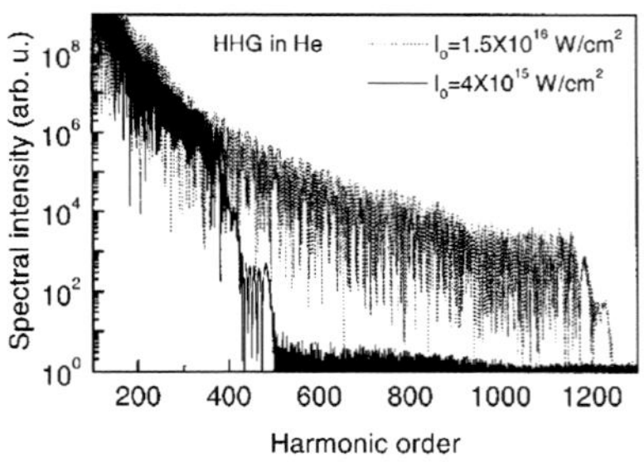

FIG. 43. Harmonic intensity spectra generated in a helium gas target at a gas pressure of 500 Torr after an interaction length of 125 μm for two different peak intensities: solid curves, $I_0 = 4\times10^{15}\,\mathrm{W/cm^2}$; dotted curve, $I_0 = 1.5\times10^{16}\,\mathrm{W/cm^2}$. The remaining laser pulse parameters are $\tau_p \approx 5$ fs and $\lambda_0 \approx 0.8\,\mu$m.

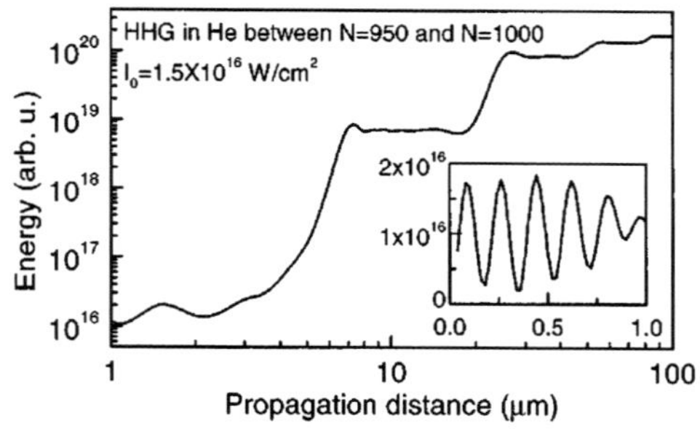

FIG. 44. Energy of the harmonic signal integrated between $N=950$ and $N=1000$ vs propagation distance for the parameters of Fig. 43 ($I_0 = 1.5\times10^{16}\,\mathrm{W/cm^2}$). The inset shows the growth of the harmonic over the first μm.

tra emerging from a He target at two peak intensity levels. For a peak intensity of $I_0 = 1.5\times10^{16}\,\mathrm{W/cm^2}$, an assessment of the influence of the magnetic component of the laser field on high harmonic generation is in order (see Appendix B). The harmonic yield is dramatically enhanced at harmonic orders higher than $N=400$ for the higher-intensity few-cycle pulse as a consequence of a more efficient nonadiabatic self-phase-matching. Owing to nonadiabatic self-phase-matching the spectrum remains nearly constant at the high-energy end, indicating that L_{fe} no longer determines the harmonic growth. A remarkable consequence of nonadiabatic self-phase-matching is that harmonics up to an order of $N=1200$ are predicted to emerge, corresponding to a photon energy in excess of 1.5 keV. The efficiency of high harmonic generation in the keV regime can be estimated by comparing calculations and experiments at lower laser intensities (Tempea *et al.*, 1999b). This suggests that $\approx 10^4 - 10^5$ photons/s in a 5% bandwidth at 1.5 keV are expected to be emitted from a He target pumped by 1-mJ sub-10-fs pulses at a 1-kHz repetition rate.

In Fig. 44 the growth of energy contained in the spectral window between the 950th and the 1000th harmonic is plotted versus interaction length for the parameters of Fig. 43. The harmonic signal grows over the coherence length $L_{fe} \approx 0.1\,\mu$m and oscillates for $\xi > L_{fe}$ over a limited propagation length, as shown in the inset. These oscillations originate from the interference between harmonic radiation generated at different positions along the propagation direction, in analogy to the Maker fringes in second-harmonic generation (Boyd, 1992). Note that the computed coherence length is a factor of 3 longer than the coherence length determined by Eq. (41). The increase in effective coherence length is related to the propagating pulse's being subject to significant modifications within one carrier oscillation cycle (nonadiabatic pulse shaping) as a consequence of the fractional ionization rate's becoming comparable to ω_0 (see Fig. 32) in intense few-cycle laser fields (Kan *et al.*, 1997). Another, even more striking, implication of nonadiabatic pulse evolution is the onset of nonadiabatic self-phase-matching for $\xi \gg L_f$. As a result, the Maker fringes disappear, and the harmonic signal exhibits a steplike increase, enhancing the energy converted into harmonic radiation by some four orders of magnitude.

Future terawatt-scale sub-10-fs laser technology is expected to allow full exploitation of nonadiabatic self-phase-matching, which is out of reach at present. Once this technology becomes available, few-cycle-driven harmonic sources emitting coherent radiation at wavelengths of a few nanometers with high average brightness and high peak brightness may open up new fields of research. High average brightness may open the way to biological microscopy or interferometric testing of x-ray optics using a compact laboratory source. This is a relative value. The expected high peak brightness even promises to offer an absolute value: two-photon-induced transitions may become measurable, indicating the advent of nonlinear x-ray optics. Intriguing consequences may include the feasibility of time-resolved spectroscopy of inner-shell electron relaxation processes and the temporal characterization of few-cycle-driven harmonic x-ray pulses, which can be of subfemtosecond duration under specific conditions to be discussed in the next subsection. Once extended into the keV regime, few-cycle-driven harmonic sources may become a powerful tool for studying the ultrafast structural dynamics of molecules and solids by means of time-resolved x-ray diffraction or absorption (Rischel *et al.*, 1997; Barty, Ben-Nun *et al.*, 1998; Chin *et al.*, 1998; Jimenez *et al.*, 1998) with never-before-achieved resolution.

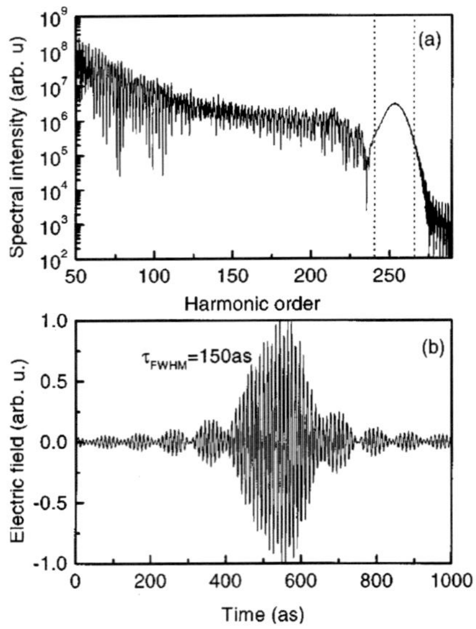

FIG. 45. Harmonic spectrum in helium (500 Torr) after a propagation distance of 9 μm for $\lambda_0=0.8\,\mu$m, $\tau_p=5$ fs, $I_0=2\times10^{15}$ W/cm^2, and $\varphi_0=0$. (a) Computed harmonic spectrum; dotted lines, harmonic orders $N=240$ and $N=265$; (b) Fourier transform of the spectral band between $N=240$ and $N=265$.

E. Attosecond x-ray pulse generation

In the time domain, harmonic emission is, within a laser oscillation period, confined to a small fraction of $T_0/2$ (within a limited frequency band near the cutoff), giving rise to a train of subfemtosecond or even attosecond bursts (1 as$=10^{-18}$ s) of short-wavelength radiation (Antoine *et al.*, 1996). This temporal structure was first proposed by Farkas and Toth (1992). Other proposals for pulse generation in this, thus far inaccessible, regime include Fourier synthesization (Hänsch, 1990) and stimulated Raman scattering (Yoshikawa and Imasaka, 1993; Kaplan, 1994; Harris and Sokolov, 1998). Although the conversion efficiency of high harmonic generation is lower than that of other schemes for attosecond pulse generation (Harris and Sokolov, 1998), the advantage of high harmonic generation is that there exist several possibilities for cutting a single attosecond pulse out of the pulse train, which are currently under intense investigation. The availability of single attosecond pulses will pave the way towards xuv/x-ray pump-probe experiments on a subfemtosecond time scale, the time scale in which, for example, inner-shell electron relaxation processes in medium- and high-Z atoms take place. Potential methods to isolate a single attosecond pulse include combining two perpendicularly polarized pulses with slightly different frequencies, known as the "atomic Pockels cell" (Corkum *et al.*, 1994; Corkum, 1996), and utilization of a time-dependent degree of ellipticity of the laser polarization (Antoine *et al.*, 1997).

A more direct method for the generation of a single-attosecond pulse is high harmonic generation with few-cycle laser pulses, as was revealed by the theoretical analysis of Christov *et al.* (1997). Under few-cycle excitation conditions, emission of the high-energy end of the harmonic spectrum can be confined to one half-oscillation period near the pulse peak, if the maximum ponderomotive potential U_p in Eq. (40) is reached for $\tau=0$. This is the case in the limit of weak ground-state depletion, resulting in the emission of a single x-ray pulse of attosecond duration within a 10% spectral window near the cutoff, as shown in Fig. 45. By contrast, in the regime of nonadiabatic self-phase-matching the emergence of a single-attosecond pulse is predicted to be feasible also in the plateau region of the harmonic spectrum.

So far the only experimental indication (there is no real evidence!) for the possible emergence of xuv emission on a subfemtosecond time scale is the coalescence of harmonics in the xuv spectrum of Ne driven by 7-fs pulses, as revealed by Fig. 35. Clearly, before exciting applications such as inner-shell spectroscopy can be tackled, more reliable techniques for the temporal characterization of these ultrashort x-ray pulses need to be developed. Cross-correlation techniques based on ionization by a harmonic photon in the presence of a strong laser field (Kálmán *et al.*, 1993) have been successful in the sub-100-fs range (Glover *et al.*, 1996; Schins *et al.*, 1996) and more recently in the sub-10-fs range (Murnane and Kapteyn, 1999). However, the resolution of this method is limited by the fundamental pulse duration to a few femtoseconds. This limitation does not appear in autocorrelation measurements utilizing two-photon ionization of atoms, which was demonstrated for the ninth harmonic of the Ti:sapphire laser at around 90 nm (Kobayashi *et al.*, 1998). Unfortunately, the two-photon transition cross section scales with λ^6 (Faisal, 1986; Kálmán, 1989), making implementation of this technique even more difficult for shorter wavelengths.

Recently, it was shown that few-cycle-driven harmonics may extend nonlinear autocorrelation techniques to the border of the soft-x-ray regime (Schnürer *et al.*, 1999b). The harmonic peak intensity predicted from the measured conversion efficiency in Fig. 39 for $\lambda \approx 13.4$ nm should be sufficient to create K-shell vacancies in boron by two-photon-induced photoemission. Before autocorrelation measurements on subfemtosecond xuv/x-ray pulses can be tackled, precise control of the the transverse beam profile of the driving laser pulse will be required. This is because the position and the phase of the attosecond harmonic burst within a selected frequency band is subject to variation across the pump laser beam, impairing the quality and duration of the focused harmonic pulse. Creating a near-flat-top laser beam profile in the interaction volume is one possible solution to this problem. Finally, the spectrum and hence the temporal structure of radiation emitted by a few-cycle-driven harmonic source depends sensitively on the absolute phase of the laser pulse φ_0 [see Eq. (14)]. The random variation of φ_0 translates into a timing jitter and fluctuations of the energy and duration of the harmonic pulse. Hence stable subfemtosecond x-ray harmonic pulse generation, like any other strong-field light-atom interaction process, calls for access to and control of φ_0 in few-cycle wave packets, a problem which is addressed in the next section.

VIII. PHASE SENSITIVITY OF STRONG-FIELD PHENOMENA

This section will show that the phase sensitivity of strong-field processes offers routes to measuring and subsequently controlling the absolute phase φ_0 of light pulses. The interactions to be addressed require intensities of the order of 10^{14} W/cm^2 or higher. Until recently, this intensity range was out of the reach of cw mode-locked laser oscillators. Advances in femtosecond technology reviewed in Sec. III have recently made it possible to focus the output of a 1-MW, sub-10-fs MDC KLM Ti:sapphire oscillator to peak intensities in excess of 5×10^{13} W/cm^2 (Xu *et al.*, 1998). An improved version of this system now delivers sub-10-fs pulses with peak powers approaching 3 MW (Poppe, Lenzner *et al.*, 1999), which should be focusable to intensities beyond the 10^{14}-W/cm^2 level. These intensities are sufficient for triggering the phase-sensitive strong-field processes discussed below, opening up ways to measure and eventually control the phase of few-cycle light pulses delivered by mode-locked laser oscillators.

A. Is few-cycle pulse evolution phase sensitive?

Before we address the influence of φ_0 on nonlinear interactions, let us consider the principal question for the design of few-cycle pulse experiments: how does the pulse phase behave during (linear) propagation? A few-cycle pulse envelope with a cosine carrier can have a finite spectral dc component, whereas the dc component of a sine wave carrying a symmetric amplitude envelope must be zero for reasons of symmetry. As the dc component does not constitute a solution to the wave equation, i.e., it cannot propagate, it is important to know which kind of pulses are capable of interacting with an extended medium without suffering significant distortions due to linear propagation. In order to answer this question, we have calculated the evolution of few-cycle pulses in vacuum, which is most conveniently described by Eq. (26) in the frequency domain and in a coordinate system moving at the speed of light in vacuum:

$$\partial_\xi \tilde{E}(r,\xi,\omega)=\frac{ic}{2\omega}\nabla_\perp^2 \tilde{E}(r,\xi,\omega). \qquad (47)$$

Equation (47) is solved by starting from a Gaussian initial spatial pulse profile

$$\tilde{E}(r,\xi=0,\omega)=E_0\exp\left[-\left(\frac{r}{a}\right)^2\right]\tilde{F}(\omega). \qquad (48)$$

Here, E_0 is the peak electric-field amplitude, $\tilde{F}$ denotes the complex Fourier spectrum of a pulse of arbitrary temporal shape, a is the radius at the beam waist, and $r=|\mathbf{r}_\perp|=\sqrt{x^2+y^2}$ is the radial coordinate, assuming cylindrical symmetry.

At the beam center $r=0$ and at propagation distances $\xi/z_0\gg1$, where $z_0=\pi a^2/\lambda_0$ is the confocal parameter, Eq. (47) can be solved analytically, yielding

$$\tilde{E}(\xi,\omega,r=0)=\frac{\omega a^2}{2ic\xi}\tilde{F}(\omega). \qquad (49)$$

Choosing the initial electric field in the time domain to be $E(\xi=0,\tau)=E_0\,\mathrm{sech}(\tau/\tau_p)\cos(\omega_0\tau+\varphi_0)$, we obtain from a Fourier transform of Eq. (49)

$$E(\xi,\tau,r=0)\approx\frac{z_0}{\xi}\sqrt{1+\frac{1}{(\omega_0\tau_0)^2}\tanh^2\left(\frac{\tau}{\tau_p}\right)}\times\mathrm{sech}\left(\frac{\tau}{\tau_p}\right)\cos[\omega_0\tau+\varphi(\tau)], \qquad (50)$$

where the phase is given by

$$\varphi(\tau)=\varphi_0-\frac{\pi}{2}+\frac{1}{\omega_0\tau_p}\tanh\left(\frac{\tau}{\tau_p}\right) \qquad (51)$$

and $\varphi_0=\varphi(\tau=0,\xi=0)$ is the initial pulse phase at the pulse center. Our analysis reveals that during propagation the pulse envelope remains symmetric, so that the pulse center remains at $\tau=0$. As the pulse center does not change in the far field, the pulse phase can be determined at $\tau=0$. The pulse phase in the far field, $\varphi(\tau=0)=\varphi_0-\pi/2$, undergoes a phase shift which comes from the Gouy phase shift $\lim_{\xi\to\infty}-\tan^{-1}(\xi/z_0)\to-\pi/2$.

Note that the Guoy phase shift and all the other changes experienced by the pulse during propagation do not depend on the initial pulse phase. Therefore it can be concluded that the evolution of few-cycle pulses is independent of the initial carrier phase φ_0. The low-

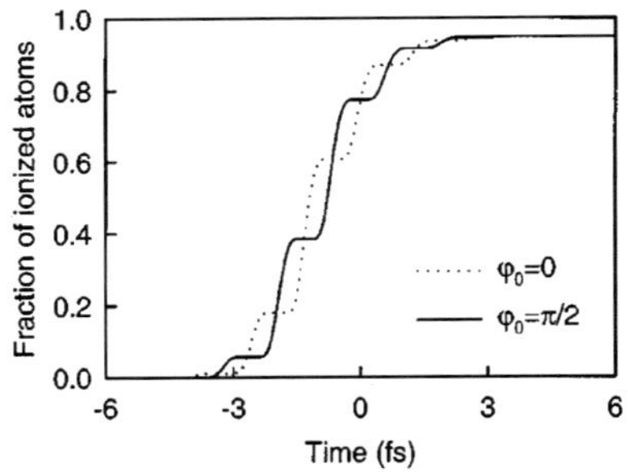

FIG. 46. Fraction of ionized helium atoms vs time. The laser pulse parameters are $\tau_p=5$ fs, $I_0=4\times10^{15}$ W/cm^2, and $\lambda_0=0.8\,\mu$m. Dotted lines, pulse phase $\varphi_0=0$; solid line, $\varphi_0=\pi/2$. These fundamental phases correspond approximately to a cosine and a sine carrier wave, respectively. The electric field is defined by Eq. (33).

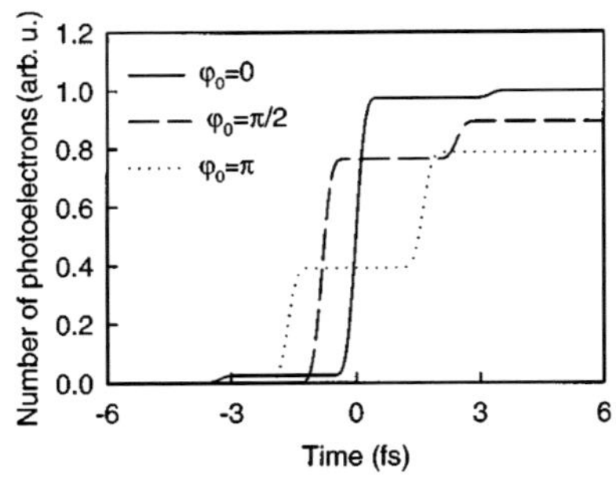

FIG. 47. Number of electrons ionized from a gold surface ($W_b=4.9$ eV) for a pulse impinging at an angle of incidence of 45°. The pulse parameters are $\lambda_0=0.8\,\mu$m, $I_0\approx2\times10^{13}$ W/cm^2, and $\tau_p=5$ fs. Pulse phases: solid curve, $\varphi_0=0$; dashed curve, $\varphi_0=\pi/2$; dotted curve, $\varphi_0=\pi$.

frequency (dc) components of the pulse spectrum are suppressed by diffraction, which manifests itself mathematically in a multiplication of the laser spectrum with ω in Eq. (49). This distorts the pulse envelope; however, the pulse phase φ_0, which is defined at the pulse center, is invariant during free-space vacuum pulse propagation. This suggests that there must exist a class of envelopes that do not have a dc component irrespective of the initial phase. This class of pulses is obtained by defining the electric field via the vector potential, $E=-\partial A/\partial t$ [see, for example, Eq. (33)]. The temporal derivative corresponds to a multiplication with ω in the frequency domain and ensures that the spectrum contains no dc components. A thorough treatment of pulse propagation in the single-cycle regime was recently presented by Shvartsburg (1996).

B. Phase sensitivity of optical field ionization

The quasistatic ionization rate $w(E)$ depends on the instantaneous electric-field strength [see Eq. (23)]. As the electric field of a few-cycle pulse depends on its phase, strong-field ionization becomes phase dependent. In Fig. 46, the ionization profile in He is depicted for the linearly polarized pulses of $\tau_p=5fs$ with $\varphi_0=0$ and $\varphi_0=\pi/2$. Ionization was calculated by using Eq. (23) together with the exact static ionization rates of helium. Whereas the ionization profile is phase sensitive, the final fraction of ionized atoms is, somewhat surprisingly, independent of φ_0. This is a general feature of strong-field ionization in the gaseous phase, as long as ionization is calculated in the quasistatic approximation. Recently, a full numerical solution of the time dependent Schrödinger equation showed some phase sensitivity (Christov, 1999).

Alternatively, there exist various other routes that provide access to φ_0 in few-cycle light pulses. Tunnel ionization at the surface of a solid occurs only during positive or negative half cycles. This symmetry breaking makes the total number of photoelectrons released by a p-polarized laser pulse impinging at oblique incidence on the photocathode phase dependent. Figure 47 shows the number of photoelectrons emitted upon irradiation of a gold photocathode with 5-fs pulses of various phases (Poppe, Fürbach, *et al.*, 1999). The ionization rate in Fig. 47 was calculated using the quasistatic ionization theory for metal surfaces (Sommerfeld, 1967). The predicted variation of the photoelectron yield versus φ_0 should be easily detectable. Another approach to determining φ_0 is the measurement of the angular distribution of electrons set free by circularly polarized light (Dietrich *et al.*, 1999), which also varies with φ_0.

C. Phase effects in high harmonic generation

Because the high-frequency atomic dipole moment $\tilde{d}_h$ given by Eq. (34), which is responsible for high harmonic generation, is also driven directly by the electric field, the emission characteristics exhibit a pronounced dependence on the phase of a few-cycle driver pulse (de Bohan *et al.*, 1998; Krausz *et al.*, 1998; Salières *et al.*, 1998; Tempea *et al.*, 1999a). This can be seen in Fig. 48, where the harmonic intensity spectra emitted from He pumped by 5-fs pulses of different phases, $\varphi_0=0$ (dotted line) and $\varphi_0=\pi/2$ (solid line), are plotted for various propagation distances. Figure 48(a) shows the harmonic spectrum after a short interaction length, reflecting the single-atom dipole response. In the cutoff region, a pulse with a cosine carrier yields a significantly stronger signal and the cutoff is shifted towards shorter wavelengths as compared to that obtained with a sine-carrier pulse.

A reliable deduction of φ_0 from high harmonic generation experiments calls for phase-dependent characteristics invariant to propagation. Figures 48(b)–(d) show the evolution of the harmonic radiation over distances comparable to the coherence length. The coherence length at the cutoff ($N=255$) is $\approx11\,\mu$m. The phase dependence of plateau harmonics is subject to strong variations with propagation. By contrast, the qualitative behavior in the cutoff region, as found in graph (a), is preserved for longer propagation distances. Hence the detection of cutoff harmonics may offer another route to measuring the phase of few-cycle light pulses.

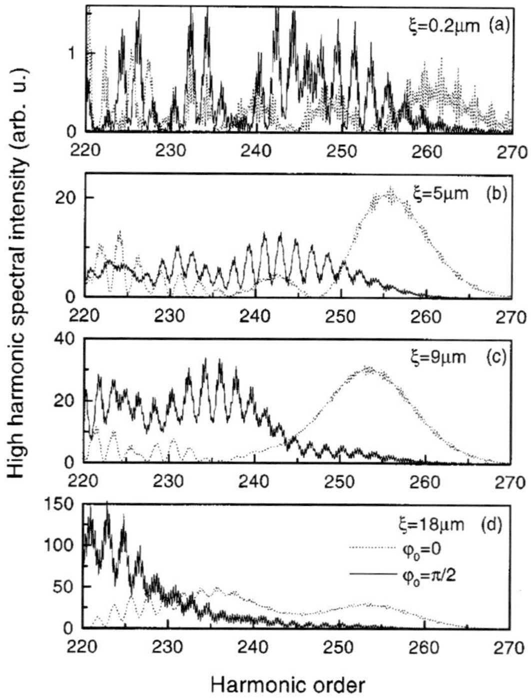

FIG. 48. Spectral intensity of the harmonic radiation in helium at a pressure of 500 Torr for the fundamental pulse phases: dotted curve, $\varphi_0=0$; solid curve, $\varphi_0=\pi/2$, after propagation distances of (a) 0.2, (b) 5, (c) 9, and (d) 18 μm. The laser parameters are $I_0=2\times10^{15}$ W/cm^2, $\lambda_0=0.8\,\mu$m, and $\tau_p=5$ fs. The electric field is defined by Eq. (33).

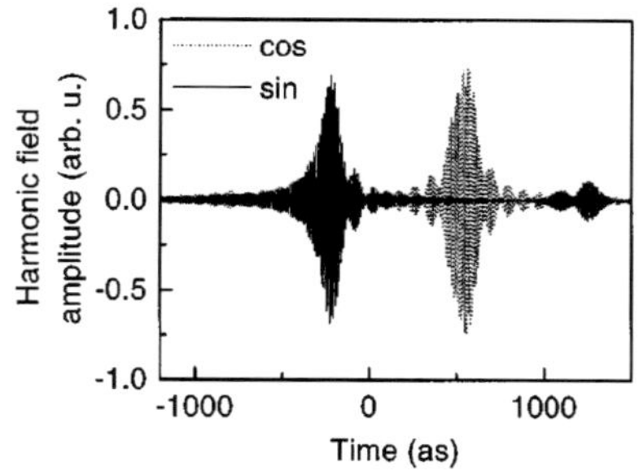

FIG. 49. Fourier transform of the harmonic amplitude spectrum between the orders 230 and 260 for the parameters of Fig. 48(c). Pulse phases: dotted line, $\varphi_0=0$; solid line, $\varphi_0=\pi/2$.

The origin of the characteristic phase dependence near the cutoff becomes clear by inspection of the cutoff harmonic signal in the time domain. Figure 49 shows the evolution of the temporal harmonic signal as obtained by an inverse Fourier transform of the harmonic spectra depicted in Fig. 48(c) transmitted through a bandpass between the 230th and the 260th harmonic order. For $\varphi_0=0$, the cutoff harmonics are generated at a single instant during the 5-fs driver pulse, whereas for $\varphi_0=\pi/2$ they emerge at two instants approximately half a cycle apart, resulting in a couple of x-ray bursts during the driver pulse.

The single attosecond pulse for $\varphi_0=0$ has a duration of approximately 160 as and presents a promising tool for time-resolved attosecond x-ray experiments. However, it is obvious that the timing jitter and the intensity fluctuations introduced by the phase sensitivity of few-cycle-driven harmonic emission impairs the utility of such a source for pump-probe experiments. Therefore the birth of attosecond science is closely linked to the development of techniques for measuring and controlling the phase of few-cycle light pulses.

IX. OUTLOOK

Intense few-cycle light pulses open up never-before-accessed parameter ranges in high-field physics. Strong-field processes induced by few-cycle laser fields hold out the promise of gaining access to the phase of the carrier wave and hence to the light fields for the first time. Presumably only a few of the many implications can be assessed at the present time. The investigations presented in this work suggest that phase-controlled light pulses will allow control of high-intensity light-matter interactions on a subcycle time scale. Consequences include the injection of an electron jet into a plasma with a subfemtosecond rise time and timing precision as well as the generation of single, well-controlled x-ray bursts of less than 100 attoseconds in duration. These pulse durations approach the atomic time scale and may allow the tracing of quantum dynamics in bound states characterized by low principal quantum numbers for the first time.

Beyond these long-term prospects, intense few-cycle laser pulses are likely to become the most efficient pump sources of ultrashort-pulsed short-wavelength radiation all the way from the xuv to the hard-x-ray regime, which may be produced at practically useful photon flux levels in the foreseeable future. At xuv and soft x-ray wavelengths few-cycle-driven high harmonics have been demonstrated to reach peak brightness levels that may allow two-photon excitations to be detected. Once achieved, nonlinear x-ray optics will become reality, with numerous intriguing implications including the autocorrelation measurement of subfemtosecond or attosecond xuv/x-ray pulses. When focused to relativistic intensities, few-cycle pulses will create ponderomotive forces as high as 0.1 MeV/μm at moderate (few-millijoule) energy levels, promising efficient conversion of light energy into kinetic energy of a collimated relativistic electron jet (Gahn *et al.*, 1999) with a duration comparable to that of the driver pulse. This electron jet could produce characteristic or channeling radiation (Kálmán, 1993, 1994), giving rise to hard x rays with similar pulse durations.

The search for laser-driven ultrafast x-ray sources is motivated by a number of important processes in the microscopic world that are very hard to gain insight into by conventional spectroscopy. Figure 50 depicts the relevant time scales for some important physical, chemical, and biological processes taking place in the microcosm. Whereas relaxation processes of weakly bound (outer-shell or valence-band) electrons can be accessed directly by pump-probe laser spectroscopy, only indirect and in

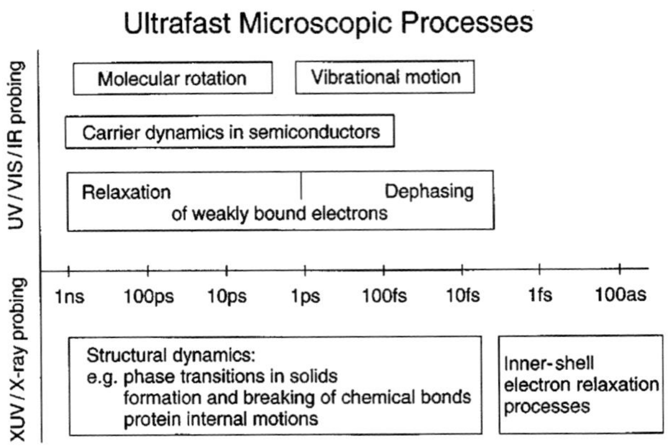

FIG. 50. Characteristic time scales of some elementary atomic, molecular, and collective processes. The boundaries of the presented time range are not intended to imply upper or lower limits on some of the processes.

many cases insufficient information about the motion of nuclei and the dynamics of inner-shell electronic transitions can be obtained by conventional ultrafast laser spectroscopy. Laser-produced x-ray pulses of a few femtoseconds or less in duration will allow the tracing of nuclear motion in a wide range of phenomena including phase transitions in the solid phase, the formation and breaking of chemical bonds in the gaseous and liquid phases, and biological processes as important as protein unfolding by means of time-resolved x-ray diffraction or absorption. Also, the wavelength and duration of these pulses will open the way to probing inner-shell atomic electron relaxation processes directly in the time domain.

In conclusion, ultrafast optics and high-field physics are likely to continue rapidly evolving over the coming years. Approaching the light oscillation cycle in intense ultrashort pulse generation issues new challenges to the research community but also promises great rewards, e.g., access to the electric and magnetic fields of light or the birth of attosecond metrology. The expected significant impact on numerous fields provides the major driving force for further progress.

ACKNOWLEDGMENTS

This review is dedicated to Professor A. J. Schmidt, who directed the authors' attention to this fascinating field. The authors are deeply indebted to Z. Cheng, M. Geissler, M. Hentschel, A. Poppe, M. Schnürer, A. Scrinzi, Ch. Spielmann, and G. Tempea for making their experimental and numerical data available for this work, as well as for fruitful discussions. We thank U. Keller for providing the data in Fig. 17. This work was sponsored by the Austrian Science Fund, Grant Nos. P-12631-PHY, Y44-PHY, and the special research program F016.

APPENDIX A: CONVERSION BETWEEN ATOMIC AND SI UNITS

Atomic units are used for convenience in quantum mechanics (Bethe and Salpeter, 1997), as by setting the fundamental constants $\hbar=m=e^2=1$ calculations are simplified substantially. In the following, conversion factors between atomic units and SI units are given.

(1) 1 atomic charge unit=charge of the electron$=1.602\times10^{-19}$ C
(2) 1 atomic mass unit=mass of the electron$=9.109\times10^{-31}$ kg
(3) 1 atomic length unit=radius of the first Bohr orbit $=5.2917\times10^{-11}$ m
(4) 1 atomic velocity unit=electron velocity in the first Bohr orbit$=2.1877\times10^{6}$ m/s
(5) 1 atomic momentum unit=electron momentum in the first Bohr orbit$=1.9926\times10^{-24}$ kg m/s
(6) 1 atomic energy unit=twice the ionization potential of hydrogen$=4.359\times10^{-18}$ J
(7) 1 atomic time unit$=2.4189\times10^{-17}$ s
(8) 1 atomic frequency unit$=4.1341\times10^{16}\,\mathrm{s}^{-1}$
(9) 1 atomic unit of electric potential=27.210 V
(10) 1 atomic unit of electric-field strength$=5.142\times10^{11}$ V/m.

APPENDIX B: DERIVATION OF THE FIRST-ORDER PROPAGATION EQUATION

The derivation of the first-order propagation equation starts from the scalar wave equation in the frequency domain

$$[\partial_z^2+\nabla_\perp^2+k^2(\omega)]E(\mathbf{r},\omega)=\frac{\omega^2}{\varepsilon_0 c^2}\hat{F}[P_{nl}(\mathbf{r},t)], \tag{B1}$$

which can be obtained from the time-domain scalar wave equation by replacing the operations ∂_t and $\int dt$ by multiplications with $-i\omega$ and i/ω in the frequency domain. The scalar wave equation is appropriate as long as the electric field and the induced polarization in the transversal dimensions (x,y) exhibit little variation over a distance comparable to the center wavelength λ_0 (Yariv and Yeh, 1984). The derivation of the first-order propagation equation for the full electric field is performed by choosing the ansatz $\tilde{E}=\tilde{U}\exp[ik(\omega)z]$. Inserting this into Eq. (B1), we obtain

$$[2ik(\omega)\partial_z+\nabla_\perp^2]\tilde{U}(\omega)=\frac{\omega^2}{\varepsilon_0 c^2}\exp[-ik(\omega)z]\hat{F}[P_{nl}], \tag{B2}$$

where the second space derivative is neglected. This approximation has been termed the slowly-evolving-wave approximation (Brabec and Krausz, 1997) and is applicable as long as changes in the electric field induced by the polarization of the medium over a distance comparable to one center wavelength λ_0 are small, i.e., $|\partial_z\tilde{E}|\ll\beta_0\tilde{E}$, with $\beta_0=k(\omega_0)=2\pi/\lambda_0$. Transforming back to $\tilde{E}$ yields

$$[\partial_z-ik(\omega)]\tilde{E}(\omega)=\frac{i}{2k(\omega)}\nabla_\perp^2\tilde{E}(\omega) -\frac{i\omega}{2\varepsilon_0 n(\omega)c}\hat{F}[P_{nl}]. \tag{B3}$$

Although this equation is not directly usable, it is the starting point for the derivation of first-order propagation equations governing the evolution of laser pulses in various nonlinear media. A detailed analysis is given in Sec. V.

APPENDIX C: INFLUENCE OF THE LORENTZ FORCE ON HIGH HARMONIC GENERATION

Our model of high-order harmonic generation Eqs. (34)–(39), does not take into account the $\mathbf{v}\times\mathbf{B}$ (Lorentz) force that tends to become significant at high laser intensities. Here, $\mathbf{v}$ is the velocity of the electron and $\mathbf{B}$ is the magnetic field of the laser pulse. The Lorentz force shifts the electron trajectory in the direction of the laser wave vector, which presents a potential limitation to high harmonic generation. To estimate its effect on our predictions presented in Sec. VII, the electron trajectory in the laser field was calculated by solving the classical equation of motion. For an intensity of $I \approx 10^{16}\,\mathrm{W/cm^2}$, we find that the electron is offset by $z_o \approx 1$ nm when it returns to the nucleus in the linearly polarized field. This shift must be compared to the transversal width of the returning electron wave function, $z_r = v_b(\tau_r - \tau_b)$, in order to determine the influence on high harmonic generation. Here, $\tau_r - \tau_b \approx 2.2$ fs is the interval between the birth and the return time of the electron and v_b is the width of the velocity distribution of the electron wave packet at its birth. As for the intensities under consideration, the electrons are set free by above-barrier ionization, and the width of the spatial distribution of the electron wave packet at its birth, x_b, cannot be determined by tunneling theory relying on the presence of a barrier (Delone and Krainov, 1991). Therefore, to estimate a lower limit for x_r, it is assumed that the initial spatial distribution is equal to the Bohr radius, $a_B \approx = 5.3\times10^{-9}$ cm/s. By virtue of the uncertainty relation $[a_B v_b \geqslant \hbar/(2m)]$ we obtain $v_b \approx 1.1$ nm/fs. This leads to $z_r \geqslant 2.5$ nm, i.e., the extension of the electron wave packet is expected to be larger than the offset z_r. This justifies neglect of the Lorentz force in our calculations.

REFERENCES

Agnesi, A., 1995, IEEE J. Quantum Electron. **30**, 1115.

Agostini, P., F. Fabre, G. Mainfray, G. Petite, and N. K. Rahman, 1979, Phys. Rev. Lett. **42**, 1127.

Agrawal, G. P., 1995, *Nonlinear Fiber Optics*, 2nd ed. (Academic, San Diego).

Akhmanov, S. A., V. A. Vysloukh, and A. S. Chirkin, 1992, *Optics of Femtosecond Laser Pulses* (American Institute of Physics, New York).

Alfano, R. R., and S. L. Shapiro, 1970, Phys. Rev. Lett. **24**, 592.

Ammosov, M. V., N. B. Delone, and V. P. Krainov, 1986, Zh. Eksp. Teor. Fiz. [Sov. Phys. JETP **64**, 1191 1986].

Antoine, P., A. L'Huillier, and M. Lewenstein, 1996, Phys. Rev. Lett. **77**, 1234.

Antoine, P., D. B. Milosevic, A. L'Huillier, M. B. Gaarde, P. Salières, and M. Lewenstein, 1997, Phys. Rev. A **56**, 4960.

Armstrong, J. A., 1967, Appl. Phys. Lett. **10**, 16.

Asaki, M. T., C.-P. Huang, D. Garvey, J. Zhou, H. C. Kapteyn, and M. M. Murnane, 1993, Opt. Lett. **18**, 977.

Augst, S., D. Strickland, D. D. Meyerhofer, S. L. Chin, and J. H. Eberly, 1989, Phys. Rev. Lett. **63**, 2212.

Augst, S., A. Talebpour, S. L. Chin, Y. Beaudoin, and M. Chaker, 1995, Phys. Rev. A **52**, R917.

Backus, S., C. G. Durfee III, G. Mourou, H. C. Kapteyn, and M. M. Murnane, 1997, Opt. Lett. **22**, 1256.

Backus, S., C. G. Durfee III, M. M. Murnane, and H. C. Kapteyn, 1998, Rev. Sci. Instrum. **69**, 1207.

Baltuska, A., Z. Wei, M. S. Pshenichnikov, and D. A. Wiersma, 1997a, Opt. Lett. **22**, 102.

Baltuska, A., Z. Wei, R. Szipöcs, M. S. Pshenichnikov, and D. A. Wiersma, 1997b, Appl. Phys. B: Lasers Opt. **65**, 175.

Bandrauk, A. D., S. Chelkowski, and H. Yu, 1997, Phys. Rev. A **56**, R2537.

Bandrauk, A. D., and H. Yu, 1999, Phys. Rev. A **59**, 539.

Barbara, P. F., J. G. Fujimoto, W. H. Knox, and W. Zinth, 1996, Eds., *Ultrafast Phenomena X* (Springer, Berlin).

Barbara, P. F., W. H. Knox, G. A. Mourou, and A. H. Zewail, 1994, Eds., *Ultrafast Phenomena IX* (Springer, Berlin).

Barr, J. R. M., D. C. Hanna, and D. W. Hughes, 1991, *Digest of Conference on Lasers and Electro-optics* (Optical Society of America, Washington, DC), Paper FQ5.

Barty, C. P. J., M. Ben-Nun, T. Guo, F. Raksi, Ch. Rose-Petruck, J. Squier, K. R. Wilson, V. V. Yakovlev, P. M. Weber, Z. Jiang, A. Ikhlef, and J. C. Kieffer, 1998, in *Time-Resolved X-Ray Diffraction*, edited by J. Helliwell and P. M. Rentzepis (Oxford University, New York), Chap. 2.

Barty, C. P. J., C. L. Gordon III, and B. E. Lemoff, 1994, Opt. Lett. **19**, 1442.

Barty, C. P. J., T. Guo, C. Le Blanc, F. Raksi, C. Rose-Petruck, J. Squier, K. R. Wilson, V. V. Yakovlev, and K. Yamakawa, 1996, Opt. Lett. **21**, 668.

Barty, C. P. J., W. White, W. Sibbett, and R. Trebino, 1998, Eds., *Ultrafast Optics*, IEEE J. Sel. Top. Quantum Electron. **4**, p. 157.

Beaud, P., M. Richardson, E. J. Miesak, and B. H. T. Chai, 1993, Opt. Lett. **18**, 1550.

Becker, U., and D. A. Shirley, 1996, *Vacuum uv and Soft X-Ray Photoionization* (Plenum, New York).

Becker, W., S. Long, and J. K. McIver, 1994, Phys. Rev. A **50**, 1540.

Beddard, T., W. Sibbett, D. T. Reid, J. Garduno-Mejia, N. Jamasbi, and M. Mohebi, 1999, Opt. Lett. **24**, 163.

Ben-Tal, N., N. Moiseyev, R. Kosloff, and C. Cerjan, 1993, J. Phys. B **26**, 1445.

Bethe, H. A., and E. E. Salpeter, 1977, *Quantum Mechanics of One- and Two-Electron Atoms* (Plenum, New York).

Blanchot, N., C. Rouyer, C. Sauteret, and A. Migus, 1995, Opt. Lett. **20**, 395.

Bloembergen, N., 1965, *Nonlinear Optics* (Addison-Wesley, Reading, MA).

Blow, K. J., and B. P. Nelson, 1988, Opt. Lett. **13**, 1026.

Blow, K. J., and D. Wood, 1988, J. Opt. Soc. Am. B **5**, 629.

Borodin, N. L., A. G. Okrimchuk, and A. V. Shestakov, 1992, in *OSA Proceedings on Advanced Solid-State Lasers*, Vol. 13, edited by Lloyd L. Chase and Albert A. Pinto (Optical Society of America, Washington, DC), p. 42.

Boyd, R. W., 1992, *Nonlinear Optics* (Academic, San Diego).

Brabec, T., M. Yu. Ivanov, and P. Corkum, 1996, Phys. Rev. A **54**, R2551.

Brabec, T., S. M. J. Kelly, and F. Krausz, 1995, in *Compact Sources of Ultrashort Pulses*, edited by I. N. Duling III (Cambridge University, Cambridge, England), p. 57.

Brabec, T., and F. Krausz, 1997, Phys. Rev. Lett. **78**, 3282.

Brabec, T., Ch. Spielmann, P. F. Curley, and F. Krausz, 1992, Opt. Lett. **17**, 1292.

Brabec, T., Ch. Spielmann, and F. Krausz, 1991, Opt. Lett. **16**, 1961.

Brabec, T., Ch. Spielmann, and F. Krausz, 1992, Opt. Lett. **17**, 748.

Bradley, D. J., and G. H. C. New, 1974, Proc. IEEE **62**, 313.

Braun, A., G. Korn, X. Liu, D. Du, J. Squier, and G. Mourou, 1995, Opt. Lett. **20**, 73.

Brunel, F., 1990, J. Opt. Soc. Am. B **7**, 521.

Budil, K. S., P. Salières, A. L'Huillier, T. Ditmire, and M. D. Perry, 1993, Phys. Rev. A **48**, R3437.

Burke, P. G., and V. M. Burke, 1997, J. Phys. B **30**, L383.

Burnett, N. H., C. Kan, and P. B. Corkum, 1995, Phys. Rev. A **51**, R3418.

Campillo, A. J., S. L. Shapiro, and B. R. Suydam, 1973, Appl. Phys. Lett. **23**, 628.

Cerullo, G., S. De Silvestri, V. Magni, and L. Pallaro, 1994, Opt. Lett. **19**, 807.

Chambaret, J. P., C. Le Blanc, G. Cheriaux, P. Curley, G. Darpentigny, P. Rousseau, G. Hamoniaux, A. Antonetti, and F. Salin, 1996, Opt. Lett. **21**, 1921.

Chang, Z., A. Rundquist, H. Wang, M. Murnane, and H. C. Kapteyn, 1997, Phys. Rev. Lett. **79**, 2967.

Chelkowski, S., A. Conjusteau, T. Zuo, and A. D. Bandrauk, 1996, Phys. Rev. A **54**, 3235.

Chen, S., A. Maksimchuk, and D. Umstadter, 1998, Nature (London) **396**, 653.

Cheng, Z., A. Fürbach, S. Sartania, M. Lenzner, Ch. Spielmann, and F. Krausz, 1999, Opt. Lett. **24**, 247.

Cheng, Z., G. Tempea, T. Brabec, K. Ferencz, C. Spielmann, and F. Krausz, 1998, in *Ultrafast Phenomena XI*, edited by T. Elsässer, J. G. Fujimoto, D. A. Wiersma, and W. Zinth (Springer, Berlin), p. 8.

Cheriaux, G., P. Rousseau, F. Salin, J. Chambaret, B. Walker, and L. Dimauro, 1996, Opt. Lett. **21**, 414.

Chilla, J. L. A., and O. E. Martinez, 1993, J. Opt. Soc. Am. B **10**, 638.

Chin, A. H., R. W. Schoenlein, T. E. Glover, P. Balling, W. P. Leemans, and C. V. Shank, 1998, in *Ultrafast Phenomena XI*, edited by T. Elsässer, J. G. Fujimoto, D. A. Wiersma, and W. Zinth (Springer, Berlin), p. 401.

Cho, S. H., B. E. Bouma, E. P. Ippen, and J. G. Fujimoto, 1999, Opt. Lett. **24**, 417.

Christov, I. P., 1999, Opt. Lett. **24**, 1425.

Christov, I. P., H. C. Kapteyn, and M. M. Murnane, 1998, Opt. Express **3**, 360.

Christov, I. P., M. M. Murnane, and H. C. Kapteyn, 1997, Phys. Rev. Lett. **78**, 1251.

Christov, I. P., M. M. Murnane, and H. C. Kapteyn, 1998, Phys. Rev. A **57**, R2285.

Christov, I. P., J. Zhou, J. Peatross, A. Rundquist, M. M. Murnane, and H. C. Kapteyn, 1996, Phys. Rev. Lett. **77**, 1743.

Constant, E., D. Garzella, P. Breger, E. Mevel, C. Dorrer, C. Le Blanc, F. Salin, and P. Agostini, 1999, Phys. Rev. Lett. **82**, 1668.

Corkum, P. B., N. H. Burnett, and F. Brunel, 1989, Phys. Rev. Lett. **62**, 1259.

Corkum, P. B., 1993, Phys. Rev. Lett. **71**, 1994.

Corkum, P. B., 1996, Nature (London) **384**, 118.

Corkum, P. B., N. H. Burnett, and M. Yu. Ivanov, 1994, Opt. Lett. **19**, 1870.

Crane, J. K., M. D. Perry, S. Hermann, and R. W. Falcone, 1992, Opt. Lett. **17**, 1256.

Curley, P. F., Ch. Spielmann, T. Brabec, F. Krausz, E. Wintner, and A. J. Schmidt, 1993, Opt. Lett. **18**, 54.

Dahlström, L., 1972, Opt. Commun. **5**, 157.

de Bohan, A., P. Antoine, D. B. Milosevic, and B. Piraux, 1998, Phys. Rev. Lett. **81**, 1837.

de Groot, S., W. van Leeuwen, and C. van Weert, 1980, *Relativistic Kinetic Theory, Principles and Applications* (North-Holland, Amsterdam).

Delone, N. B., and V. P. Krainov, 1991, J. Opt. Soc. Am. B **8**, 1207.

DeMaria, A. J., W. H. Glenn, M. J. Brienza, and M. E. Mack, 1969, Proc. IEEE **57**, 2.

DeMaria, A. J., D. A. Stetser, and H. Heynau, 1966, Appl. Phys. Lett. **8**, 174.

De Silvestri, S., P. Laporta, and O. Svelto, 1984, IEEE J. Quantum Electron. **20**, 533.

Deutsch, T., 1965, Appl. Phys. Lett. **7**, 80.

DiDomenico, M., 1964, J. Appl. Phys. **35**, 2870.

DiDomenico, M., H. M. Marcos, J. E. Geusic, and R. E. Smith, 1966, Appl. Phys. Lett. **8**, 180.

Diels, J.-C., W. Dietel, J. J. Fontaine, W. Rudolph, and B. Wilhelmi, 1985, J. Opt. Soc. Am. B **2**, 680.

Diels, J.-C., J. J. Fontaine, I. C. McMichael, and F. Simoni, 1985, Appl. Opt. **24**, 1270.

Diels, J.-C., and W. Rudolph, 1996, *Ultrashort Laser Pulse Phenomena* (Academic, New York).

Diels, J.-C., E. Van Stryland, and G. Benedict, 1978, Opt. Commun. **25**, 93.

Dietrich, P., N. H. Burnett, M. Ivanov, and P. B. Corkum, 1994, Phys. Rev. A **50**, R3583.

Dietrich, P., F. Krausz, and P. B. Corkum, 1999, "Determining the absolute carrier phase of a few-cycle laser pulse," in press.

Ditmire, T., J. K. Crane, H. Nguyen, L. B. DaSilva, and M. D. Perry, 1995, Phys. Rev. A **51**, R902.

Ditmire, T., K. Kulander, J. K. Crane, H. Nguyen, and M. D. Perry, 1996, J. Opt. Soc. Am. B **13**, 406.

Ditmire, T., and M. D. Perry, 1993, Opt. Lett. **18**, 426.

Donelly, T. D., T. Ditmire, K. Neumann, M. D. Perry, and R. W. Falcone, 1996, Phys. Rev. Lett. **76**, 2472.

Dorner, R., H. Brauning, O. Jagutzki, V. Mergel, M. Achler, R. Moshammer, J. M. Feagin, T. Osipor, A. Brauning-Demain, L. Spielberger, T. M. McGuire, M. M. Prior, N. Berrah, J. D. Bozek, C. L. Cocke, H. Schmidt-Böcking, 1998, Phys Rev. Lett. **81**, 5776.

Du, D., X. Liu, G. Korn, J. Squier, and G. Mourou, 1994, Appl. Phys. Lett. **64**, 3071.

Duling, I. N. III, 1991, Opt. Lett. **16**, 539.

Duling III, I. N., and M. L. Dennis, 1995, in *Compact Sources of Ultrashort Pulses*, edited by I. N. Duling III (Cambridge University, Cambridge, England), p. 140.

Durfee, C. G. III, S. Backus, M. M. Murnane, and H. C. Kapteyn, 1998, IEEE J. Sel. Top. Quantum Electron. **4**, 395.

Eberly, J. H., Q. Su, and J. Javanainen, 1989a, Phys. Rev. Lett. **62**, 881.

Eberly, J. H., Q. Su, and J. Javanainen, 1989, J. Opt. Soc. Am. B **6**, 1289.

Elsässer, T., J. G. Fujimoto, D. A. Wiersma, and W. Zinth, 1998, Eds., *Ultrafast Phenomena XI* (Springer, Berlin).
Esarey, E., G. Joyce, and P. Sprangle, 1991, Phys. Rev. A **44**, 3908.
Faisal, F. H. M., 1986, *Theory of Multiphoton Processes* (Plenum, New York).
Faldon, M. E., M. H. R. Hutchinson, J. P. Marangos, J. E. Muffett, R. A. Smith, J. W. G. Tisch, and C. G. Wahlström, 1992, J. Opt. Soc. Am. B **9**, 2094.
Farkas, Gy. and Cs. Toth, 1992, Phys. Lett. A **168**, 447.
Feldmann, J., J. Sacher, and E. Göbel, 1991, Opt. Lett. **16**, 241.
Fermann, M. E., 1995, in *Compact Sources of Ultrashort Pulses*, edited by I. N. Duling III (Cambridge University, Cambridge, England), p. 179.
Fermann, M. E., L. M. Yang, M. L. Stock, and M. J. Andrejco, 1994, Opt. Lett. **19**, 43.
Feurer, T., and R. Sauerbrey, 1997, in *Atomic, Molecular, and Optical Physics: Electromagnetic Radiation; Experimental Methods in the Physical Sciences, Vol. 29*, edited by F. B. Dunning and R. G. Hulet (Academic, New York), p. 193.
Fibich, G. and G. C. Papanicolaou, 1997, Opt. Lett. **18**, 1379.
Figueria-de-Morisson-Faria, C., M. Dörr, and W. Sandner, 1998, Phys. Rev. A **58**, 2990.
Fisher, R. A., P. L. Kelly, and T. K. Gustafson, 1969, Appl. Phys. Lett. **14**, 140.
Fittinghoff, D. N., P. R. Bolton, B. Chang, and K. C. Kulander, 1992, Phys. Rev. Lett. **69**, 2642.
Fork, R. L., C. H. Cruz, P. C. Becker, and C. V. Shank, 1987, Opt. Lett. **12**, 483.
Fork, R. L., B. I. Greene, and C. V. Shank, 1981, Appl. Phys. Lett. **38**, 71.
Fork, R. L., O. E. Martinez, and J. P. Gordon, 1984, Opt. Lett. **9**, 150.
Fork, R. L., C. V. Shank, and R. Yen, 1982, Appl. Phys. Lett. **41**, 223.
Fork, R. L., C. V. Shank, R. Yen, and C. A. Hirlimann, 1983, IEEE J. Quantum Electron. **19**, 500.
Freeman, R. R., P. H. Bucksbaum, W. E. Cooke, G. Gibson, T. J. McIlrath, and L. D. Van Woerkom, 1992, in *Atoms in Intense Laser Fields*, edited by M. Gavrila (Academic, San Diego), p. 43.
French, P. M., 1985, Rep. Prog. Phys. **58**, 169.
French, P. M. W., 1996, Contemp. Phys. **37**, 283.
French, P. M., M. M. Opalinska, and J. R. Taylor, 1989, Opt. Lett. **14**, 217.
French, P. M., and J. R. Taylor, 1988, Opt. Lett. **13**, 470.
French, P. M., J. A. R. Williams, and J. R. Taylor, 1987, Opt. Lett. **12**, 684.
French, P. M., J. A. R. Williams, and J. R. Taylor, 1989, Opt. Lett. **14**, 686.
Gahn, C., G. D. Tsakiris, A. Pukhov, J. Meyer-ter-Vehn, G. Pretzler, P. Thirolf, D. Habs, and K. J. Witte, 1999, Phys. Rev. Lett. **83** (in press).
Gallmann, L., D. H. Sutter, N. Matuschek, G. Steinmeyer, U. Keller, C. Iaconis, and I. A. Walmsley, 1999, Opt. Lett. **24**, 1314.
Geissler, M., G. Tempea, M. Schnürer, F. Krausz, and T. Brabec, 1999a, Phys. Rev. Lett. **83**, 2930.
Geissler, M., G. Tempea, and T. Brabec, 1999b, "Quasi-phase-matching of high harmonic generation," submitted to Phys. Rev. A.
Georgiev, D., J. Herrmann, and U. Stamm, 1992, Opt. Commun. **92**, 368.
Gildenburg, V. B., A. V. Kim, and A. M. Sergeev, 1990, Pis'ma Zh. Eksp. Teor. Fiz. [JETP Lett. **51**, 104 1990].
Gildenburg, V. B., T. A. Petrova, and A. D. Yunakovsky, 1995, Physica D **87**, 335.
Glover, T. E., R. W. Schoenlein, A. H. Chin, and C. V. Shank, 1996, Phys. Rev. Lett. **76**, 2468.
Glownia, J. H., J. Misewich, and P. P. Sorokin, 1987, J. Opt. Soc. Am. B **4**, 1061.
Greenhow, R. C., and A. J. Schmidt, 1974, in *Advances in Quantum Electronics Vol. 2*, edited by D. W. Goodwin (Academic, London), p. 158.
Grischkowski, D., and A. C. Balant, 1981, Appl. Phys. Lett. **41**, 1.
Haight, R. and P. F. Seidler, 1994, Appl. Phys. Lett. **65**, 517.
Hargrove, L. E., R. L. Fork, and M. A. Pollack, 1964, Appl. Phys. Lett. **5**, 4.
Harris, S. E. and A. V. Sokolov, 1998, Phys. Rev. Lett. **81**, 2894.
Haus, H. A., 1975a, IEEE J. Quantum Electron. **11**, 736.
Haus, H. A., 1975b, J. Appl. Phys. **46**, 3049.
Haus, H. A., 1976, IEEE J. Quantum Electron. **12**, 169.
Haus, H. A., 1984, *Waves and Fields in Optoelectronics* (Prentice-Hall, Englewood Cliffs, NJ), p. 99.
Haus, H. A., J. G. Fujimoto, and E. P. Ippen, 1991, J. Opt. Soc. Am. B **8**, 2068.
Haus, H. A., J. G. Fujimoto, and E. P. Ippen, 1992, IEEE J. Quantum Electron. **28**, 2086.
Hänsch, T. W., 1990, Opt. Commun. **80**, 71.
Heinz, P., and A. Lauberau, 1989, J. Opt. Soc. Am. B **6**, 1574.
Heritage, J. P., and M. Nuss, 1992, Eds., *Special Issue on Ultrafast Optics and Electronics*, IEEE J. Quantum Electron. **28**, p. 2084.
Herrmann, J., 1994, J. Opt. Soc. Am. B **11**, 498.
Hofer, M., M. E. Fermann, F. Haberl, M. H. Ober, and A. J. Schmidt, 1991, Opt. Lett. **16**, 502.
Hofer, M., M. H. Ober, F. Haberl, and M. E. Fermann, 1992, IEEE J. Quantum Electron. **28**, 720.
Hooker, S. M., and S. E. Harris, 1995, Opt. Lett. **20**, 1994.
Hu, S. X., and Z. Z. Xu, 1997, Appl. Phys. Lett. **71**, 2605.
Hunsche, S., S. Feng, H. G. Winful, A. Leitenstorfer, M. C. Nuss, and E. P. Ippen, 1999, J. Opt. Soc. Am. A **16**, 2025.
Iaconis, C., and I. Walmsley, 1998, Opt. Lett. **23**, 792.
Ippen, E. P., H. A. Haus, and L. Y. Liu, 1989, J. Opt. Soc. Am. B **6**, 1736.
Ippen, E. P., L. Y. Liu, and H. A. Haus, 1990, Opt. Lett. **15**, 183.
Ippen, E. P., C. V. Shank, and A. Dienes, 1972, Appl. Phys. Lett. **21**, 348.
Ivanov, M. Y., T. Brabec, and N. Burnett, 1996, Phys. Rev. A **54**, 742.
Ivanov, M. Y., and P. B. Corkum, 1993, Phys. Rev. A **48**, 580.
Jasapara, J., and W. Rudolph, 1999, Opt. Lett. **24**, 777.
Jimenez, R., C. Rose-Petruck, T. Guo, K. R. Wilson, and C. P. J. Barty, 1998, in *Ultrafast Phenomena XI*, edited by T. Elsässer, J. G. Fujimoto, D. A. Wiersma, and W. Zinth (Springer, Berlin), p. 404.
Joachain, C. J., M. Dörr, and N. J. Kylstra, 2000, Adv. At., Mol., Opt. Phys. **42**, 225.
Johnson, A. M., W. M. Simpson, and R. H. Stolen, 1984, Appl. Phys. Lett. **44**, 729.
Kálmán, P., 1989, Phys. Rev. A **39**, 3200.
Kálmán, P., 1993, Phys. Rev. A **48**, R42.
Kálmán, P., 1994, Phys. Rev. A **49**, 620.

Kálmán, P., F. Krausz, and E. Wintner, 1993, Phys. Rev. A **47**, 729.

Kalosha, V. P., M. Müller, J. Herrmann, and S. Gatz, 1998, J. Opt. Soc. Am. B **15**, 535.

Kan, C., N. H. Burnett, C. Capjack, and R. Rankin, 1997, Phys. Rev. Lett. **79**, 2971.

Kan, C., C. E. Capjack, R. Rankin, T. Brabec, and N. H. Burnett, 1996, Phys. Rev. A **54**, R1026.

Kane, D. J., and R. Trebino, 1993, IEEE J. Quantum Electron. **29**, 571.

Kaplan, A. E., 1994, Phys. Rev. Lett. **73**, 1243.

Kärtner, F. X., I. D. Jung, and U. Keller, 1996, IEEE J. Sel. Top. Quantum Electron. **2**, 540.

Kärtner, F. X., and U. Keller, 1995, Opt. Lett. **20**, 16.

Kärtner, F. X., N. Matuschek, T. Schibli, U. Keller, H. A. Haus, C. Heine, R. Morf, V. Scheuer, M. Tilsch, and T. Tschudi, 1997, Opt. Lett. **22**, 831.

Kasper, A., and K. J. Witte, 1996, Opt. Lett. **21**, 360.

Kean, P. N., X. Zhu, D. W. Crust, R. S. Grant, N. Langford, and W. Sibbett, 1989, Opt. Lett. **14**, 39.

Keldysh, L. V., 1965, Zh. Eksp. Teor. Fiz. [Sov. Phys. JETP **20**, 1307 1965].

Keller, U., 1994, Appl. Phys. B: Lasers Opt. **58**, 347.

Keller, U., 1997, Ed., *Special Issue on Ultrafast Lasers*, Appl. Phys. B: Lasers Opt. **65**, 2.

Keller, U., 1999, in *Nonlinear Optics in Semiconductors*, edited by A. Kost and E. Garmire (Academic, Boston), Vol. 59, p. 211.

Keller, U., W. H. Knox, and H. Roskos, 1990, Opt. Lett. **15**, 1377.

Keller, U., W. H. Knox, and G. W. 't Hooft, 1992, IEEE J. Quantum Electron. **28**, 2123.

Keller, U., D. A. B. Miller, G. D. Boyd, T. H. Chiu, J. F. Ferguson, and M. T. Asom, 1992, Opt. Lett. **17**, 505.

Keller, U., G. W. 't Hooft, W. H. Knox, and J. E. Cunningham, 1991, Opt. Lett. **16**, 1022.

Keller, U., K. J. Weingarten, F. X. Kärtner, D. Kopf, B. Braun, I. D. Jung, R. Fluck, C. Hönninger, N. Matuschek, and J. Aus der Au, 1996, IEEE J. Sel. Top. Quantum Electron. **2**, 435.

Kim, D.-E., Cs. Tóth, and C. P. J. Barty, 1999, Phys. Rev. A **59**, R4129.

Knox, W. H., 1988, IEEE J. Quantum Electron. **24**, 388.

Knox, W. H., M. C. Downer, R. L. Fork, and C. V. Shank, 1985, Opt. Lett. **9**, 552.

Kobayashi, Y., T. Seikikawa, Y. Nabekawa, and S. Watanabe, 1998, Opt. Lett. **23**, 64.

Koechner, W., 1996, *Solid-State Laser Engineering* (Springer, Berlin).

Kogelnik, H. W., E. P. Ippen, A. Dienes, and C. V. Shank, 1972, IEEE J. Quantum Electron. **8**, 373.

Kondo, K., N. Sarukura, K. Sajiki, and S. Watanabe, 1993, Phys. Rev. A **47**, R2480.

Koroshilov, E. V., I. V. Kryukov, P. G. Kryukov, and A. V. Sharkov, 1984, in *Ultrafast Phenomena VI*, edited by T. Yajima, K. Yoshihara, C. B. Harris, and S. Shionoya (Springer, Berlin), p. 22.

Krainov, V. P., 1997, J. Opt. Soc. Am. B **14**, 425.

Krause, J. L., K. J. Schafer, and K. C. Kulander, 1992, Phys. Rev. Lett. **68**, 3535.

Krausz, F., T. Brabec, M. Schnürer, and C. Spielmann, 1998, Opt. Photonics News **9**, 46.

Krausz, F., T. Brabec, and Ch. Spielmann, 1991, Opt. Lett. **16**, 235.

Krausz, F., M. E. Fermann, T. Brabec, P. F. Curley, M. Hofer, M. H. Ober, Ch. Spielmann, E. Wintner, and A. J. Schmidt, 1992, IEEE J. Quantum Electron. **28**, 2097.

Krausz, F., and E. Wintner, 1994, Eds., *Special Issue on Ultrashort Pulse Generation*, Appl. Phys. B: Lasers Opt. **58**, 3.

Kühlke, D., W. Rudolph, and B. Wilhelmi, 1983, IEEE J. Quantum Electron. **19**, 526.

Kuizenga, D. J., and A. E. Siegman, 1970a, IEEE J. Quantum Electron. **QE-6**, 694.

Kuizenga, D. J., and A. E. Siegman, 1970b, IEEE J. Quantum Electron. **QE-6**, 709.

Kulander, K. C., K. J. Schafer, and J. L. Krause, 1992, in *Atoms in Intense Laser Fields*, edited by M. Gavrila (Academic, San Diego), p. 247.

Kulander, K. C., K. J. Schafer, and J. L. Krause, 1993, in *Proceedings of the Workshop on Super Intense Laser Atom Physics (SILAP III)*, edited by B. Piraux (Plenum Press, New York), Vol. 316, p. 95.

Kulander, K. C., and B. W. Shore, 1989, Phys. Rev. Lett. **63**, 524.

Lablanquie, P., J. Mazeau, L. Andric, P. Selles, and A. Huetz, 1995, Phys. Rev. Lett. **74**, 2192.

Landau, L. D., and E. M. Lifshitz, 1977, *Quantum Mechanics* (Pergamon, Oxford).

Lange, H. R., A. Chrion, J.-F. Ripoche, A. Mysyrewicz, P. Breger, and P. Agostini, 1998, Phys. Rev. Lett. **81**, 1611.

Laporta, P., and V. Magni, 1985, Appl. Opt. **24**, 2014.

Laubereau, A., 1969, Phys. Lett. **29A**, 539.

Laubereau, A., and D. von der Linde, 1970, Z. Naturforsch. A **25a**, 1626.

Lemoff, B. E., and C. P. J. Barty, 1993a, Opt. Lett. **18**, 57.

Lemoff, B. E., and C. P. J. Barty, 1993b, Opt. Lett. **18**, 1651.

Lemoff, B. E., G. Y. Yin, C. L. Gordon III, C. P. J. Barty, and S. E. Harris, 1995, Phys. Rev. Lett. **74**, 1574.

Lenzner, M., J. Krüger, S. Sartania, Z. Cheng, Ch. Spielmann, G. Mourou, W. Kautek, and F. Krausz, 1998, Phys. Rev. Lett. **80**, 4076.

Lewenstein, M., Ph. Balcou, M. Yu. Ivanov, A. L'Huillier, and P. B. Corkum, 1994, Phys. Rev. A **49**, 2117.

L'Huillier, A., and P. Balcou, 1993, Phys. Rev. Lett. **70**, 774.

L'Huillier, A., P. Balcou, S. Candel, K. J. Schafer, and K. C. Kulander, 1992, Phys. Rev. A **46**, 2778.

L'Huillier, A., X. F. Li, and L. A. Lompré, 1990, J. Opt. Soc. Am. B **7**, 527.

L'Huillier, A., L.-A. Lompré, G. Mainfray, and C. Manus, 1992, in *Atoms in Intense Laser Fields*, edited by M. Gavrila (Academic, New York), p. 139.

L'Huillier, A., K. J. Schafer, and K. C. Kulander, 1991, Phys. Rev. Lett. **66**, 2200.

Li, X. F., A. L'Huillier, M. Ferray, L. A. Lompré, and G. Mainfray, 1989, Phys. Rev. A **39**, 5751.

Liang, Y., S. August, S. L. Chin, Y. Beaudoin, and M. Chaker, 1994, J. Phys. B **27**, 5119.

Libertun, A. R., R. Shelton, H. C. Kapteyn, and M. Murnane, 1999, in *Conference on Lasers and Electro-Optics*, OSA Technical Digest (Optical Society of America, Washington, DC), p. 469.

Lin, K.-H., Y. Lai, and W.-F. Hsieh, 1995, J. Opt. Soc. Am. B **12**, 468.

Long, S., W. Becker, and J. K. McIver, 1995, Phys. Rev. A **52**, 2262.

Macklin, J. J., J. D. Kmetec, and C. L. Gordon III, 1993, Phys. Rev. Lett. **70**, 766.

Maier, M., W. Kaiser, and J. A. Giordmaine, 1966, Phys. Rev. Lett. **17**, 1275.

Maine, P., D. Strickland, P. Bado, M. Pessot, and G. Mourou, 1988, IEEE J. Quantum Electron. **24**, 398.

Marburger, J. H., 1975, Prog. Quantum Electron. **4**, 35.

Marcatili, E. A. J., and R. A. Schmeltzer, 1964, Bell Syst. Tech. J. **43**, 1783.

Mark, J., L. Y. Liu, K. L. Hall, H. A. Haus, and E. P. Ippen, 1989, Opt. Lett. **14**, 48.

Martin, J.-L. A. Migus, G. A. Mourou, and A. H. Zewail, 1993, Eds., *Ultrafast Phenomena VIII* (Springer, Berlin).

Martinez, O. E., 1987, IEEE J. Quantum Electron. **23**, 1385.

Martinez, O. E., R. L. Fork, and J. P. Gordon, 1984b, Opt. Lett. **9**, 156.

Martinez, O. E., R. L. Fork, and J. P. Gordon, 1985, J. Opt. Soc. Am. B **2**, 753.

Martinez, O. E., J. P. Gordon, and R. L. Fork, 1984, J. Opt. Soc. Am. A **1**, 1003.

Matuschek, N., F. X. Kärtner, and U. Keller, 1998, IEEE J. Sel. Top. Quantum Electron. **4**, 414.

McPherson, A., G. Gibson, H. Jara, U. Johann, T. S. Luk, I. A. McIntyre, K. Boyer, and C. K. Rhodes, 1987, J. Opt. Soc. Am. B **4**, 595.

Milchberg, H. M., C. G. Durfee III, and T. J. McIlrath, 1995, Phys. Rev. Lett. **75**, 2494.

Milosevic, D. B., and A. F. Starace, 1998, Phys. Rev. Lett. **81**, 5097.

Milosevic, D. B., and A. F. Starace, 1999, Phys. Rev. A **60**, 3943.

Milosevic, N., G. Tempea, and T. Brabec, 1999, "Optical pulse compression: bulk media versus hollow waveguides," in press.

Mitschke, F. M., and L. F. Mollenauer, 1986, IEEE J. Quantum Electron. **22**, 2242.

Mitschke, F. M., and L. F. Mollenauer, 1987, Opt. Lett. **12**, 407.

Mocker, H. W., and R. J. Collins, 1965, Appl. Phys. Lett. **7**, 270.

Modena, A., Z. Najmudin, A. E. Dangor, C. E. Clayton, K. A. Marsh, C. Joshi, V. Malka, C. B. Darrow, C. Danson, D. Neely, and F. N. Walsh, 1995, Nature (London) **377**, 606.

Moiseyev, N., and F. Weinhold, 1997, Phys. Rev. Lett. **78**, 2100.

Mollenauer, L. F., and R. H. Stolen, 1984, Opt. Lett. **9**, 13.

Morgner, U., F. X. Kärtner, S. H. Cho, Y. Chen, H. A. Haus, J. G. Fujimoto, E. P. Ippen, V. Scheuer, G. Angelow, and T. Tschudi, 1999, Opt. Lett. **24**, 411.

Morin, M., and M. Piché, 1990, Opt. Lett. **14**, 1119.

Mossavi, K., Th. Hoffmann, F. K. Tittel, and G. Szabó, 1993, Appl. Phys. Lett. **62**, 1203.

Moulton, P. F., 1982, Opt. News **8**, 9.

Moulton, P. F., 1986, J. Opt. Soc. Am. B **3**, 125.

Mourou, G., 1997, Appl. Phys. B: Lasers Opt. **65**, 205.

Muller, H. G., P. Agostini, and G. Petite, 1992, in *Atoms in Intense Laser Fields*, edited by M. Gavrila (Academic, San Diego), p. 1.

Murnane, M., and H. C. Kapteyn, 1999, private communication.

Myazaki, K., and H. Takada, 1995, Phys. Rev. A **52**, 3007.

Nabekawa, Y., Y. Kuramoto, T. Togashi, T. Sekikawa, and S. Watanabe, 1998, Opt. Lett. **23**, 1384.

Nees, J., S. Biswal, F. Druon, J. Faure, M. Nantel, G. A. Mourou, A. Nishimura, H. Takuma, J. Itatani, J.-C. Chanteloup, and C. Hönninger, 1998, IEEE J. Sel. Top. Quantum Electron. **4**, 376.

New, G. H. C., 1972, Opt. Commun. **6**, 188.

New, G. H. C., 1974, IEEE J. Quantum Electron. **10**, 115.

Nibbering, E. T., P. F. Curley, G. Grillon, B. S. Prade, M. A. Franco, F. Salin, and A. Mysyrowicz, 1996, Opt. Lett. **21**, 62.

Nibbering, E. T., J. O. Duhr, and G. Korn, 1997, Opt. Lett. **22**, 1355.

Nisoli, M., S. De Silvestri, and O. Svelto, 1996, Appl. Phys. Lett. **68**, 2793.

Nisoli, M., S. De Silvestri, O. Svelto, R. Szipöcs, K. Ferencz, Ch. Spielmann, S. Sartania, and F. Krausz, 1997, Opt. Lett. **22**, 522.

Nisoli, M., S. Stagira, S. De Silvestri, O. Svelto, S. Sartania, Z. Cheng, M. Lenzner, Ch. Spielmann, and F. Krausz, 1997, Appl. Phys. B: Lasers Opt. **65**, 175.

Nisoli, M., S. Stagira, S. De Silvestri, O. Svelto, S. Sartania, Z. Cheng, G. Tempea, Ch. Spielmann, and F. Krausz, 1998, IEEE J. Sel. Top. Quantum Electron. **4**, 414.

Nuss, M. C., R. Leonhart, and W. Zinth, 1985, Opt. Lett. **10**, 16.

Oppenheimer, J. R., 1928, Phys. Rev. **13**, 66.

Osterink, L. M., and J. D. Foster, 1968, J. Appl. Phys. **39**, 4163.

Ouellette, F., and M. Piché, 1986, Opt. Commun. **60**, 99.

Patterson, F., J. Bonlie, D. Price, B. White, and P. Springer, 1999, in *Conference on Lasers and Electro-Optics*, OSA Technical Digest (Optical Society of America, Washington, DC) Post-Deadline Paper CPD6.

Patterson, F. G., and M. D. Perry, 1991, J. Opt. Soc. Am. B **8**, 2384.

Peatross, J., S. Voronov, and I. Prokopovich, 1997, Opt. Express **1**, 114.

Perelomov, A. M., V. S. Popov, and M. V. Terent'ev, 1966, Zh. Eksp. Teor. Fiz. [Sov. Phys. JETP **23**, 924 1966].

Perry, M. D., and J. K. Crane, 1993, Phys. Rev. A **48**, R4051.

Perry, M. D., and G. Mourou, 1994, Science **264**, 917.

Perry, M. D., D. Pennington, B. C. Stuart, G. Tietbohl, J. A. Britten, C. Brown, S. Herman, B. Golick, M. Kartz, J. Miller, H. T. Powell, M. Vergino, and V. Yanovski, 1999, Opt. Lett. **24**, 160.

Peterson, O. G., S. A. Tuccio, and B. B. Snavely, 1970, Appl. Phys. Lett. **17**, 245.

Petricevic, V., S. K. Gayen, and R. R. Alfano, 1989, Opt. Lett. **14**, 612.

Piché, M., 1991, Opt. Commun. **86**, 156.

Pont, M., and M. Gavrila, 1990, Phys. Rev. Lett. **65**, 2362.

Poppe, A., A. Fürbach, Ch. Spielmann, and F. Krausz, 1999, "Electronic on the time scale of the light oscillation period?" in *Ultrafast Electronics and Optoelectronics*, Trends in Optics and Photonics No. 28, edited by John E. Bowers and Wayne H. Knox (Optical Society of America, Washington, D.C.), p. 31.

Poppe, A., M. Lenzner, F. Krausz, and Ch. Spielmann, 1999, "A sub-10fs 2.5-MW Ti:sapphire oscillator," *Technical Digest of the Ultrafast Optics Conference*, to be published.

Poppe, A., L. Xu, F. Krausz, and Ch. Spielmann, 1998, IEEE J. Sel. Top. Quantum Electron. **4**, 179.

Potvliege, R. M., and R. Shakeshaft, 1989, Phys. Rev. A **40**, 3061.

Potvliege, R. M., and R. Shakeshaft, 1992, in *Atoms in Intense Laser Fields*, edited by M. Gavrila (Academic, San Diego), p. 373.

Proctor, B., and F. Wise, 1993, Appl. Phys. Lett. **62**, 470.

Pshenichnikov, M. S., A. Baltuska, R. Szipöcs, and D. A. Wiersma, 1998, in *Ultrafast Phenomena XI*, edited by T. Elsässer, J. G. Fujimoto, D. A. Wiersma, and W. Zinth (Springer, Berlin), p. 3.

Pshenichnikov, M. S., W. P. de Boeij, and D. A. Wiersma, 1994, Opt. Lett. **19**, 572.

Pukhov, A., and J. Meyer-ter-Vehn, 1996, Phys. Rev. Lett. **76**, 3975.

Rae, S. C., K. Burnett, and J. Cooper, 1994, Phys. Rev. A **50**, 3438.

Ramaswamy, M., M. Ulman, J. Paye, and J. G. Fujimoto, 1993, Opt. Lett. **18**, 1822.

Ranka, J. K., and A. L. Gaeta, 1998, Opt. Lett. **23**, 534.

Reichert, J., R. Holzwarth, Th. Udem, and T. W. Hänsch, 1999, "Measuring the frequency of light with mode-locked lasers," Opt. Commun. (in press).

Reiss, H. R., 1980, Phys. Rev. A **22**, 1786.

Reiss, H. R., 1992, Prog. Quantum Electron. **16**, 1.

Richardson, D. J., R. I. Laming, D. N. Payne, V. Matsas, and M. W. Phillips, 1991, Electron. Lett. **27**, 542.

Rischel, Ch., A. Rousse, I. Uschmann, P. A. Albouy, J. P. Geindre, P. Audebert, J. C. Gauthier, E. Förster, J. L. Martin, and A. Antonetti, 1997, Nature (London) **390**, 490.

Rizvi, N. H., P. M. W. French, J. R. Taylor, P. J. Delfyett, and L. T. Florez, 1993, Opt. Lett. **18**, 983.

Rolland, C., and P. B. Corkum, 1988, J. Opt. Soc. Am. B **5**, 641.

Rosen, C., M. Dörr, U. Eichmann, and W. Sandner, 1999, Phys. Rev. Lett. **83**, 4514.

Rouyer, C., E. Mazataud, I. Allais, A. Pierre, S. Seznec, C. Sauteret, G. Mourou, and A. Migus, 1993, Opt. Lett. **18**, 214.

Ruddock, I. S., and D. J. Bradley, 1976, Appl. Phys. Lett. **29**, 296.

Rundquist, A., C. G. Durfee III, Z. Cheng, C. Herne, S. Backus, M. M. Murnane, and H. C. Kapteyn, 1998, Science **280**, 1412.

Sala, K., M. C. Richardson, and N. R. Isenor, 1977, IEEE J. Quantum Electron. **13**, 915.

Salières, P., A. l'Huillier, P. Antoine, and M. Lewenstein, 1999, Adv. At., Mol., Opt. Phys. **41**, 83.

Salières, P., P. Antoine, A. de Bohan, and M. Lewenstein, 1998, Phys. Rev. Lett. **81**, 5544.

Salières, P., A. L'Huillier, and M. Lewenstein, 1995, Phys. Rev. Lett. **74**, 3776.

Salin, F., P. Grangier, P. Georges, and A. Brun, 1990, Opt. Lett. **15**, 1374.

Salin, F., J. Squier, and M. Piché, 1991, Opt. Lett. **16**, 1674.

Sanpera, A., P. Jönsson, J. B. Watson, and K. Burnett, 1995, Phys. Rev. A **51**, 3148.

Sarachick, E., and G. Schappert, 1970, Phys. Rev. D **1**, 2738.

Sartania, S., Z. Cheng, M. Lenzner, G. Tempea, Ch. Spielmann, F. Krausz, and K. Ferencz, 1997, Opt. Lett. **22**, 1562.

Sarukura, N., K. Hata, T. Adachi, R. Nodomi, M. Watanabe, and S. Watanabe, 1991, Phys. Rev. A **43**, 1669.

Sauteret, C., D. Husson, G. Thiell, S. Seznec, G. Gray, A. Migus, and G. Mourou, 1991, Opt. Lett. **16**, 238.

Schäfer, F. P., F. P. W. Schmidt, and J. Volze, 1966, Appl. Phys. Lett. **9**, 306.

Schafer, K. J., and K. C. Kulander, 1997, Phys. Rev. Lett. **78**, 638.

Schins, J. M., P. Breger, P. Agostini, R. C. Constantinescu, H. G. Muller, A. Bouhal, G. Grillon, A. Antonetti, and A. Mysyrowicz, 1996, J. Opt. Soc. Am. B **13**, 197.

Schnürer, M., Z. Cheng, M. Hentschel, F. Krausz, T. Wilhein, D. Hambach, G. Schmahl, M. Drescher, Y. Lim, U. Heinzmann, 1999a, "Few-cycle-driven XUV harmonics: generation and focusing," Appl. Phys. B. (in press).

Schnürer, M., Z. Cheng, M. Hentschel, G. Tempea, P. Kalman, T. Brabec, and F. Krausz, 1999b, Phys. Rev. Lett. **83**, 722.

Schnürer, M., Ch. Spielmann, P. Wobrauschek, C. Streli, N. H. Burnett, C. Kan, K. Ferencz, R. Koppitsch, Z. Cheng, T. Brabec, and F. Krausz, 1998, Phys. Rev. Lett. **80**, 3236.

Schnürer, M., C. Streli, P. Wobrauschek, M. Hentschel, R. Kienberger, Ch. Spielmann, and F. Krausz, 1999, "Femtosecond X-ray fluorescence," submitted for publication.

Schoenlein, R. W., W. P. Leemans, A. H. Chin, P. Volfbeyn, T. E. Glover, P. Balling, M. Zolotorev, K.-J. Kim, S. Chattopadhay, and C. V. Shank, 1996, Science **274**, 236.

Scrinzi, A., M. Geissler, and T. Brabec, 1999, Phys. Rev. Lett. **83**, 706.

Seidemann, T., M. Yu. Ivanov, and P. B. Corkum, 1995, Phys. Rev. Lett. **75**, 2819.

Shakeshaft, R., R. M. Potvliege, M. Dörr, and W. E. Cooke, 1990, Phys. Rev. A **42**, 1656.

Sheik-Bahae, M., A. A. Said, D. J. Hagan, M. J. Soileau, and W. W. Stryland, 1984, Opt. Eng. (Bellingham) **30**, 1228.

Shank, C. V., 1988, in *Ulrashort Laser Pulses and Applications*, Topics in Applied Physics No. 60, edited by W. Kaiser (Springer, New York), p. 5.

Shen, Y. R., 1984, *The Principles of Nonlinear Optics* (Wiley, New York).

Shimizu, F., 1967, Phys. Rev. Lett. **19**, 1097.

Shirakawa, A., I. Sakane, M. Takasaka, and T. Kobayashi, 1999, Appl. Phys. Lett. **74**, 2268.

Shkolnikov, P. L., A. E. Kaplan, and A. Lago, 1996, J. Opt. Soc. Am. B **13**, 412.

Shvartsburg, A. B., 1996, *Time-Domain Optics of Ultrashort Waveforms* (Clarendon, Oxford).

Sizer, T. II, J. D. Kafka, I. N. Duling III, C. W. Gabel, and G. A. Mourou, 1983, IEEE J. Quantum Electron. **19**, 506.

Smith, L. K., S. A. Payne, W. L. Kway, L. L. Chase, and B. H. T. Chai, 1992, IEEE J. Quantum Electron. **28**, 2612.

Smith, P. W., M. A. Duguay, and E. P. Ippen, 1974, Prog. Quantum Electron. **3**, 109.

Sommerfeld, A., 1967, *Elektronentheorie der Metalle* (Springer, Berlin), p. 109.

Sorokin, P. P., and J. R. Lankard, 1966, IBM J. Res. Dev. **10**, 162.

Spence, D. E., P. N. Kean, and W. Sibbett, 1991, Opt. Lett. **16**, 42.

Spielmann, Ch., N. H. Burnett, S. Sartania, R. Koppitsch, M. Schnürer, C. Kan, M. Lenzner, P. Wobrauschek, and F. Krausz, 1997, Science **278**, 661.

Spielmann, Ch., P. F. Curley, T. Brabec, and F. Krausz, 1994, IEEE J. Quantum Electron. **30**, 1100.

Spielmann, Ch., C. Kan, N. H. Burnett, T. Brabec, M. Geissler, A. Scrinzi, M. Schnürer, and F. Krausz, 1998, IEEE J. Sel. Top Quantum Electron. **4**, 249.

Spinelli, L., B. Couillaud, N. Goldblatt, and D. K. Negus, 1991, in *Conference on Lasers and Electro-Optics* (Optical Society of America, Washington, D.C.), Paper CPDP7.

Stankov, K. A., and J. Jethwa, 1988, Opt. Commun. **66**, 41.

Stingl, A., M. Lenzner, Ch. Spielmann, F. Krausz, and R. Szipöcs, 1995, Opt. Lett. **20**, 602.

Stix, M. S., and E. P. Ippen, 1983, IEEE J. Quantum Electron. **19**, 520.

Strickland, D., and G. Mourou, 1985, Opt. Commun. **56**, 219.

Stuart, B. C., M. D. Feit, A. M. Rubenchik, B. W. Shore, and M. D. Perry, 1995, Phys. Rev. Lett. **74**, 2248.

Sullivan, A., H. Hamster, H. C. Kapteyn, S. Gordon, W. White, H. Nathel, R. J. Blair, and R. W. Falcone, 1991, Opt. Lett. **16**, 1406.

Sutter, D. H., G. Steinmeyer, L. Gallmann, N. Matuschek, F. Morier-Genoud, U. Keller, V. Scheuer, G. Angelow, and T. Tschudi, 1999, Opt. Lett. **24**, 631.

Szatmári, S., 1994, Appl. Phys. B: Lasers Opt. **58**, 211.

Szatmári, S., B. Rácz, and F. P. Schäfer, 1987, Opt. Commun. **62**, 271.

Szipöcs, R., K. Ferencz, Ch. Spielmann, and F. Krausz, 1994, Opt. Lett. **19**, 201.

Szipöcs, R., A. Stingl, Ch. Spielmann, and F. Krausz, 1995, Opt. Photonics News **6**, 16.

Tamura, K., J. Jacobson, E. P. Ippen, H. A. Haus, and J. G. Fujimoto, 1993, Opt. Lett. **18**, 220.

Taylor, A. J., T. R. Gosnell, and J. P. Roberts, 1990, Opt. Lett. **15**, 39.

Telle, H. R., G. Steinmeyer, A. E. Dunlop, J. Stenger, D. H. Sutter, and U. Keller, 1999, Appl. Phys. B: Lasers Opt. **69**, 327.

Tempea, G., and T. Brabec, 1998a, Opt. Lett. **23**, 762.

Tempea, G., and T. Brabec, 1998b, Opt. Lett. **23**, 1286.

Tempea, G., M. Geissler, and T. Brabec, 1999a, J. Opt. Soc. Am. B **16**, 669.

Tempea, G., M. Geissler, and T. Brabec, 1999b, "Self-phase-matched high harmonic generation," Phys. Rev. Lett., in press.

Tempea, G., F. Krausz, Ch. Spielmann, and K. Ferencz, 1998, IEEE J. Sel. Top. Quantum Electron. **4**, 193.

Theobald, W., R. Häßner, C. Wülker, and R. Sauerbrey, 1996, Phys. Rev. Lett. **77**, 298.

Tien, A. C., S. Backus, H. Kapteyn, and M. Murnane, 1999, Phys. Rev. Lett. **82**, 3883.

Tisch, J. W. G., R. A. Smith, J. E. Muffett, M. Ciarrocca, J. P. Marangos, and M. H. R. Hutchinson, 1994, Phys. Rev. A **49**, R28.

Tomlinson, W. J., R. H. Stolen, and C. V. Shank, 1984, J. Opt. Soc. Am. B **1**, 139.

Toth, Cs., G. Farkas, and K. L. Vodopyanov, 1991, Appl. Phys. B: Photophys. Laser Chem. **53**, 221.

Treacy, E. B., 1969, Phys. Lett. **28A**, 34.

Trebino, R., K. W. DeLong, D. N. Fittinghoff, J. N. Sweetser, M. A. Krumbügel, B. A. Richman, and D. J. Kane, 1997, Rev. Sci. Instrum. **68**, 3277.

Tsuda, S., W. H. Knox, S. T. Cundiff, W. Y. Jan, and J. E. Cunningham, 1996, IEEE J. Sel. Top. Quantum Electron. **2**, 435.

Tsuda, S., W. H. Knox, E. A. de Souza, W. Y. Jan, and J. E. Cunningham, 1995, Opt. Lett. **20**, 1406.

Udem, Th., J. Reichert, R. Holzwarth, and T. W. Hänsch, 1999a, Phys. Rev. Lett. **82**, 3568.

Udem, Th., J. Reichert, R. Holzwarth, and T. W. Hänsch, 1999b, Opt. Lett. **24**, 881.

Umstadter, D., S.-Y. Chen, A. Maksimchuk, G. Mourou, and R. Wagner, 1996, Science **273**, 472.

Valdmanis, J. A., and R. L. Fork, 1986, IEEE J. Quantum Electron. **22**, 112.

Valdmanis, J. A., R. L. Fork, and J. P. Gordon, 1985, Opt. Lett. **10**, 131.

Wahlström, C. G., J. Larsson, A. Persson, T. Starczewski, S. Svanberg, P. Salières, Ph. Balcou, and A. L'Huillier, 1993, Phys. Rev. A **48**, 4709.

Walker, B., B. Sheehy, L. F. DiMauro, P. Agostini, K. J. Schafer, and K. C. Kulander, 1994, Phys. Rev. Lett. **73**, 1227.

Walmsley, I., and V. Wong, 1996, J. Opt. Soc. Am. B **13**, 2453.

Watanabe, S., A. Endoh, M. Watanabe, and N. Sakura, 1988, Opt. Lett. **13**, 580.

Weber, H. P., 1967, J. Appl. Phys. **38**, 2231.

Weihe, F. A., S. K. Dutta, G. Korn, D. Du, P. H. Bucksbaum, and P. L. Shkolnikov, 1995, Phys. Rev. A **51**, R3433.

Wood, Wm. M., C. W. Siders, and M. C. Downer, 1991, Phys. Rev. Lett. **67**, 3523.

Wöste, L., C. Wedekind, H. Wille, P. Rairoux, B. Stein, S. Nikolov, C. Werner, S. Niedermeier, F. Ronneberger, H. Schillinger, and R. Sauerbrey, 1997, *Laser Optoelektron.* **29**, 51 (October issue).

Xu, H., X. Tang, and P. Lambropoulos, 1992, Phys. Rev. A **46**, R2225.

Xu, L., Ch. Spielmann, A. Poppe, T. Brabec, F. Krausz, and T. Hánsch, 1996, Opt. Lett. **21**, 2008.

Xu, L., G. Tempea, A. Poppe, M. Lenzner, Ch. Spielmann, F. Krausz, A. Stingl, and K. Ferencz, 1997, Appl. Phys. B: Lasers Opt. **65**, 151.

Xu, L., G. Tempea, Ch. Spielmann, F. Krausz, A. Stingl, K. Ferencz, and S. Takano, 1998, Opt. Lett. **23**, 789.

Yamakawa, K., H. Siraga, Y. Kato, and C. P. J. Barty, 1991, Opt. Lett. **16**, 1593.

Yamakawa, K., M. Aoyama, S. Matsuoka, H. Takuma, C. P. J. Barty, and D. Fittinghoff, 1998a, Opt. Lett. **23**, 525.

Yamakawa, K., M. Aoyama, S. Matsuoka, T. Kase, Y. Akahane, H. Takuma, D. Fittinghoff, and C. P. J. Barty, 1998b, in *Ultrafast Phenomena XI*, edited by T. Elsässer, J. G. Fujimoto, D. A. Wiersma, and W. Zinth (Springer, Berlin), p. 44.

Yariv, A., 1965, J. Appl. Phys. **36**, 388.

Yariv, A., and P. Yeh, 1984, *Optical Waves in Crystals* (Academic, New York).

Yoshikawa, S. and T. Imasaka, 1993, Opt. Commun. **96**, 94.

Zakharov, V. E., and A. B. Shabat, 1972, Zh. Eksp. Teor. Fiz. [Sov. Phys. JETP **34**, 62 (1972)].

Zhang, X.-C., B. H. Kolner, and K. Leo, 1996, Eds., *Ultrafast Electronics, Photonics, and Optoelectronics*, IEEE J. Sel. Top. Quantum Electron. **2**, p. 433.

Zhou, J. P., C. P. Huang, M. M. Murnane, and H. C. Kapteyn, 1995, Opt. Lett. **20**, 64.

Zhou, J., J. Peatross, M. M. Murnane, H. C. Kapteyn, and I. P. Christov, 1996, Phys. Rev. Lett. **76**, 752.

Zhou, J., G. Taft, C. P. Huang, M. M. Murnane, H. C. Kapteyn, and I. Christov, 1994, Opt. Lett. **19**, 1149.

Ziolkowski, R. W., and J. B. Judkins, 1993, J. Opt. Soc. Am. B **10**, 186.

Zozulya, A. A., S. A. Diddams, A. G. Van Engen, and T. S. Clement, 1999, Phys. Rev. Lett. **82**, 1430.

Zuo, T., A. D. Bandrauk, M. Ivanov, and P. B. Corkum, 1995, Phys. Rev. A **51**, 3991.

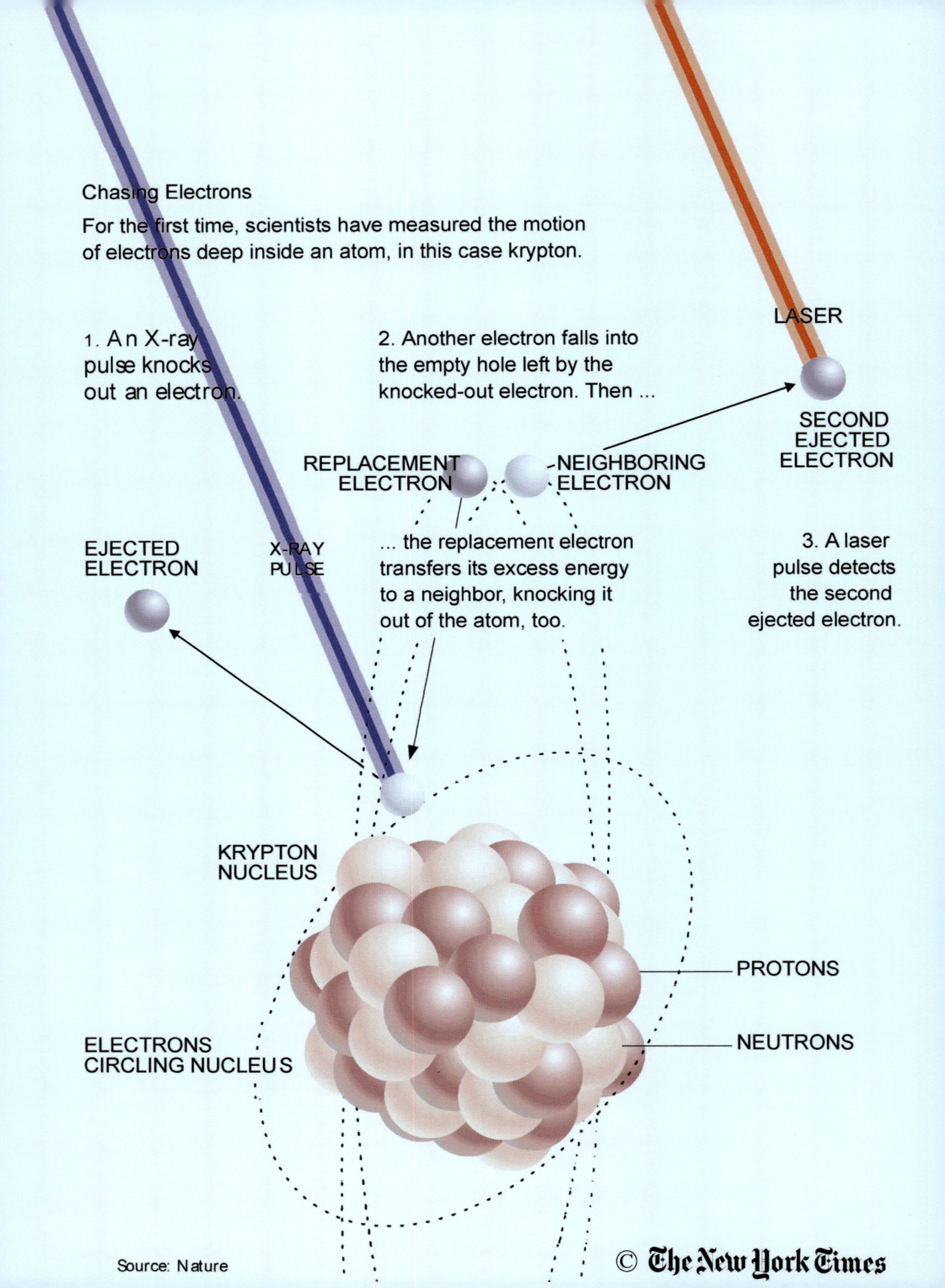

Chasing Electrons
For the first time, scientists have measured the motion of electrons deep inside an atom, in this case krypton.
1. An X-ray pulse knocks out an electron.
2. Another electron falls into the empty hole left by the knocked-out electron. Then ...
LASER
SECOND EJECTED ELECTRON
REPLACEMENT ELECTRON
NEIGHBORING ELECTRON
EJECTED ELECTRON
X-RAY PULSE
... the replacement electron transfers its excess energy to a neighbor, knocking it out of the atom, too.
3. A laser pulse detects the second ejected electron.
KRYPTON NUCLEUS
PROTONS
ELECTRONS CIRCLING NUCLEUS
NEUTRONS
Source: Nature
© The New York Times

Measurements at the Attosecond Frontier

4

Real-time observation of an electron undergoing an inner-shell atomic transition following ionization by a sub-femtosecond XUV pulse [21]. Reprinted from the New York Times with permission of the publisher.

By the turn of the millennium, the temporal confinement of atomically strong fields to the time scale of the optical period along with the exploitation of extreme nonlinearities advanced ultrafast science to the verge of the attosecond time scale [18]. Markus Drescher and his co-workers drove tunneling ionization and concomitant high-order harmonic emission (HHG) with few-cycle light and observed 1-fs-scale coherent emission of extreme-ultraviolet/soft-X-ray (XUV) light, as reported in [19]. Theoretical modeling described in [16-17] predicted that HHG driven by few-cycle light should be capable of spawning much shorter, several-100-attosecond pulses from the same interaction under the experimental conditions described in [19]. But why did we not see these sub-femtosecond pulses? There is (at least) one obvious reason.

As in all previous femtosecond measurements, the duration of XUV emission was measured, or more precisely, attempts were made to measure the emission by cross-correlating the process with the amplitude envelope $a_L(t)$ of the laser pulse. The several-femtosecond duration of this envelope limited the temporal resolution of the measurement. The solution could not have been more obvious: replace the femtosecond-varying $a_L(t)$ with the attosecond-varying electric field of the laser pulse $E_L(t)$ in the same cross correlation experiment.

A few months after the experiment reported in [19], this step indeed permitted the breaking of the attosecond barrier by demonstrating the generation and measurement of sub-femtosecond temporal confinement of coherent radiation. The editors of *Nature* accepted the work for publication 18 days following its submission and published it on November 29, 2001

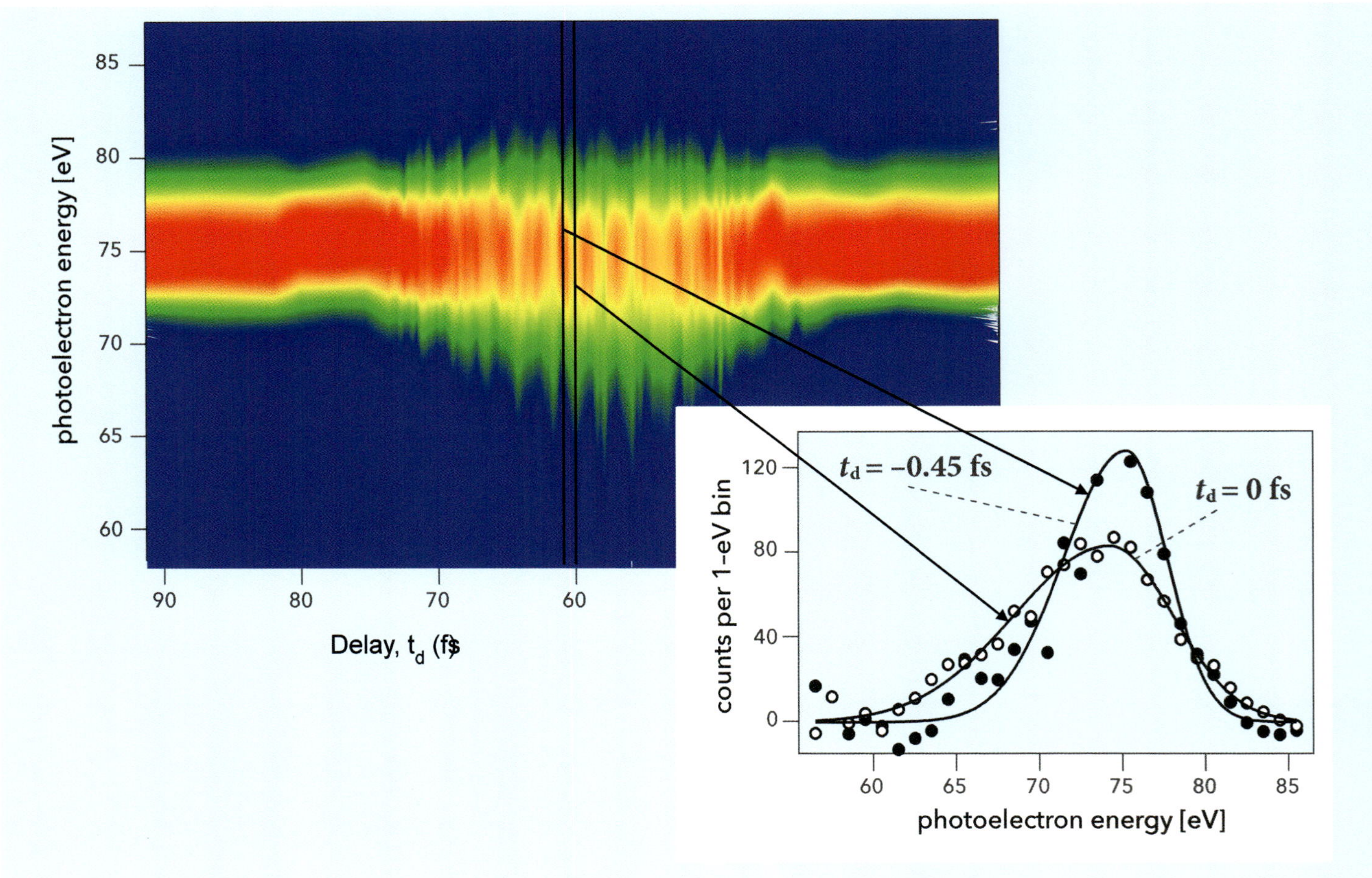

Figure 4-1 First attosecond streaking spectrogram recorded with a sub-femtosecond XUV pulse releasing photoelectrons from neon atoms and an orthogonally-polarized few-cycle near-infrared laser pulse deflecting the photoelectrons laterally, as a function of delay between the XUV pulse and the streaking laser field. Varying the delay by merely 450 as results in a dramatic change of the photoelectrons' energy distribution, providing evidence for the sub-femtosecond temporal confinement of the photoemission process [20].

[20]. The cross correlation between the oscillating few-cycle laser electric field and the XUV pulse was implemented by having the XUV pulse release photoelectrons. Their momentum was altered by $E_L(t)$ by an amount varying with the moment of release due to the rapid variation of $E_L(t)$ on the same (attosecond) time scale (Fig. 4-1). Hence, light-field-driven attosecond streaking, first proposed by Paul Corkum in Ottawa and experimentally demonstrated by Michael Hentschel, Reinhard Kienberger and their co-workers in Vienna [20], utilizes the steep, sub-femtosecond temporal gradient of the oscillating field for attosecond chronoscopy.

Retrieving the duration of the XUV pulse from the resultant cross-correlation trace relied on the assumption of an instantaneous emission of photoelectrons from atoms upon exposure to the XUV pulse. If the XUV photons are energetic enough to create an inner-shell vacancy, Auger electrons are emitted, the duration of which is dictated by the lifetime of the inner-shell vacancy rather than by the XUV excitation (provided that the latter is shorter than the former). By having the emitted Auger electrons deflected by $E_L(t)$ in the same way that the photoelectrons were previously deflected in the measurement of the duration of the XUV pulse, Markus Drescher and his collaborators were able to observe electron motion deep inside atoms – the filling of an inner-shell vacancy by an electron from an outer shell – in real time [21].

Alignment of the polarization of the strong laser field $E_L(t)$ parallel to the direction of the photoelectron emission induced by a sub-femtosecond XUV pulse permitted Reinhard Kienberger and his co-workers to accelerate and decelerate the emitted photoelectron wavepacket with the electric field of light [22]. The new interaction geometry revealed a more efficient scheme for the precise characterization of the emerging tools of attosecond technology, namely the attosecond-varying laser waveform and synchronized attosecond-duration XUV pulses (Chapter 6). Even more importantly, the experiment also conveyed the message that phase-stabilized few-cycle light (Chapter 5) is a prerequisite for exploiting the full potential of these attosecond diagnostics.

The experiments described in [20-22] include (i) the creation and observation of the first sub-femtosecond tool, (ii) its measurement and (iii) first spectroscopic application for capturing intra-atomic electron motion in real time. These achievements heralded a new field of research: attophysics as attested by Yaron Silberberg [23], Louis F. DiMauro [24], and Maciej Lewenstein [25] in their commentaries. *Nature* and *Science* celebrated these advances among the *Breakthroughs of the Year*[7,8]. However, attosecond physics was still in its infancy. Precision attosecond metrology would come of age only through the advances described in the following two chapters.

[18] F. Krausz. From femtochemistry to attophysics. *Physics World* **14**, 41 (2001). ***(Page: 174)***

[19] M. Drescher, M. Hentschel, R. Kienberger, G. Tempea, C. Spielmann, G. A. Reider, P. Corkum, F. Krausz. X-ray pulses approaching the attosecond frontier. *Science* **291**, 1923 (2001). ***(Page: 180)***

[20] M. Hentschel, R. Kienberger, C. Spielmann, G. A. Reider, N. Milosevic, T. Brabec, P. Corkum, U. Heinzmann, M. Drescher, F. Krausz. Attosecond metrology. *Nature* **414**, 509 (2001). ***(Page: 186)***

[21] M. Drescher, M. Hentschel, R Kienberger, M. Uiberacker, V. Yakovlev, A. Scrinzi, T. Westerwalbesloh, U. Kleineberg, U. Heinzmann, F. Krausz. Time-resolved atomic inner-shell spectroscopy. *Nature* **419**, 803 (2002). ***(Page: 191)***

[22] R. Kienberger, M. Hentschel, M. Uiberacker, C. Spielmann, M. Kitzler, A. Scrinzi, M. Wieland, T. Westerwalbesloh, U. Kleineberg, U. Heinzmann, M. Drescher, F. Krausz. Steering attosecond electron wave packets with light. *Science* **297**, 1144 (2002). ***(Page: 196)***

[23] Y. Silberberg. Physics at the attosecond frontier. *Nature* **414**, 494 (2001). ***(Page: 201)***

[24] L. F. DiMauro. Atomic photography. *Nature* **419**, 789 (2002). ***(Page: 203)***

[25] M. Lewenstein. Resolving physical processes on the attosecond time scale. *Science* **297**, 1131 (2002). ***(Page: 205)***

[7] The Editors. Fast moves. *Nature* **420**, 737 (19/26 December 2002).
[8] J. Giles. The fast show. *Science* **298**, 2299 (20 December 2002).

EXTREME OPTICS

Physicists routinely exploit femtosecond light pulses to track the movement of atoms in molecules, and are now developing sources that emit even shorter bursts of X-rays to study and control the motion of electrons inside atoms

From femtochemistry to attophysics

Ferenc Krausz

MANY amateur photographers are disappointed when they discover that their shots of fast-moving objects are blurred beyond recognition. The most likely reason for this is that the exposure time was not short enough to freeze the motion. In contrast, modern ultrahigh-speed cameras can take up to a million images every second and can capture motion that is normally imperceptible to the human eye. By projecting the photographs on a screen in sequence, the action can be replayed in slow motion. While these techniques are ideal for studying macroscopic objects, how can we possibly follow the motion of atoms and electrons?

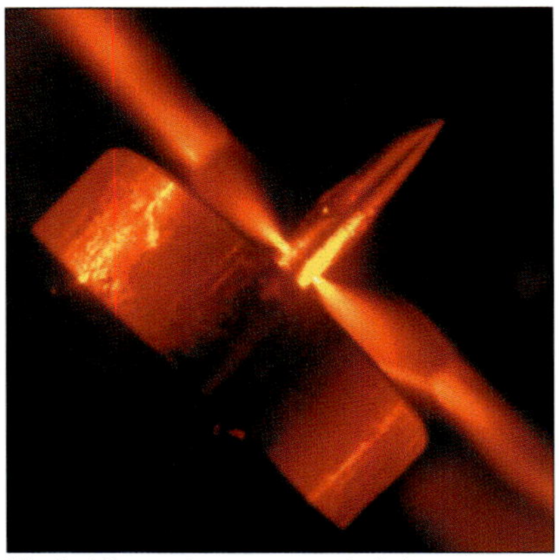

Ultrafast X-ray pulses are generated when neon atoms contained in this metal tube are irradiated with a femtosecond laser

The impetus to be able to track the movement of atoms and electrons comes from many areas of science and technology. The ability to look at chemical or biochemical processes is a prerequisite for steering reactions, while insights into the dynamics of electrons and holes in semiconductor structures are crucial for speeding up electronic devices. At a more fundamental level, following the motion of electrons in atoms is essential if we want to understand what happens inside excited atoms and exploit these processes in applications such as X-ray lasers.

So how short does the exposure have to be in order to capture the dynamics of atoms and electrons? Surprisingly, the answer is rooted in classical physics and is determined by the mass of the atom and the Coulomb force exerted on it by its neighbours. The movement of nuclei in a molecular or crystal-lattice structure, for example, can be traced with a probe lasting 10–100 femtoseconds ($1\ \text{fs} = 10^{-15}$ s). Meanwhile, the smaller mass of electrons makes them rather snappy movers. Indeed, the motion of bound electrons in excited atoms and molecules evolves on timescales of 10–1000 attoseconds ($1\ \text{as} = 10^{-18}$ s) and so requires attosecond probes.

Taking freeze-frame shots of atoms calls for exposures more than a million times shorter than those offered by the fastest high-speed cameras. This enormous gap has been bridged in the last decade thanks to the invention of lasers that produce flashes of light lasting just a few femtoseconds.

Atoms in slow motion

Physicists can now take snapshots of evolving atomic systems with pinpoint accuracy using ultrashort flashes of light to *both* trigger the dynamics *and* illuminate the system. This can be done by splitting each laser pulse with a partially transmitting mirror and delaying the less energetic "probe" pulse with respect to the stronger excitation or "pump" pulse (figure 1). In this way a powerful femtosecond laser pulse can initiate the same microscopic process in millions of molecules or sites in a crystal lattice. A weaker portion of the pulse (or a frequency-shifted replica) can then probe the dynamics by measuring changes in the optical properties of the system (e.g. the absorption) at a later instant.

We can replay the atomic or molecular dynamics in slow motion by using a series of femtosecond pulses and increasing the delay between successive pump and probe pulses. This approach is known as time-resolved or pump–probe spectroscopy and is currently the only way to study microscopic dynamics. The time resolution is only limited by the duration of the pump and probe pulses.

Physicists can now routinely generate pulses shorter than 10 fs using self-mode-locking lasers, which were invented in 1990 by Wilson Sibbett of St Andrews University in Scotland. In a mode-locked laser, the electric fields associated with all the different frequencies or modes that can exist inside the laser cavity add constructively at one point and destructively elsewhere to create a high-intensity spike.

The early mode-locked lasers developed in the 1970s were difficult to construct and operate. In contrast, self-mode locking happens naturally in carefully designed laser cavities that incorporate a solid-state amplifier such as titanium-doped sapphire – a laser medium developed in the 1980s by Peter Moulton at the Massachusetts Institute of Technology. Meanwhile

EXTREME OPTICS

1 High-speed photography at the microscale

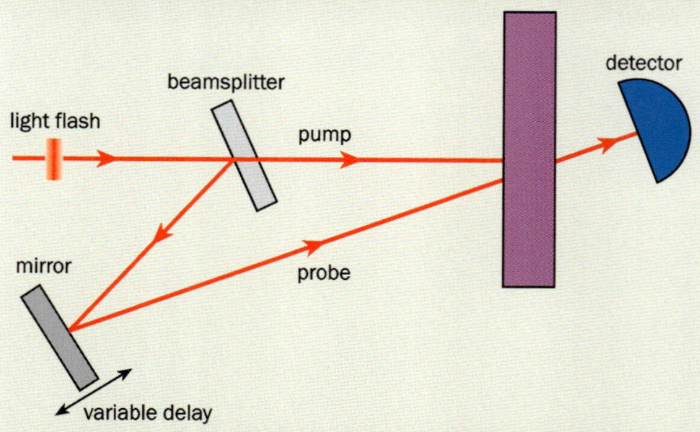

To extend high-speed photography to microscopic processes, an ultrashort laser pulse is split in two at a beamsplitter. The excitation or pump pulse initiates the dynamics, e.g. the breaking of a chemical bond, while the probe pulse captures any change in the optical properties at a later time. A sequence of measurements can by made by varying the delay between the pump and probe pulses using a movable mirror. From the change of the optical properties (e.g. absorption) versus delay, the dynamics (e.g. atomic motion) can be reconstructed. The duration of the light flashes determines the resolution of this "time microscope".

specialized "chirped" mirrors, devised in 1993 by Robert Szipöcs of the Research Institute for Solid State Physics and Optics in Budapest and the current author, can further compress the pulses (see Szipöcs *et al.* in further reading).

Using these techniques, Ursula Keller at the ETH in Zurich and, independently, Franz Kärtner and colleagues at MIT, have developed oscillators that produce pulses that last just 5 fs and have a wavelength of 800 nm. In other words, these pulses contain only two cycles of the laser field. Since light is an electromagnetic wave, the laser pulses cannot be shorter than the "carrier wavelength", λ, which therefore limits the duration of the pulse to λ/c, where c is the speed of light.

This means that the wavelength of light limits the resolution with which atoms can be observed in time, as well as limiting the spatial resolution (see the article by Vahid Sandoghdar on page 29). Interestingly, physicists can resolve atoms more readily when they are moving than when they are fixed. While it is impossible to view stationary atoms with visible light due to the diffraction limit, femtosecond pulses are capable of tracking moving atoms that are displaced by as little as 0.01 nm in real time due to the fact that the atoms move slowly compared with the speed of light.

Femtosecond chemistry and semiconductor dynamics

The formation and breaking of chemical bonds between atoms in molecules is one of the most important microscopic processes that affect our lives. Ultrafast laser pulses allow physicists and chemists to follow these femtosecond processes by tracing the displacements of the atoms, as demonstrated for the first time in the late 1980s by Ahmed Zewail at the California Institute of Technology. These displacements induce changes in the optical properties of the weakly bound valence electrons that can be revealed instantly by a visible-light probe. In complex systems, however, the atomic motion can be more accurately determined by studying core electrons close to the atomic nucleus, but this requires X-ray wavelengths.

Ultrafast laser pulses also allow us to measure the frequencies with which the chemical bonds stretch and contract. However, molecular vibrations can be measured more easily in the frequency domain using existing techniques, for example by measuring the absorption of infrared radiation as a function of frequency. So what are the advantages of ultrafast lasers?

Unlike frequency-domain techniques, ultrafast lasers provide access to the relative phase of the vibrational modes, which is crucial for reconstructing the dynamical changes in the structure of the molecule. These time-domain measurements can identify which chemical bond broke first, for instance, and the course of the chemical reaction. Moreover, femtosecond pulses can even control chemical reactions, as demonstrated recently by Gustav Gerber and Thomas Baumert at the University of Würzburg in Germany.

Ever since they became widely available, picosecond and femtosecond lasers have been used to investigate the dynamics of charge carriers in semiconductors. New insights into the basic phenomena that limit the speed of integrated circuits allow physicists to extend the limits of high-speed electronics. Ultrashort pulses of light can be used to explore the frontiers of electronics because they are far faster than the fastest electronic devices available.

Moreover, the terahertz electric fields that are produced by ultrashort optical pulses can induce and control currents in integrated circuits on timescales that are inaccessible to electronic instruments. This capability will help to speed up computers by a factor of thousand compared with the current state of the art. And even when terahertz electronics becomes a reality, femtosecond laser pulses will still be fast enough to keep pushing the frontiers of semiconductor technology.

Beyond the femtosecond barrier

So can we gain anything more by improving the resolution of pump–probe spectroscopy beyond 1 fs? The answer is clearly yes – sub-femtosecond laser pulses will enable us to track processes *inside* atoms for the first time. The first step in this direction is to generate and measure sub-femtosecond or attosecond light pulses. These pulses will have to be substantially shorter than the wave cycle of visible light, which lasts about 2 fs in the case of red light. As a result, shorter wavelengths or higher "carrier frequencies" are needed to generate sub-femtosecond light pulses.

In principle, a sequence of light pulses that are shorter than λ/c can be produced simply by adding together waves that oscillate with an angular frequency of $\omega_0 + N\Delta\omega$, where $\Delta\omega$ is a fixed shift with respect to the fundamental laser wave, $\omega_0 = 2\pi c/\lambda$, and N is an integer (figure 2). The result is a series of intense spikes separated in time by $1/\Delta\nu = 2\pi/\Delta\omega$.

The duration of these spikes is inversely proportional to both the frequency shift, $\Delta\nu$, and the number of waves that add together. This means that in order to produce the sub-femtosecond pulses using a limited number of waves, $\Delta\nu$ must be similar to the frequency of visible light. Conceptually, this technique is closely related to the mode-locking method that is generally used to generate femtosecond pulses in laser resonators (see Kapteyn and Murnane in further reading). The significant difference, however, is that $\Delta\nu$ for sub-femtosecond pulse generation must be several orders of magnitude greater than the frequency spacing between the adjacent modes of a laser resonator. Indeed, $\Delta\nu$ must be so large that no laser can amplify all these frequency-shifted waves. The only way that these waves can be produced is using nonlinear

2 How to generate sub-femtosecond pulses

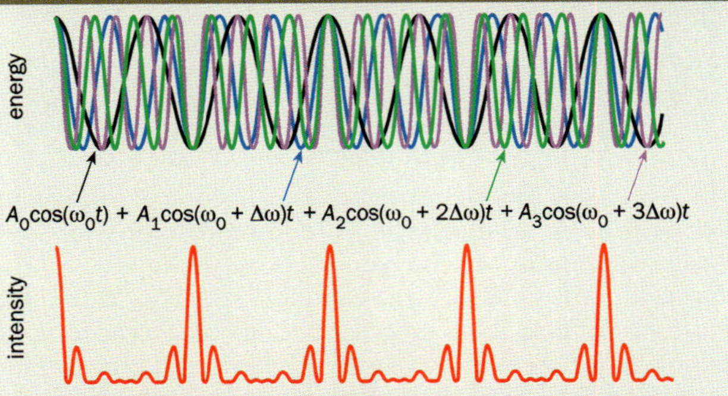

The superposition of several light waves at equidistant frequencies (top) in the ultraviolet region can give rise to a sequence of sub-femtosecond spikes (bottom) if the phases of the waves are adjusted appropriately. The repetition rate of the spikes is $\Delta\nu = \Delta\omega/2\pi$, where $\Delta\omega$ is the angular frequency difference between adjacent components.

3 Raman sidebands and high harmonics

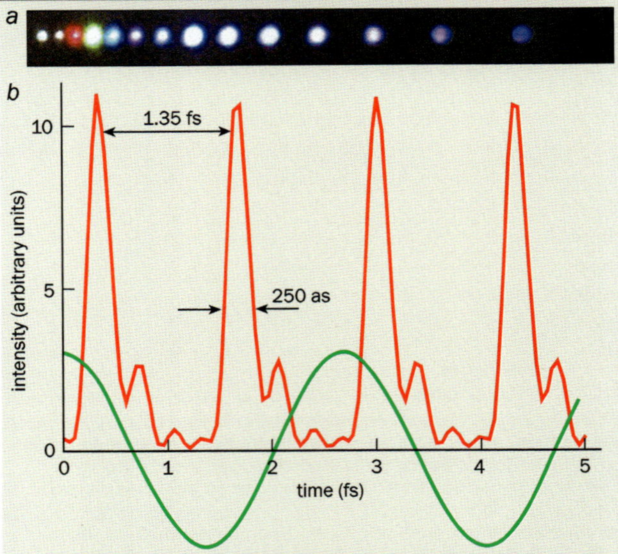

(*a*) The spectrum produced by Alexi Sokolov and co-workers at Stanford using a collinear Raman generator. From the left: the first two sidebands are at infrared wavelengths, while the next four are in the visible spectrum from red, green, blue to violet. The subsequent sidebands are in the ultraviolet and are only visible from the fluorescence they generate. The sidebands produced by exciting molecular vibrations at the difference of two laser frequencies are coherent and range in wavelength from 2.94 μm to 195 nm (see A Sokolov *et al.* 2000 *Phys. Rev. Lett.* **85** 562). (*b*) The train of attosecond pulses (red) that is predicted to emerge when a jet of argon gas is exposed to intense 40 fs laser pulses with a wavelength of 800 nm. The 11th, 13th, 15th, 17th and 19th harmonics that are generated add together to produce the ultrashort pulses. The carrier wavelength of the attosecond-pulse train is 53 nm and the train length is of the order of 10 fs. The green line represents the electric field of the driving laser (P Paul *et al.* 2001 *Science* **292** 1689).

optical techniques that are not part of the femtosecond laser oscillator itself.

In general, two nonlinear optical phenomena can be exploited to produce waves that are phase-coherent with the incident laser beam but have much higher frequencies – stimulated Raman scattering and high-order harmonic generation. In order to keep the dispersion of ultraviolet light and the absorption of shorter wavelengths as low as possible, both processes are carried out in a gas.

Raman scattering occurs when light passes through a gas of molecules. The light can excite vibrational or rotational energy levels in the molecules, which subsequently modulate the laser radiation. The potential of stimulated Raman scattering for generating trains of sub-femtosecond pulses has been demonstrated by several groups, including Alexi Sokolov and Stephen Harris at Stanford University in the US. They excited vibrational oscillations in deuterium molecules using two narrow-band lasers, which were chosen so that the difference in frequency between the two beams matched the oscillation frequency of the molecules. The periodic movement of the deuterium nuclei affects the extent to which the electrons are displaced by the laser. This periodic nuclear motion leads to phase-coherent sidebands that are frequency shifted with respect to the incident laser waves by the vibrational frequency ($\Delta\nu \approx 90$ THz). The spectrum of these coherent collinear Raman modes extends from the infrared to the ultraviolet (figure 3*a*). Only a fraction of this spectrum – comprising the blue, violet and ultraviolet lines – is needed to generate a train of sub-femtosecond pulses separated by $1/\Delta\nu \approx 11$ fs, provided the relative phase of the Raman modes is properly adjusted.

Even shorter pulses can be produced by shifting the laser frequency into the extreme ultraviolet using high-order harmonic generation. This phenomenon occurs when a high-power beam of linearly polarized femtosecond laser pulses is passed through an atomic gas. The intense electric field of the light pulse can rip an electron away from an atom and then accelerate it. When the light field reverses in the next half-cycle, the electron can slam back into the ion and give up its energy in the form of an X-ray photon. This process is repeated every half cycle of the laser field, resulting in the emergence of discrete lines in the extreme ultraviolet and X-ray spectrum. These "high-order harmonics" are separated in frequency by twice the frequency of the driving laser field and combine with each other to produce a train of attosecond pulses (figure 3*b*).

Conclusive experimental evidence for the existence of such a pulse train was obtained only recently by a collaboration between French and Dutch physicists, led by Harm Geert Muller at the FOM Institute for Atomic and Molecular Physics in Amsterdam and Pierre Agostini at the Centre d'Etudes de Saclay near Paris. The experiment indicated the generation of a train of 250 as pulses using 40 fs pump pulses from a Ti:sapphire laser.

Sub-femtosecond spectroscopy

Stimulated Raman scattering and high-harmonic generation unfortunately have their drawbacks for pump–probe spectroscopy. The time interval between the pulses – 11 fs in the Stanford experiment and 1.3 fs in the Amsterdam–Paris work – is so short that the physical, chemical or biological processes under study do not have time to completely decay and return to their initial state before the next pulse arrives. The high repetition rate of the pulses therefore imposes an unacceptably severe restriction on the processes that can be studied.

Clearly pump–probe spectroscopy can only benefit from these ultrafast sources if a single pulse can be selected from the train, as Paul Corkum of the Steacie Institute of Molecular Sciences in Ottawa, Canada, pointed out in his pioneering proposal for attosecond optics. The most promising way of isolating such a pulse is to drive the molecules that produce the

EXTREME OPTICS

4 Intense few-cycle laser pulses

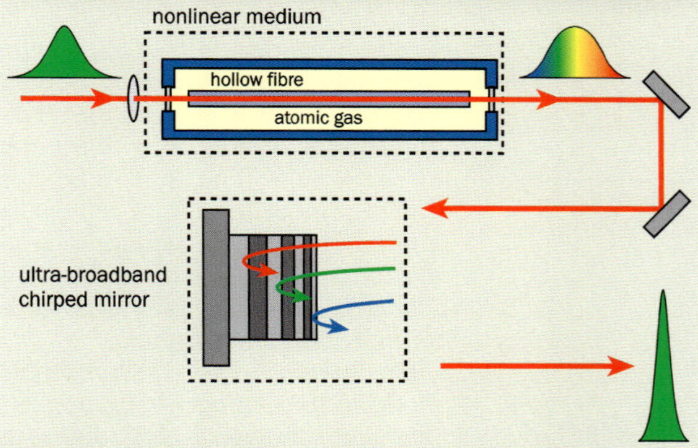

To produce high-energy light pulses comprising less than three wave cycles, 20 fs pulses from a Ti:sapphire laser are first propagated through a hollow-core waveguide filled with neon. The intensity-dependent refractive index of the gas modulates the pulse, resulting in the emergence of new frequency components. The leading edge of the pulse is "red-shifted" to lower frequencies while the trailing edge is "blue-shifted" to higher frequencies. When these pulses impinge on a specially designed "chirped" multilayer mirror, the red-shifted components at the front edge penetrate deeper before being reflected than the blue-shifted components at the trailing edge. This mirror compresses the frequency-broadened pulse in time. The system is capable of delivering 5 fs pulses with a peak power of 0.1 terawatt (S Sartania *et al.* 1997 *Opt. Lett.* **22** 1526–1528).

Raman lines – or the atoms that emit the high harmonics – for such a short time that a single pulse naturally arises from the generation process. This avoids the formidable task of having to select individual pulses from a multi-terahertz train.

This approach requires that pulses from a Ti:sapphire laser must last no longer than 5–10 fs and must have about a millijoule of energy to produce a single sub-femtosecond burst of ultraviolet or X-ray radiation. But the 5–10 fs pulses that emerge directly from a Ti:sapphire laser oscillator have nanojoules of energy, rather than millijoules. Amplifying this energy by a factor of a million comes at the cost of stretching the pulses in time to about 20 fs.

One way around this problem is to take the intense 20 fs pulses and pass them through a hollow waveguide filled with a gas, a device that was invented in 1995 by Orazio Svelto and Sandro De Silvestri of Politecnico di Milano in Italy. Since the refractive index of the gas varies with intensity, the pulse broadens in frequency – the leading edge of the pulse is shifted to a lower frequency while the trailing edge is shifted to a higher frequency. These frequency-broadened pulses are next compressed in time using ultra-broadband "chirped multilayer mirrors" that were developed by my group in Vienna and researchers at Budapest (figure 4). These mirrors are specially designed so that the red-shifted light at the leading edge of the pulse penetrates further before being reflected than the blue-shifted light at the trailing edge. The emerging pulse is less than 10 fs in duration and has a peak intensity in excess of 100 gigawatts. Pulses with these characteristics are opening the way to the generation of single sub-femtosecond pulses.

The first successful steps in this direction were taken recently by exploiting both Raman scattering and high-harmonic generation. Georg Korn and co-workers at the Max Born Institute in Berlin passed 15 fs pulses of violet light through pre-excited sulphur-fluoride molecules with a vibrational frequency of 23 THz. Continuous sidebands were generated, which resulted in the emergence of single pulses that had a wavelength of 400 nm and that last just 4 fs.

In another recent experiment, my group – in collaboration with researchers at the University of Bielefeld in Germany and the National Research Council of Canada in Ottawa – generated single soft X-ray pulses with a wavelength of 14 nm. These isolated X-ray bursts emerged as a collimated laser-like beam from high-harmonic generation in a gas of neon atoms exposed to intense pulses containing a few cycles of the laser field. We measured an upper limit of some 2 fs on the pulse duration by using the X-ray burst as a "pump" pulse to photo-ionize a gas of krypton atoms, and a second time-delayed laser as a "probe" (see Drescher *et al.* in further reading).

The techniques we developed offer the potential for generating and measuring single attosecond X-ray pulses. Theoretical calculations predict that the X-ray pulses emerging from our experiment lasted 600 as, but to verify this value experimentally requires either higher-energy X-ray pulses or more efficient measurement apparatus (figure 5*a*).

X-rays are not dispersed as widely in materials as ultraviolet pulses, which means that X-ray pulses lasting 100 as are quite robust. However, their production efficiency is far inferior to that of ultraviolet pulses. Now that there are suitable techniques for generating and measuring single X-ray pulses lasting 100–200 as, research must focus on enhancing the X-ray yield from high-harmonic generation for reliable attosecond pump–probe spectroscopy.

As a first application, we used a series of our 1.8 fs pulses to capture snapshots of the 7 fs laser that was used to generate the X-rays (figure 5*b*). This allowed us to resolve the intensity profile (i.e. the shape) of a light pulse lasting several femtoseconds for the first time.

Towards attophysics

Physicists are now close to controlling the motion of electrons on a timescale that is substantially shorter than the oscillation period of visible light. Indeed, in our experiment the electric dipoles of the neon atoms may have been set oscillating at the highest frequencies for just a few hundred attoseconds by an intense ultrashort laser. We can also use pulses containing a few cycles of the light field to rip an electron wavepacket from the core of an atom and set it free with similar temporal precision.

The specific time structure of both these processes depends on the evolution of the laser field within the optical cycle. This evolution is, in turn, affected by the relative phase of the carrier wave with respect to the amplitude envelope, the so-called carrier-envelope phase. For example, if the carrier-envelope phase of the laser pulse in figure 5*a* is shifted by $\pi/2$, then the electric field would reach a maximum at the centre of the pulse, rather than zero. In a pulse containing less than three oscillations, this phase shift can significantly modify the evolution of the electric and magnetic fields, and thereby affect the electron trajectories.

The first experimental indications of this phase dependence were observed recently by Gerhard Paulus and Herbert Walther at the Max Planck Institute of Quantum Optics near Munich and collaborators from the Politecnico di Milano. Meanwhile, Stephen Cundiff and co-workers at the JILA laboratory in Boulder, Colorado, and, independently, a collaboration between researchers at Vienna and Munich, have drawn on an ingenious idea of Theodor Hänsch at

MPI Munich to stabilize the carrier-envelope phase of ultrashort low-energy pulses.

Once amplified high-intensity pulses with a stabilized phase become available, strong-field processes – such as optical-field ionization or high-order harmonic generation – will allow us to measure the carrier-envelope phase directly. Access to this parameter – together with a measurement of the variation in amplitude and the frequency of the laser field using standard techniques – will allow us to determine and control the evolution of the electromagnetic field in a light wave, and even to synthesize arbitrary waveforms of light containing a few cycles.

Such waveforms might allow us to control the motion of an electron after it has been detached from an atom or a molecule. Direct consequences include the reliable generation and control of attosecond pulses at extreme ultraviolet and soft X-ray wavelengths via high-order harmonic generation. And at much higher intensities it may be possible to accelerate attosecond electron pulses up to several mega-electron-volts in energy.

Since these X-ray and electron pulses are synchronized to the laser field with attosecond precision, we will be able to take snapshots of different stages of atomic, molecular and plasma dynamics. These processes will be triggered and steered by the intense field-controlled waveforms of light and probed by X-ray pulses, enabling us to both control and reconstruct the motion of electrons at atomic length scales and with attosecond resolution. In this way, for example, the trajectory of an electron wavepacket might be measured from the instant it is ejected from the atom and could be tailored for different applications. These applications might include the development of more efficient and shorter-wavelength X-ray lasers, or sources of ultrafast, monoenergetic relativistic electron pulses for the next-generation of particle accelerators.

5 Attosecond pulses in action

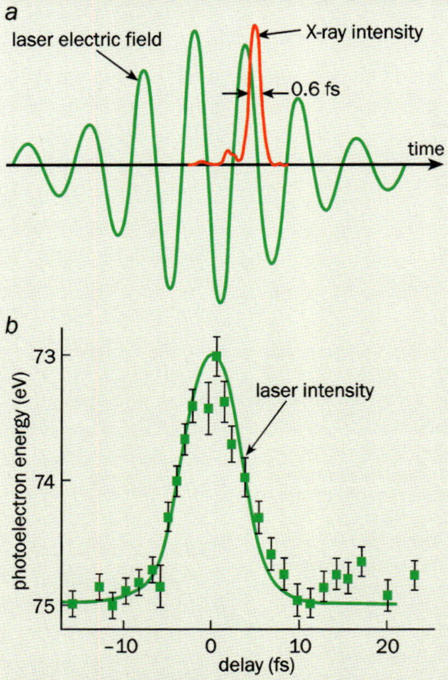

(*a*) The predicted intensity envelope of the X-ray pulse (red) generated in a high-order-harmonic experiment with neon gas and near-infrared laser pulses lasting 7 fs (green). The isolated X-ray pulse is calculated to last 600 as and have a wavelength of 14 nm. Measuring the duration of such short pulses is challenging. The first measurement reported by a collaboration of Austrian, German and Canadian physicists set an upper limit of approximately 2 fs (see Drescher *et al.* in further reading). It should be possible to further improve the measurements. (*b*) The intensity versus time profile of a 7 fs laser pulse was recorded using 1.8 fs X-ray pulses to knock an electron out of an atom. The energy shift of the electron is proportional to the intensity of the longer laser pulse at the instant of X-ray absorption. By measuring this energy shift as a function of delay between the laser and X-ray pulse (points), the intensity (line) was reconstructed with a resolution dictated by the X-ray pulse duration. This is the first "time-microscope" measurement with a resolution approaching 1 fs.

Looking into atoms

Attosecond X-ray pulses will also pave the way towards watching the motion of electrons *inside* atoms. For example, we expect to be able to track the motion of an electron within a femtosecond of it being set free from a bound state by an intense laser field. We could also follow the relaxation of the remaining bound electrons to their new equilibrium states with attosecond resolution.

Once we can generate attosecond pulses with much higher energies, we will be able to use them as both pumps and probes in pump–probe spectroscopy. We could then selectively remove electrons from the inner shells with a strong attosecond X-ray pulse and probe the filling of the inner-shell vacancy to an accuracy better than 1 fs using a weaker time-delayed replica of the same pulse. Is there any need for this? Why not simply depend on the relaxation times derived from linewidth measurements? The simple answer is that the influence of an ultrashort intense field on the bound-electron dynamics – important for applications such as X-ray lasers – cannot be studied in the frequency domain.

Finally, back to basics. Bohr's simple model of the hydrogen atom predicts that the electron takes about 150 as to orbit around the proton. According to the laws of quantum mechanics, no observable dynamics takes place as long as the electron resides in its ground state (1s). However, this situation changes if a fraction of the electron wavefunction is excited into the first excited state (2s). The time-dependent Schrödinger equation predicts that the two components ($|1s\rangle$ and $|2s\rangle$) of the new composite wavefunction will beat at a frequency equal to the transition frequency between the 1s and 2s energy levels (figure 6).

This beating gives rise to a cyclic change in the radial probability distribution of the electron with a period of about 400 as. This "breathing" of the hydrogen atom may be regarded as one of the most fundamental quantum dynamic phenomena in nature. But can it be observed? According to Armin Scrinzi and Thomas Brabec from the Vienna University of Technology, it might be detectable by exploiting the fact that the electron can be detached more easily when it is closer to the atomic core. They predict that measuring the photoionization yield with X-ray pulses shorter than 200 as in a pump–probe experiment might provide direct insight into this quantum dynamics. Although it is only of academic interest at the moment, inner-shell quantum beats may eventually be exploited, for example as clocks for future ultrahigh-speed molecular electronics operating beyond 1 THz.

Outlook

Ultrashort flashes of light allow us to take snapshots of microscopic particles and let us reconstruct their motion. State-of-the-art femtosecond lasers emit pulses that are short enough to resolve the motion of atoms in molecules and solids, as well as charge-carrier dynamics in semiconductors. Meanwhile, specially tailored femtosecond pulses also allow these processes to be controlled and will lead to numerous applications in chemistry, biochemistry and electronics.

The frontiers of both time-resolved spectroscopy and the control of microscopic dynamics are about to be radically

EXTREME OPTICS

6 Attosecond breathing of the hydrogen atom

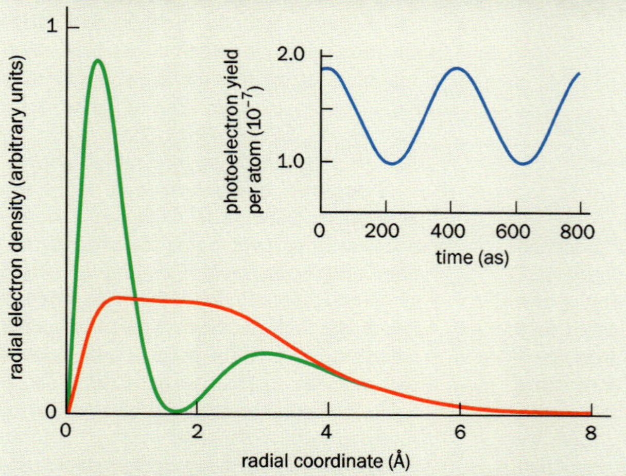

The radial electron density of a 1s–2s superposition state of hydrogen. The time-dependent Schrödinger equation predicts that the wavefunction oscillates between states of maximal (green) and minimal (red) electron density with a period of about 400 as near the nucleus. According to theoretical calculations, this fundamental quantum phenomenon may be observable using attosecond X-ray pulses that ionize the hydrogen atom. Theory predicts that the ionization yield will be modulated at the "breathing" period as a function of the time delay between the attosecond X-ray pulse and the excitation pulse (see insert).

extended due to the emerging technical capability that will allow us to synthesize intense pulses containing a few cycles of the laser field that evolve in a precisely determined way. With this technique we will soon be able to control the motion of electron wavepackets on attosecond timescales, just as we can currently control the motion of nuclear wavepackets within a few femtoseconds. The single attosecond X-ray pulses and attosecond electron bunches that will arise from our ability to control electron wavepackets will enable us to probe and steer electron motion with unprecendented precision.

The availability of these tools will herald a new age of experimental physics that will be so radical and so widespread that it will demand a new name. Get ready to welcome the era of attophysics.

Further reading

T Brabec and F Krausz 2000 Intense few-cycle laser fields: frontiers of nonlinear optics *Rev. Mod. Phys.* **72** 545–592

P Corkum 1995 Breaking the attosecond barrier *Optics and Photonics News* **6** March pp18–22

M Drescher *et al.* 2001 X-ray pulses approaching the attosecond frontier *Science* **291** 1923–1927

H Kapteyn and M Murnane 2000 Ultrashort light pulses: life in the fast lane *Physics World* January pp31–35

G Steinmeyer *et al.* 1999 Frontiers in ultrashort pulse generation *Science* **286** 1507–1512

R Szipöcs *et al.* 1995 Pushing the limits of femtosecond technology *Optics and Photonics News* **6** June pp17–20

A Zewail 2000 Femtochemistry: atomic-scale dynamics of the chemical bond (adapted from the Nobel Lecture) *J. Phys. Chem.* **A 104** 5660–5694

www.femtolasers.com

Ferenc Krausz is the director of the Centre for Advanced Light Sources at the Photonics Institute, Vienna University of Technology, Gusshausstrasse 27, A-1040 Vienna, Austria, e-mail krausz@tuwien.ac.at

RESEARCH ARTICLES

X-ray Pulses Approaching the Attosecond Frontier

Markus Drescher,[1,2] Michael Hentschel,[1] Reinhard Kienberger,[1] Gabriel Tempea,[1] Christian Spielmann,[1] Georg A. Reider,[1] Paul B. Corkum,[3] Ferenc Krausz[1]*

Single soft-x-ray pulses of ~90–electron volt (eV) photon energy are produced by high-order harmonic generation with 7-femtosecond (fs), 770-nanometer (1.6 eV) laser pulses and are characterized by photoionizing krypton in the presence of the driver laser pulse. By detecting photoelectrons ejected perpendicularly to the laser polarization, broadening of the photoelectron spectrum due to absorption and emission of laser photons is suppressed, permitting the observation of a laser-induced downshift of the energy spectrum with sub-laser-cycle resolution in a cross correlation measurement. We measure isolated x-ray pulses of 1.8 (+0.7/−1.2) fs in duration, which are shorter than the oscillation cycle of the driving laser light (2.6 fs). Our techniques for generation and measurement offer sub-femtosecond resolution over a wide range of x-ray wavelengths, paving the way to experimental attosecond science. Tracing atomic processes evolving faster than the exciting light field is within reach.

Ultrashort-pulse lasers constitute the fastest probes available for tracing transitions between different states of matter. They allow measuring time intervals on a femtosecond (1 fs = 10^{-15} s) time scale and provide access to fundamental physical, chemical, and biological processes on a microscopic scale in the time domain (*1*). A delayed probe pulse permits taking snapshots of atomic, molecular, or condensed-matter dynamics after excitation by a pump pulse (*2*). Any variation of the state of the object under study occurring during excitation by the pump or illumination by the probe remains unresolved; hence, the fastest speed at which events can be followed is limited by the laser pulse duration.

More than 30 years of evolution of ultrashort-pulse laser technology recently culminated in the generation of laser pulses shorter than 10 fs in the visible and near-infrared spectral ranges (*3–9*). These pulses comprise only a few oscillation cycles of the electromagnetic field and approach the ultimate limit set by the laser oscillation period $T_0 = \lambda_L/c$, where λ_L is the laser (carrier) wavelength and c is the speed of light in vacuum. Several fundamental atomic processes such as inner-shell electronic relaxation or ionization by optical tunneling, however, take place within a fraction of the oscillation period of visible or near-infrared radiation and require very short probes for being investigated. Although trains of attosecond (1 as = 10^{-18} s) pulses at ultrahigh (T_0^{-1} or $2T_0^{-1}$) repetition rates can result from the superposition of frequency-converted coherent radiation (*10–12*) (most simply by superimposing on the driving field its second and third harmonic), they are of limited use for this purpose because of the very short (~1 to 2 fs) pulse intervals within the train. Isolated bursts of radiation on the order of 1 fs or shorter are required to reliably trace these dynamics never before accessed in the time domain. These single bursts need to be carried at substantially shorter wavelengths, preferably in the extreme ultraviolet (XUV) or x-ray regime, if bound atomic electrons are to be accessed in ultrafast spectroscopy.

High-harmonic generation (*13–15*) using few-cycle pulses (*16–18*) is ideal for this purpose. High harmonics are generated by electrons removed from the atom by the high laser field and then driven back to their ion of origin. Because their recollision energy is greatest when the laser field is largest, the most energetic "harmonic" radiation is produced at the peak of the laser pulse. We select radiation produced only within a fraction of the laser cycle near the peak of the pulse by passing the harmonics through a filter that transmits only the highest frequencies. This method has been predicted to be capable of generating single XUV/x-ray pulses of ~100-as duration emitted in a collimated, laser-like beam (*19–21*).

Whereas the key techniques for generation have been available for several years (*16*, *22*), methods for temporal characterization of XUV and x-ray harmonics have been restricted to bandwidths that are small compared with the laser photon energy $\hbar\omega_L$. The key to the progress reported here is a newly developed measurement technique. It allows measuring the duration of broadband ($>\hbar\omega_L$) x-ray pulses with sub-laser-cycle resolution. The method is intuitive, can be generalized to a wide range of x-ray wavelengths, and offers sub-femtosecond resolution. Using this technique, we set an upper limit of 2.5 fs on the duration of isolated 90-eV x-ray pulses by our 7-fs laser-driven high-harmonic source.

Temporal characterization of high-order laser harmonics. There are currently two general approaches to pulse characterization being investigated. One aims to extend autocorrelation, developed for measuring visible pulses, into the x-ray region. Autocorrelation requires that the pulse is split into two replica pulses and then recombined in a nonlinear medium where the nonlinear process is sensitive to their temporal overlap. Unfortunately, except at the longest wavelengths (*23*), the XUV/x-ray pulses generated by high harmonics are currently too weak to induce measurable nonlinearities of atomic media. Thus, although autocorrelation is conceptually simple and well understood, so far it has been impossible to implement in the x-ray regime. Recently, Papadogiannis and co-workers (*24*) proposed an interesting variation of this concept by splitting the driving pulse rather than the harmonic pulse. The idea is that the ionizing gas, which produces the harmonics, can also be used as the nonlinear medium to measure their duration (*25*). They observed attosecond sub-structures on a much longer (>10 fs) autocorrelation feature. However, although this approach is experimentally simple, it is conceptually complex because production and measurement are intertwined (*26*).

Our approach falls within a second class of experiments that began with the work of Schins and co-workers (*27–31*). Here the x-ray pulse is "cross correlated" against a short visible pulse. An x-ray photon excites a bound atomic electron into a positive-energy state in the presence of a laser field. Laser-induced transitions from this state or a laser-induced shift in its energy (Stark shift) provide the nonlinearity linking the x-ray to the laser pulse, enabling us to compare the x-ray pulse duration with that of the visible pulse. In the experiments reported so far (*27–31*), a number of laser photons increasing rapidly with the laser intensity are absorbed and emitted by the freed electron (forced to oscillate at ω_L). To identify the Stark shift amidst the resulting sidebands in the photoelectron energy spectrum or to measure the amplitude of the sidebands, the x-ray spectral width must be kept well below the laser photon energy $\hbar\omega_L \approx 1.6$ eV to avoid overlap of the sidebands with the main peak.

Without modification, Schins's approach cannot be extended to sub-laser-cycle x-ray pulses for two reasons. (i) If we are to measure x-ray pulses with durations of $\tau_x < T_0$,

[1]Institut für Photonik, Technische Universität Wien, Gusshausstr. 27, A-1040 Wien, Austria. [2]Fakultät für Physik, Universität Bielefeld, D-33615 Bielefeld, Germany. [3]National Research Council of Canada, Ottawa, Ontario, Canada K1 A 0R6.

*To whom correspondence should be addressed. E-mail: ferenc.krausz@tuwien.ac.at

the photoelectron energy distribution must become broader than the laser photon energy so that it is no longer possible to identify individual sidebands and measure their amplitudes or determine the Stark shift. (ii) If the field is increased to increase the Stark shift so it becomes measurable, absorption and emission of laser photons tend to smear the photoelectron energy distribution beyond redemption.

In what follows, we show that choosing the geometry of observation carefully and confining the laser field to a few oscillation cycles opens the door to broadband ($>\hbar\omega_L$) x-ray pulse measurement with sub-femtosecond resolution.

Laser-assisted lateral x-ray photoionization. To gain insight into the role of the observation geometry, consider the laser-assisted photoelectric effect in a semi-classical analysis (*32–35*). The electron is set free by an x-ray photon of energy $\hbar\omega_x$ at a delay time t_d with respect to the peak of the visible laser pulse having an initial kinetic energy $W_0 = mv_i^2/2 = \hbar\omega_x - W_b$, where W_b is the ionization potential of the atom, in an arbitrary direction. For $\hbar\omega_x \gg W_b$, the strong, linearly polarized "probe" laser field $E_L(t) = E_a(t)\cos(\omega_L t + \varphi)$ is predominant in governing electron motion after ionization. In the adiabatic limit $dE_a/dt \ll E_a\omega_L$ (which is a remarkably good approximation even for our 7-fs pulses) the equation of motion can then be readily integrated, yielding the connection

$$v_{\|,f} = \sqrt{\frac{4U_p(t_d)}{m}}\sin(\omega_L t_d + \varphi) + v_{\|,i} \quad (1)$$

between the initial ($v_{\|,i}$) and final value ($v_{\|,f}$) of the velocity component parallel to the laser polarization. Here $U_p(t) = e^2E_a^2(t)/4m\omega_L^2$, which, in the adiabatic limit, represents the cycle-averaged quiver energy of the electron (mass m and charge $-e$) and is commonly referred to as the ponderomotive energy (*36*). For the transverse component, $v_{\perp,f} = v_{\perp,i}$. Using these relations along with the above constraint on the initial kinetic energy, the final kinetic energy (after the laser pulse) can, in the limit of $\omega_x \gg \omega_L$, be expressed as

$$\begin{aligned} W_f \approx W_0 &- U_p(t_d) + U_p(t_d)\cos 2\omega_L t_d \\ &+ 4U_p(t_d)\cos^2\theta\sin^2\omega_L t_d \\ &+ \sqrt{8W_0U_p(t_d)}\cos\theta\sin\omega_L t_d \end{aligned} \quad (2)$$

where θ is the angle of the final electron momentum $\mathbf{p}_f$ with respect to the laser polarization (Fig. 1). Here, for the sake of brevity we assumed $\varphi = 0$. To account for x-ray excitation of nonzero temporal and spectral width, we expanded our simple analysis by a numerical simulation. It allows us to calculate the electron energy spectrum observed within a given range of observation angles θ. In spite of its deficiencies, however, Eq. 2 provides insight into the basic physics and leads us to the optimum design of the experiment for x-ray pulse measurement.

The second term on the right-hand side of Eq. 2 lowers the final kinetic energy of the electron by its quiver energy at the instant of its "birth." This energy is transferred to the laser field by blue-shifting its carrier frequency for fs pulse durations (*37, 38*). The third to fifth terms correspond to the absorption and emission of laser photons in a quantum description. For a brief analysis, consider two limiting cases represented by an x-ray bandwidth much smaller or much greater than the laser photon energy, respectively. The former condition implies $\tau_x \gg T_0$, resulting in the appearance of discrete sidebands spaced by $\hbar\omega_L$ around the main peak in the photoelectron energy spectrum. This is the regime of previous studies (*27–31*). For $W_0 \gg U_p$, the overall width of the sideband spectrum can be evaluated from Eq. 2 as $2\sqrt{8W_0U_p(t_d)}\cos\theta$, and its center of gravity is upshifted by $2U_p(t_d)\cos^2\theta$ with respect to $W_0 - U_p(t_d)$ (the position of the main peak) due to a higher amplitude of Stokes (photon absorption) as compared with anti-Stokes sidebands (photon emission). In this limit, both the ponderomotive shift $U_p(t_d)$ and the amplitude of sidebands can be accurately measured.

In the opposite limit, for an x-ray spectral width larger than $\hbar\omega_L$, the sidebands completely coalesce (due to a distribution of W_0 over a range broader than $\pm\hbar\omega_L$), forming a single broad spectral feature with a width and center-of-gravity position as given above. Consequently, neither the individual sideband amplitudes nor the ponderomotive shift can be reliably acquired from the photoelectron spectrum any longer. The ponderomotive shift tends to be fully obscured by an upshift in the center-of-gravity of the overall spectral feature (fourth term in Eq. 2). This upshift means a net absorption of laser photons and is most pronounced for $\theta \approx 0$. Moreover, $\hbar\omega_x \gg W_b > U_p$ implies $\sqrt{8W_0U_p} \gg U_p$, i.e., the photoelectron energy spectrum is "smeared out" over an energy range much broader than U_p, impeding a reliable measurement of its shift of the order of $U_p(t_d)$. Electrons ejected at $\theta \approx 90°$ are special in that their energy spectrum is hardly distorted by the absorption and emission of laser photons, but is just shifted by the oscillating laser field. This central finding from our simple analysis is the basis for our broadband x-ray pulse measurement.

Choosing the observation angle as $\theta \approx 90°$ (orthogonal to the laser polarization) eliminates the last two terms in Eq. 2 and thereby makes the ponderomotive shift $-U_p(t_d)$ clearly observable. Because $U_p(t) \propto E_a^2(t)$, measurement of the ponderomotive shift as a function of the delay t_d yields a cross correlation between the temporal intensity profiles of the x-ray and laser pulses. The predicted oscillations in the electron energy shift (third term in Eq. 2) offer a sub-$T_0/4$ probe for measuring x-ray emission characteristics on an attosecond time scale, reminiscent of the concept of an optically driven streak camera (*25, 39*). These oscillations average out for $\tau_x \geq T_0/2$.

Intuitively, the simplified dynamics (oscillating downshift) can be understood as an impact of the laser field on the electron momentum distribution (Fig. 1). The final x-ray photoelectron momentum distribution $\mathbf{p}_f(\theta)$ gets shifted along the laser polarization, owing to a momentum transfer that reverses its sign twice within a laser oscillation period. This momentum transfer translates into a downshift in W_f for detection angles near $\theta = 90°$, oscillating at $2\omega_L$ with its envelope following $U_p(t_d)$, i.e., the temporal intensity profile of the laser pulse.

In this observation geometry (Fig. 1), we confine detection to a small solid angle around $\theta = 90°$ instead of collecting all electrons ejected in 2π (or 4π), as has been

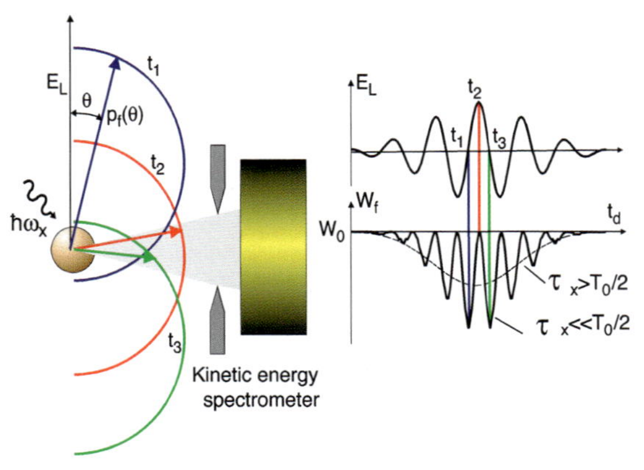

Fig. 1. Semiclassical modeling of laser-assisted x-ray photoionization. The electrons set free at different instants (t_1, t_2, t_3) by an x-ray photon are ejected with different angle distributions of their final momenta $\mathbf{p}_f(\theta)$ in the presence of a laser field. The momentum transferred along the laser polarization translates into a downshift of the kinetic energy W_f of the electrons observed (within a small solid angle) orthogonally to the laser polarization. The variation of W_f with the time delay between the instant of x-ray absorption and the peak of the laser pulse (assuming an x-ray pulse duration $\tau_x \ll T_0/2$ and observation near $\theta = 90°$) exhibits a modulation at twice the laser frequency with an envelope following the temporal evolution of the laser intensity. For $\tau_x > T_0/2$, the oscillations average out (dashed line).

RESEARCH ARTICLES

the case in all relevant experiments so far (*27–31*). In addition to the benefits discussed above, this geometry also helps suppress undesirable background resulting from electrons produced by multiphoton above-threshold ionization (ATI) by the laser field alone. Other than the particular observation geometry, confinement of the interaction to a few-laser oscillation cycles is important (i) for enhancing the ponderomotive downshift at low levels of ATI background and (ii) for a steep temporal intensity gradient, yielding a high "streaking" speed for measuring x-ray transients down to the 1-fs range, where the sub-$T_0/4$ probe (see third term in Eq. 2) comes into play.

Apparatus for laser harmonic generation and measurement. To exploit the potential of the above described technique for measuring x-ray bursts with high temporal resolution, jitter between the x-ray pulse and the laser pulse probing the x-rays needs to be minimized. This requirement can be most efficiently fulfilled by using the same laser pulse for generation and measurement, which becomes feasible because of the low levels of ionization at which few-cycle drivers are able to efficiently generate harmonics up to the 100-eV region (*40*). As a result, an enhanced conversion efficiency is combined with a nearly undistorted propagation of the laser pulse through the weakly ionizing harmonic source. This enables us to record high-resolution cross-correlations with the co-propagating, perfectly synchronized harmonic pulse by using a compact, highly stable delay stage.

We focus laser pulses carried at $\lambda_L \approx 770$ nm ($\hbar\omega_L \approx 1.6$ eV) with a duration of 7 fs and pulse energy of 0.5 mJ delivered at a repetition rate of 1 kHz (*22*) into a quasi-static neon gas cell (Fig. 2). Harmonics in the same geometry and under similar experimental conditions as described in (*40*) are produced up to photon energies of ~120 eV with a divergence angle of ~0.7 mrad in the near-cutoff region. The co-propagating x-ray and laser pulses are passed through a pellicle with a central circular zone formed by a zirconium foil that is transparent for the x-rays but blocks laser radiation. The resulting annular laser beam and the concentric harmonic beam hit a Mo/Si spherical multilayer mirror ($R_{peak} \approx 60\%$ at 90 eV, bandwidth ≈ 5 eV) consisting of two piezo-controlled concentric parts. The two-component mirror forms a compact, virtually jitter-free delay stage for the laser and the harmonic pulses with a maximum scan range of ~50 fs and a resolution and reproducibility of better than 0.1 fs. The folding angle is kept below 5° to minimize astigmatism. Spherical aberration gives rise to a time smearing of <0.05 fs for the harmonic and <1 fs for the laser beam. The harmonics are focused to a diameter of ~2 μm (*41*), which is small compared with the ~20-μm diameter of the focused laser beam, ensuring a nearly constant laser intensity in the interaction volume. The target atoms effuse from a 100-μm-long, 20-μm-wide slit aligned parallel to the propagation direction of the beams. The entrance of a 0.4-m time-of-flight (TOF) photoelectron spectrometer (resolution ≈ 0.5 eV at 75 eV) with an acceptance angle of 0.3 srad is brought in close proximity (5 mm) to the interaction volume. The spectrometer axis is aligned perpendicularly to the laser beam direction and to the laser polarization (θ = 90°).

X-ray/laser-pulse cross correlation in the few-cycle regime. The lowest plot (full line) in Fig. 3 depicts the 4p-photoelectron energy spectrum of krypton obtained in the absence of the laser field. Krypton has been chosen as a target because of the nearly isotropic momentum distribution of 4p electrons at $\hbar\omega_x \approx 90$ eV photon energy (*42*, *43*). The shape of the 4p spectrum is determined by the high-harmonic spectrum after reflection off the Mo/Si multilayer (the dashed line showing the spectral reflectance shifted by the ionization potential W_b = 14.0 eV of Kr). The mirror reflects harmonics H55 and H57, which can be barely discriminated due to their coalescence. Assuming no spectral phase modulation, Fourier transformation of the spectrum yields a pulse duration of 0.4 fs (full width at intensity half maximum).

To measure the x-ray pulse duration, we performed x-ray photoionization experiments in the presence of the few-cycle laser field. The laser intensity was adjusted to result in a maximum ponderomotive energy of $U_p \approx 2$ eV, ensuring negligible ATI for our pulse duration. Figure 3 depicts x-ray

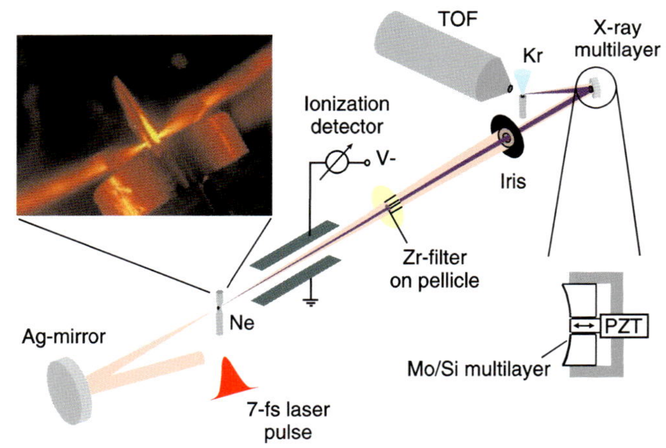

Fig. 2. Schematic diagram of the experimental setup. The focused 7-fs laser beam fights its way through the walls of a metallic tube (diameter ≈ 3 mm) by ablation and interacts with neon atoms inside the tube to produce high-harmonic radiation. The ionizing neon gas streaming out of the tube can be traced because of its fluorescence emission (see photograph). The laser and the highly collimated x-ray beam emerging from the interaction region within the tube co-propagate collinearly (along the axis of the emission cone) through a 2-m beamline toward the measurement. Differential pumping stages reduce the pressure from ~4 × 10^{-2} mbar in the source chamber to the range of ~10^{-5} mbar in the experimental chamber. An ionization detector in the higher-pressure part of the beamline serves for monitoring the spectrally integrated high-harmonic flux. In the beamline, the laser and harmonic beams pass through a 200-nm-thick, 3-mm-diameter zirconium foil placed on a 5-μm-thick nitrocellulose pellicle to cover a hole of 2-mm diameter. The energy transported by the resulting annular laser beam can be adjusted with a motorized iris between a fraction of a microjoule and a few tens of microjoules. The Mo/Si multilayer consists of an annular part that has an outer diameter of 10 mm with a concentric hole of 3-mm diameter hosting a miniature mirror of slightly smaller diameter. Both parts originate from the same substrate, ensuring identical radii of curvature (R = 70 mm). The miniature central mirror is mounted on a quadrant piezo stage (PZT), allowing alignment and translation with respect to the external part.

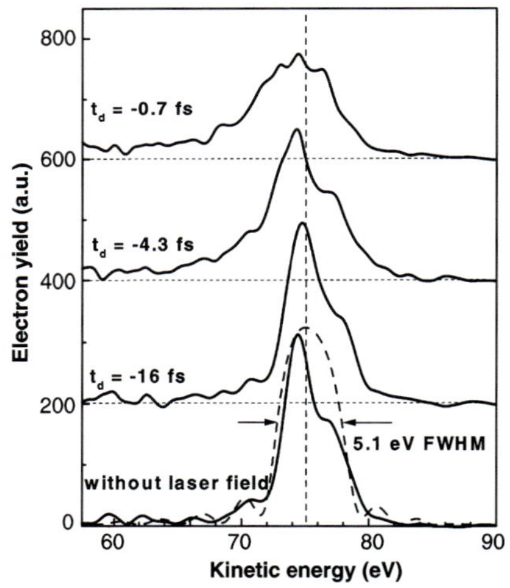

Fig. 3. X-ray photoelectron spectra produced by 90-eV harmonics of 7-fs laser pulses in Kr in the presence of the laser field for different delays t_d of the x-ray pulse with respect to the laser pulse. The laser peak intensity at target is calculated from the maximum ponderomotive shift as ~4 × 10^{13} W/cm^2. The spectra were recorded over an acquisition period of 200 s by collecting ~1500 signal counts with a minor background from ATI electrons, which were subtracted. The dashed line depicts the reflectivity (peak ≈ 60%, bandwidth ≈ 5.1 eV) of the Mo/Si multilayer shifted by the ionization potential.

RESEARCH ARTICLES

photoelectron spectra recorded at various delays t_d of the x-ray pulse with respect to the laser pulse. Comparison of the spectra recorded in the absence of the laser field and for $t_d = -16$ fs (no temporal overlap) shows little influence of space-charge effects by ATI electrons.

A contour plot (Fig. 4) illustrates the evolution of the 4p spectrum of Kr for increasing delay of the x-ray pulse (recorded from $t_d = -16$ fs to $t_d = 25.3$ fs in 27 steps). In spite of the high laser peak intensity, the x-ray photoelectron spectrum remains "intact" throughout the region of temporal overlap between the laser pulse and the x-ray pulse. Eq. 2 reveals that this is a direct consequence of our specific observation geometry. Under our experimental conditions, the photoelectron spectrum would be broadened to cover a range of $2\sqrt{8W_0U_p(t_d = 0)} \approx 60$ eV in a 2π (or 4π) detection geometry (*27–31*). By contrast, in our lateral ($\theta \approx 90°$) observation geometry the effective broadening amounts to merely a few eV (caused by the finite detection solid angle). As a result, a downshift on the order of $U_p \approx 2$ eV is clearly discernible. A notable shift of the center of gravity is confined to a delay range of less than 10 fs and accompanied by some (asymmetric) broadening. This becomes evident from the white iso-spectral-density lines or from a comparison of the broadened spectra recorded at the slope ($t_d = -4.3$ fs) and at the peak ($t_d = -0.7$ fs), respectively, with a spectrum well before the arrival of the laser pulse ($t_d = -16$ fs) in Fig. 3. These observations are in excellent agreement with simulations based on our semiclassical model and performed for finite x-ray pulse duration and bandwidth.

X-ray pulses with sub-laser-cycle duration. For a quantitative analysis, we evaluated the first-order moments (center of gravity) of the measured photoelectron spectra as a function of t_d (Fig. 5). According to Eq. 2, the center-of-gravity shift with varying delay t_d displays directly the evolution of the ponderomotive energy U_p during the laser pulse. Assuming the laser pulse as bandwidth-limited at the entrance of the high-harmonic target, we simulated the propagation of the 7-fs pulses with an initial peak intensity of 7×10^{14} W/cm^2 through the neon target (*44*). The initial pulse shape is obtained by inverse Fourier-transformation from the measured pulse spectrum. The ionizing target becomes somewhat dispersive for the peak and trailing edge of the pulse, resulting in a pulse-broadening to ~7.5 fs. However, because ionization is virtually confined to a single cycle near the pulse peak, the leading edge is hardly affected by dispersion in the harmonic source, constituting a steep probe (rise time $\tau_r \approx 5$ fs) for the x-ray temporal intensity profile.

The laser "probe" obtained from this numerical analysis is idealized in that it has a rise time τ_r equal to its theoretical limit. To obtain a conservative upper limit on the x-ray pulse duration, we first convolve this idealized probe with a hypothetical x-ray pulse obtained by inverse Fourier transformation of the lowest spectrum shown in Fig. 3, allowing for an adjustable quadratic spectral phase (as a fit parameter) that determines the x-ray pulse duration τ_x. The solid line in Fig. 5 represents the best fit to the measured data, obtained from a convolution of a 2.4-fs x-ray pulse with the idealized probe. For a realistic rise time of the laser pulse front increased by $\delta\tau_r$, this value is to be corrected by $\delta\tau_x \approx -(\tau_r/\tau_x)\delta\tau_r$. With a conservative estimate of $\delta\tau_r \geq 0.3$ fs from inspection of the most prominent pulse-broadening effects (spherical aberration and nonlinear spectral phase) and an F-stat goodness-of-fit analysis of the cross-correlation (Fig. 5), we obtain $\tau_x = 1.8$ (+0.7/−1.2) fs full width at intensity half maximum (FWHM) (*45*).

This result has been corroborated by tracing the evolution of the low-energy electron yield as a function of the delay t_d between the x-ray pulse and the laser pulse. Numerical simulations based on our semi-classical analysis of laser-assisted x-ray photoionization predict this signal to be "switched on" somewhat faster than the rise time of the front edge of the laser pulse because of a finite threshold intensity needed for the production of these electrons. The shorter rise time offers a probe with enhanced resolution for x-ray pulse measurements. The evaluation of the electron yield (Fig. 6) with $W_f \leq 71$ eV as a function of t_d by fitting an error function to the data results in rise times of 3.1 ± 0.6 fs and 2.9 ± 0.5 fs, respectively. The error-function fit is consistent with a Gaussian-like x-ray pulse profile and a steep probe function with a rise time comparable to or shorter than the pulse duration. Deconvolution yields an upper limit of ~2 fs on the x-ray pulse duration, in fair agreement with the above result.

Current limitations and future challenges. There are several predictions suggesting that isolated XUV/x-ray pulses in the range of 100 as can be generated (*18–21, 46*). However, before these pulses become available for experiments, a few problems need solving.

First, we need a reliable and simple method of measuring the temporal structure of broadband x-ray emission with a resolution of approximately 0.1 fs. We believe that the technique introduced in this paper is capable of solving this problem with the sub-$T_0/4$ probes that originate from the predicted oscillations in the electron energy shift (Fig. 1, Eq. 2). However, both the measurement (*47*)

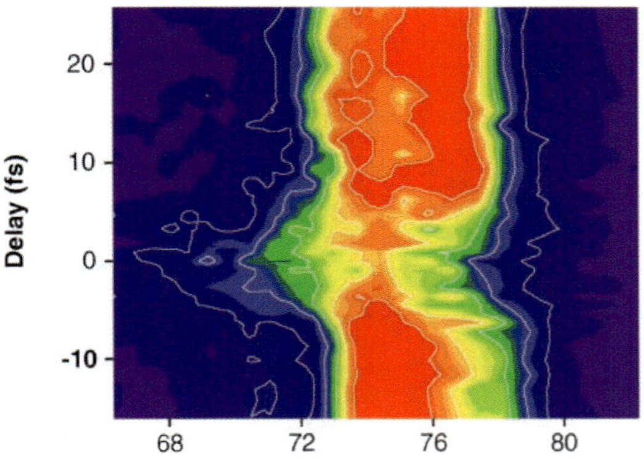

Fig. 4. Contour plot of the evolution of the 4p spectrum of Kr comprising 28 spectra recorded within the depicted range of delay time between the x-ray and laser pulse. Blue, low-spectral electron density; red, high-spectral electron density; white, iso-spectral-density lines. Typically, 1000 to 1500 counts contributed to the spectra, which were renormalized to a constant area for the contour diagram. It took approximately 2 hours to acquire the data used for this plot. For delays $t_d > 5$ fs, a gradual spectral redistribution of x-ray energy in favor of H57 occurs due to laser energy drift-induced minor shift in the harmonic peak positions with respect to the (steep) edges of the Mo/Si mirror reflectance curve (Fig. 3). This is caused by a drift of the laser pulse energy as small as a few percent, which can not be avoided on a time scale of hours.

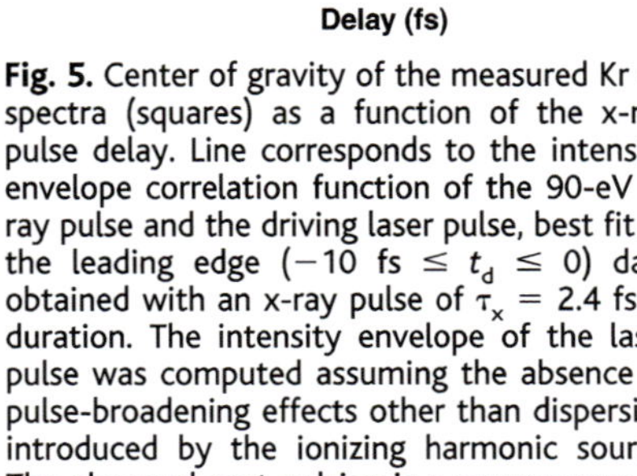

Fig. 5. Center of gravity of the measured Kr 4p spectra (squares) as a function of the x-ray pulse delay. Line corresponds to the intensity envelope correlation function of the 90-eV x-ray pulse and the driving laser pulse, best fit to the leading edge (-10 fs $\leq t_d \leq 0$) data obtained with an x-ray pulse of $\tau_x = 2.4$ fs in duration. The intensity envelope of the laser pulse was computed assuming the absence of pulse-broadening effects other than dispersion introduced by the ionizing harmonic source. The observed post-pulsing is a consequence of a small residual spectral phase carried by the intial laser pulse (*54*), which is neglected in our modeling.

and the generation (*48*) technique demonstrated here need to be improved to permit reliable attosecond metrology. Lastly, improved characterization and control of strong laser pulses is required for achieving this objective. Because attosecond pulses are generated by sub-period dynamics from a visible pulse, the optical pulse field must be controlled on a sub-period scale. That is, we require the optical oscillations to fit reproducibly (with attosecond precision) within the pulse envelope for the generation of single attosecond pulses with reproducible amplitude and timing. Technically, the carrier-envelope phase [denoted by φ in the expression of $E_L(t)$] must be controlled, or at least known. Carrier-envelope phase control (*49-51*) with sub-cycle precision (*52*) has been demonstrated for low-energy pulse trains. Amplification of these pulses with low jitter of φ and subsequent accurate single-shot measurement of φ for the intense laser pulses (*53*) is another prerequisite for the development of attosecond metrology.

In conclusion, we believe that the experimental techniques are now available for measuring sub-light-cycle dynamics. The single attosecond x-ray pulses that will be generated will open a new chapter in ultrafast spectroscopy. Just as femtosecond pulses allow the observation of the motion of atoms in molecules, attosecond pulses will allow us to trace electronic motion and electronic relaxation (such as inner-shell dynamics) in atoms and molecules. In addition, attosecond science will have strong, phased laser fields perfectly synchronized with the attosecond pulse. The laser fields will be strong enough and well-enough controlled to govern electron dynamics (in fact, this is how high harmonics are produced), and well-enough synchronized to attosecond pulses to allow sub-laser-cycle measurements. Studying strong-field electron dynamics (such as optical-field ionization) and controlling these processes will be a forefront of attosecond science.

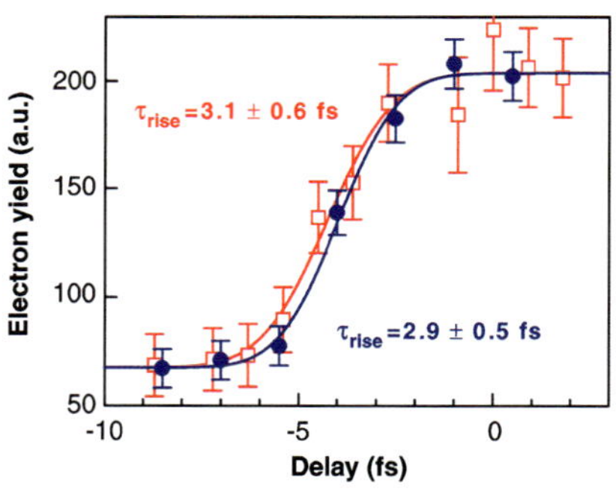

Fig. 6. Electron yield at the low-energy tail of the 4p spectrum obtained on different days and under slightly different conditions within the energy interval of 68 to 71 eV (open squares) and of 64 to 71 eV (solid circles) versus delay time t_d between x-ray and laser pulse. The corresponding lines represent error-function fits to the measured data, yielding a rise time of 3.1 ± 0.6 fs and 2.9 ± 0.5 fs, respectively.

References and Notes

1. J.-C. Diels, W. Rudolph, *Ultrashort Laser Pulse Phenomena* (Academic Press, New York, 1996).
2. Fast heavy-ion impact can trigger atomic processes (e.g., ionization) on an unprecedented, attosecond time scale as demonstrated by R. Moshammer *et al.* [*Phys. Rev. Lett.* **79**, 3621 (1997)]. However, the lack of a synchronized probe pulse of comparable duration does not allow tracing of dynamics after the impact.
3. A. Baltuska *et al.*, *Appl. Phys. B* **65**, 175 (1997).
4. M. Nisoli *et al.*, *Appl. Phys. B* **65**, 189 (1997).
5. U. Morgner *et al.*, *Opt. Lett.* **24**, 411 (1999).
6. D. H. Sutter *et al.*, *Opt. Lett.* **24**, 631 (1999).
7. A. Shirakawa *et al.*, *Appl. Phys. Lett.* **74**, 2268 (1999).
8. G. Steinmeyer *et al.*, *Science* **286**, 1507 (1999).
9. T. Brabec, F. Krausz, *Rev. Mod. Phys.* **72**, 545 (2000).
10. T. W. Hänsch, *Opt. Commun.* **80**, 71 (1990).
11. G. Farkas, C. Toth, *Phys. Lett. A* **168**, 447 (1992).
12. S. E. Harris *et al.*, *Opt. Commun.* **100**, 487 (1993).
13. A. L'Huillier, Ph. Balcou, *Phys. Rev. Lett.* **70**, 774 (1993).
14. J. J. Macklin, J. D. Kmetec, C. L. Gordon III, *Phys. Rev. Lett.* **70**, 766 (1993).
15. M. Lewenstein *et al.*, *Phys. Rev. A* **49**, 2117 (1994).
16. Ch. Spielmann *et al*, *Science* **278**, 661 (1997).
17. M. Schnürer *et al.*, *Phys. Rev. Lett.* **80**, 3236 (1998).
18. Ch. Spielmann *et al.*, *IEEE J. Sel. Top. Quantum Electr.* **4**, 249 (1998).
19. K. J. Schafer, K. C. Kulander, *Phys. Rev. Lett.* **78**, 638 (1997).
20. I. P. Christov, M. M. Murnane, H. C. Kapteyn, *Phys. Rev. Lett.* **78**, 1251 (1997).
21. C. Kan *et al.*, *Phys. Rev. Lett.* **79**, 2971 (1997).
22. S. Sartania *et al.*, *Opt. Lett.* **22**, 1562 (1997).
23. Y. Kobayashi *et al.*, *Opt. Lett.* **23**, 64 (1998).
24. N. A. Papadogiannis *et al.*, *Phys. Rev. Lett.* **83**, 4289 (1999).
25. E. Constant *et al.*, *Phys. Rev. A* **56**, 3870 (1997).
26. The signal detected by Papadogiannis *et al.* (*24*) can be regarded as a high-order autocorrelation of the driver laser pulse. The extent to which this provides insight into the temporal structure of the generated high harmonics (autocorrelation signal) is not quite clear at present.
27. J. M. Schins *et al.*, *Phys. Rev. Lett.* **73**, 2180 (1994).
28. J. M. Schins *et al.*, *J. Opt. Soc. Am. B* **13**, 197 (1996).
29. T. E. Glover *et al.*, *Phys. Rev. Lett.* **76**, 2468 (1996).
30. A. Bouhal *et al.*, *J. Opt. Soc. Am. B* **14**, 950 (1997).
31. E. S. Toma *et al.*, *Phys. Rev. A* **62**, 061801(R) (2000).
32. T. F. Gallagher, *Phys. Rev. Lett.* **61**, 2304 (1988).
33. H. B. van Linden van den Heuvell, H. G. Muller, in *Multiphoton Processes*, S. J. Smith, P. L. Knight, Eds. (Cambridge Univ. Press, Cambridge, 1988), pp. 26–44.
34. P. B. Corkum, N. H. Burnett, F. Brunel, *Phys. Rev. Lett.* **62**, 1259 (1989).
35. P. B. Corkum, *Phys. Rev. Lett.* **71**, 1994 (1993).
36. Note that within the adiabatic approximation, the cycle-averaged quiver energy is a meaningful physical quantity even if $\tau_x \ll T_0$ because the velocity and, hence, the kinetic energy can be unambiguously decomposed into a quiver (U_p) and drift component.
37. R. R. Freeman *et al.*, *Phys. Rev. Lett.* **59**, 1092 (1987).
38. E. Yablonovitch, *Phys. Rev. Lett.* **60**, 795 (1988).
39. A. Scrinzi, M. Geissler, T. Brabec, *Phys. Rev. Lett.* **86**, 412 (2001).
40. M. Schnürer *et al.*, *Phys. Rev. Lett.* **83**, 722 (1999).
41. M. Schnürer *et al.*, *Appl. Phys. B* **70**, S227 (2000).
42. U. Becker, D. A. Shirley, *VUV and Soft X-Ray Photoionization* (Plenum, New York, 1996), p. 152.
43. The photoelectrons originating from the 4s level (kinetic energy ≈ 62 eV) are suppressed in this geometry due to their anisotropic momentum distribution, (*42*).
44. Laser pulse propagation and generation of high-order harmonics in our neon target were modeled from the quantum theory of M. Lewenstein *et al.* (*15*), expanded to account for nonadiabatic effects, ground-state depletion, and the influence of the atomic Coulomb field on ionization (*9*) and from Maxwell's wave equation. The latter also includes the refractive index of the neutral atoms and the absorption coefficient at high-harmonic frequencies (*40*). The laser and target parameters used are as follows: input laser pulse shape, as calculated from the measured spectrum; initial duration, 7 fs; target length, 2.5 mm; and pressure, 200 mbar.
45. Due to a sensitivity of τ_x to minor changes in τ_r our error propagation analysis yields a lower boundary of the confidence interval of τ_x that is inconsistent with the measured x-ray spectrum (Fig. 3). Therefore, the lower boundary has been replaced by the limit obtained from our numerical analysis (*44*), $\tau_{x,min} \approx 0.6$ fs.
46. P. B. Corkum, N. H. Burnett, M. Yu. Ivanov, *Opt. Lett.* **19**, 1870 (1994).
47. Our simulations indicate that exploitation of the attosecond potential of our measurement technique calls for smaller solid angles of detection than used in our experiments. In our attempts to resolve sub-femtosecond x-ray emission structure, we have been able to observe first indications of the predicted oscillations (third term in Eq. 2) over a delay range of several laser cycles near $t_d \approx 0$. Currently, however, the signal-to-noise ratio is strongly compromised by the reduced solid angle of detection. Somewhat higher harmonic photon yields appear to be necessary for performing attosecond-resolution measurements.
48. To provide more x-ray photons for measurement and to shorten the x-ray pulses, the generation process needs to be improved. Our simulations (*44*) confirm the sub-laser-cycle duration of the 90-eV harmonic emission for the parameters used in our experiments but indicate a sensitive dependence of its "sub-structure" on the laser peak intensity I_p. The ability to keep I_p at ~5 × 10^{14} W/cm^2 across the entire interaction volume would allow us to confine 90-eV harmonic emission to a fraction of a half cycle near the pulse peak, with $\tau_x \approx 0.6$ fs (limited by the Mo/Si mirror bandwidth). However, an increase of I_p to ~7 × 10^{14} W/cm^2 at the center of our Gaussian beam profile is predicted to give rise to a double-peaked emission of ~2-fs duration for near-axis rays. Reducing the on-axis intensity to ~5 × 10^{14} W/cm^2 would remove this broadening, but only at the (currently unacceptable) expense of 90-eV photon yield (due to a lateral confinement of the emission volume). Moreover, increasing the bandwidth of the multilayer for supporting shorter x-ray pulses tends to further compromise the photon flux on target. Increasing the power of few-cycle driven x-ray harmonic sources and/or developing techniques for efficient conversion of a Gaussian into a flat-top beam profile, therefore, will be of central importance for pushing the limits of the demonstrated techniques for generation and measurement into the 100-as range.
49. L. Xu *et al.*, *Opt. Lett.* **21**, 2008 (1996).
50. D. J. Jones *et al.*, *Science* **288**, 635 (2000)
51. A. Apolonski *et al.*, *Phys. Rev. Lett.* **85**, 740 (2000).
52. A. Poppe *et al.*, *Appl. Phys. B* **72**, 373 (2001). Published online 13 December 2000; 10.1007/ s003400000526.
53. P. Dietrich, F. Krausz, P. B. Corkum, *Opt. Lett.* **25**, 16 (2000).
54. Z. Cheng *et al.*, *Opt. Lett.* **24**, 247 (1999).

55. U. Heinzmann from the University of Bielefeld and his colleagues, U. Kleineberg and Y. Lim, provided invaluable x-ray optical components and support for the experiments. The Zr filters were provided by G. Schmahl from the University of Göttingen. The computer code used for simulating high-harmonic generation was developed under the guidance of T. Brabec. All contributions are gratefully acknowledged. Sponsored by the Austrian Science Fund (grants Y44-PHY and F016) of the Federal Ministry of Education, Science, and Culture. M.D. acknowledges a fellowship in the frame of the European ATTO network and support from the Ministry of Education, Science and Research NRW.

22 December 2000; accepted 2 February 2001
Published online 15 February 2001;
10.1126/science.1058561
Include this information when citing this paper.

CORRECTIONS AND CLARIFICATIONS

ERRATUM

post date 20 April 2001

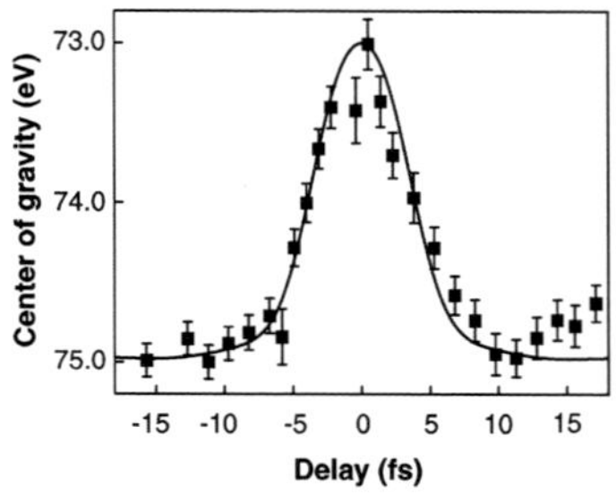

Reports: "X-ray pulses approaching the attosecond frontier" by M. Drescher *et al.* (9 Mar. 2001, p. 1923). In Figure 5, the number "76.0" at the base of the *y* axis should have read "75.0." The correct figure is shown here.

articles

Attosecond metrology

M. Hentschel*†, R. Kienberger*†, Ch. Spielmann*, G. A. Reider*, N. Milosevic*, T. Brabec*, P. Corkum‡, U. Heinzmann§, M. Drescher§ & F. Krausz*

* *Institut für Photonik, Technische Universität Wien, Gusshausstr. 27, A-1040 Wien, Austria*
‡ *Steacie Institute of Molecular Sciences, NRC Canada, Ottawa, Canada K1A 0R6*
§ *Fakultät für Physik, Universität Bielefeld, D-33615 Bielefeld, Germany*
† *These authors contributed equally to this work*

The generation of ultrashort pulses is a key to exploring the dynamic behaviour of matter on ever-shorter timescales. Recent developments have pushed the duration of laser pulses close to its natural limit—the wave cycle, which lasts somewhat longer than one femtosecond (1 fs = 10^{-15} s) in the visible spectral range. Time-resolved measurements with these pulses are able to trace dynamics of molecular structure, but fail to capture electronic processes occurring on an attosecond (1 as = 10^{-18} s) timescale. Here we trace electronic dynamics with a time resolution of ≤ 150 as by using a subfemtosecond soft-X-ray pulse and a few-cycle visible light pulse. Our measurement indicates an attosecond response of the atomic system, a soft-X-ray pulse duration of 650 ± 150 as and an attosecond synchronism of the soft-X-ray pulse with the light field. The demonstrated experimental tools and techniques open the door to attosecond spectroscopy of bound electrons.

Experimental access to fast-evolving microscopic processes requires a short excitation ('pump') pulse, which sets the process going, and a short probe pulse for taking snapshots of the evolution of the process[1]. Recent advances in ultrafast optics have made available laser pulses as short as a few femtoseconds for this time-resolved (or pump–probe) measurement technique[2,3]. These pulse durations are practically at the limit set by the laser field oscillation cycle in the visible spectral range. They allow the tracking of changes in the nuclear structure of molecules—such as vibrations, or the breaking and formation of chemical bonds—because the characteristic time-scale for atomic motion on an atomic length scale (0.1 nm) extends from a few femtoseconds to a few thousand femtoseconds (refs 4, 5). But time-domain access to a wide range of electron dynamics in the atomic shells has been frustrated so far by the speed of electronic relaxation processes. The wavefunction of bound electrons, following an energetic excitation—for example, ionization or the creation of an inner-shell vacancy—tends to evolve on an attosecond time-scale. Until now, insight into these processes could be gained only indirectly from frequency-domain measurements of transition linewidths[6].

Coherent extreme-ultraviolet (XUV) and X-ray sources are not only important tools for atomic spectroscopy but, owing to their short wave cycle, also offer the potential for producing electromagnetic radiation of attosecond duration. High-order harmonics of femtosecond laser radiation[7,8] have been recently shown to be[9,10] sources of trains of attosecond XUV pulses, separated by one half-cycle of the laser period with the train extending over some 10 fs. Yet, attosecond time-resolved measurements still appeared to be out of reach, because straightforward interpretation of spectroscopic data requires isolated, single pulses[11,12]. Moreover, the XUV fluences available from current high-harmonic sources are many orders of magnitude lower than required by XUV-pump/XUV-probe spectroscopy, while conventional schemes of visible-light/XUV cross-correlation, yielding a convolution of the respective pulse envelopes, do not provide subfemtosecond time resolution. Using few-cycle visible laser pulses[13] for the generation of isolated soft-X-ray pulses[14], together with a special geometry for detecting photoelectrons originating from atoms exposed simultaneously to both of these pulses, our work offers a way to overcome these limitations.

Here we report the observation of light-field-induced petahertz frequency (1 PHz = 10^{15} Hz) modulation of the width of the kinetic energy distribution of photoelectrons; these photoelectrons were derived from an atomic gas (krypton, in our experiments) excited by an ultrashort 90-eV soft-X-ray (henceforth briefly X-ray) pulse. Our measurements yield an X-ray pulse duration of τ_x = 650 ± 150 as, and indicate that more than 90% of the X-ray photons, which are spectrally filtered by a bandpass at ~90 eV, are confined within a single, isolated pulse of this duration. The experimental results bear direct evidence of the X-ray pulse being synchronized to the field oscillations of the visible light pulse with attosecond precision; they also indicate bound–free electronic transitions from the 4*p* state of krypton responding to 90-eV excitation on an attosecond timescale. With these tools we have traced the electric field oscillations in a visible light wave with a resolution of better than 150 as. These measurements constitute, to our knowledge, the first demonstrations of attosecond metrology. Beyond measuring the duration of X-ray pulses with attosecond resolution, our approach will permit attosecond spectroscopy of bound electron dynamics in atoms and molecules using a low-fluence subfemtosecond X-ray pulse and a strong visible light field.

X-ray photoemission controlled by a strong light field

Angle-resolved X-ray-induced photoemission assisted by visible light has been recently used to measure the cross-correlation of the intensity envelope of a 7-fs visible light pulse with that of an ultrashort soft-X-ray pulse[14]. Drawing on this previous work, we show here control of the final kinetic energy of X-ray-generated photoelectrons by the oscillating light field rather than by its cycle-averaged intensity. This enables us to probe, on an attosecond timescale, the duration of the X-ray pulse and its timing jitter with respect to the light field. The key to this progress has been (1) an approach that allows us to combine an enhanced solid angle of detection with attosecond temporal resolution, and (2) matching the X-ray pulse front to the light wave front within the volume of their interaction with nanometre-scale precision and stability. Photoelectrons from atoms exposed simultaneously to a strong light field and an X-ray pulse were detected within a cone aligned orthogonally to the beam direction and electric field vector of the linearly polarized light field. This orthogonal detection geometry was chosen to suppress a background in the photoelectron spectrum produced primarily along the electric field vector of the light pulse by above-threshold ionization (ATI).

The effect of the light field on the X-ray photoelectron energy spectrum is now well understood in a quantum-mechanical framework[15,16]. In the strong-field limit, the most conspicuous features can also be accounted for by a quasi-classical model treating the interaction as a two-step process[14]: first, the photoelectron is ejected by a short X-ray pulse with a distribution of initial momenta known from conventional photoionization studies[6]; and second, the photoelectron is subsequently accelerated as a classical particle by the electric field of a femtosecond light pulse. In accordance with the quantum treatment, the model predicts that, depending on the oscillation phase of the light field at the instant of 'birth' of the electron, a momentum component along the electric field vector, Δp_x, is added to the initial momentum of the electron, resulting in a shift of the photoelectron distribution up or down in momentum space (Fig. 1). The momentum transfer gives rise to a spread of the final kinetic energy of photoelectrons collected within a finite solid angle as revealed by the top and bottom of Fig. 1. The effect increases with increasing detection cone and vanishes for monodirectional detection.

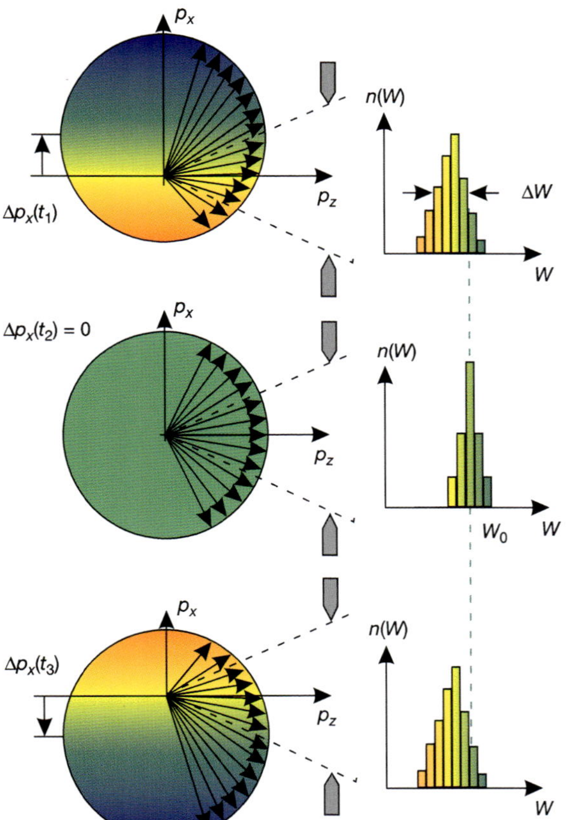

Figure 1 The principle of measuring cross-correlation between light electric field and X-ray intensity with attosecond resolution by means of 'two-colour' photoionization. The photoelectrons created initially with isotropic momentum distribution by the absorption of X-ray photons pick up momentum from the strong femtosecond laser field. In the slowly varying wave approximation, the field-induced momentum change Δp_x depends on the phase, amplitude and oscillation frequency of the light electric field at the instant of birth and is independent of the subsequent evolution of the light field. At birth instants t_1, $t_2 = t_1 + T_0/4$ and $t_3 = t_1 + T_0/2$ chosen such that the light field crosses zero at t_1 and t_3, the light-induced momentum change Δp_x deforms the final photoelectron momentum and energy distribution as shown. Different colours represent electrons of different final kinetic energy, with the energy increasing from the red to the blue. The momentum transfer and hence the change in the energy spectrum is largest if the electric field of the light pulse crosses zero at the moment of birth of the photoelectron (top and bottom). For a birth instant coinciding with the peak of the light electric field (centre) the final energy spectrum is unaffected by the light field and is determined by the spectrum of the X-ray pulse (represented by a single colour and a well-defined radius of the sphere in momentum space for simplicity).

Scanning the instant of birth of the photoelectron through the light field oscillations by changing the relative delay t_d between the light pulse and X-ray pulse results in a modulation of the spectral width $\Delta W(t_d)$ with a period equal to one-half of the light oscillation period T_0. Figure 1 reveals that a confined detection cone aligned orthogonally to the light field vector also gives rise to a downshift of the centre of gravity of the photoelectron spectrum. This energy shift was used in our earlier study as a primary observable[14]. The key advantage of exploiting instead the broadening of the spectrum is that this effect is enhanced by increasing the detection aperture, and thus allows for a significantly enhanced signal yield and modulation amplitude. By contrast, the modulation in the centre of gravity of the photoelectron spectrum gets increasingly washed out with increasing detection angle, resulting in deterioration of the temporal resolution.

In the above description, we tacitly assumed an X-ray pulse duration τ_x that was very short compared to $T_0/2$, as well as an instantaneous response of the electronic transition, defining precisely the instant of birth of the photoelectron on a timescale of $T_0/2$. A finite X-ray pulse duration and timing jitter of the X-ray pulse (relative to the phase of the light field) of any origin results in a reduced depth of the above-discussed modulation of $\Delta W(t_d)$. In fact, an X-ray pulse duration approaching $T_0/2$ smears out the modulation completely. Our results rely on this line of argument: the modulation depth of $\Delta W(t_d)$ sets a reliable upper limit, on an attosecond timescale, on the X-ray pulse duration and timing jitter.

Cross-correlation of light electric field and X-ray intensity

The experiments reported below have been performed with a modified version of the apparatus described previously[14]. Major modifications included an enhanced detection angle and the installation of a feedback-controlled piezo delay stage as well as X-ray beam diagnostics. Figure 2 illustrates schematically the main steps in creating and processing the X-ray and light pulses for the photoemission experiment discussed in the preceding section.

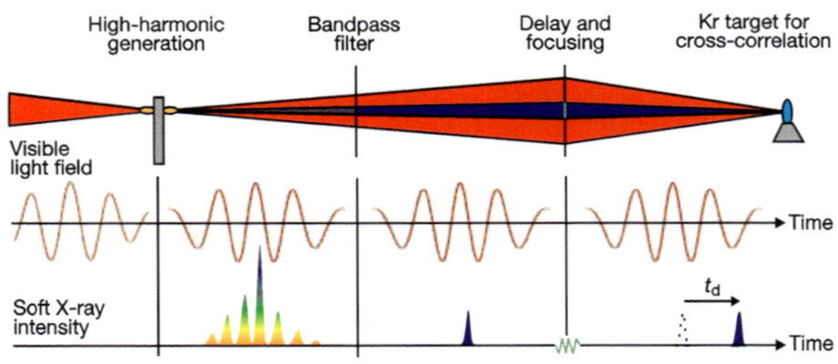

Figure 2 Diagram of the principal physical effects and processes preceding our light-field-controlled X-ray photoemission experiment. First, the few-cycle visible light pulse interacts with a jet of neon atoms to produce a train of pulses in the XUV and soft-X-ray spectral range. The rainbow colours of the individual pulses symbolize the respective spectral energy content: the highest photon energy components are contained only in a single pulse generated near the peak of the driver pulse. Note the increase of the instantaneous frequency in the centre of the visible light pulse (addressed as a dynamic blue shift in the text) past the interaction due to self phase modulation in the laser-generated plasma. The collinear laser and soft-X-ray beams are passed through a small circular Zr filter matched in size to the diameter of the harmonic beam. It stops the central portion of the laser beam, and transmits the low-divergence harmonic beam near 90 eV along with an annular laser beam surrounding it. The two co-axial beams then are focused onto the Kr atomic gas target by a concentric piezo-controlled double mirror unit[14] (rays actually reflected at this point are unfolded in our schematic illustration for convenience). The central Mo/Si reflector serves simultaneously as a focusing mirror for the 90-eV soft-X-ray harmonics, and as a bandpass filter that selects the highest-energy photons contained only in the central harmonic pulse. Furthermore, this multifunctional unit provides the necessary adjustable delay between the visible light and soft-X-ray harmonic pulse.

The kinetic energy spectrum of 4p photoelectrons ejected from krypton atoms under simultaneous irradiation by the 90-eV X-ray pulse ($\lambda \approx 14$ nm) and the light pulse ($\lambda \approx 750$ nm) driving the harmonic radiation was measured as a function of t_d, which is the delay between the X-ray and the light pulse, both polarized along the x direction in Fig. 1. The Kr 4p electrons were chosen because of their favourable yield along the z direction at our X-ray photon energies[6]. The light pulse (peak intensity, $\sim 5\times10^{13}$ W cm^{-2}) overlapped with the more tightly focused X-ray pulse in a volume of krypton gas kept at a pressure of a few times 10^{-2} mbar, limited to this value to prevent space-charge effects. The detection cone of our time-of-flight spectrometer is defined by the aperture of the spectrometer and its distance from the krypton target, with its full angle limited to $\sim 40°$ by the numerical aperture of the electron lens system employed.

The delay was varied in steps of $\Delta t_d = 150$ as in the central range of temporal overlap (-5 fs $\leq t_d \leq 5$ fs, with $t_d = 0$ near the peak of the light pulse), and in larger steps outside this range. Figure 3 shows two representative Kr 4p spectra corrected for an ATI background and recorded at $t_d = -450$ as and $t_d = 0$, along with respective asymmetric gaussian fits. This model function yields excellent fits to our data, characterized by a fidelity parameter $R^2 > 0.95$. The noise increasing to lower energies can be attributed to ATI electrons. The contour plot in Fig. 4 depicts a series of Kr 4p photoelectron spectra (processed as described above) as a function of delay t_d. The data clearly bring to light a quasi-periodic evolution of the photoelectron energy spectrum. Both the width and, as a consequence of normalization to constant area, the peak height of the photoelectron energy spectrum are subject to a modulation versus t_d at a period of $\sim T_0/2$. Coupled to a modulation of the spectral width, we observe a periodic downshift of the centre of gravity of the spectrum, in agreement with our model's (Fig. 1) prediction for a narrow detection solid angle. However, the spectral shift is found to be contaminated by much larger uncertainties than is the variation of the spectral width.

For a quantitative analysis, we study the width of the photoelectron spectrum ΔW as a function of t_d. The upper panel of Fig. 4 shows the variation of the full-width at half-maximum $\Delta W(t_d)$ of the gaussian fits versus delay between the X-ray and light pulse. In what follows we focus on the oscillatory part of $\Delta W(t_d)$, which was obtained by subtracting the cycle-averaged 'd.c.' component of $\Delta W(t_d)$. The result is depicted by dots in Fig. 5 along with the respective error bars. From an inspection of the error bars at points of maximum slope in Fig. 5, we infer a time resolution of our sampling scheme of ≤ 150 as. The modulation period of the data is close to 1.25 fs in the wings of the correlation function $\Delta W(t_d)$, in agreement with a carrier wavelength of ~ 750 nm of the light pulse. However, it decreases to ~ 0.9 fs near zero delay, indicative of a pronounced blue shift of the light pulse at its peak, the origin of which we shall discuss below.

Subfemtosecond X-ray pulse

To evaluate the X-ray pulse duration from our light-field/X-ray-intensity cross-correlation data (dots in Fig. 5), we simulated the light-field-induced variation of the Kr 4p photoelectron spectrum versus t_d based on the quasi-classical model described above. The duration τ_x of our gaussian model X-ray pulse was used as the only fit parameter. In the analytic representation of the light pulse, the pronounced frequency upshift at the pulse centre and a small residual chirp apparent in the measured correlation function $\Delta W(t_d)$ was taken into account as described in the caption of Fig. 5. Under the assumption of an isotropic initial momentum distribution of X-ray photoelectrons and a uniform detection efficiency within the aperture of the spectrometer, our simple model reproduces well all observed key features (oscillating broadening and downshift of the photoelectron spectrum) except for a pronounced asymmetry of the broadened spectra.

The red line in Fig. 5 shows the results of our model calculations for an X-ray pulse length of $\tau_x = 650$ as, which—using no fit parameter other than τ_x—constitutes the best fit to the measured data in the range of maximum modulation depth (-1 fs $\leq t_d \leq 2$ fs). The inset displays, in addition, the computed correlation function $\Delta W(t_d)$ for $\tau_x = 500$ as and $\tau_x = 800$ as, revealing modulations respectively significantly enhanced and suppressed with respect to the measured data. These data provide a safe estimate for the duration of our 90-eV X-ray pulse within the range of $\tau_x = 650 \pm 150$ as, with the lower limit applying under the assumption of zero timing jitter of the X-ray pulse relative to the light wave.

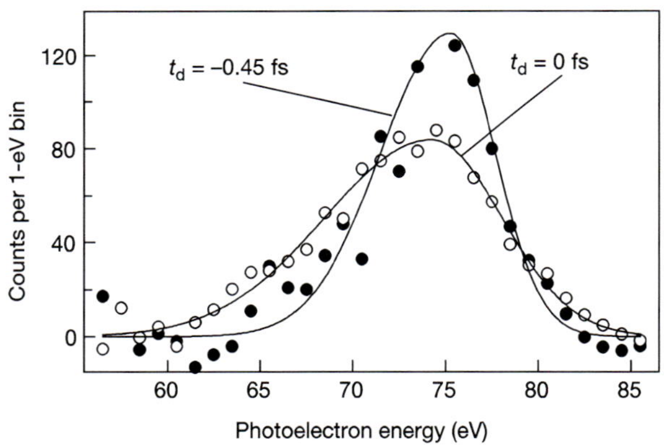

Figure 3 Kr 4p photoelectron spectra produced by $\sim$90-eV soft-X-ray pulses in the presence of a strong visible light field at two different delays t_d of the X-ray pulse. (Positive t_d corresponds to a retardation of the X-ray pulse.) The dots represent results corrected for ATI background. The lines show asymmetric gaussian fits to the data. The spectra were acquired over a measurement time of 2 minutes each.

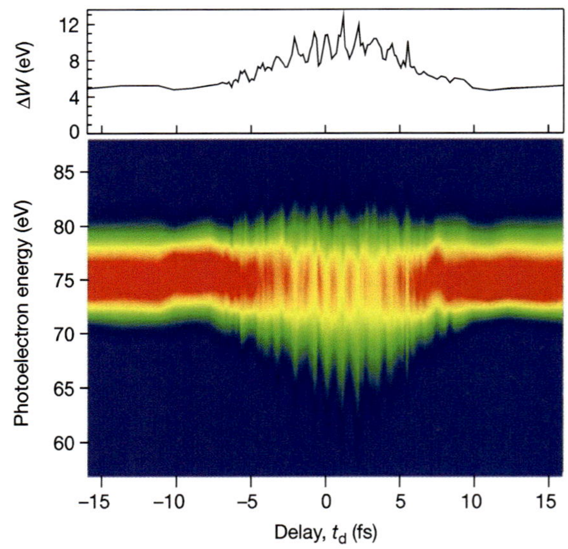

Figure 4 Cross-correlation of X-ray pulse with few-cycle laser pulse. Lower panel, contour plot of the Kr 4p photoelectron spectra (vertical axis) as a function of delay t_d of the soft-X-ray pulse with respect to the light pulse (horizontal axis). The plot displays the asymmetric gaussian fits introduced in Fig. 3, and comprises some 120 fitted spectra of the 4p feature, each normalized to the same number of counts, resulting in a constant area under the spectral distribution functions. Upper panel, spectral width ΔW evaluated from the gaussian fits as a function of the delay t_d. The spectra recorded at large delays, that is, those unaffected by the light field, are consistent with the X-ray spectrum centred at $\sim$90 eV and with a binding energy of $\sim$14 eV of the Kr 4p electron.

articles

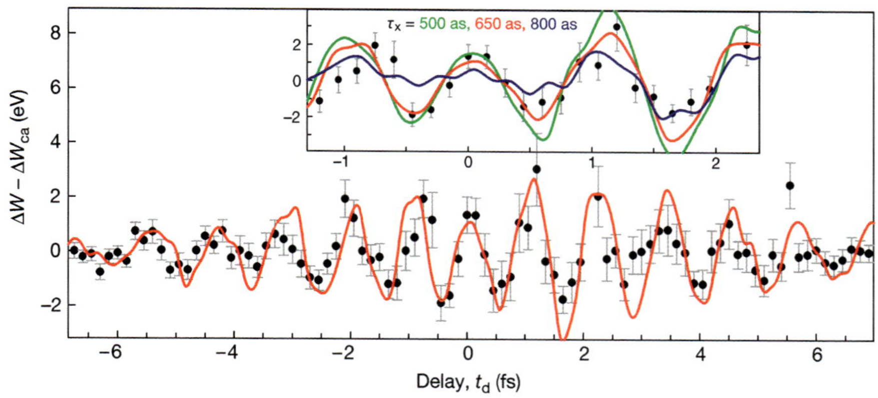

Figure 5 Spectral width ΔW of the Kr 4p photoelectron spectra as a function of t_d. The dots represent the oscillating component of the spectral width, $\Delta W - \Delta W_{ca}(t_d)$, where $\Delta W_{ca}(t_d)$ stands for the cycle-averaged value of $\Delta W(t_d)$. The full red line is the result of simulations based on the quasi-classical theory of two-colour X-ray photoionization by assuming a 7.5-fs, 750-nm linearly polarized light field of 5×10^{13} W cm^{-2} peak intensity and a 650-as, 90-eV gaussian X-ray pulse. The model light pulse includes a gaussian frequency sweep at the pulse peak superimposed on a weak quadratic component. Inset, coloured lines show the result of the same simulation for different pulse durations near $t_d = 0$. The conspicuous asymmetry in the calculated and measured $\Delta W(t_d)$ about $t_d = 0$ is due to a change of the light carrier frequency occurring on a timescale T_0. Consequently, the slowly varying wave approximation breaks down and the final state of the electron becomes dependent on the actual evolution of the light field. The substructure in the curves is an artefact, resulting from the statistical initial conditions in our simulations.

The emergence of a subfemtosecond pulse from high harmonic generation in the few-cycle regime was corroborated by numerical simulations of our X-ray source with a computer code[17]. This code solved Maxwell's wave equations in three dimensions and computed the radiation of the strongly-driven atomic dipoles using a generalization of the quantum treatment of ref. 18. These simulations yield a 530-as X-ray pulse near the propagation axis in the far field within a 5-eV spectral range near 90 eV (full line in Fig. 6), accompanied by a few small satellite pulses. The appearance of satellites separated by $\sim T_0/2$ from the central pulse leads to a spectral modulation with a period of twice the laser photon energy, as revealed by the calculated spectrum (full line) in the inset of Fig. 6. The depth of this modulation provides a sensitive measure of satellite content. The measured spectrum of the harmonic X-ray pulse reflected by our Mo/Si multilayer (dotted line in the inset of Fig. 6) sets a safe upper limit of 10% to the total satellite fluence relative to the total X-ray fluence within a 5-eV range around 90 eV.

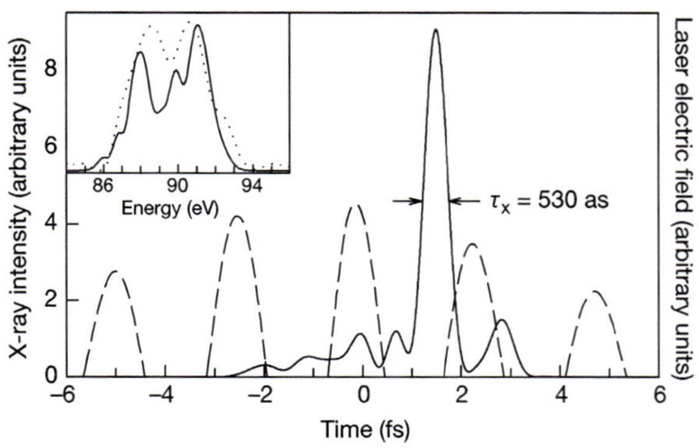

Figure 6 Calculated far-field, near-axis temporal intensity profile of a soft-X-ray pulse. The X-ray pulse is produced in a 3-mm-long 200-mbar neon gas volume by a 7-fs, 750-nm gaussian laser pulse with an on-axis peak intensity of 9×10^{14} W cm^{-2}. For the electric field of the laser pulse, $E(t) \propto \exp(-t^2/\tau_L^2)\cos(\omega_0 t + \varphi)$ with $\varphi = 0$ (cosine pulse), where τ_L is the pulse duration, ω_0 is the angular carrier frequency and φ is the 'absolute' phase. The dashed line shows the on-axis electric field of the laser pulse leaving the interaction region. The calculated X-ray radiation is selected within a 5-eV spectral range near 90 eV. Inset, calculated (full line) and measured (dotted line) X-ray pulse spectrum selected by the Mo/Si reflector, showing that about 90% of the total fluence is within a 5-eV range around 90 eV.

The agreement of the measured X-ray pulse duration with that obtained from our numerical calculations[16] (within the experimental error) shows that the timing jitter of the X-ray pulse relative to the visible light field is much smaller than 1 fs. In other words, our experiment shows that the subfemtosecond X-ray pulse is locked to the carrier wave of its generating few-cycle light pulse with attosecond precision. This is surprising, given that the 'absolute' phase φ of the wave (defined in the caption of Fig. 6) with respect to the amplitude envelope is random in our laser pulses[3], slightly modifying the evolution of the electric field from pulse to pulse. The attosecond timing stability of our subfemtosecond X-ray pulse to its few-cycle driver is due either to a (surprising) robustness of the high-harmonic generation process against these modifications, or (more likely) to a substantially reduced X-ray yield for values of φ other than the optimum assumed in Fig. 6. In the latter case, only a fraction of all laser pulses would provide an appreciable contribution to the detected photoelectron yield. If this is true, few-cycle driver pulses with stabilized absolute phase would markedly improve the data acquisition rate in the present and related experiments.

Probing light-field oscillations

With its duration extracted from the central part of $\Delta W(t_d)$, the X-ray pulse can now be used, as a first application, to probe the evolution of the electric field in the light pulse. The sweep of the instantaneous carrier frequency ν_{inst} (or wavelength λ_{inst}) in the visible light pulse can be evaluated from $\Delta W(t_d)$ by fitting a sinusoidal oscillation of adjustable period to the data in Fig. 5. The dots in Fig. 7 show the carrier frequency sweep evaluated in this manner, revealing a dynamic blue shift from a carrier wavelength of ~750 nm to some 550 nm. To understand this phenomenon, we have to remember that the light pulse used here is derived from the one generating the X-ray pulse (Fig. 2). Also shown in Fig. 7 is the calculated frequency sweep of the laser pulse leaving the neon-gas X-ray source.

The measured dynamic frequency shift at the centre of the pulse is larger than predicted by our numerical study, but reflects qualitatively the predicted behaviour. The qualitative agreement suggests that the observed blue shift can be attributed to ionization-induced

articles

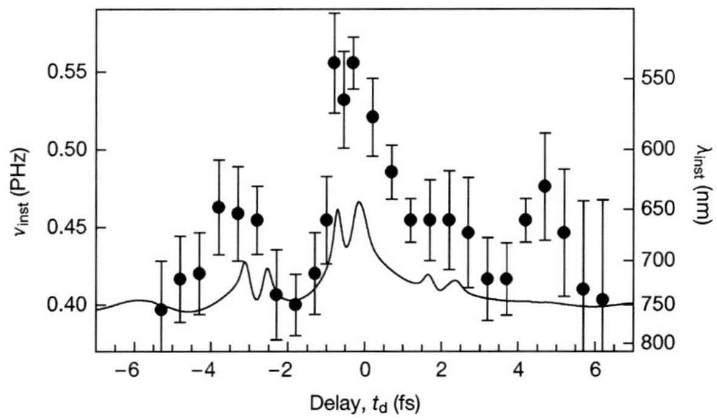

Figure 7 Calculated (line) and measured (dots) instantaneous frequency of the few-cycle light field. The experimental data are derived from $\Delta W(t_d)$ shown in Fig. 5. The computed frequency sweep originates from self-phase modulation of the light pulse upon its interaction with the neon gas X-ray source.

self-phase modulation in the X-ray generation process. These results provide evidence of optical field ionization driven by a few-cycle pulse being confined to less than one single oscillation cycle. The direct probing of the field oscillations in a light wave as implemented here is another demonstration of attosecond metrology. It should permit complete measurement of the electric field of absolute-phase-stabilized few-cycle light[19,20]. The presence of a substantial dynamic blue shift in the light pulse used to control X-ray photoemission provides a further strong argument for the isolated nature of our X-ray pulse: the depth of modulation of $\Delta W(t_d)$ near $t_d = 0$, where the modulation period drops to ~0.9 fs, would rapidly decrease with increasing satellites spaced by ~1.25 fs from the main pulse (Fig. 6).

Attosecond spectroscopy

With the availability of single subfemtosecond X-ray pulses, it is now possible to extend time-resolved spectroscopy into the attosecond domain by using these pulses for both triggering and probing bound–bound or bound–free electronic transitions in atoms or molecules. However, the subfemtosecond X-ray pulses currently available do not yet have sufficient fluence for X-ray-pump/X-ray-probe spectroscopy. Generalization of the present concept of light-field-controlled X-ray photoemission should allow the substitution of either the X-ray pump or the X-ray probe pulse by a strong few-cycle laser field without losing attosecond resolution. This approach vastly relaxes requirements on the X-ray pulse fluence, holding promise for extending the technique up to photon energies approaching one kiloelectron volt (keV) with next-generation high-harmonic sources.

In one class of proposed experiments, an absolute-phase-stabilized[19] quasi-single-cycle visible laser pulse (for example, τ_p = 4 fs at λ = 750 nm, as shown by Fig. 20 in ref. 3) sets free a subfemtosecond wavepacket by optical field ionization; this wavepacket is driven back to its parent ion an optical cycle later with energies of up to a few keV and produce collisional excitations. The relaxation processes following subfemtosecond collisional excitation are important for a number of phenomena, such as X-ray lasing and non-sequential photoionization. These processes could be traced in time by measuring the energy of photoelectrons detached from the ion by a delayed, subfemtosecond X-ray probe pulse. In another class of studies, the subfemtosecond X-ray pulse might serve as a pump pulse to create inner-shell vacancies. The temporal evolution of the subsequent inner-shell relaxation processes could be traced by time-resolved spectroscopy of the emitted Auger electron(s). The Auger spectrum, modified by a few-cycle light field and measured as a function of t_d, would provide access to the (few-femtosecond to attosecond) inner-shell relaxation times in the range of several hundred eV to keV in the same way as the correlation functions in Figs 4 and 5 yield the X-ray pulse duration. The experimental tools are now becoming available for attosecond spectroscopy. □

Received 1 October; accepted 19 October 2001.

1. Krausz, F. From femtochemistry to attophysics. *Phys. World* **14**(9), 41–46 (2001).
2. Steinmeyer, G., Sutter, D. H., Gallmann, L., Matuschek, N. & Keller, U. Frontiers in ultrashort pulse generation: pushing the limits in linear and nonlinear optics. *Science* **286**, 1507–1512 (1999).
3. Brabec, T. & Krausz, F. Intense few-cycle laser fields: frontiers of nonlinear optics. *Rev. Mod. Phys.* **72**, 545–591 (2000).
4. Zewail, A. Femtochemistry: atomic-scale dynamics of the chemical bond (adapted from the Nobel Lecture). *J. Phys. Chem. A* **104**, 5660–5694 (2000).
5. Bhattacharjee, Y. Measuring the immeasurable. *Nature* **412**, 474–476 (2001).
6. Becker, U. & Shirley, D. A. *VUV and Soft X-Ray Photoionization* 152 (Plenum, New York, 1996).
7. L'Huillier, A. & Balcou, P. High-order harmonic generation in rare gases with a 1-ps 1053-nm laser. *Phys. Rev. Lett.* **70**, 774–777 (1993).
8. Macklin, J. J., Kmetec, J. D. & Gordon, C. L. High-order harmonic generation using intense femtosecond pulses. *Phys. Rev. Lett.* **70**, 766–769 (1993).
9. Papadogiannis, N. A., Witzel, B., Kalpouzos, C. & Charalambidis, D. Observation of attosecond light localization in higher order harmonic generation. *Phys. Rev. Lett.* **83**, 4289–4292 (1999).
10. Paul, P. *et al.* Observation of a train of attosecond pulses from high harmonic generation. *Science* **292**, 1689–1692 (2001).
11. Ivanov, M., Corkum, P. B., Zuo, T. & Bandrauk, A. Routes to control of intense-field atomic polarizability. *Phys. Rev. Lett.* **74**, 2933–2936 (1995).
12. Christov, I. P., Murnane, M. M. & Kapteyn, H. C. High-harmonic generation of attosecond pulses in the "single-cycle" regime. *Phys. Rev. Lett.* **78**, 1251–1254 (1997).
13. Spielmann, Ch. *et al.* Generation of coherent X-rays in the water window using 5-femtosecond laser pulses. *Science* **278**, 661–664 (1997).
14. Drescher, M. *et al.* X-ray pulses approaching the attosecond frontier. *Science* **291**, 1923–1927 (2001). Published online 15 February 2001; 10.1126/science.1058561.
15. Kitzler, M., Milosevic, N., Scrinzi, A. & Brabec, T. Theory of attosecond streak cameras. *Phys. Rev. Lett.* (submitted).
16. Itatani, J. *et al.* Attosecond streak camera. *Phys. Rev. Lett.* (submitted).
17. Milosevic, N., Scrinzi, A. & Brabec, T. Ab initio numerical calculation of attosecond pulse generation. *Phys. Rev. Lett.* (submitted).
18. Lewenstein, M., Balcou, Ph., Ivanov, M. Yu., L'Huillier, A. & Corkum, P. B. Theory of high-harmonic generation by low-frequency laser fields. *Phys. Rev. A* **49**, 2117–2132 (1994).
19. Poppe, A. *et al.* Few-cycle optical waveform synthesis. *Appl. Phys. B* **72**, 373–376 (2001). Published online 13 December 2000; 10.1007/s003400000526.
20. Paulus, G. G. *et al.* Evidence of "absolute-phase" phenomena in photoionization with few cycle laser pulses. *Nature* **414**, 182–184 (2001).

Acknowledgements

We thank M. Uiberacker for assistance with measurements, and Y. Lim and U. Kleineberg for manufacturing the X-ray multilayer mirror. Discussions with M. Ivanov are gratefully acknowledged. This work was supported by the Austrian Science Fund and by the European ATTO network.

Competing interests statement

The authors declare that they have no competing financial interests.

Correspondence and requests for materials should be addressed to F.K. (e-mail: ferenc.krausz@tuwien.ac.at).

articles

Time-resolved atomic inner-shell spectroscopy

M. Drescher*†, M. Hentschel*, R. Kienberger*, M. Uiberacker*, V. Yakovlev*, A. Scrinzi*, Th. Westerwalbesloh†, U. Kleineberg†, U. Heinzmann† & F. Krausz*

* *Institut für Photonik, Technische Universität Wien, Gusshausstrasse 27, A-1040 Wien, Austria*
† *Fakultät für Physik, Universität Bielefeld, Universitätsstrasse 25, D-33615 Bielefeld, Germany*

The characteristic time constants of the relaxation dynamics of core-excited atoms have hitherto been inferred from the linewidths of electronic transitions measured by continuous-wave extreme ultraviolet or X-ray spectroscopy. Here we demonstrate that a laser-based sampling system, consisting of a few-femtosecond visible light pulse and a synchronized sub-femtosecond soft X-ray pulse, allows us to trace these dynamics directly in the time domain with attosecond resolution. We have measured a lifetime of $7.9^{+1.0}_{-0.9}$ fs of M-shell vacancies of krypton in such a pump–probe experiment.

Many-body systems with excess internal energy relax towards states of lower energy by rearrangement of molecular, atomic or nuclear structure. Observing these processes in real time requires a pump pulse for initiating the microscopic dynamics and a delayed probe pulse for detecting transition states of the evolving system. Time-resolved spectroscopy based upon this pump–probe approach[1] is now routinely used for tracking atomic motion in molecules with femtosecond laser pulses[2].

Electronic motion in weakly bound (Rydberg) states can also be captured with optical or infrared pulses[3]. However, the relaxation dynamics of core-excited atoms is out of the reach of femtosecond optical techniques. Excitation of a strongly bound electron from an atomic inner shell is followed by an ultrafast rearrangement of the electronic system, resulting in a disappearance of the inner-shell vacancy (henceforth briefly referred to as core hole). Characteristic time constants of the concomitant electronic dynamics range from a few attoseconds (1 as $= 10^{-18}$ s) to a few femtoseconds depending on the energy of the core hole and the strength of electronic coupling.

The lifetime of core holes can be inferred from the energy spectral width of the photons or electrons emitted upon the decay of the excited atomic state. However, except for specific cases of isolated resonances without incoherent background, energy-domain measurements on their own are—in general—unable to provide detailed and accurate insight into the evolution of multi-electron dynamics. As a complementary technique, time-domain access would be desirable. Time-resolved spectroscopy of core-excited atoms calls for several new tools and techniques. The core hole must be created within a time interval that is short compared to its decay time: that is, within typically less than 1 femtosecond. Sampling of the subsequent atomic relaxation calls for a synchronized probe pulse and for a suitable probing technique.

The recent demonstration of isolated sub-femtosecond soft-X-ray pulses[4,5] and trains of pulses[6,7] along with attosecond sampling of photo-electron wave packets with a few-cycle optical wave[5,8] opened the door to attosecond atomic spectroscopy[9]. Here we use these tools and techniques to track rearrangement of the electronic system of a core-excited atom. The core hole is created within less than a femtosecond and its subsequent decay can be observed on a few-femtosecond timescale. The applied technique offers attosecond resolution and will allow time-resolved spectroscopy of a wide range of atomic processes.

Decay of atomic core holes

Excitation of an electron from an atomic shell other than the outermost valence orbital by the impact of an energetic particle—for example, a photon or electron (here we use photons)—creates a transient hole state of energy W_h (see process a in Fig. 1a). This core hole is not stable, so the system tends to minimize its energy by filling the vacancy with an electron from an outer shell (see process b in Fig. 1a). The excess binding energy $W_h - W_1$ is either carried away by an extreme ultraviolet or X-ray fluorescence photon or transferred via electrostatic forces to another (Auger) electron, which subsequently escapes from the atomic binding (see process c in Fig. 1a). The latter relaxation channel is most probable for light atoms and moderate binding energies ($W_h < 1$ keV) but remains intact for $W_h > 1$ keV (ref. 10).

Owing to several Auger relaxation channels, the energy spectrum of the ejected electrons usually contains a number of Auger lines. Further, several photo lines originating from atomic shells with binding energy smaller than the energy of the exciting X-ray photon, $\hbar\omega_X$, add to the spectrum. In Fig. 1a just a single Auger line and a couple of photo lines represent this manifold. The mean kinetic energy of the Auger electron $W_{kin} = W_h - W_1 - W_2$ and its spectral width Γ are fully determined by intrinsic atomic properties. By contrast, the photo electrons carry the energy of the absorbed photon diminished by their atomic binding energy and hence—in the absence of resonances—their energy distribution reflects that of the exciting radiation.

The feasibility of time-resolved inner-shell spectroscopy relies on X-ray pulses with a duration τ_X much shorter than the lifetime of the generated core hole τ_h. Under these circumstances the temporal evolution of the production rate dN/dt or—in the language of quantum mechanics—the temporal shape $|\psi(t)|^2$ of the ejected wave packet is very different for photo and Auger electrons (Fig. 1b). If the photo-ionization cross-section is nearly constant within the bandwidth of the X-ray pulse, $|\psi_{photo}(t)|^2$ follows the temporal profile of the exciting X-ray pulse $|E_X(t)|^2$ and thus the production rate of core holes. For $\tau_X \ll \tau_h$, the front edge of $|\psi_{Auger}(t)|^2$ rises within τ_X, whereas its trailing edge tracks the decay of the core holes. Thus, sampling the time evolution of photo and Auger electron emission explores the generation and decay of core-excited atoms in real time.

Sampling electron emission with light

Strong-field physics provides the key to trace electron emission from atoms. The presence of a strong light field affects the ejected electrons' motion and can be used to probe the emission. Assuming a linearly polarized few-cycle probe light field $E_L(t) = E_a(t)\cos(\omega_L t + \varphi)$, the final momentum (reached after the light pulse left the interaction volume) of an electron released at the

articles

instant t_r is changed by $\Delta p(t_r) = [eE_a(t_r)/\omega_L]\sin(\omega_L t_r + \varphi)$ along the light electric field vector according to a simple classical analysis[4] (Fig. 2a). E_a is the amplitude envelope, and ω_L and φ stand for the carrier frequency and carrier-envelope phase of the laser pulse, respectively. The final energy is given by:

$$W_f(t_r) = W_i + 2U_p(t_r)\sin^2(\omega_L t_r + \varphi)\cos 2\theta + [8W_i U_p(t_r)]^{1/2}\sin(\omega_L t_r + \varphi)\cos\theta \quad (1)$$

where $U_p(t) = e^2E_a^2(t)/4m_e\omega_L^2$ is the electron's cycle-averaged quiver energy in the probe light field, e, m_e and W_i are the electron's charge, mass and initial energy, and θ is the angle between its final momentum vector and the light electric field vector. For $\theta \approx 0°$ (Fig. 2b) and $W_i \gg \hbar\omega_L$, a moderate field strength ($U_p \approx \hbar\omega_L$) is able to shift the electron energy by many electronvolts. The energy shift varies from zero to its maximum value within a quarter-wave cycle $T_0/4 = \pi/2\omega_L$ ($\approx$0.6 fs at $\lambda_L = 750$ nm). This time-to-energy mapping permits sampling of electron emission with attosecond resolution. The final energy spectrum for an initial energy distribution $dN/dW = f_W(W_i)$ and time structure $dN/dt = f_t(t_r - t_{r0})$, where t_{r0} represents the instant of the emission peak, can be calculated for any detection direction θ and timing t_{r0} by using equation (1). Experimentally, t_{r0} can be varied by varying the arrival time, that is, the delay Δt (Fig. 2b), of the probe light pulse relative to the X-ray pump pulse that triggers the electron emission. For large initial kinetic energy, $W_i \gg \hbar\omega_L$, the validity of this simple classical treatment of the influence of a strong light field on electrons emitted within a fraction of the light wave cycle T_0 has been corroborated by several photo-emission experiments[5,8] as well as by quantum mechanical analyses[11,12].

We have generalized one (ref. 12) of these quantum theories addressing photo-electron emission in the presence of a strong laser field to describe the behaviour of Auger electrons emitted under similar conditions (for more details see Supplementary Information). Our theory yields the temporal evolution of hypothetical single-line Auger electron spectra probed by strong few-cycle light for different decay times τ_h as shown in Fig. 3. In our modelling we assumed $\tau_X = 0.5$ fs (at $\hbar\omega_X = 100$ eV) and $\tau_L = 5$ fs (at $\lambda_L = 750$ nm, $\hbar\omega_L \approx 1.6$ eV), corresponding to the current state of the art. For $\tau_h < T_0/2$ (Fig. 3a–c), the energy distribution of the Auger electrons exhibits pronounced variation as the delay Δt is varied by as little as $T_0/4$ (for $\tau_h \leq T_0/5$ in close agreement with the results of the above classical treatment). From the energy distributions recorded for different values of Δt, the temporal evolution

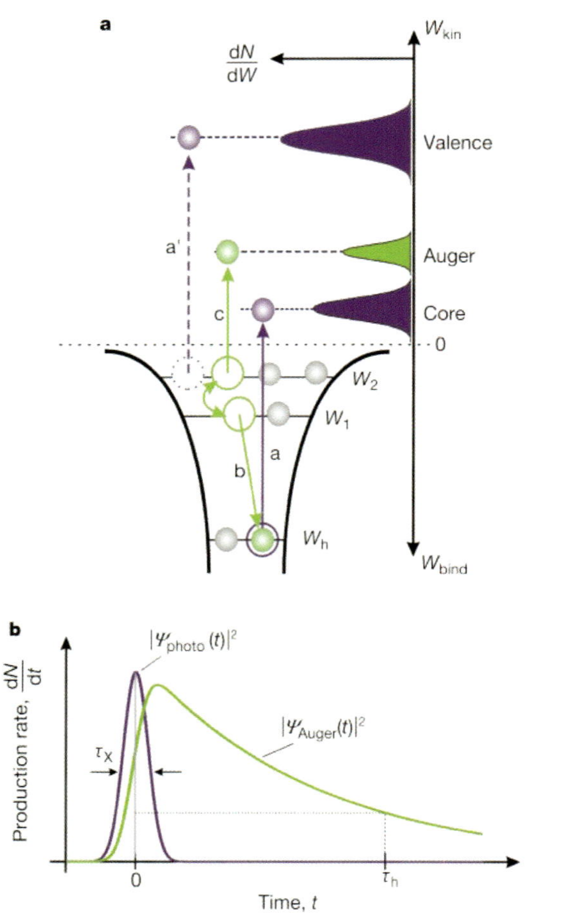

Figure 1 Schematic illustration of atomic excitation and relaxation processes following exposure to an ultrashort X-ray pulse. **a**, In the absence of resonances, the atom instantly responds by ejecting photo electrons (processes a, a') with $|\psi_{photo}(t)|^2$ following the temporal intensity profile of the exciting X-ray pulse. The inner-shell vacancy (W_h) is subsequently filled (process b) by an electron from an outer shell upon emission (process c) of a secondary (Auger) electron with $|\psi_{Auger}(t)|^2$ tracing the decay of the inner-shell vacancy. **b**, Probing the emitted photo and Auger electrons reveals the excitation and relaxation dynamics of core-excited atoms. W_{kin}, kinetic energy; W_{bind}, binding energy; W_h, $W_{1,2}$, binding energies of core and valence electrons, respectively; N, number of freed electrons. τ_X, duration of X-ray pulse; τ_h, decay time of core hole. The photo and Auger electrons are represented here (as in Fig. 5) in violet and green, respectively.

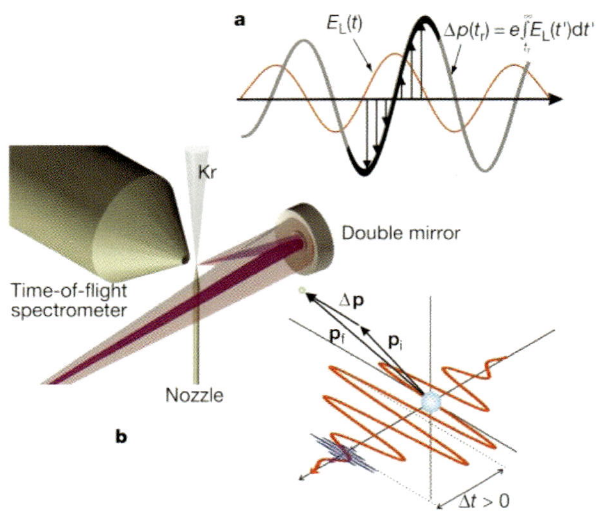

Figure 2 Attosecond two-colour sampling technique for probing electron emission from atoms. An extreme ultraviolet or X-ray pulse excites the atomic target and induces electron emission (see Fig. 1a). A delayed probe light pulse transfers a momentum Δp to the ejected electron after its release. p_i and p_f represent the electron's initial and final momentum, respectively. **a**, The transferred momentum sensitively depends on the phase and amplitude of the light field vector $E_L(t)$ at the instant of release resulting in a time-to-energy mapping on an attosecond timescale. For processes lasting less than the light cycle the oscillating light field constitutes a sub-femtosecond probe, whereas processes lasting longer than the light-wave cycle are sampled by the amplitude envelope of the laser pulse. In both cases, a sequence of light-affected electron energy spectra are recorded at different delays Δt, from which the time evolution of electron emission is reconstructed. **b**, In our experiments, we used a 97-eV, sub-femtosecond soft-X-ray pulse for excitation and a 750-nm (1.6 eV), sub-7-fs few-cycle light pulse for probing electron emission. The two pulses are collinearly focused into a krypton gas target by a two-component mirror similar to that used in ref. 5 but designed to reflect photons with higher energies. The kinetic energy distribution of the ejected photo and Auger electrons has been measured with a time-of-flight spectrometer aligned parallel to the polarization direction of the light pulse and the X-ray pulse.

of both the emission rate and a possible energy sweep of the electrons may be retrieved with a resolution of about $T_0/20$ ($\approx$100 as at $\lambda_L = 750$ nm) in future studies of sub-femtosecond inner-shell dynamics.

For τ_h approaching and exceeding T_0 (Fig. 3c–e), which is relevant to our current experiment, variation of the energy distribution with Δt within the laser cycle gradually disappears. The energy spectrum is broadened and broken up into sidebands spaced by $\hbar\omega_L$ at any Δt within the range of temporal overlap of electron emission and the laser pulse[13–16]. The sidebands originate from quantum interferences between different portions of the ejected electron wave packet experiencing the same momentum shift by the laser field. Although the overall width of the broadened energy distribution can still be estimated from the classical treatment, the emergence of the sideband structure can only be accounted for by quantum models. For electron emission times comparable to or longer than T_0 (Fig. 3d, e) the number of electrons scattered into sidebands is predicted to be proportional to the convolution of electron emission and the laser pulse. Hence for dynamics extending over T_0 or longer, evaluation of the X-ray-pump/visible-light-probe experiments does not rely on the energy–time mapping (in equation (1)) required for sub-cycle dynamics. Rather, the sideband area versus Δt directly yields the temporal evolution of electron emission (by deconvolution), provided that the laser pulse is sufficiently short (a few femtoseconds) in duration.

Recently we applied this general concept to sampling photo-electron emission with a few-cycle probe light pulse. From the measured photo-electron emission time we obtained the duration of the ionizing soft-X-ray pulse[5,8]. The measurements revealed that few-cycle light-driven atoms are capable of emitting isolated soft-X-ray pulses ($\hbar\omega_X \approx 100$ eV) with sub-femtosecond duration ($\tau_X < 0.5$ fs) in excellent synchrony ($\Delta t_{jitter} < 0.2$ fs) with the phase of its few-cycle ($\tau_L \approx 7$ fs) driver light wave[8]. These pulses, along with the above concept for electron sampling, combine to give an attosecond-resolution pump–probe technique for tracing electron dynamics after core-level excitation of atoms (or molecules).

Tracing the decay of an M-shell vacancy in krypton

For the first proof-of-principle experiments we selected one of the most comprehensively studied atomic inner-shell relaxation processes, the MNN Auger decay in krypton[17–20]. The Lorentz-type line profiles of the resultant Auger electrons imply an exponential decay of the core hole with a few-femtosecond lifetime. Using the notation of Fig. 1, in this case W_h, W_1 and W_2 correspond to the 3d, 4s and 4p sub-shells, respectively.

The pump–probe apparatus used here has been described elsewhere[5]. Briefly, the few-cycle light pulse ($\tau_L \approx 7$ fs, $\lambda_L = 750$ nm, $T_0 = 2.5$ fs) is delivered in an annular beam and hits the outer part of the two-component mirror shown in Fig. 2b. The soft-X-ray pulse with a continuum spectrum extending to photon energies of around 100 eV (ref. 8) propagates in a collinear beam matching the diameter of the inner part of the two-component mirror. For the current experiments, a dedicated Mo/Si multilayer mirror with a reflectance curve peaking at 97 eV has been manufactured. The energy spectrum of the X-ray pulse filtered by the mirror was upshifted as compared to previous experiments[5,8] to avoid the excitation of 3d electrons into Rydberg states, which comes into play at $\hbar\omega_X < 95$ eV. The energy upshift comes at the expense of a narrower bandwidth of the Mo/Si multilayer (full-width at half-maximum, FWHM $\approx$ 3 eV), resulting in an increased pulse duration, $\tau_X \approx 0.9$ fs, as compared to recent experiments[5,8].

Both the X-ray and the light pulse are polarized along the axis of the time-of-flight electron energy spectrometer (Fig. 2b) to collect the electrons with a final momentum directed nearly parallel to the laser polarization. This parallel detection geometry is characterized by $\theta \approx 0°$ and 'activates' the last term in equation (1). A comparison

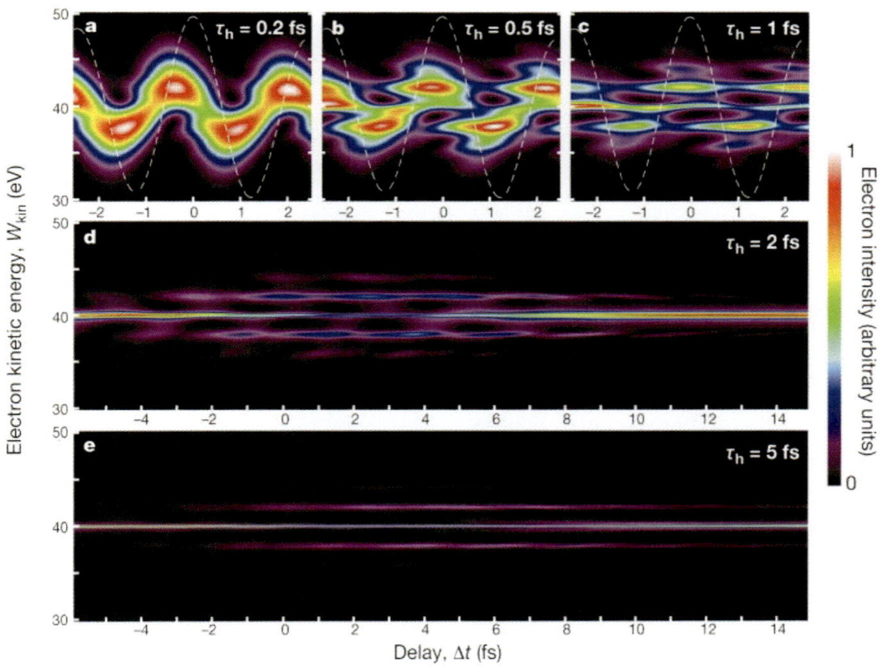

Figure 3 Laser sampling of Auger electron emission with attosecond resolution. Simulated electron spectra of a single isolated Auger line resulting from a hypothetical atomic core hole decay of different decay time constants τ_h in the presence of a few-cycle 750-nm laser field ($T_0 = 2.5$ fs) with $\varphi \approx 0$ using our quantum mechanical model (Supplementary Information). The spectra are shown in false-colour representation at small delays near $\Delta t = 0$ for sub-femtosecond decays (**a**, **b**, **c**) and over an extended delay range for decays lasting longer than the wave cycle (**d**, **e**). The dashed line represents the laser electric field. The durations of the exciting X-ray and probing laser pulse and the peak intensity of the latter were assumed to be $\tau_x = 0.5$ fs, $\tau_L = 5$ fs (FWHM) and $I_p = 5\times10^{11}$ W cm^{-2}, respectively.

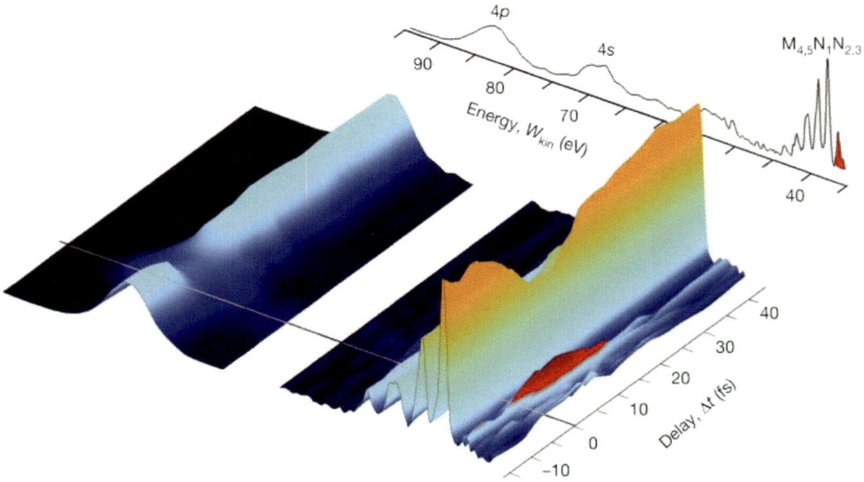

Figure 4 Evolution of electron spectra following core excitation. The spectra were recorded after the exposure of krypton atoms to a 97-eV sub-femtosecond X-ray pulse and a sub-7-fs, 750-nm laser pulse at different delays between the pulses. The surface plot shows the evolution of the 4*p* photo line and the $M_{4,5}N_1N_{2,3}$ Auger lines with Δt. The spectrum highlighted in the background was recorded at $\Delta t = 5$ fs. According to equation (1) the kinetic energy spectrum of the sub-laser-cycle-duration photo electrons (4*s* and 4*p* lines) should be periodically shifted versus Δt. However, for τ_X comparable to $T_0/2$ this periodic shift merges to a continuous broadening (probing the laser amplitude envelope $E_a(t)$) owing to the absence of stabilization of the absolute phase φ in our laser pulses, because light pulses with $\varphi \approx 0$ and $\varphi \approx \pi$ induce shifts of opposite sign and turn asymmetric shifts into symmetric broadening. The laser peak intensity was evaluated from the broadening of photo lines as $I_p \approx 5 \times 10^{11}\,\mathrm{W\,cm^{-2}}$. The evolution of the first-order sideband (highlighted in red) of the lowest-energy member of the $M_{4,5}N_1N_{2,3}$ Auger group reflects the delayed decay dynamics of the krypton 3*d* core hole. A pronounced positive temporal shift of the side-band maximum with respect to the photo-line minimum is clearly discernible (see also Fig. 5).

of the second and third terms in equation (1) reveals that, for $W_i \gg \hbar\omega_L$, this results in a detectable change of the electron energy at substantially lower light-field strengths compared to the orthogonal detection geometry ($\theta \approx 90°$). The low field strength helps to avoid both multiphoton ionization and a perturbation to the studied electronic relaxation. The experimental results are summarized in a surface plot compiled from 20 spectra recorded at a sequence of pump–probe delays Δt (defined in Fig. 2b), each one integrated over 300,000 laser shots and normalized to constant integral electron counts (Fig. 4). By definition, $\Delta t = 0$ at the coincidence of the peaks of the X-ray and laser pulses, which has been experimentally determined by observing the maximum broadening of the 4*p* and 4*s* photo lines.

The improved transmittivity and spectral resolution (~0.5 eV) of the time-of-flight spectrometer at lower kinetic energies favours the $M_{4,5}N_1N_{2,3}$ Auger group around 40 eV for an in-depth analysis. Before doing this we can make a few instructive observations. The photo lines are broadened smoothly (without the appearance of sidebands) in the range of temporal overlap between the laser and the X-ray pulse, indicating the sub-T_0 duration of the emitted electron wave packet (for more detailed comments, see legend of Fig. 4). In strong contrast, the Auger lines do not exhibit notable broadening. Instead, they are redistributed into sidebands spaced by $\hbar\omega_L \approx 1.6$ eV under the influence of the laser field. This is indicative of an Auger emission time comparable to or longer than T_0 according to our above discussion. The conspicuous suppression of the most prominent Auger peak and the simultaneous appearance of its first-order sideband (highlighted in red; other sidebands are hidden in the MNN group) indeed survive the broadening of the photo lines (for $\Delta t > 0$), confirming prolonged emission of the Auger electrons compared to that of photo-electrons.

For a quantitative analysis of the Auger emission we evaluated the sideband area A_{sb} (highlighted in red in Fig. 4) from the normalized electron spectra by combined gaussian peak-fitting to the whole $M_{4,5}N_1N_{2,3}$ group. The evolution of $A_{sb}(\Delta t)$ is shown by the circles in Fig. 5a. Figure 5b depicts the broadening $\delta w(\Delta t) = [\Delta w(\Delta t)^2 - \Delta w(\infty)^2]^{1/2}$ of the width Δw of the 4*p* photo line (circles) recorded

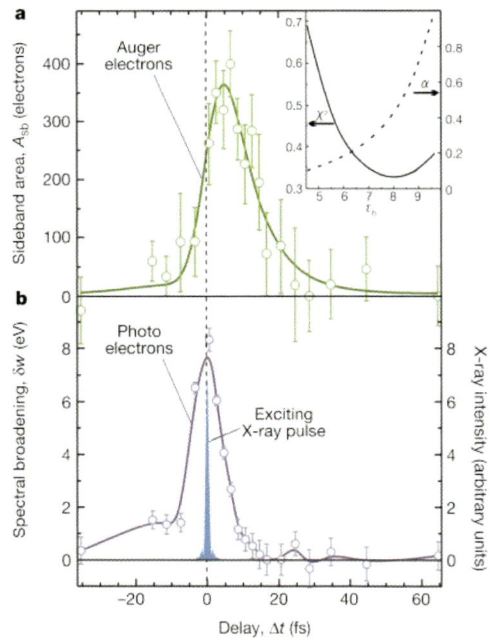

Figure 5 Probing the temporal evolution of Auger electron emission. **a**, Sideband area (circles) of the first-order sideband of the lowest-energy Kr $M_{4,5}N_1N_{2,3}$ Auger line extracted from the spectra of Fig. 4 and a fitted convolution of the exponential decay curve $R_0 \int \exp[-(t-t')/\tau_h] E_X^2(t')\mathrm{d}t'$ with a power 2α of the amplitude envelope of the laser field. Because $\tau_X \ll \tau_L$, the latter is proportional to the spectral broadening of the 4*p* photoline from the same electron spectrum, presented in **b**. Panel **b** also displays the temporal intensity profile of the X-ray pulse computed from its measured energy distribution (FWHM ≈ 3 eV) assuming the absence of chirp, which applies to cut-off harmonic radiation under our experimental conditions as was verified in a recent experiment[5]. Inset, the confidence band for τ_h is confined by the merit function χ^2 and the valid range of α to $\tau_h = 7.9^{+1.0}_{-0.9}$ fs.

at the same instants. For $\tau_{el} \ll \tau_L$ and $\delta w(0)^2 \gg \Delta w(\infty)^2$, according to equation (1), the broadening displays the amplitude envelope $E_a(t)$ of the light field with the leading edge of the pulse being sampled at positive delays $\Delta t > 0$. A spline fit to the measured data (line in Fig. 5b) yields a FWHM duration of the field envelope $E_a(t)$ of slightly less than 10 fs, in excellent agreement with the result of a simultaneous autocorrelation measurement of the laser pulse ($\tau_L \approx 7$ fs). From $\delta w(\Delta t)$ we can also determine the start of the Auger emission ($\Delta t = 0$) with an accuracy of ± 0.5 fs.

In the low-intensity (perturbative) limit, the transition rate into the first-order sideband is proportional to the instantaneous laser intensity $E_a^2(t)$. However, for increasing intensities, depletion of the main peak and transition from the first-order into the second-order sideband causes deviation from this scaling. A modified scaling law of the form $E_a^{2\alpha}(t)$ with $0 < \alpha < 1$ represents a convenient approximation[16]. We confirmed the validity of this model for the relevant intensity range (10^{11}–10^{12} W cm^{-2}) experimentally by measuring A_{sb} as a function of the laser-pulse energy at a fixed delay Δt. We obtained $\alpha = 0.5 \pm 0.2$, in close agreement with previous findings[16] about the effective sublinear scaling of A_{sb} with laser intensity. We can now obtain quantitative dynamical information about the decay of the krypton M-shell vacancies by fitting the convolution of the Auger emission rate $R_0 \int \exp[-(t - t')/\tau_h] E_X^2(t')\,dt'$ and the effective sampling function $E_a^{2\alpha}(t - \Delta t)$ to the measured sideband area $A_{sb}(\Delta t)$. With $\Delta t = 0$ and $E_a(t)$ known from the photoelectron data and $E_X^2(t)$ estimated (Fig. 5b) to be gaussian with $\tau_X = 0.9$ fs, the only fit parameter that remains to be determined is the lifetime of the core hole, τ_h.

To double check the validity of our effective sampling function we have computed the best fit for several values of α between 0 and 1. The inset in Fig. 5a shows the optimum value of τ_h as a function of α along with the corresponding value of the merit function χ^2. The best-quality fit (represented by the line in Fig. 5a) has been obtained for $\alpha \approx 0.4$, which is within the above confidence interval acquired from an independent measurement. This analysis yields $\tau_h = 7.9^{+1.0}_{-0.9}$ fs for the lifetime of $M(d_{5/2})$ vacancies in krypton. The measured lifetime yields an Auger linewidth of $\Gamma = \hbar/\tau_h = 84 \pm 10$ meV (not resolved by our spectrometer). This is in good accordance with 88 ± 4 meV, the result of recent energy-domain measurements[19]. The good agreement also indicates that post-collision interactions[21] have little, if any, influence on our time-resolved Auger spectra in spite of the near-threshold (97 eV) excitation.

Exploring atomic dynamics by time-domain studies

The exponential decay of krypton M-shell vacancies served as a benchmark process for testing the feasibility of time-resolved atomic inner-shell spectroscopy. The measured decay time can also be inferred from energy-domain measurements.

We believe we may nevertheless expect new insight into atomic dynamics from time-domain experiments. Whenever continuum states are involved in the atomic relaxation dynamics, several interfering relaxation pathways may connect the same initial and the same final state, resulting in a complex non-exponential temporal behaviour. For example, in the photoionization process, Fano-type[22] or shape[23] resonances are indicative of the outgoing electron being captured by its parent ion in a transient state for a short while. In the Auger process, post-collision interactions[21] tend to give rise to complications. Measuring the energy sweep and (non-exponential) envelope of the emitted electron wave packets with attosecond resolution will provide comprehensive insight into the transient states involved. Full exploration of the dynamical behaviour of multi-electron systems will therefore require time-domain spectroscopy to complement energy-domain measurements.

Atomic inner-shell processes have so far been studied only in the absence of external perturbations (other than the particle initiating the process). However, in high-field interactions atoms are often exposed to light electric-field strengths comparable to that of the effective Coulomb fields acting on inner-shell electrons. Atoms exposed to atomic-scale light fields (relevant in, for example, X-ray laser research) may exhibit a dynamical behaviour significantly different from that of unperturbed atoms. Because of the short-lived optical perturbation this can be investigated exclusively by time-domain techniques.

The evolution of time-resolved atomic physics, what might be dubbed attophysics, and its impact on other fields will—beyond experimental advances—critically depend on progress in the theoretical understanding of the dynamics of excited multi-electron systems and their interaction with strong fields. □

Received 5 August; accepted 20 September 2002; doi:10.1038/nature01143.

1. Töpler, A. Vibroskopische Beobachtungen über die Schwingungsphasen singender Flammen (der chemischen Harmonica) mit Benutzung des Schlierenapparates. *Ann. Phys. Chem.* **128,** 126–139 (1866).
2. Zewail, A. Femtochemistry: atomic-scale dynamics of the chemical bond (adapted from the Nobel Lecture). *J. Phys. Chem. A* **104,** 5660–5694 (2000).
3. Bromage, J. & Stroud, C. R. Jr Excitation of three-dimensionally localized atomic electron wave packet. *Phys. Rev. Lett.* **83,** 4963–4966 (1999).
4. Drescher, M. *et al.* X-ray pulses approaching the attosecond frontier. *Science* **291,** 1923–1927 (2001) published online 15 February 2001; 10.1126/science.1058561.
5. Hentschel, M. *et al.* Attosecond metrology. *Nature* **414,** 509–513 (2001).
6. Papadogiannis, N. A., Witzel, B., Kalpouzos, C. & Charalambidis, D. Observation of attosecond light localization in higher order harmonic generation. *Phys. Rev. Lett.* **83,** 4289–4292 (1999).
7. Paul, P. *et al.* Observation of a train of attosecond pulses from high harmonic generation. *Science* **292,** 1689–1692 (2001).
8. Kienberger, R. *et al.* Steering attosecond electron wave packets with light. *Science* **297,** 1144–1148 (2002) published online 11 July 2002; 10.1126/science.1073866.
9. Krausz, F. Attosecond spectroscopy comes of age. *Opt. Photon. News* **13,** 62–68 (May 2002).
10. Krause, M. O. Atomic radiative and radiationless yields for K and L shells. *J. Phys. Chem. Ref. Data* **8,** 307–328 (1979).
11. Itatani, J. *et al.* Attosecond streak camera. *Phys. Rev. Lett.* **88,** 173903 (2002).
12. Kitzler, M., Milosevic, N., Scrinzi, A., Krausz, F. & Brabec, T. Quantum theory of attosecond XUV pulse measurement by laser-dressed photoionization. *Phys. Rev. Lett.* **88,** 173904 (2002).
13. Freeman, R. R. & Bucksbaum, P. H. Investigation of above-threshold ionization using subpicosecond pulses. *J. Phys. B* **24,** 325–347 (1991).
14. Schins, J. M. *et al.* Observation of laser-assisted Auger decay in argon. *Phys. Rev. Lett.* **73,** 2180–2183 (1994).
15. Glover, T. E., Schoenlein, R. W., Chin, A. H. & Shank, C. V. Observation of laser assisted photoelectric effect and femtosecond high order harmonic radiation. *Phys. Rev. Lett.* **76,** 2468–2471 (1996).
16. Bouhal, A. *et al.* Cross-correlation measurement of femtosecond noncollinear high-order harmonics. *J. Opt. Soc. Am. B* **14,** 950–956 (1997).
17. Aksela, H., Aksela, S. & Pulkkinen, H. Correlation effects in $4s^0 4p^6$ and $4s^1 4p^5$ configurations of krypton studied by the M-NN Auger decay. *Phys. Rev. A* **30,** 2456–2461 (1984).
18. Carlson, T. A. *et al.* Angular distribution of ejected electrons in resonant Auger processes of Ar, Kr, and Xe. *Phys. Rev. A* **39,** 1170–1185 (1988).
19. Jurvansuu, M., Kivimäki, A. & Aksela, S. Inherent lifetime widths of Ar $2p^{-1}$, Kr $3d^{-1}$, Xe $3d^{-1}$, and Xe $4d^{-1}$ states. *Phys. Rev. A* **64,** 012502 (2001).
20. Schmidtke, B. *et al.* The Kr $M_{4,5}N_1N_{2,3}$ ($^1P_1$) Auger decay: measurement of the transferred spin polarization and analysis of Auger amplitudes. *J. Phys. B* **34,** 4293–4310 (2001).
21. Borst, M. & Schmidt, V. Vanishing postcollision interaction in inner-shell photoionization. *Phys. Rev. A* **33,** 4456–4458 (1986).
22. Fano, U. Effects of configuration interaction on intensities and phase shifts. *Phys. Rev.* **124,** 1866–1878 (1961).
23. Becker, U. *et al.* Subshell photoionization of Xe between 40 and 1000 eV. *Phys. Rev. A* **39,** 3902–3911 (1989).

Supplementary Information accompanies the paper on *Nature*'s website (**http://www.nature.com/nature**).

Acknowledgements Sponsored by the Fonds zur Förderung der wissenschaftlichen Forschung (Austria), the Deutsche Forschungsgemeinschaft (Germany) and by the European Atto Network.

Competing interests statement The authors declare that they have no competing financial interests.

Correspondence and requests for materials should be addressed to M.D. (e-mail: drescher@physik.uni-bielefeld.de) or F.K. (e-mail: ferenc.krausz@tuwien.ac.at).

RESEARCH ARTICLE

Steering Attosecond Electron Wave Packets with Light

R. Kienberger,[1] M. Hentschel,[1] M. Uiberacker,[1] Ch. Spielmann,[1,2] M. Kitzler,[1] A. Scrinzi,[1] M. Wieland,[3] Th. Westerwalbesloh,[4] U. Kleineberg,[4] U. Heinzmann,[4] M. Drescher,[4] F. Krausz[1]*

Photoelectrons excited by extreme ultraviolet or x-ray photons in the presence of a strong laser field generally suffer a spread of their energies due to the absorption and emission of laser photons. We demonstrate that if the emitted electron wave packet is temporally confined to a small fraction of the oscillation period of the interacting light wave, its energy spectrum can be up- or downshifted by many times the laser photon energy without substantial broadening. The light wave can accelerate or decelerate the electron's drift velocity, i.e., steer the electron wave packet like a classical particle. This capability strictly relies on a sub-femtosecond duration of the ionizing x-ray pulse and on its timing to the phase of the light wave with a similar accuracy, offering a simple and potentially single-shot diagnostic tool for attosecond pump-probe spectroscopy.

Interaction of intense laser light with a free electron in the vicinity of an atom has been found to give rise to a series of side bands in the electron energy spectrum (*1–4*). The side bands are a manifestation of the absorption and emission of a discrete number of photons by the electron. In the specific case of the free electron arising from photoionization by an extreme ultraviolet or x-ray photon, the electron energies are spread due to simultaneous absorption and emission of laser photons without causing any appreciable change in the electron's average drift velocity (*5–7*). We demonstrate that temporal confinement of the photoelectron wave packet to a small fraction of the half wave cycle of the interacting laser light, i.e., to less than 1 femtosecond (1 fs = 10^{-15} s), and its timing to the light phase with similar precision allow light to substantially change the electron's drift velocity.

The temporal structure of the photoelectron wave packet is determined by that of the ionizing radiation. We produce a soft x-ray (henceforth briefly x-ray) burst by high-order harmonic generation in an atomic gas (*8, 9*) and control its temporal extension by spectral bandpass filtering and by varying the intensity of the few-cycle 750-nm driver laser pulse (pulse duration: τ_L = 7 fs, wave period: T_0 = 2.5 fs). Selecting the highest photon-energy ("cut-off") radiation from the harmonic spectrum results in an isolated sub-femtosecond x-ray pulse (*9*). Using this x-ray pulse to ionize an atomic gas in the presence of a strong few-cycle light field, we demonstrate that the photoelectron spectrum can be up- and downshifted by many times the laser photon energy and by many times the time-averaged oscillatory energy of the electron in the laser field.

The shift of photoelectron energy by up to tens of electronvolts turns into a spread of electron energies over a similar range, if either the duration of the x-ray pulse or its timing jitter becomes comparable to or longer than $T_0/2$, which provides a simple and sensitive attosecond diagnostic tool (*10, 11*). Our experiments yielded evidence for isolated sub–0.5-fs x-ray pulses emerging with a timing stability to within ±0.2 fs with respect to the phase of the driving light from few-cycle–driven high harmonic generation. The x-ray pulse is found to emerge near the zero transition of the laser electric field. The light-induced energy shift of the photoelectron wave packet excited by this x-ray pulse scales linearly with the instantaneous electric field amplitude up to field strengths in excess of 10^8 V/cm, allowing direct and accurate measurement of the electric field strength of intense laser light as well as monitoring of the stability of the absolute phase of few-cycle light pulses. Previous attosecond time-resolved studies required measurement of electron energy spectra for a number of different delays between the x-ray pump pulse and a visible probe laser (*8, 9, 12*). By contrast, the approach presented here provides access to the key characteristics of an attosecond sampling (pump-probe) apparatus consisting of a sub-fs x-ray pulse and a few-cycle visible light pulse (*13*) without any external delay between the two pulses which can be implemented—in principle—in a single laser shot.

Strong Electron-Light Coupling

Energy exchange between light and an electron is most efficient if the electron is moving along the electric field vector of linearly polarized radiation. We studied the electrons emitted in this direction (*11*). In a semiclassical treatment, the electron is detached by absorption of an x-ray photon of energy $\hbar\omega_x$ from an atom and released with an initial energy of $W_0 = m_e v_i^2/2 = \hbar\omega_x - W_b$ in the direction of the electric field $E_L(t) = E_a(t)\cos(\omega_L t + \varphi)$ of the light pulse. Here, ω_x and ω_L stand for the x-ray and light frequency, respectively; W_b is the electron's atomic binding energy; $E_a(t)$ represents the amplitude envelope of the light pulse; and the phase φ adjusts the carrier wave's timing to the envelope, which has been referred to as absolute or carrier-envelope phase.

Solution of the classical equation of motion in the adiabatic limit, $dE_a(t)/dt \ll E_a(t)\omega_L$, and in the dipole approximation (the electron remains localized within a fraction of the wavelength during the interaction) yields the final velocity $v_f = v_i + [eE_a(t_r)/m_e\omega_L]\sin(\omega_L t_r + \varphi)$. Energy gain or loss is maximal for release times at which $E_L(t_r) = 0$, i.e., $\omega_L t_r + \varphi = \pi/2 + m\pi$, resulting in the minimum and maximum final electron energy of $W_f = W_0 + 2U_p(t_r) \pm \Delta W$ for m as an odd or even integer, respectively. Here, $\Delta W = [8W_0U_p(t_r)]^{1/2}$ and $U_p < W_0/2$ has been assumed. Furthermore, $U_p(t_r) = e^2E_a^2(t_r)/4m_e\omega_L^2$ is the electron's quiver energy averaged over an optical cycle and, hence, $\Delta W \propto E_a(t_r)$. If $U_p \ll W_0$, the energy shift is dominated by ΔW and is therefore linearly proportional to the laser electric field strength. For a large initial velocity along the laser polarization ($m_e v_i^2/2 \gg \hbar\omega_L$), this energy gain or loss may far exceed the laser photon energy even for moderate field strengths characterized by $U_p < \hbar\omega_L$. The momentum transfer associated with this energy exchange permits efficient steering of the electron's drift motion (*14*). Why could this light-control of electronic motion not be observed experimentally until now (*15*)? An obvious reason is the lack of sufficiently accurate and reproducible timing of the instant of release to the phase of the light field. If an x-ray pulse with a temporal extension comparable to or longer than T_0 excites the atom, the electron's release cannot be timed to any specific phase of the light field.

We have used a quantum mechanical treatment (*11*) to analyze photoelectron motion in the laser field for ionizing x-ray pulses with durations ranging between $\tau_x > T_0$ and $\tau_x \ll T_0$ and timed to ensure maximum energy gain from the light field in the limit of $\tau_x \ll T_0$ (Fig. 1A). The relevant parameters of the interaction are given in the caption of

[1]Institut für Photonik, Technische Universität Wien, Gusshausstr. 27, A-1040 Wien, Austria. [2]Physikalisches Institut EP1, Universität Würzburg, Am Hubland, D-97074 Würzburg, Germany. [3]Fachhochschule Koblenz, RheinAhrCampus Remagen, Südallee 2, D-53424 Remagen, Germany. [4]Fakultät für Physik, Universität Bielefeld, D-33615 Bielefeld, Germany.

*To whom correspondence should be addressed. E-mail: ferenc.krausz@tuwien.ac.at

RESEARCH ARTICLE

Fig. 1. The narrow spectral peaks that are spread over a range of $2\Delta W$ for $\tau_x \geq T_0$ gradually disappear as τ_x, the full width at half maximum (FWHM) of the intensity profile of the wave packet, is shortened to less than $T_0/2$. For $\tau_x < T_0/5$, they turn into a single smooth spectrum upshifted by ΔW. This behavior occurs regardless of whether temporal broadening of the electron wave packet is caused by spectral narrowing or by spectral phase modulation (Fig. 1, B and C, respectively). Strong electron-light coupling (high v_i parallel to E_L) along with sub-$T_0/5$ temporal confinement is predicted to give rise to a striking nonlinear optical effect: low-order transitions (absorption or emission of a single or a few photons) can be completely "switched off" upon simultaneous activation of a high-order transition (Fig. 2). We demonstrate this phenomenon experimentally and address its implications for attosecond metrology.

Controlling the Duration of Photoemission

The temporal extension of the electron wave packet can be affected by shaping the ionizing x-ray pulse. The briefest emission of extreme ultraviolet or x-ray radiation has so far been demonstrated by the generation of high-order harmonics of intense femtosecond light pulses in an atomic gas (*16*, *17*). High-harmonic radiation driven with sub–100-fs laser pulses typically lasts a few tens of femtoseconds (*18*, *19*) and exhibits a quasi-periodic substructure of attosecond spikes at the highest (cut-off) harmonic photon energies (*12*, *20*). The spectrum of this radiation consists of discrete lines at odd multiples of the laser frequency as a spectral manifestation of temporal periodicity with a period of $T_0/2$. In the absence of a notable frequency dependence of the spectral phase, the overall emission time is related to the line width of the individual harmonics, whereas the duration of individual spikes is determined by the overall spectral width of the band of lines filtered out of the harmonic spectrum. The cut-off harmonics of a few-cycle light pulse were predicted to merge into a continuum, forming an isolated sub-femtosecond pulse (*21–24*). This was confirmed in a recent experiment (*9*). We demonstrate that the transition from a quasi-periodic emission with a temporal extension $\geq T_0$ to a single-pulsed sub-femtosecond emission can be continuously controlled with the pump intensity in a few-cycle–driven high-harmonic source.

Figure 3 depicts high-harmonic spectra emitted from a neon gas source pumped by 7-fs, 750-nm linearly polarized laser pulses (*8*, *9*, *25*) for a range of laser peak intensities. For the highest peak intensity the spectrum consists of well-resolved harmonic lines. Their line width implies an emission time of $\geq T_0$, i.e., the appearance of at least a couple of satellite pulses spaced by $T_0/2$ from the main pulse. Reducing the pump intensity results in harmonics merging to a continuum within the high-reflectance band (85 to 100 eV) of our Mo/Si multilayer (dashed line in Fig. 3), implying emission in a single pulse of potentially sub-femtosecond duration. The continuum portion of the few-cycle–generated harmonic spectrum is predicted to originate from driver pulses with $\varphi \approx 0$ or π, which give rise to a substantially higher peak tunneling current and, hence, dipole acceleration in the cut-off region as compared to $\varphi \approx \pm\pi/2$ for laser pulse durations approaching T_0 (*13*, *23*, *24*).

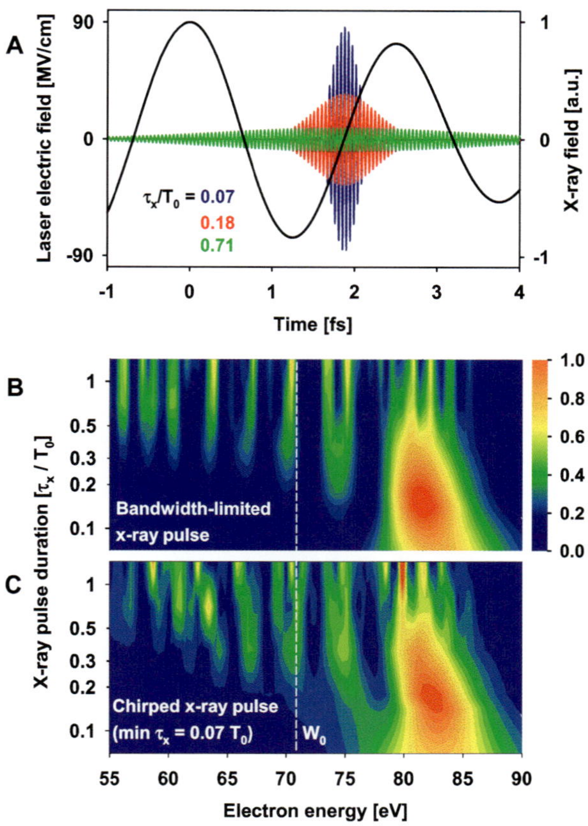

Fig. 1. Computed energy distributions of 2p photoelectrons detached from neon atoms (W_b = 21.5 eV) by a 93-eV, soft x-ray pulse in the presence of a strong 7-fs, 750-nm few-cycle light pulse for different x-ray pulse durations τ_x (contour plots; the electron yield increases from the blue to the red) using the quantum theory presented in (*11*). The electrons are detected within an infinitesimally small cone with its axis aligned parallel to the electric field vectors of the linearly polarized x-ray and laser fields (**A**). The x-ray pulse is timed to peak at the zero transition of the laser electric field. The duration of the Gaussian x-ray pulse has been varied by changing the spectral width (**B**) or by imposing a quadratic phase of variable magnitude on the spectrum (**C**). The white dashed line at $W_0 = \hbar\omega_x - W_b \approx 71$ eV indicates the position of the peak of the photoelectron spectrum in the absence of the laser field.

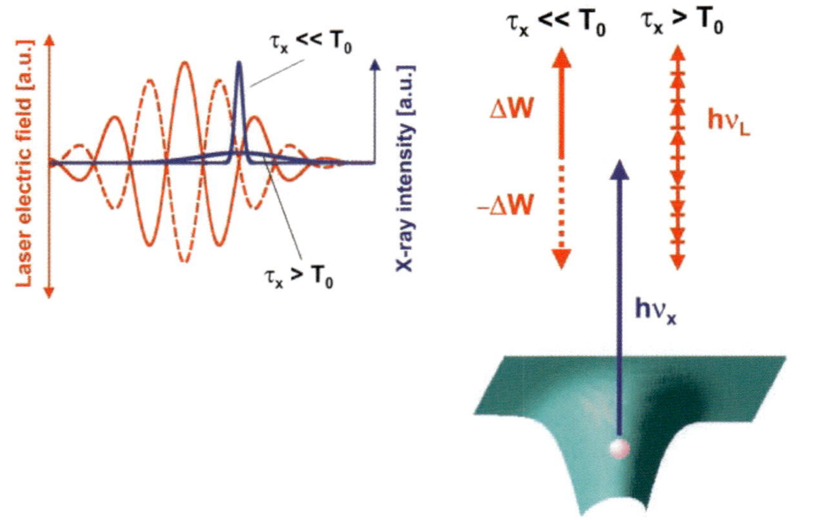

Fig. 2. Influence of a strong, linearly polarized laser field on the final energy distribution of x-ray–induced photoelectrons detected along the laser polarization as a function of the x-ray pulse duration τ_x relative to the wave cycle of the laser T_0.

RESEARCH ARTICLE

Acceleration of Electron Wave Packets with Light

The harmonic beam and its pump laser beam propagate collinearly down a 2-m-long evacuated beamline to hit a concave spherical Mo/Si multilayer mirror (radius of curvature, 7 cm) at near normal incidence. The mirror has high reflectivity within a 15-eV band centered at 93 eV (see dashed line in Fig. 3) and over the entire spectrum of the sub–10-fs laser pulse, and it focuses the two collinear beams into a neon gas jet effusing from a metal-coated glass tube (inner diameter, 50 μm). The laser pulse intensity was restricted to values insufficient for laser-induced ionization. The photoelectrons produced by x-ray photons were collected within a narrow cone aligned parallel to the laser and x-ray polarization. The detection cone was defined by the 5-cm-diameter microsphere-plate detector placed 40 cm from the neon target to allow a time-of-flight analysis of the photoelectrons. The x-ray beam spot size in the target was more than an order of magnitude smaller than that of the laser beam, ensuring that all the x-ray photoelectrons were exposed to the same light field.

Before hitting the Mo/Si mirror, the beams are passed through an aperture of adjustable diameter (iris), which was used for fine adjustment of the laser pulse intensity in the neon target. The iris selected near-axis rays within a diameter of ≤3 mm of the laser beam of several centimeters in diameter. The transmitted laser beam had a nearly flat-top intensity profile, implying a quadratic and linear scaling of focused intensity and electric field E_L, respectively, with transmitted pulse energy in the photoionization region. By varying the iris diameter and monitoring the transmitted pulse energy, we could therefore accurately control relative changes of E_L in the interaction volume.

Figures 4 and 5 display electron energy spectra as a function of the measured relative change of E_L in a false-color representation. In the absence of a light field, the photoelectron energy spectrum (lowest plot on the right of Fig. 5) mimics that of the x-ray pulse downshifted by the ionization potential of neon (W_b = 21.5 eV). Figure 4 reveals a spread of electron energies rapidly increasing with increasing laser field strength when photoionization is performed using a train of x-ray pulses with spectrum a in Fig. 3. The data depicted in Fig. 5 have been collected with x-ray pulses having a continuous spectrum. In contrast to the behavior shown in Fig. 4, with increasing field strength the electron energy spectrum is upshifted as a whole without a dramatic distortion of its shape. This indicates that the x-ray pulse is of sub-femtosecond duration and is emitted at the zero transition of the driving laser electric field from the harmonic source (*26*), in accordance with numerical simulations (*9*). The simultaneous appearance of an up- and down-shifted spectral feature is the consequence of pump pulses with $\varphi \approx 0$ and $\varphi \approx \pi$ contributing to the electron spectra accumulated over many laser shots, i.e., of the lack of phase stabilization of our sub–10-fs laser pulses.

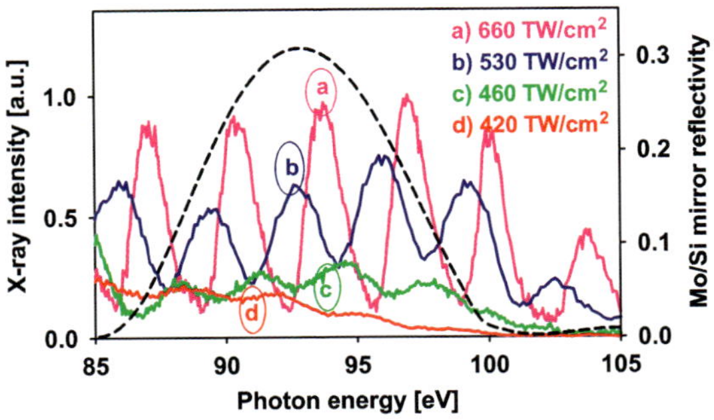

Fig. 3. High-order harmonic spectra emitted from a 2-mm-long, 110-mbar neon gas source exposed to linearly polarized 7-fs, 750-nm laser pulses with peak intensities given in the figure. Dashed line, reflectance curve of the broadband Mo/Si multilayer mirror used to select radiation for the photoionization experiments, as calculated using data from in situ measurements of the roughness and thickness of the Mo and Si layers. The reflectivity band extends over some 15 eV with a FWHM of ≈ 9 eV.

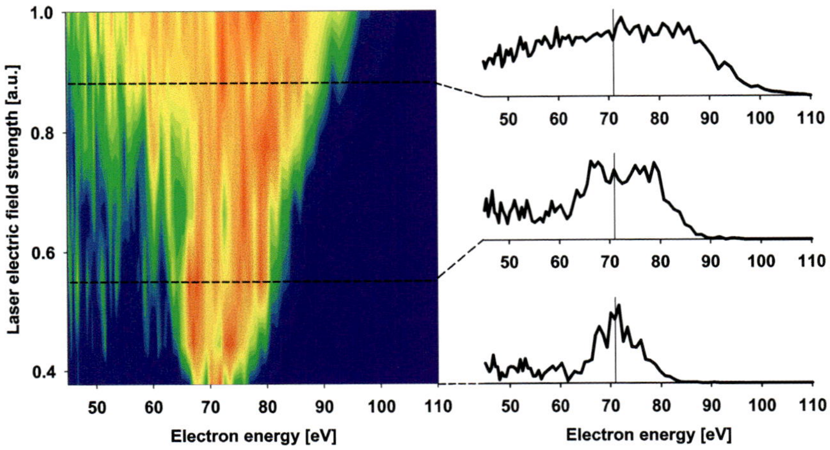

Fig. 4. Photoelectron spectra generated at different light field strengths by a sequence of sub-femtosecond soft x-ray pulses. The peak intensity of the laser pulse driving the harmonics was adjusted to 650 to 700 TW/cm² to produce a discrete harmonic spectrum (see Fig. 3) within the range of 85 to 100 eV with a duration of the pulse train of $\geq T_0$.

Figure 5 reveals the linear dependence of the energy shift on E_L, corroborating the prediction of our simple classical analysis. The measured spectra can be well reproduced by simulating the electron wave packet as a classical ensemble of electrons distributed across initial energies and release times whose motions are governed by the laws of classical physics. Approximating the initial energy distribution with a Gaussian function (bottom red line in Fig. 5) and assuming a bandwidth-limited x-ray pulse ($\tau_x \approx 0.2$ fs) that peaks at $E_L(t_r) = 0$, the classical and the quantum analysis delivers results (central and upper red lines in Fig. 5) in good agreement with the spectra measured for different strengths of $E_a(t_r)$. Attosecond confinement appears to mask quantum effects in the interaction of electrons with light.

Attosecond Diagnostics

The generation of isolated sub-femtosecond x-ray pulses synchronized to a strong few-cycle light pulse (*9*) opens up a route to time-resolved (pump-probe) inner-shell atomic spectroscopy by using the x-ray pulse as a pump and the light pulse as a probe (*13*). Indispensable to the routine application of this new tool is a reliable, simple, and fast diagnostic test of its key characteristics. These include the strength and phase-stability of the few-cycle probe field, the timing jitter between the x-ray pulse and light wave, and the duration of the x-ray pulse. The sensitivity of the energy spectrum of the x-ray–induced, light-accelerated photoelectrons to these parameters makes it an ideal diagnostic tool for attosecond spectroscopy.

Strength and phase stability of the few-cycle probe light field. Because of the favorable timing of the sub-femtosecond harmonic pulse to its few-cycle driver, which satisfies $E_L(t_r) \approx 0$, the photoelectron energy shift in Fig. 5 directly probes the instantaneous electric field amplitude $E_a(t_r)$ with an accuracy of better than 10%. This provides direct experimental access to the peak electric field strength of an intense light pulse. Moreover, the light-accelerated electron

RESEARCH ARTICLE

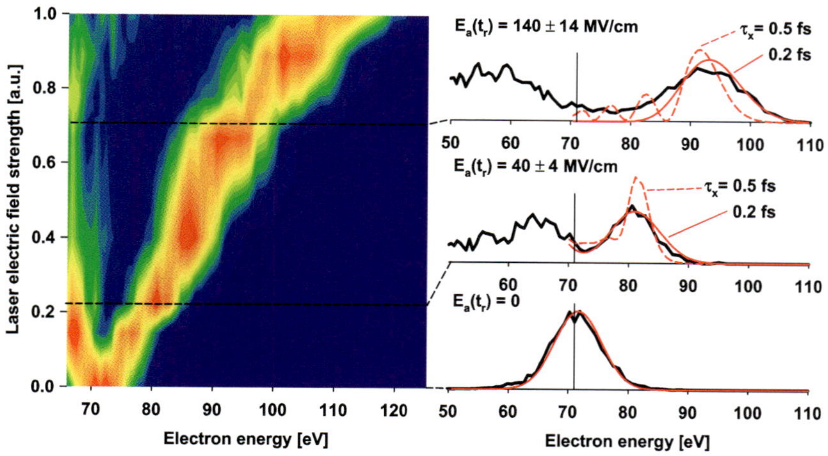

Fig. 5. Photoelectron spectra generated at different light field strengths by an isolated sub-femtosecond x-ray pulse. The peak intensity of the laser pulse driving the harmonics was adjusted in the range of 400 to 450 TW/cm² to generate a continuous harmonic spectrum in the 85- to 100-eV spectral range (Fig. 3). $E_L(t_r)$ has been evaluated from the observed energy shift. The bottom red line is the model x-ray spectrum (downshifted by the electron binding energy of 21.5 eV) used in the simulations resulting in the central and upper red lines for bandwidth-limited 0.2-fs (solid lines) and 0.5-fs soft x-ray pulses carrying a quadratic spectral phase (dashed lines).

proves a sensitive probe of the absolute-phase stability of intense few-cycle light. If φ is firmly locked to zero, the photoelectrons would be exclusively accelerated (with the convention of $E_L > 0$ implying a field vector that points toward the electron detector). This would imply a disappearance of the downshifted spectral feature, offering a reliable diagnostics of the phase stability of intense few-cycle pulses (*27*) for future phase-stabilized systems.

Timing jitter between the x-ray pulse and the probe light phase. Jitter affects the temporal resolution to the same extent as τ_x and T_0 (*13*); therefore, its measurement is imperative for attosecond spectroscopy. The periodic harmonic structure can be perceived in several downshifted spectra (*28*), e.g., in the central plot in Fig. 5 (the structure tends to get blurred in the upshifted features because of time-of-flight resolution worsening for increasing electron energy). Resolving the harmonic structure in energy-shifted spectra results in an upper limit of ±10% of the energy shift on smearing of the electron spectrum (accumulated over many laser shots) due to fluctuations in $E_L(t_r)$ by either laser intensity variations or x-ray pulse timing jitter. This yields a shot-to-shot timing stability of the x-ray pulse to the phase of its few-cycle driver of better than ±0.2 fs (*29*).

X-ray pulse duration. Figures 4 and 5 confirm that the energy spectrum of the light-accelerated x-ray photoelectron responds sensitively to temporal broadening of the ionizing x-ray pulse. Although a conspicuous energy spread is predicted to come into play only as τ_x becomes longer than $T_0/3$ (Fig. 1C), a considerable narrowing of the energy-shifted spectral feature already sets in for $\tau_x > T_0/10$. We calculated the light-accelerated photoelectron spectra for chirped x-ray pulses by assuming a quadratic as well as higher-order spectral phase. The root-mean-square deviation of the calculated spectra from the measured ones was a minimum for near-bandwidth–limited pulses of $\tau_x \approx 0.2$ fs (solid red lines in Fig. 5) and was increased by more than a factor of three for chirped pulses with $\tau_x \geq 0.5$ fs (*30*). Considering the minor role of spectral smearing effects, our analysis yields a safe upper limit of $\tau_x \leq 0.5$ fs for the x-ray pulse duration.

Conclusions

Photoelectron wave packets excited by an x-ray pulse within a time interval of several hundred attoseconds can be efficiently accelerated or decelerated by the electric field of light. The sensitivity of the final energy distribution of the electron to (i) the strength and (φ dependent) sub-cycle evolution of the light field, (ii) the timing of the x-ray pulse with respect to the light phase, and (iii) the sub-femtosecond temporal structure of the x-ray pulse permits reliable measurement of the key parameters of an attosecond pump-probe apparatus based on a sub-femtosecond x-ray pulse synchronized to a strong few-cycle light wave (*13*). These measurements consume orders of magnitude less time and can be performed with a much simpler apparatus as compared to the previous implementation of time-domain attosecond metrology (*9*). The substantially enhanced electron yield and larger variation of the electron spectrum owing to detection parallel to the x-ray and laser polarization (*11*), respectively, hold out the promise of sub–100-as resolution in the measurement of the x-ray pulse duration and timing jitter. However, the potential of this attosecond diagnostics can be fully exploited only with phase-stabilized few-cycle light. With φ stabilized, both the fractional energy carried by possible satellites (*11*) and the chirp (*10*) of the sub-femtosecond x-ray pulse can be determined. Accurate chirp measurements will be a key to the reliable generation of sub–100-as x-ray pulses.

References and Notes

1. A. Weingartshofer, J. K. Holmes, G. Caudle, E. M. Clarke, H. Krüger, *Phys. Rev. Lett.* **39**, 269 (1977).
2. R. R. Freeman, P. H. Bucksbaum, *J. Phys. B.* **24**, 325 (1991).
3. J. M. Schins *et al.*, *Phys. Rev. Lett.* **73**, 2180 (1994).
4. T. E. Glover, R. W. Schoenlein, A. H. Chin, C. V. Shank, *Phys. Rev. Lett.* **76**, 2468 (1996).
5. This applies in the nonrelativistic limit for interaction times short enough to prevent the electron from traveling distances comparable to or larger than the light wavelength during the interaction (dipole approximation). These conditions are fulfilled by femtosecond pulses of visible light at intensities below 10^{18} W/cm². In this intensity regime of interactions, ponderomotive effects provide the only means of affecting electron motion by light. For ultrashort (<1 ps) pulse durations, electrons released into a strong field lose energy equal to the ponderomotive potential of the laser, which is transferred into a blue shift of the radiation (*6*). The ponderomotive potential of long pulses, on the other hand, is capable of modifying the momentum of free electrons (*7*).
6. E. Yablonovits, *Phys. Rev. Lett.* **60**, 795 (1988).
7. P. H. Bucksbaum, M. Bashkansky, T. J. McIlrath, *Phys. Rev. Lett.* **58**, 439 (1987).
8. M. Drescher *et al.*, *Science* **291**, 1923 (2001).
9. M. Hentschel *et al.*, *Nature* **414**, 509 (2001).
10. J. Itatani *et al.*, *Phys. Rev. Lett.* **88**, 173903 (2002).
11. M. Kitzler, N. Milosevic, A. Scrinzi, F. Krausz, T. Brabec, *Phys. Rev. Lett.* **88**, 173904 (2002).
12. P. M. Paul *et al.*, *Science* **292**, 1689 (2001).
13. F. Krausz, *Opt. Photon* **13**, 62 (2002).
14. In our detection geometry, only electrons that are ejected with zero velocity in the direction orthogonal to the detection cone can be observed. These electrons experience only acceleration or deceleration by the field without any change in the direction of their momentum. However, electrons with nonzero transversal velocity component are also ejected. These electrons are indeed steered because the laser field changes the direction of their motion. They could be observed by aligning the detection cone so that it forms a nonzero angle with the laser electric field vector.
15. Control of the electron's drift velocity could be previously accomplished by using a microwave field [D. A. Tate, D. G. Papaioannou, T. F. Gallagher, *Phys. Rev. A* **42**, 5703 (1990)].
16. A. L'Huillier, P. Balcou, *Phys. Rev. Lett.* **70**, 774 (1993).
17. J. J. Macklin, J. D. Kmetec, C. L. Gordon III, *Phys. Rev. Lett.* **70**, 766 (1993).
18. Y. Kobayashi, T. Sekikawa, Y. Nabekawa, S. Watanabe, *Opt. Lett.* **23**, 64 (1998).
19. E. S. Toma *et al.*, *Phys. Rev. A* **62**, 061801(R) (2000).
20. N. A. Papadogiannis, B. Witzel, C. Kalpouzos, D. Charalambidis, *Phys. Rev. Lett.* **83**, 4289 (1999).
21. I. P. Christov, M. M. Murnane, H. C. Kapteyn, *Phys. Rev. Lett.* **78**, 1251 (1997).
22. C. Kan, N. H. Burnett, C. E. Capjack, R. Rankin, *Phys. Rev. Lett.* **79**, 2971 (1997).
23. A. de Bohan, P. Antoine, D. B. Milosevic, B. Piraux, *Phys. Rev. Lett.* **81**, 1837 (1998).
24. T. Brabec, F. Krausz, *Rev. Mod. Phys.* **72**, 545 (2000).
25. M. Schnürer *et al.*, *Phys. Rev. Lett.* **83**, 722 (1999).
26. The harmonic spikes in the train corresponding to spectrum a in Fig. 3 are also predicted to emerge near the zero transition of the driving laser electric field. As a consequence, only an up- and downshifted spectral feature would be expected, if the spikes were not longer than $T_0/5$ and the probe field $E_L(t)$ would be perfectly periodic. A blue shift near the laser pulse peak [induced by ionization in the harmonic generation process; see (*9*)], however, impairs synchronism between the x-ray pulse sequence and the zero transitions of the probe laser field. In addition, the spikes

RESEARCH ARTICLE

are predicted to carry a substantial chirp and, hence, to be broadened in the "plateau" region.

27. G. G. Paulus *et al.*, *Nature* **414**, 182 (2001).
28. Note that tiny variations in the peak laser intensity can make the harmonic structure more or less pronounced as revealed by spectra c and d in Fig. 3. However, the modulation depth was kept below 25% in all measurements shown in Fig. 5, resulting in negligible fractional satellite energy content.
29. It is important to emphasize that this timing stability may not be valid for arbitrary values of the absolute phase φ of the driver pulse because laser shots characterized by a φ considerably deviating from 0 or π are expected to deliver only a small contribution to energy-shifted spectral features. As a consequence, the evaluated upper limit on the timing jitter applies to laser pulses with $\varphi \approx 0$ or π only.
30. This finding is in accordance with our recent measurement (*9*) resulting in almost zero phase distortion over the 5-eV bandwidth (FWHM) of the sub-femtosecond harmonic pulse generated under similar experimental conditions in the same spectral range.
31. The theory used extensively in this work (*11*) was developed under the guidance of T. Brabec. The transmission-grating x-ray spectrograph was provided by T. Wilhein and by G. Schmahl. The computer-controlled data acquisition system was significantly improved by M. Hornung. We had illuminating discussions with G. Reider. All contributions are gratefully acknowledged. Sponsored by the Fonds zur Förderung der Wissenschaftlichen Forschung in Österreich (Austria, grants Y44-PHY and F016), the Deutsche Forschungsgemeinschaft (Germany, grants SPP1053, HE1049/9, and KL1077/1), and the European Union's Human Potential Programme under contract HPRN-2000-00133 (Atto).

13 May 2002; accepted 1 July 2002
Published online 11 July 2002;
10.1126/science.1073866
Include this information when citing this paper.

REPORTS

Imaging Quasiparticle Interference in $Bi_2Sr_2CaCu_2O_{8+\delta}$

J. E. Hoffman,[1] K. McElroy,[1] D.-H. Lee,[1,6] K. M Lang,[1,2] H. Eisaki,[3,4,5] S. Uchida,[5] J. C. Davis[1,6]*

Scanning tunneling spectroscopy of the high-T_c superconductor $Bi_2Sr_2CaCu_2O_{8+\delta}$ reveals weak, incommensurate, spatial modulations in the tunneling conductance. Images of these energy-dependent modulations are Fourier analyzed to yield the dispersion of their wavevectors. Comparison of the dispersions with photoemission spectroscopy data indicates that quasiparticle interference, due to elastic scattering between characteristic regions of momentum-space, provides a consistent explanation for the conductance modulations, without appeal to another order parameter. These results refocus attention on quasiparticle scattering processes as potential explanations for other incommensurate phenomena in the cuprates. The momentum-resolved tunneling spectroscopy demonstrated here also provides a new technique with which to study quasiparticles in correlated materials.

In an ideal metal, the Landau-quasiparticle eigenstates are Bloch wavefunctions characterized by wavevector $\vec{k}$ and energy ε. Their dispersion relation, $\varepsilon(\vec{k})$, can be measured with momentum-resolved techniques such as angle-resolved photoemission spectroscopy (ARPES). By contrast, real space imaging techniques, such as scanning tunneling microscopy (STM), cannot be used to measure $\varepsilon(\vec{k})$. This is because the local-density-of-states LDOS(E) spectrum at a single location $\vec{r}$ is related to the $\vec{k}$-space eigenstates $\psi_k(\vec{r})$ by

$$\mathrm{LDOS}(E,\vec{r}) \propto \sum_k |\psi_k(\vec{r})|^2\delta(E-\varepsilon(\vec{k})) \quad (1)$$

and substitution of a Bloch wavefunction into Eq. 1 shows LDOS(E) to be spatially uniform.

When sources of disorder such as impurities or crystal defects are present, elastic scattering mixes eigenstates that have different $\vec{k}$ but are located on the same quasiparticle contour of constant energy (CCE) in $\vec{k}$-space. When scattering mixes states $\vec{k}_1$ and $\vec{k}_2$, an interference pattern with wavevector $\vec{q} = \vec{k}_2 - \vec{k}_1$ appears in the norm of the quasiparticle wavefunction and LDOS modulations with wavelength $\lambda = 2\pi/|\vec{q}|$ appear. These phenomena can be observed by STM as modulations of the differential tunneling conductance, which are often (imprecisely) referred to as "Friedel oscillations." STM studies of such conductance modulations have allowed the first direct probes of the quantum interference of electronic eigenstates in metals and semiconductors (*1–5*).

The Bogoliubov-quasiparticles in a Bardeen-Cooper-Schrieffer superconductor are also Bloch states but with dispersion

$$E_\pm(\vec{k}) = \pm\sqrt{\varepsilon(\vec{k})^2 + \Delta(\vec{k})^2} \quad (2)$$

where $|\Delta(\vec{k})|$ is the $\vec{k}$-dependent magnitude of the energy gap at the Fermi surface (CCE for $\varepsilon(\vec{k}) = 0$ in the normal state). Elastic scattering of Bogoliubov-quasiparticles can also result in conductance modulations. Quasiparticle standing waves have been imaged by STM in the conventional superconductor Nb (*6*) and in the CuO chains of YBCO (*7*). For the cuprates in general, it has long been proposed that conductance modulations due to quasiparticle scattering should occur, and that both the homogeneous electronic structure and superconducting gap anisotropy could be extracted from measurement of their properties (*8*).

When LDOS modulations are detected by STM in a given sample, certain CCE in $\vec{k}$-space can be reconstructed by analyzing the Fourier transform of the real-space LDOS(E) image (*3, 9, 10*). This is potentially a powerful technique because, for any

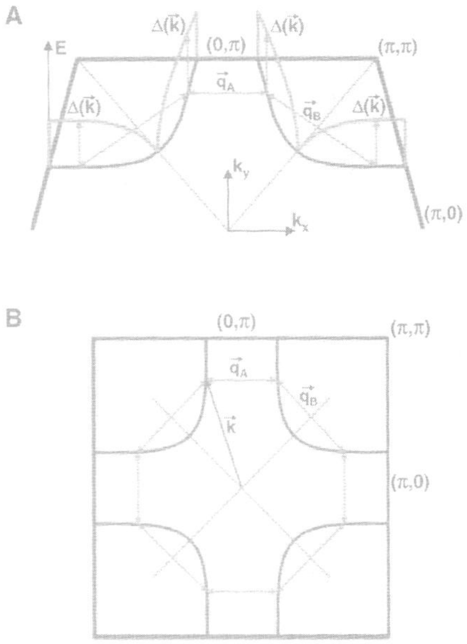

Fig. 1. (A) A perspective view of the superconducting energy gap Δ (green) as a function of the location along the Fermi surface (black). (B) Schematic Fermi surface of Bi-2212. Vectors connecting the eight areas of the Fermi surface with identical $|\Delta|$ are shown and labeled by blue and red arrows depending on the type of elastic scattering process at $E = |\Delta|$ that connects them.

[1]Department of Physics, University of California, Berkeley, CA 94720–7300, USA. [2]Electromagnetic Technology Division, National Institute of Standards and Technology, Boulder, CO 80305–3328, USA. [3]Department of Applied Physics, Stanford University, Stanford, CA 94305–4090, USA. [4]Agency of Industrial Science and Technology, 1-1-1 Central 2, Umezono, Tsukuba, Ibaraki, 305-8568 Japan. [5]Department of Superconductivity, University of Tokyo, Tokyo, 113-8656 Japan. [6]Materials Sciences Division, Lawrence Berkeley National Laboratory, Berkeley, CA 94720, USA.

*To whom correspondence should be addressed. E-mail: jcdavis@socrates.berkeley.edu

news and views

Laser science

Physics at the attosecond frontier

Yaron Silberberg

Ultrashort laser pulses allow physicists and chemists to watch fast molecular motion as it happens. But many fundamental atomic processes are even faster and require the shortest pulses ever created.

As every photographer knows, a flash of light can stop the action. Just as a fast flash lamp can freeze the image of a bullet in mid-flight, so short laser pulses can be used to probe fast molecular motion. It is no surprise, then, that laser scientists have been pushing for ever-shorter pulses of light in order to follow ever-more rapid processes. The quest has taken us from the first sub-picosecond (1 picosecond is 10^{-12} s) pulses, more than a generation ago, to the development of femtosecond optics (1 femtosecond is 10^{-15} s). These time periods are hard to imagine, but a femtosecond is to a minute what a minute is to the age of the Universe.

Femtosecond pulses led to femtochemistry, the experimental study of fast chemical reactions and molecular dynamics. Even the fastest molecular vibrations appear completely still when probed with a pulse lasting a few femtoseconds. Now, we are entering the era of attosecond pulses (1 attosecond is 10^{-18} s). On page 509 of this issue Hentschel *et al.* describe[1] the generation and use of pulses lasting 650 attoseconds, in what might be the dawn of attophysics — the study of dynamics on timescales fast enough to follow electronic motion within atoms.

The road from picosecond to femtosecond light pulses has seen laser technology evolve towards lasers that emit light with greater 'spectral' width — that is, covering a wider range of wavelengths. A short pulse results when all the spectral components in the light beam interfere in a way that adds up to a single burst of light. The duration of this pulse is inversely proportional to the spectral width — so a wider spectral

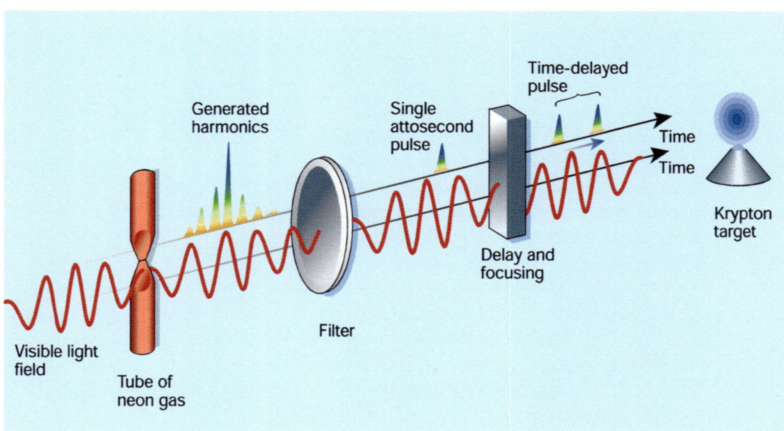

Figure 1 Generating and using attosecond laser pulses. In their experiment, Hentschel *et al.*[1] shine an ultrashort visible light pulse on a gas of neon atoms to produce higher frequency 'harmonic' radiation (rainbow-coloured pulses) at ultraviolet and X-ray wavelengths. The visible and harmonic pulses pass together through a zirconium filter, which reduces the train of harmonic pulses to a single attosecond pulse (1 attosecond is 10^{-18} s). Both the attosecond pulse and the optical beam are focused onto the krypton target where they accomplish the very first attosecond measurement. The authors monitor the attosecond dynamics of photoelectrons emitted by the krypton gas by varying the delay between the attosecond and visible pulses.

span means a shorter pulse. This approach reached its natural limit when the spectral span of the laser became a significant portion of the visible spectrum. This is because the pulse cannot be shorter than the period of the optical wave, which is a few femtoseconds in the visible range. So it was obvious that a different approach would be needed to break into the attosecond range.

The latest advances are based on research into 'high-order harmonics' of femtosecond laser pulses. Nearly a decade ago[2,3] physicists used intense femtosecond pulses to ionize a rare gas (such as neon) and found that new electromagnetic waves were generated at much higher frequencies (and shorter wavelengths) alongside the original optical pulse (at visible and short infrared wavelengths). It is fairly straightforward to produce these high-order harmonics, whose wavelengths extend well into the ultraviolet and X-ray range, but measuring and controlling them to produce ultrashort pulses is much harder.

Precise analysis of the way the gas atoms interact with the laser field requires careful application of quantum mechanics, but a simple model developed in 1993 by Paul Corkum[4] (one of the authors of the Hentschel *et al.* paper) provides some insight. In this model, the original laser pulse tears an electron away from an atom, and the freed electron moves in response to the laser field, being accelerated and decelerated as the electromagnetic wave oscillates. The new harmonics are generated when the electron collides with the ion it left behind. This radiation is termed harmonic because its frequencies are multiples of the original laser frequency.

Corkum argued that electrons that are ejected precisely at the peaks or the crests of the optical pulse are much more likely to radiate. So the radiation is produced in very short bursts, which occupy just a fraction of the optical cycle. These short bursts explain the much wider spectral span of the high-order harmonics. The bursts are estimated to last some 100 attoseconds, and are expected to appear twice in each optical cycle of the original pulse. Last year, two experiments confirmed that the duration of these pulses is in the attosecond range[5,6].

Attosecond pulses produced in this way appear as a train of pulses separated from each other by half an optical wavelength. However, most experiments require isolated pulses. The key to isolating a single attosecond pulse is to start with a very short pulse of just a few optical periods. Such a pulse produces only a few attosecond bursts, and it is possible to select just one by appropriate filtering (Fig. 1). So Hentschel *et al.*[1] start with a 7-femtosecond optical pulse and estimate that, after filtering, more than 90% of the energy of the new radiation they produce is contained in a single 650-attosecond pulse.

But that's not all. After generating an isolated attosecond pulse, Hentschel *et al.* shine it, together with the optical pulse that generated it, onto krypton gas, to perform the first attosecond measurements of an electronic process. While varying the delay between the two pulses with attosecond accuracy they monitor the energy spectrum of any photo-electrons ejected from the gas. (Changing the difference between the paths taken by the two pulses by 0.3 micrometres changes the delay between the two pulses by 1 femtosecond, so it is easy to control the pulse timing in this way.) With this measurement Hentschel *et al.* achieve three important things: they characterize the attosecond pulse; they measure the detailed structure of the optical field; and they perform the first sub-femtosecond measurement of electron dynamics associated with photo-ionization.

What sort of experiments are possible with attosecond pulses? Should we expect an explosion of new results? At these timescales, chemistry is essentially frozen in time, so the only dynamics to be studied are those of electrons, as they are much lighter and faster than nuclei. Most femtochemistry measurements are based on the pump–probe method: one pulse sets the system into motion, and a second one probes it after a controlled delay. But this approach is not possible with attosecond pulses: they are just too weak. Moreover, manipulating such pump and probe pulses is harder to do because their wavelength is in the X-ray range, for which traditional optical tools (such as lenses and beamsplitters) do not exist.

The work of Hentschel *et al.* hints at the likely direction that attophysics will take in the near future. They achieve attosecond measurements by carefully timing the delay between an attosecond pulse and an optical pulse. Yet the experiment also demonstrates the difficulties associated with this approach, because the information in the attosecond pulse, the optical field and the physics of the interaction are all interwoven. The interpretation of such experiments will necessarily be complex, and any physical information will have to be derived indirectly. We have now entered the attosecond world, but we will need a better guidebook to help us find our way around it. ■

Yaron Silberberg is in the Department of Physics of Complex Systems, Weizmann Institute of Science, 76100 Rehovot, Israel.
e-mail: yaron.silberberg@weizmann.ac.il

1. Hentschel, M. *et al. Nature* **414**, 509–513 (2001).
2. L'Huillier, A. & Balbou, P. *Phys. Rev. Lett.* **70**, 774–777 (1993).
3. Macklin, J. J., Kmetec, J. D. & Gordon, C. L. III *Phys. Rev. Lett.* **70**, 766–769 (1993).
4. Corkum, P. B. *Phys. Rev. Lett.* **71**, 1994–1997 (1993).
5. Papadogiannis, N. A. *et al. Phys. Rev. Lett.* **83**, 4289–4292 (1999).
6. Paul, P. *et al. Science* **292**, 1689–1692 (2001).

Oceanography

Sea snow microcosms

Farooq Azam and Richard A. Long

Marine bacteria can respond to organic particles in sea water, creating hotspots of bacterial growth and carbon cycling. This microscale behaviour should be included in models of the oceanic carbon cycle.

Oceanographers exploring the ocean's carbon cycle, and its role in climate change, do so at a grand scale. Satellite measurements of ocean colour are used to infer carbon dioxide (CO_2) assimilation into organic matter by photosynthetic organisms — a process known as carbon fixation. Instruments on ships and buoys assess how fast carbon is being fixed, what fraction is respired back by marine organisms to CO_2, and what fraction of organic matter sinks into the deep ocean, isolating its carbon from air–sea exchange for millennia. These studies paint a broad-brush picture of carbon cycling in the oceans, but they cannot predict future changes — in response to increases in atmospheric CO_2, for example.

Without looking at the ocean at the finer scale we cannot hope to understand or predict these changes[1,2]. It has long been recognized that bacteria are responsible for consuming much of the organic matter in marine systems. But the current climate-change models ignore how bacteria in the ocean respond to the patchy distribution of organic matter. This may well be a serious omission. A simulation, reported in *Limnology and Oceanography* by Kiørboe and Jackson[3], greatly reinforces our understanding of how marine bacteria interact with the patchy distribution of organic matter in the ocean. The results indicate that the response of bacteria to sinking organic particles has a big influence on the oceanic carbon cycle.

Kiørboe and Jackson[3] model the interactions of bacteria with marine snow (Fig. 1, overleaf). A cubic metre of sea water contains thousands of these rapidly sinking flakes[4]. They are the main vehicle for transporting organic matter, and its associated carbon, from the upper ocean[4] (where it readily exchanges with the atmosphere) to the ocean floor (where it eventually settles in sediments). The fate of marine snow, and the depth at which its carbon is converted back to CO_2 by bacteria, is critical for under-news and views

Atomic photography

Louis F. DiMauro

No ordinary camera can capture the motion of electrons inside an atom. But the advent of ultrafast laser pulses brings the necessary 'shutter speed' for snapping them as they tumble between energy levels close to the nucleus.

In stop-action photography, a moving body is frozen on film by the speed of the shutter or the flash of a strobe light. In this photograph (Fig. 1) of the great American pastime baseball, the shutter speed was sufficient to freeze the motion of the crowd, and the batter's anxious expression, as the pitcher throws the ball towards him. But there is one detail that the camera failed to capture — the motion of the ball. The ball's speed was so high (around 150 kilometres per hour) that it is blurred in the photograph, masking such details as the stitching on its surface. So there are different timescales for the action on the ball field, some too fast to register easily. The same limitation applies at the atomic scale, but the paper by Drescher *et al.*[1], on page 803 of this issue, is set to change this. The authors have introduced a new type of 'photography' that freezes an atom in motion and heralds a new field of research — attophysics.

Consider a water molecule moving through a solution on a picosecond timescale ($1\,\text{ps} = 10^{-12}$ s). On a faster timescale (down to 10^{-14} s), the atomic nuclei of hydrogen and oxygen inside the water molecule also vibrate and bend relative to each other. Scientists have studied this motion since the late nineteenth century by unravelling the mysteries contained in the spectrum of radiation that can be emitted by such a molecule. But the advent of ultrafast laser pulses has revolutionized time-resolved spectroscopy over the past quarter-century.

Using the technique of 'pump–probe' laser spectroscopy, the experimentalist can initiate (pump) and watch (probe) the motion of molecules in real time with a resolution of a few femtoseconds ($1\,\text{fs} = 10^{-15}$ s). This has not only resulted in a direct measure of how molecules move during a chemical reaction but, perhaps more importantly, it has opened up a new school of thought that views the quantum world in terms of moving 'wave packets' representing the evolution of quantum-mechanical amplitudes and phases. In 1999, Ahmed Zewail received the Nobel prize in chemistry for his seminal contributions to this field of femtochemistry.

Although femtosecond light pulses are still an important scientific tool, they cannot freeze every aspect of atomic motion that is relevant in the structure of matter, just as the photograph fails to capture the details of the baseball. The frontier in ultrafast spectroscopy is the study of the motion of electrons bound inside the atom, in orbits close to the nucleus. This new scale is defined by the time it takes the electron in the innermost orbit of the hydrogen atom to complete one turn around the proton nucleus. The period of this orbit — 24×10^{-18} s, or 24 attoseconds — is at least 100 times shorter than the duration of the shortest laser pulse.

L. F. DIMAURO

Figure 1 **Spot the ball — the limitations of action photography.**

But the period of the laser-pulse cycle at or near visible wavelengths imposes a fundamental limit on the resolution that can be achieved, at approximately a few femtoseconds. Consequently, researchers trying to make light pulses of attosecond duration have looked to other wavelengths, particularly those around the border between the X-ray and ultraviolet (XUV) regimes. One successful approach, using the strong-field phenomenon known as high-harmonic generation, produced the first measurable light pulses in the attosecond regime[2,3].

But forming attosecond light pulses is only one of the scientific challenges to be faced in making attophysics a reality. Equally challenging is the detection and propagation of these fragile pulses. Drescher *et al.*[1], however, have solved all of these problems for the specific case of measuring the decay time

news and views

of an inner-shell electron excitation, which until now had been studied only indirectly through the emission spectrum.

The authors used an attosecond XUV pulse to initiate the excitation process in an atom of krypton. This pulse frequency is high enough to ionize the atom, knocking out an inner-shell electron and leaving behind an unstable 'hole'. The decay path from this excited state involves two electrons; one drops from a higher shell to fill the hole, and another — the 'Auger electron' — is shaken out of the atom. The Auger decay process occurs between 10^{-14} and 10^{-16} s after ionization, and by sending in a second, longer (femtosecond) optical pulse, Drescher *et al.* could pick up the emission of the Auger electron. The time of release and the final energy of the Auger electron in the intense optical field are related. So by measuring the Auger electron's energy as a function of the time delay between the XUV and optical pulses, Drescher *et al.* were able to measure the Auger decay time with attosecond resolution.

The emergence of new areas with broad scientific impact is usually the result of creative contributions from a large community of researchers over a long period of time. But there are a few papers that announce the beginning of a new era, and the paper by Drescher *et al.*[1] falls into this category. We are entering a new realm of hyperfast measurement; the age of attophysics has begun. ■

Louis F. DiMauro is at the Brookhaven National Laboratory, Upton, New York 11973, USA.
e-mail: dimauro@bnl.gov

1. Drescher, M. *et al. Nature* **419,** 803–807 (2002).
2. Paul, P. *et al. Science* **292,** 1689–1692 (2001).
3. Hentschel, M. *et al. Nature* **414,** 509–513 (2001).

Cell biology

Survival in three dimensions

Kenneth M. Yamada and Katherine Clark

Whether a cell lives or dies depends on various local cues. New work reveals that those cues include a cell's spatial relationship with its neighbours and polarized interactions with the adjacent extracellular matrix.

Successful cancer treatment requires both the efficient killing of tumour cells and the survival of normal cells. Because cancer chemotherapy involves drug-induced apoptosis (programmed cell death)[1], a crucial question is what actually determines susceptibility to this process or protection from it. As described in *Cancer Cell*, Weaver *et al.*[2] have identified a mechanism that is relevant to both normal and cancer cells, in which — surprisingly — the determinants are the three-dimensional organization of the cells and their interactions with the surrounding extracellular matrix.

Most human cancers arise from epithelial cells, which cover body surfaces and the interior of organs such as the breast, lung and intestine. Epithelial cells are normally anchored to an extracellular matrix called the basement membrane, which gives them polarity. They are organized into three-dimensional structures such as glands. This tissue organization is disrupted when malignant cancer cells invade across basement membranes and into other tissues. Indeed, the ability of tumour cells to thrive in the absence of contact with basement membranes or other such substrates is a well-known characteristic of malignancy[3]. Weaver *et al.*[2] now show that these features also make cancer cells vulnerable to drug-induced apoptosis.

In a three-dimensional cell-culture model system, there is a marked difference in tissue organization between a non-malignant breast epithelial cell line and a malignant cell line derived from it. When grown with components of basement membrane, the non-cancerous cells form small, gland-like structures surrounded by locally secreted basement membrane (Fig. 1a, d). In contrast, the tumour-cell counterpart fails to form glandular structures (Fig. 1b). Weaver *et al.* found that treatment with a variety of drugs or immunological inducers of apoptosis killed the tumour cells, but not the well-organized normal cells.

The authors then tested whether three-dimensional organization was crucial to these outcomes. They disrupted the organization of epithelial tissue by growing the cells on flat tissue-culture substrates, producing a single layer (a monolayer). Alternatively, cell-to-cell adhesion could also be disrupted by inhibiting the function of the adhesion molecule known as E-cadherin.

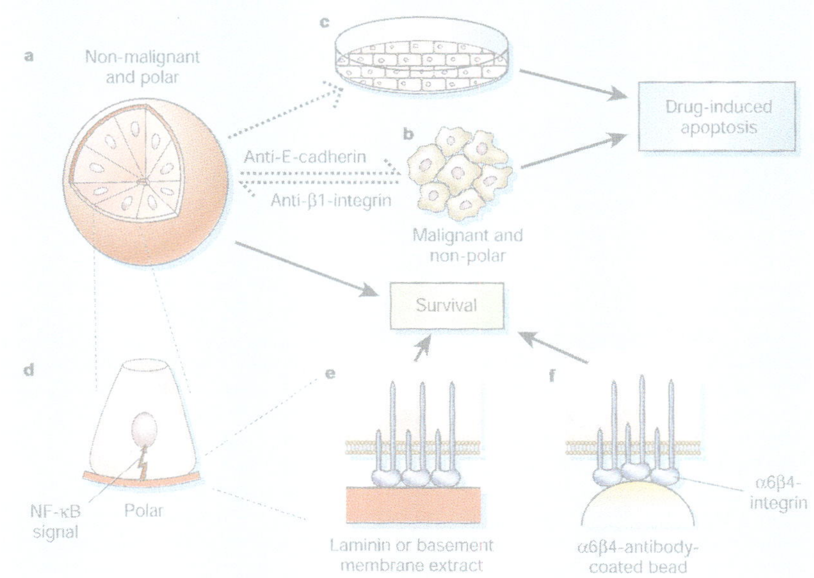

Figure 1 **Three-dimensional cell organization and polarity regulate sensitivity to drug-induced apoptosis — a summary of the results of Weaver *et al.*[2].** a, Normal, non-malignant breast epithelial cells have polarity and are anchored to a basement membrane (red); they are resistant to drug-induced apoptosis. b, Malignant breast cells lack polarity and are susceptible to drug-induced apoptosis. c, Converting normal cells to a monolayer by growing them on a flat collagen-coated plastic dish, or disrupting epithelial organization using anti-E-cadherin antibodies, produces cells that are sensitive to apoptosis. Conversely, treatment of breast tumour cells with anti-β1-integrin antibodies induces polarized organization and protects the cells against apoptosis. d, Protection is mediated by the NF-κB signalling pathway, which is active in epithelial cells that possess polarity. e, f, Key molecules involved in this process are laminin in basement membranes and an adhesion receptor, the α6β4-integrin. Cells are also protected from apoptosis by the addition of laminin or a basement-membrane extract (e) or by binding to beads coated with an antibody that mimics laminin by binding to the α6β4-integrin (f).

tions, whether these transitions be developmental or adaptive. If stochasticity is a fact of life, states are by definition metastable, and fluctuations can cause transitions between them. We are comfortable with the notion that antigenic phase variation in bacteria is stochastic (*10*), but might not stochastic mechanisms equally underlie the well-orchestrated, seemingly deterministic progression of states we call organismal development? We recognize the importance of the stochastic genetic mechanisms that generate diversity in the immune system. When coupled to suitable feedback mechanisms, such as the clonal amplification of cells expressing a particular antigen, these constitute a powerful means of learning, of crafting appropriate responses to unforeseen situations. Will we discover an analogous role for stochastic gene activation?

Stochasticity is inherent in all biological processes and it can be argued that the proliferation of both noise and noise-reduction systems is a hallmark of organismal evolution. One of us (*11*) has suggested that the reason it took so long to figure out that most genomes, including our own, are stuffed with transposons is because they "jump" from one chromosomal site to another so rarely that they are almost invisible genetically. Genome growth, dependent on the stochastic processes of gene duplication and transposition, may well have necessitated the prior invention of ways to suppress the inevitable countervailing deletions and genome scrambling caused by homologous recombination. Adrian Bird (*12*) has suggested that the inherent imprecision of gene regulation also sets an upper limit upon gene numbers. He has argued that the jump in gene numbers that accompanied the prokaryotic-eukaryotic transition was made possible by the accretion of multiple transcriptional noise-reduction mechanisms including chromatin, DNA methylation, the separation of the transcriptional from the translational apparatus, and the introduction of a complex quality control machinery into the production of mRNAs. And so the question is this: To what extent is the seemingly inexorable increase in complexity that we call evolution driven by the counterpoint of noise and noise reduction, of chance and the necessity of inventing and accumulating mechanisms to render coherent its random gifts?

References

1. E. Palsson, E. C. Cox, *Proc. Natl. Acad. Sci. U.S.A.* **93**, 1151 (1996).
2. H. H. McAdams, A. Arkin, *Proc. Natl. Acad. Sci. U.S.A.* **94**, 814 (1997).
3. H. H. McAdams, A. Arkin, *Trends Genet.* **15**, 65 (1999).
4. M. Thattai, A. van Oudenaarden, *Proc. Natl. Acad. Sci. U.S.A.* **98**, 8614 (2001).
5. E. M. Ozbudak *et al.*, *Nature Genet.* **31**, 69 (2002).
6. M. B. Elowitz, A. J. Levine, E. D. Siggia, P. S. Swain, *Science* **297**, 1183 (2002).
7. J. von Neumann, in *Automata Studies*, C. E. Shannon, J. E. McCarthy, Eds. (Princeton Univ. Press, Princeton, NJ, 1956), pp. 43–98.
8. A. Wagner, *Nature Genet.* **24**, 355 (2000).
9. A. Becskei, L. Serrano, *Nature* **405**, 590 (2000).
10. P. van de Putte, N. Goosen, *Trends Genet.* **8**, 457 (1992).
11. N. V. Fedoroff, *Proc. Natl. Acad. Sci. U.S.A.* **97**, 7002 (2000).
12. A. Bird, S. Tweedie, *Philos. Trans. R. Soc. London Ser B.* **349**, 249 (1995).
13. "Is uncertainty the only thing that is certain?", *New York Times*, 17 July 2002, page 1.

PERSPECTIVES: PHYSICS

Resolving Physical Processes on the Attosecond Time Scale

Maciej Lewenstein

Many molecular and atomic processes occur on a femtosecond time scale (1 fs = 10^{-15} s), which can be probed with coherent laser excitation. Now physicists are reaching for the next frontier. On page 1144 of this issue, Kienberger *et al.* (*1*) lead the way toward resolving physical processes on the even faster attosecond scale (1 as = 10^{-18} s).

To achieve fast time resolution, the duration of the laser pulse must be controlled. With two short laser pulses, one can perform "pump-probe" experiments, in which the first pulse excites the system and the second pulse probes the resulting dynamics. For this method to work, the pulses must be shorter than the characteristic time scale of the dynamics in question.

A laser pulse cannot be shorter than the oscillation period of the electromagnetic field in the pulse. Hence, for a laser with a wavelength λ of ~800 nm, the shortest pulse duration is ~4 to 5 fs. In time-resolved spectroscopy, such femtosecond pulses are used, for instance, to monitor the dynamics of molecular reactions. However, tightly bound electrons in atoms and ions cannot be observed with femtosecond techniques, because their dynamics occur typically on the attosecond time scale.

To decrease the pulse duration beyond the femtosecond time scale requires coherent sources of ultraviolet and soft x-ray radiation. A method called high order harmonic generation (HHG) (*2*) offers a possibility of realizing this task. When an atomic gas is irradiated with an intense femtosecond pulse, the gas produces harmonics of the laser field. The frequencies of these harmonics are odd integers of the laser frequency (ω_L) and extend up to few hundred ω_L (*3*). They reach wavelengths of a few nanometers and periods in the attosecond regime.

In a macroscopic medium, HHG requires constructive interference (phase-matching) between the contributions of individual atoms. According to the "simple man's model" (*4*, *5*), the HHG process occurs in three steps. First, the laser field causes an electron to tunnel to those regions in space where interactions with its nucleus are practically negligible. It then

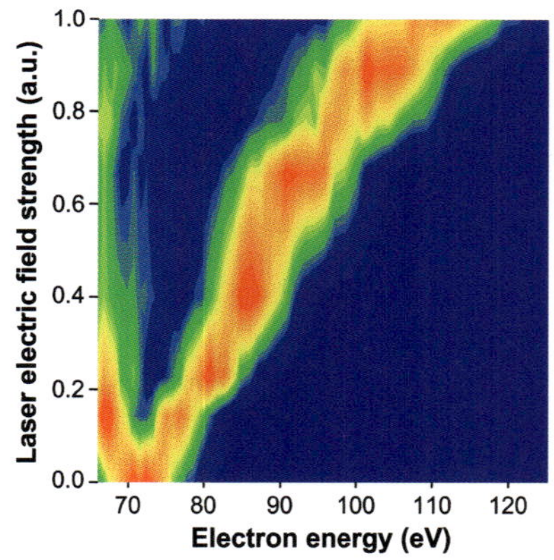

How to steer electrons. The shift of this photoelectron energy spectrum toward higher energies with increasing laser field represents "steering" of the electronic wavepacket. [From (*7*)]

oscillates in the laser field as a free charge. If it comes back to the nucleus, it may recombine, emitting harmonics. This model has a solid quantum mechanical basis (*6*); in particular, it incorporates quantum mechanical interferences between various electronic trajectories.

Antoine *et al.* predicted (*7*) that the harmonics, generated in a macroscopic medium under phase-matching conditions, are locked in phase. Twice in a laser period T_L, groups of neighboring harmonics interfere constructively for a very short time interval, producing a train of attosecond pulses separated by $T_L/2$. The first indirect evidence that HHG generates at-

The author is at the Institute for Theoretical Physics, Universität Hannover, 30167 Hannover, Germany. E-mail: lewen@itp.uni-hannover.de

SCIENCE'S COMPASS

tosecond pulse trains was reported by Papadogiannis *et al.* (*8*). Last year, de Paul *et al.* measured a train of 250-as pulses directly in the time domain (*9*).

Can such pulse trains be used for time-resolved attosecond spectroscopy? Salières *et al.* proposed in 1996 to use them for monitoring processes that occur with the period $T_L/2$ of the train. Candidates for such processes are harmonic generation itself or above threshold ionization (*2*), a laser-induced ionization process in which electrons absorb more photons than they need to be released. All of these processes can be described by "simple man's models" (*10*). Unfortunately, today's attosecond pulses are not sufficiently intense to realize time-resolved attosecond spectroscopy of these processes.

Although HHG is now an established source for attosecond pulse trains, it has one major limitation. Applications of time-resolved spectroscopy to dynamics that do not occur with the $T_L/2$ period require single isolated attosecond pulses. But there may be a way around this problem. For laser pulses shorter than 10 fs, the resulting individual harmonics fall below the femtosecond limit. Because the laser pulse lasts for only a few T_L, the harmonics cannot develop; instead, a soft x-ray attosecond pulse should be generated.

Krausz and co-workers (*11*) generated such an isolated attosecond x-ray pulse ($\lambda \approx 14$ nm) by irradiating a very short laser pulse ($\lambda \approx 750$ nm) of ~5-fs duration on a krypton gas sample, and filtering the outgoing radiation to a 5-eV range around 90 eV. They then irradiated the target krypton sample simultaneously with the x-ray pulse and a laser pulse of visible light of a few laser cycles duration. The x-ray pulse ionized the krypton atoms. The energy spectrum of the photoelectrons depended on the phase of the laser pulse at the moment of the electron's detachment. When the authors changed the relative delay between the laser and the soft x-ray pulse, a modulation of the spectral width appeared, allowing the duration of the x-ray pulse to be estimated as ~650 as.

Krausz and co-workers now describe (*1*) the first genuine application of isolated attosecond pulses for time-resolved attosecond spectroscopy. They study the absorption and emission of laser photons by electronic wavepackets created by soft x-ray radiation. Normally, the photoelectron energy spreads as a result of the photon absorption or emission (*12*). If, however, the emitted electron wavepacket is temporally confined to a fraction of T_L, its energy spectrum may be up- or down-shifted by several laser photon energies without broadening. The laser light can then "steer" the electron wavepacket like a classical particle. The results of such "steering" depend on the timing of the attosecond x-ray pulse relative to the absolute phase of the laser (see the figure), offering a simple, single-shot tool for time-resolved attosecond spectroscopy.

Attophysics has moved from dream to reality. One can expect fruitful applications of time-resolved attosecond spectroscopy to HHG or to above threshold ionization processes induced by ultrashort laser pulses, in which the absolute phase of the laser pulse plays a crucial role (*13*). Attosecond spectroscopy will provide diagnostics and perhaps new ways of controlling these processes, in particular to obtain better ways of short x-ray coherent pulse generation.

References and Notes

1. R. Kienberger *et al., Science* **297**, 1144 (2002); published online 11 July 2002 (*Science* 10.1126/science.1073866).
2. C. J. Joachain, M. Dörr, N. Kylstra, *Adv. Atom. Mol. Opt. Phys.* **42**, 225 (2000).
3. Ch. Spielmann *et al., Science* **278**, 661 (1997).
4. P. B. Corkum, *Phys. Rev. Lett.* **71**, 1994 (1993).
5. K. C. Kulander, K. J. Schafer, J. L. Krause, in *Super-Intense Laser-Atom Physics*, B. Piraux, A. L'Huillier, K. Rzcàzẽwski, Eds. (Plenum, New York, 1993).
6. M. Lewenstein *et al., Phys. Rev. A* **49**, 2117 (1994).
7. P. Antoine, A. L'Huillier, M. Lewenstein, *Phys. Rev. Lett.* **77**, 1234 (1996).
8. N. Papadogiannis *et al., Phys. Rev. Lett.* **83** 4289 (1999).
9. P. M. Paul *et al., Science* **292**, 1689 (2001).
10. P. Salières *et al., Science* **292**, 902 (2001).
11. M. Hentschel *et al., Nature* **414**, 509 (2001).
12. N. M. Kroll, K. M. Watson, *Phys. Rev. A* **8**, 804 (1973).
13. G. G. Paulus *et al., Nature* **414**, 182 (2001).

14. I thank the Deutsche Forschungsgemeinschaft, Alexander von Humboldt-Foundation, the VolkswagenStiftung, the ESF, and European Union Research Training Network "Cold Quantum Gases."

PERSPECTIVES: NEUROSCIENCE

It Takes More Than Two to Nogo

Clifford J. Woolf and Stefan Bloechlinger

The environment of the adult mammalian central nervous system (CNS) is hostile to the growth of axons and is a major contributor to the inability of injured neurons to regenerate. Much of this inhibition is caused by myelin, the insulating lipid and protein material that is wrapped around axons, ensuring rapid transmission of electrical signals along central nerve fibers. In the CNS, myelin is produced by supporting glial cells called oligodendrocytes. These cells also make growth-inhibitory proteins that become embedded within the myelin sheath.

Enhanced online at www.sciencemag.org/cgi/content/full/297/5584/1132

The authors are in the Neural Plasticity Research Group, Department of Anesthesia, Massachusetts General Hospital and Harvard Medical School, Boston, MA 02129, USA. E-mail: woolf.clifford@mgh.harvard.edu

Injured nerve fibers that make contact with CNS myelin cease to regenerate. At least three growth-inhibitory proteins have been identified so far: Nogo-A, named for its inhibitory action on axonal growth; myelin-associated glycoprotein (MAG); and oligodendrocyte myelin glycoprotein (OMgp). Although Nogo-A is known to bind to the Nogo receptor (NgR), the receptors for MAG and OMgp have remained elusive. Now, in an extraordinary and unexpected convergence reported by several groups including Liu *et al.* (*1*) on page 1190 of this issue, all three molecules appear to bind to the same receptor, NgR (*2–4*). This discovery opens up exciting new possibilities for overcoming axonal growth inhibition, a vital step in neuronal regrowth after brain or spinal cord injury.

During the development of the nervous system and in neurons grown in culture, the extension of axons from the cell body begins with the formation of small processes whose active tips have a specialized structure called the axonal growth cone. The growth cone interacts with the environment to determine the direction and rate of axon elongation. When the growth cone contacts CNS myelin, its cytoskeletal structure is altered, causing it to collapse and resulting in cessation of axonal growth. Nogo-A, MAG, and OMgp all contribute to the inhibitory action of CNS myelin on axonal growth and regeneration.

Nogo-A, a member of the reticulon family of proteins, has two inhibitory domains: a cell surface domain called Nogo-66 (*5*), and a long amino-terminal region (*6*) (see the figure). The Nogo-66 domain on the oligodendrocyte surface binds to NgR, a leucine-rich repeat protein that is attached to the extracellular surface of the neuronal membrane by glycophosphatidylinositol (GPI) (*4*). The location of the inhibitory amino-terminal domain of Nogo-A may be cytoplasmic, although this remains unclear (*5, 6*). If this is the case, then the amino-terminal domain of Nogo-A can inhibit axonal growth only when myelin is disrupted by injury. No receptor

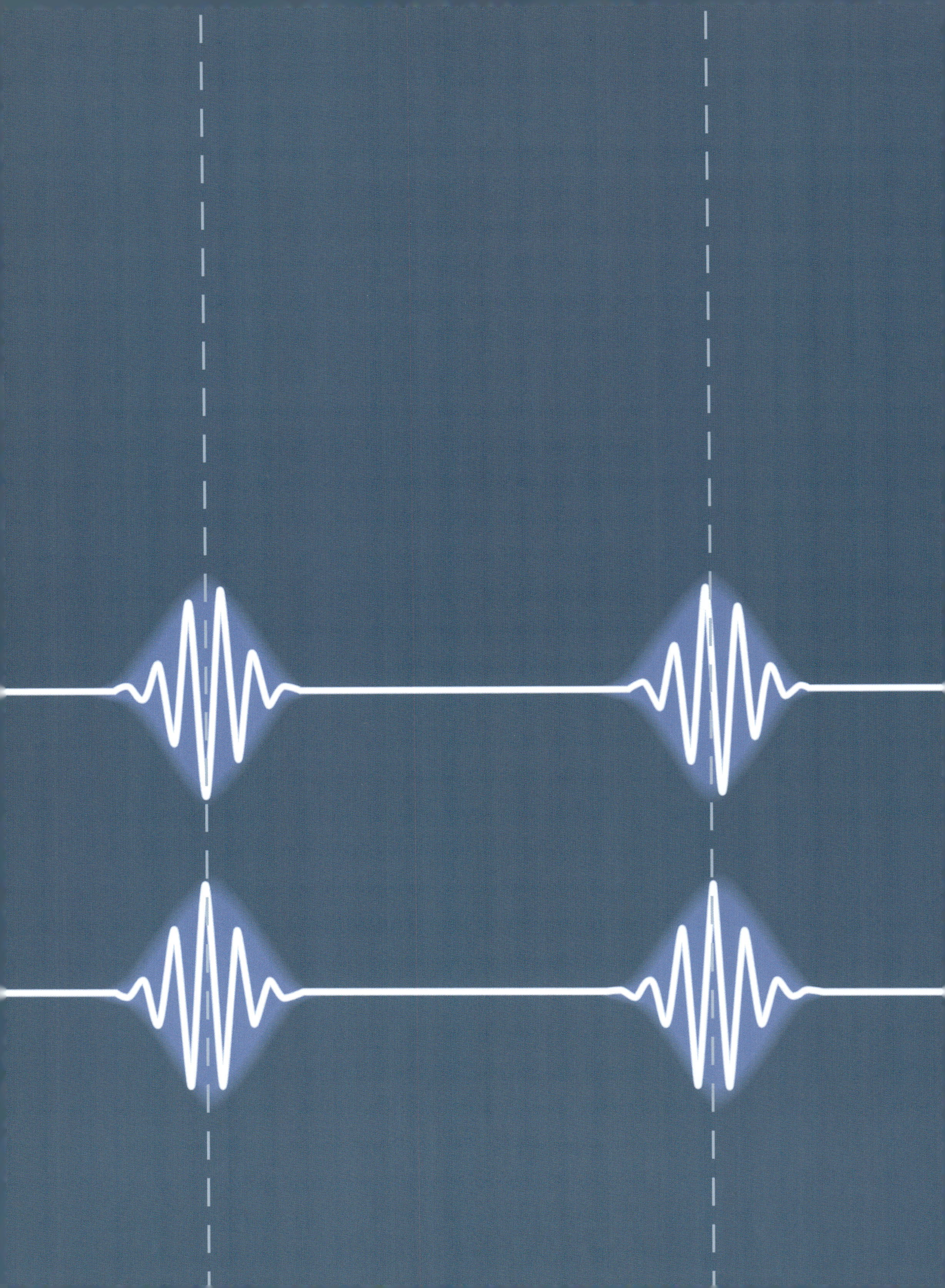

Controlled Attosecond-Varying Force Fields

5

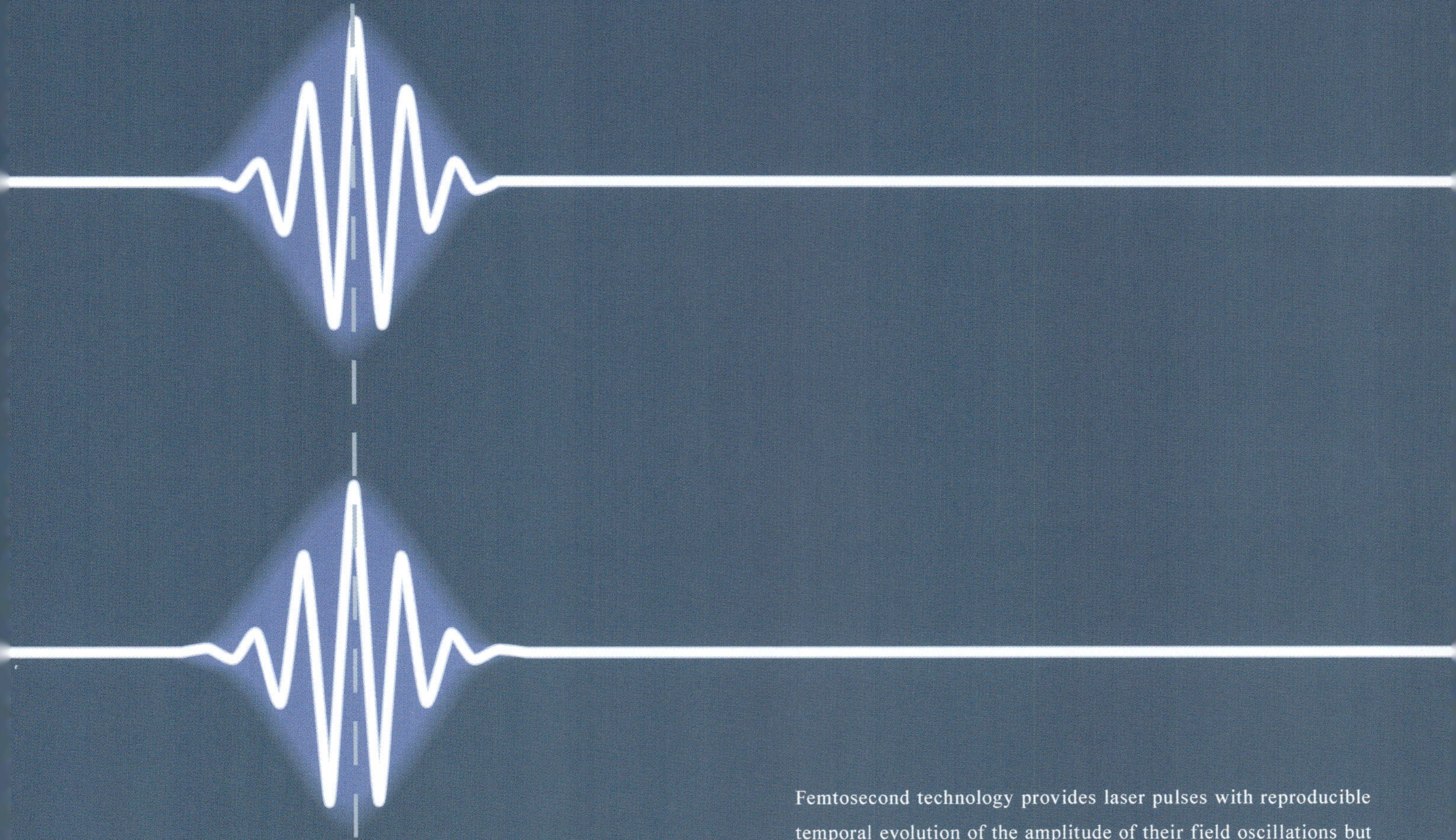

Femtosecond technology provides laser pulses with reproducible temporal evolution of the amplitude of their field oscillations but with randomly-varying waveforms (upper pulse train). Control of the carrier-envelope phase of ultrashort laser pulses allowed their production with reproducible waveform (lower pulse train), providing a controlled attosecond-varying force for attosecond metrology.

The extreme sensitivity of the attosecond-scale temporal evolution of cut-off high-order harmonic radiation to the sub-cycle evolution of the driving electric field is shown below (Fig. 5-1). Changing the carrier-envelope phase (CEP) φ from zero to $\pi/2$ modifies the peak field strength by only a few percent. This miniscule change implies a dramatic change in the predicted temporal profile of the XUV burst. In fact, the single attosecond pulse for $\varphi = 0$ (Fig. 5-1, red line) collapses into a much longer pulse of greatly reduced energy (Fig. 5-1, blue line). Consequently, the lack of reproducible φ in the experiments described in the previous chapter corresponds to the generation of XUV emission bursts with varying temporal distribution and timing with respect to the generating optical field in consecutive pulses of the driving laser light. This variation, in turn, implies a temporal smearing of any physical measurable recorded as a function of delay between the XUV burst and the driving laser field in time-resolved spectroscopic experiments relying on multiple pump-probe exposures for each delay setting.

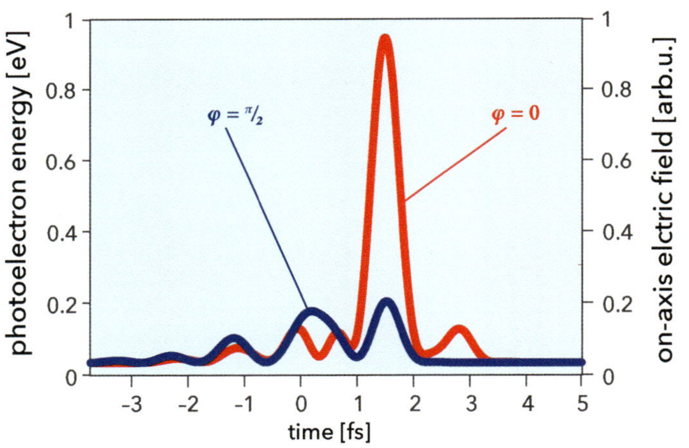

Figure 5-1 Computed temporal evolution of high-order harmonic emission from neon atoms driven by linearly-polarized 800-nm, 7-fs laser pulses, using the modelling described in [17]. The red and blue lines show the intensity evolution of the corresponding high-harmonic emission near cutoff as transmitted through a bandpass filter centered at a photon energy of 90 eV for two different values of the carrier-envelope phase, φ.

Hence, the development of techniques for the measurement and control of the phase of few-cycle light pulses is directly connected to the birth of attosecond science [17].

A pulse-to-pulse change in CEP, $\Delta\varphi$, was first accessed and observed experimentally in [1]. A nonlinear self-referencing technique invented by Theodor Hänsch permitted instantaneous measurement of $\Delta\varphi$ as demonstrated by Alexander Apolonskiy and his collaborators in Vienna [26] and independently, by John Hall and his co-workers in Boulder[9]. These two advances led to the control of the CEP in mode-locked laser oscillators. The breakthrough relied on a combination of self-referencing [26] with an effect discovered in [1]. $\Delta\varphi$ is dependent on the peak power of the laser pulse circulating in the laser cavity and can hence be affected by the power used for pumping the gain medium. This effect has been utilized for controlling the CEP evolution in mode-locked lasers ever since the first demonstrations of CEP control. Derivation of an error signal from electronic referencing of $\Delta\varphi$ to a preset value of the desired intracavity round-trip CEP shift and controlling the pump power with this error signal using an electro-optic modulator allowed the fine-tuning of the intra-cavity round-trip CEP shift, $\Delta\varphi$. One of the first implementations of this CEP control technology, which is now commercially available and widely used all over the world, was reported by Andreas Poppe and his collaborators [27].

Through the approach described, one is able to control the evolution of the CEP but is unable to determine the value of φ in any pulse of the train delivered by the mode-locked laser. Measurement of φ calls for a field-matter interaction that yields a physical observable dependent on φ in a well-understood and predictable manner. Peter Dietrich and his co-workers proposed such a process called strong-field-induced above-threshold ionization [28]. As a result of pioneering work by Gerhard Paulus and his co-workers, the phenomenon is now widely utilized (in a slightly different interaction geometry) for determining the absolute value of CEP of intense few-cylce laser pulses in a single shot, without the need for any control of the CEP, as described in [29].

The above-mentioned work underlines how through pushing the frontiers of femtosecond technology to octave-spanning bandwidth and to the few-cycle regime, one was able to render

[9] David J. Jones *et al.* Carrier-envelope phase control of femtosecond mode-locked lasers and direct optical frequency synthesis. *Science* **288**, 635 (2000).

the CEP and thereby the oscillating light field both *controllable* and *accessible* to measurement.

The CEP-controlled few-cycle laser pulses delivered by the oscillator [27] must be further amplified to become intense enough for attosecond XUV pulse generation and measurement. This amplification process introduces random shifts to the CEP, which need to be measured in individual laser shots and adequately compensated for, via a feedback loop. The first implementation of this important step was reported by Andrius Baltuška, Thomas Udem and their collaborators in [30]. This technological development resulted in the first demonstration of strong few-cycle light fields with controlled waveforms, the impact of which has been addressed by Phil Bucksbaum in his commentary [31].

The next chapter reviews how, shortly after coming into existence, waveform-controlled few-cycle laser light has permitted the reproducible generation of isolated attosecond pulses as well as their precise measurement. The very same measurement also yields complete information about the strength and temporal evolution of the laser waveform itself, which provides a controlled attosecond force not only for reproducible attosecond pulse generation but more generally for triggering, steering and/or tracking atomic-scale electron motion.

[26] A. Apolonskiy, A. Poppe, G. Tempea, C. Spielmann, T. Udem, R. Holzwarth, T. Hänsch, F. Krausz. Controlling the phase evolution of few-cycle light pulses. *Physical Review Letters* **85**, 740 (2000).
(Page: 212)

[27] A. Poppe, R. Holzwarth, A. Apolonskiy, G. Tempea, C. Spielmann, T. Hänsch, F. Krausz. Few-cycle optical waveform synthesis. *Applied Physics B* **72**, 373 (2001).
(Page: 216)

[28] P. Dietrich, F. Krausz, P. Corkum. Determining the absolute carrier phase of a few-cycle laser pulse. *Optics Letters* **25**, 16 (2000).
(Page: 220)

[29] G. Paulus, F. Lindner, H. Walther, A. Baltuška, E. Goulielmakis, M. Lezius, F. Krausz. Measurement of the phase of few-cycle laser pulses. *Physical Review Letters* **91**, 253004 (2003).
(Page: 223)

[30] A. Baltuška, T. Udem, M. Uiberacker, M. Hentschel, E. Goulielmakis, C. Gohle, R. Holzwarth, V. Yakovlev, A. Scrinzi, T. Hänsch, F. Krausz. Attosecond control of electronic processes by intense light fields. *Nature* **421**, 611 (2003).
(Page: 227)

[31] P. Bucksbaum. Ultrafast control. *Nature* **421**, 593 (2003).
(Page: 232)

VOLUME 85, NUMBER 4 PHYSICAL REVIEW LETTERS 24 JULY 2000

Controlling the Phase Evolution of Few-Cycle Light Pulses

A. Apolonski,[1,2] A. Poppe,[1] G. Tempea,[1] Ch. Spielmann,[1] Th. Udem,[3] R. Holzwarth,[3] T. W. Hänsch,[3] and F. Krausz[1]

[1]*Institut für Photonik, Technische Universität Wien, Gusshausstrasse 27, A-1040 Wien, Austria*
[2]*Institute of Automation and Electrometry, SB RAS, Novosibirsk 630090, Russia*
[3]*Max-Planck-Institut für Quantenoptik, Hans-Kopfermann-Strasse 1, D-85748 Garching, Germany*

(Received 13 April 2000)

Using a coherent nonlinear optical technique, slipping of the carrier through the envelope of 6-fs light wave packets emitted from a mode-locked-oscillator/pulse-compressor system has been measured, permitting the generation of intense, few-cycle light with precisely reproducible electric and magnetic fields. These pulses open the way to controlling the evolution of strong-field interactions on the time scale of the light oscillation cycle and are indispensable to reproducible attosecond x-ray pulse generation.

PACS numbers: 42.65.Re, 32.80.–t, 42.25.Bs, 42.65.Ky

Exposing matter to a few oscillation cycles of intense optical radiation permits bound electrons to survive in the vicinity of the nucleus up to unprecedented intensity levels, allowing them to enter previously inaccessible, extreme regimes of nonlinear optics [1]. Observed and predicted consequences include optical-field ionization rates comparable to the light frequency [2], making atoms radiate high-order harmonics of the driving laser up to photon energies exceeding by more than a factor of 300 the energy of the laser photons [3], and the possibility of generating *isolated* x-ray pulses of attosecond duration [4]. These phenomena emerge in the strong-field, low-frequency regime of nonlinear optics, where the electronic motion is directly controlled by the electric and (at relativistic intensity levels also) magnetic fields. In a few-cycle laser pulse, with an electric field

$$E(t) = A(t)\cos(\omega_0 t + \varphi), \quad (1)$$

the fields sensitively depend on the parameter φ, determining the position of the carrier wave (oscillating at frequency ω_0) with respect to the amplitude envelope $A(t)$. The resultant sensitivity of high-field phenomena to the absolute phase φ is beautifully demonstrated by the dependence of the predicted attosecond time structure of few-cycle-driven high harmonics on φ, as shown in Fig. 1. As a consequence, one of the greatest challenges in ultrafast laser physics and nonlinear optics, the *reproducible* generation of isolated attosecond x-ray pulses, relies on light pulses with reproducible absolute phase.

Rapid changes of φ in the pulse train delivered by mode-locked oscillators [5] and failure of gaining access to this parameter prevented precise control over strong-light-field-driven processes thus far. In this Letter we report on precision measurement of the variation of φ in sub-10-fs light pulses. The demonstrated technique allows selection of pulses with a fixed (but not known) absolute phase from the high-repetition-rate pulse train for the generation of high-intensity few-cycle light pulses with precisely reproducible electric and magnetic fields. Recent measurement of the interferometric cross correlation of successive sub-10-fs laser pulses revealed that φ suffers a shift upon each round trip of the pulse in the resonator due to a difference in the phase and group velocities in the laser components [5]. This round-trip phase shift typically accumulates to several hundred times 2π plus a physically relevant component $\Delta\varphi$, which obeys $0 < \Delta\varphi < 2\pi$ and accounts for a corresponding shift of the carrier with respect to the envelope. As a result, the carrier is offset at the carrier-envelope-offset frequency $f_{ceo} = (\Delta\varphi/2\pi)f_r$ (f_r is the pulse repetition rate), which leads to an offset of the phase-locked comb of laser modes, yielding a frequency of the nth mode $f_n = f_{ceo} + nf_r$ [6,7].

In Ref. [5], which addressed φ and possible ways of its control for the first time, $\Delta\varphi$ was shown to be also affected by nonlinear effects, which translate pulse energy fluctuations into a jitter of $\Delta\varphi$. Hence, precise measurement of the temporal evolution of φ in the output pulse train is required for selecting pulses with a fixed absolute phase, i.e., pulses having reproducible waveforms. Triggered by

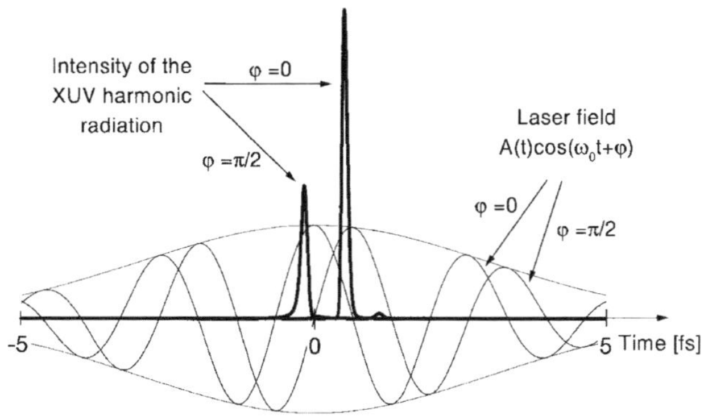

FIG. 1. Calculated temporal evolution of the cycle-averaged intensity of high-harmonic emission within 10% bandwidth at 3.2 nm from helium (interaction length 18 μm, pressure 500 Torr) driven by linearly polarized 5-fs light pulses ($\lambda_0 = 750$ nm, $I_{peak} = 2 \times 10^{15}$ W/cm^2) for two different values of the absolute phase φ. For the underlying physics, see Ref. [1].

the recent availability of intense few-cycle laser pulses [8] down to pulse durations as short as 4 fs [1,9] several methods have been proposed for gaining access to either φ or $\Delta\varphi$ [1]. These include techniques drawing on strong-field processes, such as optical field ionization [10–12] or high harmonic generation [13], and on heterodyning different parts of the mode-locked spectrum [6,7,14].

Locking the equidistant frequency comb to stabilized external harmonic frequencies, as recently demonstrated for precision frequency-domain metrology [15–17], constitutes one possible means of implementing the latter concept. In our work, we have opted for another implementation of this approach. By frequency converting *directly* a portion of the spectrally broadened laser spectrum and coherently mixing this converted radiation with the fundamental laser field at the shifted frequency we are able to precisely measure and control the evolution of φ in the high-repetition-rate train of few-cycle pulses emerging from a mode-locked-oscillator/pulse-compressor system. The phase-locked frequency comb of the laser is first broadened by self-phase modulation to span a full octave. In this process, f_r is preserved [17] but f_{ceo} may be affected, as discussed below. The low-frequency wing of the spectrum, $f_{\mathrm{ceo}} + n_{\mathrm{low}} f_r$, is then frequency doubled and recombined with the high-frequency wing, $f_{\mathrm{ceo}} + n_{\mathrm{high}} f_r$. Beating of the modes $n_{\mathrm{high}} = 2n_{\mathrm{low}}$ with the second harmonic of modes n_{low} gives rise to a beat note at $2 \times (f_{\mathrm{ceo}} + n_{\mathrm{low}} f_r) - (f_{\mathrm{ceo}} + n_{\mathrm{high}} f_r) = f_{\mathrm{ceo}}$ [18], yielding direct information about the round-trip phase shift $\Delta\varphi$ and thereby the temporal evolution of φ in the output pulse train. Beat notes of the same origin are also expected at $nf_r \pm f_{\mathrm{ceo}}$, for $n = 1, 2, \ldots$.

The cascaded nonlinear processes (second harmonic generation of a self-phase-modulated pulse) call for high peak powers and very low pulse energy fluctuations. To this end, a specially designed Kerr-lens-mode-locked, mirror-dispersion-controlled Ti:sapphire laser has been developed. The resonator is made up of broadband chirped mirrors and incorporates a 1:1 imaging telescope consisting of two 1-m-radius mirrors and several plane folding mirrors (not shown in Fig. 2) to extend its length to $\approx$6 m. Pumped with a single-frequency, frequency-doubled $\mathrm{Nd{:}YVO_4}$ laser ($P_{\mathrm{pump}} \approx 4.5$ W), the laser generates a highly stable train of 9-fs, 20-nJ pulses at a repetition rate of 24 MHz (Fig. 2). These megawatt pulses produce substantial spectral broadening by self-phase modulation in a standard single-mode optical fiber (FS-SC-3314, Thorlabs). Figure 3 depicts the broadened spectrum, which spans from less than 500 nm to more than 1100 nm. A 1-mm-thick BBO doubling crystal oriented for type-I phase matching at $\approx$1080 nm is placed in close proximity of the fiber output to maximize frequency doubling efficiency.

The spectrally broadened pulses are passed through a chirped-mirror compressor, yielding sub-6-fs, 6-nJ pulses in a diffraction-limited beam (see inset of Fig. 3). Chirped

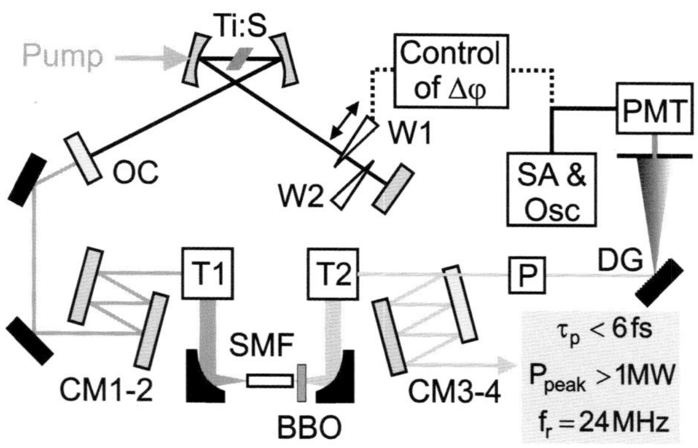

FIG. 2. Schematic of the experimental setup. Ti:S, highly doped ($\alpha_{514} = 5\ \mathrm{cm}^{-1}$) 3-mm-long titanium-doped sapphire crystal; OC, 25% output coupler; W1-2, thin (0.1–0.7 mm) fused-silica wedges; CM1-4, ultrabroad-band chirped mirrors providing precise dispersion control over 630–950 nm; T1-2, 1:5 magnifying and demagnifying telescope, respectively; SMF, 3-mm-long single-mode fiber; P, polarizer; DG, diffraction grating; PMT, photomultiplier; SA & Osc, rf spectrum analyzer and digital oscilloscope.

mirror CM4 transmits some 30% of incident radiation near 540 nm. The spectral components leaking through this mirror near 540 nm are dispersed by a diffraction grating and directed through a polarizer and a slit onto a photomultiplier. The polarizer has been aligned to permit interference between the orthogonally polarized fundamental and frequency-doubled laser fields. The upper diagram in Fig. 4 shows a typical rf power spectrum of the combined fundamental and frequency-doubled laser radiation entering the detector within the wavelength range of (540 ± 0.8) nm. The sharp features showing up in the spectrum at $\approx$4 MHz and $\approx f_r - 4$ MHz disappear as the polarizer is rotated to have its principal axes aligned along the polarization directions of the fundamental and

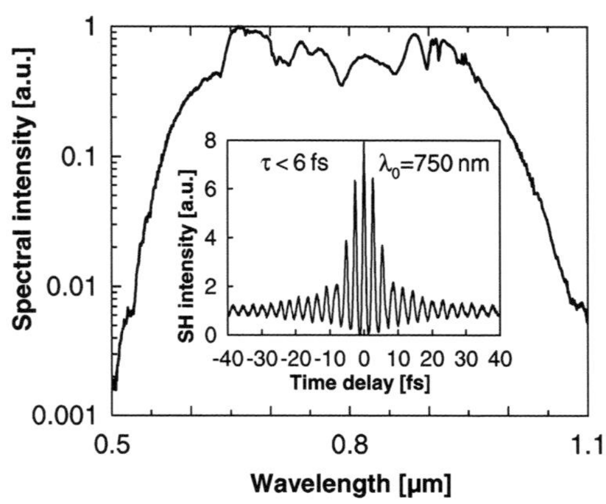

FIG. 3. Spectrum and interferometric autocorrelation (inset) of the pulses exiting the pulse compressor consisting of the single-mode fiber SMF and chirped mirrors CM3-4, respectively (see Fig. 2). The high fringe visibility over the entire autocorrelation trace indicates a pulse close to its bandwidth limit, which would allow a pulse duration of $\approx$5 fs.

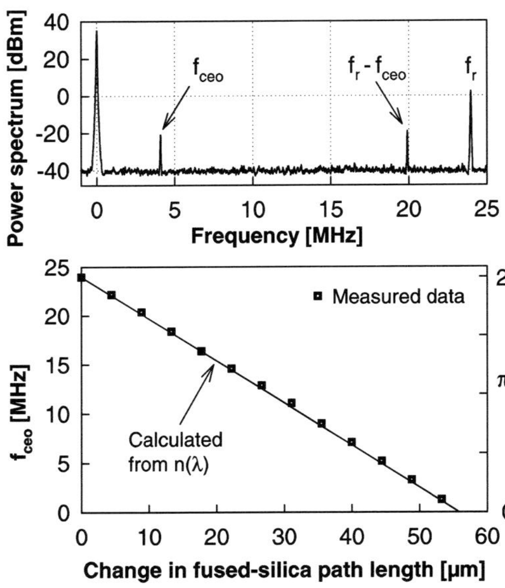

FIG. 4. Upper diagram, rf power spectrum of the combined frequency-doubled and fundamental laser radiation at 540 nm within an ≈1.5-nm bandwidth, as recorded using a 30-kHz rf resolution bandwidth. Lower diagram, carrier-envelope-offset frequency versus change in round-trip path length through the fused silica wedge W1. The accuracy of measuring (and setting) f_{ceo} is merely limited by random shifts of the beat within some ±15 kHz over a convenient measurement time of some 10–15 seconds.

frequency-doubled radiation, indicating that the features originate from a beat between these two fields. With the polarizer aligned for maximum beating signal, the power of these beat notes is some 20 dB below that emerging at f_r, a difference rapidly increasing as the detection bandwidth is increased beyond 2 nm as a consequence of the temporal offset between the frequency-doubled and fundamental wave packet at 540 nm.

To provide additional evidence for the origin of the observed rf spectral features we have varied the round-trip phase shift $\Delta\varphi$ by translating a thin wedged fused silica plate in the laser cavity and measured the frequency of the beat signal as a function of the (round-trip) path length in the fused silica plate. The results are depicted by squares in Fig. 4. The absolute phase φ in Eq. (1) is defined in a retarded frame (traveling at the group velocity v_g), hence it is varying as $\varphi(z) = \varphi_0 + (\beta_0 - \omega_0 v_g^{-1})z$ during propagation through a dispersive transparent medium [5]. Here $\beta_0 = \beta(\omega_0) = \omega_0 n(\omega_0)/c$ is the propagation constant at ω_0 and $v_g = (\partial\beta/\partial\omega)_{\omega_0}^{-1}$ is the group velocity. From these expressions straightforward algebra yields $\delta\varphi = 2\pi(\partial n/\partial\lambda)_{\lambda_0}\delta z$, where $\partial n/\partial\lambda = -0.018\ \mu\mathrm{m}^{-1}$ for fused silica at $\lambda_0 = 790$ nm, the center laser wavelength. The change in f_{ceo} is then predicted as $f_{ceo}(\delta z) = f_{ceo,0} - 0.018 f_r \delta z$ [μm] and depicted by the full line in the lower diagram of Fig. 4 for the starting position ($\delta z = 0$) set to yield $f_{ceo} = f_r$. In contrast with previously used techniques (tilting and translating the cavity end mirror [15,16]), the present approach allows precise determination of a change in $\Delta\varphi$ and f_{ceo} *independently* and thereby experimental verification of the relationship between these two quantities for the first time. The excellent agreement between the *calculated* slope $\partial f_{ceo}/\partial z$ and the *measured* variation of f_{ceo} provides conclusive evidence of the origin of the observed beat signal and our ability to measure and control the pulse-to-pulse shift in φ with a high precision.

The stability of f_{ceo} invites to "slowing down" the evolution of φ to frequencies as low as a few kHz and observing it directly in the time domain. To this end, we have tuned the carrier-slip beat to $f_{ceo} < 10$ kHz and analyzed the photomultiplier signal amplified by a low-pass amplifier ($f_T = 100$ kHz). Figure 5 shows millisecond snapshots of the temporal evolution of the intensity of 540-nm light composed of fundamental laser radiation and frequency-doubled 1080-nm light (with the dc component eliminated), reflecting the evolution of φ in the laser pulse train [19]. The nearly periodic evolution of the carrier-slip beat signal over several cycles in Fig. 3 implies that the round-trip phase shift can be kept "frozen" at values as low as $\Delta\varphi = 2\pi f_{ceo}/f_r < 10^{-3}$ rad for extended periods as long as milliseconds without active stabilization; i.e., random phase jitter remains less than this value over milliseconds. As a consequence, full electronic control or stabilization of φ can be accomplished with a sub-millisecond-response servo loop adjusting either the path length in the fused silica wedge or the intracavity pulse energy (upon controlling the pump power) [5] such that f_{ceo} is phase locked to a user-defined value.

It is important to note that the generation of phase-stable few-cycle pulses at the output of the fiber/chirped-mirror compressor by locking f_{ceo} to zero in the present scheme does not strictly imply phase stabilization *directly* at the output of the mode-locked oscillator as well. This

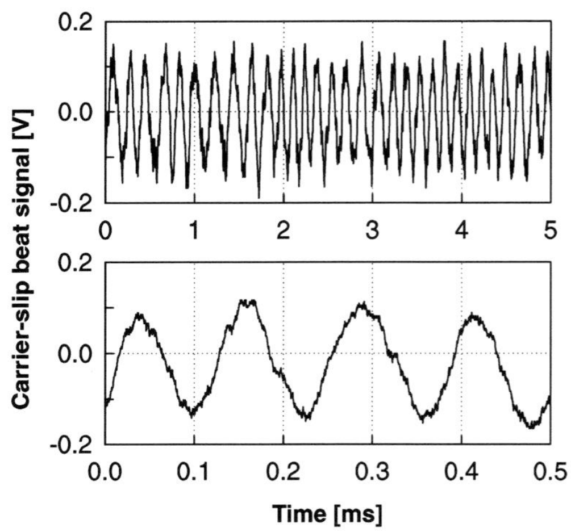

FIG. 5. Snapshots of the temporal evolution of the carrier-slip beat signal reflecting slow quasiperiodic variation of the absolute phase φ in the sub-6-fs pulse train.

VOLUME 85, NUMBER 4 PHYSICAL REVIEW LETTERS 24 JULY 2000

is because drift or fluctuations in the pulse parameters at the fiber input and in the fiber parameters (e.g., temperature, refractive index) may be translated into a corresponding change in φ by linear and nonlinear effects [5]. The resultant time-dependent shift of φ then emerges with opposite sign at the output of the oscillator, if φ is stabilized at the fiber output. Nevertheless, in the present case these external fluctuations and drifts in φ are estimated as very low ($<\pi/50$) owing to the high pulse energy stability, moderate spectral broadening in the fiber, and the short fiber length.

Whereas locking both f_{ceo} and f_r is imperative for applications in frequency-domain metrology [7,15–17], neither of them needs to be stabilized if intense phase-stabilized few-cycle light pulses are to be generated. This is a consequence of (i) the very low jitter of $\Delta\varphi$ observed in our experiments and (ii) the fact that the pulse energy can be significantly boosted (by extracavity amplification) only at substantially reduced repetition rates (≤ 1 kHz). Setting f_{ceo} equal to a few hundred kHz and using a tracking oscillator will allow one to produce a virtually noise-free quasisinusoidal signal reflecting the variation of φ in the 24-MHz pulse train even more accurately than do the traces shown in Fig. 5. Triggering the pulse picker to, say, the positive-slope zero transition of the beat signal will select pulses with φ fixed to within $2\pi(f_{\text{ceo}}/f_r) < \pi/10$ for external amplification (and moderate pulse compression). The phase of the resultant intense few-cycle pulses can be fine tuned externally by the same technique used for adjusting $\Delta\varphi$ inside the cavity.

In conclusion, we have demonstrated a sub-10-fs source equipped with a phase-tracing system capable of delivering seed pulses with a locked pulse-to-pulse phase for kHz-rate amplifier/compressor systems without any active stabilization. The presented results open the way to generating few-cycle light pulses focusable to intensities up to 10^{18} W/cm^2 and beyond with precisely reproducible electric and magnetic fields. Exposing matter to these ultraintense electromagnetic transients will allow unprecedented control of optical field ionization of atoms and subsequent motion of the freed electron wave packets in strong fields. Anticipated impacts of this new experimental capability include the controlled generation of isolated attosecond bursts of coherent soft-x-ray radiation and highly collimated relativistic electron pulses with sublaser-cycle duration.

The authors gratefully thank T. Brabec and P. Corkum for many fruitful discussions. This research has been sponsored by the Austrian Science Fund, within the framework of the Special Research Program F016 and Grant No. Y44-PHY, as well as by INTAS, Grant No. 97-1058. C. S. acknowledges support by APART.

Note added.—Since the submission of this paper, Jones *et al.* [20] reported full electronic control of f_{ceo} and f_r and similar results have been achieved independently by Holzwarth *et al.* [21], yielding unprecedented frequency rulers for metrology.

Note added.—With a somewhat improved carrier-slip beat signal we have achieved full control of the evolution of φ by phase locking f_{ceo} to an rf local oscillator by fine control of the pump power as described in Ref. [21].

[1] T. Brabec and F. Krausz, Rev. Mod. Phys. **72**, 545–591 (2000).

[2] C. Kan *et al.*, Phys. Rev. Lett. **79**, 2971 (1997).

[3] Z. Chang *et al.*, Phys. Rev. Lett. **79**, 2967 (1997); Ch. Spielmann *et al.*, Science **278**, 661 (1997); M. Schnürer *et al.*, Phys. Rev. Lett. **80**, 3236 (1998).

[4] I. P. Christov *et al.*, Phys. Rev. Lett. **78**, 1251 (1997).

[5] L. Xu *et al.*, Opt. Lett. **21**, 2008 (1996).

[6] H. R. Telle *et al.*, Appl. Phys. B **69**, 327 (1999).

[7] J. Reichert *et al.*, Opt. Commun. **172**, 59 (1999).

[8] M. Nisoli *et al.*, Opt. Lett. **22**, 522 (1997); S. Sartania *et al.*, Opt. Lett. **22**, 1526 (1997).

[9] Z. Cheng *et al.*, in *Ultrafast Phenomena XI,* edited by T. Elsässer, J. G. Fujimoto, D. A. Wiersma, and W. Zinth (Springer, Berlin, 1998), p. 8.

[10] A. Poppe *et al.*, in *Trends in Optics and Photonics Vol. 28*, edited by J. E. Bowers and W. H. Knox (Optical Society of America, Washington, DC, 1999), p. 31.

[11] I. P. Christov, Opt. Lett. **24**, 1425 (1999).

[12] P. Dietrich *et al.*, Opt. Lett. **25**, 16 (2000).

[13] F. Krausz *et al.*, Opt. Photonics News **9**, 46 (1998); A. de Bohan *et al.*, Phys. Rev. Lett. **81**, 1837 (1998); G. Tempea *et al.*, J. Opt. Soc. Am. B **16**, 669 (1999).

[14] P. B. Corkum (private communication); F. Kärtner (private communication); M. S. Pshenichnikov (private communication).

[15] T. Udem *et al.*, Phys. Rev. Lett. **82**, 3568 (1999).

[16] S. A. Diddams *et al.*, Opt. Lett. **25**, 186 (2000).

[17] J. Reichert *et al.*, Phys. Rev. Lett. **84**, 3232 (2000).

[18] Many modes within the detection bandwidth may contribute to this signal.

[19] Although this signal, in principle, allows determination of φ of any pulse in the output train, in practice, the lack of precise knowledge of all the phase and group retardations introduced by linear and nonlinear effects in the fiber and the frequency-doubling crystal prevents a reliable measurement of the absolute phase by this technique.

[20] D. J. Jones *et al.*, Science **288**, 635 (2000).

[21] R. Holzwarth *et al.*, "An optical frequency synthesizer for precision spectroscopy" (to be published).

Appl. Phys. B 72, 373–376 (2001) / Digital Object Identifier (DOI) 10.1007/s003400000526

Applied Physics B
Lasers
and Optics

Rapid communication

Few-cycle optical waveform synthesis

A. Poppe[1,*], R. Holzwarth[2], A. Apolonski[1,3], G. Tempea[1], Ch. Spielmann[1], T.W. Hänsch[2], F. Krausz[1]

[1] Institut für Photonik, Technische Universität Wien, Gusshausstrasse 27, 1040 Wien, Austria
[2] Max-Planck-Institut für Quantenoptik, Hans-Kopfermann-Strasse 1, 85748 Garching, Germany
[3] Institute of Automation and Electrometry, SB RAS, Novosibirsk 630090, Russia

Received: 23 November 2000/Published online: 13 December 2000 – © Springer-Verlag 2000

Abstract. We demonstrate cycle-slip-free electronic control of the carrier-envelope phase evolution of few-femtosecond light pulses, optical access to its sub-cycle fluctuations and suppression of the amplitude noise of a mode-locked laser by phase control. As a result, few-cycle light waveforms can now be synthesized with electric and magnetic fields reproducible to within a known phase error (< 0.3 rad) and an unprecedentedly low amplitude error ($< 3\%$). The exploration of phase-sensitive light–matter interactions can be tackled.

PACS: 42.65.Re; 42.65.Ky; 42.25.Bs; 32.80.-t; 52.35.Mw

The synthesization of electromagnetic waveforms can be routinely accomplished by radio-frequency (rf) signal generators but has not been feasible in the optical region so far. Extending this capability to light frequencies by generating few-cycle light waves with a reproducible evolution of the electric and magnetic fields would, beyond its fundamental significance, open the door for unprecedented control of the evolution of ultrafast microscopic processes on a time scale of the light oscillation period [1].

Phase-coherent locking of beat notes of a mode-locked laser to radio-frequency (rf) local oscillators has recently allowed direct electronic control of both the repetition rate f_{r} and the carrier-envelope-offset frequency f_{ceo}, which not only resulted in a simple and powerful 'frequency ruler' with lines at $f_{\mathrm{ceo}} + n f_{\mathrm{r}}$ spanning a full octave [2–4] but also offered the potential for reproducible optical waveform synthesis in the time domain. In this work we exploit this potential for the first time.

Whereas laser mode locking is capable of generating a pulse train $E_n(t) = A_n(t)\exp(-\mathrm{i}\omega_0 t + \mathrm{i}\varphi_n) + \text{c.c.}$ with a nearly identical complex envelope $A_n(t)$[1] of the electric field $E_n(t)$, i.e. $A_{n+k}(t) \approx A_n(t)$, it is not able to stabilize the carrier-envelope phase φ_n [5]. This is because a pulse-to-pulse shift $\Delta\varphi_n$, arises from a difference between the round-trip phase and group delay in the laser, which is subject to minor variations, $\Delta\varphi_n = \Delta\varphi_0 + \delta_n$, resulting in

$$\varphi_n = \varphi_0 + \sum_{k=1}^{n} \Delta\varphi_k = \varphi_0 + n\Delta\varphi_0 + \sum_{k=1}^{n} \delta_k. \tag{1}$$

The deviation of φ_n from the predictable value $\varphi_0 + n\Delta\varphi_0$ may accumulate to many times 2π over extended measurement times $T_{\mathrm{m}} \gg f_{\mathrm{r}}^{-1}$. Owing to the relationship $f_{\mathrm{ceo}} = (\Delta\varphi/2\pi) f_{\mathrm{r}}$ [6–8], phase locking of f_{ceo} can substantially reduce this random deviation. Whereas the avoidance of cycle slips of the stabilized beat notes with respect to rf clocks warrants stable optical frequency synthesis, the requirement for reproducible waveform synthesis is much more stringent: the carrier-envelope phase jitter

$$\sigma_\varphi(T_{\mathrm{m}}) = \langle(\varphi_n - \varphi_0 - n\Delta\varphi_0)^2\rangle_N^{1/2} \tag{2}$$

(where the angle brackets denote the average over $N = f_{\mathrm{r}} T_{\mathrm{m}}$) must not exceed a tiny fraction of π. *This requirement was not addressed before* and is the focus of this paper. The extent to which it can be fulfilled determines the feasibility of generating reproducible light waveforms at predictable instants.

In this paper, we report the emergence of this new experimental capability owing to: (i) full electronic control of the evolution of φ_n without cycle slips over extended periods of time in a few-femtosecond pulse train; (ii) measurement of σ_φ with a sub-cycle precision; and (iii) suppression of the amplitude noise of a mode-locked laser by means of phase control. This is achieved by exploiting optical nonlinearities for both closing the servo loop and as a diagnostic tool, for determining the sub-cycle jitter of φ_n, which is indispensable for a reproducible waveform generation. As a result, we demonstrate the generation of 5.8-fs light pulses (at $\lambda_0 \approx 0.8\,\mu$m, where the field oscillation period is ≈ 2.7 fs) with their carrier-envelope phase locked to within 0.3 rad of an electronic oscillator and their field amplitude stabilized to within 3% for extended periods of time. Although φ_0 is still unknown, few-cycle light waveforms can be reproducibly synthesized for the first time, creating ideal conditions for measuring φ_0 by phase-sensitive nonlinear effects [1].

*Corresponding author.

[1] $A_n(t) = |A_n(t)|e^{-\mathrm{i}\gamma(t)}$ accounts for amplitude and possibly frequency modulation with the chirp $\gamma(t)$ being a nonlinear function of t.

Figure 1 shows the schematic of our optical pulse generator. In contrast with previous work [2–4], the mode-locked spectrum emitted by a Ti:sapphire oscillator is broadened in a standard (rather than photonic-crystal) single-mode fiber to span more than an octave (500–1100 nm). Chirped mirrors compress the pulses exiting the fiber [9], hence the system delivers sub-6-fs pulses (see right inset in Fig. 4 and left inset in Fig. 6) with a peak power exceeding 1 MW at a repetition rate of 24 MHz. The wings of the spectrum (left inset in Fig. 4) extend from 500 to 1100 nm, spanning a full octave. A dichroic chirped mirror in the compressor (CM4 in Fig. 2) transmits the spectral components near 1080 nm and 540 nm towards a nonlinear interferometer generating the carrier-slip beat signal, which is schematically illustrated in Fig. 2. A dichroic mirror (DM) transmits the infrared components at 1080 nm, which are frequency-doubled in a 1-mm-thick BBO crystal and recombined with the 540-nm beam propagating through the other arm of the Mach–Zender interferometer at the beam splitter (BSP; SM represents the silver-coated steering mirrors). The spatial and temporal overlap of the two wave packets are ensured by the telescope formed by the lenses L1–2 and by the wedged glass plates W3–4, respectively. Interference between the orthogonally polarized 540-nm wave packets is enforced by a polarizer (P) aligned to yield a maximum beat signal at f_{ceo}. The optimized carrier-slip beat signal exhibits an improved signal-to-noise ratio (> 30 dB in a 100 kHz bandwidth), as compared to our previously used setup [9].

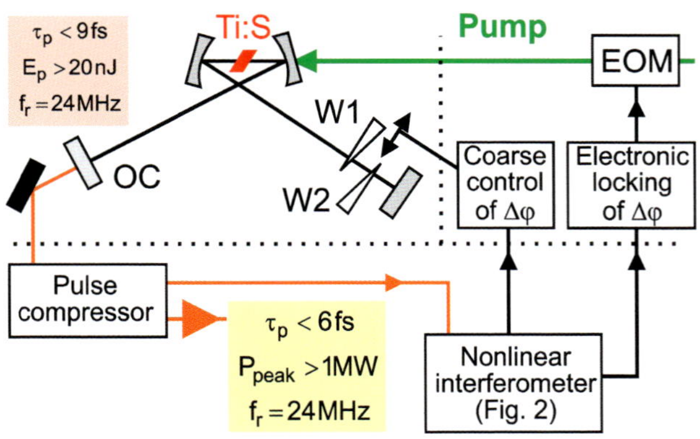

Fig. 1. Schematic of the few-cycle optical pulse generator and the servo loop used for carrier-envelope phase locking

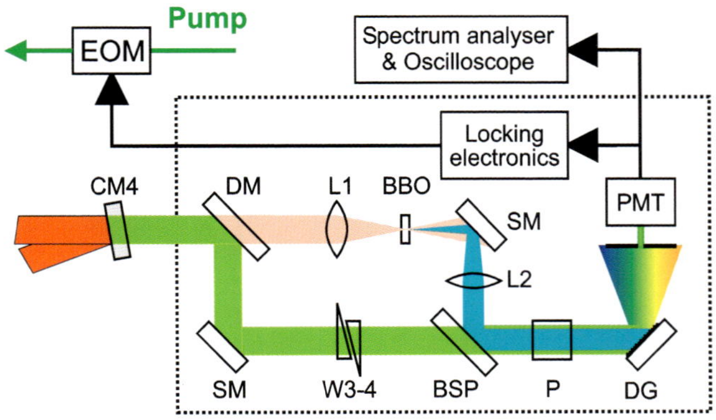

Fig. 2. Schematic of the optical setup used for generating the carrier-slip beat signal at ≈ 540 nm (for explanations see text)

We have phase locked the carrier-slip beat note to $f_{\mathrm{local}} =$ 1 MHz (synthesized by a stabilized local oscillator) by adjusting the pump power with an electro-optic modulator (EOM) [4]. This servo loop relies on nonlinearities (Kerr phase shift, Raman frequency shift [5]) translating a minor change of W into a significant change of $\Delta\varphi$ and hence that of f_{ceo}, as shown in Fig. 3. Once f_{ceo} is manually adjusted by translating a thin fused-silica wedge (W2 in Fig. 1) to be within $f_{\mathrm{local}} \pm 100$ kHz, the servo loop pulls f_{ceo} to f_{local} and phase-locks the carrier-slip beat note to the signal of the local oscillator. The rf spectrum of the in-lock carrier-slip beat note is shown in Fig. 4, depicting the average of 20 scans. With the loop interrupted, f_{ceo} fluctuates over the full frequency range shown in Fig. 4 during the same measurement time (≈ 10 s).

Measurement of the ratio of the frequency of the in-lock carrier-slip beat signal to that of the local oscillator with a two-channel frequency counter yields no deviation of f_{ceo} from f_{local} to within the gate-time-limited resolution of 100 mHz up to the maximum gate time of $T_{\mathrm{m}} = 10$ s allowed by our frequency counter. This implies that the servo loop suppresses the carrier-envelope phase jitter to $\sigma_{\varphi} < 2\pi$ for $T_{\mathrm{m}} \geq 10$ s, i.e. φ_n is locked to the signal of the local oscillator firmly enough to prevent uncontrolled cycle slips from occur-

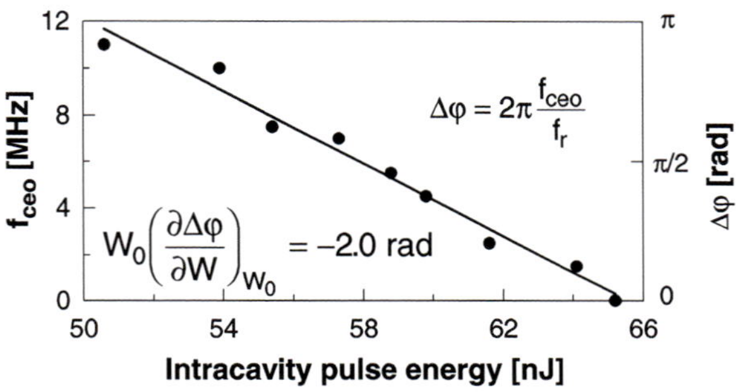

Fig. 3. Measured dependence of the carrier-envelope offset frequency f_{ceo} (and the pulse-to-pulse phase shift $\Delta\varphi$) on the intracavity pulse energy W (*dots*) and a linear regression to the depicted experimental data

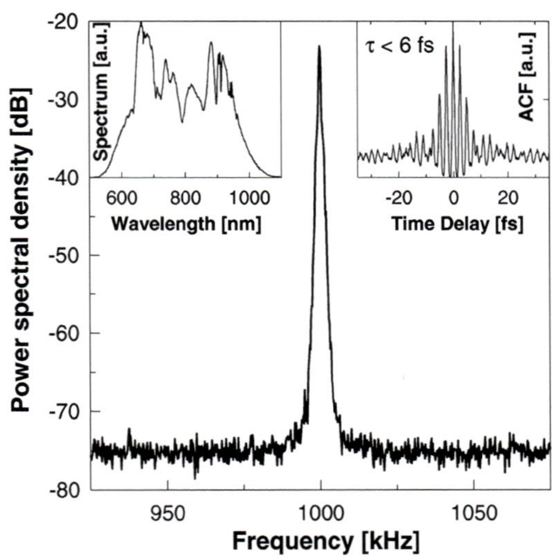

Fig. 4. Power spectral density of the in-lock carrier-slip beat signal at 540 nm recorded with a 1-kHz-resolution bandwidth. Spectrum (*left inset*) and interferometric autocorrelation (*right inset*) of the pulses exiting the pulse compressor consisting of the single-mode fiber SMF and chirped mirrors

ring. It is important to note that this degree of control can be achieved *without stabilizing f_r to an rf oscillator.* Minor variations in f_r merely shift the instants at which a pulse exits the laser without affecting phase locking, because the latter is performed at $f_{local} \ll f_r$ and is hence incapable of resolving this event. With f_r 'freely floating' merely the absence of synchronism of the pulse train with f_{local} has some negative implication. The selection of phase-stable pulses from the train by triggering the pulse picker using the local oscillator is tainted with a 'sampling' uncertainty of $\delta\varphi_{error} \leq 2\pi(f_{ceo}/f_r)$ due to desynchronism. This source of carrier-envelope phase uncertainty can be fully eliminated by simply deriving f_{local} from f_r via frequency division, *without the need for stabilizing f_{ceo} and f_r separately.*

As discussed above, cycle-slip-free operation is necessary but not sufficient for synthesizing reproducible waveforms in the time domain. To gain direct optical access to the sub-cycle jitter of φ_n, we utilize the fact that the parameter primarily controlled by the servo loop, namely the pulse energy W_n: (i) can be directly and permanently monitored; and (ii) is directly connected to $\Delta\varphi_n$ (see Fig. 3). As a consequence, the carrier-envelope phase jitter $\sigma_{\varphi(W)}$ that accumulates over T_m due to (controlled or random) variation of W_n can be determined from the power spectral density $S_W(f)$ of pulse energy variations [5] as

$$\sigma_{\varphi(W)}(T_m) = \frac{W_0}{\sqrt{2\pi}} \left|\frac{\partial\Delta\varphi}{\partial W}\right|_{W_0} \left(\int_{1/T_m}^{f_r/2} S_W(f) \frac{f_r^2}{f^2} \mathrm{d}f \right)^{1/2}, \quad (3)$$

where W_0 is the average intracavity pulse energy and we have made use of the approximately linear dependence of $\Delta\varphi$ on $\Delta W/W$ with a slope of $W_0|\partial\Delta\varphi/\partial W|_{W_0} = 2$ rad at $W_0 = 60$ nJ [2]. The higher weight of low-frequency contributions to $\sigma_{\varphi(W)}$ is due to the increased time over which a deviation with frozen sign can accumulate.

Figure 5 depicts the measured $S_W(f)$ at low frequencies and the calculated $\sigma_{\varphi(W)}(T_m)$ for phase-locked as well as unlocked operation. The increase of $\sigma_{\varphi(W)}$ with T_m well beyond 2π for $T_m < 1$ s can be reconciled with the observation of cycle-slip-free operation (over periods of $T_m \geq 10$ s) by assuming substantial carrier-envelope phase jitter originating from effects other than pulse energy fluctuations below frequencies of several tens of Hz. Hence, in this range a significant fraction of the pulse energy fluctuations have to be 'rephased' by the control loop to compensate a jitter from other sources and $\sigma_{\varphi(W)}$ mirrors this jitter.

Although the phase jitter originating from effects other than energy noise rapidly increases with T_m up to our longest measurement times (see upper thick grey line in Fig. 5), energy-noise-induced phase jitter becomes vastly dominant for $T_m > 1$ s, as indicated by the dramatic noise reduction under locked conditions at frequencies below 1 Hz. This leads to a root mean square of the residual pulse energy fluctuations of $\sigma_W \approx 0.01\%$ over the spectral range 0.1 Hz – 0.1 MHz at the output of the carrier envelope-phase-locked oscillator,

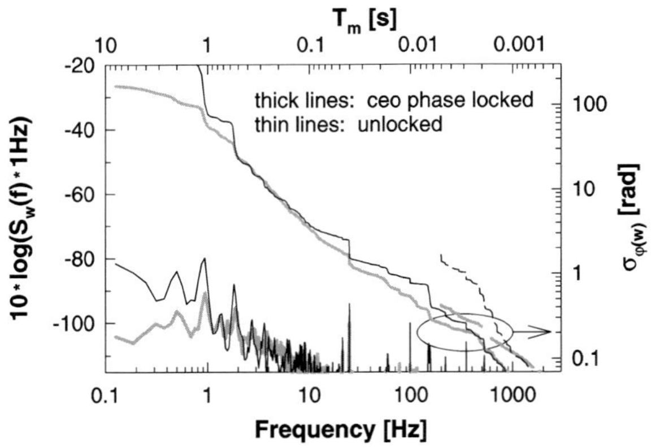

Fig. 5. *Lower traces*: Power spectral density $S_W(f)$ of the pulse energy fluctuations or its controlled variations at the output of the sub-10-fs Ti:sapphire laser with the servo loop interrupted (*thin black line*) and closed (*thick grey line*), shown for $f < 10$ Hz, where the two spectra conspicuously deviate. *Upper traces*: Carrier-envelope phase jitter $\sigma_{\varphi(W)}$ as a function of T_m introduced by pulse energy variations, with the servo loop interrupted (*thin black full and dashed lines*) and closed (*thick grey full and dashed lines*), as calculated from (3) by using the respective $S_W(f)$. The *dashed lines* represent $\sigma_{\varphi(W)}(T_m)$ obtained from a different data set, recorded with a somewhat 'noisier' laser. Nevertheless, the servo loop suppresses the phase jitter (by dampening the energy fluctuations) to a comparable level as in the 'quiet' laser (*full lines*), providing evidence for the robustness of phase locking. The arrow indicates the phase jitter level (≈ 0.2 rad) at which the servo loop starts responding

which is, to our knowledge, an unprecedented value. This is enhanced by somewhat less than a factor of 10 at the fiber output (due presumably to mechanically induced beam pointing instabilities at the fiber input), resulting in a reproducibility of the field amplitudes to within less than 3% in the sub-6-fs pulse train.

As compared to the sub-Hz regime, smaller but clearly notable and reproducible dampening of pulse energy fluctuations is caused by the servo loop in the sub-kHz range, as is evident from a corresponding suppression of $\sigma_{\varphi(W)}$, indicating the dominance of energy noise in this range. In repeated measurements (see also the dashed lines), $\sigma_{\varphi(W)}$ gets reproducibly suppressed at frequencies where it approaches ≈ 0.3 rad for increasing T_m. The highest frequency at which the suppression of $\sigma_{\varphi(W)}$ sets in has ranged from 500 Hz to 1000 Hz in our experiments, depending on the relevant noise level (with the loop opened), which has varied somewhat in repeated measurements. Because energy-noise-induced jitter appears to dominate for $T_m < 10$ ms, from these observations we may infer that the jitter of the carrier-envelope phase is kept safely below 0.3 rad by the servo loop. This value is comparable to $\delta\varphi_{error} \approx \pi/12$ and it will be interesting to see, whether the carrier-envelope phase jitter can be further lowered by deriving f_{local} from f_r in the future.

Using the SPIDER technique [10] for determining the amplitude envelope and frequency chirp, we can determine the range of waveforms our light generator can 'synthesize' reproducibly at regular times. The full line depicts such a waveform in Fig. 6, which is reproduced within the thickness of the line in each pulse following the trigger edge of a clock signal from the local oscillator. Shifting the phase of this trigger edge with respect to the signal controlling the servo loop results in a corresponding shift of the carrier with respect to the amplitude enve-

---

[2] $\sigma_{\varphi(W)}$ here is defined in the same way as σ_φ in (2) except that perturbations to φ here results, by definition, from pulse energy fluctuations only. Expression (3) is obtained by replacing in (3) of [5] the lower boundary of integration with $1/T_m$, rather than the higher boundary, as mistakenly stated in the context of Fig. 3 in [5].

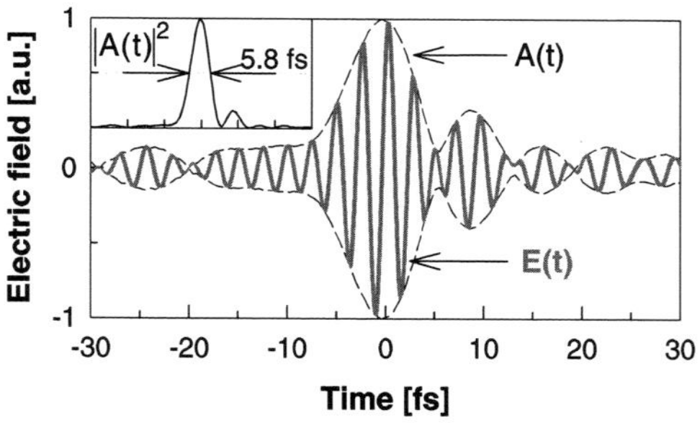

Fig. 6. Electric field $E(t) = A(t)\exp[-i\omega_0 t + i\varphi] + \text{c.c.}$ of a waveform (*full line*) emitted by our carrier-envelope-phase-locked few-cycle pulse generator at a particular setting of the trigger phase from the local oscillator yielding $\varphi = 1.25$ rad $[\gamma(0) = 0]$, as obtained from a SPIDER measurement. The intensity envelope $I(t) \propto A(t)^2$ in the *inset* has a half-width of 5.8 fs. Contrasting this with the computed 3.5-fs duration of the pulse in the absence of phase errors indicates the presence of a significant chirp $\gamma(t)$ carried by the electric field oscillations. The waveform is reproduced with an accuracy represented by its thickness at a repetition rate of $f_{\text{ocal}} = 1$ MHz

lope (and chirp). Although the absolute phase is still unknown and hence no definite relation between the trigger phase and a particular waveform can be established, the presented optical waveform generation, due to its unique reproducibility, opens the door for exploring nonlinear optical effects sensitive to the absolute phase φ [1]. In spite of the similarities in the techniques applied *no previous experimental work warrants these claims,* indicating the need for a different approach to synthesizing frequencies and waveforms.

Upon their observation, phase-sensitive nonlinear optical effects can immediately be exploited for determining the absolute value of φ and assigning it to the corresponding trigger phase in the demonstrated few-cycle waveform generator. This step will, in combination with shaping the amplitude envelope and tailoring the chirp using frequency-domain techniques [11], allow the synthesis of few-cycle light with arbitrary waveforms. Extending this capability from radio frequencies to the optical regime opens up new prospects in a number of fields where light is used for triggering, tracing and controlling microscopic processes.

Acknowledgements. The SPIDER measurements of M. Laubscher, A. Müller and B. Bacovic are gratefully acknowledged. This research has been sponsored by the Austrian Science Fund, Grants Y44-PHY and F016, as well as by INTAS, Grant 97-1058.

References

1. T. Brabec, F. Krausz: Rev. Mod. Phys. **72**, 545 (2000)
2. S.A. Diddams D.J. Jones, J. Ye, S.T. Cundiff, J.L. Hall, J.K. Ranka, R.S. Windeler, R. Holzwarth, T. Udem, T.W. Hänsch: Phys. Rev. Lett. **84**, 5102 (2000)
3. D.J. Jones, S.A. Diddams, J.K. Ranka, A. Stentz, R.S. Windeler, J.L. Hall, S.T. Cundiff: Science **288**, 635 (2000)
4. R. Holzwarth, T. Udem, T.W. Hänsch, J.C. Knight, W.J. Wadsworth, P.S.J. Russel: Phys. Rev. Lett. **85**, 2264 (2000)
5. L. Xu, C. Spielmann, A. Poppe, T. Brabec, F. Krausz, T.W. Hänsch: Opt.Lett. **21**, 2008 (1996)
6. J.N. Eckstein: Ph.D. dissertation (Stanford University 1978)
7. H.R. Telle, G. Steinmeyer, A.E. Dunlop, J. Stenger, D.H. Sutter, U. Keller: Appl. Phys. B **69**, 327 (1999)
8. J. Reichert, R. Holzwarth, T. Udem, T.W. Hänsch: Opt. Commun. **172**, 59 (1999)
9. A. Apolonski, A. Poppe, G. Tempea, Ch. Spielmann, T. Udem, R. Holzwarth, T.W. Hänsch, F. Krausz: Phys. Rev. Lett. **85**, 740 (2000)
10. I. Walmsley, V. Wong: J. Opt. Soc. Am. B **13**, 2453 (1996)
11. A.M. Weiner: Rev. Sci. Instrum. **71**, 1929 (2000)

OPTICS LETTERS / Vol. 25, No. 1 / January 1, 2000

Determining the absolute carrier phase of a few-cycle laser pulse

P. Dietrich

Institut für Experimentalphysik, Freie Universität Berlin, Arnimallee 14, 14195 Berlin, Germany

F. Krausz

Institut für Angewandte Elektronik und Quantenelektronik, Technische Universität Wien, 1040 Vienna, Austria

P. B. Corkum

Steacie Institute for Molecular Sciences, National Research Council, Ottawa, Ontario K1A 0R6, Canada

Received September 20, 1999

In a strong laser field, electrons tunnel from an atom at a rate determined by the instantaneous field. If the pulse is only a few cycles in duration, the highly nonlinear nature of tunnel ionization ensures that the resultant electron wave packet is primarily formed in less than one period. Measuring the direction of above-threshold-ionization electrons produced by circularly polarized light provides a direct method of measuring the absolute carrier phase of a single pulse. The method is robust, surviving spatial and temporal integration as well as intensity fluctuations. © 2000 Optical Society of America

OCIS codes: 270.1670, 270.4180, 270.6620.

Although it is possible to produce high-power laser pulses with durations of a few periods both directly from oscillators[1] and by optical pulse compression,[2] so far the absolute carrier phase is undetermined. Changes in the carrier phase from one oscillator pulse to the next can be minimized by introduction of dispersive material inside the cavity,[3] but, without a method for measuring the absolute carrier phase, it is not possible to know the phase or to prevent its slow drift. We show that the absolute carrier phase can be measured by use of the angular distribution of above-threshold electrons.

Quantum mechanically, phase dependence (including sensitivity to the absolute carrier phase) arises from interference of multiple pathways to the same final state.[4] Only short pulses, or high-order nonlinear phenomena, permit this kind of interference from within the same pulse. High harmonic generation (HHG) is a high-order process, and its use for measuring the absolute carrier phase has been proposed.[5] However, HHG requires the coherent response of many atoms in a complex, ionizing environment. Therefore a dependence on the absolute carrier phase is found only for the highest harmonics and is highly sensitive to intensity fluctuations, making its use difficult.

If the pulse is short enough, high-order nonlinearities are not necessary. By selecting electrons that can be produced by two or more low-order pathways, one can observe a dependence on the absolute carrier phase.[6] Compared with HHG, multiphoton ionization is less complex because it is a single-atom process. Such low-order processes, however, place severe constraints on the electron spectral resolution unless the pulse is short indeed.

Here we concentrate on the single-atom above-threshold-ionization (ATI) response, using circularly polarized light. We study processes in which more than 100 photons are absorbed, so a classical description of strong-field ionization is appropriate. We use circularly polarized light for several reasons. First, the amplitude of the electric field varies smoothly with time (Fig. 1). Therefore ionization is not modulated by the rapidly changing electric field as for linearly polarized light. Second, once it is ionized, circularly polarized light prevents electron rescattering off the parent ion.[7] Consequently, the direction of the above-threshold electrons is uniquely determined by the absolute carrier phase of the laser at the time of ionization.

We apply the quasi-static procedure[8] to describe ATI. It makes our approach intuitively clear because it naturally follows the oscillations of the electric field. The quasi-static procedure can be derived from a quantum-mechanical treatment[9] and has been

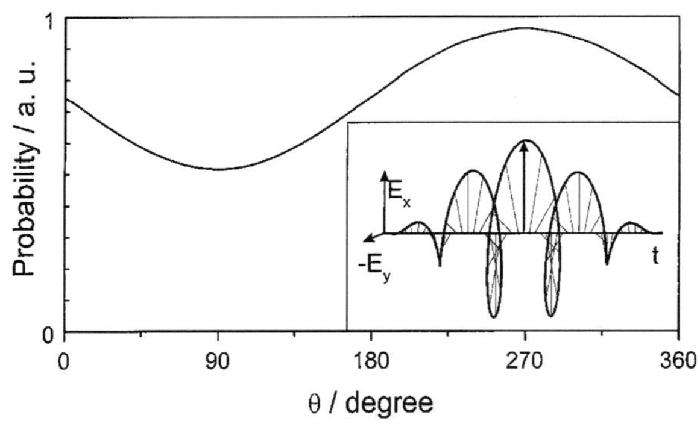

Fig. 1. Angular electron distribution for tunnel ionization of He with circularly polarized light at 800 nm with 4.8-fs duration. The absolute phase ϕ was set to 0. The distribution peaks in the direction $270° \triangleq -90°$. Inset, the time-dependent electric field vector rotating around the direction of propagation. Changing the absolute phase of the pulse will rotate the electric field so the direction at $t = 0$ is changed.

0146-9592/00/010016-03$15.00/0

January 1, 2000 / Vol. 25, No. 1 / OPTICS LETTERS

confirmed by measurements of the ATI[10] and the HHG.[11] First we determine the probability of ionization as function of the instantaneous laser field, using Ammosov–Delone–Krainov tunneling rates.[12] Second, the motion of the electrons after ionization is calculated from Newton's equations.

The final direction of the electron rotates with the electric field vector. This motion consists of a large drift in the direction 90° ahead of the electric field direction at the time of ionization and a circular component whose velocity goes to zero as the field goes to zero. For a long pulse the angular electron distribution is isotropic because tunneling is equally likely for any phase, but for short pulses significant ionization occurs during less than an optical cycle. Thus the absolute carrier phase determines the direction of the field at the moment of maximum ionization and therefore the direction of the electrons (but not the total ionization yield). Although changes in the fluence will change the total yield, for circular polarization it is possible to find conditions in which the phase dependence of the angular distribution is robust.

Figure 1 shows the angular distribution of the ATI electrons from He for a 4.8-fs pulse and a peak field of $6 \times 10^{10}\,\mathrm{V\,m^{-1}}$. We calculate the number of electrons per time step (less than 10^{-3} of the optical period) brought into the continuum. Then the equations of motion for the electrons in the time-varying field $\mathbf{E}(t) = E_{\mathrm{env}}(t)[\mathbf{e}_x \cos(\omega t + \phi) + \mathbf{e}_y \sin(\omega t + \phi)]$ are integrated analytically with the initial condition $v_0 = 0$. The final velocities $v_{x,y}(\infty) = -(e/m)\int_{t_0}^{\infty} E_{x,y}(t)\mathrm{d}t = (e/m)[A_{x,y}(\infty) - A_{x,y}(t_0)]$ are given by the vector potential A at the time of the ionization t_0 and the end of the pulse. The final angle θ is given by $\tan\theta = v_y(\infty)/v_x(\infty)$ (i.e, $\theta = 0$ corresponds to the x direction), from which we obtain the angular distribution by summing all trajectories with their respective weights that are due to the ionization probability. We use $\phi = 0$ throughout these calculations. A change $\Delta\phi$ of the absolute phase rotates the angular distribution by $\Delta\phi$. The envelope of the electric field is assumed to be $E_{\mathrm{env}}(t) = E_0 \cos^2(\pi t/\tau)$, with $t \in [-\tau/2, \tau/2]$. The difference from a Gaussian profile in the wings of the profile is not significant because ionization occurs in the central part of the pulse. The half-width of the intensity is $\Delta\tau \approx 0.3640\tau$, the peak intensity is $I_0 = \epsilon_0 c E_0^2/2$, and the fluence of the pulse is $F = I_0 3\tau/8$. We characterize the angular distributions by the direction θ_0 of the angular distribution $S(\theta)$ and by the contrast $R_c = S_{\max}/S_{\min}$. In Fig. 1 the direction is $\theta_0 = 270°$ or $\theta_0 = -90°$ and the contrast is $R_c = 1.9$. No spatial intensity distribution was taken into account in this example.

The solid curves in Fig. 2 represent results without spatial integration. Figure 2(a) shows the angle of the peak of the electron distribution as a function of the peak electric field for ionization of He with pulses of 4.8-fs duration (FWHM) and a center wavelength of $\lambda = 800$ nm. For low fields the direction is $-90°$ and the contrast R_c is high because ionization occurs mainly during a small fraction of the pulse about the peak. With increasing intensity, ionization occurs preferentially on the rising part of the pulse, and the direction changes [solid curve in Fig. 2(a)]. The direction of the electrons is thus a measure of the phase at which most ionization occurred. With increasing field the angular distribution also becomes more isotropic, with minimum contrast at a field where ionization starts to saturate [Fig. 2(a)]. Increasing the field further increases the contrast again because ionization is confined to an increasingly shorter part of the rising edge [solid curve in Fig. 2(b)]. From Fig. 2 we conclude that the angular electron distribution does not change with changes in the peak field for low intensities.

It is experimentally difficult to place He atoms only at the center of the laser focus, so we must treat the spatial field distribution. First, the pulses have a transverse intensity distribution that is assumed to be Gaussian. At the spatial center of a high-intensity pulse, ionization will occur before the temporal peak of the pulse. In contrast, in the wings of the radial profile, ionization occurs mainly at the temporal peak. Inasmuch as the direction of the angular distribution depends on the direction of the electric field at the time of ionization, integration over the radial intensity distribution will smear the angular distribution, thus reducing the contrast. The dashed curves in Fig. 2 show the results for the direction and the contrast after radial integration. The results are unchanged as long as the intensity is below saturation because ionization occurs only in the center of the beam.

The second effect that reduces the contrast is the intensity variation in the propagation direction. The effects are similar to the influence of a radial variation. In addition, there are phase variations of the pulse in the radial direction owing to the curvature of the focused wave fronts and axially owing to the Guoy phase. Both variations further smear the angular distribution at high intensities, and the resultant anisotropy is small (dotted curves in Fig. 2).

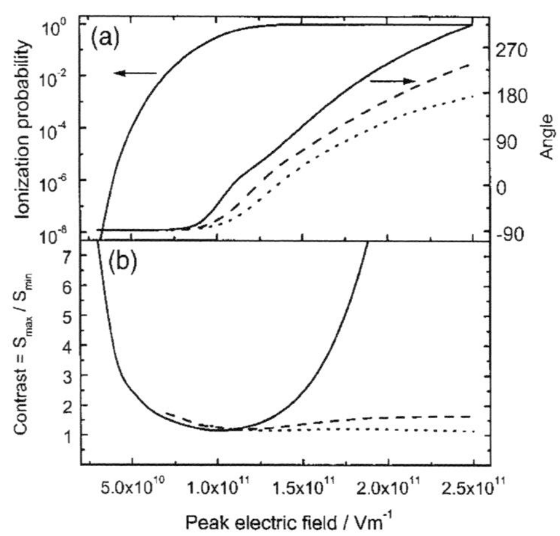

Fig. 2. (a) Direction (right-hand axis) and (b) contrast of the angular electron distribution as a function of the peak electric field for strong-field ionization of He by circularly polarized light (800 nm, 4.8 fs). The ionization probability is given in (a), left axis, without spatial integration. Solid curves, no spatial integration; dashed curves, radial integration; dotted curves, full spatial integration.

OPTICS LETTERS / Vol. 25, No. 1 / January 1, 2000

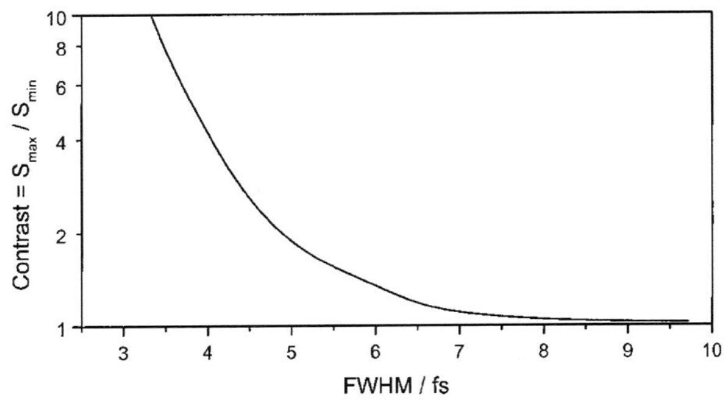

Fig. 3. Contrast of the angular electron distribution as a function of pulse duration calculated for a constant ionization yield of He (10^{-3}).

These results suggest that the absolute carrier phase can be measured on a single-shot basis. Comparing the results in Fig. 2 shows that the optimum regime is at low fields [see Fig. 2(a)], where ionization occurs only at the peak of the field. The electron distribution is then unambiguously related to the absolute carrier phase, and amplitude fluctuations do not change the direction.

So far we have assumed state-of-the-art pulses. For longer pulses the directionality decreases because ionization occurs during a larger time interval. Figure 3 shows the dependence of the contrast on the pulse duration, calculated for an ionization yield of 0.1%. The contrast increases as the laser fluence decreases [Fig. 2(b)]. Thus for longer pulses one could use lower peak fields, but this would reduce the signal (i.e., the number of electrons). For an ionization probability of 2×10^{-6}, the contrast is 1.44 for a 6.3-fs pulse.

Our approach, although it is well founded, ignores some quantum-mechanical effects. We treat the electrons classically, neglecting the spatial extent of the wave function and the distribution of initial velocities. Inasmuch as we are measuring the electron distribution at macroscopic distances, the velocity distribution can lead to a blurring of the angular distribution. For small variations of the velocities Δv the corresponding change of the angle is $\Delta\theta \approx \Delta v/v$, where v is the final velocity, which scales with the electric field. The typical spreading of the electron wave function Δv for these cases is $\Delta v \simeq 0.7$ nm/fs.[7,9,13] Using $v \approx 7.5 \times 10^6\ \mathrm{m\,s^{-1}}$ at $E = 10^{11}\ \mathrm{V\,m^{-1}}$, we obtain $\Delta\theta \approx 0.1$, which is much smaller than the width of the angular distribution. The results in Fig. 2 are almost unaffected.

We emphasize the practicability of our approach. Using parameters from Ref. 14 (5-fs duration, 100-μJ energy), we can obtain a focal volume of $6 \times 10^{-12}\ \mathrm{cm}^3$ (FWHM focal diameter, $\approx 30\ \mu$m) while maintaining a peak intensity of $10^{15}\ \mathrm{W\,cm^{-2}}$. Assuming a gas density of 10^{-3} Torr (achievable in a supersonic jet expansion), there are 10^7 atoms in the focal volume. We can limit ionization probability to 10^{-4} and still produce 1000 electrons per pulse, sufficient for single-shot electron imaging.[15]

In conclusion, femtosecond science aims to produce intense electric fields with almost arbitrary time structure (consistent with Maxwell's equations). Doing so requires measurement and control of the absolute carrier phase. Multiphoton ionization with circularly polarized light offers a realistic single-shot measurement method.

P. Corkum's e-mail address is paul.corkum@nrc.ca.

References

1. J. Zhou, G. Taft, C.-P. Huang, M. M. Murnane, H. C. Kapteyn, and I. P. Christov, Opt. Lett. **19,** 1149 (1994); I. D. Jung, F. X. Kärtner, N. Matuschek, D. H. Sutter, F. Morier-Genoud, G. Zhang, U. Keller, V. Scheuer, M. Tilsch, and T. Tschudi, Opt. Lett. **22,** 1009 (1997); L. Xu, G. Tempea, Ch. Spielmann, and F. Krausz, Opt. Lett. **23,** 789 (1998).
2. M. Nisoli, S. DeSilvestri, O. Svelto, R. Szipöcs, K. Ferencz, Ch. Spielmann, S. Sartania, and F. Krausz, Opt. Lett. **22,** 522 (1997); A. Baltuka, Z. Wei, M. S. Pshenichnikov, and D. A. Wiersma, Opt. Lett. **22,** 102 (1997).
3. L. Xu, Ch. Spielmann, A. Poppe, T. Brabec, F. Krausz, and T. W. Hänsch, Opt. Lett. **21,** 2008 (1996).
4. P. Brumer and M. Shapiro, Chem. Phys. Lett. **126,** 541 (1986); Annu. Rev. Phys. Chem. **43,** 257 (1992).
5. F. Krausz, T. Brabec, M. Schnürer, and Ch. Spielmann, Opt. Photon. News **9**(7), 46 (1998); A. de Bohan, Ph. Antoine, D. B. Miloević, and B. Piraux, Phys. Rev. Lett. **81,** 1837 (1998); G. Tempea, M. Geissler, and T. Brabec, J. Opt. Soc. Am. B **16,** 669 (1999).
6. E. Cormier and P. Lambropoulos, Eur. Phys. J. D **2,** 15 (1998).
7. P. Dietrich, N. H. Burnett, M. Yu. Ivanov, and P. B. Corkum, Phys. Rev. A **50,** R3585 (1994).
8. P. B. Corkum, Phys. Rev. Lett. **71,** 1994 (1993).
9. M. Lewenstein, Ph. Balcou, M. Yu. Ivanov, A. L'Huillier, and P. B. Corkum, Phys. Rev. A **49,** 2117 (1994).
10. G. G. Paulus, W. Nicklich, H. Xu, P. Lambopoulos, and H. Walther, Phys. Rev. Lett. **72,** 2851 (1994).
11. A. L'Huillier, M. Lewenstein, P. Salières, Ph. Balcou, M. Yu. Ivanov, J. Larsson, and C. G. Wahlström, Phys. Rev. A **48,** R3433 (1993).
12. M. V. Ammosov, N. B. Delone, and V. Kraĭnov, Sov. Phys. JETP **64,** 1191 (1986).
13. P. B. Corkum, N. H. Burnett, and M. Y. Ivanov, Opt. Lett. **19,** 1870 (1994).
14. S. Sartania, Z. Cheng, M. Lenzner, G. Tempea, Ch. Spielmann, F. Krausz, and K. Ferencz, Opt. Lett. **22,** 1562 (1997).
15. H. Helm, N. Bjerre, M. J. Dyer, D. L. Huestis, and M. Saeed, Phys. Rev. Lett. **70,** 3221 (1993).

VOLUME 91, NUMBER 25 PHYSICAL REVIEW LETTERS week ending 19 DECEMBER 2003

Measurement of the Phase of Few-Cycle Laser Pulses

G. G. Paulus,[1,2,3] F. Lindner,[2] H. Walther,[2,3] A. Baltuška,[4] E. Goulielmakis,[4] M. Lezius,[4,5] and F. Krausz[2,4]

[1]*Department of Physics, Texas A&M University, College Station, Texas 77843-4242, USA*
[2]*Max-Planck-Institut für Quantenoptik, 85748 Garching, Germany*
[3]*Ludwig-Maximilians-Universität München, 85748 Garching, Germany*
[4]*Institut für Photonik, Technische Universität Wien, Gusshausstrasse 27, A-1040 Wien, Austria*
[5]*Institut für Ionenphysik, Universität Innsbruck, Technikerstrasse 25, A-6020 Innsbruck, Austria*
(Received 17 June 2003; published 18 December 2003)

For the shortest pulses generated to date, the amplitude of the electromagnetic wave changes almost as rapidly as the field oscillates. The temporal variation of the field, which directly governs strong-field interactions, therefore depends on whether the maximum of the pulse amplitude coincides with that of the wave cycle or not, i.e., on the phase of the field with respect to the pulse envelope. It is demonstrated that the direction of electron emission from photoionized atoms can be controlled by varying the phase of the field, providing for the first time a tool for its accurate determination. Directing fast electron emission to the right or to the left with the light phase constitutes a new kind of coherent control.

DOI: 10.1103/PhysRevLett.91.253004 PACS numbers: 32.80.Rm, 42.50.Hz, 42.65.Re

Generation of laser pulses with a pulse energy around 500 μJ and duration of 5 fs (measured at full width at half maximum) has become routine in past few years [1,2]. One of the most intriguing features of such pulses is that their envelope varies almost as fast as the electromagnetic field itself. Considering a typical wavelength of 760 nm, an optical cycle has a duration of 2.5 fs; i.e., the pulses consist of very few optical cycles only (few-cycle pulses). One of the most momentous consequences is that the time variation of the laser field will depend on the phase of the carrier frequency with respect to the envelope, the so-called carrier-envelope (CE) phase. Although few-cycle pulses have hitherto mainly been used for generating coherent soft x rays [3,4] and attosecond pulses [5], there is a high potential for applications in many other fields, ranging from coherent control of molecular dynamics to particle acceleration. All the processes involved depend on the time variation of the field and thus on the CE phase. The ability to measure and precisely stabilize the CE phase is therefore a crucial point for all applications. So far, clear evidence of phase effects has been seen in photoionization without phase control for the few-cycle pulses [6], in soft x-ray generation with phase-stabilized pulses [7], and virtually at the same time as the present experiment in photoemission from surfaces with non-amplified few-cycle pulses [8]. Phase-dependent effects in semiconductors have been predicted [9]. Here we present experimental results of nonlinear photoionization with intense phase-stabilized few-cycle visible light pulses. The spatially asymmetric field of few-cycle laser pulses leads to different photoionization rates for opposite directions. By detecting this difference it was possible for the first time to measure the CE phase of powerful few-cycle laser pulses.

The electric part $\mathcal{E}(t)$ of the field of a laser pulse can be written as

$$\vec{\mathcal{E}}(t) = \vec{e}_x \mathcal{E}_0(t) \cos(\omega t + \varphi), \quad (1)$$

where $\vec{e}_x$ denotes the axis of polarization, $\mathcal{E}_0(t)$ denotes the envelope of the pulse, $\omega/2\pi$ denotes the carrier frequency of the laser, and φ denotes the phase. The latter describes a general property of the pulse and is therefore also commonly referred to as the "absolute" phase. For a few-cycle pulse, the width of $\mathcal{E}_0(t)$ (measured at half the maximum) is about 2 times the optical period $2\pi/\omega$. With respect to φ, the convention of choosing $t = 0$ at the maximum of the envelope is used. Then, $\varphi = 0$ corresponds to a "cosine-like" pulse with the absolute maximum of the electric field pointing to the right. Accordingly, we speak of a sine-like pulse for $\varphi = -\pi/2$ (see Fig. 1). Evidently, it is possible to tailor the field of ultrashort pulses by controlling their absolute (CE) phase.

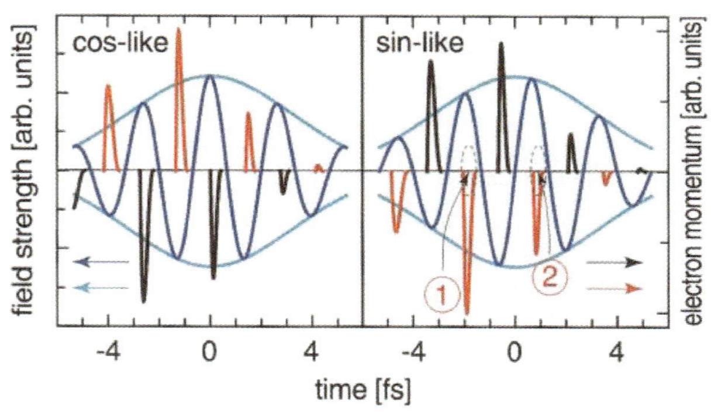

FIG. 1 (color). Electric field (blue) and envelope (light blue) of a 5 fs laser pulse for CE phase 0 (left panel) and $-\pi/2$ (right panel). The drift momentum of high-energy photoelectrons as a function of the ionization time t_0 is shown in black (emission to the right) and red (emission to the left). It was calculated with a classical model. Note that highest momentum does not necessarily coincide with highest electron yield: In the right panel the ionization probability at times labeled with 1 is considerably lower than those labeled with 2 because of lower field strength at t_0.

0031-9007/03/91(25)/253004(4)$20.00

VOLUME 91, NUMBER 25 PHYSICAL REVIEW LETTERS week ending 19 DECEMBER 2003

Conventional femtosecond lasers, however, generate pulses with more or less randomly changing phase. Its stabilization is routinely performed for femtosecond laser oscillators and has led to a breakthrough in frequency metrology [10–12]; for a recent review, see [13]. In contrast, phase stabilization for amplified pulses is a very recent achievement [7]. Considering the asymmetry of the field of few-cycle pulses [14], it appears that an obvious approach is to look for the signatures of the CE phase in nonlinear photoionization. Particular emphasis is given to noninversion symmetric emission of the photoelectrons, i.e., a different count rate for electrons emitted to the right and to the left. A number of theoretical investigations have been devoted to the problem in the past years [15–20]. Most of them concentrate on total yields of photoelectrons.

In our experiments xenon atoms were ionized with 5 fs long laser pulse of peak intensities of nearly 10^{14} W/cm^2. Absorption of eight photons is necessary in order to subdue the ionization threshold. As a matter of fact, processes of much higher order, i.e., absorption of photons in excess of those necessary for ionization, can be observed, and thus correspondingly high photoelectron energies. This phenomenon is known as above-threshold ionization (ATI) [21]. Our experimental approach to analyzing this phenomenon relies on time-of-flight spectroscopy; see Fig. 2.

Figure 3(a) shows a series of ATI spectra corresponding to different CE phases. Pulses with known phase differences $\Delta\varphi$ can be realized by delaying the envelope with respect to the carrier. Since the envelope propagates with the group velocity and the carrier with the phase velocity of light, glass dispersion can be used to shift the CE phase. At a central wavelength of 760 nm adding 52 μm of fused silica will change φ by 2π without distinctly affecting the pulse duration. Thus, two glass wedges which can be shifted with respect to each other allow any phase to be adjusted (see Fig. 2). Simply by inspecting the high-energy part (> 20 eV) of the spectra one can verify that $\Delta\varphi = \pi$ corresponds to a change from left to right, while $\Delta\varphi = 2\pi$ faithfully reproduces the spectra. For high electron energies, the ratio of left/right electron yield approaches 1 order of magnitude and can be controlled by adjusting the CE phase. This constitutes a new kind of *coherent control* and, to our knowledge, is the highest contrast observed so far.

Figure 3(b) displays the ratio of photoemission to the left and to the right for different spectral regions. The ratio of high- and low-energy electrons oscillates with a periodicity of $\Delta\varphi$ in excellent agreement with the periodicity expected from glass dispersion. Besides the much higher contrast for high-energy electrons, a phase shift of both sets of data is also apparent. To determine the CE phase we have to establish the connection between the phase of the field and the ATI signal. This is done by using the high-energy (> 20 eV) electrons. One reason certainly is their higher sensitivity to the CE phase. More importantly, the dominant overall features in this spectral region are known not to depend on specific properties of the atomic species or on intensity. Figure 4 shows a comparison between measurements and results from classical calculations, the latter taking into account only high-energy electrons. The concept of the calculation will be explained below. In the experimental as well as in the theoretical data the ratio of emission to the left and right depends in a characteristic way on electron energy and the CE phase. This leads to the inclined stripes in the false-color representations. Another common feature is that the asymmetry of left-right emission increases for increasing electron energy. It should be noted that, unlike the contrast of asymmetry, these characteristic features do *not* depend critically on pulse shape or pulse duration. The sharp edges visible in the calculation originate from classical energy conservation. This artifact is removed by energy-time uncertainty in quantum mechanics. Matching the characteristic features of the theoretical and experimental data leads

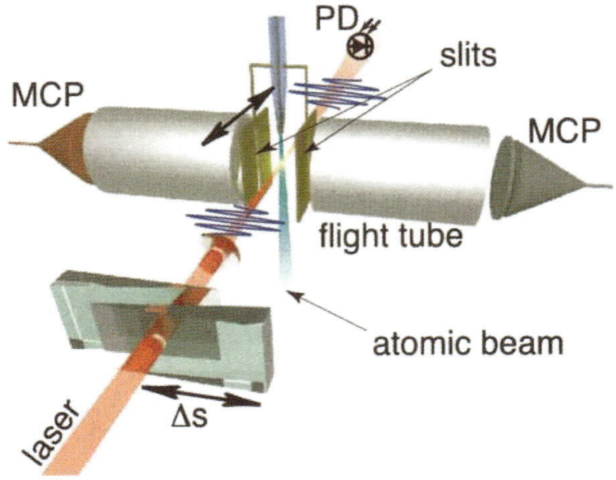

FIG. 2 (color). "Stereo-ATI" spectrometer. Two opposing electrically and magnetically shielded time-of-flight spectrometers are mounted in an ultrahigh vacuum apparatus. Xenon atoms fed in through a nozzle from the top are ionized in the focus of a few-cycle laser beam. The focal length is 250 mm (the lens shown in the sketch is in reality a concave mirror), and the pulse energy was attenuated to 20 μJ. The laser polarization is linear and parallel to the flight tubes. Note that the laser field changes sign while propagating through the focus. Slits with a width of 250 μm are used to discriminate electrons created outside the laser focus region chosen. The slits can be moved from outside the vacuum system. A photodiode (PD) and microchannel plates (MCP) detect the laser pulses and photoelectrons, respectively. Throughout this Letter, spectra corresponding to electrons detected with the left red-colored (right black-colored) detector are colored in red (black). The laser repetition rate is 1 kHz and 50 electrons per pulse are recorded at each MCP. A pair of glass wedges (apex angle 2.8°) is used to optimize dispersion and adjust the CE phase. Therefore, movement of one wedge by Δs introduces $\Delta x = \Delta s \tan(2.8°)$ glass.

VOLUME 91, NUMBER 25 PHYSICAL REVIEW LETTERS week ending 19 DECEMBER 2003

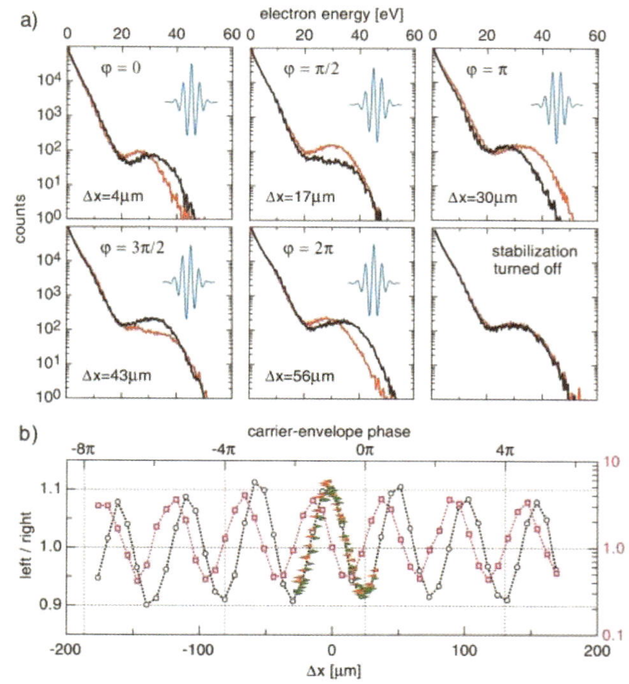

FIG. 3 (color). (a) Photoelectron spectra for different CE phases controlled by fine movement of one of the wedges. Δx indicates the glass added hereby. The black curves correspond to emission to the right (positive direction), the red ones to the opposite direction. The insets show the corresponding real time variation of the electric field, as deduced from the phase assignment shown in Fig. 4. Only without phase stabilization were identical spectra measured left and right as expected. (b) Left-right ratio of the total electron yield (circles) and high-energy electrons (squares) as a function of glass thickness Δx added or subtracted by moving one of the wedges. $\Delta x = 0$ corresponds to optimal dispersion compensation, i.e., the shortest pulses. Maximal left/right ratio for the total yield does not coincide with that for high-energy electrons. Note the different scales for low- and high-energy electrons. The upper x scale indicates the carrier-envelope phase of the pulse, as deduced from the comparison with theory shown in Fig. 4. The measurement of the CE phase drift was performed by a scan forth (triangles pointing to the right) and back (triangles pointing to the left) with finer variation of the glass thickness. The mismatch at the start and the end point indicates a phase drift of ≈ 50 mrad/min.

to an unambiguous determination of the CE phase with an estimated error of $\pi/10$.

The calculation used to determine the CE phase is based on the insight that, at high intensities, ATI can be understood in terms of quasistatic field ionization at some time t_0 and subsequent classical evolution of the electron in the laser field [22,23]. For a recent review, see [24]. A considerable fraction of the photoelectrons generated when atoms or molecules are exposed to intense laser pulses do not escape directly. Rather they return to the ion core within times of typically less than an optical cycle and in precise synchronization with the electric field of the laser pulse. The characteristics of returning electrons can be observed in the kinetic energy spectra of photoelectrons. The underlying mechanism is elastic scattering of the electrons when they return to the ion core at time $t = t_1$ and are subsequently accelerated by the laser field. As a consequence, photoelectrons with high kinetic energy can be observed. In the photoelectron spectra, they manifest themselves in the high-energy plateaulike annex to the exponentially decreasing count rate at low electron energies [25,26].

The kinetic energy spectrum of the plateau electrons is sensitive to the CE phase for several reasons. First of all, high-energy electrons returning to the ion core can be created only in subfemtosecond time intervals close to peaks of the electric field of the laser pulse (see Fig. 1). However, the probability that they tunnel through the atomic potential at t_0 depends exponentially on the field strength $\mathcal{E}(t_0)$ and as few-cycle pulses are involved it is likely only for those very few optical half-cycles close to the pulse maximum. Generally, the highest kinetic energies are reached for electrons returning to the core when the electric field becomes nearly zero [$\mathcal{E}(t_1) = 0$]. For few-cycle pulses in addition, the field amplitude $\mathcal{E}_0$ needs to be as high as possible for $t > t_1$ in order to allow efficient acceleration after rescattering. Since the start time t_0 and return time t_1 differ by almost one optical cycle, both conditions, namely, the highest possible field strength at t_0 and highest possible amplitude after t_1 are hard to meet and result in a strong dependence of photoionization on the CE phase. Number, strength, and timing of the wave packets lead to distinctive structures in the ATI spectra. Their analysis therefore provides detailed information about the key processes of attosecond science. Quantum mechanical calculations are in very good qualitative agreement with this classical treatment [20].

Having established that ATI can be used for precise determination of the CE phase, an example of the sensitivity of the instrument should be given. In Fig. 3(b) also a measurement with small steps $\Delta\varphi$ is shown. This measurement reveals small phase drifts, which have to be attributed to the laser system and its stabilization scheme. The phase drifts are of the order of 50 mrad/min, measured over more than 10 min. As the stereo-ATI phase meter not only can measure the CE phase and also detect small phase drifts, it clearly has the potential to be used in a feedback loop. In this respect it should be emphasized that the target gas pressure is so low ($< 10^{-4}$ mbar) that the ATI experiment does not affect the laser beam; i.e., a stereo-ATI phase meter can be placed anywhere in a laser beam line.

In summary, we have shown that strong-field photoionization provides a very efficient means of measuring the CE phase of few-cycle femtosecond laser pulses with *no* $\pm\pi$ phase ambiguity inherent in many other schemes. Conversely, the emission direction of photoelectrons can be steered by choosing the CE phase, establishing a new kind of coherent control scheme. Control of the CE phase allows one to tailor the field of a few-cycle pulse. A

VOLUME 91, NUMBER 25 PHYSICAL REVIEW LETTERS week ending 19 DECEMBER 2003

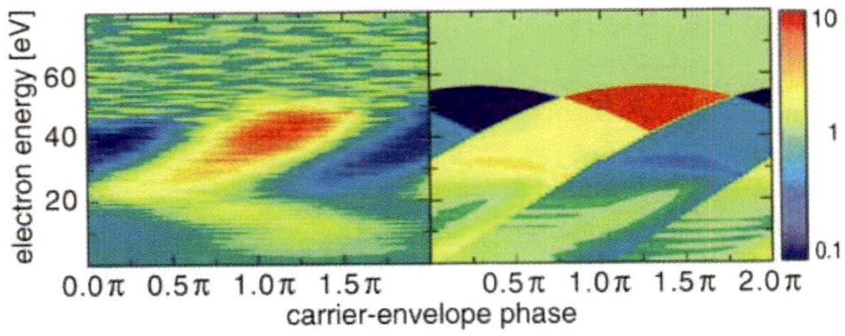

FIG. 4 (color). Determination of the CE phase. False-color representation (logarithmic scale) of the left-right ratio of photoelectrons as a function of the electron energy and the CE phase. Positive values (red) indicate dominant left emission, negative values (blue) dominant right emission. For the theoretical part (right panel), only rescattered electrons were taken into account. The parameters used are the same as in the experiment. This comparison allows precise carrier-envelope phase determination. With the convention of Eq. (1), the electric field is now uniquely characterized.

tailored field variation would deform molecular potentials in a virtually arbitrary way. This could help to induce or analyze exotic chemical reactions. In fact, it is possible to choose conditions such that the field vanishes upon the return of the wave packet. The very high current densities corresponding to the returning wave packets should facilitate electron-ion scattering experiments, hitherto hardly possible. Furthermore, investigation of interference patterns in ATI spectra will allow ionization to be studied in real time.

This work has been supported by the Austrian Science Fund (Grants No. F016, No. Z63, and No. P15382) and by The Welch Foundation (Grant No. A-1562).

[1] M. Nisoli, S. De Silvestri, O. Svelto, R. Szipöcs, K. Ferencz, Ch. Spielmann, S. Sartania, and F. Krausz, Opt. Lett. **22**, 522 (1997).

[2] G. Steinmeyer, D. Sutter, L. Gallmann, N. Matuschek, and U. Keller, Science **286**, 1507 (1999).

[3] Ch. Spielmann, N. H. Burnett, S. Sartania, R. Koppitsch, M. Schnürer, C. Kan, M. Lenzner, P. Wobrauschek, and F. Krausz, Science **278**, 661 (1997).

[4] P. Villoresi, P. Ceccherini, L. Poletto, G. Tondello, C. Altucci, R. Bruzzese, C. de Lisio, M. Nisoli, S. Stagira, G. Cerullo, S. De Silvestri, and O. Svelto, Phys. Rev. Lett. **85**, 2494 (2000).

[5] M. Hentschel, R. Kienberger, Ch. Spielmann, G. A. Reider, N. Milosevic, T. Brabec, P. Corkum, U. Heinzmann, M. Drescher, and F. Krausz, Nature (London) **414**, 509 (2001).

[6] G. G. Paulus, F. Grasbon, H. Walther, P. Villoresi, M. Nisoli, S. Stagira, E. Priori, and S. De Silvestri, Nature (London) **414**, 182 (2001).

[7] A. Baltuška, Th. Udem, M. Uiberacker, M. Hentschel, E. Goulielmakis, Ch. Gohle, R. Holzwarth, V. S. Yakovlev, A. Scrinzi, T. W. Hänsch, and F. Krausz, Nature (London) **421**, 611 (2003).

[8] A. Apolonski, P. Dombi, G. G. Paulus, M. Kakehata, R. Holzwarth, Th. Udem, Ch. Lemell, J. Burgdörfer, T.W. Hänsch, and F. Krausz, Phys. Rev. Lett. (to be published).

[9] O. D. Mücke, T. Tritschler, M. Wegener, U. Morgner, and F. X. Kärtner, Phys. Rev. Lett. **89**, 127401 (2002).

[10] J. Reichert, R. Holzwarth, Th. Udem, and T.W. Hänsch, Opt. Commun. **172**, 59 (1999).

[11] D. J. Jones, S. A. Diddams, J. K. Ranka, A. Stentz, R. S. Windeler, J. L. Hall, and S. T. Cundiff, Science **288**, 635 (2000).

[12] H. R. Telle, G. Steinmeyer, A. E. Dunlop, J. Stenger, D. H. Sutter, and U. Keller, Appl. Phys. B **69**, 327 (1999).

[13] S. T. Cundiff and Jun Ye, Rev. Mod. Phys. **75**, 325 (2003)

[14] A relationship of few-cycle pulses with bichromatic fields with commensurate frequencies is obvious: A. N. Chudinov, Yu. E. Kapitzky, A. A. Shulginov, and B. Ya. Zel'dovich, Opt. Quantum Electron. **23**, 1055 (1991); D.W. Schumacher, F. Weihe, H. G. Muller, and P. H. Bucksbaum, Phys. Rev. Lett. **73**, 1344 (1994).

[15] E. Cormier and P. Lambropoulos, Eur. Phys. J. D **2**, 15 (1998).

[16] P. Dietrich, F. Krausz, and P. B. Corkum, Opt. Lett. **25**, 16 (2000).

[17] I. P. Christov, Appl. Phys. B **70**, 459 (2000).

[18] R. M. Potvliege, N. J. Kylstra, and C. J. Joachain, J. Phys. B **33**, L743 (2000).

[19] S. Chelkowski and A. D. Bandrauk, Phys. Rev. A **65**, 061802(R) (2002).

[20] D. B. Milošević, G. G. Paulus, and W. Becker, Opt Express **11**, 1418 (2003)

[21] P. Agostini, F. Fabre, G. Mainfray, G. Petite, and N. K. Rahman, Phys. Rev. Lett. **42**, 1127 (1979).

[22] P. B. Corkum, Phys. Rev. Lett. **71**, 1994 (1993).

[23] K. C. Kulander, K. J. Schafer, and K. L. Krause, in *Super-Intense Laser-Atom Physics*, edited by B. Piraux, A. L'Huillier, and K. Rzazweski, NATO Advanced Study Institutes, Ser. B, Vol. 316 (Plenum, New York, 1993), p. 95.

[24] W. Becker, F. Grasbon, R. Kopold, D. B. Milošević, G. G. Paulus, and H. Walther, Adv. At. Mol. Opt. Phys. **48**, 35 (2002).

[25] G. G. Paulus, W. Nicklich, H. Xu, P. Lambropoulos, and H. Walther, Phys. Rev. Lett. **72**, 2851 (1994).

[26] G. G. Paulus, W. Becker, W. Nicklich, and H. Walther, J. Phys. B **27**, L703 (1994).

dispersion of the Karin family requires that its parent body was pre-fragmented or reaccumulated. This supports the overall picture that all large members of asteroid families are reaccumulated bodies, given that the Karin cluster is at least a second-generation family in the lineage of the older Koronis parent body.

Our simulations can also be used to determine the impact energy needed to produce a given degree of disruption as a function of the internal structure of the parent body. From Table 1, we see that disrupting a pre-shattered target requires less energy per unit mass than for a monolithic body. This, at first glance a surprising result, is related to the fact that the fractures as modelled here (no porosity and no material discontinuities except damage) do not affect shock waves but only tensile waves. Hence, fragments can be set in motion immediately upon being hit by the shock wave without having to wait for fracture to occur in a following tensile wave. Thus, transfer of momentum is more efficient and disruption facilitated. Note that the presence of large voids such as those in rubble piles affect the propagation of the shock wave, thus reversing this trend. A more detailed study of these properties will be presented in another paper. Nevertheless, it is already clear that the internal structure of an asteroid plays an important role in the determination of its response to impacts (see also ref. 18). This is not only relevant for estimating the collisional lifetime of a body in the asteroid belt, but also for developing strategies to deflect a potential Earth impactor.

A last important implication of our simulations concerns the properties of family members with sizes smaller than can currently be observed. The break-up of a pre-shattered Karin parent body produces several thousand fragments with sizes ranging from a few kilometres down to our resolution limit of 0.2 km (Fig. 1). Their ejection speeds can allow some of them to be injected into efficient transport mechanisms leading to Earth-crossing orbits. Therefore, we conclude that the Karin break-up event may well have produced some near-Earth asteroids and even some meteorities. According to our simulations, the potential near-Earth asteroids with size at least above 0.2 km result from the gravitational reaccumulation of smaller fragments, and are therefore all aggregates. □

Received 4 October; accepted 5 December 2002; doi:10.1038/nature01364.

1. Milani, A. & Knežević, Z. Asteroid proper elements and the dynamical structure of the asteroid main belt. *Icarus* **107,** 219–254 (1994).
2. Zappalà, V., Bendjoya, P., Cellino, A., Farinella, P. & Froeschlé, C. Asteroid families: Search of a 12,487-asteroid sample using two different clustering techniques. *Icarus* **116,** 291–314 (1995).
3. Farinella, P., Davis, D. R. & Marzari, F. in *Completing the Inventory of the Solar System* (eds Rettig, T. W. & Hahn, J. M.) 45–55 (ASP Conference Series 107, Astronomical Society of the Pacific, San Francisco, 1996).
4. Marzari, F., Farinella, P. & Davis, D. R. Origin, aging, and death of asteroid families. *Icarus* **142,** 63–77 (1999).
5. Bottke, W. F., Vokrouhlický, D., Brož, M., Nesvorný, D. & Morbidelli, A. Dynamical spreading of asteroid families via the Yarkovsky effect. *Science* **294,** 1693–1696 (2001).
6. Nesvorný, D., Bottke, W. F., Dones, L. & Levison, H. F. The recent breakup of an asteroid in the main-belt region. *Nature* **417,** 720–722 (2002).
7. Michel, P., Benz, W., Tanga, P. & Richardson, D. C. Collisions and gravitational reaccumulation: Forming asteroid families and satellites. *Science* **294,** 1696–1700 (2001).
8. Michel, P., Benz, W., Tanga, P. & Richardson, D. C. Formation of asteroid families by catastrophic disruption: Simulations with fragmentation and gravitational reaccumulation. *Icarus* **160,** 10–23 (2002).
9. Melosh, J. G. & Ryan, E. V. Asteroids: shattered but not dispersed. *Icarus* **129,** 562–564 (1997).
10. Tillotson, J. H. *Metallic Equations of State for Hypervelocity Impact* (General Atomic Report GA-3216, San Diego, 1962).
11. Thompson, S. L. & Lauson, H. F. Improvement in the chart D radiation hydrodynamic code III: revised analytic equation of state. (Report SC-RR-71 0714, Sandia National Laboratory, Albuquerque, 1972).
12. Belton, B. *et al.* The bulk density of asteroid 243 Ida from Dactyl's orbit. *Nature* **374,** 785–788 (1995).
13. Yeomans, D. K. *et al.* Radio science results during the NEAR-Shoemaker spacecraft rendezvous with Eros. *Science* **289,** 2085–2088 (2000).
14. Flynn, G. J., Moore, L. B. & Klöck, W. Density and porosity of stone meteorites: implications for the density, porosity, cratering, and collisional disruption of asteroids. *Icarus* **142,** 97–105 (1999).
15. Bottke, W. F., Nolan, M. C., Greenberg, R. & Kolvoord, R. A. Velocity distributions among colliding asteroids. *Icarus* **107,** 255–268 (1994).
16. Benz, W. & Asphaug, E. Catastrophic disruptions revisited. *Icarus* **142,** 5–20 (1999).
17. Richardson, D. C., Quinn, T., Stadel, J. & Lake, G. Direct large-scale *N*-body simulations of planetesimal dynamics. *Icarus* **143,** 45–59 (2000).
18. Asphaug, E. *et al.* Disruption of kilometre-sized asteroids by energetic collisions. *Nature* **393,** 437–440 (1998).

Acknowledgements P.M. acknowledges financial support from the Action Thématique Innovante 2001 of the French INSU, the programme 'Bonus-Qualité-Recherches 2001' and the Cassini laboratory of the Côte d'Azur Observatory (OCA). W.B. and D.C.R. acknowledge support respectively from the Swiss National Science Foundation and NASA through the Office of Space Science. Simulations were carried out on 4-processor Compaq DEC Alpha workstations thanks to the SIVAM project and the ILGA team of the OCA and on a Beowulf installed by the society Alineos.

Competing interests statement The authors declare that they have no competing financial interests.

Correspondence and requests for materials should be addressed to P.M. (e-mail: michel@obs-nice.fr).

Attosecond control of electronic processes by intense light fields

A. Baltuška*, Th. Udem†, M. Uiberacker*, M. Hentschel*, E. Goulielmakis*, Ch. Gohle†, R. Holzwarth†, V. S. Yakovlev*, A. Scrinzi*, T. W. Hänsch† & F. Krausz*

* *Institut für Photonik, Technische Universität Wien, Gusshausstrasse. 27, A-1040 Wien, Austria*

† *Max-Planck-Institut für Quantenoptik, Hans-Kopfermann-Strasse 1, D-85748 Garching, Germany*

The amplitude and frequency of laser light can be routinely measured and controlled on a femtosecond (10^{-15} s) timescale[1]. However, in pulses comprising just a few wave cycles, the amplitude envelope and carrier frequency are not sufficient to characterize and control laser radiation, because evolution of the light field is also influenced by a shift of the carrier wave with respect to the pulse peak[2]. This so-called carrier-envelope phase has been predicted[3–9] and observed[10] to affect strong-field phenomena, but random shot-to-shot shifts have prevented the reproducible guiding of atomic processes using the electric field of light. Here we report the generation of intense, few-cycle laser pulses with a stable carrier envelope phase that permit the triggering and steering of microscopic motion with an ultimate precision limited only by quantum mechanical uncertainty. Using these reproducible light waveforms, we create light-induced atomic currents in ionized matter; the motion of the electronic wave packets can be controlled on timescales shorter than 250 attoseconds (250×10^{-18} s). This enables us to control the attosecond temporal structure of coherent soft X-ray emission produced by the atomic currents—these X-ray photons provide a sensitive and intuitive tool for determining the carrier-envelope phase.

Matter exposed to intense laser light undergoes ionization, which gives rise to a broad range of phenomena in atoms, molecules and plasmas. An electronic wave packet is set free around each oscillation peak of a laser electric field that is strong enough to overcome the effective binding potential (Fig. 1a). The ensuing motion of the wave packets released by optical-field ionization depends on the subsequent evolution of the driving laser field[11,12]. A laser pulse consisting of many wave cycles launches a number of wave packets at different instants. Each of these follows a different trajectory because of the differences in the initial conditions of their motion, preventing a precise control of strong-field-induced electronic dynamics. Intense few-cycle light pulses with adjustable carrier-envelope (C-E) phase hold promise for a marked improvement. With the C-E phase set to make the peak of the oscillating electric field coincide with the pulse peak (blue line in Fig. 1c) and the strength of the field just sufficient to reach the ionization threshold at the pulse centre, a single, isolated electronic wave packet can be formed. The motion of this wave packet launched into an accurately controlled electric (and magnetic) field at a well-defined instant is determined with the highest possible precision permitted by quantum mechanics. Hence, few-cycle light with controlled electric-field waveforms offers the ultimate control of strong-field-driven atomic, molecular and plasma dynamics. This prospect constitutes the primary motivation for the work reported here.

In the present experiments, we adjust the peak intensity of linearly polarized few-cycle laser pulses such that several electronic wave packets in the vicinity of the pulse peak are set free. First they are removed from their parent ion, but within a laser period they are pulled back by the laser electric field (Fig. 1b). The highest-energy portion of the wave packet re-collides with the ion near the second (counting from the wave packet's 'birth') zero transition of the laser electric field, and results in the emission of an energetic (soft-X-ray) photon[11,13]. Being repeated many times in a multi-cycle laser field, this elementary process results in a series of equidistant emission lines corresponding to high-order odd harmonics of the driver laser light[14] and a sequence of sub-femtosecond bursts in the time domain[15]. By contrast, our few-cycle pulses release only a few wave packets near the pulse centre. The grey arrows in Fig. 1c depict

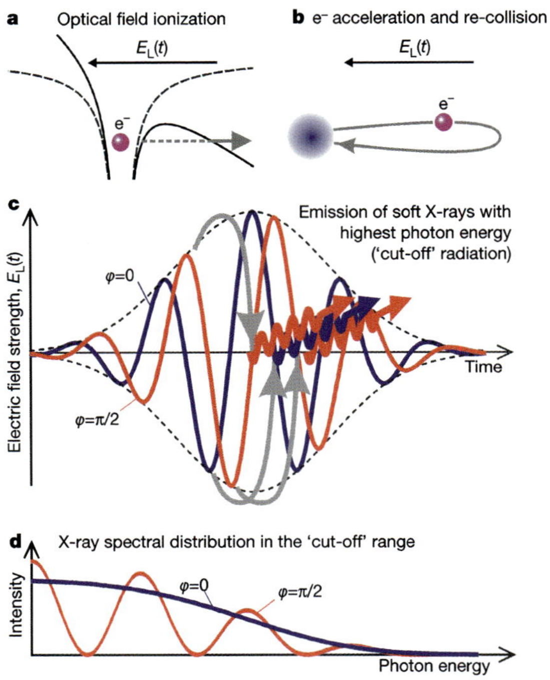

Figure 1 Optical-field ionization and generation of coherent extreme ultraviolet and soft-X-ray radiation from an atom exposed to a strong, linearly polarized, few-cycle light pulse. **a**, The effective Coulomb potential binding valence electrons to the atomic core (dashed curve) is temporarily suppressed around the oscillation peak of the laser electric field. A valence electron can tunnel through or escape above the potential barrier formed by the superposition of the atomic Coulomb field and the instantaneous laser field (solid curve). The electron wave packet that is set free within one-half of the laser period ($T_0 = 2.5$ fs at a wavelength of 750 nm) is temporally confined to less than $T_0/10$ owing to a nonlinear dependence of the ionization probability on the strength of the laser electric field E_L. Atoms exposed to a few-cycle pulse emit one or several wave packets near the pulse peak, depending on the peak intensity (relative to the ionization threshold) and the C-E phase φ. **b**, The freed electron is moved away from the atomic core and then pulled back to it by a linearly polarized field. Re-collision of the electron with its parent ion may trigger several processes, including secondary electron emission, excitation of bound electrons and emission of an energetic (soft-X-ray) photon. **c**, The highest-energy ('cut-off') X-ray photons are emitted near the zero transitions of the laser electric field around the pulse peak. With $\varphi \approx 0$ (cosine waveform, blue line) they are temporally confined to one single burst, whereas $\varphi \approx \pi/2$ (sine waveform, red line) yields two bursts of comparable amplitude separated by $\sim T_0/2$. **d**, As a result, the spectral distribution of the emitted 'cut-off' X-rays is continuous or modulated quasi-periodically, respectively.

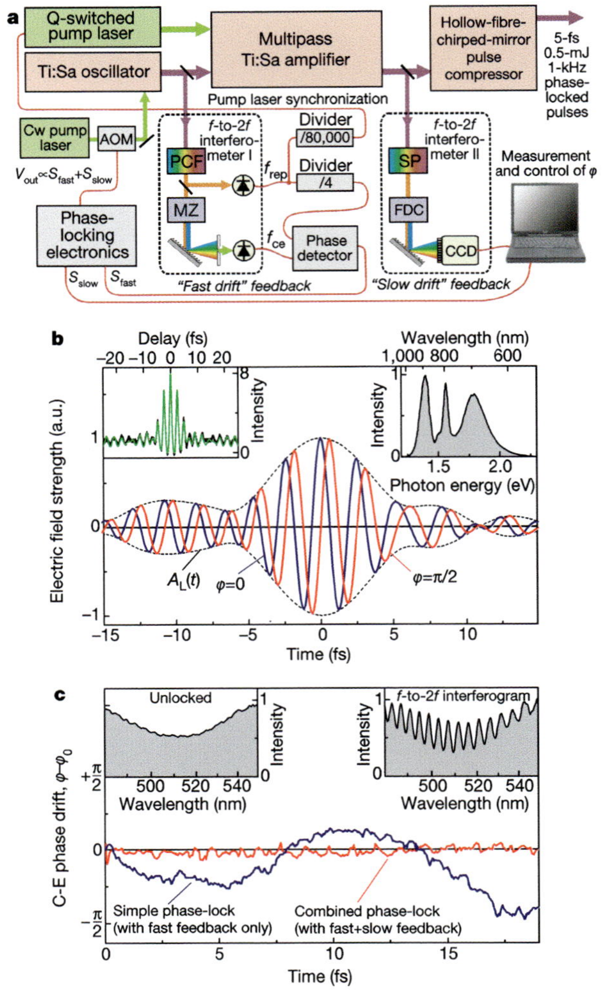

Figure 2 Overview of C-E-phase-stabilized high-power laser system. **a**, Schematic of the laser set-up. AOM, acousto-optical modulator; PCF, photonic crystal fibre; MZ, Mach–Zehnder set-up; SP, 2-mm sapphire plate; FDC, frequency-doubling crystal. The C-E phase of the pulses delivered by the titanium:sapphire (Ti:Sa) oscillator is controlled by tracking the *f*-to-2*f* signal in interferometer I and controlling the pump power (Verdi, Coherent) through a feedback based on the AOM. Frequency dividers /4 and /80,000 are used to derive, respectively, the reference frequency for the stabilization of the phase slip behind interferometer I and the repetition rate of pulses amplified in a multipass amplifier (Femtopower Compact Pro, Femtolasers; pump source: Corona, Coherent). With this technique the C-E phase recurrence of the Ti:Sa oscillator is set such that every fourth pulse (and therefore also every 80,000th pulse) exhibits the same electric-field waveform. The residual drift of the C-E phase behind the laser amplifier is monitored with interferometer II and pre-compensated by shifting accordingly the C-E phase of the oscillator. See Methods for further details. **b**, Pulse properties behind the hollow-fibre-chirped-mirror compressor. The blue and red curves depict possible electric-field waveforms compatible with $A_L(t)$ and $\omega_L(t)$ retrieved from the interferometric autocorrelation (left inset) and the pulse spectrum (right inset) for different settings of the C-E phase; **c**, C-E phase drift versus time with and without slow-drift feedback (compare **a**); insets show the *f*-to-2*f* interference spectrograms recorded with interferometer II. a.u., arbitrary units; f_{rep}, repetition rate of the oscillator; f_{ce}, *f*-to-2*f* beat signal; S_{fast} and S_{slow}, electric signals from the 'fast' and 'slow' feedback loops; V_{out}, amplitude of AOM control voltage.

Erratum: In Fig. 2c of this Letter, the units of time on the x axis should be seconds, not femtoseconds.

those that have the highest kinetic energy at re-collision and, consequently, produce the highest-energy (cut-off) X-ray photons. The spectral distribution of these photons depends sensitively on whether they are emitted in a single sub-femtosecond X-ray burst or in a couple of bursts, which can be controlled with the C-E phase of the few-cycle laser pulse (Fig. 1d). Such sub-femtosecond X-ray bursts were recently produced and measured using randomly phased few-cycle light pulses[16,17]. As the emission of two bursts of comparable energy can only occur within a narrow range of C-E phases, statistically, the highest-energy X-ray photons were carried in one dominant burst. Yet, optimization of the excitation conditions for maximum concentration of X-ray emission energy in a single pulse and their accurate reproduction has not been possible with randomly phased laser pulses. Here we report the generation of this radiation—which we do not refer to as high harmonics for the reasons explained below—with phase-stabilized light pulses. The spectral distribution of the highest-energy X-ray photons emerging from the few-cycle-driven ionization process enables us to determine the C-E phase. In combination with standard pulse diagnostics, this approach permits complete characterization of the temporal structure of few-cycle light waveforms (with an ambiguity in the direction of the field).

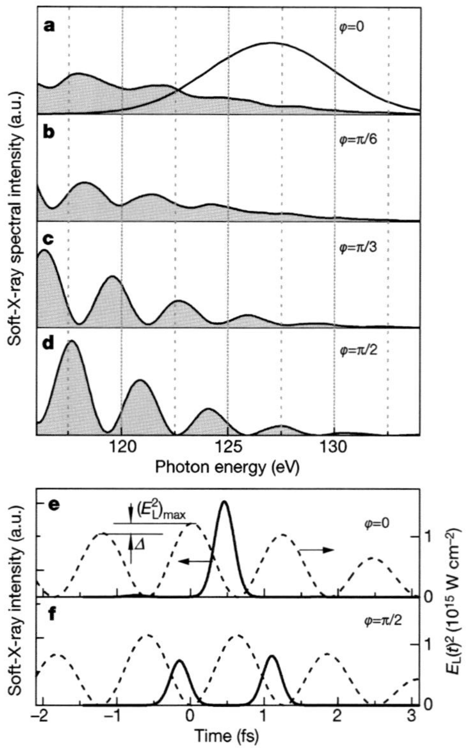

Figure 3 Numerical simulations of few-cycle-driven coherent soft-X-ray emission from ionizing atoms. **a–d**, Cut-off-range spectra for different C-E-phase settings of the driving laser pulse. **e**, **f**, The solid curves depict the temporal intensity profile of the cut-off harmonic radiation filtered through a gaussian bandpass filter with a full-width at half-maximum of 7 eV (solid line in **a**), whereas the dashed curves plot $E_L(t)^2$. The parameters of the calculations are as follows: laser pulse duration, $\tau_L = 5$ fs; pulse energy, 0.2 mJ; beam diameter, $2w_L = 122\,\mu$m; medium, neon; pressure, 100 mbar; length, 2 mm. These results, along with previous numerical studies[4,9], fully support our intuitive analysis: the few-cycle cosine wave is predicted to produce a single soft-X-ray burst (filtered in the cut-off, **e**). Deviation of φ from zero gradually suppresses the magnitude of the main burst and gives rise to a satellite spike. The latter becomes most prominent for $|\varphi| \rightarrow \pi/2$ (**f**). In the frequency domain, the isolated pulse emerging for $\varphi = 0$ implies a continuous spectrum (**a**), which gets increasingly modulated with the appearance of the second burst for $|\varphi| \rightarrow \pi/2$ (**b–d**). Detailed numerical simulations show that the values of φ corresponding to the smoothest and the most deeply modulated cut-off spectra are shifted by about 20° from the intuitively anticipated values of 0 and $\pi/2$. The relative bandwidth of the cut-off continuum linearly scales with the ratio $\Delta/(E_L^2)_{max}$ (compare **e**), which, in turn, is related to the number of cycles in the optical field, τ_L/T_0. In fact, the continuum bandwidth is as broad as ~20 eV for $\tau_L/T_0 = 2$ and a cut-off photon energy of 125 eV, and shrinks to only 3.8 eV for $\tau_L/T_0 = 5$.

The electric field of a linearly polarized light pulse can be generally written as $E_L(t) = A_L(t)\cos[\omega_L(t)t + \varphi]$. This description includes three physical quantities: the amplitude envelope $A_L(t)$, where t is time, the instantaneous frequency of the field oscillations $\omega_L(t) = \omega_0 + \beta(t)$, where $\beta(t)$ is a possible frequency sweep across the pulse, and the C-E phase φ. In few-cycle pulses, a change in φ affects $E_L(t)$ considerably[2]. Whereas $A_L(t)$ and $\omega_L(t)$ can be obtained from standard pulse diagnostic techniques and flexibly modified in pulse shapers, φ has remained immeasurable so far. In general, pulses delivered by femtosecond lasers have a reproducible envelope and carrier frequency but carry a random C-E phase[2]. It was demonstrated recently that the pulse-to-pulse shift of φ can be tracked in real time and stabilized using a so-called f-to-$2f$ interferometer[18–21]. In such a scheme, the laser spectrum is broadened to a full octave and the beat between the frequency-doubled low-frequency spectral components and the high-frequency components is detected and used for active phase stabilization.

In our experiments, we amplify phase-stabilized pulses from a sub-10-fs titanium:sapphire (Ti:Sa) laser (seed oscillator) in a multipass chirped-pulse Ti:Sa amplifier operating at a 1-kHz repetition rate (Fig. 2a). Amplification and temporal recompression yield 20-fs, 1-mJ pulses, which are subsequently spectrally broadened by self-phase modulation in a hollow-core waveguide filled with neon gas, and finally compressed on reflection off chirped multilayer mirrors. This system delivers 5-fs, 0.5-mJ pulses with a carrier wavelength of 750 nm (Fig. 2b)[22]. The C-E phase jitter emerging in the waveguide has been measured by comparing the output beam with a reference split off the input beam in a linear interferometer. The jitter is found to be less than 50 mrad (root-mean-square, r.m.s.) owing to the excellent energy stability of the amplified pulses (<1% r.m.s.). The C-E phase can therefore be monitored directly at the output of the amplifier with a broadband f-to-$2f$ interferometer[23,24].

The drift of the C-E phase can be retrieved from the spectral shift of the recorded pattern of the f-to-$2f$ spectral interference (right inset in Fig. 2c). In the absence of phase stabilization of the seed oscillator, the f-to-$2f$ interferogram is completely blurred after approximately 200 laser shots (left inset of Fig. 2b). In contrast, seeding the amplifier with phase-stabilized pulses preserves good fringe visibility over thousands of laser shots. The phase evolution derived from the change of the interference pattern (blue curve in Fig. 2c) appears to be frozen for periods as long as several seconds. Considering the fact that some extra phase noise might be introduced in the measurement process, the blue curve in Fig. 2c accounts for the maximal possible drift of the C-E phase present in the laser system. Due to the slowness of this drift, it can be readily tracked by computer analysis of the f-to-$2f$ interferogram, and pre-compensated by a phase offset of opposite sign through a 'slow-drift' feedback loop (Fig. 2a). The simultaneous use of both feedback loops allows stabilizing the C-E phase for extended periods (≫1 minute). The residual variation of the C-E phase retrieved using the in-loop f-to-$2f$ interferometer is shown by red curve in Fig. 2c. In this case, the upper limit of the C-E phase drift at the output of the laser system can be obtained with an additional, out-of-loop interferometer.

With these intense phase-controlled few-cycle pulses, we can now set out to control electronic motion within the field oscillation cycle of visible light and exploit this technical capability to measure the value of the C-E phase by gauging it with a sensitive strong-field interaction. According to the intuitive insight into the underlying physics, outlined above, the energetic radiation from an atom ionized by a linearly polarized few-cycle pulse constitutes an excellent candidate for such a C-E-phase calibration. Owing to the fact that the emission of the highest-energy (cut-off) photons is confined to about a single light oscillation period, T_0, the time structure of the generated X-rays is expected to depend sensitively on the C-E phase of the few-cycle driver light. In fact, for a cosine waveform featuring a single strongest half-cycle ($\varphi \approx 0 \pm n\pi$, where $n = 0,1,..$), the highest-energy photons are emitted within a single burst resulting from the wave packet that was 'born' near $t = -T_0/2$ and re-collided with the parent ion in the vicinity of $t = T_0/4$ (Fig. 1c). For a sine waveform ($\varphi \approx \pi/2 \pm n\pi$) there are two most intense half cycles resulting in a pair of most energetic re-colliding wave packets and, correspondingly, two cut-off X-ray bursts emitted near $t = 0$ and $T_0/2$ (Fig. 1c). Therefore, for these two different values of φ we expect distinctly different spectral distributions of the highest energy photons (Fig. 1d): a continuous distribution for $\varphi \approx 0 \pm n\pi$ (where $n = 0,1,..$), and a distribution exhibiting deep (quasi-) periodic modulation in the case of $\varphi \approx \pi/2 \pm n\pi$. These intuitive considerations have been fully confirmed by our numerical simulations[25,26] for the realistic experimental conditions of a macroscopic generation medium. Figure 3 depicts representative results. Our investigations reveal that $\tau_L/T_0 \leq 2.5$ (where τ_L is the pulse width, full-width at half-maximum) is required to produce a soft-X-ray continuum of sizeable relative bandwidth (>10%; that is, >10 eV in the 100-eV range). For pulses that meet this requirement and have a duration close to their bandwidth limit, the above-described dependence of the spectral structure of the cut-off radiation on φ is found to be robust within a broad range of parameters. Consequently, it can be used for calibrating the value of the C-E phase.

We have generated coherent soft X-rays by gently focusing our phase-stabilized 5-fs pulses into a 2-mm-long sample of neon gas. Figure 4 shows a series of soft-X-ray spectra produced under the conditions that are described in Fig. 4 legend for different values of the C-E phase of the 5-fs pump pulses. For a setting of $\varphi = \varphi_0$ (Fig. 4b), a broad structureless continuum appears in the cut-off region ($\hbar\omega > 120$ eV). Notably, with a change of the phase, the continuous spectral distribution of the cut-off radiation gradually transforms into discrete harmonic peaks, with the maximum modulation depth appearing for the settings of $\varphi = \varphi_0 \pm \pi/2$. This behaviour is in agreement with the intuitive picture presented above, and allows us to identify φ_0 as zero with a residual ambiguity of $n\pi$, where n is an integer. This ambiguity in the determination of φ relates to the inversion symmetry of the interaction with the atomic gas medium. In fact, a π-shift in φ results in no change of the light waveform other than reversing the direction of the electromagnetic field vectors. This phase flip does not modify the intensity of the radiated X-ray photons, but it might become observable in photoelectron experiments[8,10].

Another noteworthy feature that has emerged from the soft-X-ray experiments with phase-stabilized pulses is the frequency shift of the harmonic peaks as a function of φ. This effect has been predicted by previous[27] simulations, as well as by our current simulations, and clearly manifests itself in the spectra shown in Fig. 4. This shift is due to a phase shift of the satellite burst with respect to the main X-ray burst, presumably as a consequence of the variation of the field amplitude within T_0 induced by that of the C-E phase. This behaviour reveals that the 'harmonic' fringes visible in the cut-off region of few-cycle-driven high-harmonic spectra do not represent genuine harmonics, because the latter should remain invariant to the change in the C-E phase as long as the laser frequency ω_L is left unchanged. This shift, induced by C-E phase, completely smears the harmonic structure of the near-cut-off soft X-rays generated by our 5-fs pulses in the absence of phase stabilization (Fig. 4e). Real high-order harmonics (invariant to shifts in φ) appear at photon energies some 40% below cut-off (plateau harmonics, not shown in Fig. 4), which are also observed in the absence of phase stabilization. The invariance of these plateau harmonics to shifts in φ confirms the fact that $\omega_L(t)$ is not affected by varying the C-E phase in our experiment.

The periodic spectral modulation present for $\varphi \approx \pi/2$ up to the highest photon energies observed, $\hbar\omega_{\text{cut-off}} \approx 130$ eV (Fig. 4d), completely disappears over a band of ~16 eV in the cut-off range as φ is shifted to zero (Fig. 4b). This is in fair agreement with the 20-eV continuum bandwidth predicted by our simple considerations for a 5-fs, 750-nm ($\tau_L/T_0 = 2$) laser pulse. The C-E phase can be evaluated with the highest possible accuracy by recording pairs of soft-X-ray harmonic spectra at $\varphi_1 = \varphi$ and $\varphi_2 = \varphi + \pi/2$, where φ is varied in small steps. In this method, $\varphi = 0$ (or, equivalently, π) can be identified from a pair of spectra that exhibit, respectively, the smallest and the largest modulation depth in the cut-off range. The current phase and intensity stability of our few-cycle pulses permits identification of a 'cosine' pulse with an accuracy better than $\pi/5$. The obtained value of φ is subsequently used to calibrate the interference pattern recorded in the second f-to-$2f$ interferometer. With this calibration done, we can change φ with an accuracy of better than $\pi/10$ using the interferometer.

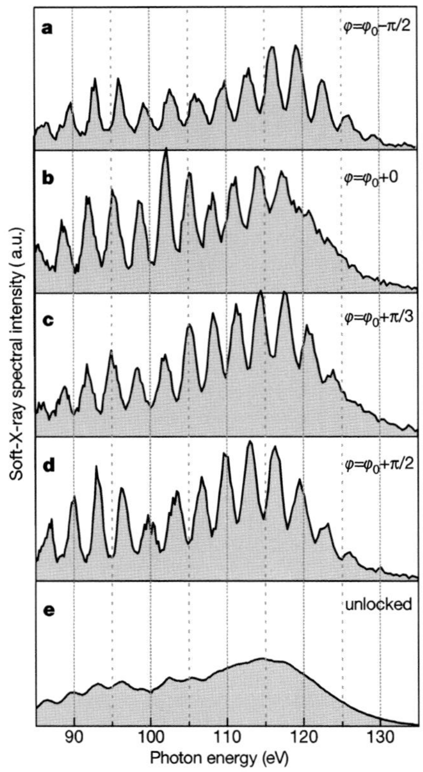

Figure 4 Measured spectral intensity of few-cycle-driven soft-X-ray emission from ionizing atoms. **a–d**, Data obtained with phase-stabilized pulses for different C-E-phase settings. **e**, Spectrum measured without phase stabilization. CCD exposure time was 0.5 s in all cases. The coherent energetic radiation was generated by gently focusing 5-fs, 0.2-mJ laser pulses into a 2-mm-long neon gas medium. The on-axis peak intensity of the pump pulse was estimated to be 7×10^{14} W cm^{-2}. The neon gas was supplied in a thin-walled metal tube with a backing pressure of 160 mbar. The pressure in the interaction region, however, was somewhat lower owing to the gas expansion into the surrounding vacuum chamber. Thin zirconium foils were installed in the pathway of the generated soft-X-ray radiation in order to block the laser light as well as low-order harmonics. The high-energy part of the spectrum (above 80 eV) was spectrally analysed with a simple spectrometer consisting of a 10,000 lines mm^{-1} transmission grating and a back-illuminated soft-X-ray CCD camera (Roper Scientific).

In this work we have demonstrated the generation of intense few-cycle light pulses with reproducible temporal evolution of the electromagnetic field. With these controlled intense light waveforms, we have explored the sensitivity of microscopic atomic currents to the timing of light field oscillations with respect to the pulse peak. The spectral distribution of the soft-X-ray radiation emitted by the atomic currents change as this timing is changed by less than 250 as, allowing a calibration of the C-E phase with a precision better than $\pi/5$. Intense few-cycle light pulses with a reproducible and known waveform allow precise control of the emission of an electron wave packet from an atom (or molecule) and its subsequent motion. The emerging field of attosecond physics is expected to benefit from this capability in several ways. First, the re-colliding electron wave packet may serve as a sub-femtosecond excitation pulse for atomic electron dynamics, or as a source of well-controlled, isolated sub-femtosecond X-ray pulses (Fig. 3e). Second, measurement of the shape and chirp[28,29] of sub-femtosecond pulses is becoming feasible by recently demonstrated attosecond diagnostic techniques[16,17]. Third, time-resolved inner-shell spectroscopy[30] could now be extended to tracing atomic processes at an attosecond timescale. Last, extending intense light waveform control to ultrahigh intensities would allow the motion of relativistic electrons to be steered with unprecedented accuracy. □

Methods

C-E phase stabilization loop

The high-power laser set-up (Fig. 2a) incorporates a phase-stabilized 10-fs laser system (FS 800, Menlo Systems). It consists of a dispersive-mirror-controlled Kerr-lens mode-locked Ti:sapphire oscillator (Femtosource Compact Pro, Femtolasers), an f-to-$2f$ unit ('interferometer I' in Fig. 2a), and phase-locking electronics driving an acousto-optic modulator (AOM in Fig. 2a). About 50% of the output of the oscillator is directed into a 2-cm-long photonic crystal fibre (PCF in Fig. 2a) to broaden the spectrum of the pulses beyond an octave. Subsequently, the low-frequency components f_{low} of the broadened spectrum are frequency-doubled to produce an interference beat with the fundamental light at the high-frequency end of the spectrum, that is, $f_{\mathrm{high}} = 2f_{\mathrm{low}}$. The optical set-up producing this beat signal is referred to as an f-to-$2f$ interferometer. The phase of the two interfering quasi-monochromatic wave packets differs by $2\varphi - \varphi + \phi$, where ϕ is an unknown constant phase shift that prevents access to the value of φ in this type of experiment. For practical reasons, ϕ cannot be reliably calibrated from an f-to-$2f$ set-up. Despite the fact that the value of the C-E phase remains unknown, the information on its drift is sufficient to phase-stabilize the oscillator. In our laser system, the spectrally broadened oscillator pulses are fed into a Mach–Zehnder set-up (MZ in Fig. 2a) that has a frequency-doubling crystal in one of the arms[2,20,21]. The interference with the original spectrum leads to the f-to-$2f$ beat signal $f_{\mathrm{ce}} = \Delta\varphi f_{\mathrm{rep}}/2\pi$ at around 530 nm, where $\Delta\varphi$ is a pulse-to-pulse shift of the C-E phase. The phase detector compares the electric signal detected at the output of the MZ with a reference, the frequency of which is derived by dividing the repetition rate of the oscillator ($f_{\mathrm{rep}} \approx 80$ MHz) by a factor of 4. Forcing the two signals, $1/4 f_{\mathrm{rep}}$ and ω_{CE} to oscillate in phase yields $\Delta\varphi = 1/4 \times 2\pi$ so that every fourth pulse looks alike. The rough adjustment of $\Delta\varphi$ is performed by changing the optical path length through a thin intracavity fused-silica plate. At this stage, φ is approximately reproduced in every fourth round trip in the laser cavity, which results in a quasi-periodic modulation of the f-to-$2f$ beat signal at $f_{\mathrm{ce}} \approx f_{\mathrm{rep}}/4 \approx 20$ MHz. Next, to enhance the accuracy of the phase reproducibility, the f-to-$2f$ beat signal is phase-locked to the $f_{\mathrm{rep}}/4$ reference. This phase-locking is achieved by controlling $\Delta\varphi$ via nonlinear effects in the Ti:Sa crystal through variation of the pump intensity[2]. For this purpose, the beam of the continuous-wave (c.w.) pump laser is passed through the AOM, which varies the power of the transmitted beam proportionately to the phase error signal described above. As a result, every fourth pulse, in the 80-MHz pulse train, and therefore also every 80,000th pulse picked by the multipass amplifier, carries the same C-E phase, giving rise to an accurate optical field reproduction in the radiation seeded into the amplifier. To correct additional C-E phase drift emerging during the process of amplification, the spectral interferograms detected behind the amplifier in the second f-to-$2f$ unit ('interferometer II' in Fig. 2a) are analysed in a personal computer by a Fourier-transform algorithm. The resultant (slowly varying) phase error signal is subsequently combined with the feedback from the phase detector in the first f-to-$2f$ interferometer and used to control the AOM.

Measurement of phase jitter introduced by the hollow waveguide

The method of (linear) spectral interference was used to assess the C-E phase jitter added by the hollow-fibre chirped-mirror pulse compressor (Fig. 2a). The measurement was performed by splitting off a reference beam behind the multipass amplifier, and interfering it with the spectrally broadened and temporally compressed light. The resultant spectral interferograms, recorded with a CCD (charge coupled device)-based spectrometer, reflected the shot-to-shot phase jitter that has two distinct contributions. The first arises from the mechanical path-length fluctuations of the interferometer arms, and corresponds to a linear variation of the spectral phase. The second represents the shift of the C-E phase resulting from the beam pointing and intensity instabilities of the light coupled into the hollow waveguide. As the C-E phase jitter is frequency-independent, it can be clearly separated from the contribution added by the mechanical arm length drift.

Numerical simulations of the generation and propagation of soft X-rays

We used an improved version of the code first presented in refs 25 and 26. The code solves the wave propagation equations in three dimensions of space, where cylindrical symmetry is assumed for the transverse beam profile, and the slowly evolving wave approximation[9] is used. The atomic dipole response to the laser field was calculated by an extension of the model of Lewenstein *et al.*[14,25], using accurate static field ionization rates obtained from numerical fully quantum-mechanical calculations. Harmonic absorption and diffraction is taken into account, as well as the geometrical (Gouy) phase shift, dispersion introduced by free electrons, ionization energy losses, and the delay due to the refractive index of neutral atoms.

Received 8 November 2002; accepted 3 January 2003; doi:10.1038/nature01414.

1. Steinmeyer, G., Sutter, D., Gallmann, L., Matuschek, N. & Keller, U. Frontiers in ultrashort pulse generation: Pushing the limits in linear and nonlinear optics. *Science* **286**, 1507–1512 (1999).
2. Xu, L., Spielmann, C., Poppe, A., Brabec, T. & Krausz, F. Route to phase control of ultrashort light pulses. *Opt. Lett.* **21**, 2008–2010 (1996).
3. Krausz, F., Brabec, T., Schnürer, M. & Spielmann, C. Extreme nonlinear optics: Exploring matter to a few periods of light. *Opt. Photon. News* **9**, 46–51 (1998).
4. de Bohan, A., Antoine, P., Milošević, D. B. & Piraux, B. Phase-dependent harmonic emission with ultrashort laser pulses. *Phys. Rev. Lett.* **81**, 1837–1840 (1998).
5. Cormier, E. & Lambropoulos, P. Effect of the initial phase of the field in ionization by ultrashort laser pulses. *Eur. Phys. J. D* **2**, 15–20 (1998).
6. Tempea, G., Geissler, M. & Brabec, T. Phase sensitivity of high-order harmonic generation with few-cycle laser pulses. *J. Opt. Soc. Am. B* **16**, 669–673 (1999).
7. Christov, I. P. Phase-dependent loss due to nonadiabatic ionization by sub-10-fs pulses. *Opt. Lett.* **24**, 1425–1427 (1999).
8. Dietrich, P., Krausz, F. & Corkum, P. B. Determining the absolute carrier phase of a few-cycle laser pulse. *Opt. Lett.* **25**, 16–18 (2000).
9. Brabec, T. & Krausz, F. Intense few-cycle laser fields: Frontiers of nonlinear optics. *Rev. Mod. Phys.* **72**, 545–591 (2000).
10. Paulus, G. G. *et al.* Absolute-phase phenomena in photoionization with few-cycle laser pulses. *Nature* **414**, 182–184 (2001).
11. Corkum, P. B. Plasma perspective on strong-field multiphoton ionization. *Phys. Rev. Lett.* **71**, 1994–1997 (1993).
12. Niikura, H. *et al.* Using correlated pairs for attosecond-resolution molecular wave packet measurements. *Nature* (in the press).
13. Schafer, K. J., Yang, B., DiMauro, L. F. & Kulander, K. C. Above threshold ionization beyond the high harmonic cutoff. *Phys. Rev. Lett.* **70**, 1599–1602 (1993).
14. Lewenstein, M., Balcou, P., Ivanov, M. Y., L'Huillier, A. & Corkum, P. B. Theory of high-harmonic generation by low-frequency laser fields. *Phys. Rev. A* **49**, 2117–2132 (1994).
15. Paul, P. M. *et al.* Observation of a train of attosecond pulses from high harmonic generation. *Science* **292**, 1689–1692 (2001).
16. Hentschel, M. *et al.* Attosecond metrology. *Nature* **414**, 509–513 (2001).
17. Kienberger, R. *et al.* Steering attosecond electron wave packets with light. *Science* **297**, 1144–1148 (2002).
18. Udem, T. *Phasenkohärente optische Frequenzmessungen am Wasserstoffatom* Thesis, Ludwig-Maximilians-Univ. (1997).
19. Telle, H. R. *et al.* Carrier-envelope offset phase control: A novel concept for absolute optical frequency measurement and ultrashort pulse generation. *Appl. Phys. B.* **69**, 327–332 (1999).
20. Jones, D. J. *et al.* Carrier-envelope phase control of femtosecond mode-locked lasers and direct optical frequency synthesis. *Science* **288**, 635–639 (2000).
21. Apolonski, A. *et al.* Controlling the phase evolution of few-cycle light pulses. *Phys. Rev. Lett.* **85**, 740–743 (2000).
22. Sartania, S. *et al.* Generation of 0.1-TW 5-fs optical pulses at a 1-kHz repetition rate. *Opt. Lett.* **22**, 1562–1564 (1997).
23. Kakehata, M. *et al.* Single-shot measurement of carrier-envelope phase changes by spectral interferometry. *Opt. Lett.* **26**, 1436–1438 (2001).
24. Baltuška, A., Fuji, T. & Kobayashi, T. Self-referencing of the carrier-envelope slip in a 6-fs visible parametric amplifier. *Opt. Lett.* **27**, 1241–1243 (2002).
25. Milosevic, N., Scrinzi, A. & Brabec, T. Numerical characterization of high harmonic attosecond pulses. *Phys. Rev. Lett.* **88**, 093905 (2002).
26. Yakovlev, V. S. & Scrinzi, A. High harmonic imaging of few-cycle laser pulses. *Phys. Rev. Lett.* (submitted).
27. Priori, E. *et al.* Nonadiabatic three-dimensional model of high-order harmonic generation in the few-optical-cycle regime. *Phys. Rev. A* **61**, 063801 (2000).
28. Itatani, J. *et al.* Attosecond streak camera. *Phys. Rev. Lett.* **88**, 173903 (2002).
29. Kitzler, M., Milosevic, N., Scrinzi, A., Krausz, F. & Brabec, T. Quantum theory of attosecond XUV pulse measurement by laser dressed photoionization. *Phys. Rev. Lett.* **88**, 173904 (2002).
30. Drescher, M. *et al.* Time-resolved atomic inner-shell spectroscopy. *Nature* **419**, 803–807 (2002).

Acknowledgements We thank H. A. Haus, M. S. Pshenichnikov, G. F. Tempea and R. Kienberger for discussions, and Ch. Warmuth, Z. Cheng and M. Wieland for technical assistance. This work was sponsored by the Fonds zur Förderung der wissenschaftlichen Forschung (Austria) and the European ATTO network. We also thank Femtolasers GmbH and Menlo Systems GmbH for support.

Competing interests statement The authors declare competing financial interests: details accompany the paper on *Nature*'s website (▸ **http://www.nature.com/nature**).

Correspondence and requests for materials should be addressed to F.K. (e-mail: ferenc.krausz@tuwien.ac.at).

an impediment to light supply would be a disadvantage for the host plant, but the effects of lichen cover on leaf surfaces have never been quantified. Affected leaves might be able to adjust to the new conditions of shading, by biochemical means and by structural alterations, such as the movement of chloroplasts to new locations within the cell[5]. For most tropical rainforest leaves, the lichen coverage increases gradually as they age. So leaves might be able to adapt to the new light conditions and retain their productive potential.

Anthony and colleagues[1] have studied lichens on the leaves of the palm *Calamus australis* in northern Queensland, Australia. Their first concern was to determine precisely how light conditions alter for a host leaf as a consequence of developing lichen cover. They found that light interception values for lichens had a mean of 48%, so the lichens effectively halved the amount of light reaching the host leaves. But did the lost light lie within the wavelengths required for photosynthesis by the host? It did — the light absorbed by the lichens had a similar spectral composition to that absorbed by the host leaf.

However, the possibility remained that compensation could occur by adjustment of a leaf's photosynthetic systems. Anthony *et al.* investigated this by examining the relationship between light intensity and carbon gain by photosynthesis in parts of the leaf either covered by or clear of lichens. They found that the two areas behaved quite differently. The part covered by lichens showed characteristics of a classic 'shade leaf', having a capacity to fix carbon effectively even at very low levels of irradiance. Respiration rates and the light compensation point (the irradiance level at which carbon fixation balances respiratory demand) remained unaltered, but the more efficient use of low light levels showed that lichen-covered parts of the leaf had acclimated to the new conditions. Checking the chlorophyll levels in different parts of the leaves revealed that the shaded portions had increased concentrations, although the ratio of the two main types of chlorophyll (*a*/*b*) remained the same. The gradual accretion of lichen cover during the lifetime of a leaf evidently allows this adjustment to take place, and local photosynthetic efficiency can thereby be maintained. Modelling the total carbon gain for entire leaves, Anthony *et al.* found that the general carbon economy of the host was not greatly affected by the presence of the epiphylls.

These findings would appear to resolve the physiological conundrum, first described for coffee plants[3], in which increased epiphyte shading resulted in no loss of productivity. But other questions remain, such as those relating to coevolution of organisms within rainforest and the consequent high biodiversity of this habitat. The leaf can adapt to the lichen epiphyte, but has the lichen also adapted in a way that will ensure its host's survival? For example, the fact that the lichen steals only half of the available light could be the result of a trade-off between further productivity by the epiphyll and the consequent danger of reducing the productivity of the host leaf and possibly destroying it. Then there is the question of why leaves with a particularly long life-span generally have low levels of epiphylls[4]. Is the development of some chemical or structural defence mechanism by these highly persistent leaves an alternative strategy to that of appeasement? ■

Peter D. Moore is in the Department of Life Sciences, King's College London, London SE1 9NN, UK.
e-mail: peter.moore@kcl.ac.uk

1. Anthony, P. A., Holtum, J. A. M. & Jackes, B. R. *Funct. Ecol.* 16, 808–816 (2002).
2. Richards, P. W. *The Tropical Rainforest* (Cambridge Univ. Press, 1957).
3. Roskoski, J. P. *Oikos* 37, 252–256 (1981).
4. Coley, P. D., Kursar, T. A. & Machado, J.-L. *Ecology* 74, 619–623 (1993).
5. Kasahara, M. *et al. Nature* 420, 829–832 (2002).

Attophysics

Ultrafast control

Philip H. Bucksbaum

The successes of two pioneering groups are now brought together to create trains of identical ultrashort laser pulses that can control what's happening inside an atom.

One attosecond is one-millionth of one-millionth of one-millionth of a second. If the next second of your life were expanded to fill the history of the Universe, then one attosecond would correspond to less than one second of that history. In one attosecond (10^{-18} s), light travels slightly further than the length of a single water molecule, and the molecule itself appears utterly frozen. The natural vibrations of a molecule take place over tens of thousands of attoseconds, and rotations take millions of attoseconds — a relative eternity. Even the rapid motion of atomic electrons is typically measured in hundreds to thousands of attoseconds; and a single oscillation of a wave of visible light lasts nearly 2,000 attoseconds.

The title, then, of the report by Baltuška *et al.*[1] on page 611 of this issue — "Attosecond control of electronic processes by intense light fields" — is impressive indeed. The authors claim to have harnessed light on an attosecond scale, and used the light to control electrons with similar precision. This work combines the efforts of two creative research groups in optical science: the precision measurements of Theodor Hänsch's group at the Max Planck Institute for Quantum Optics, and the ultrafast-laser expertise of Ferenc Krausz's group at the Technical University of Vienna. Krausz, Hänsch and their colleagues have achieved this dizzying feat by delicately stacking together two techniques that have grown out of their research into ultrafast-laser technology: the production of subfemtosecond soft X-ray pulses[2], and the production of an all-optical-frequency standard[3].

Both of these achievements are based on the Kerr-lens mode-locked laser, the workhorse of ultrafast science, invented only a dozen years ago and now found in most of the world's ultrafast-research laboratories[4]. It produces a continuous train of 10–100-femtosecond pulses (10,000–100,000 attoseconds) in the near-infrared region of the optical spectrum, through a fortuitous nonlinear effect known as the Kerr lens (named

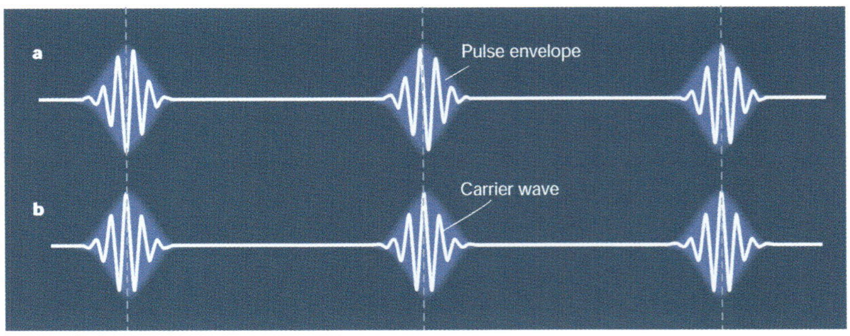

Figure 1 Locking onto the laser frequency. a, The output of a typical ultrafast laser, based on the Kerr lens, forms a train of non-identical pulses: the optical carrier wave inside each pulse is shifted with respect to the pulse envelope, because its frequency is not an integer multiple of the pulse repetition frequency. b, If, however, the laser frequency is stabilized (using an optical comb generator[3]), truly identical pulses are created.

news and views

after the Scottish physicist John Kerr, 1824–1907). The lens is created by the light itself, and forces the photons in a laser cavity to bunch up into extremely short pulses. These pulses can be thought of as electromagnetic waves confined in a single short envelope, bouncing back and forth between the end mirrors of the laser cavity. If one of the end mirrors is partly transmitting, a small portion of the pulse leaks out on each round trip, creating a train of output pulses.

But, although transmitted through the mirror on successive round trips, the pulses in the train do not look identical. The optical carrier wave appears shifted by some fraction of a cycle on every bounce (Fig. 1a). This is because the carrier frequency is not commensurate with the pulse repetition frequency — that is, the ratio of the optical frequency to the repetition frequency is not an integer. If the ratio were an integer, then the optical frequency could be simply related to the round-trip frequency.

Several years ago, Hänsch realized that if this defect could be overcome, it would be possible to build all-optical-frequency standards, which would significantly simplify and improve the precision of time measurement. His group found a way to do this, and the result, known as the 'optical comb generator', is revolutionizing the time standard. The details involve 'holy fibres', self-phase modulation and second-harmonic generation, a fascinating story which has been told in a previous issue of *Nature*[3]. Suffice to say here that Hänsch learned how to make a Kerr-lens laser in which every pulse in the train is truly identical (Fig. 1b).

Meanwhile, Krausz's group was pursuing a different application of Kerr-lens ultrafast lasers. They amplified light pulses only a few cycles long to peak powers of 100 billion watts, to create bursts of short-wavelength radiation. The generation of energetic ultrashort pulses with these lasers is not trivial. Straightforward laser amplification will destroy both the laser and its light because the intensities are high enough to rip atoms apart. This catastrophe can be averted using exotic techniques with equally exotic names: chirped-pulse amplification, hollow-fibre self-phase modulation and chirped mirror compression. This is ultrafast optics raised to a high art, and Krausz is a virtuoso.

When the high-powered light is focused into atoms in a vapour, the atoms react in a number of ways. Some of them ionize rapidly, creating a shower of energetic electrons through what is called 'above-threshold ionization'[5]. Other atoms absorb laser light and re-radiate at very high harmonics (or multiples) of the laser frequency[6]. This high-harmonic radiation constitutes a unique source of 'soft' X-rays, with pulse durations that can be less than a femtosecond[2].

Krausz's group and others have studied the mechanism for high-harmonic generation. Theories suggest that the high harmonics are due to coherent radiation created by the recollisions of ionized electrons with their parent atoms[7,8]. If so (and experiments seem to support this), then the soft X-rays should arrive in attosecond bursts, occurring on every cycle of the laser field. Thus, the pattern of X-ray bursts matches the pattern of the optical carrier wave within the pulse envelope — and this is where the work of Hänsch's group and of Krausz's group now comes together[1].

When Krausz's laser is stabilized by Hänsch's technique, each 100-billion-watt pulse from the amplifier looks identical. This means that electrons ripped from the atoms in the focus of the laser beam follow identical paths, recollide with their parent atoms at identical stages of the pulse, and produce optical and electron spectra that record this violent motion. All previous experiments have been forced to average over non-identical laser pulses, and so many special characteristics of the atomic-scale process were washed out. In particular, the harmonics spectrum is now revealed to be not always 'harmonic' (an integer multiple of the pump frequency) after all. Rather, the spectrum is modulated by the coherent interference between successive X-ray bursts initiated at different parts of successive cycles of the driving light field.

The phase-locking to produce identical pulses is successful down to a small fraction of an optical period, amounting to a temporal smearing of less than 200 attoseconds. Now, 200 attoseconds is not one attosecond, so the title of Baltuška and colleagues' paper[1] is perhaps rushing things a bit. Nonetheless, this is a remarkable achievement. This level of precision can capture the motion of photo-excited electrons as they move around their parent ion. The success of Hänsch, Krausz and their colleagues in creating frequency-locked high-field laser pulses marks the beginning of the era of 'attophysics' — the study of physical processes on the attosecond timescale. ■

Philip H. Bucksbaum is at the FOCUS Center, University of Michigan, Ann Arbor, Michigan 48109-1120, USA.
e-mail: phb@umich.edu

1. Baltuška, A. *et al. Nature* **421**, 611–615 (2003).
2. Hentschel, M. *et al. Nature* **414**, 509–513 (2001).
3. Udem, T., Holzwarth, R. & Hänsch, T. W. *Nature* **416**, 233–237 (2002).
4. Spence, D. E., Kean, P. N. & Sibbett, W. *Opt. Lett.* **16**, 42–44 (1991).
5. Agostini, P., Fabre, F., Mainfray, G., Petite, G. & Rahman, N. K. *Phys. Rev. Lett.* **42**, 1127–1130 (1979).
6. Ferray, M. *et al. J. Phys. B* **21**, L31–L35 (1988).
7. Corkum, P. B. *Phys. Rev. Lett.* **71**, 1194–1197 (1993).
8. Krause, J. L., Schafer, K. J. & Kulander, K. C. *Phys. Rev. Lett.* **68**, 3535–3538 (1992).

Cancer

More than skin deep

Diana Bolotin and Elaine Fuchs

Activation of the nuclear factor NF-κB has been linked with many human cancers. But in tumours of some tissues this molecule is inactive. New work examines how turning off NF-κB promotes cancer in one tissue, the skin.

What do cell death, cell proliferation, immune responses and tumour development have in common? The answer is that they are just a few of the many biological processes in mammals that are regulated by proteins of the nuclear factor NF-κB family[1,2]. Widely expressed throughout an organism, the NF-κB subunits are also exquisitely regulated in terms of their ability to enter the cell nucleus and activate key genes. However, although much is known about these proteins, many issues remain unexplored. For instance, studies of the immune system and several cell lines have indicated a role for NF-κB in promoting cell proliferation and protecting against programmed cell death (apoptosis)[2,3]. So it is puzzling that, in the skin, these proteins seem to oppose proliferation[4,5]. On page 639 of this issue[6], Dajee *et al.* address the paradox that inhibition of NF-κB in the skin can lead to increased cell proliferation — and cancer. Their findings may be important not only for understanding how skin tumours develop, but also for exploring the design and use of NF-κB inhibitors in treating cancer.

Our cells regulate the activity of NF-κB proteins — which exist as dimers — largely by controlling their intracellular localization, using the IκB molecule[7]. When bound to IκB, NF-κB proteins are excluded from the nucleus and so are prevented from activating their target genes. However, certain stimuli can induce an enzyme complex, the IκB kinase, to modify IκB, targeting it for degradation. The NF-κB dimer is thereby released and moves into the nucleus, resulting in the expression of genes with functions appropriate to the stimuli.

For many years, the function of NF-κB in the organism seemed relatively straightforward, fitting a profile for the protein as the cell's protector against apoptosis and an enhancer of proliferation. So, in several organs and tissues, high levels of NF-κB can prevent apoptosis by activating genes for anti-apoptotic proteins such as Bcl-xL and cellular inhibitors of apoptosis[2]. And in the immune system, where these proteins have

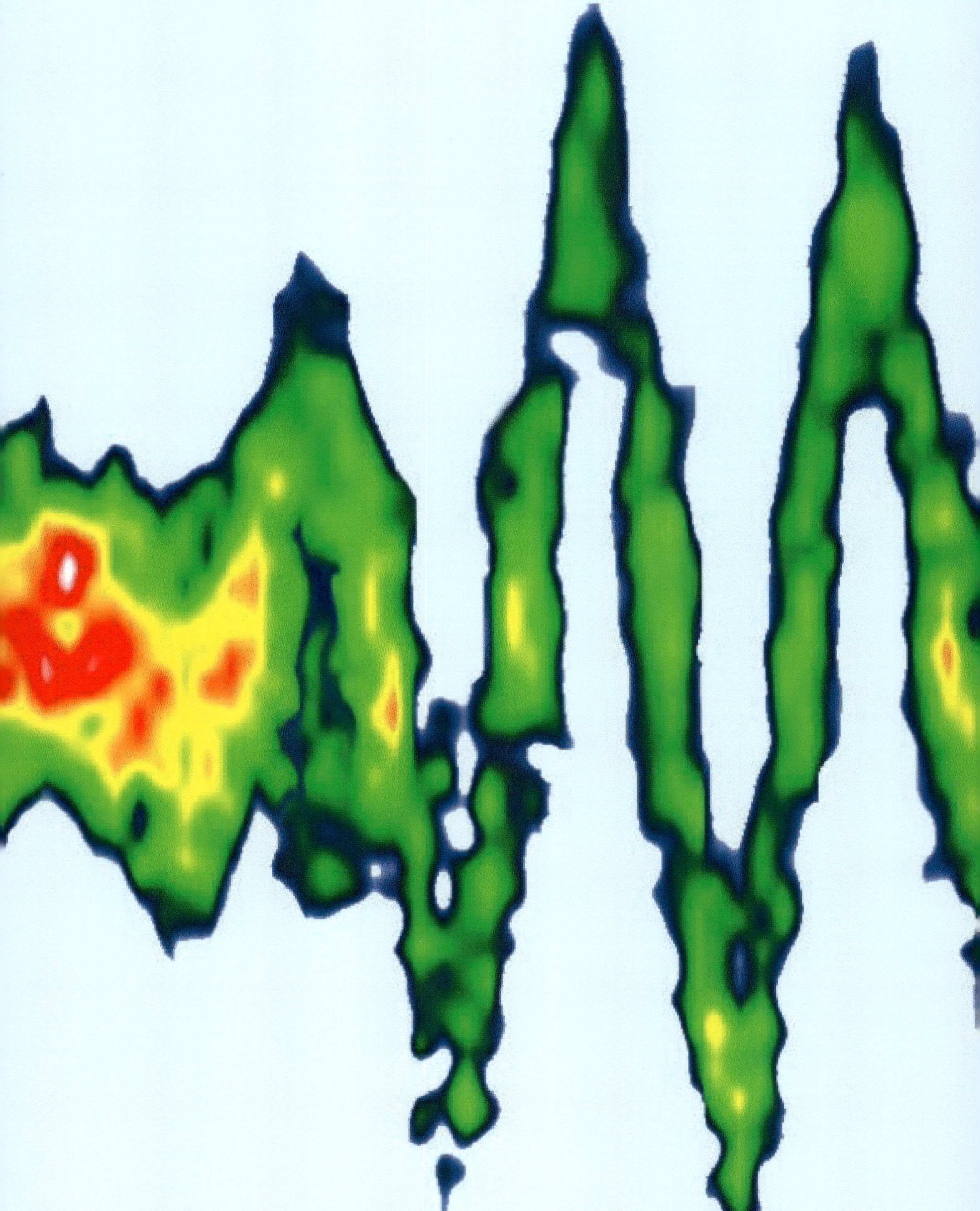

Attosecond Metrology of the First Generation: AM 1.0

6

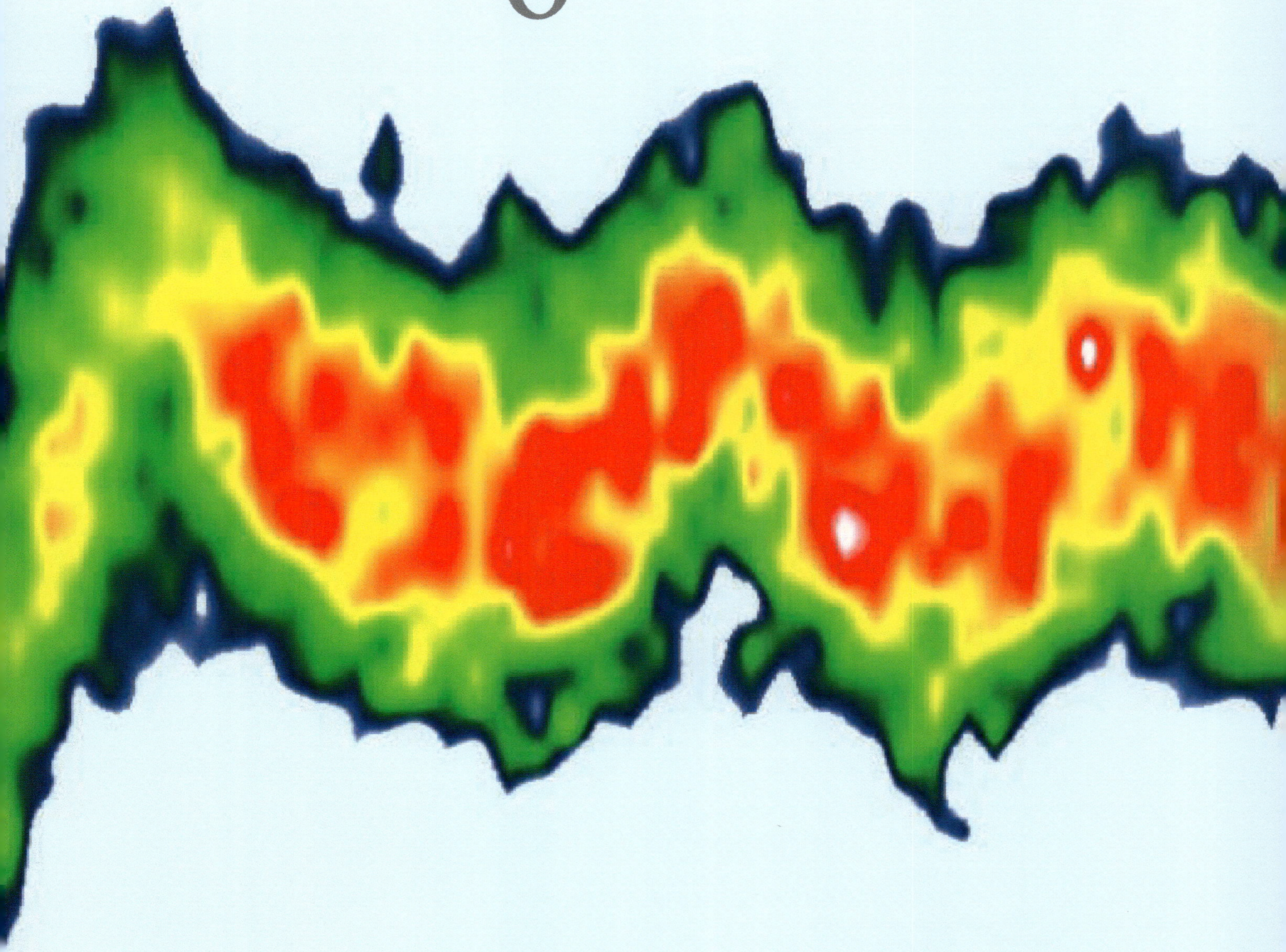

The first attosecond-resolution „photograph" of a light wave recorded in Vienna in 2004. The trace serves as a basis of the logo of the Laboratory of Attosecond Physics, www.attoworld.de, jointly operated by the Ludwig-Maximilians-Universität München and the Max Planck Institute of Quantum Optics. Attosecond XUV pulses have been used to capture the waving of red light [36]. © Eleftherios Goulielmakis

The reliable generation, measurement and spectroscopic applications of isolated attosecond pulses critically rely on strong[10] laser fields with precisely controlled temporal evolution. Although these controlled, strong laser fields first became available in 2003 [30], Paul Corkum and his co-workers in Ottawa and Thomas Brabec and his co-workers in Vienna had already worked out a theoretical concept in 2002 of how the resultant attosecond-controlled force field could be utilized for the temporal characterization of attosecond XUV pulses [32-33].

Their concept, which was inspired by [20, 22] and a prior idea from Paul Corkum, was based on the assumption that energetic (XUV) photons release photoelectrons with either negligible delay or with a delay whose dependence on the excitation frequency is negligible within the frequency band of the photons. Under these circumstances the incident XUV pulse is expected to launch photoelectrons from the atomic ensemble exposed to the pulse with an overall temporal profile that mimicks that of the incident XUV pulse. In the presence of a strong laser field, photoelectrons released at slightly differing instants within this emission profile accelerate slightly differently, due to the variation of the laser field on the same time scale. This attosecond-varying force modifies (streaks) the distribution of the photoelectrons' final momentum (Fig. 6-1).

Measuring the final momentum distribution of XUV-induced laser-streaked photoelectrons vs. delay between the XUV and the laser pulse (referred to as an attosecond streaking spectrogram) provides access to both the temporal intensity profile (and possible frequency chirp) of the XUV pulse and the evolution of the laser field used for streaking [32-33]. Light-field-driven attosecond streaking was later shown by Vladislav Yakovlev and his co-workers to measure the quantum phase of the emitted electron wavepacket [34].

The birth of intense controlled few-cycle optical waveforms was soon followed by putting these concepts into practice. The first step towards this end was the generation of isolated attosecond pulses with a reproducible temporal profile and reproducible timing with respect to the controlled strong driving field [30]. Owing to the attosecond synchronism between the two pulses, the very same laser waveform, at somewhat reduced field strength, could be used for precision

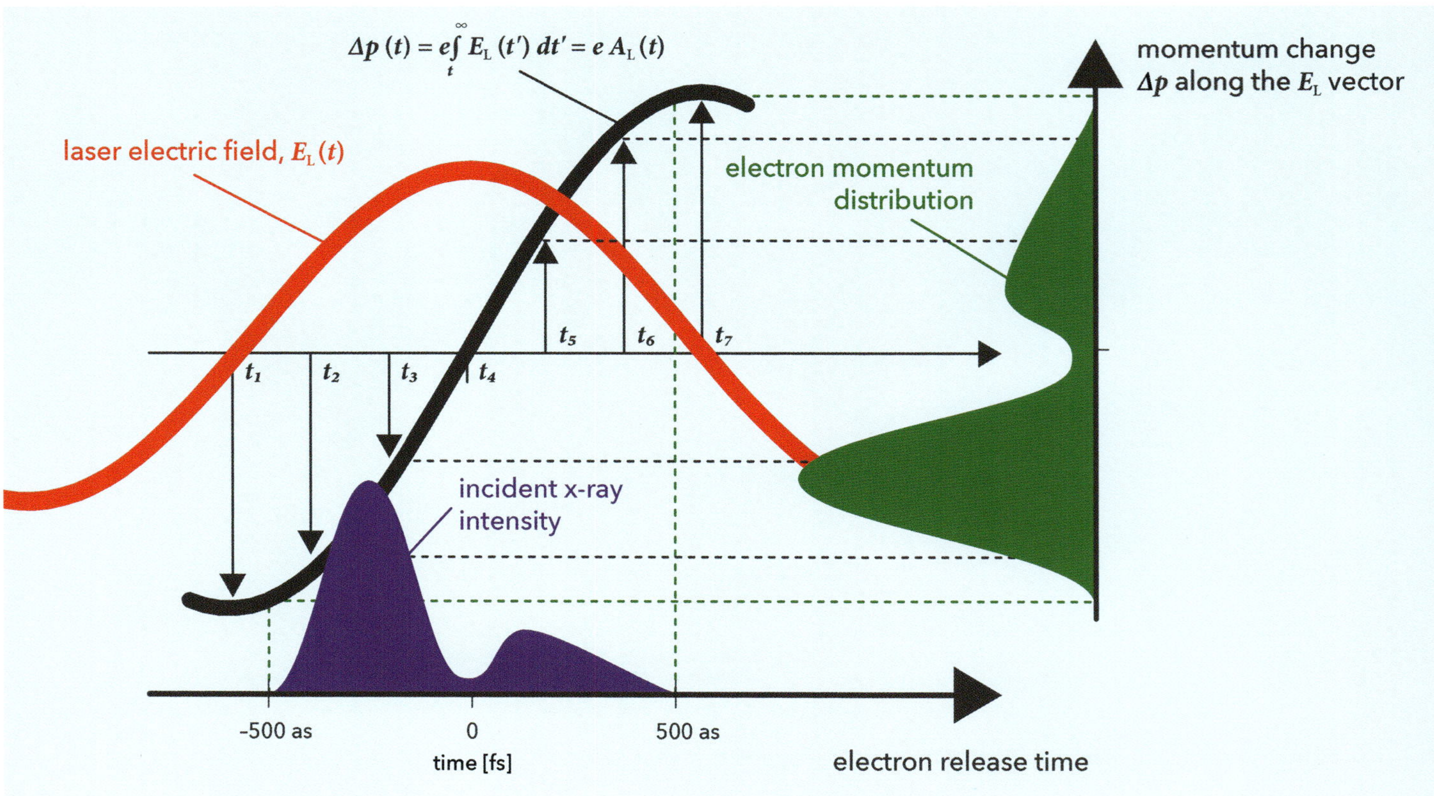

Figure 6-1 Light-field-driven attosecond streak camera.

[10] Here and throughout this text: rivalling intra-atomic Coulomb fields.

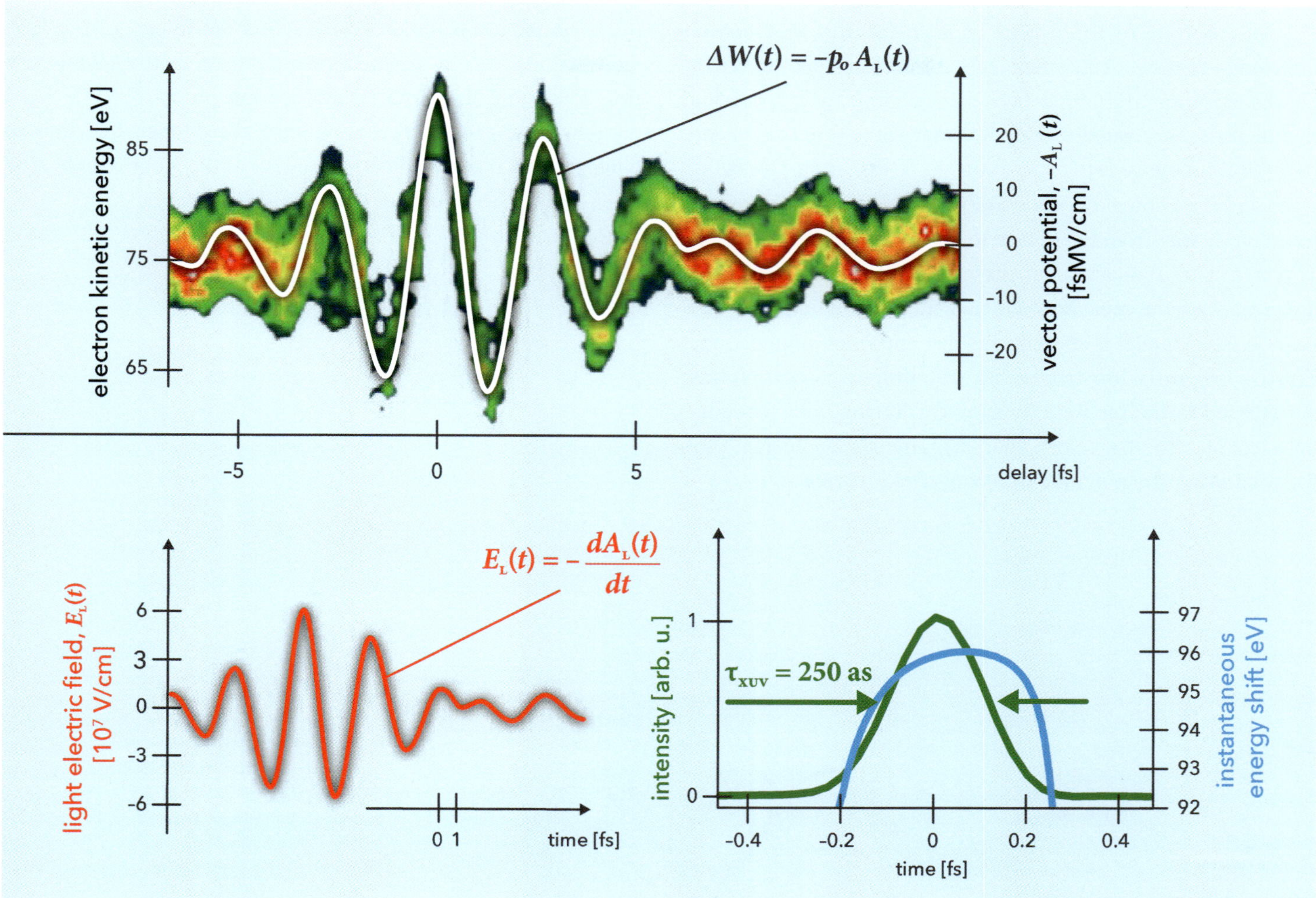

Figure 6-2 First attosecond streaking spectrogram (upper panel) measured in 2004 [36]. Lower panels: retrieved streaking field evolution (left) and temporal profile and chirp of the attosecond XUV pulse releasing the photoelectrons (right).

temporal characterization of the attosecond pulse in the first light-field-driven streaking measurement, performed by Reinhard Kienberger and his collaborators [35].

The very same method, implemented over an extended delay range by scanning the XUV pulse across the laser waveform, allowed Eleftherios Goulielmakis and his co-workers to acquire complete information about the strength and temporal evolution of the oscillating laser field [36]. The measurement of the first attosecond streaking spectrogram, along with the streaking field evolution and the attosecond pulse profile and chirp retrieved from the measurement are shown above (Fig. 6-2). Subsequently, advances in few-cycle femtosecond technology permitted waveform-controlled light to be confined to a single wave cycle [37], which led to the generation of isolated sub-100-as pulses for the first time [38] (Fig. 6-3).

High-fidelity measurement of an isolated attosecond pulse and its driver laser waveform via attosecond streaking also allowed precision tests of the basic models of strong-field interactions to be performed for the first time. The isolation of a single attosecond pulse implies the isolation of a single tunneling and recollision event and wavepacket motion between the instants of ionization and recollision. All this happens within a well-defined and precisely-known oscillation period of the strong laser field driving the interaction. Analysis of classical electron trajectories between the ionization and recollision within the framework of Paul Corkum's semiclassical model yields remarkable findings. The emitted XUV pulses are found to either have a positive or negative chirp for trajectories shorter and or longer than the one resulting in the highest-energy emission, respectively (Fig. 6-4).

An example of the power of high-fidelity attosecond metrology is the measurement of the intrinsic spectral chirp, i.e. the variation of group delay versus photon energy, carried by the attosecond pulse during its emergence. This chirp can be unambiguously retrieved from the attosecond streaking trace and the known dispersion of the system traversed by the pulse on its way from the source to the measurement (Fig. 6-4). The error bars indicate an accuracy of better than 25 as. The agreement of the measured chirp with the prediction of the semiclassical model is striking and has important implications. Firstly, the emission from long trajectories is completely suppressed in the far field by spatial filtering, and secondly, classical electron trajectory analysis provides a powerful tool for predicting the outcome of strong-field interactions.

The resultant tools of attosecond metrology of the first generation, AM 1.0, are the strong controlled attosecond-varying force and the isolated attosecond pulse of XUV light synchronized to the force, which have opened the door to steering and tracking electron motion at the atomic scale. The next chapter reviews the first implementations.

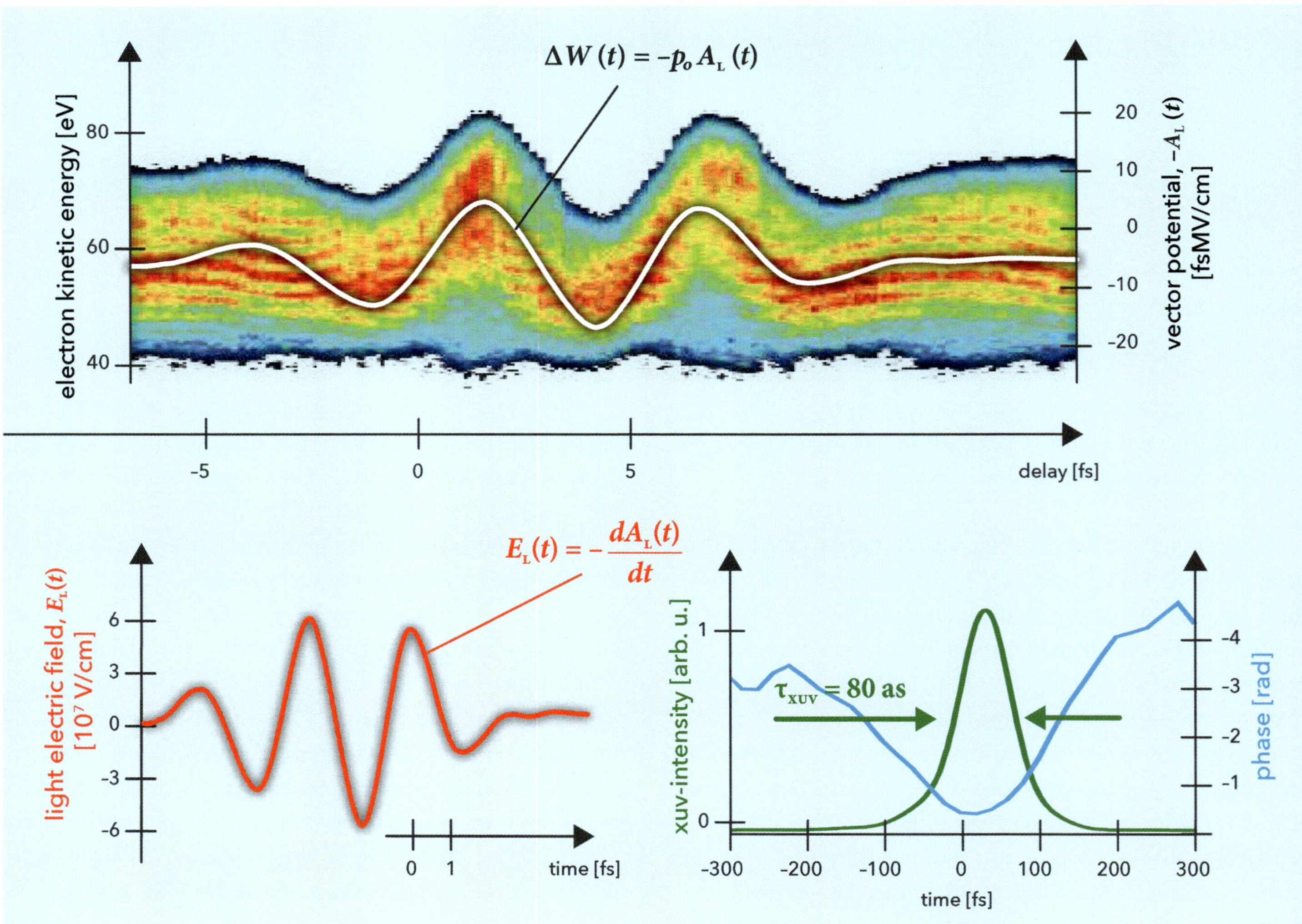

Figure 6-3 Attosecond streaking spectrogram (upper panel) measured in 2008 [38]. Lower panels: retrieved streaking field evolution (left) and temporal profile and chirp of the attosecond XUV pulse releasing the photoelectrons (right).

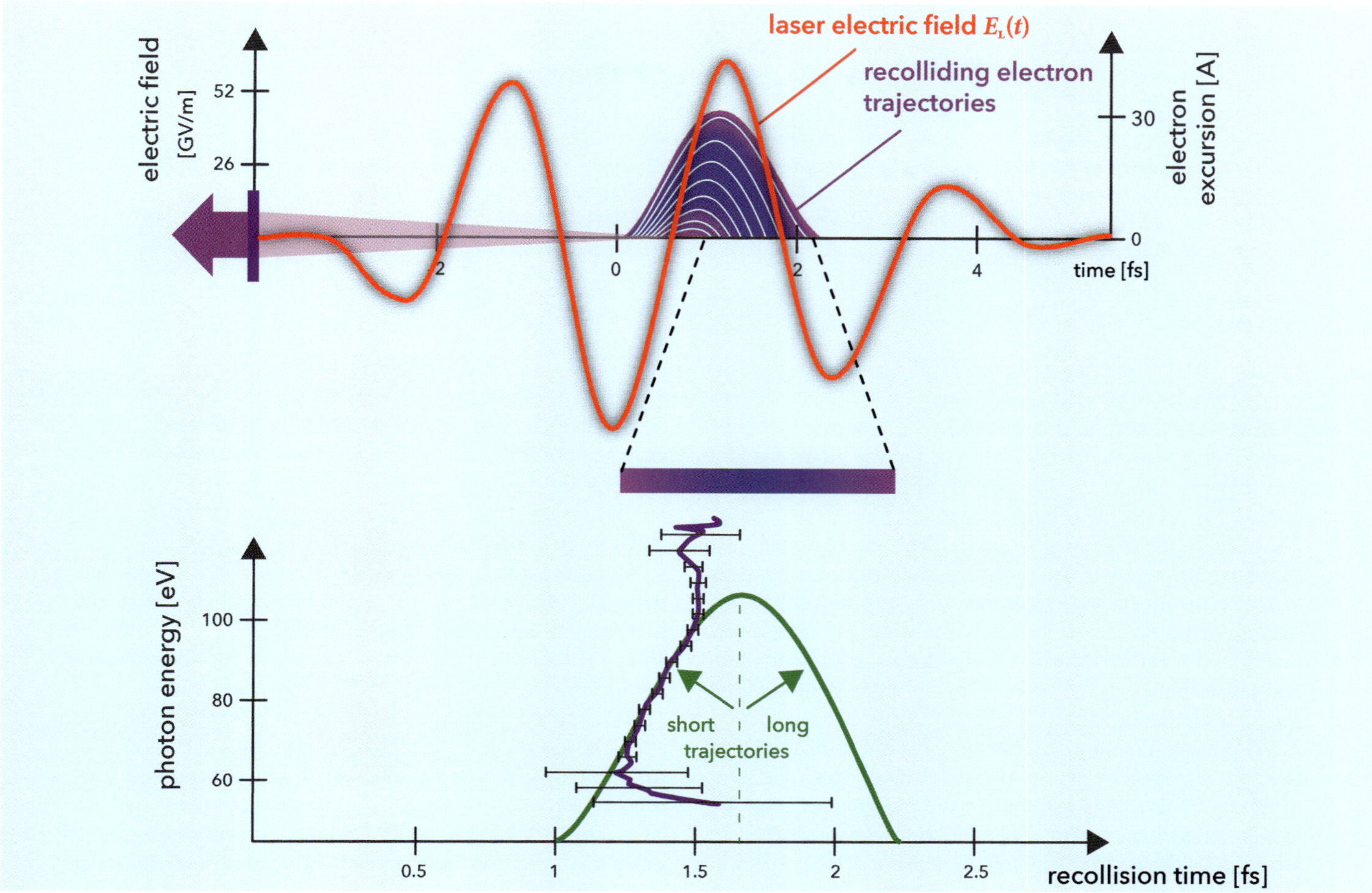

Figure 6-4 Attosecond real-time insight into electron recollision dynamics during high-order harmonic generation with a near-single-cycle laser pulse.

[32] J. Itatani, F. Quéré, G. L. Yudin, M. Ivanov, F. Krausz, P. Corkum. Attosecond streak camera. *Physical Review Letters* **88**, 173903 (2002). ***(Page: 240)***

[33] M. Kitzler, N. Milosevic, A. Scrinzi, F. Krausz, T. Brabec. Quantum theory of attosecond XUV pulse measurement by laser dressed photoionization. *Physical Review Letters* **88**, 173904 (2002). ***(Page: 244)***

[34] V. Yakovlev, J. Gagnon, N. Karpowicz, F. Krausz. Attosecond streaking enables the measurement of quantum phase. *Physical Review Letters* **105**, 073001 (2010). ***(Page: 248)***

[35] R. Kienberger, E Goulielmakis, M. Uiberacker, A. Baltuška, V. Yakovlev, F. Bammer, A. Scrinzi, T. Westerwalbesloh, U. Kleineberg, U. Heinzmann, M. Drescher, F. Krausz. Atomic transient recorder. *Nature* **427**, 817 (2004). ***(Page: 252)***

[36] E. Goulielmakis, M. Uiberacker, R. Kienberger, A. Baltuška, V. Yakovlev, A. Scrinzi, T. Westerwalbesloh, U. Kleineberg, U. Heinzmann, M. Drescher, F. Krausz. Direct measurement of light waves. *Science* **305**, 1267 (2004). ***(Page: 257)***

[37] E. Goulielmakis, V. Yakovlev, A. Cavalieri, M. Uiberacker, V. Pervak, A. Apolonski, R. Kienberger, U. Kleineberg, F. Krausz. Attosecond control and measurement: lightwave electronics. *Science* **317**, 769 (2007). ***(Page: 260)***

[38] E. Goulielmakis, M. Schultze, M. Hofstetter, V. Yakovlev, J. Gagnon, M. Uiberacker, A. L. Aquila, E. M. Gullikson, D. T. Attwood, R. Kienberger, F. Krausz, U. Kleineberg. Single-cycle nonlinear optics. *Science* **320**, 1614 (2008). ***(Page: 267)***

VOLUME 88, NUMBER 17 PHYSICAL REVIEW LETTERS 29 APRIL 2002

Attosecond Streak Camera

J. Itatani,[1] F. Quéré,[1] G. L. Yudin,[1] M. Yu. Ivanov,[1] F. Krausz,[2] and P. B. Corkum[1]

[1]Steacie Institute for Molecular Sciences, National Research Council of Canada, Ottawa, Ontario, Canada K1A 0R6

[2]Institute für Photonik, Technische Universität Wien, Gusshausstrasse 27, A-1040 Wien, Austria

(Received 29 August 2001; published 16 April 2002)

An electron generated by x-ray photoionization can be deflected by a strong laser field. Its energy and angular distribution depends on the phase of the laser field at the time of ionization. This phase dependence can be used to measure the duration and chirp of single sub100-attosecond x-ray pulses.

DOI: 10.1103/PhysRevLett.88.173903 PACS numbers: 42.65.Ky, 41.50.+h

Historically, advances in our ability to measure fast phenomena have led to corresponding scientific advances [1]. It is now possible to use the physics of high harmonic generation for producing single attosecond (10^{-18} sec, asec) pulses in the extreme ultraviolet to soft x-ray regions [2,3]. The next frontier will be to utilize such ultrashort pulses to probe the electronic dynamics of atoms and molecules which occurs on the attosecond time scale [4]. However, one impediment for experimentally demonstrating attosecond pulses has been the problem of finding a good method for their characterization [5–7].

We present a method to determine the duration and chirp of a single attosecond x-ray pulse, based on the ionization of atoms by the x-ray photons in the presence of a strong low-frequency laser field. This method relies on two basic ideas. (i) The subcycle oscillation of the laser electric field is used as a time reference to determine the duration of the x-ray pulse. This can be done only if the x-ray pulse is shorter than the period of the oscillating field. (ii) The photoelectron signal generated by the x-ray pulse is resolved in energy and angle simultaneously. Depending on the laser-field polarization, the information on the pulse duration is contained in the width of the energy spectrum at a given observation angle (linear polarization), or in the width of the angular distribution at a given energy (circular polarization). We show that the resolution limit depends on the photon energy, bandwidth, and chirp of the x-ray pulse. The resolution improves with increasing photon energy, and is on the order of 70 asec at 100 eV for transform-limited pulses.

A group including two of the authors has implemented a closely related configuration, where the photoelectrons are collected in the direction perpendicular to the laser polarization, and successfully characterized x-ray pulses shorter than one oscillation period of the laser field used for their generation [3,8].

The ionization of atoms by x rays in the presence of laser fields has been used for several years to measure the duration of x-ray pulses longer than the optical period of the laser field. In these measurements, Auger electrons [9] or x-ray photoelectrons [10–12] are dressed by the laser field to cross correlate the x-ray pulse and the envelope of the laser pulse. Two effects are used for this cross correlation. (i) The appearance of sidebands in the photoelectron energy spectrum due to the emission and absorption of laser photons by the x-ray generated photoelectrons. (ii) The ponderomotive shift of these peaks, due to the ac Stark shift of the continuum induced by the laser field.

As shown later, as the pulse duration falls below one period of the laser radiation, the sidebands broaden and merge, and the ponderomotive shift is no longer observable. Therefore, these cross-correlation methods cannot be directly extended to subcycle x-ray pulses.

By analyzing the dynamics of the electrons, we will show how the laser field affects the photoelectron distribution in the case of subcycle x-ray pulses, and how this effect can be used to characterize these pulses. The large difference in frequencies between the x-ray pulse Ω_X and the laser pulse ω_L naturally divides the two-color ionization process in two steps: absorption of an x-ray photon followed by acceleration in the laser field [13]. The x ray produces an electron with a kinetic energy $W_0 = m_e v_0^2/2 = \hbar\Omega_X - I_p$, where I_p is the ionization potential and $\mathbf{v}_0$ is the initial velocity. If the $U_p \gg \hbar\omega_L$ and $\hbar\Omega_X \gg I_p$, classical mechanics can be used to determine the time-dependent velocity $\mathbf{v}(t)$ in the field [14,15]. For the laser electric field $\mathbf{E}_L(t) = E_0(t)/\sqrt{1+\varepsilon^2}\{\cos(\omega_L t + \varphi)\mathbf{e}_x + \varepsilon \sin(\omega_L t + \varphi)\mathbf{e}_y\}$ with ellipticity ε, this velocity is expressed as

$$\mathbf{v}(t) = -\frac{e}{m_e}\mathbf{A}(t) + \left[\mathbf{v}_0 + \frac{e}{m_e}\mathbf{A}(t_i)\right], \tag{1}$$

where $\mathbf{A}(t)$ is the vector potential of the field, $\mathbf{E}_L = -\partial\mathbf{A}/\partial t$, and t_i is the time of ionization.

In Eq. (1) the first term describes the electron's quiver motion and goes to zero as the laser pulse ends. The second term is determined by the initial condition $\mathbf{v}(t_i) = \mathbf{v}_0$ at the time of ionization. This term is the final drift velocity $\mathbf{v}_f = \mathbf{v}_0 + (e/m_e)\mathbf{A}(t_i)$ measured after the laser pulse. The fact that $\mathbf{v}_f$ is different from $\mathbf{v}_0$ is the classical equivalent to the absorption, emission, and scattering of laser photons by the photoelectron [13,16].

The dashed circle of radius v_0 in Fig. 1 shows the electron velocity distribution in the xy plane ($v_{0,z} = 0$) produced by the x rays alone. The solid circle in Fig. 1 represents the drift velocity distribution in the presence of

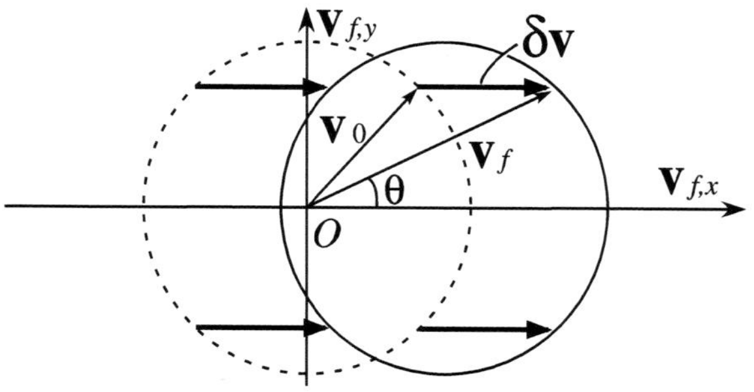

FIG. 1. Effect of a strong laser field on the photoelectrons ionized by the x-ray pulse at a given phase ϕ_i. Dashed circle—drift velocity distribution (v_{0x}, v_{0y}) of the photoelectrons without the laser field; solid circle—distribution with the laser field. Note that neither the uncertainty in $\mathbf{v}_0$ due to the bandwidth of the x-ray pulse nor the angular dependence of the emission probability are represented for simplicity.

the laser field. It is a translation of the dashed circle by $\delta\mathbf{v} = (e/m_e)\mathbf{A}(t_i)$. For linearly polarized light ($\varepsilon = 0$), the translation is parallel to the electric field, and the electron distribution sweeps back and forth along the field direction as t_i varies. For circularly polarized light ($\varepsilon = \pm 1$), the electron distribution is displaced perpendicular to the laser electric field at the moment of birth, and it rotates around the dashed circle as the time of ionization is varied.

For all polarizations, electrons that are freed at slightly different times t_i will have different final velocities because of the dependence of $\mathbf{v}_f$ on the optical phase $\phi_i = \omega_L t_i + \varphi$. It is this phase dependence that offers an opportunity for the measurement of subcycle duration. To evaluate the temporal resolution, it is convenient to reformulate Eq. (1) in terms of the drift kinetic energy $K = m_e v_f^2/2$ for electrons moving in the xy plane:

$$
\begin{aligned}
K_L &= W_0 + 2U_p \cos 2\theta \sin^2\phi_i \pm \alpha_L\sqrt{8W_0U_p}\,\cos\theta\,\sin\phi_i\,,\\
K_C &= W_0 - U_p \cos 2(\phi_i - \theta) \pm \alpha_C\sqrt{4W_0U_p}\,\sin(\phi_i - \theta), \qquad (2)\\
\alpha_L &= \{1 - (2U_p/W_0)\sin^2\theta\sin^2\phi_i\}^{1/2}, \qquad \alpha_C = \{1 - (U_p/2W_0) - (U_p/2W_0)\cos[2(\phi_i - \theta)]\}^{1/2}.
\end{aligned}
$$

Here K_L, K_C are the kinetic energies for linear and circular polarization of the laser field, respectively; $U_p = e^2E_0^2(t_i)/4m_e\omega_L^2$ is the ponderomotive potential of the laser field at the time of ionization; θ is the angle of observation measured from the x axis as shown in Fig. 1; α_L and α_C are correction terms to the expression given in [8], and deviate from 1 when U_p becomes comparable to or higher than W_0. If the field is increased to $U_p > W_0/2$ (linear polarization) or $U_p > W_0$ (circular polarization), some of the photoelectrons are deflected into the reverse direction from $\mathbf{v}(t_i) = \mathbf{v}_0$. The branches of negative signs account for these backscattered electrons, and are required only for this high intensity regime. In this case, another peak appears on the lower energy side of photoelectron spectra, which makes their interpretation more complex. Hereafter, we will restrict the discussion to $U_p < W_0/2$ (linear polarization) and $U_p < W_0$ (circular polarization).

There are many possibilities for measurement because the photoelectron spectra depend on the observation angle and the laser polarization. We will concentrate on three configurations. For a linearly polarized field, if we observe only electrons with $\mathbf{v}_f$ parallel to the laser polarization, Eq. (2) shows that the drift energy will sweep above and below W_0 as the phase of birth varies from 0 to 2π. If we observe only electrons with $\mathbf{v}_f$ perpendicular to the laser polarization, Eq. (2) shows that it will sweep only below W_0 twice per laser period. Measurements that are angle integrated will have a degraded resolution. In circularly polarized light, if we observe only electrons of a given energy, then their direction will sweep. For any polarization, measurements that are energy integrated [5] will have a degraded resolution.

With a linearly polarized field, the width ΔE of the angle-resolved photoelectron spectrum is determined by the following factors: (i) the variation of the moment of birth of photoelectrons due to the x-ray pulse duration, and (ii) the bandwidth and the chirp of the x-ray pulse. Thus, the duration and chirp can be determined by comparing the spectra with and without the laser field.

For a transform-limited x-ray pulse, the width of a photoelectron spectrum is given as $\Delta E[\tau_X, \Delta(\hbar\Omega_X)] = \{(|\partial K/\partial\phi_i|\omega_L\tau_X)^2 + [|\partial K/\partial W_0|\Delta(\hbar\Omega_X)]^2\}^{1/2}$ provided that the streaking speed $\partial K/\partial\phi_i$ is constant within the x-ray pulse. Within one optical period, the best temporal resolution is achieved at the phase that maximizes the resolving parameter $\beta_L \equiv (|\partial K/\partial\phi_i|\omega_L\tau_X)/\{|\partial K/\partial W_0|\Delta(\hbar\Omega_X)\}$. It occurs at $\phi_i = m\pi$ for $\theta = 0$ or $\phi_i = (m/2 + 1/4)\pi$ for $\theta = \pi/2$, respectively. β_L scales as $v_0E_0(t_i)\tau_X^2$ for $\theta = 0$ and $E_0^2(t_i)\tau_X^2$ for $\theta = \pi/2$. Physically, as the x-ray pulse duration decreases, its bandwidth increases as $\propto 1/\tau_X$ and the contribution of sweeping to the photoelectron bandwidth ΔE decreases as $\propto \tau_X$.

We set the condition for the resolution limit as $\beta_L = 1$, i.e., the x-ray bandwidth is equal to the broadening by the laser streak. This is the criterion used in traditional streak cameras. For transform-limited x-ray pulses at 100 eV, the resolution limit is 70 asec for $\theta = 0$ and 100 asec for $\theta = \pi/2$. Here we assumed helium as an ionizing atom ($I_p = 24.6$ eV), laser wavelength $\lambda_L = 0.8\ \mu$m, and the intensity of the laser field set by $U_p = W_0/2$ ($I = 6.3 \times 10^{14}$ W/cm^2).

One drawback of using a linearly polarized laser field is that the streaking speed $\partial K/\partial\phi_i$ varies within the laser period, as seen in Eq. (2). If the x-ray pulse duration

VOLUME 88, NUMBER 17 PHYSICAL REVIEW LETTERS 29 APRIL 2002

becomes close to the streaking period, the streaking speed will change within the x-ray pulse, which will make the data interpretation more complicated. This difficulty can be removed by using a circularly polarized laser field.

With circularly polarized light, a photoelectron at a given kinetic energy streaks in angle as shown in Eq. (2). This angular spread $\Delta\theta$ of photoelectron distribution in the xy plane can be used to measure the duration of x-ray pulses. The main advantage of this configuration is that the streaking speed is constant over the whole optical cycle, and is automatically calibrated by the frequency of the laser field. This method generalizes the concept of the attosecond streak camera [5], which is based on the measurement of the angular distribution of photoelectrons without resolving the kinetic energy. Energy-integrated measurements require $U_p \gg W_0$, a condition difficult to achieve experimentally. The temporal resolution of this configuration is again limited by the bandwidth of the x-ray pulse and is similar to that for linear polarization (70 asec for 100 eV photons).

The electron spectrum can also be calculated quantum mechanically [atomic units are used for compactness in Eqs. (3) and (4)]. Following [17], the amplitude of populating a state $|\mathbf{v}\rangle$ with kinetic momentum $\mathbf{v}$ at a moment T after the end of both the laser and the x-ray pulses is

$$a_{\mathbf{v}}(T) = -i \int_{-\infty}^{T} dt\, \mathbf{d}_{\mathbf{p}(t)} \mathbf{E}_X(t) e^{-i \int_t^T dt' (1/2[\mathbf{p}(t')]^2 + I_p)}, \tag{3}$$

where $\mathbf{p}(t) = \mathbf{v} + \mathbf{A}(t)$ is the instantaneous kinetic momentum, and $\mathbf{d}_{\mathbf{p}(t)}$ is the dipole transition matrix element from the ground state to the continuum with kinetic momentum $\mathbf{p}(t)$. The x-ray field $\mathbf{E}_X(t)$ includes the fast oscillations of the carrier, chirp, and the pulse envelope.

We assume that the x-ray pulse is a linearly chirped Gaussian, with dimensionless chirp ξ defined in the spectral domain as $\tilde{E}_X(\Omega) \sim \exp[-(\Omega - \Omega_X)^2\tau^2(1 - i\xi)/2]$. In time domain, positive chirp $\xi > 0$ corresponds to the instantaneous frequency increasing with time: $E_X(t) \sim \exp[-(t - t_0)^2(1 + i\xi)/2\tau^2(1 + \xi^2) - i\Omega_X t]$, where t_0 is the peak of the x-ray pulse and its duration is $\tau_X \sim \tau\sqrt{1 + \xi^2}$. For a sufficiently short x-ray pulse, Eq. (3) can be approximated using the saddle point method, so we have

$$|a_{\mathbf{v}}|^2 \propto \frac{|\mathbf{d}_{\mathbf{p}(t_0)} \mathbf{E}_X(t_0)|^2}{\mu} \exp\left[-\left(\frac{\mathbf{p}^2(t_0)}{2} - W_0\right)^2 \frac{\tau^2}{\mu^2}\right],$$
$$\mu = \sqrt{(1 + \eta\xi)^2 + \eta^2}, \qquad \eta = \mathbf{E}_L(t_0)\mathbf{p}(t_0)\tau^2. \tag{4}$$

Equations (3) and (4) apply for all directions of observation and laser polarizations. In the field-free case $\mu = 1$ and Eq. (4) corresponds to the spectrum of the x-ray pulse shifted by $-I_p$. Equation (4) shows that, in the presence of the laser pulse, the width of the electron spectrum depends on $\mathbf{E}_L(t_0)$ at the peak of the x-ray pulse. To measure the x-ray pulse duration, one must produce an observable change in μ.

Figure 2a shows photoelectron spectra in the direction parallel to the laser polarization produced by 70 asec (full width at half maximum, FWHM) transform-limited x-ray pulses for the same condition that we used to derive the resolution limit above. The center of the x-ray pulse is set at the maximum of the laser field ($\omega_L t_0 + \varphi = 0$). The solid curve corresponds to the field-free spectrum. In the presence of the laser field, classical [open circles, from Eq. (2)] and quantum-mechanical [dotted curve, from Eq. (3)] calculations agree well. The laser field broadens the spectrum by $\sqrt{2}$ as predicted in the semiclassical analysis above.

In the case of a chirped x-ray pulse, substantial modification of the spectrum can be achieved when $\eta\xi \sim 1$, which for large chirps allows one to deal with larger bandwidth $\Delta\Omega_X \sim 1/\tau$. Physically, if the photoelectrons are positively streaked ($\partial K/\partial\phi_i > 0$), the width of the photoelectron spectrum will be larger than the transform-limited case for positive chirp of the x-ray pulse, and narrower for negative chirp. Equation (4) allows us to determine both the sign and the magnitude of the chirp.

Figure 2b shows the photoelectron spectra produced by chirped x-ray pulses (70 asec FWHM, $\xi = \pm\sqrt{3}$). The spectra are calculated using Eq. (3) [Eq. (4) gives almost identical results]. Other conditions are the same as in

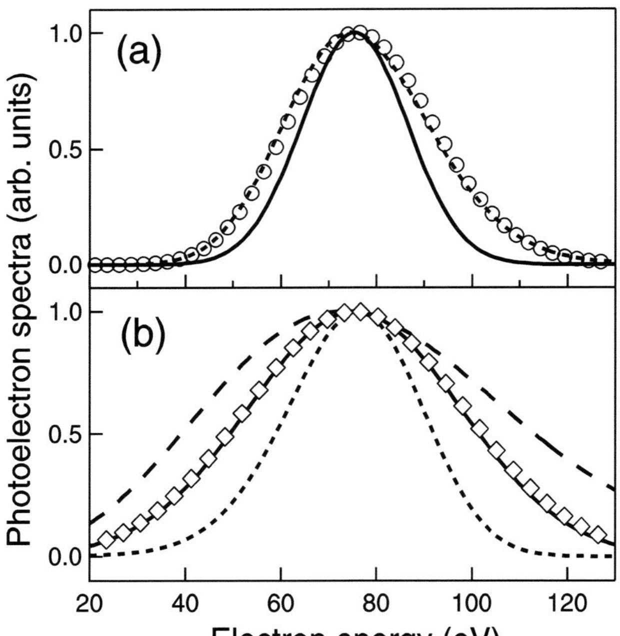

FIG. 2. (a) The photoelectron spectra produced by a transform-limited x-ray pulse (70 asec FWHM). Solid curve—field-free case; dotted curve—with a laser field ($I = 6.3 \times 10^{14}$ W/cm^2). The classical result is shown by open circles. (b) The photoelectron spectra produced by chirped x-ray pulses (70 asec FWHM, $\xi = \pm\sqrt{3}$). Solid curve—field-free case; dashed curve is for the positive chirp; dotted curve is for the negative chirp. Open diamonds are for the transform-limited case (35 asec FWHM, $\xi = 0$). The field intensity is the same as in (a).

Fig. 2a. The solid line shows the field-free photoelectron spectrum. Depending on the sign of the chirp, we can clearly see the broadening or narrowing of the spectrum. For a transform-limited pulse (35 asec FWHM, $\xi = 0$), the broadening can hardly be seen (open diamonds in Fig. 2b).

In the above analysis, we have made a number of implicit assumptions that might influence experiments. Shot-to-shot fluctuation of the streaking phase ϕ_i will increase the width of the photoelectron spectra and degrade the temporal resolution. In high harmonic generation, the most likely source of attosecond pulses, the x-ray pulses are naturally phase locked to the driving laser pulses. Therefore, the combination of these x-ray and laser pulses is ideal for the attosecond streak camera. Fluctuation of the field intensity ($\propto U_p$) will also degrade the resolution. However, with Ti:sapphire lasers we can expect a good stability of $\sim$5%.

The maximum temporal resolution requires the highest feasible intensity of the laser field. Above-threshold ionization (ATI) by the laser field alone can produce a significant number of background electrons which limit the intensity that can be used. There are three ways to minimize the influence of ATI electrons. (i) A few-cycle laser pulse reduces the total ionization probability at a given intensity [8]. (ii) ATI tends to produce low energy electrons directed along the laser polarization. The energy and angle of observation can be chosen to reduce their influence [8]. (iii) For the linearly polarized case, the small background of rescattered electrons can be suppressed by introducing a slight ellipticity to the field.

Angular dependence of the photoionization cross section may seem to present a problem for correctly interpreting the observed photoelectron spectrum. However, it is technically feasible to cancel this anisotropy by integrating the photoelectron signal while rotating the polarization of the x-ray pulse with respect to the laser polarization. In the case of high harmonics, this can be done by rotating the polarization direction of the laser pulses used for x-ray generation.

Before concluding, we revisit the multicycle measurements from a subcycle perspective. As we have seen, the strong laser field distorts the photoelectron spectrum on the subcycle time scale. This effect corresponds to a frequency modulation of the photoelectron wave function within the laser optical cycle. If the x-ray pulse is longer than the laser period τ_L, this modulation is repeated identically every optical cycle. This leads to a temporal periodicity of the wave function, which is responsible for the appearance of sidebands spaced by $\hbar\omega_L$ in the energy spectrum.

The positions of the sidebands are determined by the phase shift of the wave function that occurs from one optical cycle to the following one. This is analogous to the frequency comb spectrum of a pulse train from a mode-locked laser. The position of the comb is determined by the shift of the carrier-envelope optical phase from pulse to pulse [18,19]. For the electrons, a similar phase shift occurs because an electron born at time t makes one oscillation more in the laser field than one born at $t + \tau_L$. The quantum phase acquired during this extra oscillation is $2\pi U_p/\hbar\omega_L$, and accounts for the ponderomotive shift of the sidebands by $-U_p$.

In conclusion, the attosecond streak camera will allow resolution $\lesssim$100 asec. To realize this resolution experimentally, any phase variation of the laser field with respect to the x-ray pulse must be less than $2\pi(\tau_X/\tau_L)$ over the interaction volume and between laser shots. This criterion appears achievable for attosecond x-ray pulses produced by high harmonic generation.

This work is supported in part by Photonics Research Ontario.

Note added.—Recently isolated attosecond pulses [3] as well as a train of pulses [20] were successfully characterized using x-ray photoionization of atoms in a laser field. The former [3] is closely related to our proposed method, but utilizes the broadening of photoelectron spectra due to a large angle of collection. The latter [20] is not directly related to our method, but, by using Eq. (3), this experiment can be understood as the interference of photoelectron wave packets with laser-induced phase shifts (see also [21]).

[1] A. H. Zewail, J. Phys. Chem. A **104**, 5660 (2000).
[2] T. Brabec and F. Krausz, Rev. Mod. Phys. **72**, 545 (2000), and references therein.
[3] M. Hentschel *et al.*, Nature (London) **414**, 509 (2001).
[4] F. Krausz, Phys. World **14**, 41 (2001).
[5] E. Constant *et al.*, Phys. Rev. A **56**, 3870 (1997).
[6] Y. Kobayashi *et al.*, Opt. Lett. **23**, 64 (1998).
[7] A. Scrinzi, M. Geissler, and T. Brabec, Phys. Rev. Lett. **86**, 412 (2001).
[8] M. Drescher *et al.*, Science **291**, 1923 (2001).
[9] J. M. Schins *et al.*, Phys. Rev. Lett. **73**, 2180 (1994).
[10] T. E. Glover *et al.*, Phys. Rev. Lett. **76**, 2468 (1996).
[11] A. Bouhal *et al.*, J. Opt. Soc. Am. B **14**, 950 (1997).
[12] E. S. Toma *et al.*, Phys. Rev. A **62**, 061801(R) (2000).
[13] A. M. Dykhne and G. L. Yudin, Usp. Fiz. Nauk **121**, 157 (1977) [Sov. Phys. Usp. **20**, 80 (1977)]; Usp. Fiz. Nauk **125**, 377 (1978) [Sov. Phys. Usp. **21**, 549 (1978)].
[14] T. F. Gallagher, Phys. Rev. Lett. **61**, 2304 (1988).
[15] P. B. Corkum, N. H. Burnett, and F. Brunel, Phys. Rev. Lett. **62**, 1259 (1989).
[16] E. Yablonovitch, Phys. Rev. Lett. **60**, 795 (1988).
[17] M. Lewenstein *et al.*, Phys. Rev. A **49**, 2117 (1994).
[18] D. J. Jones *et al.*, Science **288**, 635 (2000).
[19] A. Apolonski *et al.*, Phys. Rev. Lett. **85**, 740 (2000).
[20] P. M. Paul *et al.*, Science **292**, 1689 (2001).
[21] F. Quéré *et al.* (to be published).

VOLUME 88, NUMBER 17 PHYSICAL REVIEW LETTERS 29 APRIL 2002

Quantum Theory of Attosecond XUV Pulse Measurement by Laser Dressed Photoionization

Markus Kitzler, Nenad Milosevic, Armin Scrinzi, Ferenc Krausz, and Thomas Brabec*

Institut für Photonik, Technische Universität Wien, Gusshausstrasse 27/387, A-1040 Wien, Austria

(Received 31 August 2001; published 16 April 2002)

The first reported measurements of single attosecond pulses use laser dressed single-photon extreme ultraviolet (XUV) ionization of gas atoms. The determination of XUV pulse duration from the electron spectrum is based on a classical theory. Although classical models are known to give a qualitatively correct description of strong laser atom interaction, the validity must be scrutinized by a quantum-mechanical analysis. We establish a theoretical framework for the accurate temporal characterization of attosecond XUV pulses. Our analysis reveals an improved scheme that allows for direct experimental discrimination between single and multiple attosecond pulses.

DOI: 10.1103/PhysRevLett.88.173904 PACS numbers: 42.65.Re, 32.80.Rm, 32.80.Wr, 42.65.Ky

Extreme ultraviolet (XUV) pulse durations are usually determined by a laser-XUV pulse cross-correlation measurement based on the laser-induced shift of the photoelectron spectrum [1–5]. The time resolution of this method is limited by the duration of the laser pulse and, therefore, cannot be used in the attosecond (10^{-18} sec, asec) time domain. Recently, several cross-correlation methods [6–9] were proposed that have the potential to resolve asec pulse durations. The most promising concept is based on the laser-induced shift of the photoelectron spectrum, that is produced by laser dressed single-photon ionization of gas atoms by an XUV asec pulse. Two modifications of this method were used to measure an asec pulse train [9] and an isolated XUV pulse with a duration of 650 asec [8]. Our investigation focuses on the scheme realized in Ref. [8], which has the advantage that single asec pulses can be measured. The difference to earlier experiments [1–5] is that the sublaser cycle resolution is achieved by measuring the electron spectrum only over a limited solid angle [8]. The XUV pulse duration was determined by using a classical model that relates the shift and the broadening of the electron spectrum to the XUV duration. The classical model allows efficient numerical implementation making this method a particularly attractive method for asec pulse measurement. Although classical models in strong laser-field physics are known to give a qualitatively correct description, quantitative agreement between a classical model and a full quantum-mechanical description is rarely the case. Therefore, a reliable determination of sub-fs pulse durations with laser dressed single XUV photoionization must depend on a thorough quantum-mechanical analysis.

We perform such an analysis using exact integration of the Schrödinger equation and analytical integration under the strong field approximation (SFA), where the influence of the atomic Coulomb potential on the free electrons is neglected. Based on the SFA we derive a quantum-mechanical and a semiclassical equation for the electron spectrum that can be integrated orders of magnitude faster than the Schrödinger equation. The semiclassical theory presents a generalization of the classical model used in Ref. [8] and, in special cases, becomes identical with the classical model. The validity of the semiclassical approach is verified by a comparison to exact results, which in a certain parameter range yields good agreement. Finally, our analysis is used to identify an improved configuration for asec pulse measurement. This setup is more than 1 order of magnitude more efficient than the one used so far [8] and allows for direct experimental discrimination between single and multiple asec pulses.

Our analysis starts from the three-dimensional Schrödinger equation in atomic units,

$$i\partial_t\Psi(\mathbf{r}) = \left[-\frac{1}{2}\nabla^2 - \frac{1}{r} - \mathbf{r}E(t)\right]\Psi(\mathbf{r}), \quad (1)$$

where the electron is coupled to the classical electromagnetic field in dipole approximation and in the length gauge. Here, $\mathbf{r} = (x, y, z)$ denotes the space coordinates, ∇ is the corresponding gradient operator, ∂_t is the time derivative, and Ψ represents the electron wave function. The electric field $\mathbf{E}(t) = \mathbf{E}_l + \mathbf{E}_x$ comprises a laser, $\mathbf{E}_l$, and an XUV, $\mathbf{E}_x$, contribution. The laser and XUV components, $\mathbf{E}_{l,x} = \hat{\mathbf{e}}_{l,x}E_{l,x}(t - t_{l,x})\cos[\omega_{l,x}(t - t_{l,x})]$, are characterized by the polarization vector $\hat{\mathbf{e}}_{l,x}$, envelope $E_{l,x}(t)$, and by the carrier frequency $\omega_{l,x}$. Further, we assume that the laser pulse peak is at $t_l = 0$ and that the XUV pulse peak is delayed by a time $t_x = t_d$. Finally, the electric field is related to the vector potential by $\mathbf{E} = -(1/c)\partial_t\mathbf{A}$, where c denotes the velocity of light.

In the following, Eq. (1) is solved analytically by applying the SFA [10], i.e., the Coulomb potential is neglected as compared to the laser field. As a result, the Coulomb continuum eigenfunctions may be substituted by plane waves, and Eq. (1) is integrated by using the ansatz $\Psi = |0\rangle\exp(iI_pt) + \int d^3p\, b(\mathbf{p},t)\,|\mathbf{p}\rangle$. Here, $|0\rangle$ denotes the ground state with ionization potential I_p, the plane wave is given by $|\mathbf{p}\rangle = \exp(i\mathbf{pr})$ with $\mathbf{p}$ the electron momentum, and $b(\mathbf{p},t)$ is the momentum space wave function of free electrons. Within the SFA we obtain the electron spectrum as [11]

0031-9007/02/88(17)/173904(4)$20.00

VOLUME 88, NUMBER 17 PHYSICAL REVIEW LETTERS 29 APRIL 2002

$$b(\mathbf{p}) = i\int_{-\infty}^{\infty} \mathbf{E}(t')\mathbf{d}[\mathbf{p} - \mathbf{A}(t')]\exp\left[-i\int_{t'}^{\infty} \frac{1}{2}[\mathbf{p} - \mathbf{A}(t'')]^2\, dt'' + iI_p t'\right] dt'. \tag{2}$$

The dipole moment is determined by $\mathbf{d}(\mathbf{p}) = \langle \mathbf{p}|\mathbf{r}|0\rangle$. For the special case of continuum transitions from hydrogen bound s states, $\mathbf{d} \propto \mathbf{p}/(\mathbf{p}^2 + 2I_p)^3$. For calculations in noble gases we have derived dipole moments from cross sections tabulated in Ref. [12].

The integral in Eq. (2) is calculated in two ways, by a Fourier-Bessel expansion of the exponential function in harmonics of the laser field [13] and by the stationary phase method [14]. The first route results in

$$|b(\mathbf{p})|^2 = \left| \sum_{n=-\infty}^{+\infty} i^n J_n(a,b) F_n(p) \right|^2, \tag{3}$$

where $F_n(p) = \int_{-\infty}^{+\infty} \mathbf{d} \cdot \mathbf{E}_x(t - t_d)\exp\{i[p^2/2 + I_p + U_p(t_d) + n\omega_l]t\}\, dt$, and $U_p(t) = E_l^2(t)/(4\omega_l^2)$ is the laser ponderomotive potential. The contributions F_n from the nth sideband are weighted by the generalized Bessel functions $J_n(a,b) = [\mathrm{sgn}(-a)]^n \sum_{l=-\infty}^{+\infty} J_{n+2l}(|a|)J_{-l}(b)$ with $a = \mathbf{p} \cdot \hat{\mathbf{e}}_l E_l(t_d)/\omega_l^2$ and $b = E_l^2(t_d)/(8\omega_l^3)$. Note that Eqs. (3) and (2) give identical results as long as the XUV pulse duration is shorter than the laser optical cycle.

A simpler solution of Eq. (2) can be obtained in the semiclassical limit by stationary phase integration. This results in

$$|b(\mathbf{p})|^2 \approx \left| \sum_{t_s} \sqrt{\frac{\pi}{2i\ddot{S}(t_s)}}\, \mathbf{E}_x(t_s - t_d)\mathbf{d}[\mathbf{p} - \mathbf{A}(t_s)] \times \exp[-iS(t_s)] \right|^2, \tag{4}$$

where $S(t_s) = (1/2)\int_{t_s}^{\infty} [\mathbf{p} - \mathbf{A}(t'')]^2\, dt'' + (\omega_x - I_p)t_s$ is the classical action, and $\ddot{S}(t_s) = -E(t_s)[p\cos\theta - A(t_s)]$ is the second time derivative of the classical action. Here, θ denotes the angle between the momentum $\mathbf{p}$ and the polarization axis of the laser electric field. The stationary phase points t_s are determined by solution of the stationary phase equation $(1/2)[\mathbf{p} - \mathbf{A}_l(t_s)]^2 = \omega_x - I_p$.

The SFA equation (3) and the semiclassical Eq. (4) reduce the computational effort for calculating the electron spectrum by 4 to 5 orders of magnitude as compared to the integration of Eq. (1). This makes the accurate determination of the XUV pulse duration including quantum effects practically possible. It will be shown below that Eq. (4) has a more limited validity range than Eq. (3); however, it has the advantage that contact can be made with the classical analysis of Ref. [8].

The asec pulse measurement method in Ref. [8] relies on the fact that the electron spectrum is measured in a limited solid angle θ, φ. The XUV pulse duration is determined by measuring the nth moment of the electron energy spectrum as a function of the XUV pulse delay t_d, which is given by

$$\langle \Omega^n(t_d)\rangle = \frac{\int_{\theta_0}^{\theta_1}\int_0^{\infty} \Omega^n |b(\Omega,\theta)|^2 \sqrt{\Omega}\, d\Omega \sin\theta d\theta}{\int_{\theta_0}^{\theta_1}\int_0^{\infty} |b(\Omega,\theta)|^2 \sqrt{\Omega}\, d\Omega \sin\theta d\theta}, \tag{5}$$

where we denote the electron energy by $\Omega = p^2/2$. The angle θ to the polarization axis is confined between θ_0 and θ_1. We focus here on configurations with cylindrical symmetry, where the linear laser and XUV polarizations coincide, for which the integral over φ drops out of Eq. (5). Usually, the center of gravity $\langle\Omega\rangle$ or the rms width of the electron spectrum $\Delta\Omega = \sqrt{\langle\Omega^2\rangle - \langle\Omega\rangle^2}$ is used for asec pulse measurement [8].

Equations (3) or (4) in combination with Eq. (5) allow an efficient numerical evaluation of $\langle\Omega^n\rangle$. With the use of Eq. (4), $\langle\Omega^n\rangle$ can be further simplified and yields a simple semiclassical formula. For that we insert Eq. (4) and transform from (Ω, θ) to (t_s, θ), where $\Omega(t_s)$ is defined by the stationary phase equation. This yields

$$\langle \Omega^n(t_d)\rangle = \frac{\int_{\theta_0}^{\theta_1}\int_{-\infty}^{\infty} \Omega^{n+1/2}(t_s)E_x^2(t_s - t_d)[\hat{\mathbf{e}}_x \mathbf{d}(t_s)]^2\, dt_s \sin\theta d\theta}{\int_{\theta_0}^{\theta_1}\int_{-\infty}^{\infty} \Omega^{1/2}(t_s)E_x^2(t_s - t_d)[\hat{\mathbf{e}}_x \mathbf{d}(t_s)]^2\, dt_s \sin\theta d\theta}. \tag{6}$$

Equation (6) puts the classical derivation of Ref. [8] on a more general and rigorous theoretical footing. In the special case of transitions to spherically symmetric s continuum waves (angular momentum $l = 0$), the result of Ref. [8] is exactly recovered. This is because, by virtue of the stationary phase condition, $\mathbf{d}$ becomes independent of the integration variables and drops out of Eq. (6).

The quality of the approximations utilized in our derivation was tested by comparison to an exact numerical solution of Eq. (1) based on a discretization of the Schrödinger equation in momentum space, as described in Ref. [15]. For Fig. 1 we calculated the rms width $\Delta\Omega$ in various ways. The modulation of $\Delta\Omega$ appears for pulses shorter than one laser half-cycle and gives a sensitive measurement of the XUV pulse duration. The modulation increases with decreasing XUV pulse duration. The filled circles, solid line, and dotted line refer to the exact numerical solution of Eq. (1), to the SFA equation (3) in combination with Eq. (5), and to the integration of Eq. (6), respectively. The exact solution agrees well with the SFA solution, which demonstrates the validity of the SFA. The agreement with Eq. (6) is best around the laser pulse center. The reason is that the semiclassical calculation of the electron spectrum based upon the stationary phase method fails for vanishing laser intensities. In this limit, the electron spectrum becomes a delta peak at energy $\omega_x - I_p$. As a result $\Delta\Omega \to 0$ for large t_d, whereas in reality the width never drops below the spectral width of the XUV pulse. The calculations in Fig. 1 were repeated by varying the laser and XUV pulse parameters in the range relevant for asec pulse measurement (XUV/laser duration between

VOLUME 88, NUMBER 17 PHYSICAL REVIEW LETTERS 29 APRIL 2002

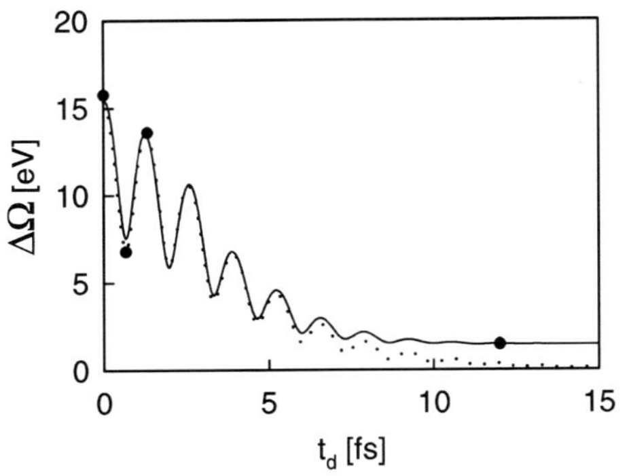

FIG. 1. rms width $\Delta\Omega$ of the electron spectrum for a hydrogen (1s) ground state as a function of delay time, t_d, between the laser and XUV pulse. The pulse parameters are laser wavelength $\lambda_l = 800$ nm, laser FWHM pulse duration 5 fs, sech laser pulse envelope, laser intensity $I_l = 3 \times 10^{13}$ W/cm^2, $\omega_x = 90$ eV, XUV pulse duration 500 asec, XUV pulse intensity $I_x = 10^{12}$ W/cm^2. Laser and XUV polarization are chosen parallel to the direction of observation, $\hat{\mathbf{z}} = \hat{\mathbf{e}}_l \parallel \hat{\mathbf{e}}_x \parallel \mathbf{p}$, and $\theta_1 = 2°$. $\Delta\Omega$ was calculated in various ways: exact numerical solution of Eq. (1) (filled circles), evaluation of the SFA equation (3) (solid line), and integration of the semiclassical Eq. (6) (dotted line).

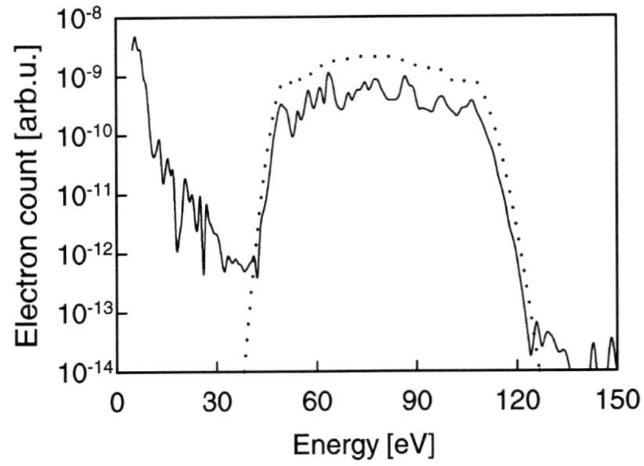

FIG. 2. Photoelectron spectrum for the parameters of Fig. 1 and delay $t_d = 0$. The solid and dotted lines denote the electron spectrum as determined by numerical integration of the Schrödinger equation (1) and the SFA equation (3), respectively. Note that, in the derivation of Eq. (3), laser-induced ionization is not included, which is responsible for the low-frequency part (<30 eV) in the exact calculation.

0.1/5 and 1/15 fs, XUV/laser peak intensity between 10^{12} and 10^{14} W/cm^2, sech, and Gaussian pulse shape). The only parameter critical for the applicability of the semiclassical model is the laser peak intensity, for the reasons discussed above. For intensities well below 10^{13} W/cm^2 the semiclassical theory predicts wrong rms widths (see Fig. 1) and the SFA equation must be used. Finally, we did not plot the classical result, which is identical with the semiclassical calculation for the parameters chosen here. This is because for $\theta \approx 0°$, $\mathbf{d}$ becomes again independent of the integration variables and drops out of Eq. (6). For $\theta \approx 90°$ the effect of $\mathbf{d}$ is maximum and introduces a factor of 2 in the modulation depth of the rms width.

To further corroborate the applicability of the SFA, we have compared electron spectra in Fig. 2, which were calculated by an exact solution of Eq. (1) (solid line) and by Eq. (3) (dotted line). The agreement is excellent, justifying the omission of the Coulomb potential in the derivation of Eq. (3). Figure 2 reveals another important result. Except for the XUV photoelectrons the spectrum at low energies contains a contribution of ATI (above-threshold ionization) electrons generated by the laser directly. The experimental setup must be chosen in such a way that the two contributions fall into well-separated spectral ranges. Otherwise the laser dressed photoionization signal, carrying the information on the asec pulse duration, is covered by the ATI electrons. So far, in order to minimize ATI, electrons were observed perpendicular to the laser polarization axis [8]. The spectrum in Fig. 2 shows that the level of ATI electrons is less than anticipated. Even for $\hat{\mathbf{e}}_l \parallel \mathbf{p}$, where ATI reaches the maximal energies, the laser induced part of the ionization spectrum is well separated from the x-ray photo ionization spectrum. This can be understood from a simple estimate. The laser-induced part of the ATI spectrum consists of a main part extending to $2U_p$ and a smaller contribution coming from rescattering, with electron energies up to $10U_p$ [16]. Only the $2U_p$ contribution is strong enough to cover the XUV-induced electron signal. As a result, asec-pulse measurement is feasible, as long as the lower cutoff of the XUV photoelectron spectrum does not overlap with the $2U_p$ part of the laser-induced ATI spectrum. Based on the stationary phase equation derived above the simple condition $8U_p \leq \omega_x - I_p$ is found, determining the parameter range in which the two spectral contributions remain seperate. Laser-induced ionization becomes dominant and buries the single XUV photoionization signal only for laser peak intensities in excess of 10^{14} W/cm^2. This finding adds an additional degree of freedom to the realization of an optimum setup for asec pulse measurement.

A major problem in asec-pulse measurements is the low efficiency of harmonic sources and the resulting long data collection times and poor signal-to-noise ratio. So far the asec measurement was performed with Kr 4p electrons, for which the single XUV photon ionization is dominated by a transition to a spherically symmetric s continuum wave [12]. We have compared various experimental setups with respect to efficiency in electron yield. We find an optimal setup when laser and x-ray polarization are chosen parallel to the direction of observation of the electron spectrum, i.e., $\hat{\mathbf{e}}_l \parallel \hat{\mathbf{e}}_x \parallel \mathbf{p}$. Instead of Kr 4$p$ we propose to use Ne 2p electrons. The overall gain in the XUV electron yield as compared to the setup in Ref. [8] is a factor of 30 for the same opening angle of 40°. This enhancement can be attributed to two reasons. First, the Ne 2p transition is an order of magnitude more efficient than the Kr 4p transition. Second, the XUV photoionization (90 eV) of Ne 2p electrons is dominated by a transfer into a d continuum wave, which extends in the direction

VOLUME 88, NUMBER 17 PHYSICAL REVIEW LETTERS 29 APRIL 2002

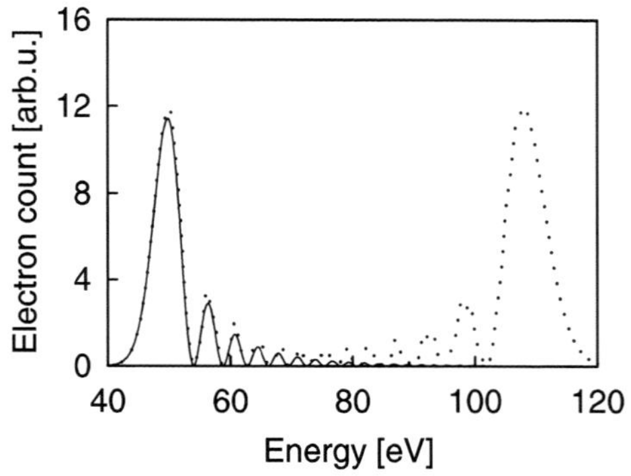

FIG. 3. Signature of two attosecond pulses with delay times $t_d = \pm 1/4$ laser optical cycles compared to a single pulse with delay $t_d = 1/4$. The remaining pulse parameters are as in Fig. 1. Electrons released at $t_d = \pm 1/4$ experience maximal acceleration and maximum deceleration by the laser, creating two well-separated electron peaks (dotted line). A single pulse creates only a single peak (solid line).

of the x-ray polarization. Therefore, a measurement along the x-ray polarization direction captures a larger part of the photoelectrons, which gives an enhancement by a factor of 3. The gain enhancement by more than 1 order of magnitude in our improved setup makes a more efficient measurement of asec pulses possible and allows an extension of the method towards shorter wavelengths.

Higher harmonic generation with few cycle laser pulses tends to generate a single asec pulse with satellites at a repetition rate of twice the laser frequency. For practical applications it is essential to discriminate between a single asec pulse and a pulse train. Recently, indirect evidence of single asec pulses was obtained by using the harmonic spectrum [8]. Measurement of the electron spectrum in the laser polarization direction will allow for the first time to directly distinguish a single asec pulse from a pulse train (see Fig. 3). This is because the shift of electron energies becomes sensitive to the sign of the laser vector potential at the time of XUV ionization. Depending on the sign, electrons are either accelerated or decelerated by the laser, leading to two well-separated peaks in the electron spectrum. For one main and one or two smaller satellite pulses, the ratio of the peaks in the electron spectrum reflects the relative energy content carried in the satellites. In an experiment, data are usually collected over several shots. For the observation of a single peak, it is essential that the absolute phase of the laser field is kept constant during data acquisition.

In conclusion, we have developed a quantum-mechanical model of laser dressed single x-ray photoionization. It was shown that laser dressed photoionization can be quantitatively modeled by using the strong field approximation, i.e., by neglecting the Coulomb potential during the continuum evolution of the electron. Based on the strong field approximation, we derived a quantum-mechanical and a semiclassical expression that can be evaluated efficiently, making an accurate determination of the asec pulse duration including quantum effects practicable. The validity range of the previously used classical model was established. We identified an optimum setup for an asec pulse measurement that is a factor of 30 more efficient than existing schemes. Finally, the proposed setup discriminates between single and multiple attosecond pulses. The results revealed by our quantum-mechanical investigation open the way towards an accurate and more efficient measurement of asec pulses.

The authors gratefully acknowledge the support of A. J. Schmidt, and invaluable discussions with P. Corkum, and G. Reider. This work was supported by the Austrian Science Fund, Special Research Program F016 and project Y142-TPH.

*Email address: brabec@tuwien.ac.at

[1] J. M. Schins *et al.,* Phys. Rev. Lett. **73**, 2180 (1994).
[2] J. M. Schins *et al.,* J. Opt. Soc. Am. B **13**, 197 (1996).
[3] T. E. Glover, R. W. Schoenlein, A. H. Chin, and C. V. Shank, Phys. Rev. Lett. **76**, 2468 (1996).
[4] A. Bouhal *et al.,* J. Opt. Soc. Am. B **14**, 950 (1997).
[5] E. S. Toma *et al.,* Phys. Rev. A **62**, 061801(R) (2000).
[6] E. Constant, V. D. Taranukhin, A. Stolow, and P. B. Corkum, Phys. Rev. A **56**, 3870 (1997).
[7] A. Scrinzi, M. Geissler, and T. Brabec, Phys. Rev. Lett. **86**, 412 (2001).
[8] M. Drescher *et al.,* Science **291**, 1923 (2001); M. Hentschel *et al.,* Nature (London) **414**, 509 (2001).
[9] P. M. Paul *et al.,* Science **292**, 1689 (2001).
[10] L. V. Keldysh, Sov. Phys. JETP **20**, 1307 (1965).
[11] T. Brabec and F. Krausz, Rev. Mod. Phys. **72**, 545 (2000).
[12] U. Becker and D. A. Shirley, *VUV and Soft X-ray Photoionization* (Plenum, New York, 1996).
[13] M. Abramowitz and I. A. Stegun, *Handbook of Mathematical Functions* (Dover, New York, 1972).
[14] C. M. Bender and S. A. Orszag, *Advanced Mathematical Methods for Scientists and Engineers* (McGraw Hill, New York, 1978).
[15] A. Scrinzi, T. Brabec, and M. Walser, in *Proceedings of the Workshop on Super-Intense Laser-Atom Physics,* edited by B. Piraux and K. Rzaszewski (Kluwer, Dordrecht, 2001), p. 313.
[16] D. B. Milosevic and F. Ehlotzky, Phys. Rev. A **58**, 3124 (1998).

PRL 105, 073001 (2010) PHYSICAL REVIEW LETTERS week ending 13 AUGUST 2010

Attosecond Streaking Enables the Measurement of Quantum Phase

V. S. Yakovlev,[1,2,*] J. Gagnon,[1,2] N. Karpowicz,[2] and F. Krausz[1,2]

[1]*Department für Physik, Ludwig-Maximilians-Universität, Am Coulombwall 1*

[2]*Max-Planck-Institut für Quantenoptik, Hans-Kopfermann-Straße 1, D-85748 Garching, Germany*

(Received 7 May 2010; published 10 August 2010)

Attosecond streaking, as a measurement technique, was originally conceived as a means to characterize attosecond light pulses, which is a good approximation if the relevant transition matrix elements are approximately constant within the bandwidth of the light pulse. Our analysis of attosecond streaking measurements on systems with a complex response to the photoionizing pulse reveals the relation between the momentum-space wave function of the outgoing electron and the result of conventional retrieval algorithms. This finding enables the measurement of the quantum phase associated with bound-continuum transitions.

DOI: 10.1103/PhysRevLett.105.073001 PACS numbers: 32.80.−t

The absorption of an energetic photon by an atom or molecule starts a sequence of events which may result in the emission of one or several electrons. Recent advances in attosecond science allow time-resolved measurements of such dynamics [1–3]. In general, the resulting electron wave packets carry valuable information about the processes that produced them. Until recently, the full characterization of an electron wave packet was impossible—while the spectrum of a wave packet can easily be measured, the phase information was inaccessible. In this Letter, we show rigorously that attosecond streaking [4,5] is an appropriate tool to measure the energy dependence of the phase associated with a particular bound-free transition.

Attosecond streaking consists in recording a set of photoelectron spectra over a range of delays between an ionizing extreme ultraviolet (XUV) pulse and an optical waveform, the intensity of which should be too weak to affect bound electrons, but strong enough to significantly accelerate or decelerate free electrons. This interaction with free electrons is referred to as *streaking*, and the role of the streaking waveform is usually played by a few-cycle near-infrared laser pulse with a stabilized carrier-envelope phase. A set of such laser-dressed electron spectra comprises a *spectrogram*. For a comprehensive review of attosecond streaking measurements and reconstruction techniques, we refer the reader to the original papers [4–9] and focus on the most important concepts. In the following, the photoionizing radiation is referred to as an "XUV pulse", although this radiation may consist of several pulses, and it does not have to be strictly in the XUV spectral range. Let us consider the interaction of a linearly-polarized XUV pulse with a quantum-mechanical system, which is initially found in a stationary bound state $|\Psi_0\rangle$ of its Hamiltonian $\hat{H}_0$. The XUV pulse launches a single-electron wave packet $|\Psi(t)\rangle$, the propagation of which in the ionic potential is conveniently described in a basis formed by continuum eigenstates of $\hat{H}_0$. In the following, $|\mathbf{p}\rangle$ will represent such an eigenstate, so that $|\langle\mathbf{p}|\Psi(t)\rangle|^2$ is the probability density of detecting a photoelectron with an asymptotic momentum $\mathbf{p}$. In this basis, the motion of the photoelectron is determined by $\langle\mathbf{p}|\Psi(t)\rangle = \tilde{\chi}(\mathbf{p})e^{-i(p^2/2)t}$ (atomic units are used throughout this Letter). The probability amplitudes $\tilde{\chi}(\mathbf{p})$ fully describe the properties of the electron wave packet. For each direction of observation, we define a *time-domain wave packet* as

$$\chi(t) = \frac{i}{\pi}\int_0^\infty \tilde{\chi}(\sqrt{2\epsilon})e^{-i(\epsilon-\epsilon_0)t}d\epsilon = \frac{i}{\pi}\int_0^\infty \langle\mathbf{p}|\Psi(t)\rangle e^{i(p_0^2/2)t}pdp, \tag{1}$$

where $\epsilon = p^2/2$ stands for the energy of the electron infinitely far from the ion, and $\epsilon_0 = p_0^2/2$ is a central energy of the wave packet. As we show below, $\chi(t)$ is the quantity that is recovered by analyzing the attosecond streaking spectrogram. This provides much needed rigor to what has been vaguely referred to as the reconstructed "wave packet."

If an electron is freed as a result of single-photon ionization, the properties of the electron wave packet derive from the spectral components of the incoming light pulse multiplied by the respective complex-valued transition matrix element $D(\mathbf{p})$ [10]. Let us consider an XUV pulse with the electric field $E_{\mathrm{XUV}}(t) = \mathrm{Re}[\mathcal{E}_{\mathrm{XUV}}(t)e^{-i\Omega t}]$, where Ω is the central frequency, and $\mathcal{E}_{\mathrm{XUV}}(t)$ is the complex envelope of the pulse. First-order perturbation theory combined with the dipole and rotating-wave approximations yields

$$\tilde{\chi}(\mathbf{p}) = -\frac{i}{2}\tilde{\mathcal{E}}_{\mathrm{XUV}}\left(\frac{p^2}{2} - \frac{p_0^2}{2}\right)D(\mathbf{p}), \tag{2}$$

where

$$\tilde{\mathcal{E}}_{\mathrm{XUV}}(\omega) = \int_{-\infty}^{\infty} \mathcal{E}_{\mathrm{XUV}}(t)e^{i\omega t}dt \tag{3}$$

is the Fourier transform of the complex XUV envelope. The central momentum p_0 is related to the central frequency Ω of the XUV pulse and the ionization potential W by energy conservation: $p_0^2/2 = \epsilon_0 = \Omega - W$.

In the absence of the streaking field, an explicit expression for the photoelectron spectrum $S_0(\mathbf{p}) = |\langle \mathbf{p}|\Psi(t)\rangle|^2 = |\tilde{\chi}(\mathbf{p})|^2$ is

$$S_0(\mathbf{p}) = \left|\frac{1}{2}\int_{-\infty}^{\infty} dt \mathcal{E}_{\mathrm{XUV}}(t) D(\mathbf{p}) e^{i((p^2/2)-(p_0^2/2))t}\right|^2. \quad (4)$$

As it was originally shown in [5], the presence of a streaking laser pulse $\mathbf{E}_L(t) = -\partial \mathbf{A}_L/\partial t$ delayed by τ with respect to the XUV pulse is accounted for by the following generalization of Eq. (4):

$$S(\mathbf{p}, \tau) = \left|\frac{1}{2}\int_{-\infty}^{\infty} dt \mathcal{E}_{\mathrm{XUV}}(t + \tau) G_0(\mathbf{p}, t) e^{i((p^2/2)-(p_0^2/2))t}\right|^2. \quad (5)$$

Here, $G_0(\mathbf{p}, t) = D(\mathbf{p} + \mathbf{A}_L(t)) e^{i\Phi(\mathbf{p},t)}$ with $\Phi(\mathbf{p}, t)$ being the Volkov phase:

$$\Phi(\mathbf{p}, t) = -\int_t^{\infty}\left(\mathbf{p}\mathbf{A}_L(t') + \frac{1}{2}A_L^2(t')\right)dt'. \quad (6)$$

In the original derivation [5], $D(\mathbf{p}) = \langle \mathbf{p}|z|\Psi_0\rangle$ was evaluated with $\langle \mathbf{p}|$ as a plane-wave state, which is equivalent to the so-called strong-field approximation. This approximation was improved [11] by taking $\langle \mathbf{p}|$ as a continuum eigenstate of $\hat{H}_0$, which is known as the Coulomb-Volkov approximation. We use a variant of this approximation [12].

Starting from the work [7,13] of Y. Mairesse and F. Quéré, the analysis of attosecond spectrograms is based on the striking similarity between Eq. (5) and the definition of a spectrogram in the context of frequency-resolved optical gating (FROG):

$$S_{\mathrm{FROG}}(\omega, \tau) = \left|\frac{1}{2}\int_{-\infty}^{\infty} dt P(t + \tau) G(t) e^{i\omega t}\right|^2, \quad (7)$$

where $P(t)$ and $G(t)$ are interpreted as a *pulse* [14] and a *gate*, respectively. Powerful algorithms were developed for retrieving both the pulse and the gate from a FROG spectrogram [15,16]. The application of these algorithms to the analysis of streaking spectrograms is hindered by the fact that $G_0(\mathbf{p}, t)$, unlike $G(t)$, depends on $\mathbf{p}$. Fortunately, this dependence is often weak, especially if the matrix element $D(\mathbf{p})$ is almost constant within the bandwidth of the XUV pulse. By replacing $G_0(\mathbf{p}, t)$ with

$$G_1(t) = D(\mathbf{p}_0 + \mathbf{A}_L(t)) e^{i\Phi(\mathbf{p}_0,t)}, \quad (8)$$

which is known as the central momentum approximation, the streaking spectrogram is considered as a FROG spectrogram with $\omega = p^2/2 - p_0^2/2$. Within this framework, the pulse retrieved by a FROG algorithm was expected [8,17] to be

$$P_1(t) = \mathcal{E}_{\mathrm{XUV}}(t). \quad (9)$$

In other words, FROG was thought to retrieve the complex envelope of the XUV pulse. We find that this is *generally not correct*. Given a spectrogram defined by Eq. (5), FROG will rather retrieve a pulse

$$P_2(t) = \chi(t) \quad (10)$$

and a gate given by Eq. (15) below, which is approximately represented by

$$G_2(t) \approx e^{i\Phi(\mathbf{p}_0,t)}. \quad (11)$$

Before we provide the mathematical background for this fact, let us illustrate it by an example where the central momentum approximation (8) spectacularly breaks down. We take an artificial matrix element that roughly models the $3s$ Cooper minimum in argon [18]:

$$D(p) = p^2/2 - p_0^2/2 \quad (12)$$

with $p_0 = 1.77$ at.u. (choosing the central energy to coincide with the Cooper minimum at 42.5 eV). Although $D(p)$ is a real quantity in this example, it can be considered having a piecewise constant phase with a discontinuity of π at $p = p_0$. The XUV and laser pulses are assumed to be polarized along the direction of photoelectron detection. Both pulses are bandwidth limited with Gaussian envelopes: $\mathcal{E}_{\mathrm{XUV}}(t) \propto \exp[-0.036t^2]$ (shown in Fig. 2) and $A_L(t) = 0.1\exp[-3.24 \times 10^{-5}t^2]\cos(0.0608t)$. The full width at half maximum (FWHM) of the XUV intensity is equal to 150 as. The FWHM of the laser pulse is 5 fs; its central wavelength is 750 nm.

Figure 1(a) shows a spectrogram calculated using Eq. (5). The FROG reconstruction [8] yields a very similar spectrogram, Fig. 1(b). On the other hand, the conventional central momentum approximation, defined by Eqs. (7)–(9), yields a spectrogram, shown in Fig. 1(c), which has very little resemblance to the original spectrogram. The approximation breaks down in this example because $D(p_0) = 0$. However, the pulse and gate pair given by Eqs. (10) and (11) yield a spectrogram that is very close to the original one, Fig. 1(d).

In Fig. 2, we compare the XUV envelope $\mathcal{E}_{\mathrm{XUV}}(t)$, the time-domain wave packet $\chi(t)$, and the pulse $P(t)$ retrieved by FROG from the spectrogram shown in Fig. 1(a). The retrieved pulse nearly coincides with the wave packet $\chi(t)$. In the momentum space, we see that the wave packet and the retrieved pulse only differ in a narrow spectral region near the phase discontinuity, as shown in Fig. 3.

To gain insight as to why the time-domain wave packet (1) plays the role of the pulse, we use Eqs. (2) and (3) to express $\mathcal{E}_{\mathrm{XUV}}(t + \tau)$ in (5) via $\tilde{\chi}(p)$ and then expand $[D(p)]^{-1}$ in a Taylor series, which eventually allows us to rewrite the master equation (5) in a form where $\chi(t)$ plays the role of the pulse. In the case when the XUV and

PRL **105,** 073001 (2010) PHYSICAL REVIEW LETTERS week ending 13 AUGUST 2010

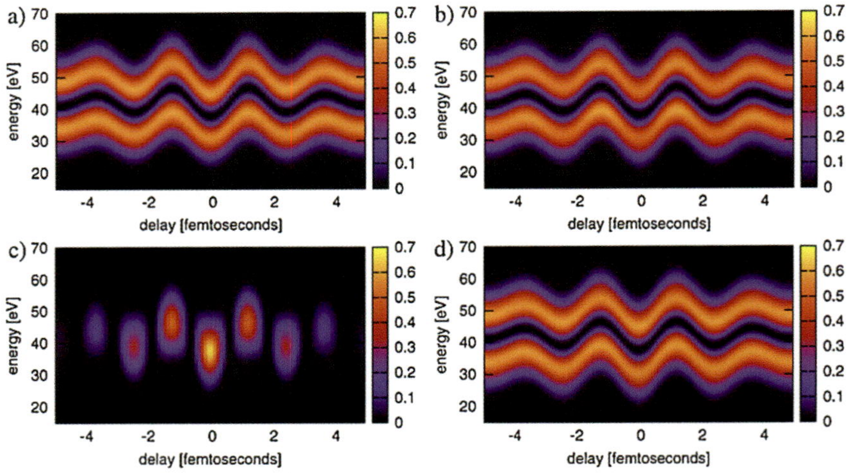

FIG. 1 (color online). (a) A streaking spectrogram calculated with the aid of Eq. (5) using a model transition matrix element (12). (b) The spectrogram reconstructed using a FROG algorithm. (c) The spectrogram calculated using the FROG ansatz (7) with the pulse $P_1(t) = \mathcal{E}_{\rm XUV}(t)$ and the gate $G_1(t)$. (d) The FROG spectrogram evaluated with the pulse being the time-domain wave packet $\chi(t)$ and the gate $G_2(t) = \exp\{i\Phi(\mathbf{p}_0, t)\}$.

streaking pulses are polarized in the direction where photoelectrons are detected, $S(p, \tau)$ can be rewritten, without any approximations, as

$$S(p,\tau) = \left| \frac{1}{2} \int_{-\infty}^{\infty} dt \chi(t+\tau) \sum_{n=0}^{\infty} a_n \frac{\partial^n g(p,t)}{\partial t^n} \right|^2, \quad (13)$$

where $a_n = \frac{(-i)^n}{n!} \frac{\partial^n}{\partial \omega^n} \frac{1}{D(\sqrt{2\omega + p_0^2})}|_{\omega=0}$ and

$$g(p,t) = D(p + A_L(t)) e^{i\Phi(p,t) + i((p^2/2) - (p_0^2/2))t}. \quad (14)$$

Consequently, the FROG gate associated with the time-domain wave packet is given by

$$G_2(t) = \left(\sum_{n=0}^{\infty} a_n \frac{\partial^n g(p,t)}{\partial t^n} \right) \Bigg|_{p=p_0}. \quad (15)$$

Applying the central momentum approximation to Eq. (15), rather than Eq. (5), yields a spectrogram that is closer to the exact one. To show this analytically, we neglect all derivatives of $A_L(t)$ in the expression for $\partial^n g(p, t)/\partial t^n$, which is justified if the exponential function in Eq. (14) oscillates with a period that is much shorter than the optical cycle of the streaking field. Once this approximation is made, we recognize that the sum over n in Eq. (13) becomes the Taylor expansion of $[D(p + A_L(t))]^{-1}$, which simplifies Eq. (13) to

$$S(p,\tau) \approx \left| \frac{1}{2} \int_{-\infty}^{\infty} dt \chi(t+\tau) e^{i\Phi(p,t)} e^{i((p^2/2)-(p_0^2/2))t} \right|^2. \quad (16)$$

Thus, $G_2(t) \approx \exp\{i\Phi(\mathbf{p}_0, t)\}$. Importantly, the central momentum approximation applied this way does not affect the transition matrix element $D(p)$, while the conventional expression for the FROG gate (8) approximates $D(p + A_L(t))$ with $D(p_0 + A_L(t))$. In the example presented above this was a very poor approximation.

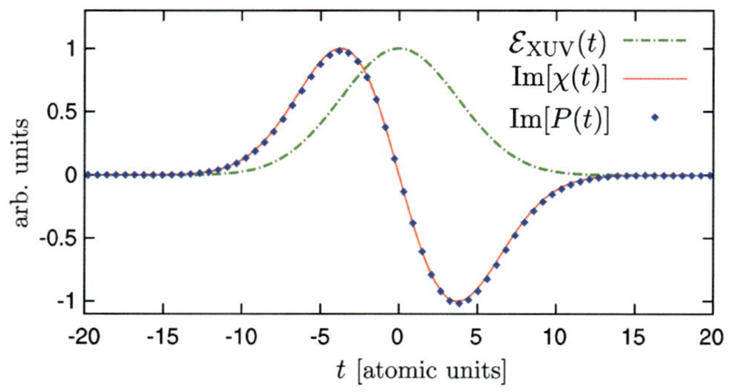

FIG. 2 (color online). The envelope of the XUV pulse $\mathcal{E}_{\rm XUV}(t)$ (dash-dotted line), the time-domain wave packet $\chi(t)$ defined by Eq. (1) (solid line), and the pulse $P(t)$ retrieved by FROG from the spectrogram shown in Fig. 1(a) (diamonds). In this example, the real part of the wave packet is equal to zero, so we compare the imaginary parts of $\chi(t)$ and $P(t)$.

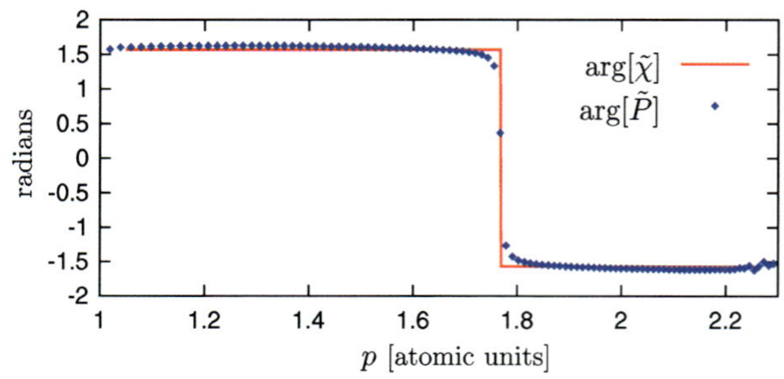

FIG. 3 (color online). The phase of the momentum-space wave packet $\arg[\tilde{\chi}(p)]$ (solid line) in comparison with the phase of the retrieved wave packet $\arg[\tilde{P}(\frac{p^2}{2} - \frac{p_0^2}{2})]$ (diamonds).

As a remark, we were able to find cases where Eq. (16) was a seemingly worse approximation to the true spectrogram (5) than the conventional central momentum approximation [19]. Nevertheless, FROG retrieved the time-domain wave packet. This confirms that the central momentum approximation is *more generally applicable* if the spectrogram is expressed via $\chi(t)$ according to Eq. (13).

In conclusion, we have established that the time-domain wave packet $\chi(t)$ defined by Eq. (1) determines the output of attosecond streaking measurements processed with conventional retrieval algorithms, such as "frequency-resolved optical gating for complete reconstruction of attosecond bursts" (FROG CRAB) [13]. The analysis presented in this Letter was focused on direct single-photon ionization with a complex transition matrix element, using a Cooper minimum as an example. However, it is known that Eq. (5) formally describes more sophisticated streaking measurements, such as those of the Fano resonances and autoionization decay [20,21]. Therefore, our analysis is directly applicable to these situations as well, Eq. (1) being a fundamental relation between the momentum-space wave function of a photoelectron and the pulse retrieved from a streaking spectrogram.

Having established this relation, we claim that the quantum phase associated with bound-free transitions can be measured (up to a constant phase) by means of attosecond streaking. According to Eq. (2), this is possible if the spectral phase of an attosecond pulse is known. An XUV pulse can be fully characterized in a streaking measurement performed on an atom where the accurate matrix elements are known from a reliable model or their phases are known to be negligible. For example, helium can be used for such a calibration. Then another streaking measurement on the system under scrutiny will provide all the necessary information to retrieve the unknown phases of quantum transitions and thus completely characterize photoionization dynamics. Such a measurement would be a test and, possibly, a challenge for many-electron theories of photoionization, especially if the ionization involves a resonance state or a Cooper minimum.

Supported by the DFG Cluster of Excellence: Munich-Centre for Advanced Photonics.

*vladislav.yakovlev@physik.uni-muenchen.de

[1] M. Drescher, M. Hentschel, R. Kienberger, M. Uiberacker, V. Yakovlev, A. Scrinzi, T. Westerwalbesloh, U. Kleineberg, U. Heinzmann, and F. Krausz, Nature (London) **419**, 803 (2002).

[2] M. Swoboda, T. Fordell, K. Klünder, J. M. Dahlström, M. Miranda, C. Buth, K. J. Schafer, J. Mauritsson, A. L'Huillier, and M. Gisselbrecht, Phys. Rev. Lett. **104**, 103003 (2010).

[3] M. Schultze, M. Fieß, N. Karpowicz, J. Gagnon, M. Korbman, M. Hofstetter, S. Neppl, A. L. Cavalieri, Y. Komninos, Th. Mercouris, C. A. Nicolaides, R. Pazourek, S. Nagele, J. Feist, J. Burgdörfer, A. M. Azzeer, R. Ernstorfer, R. Kienberger, U. Kleineberg, E. Goulielmakis, F. Krausz, and V. S. Yakovlev, Science **328**, 1658 (2010).

[4] J. Itatani, F. Quéré, G. L. Yudin, M. Y. Ivanov, F. Krausz, and P. B. Corkum, Phys. Rev. Lett. **88**, 173903 (2002).

[5] M. Kitzler, N. Milosevic, A. Scrinzi, F. Krausz, and T. Brabec, Phys. Rev. Lett. **88**, 173904 (2002).

[6] R. Kienberger, E. Goulielmakis, M. Uiberacker, A. Baltuska, V. Yakovlev, F. Bammer, A. Scrinzi, T. Westerwalbesloh, U. Kleineberg, U. Heinzmann, M. Drescher, and F. Krausz, Nature (London) **427**, 817 (2004).

[7] F. Quere, Y. Mairesse, and J. Itatani, J. Mod. Opt. **52**, 339 (2005).

[8] J. Gagnon, E. Goulielmakis, and V. S. Yakovlev, Appl. Phys. B **92**, 25 (2008).

[9] J. Gagnon and V. S. Yakovlev, Opt. Express **17**, 17 678 (2009).

[10] J. Mauritsson, M. B. Gaarde, and K. J. Schafer, Phys. Rev. A **72**, 013401 (2005).

[11] G. L. Yudin, S. Patchkovskii, and A. D. Bandrauk, J. Phys. B **41**, 045602 (2008).

[12] A. S. Kornev and B. A. Zon, J. Phys. B **35**, 2451 (2002).

[13] Y. Mairesse and F. Quéré, Phys. Rev. A **71**, 011401(R) (2005).

[14] In the literature on frequency-resolved optical gating, $P(t)$ is usually referred to as probe.

[15] R. Trebino, K. W. DeLong, D. N. Fittinghoff, J. N. Sweetser, M. A. Krumbugel, B. A. Richman, and D. J. Kane, Rev. Sci. Instrum. **68**, 3277 (1997).

[16] D. J. Kane, IEEE J. Quant. Electron. **35**, 421 (1999).

[17] G. Sansone, E. Benedetti, F. Calegari, C. Vozzi, L. Avaldi, R. Flammini, L. Poletto, P. Villoresi, C. Altucci, R. Velotta, S. Stagira, S. De Silvestri, and M. Nisoli, Science **314**, 443 (2006).

[18] J. W. Cooper, Phys. Rev. **128**, 681 (1962).

[19] For example, if we use $D(p) = \exp\{4i(p^2 - p_0^2)^2\}$ instead of Eq. (2).

[20] M. Wickenhauser, J. Burgdörfer, F. Krausz, and M. Drescher, Phys. Rev. Lett. **94**, 023002 (2005).

[21] Z. X. Zhao and C. D. Lin, Phys. Rev. A **71**, 060702(R) (2005).

Strömgren sphere radii for the observed SDSS quasars. This is in contrast to the simplified monolithic model of quasar formation[16,1], for which the H II region has zero volume at the time when the quasar turns on $[R(t_{age}=0)=0]$ and the distribution extends down to $R_p = 0$.

To constrain the neutral fraction, we have computed the likelihood of observing $R_{obs} = 4.5$ Mpc around SDSS J1030+0524 and $R_{obs} = 4.7$ Mpc around SDSS J1148+5251 as a function of x_{HI} and f_{lt}. The results are plotted in Fig. 2. Figure 2a shows the locus of most likely values (thick line) as well as likelihood contours at 0.1 of the peak value (dashed lines). The fiducial lifetime (f_{lt}) favours $x_{HI} \approx 1$, while with $f_{lt} \ll 1$ the distributions for R_p lie substantially below the observed value of 4.5 Mpc, making smaller values of x_{HI} more likely. Extrapolation of the most likely contour for the Elvis *et al.*[19] template yields $f_{lt} \approx 2x_{HI}$. This implies that $x_{HI} \approx 10^{-3}$ would require a lifetime as short as 2×10^4 years, which is ruled out by the variability properties of quasars in SDSS[26]. The a posteriori probability distributions for x_{HI} given f_{lt} are plotted in Fig. 2b and robustly yield the constraint $x_{HI} \gtrsim 0.01$. For $f_{lt} \gtrsim 0.3$, we find $x_{HI} \gtrsim 0.1$ and $x_{HI} \gtrsim 0.4$ at the 90% level for the Elvis *et al.*[19] and Telfer *et al.*[18] spectra, respectively. The fiducial model with $f_{lt} = 1$ yields corresponding constraints of $x_{HI} \gtrsim 0.3$ and $x_{HI} \gtrsim 0.6$.

The inference of a large neutral fraction at $z \approx 6.3$ presents a challenge to theories of cosmological reionization when combined with the large optical depth[3] to electron scattering after cosmological recombination, $\tau_{es} = 0.17 \pm 0.04$. Consider a toy model in which the Universe was partially reionized at a high redshift $z_{reion,I}$, leaving a neutral fraction x_{HI} until complete reionization was reached at $z_{reion,II} \approx 6.25$. The optical depth is then $\tau_{es} = 0.04 + 0.002(1 - x_{HI})\,[(1 + z_{reion,I})^{3/2} - 19.5]$. If the IGM ionization fraction increased monotonically, then given $\tau_{es} > 0.13$, the Universe needed to be reionized earlier than $z_{reion,I} \approx 18$ or $z_{reion,I} \approx 24$, assuming the Elvis *et al.*[19] and Telfer *et al.*[18] spectra, respectively. If $\tau_{es} = 0.17$ and $x_{HI} > 0.7$, as implied by the Telfer *et al.*[18] spectrum, then a monotonic reionization history requires significant reionization at an implausibly high redshift ($z \gtrsim 30$). In this case, a more plausible alternative is a non-monotonic history with an early reionization peak, possibly due to the formation of massive population-III stars[27,28]. □

Received 14 October 2003; accepted 15 January 2004; doi:10.1038/nature02336.

1. White, R. L., Becker, R. H., Fan, X. & Strauss, M. A. Probing the ionization state of the universe at $z > 6$. *Astron. J.* **126**, 1–14 (2003).
2. Fan, X. *et al.* Evolution of the ionizing background and the epoch of reionization from the spectra of $z \sim 6$ quasars. *Astron. J.* **123**, 1247–1257 (2002).
3. Kogut, A. *et al.* First-year Wilkinson microwave anisotropy probe (WMAP) observations: temperature-polarization correlation. *Astron. J. Suppl.* **148**, 161–173 (2003).
4. Hu, E. M. *et al.* A redshift $z = 6.56$ galaxy behind the cluster Abel 370. *Astrophys. J.* **568**, L75–L79 (2002).
5. Haiman, Z. The detectability of high-redshift Lyα emission lines prior to the reionization of the universe. *Astrophys. J.* **576**, L1–L4 (2002).
6. Willott, C. J., McLure, R. J. & Jarvis, M. J. A $3 \times 10^9 M_\odot$ black hole in the quasar SDSS J1148+5251 at $z = 6.41$. *Astrophys. J.* **587**, L15–L18 (2003).
7. Walter, F. *et al.* Molecular gas in the host galaxy of a quasar at redshift $z = 6.42$. *Nature* **424**, 406–408 (2003).
8. Richards, G. T. *et al.* Broad emission-line shifts in quasars: an orientation measure for radio-quiet quasars? *Astron. J. Suppl.* **124**, 1–17 (2002).
9. Barkana, R. & Loeb, A. In the beginning: the first sources of light and the reionization of the universe. *Phys. Rep.* **349**, 125–238 (2001).
10. Gunn, J. E. & Peterson, B. A. On the density of neutral hydrogen in intergalactic space. *Astrophys. J.* **142**, 1633–1641 (1965).
11. Pentericci, L. *et al.* VLT optical and near-infrared observations of the $z = 6.28$ quasar SDSS J1030+0524. *Astron. J.* **123**, 2151–2158 (2002).
12. Martini, P. QSO lifetimes. Preprint at ⟨http://arXiv.org/astro-ph/0304009⟩ (2003).
13. Miralda-Escude, J., Haehnelt, M. & Rees, M. Reionization of the inhomogeneous universe. *Astrophys. J.* **530**, 1–16 (2000).
14. Barkana, R. & Loeb, A. GRBs versus Quasars: Lyman-α signatures of reionization verses cosmological infall. *Astrophys. J.* (in the press); preprint at ⟨http://arXiv.org/ astro-ph/0305470⟩ (2003).
15. Shapiro, P. R. & Giroux, M. L. Cosmological, H II regions and the photoionization of the intergalactic medium. *Astrophys. J.* **321**, L107–L112 (1987).
16. Cen, R. & Haiman, Z. Quasar Stromgren spheres before cosmological reionization. *Astrophys. J.* **542**, L74–L78 (2000).
17. Madau, P. & Rees, M. J. The earliest luminous sources and the damping wing of the Gunn-Peterson trough. *Astrophys. J.* **542**, L69–L73 (2000).
18. Telfer, R. C., Zheng, W., Kriss, G. A. & Davidsen, A. F. The rest-frame extreme-ultraviolet spectral properties of quasi-stellar objects. *Astron. J.* **565**, 773–785 (2002).
19. Elvis, M. *et al.* Atlas of quasar energy distributions. *Astron. J. Suppl.* **95**, 1–68 (1994).
20. Shields, G. A. *et al.* The black hole-bulge relationship in quasars. *Astrophys. J.* **583**, 124–133 (2003).
21. Ferrarese, L. Beyond the Bulge: a fundamental relationship between supermassive black holes and dark matter halos. *Astrophys. J.* **587**, 90–97 (2001).
22. Volonteri, M., Haardt, F. & Madau, P. The assembly and merging history of supermassive black holes in hierarchical models of galaxy formation. *Astrophys. J.* **582**, 559–573 (2003).
23. Yu, Q. & Tremaine, S. Observational constraints on growth of massive black holes. *Mon. Not. R. Astron. Soc.* **335**, 965–976 (2002).
24. Martini, P. & Weinberg, D. H. Quasar clustering and the lifetime of quasars. *Astrophys. J.* **547**, 12–26 (2001).
25. Jakobsen, P., Jansen, R. A., Wagner, S. & Reimers, D. Caught in the act: a helium-reionizing quasar near the line of sight to Q0302003. *Astron. Astrophys.* **397**, 891–898 (2003).
26. Martini, P. & Schneider, D. P. Preprint at ⟨http://arXiv.org/astro-ph/0309650⟩ (2003).
27. Wyithe, J. S. B. & Loeb, A. Reionization of hydrogen and helium by early stars and quasars. *Astrophys. J.* **586**, 693–708 (2003).
28. Cen, R. The implications of Wilkinson microwave anisotropy probe observations for population III star formation processes. *Astrophys. J.* **591**, L5–L8 (2003).
29. Miralda-Escude, J. Reionization of the intergalactic medium and the damping wing of the Gunn-Peterson trough. *Astrophys. J.* **501**, 15–22 (1998).
30. Spergel, D. N. *et al.* First-year Wilkinson microwave anisotropy probe (WMAP) observations: determination of cosmological parameters. *Astron. J. Suppl.* **148**, 175–194 (2003).

Acknowledgements This work was supported in part by grants from ARC, NSF and NASA.

Competing interests statement The authors declare that they have no competing financial interests.

Correspondence and requests for materials should be addressed to A.L. (aloeb@cfa.harvard.edu).

Atomic transient recorder

R. Kienberger[1]*, E. Goulielmakis[1]*, M. Uiberacker[1]*, A. Baltuska[1], V. Yakovlev[1], F. Bammer[2], A. Scrinzi[1], Th. Westerwalbesloh[3], U. Kleineberg[3], U. Heinzmann[3], M. Drescher[3] & F. Krausz[1,4]

[1] *Institut für Photonik, Technische Universität Wien, Gusshausstraße 27, A-1040 Wien, Austria*
[2] *Institut für Spanlose Fertigung und Hochleistungslasertechnik, Technische Universität Wien, Franz-Grillstr. 1, Arsenal Obj. 207, A-1030 Wien, Austria*
[3] *Fakultät für Physik, Universität Bielefeld, D-33615 Bielefeld, Germany*
[4] *Max-Planck-Institut für Quantenoptik, Hans-Kopfermann-Straße 1, D-85748 Garching, Germany*

* These authors contributed equally to this work

In Bohr's model of the hydrogen atom, the electron takes about 150 attoseconds (1 as = 10^{-18} s) to orbit around the proton, defining the characteristic timescale for dynamics in the electronic shell of atoms. Recording atomic transients in real time requires excitation and probing on this scale. The recent observation of single sub-femtosecond (1 fs = 10^{-15} s) extreme ultraviolet (XUV) light pulses[1] has stimulated the extension of techniques of femtochemistry[2] into the attosecond regime[3,4]. Here we demonstrate the generation and measurement of single 250-attosecond XUV pulses. We use these pulses to excite atoms, which in turn emit electrons. An intense, waveform-controlled, few cycle laser pulse[5] obtains 'tomographic images' of the time-momentum distribution of the ejected electrons. Tomographic images of primary (photo)electrons yield accurate information of the duration and frequency sweep of the excitation pulse, whereas the same measurements on secondary (Auger) electrons will provide insight into the relaxation dynamics of the electronic shell following excitation. With the current ~750-nm laser probe and ~100-eV excitation, our transient recorder is capable of resolving atomic electron dynamics within the Bohr orbit time.

In electron-optical chronoscopy the rapid deflection of a bunch of photoelectrons maps their temporal profile to a spatial distribution[6,7]. From the electrons' streaked image the temporal structure of the light pulse triggering the photoemission can be inferred with sub-picosecond resolution. The resolution of streak cameras is limited by the spread of the electron transit time owing to the spread of their initial momenta. In this work we report sampling of sub-femtosecond electron emission from atoms by drawing on the same basic concept. The improved time resolution results from several essential modifications of image-tube streak cameras: the oscillating electric field of light is used for affecting the electrons' motion (1), the field is virtually jitter-free (2), and is applied along the direction of electron motion, which implies their acceleration or deceleration instead of their deflection (3), directly at the location and instant of emission (4). Whereas (1) implies 'only' a dramatically enhanced streaking speed, the consequences of (2)–(4) are more profound: (2) allows us to systematically vary the timing of the probing field with an accuracy within the electron bunch length, and because of (3) this capability results in the 'projection' of the initial time–momentum distribution of electron emission into a series of distributions of the electrons' final momentum, while (4) prevents the initial momentum spread from introducing any measurement error. As a result of (2)–(4), the spread of initial electron momenta no longer limits the time resolution. Moreover, a possible temporal variation of this initial momentum spread during the emission can be captured just as can the temporal variation of the emission intensity.

The final momentum distribution of electrons detached from atoms by an impulsive excitation in the presence of an intense, linearly polarized laser field $E_L(t) = E_0(t)\cos(\omega_L t + \varphi)$ and observed within a narrow cone aligned parallel to the laser electric field vector is given by:

$$\sigma(p) = \int_{-\infty}^{\infty} n_e(p - eA_L(t), t)\mathrm{d}t \qquad (1)$$

where $n_e(p_i, t)$ is the initial time–momentum distribution of emitted electrons and e is the electron charge. The vector potential in the Coulomb gauge $A_L(t) = \int_t^{\infty} E_L(t')\mathrm{d}t'$ accounts for the

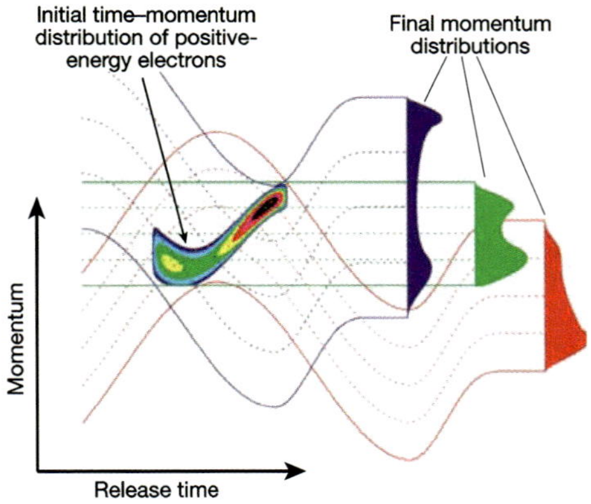

Figure 1 Principle of the atomic transient recorder. The initial time–momentum distribution of positive-energy (photo or Auger) electrons emitted from atoms excited by an attosecond pulse carries direct time-domain information about the excitation and relaxation dynamics of the electronic shell. A linearly polarized, few-cycle laser pulse, with its electric field directed parallel to the direction of observation of the ejected electrons, creates at different delays different tomographic projections of the time–momentum distribution $n_e(p_i, t)$ of atomic electron emission on momentum space. These final momentum distributions can readily be measured after the laser pulse left the interaction region. In the absence of the laser field, the electron momenta do not change after detachment and accumulation of electrons with a given final momentum (mathematically: time-integration) follows lines of constant momentum (green lines) yielding the field-free final momentum distribution: $\sigma_0(p) = \int_{-\infty}^{\infty} n_e(p, t)\mathrm{d}t$ (green profile). In the presence of a probing field, accumulation of electrons with a given final momentum occurs along the lines of constant canonical momentum $p_i + eA_L(t)$, (blue and red lines, see equation (1)). The corresponding projections (streaked spectra) are represented in red and blue. From a suitable set of such tomographic projections the time–momentum distribution of electron emission can be retrieved, providing direct time-domain insight into atomic dynamics triggered by an attosecond excitation (in our experiments XUV pulse) synchronized to the probing laser field.

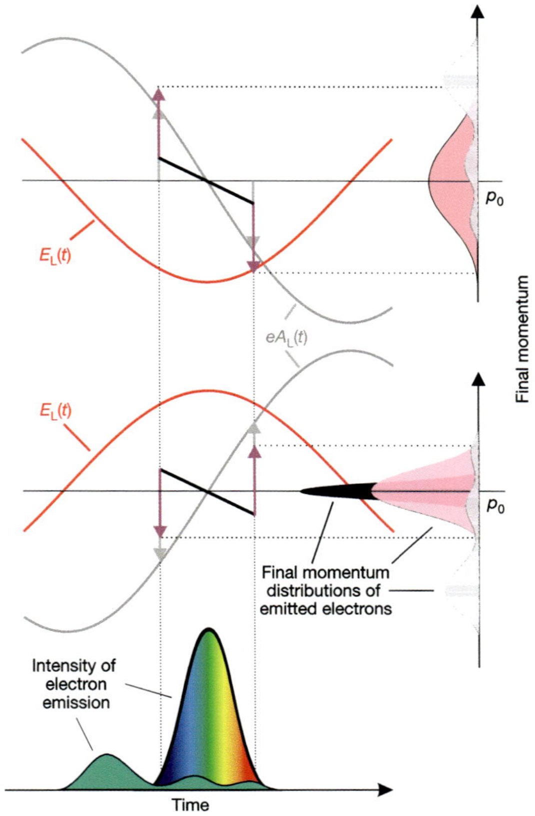

Figure 2 Electron streak records in specific cases. The electric field vector of the linearly polarized probe laser pulse points towards the electron detector for $E_L > 0$. In the absence of a temporal sweep of initial momenta of the emitted atomic electrons, described by $n_e(p_i, t) = f(p_i)g(t)$, the momentum transfer $\Delta p = eA_L(t_r)$ (grey line), maps the temporal emission profile $g(t)$ (green profile) uniquely into a similar distribution of final momenta (grey profiles) around the mean initial momentum p_0, provided that the emission is temporally confined between adjacent zero transitions of the probing laser field $E_L(t)$. In this specific case, the temporal intensity profile of electron emission is directly mirrored by the streak record just as in a conventional streak camera. However, any sweep of the electrons' initial momentum (for example, a linear one as indicated by the black line for the single-peaked electron emission) revokes the unique correspondence between an electron's final momentum and its release time, preventing the retrieval of accurate temporal information from single streak records. In this case, at least two streak records (pink profiles)—in addition to the field-free spectrum (black profile)—are required. In the absence of a nonlinear temporal momentum sweep the streak records obtained near the zero transitions of $A_L(t)$ with opposite slopes (pink profiles) together with the field-free spectrum (black profile) allow determination of all relevant characteristics: the temporal profile, duration and momentum chirp of emission. Note that a linear momentum sweep leads to an asymmetric broadening of the streaked spectra at these delays and it is this asymmetry that makes the measurement highly sensitive to the momentum sweep, that is, highly sensitive to deviations of the pulse duration from the Fourier limit.

momentum boost imparted to the free electron by the laser field from the time of its release t until the pulse has passed. In the classical description of the freed electrons' motion in the strong laser field, the final momentum spectrum can be viewed as a projection of $n_e(p_i, t)$ along the lines where $p = p_i + eA_L(t)$ is constant (see Fig. 1); p_i is the initial kinetic momentum of the electron. By delaying the laser field with respect to the excitation that triggers electron emission, a set of tomographic records is obtained, from which the initial time–momentum distribution $n_e(p_i, t)$ can be uniquely reconstructed if the emission terminates within $T_0/2 = \pi/\omega_L$. The method is closely related to frequency-resolved optical gating[8,9] with the oscillating laser field (rather than its envelope) being employed as a gate. In the simplest cases one or two streaked spectra are sufficient for retrieving $n_e(p_i, t)$, as discussed in Fig. 2.

For our experiments the XUV excitation pulses have been generated from neon atoms ionized by an intense linearly polarized few-cycle laser pulse (full-width at intensity half-maximum (FWHM): $\tau_L = 5$ fs, carrier wavelength: $\lambda_L = 750$ nm, $T_0 = 2.5$ fs) with controlled waveform[5] by a process that leads to the emission of high-order harmonics for multi-cycle drivers[10,11]. Proper adjustment of the laser peak intensity yielded highest-energy (cutoff) XUV emission in the high-reflectivity band of our Mo/Si multilayer mirror, which was used for focusing the XUV as well as laser pulses into another neon gas jet for recording atomic transients. Both an intuitive model[12,13] and extensive numerical simulations[14–18] predict the cutoff emission to be confined to the vicinity of zero transition(s) of the laser electric field after the most intense half-cycle(s) in a few-cycle driver.

Figure 3 summarizes representative streaked neon photoelectron spectra recorded with the XUV and laser pulse impinging with a fixed relative timing set in the XUV generation process. For a cosine driver waveform ($\varphi = 0$), cutoff radiation (filtered by the Mo/Si multilayer) is predicted to be emitted in a single bunch at the zero transition of $E_L(t)$ following the pulse peak. The photoelectrons knocked off in the direction of the peak electric field at this instant should gain the maximum increase of their momentum and energy. Figure 3d corroborates this prediction. The clear upshift is consistent with the XUV burst coinciding with the zero transition of the laser electric field (see Fig. 2). Possible satellites would appear at the adjacent zero transitions of $E_L(t)$ and suffer an energy downshift. The absence of a downshifted spectral peak of substantial intensity indicates a clean single sub-fs pulse generation. With the phase

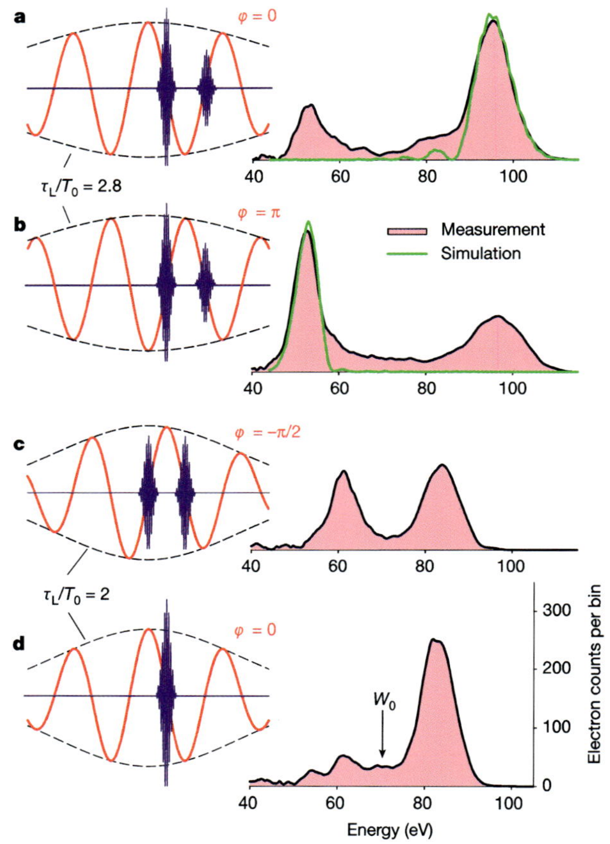

Figure 3 Streaked photoelectron spectra recorded at a fixed delay of probe laser light. Energy distribution of photoelectrons emitted from neon atoms excited by a sub-fs XUV pulse carried at a photon energy of $\hbar\omega_{XUV} \approx 93.5$ eV (selected by the Mo/Si mirror). The photoelectron spectrum peaks at $W_0 = \hbar\omega_{XUV} - W_b \approx 72$ eV in the absence of $E_L(t)$, where $W_b = 21.5$ V is the binding energy of the most weakly bound valence electrons in Ne. The spectrally filtered cutoff XUV bursts and the 5-fs, 750-nm driver laser pulses are depicted by blue and red lines, respectively. **a**, **b**, Streaked spectra obtained with respectively 'cosine' and '–cosine' laser pulses of a normalized duration of $\tau_L/T_0 = 2.8$ and of a peak electric field of $E_0 = 140$ MV cm^{-1}. The green lines on the right-hand side depict spectra computed with an XUV burst derived from the measured asymmetric XUV radiation filtered by the mirror (see Fig. 5a) under the assumption of zero spectral phase. The satellite pulse is not modelled in this way, because the corresponding modulation of the spectrum was not considered in the calculation. The difference in broadening of the up- and down-shifted spectral features appears to be a consequence of the quadratic temporal frequency sweep resulting from the asymmetric spectral distribution of the XUV burst (see Fig. 5a). **c**, **d**, Streaked spectra obtained with respectively 'sine' and 'cosine' laser pulses characterized by $\tau_L/T_0 = 2$ and $E_0 = 75$ MV cm^{-1}.

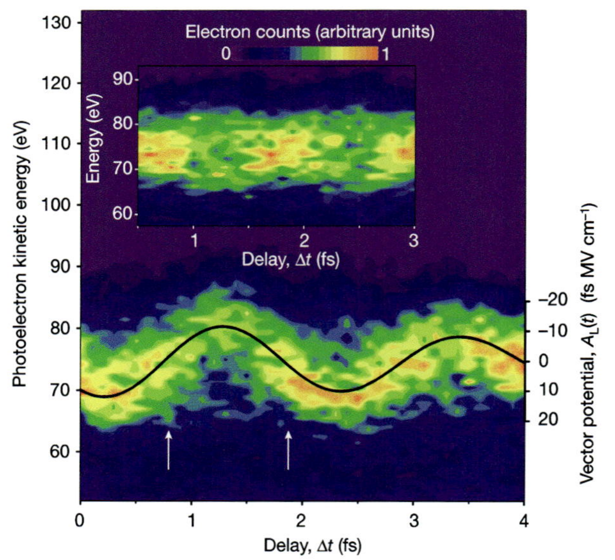

Figure 4 ATR measurement: a series of tomographic projections (streaked kinetic energy spectra) of the initial time–momentum distribution of photoelectrons knocked out by a single sub-fs XUV pulse (in false-colour representation). A few-cycle laser pulse with a cosine waveform and a normalized duration of $\tau_L/T_0 = 2$ was used for both generating the single 93-eV sub-fs excitation pulse and for probing photoelectron emission in the atomic transient recorder. Black line, $A_L(t)$ of the probing field evaluated from the peak shift of the streaked spectra (see scale on the right hand side). From $A_L(t)$ the electric field of the light wave can also be determined by using the relationship $E_L(t) = -dA_L/dt$. The electric field vector points towards the electron detector for $E_L > 0$. Inset, streaked spectra obtained under the same experimental conditions except for the carrier-envelope phase of the few-cycle pulse, which was left unstabilized.

letters to nature

adjusted to yield a sine waveform ($\varphi = -\pi/2$), cutoff emission is predicted to come in twin pulses (Fig. 3c). The double-peaked streaked spectrum (Fig. 3c) clearly reflects this time structure. High-energy XUV photons are now distributed in two bursts, each of which is less than half as intense as the isolated burst produced by the cosine waveform (Fig. 3d). These measurements demonstrate how light waveform control allows shaping XUV emission on a sub-fs timescale.

Other than φ the normalized pulse duration τ_L/T_0 plays a crucial role in the generation of a single XUV pulse, as demonstrated by experiments with slightly longer, 7-fs laser pulses ($\tau_L/T_0 = 2.8$). Under these conditions a satellite pulse is unavoidable for any setting of φ and most careful filtering of the cutoff photons (Fig. 3a, b). A reduced difference in intensity of the adjacent half-cycles of the few-cycle wave implies that the highest-energy spectral components of adjacent XUV bursts reach the energy band selected by our 15%-bandwidth bandpass filter.

With isolated sub-fs XUV pulses at our disposal atomic transients can now be triggered and their subsequent evolution be captured by probing electron emission with a synchronized wave of laser light in the atomic transient recorder (ATR). In the first ATR measurements presented here the objects of scrutiny were photoelectrons. Figure 4 shows a series of streaked spectra of photoelectrons emitted from neon at different delays of $E_L(t)$. If the electrons' initial kinetic energy is much larger than their average quiver energy, their laser-induced energy shift $\Delta W(t) \approx (p_0/m)\Delta p(t) = (ep_0/m)A(t)$, where m is the electron mass, probes $A_L(t)$ directly at the instant t of electron release and allows its determination without having to analyse the structure of the streaked spectra. The evaluated $A_L(t)$ is depicted by the black line in Fig. 4, revealing $T_0/2 \approx 1$ fs at the pulse peak, which is indicative of a strong blue shift. This was observed previously and found to originate from ionization-induced self-phase modulation in the XUV source[1]. With $A_L(t)$ known, we can set out to characterize the electron emission. Waveform control is as essential here as for the reproducible generation of sub-fs bursts (see Fig. 3); its absence smears the streak records beyond redemption (inset in Fig. 4).

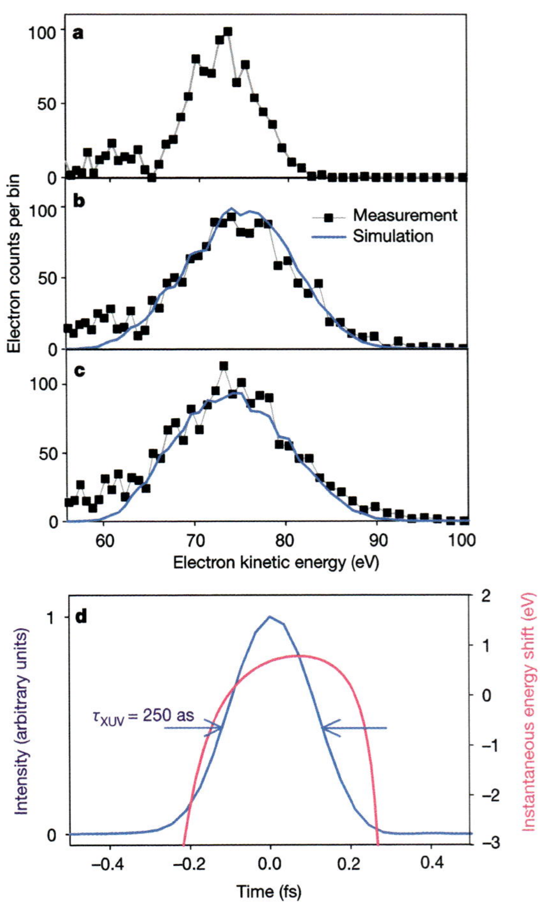

Figure 5 Selected streaked spectra from the ATR measurement of photoelectron emission from neon excited with a 93-eV sub-fs pulse (Fig. 4). **a**, Spectrum in the absence of laser field. **b**, **c**, Streaked spectra recorded at delays that imply coincidence of electron emission with (adjacent) zero transitions of $A_L(t)$, see arrows in Fig. 4. **d**, Temporal intensity profile and energy sweep of the sub-femtosecond XUV excitation pulse evaluated from the ATR measurements.

Figure 5 depicts the photoelectron spectra in the absence of the probe laser field (Fig. 5a) and the streaked spectra (Fig. 5b, c) recorded at adjacent zero transitions of $A_L(t)$, which are most sensitive to a possible quadratic spectral phase (linear chirp) carried by the XUV pulse. The best fits to these data (blue lines in Fig. 5) yield $\tau_{XUV} = 250(-5/+30)$ as and $\tau_{jitter} = 260(\pm 80)$ as for the duration and jitter of the XUV pulse, respectively. The remarkable accuracy of τ_{XUV} derives from the use of several tomographic projections of the time–momentum distribution of photoelectrons (spectra shown in Fig. 5). The inspection of other streaked spectra justified the neglect of higher-order spectral phase in the above analysis. The large jitter appears to be extrinsic (vibrations due to the vacuum pumps) because it is virtually absent in the absence of a delay between XUV and laser pulse (Fig. 3). The temporal intensity profile and chirp of the XUV pulse obtained from the ATR measurements are shown in Fig. 5d. The pulse is Fourier-limited with a weak temporal chirp resulting from the asymmetric shape of the pulse spectrum (Fig. 5a).

The briefest time interval within which two different atomic events can be recognized as different by our apparatus is ultimately limited by quantum mechanical uncertainty. This dictates that any short time structure comes with a broad energy spectrum. In order that the streaks of two events separated by δt in time can be resolved they must be shifted with respect to each other by at least as much as their own spectral width $\delta W \approx \hbar/\delta t$. From this requirement we obtain the ATR resolving power as:

$$\delta t = \frac{T_0}{2\pi}\sqrt{\frac{\hbar\omega_L}{\Delta W_{max}}} \qquad (2)$$

where ΔW_{max} stands for the energy shift suffered by the electron ejected at the peak of $A_L(t)$. Under our current experimental conditions ΔW_{max} can approach 25 V before the onset of laser-induced ionization, yielding $\delta t < 100$ as for $T_0 = 2$ fs. The atomic transient recorder driven by red light and probing electrons with a kinetic energy near 100 V is able to distinguish events within 100 as, constituting what is to our knowledge the shortest interval of time directly measurable to date.

Extension of the presented experiments to simultaneously exploring the temporal variation of emission intensity and momentum distribution of primary (photo) and secondary (Auger) electron emission will provide unprecedented insight into the excitation and relaxation dynamics of the electronic shell of atoms and molecules. The currently used time window of $T_0/2 \approx 1$ fs can be extended to several 10 fs by difference frequency generation with the few-cycle laser pulse while keeping the resolution of the ATR in the sub-fs regime, owing to $\delta t \propto \sqrt{T_0}$. With an extended time window a detailed characterization of trains of attosecond bursts (in contrast with the measurement of parameters averaged over the train[19–21]) will also become feasible. At near-keV excitation energies, the atomic transient recorder will allow time-domain metrology with a resolution approaching the atomic unit of time (24 as). □

Methods

We sample electron emission from atoms using the experimental set-up that was described in detail[1] and employed recently for probing Auger electrons on a few-fs timescale[22]. The essential innovation here is that waveform-controlled few-cycle light now provides a reproducible excitation burst for accurate triggering and a reproducible streaking field for capturing sub-fs electron emission from atoms. For excitation, XUV bursts are produced from Ne atoms ionized by intense, few-cycle waveform-controlled light pulses[5] in a process giving rise to high-order harmonics of the incident light for periodic (multi-cycle) pumping[10,11]. The collinear XUV and laser beams are focused into a neon gas jet and delayed with respect to each other for the ATR measurements with a two-component Mo/Si broadband multilayer mirror[1] (radius of curvature $= -70$ mm) placed 2.5 m downstream from the source. The internal part of the laser beam (~3 mm in diameter) was blocked with a zirconium filter (transmitting the XUV pulse) in front of the Mo/Si mirror. The resultant annular laser beam (with a dark spot 3 mm in diameter in its centre) and the XUV beam (~3 mm in diameter) confined by the annular laser beam were focused by the external and internal sections of the two-component Mo/Si mirror[1], respectively. The two components can be aligned separately and translated with respect to each other along the optical axis of the mirror.

The reflectivity band of the multilayer extends from 85 V to 100 V with a peak reflectance of ~30% and a FWHM of ~9 V. Electrons ejected following the XUV excitation are collected within a narrow cone (<4°) aligned parallel to the laser and XUV polarization and analysed with a time-of-flight spectrometer. Two types of experiments have been implemented. First, by removing the zirconium filter we used the internal part of the Mo/Si mirror to focus both the XUV and the laser beam to eliminate any external source of fluctuations in the timing between the excitation and probing pulses (results summarized in Fig. 3). In the second type of studies, the XUV and the laser beam were reflected by different sections of the focusing mirror as described above. In these investigations the probing laser field could be delayed with respect to the XUV excitation pulse by the translation of the internal part of the mirror on a nanometre scale, yielding the data shown in Figs 4 and 5.

Received 24 October; accepted 5 December 2003; doi:10.1038/nature02277.

1. Hentschel, M. *et al.* Attosecond metrology. *Nature* **414**, 509–513 (2001).
2. Zewail, A. Femtochemistry: atomic-scale dynamics of the chemical bond (adapted from the Nobel Lecture). *J. Phys. Chem. A* **104**, 5660–5694 (2000).
3. Itatani, J. *et al.* Attosecond streak camera. *Phys. Rev. Lett.* **88**, 173903 (2002).
4. Kitzler, M., Milosevic, N., Scrinzi, A., Krausz, F. & Brabec, T. Quantum theory of attosecond XUV pulse measurement by laser-dressed photoionization. *Phys. Rev. Lett.* **88**, 173904 (2002).
5. Baltuska, A. *et al.* Attosecond control of electronic processes by intense light fields. *Nature* **421**, 611–615 (2003).
6. Bradley, D. J., Liddy, B. & Sleat, W. E. Direct linear measurement of ultrashort light pulses with a picosecond streak camera. *Opt. Commun.* **2**, 391 (1971).
7. Schelev, M. Ya., Richardson, M. C. & Alcock, A. J. Image-converter streak camera with picosecond resolution. *Appl. Phys. Lett.* **18**, 354 (1971).
8. Kane, D. J. & Trebino, R. Characterization of arbitrary femtosecond pulses using frequency-resolved optical gating. *IEEE J. Quantum Electron.* **29**, 571–579 (1993).
9. Sekikawa, T., Katsura, T., Miura, S. & Watanabe, S. Measurement of the intensity-dependent atomic dipole phase of a high harmonic by frequency-resolved optical gating. *Phys. Rev. Lett.* **88**, 193902 (2002).
10. L'Huillier, A. & Balcou, Ph. High-order harmonic generation in rare gases with a 1-ps 1053-nm laser. *Phys. Rev. Lett.* **70**, 774–777 (1993).
11. Macklin, J. J., Kmetec, J. D. & Gordon, C. L. III High-order harmonic generation using intense femtosecond pulses. *Phys. Rev. Lett.* **70**, 766–769 (1993).
12. Schafer, K. J., Yang, B., DiMauro, L. F. & Kulander, K. C. Above threshold ionization beyond the high harmonic cutoff. *Phys. Rev. Lett.* **70**, 1599–1602 (1993).
13. Corkum, P. B. Plasma perspective on strong-field multiphoton ionization. *Phys. Rev. Lett.* **71**, 1994–1997 (1993).
14. Lewenstein, M., Balcou, Ph., Ivanov, M. Yu., L'Huillier, A. & Corkum, P. B. Theory of high-harmonic generation by low-frequency laser fields. *Phys. Rev. A* **49**, 2117–2132 (1994).
15. Christov, I. P., Murnane, M. M. & Kapteyn, H. C. High-harmonic generation of attosecond pulses in the "single-cycle" regime. *Phys. Rev. Lett.* **78**, 1251–1254 (1997).
16. Kan, C., Burnett, N. H., Capjack, C. E. & Rankin, R. Coherent XUV generation from gases ionized by several cycle optical pulses. *Phys. Rev. Lett.* **79**, 2971–2974 (1997).
17. de Bohan, A., Antoine, P., Milosevic, D. B. & Piraux, B. Phase-dependent harmonic emission with ultrashort laser pulses. *Phys. Rev. Lett.* **81**, 1837–1840 (1998).
18. Tempea, G., Geissler, M. & Brabec, T. Phase sensitivity of high-order harmonic generation with few-cycle laser pulses. *J. Opt. Soc. Am. B* **16**, 669–674 (1999).
19. Paul, P. M. *et al.* Observation of a train of attosecond pulses from high harmonic generation. *Science* **292**, 1689–1692 (2001).
20. Mairesse, Y. *et al.* Attosecond synchronisation of high-harmonic soft X-rays. *Science* **302**, 1540–1543 (2003).
21. Tzallas, P., Charalambidis, D., Papadogiannis, N. A., Witte, K. & Tsakiris, G. D. Direct observation of attosecond light bunching. *Nature* **426**, 267–271 (2003).
22. Drescher, M. *et al.* Time-resolved atomic inner-shell spectroscopy. *Nature* **419**, 803–807 (2002).

Acknowledgements This work was sponsored by the Fonds zur Förderung der wissenschaftlichen Forschung (Austria), the Deutsche Forschungsgemeinschaft and the Volkswagenstiftung (Germany) and by the European Union's Human Potential Programme.

Competing interests statement The authors declare that they have no competing financial interests.

Correspondence and requests for materials should be addressed to F.K. (ferenc.krausz@tuwien.ac.at).

Giant magnetoresistance in organic spin-valves

Z. H. Xiong, Di Wu, Z. Valy Vardeny & Jing Shi

Department of Physics, University of Utah, Salt Lake City, Utah 84112, USA

A spin valve is a layered structure of magnetic and non-magnetic (spacer) materials whose electrical resistance depends on the spin state of electrons passing through the device and so can be controlled by an external magnetic field. The discoveries of giant magnetoresistance[1] and tunnelling magnetoresistance[2] in metallic spin valves have revolutionized applications such as magnetic recording and memory, and launched the new field of spin electronics[3]—'spintronics'. Intense research efforts are now devoted to extending these spin-dependent effects to semiconductor materials. But while there have been noteworthy advances in spin injection and detection using inorganic semiconductors[4–6], spin-valve devices with semiconducting spacers have not yet been demonstrated. π-conjugated organic semiconductors may offer a promising alternative approach to semiconductor spintronics, by virtue of their relatively strong electron–phonon coupling[7] and large spin coherence[8]. Here we report the injection, transport and detection of spin-polarized carriers using an organic semiconductor as the spacer layer in a spin-valve structure, yielding low-temperature giant magnetoresistance effects as large as 40 per cent.

π-conjugated organic semiconductors (OSEs) are a relatively new class of electronic materials that are revolutionizing important technological applications including information display[9] (ref. 10 and references therein) and large-area electronics (ref. 11 and references therein), owing to their ability to be economically processed in large areas, their compatibility with low-temperature processing, the tunability of their electronic properties, and the simplicity of thin-film device fabrication. The virtually limitless flexibility of synthetic organic chemistry allows the fabrication of π-conjugated OSE structures with a degree of control unattainable with the conventional inorganic semiconductors. In addition, the OSEs have extremely weak spin-orbit interaction and weak hyperfine interaction, so that electron spin diffusion length is especially long[8]. These properties make them ideal for spin-polarized electron injection and transport applications.

Here we have chosen the small π-conjugated molecule 8-hydroxy-quinoline aluminium (Alq_3), most commonly used in organic light-emitting diodes (OLEDs)[10], to serve as an OSE spacer in organic spin-valves, because it can easily be deposited as thin films and integrated with a variety of metallic electrodes. As shown in Fig. 1a, the vertical organic spin-valves that we fabricated consist of three layers: two ferromagnetic electrode films (FM_1 and FM_2, respectively) and the OSE spacer. By engineering the two FM electrodes to have different coercive fields (H_{c1} and H_{c2}, respectively), their magnetization directions can have either a parallel or anti-parallel alignment configuration when sweeping the external magnetic field, H; this is essential for proving the spin-valve effect.

Intrigued by the possibility of spin-injection involving a π-conjugated OSE oligomer (sexi-thiophene, T_6)[12], where both the opposite FM electrodes were $La_{0.67}Sr_{0.33}MnO_3$ (LSMO), we have chosen LSMO for the bottom electrode (FM_1) and cobalt as the top electrode (FM_2) in our devices (Fig. 1a, b). LSMO is believed to be a half-metallic ferromagnet that possesses near-100% spin-polarization[13]. We note that unlike metallic FM films such as cobalt, nickel, iron or their alloys, the LSMO films are already stable against oxidation. In fact, our LSMO films have been cleaned and re-used

REPORTS

Direct Measurement of Light Waves

E. Goulielmakis,[1*] M. Uiberacker,[1*] R. Kienberger,[1] A. Baltuska,[1] V. Yakovlev,[1] A. Scrinzi,[1] Th. Westerwalbesloh,[2] U. Kleineberg,[2] U. Heinzmann,[2] M. Drescher,[2] F. Krausz[1,3]†

The electromagnetic field of visible light performs ~10^{15} oscillations per second. Although many instruments are sensitive to the amplitude and frequency (or wavelength) of these oscillations, they cannot access the light field itself. We directly observed how the field built up and disappeared in a short, few-cycle pulse of visible laser light by probing the variation of the field strength with a 250-attosecond electron burst. Our apparatus allows complete characterization of few-cycle waves of visible, ultraviolet, and/or infrared light, thereby providing the possibility for controlled and reproducible synthesis of ultrabroadband light waveforms.

Although the wave nature of light has long been known, it has not been possible to measure directly the oscillating field of light. Radiation in the visible and higher frequency spectral ranges can so far only be characterized in terms of physical quantities averaged over the wave period. Nonlinear optical techniques now allow measurement of $\varepsilon_L(t)$, the amplitude envelope, and $\omega_L(t)$, the carrier frequency, as a function of time t, for light pulses with durations that approach the wave cycle (*1*, *2*). The carrier-envelope phase φ, which determines the timing between $\varepsilon_L(t)$ and $\omega_L(t)$, can also be measured (*3*). These measurements rely on carrier-envelope decomposition, which is physically meaningful only as long as the frequency spectrum of the wave is confined to less than one octave (*4*). If the radiation is composed of frequencies spanning a broader range (*5–17*), direct access to the field is required. Attosecond pulses of extreme ultraviolet (XUV) light were predicted to suit for this purpose (*18*, *19*). We report the direct measurement of the buildup and disappearance of the electric field of a light pulse through the use of an attosecond probe.

The electric field is defined as the force exerted on a point charge of unit value. Its conceptually most direct measurement must therefore rely on measurement of this force. In a light wave, the electric field E_L, and hence the force $F = qE_L$ it exerts on a particle with charge q, are subject to rapid variations. Access to this force is possible only if the probe charge is instantly placed in the field, i.e., within a time interval τ_{probe} over which the temporal variation of the force is "frozen", i.e., $\tau_{probe} \ll T_0 = (2\pi)/\omega_L$, where T_0 is the wave period. The probe charge can be launched into the field by knocking electrons free from atoms or ions instantly. In a linearly polarized wave, the change of the electrons' momentum $\Delta p(\vec{r}, t)$ at location $\vec{r}$ and time t along the direction of the electric field is given by

$$\Delta p(\vec{r},t) = e\int_t^\infty E_L(\vec{r},t')\mathrm{d}t' = eA_L(\vec{r},t) \quad (1)$$

where e is the electron charge and $A_L(\vec{r},t)$ is the vector potential of the electric field $E_L(\vec{r},t) = E_0\varepsilon_L(\vec{r}, t)\cos(kz - \omega_L t + \varphi)$, where E_0 is the maximum field amplitude, and k is the wave vector. In our analysis, we assumed the wave to propagate along the z direction, and $t = t_{real} - z/v_g$ was defined in a retarded frame to yield $t = 0$ as locked to the peak of the pulse travelling at the group velocity v_g.

The relation $E_L(\vec{r},t) = -\partial A_L(\vec{r},t)/\partial t$ implies that measuring the momentum boost $\Delta p(\vec{r},t)$ imparted to the freed electrons by the field at the location $\vec{r}$ at two instants differing in time by $\delta t \ll T_0/4$ will yield the electric field strength and direction directly as $E_L(\vec{r},t) = [\Delta p(\vec{r},t - \delta t/2) - \Delta p(\vec{r},t + \delta t/2)]/e\delta t$. This measurement procedure relies on a momentary release of the electrons within $\tau_{probe} \leq T_0/4$. For near infrared, visible, and ultraviolet light, this condition dictates that $\tau_{probe} <$ 1 fs. Varying the timing of such a subfemtosecond electron probe across the laser pulse provides complete information on the electric field of the light wave.

These considerations suggest that the electron probe needs to be localized not only in time to a tiny fraction of the wave period T_0, but also in space to a tiny fraction of the wavelength λ_L of the light wave to be measured. The latter requirement can be substantially relaxed if we trigger the electron release with an energetic photon pulse that copropagates with the laser wave in a collinear beam (Fig. 1). Because the timing of the probe electrons relative to the light field is invariant to space in this case, in a gently focused laser beam they can be released and are subsequently allowed to move over distances substantially larger than λ_L, in a volume within which the spatial variation of the field amplitude $\varepsilon_L(\vec{r},t)$ is negligibly small for a fixed value of t.

Putting the above concept into practice requires the electron probe to be scanned through the entire laser pulse. For each newly set timing t, measurement of the momentum shift $\Delta p(t)$ of the probing electrons requires the laser pulse to pass through the measurement apparatus again. Full characterization of the light waveform is therefore only feasible if it can be reproducibly generated for repeated measurements. Another equally important prerequisite for implementation of the above concept is the availability of an energetic instantaneous excitation (for launching the probing electrons) that is not only confined temporally to a fraction of 1 fs but is also synchronized to the light wave with similar

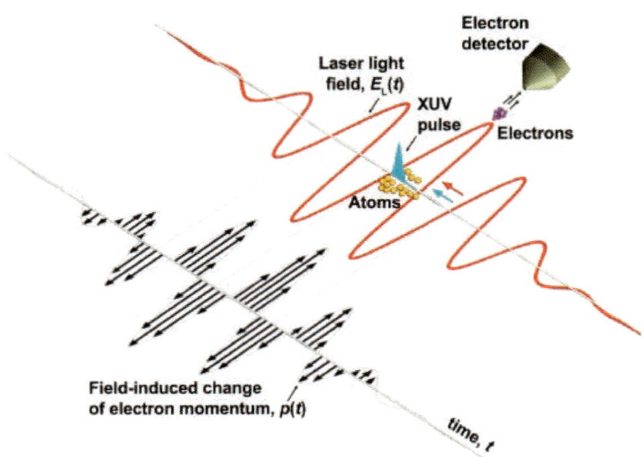

Fig. 1. Schematic of the measurement principle. A few-cycle pulse of laser light, together with a synchronized subfemtosecond XUV burst, is focused into an atomic gas target. The XUV pulse knocks electrons free by photoionization. The light electric field $E_L(t)$ to be measured imparts a momentum change to the electrons (black arrows), which scales as the instantaneous value of the vector potential $A_L(t)$ at the instant of release of the probing electrons. The momentum change is measured by an electron detector, which collects the electrons ejected along the direction of the linearly polarized $E_L(\vec{r},t)$.

[1]Institut für Photonik, Technische Universität Wien, Gusshausstraβe 27, A-1040 Wien, Austria. [2]Fakultät für Physik, Universität Bielefeld, D–33615 Bielefeld, Germany. [3]Max-Planck-Institut für Quantenoptik, Hans-Kopfermann-Straβe 1, D–85748 Garching, Germany.

*These authors contributed equally to this work.
†To whom correspondence should be addressed. E-mail: ferenc.krausz@tuwien.ac.at

accuracy. With the generation of waveform-controlled, intense, few-cycle light pulses (*20*) and their successful application to producing single 250-as XUV pulses synchronized to the driver light wave (*21*), these preconditions are now fulfilled. The waveform-controlled pulses—after having produced the attosecond photon probe—allow through nonlinear optical frequency conversion the synthesis of reproducible, synchronized, ultrabroadband, few-cycle waveforms (*5–17*). These can be repeatedly sent into the measurement apparatus with exactly the same waveform, and the subfemtosecond XUV pulse is able to produce the electrons by photoionization for probing the oscillating light field with sufficient temporal resolution.

The electrons knocked free from the atoms by the XUV pulse can be most conveniently detected if the direction of their movement is left unchanged by the light field. This applies if electrons are detected within a narrow cone aligned with the electric field vector of the linearly polarized laser wave along the x direction and are ejected with a large-enough initial momentum p_i to fulfill $|p_i| > |\Delta p_{max}|$, where Δp_{max} is the maximum momentum shift induced by the field. A large initial momentum also benefits the measurement by enhancing the change of the electrons' kinetic energy ΔW, according to $\Delta W \approx (p_i/m)\Delta p$, and m is the electron's mass. This expression, together with Eq. 1, implies that the energy shift scales linearly with both the electric field and the wavelength of the light field to be probed (*22*). The importance of a large ΔW lies in the facts that the probing electrons are emitted with an inherent uncertainty $\delta W_{probe} \approx \hbar/\tau_{probe}$ (where $\hbar$ is Planck's constant h divided by 2π) and that the dynamic range over which the light field strength can be reliably measured scales with $\Delta W_{max}/\delta W_{probe}$ (ΔW_{max} is the maximum shift in the pulse).

Measurement of $E_L(t)$ over a substantial dynamic range requires a ΔW_{max} of several tens of electron volts. For an initial kinetic energy of $W_i \approx 100$ eV, this condition is satisfied for $E_0 < 10^8$ V/cm for near-infrared light and requires $E_0 \approx 3 \times 10^8$ V/cm for ultraviolet light (*22*). Noble gases with a low atomic number (such as helium and neon) safely resist ionization by a few-cycle field at these field strengths (*23*). The accuracy of definition of the location $\vec{r}$ is dictated by the size of the volume within which $\varepsilon(\vec{r},t)$ is approximately independent of $\vec{r}$. If the field is probed in the beam focus, this condition requires the probing electrons to be confined—during their interaction with the laser field—laterally (xy) and longitudinally (z) to a small fraction of the diameter and to the confocal parameter of the beam, respectively.

In a proof-of-concept experiment, we directly measured the $E_L(t)$ of the few-cycle laser pulse used for producing the attosecond photon probe (Fig. 1). Linearly polarized, waveform-controlled, <5-fs, 0.4-mJ, 750-nm ($T_0 = 2.5$ fs) laser pulses (*20*), with carefully optimized values of φ and E_0, produce single 250-as XUV pulses at $(\hbar\omega_{xuv})_{mean} = 93$ eV in a gas of neon atoms (*21*). The XUV pulse copropagates with the laser pulse in a collinear, laserlike beam to a second neon target placed in the focus of a spherical, two-component, Mo/Si multilayer mirror (*21*). The mirror, of 120-mm focal length, reflects XUV radiation over a band of ~9 eV, centered at ~93 eV. Consequently, the XUV pulse sets electrons free by photoionization with an initial kinetic energy of $p_i^2/2m = \hbar\omega_{xuv} - W_b$, (where W_b is the electron's binding energy) spread over an ~9-eV band, implying that $\delta W_{probe} \approx 9$ eV. The electrons' energy shift $\Delta W(t) \approx e(p_i/m)A_L(t)$ probes the laser vector potential. The volume of light-field probing is defined laterally by the <10-μm diameter of the XUV beam at its waist and longitudinally by the <50-μm size of the neon jet, which is well confined within the focal volume of the laser beam (diameter, >60 μm; confocal parameter, >5 mm). For $p_i^2/2m \approx 100$ eV, the electrons traveled less than 1 μm within 100 fs and hence remained safely confined to the region of constant laser field amplitude.

The field-induced variation of the final energy spectrum of the probe electrons versus delay between the XUV burst and the laser pulse (Fig. 2) reveal, without the need of any detailed analysis, that probing is implemented by a single burst of subfemtosecond duration that is synchronized with subfemtosecond accuracy to the measured laser field. $E_L(t)$ can now be directly (i.e., without any iterative steps) obtained through the procedure outlined above (Fig. 3). From the measured spectrum of the few-cycle laser pulse (Fig. 3, inset), we calculated $E_L(t)$ by a simple Fourier transformation on the assumption of absence of spectral phase variations. The result, with E_0 and φ chosen to yield the best match to the measured values, is shown in gray. The excellent fit to the measured field evolution indicates a near-transform-limited pulse. Its duration was evaluated as 4.3 fs, in good agreement with the result of an autocorrelation measurement.

It has been predicted by theory that the few-cycle pulse pumping the XUV source has a "cosine" waveform ($\varphi \approx 0$) if a single subfemtosecond pulse emerges from the ionizing atoms (*24*). Our results (Fig. 3) yield the exper-

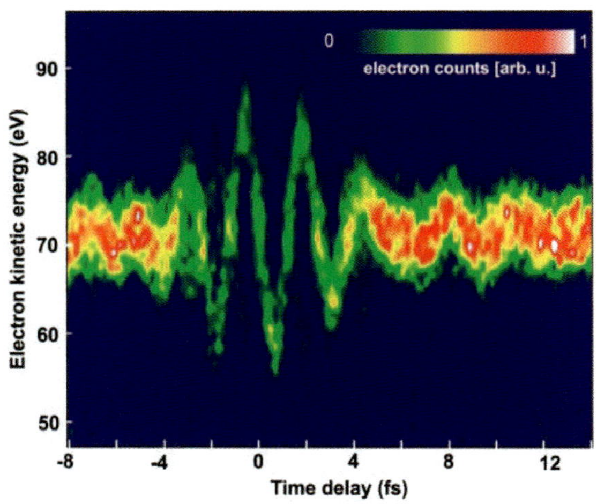

Fig. 2. A series of kinetic energy spectra of electrons detached by a 250-as, 93-eV XUV pulse from neon atoms in the presence of an intense <5-fs, 750-nm laser field, in false-color representation. The delay of the XUV probe was varied in steps of 200 as, and each spectrum was accumulated over 100 s. The detected electrons were ejected along the laser electric field vector with a mean initial kinetic energy of $p_i^2/2m \approx \hbar\omega_{xuv} - W_b = 93$ eV $- 21.5$ eV $= 71.5$ eV. The energy shift of the electrons versus the timing of the XUV trigger pulse that launches the probing electrons directly represents $A_L(t)$. arb. u., arbitrary units.

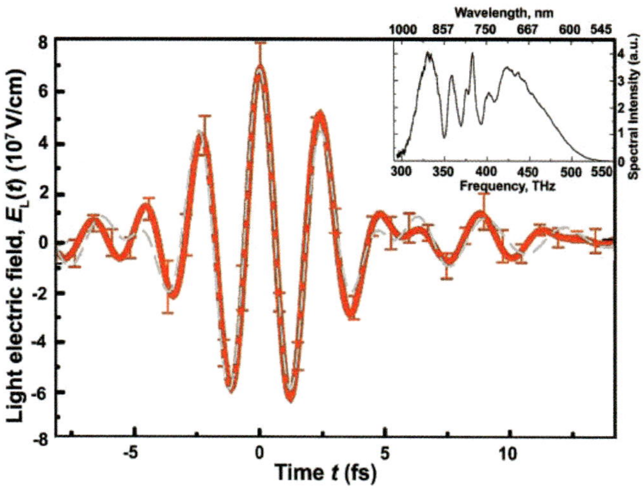

Fig. 3. $E_L(t)$ reconstructed (red line) from the data depicted in Fig. 2 and calculated (gray line) from the measured pulse spectrum (inset) with the assumed absence of a frequency-dependent phase and with E_0 and φ chosen so as to afford optimum matching to the measured field evolution. a.u., arbitrary units.

imental evidence. From this measurement, we also learn that the electric field points toward the electron detector at the pulse peak and that its strength is ~7×10^7 V/cm. With the temporal evolution, strength, and direction of $E_L(t)$ measured, we have performed a complete characterization of a light pulse in terms of its classical electric field.

Direct probing of light-field oscillations represents what we believe to be a substantial extension of the basic repertoire of modern experimental science. The door to practical applications is opened by the creation of the key element of the demonstrated light-field detector, the synchronized attosecond electron probe, in a noninvasive manner. In fact, our intense <5-fs laser pulse appears to be capable of producing the necessary XUV trigger burst without suffering any noticeable back-action to its own temporal shape (Fig. 3). After having produced the attosecond photon probe, this powerful few-femtosecond pulse is ideally suited for the synthesis of ultrabroadband, few-cycle, optical waveforms (*5–17*). Being composed of radiation extending from the infrared through the visible to the ultraviolet region, the resultant few-cycle, monocycle, and conceivably even subcycle waveforms will offer a marked degree of control over the temporal variation of electric and magnetic forces on molecular and atomic time scales.

These light forces, in turn, afford the promise of controlling quantum transitions of electrons in atoms and molecules and—at relativistic intensities—their center-of-mass motion. Reproducible ultrabroadband light wave synthesis, a prerequisite for these prospects to materialize, is inconceivable without subfemtosecond measurement of the synthesized waveforms. Beyond providing the subfemtosecond electron probe for these measurements, the substantial experimental efforts associated with the construction and reliable operation of a subfemtosecond photon source will pay off in yet another way. The envisioned control of electronic motion with light forces can only be regarded as accomplished once it has been measured. Owing to their perfect synchronism with the synthesized light waveforms, the subfemtosecond photon probe will allow us to test the degree of control achieved by tracking the triggered (and hopefully steered) motion in a time-resolved fashion.

References and Notes

1. R. Trebino *et al.*, *Rev. Sci. Instrum.* **68**, 3277 (1997).
2. C. Iaconis, I. A. Walmsley, *Opt. Lett.* **23**, 792 (1998).
3. G. G. Paulus *et al.*, *Phys. Rev. Lett.* **91**, 253004 (2003).
4. T. Brabec, F. Krausz, *Phys. Rev. Lett.* **78**, 3282 (1997).
5. T. W. Hänsch, *Opt. Commun.* **80**, 71 (1990).
6. S. Yoshikawa, T. Imasaka, *Opt. Commun.* **96**, 94 (1993).
7. A. E. Kaplan, P. L. Shkolnikov, *Phys. Rev. Lett.* **73**, 1243 (1994).
8. K. Shimoda, *Jpn. J. Appl. Phys.* **34**, 3566 (1995).
9. S. E. Harris, A. V. Sokolov, *Phys. Rev. Lett.* **81**, 2894 (1998).
10. A. Nazarkin, G. Korn, *Phys. Rev. Lett.* **83**, 4748 (1999).
11. O. Albert, G. Mourou, *Appl. Phys. B* **69**, 207 (1999).
12. M. Wittman, A. Nazarkin, G. Korn, *Phys. Rev. Lett.* **84**, 5508 (2000).
13. Y. Kobayashi, K. Torizuka, *Opt. Lett.* **25**, 856 (2000).
14. A. V. Sokolov, D. R. Walker, D. D. Yavuz, G. Y. Yin, S. E. Harris, *Phys. Rev. Lett.* **87**, 033402 (2001).
15. K. Yamane *et al.*, *Opt. Lett.* **28**, 2258 (2003).
16. M. Y. Shverdin, D. R. Walker, D. D. Yavuz, G. Y. Yin, S. E. Harris, in *OSA Trends in Optics and Photonics Series (TOPS) Vol. 96, Conference on Lasers and Electro-Optics (CLEO)* (Optical Society of America, Washington, DC, 2004), Postdeadline paper CPDC1.
17. K. Yamane, T. Kito, R. Morita, M. Yamashita, in *OSA Trends in Optics and Photonics Series (TOPS)*, vol. 96, *Conference on Lasers and Electro-Optics (CLEO)*, (Optical Society of America, Washington, DC, 2004), Postdeadline paper CPDC2.
18. R. Kienberger *et al.*, *Science* **297**, 1144 (2002).
19. A. D. Bandrauk, Sz. Chelkowski, N. H. Shon, *Phys. Rev. Lett.* **89**, 283903 (2002).
20. A. Baltuska *et al.*, *Nature* **421**, 611 (2003).
21. R. Kienberger *et al.*, *Nature* **427**, 817 (2004).
22. In the limit of $|\Delta p_{max}| \ll |p_i|$, the change in the electrons' final kinetic energy is given by $\Delta W_{max} \approx [8W_i U_{p,max}]^{1/2}$, where $U_{p,max} = e^2 E_0^{\,2}/4m_e\omega_L^{\,2}$ is the electrons' quiver energy averaged over an optical cycle at the peak of the light pulse.
23. Increase of the excitation energy $\hbar\omega_{xuv}$ tends to reconcile the conflicting requirements of avoiding field ionization and ensuring a high dynamic range.
24. T. Brabec, F. Krausz, *Rev. Mod. Phys.* **72**, 545 (2000).

25. We are grateful to B. Ferus for creating the artwork. Sponsored by the fonds zur Förderung der Wissenschafllichen Forschung (Austria, grant nos. Y44-PHY, P15382, and F016), the Deutsche Forschungsgemeinschaft and the Volkswagenstiftung (Germany), the European ATTO and Ultrashort XUV Pulses for Time-Resolved and Non-Linear Applications networks, and an Austrian Programme for Advanced Research and Technology fellowship to R.K. from the Austrian Academy of Sciences.

28 May 2004; accepted 20 July 2004

Nanoribbon Waveguides for Subwavelength Photonics Integration

Matt Law,[1,2*] Donald J. Sirbuly,[1,2*] Justin C. Johnson,[1] Josh Goldberger,[1] Richard J. Saykally,[1] Peidong Yang[1,2†]

Although the electrical integration of chemically synthesized nanowires has been achieved with lithography, optical integration, which promises high speeds and greater device versatility, remains unexplored. We describe the properties and functions of individual crystalline oxide nanoribbons that act as subwavelength optical waveguides and assess their applicability as nanoscale photonic elements. The length, flexibility, and strength of these structures enable their manipulation on surfaces, including the optical linking of nanoribbon waveguides and other nanowire elements to form networks and device components. We demonstrate the assembly of ribbon waveguides with nanowire light sources and detectors as a first step toward building nanowire photonic circuitry.

Photonics, the optical analog of electronics, shares the logic of miniaturization that drives research in semiconductor and information technology. The ability to manipulate pulses of light within sub-micrometer volumes is vital for highly integrated light-based devices, such as optical computers, to be realized. Recent advances in the use of photonic band gap (*1, 2*) and plasmonic (*3, 4*) phenomena to control the flow of light are impressive in this regard. One alternative route to integrated photonics is to assemble photonic circuits from a collection of nanowire elements that assume different functions, such as light creation, routing, and detection. Chemically synthesized nanowires have several features that make them good photonic building blocks, including inherent one-dimensionality, a diversity of optical and electrical properties, good size control, low surface roughness, and, in principle, the ability to operate above and below the diffraction limit. The toolbox of nanowire device elements already includes various types of transistors (*5*), light-emitting diodes (*6*), lasers (*7, 8*), and photodetectors (*9*). An important step toward nanowire photonics is to develop a nanowire waveguide that can link these various elements and provide the flexibility in interconnection patterns that is needed to carry out complex tasks such as logic operations (*10*). Our demonstration of nanowire-based photonics complements and expands upon recent work on optical beam steering in mesostructured silica cavities (*11*) and on subwavelength structures made lithographically (*12, 13*) and by the drawing of silica microfibers (*14*).

Nanoscale ribbon-shaped crystals of binary oxides exhibit a range of interesting properties including extreme mechanical flexibility, surface-mediated electrical conductivity (*15*), and lasing (*16*). As part of a recent study of the photoluminescence (PL) of SnO_2 nanoribbons, we noted that ribbons with high

[1]Department of Chemistry, University of California, Berkeley, CA 94720, USA. [2]Materials Science Division, Lawrence Berkeley National Laboratory, 1 Cyclotron Road, Berkeley, CA 94720, USA.

*These authors contributed equally to this work.
†To whom correspondence should be addressed. E-mail: p_yang@uclink.berkeley.edu

SPECIALSECTION

that is responsible for such quantum-mechanical phenomena as electron diffraction through crystals. Likewise, the bound electron state possesses a quantum phase, which can change sign from place to place in the molecule. The quantum phases of the returning electron wave interfere with the different parts of the molecular bound state wave function, creating attosecond modulation of the electronic structure that is recorded in the shape of the HHG spectrum (Fig. 2).

The physics of HHG is still under active discussion, and this coherent scattering picture has many subtleties that are still unresolved. Nonetheless, the effect of molecular shape on the HHG spectrum is genuine and dramatic. For example, two simple molecules O_2 and N_2 are similar enough in their gross structure, each consisting of two atoms separated by about an angstrom, and a difference of only two electrons out of more than a dozen; but their highest occupied orbitals, the ones that will field-ionize in an intense laser field, have very different symmetries. HHG spectra reveal this difference, and recent experiments have used the spectra to perform tomographic reconstructions of the wave function responsible for ionization.

Conclusion

Only a few years ago, the direct measurement of transient phenomena lasting less than an optical cycle seemed a supreme challenge. Now scientists have brought together techniques from atomic and laser physics to make these measurements possible, if not yet routine. Attoscience is a new field, with the promise to reveal some of the fastest processes of chemistry and atomic physics, or to freeze motion that will allow us to view the structure of matter under extreme conditions. The subject is still dominated by attempts to understand and improve the sources, and to interpret the data, and there are still frontiers in source development: For example, attosecond pulses at subangstrom wavelengths have not been demonstrated. There are even more challenges in the theory of attosecond molecular dynamics: One of the most difficult issues yet to be resolved in attoscience is how we should describe the structure and motion of electrons and nuclei in physical chemistry and molecular physics to accommodate processes that are so fast that cherished concepts such as potential energy surfaces or the Born Oppenheimer approximation are no longer valid. Despite these difficulties, the rapid progress in attoscience over the past few years is not abating, and we may anticipate new physical insights in the coming decade.

References and Notes

1. W.-M. Yao *et al.*, *J. Phys. G: Nucl. Part. Phys.* **33**, 1 (2006); http://meetings.aps.org/link/BAPS.2007.APR.B2.1.
2. D. E. Spence, P. N. Kean, W. Sibbett, *Opt. Lett.* **16**, 42 (1991).
3. M. Nisoli, S. De Silvestri, O. Svelto, *Appl. Phys. Lett.* **68**, 2793 (1996).
4. C. P. Hauri *et al.*, *Appl. Phys. B* **79**, 673 (2004).
5. R. L. Fork, C. H. Brito Cruz, P. C. Becker, C. V. Shank, *Opt. Lett.* **12**, 483 (1987).
6. E. Matsubara *et al.*, *J. Opt. Soc. Am. B* **24**, 985 (2007).
7. M. Hentschel *et al.*, *Nature* **414**, 509 (2001).
8. Ahmed Zewail received the 1999 Nobel Prize in Chemistry for his studies of the transition states of chemical reactions using femtosecond spectroscopy.
9. S. Augst *et al.*, *Phys. Rev. Lett.* **63**, 2212 (1989).
10. P. B. Corkum, *Phys. Rev. Lett.* **71**, 1994 (1993).
11. X. F. Li *et al.*, *Phys. Rev. A.* **39**, 5751 (1989).
12. J. L. Krause *et al.*, *Phys. Rev. A.* **45**, 4998 (1992).
13. K. J. Schafer, K. C. Kulander, *Phys. Rev. Lett.* **78**, 638 (1997).
14. R. López-Martens *et al.*, *Phys. Rev. Lett.* **94**, 033001 (2005).
15. N. M. Naumova, J. A. Nees, G. A. Mourou, *Phys. Plasmas* **12**, 056707 (2005).
16. A. A. Zholents, G. Penn, *Phys. Rev. ST Accel. Beams* **8**, 050704 (2005).
17. M. Lein, N. Hay, R. Velotta, J. P. Marangos, P. L. Knight, *Phys. Rev. A.* **66**, 023805 (2002).
18. J. Itatani *et al.*, *Nature* **432**, 867 (2004).
19. T. Kanai, S. Minemoto, H. Sakai, *Nature* **435**, 470 (2005).
20. This paper was written at the Stanford PULSE Center, with support from the NSF and from the Stanford Linear Accelerator Center, a national laboratory operated by Stanford University on behalf of the U.S. Department of Energy, Office of Basic Energy Sciences. We acknowledge useful discussions with M. Guehr, who also provided Fig. 2 for this article.

10.1126/science.1142135

REVIEW

Attosecond Control and Measurement: Lightwave Electronics

E. Goulielmakis,[1] V. S. Yakovlev,[2] A. L. Cavalieri,[1] M. Uiberacker,[2] V. Pervak,[2] A. Apolonski,[2] R. Kienberger,[1] U. Kleineberg,[2] F. Krausz[1,2]*

Electrons emit light, carry electric current, and bind atoms together to form molecules. Insight into and control of their atomic-scale motion are the key to understanding the functioning of biological systems, developing efficient sources of x-ray light, and speeding up electronics. Capturing and steering this electron motion require attosecond resolution and control, respectively (1 attosecond = 10^{-18} seconds). A recent revolution in technology has afforded these capabilities: Controlled light waves can steer electrons inside and around atoms, marking the birth of lightwave electronics. Isolated attosecond pulses, well reproduced and fully characterized, demonstrate the power of the new technology. Controlled few-cycle light waves and synchronized attosecond pulses constitute its key tools. We review the current state of lightwave electronics and highlight some future directions.

Quantum mechanics predicts the characteristic time scale for the rapidity of microscopic dynamics as $\Delta t \sim \hbar/\Delta W$, where ΔW is the spacing between the relevant energy levels of the microscopic system and $\hbar$ is Planck's constant. The milli–electron volt and multi–electron volt energy spacing of vibrational and electronic energy levels, respectively, imply that structural dynamics of molecules and solids as well as related chemical reactions and phase transitions evolve on a femtosecond time scale, whereas electronic motion on the atomic scale is to be clocked in attoseconds.

Before the invention of the laser, the resolution of time-resolved spectroscopy was limited by the nanosecond duration of pulses of incoherent light. The laser and the successive technological developments for the generation and measurement of ultrashort laser pulses improved the resolving power of pump-probe spectroscopy from several nanoseconds to several femtoseconds (*1*). The birth of femtosecond technology permitted real-time observation of the breakage and formation of chemical bonds (*2*).

We review the recent developments in the optical technology that have led to the breaking of the femtosecond barrier and provided real-time access to intra-atomic electron dynamics. Consequences include the observation of electronic motion deep inside (i.e., in inner shells of) atoms (*3*) and its control in real time (*4*). We address the underlying physical concepts and highlight the current status as well as future prospects of attosecond technology (*5*).

Femtosecond Technology: Control and Measurement with the Amplitude and Frequency of Light

Control and measurement of dynamics are intertwined. Time-resolved measurement relies on a physical quantity varying in a controlled, reproducible fashion on the relevant time scale.

[1]Max-Planck-Institut für Quantenoptik (MPQ), Hans-Kopfermann-Straße 1, D-85748 Garching, Germany. [2]Department für Physik, Ludwig-Maximilians-Universität, Am Coulombwall 1, D-85748 Garching, Germany.

*To whom correspondence should be addressed. E-mail: krausz@lmu.de

Attosecond Spectroscopy

Femtosecond technology is the result of controlling the nonlinear polarization of matter with the amplitude of light. Controlling the absorption and/or refractive index in this way has yielded—along with group delay dispersion control—femtosecond pulses from laser oscillators by means of passive mode locking (*1*). Measurement of the pulses relies on the same physical effect: control of the nonlinear polarization of matter by a replica of the pulse in the presence of the pulse to be characterized (*6*, *7*).

The controlled, well-characterized evolution of the amplitude envelope and carrier-frequency sweep (chirp) of ultrashort laser pulses permits measurement (*2*) and control (*8*) of quantum transitions on a femtosecond time scale. Reliance on these cycle-averaged quantities implies that measurement resolution and control speed are ultimately limited by the carrier wave cycle. The carrier wave cycle period is about 3 fs in the near infrared, where low dispersion favors the generation of the shortest laser pulses.

Toward Attosecond Technology: X-ray–Pump/X-ray–Probe Spectroscopy?

Femtosecond measurement and control techniques utilizing nonlinear material response could, in principle, be extended into the attosecond time domain by using intense attosecond pulses of extreme ultraviolet (XUV) or x-ray light. Unfortunately, in these regions of the optical spectrum, the probability of two-photon absorption is prohibitively low. X-ray–pump/x-ray–probe spectroscopy and x-ray quantum control therefore rely on x-ray intensities that can be attained only with large-scale free-electron lasers (*9*). Even though these sources are expected to eventually deliver their radiation in subfemtosecond pulses (*10*) and XUV sources pumped by large-scale, high energy lasers have already pushed the frontiers of nonlinear optics to the range of several tens of electron volts (*11*, *12*), proliferation of attosecond technology and its widespread applications call for another approach—one that relies on light sources suitable for small laboratories.

An Alternative Route: Light-Induced Electronic Motion Within the Wave Cycle of Light

The electric field of visible, near-infrared or infrared (henceforth, referred to collectively as NIR) laser light, $E_L(t)$, exerts a force on electrons that varies on a subfemtosecond scale. The use of this gradient for initiating and probing the subsequent dynamics with attosecond timing precision and resolution has led to the emergence of an attosecond technology that does not rely on the existence of intense x-ray pulses.

According to theory, strong-field–induced ionization of atoms is confined to subfemtosecond intervals near the peaks of the oscillating NIR field (Fig. 1A), setting a subfemtosecond electron wave packet $\psi_{free}(t)$ free; this prediction was recently confirmed by real-time observation (*13*). The subfemtosecond ionization process, locked to the peak of NIR field oscillation with corresponding precision, may serve as a subfemtosecond "starter gun" for a wide range of dynamics (including electronic and subsequent nuclear motion and related dynamics of molecular structure).

The energetic radiation concurrent with optical-field ionization in a linearly polarized laser field (*14*) provides a means of triggering motion in a more gentle way, without applying an intense field. The broadband XUV light building up during the propagation of the driving pulse through an ensemble of atoms was predicted to emerge in a subfemtosecond burst within every half optical cycle that is sufficiently intense for ionization (*15*). In this Review, we focus on these photonic tools of attosecond technology, noting that the recolliding electron wave packet itself can also be used to explore dynamics (*16*–*18*).

The electronic (and concurrent) phenomena following the attosecond trigger can be tracked with comparable resolution with the use of the generating NIR field, the oscillations of which are synchronized with the trigger event (tunnel ionization, recollision, or XUV burst from recollision). Let us consider an electron ejected during the unfolding evolution of the system under scrutiny. In the presence of a linearly polarized NIR field, outgoing electrons collected along the direction of the laser electric field $E_L(t)$ (Fig. 1B) suffer a momentum change $\Delta p(t) = e\int_t^\infty E_L(t')dt' = eA_L(t)$. Here, e is the charge of the electron, t is the instant of its emission, and $A_L(t)$ is the vector potential of the laser field. This change varies monotonically within a half wave cycle of the laser field, mapping the temporal evolution of the emitted subfemtosecond electron wave packet to a corresponding final momentum and energy distribution of the emitted electrons. We recognize here a new embodiment of the basic concept of streak imaging, with the streaking controlled by the NIR field in this case. The electrons can gain or lose an energy of several electron volts depending on whether they have been emitted some hundred attoseconds sooner or later than the peak of the field cycle (Fig. 1B), endowing this light-field–driven streak camera with attosecond resolution (*19*, *20*).

The physical concepts and mechanisms outlined above have opened a realistic prospect of measuring and controlling electron dynamics on an attosecond time scale. However, full exploitation of the potential offered by this conceptual framework requires precise control over the applied laser field.

Steering Electrons with the Electric Field of Controlled Light Waves

In a laser pulse comprising many field cycles, the subfemtosecond electron and photon bursts born in the ionizing field emerge every half cycle, resulting in a train of subfemtosecond bursts (*21*). Its spectrum consists of equidistant lines that represent high-order odd harmonics of the driving field (*22*). The subfemtosecond pulse train and its multicycle driver constitute powerful tools for controlling and tracking electron dynamics that terminate within the NIR field oscillation cycle (*23*–*25*). The characteristics of the individual bursts in the train can be difficult to control and measure. Extension of attosecond control and metrology to a wide range of electronic phenomena (including all those extending in time over more than half an optical cycle) calls for the isolation of one to several attosecond pulses and precise control of their properties (*26*).

Isolation was predicted to be achievable by manipulating the polarization state (polarization gating) of the driving field (*27*) or shortening its duration to nearly a single cycle of the carrier wave (*28*). In fact, intense few-cycle laser pulses lasting several femtoseconds (*1*) led to the first observation of light lasting for less than 1 fs (*29*). Full control of the number and properties of bursts isolated, however, requires precise control of the driving electric-field waveform $E_L(t)$.

Intense few-cycle laser pulses with controlled waveform (*4*)—along with careful bandpass filtering of the highest-energy (cutoff) harmonics—have indeed permitted the reproducible generation of single and twin subfemtosecond pulses with cosine- and the sine-shaped electric field waveforms, respectively, as well as their reliable measurement by means of attosecond streaking (*30*, *31*) (fig. S1). The series of streaked electron spectra shown in Fig. 1C yields complete information about the driving laser field and allows determination of the temporal shape, duration, and a possible chirp of the emitted subfemtosecond XUV pulse (Fig. 1D) as well as the degree of its synchronism with its driver (*31*–*33*). These experiments reveal that few-cycle light with controlled waveform permits reproducible generation and complete characterization of subfemtosecond light. Owing to the attosecond synchronism between them, these tools allow attosecond metrology without the need for high-intensity x-rays.

Lightwave Electronics: Versatile Technology for Control and Chronoscopy on the Electronic Time Scale

Controlled light fields permit control of microscopic electric currents at the atomic scale just as synthesized microwave fields permit control of currents at the mesoscopic scale in semiconductor chips. By analogy to microwave electronics, we propose to name this new technology lightwave electronics. In marked contrast with previous quantum control through controlling transitions between quantum states with the amplitude and frequency of light (*8*), lightwave electronics gives way to controlling dynamics directly by the force that the electric field of intense light exerts on electrons. This new approach is powerful in that (i) it provides a direct way of affecting the position and momentum of electrons and (ii) control is matched in speed to the electronic time scale.

The first notable manifestations of the power of lightwave electronics include controlling subfemtosecond XUV emission (*4*, *25*, *31–35*); molecular dissociation (*36*); measuring subfemtosecond XUV and electron pulses (*21*, *31–33*, *35*); and imaging dynamic changes in molecular structure (*16–18*) by means of steering electron wave packets with light fields. Sub–wave cycle (i.e., subfemtosecond) control of electron wave packet motion can be accomplished by mixing multicycle fields (*25*, *35*). However, full control over the entire system evolution from attosecond to femtosecond time scales—and, consequently, precision attosecond measurements—relies on optical waveforms that can be accurately controlled in shape and polarization state (*4*, *37*) and confined to several cycles. Exploitation of the full potential of lightwave electronics for attosecond control and metrology requires waveform-controlled broadband light.

The attosecond streaking spectrogram shown in Fig. 1C yields the complete history of photoelectron emission induced by a subfemtosecond XUV pulse as well as of the streaking optical field (*38*). Similar spectrograms can also be recorded with secondary (Auger) electrons, which uncover the history of a quantum transition of an electron deep inside an atom in real time (*3*). Attosecond streaking is not the only way of observing the temporal evolution of atomic-scale electron motion. Energetic excitation of atoms, molecules, or solids is often accompanied by the promotion of valence electrons to excited states (referred to as a shake up). The transient population of these states provides information about the instantaneous (electronic) state of the system. This population can be probed by means of attosecond tunneling induced by the time-delayed waveform-controlled few-cycle laser field (*13*).

In attosecond streaking and tunneling spectroscopy, the subfemtosecond XUV pulse serves as a pump and the controlled optical wave as the probe. Their roles can be interchanged: The optical field may—by means of tunnel ionization (*13*)—initiate an electronic process and control the unfolding dynamics (*36*), and the subfemtosecond XUV pulse may probe this process by, for example, photoelectron spectroscopy (*39*).

Attosecond technology based on a controlled optical wave and a synchronized subfemtosecond

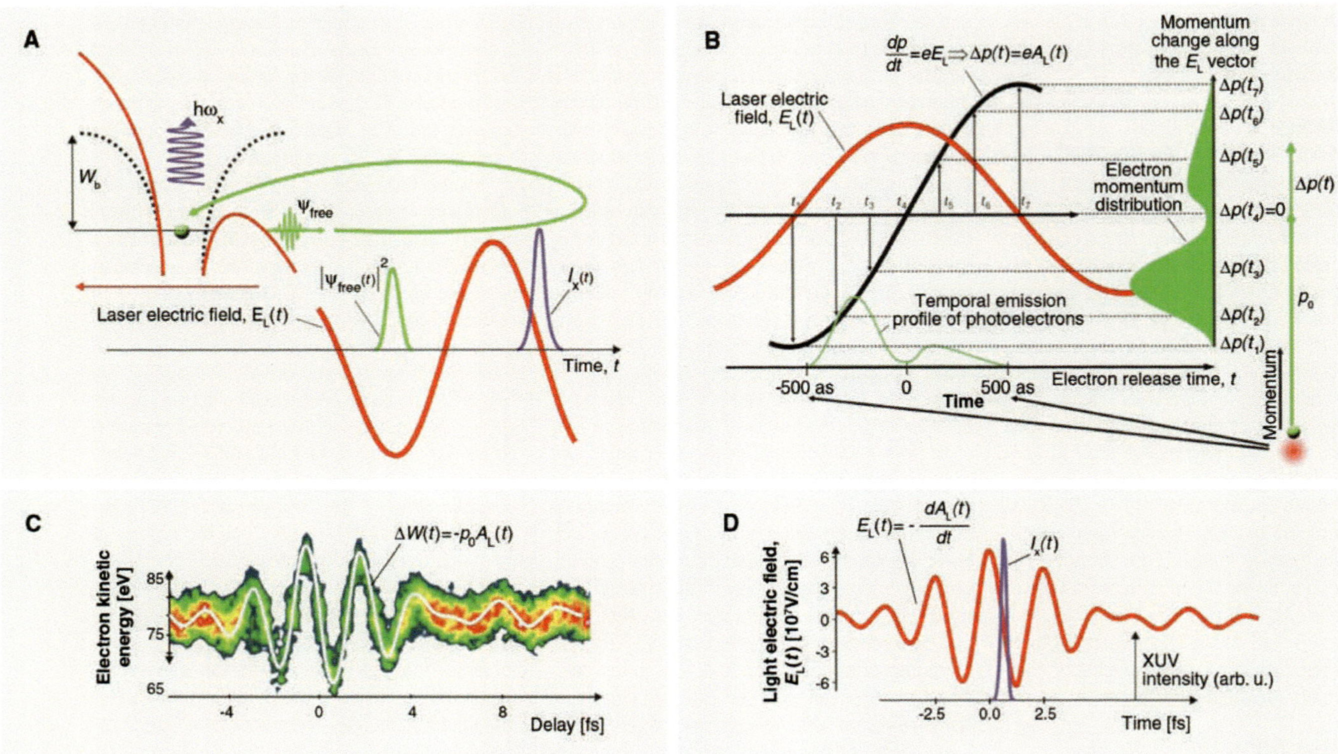

Fig. 1. The birth and measurement of a subfemtosecond (XUV) light pulse. (**A**) The field of a femtosecond NIR laser pulse, $E_L(t)$, is able to suppress the Coulomb potential in atoms sufficiently to allow a valence electron to tunnel through the narrow barrier and release a subfemtosecond wave packet, $\psi_{free}(t)$, near the peaks of its most intense oscillations. The wave packet is subsequently removed from the vicinity of the atomic core and less than a period later pushed back by the reversed field (green trajectory). Upon recollision, it interferes with the bound-state portion of the electron wave function. This interference results in high-frequency oscillations, emitting broadband XUV light. The highest-frequency portion, with intensity $I_X(t)$, is temporally confined to a small fraction of the optical period. (**B**) Concept of optical-field–driven streak imaging of electron emission from atoms. Electrons released by an XUV pulse parallel to the direction of electric field (red line) suffer a change in their initial momenta that is proportional to the vector potential of the field (black line) at the instant of release, mapping the intensity profile of the emitted electron and hence of the ionizing subfemtosecond XUV pulse into a corresponding final momentum and energy distribution of electrons. (**C**) Streaked spectra of photoelectrons released from neon atoms by a single subfemtosecond XUV pulse ($\hbar\omega_X \approx 95$ eV) recorded for a series of delays between the XUV pulse and NIR field (streaking spectrogram). The laser field causes only a moderate broadening (streaking) of the electron spectra; its main effect is the shift of the center of mass of the electron spectrum. In the limit of large initial kinetic energy of the electrons, this shift is linearly proportional to the vector potential of the field at the instant of impact of the XUV pulse (see white line). (**D**) Electric field of the NIR wave and intensity of the XUV pulse as retrieved from the streaking spectrogram shown in (C). The measurement confirms that the few-cycle waveform must be near cosine-shaped to warrant the production of a single subfemtosecond pulse and reveals a near-Fourier–limited XUV pulse of a duration of 250 attoseconds. arb. u., arbitrary units.

Attosecond Spectroscopy

pulse is both versatile and sufficiently compact and affordable to allow proliferation in small laboratories. The core infrastructure required is sketched in Fig. 2. The scope of the technology largely depends on the characteristics and flexibility of its key tools: synthesized fields of laser light and attosecond pulses synchronized to them.

From Controlled Light Waveforms Toward Optical Waveform Synthesis

Lightwave electronics benefits from ever-broader optical bandwidth in several ways. Superposition of spectral components beyond an octave permits subcycle shaping of the light waveform and hence sculpting of the electric force on the electronic time scale for steering electrons in atomic systems. The increased bandwidth and improved dispersion control also lead to shorter optical pulses, which allow advancing subfemtosecond XUV pulse generation in terms of all relevant parameters: duration, intensity, and photon energy.

Intense few-cycle optical pulses (~5 fs) were first generated (*40*) by spectral broadening of femtosecond pulses (initially ~25 fs) in a hollow-core waveguide, and the resultant supercontinuum was compressed by reflection off of chirped multilayer dielectric mirrors (*41*) (Fig. 3A). Combined with waveform control (*4*), the technology matured to become a workhorse for attosecond science. Its current state is represented by the results summarized in Fig. 3, B and C. The spectrum of near–20-fs, 800-nm pulses is first broadened by self-phase modulation in a gas-filled capillary to a supercontinuum reaching from 400 to 1000 nm, followed by compression of the ~400-μJ pulses to ~3.5 fs with octave-spanning chirped mirrors (*42*). These pulses constitute the shortest controlled optical waveforms demonstrated to date.

Figure 3D shows how a single subfemtosecond XUV pulse can emerge from an atom ionized by a linearly polarized few-cycle light field. The green lines depict the kinetic energy, W_{kin}, with which different portions of the freed electron wave packet (ψ_{free} in Fig. 1B) return to the core as a function of return time, determining the energy of the emitted harmonic photon, $\hbar\omega = W_{kin} + W_b$, where W_b is the binding energy of the electron. For the optimum (near-cosine) waveform (Fig. 3D), we can estimate—in the limit of $\hbar\omega_X >> W_b$—the bandwidth over which photons of the highest energy (near the cutoff of the emitted spectrum depicted in violet) are emitted from one recollision only as

$$\frac{\Delta W_{single-pulse}}{\hbar\omega_X} \approx 0.6\left(\frac{T_{osc}}{T_{FWHM}}\right)^2 \qquad (1)$$

where T_{osc} and T_{FWHM} stand for the oscillation period (osc) of the carrier wave and the full width at half maximum (FWHM) duration of the laser pulse ionizing the atoms, respectively (*43*). Equation 1 underscores the importance of minimizing the number of cycles within the driver pulse. The intense sub–1.5 cycle ($T_{FWHM} < 1.5T_{osc}$) laser pulses that can now be produced with commercially available ingredients of ultrafast laser technology (Fig. 3, A to C) constitute a promising tool for pushing the frontiers of attosecond-pulse technology.

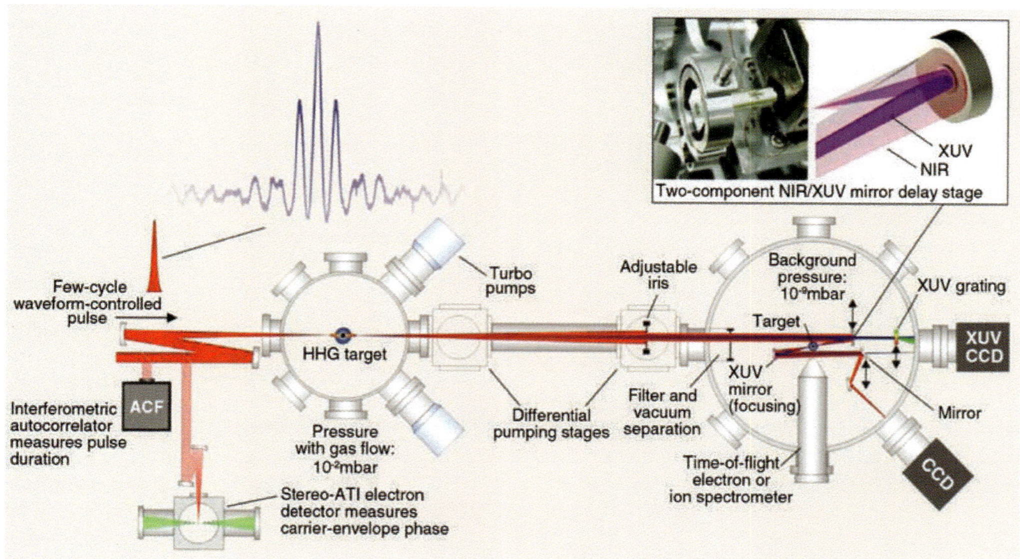

Fig. 2. Schematic of an experimental setup for attosecond-pulse generation and attosecond metrology as well as spectroscopy: the AS-1 attosecond beamline at MPQ. The intense, waveform-controlled few-cycle NIR laser pulse generates a subfemtosecond XUV pulse in the first interaction medium (jet of noble gas). The collinear XUV and NIR beams then propagate into a second vacuum chamber, where they are focused by a two-component XUV multilayer mirror into a gas target. The inner and outer part of the two-component mirror reflects and focuses the XUV and the (more divergent) NIR beam, respectively. By positioning the internal mirror with a nanometer-precision piezotranslator, the XUV pulse can be delayed with respect to the NIR pulse with attosecond accuracy. Analysis of the generated electrons and/or ions as a function of delay between the subfemtosecond XUV and the waveform-controlled NIR pulse permits characterization of the attosecond tools (Fig. 1) as well as real-time observation of atomic-scale electron dynamics in all forms of matter by means of attosecond streaking spectroscopy (*3*) and attosecond tunneling spectroscopy (*13*) with the same apparatus. In the former case, the final energy distribution of the outgoing electron is analyzed with a time-of-flight spectrometer; in the latter case, the observables measured as a function of delay are multiple-charged ions. ACF, auto-correlation function; CCD, charged-coupled device.

Advancing femtosecond technology to its ultimate limit is the key to extending the frontiers of both measuring and controlling dynamics with attosecond precision. Metrology benefits from broader spectral coverage and shorter duration of subfemtosecond pulses. The scope of control critically depends on the variety of waveforms, which is determined by the available bandwidth of intense coherent radiation. The shortest intense laser pulses also provide ideal conditions for generating the broadest supercontinua of coherent light, extending over more then three octaves (*44*). Advanced chirped multilayer mirror technology holds promise for versatile and scalable (multichannel) implementation of optical waveform synthesis. A prototypical three-channel synthesizer operational over the 1- to 3-eV band indicates the vast variety of possibilities for steering electrons in atomic systems with synthesized optical waves (fig. S2).

From Subfemtosecond Toward Attosecond Pulses

Equation 1 suggests that femtosecond technology must approach the monocycle limit to push the frontiers of XUV pulse generation toward and below 100 as. However, there are other options. Modulation of the polarization state of the ionizing NIR pulse (*33*, *37*, *45*) or of its amplitude by coherent addition of second harmonic radiation (*46*, *47*) can efficiently increase $\Delta W_{single-pulse}$ even by using several-cycle NIR driver pulses. Implementation of polarization gating with waveform-controlled, approximately two-cycle NIR pulses (*37*) has recently led to a spectacular achievement: the generation of isolated XUV

SPECIALSECTION

pulses at $\hbar\omega_{carrier} \approx 36$ eV with a bandwidth of $\Delta W_{single\text{-}pulse} \approx 15$ eV. Along with dispersion control and trajectory selection (*48*), these isolated pulses resulted in near–single-cycle XUV pulses that had a duration of 130 as (*33*) (see the low-energy streaking spectrogram in Fig. 4A).

The increased $\Delta W_{single\text{-}pulse}$ with several-cycle pulses comes, however, at the expense of efficiency because (i) only a small temporal fraction of the driver pulse participates in the generation process and (ii) the cycles preceding the generation moment pre-ionize the atoms, causing substantial depletion of the ground state before the XUV pulse can be emitted. The unprecedented confinement of electromagnetic energy into a single well-controlled oscillation of light in sub–1.5-cycle NIR pulses (Fig. 3) avoids these shortcomings and offers several more benefits. Indeed, the high contrast between the wave crests at $t = -T_{osc}/2$ and earlier ones in Fig. 3D permits—thanks to the exponential scaling of ionization probability with the field strength—adjustment of the pulse intensity so as to allow the electron to survive with a high probability in its ground state until $t = -T_{osc}/2$ (in Fig. 3D) and then to be set free with a high probability near this instant through tunneling. The resulting wave packet, ψ_{free} (Fig. 1A), is launched with unprecedented amplitude. Upon its return to the core at about $t \approx T_{osc}/4$, it interferes with the ground state portion of the wave function causing

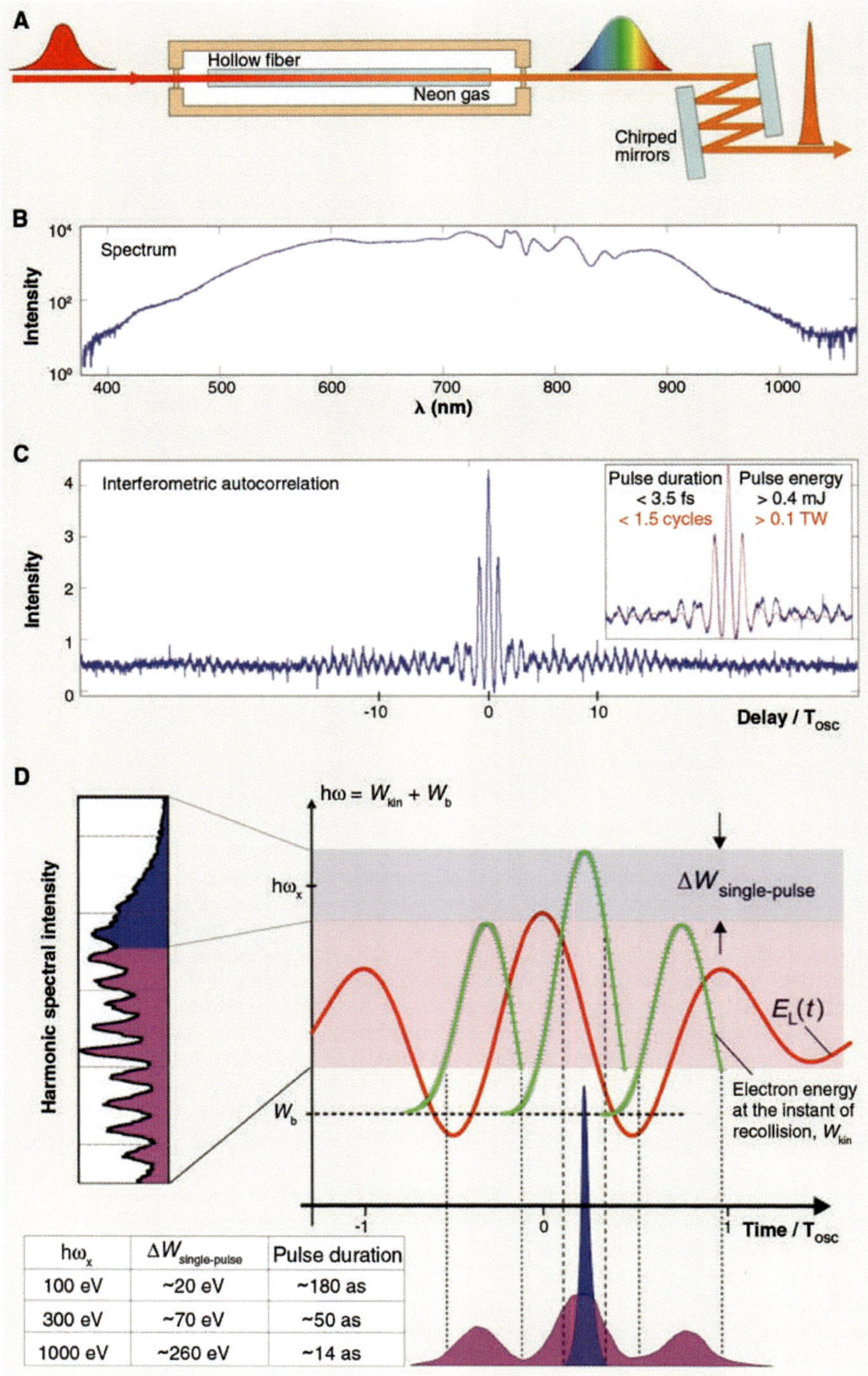

$h\omega_x$	$\Delta W_{single\text{-}pulse}$	Pulse duration
100 eV	~20 eV	~180 as
300 eV	~70 eV	~50 as
1000 eV	~260 eV	~14 as

Fig. 3. Toward intense, monocycle optical pulses. **(A)** Hollow-fiber/chirped-mirror optical pulse compressor for the generation of intense, few-cycle laser pulses. **(B)** The spectrum of millijoule-energy, near–20-fs, phase-controlled near-infrared laser pulses delivered at a repetition rate of 3 kHz from a Ti:sapphire laser (Femtopower, Femtolasers GmbH) is first broadened to more than an octave (~450 to 950 nm) in a neon-filled capillary (pressure ~ 2 bar, length ~ 1 m, bore diameter 250 μm). **(C)** The spectrally broadened pulses are compressed by octave-spanning chirped multilayer mirrors. The interferometric autocorrelation trace indicates a near–bandwidth-limited ~3.5-fs pulse carried at $\lambda_L \approx 720$ nm. **(D)** Energies of XUV photons emitted from an atom exposed to a linearly polarized 3.5-fs, 720-nm Gaussian laser pulse that is sufficiently intense for efficient tunnel ionization. The photon energies (green lines) are depicted as a function of the return time of the electron to the atomic core and have been obtained by analyzing classical electron trajectories (*43*). The width of the energy band within which only one recollision contributes to XUV emission determines the bandwidth $\Delta W_{single\text{-}pulse}$ available for the generation of an isolated attosecond pulse. Based on the classical trajectory analysis, $\Delta W_{single\text{-}pulse}$ is found to be maximum for a near–cosine-shaped waveform, $E_L(t) = E_0\varepsilon(t)\cos(\omega_L t + \varphi)$, with the carrier-envelope phase φ varying between $\pi/12$ and $\pi/6$ depending on the pulse shape and intensity. Here, E_0 and ω_L stand for the amplitude and angular frequency of the oscillations of the laser electric field, with $\varepsilon(t)$ being the amplitude envelope function. The table summarizes bandwidths over which the sub–1.5-cycle pulses are predicted to support the emergence of an isolated attosecond pulse at different XUV photon energies. The estimated pulse durations derive from the conservative assumption that no spectral components are available for the pulse synthesis outside the given spectral window.

Attosecond Spectroscopy

high-frequency (soft–x-ray) dipole oscillations with unprecedented amplitude. Hence, the intense near–single-cycle waveforms appear to constitute ideal tools for the pursuit of powerful soft–x-ray pulses with pulse durations approaching the atomic unit of time (see table in Fig. 3D) and photon energies approaching the kilo–electron volt frontier.

A first indication of the potential that near–single-cycle NIR pulses offer for attosecond-pulse generation is provided by the high-energy streaking spectrogram in Fig. 4A. It has been recorded with sub–two-cycle (near–4-fs) NIR (750-nm) pulses and XUV pulses filtered by a molybden-silicon chirped multilayer mirror with a high-reflectivity band of ~16 eV (FWHM) centered at ~93 eV (Fig. 4B). The bandwidth (FWHM) of the bandpass filtered XUV light amounts to ~13 eV. The streaking spectrogram indicates isolated sub–170-as pulses that are near bandwidth-limited (*49*). The pulses contain more than 10^6 photons, which are delivered at a repetition rate of 3 kHz, implying a photon flux of >3 × 10^9 photons per second transported in a near–diffraction-limited beam. With the Mo/Si mirror, which has a 10-cm focal length, used in the MPQ attosecond beamline AS-1 (Fig. 2), this beam can be focused to a spot diameter of several micrometers.

The recently demonstrated, waveform-controlled sub–1.5-cycle NIR pulses are capable of producing $\Delta W_{\text{single-pulse}} > 30$ eV at $\hbar\omega_X \approx 120$ eV (Fig. 4D), offering the potential for generating sub–100-as pulses at a wavelength of ~10 nm (*42*). To exploit this potential, researchers must develop ultrabroadband soft–x-ray multilayer mirrors with precisely controlled group-delay (GD) dispersion. Figure 4C plots the reflectivity and GD of the first chirped multilayer mirror design that supports sub–100-as pulses. Chirped multilayer mirror technology (both metallic and dielectric) is likely to play as important a role in advancing attosecond technology as it has played in advancing femtosecond technology to its ultimate limits.

Prospects

New research tools allow scientists to seek answers to questions that, in the absence of a means of addressing them, have never been posed before. Grand questions serve as a compass providing directions for research that promises the most extensive return. We pose a few questions across disparate areas of science, connected by the fundamental role that atomic-scale electronic motion plays in physical, chemical, and biological processes.

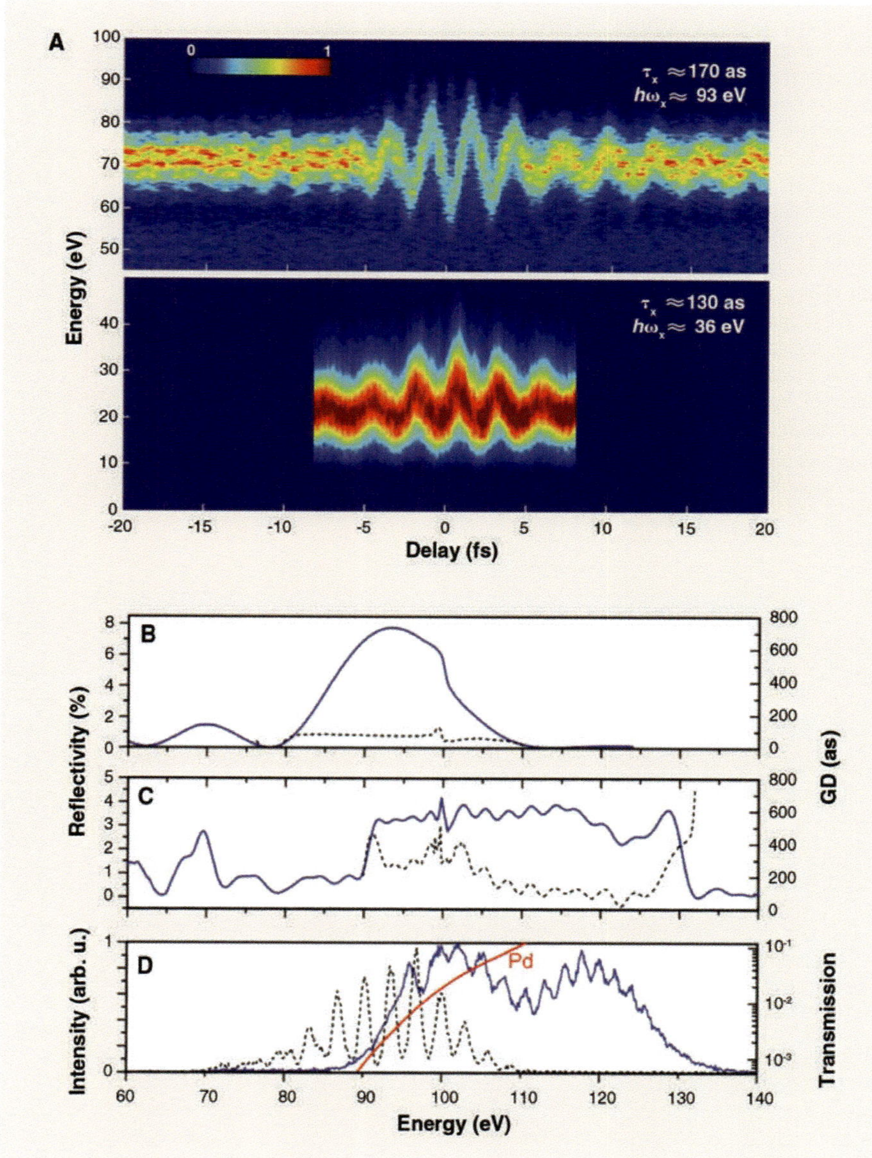

Fig. 4. Current frontiers of attosecond technology. **(A)** Attosecond streaking spectrograms recorded with few-cycle NIR pulses ($\lambda_L \approx 750$ nm). The low-energy spectrogram (bottom) shows the carrier photon energy of an XUV pulse, $\hbar\omega_X \approx 36$ eV, with argon target atoms (courtesy of M. Nisoli). The high-energy spectrogram (top) shows $\hbar\omega_X \approx 93$ eV, with neon target atoms (*49*). **(B)** Computed reflectivity and group delay of the chirped Mo/Si multilayer mirror used for filtering and focusing the isolated sub–170-as, 93-eV pulses. **(C)** Computed reflectivity and group delay of an ultrabroadband chirped multilayer Mo/Si mirror designed for compensating the chirp carried by an attosecond pulse filtered near cutoff from short-trajectory emission (group delay dispersion ~ –0.007 fs^2) and for reflecting a band sufficient for sub–100-as pulse generation (bandwidth > 35 eV). **(D)** Spectrum of high-order harmonics emitted from neon ionized by ~4-fs–duration NIR (~720-nm) pulses near cutoff as transmitted through a high-pass filter (150-nm palladium). The transmission of the Pd foil is shown by the red curve. The spectrum has been recorded with the carrier-envelope phase φ of the waveform-controlled NIR pulse set to provide the broadest continuum. Variation of φ reshapes the overall distribution of the continuum (*42*) rather than introducing a pronounced harmonic-like structure, as observed with several-cycle driver pulses for appropriate phase setting (dashed line). The smooth continuum with a bandwidth of >30 eV suggests the feasibility of sub–100-as XUV pulse generation without manipulation of the driving pulse (e.g., superimposing its second harmonic or gating its polarization).

Atomic physics and x-ray science. How is the energy of an x-ray photon distributed between electrons upon its absorption by an atom? Can electronic transitions deep inside atoms be affected by controlled ultrastrong external fields rivaling in strength the internal Coulomb fields, e.g., for opening up novel routes to efficient, compact x-ray lasers?

Physical chemistry, molecular biology, bioinformatics, and photovoltaics. Can controlled light fields offer a fundamentally new way of modifying the structure and/or composition of molecules by driving electron wave packets across molecules with synthesized optical fields? What are the microscopic mechanisms underlying biological information transport? Can charge-transfer in host-guest systems (e.g., dye-semiconductor assemblies) be exploited for developing solar cells with unprecedented efficiency?

Information technology. Can electron-based information processing and storage be downscaled to atomic dimensions and sped up to the atomic time scale (i.e., to optical frequencies)? Can these ultimate limits be reached by exploiting electric interactions (electronics) or magnetic interactions (spintronics) or collective electron motion (plasmonics)? Which incarnation of lightwave electronics will be the ultimate electron-based information technology?

The answers to these questions will rely on exploring and controlling the microscopic motion of electrons, on atomic scales in space and time. Attosecond technology now offers the tools for tackling these and many other exciting questions. The importance of the answers being sought will drive its proliferation.

References and Notes

1. T. Brabec, F. Krausz, *Rev. Mod. Phys.* **72**, 545 (2000).
2. A. H. Zewail, *J. Phys. Chem. A* **104**, 5660 (2000).
3. M. Drescher *et al.*, *Nature* **419**, 803 (2002).
4. A. Baltuska *et al.*, *Nature* **421**, 611 (2003).
5. For a comprehensive historical review, see P. Agostini, L. F. DiMauro, *Rep. Prog. Phys.* **67**, 813 (2004).
6. R. Trebino *et al.*, *Rev. Sci. Instrum.* **68**, 3277 (1997).
7. C. Iaconis, I. A. Walmsley, *IEEE J. Quantum Electron.* **35**, 501 (1999).
8. P. W. Brumer, M. Shapiro, *Principles of the Quantum Control of Molecular Processes* (Wiley, New York, 2003).
9. W. Ackermann *et al.*, *Nature Phot.* **1**, 336 (2007).
10. A. A. Zholents, W. M. Fawley, *Phys. Rev. Lett.* **92**, 224801 (2004).
11. P. Tzallas, D. Charalambidis, N. A. Papadogiannis, K. Witte, G. Tsakiris, *Nature* **426**, 267 (2003).
12. T. Sekikawa, A. Kosuge, T. Kanai, S. Watanabe, *Nature* **432**, 605 (2004).
13. M. Uiberacker *et al.*, *Nature* **446**, 627 (2007).
14. P. B. Corkum, *Phys. Rev. Lett.* **71**, 1994 (1993).
15. P. Antoine, A. L'Huillier, M. Lewenstein, *Phys. Rev. Lett.* **77**, 1234 (1996).
16. H. Niikura *et al.*, *Nature* **417**, 917 (2002).
17. T. Kanai, S. Minemoto, H. Sakai, *Nature* **435**, 470 (2005).
18. S. Baker *et al.*, *Science* **312**, 424 (2006).
19. The first implementation of the basic concept of a light-field–driven streak camera has drawn on an orthogonal detection geometry (electrons collected along a direction orthogonal to the electric field vector of the streaking NIR field (*29*).
20. The laser field needed to induce this change in electron energy is orders of magnitude less intense than that required for strong-field ionization. Hence, this streaking field has negligible influence on the atomic or molecular processes under study, unless its oscillations happen to be in resonance with a transition from an occupied to an unoccupied quantum state of the system.
21. P. M. Paul *et al.*, *Science* **292**, 1689 (2001).
22. M. Ferray *et al.*, *J. Phys. B* **21**, L31 (1988).
23. K. J. Schafer, M. B. Gaarde, A. Heinrich, J. Biegert, U. Keller, *Phys. Rev. Lett.* **92**, 023003 (2004).
24. T. Remetter *et al.*, *Nature Phys.* **2**, 323 (2006).
25. J. Mauritsson *et al.*, *Phys. Rev. Lett.* **97**, 013001 (2006).
26. Experiments that can be performed with subfemtosecond pulse trains [including those described in (*23–25*)] would benefit from using a small number of well-controlled and characterized subfemtosecond pulses.
27. P. B. Corkum, N. H. Burnett, M. Yu. Ivanov, *Opt. Lett.* **19**, 1870 (1994).
28. I. P. Christov, M. M. Murnane, H. C. Kapteyn, *Phys. Rev. Lett.* **78**, 1251 (1997).
29. M. Hentschel *et al.*, *Nature* **414**, 509 (2001).
30. J. Itatani *et al.*, *Phys. Rev. Lett.* **88**, 173903 (2002).
31. R. Kienberger *et al.*, *Nature* **427**, 817 (2004).
32. E. Goulielmakis *et al.*, *Science* **305**, 1267 (2004).
33. G. Sansone *et al.*, *Science* **314**, 443 (2006).
34. I. P. Christov, R. Bartels, H. C. Kapteyn, M. M. Murnane, *Phys. Rev. Lett.* **86**, 5458 (2001).
35. N. Dudovich *et al.*, *Nature Phys.* **2**, 781 (2006).
36. M. F. Kling *et al.*, *Science* **312**, 246 (2006).
37. I. J. Sola *et al.*, *Nature Phys.* **2**, 319 (2006).
38. F. Quéré, Y. Mairesse, J. Itatani, *J. Mod. Opt.* **52**, 339 (2005).
39. A. D. Bandrauk, S. Chelkowski, H. S. Nguyen, *Int. J. Quantum Chem.* **100**, 834 (2004).
40. M. Nisoli *et al.*, *Opt. Lett.* **22**, 522 (1997).
41. R. Szipöcs, K. Ferencz, C. Spielmann, F. Krausz, *Opt. Lett.* **19**, 201 (1994).
42. A. L. Cavalieri *et al.*, *New J. Phys.* **9**, 242 (2007).
43. The electron kinetic energies have been calculated in terms of the classical description of the center-of-mass motion of the freed electron wave packet (Fig. 2), under the assumption of a Gaussian pulse shape and by using the strong-field approximation. The factor of ~0.6 in Eq. 1 depends on the pulse shape and intensity but varies less than 15% in the relevant parameter range.
44. N. Aközbek *et al.*, *New J. Phys.* **8**, 177 (2006).
45. Z. Chang, *Phys. Rev. A.* **70**, 043802 (2004).
46. T. Pfeifer *et al.*, *Phys. Rev. Lett.* **97**, 163901 (2006).
47. Y. Oishi, M. Kaku, A. Suda, F. Kannari, K. Midorikawa, *Opt. Exp.* **14**, 7230 (2006).
48. R. Lopez-Martens *et al.*, *Phys. Rev. Lett.* **94**, 033001 (2005).
49. M. Schultze *et al.*, *New J. Phys.* **9**, 243 (2007).

50. We apologize that many original research papers could not be cited because of space limitations. This work is supported by the Max-Planck-Society and the Deutsche Forschungsgemeinschaft through the DFG Cluster of Excellence Munich-Centre for Advanced Photonics (www.munich-photonics.de). E.G. acknowledges support from a Marie Curie Intra-European Fellowship. We thank B. Ferus, M. Hofstätter, B. Horvath, and M. Schultze for their support in the preparation of this manuscript.

Supporting Online Material
www.sciencemag.org/cgi/content/full/317/5839/769/DC1
Figs. S1 and S2
References

10.1126/science.1142855

REVIEW

Harnessing Attosecond Science in the Quest for Coherent X-rays

Henry Kapteyn, Oren Cohen, Ivan Christov, Margaret Murnane*

Modern laser technology has revolutionized the sensitivity and precision of spectroscopy by providing coherent light in a spectrum spanning the infrared, visible, and ultraviolet wavelength regimes. However, the generation of shorter-wavelength coherent pulses in the x-ray region has proven much more challenging. The recent emergence of high harmonic generation techniques opens the door to this possibility. Here we review the new science that is enabled by an ability to manipulate and control electrons on attosecond time scales, ranging from new tabletop sources of coherent x-rays to an ability to follow complex electron dynamics in molecules and materials. We also explore the implications of these advances for the future of molecular structural characterization schemes that currently rely so heavily on scattering from incoherent x-ray sources.

Next year, 2008, will mark the 50th anniversary of the revolutionary paper by Schawlow and Townes that proposed the laser (*1*). This paper extended concepts first used to demonstrate the maser in the microwave region of the spectrum into the visible spectrum. Soon after the laser was demonstrated, scientists discovered how to control laser light to generate extremely short nanosecond, picosecond, and even femtosecond pulses. Given the origin of the laser, it was also natural to attempt to generate coherent light at shorter and shorter wavelengths. However, this effort proved very challenging because of the punishing power scaling inherent in lasers. Basic physics dictates that the energy required to implement a laser scales roughly as $1/\lambda^5$; that is, a laser at a 10 times shorter wavelength (λ) requires ~100,000 times the input power. Thus, the first x-ray lasers implemented in the 1980s used the building-sized Nova fusion laser at Lawrence Livermore National Laboratory as a power source to generate soft (relatively long-wavelength) x-rays. Since that initial x-ray laser, considerable progress has been made in downscaling the laser needed as the power source

JILA and the National Science Foundation Center for Extreme Ultraviolet Science and Technology, University of Colorado at Boulder, Boulder, CO 80309–0440, USA.

*To whom correspondence should be addressed. E-mail: murnane@jila.colorado.edu

REPORTS

Single-Cycle Nonlinear Optics

E. Goulielmakis,[1*] M. Schultze,[1] M. Hofstetter,[2] V. S. Yakovlev,[2] J. Gagnon,[1] M. Uiberacker,[2] A. L. Aquila,[3] E. M. Gullikson,[3] D. T. Attwood,[3] R. Kienberger,[1] F. Krausz,[1,2*] U. Kleineberg[2*]

Nonlinear optics plays a central role in the advancement of optical science and laser-based technologies. We report on the confinement of the nonlinear interaction of light with matter to a single wave cycle and demonstrate its utility for time-resolved and strong-field science. The electric field of 3.3-femtosecond, 0.72-micron laser pulses with a controlled and measured waveform ionizes atoms near the crests of the central wave cycle, with ionization being virtually switched off outside this interval. Isolated sub-100-attosecond pulses of extreme ultraviolet light (photon energy ~ 80 electron volts), containing ~0.5 nanojoule of energy, emerge from the interaction with a conversion efficiency of ~10^{-6}. These tools enable the study of the precision control of electron motion with light fields and electron-electron interactions with a resolution approaching the atomic unit of time (~24 attoseconds).

Nonlinear electron-light interactions driven by strong light fields of controlled waveform (*1*) have allowed for the control of electronic motion at light frequencies and the real-time observation of electron dynamics inside and between atoms with ~100-as resolution (*2–7*). However, time-domain access to a number of fundamental processes, such as the intra-atomic energy transfer between electrons (resulting, for example, in shake-up) (*8*), the response of an atomic electron system to external influence (e.g., to ionizing radiation) (*9*) and its rearrangement after the sudden loss of one or more electrons (*10*), the charge transfer in biologically relevant molecules (*11*) and related changes in chemical reactivity (*12*) or because of nonadiabatic tunneling (*13*, *14*), would require (or benefit from) an improved temporal resolution.

We used waveform-controlled sub-1.5-cycle near-infrared (NIR) light to demonstrate the generation of robust, energetic, isolated sub-100-as pulses of extreme ultraviolet (XUV) radiation and their precise temporal characterization. Photoionization confined to a single wave cycle results in observables (such as high-harmonic photons and electrons emitted by above-threshold ionization) that can now be related to several distinguishable subcycle ionization events and subsequent electron trajectories with a known timing with respect to the driving field, whose strength and temporal evolution is accurately known (*3*). These circumstances provide ideal conditions for testing models of strong-field control of electron motion and electron-electron interactions.

The generation of attosecond pulses benefits from the abrupt onset of ionization within a single half-cycle, which minimizes the density of free electrons and, hence, the distortion of the driving wave and its dephasing with the generated harmonic wave. As a result, the coherent build-up of the harmonic emission over an extended propagation is maximized. In addition, the order-of-magnitude variation of the ionization probability between adjacent half-cycles creates unique conditions for single sub-100-as pulse emission without the need for sophisticated gating techniques (*5*, *15*, *16*).

On the measurement side, improved resolution results from three provisions: (i) shorter

[1]Max-Planck-Institut für Quantenoptik, Hans-Kopfermann-Strasse 1, D-85748 Garching, Germany. [2]Department für Physik, Ludwig-Maximilians-Universität, Am Coulombwall 1, D-85748 Garching. [3]Center for X-Ray Optics, Lawrence Berkeley National Laboratory, Berkeley, CA 94720, USA.

*To whom correspondence should be addressed. E-mail: elgo@mpq.mpg.de (E.G.); krausz@lmu.de (F.K.); ulf.kleineberg@physik.uni-muenchen.de (U.K.)

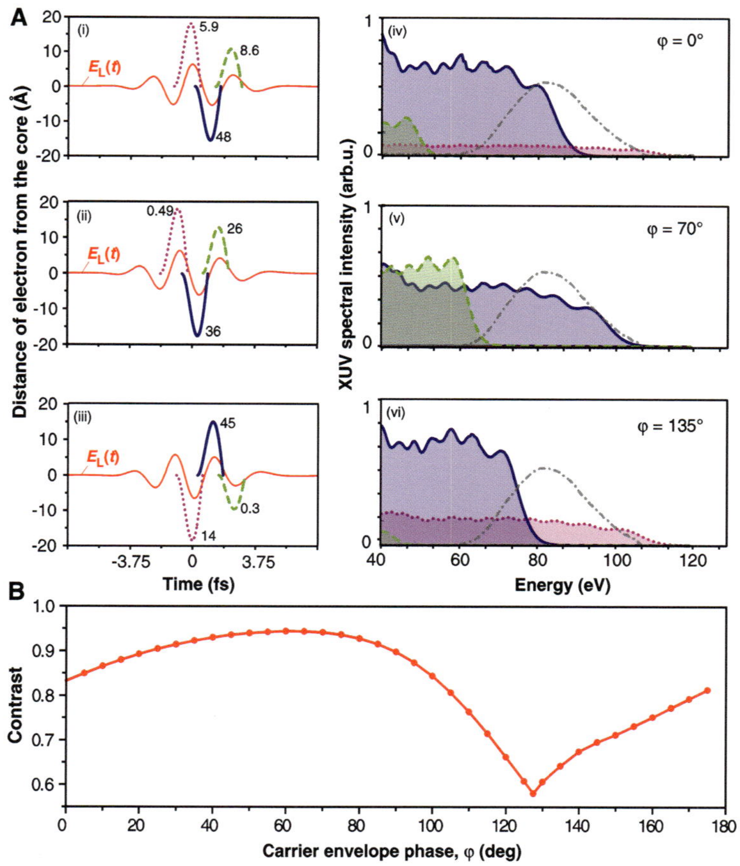

Fig. 1. Simulation of sub-femtosecond XUV emission from neon atoms ionized by a linearly polarized, sub-1.5-cycle, 720-nm laser field. E_0 and $a_L(t)$ are inferred from best agreement between the modeled (*17*) and measured (Fig. 2) spectra and the streaking spectrogram (Fig. 3), respectively. The laser field liberates electrons near its most intense wave crests. (**A**) Classical trajectories of maximum return energy (left panels) and spectra of emerging XUV emission (right panels) are shown for waveforms consistent with $a_L(t)$ inferred from Fig. 3 (correspondence established by colors and line style). The numbers in (i) to (iii) quantify, in units of 10^{-4}, the ionization probability and, hence, the squared modulus of the amplitude of the electron wave packets launched. This amplitude substantially dictates the intensity of XUV emission upon recollision: Contrast ionization probability in (i) to (iii) with the corresponding emission intensities in (iv) to (vi). The pink "dotted-line" emission is not visible in (v) because of the lower ionization probability (by two orders of magnitude) with respect to that resulting in the purple "solid-line" emission [see (ii)]. The gray dashed-and-dotted lines denote the bandpass used in our experiments (fig. S2). φ = 70° and φ = 135° yield highest contrast [see (B)] and highest XUV cutoff energy, respectively. arb.u., arbitrary units. (**B**) Contrast versus CE phase. Here, contrast is defined as the ratio of the energy of the main attosecond XUV pulse to the overall XUV emission energy transmitted through the bandpass [gray dashed-and-dotted lines in (A)].

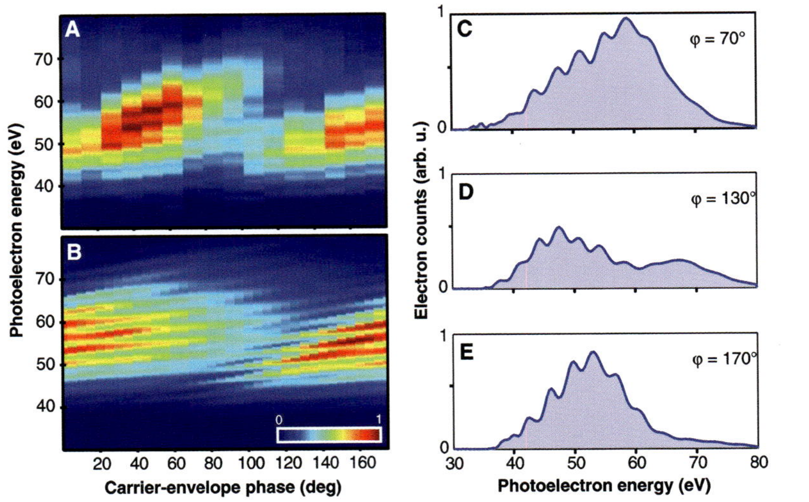

Fig. 2. Control of bandpass-filtered XUV emission with the waveform of monocycle light. Measured **(A)** and simulated **(B)** (*17*) photoelectron spectra versus CE phase, with the delay increased in steps of ~11° (π/16 rad). **(C** to **E)** Spectra measured at the CE phase setting closest to the values selected in Fig. 1A. The zero of the CE phase scale in (A) was set to yield the best agreement with the modeled spectra in (B).

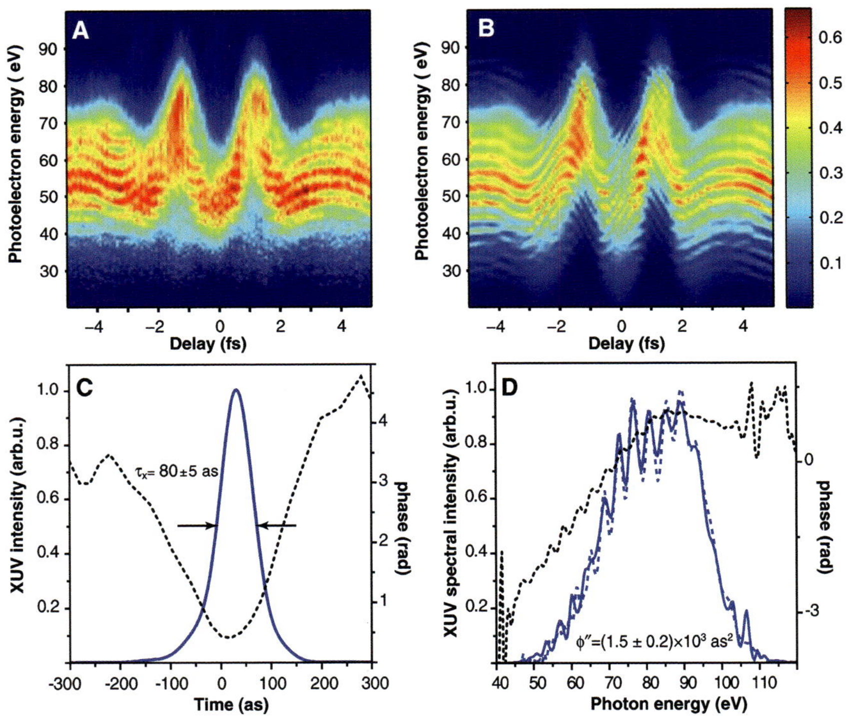

Fig. 3. Sub-100-as XUV pulse retrieval. **(A)** Measured ATR spectrogram compiled from 126 energy spectra of photoelectrons launched by an XUV pulse with a bandwidth of ~28 eV (FWHM) and recorded at delay settings increased in steps of 80 as. Here, a positive delay corresponds to the XUV pulse arriving before the NIR pulse. The high flux of the XUV source allows this spectrogram to be recorded within ~30 min. **(B)** ATR spectrogram reconstructed after ~10^3 iterations of the FROG algorithm (*17*). **(C)** Retrieved temporal intensity profile and spectral phase of the XUV pulse. The intrinsic chirp of the XUV emission (Fig. 4B) is almost fully compensated by a 300-nm-thick Zr foil introduced into the XUV beam between the attosecond source and the ATR measurement. Arrows indicate the temporal FWHM of the XUV pulse. **(D)** XUV spectra evaluated from the measurement of the XUV-generated photoelectron spectrum in the absence of the NIR streaking field (blue dashed line) and from the ATR retrieval (blue solid line). The black dotted line indicates the retrieved spectral phase.

XUV pulse duration, (ii) improved signal-to-noise (S/N) ratio due to the increased XUV photon flux, and (iii) stronger streaking before the onset of the NIR field–induced ionization in attosecond streaking (*2*) or enhanced S/N ratio due to a reduced number of tunneling steps in attosecond tunneling spectroscopy (*14*).

Figure 1 summarizes results of the modeling of the single-cycle interaction of ionizing NIR radiation with an ensemble of neon atoms (*17*). In Fig. 1A, the left panels plot possible NIR electric waveforms, $E_L(t) = E_0 a_L(t) e^{-i(\omega_L t+\varphi)} + cc$ (where *cc* stands for complex conjugate) derived from our streaking measurements (as presented in the next sections) for different settings of the carrier-envelope (CE) phase, φ. Here, E_0 is the peak electric-field strength, $a_L(t)$ is the normalized complex amplitude envelope, and ω_L is the carrier frequency. The probability of ionization outside the central cycle is more than two orders of magnitude lower than that at the field maximum and hence is negligible.

The spectra of XUV emissions originating from the individual recollisions (*18*) are predicted to differ by tens of electron volts in cut-off energy and by up to orders of magnitude in intensity as a consequence of the single-cycle nature of the driving field. The strong variation of emission energies and intensities within a single wave cycle creates ideal conditions for isolated sub-100-as pulse generation. Indeed, filtering radiation with the bandpass depicted by the dashed-and-dotted line is predicted to isolate XUV radiation with more than 90% of its energy delivered in a single attosecond pulse for a range of CE phases as broad as $30° \leq \varphi \leq 90°$ (Fig. 1B). In contrast, with few-cycle-driven harmonic generation resulting in isolated subfemtosecond pulses over only a relatively narrow range of the CE phase near $\varphi \approx 0°$ (*3*), single-cycle excitation appears to permit robust isolated attosecond pulses for a variety of driver waveforms, ranging from near-cosine– to sine-shaped ones, owing to the order-of-magnitude variation of the ionization probability within a single wave cycle.

We used phase-controlled sub-1.5-cycle laser pulses carried at a wavelength of $\lambda_L = 2\pi c/\omega_L =$ 720 nm (*19*) to generate XUV harmonics in a neon gas jet up to photon energies of ~110 eV (fig. S1). The emerging XUV pulse—following a spectral filtering through a bandpass (dashed-and-dotted line in Fig. 1A) introduced by metal foils and a Mo/Si multilayer mirror (fig. S2)—subsequently propagates, along with its NIR driver wave, through a second jet of neon atoms in which the XUV pulse ionizes the atoms in the presence of the NIR field. The freed electrons with initial momenta directed along the electric-field vector of the linearly polarized NIR field are collected and analyzed with time-of-flight spectrometry (*17*).

The variation of the measured photoelectron spectra versus CE phase shows good agreement with the predictions of our simulations (Fig. 2, A and B). Figure 2, C to E, shows plots of electron spectra corresponding to the CE phase

REPORTS

settings selected in Fig. 1A. Apart from a downshift by the ionization potential of neon (21.5 eV), they reveal close resemblance to the XUV spectra transmitted through the bandpass in Fig. 1A(v), (vi), and (iv), respectively. Figure 2C depicts the broadest filtered spectrum produced by a single recollision [full-line spectrum in Fig. 1A(v)]. Emission from the same recollision dominates also in the spectrum shown in Fig. 2E, with this spectrum red-shifted and (upon transmission through the bandpass) correspondingly narrowed, as predicted in Fig. 1A(iv). The two humps of the spectrum plotted in Fig. 2D are indicative of contributions from two recollisions, in accordance with the "solid-line" and "dotted-line" contributions to the emission spectrum in Fig. 1A(vi).

Before measuring the XUV pulse, we optimized the generation process by "fine-tuning" the NIR laser peak intensity to achieve the broadest possible XUV spectrum transmitted through the bandpass in the range of CE phase settings where the contrast is maximized (50° – 80°, according to Fig. 1B) to generate a clean single pulse with the shortest possible duration. For the temporal characterization of the generated XUV supercontinuum, we shined the NIR field into the neon atoms ionized by the XUV pulse to implement the atomic transient recorder (ATR) technique introduced in (*2*, *3*). The NIR field boosts or decreases the momenta of the photoelectrons, depending on their instant of release within the 2.4-fs period of the laser field, resulting in broadened and shifted (streaked) spectra of the electrons' final energy distribution. Figure 3A is a plot of a series of streaked spectra recorded versus delay between the XUV and NIR pulse, which we refer to as an ATR or streaking spectrogram. It is practically equivalent to the spectrogram obtained by frequency-resolved optical gating (FROG) (*20*), with the oscillating NIR field constituting an attosecond phase gate in the present case (*21*). As a consequence, a FROG retrieval algorithm (*22*) allows complete determination of both the (gated) XUV pulse and the (gating) NIR laser field (*17*). The reconstructed ATR spectrogram is plotted in Fig. 3B and reveals excellent agreement with the measured one.

The retrieved temporal intensity profile and phase of the XUV pulse are shown in Fig. 3C. The pulse duration of $\tau_x = 80 \pm 5$ as is close to its transform limit of 75 as, with a small positive chirp of $\phi'' = (1.5 \pm 0.2) \times 10^3$ as^2 being responsible for the deviation. As a further consistency check of the attosecond pulse retrieval, we compared the measured NIR field–free electron spectrum (dashed blue line in Fig. 3D) with the electron spectrum calculated from the retrieved attosecond pulse (solid line in Fig. 3D). Given that the pulse retrieval draws on streaked spectra that are strongly distorted with respect to the field-free one, the degree of agreement between the retrieved and directly measured spectrum provides yet another conclusive testimony of the reliability of the retrieved data.

The ATR retrieval algorithm indicates the presence of a satellite pulse accompanying the main attosecond pulse, containing ~1% of the energy of the main pulse. This amount of satellite is consistent with the depth of the experimentally observed modulation in the XUV spectrum (Fig. 3D). However, this result is inconsistent with our numerical modeling, which predicts a satellite energy content of some 6 to 7% for the optimum range of CE phase settings (Fig. 1B). From analysis of the streaked spectra recorded at the maximum of the NIR electric field (fig. S3), where the momentum of the electrons released by the main attosecond pulse and its satellite is shifted in opposite directions, we inferred a relative satellite energy of ~8%, which is in good agreement with the prediction of our modeling. As a consequence, the fringe visibility in the XUV spectrum is lower than was implied by the relative amplitude of the satellite pulse. The discrepancy may originate from a temporal jitter between the main pulse and the satellite pulse and/or from a different spatial amplitude distribution of the beams transporting the emission from the adjacent recollision events. The important lesson from these findings is that the fringe visibility in the XUV spectrum does not allow a reliable determination of the energy carried by the satellite(s) accompanying the main attosecond pulse.

The laser waveform evaluated from the ATR measurement is preserved between the location of the generation and measurement. Figure 4A illustrates the evaluated NIR waveform with electric-field amplitude corresponding to a peak intensity of $I_0 \sim (5.8 \pm 0.5) \times 10^{14}$ W/cm^2, as evaluated from the cut-off of our XUV spectra. The pulse duration [full width at half maximum (FWHM)] of 3.3 fs is in good agreement with the results of previous interferometric autocorrelation measurements (*19*), and the evaluated CE phase of ~50° is consistent with the optimum contrast according to our modeling (Fig. 1B).

Accurate knowledge of the attosecond XUV pulse parameters, the temporal evolution of the generating NIR field, and the emergence of the former from a single recollision permit one to perform precision tests of models of light-electron interactions underlying the ionization and subsequent attosecond pulse generation processes. As an example, we calculated the intrinsic spectral chirp (i.e., the variation of the group delay versus frequency) carried by the attosecond XUV pulse during its emergence from the chirp measured by the ATR and the known dispersion of the components traversed by the pulse on its way

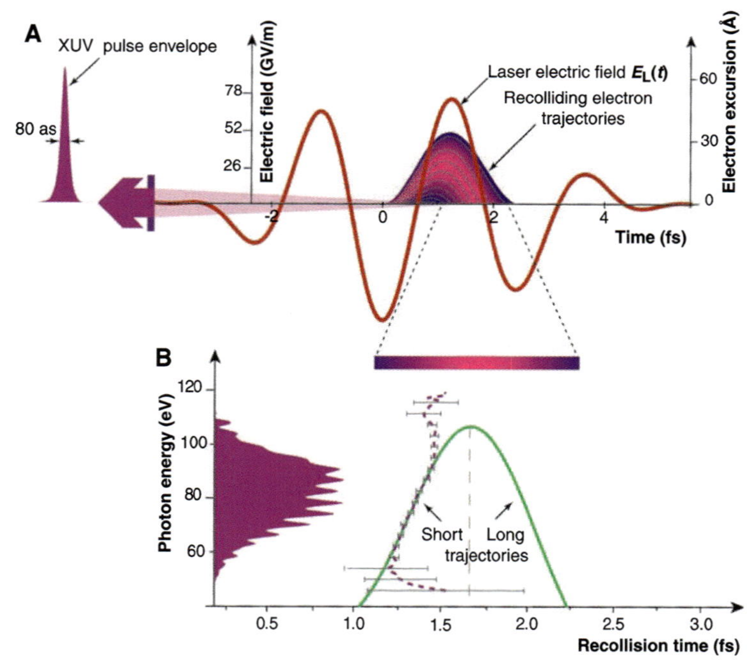

Fig. 4. **(A)** Retrieved electric field of the NIR laser pulse used for generating and measuring the attosecond XUV pulse shown in Fig. 3. The duration of the pulse (FWHM of the cycle-averaged intensity profile) is $\tau_L = 3.3$ fs, and the CE phase is evaluated as $\varphi \sim 50°$. The classical electron trajectories responsible for the emission of the filtered attosecond pulse are calculated with the plotted electric field and shown in the same panel. The color coding indicates the final return energy of the electrons. **(B)** Energy of the recolliding electron plus ionization potential (which is equal to the emitted XUV photon energy) versus recollision time evaluated from the classical trajectory analysis (solid green line), and emitted photon energy versus emission time (dashed purple line) inferred from the chirp of the measured attosecond pulse and the dispersion of the metal filter that the attosecond pulse passes through before the measurement. The basic idea for the graphical representation is borrowed from (*29*). Error bars indicate the uncertainty in the retrieved group delay.

from the source to the measurement (fig. S2). The result (purple dashed line in Fig. 4B) is compared with the intrinsic attosecond chirp (green solid line in Fig. 4B) calculated from a classical trajectory analysis (*23*, *24*). There is a notable discrepancy at the high-energy components of the wave packet, possibly because of quantum effects near cutoff. Nevertheless, the agreement with the attochirp resulting from short trajectories is stunning in the main part of the spectrum, where the S/N ratio is excellent. This agreement indicates the powerfulness of semiclassical modeling of strong-field interactions (*25*, *26*) and the negligible role of long trajectories in contributing to the XUV radiation in the far field (*27*).

In a similar way, the confinement of interaction between the ionizing field and the atom to a single wave cycle will permit accurate quantitative tests of theories of strong-field ionization. The sub-100-as XUV pulses emerging from the interaction with a flux greater than 10^{11} photons/s—along with their monocycle NIR driver wave—will push the resolution limit of attosecond spectroscopy to the atomic unit of time (~24 as) and allow for the real-time observation of electron correlations, by means of streaking (*6*), tunneling (*14*), or absorption (*28*) spectroscopy.

References and Notes

1. A. Baltuska *et al.*, *Nature* **421**, 611 (2003).
2. R. Kienberger *et al.*, *Nature* **427**, 817 (2004).
3. E. Goulielmakis *et al.*, *Science* **305**, 1267 (2004).
4. M. F. Kling *et al.*, *Science* **312**, 246 (2006).
5. G. Sansone *et al.*, *Science* **314**, 443 (2006).
6. A. L. Cavalieri *et al.*, *Nature* **449**, 1029 (2007).
7. E. Goulielmakis *et al.*, *Science* **317**, 769 (2007).
8. S. Svensson, B. Eriksson, N. Martensson, G. Wendin, U. J. Gelius, *J. Electron Spectrosc. Relat. Phenom.* **47**, 327 (1988).
9. S. X. Hu, L. A. Collins, *Phys. Rev. Lett.* **96**, 073004 (2006).
10. J. Breidbach, L. S. Cederbaum, *Phys. Rev. Lett.* **94**, 033901 (2005).
11. A. I. Kuleff, J. Breidbach, L. S. Cederbaum, *J. Chem. Phys.* **123**, 044111 (2005).
12. F. Remacle, R. D. Levine, *Proc. Natl. Acad. Sci. U.S.A.* **103**, 6793 (2006).
13. G. Yudin, M. Y. Ivanov, *Phys. Rev. A* **64**, 035401 (2001).
14. M. Uiberacker *et al.*, *Nature* **446**, 627 (2007).
15. T. Pfeifer *et al.*, *Phys. Rev. Lett.* **97**, 163901 (2006).
16. Y. Oishi *et al.*, *Opt. Express* **14**, 7230 (2006).
17. See supporting material on *Science* Online.
18. C. A. Haworth *et al.*, *Nature Phys.* **3**, 52 (2007).
19. A. L. Cavalieri *et al.*, *New J. Phys.* **9**, 242 (2007).
20. R. Trebino, D. J. Kane, *J. Opt. Soc. Am. A* **10**, 1101 (1993).
21. Y. Mairesse, F. Quéré, *Phys. Rev. A* **71**, 011401(R) (2005).
22. D. J. Kane, G. Rodriguez, A. J. Taylor, T. S. Clement, *J. Opt. Soc. Am. B* **14**, 935 (1997).
23. P. Salières *et al.*, *Science* **292**, 902 (2001).
24. V. S. Yakovlev, A. Scrinzi, *Phys. Rev. Lett.* **91**, 153901 (2003).
25. P. B. Corkum, *Phys. Rev. Lett.* **71**, 1994 (1993).
26. K. J. Schafer, B. Yang, L. F. DiMauro, K. C. Kulander, *Phys. Rev. Lett.* **70**, 1599 (1993).
27. R. López-Martens *et al.*, *Phys. Rev. Lett.* **94**, 033001 (2005).
28. Z.-H. Loh *et al.*, *Phys. Rev. Lett.* **98**, 143601 (2007).
29. K. Varjú *et al.*, *Laser Phys.* **15**, 888 (2005).

30. Supported by the Max Planck Society and the Deutsche Forschungsgemeinschaft Cluster of Excellence: Munich Centre for Advanced Photonics (www.munich-photonics.de). E.G. acknowledges a Marie-Curie fellowship (MEIF-CT-2005-02440) and a Marie-Curie Reintegration grant (MERG-CT-2007-208643). A.L.A. is supported by the NSF Extreme Ultraviolet Engineering Research Center. R.K. acknowledges support from the Sofia Kovalevskaya award of the Alexander von Humboldt Foundation.

Supporting Online Material
www.sciencemag.org/cgi/content/320/5883/1614/DC1
SOM Text
Figs. S1 to S4
References

17 March 2008; accepted 27 May 2008
10.1126/science.1157846

The Formation Conditions of Chondrules and Chondrites

C. M. O'D. Alexander,[1*] J. N. Grossman,[2] D. S. Ebel,[3] F. J. Ciesla[1]

Chondrules, which are roughly millimeter-sized silicate-rich spherules, dominate the most primitive meteorites, the chondrites. They formed as molten droplets and, judging from their abundances in chondrites, are the products of one of the most energetic processes that operated in the early inner solar system. The conditions and mechanism of chondrule formation remain poorly understood. Here we show that the abundance of the volatile element sodium remained relatively constant during chondrule formation. Prevention of the evaporation of sodium requires that chondrules formed in regions with much higher solid densities than predicted by known nebular concentration mechanisms. These regions would probably have been self-gravitating. Our model explains many other chemical characteristics of chondrules and also implies that chondrule and planetesimal formation were linked.

Chondrules make up ~20 to 80 volume % of most chondrites and formed at peak temperatures of ~1700 to 2100 K (*1*). Chondrules in the different chondrite groups have distinct physical and chemical properties (*2*), as well as distinct age ranges (*3*), indicating that they formed in relatively local environments via a process that operated at least periodically between ~1 and 4 million years after the formation of the solar system.

It was long thought that individual chondrules behaved as chemically closed systems during their formation, inheriting their compositions from their precursors (*4*, *5*). However, for likely cooling rates of 10 to 1000 K/hour (*1*) and at the low pressures (total pressure ≈ 10^{-6} to 10^{-3} bars) of the solar protoplanetary disk (nebula), experiments (*6*–*8*), natural analogs (*9*, *10*), and theoretical calculations (*11*, *12*) all show that there should be extensive evaporation of major and minor elements, in the order S > Na, K > Fe > Si > Mg.

Elemental fractionations in chondrules are generally a function of volatility (*4*, *5*). If evaporation in the nebula produced the alkali metal and Fe fractionations, the more volatile elements (such as S) should be entirely absent, which they are not. In addition, the fractionated elements should exhibit large and systematic isotopic fractionations, which they do not (*13*).

Here we demonstrate that chondrules did indeed behave as essentially closed systems during melting, at least for elements with volatilities less than or equal to that of Na. We also propose a means of resolving the apparent conflict between this result and experimental and theoretical expectations that chondrules should have suffered considerable evaporation during formation. Our conclusions have implications for mechanisms of dust concentration in the solar nebula, for chondrule formation, and for planetesimal formation.

Chondrules are dominated by olivine [$(Mg,Fe)_2SiO_4$], pyroxene [$(Mg,Fe,Ca)SiO_3$], Fe-Ni metal, and quenched silicate melt (glass). Many of the more volatile elements (such as Na) can diffuse rapidly, particularly in melts and glasses. Therefore, it is possible that volatiles were completely lost when chondrules melted, and reentered the chondrules during cooling or even after solidification. However, Na clinopyroxene/glass ratios show that the Na contents of the final chondrule melts (now glass) had approximately their observed, relatively high abundances at temperatures of ~1600 to 1200 K (*14*–*16*).

Calculations suggest that chondrule melts could have been stabilized in the nebula by substantially enriching solids (chondrule precursors or other dust) relative to gas (*11*, *12*). This also substantially increases the condensation temperatures of even highly volatile elements such as S (*11*, *12*). Even in solid-enriched systems, there is an initial phase of evaporation when a chondrule melts, but subsequent chondrule/gas re-equilibration would erase any isotopic fractionations (*12*). If the enrichment of solids is high, little evaporation may be needed to reach chondrule/gas equilibrium, and the behavior of volatile elements during cooling would resemble closed-system behavior. However, even at a high total pressure of 10^{-3} bars, with a solids enrichment of 1000 relative to the solar composition, all the Na would evaporate at near-liquidus temperatures, and substantial recondensation only begins well below 1600 K (*11*). Locally enriching chondrule-sized or smaller solids by 1000 times

[1]Department of Terrestrial Magnetism, Carnegie Institution of Washington, Washington, DC 20015, USA. [2]U.S. Geological Survey, Reston, VA 20192, USA. [3]American Museum of Natural History, New York, NY 10024, USA.

*To whom correspondence should be addressed. E-mail: alexande@dtm.ciw.edu

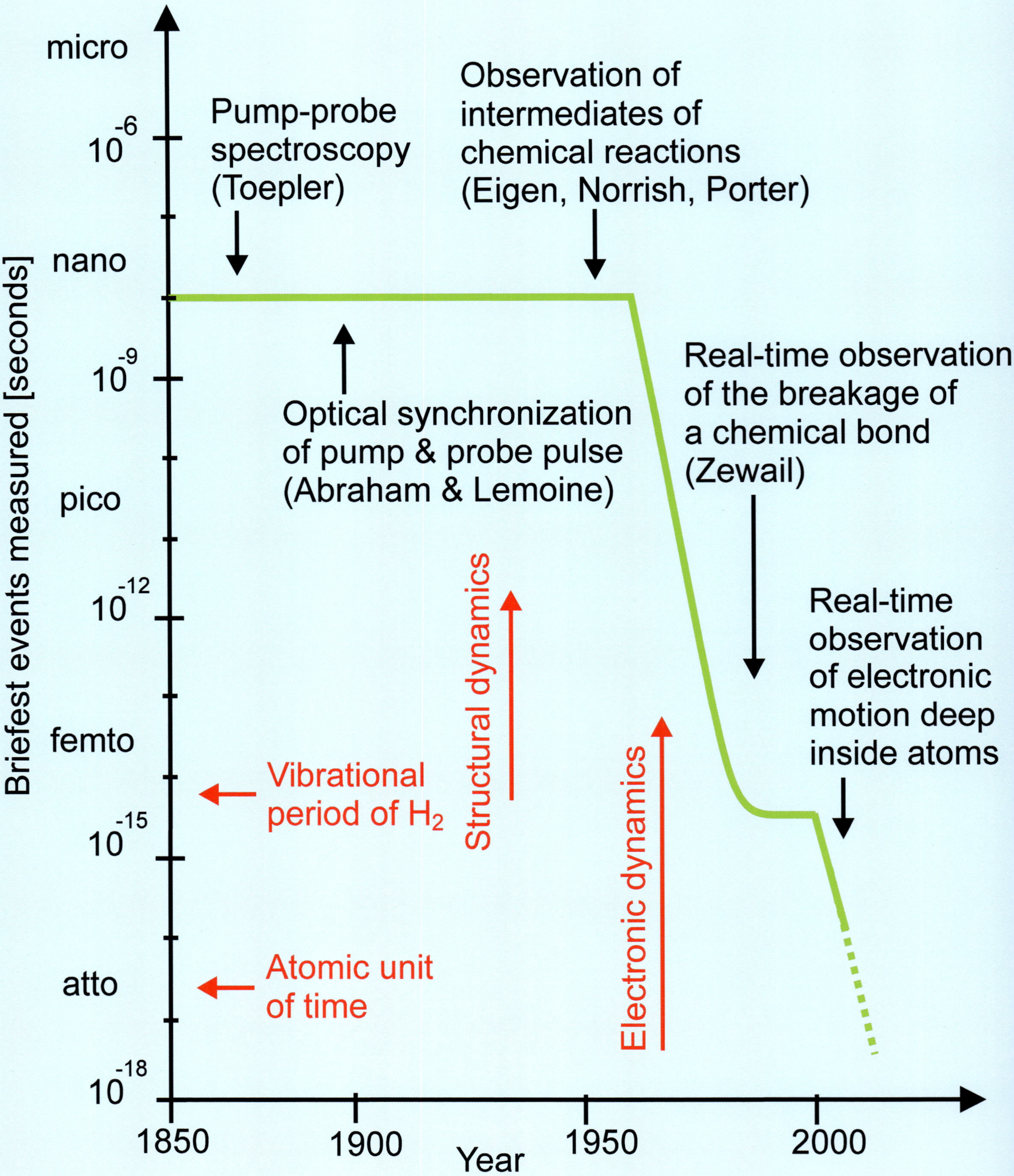

micro
10^{-6}
nano
10^{-9}
pico
10^{-12}
femto
10^{-15}
atto
10^{-18}
Briefest events measured [seconds]
1850
1900
1950
2000
Year
Pump-probe spectroscopy (Toepler)
Observation of intermediates of chemical reactions (Eigen, Norrish, Porter)
Optical synchronization of pump & probe pulse (Abraham & Lemoine)
Real-time observation of the breakage of a chemical bond (Zewail)
Real-time observation of electronic motion deep inside atoms
Vibrational period of H_2
Atomic unit of time
Structural dynamics
Electronic dynamics

Review of the Birth of the New Discipline

7

Opposite: The evolution of techniques for real-time observation of microscopic processes. Discontinuities in the slope of the briefest measured events vs. years indicate revolutions in technology.

The previous chapters have revealed how the measurement and control of the sub-cycle evolution of few-cycle light fields have spawned a radically new approach to exploring and controlling processes of the microcosm. The electric field of visible light varying on an attosecond time scale permitted the manipulation and tracking of the atomic-scale motion of electrons. Striking implications include the controlled generation and measurement of single attosecond pulses of extreme ultraviolet light (as well as trains of pulses) and the real-time observation of atomic-scale electron dynamics.

The tools and techniques for steering and tracing electronic motion in atoms, molecules, and nanostructures are now becoming available, marking the birth of attosecond physics. In this article these advances are reviewed and some of the expected implications are addressed.

[39] F. Krausz, M. Y. Ivanov. Attosecond physics. *Reviews of Modern Physics* **81**, 164 (2009).
(Page: 275)

REVIEWS OF MODERN PHYSICS, VOLUME 81, JANUARY–MARCH 2009

Attosecond physics

Ferenc Krausz

Department für Physik, Ludwig-Maximilians-Universitat, Am Coulombwall 1, D-85748 Garching, Germany and Max-Planck-Institut für Quantenoptik, Hans-Kopfermann-Strasse 1, D-85748 Garching, Germany

Misha Ivanov

Steacie Institute for Molecular Sciences, National Research Council of Canada, 100 Sussex Drive, Ottawa, Ontario, Canada K1A 0R6

(Published 2 February 2009)

Intense ultrashort light pulses comprising merely a few wave cycles became routinely available by the turn of the millennium. The technologies underlying their production and measurement as well as relevant theoretical modeling have been reviewed in the pages of *Reviews of Modern Physics* (Brabec and Krausz, 2000). Since then, measurement and control of the subcycle field evolution of few-cycle light have opened the door to a radically new approach to exploring and controlling processes of the microcosm. The hyperfast-varying electric field of visible light permitted manipulation and tracking of the atomic-scale motion of electrons. Striking implications include controlled generation and measurement of single attosecond pulses of extreme ultraviolet light as well as trains of them, and real-time observation of atomic-scale electron dynamics. The tools and techniques for steering and tracing electronic motion in atoms, molecules, and nanostructures are now becoming available, marking the birth of attosecond physics. In this article these advances are reviewed and some of the expected implications are addressed.

DOI: 10.1103/RevModPhys.81.163 PACS number(s): 42.65.Re, 32.80.−t

CONTENTS

0034-6861/2009/81(1)/163(72)

I. INTRODUCTION

Electrons have played a central role in the scientific and technological revolution of the 20th century.[1] It is widely anticipated that in the 21st century their role as a key driver of scientific and technological advances will be increasingly taken over by photons. We disagree with this assessment. Some half a century after the discovery of the transistor and the laser, our insight into and influence upon the atomic-scale motion of electrons in matter is still in its infancy. Consequently, the advance of science and technology by research on electrons is by no means over. On the contrary, it is just beginning, given the fact that the experimental tools and techniques for real-time observation and steering of electronic dynamics on the atomic scale are now becoming available.

In the microcosm, the borders between biology, chemistry, and physics tend to disappear. The gaps between these apparently so disparate fields are bridged by the microscopic motion of electrons in atoms, molecules, and nanoscale structures. Electronic motion inside atoms is behind the emission of visible, ultraviolet, and x-ray light.[2] Electronic dynamics on the molecular scale is responsible for the transport of bioinformation and initiates changes in the chemical composition and function of biological systems. The speed of information processing can be increased by ever faster switching of electronic currents in ever smaller nanometer-scale circuitry. Details of these motions often occur on time scales of tens to ten thousands of attoseconds (1 as$=10^{-18}$ s) and require commensurate measurement and control techniques for their observation and steering, respectively.

What are the most efficient ways of putting atoms into highly excited states that allow x-ray light amplification and thus creation of compact x-ray lasers? Can the function of biomolecules be manipulated and novel molecular structures be formed by steering electrons in chemical bonds? How does charge transfer occur in molecules assembled on surfaces and how can it be optimized for more efficient solar cells or for fighting radiation damage during biological imaging? What are the ultimate size and speed limits of electronic information processing and magnetic information storage, and how can we approach these limits? How can energy be most efficiently transported into high-density matter to ignite nuclear fusion? Answers to these scientific and technological questions will require insight into and possibly control of microscopic electron motion.

We address here (i) the key concepts and experimental tools which provide the means of observing and controlling the atomic-scale motion of electrons in real time, (ii) the theoretical models critical for connecting experimental observables with microscopic variables, and (iii) expected implications of this revolution in technology.[3]

[1]For a general review, see Huebener (2005).

[2]The requirement of atomic-scale electronic motion for light emission can be circumvented only by accelerating them to relativistic energies. Only relativistic electrons can radiate visible (or shorter-wavelength) light in the absence of (sub-) atomic-scale interactions.

[3]To make sure that we draw clear borderlines between facts and conclusions based upon experiments, on the one hand, and our visions, expectations and prediction (necessarily containing a speculative component) on the other hand, comments of the latter type will be systematically introduced by phrases such as "we believe," "we expect," "in our opinion," etc. In this way, we are making an attempt to write not a conventional review, but a review *and* foresight article in combination, with the two components clearly separated. We feel that, at this early stage of its evolution, the advancement of attosecond physics is best served this way.

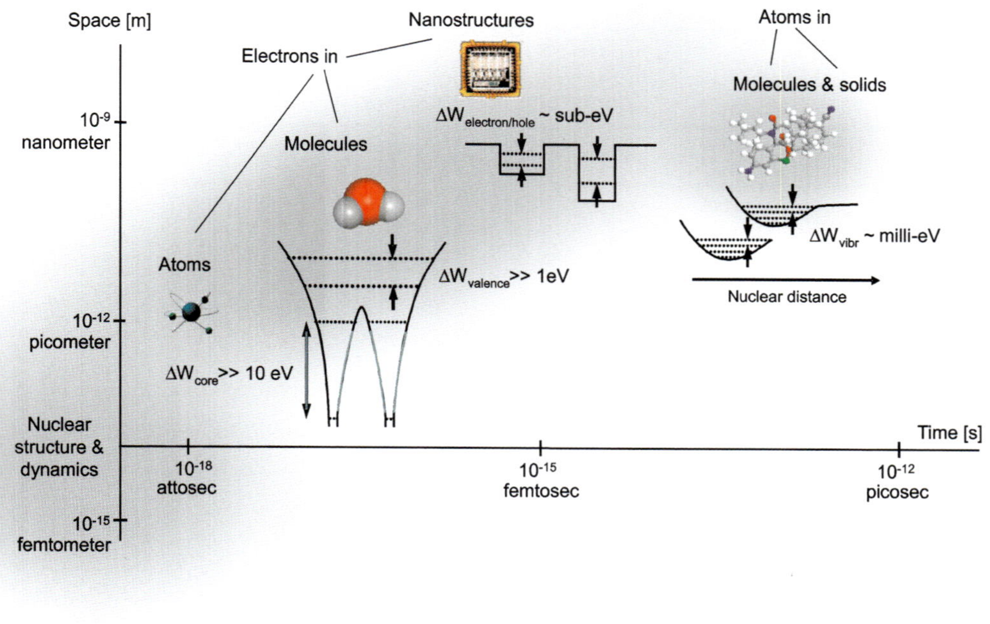

FIG. 1. (Color) Characteristic length and time scales for structure and dynamics in the microcosm, respectively.

A. Time and length scales in the microcosm

Structure and dynamics in the microcosm are inherently connected by the laws of quantum mechanics. Take, for example, a particle put in a superposition of its ground state $\phi_0(x)$ of energy W_0 and the first excited state $\phi_1(x)$ of energy W_1. Such a superposition state is referred to as a *wave packet*. Change in the position of its center of mass is the closest quantum mechanical analog of classical motion. Solution of the Schrödinger equation for the particle's wave function $\psi(x)$ yields an oscillatory motion with the oscillation period $T_{osc} = 2\pi(\hbar/\Delta W)$, where $\Delta W = W_1 - W_0$. The larger the energy separation ΔW between the two eigenstates, the faster is the particle's motion in the superposition state. This energy separation is dictated by the spatial extent of the potential confining the motion, in addition to the particle's mass. Quantum mechanics connects the rapidity of dynamics with the extension of structure.

The millielectronvolt-scale energy spacing of vibrational energy levels implies that molecular vibrations occur on a time scale of tens to hundreds of femtoseconds (Fig. 1). This defines the characteristic time scale for the motion of atoms in a molecule, including those resulting in irreversible structural changes during chemical reactions (Zewail, 2000). The motion of individual electrons in semiconductor nanostructures, molecular orbitals, and the inner shells of atoms occurs on progressively shorter intervals of time ranging from tens of femtoseconds to less than an attosecond. Motion within nuclei is predicted to unfold even faster, typically on a zeptosecond time scale.

For a wave packet made of two bound eigenstates, the oscillation period $T_{osc} = 2\pi(\hbar/\Delta W)$ also determines the response time of the particle to an external perturbation. For example, a two-level atom responds to the radiation field similar to a classical damped electron oscillator of eigenfrequency $\omega_0 = 2\pi/T_{osc}$ (Siegman, 1986; Boyd, 2003), with $T_0 = 2\pi/\omega_0$ the response time of the atom to an external perturbation. Going from two-level atoms to plasmas, we see similar behavior. The rapidity of collective motion of free electrons is dictated by the plasma oscillation period T_p, which is inversely proportional to the square root of the density of electrons N, $T_p = (2\pi/|e|)\sqrt{\varepsilon_0 m_e/N}$ (Kruer, 2003). Here e and m_e are the electron's charge and mass, respectively, and ε_0 is the dielectric constant. A characteristic length scale in a free electron gas in the absence of any static structure is the mean distance between adjacent electrons: $\ell_{mean} \approx N^{-1/3}$. For a mean distance comparable to atomic dimensions, $\ell_{mean} \approx 0.1$ nm, we obtain $T_p \approx 120$ as. This is close to the oscillation period of valence electron wave packets in bound atomic or molecular systems. Thus the motion of electrons inside atoms, in molecular orbitals, or confined in nanometer-scale potentials in semiconductors, like the collective dynamics of free electrons in high-density ionized gases, solid-density plasmas, or fusion targets, is naturally measured in attoseconds; see Fig. 2. Attosecond physics is the science of electrons in motion, both collective and individual, on atomic and molecular scales and in high-density mesoscopic systems.

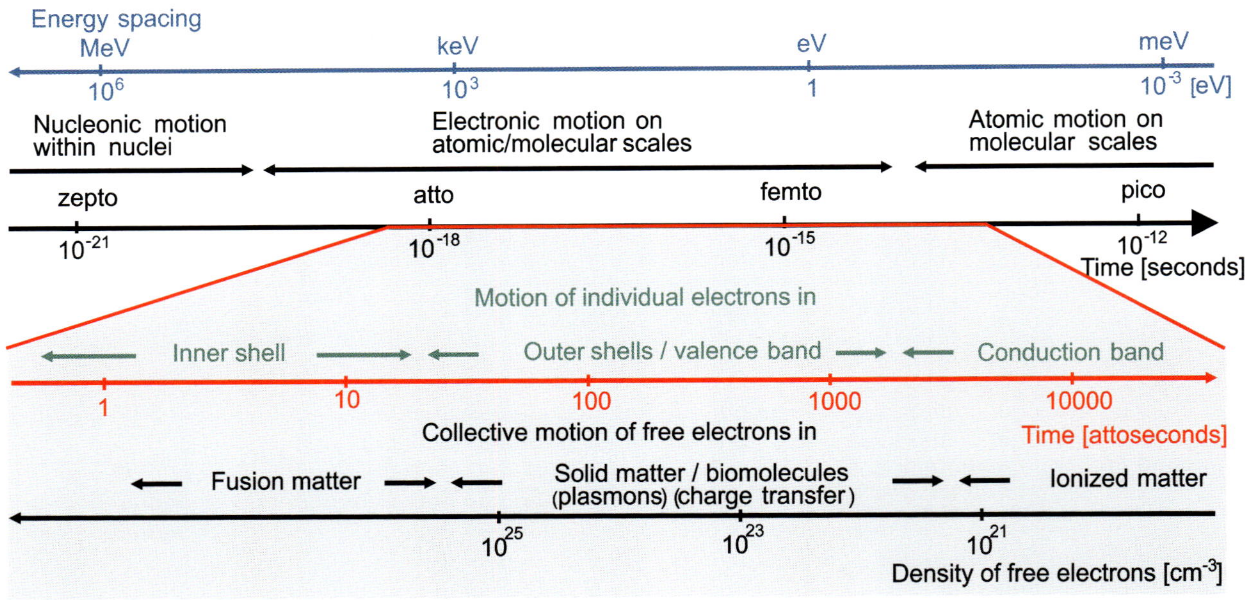

FIG. 2. (Color) Characteristic time scales for microscopic motion and its connection with energy spacing between relevant stationary states (upper panel); characteristic time scales for the motion of one or several electrons and for the collective motion of an electronic ensemble (lower panel).

B. Concepts for resolving microscopic structure and dynamics

Looking into the microcosm with high resolution in space or time requires a physical quantity with a well-controlled spatial or temporal gradient, respectively. Both types of gradients arise in waves, with the wavelength and the oscillation period determining the steepness of the spatial and temporal gradients of the oscillating fields, respectively. In waves these variations are periodic. Periodic spatial or temporal gradients can be efficiently utilized for resolving structures or dynamics (quasi)periodic on the same scale. For example, Angstrom-resolution x-ray crystallography or single-molecule imaging (Neutze *et al.*, 2000) and frequency-domain spectroscopy draw on this concept. Measurement of the frequency-domain response over a sufficiently wide range via x-ray diffraction, absorption, or scattering may, in principle, allow retrieval of the electronic structure and linear temporal response on a subatomic scale in space and time using Kramers-Kronig relationships.[4]

However, uncovering the location of an individual specimen in isolated assemblies (e.g., macromolecular structures on surfaces), or the relative timing of isolated brief events becomes increasingly challenging with increasing distance in space or time. Stationary metrology also fails when it comes to reconstructing a nonlinear response. Under these circumstances, measurements must be performed with an *isolated* temporal or spatial gradient. In general, direct space- or time-domain approaches offer transparent and intuitive measurement tools.[5] In space, the technique has become known as *microscopy*. The gradient is introduced by the radial profile of a focused light or electron beam or by a tip. In the temporal domain, a sharp gradient is introduced by a physical quantity rapidly varying in time in a controlled fashion. Its use for measuring brief time intervals and processes taking place within these intervals has been dubbed *chronoscopy*, or time-resolved metrology.

The concept of streak imaging constitutes an instructive example of using a sharp temporal gradient for chronoscopy (Fig. 3). Here the steep gradient is introduced by a voltage ramped within a fraction of a nanosecond but results in subpicosecond resolution. Temporal resolution can be generally much better than the duration Δt of the signal gradient. How much better depends on the signal-to-noise ratio S/N (Fig. 4), yielding the noise-limited resolution $\Delta t_{\text{noise}} \approx \Delta t(N/S)$. Timing jitter Δt_{jitter} also limits the temporal resolution if the measurement requires averaging: $\Delta t_{\text{resolution}} \approx \sqrt{\Delta t_{\text{noise}}^2 + \Delta t_{\text{jitter}}^2}$. The temporal resolution depends on the duration of the measuring transient, the signal-to-noise ratio, and the reproducibility of timing with respect to the process under study.

C. Revolutions in time-resolved metrology

Three radically different technologies have paved the way for advancing time-resolved metrology from the

[4]See, e.g., Abbamonte *et al.* (2004) and Wernet *et al.* (2004).

[5]By advocating direct time- (or space-) domain approaches we do not mean to discard their frequency- and momentum-domain counterparts. When linear response allows their use, they are not only valuable alternatives to direct methods, but also able to complement them to increase their precision.

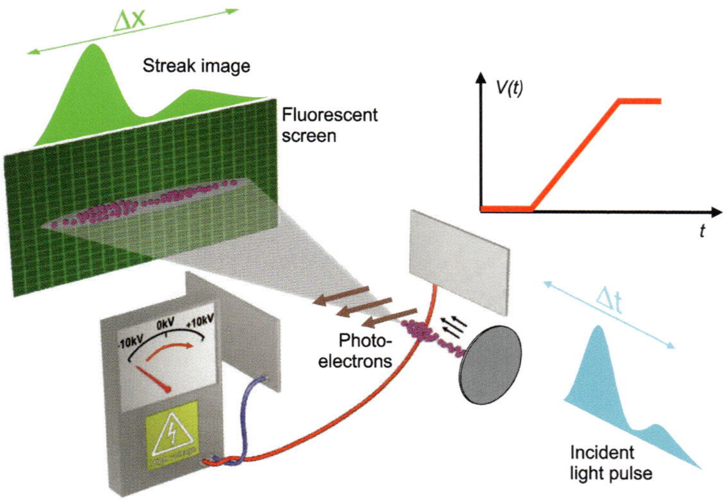

FIG. 3. (Color) Schematic of a microwave-controlled streak camera. Photoelectrons knocked off from a photocathode by a short light pulse (to be measured) are deflected on their way to a screen by a transverse microwave voltage ramp $V(t)$ to an extent depending on their instant of release (Bradley *et al.*, 1971; Schelev *et al.*, 1971). The temporally varying deflection "streaks" the location of the impact of the electrons on the screen, mapping the temporal profile of the light pulse to a spatial distribution, "streak image," of electrons on a fluorescent screen.

nanosecond to the attosecond regime: microwave electronics→picoseconds, ultrafast optics→femtoseconds, and lightwave electronics→attoseconds (Fig. 5). Technological breakthroughs triggered each of these developments: transistors for fast switching of electric current, lasers and nonlinear optics for ultrashort light pulse generation and measurement, and synthesized light waveforms for steering electrons on the atomic scale.

1. Microwave electronics

For a long time, electrical signals provided the steepest controlled gradients for measuring fast-evolving phenomena. The invention of the transistor triggered rapid evolution of high-speed electrical transients from nanoseconds toward picoseconds (Fig. 5). Their spectrum spans several octaves from radio to microwaves. However, the refractive index of any material varies substantially over this spectral range due to lattice vibrations. As a result, the distortion-free propagation length of few-picosecond electrical signals is as short as a few millimeters. This length scales quadratically with the duration of transients, preventing any substantial further improvement in the speed of practically useful electrical

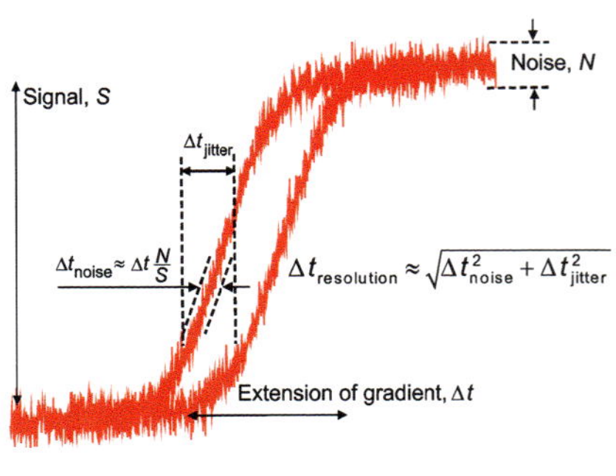

FIG. 4. (Color online) Typical reference (clocking) signal gradients (tainted with noise and jitter) for triggering or probing fast processes.

transients. Advancement of time-domain metrology beyond this frontier called for a new technology.

2. Ultrafast optics

The refractive index of transparent optical materials is nearly constant in the visible or near-infrared spectral range over relative bandwidths of several percent, creating the potential for robust femtosecond transients. This potential was exploited by coherent light amplification in a broadband laser medium and the use of nonlinear optical techniques. The resultant ultrashort laser pulses[6] approached the 1-fs barrier by the turn of the millennium. As with high-speed electronics, increasing dispersion toward the ultraviolet due to subpetahertz electronic transitions limits the fastest robust optical transients to several femtoseconds. Again, a radically new technology was required for further advance.

3. Lightwave electronics

Controlled variation of the amplitude or frequency of light pushed the frontiers of ultrafast science close the femtosecond frontier. However, light exhibits much steeper gradients than those of the cycle-averaged quantities: the subcycle slopes in the oscillating field. With the advent of controlled few-cycle light waves, a controlled force that is variable on the electronic (i.e., attosecond) time scale and rivals interatomic forces became available and opened the door for versatile attosecond control and metrology.

In nonlinear interactions, electronic motion induced by the controlled light force may exhibit even steeper gradients than the light field itself, giving way to the generation of attosecond electron and photon [extreme

[6]For recent reviews, see Brabec and Krausz (2000); Keller (2003). Early reviews include DeMaria *et al.* (1969); Bradley and New (1974); Greenhow and Schmidt (1974); Smith *et al.* (1974). Also see Akhmanov *et al.* (1992); Diels and Rudolph (1996).

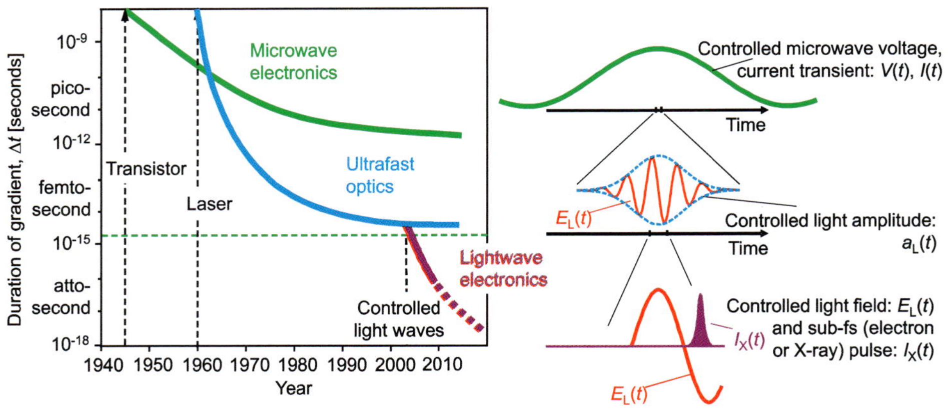

FIG. 5. (Color) Evolution of ultrafast science. The duration of the fastest controlled signal gradients dictates the speed of control and resolution of probing of microscopic processes. Microwave electronics, ultrafast optics, and lightwave electronics offer controlled transient for these purposes on successively shorter time scales, with picosecond, femtosecond, and attosecond gradients.

ultraviolet (XUV) or x-ray] pulses. The resultant pulse (see Fig. 5) is synchronized with the generating light field, allowing the sub-fs pulse to be used either as a "starter gun" (pump) for triggering ultrafast dynamics that can be referenced to the sub-fs laser field transient or as a "photofinish" (probe) taking a snapshot of microscopic motion initiated (and subsequently steered) by the few-cycle light wave or another signal derived from it. The speed of metrology and control is no longer restricted by the femtosecond pulse envelope but can utilize attosecond field oscillations (see Secs. V and VII).

By analogy with microwave electronics, this radically new technology has been dubbed *lightwave electronics* (Goulielmakis *et al.*, 2007). As microwave electronics controls electronic currents in nanoscale circuits on a picosecond time scale, lightwave electronics controls atomic-scale currents on the attosecond scale and constitutes the technological backbone of attosecond science.

As usual, this revolution was also preceded by much research, including proposals for sub-fs pulse generation[7] and measurement,[8] and the advancement of femtosecond technology and strong-field interactions, including high-order harmonic generation.[9] These advances paved the way to the birth of attosecond technology at the turn of the millennium.

[7]Hänsch, 1990; Farkas and Tóth, 1992; Harris *et al.*, 1993; Corkum *et al.*, 1994; Kaplan, 1994; Ivanov *et al.*, 1995; Kaplan and Shkolnikov, 1995; Antoine *et al.*, 1996; Protopapas *et al.*, 1996; Antoine *et al.*, 1997; Christov *et al.*, 1997; Kan *et al.*, 1997; Schafer and Kulander, 1997; de Bohan *et al.*, 1998; Harris and Sokolov, 1998; Lappas and L'Huillier, 1998; Tempea *et al.*, 1999.

[8]Constant *et al.*, 1997; Scrinzi *et al.*, 2001a, 2001b.

[9]Reviewing this early work is beyond the scope of this paper. The reader is referred to Salieres *et al.*, 1999; Joachain *et al.*, 2000; Agostini and DiMauro, 2004; Eden, 2004; Scrinzi *et al.*, 2005; Pfeifer, Spielmann, and Gerber, 2006.

D. Evolution of the study of transient phenomena

Before 1900 it was already known that short flashes of light permit the recording of rapid phenomena by spark photography. In 1864, Toepler extended the technique to study microscopic dynamics (Krehl and Engemann, 1995). He generated sound waves with a short light spark and subsequently photographed them with a second spark that was delayed electronically with respect to the first one initiating the motion. By taking pictures of the sound wave as a function of the delay time, he obtained a complete history of sound-wave phenomena. Pump-probe spectroscopy was created. Abraham and Lemoine (1899) improved the technique by deriving the pump and probe flash from the same spark with a variable optical path length between them. The resultant optical synchronism between the triggering and photographing flash opened the way to improving the resolution of pump-probe spectroscopy to the limit dictated by the flash duration.

With these milestones, the conceptual framework for studying transient microscopic phenomena was complete. Subsequent progress in time-resolved measurements was driven by developing sources of ever shorter light flashes and techniques for their measurement. The resolution of time-resolved spectroscopy was limited by the nanosecond duration of pulses of incoherent light for more than half a century (Fig. 6). Broadband lasers and nonlinear optical techniques for the generation and measurement of ultrashort laser pulses improved the resolving power of pump-probe spectroscopy from several nanoseconds to several femtoseconds, by six orders of magnitude within merely two and a half decades (Fig. 6). The creation of femtosecond technology—four decades after the first observation of intermediates of chemical reactions by Norrish and Porter (1949) and Eigen (1954)—permitted real-time observation of the breakage and formation of chemical bonds, a field triggered by Zewail (1994, 2000).

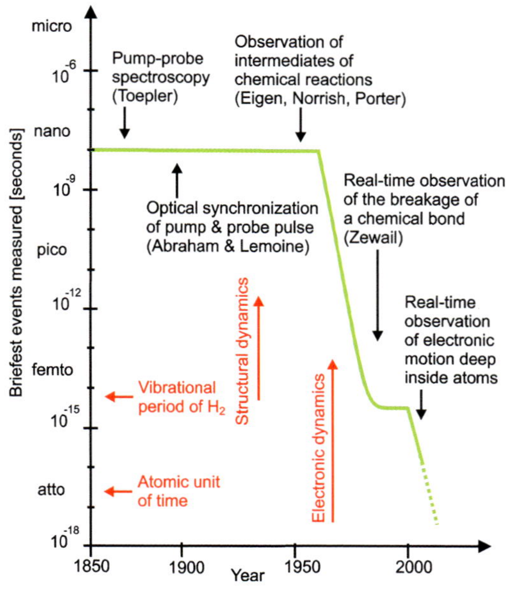

FIG. 6. (Color) Evolution of techniques for real-time observation of microscopic processes. Discontinuities in the slope of briefest measured events vs years indicate revolutions in technology.

Progress in temporal resolution was again halted in the mid-1980s, when the duration of the shortest laser pulses approached the oscillation period of the light wave carrying the pulse. Breaking the femtosecond barrier and providing thereby real-time access to intra-atomic electron dynamics came about at the turn of the millennium and permitted the observation of electronic motion deep inside (i.e., in inner shells of) atoms (Drescher *et al.*, 2002) and control of atomic-scale electron motion in real time (Baltuška, Udem, *et al.*, 2003).

II. CONCEPTS FOR ULTRAFAST MEASUREMENTS AND CONTROL

A. General connection between measurement and control

The measurement and control of ultrafast processes are inherently intertwined. In pump-probe spectroscopy, the triggering of a process must be precisely timed with respect to the observing clock: in femtosecond chemistry this capability has been exploited not only for time-resolved measurements[10] but also for control (Rice and Zhao, 2000; Brumer and Shapiro, 2003).

Any measurement process controls the appearance of a certain signal. Insight into the dynamics hinges on understanding the time-dependent changes of this signal. For instance, photoelectron spectra[11] may constitute this signal. The fact that these spectra change in a controlled and reproducible way upon varying the time delay between the pump and probe in time-resolved photoelectron spectroscopy is also a manifestation of ultrafast control over the whole process.

Attosecond technology entails the ability to control and shape not only the envelopes of the laser pulses but also the subcycle structure of the field oscillations, extending both measurement and control techniques into the new realm of light-wave engineering and lightwave electronics. Attosecond control and metrology are based on this revolution in technology, but, at the same time, they draw on concepts that have been successfully applied on longer time scales.

B. Concepts for ultrafast measurements

1. Gating, both time and frequency resolved

For sampling the evolution of the electric field $E(t)$ of a light pulse, we need a narrow temporal gate $G(t-\tau)$ with its delay τ scanned across the pulse to be measured:

$$S(\tau) \propto \int_{-\infty}^{\infty} E(t)G(t-\tau)dt. \tag{1}$$

If the gate is much shorter than the process traced, the measured signal $S(\tau)$ yields the temporal evolution of the process,[12] here $E(t)$. The fastest gate femtosecond metrology can offer is the femtosecond laser pulse itself. If we replace $G(t-\tau)$ with $E(t-\tau)$ in Eq. (1), $S(\tau)$ is the linear autocorrelation of the pulse. It allows retrieval of the pulse envelope $a_L(t)$ in Eq. (6) for transform-limited pulses only. In the more general case of chirped pulses, one has to resort to nonlinear autocorrelation. The most powerful and robust version is frequency-resolved optical gating (FROG; Trebino *et al.*, 1997), which acquires the spectrogram of the signal,

$$S(\omega,\tau) = \left| \int_{-\infty}^{\infty} E(t)G(t-\tau)e^{i\omega t}dt \right|^2. \tag{2}$$

It is the set of spectra of all gated chunks of $E(t)$ as the gate $G(t-\tau)$ is scanned across the signal by varying τ.

[10]For reviews, see Zewail, 1994; de Boeij *et al.*, 1998; Neumark, 2001; Stolow, 2003a, 2003b; Stolow *et al.*, 2004.

[11]Time-resolved photoelectron spectroscopy is key to observing dynamics in real time including structural dynamics in molecules as well as electronic dynamics on surfaces. Some early publications include Seel and Domcke, 1991; Baumert *et al.*, 1993; Assion *et al.*, 1996; Cyr and Hayden, 1996; Greenblatt *et al.*, 1997; Blanchet *et al.*, 1999; Lehr *et al.*, 1999. For reviews, see Haight, 1995; Petek and Ogawa, 1997; Neumark, 2001; Reid, 2003; Stolow, 2003a, 2003b; Stolow *et al.*, 2004; Bauer, 2005; Suzuki, 2006.

[12]Note that the pulse measurement by streak camera in Fig. 3 can also be modeled by a similar equation except that τ must be replaced with x/v_{defl}, where v_{defl} is the speed at which the electron beam's position x is scanned across the screen by the deflecting field, and instead of the field it is the cycle-averaged temporal intensity profile of the pulse, $I(t) \propto |E(t)|^2$, that is sampled.

The gate can generally be complex or even purely imaginary (phase gate). Knowledge of the spectrogram allows one to reconstruct both the gate $G(t)$ (Trebino, 2000) and the signal's complex pulse envelope $a_L(t)$; see Eq. (6).

2. Spectral interferometry

The temporal evolution of a signal can be inferred from measuring the amplitudes and relative phases of their spectral components. The latter can be accessed by converting spectral phase variations into modulation of signal amplitude. Spectral phase interferometry for direct electric field reconstruction (SPIDER; Iaconis and Walmsley, 1998) implements this concept. Let $E(\omega) = a(\omega)\exp[i\phi(\omega)]$ be the complex Fourier transform of $E(t)$ with the real spectral amplitude $a(\omega)$ and phase $\phi(\omega)$. We generate a second pulse—a "replica" $E_{\text{replica}}(t)$ of the first—delayed in time by τ and shifted (sheared) in frequency by a small amount δ. Its Fourier transform will be

$$a_{\text{replica}}(\omega)\exp[i\phi_{\text{replica}}(\omega)] = a(\omega+\delta)\exp[i\phi(\omega+\delta) - i\omega\tau]. \tag{3}$$

Spectrally resolved interference of the two pulses at the detector yields the SPIDER interferogram

$$\begin{aligned} S_{\text{SPIDER}}(\omega) &\propto |a(\omega)\exp[i\phi(\omega)] \\ &\quad + a_{\text{replica}}(\omega)\exp[i\phi_{\text{replica}}(\omega)]|^2 \\ &= |a(\omega)\exp[i\phi(\omega)] + a(\omega+\delta) \\ &\quad \times \exp[i\phi(\omega+\delta) - i\omega\tau]|^2 \\ &= a^2(\omega) + a^2(\omega+\delta) + 2a(\omega)a(\omega+\delta) \\ &\quad \times \cos\{\omega\tau - [\phi(\omega+\delta) - \phi(\omega)]\}. \end{aligned} \tag{4}$$

It allows one to find the spectral phase $\phi(\omega)$ from the variation of the phase difference $\phi(\omega+\delta)-\phi(\omega)$ across the spectrum: the interferogram maps this phase variation onto modulation of the amplitude. Since $a(\omega)$ can be measured independently, the complex envelope $a_L(t)$ of the pulse [see Eq. (6)], can be completely reconstructed. SPIDER also relies on a well-controlled temporal gradient, to impose the frequency shear. One can think of propagating the original pulse and its time-delayed replica in a medium which shifts the photon energy of the delayed pulse in a controlled way. This implies that some property of the medium must have been changed within the delay interval, requiring the presence of a controlled temporal gradient. By increasing the separation between the two pulses, the steepness of this gradient can be lowered at the expense of increased demands on the spectrometer's resolving power. Thus we conclude that the use of spectral interferometry does not obviate the need for a gradient in time, but may substantially relax the requirements on its steepness. The price to pay is the need for high spectral resolution and shot-to-shot timing stability (in averaging measurements).

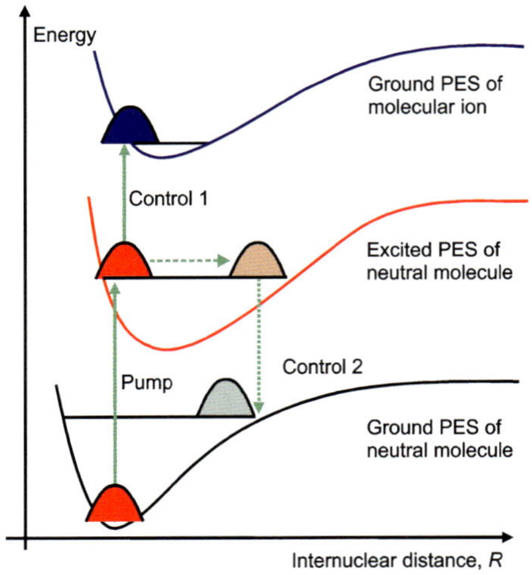

FIG. 7. (Color) Nuclear wave-packet control in a molecule with short pulses; PES, potential energy surface. A pump pulse initiates vibrational motion by exciting the molecule into an excited electronic state. Depending on its time delay with respect to the pump, a control pulse catches the vibrational wave packet at different parts of its trajectory. Control 1 promotes it up to the ground state of the molecular ion. At a different delay (control 2) the molecule will be electronically deexcited, dumping the wave packet back to an excited vibrational manifold of the ground electronic surface.

C. Evolution of ultrafast control

By analyzing ultrafast measurements, we have seen how time- and frequency-domain approaches complement each other. A similar symbiosis comes to light in the evolution of approaches to controlling microscopic dynamics. Here we show that early control approaches may be regarded as two limits of a general approach, which converge on the attosecond time scale.

1. Early approaches

The connection between time-domain measurement and control becomes transparent when considering the time-domain wave-packet control scenario in Fig. 7 (Rice and Zhao, 2000). A pump pulse promotes the molecule into an excited electronic state, initiating nuclear motion. A properly timed control pulse catches the nuclear wave packet at the right place on its trajectory along the excited electronic potential energy surface (PES) to induce a transition toward the desired outcome.

Timing between oscillations of two narrowband optical fields was shown to be an alternative approach to controlling dynamics (Brumer and Shapiro, 2003). In Fig. 8 we show an example of controlling the transition between an initial state $|i\rangle$ and a final state $|f\rangle$ with the superposition of the fields $E_\omega(t)=E_1\exp(-i\omega_L t)+\text{c.c.}$ (inducing a three-photon transition) and $E_\Omega(t) = E_3\exp(-i\Omega t+i\phi)+\text{c.c.}$ of frequency $\Omega=3\omega_L$ (inducing a one-photon transition). Our control knob is the relative phase ϕ between the two colors. In the lowest order

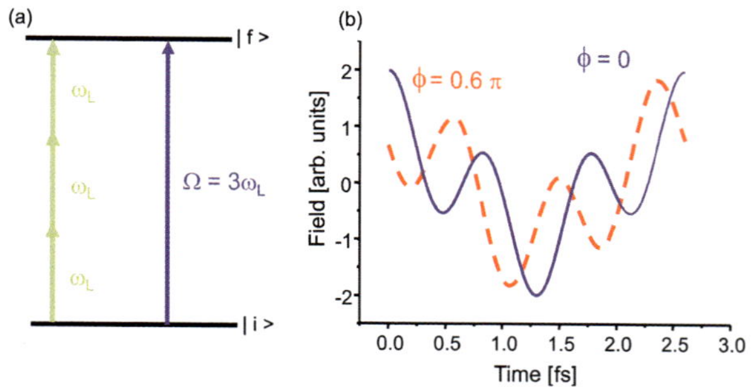

FIG. 8. (Color) Coherent control of quantum transition by changing the relative phase between two waves of different frequency. On the left panel, transition between initial and final states can proceed via two pathways, involving the absorption of one and three photons. Its probability is controlled by the relative phase ϕ between the two colors, ω_L and $3\omega_L$. The right panel shows that changing the relative phase ϕ results in shaping of the electric field oscillations on the subcycle, i.e., sub-fs, time scale, in this example for λ_L=800 nm. Solid line, ϕ=0; dashed line, ϕ=0.6π. Although timing is varied on a sub-fs scale the controlled transition usually unfolds over a much longer period.

of perturbation theory, the total transition probability is given by

$$W_{if} \propto |d^{(3)}|^2 E_1^{\ 6} + |d^{(1)}|^2 E_3^{\ 2} + 2|d^{(3)}d^{(1)}| E_1^{\ 3} E_3 \cos(\phi + \theta_{if}), \tag{5}$$

where $d^{(1)}$ and $d^{(3)}$ are the corresponding transition matrix elements and θ_{if} is the relative phase between them.[13] Changing ϕ controls the transition probability, shifting the outcome between the minima and maxima of the interference pattern. Figure 8(b) shows the total electric field $E_\Sigma(t) = E_\omega(t) + E_\Omega(t)$ for $E_1 = E_3$ and for two values of ϕ. Although the approach was conceived in the framework of quantum physics, its functioning can be traced directly to the controlled subcycle modulation of the classical field oscillations in Fig. 8(b) (Franco and Brumer, 2006). Thus the simplest manifestation of lightwave engineering, which uses only two colors, already appears here—even though the evolution of the controlled system towards its final state occurs only on a multicycle (multi-fs) scale due to the perturbative nature of the interaction.

2. Advanced control via pulse shaping

The two control scenarios described above are profoundly connected. The nuclear wave packet created in Fig. 7 is a coherent superposition of many states $|v\rangle$. The transform-limited pump pulse populates these states in phase, while the control pulse generates interference of multiple pathways, the outcome of which depends on

[13]Note that this result is based on the perturbation theory and is not applicable at high light intensities or for transitions between nonstationary (dressed) states, where the phases θ become time dependent.

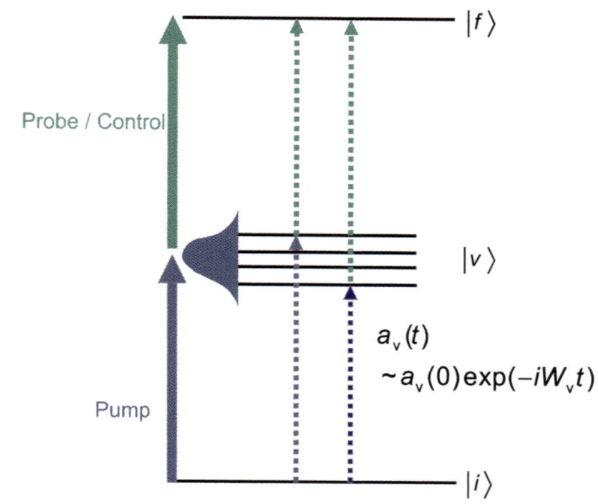

FIG. 9. (Color) Frequency-domain interpretation of wave-packet control with short pulses. Multiple pathways via various intermediate states interfere, with their interference controlled by the relative phases determined by the free wave-packet evolution.

the relative phases between the (complex) amplitudes $a_v(t) = a_v(0)\exp(-iW_v t)$. These phases are controlled via the time delay between the pump and control pulses (Fig. 9).

We do not have to rely on the free evolution of the wave packet to adjust the phases of constituent states for the desired interference in the final state. The same goal can be achieved by changing the relative phases of those colors that populate the intermediate states or complete the transition to the final state. Moreover, using strong laser fields we can also reshape laser-dressed potential energy surfaces[14] and thus alter the wave packet's temporal evolution.

State-of-the-art pulse-shaping technology (Weiner *et al.*, 1988; Weiner, 1995) allows flexible changes of phase, amplitude, and even polarization (Brixner and Gerber, 2001; Brixner *et al.*, 2004) of different spectral components that make up a femtosecond pulse. In the time domain, it controls the evolution of cycle-averaged quantities such as the amplitude vector and the instantaneous frequency of the field oscillations, on a time scale of several tens of femtoseconds. Key ingredients of this technology are feedback loops and learning algo-

[14]The idea of Born-Oppenheimer potential energy surfaces (PESs) shaped by a laser field is quite general. For the effects of resonant fields in creating "light-induced potential energy surfaces" see Bandrauk *et al.* (1994) for an overview. Various applications of this concept for molecular control (Chang *et al.*, 2003; Sola, 2004) include laser catalysis of reactive scattering (Shapiro and Zeiri, 1986). For time-dependent Born-Oppenheimer PESs shaped by nonresonant fields see Ivanov *et al.*, 1996; Corkum *et al.*, 1997; Niikura, Corkum, and Villeneuve, 2003; Sussman *et al.*, 2005, 2006.

rithms, allowing efficient optimization toward the desired outcome.[15]

D. Principles of attosecond control and measurement of electronic dynamics

In the quest for attosecond-scale control and metrology of electron dynamics, the fs-scale gradient of cycle-averaged quantities such as the amplitude and frequency of visible light could be steepened by using shorter-wavelength light. In practice, both the efficiency of coherent light generation and, not independently, the polarizability of matter diminish dramatically toward the XUV and x-ray regimes. With complex, large-scale infrastructures these limitations may be overcome, but a versatile, laboratory-scale technology for implementing control and metrology on the attosecond time scale requires a radically new approach. In our opinion, it will rely on controlled subcycle, i.e., subfemtosecond, variation of strong laser fields in the visible and nearby spectral ranges.

Subcycle light waveform control requires bandwidths of the order of an octave or more, implying "single"-cycle pulses in the Fourier limit (see Sec. III.A). Extending the bandwidth to several octaves will permit on-demand light waveform synthesis (see Sec. III.B), resulting in electric fields with strengths varying from zero to several volts per angstrom within less than 100 as. The resultant tailored force will permit control of electron motion on the atomic scale. We envision that light-wave engineering and lightwave electronics will open a new approach to controlling and exploring the microcosm: rather than controlling transitions between electronic quantum states of definite energy by such cycle-averaged quantities as the amplitude and frequency of light, electron wave packets will be steered with attosecond light forces.

III. INTENSE, CONTROLLED LIGHT FIELDS

Microwave electronics has allowed the temporal evolution of electric current and voltage to be controlled up to microwave frequencies of tens of gigahertz, creating the basis for high-speed data processing and short-distance communication.[16] Ultrafast optics has permitted controlled modulation of light up to terahertz frequencies, revolutionizing long-distance telecommunication as well as exploration and control of the motion of *atoms* in microscopic systems. Nonlinear optics and ultrafast optics allow one to convert controlled terahertz amplitude modulation of light into controlled terahertz-frequency electric fields, by means of optical rectification of femtosecond light pulses.[17] However, even these fastest electric fields synthesized are far too slow for controlling and tracing the motion of *electrons* on atomic scales, which requires an electric force variable at petahertz frequencies. Petahertz-frequency (sub-PHz to few PHz) electric fields are present in near-infrared (NIR), visible (VIS), and ultraviolet (UV) radiation (henceforth, briefly, light). Here we review the recent emergence of light waves with controlled evolution of the field (Sec. III.A) and discuss future prospects for scaling the technology of light field control to bandwidths far beyond an octave (Sec. III.B) and field strengths in the relativistic regime (Sec. III.C).

A. Light-wave control: Brief history and current state of the art

Consider the electric field of a linearly polarized light pulse. For pulses with bandwidths below one octave the light field $E_L(t)$ can be decomposed into a carrier wave and an amplitude envelope,

$$E_L(t) = a_L(t)e^{-i(\omega_L t+\varphi)} + \text{c.c.}, \tag{6}$$

in a physically meaningful manner (Brabec and Krausz, 1997).[18] Here $a_L(t)=|a_L(t)|e^{-i\beta_L(t)}$ is the complex envelope, $|a_L(t)|$ is the time-dependent (real) field amplitude, and $\beta_L(t)$ stands for a possible sweep (chirp) of the carrier frequency ω_L. Last but not least, φ determines the timing of the field oscillations with respect to the pulse peak, $\Delta t_{\text{peak}}=\varphi/\omega_L$ (see Fig. 10) and has been referred to as the carrier-envelope (CE) phase.

In relatively narrowband ($\Delta\omega/\omega_L \ll 1$) pulses, a change of the CE phase has no measurable physical consequence. Measurement of the shape of a light wave and its control (based on this measurement) rely on using pulses with bandwidth $\Delta\omega$ constituting substantial fraction of ω_L. In the absence of spectral phase variation, such pulses comprise merely a few oscillation cycles of the electromagnetic field, henceforth referred to as few-cycle pulses. A change in the CE phase changes significantly their temporal evolution (Fig. 10). In turn, this change affects the atomic-scale motion of electrons driven by these fields, which can be utilized for measuring the CE phase[19] (see Sec. IV). Along with the measurement of the complex envelope $a_L(t)$ (see Sec. II.B) it determines the waveform, opening a route to light waveform control, outlined in Fig. 11(a). With φ known, the desired waveform can be realized by passing the pulse

[15]The work of Judson and Rabitz (1992) led to a series of beautiful experiments including Assion *et al.*, 1998, 2001; Bartels *et al.*, 2000, 2004; Brixner, Damrauer, and Gerber, 2001; Brixner, Damrauer, Niklaus, and Gerber, 2001; Levis *et al.*, 2001; Brixner *et al.*, 2003; Daniel *et al.*, 2003; Wohlleben *et al.*, 2003, 2004, 2005; Vogt *et al.*, 2005.

[16]For a recent review, see Pozar (2004).

[17]For a recent review, see Xu and Zhang (2006).

[18]Note that we changed the sign convention in defining φ as compared with that used by Brabec and Krausz (2000) to conform with the more customary convention of Eq. (9).

[19]Strong-field CE phase effects have been predicted by Haljan *et al.*, 1997; Cormier and Lambropoulos, 1998; de Bohan *et al.*, 1998; Tempea *et al.*, 1999; Dietrich *et al.*, 2000; Nakajima and Watanabe, 2006.

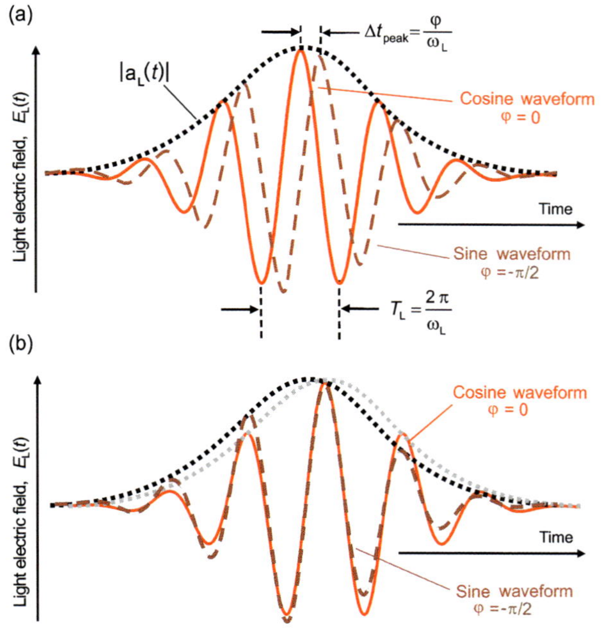

FIG. 10. (Color) Laser pulses containing two oscillation cycles within the full width at half maximum of their intensity profile, with sine- and cosine-shaped waveforms, with their (a) amplitude envelopes and (b) field oscillations coinciding.

through a thin optical medium of length L, shifting the CE phase by

$$\Delta\varphi(L) = 2\pi L(\partial n/\partial\lambda)_{\lambda_L}, \tag{7}$$

where λ_L is the carrier wavelength. The dephasing length which offsets the CE phase by π, $L_{\text{deph}} = 1/2|\partial n/\partial\lambda|^{-1}_{\lambda_L}$, amounts in transparent optical materials to tens of microns, permitting CE phase adjustments by a small translation of thin prisms (Xu *et al.*, 1996).

The above approach to light waveform control requires a single-shot measurement of φ. Stabilizing the frequency comb of femtosecond laser oscillators (Hänsch, 1997; Reichert *et al.*, 1999; Telle *et al.*, 1999) obviated the need for single-shot phase measurement by providing pulse trains with stabilized pulse-to-pulse CE phase shift.[20] Selection and amplification of pulses with constant (but unknown) φ from the pulse train yielded intense few-cycle pulses with stabilized and adjustable CE phase.[21] With these pulses in place, measurement of φ can be performed over many pulses (via signal averaging) to produce intense few-cycle pulses with controlled waveform [see Fig. 11(b)]. Figure 12 shows the first controlled and completely characterized light wave. Lightwave engineering is now approaching the single-cycle limit (Cavalieri, Goulielmakis, *et al.*, 2007; Goulielmakis, Schultze, *et al.*, 2008) and being extended to UV (Shverdin *et al.*, 2005) and IR frequencies (Fuji *et al.*, 2006; Manzoni *et al.*, 2006; Hauri *et al.*, 2007).

B. Light waveform synthesis

For atomic-scale electronic motion control we need to control the steering force on a sub-fs time scale. So far, the room for subcycle tailoring of the light waveform has been modest. Indeed, comparison of the cosine- and sine-shaped few-cycle waves in Fig. 10(b) shows merely a few percent difference in the electric field. More flexibility in subcycle shaping of the field requires a spectrum stretching substantially beyond an octave.

Self-phase-modulation (SPM) (Alfano and Shapiro, 1970) and cascaded stimulated Raman scattering (Kaplan, 1994; Harris and Sokolov, 1997; Sokolov *et al.*, 2001, 2005) of intense laser pulses proved most efficient for broadening the spectrum of an optical pulse.[22] In a gas-filled, hollow-core fiber (Nishioka *et al.*, 1995; Nisoli *et al.*, 1996, 1997; Sartania *et al.*, 1997; Oishi *et al.*, 2005; Suda *et al.*, 2005) or in a self-induced plasma wave guide (Hauri *et al.*, 2004; Couairon *et al.*, 2006; Guandalini *et al.*, 2006; Chen *et al.*, 2007), SPM permits efficient spectral broadening of powerful femtosecond pulses. These can be compressed to the few-cycle regime using dispersion control with multilayer chirped mirrors (Szipöcs *et al.*, 1994; Sting *et al.*, 1995; Jung *et al.*, 1997; Matuschek *et al.*, 1998; Tempea *et al.*, 1998). Recently, the latter approach has led to a supercontinuum stretching from <250 nm to >1000 nm, using 5-fs input pulses (Goulielmakis, Koehler, *et al.*, 2008); see Fig. 13(a). Phase modulation with twin pulses may result in further progress (Yamashita *et al.*, 2006; Matsubara *et al.*, 2007). We believe that covering the wavelength range of ~0.12–1.2 μm (~1–10 eV) with a supercontinuum generated by an intense few-fs pulse is within reach.

Control of the amplitude and phase of the generated spectral components can be realized by separating them in space (Weiner, 1995) and addressing the amplitudes and phases of the individual channels independently. Extending this control over several octaves is a challenge. Figure 13(b) outlines a possible implementation based on the work of Binhammer *et al.* (2005, 2006) using a prism-based system. Another route drawing on several discrete spectral channels separated by dichroic chirped mirrors is also being pursued (Goulielmakis *et al.*, 2007). Figure 14 shows several possible waveforms and demonstrates that superimposing waves of frequencies spanning several octaves permits efficient attosecond shaping of light waveforms. Attosecond sampling of the generated waveform (Sec. V) will allow its measurement and completes the toolbox of lightwave engineering.

[20] Apolonski *et al.*, 2000; Holzwarth *et al.*, 2000; Jones *et al.*, 2000; Kakehata *et al.*, 2001; Ye and Cundiff, 2005; Li *et al.*, 2007; Yu *et al.*, 2007. For detailed discussion, see Cundiff, 2002; Udem *et al.*, 2002; Cundiff and Ye, 2003.

[21] Baltuška, Udem, *et al.*, 2003; Baltuška, Uiberacker, *et al.*, 2003; Kakehata *et al.*, 2004, 2006; O'Keeffe *et al.*, 2004; Lee *et al.*, 2005; Gagnon *et al.*, 2006; Hong *et al.*, 2006; Li *et al.*, 2006; Imran *et al.*, 2007; Mashiko *et al.*, 2007; Wu *et al.*, 2007.

[22] For a review, see Zheltikov (2006a).

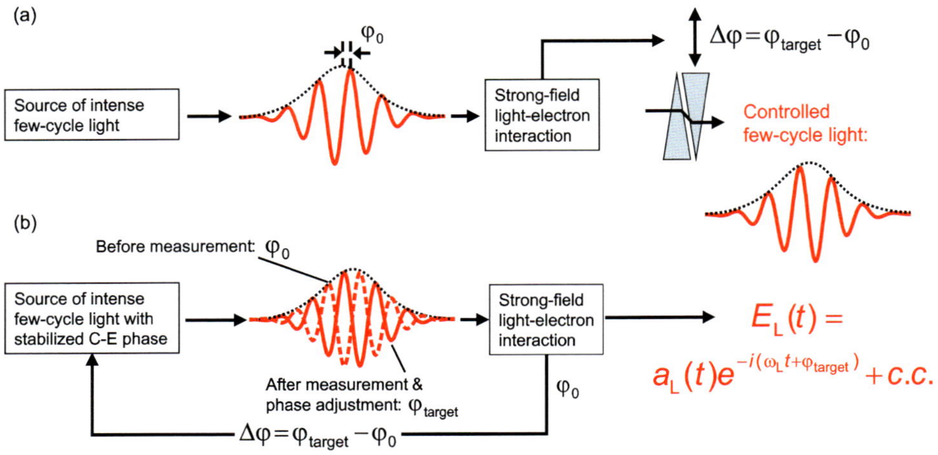

FIG. 11. (Color) Control of the waveform of few-cycle laser pulses relies on measurement of the CE phase by a strong-field light-electron interaction. (a) Once the measurement is performed (yielding φ_0), the required CE phase φ_{target} can be set by imposing an external phase shift $\Delta\varphi=\varphi_{target}-\varphi_0$, leading to the desired waveform. (b) If the ultrashort-pulse source delivers pulses with a stabilized CE phase, its adjustment to the required value can be performed inside the source via an electronic feedback loop, obviating the need for fast phase measurement.

C. Towards ultrastrong controlled light fields

Today, light waveform control has been realized at power levels where the electric field—upon suitable focusing—can rival the atomic Coulomb field. State-of-the-art attosecond technology draws its tools from the interaction of light fields with electrons, either bound or of low positive energy, at these fields strengths. There exist several proposals for extending lightwave electronics to relativistic interactions, holding promise of enriching attosecond science with high-flux, high-energy electron and photon pulses of subfemtosecond duration (see Sec. VI). A concept[23] at Max Planck Institute of Quantum Optics draws on three well-established technologies: (i) chirped-pulse amplification (CPA),[24] for providing subpicosecond to several-picosecond pulses with high energy; (ii) noncollinear optical parametric amplification (NOPA) over a near-octave-spanning bandwidth,[25], which was recently scaled to multiterawatt power levels;[26] and (iii) chirped multilayer mirrors (Szipöcs *et al.*, 1994; Steinmeyer *et al.*, 1999) for ultrawideband, high-throughput dispersion control scalable for handling ultrahigh powers. The emerging source dubbed Petawatt Field Synthesizer may extend attosecond electron steering into the highly relativistic regime.

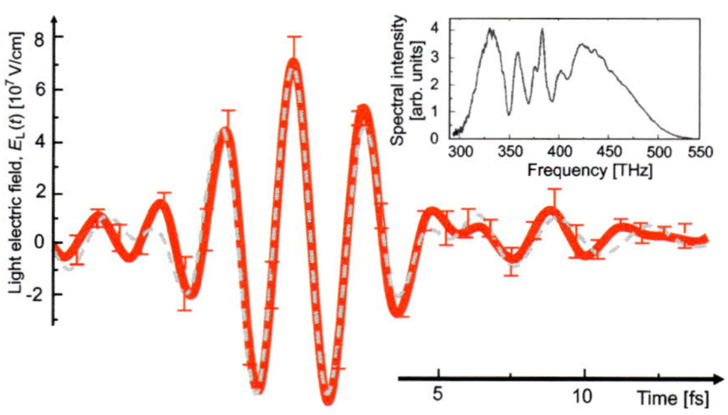

FIG. 12. (Color) Electric field (solid line) of the first completely characterized light wave (Goulielmakis *et al.*, 2004). The measurement has been performed with an attosecond streak camera, described in Sec. V, and yields the temporal evolution, direction, and strength of the field. The pulses are carried at a wavelength of $\lambda_L=750$ nm with a duration full width at half maximum of 4.3 fs. Dashed line: electric field calculated from the measured spectrum (inset) by Fourier transform, assuming that all spectral components are in phase. Adapted from Goulielmakis *et al.*, 2004.

IV. LIGHTWAVE ELECTRONICS

We use the term lightwave electronics to refer to a microscopically[27] significant response of electrons to light oscillations on the time scale of the laser cycle and below. Our discussion begins by assessing the conditions for the appearance of such a response. As shown, it oc-

[23]Proposed by Karsch (2004), for more details, see www.attoworld.de/PFS.html.

[24]Invented by Strickland and Mourou (1985), for a recent review, see Mourou, Tajima, and Bulanov (2006).

[25]Takeuchi and Kobayashi, 1994; Gale *et al.*, 1995; Cerullo *et al.*, 1997, 1998, 1999; Wilhelm *et al.*, 1997; Shirakawa and Kobayashi, 1998; Shirakawa *et al.*, 1998, 1999; Riedle *et al.*, 2000; Kobayashi *et al.*, 2001; Zavelani-Rossi *et al.*, 2001; Baltuška *et al.*, 2002a, 2002b, 2002c; Dubietis *et al.*, 2006.

[26]Tavella *et al.*, 2006, 2007; Witte *et al.*, 2006; Renault *et al.*, 2007.

[27]This attribute is meant to stress that no light propagation over a *macroscopic* distance, which is sensitive to even miniscule changes in the refractive index, is needed to "feel" this response.

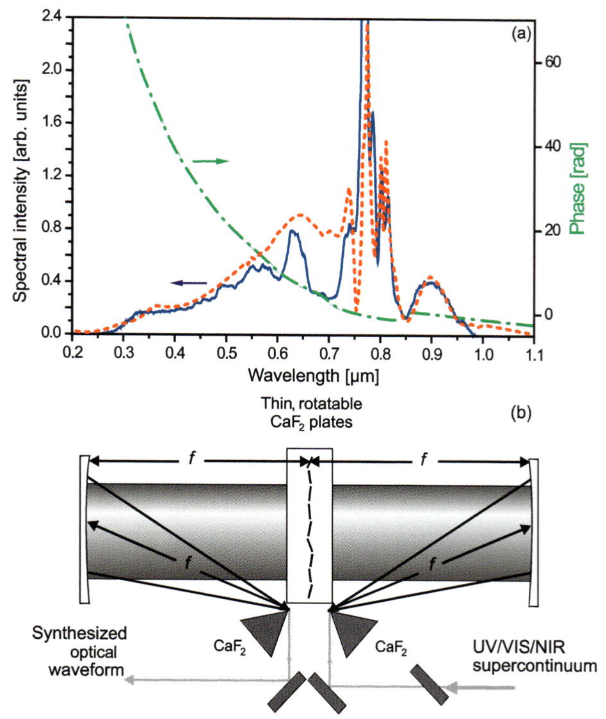

FIG. 13. (Color) Toward ultrawideband optical waveform synthesis. (a) Spectral supercontinuum generated by self-phase modulation of 5-fs, 0.2-mJ, 750-nm laser pulses in a self-induced plasma channel created in a cell filled with helium at a pressure of 25 bars (solid curve). Dotted curve, dashed-dotted curve, spectral intensity and phase, respectively, obtained from numerical simulation. Adapted from Goulielmakis, Koehler, *et al.*, 2008. (b) Possible implementation of arbitrary light waveform synthesis by adjusting the phase and amplitude of spectral components independently in separate channels. The UV-transparent CaF_2 prisms separate the spectral components of the incident continuum, the amplitude and phase of which are then controlled in the Fourier plane of a $4f$ configuration: rotation of thin CaF_2 plates will result in tuning the phases whereas translation of the plates coated with a thin metal layer of variable thickness will permit amplitude control. Courtesy of E. Goulielmakis.

curs naturally in few-cycle strong-field interactions and plays a key role in developing attosecond technology.

Below, we deal with the atomic response to light fields. Therefore we use the atomic (Hartree) system of units. The corresponding expressions are obtained from the usual CGS units by setting the electron mass m_e, the electron charge $|e|$, and the Planck constant $\hbar$ equal to unity. In these units, the natural scales derive from the physics of the hydrogen atom. The unit of energy is 27.21 eV, twice the ionization potential of hydrogen. The unit of time is 24.2 as, $1/2\pi$ times the electron orbit time around the nucleus in the first Bohr orbit. The unit of length is the radius of this orbit, equal to 0.529 Å. The unit of the electric field strength is that of the nucleus experienced by the electron in this orbit, 5.14×10^9 V/cm. A linearly polarized laser field with such electric field amplitude has the intensity of 3.5×10^{16} W/cm^2. The speed of light in these units is $c = 137$.

A. Significant variations of atomic electron density within the wave cycle of light

In atoms and small molecules transitions between ground and excited electronic states typically lie in the UV-VUV range. As discussed in Sec. I.A, this implies a near-instantaneous response of the electronic system to NIR and VIS light fields. The scale of the corresponding distortions to the electron density can be estimated from electronic polarizabilities, with the static polarizability α_{st} offering a good estimate for NIR-VIS light. The dipole moment induced by the nonresonant electric field $E_0 \cos \omega_L t$ is $d \approx \alpha_{st} E_0 \cos \omega_L t$. At the same time, $d \approx N_e \Delta R$, where N_e is the number of active electrons (typically electrons in the valence shell) and ΔR is the characteristic displacement per electron. These two equations yield $\Delta R \approx \alpha_{st} E_0 / N_e$. Given that polarizabilities typically do not exceed a few atomic units per participating electron, we see that microscopically significant distortion of the electron density can be induced only with light field strengths approaching typical interatomic field strengths. This implies the onset of ionization, which will dominate modification of the electronic cloud. Then, significant changes within one wave cycle imply ionization rates comparable to the light frequency. This is feasible only with a few-cycle laser pulse (Spielmann *et al.*, 1998); otherwise, ionization will saturate before the desired intensity is reached (Lambropoulos, 1985). The unprecedented ionization rate achievable with few-cycle light benefits the efficient generation of coherent XUV and THz radiation (Gildenburg and Vvedenskii, 2007) likewise.

Thus a microscopically significant change in the electron density of atomic systems within the wave cycle of visible light requires few-cycle light fields. The resultant atomic-scale electronic motion unfolds on the attosecond time scale, is highly sensitive to the subcycle evolution of the light field, and lies at the heart of lightwave electronics. We now examine this motion more closely.

In the dipole approximation, for a linearly polarized laser field the interaction potential energy for an electron at position $\mathbf{r} = (x, y, z)$ is (in atomic units with $|e| = m_e = \hbar = 1$)

$$V(\mathbf{r}, t) = V_{\text{atom}}(\mathbf{r}) + x E_L(t), \tag{8}$$

where we assume that the laser electric field

$$E_L(t) = E_0 f(t) \cos(\omega_L t + \varphi) \tag{9}$$

is polarized along the x axis and positive when pointing in the direction of increasing x coordinate, see Fig. 15. Here $f(t)$ is a real-valued envelope function, peaked at $t=0$, $f(0)=1$. A snapshot of the total potential at the peak of the field oscillation, $\cos(\omega_L t + \varphi) = -1$, along the direction x of laser polarization, is shown in the elliptical inset of Fig. 15. The strong electric field suppresses the barrier that keeps electrons bound, allowing them to

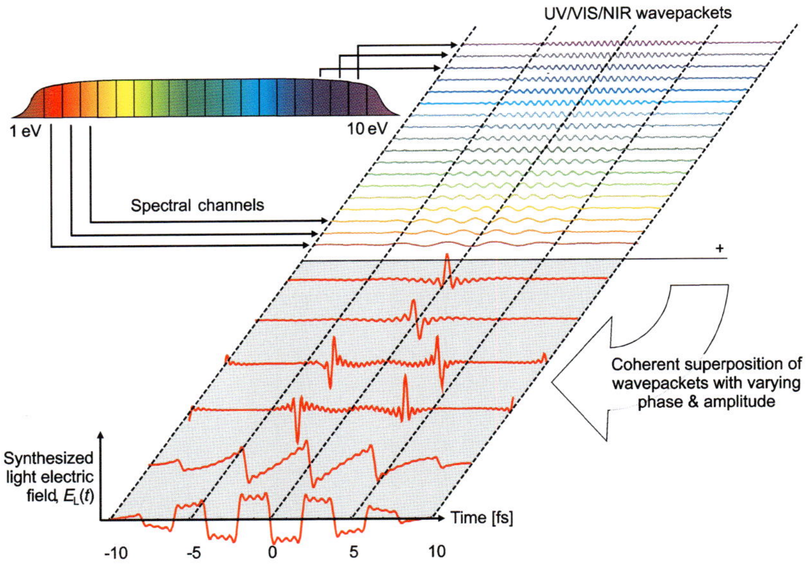

FIG. 14. (Color) Representative optical waveforms (lower panel) that have been numerically synthesized from coherent superposition of quasimonochromatic wave packets (shown on the right-hand side of the upper panel) delivered in $N=18$ spectral channels of the supercontinuum stretching over 1–10 eV (left-hand side of the upper panel). Courtesy of E. Goulielmakis.

tunnel out each time the oscillating electric field passes its maxima, twice every laser cycle.[28] As a result, the duration of the electron wave packet which emerges on the other side of the barrier is a small fraction of the half oscillation of the driving field, less than 1 fs in a NIR field.[29]

Once the electron appears at the tunnel exit, it is accelerated away from the parent ion until the field reverses its direction in about a quarter cycle. At field strengths of 10^9 V/cm, the total spatial excursion of the electron driven by a NIR laser field reaches several tens of angstroms, greatly exceeding the size of an atom or a small molecule. If the electron is freed with substantial probability within a single half cycle in a few-cycle field, such an excursion causes a dramatic change in the microscopic electron density distribution.

The electron's motion during these oscillations depends on the phase of the electric field $E_L(t)$ at which ionization has occurred, $\phi_0=\omega_L t_0+\varphi$. If the electron is returned, which may happen for the phases ϕ_0 after the peak of the oscillating field, it may recollide with the parent ion; see the inset in Fig. 15. Upon return, the electron wave packet may interact with other electrons left behind in their bound states. Many interactions are confined to a fraction of the optical cycle, giving rise to a wealth of subfemtosecond phenomena. Figure 15 summarizes the most important ones, which will be discussed.

B. Semiclassical modeling of strong field-matter interaction

Our previous discussion of intense laser-atom interaction relied on the semiclassical model introduced by Corkum (1993), Kulander *et al.* (1993), and Schafer *et al.* (1993).[30] It is based on the single active electron (SAE) approximation,[31] a key ingredient in our understanding of the interaction between an intense infrared laser field and an atom or a small molecule. This approximation assumes that only one electron actively participates in the ionization of the atom or molecule by the low-frequency laser field. Other electrons are involved only in the screening of the nucleus, creating an effective single-electron potential. The SAE approximation is justified when the energies of multielectron excitations are large compared to the laser frequency and significantly exceed those of single-electron excitations.[32]

The first building block of the semiclassical model is tunnel ionization or, more generally, optical field ionization, often modeled using the Keldysh theory or its refined versions.[33] Once the electronic wave packet emerges on the other side of the tunneling barrier at x_0 (see inset in Fig. 15), it is pulled far away within a small

[28]Recently, this process was observed in real time; see Uiberacker *et al.*, 2007, and Fig. 49 in Sec. VII.

[29]A tutorial discussion of this process can be found in Ivanov *et al.* (2005).

[30]This model had an important precursor: the so-called "atomic antenna" model (Kuchiev, 1987). Relevant early work also includes Corkum *et al.*, 1989; Krause *et al.*, 1992.

[31]Detailed discussion can be found in Kulander *et al.*, 1993; see also Schafer and Kulander, 1990. Later important work includes Muller and Kooiman, 1998.

[32]These conditions are met by noble gases, where the SAE approximation has been extremely successful. For alkaline-metal atoms, nonlocal modifications of the single-electron potential are already required but the SAE approximation still performs well (Sheehy *et al.*, 1999; Gaarde *et al.*, 2000). In small molecules, the SAE approximation is still applicable, but in polyatomic systems, especially when charge-transfer excitations are located within a few electron-volts from the ground electronic state, this approximation fails (see Lezius *et al.*, 2001; Markevitch *et al.*, 2003).

[33]For reviews, see Delone and Krainov, 1985; Faisal, 1987; Reiss, 1992; Becker and Faisal, 2005. For earlier references, see Keldysh, 1964; Perelomov *et al.*, 1966a, 1966b; Perelomov and Popov, 1967; Faisal, 1973; Reiss, 1980a, 1980b.

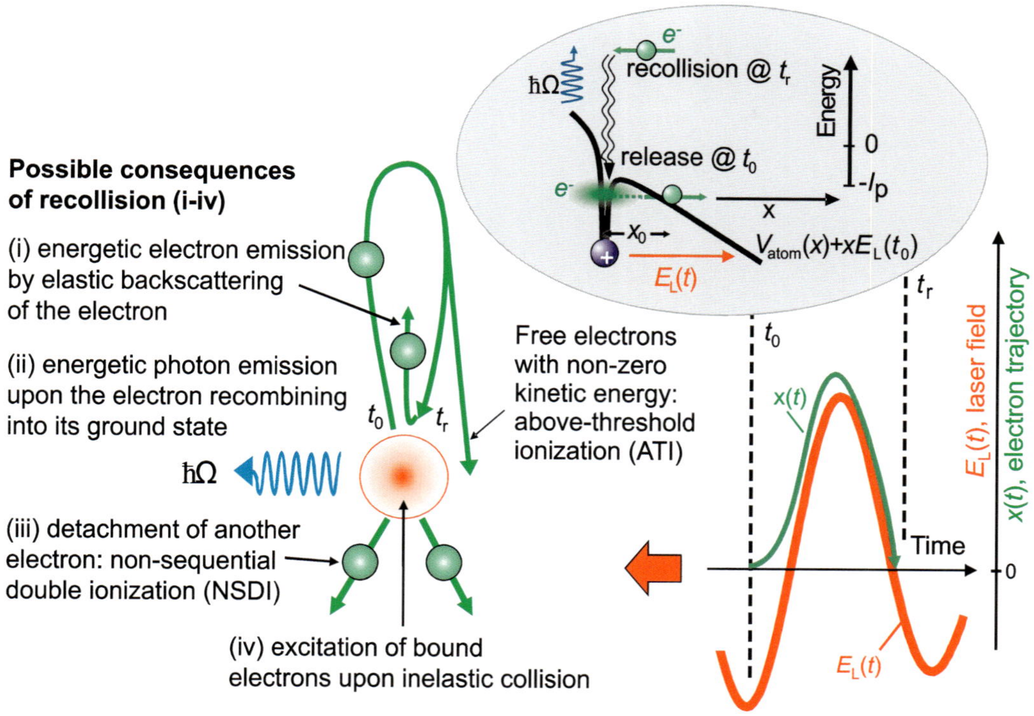

FIG. 15. (Color) Optical field ionization of an atom (at moment t_0) and subsequent recollision of the detached electron (at moment t_r) with the parent ion in a strong, linearly polarized ultrashort-pulsed laser field. (i)–(iv) Possible consequences of recollision. Inset: Release and recollision of electron that returns with highest energy to the parent ion, resulting—via process (ii)—in the highest-energy photons emitted.

fraction of the electric field oscillation cycle. Consequently, the Coulomb attraction to the parent ion quickly becomes a small perturbation compared to the laser field. The simplest approximation is to neglect the Coulomb field after ionization. This is the basis of the quantum strong-field approximation (SFA) (see footnote 32), a theoretical workhorse in strong-field physics. Once $V_{\text{atom}}(\mathbf{r})$ is dropped in Eq. (8), the center-of-mass motion of the quantum wave packet is described by

$$\ddot{x} = -E_L(t). \tag{10}$$

A further essential assumption of the semiclassical model is that the electron appears at the "exit" of the tunneling barrier x_0 with zero velocity $v(t_0)=0$. Then Eq. (10) yields for the electron's velocity [equal to its (kinetic) momentum in atomic units]

$$v(t) = v(t_0) + A_L(t) - A_L(t_0) = A_L(t) - A_L(t_0). \tag{11}$$

Here the vector potential $A_L(t)$ is defined by $E_L(t) = -dA_L/dt$. In the limit of a slowly varying envelope, $df/dt \ll \omega_L f$, we have

$$A_L(t) \approx -(E_0/\omega_L) f(t) \sin(\omega_L t + \varphi), \tag{12}$$

which is valid down to the few-cycle regime. In this limit, Eq. (11) yields

$$v(t) = A_L(t) - A_L(t_0) \approx \frac{E_0}{\omega_L} f(t_0) \sin\phi_0 - \frac{E_0}{\omega_L} f(t) \sin(\omega_L t + \varphi), \tag{13}$$

where $\phi_0 = \omega_L t_0 + \varphi$ is the phase of the laser field at the instant of ionization. Figure 16 shows $v(t)$ for different instants of release. The final momentum is

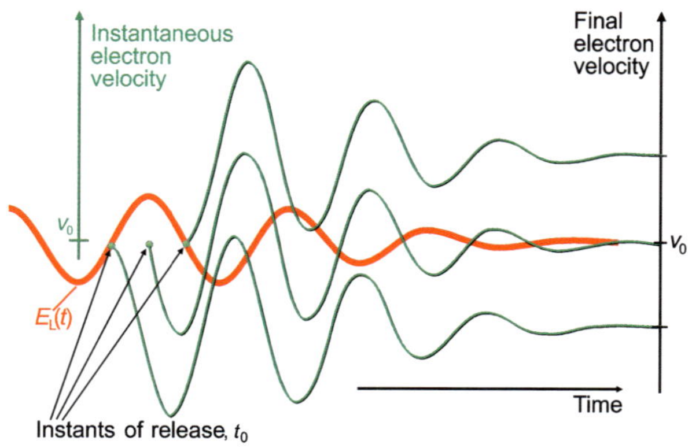

FIG. 16. (Color) Temporal variation of the electron velocity component parallel to the electric field of a linearly polarized light wave for different moments of release. The electron's final velocity v_f shown on the right-hand side sensitively depends on the instant of release.

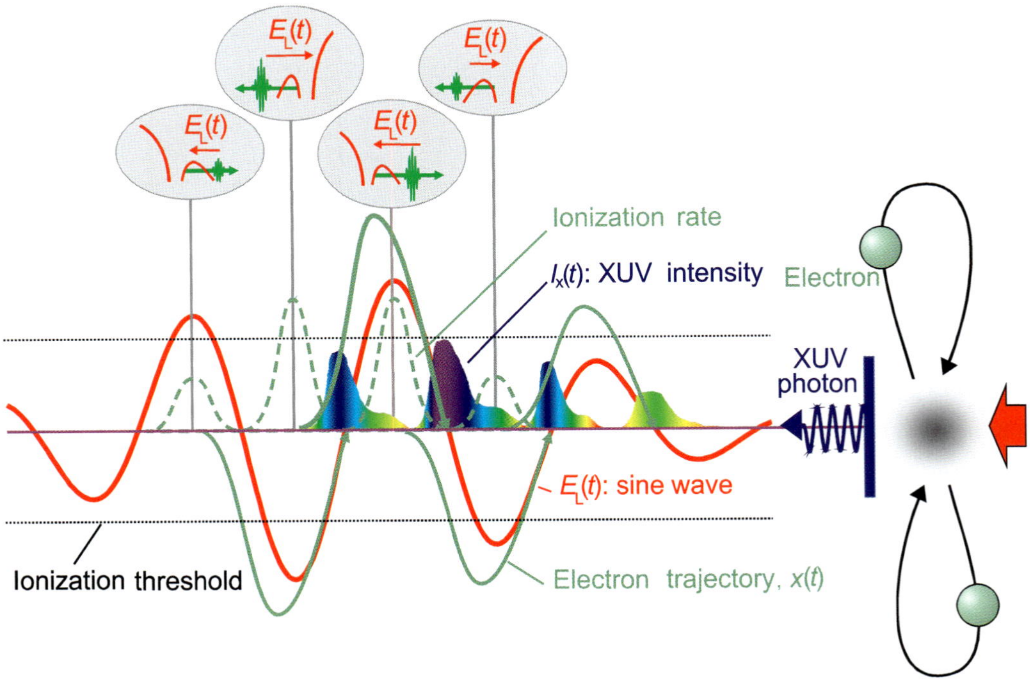

FIG. 17. (Color) Trajectories of electrons returning to the vicinity of the core with maximum kinetic energy (solid green lines) for a laser pulse consisting of two oscillation cycles within the full width at half maximum of its cycle-averaged intensity profile (e.g., a 5-fs pulse at a carrier wavelength of $\lambda_L=750$ nm). We have set $\varphi=-\pi/2$, i.e., used a "sine-shaped" electric field waveform. Dashed line sketches time dependence of ionization rate.

$$\begin{aligned} v_f &= A_L(t\rightarrow\infty)-A_L(t_0)\\ &= -A_L(t_0)\\ &\approx (E_0/\omega_L)f(t_0)\sin\phi_0. \end{aligned} \tag{14}$$

Further integration yields the electron trajectory

$$\begin{aligned} x(t) &\approx x_0+\frac{E_0}{\omega_L^2}[f(t)\cos(\omega_L t+\varphi)-f(t_0)\cos(\omega_L t_0+\varphi)]\\ &\quad+\frac{E_0}{\omega_L}f(t_0)\sin(\omega_L t_0+\varphi)(t-t_0). \end{aligned} \tag{15}$$

For an instant of release shortly after the field oscillation peak, $x(t)$ is sketched in Fig. 15. In a few-cycle, linearly polarized laser field, ionization is significant only a few times near the strongest field oscillation peaks. Figure 17 illustrates trajectories launched shortly after the oscillation peak and returning to the core with maximum kinetic energy, which are responsible for emission of the highest-energy photons (upon recombination) and electrons (upon elastic backscattering). Predictions of this simple classical model including the assumption $v(t_0)=0$ have been checked against full quantum mechanical simulations, with good agreement in the strong-field limit (de Bohan *et al.*, 2002).

Is this classical analysis useful if the final step of the strong-field interaction has to be handled quantum mechanically? For example, high-order harmonic generation (HHG; see Sec. IV.C.2) is governed by the interference of the "continuum" electron wave packet ψ_c revisiting the core with the portion of the wave function left behind in the ground atomic state ψ_g. Classical analysis enters via the quantum-classical correspondence. The wave packet ψ_c evolves as

$$\psi_c(\mathbf{r},t)\propto\sqrt{w_i(t_0)}\exp[iS(\mathbf{r},t,t_0)]. \tag{16}$$

Here S is the action along the classical trajectory that starts at the moment t_0 with zero initial velocity and arrives at the point $\mathbf{r}$ at the moment t and $w_i(t_0)$ is the ionization rate. The moment t_0 is uniquely defined by the $\mathbf{r}$ and t, i.e., $t_0=t_0(t,\mathbf{r})$. In general, the action and trajectories are quite complicated. However, in the SFA the result is relatively simple:

$$S(\mathbf{r},t,t_0)=[\mathbf{p}+\mathbf{A}_L(t)]\mathbf{r}-\tfrac{1}{2}\int_{t_0}^{t}[\mathbf{p}+\mathbf{A}_L(t')]^2dt', \tag{17}$$

where $\mathbf{A}_L(t)$ is the vector potential of the laser field, $\mathbf{p}$ is the electron's drift (canonical) momentum, and $\mathbf{v}(t)=\mathbf{p}+\mathbf{A}_L(t)$ is the instantaneous (kinetic) momentum. The relevant canonical momentum $\mathbf{p}$ is defined by the assumption that, at the moment of ionization t_0, $\mathbf{v}(t_0)=\mathbf{p}+\mathbf{A}_L(t_0)=0$. The first term in Eq. (17) gives the spatial dependence of the wave function, which in the SFA is a plane wave. The second term describes the energy-dependent contribution to its phase—the integral of the instantaneous kinetic energy of the electron along the trajectory. Equations (16) and (17) imply that the wave packet's instantaneous wavelength and instantaneous carrier frequency are determined by the instantaneous values of the electron's momentum and kinetic energy, respectively. Equations (16) and (17) establish the link between the quantum wave function and the underlying classical evolution.[34]

Another key aspect of ψ_c is the dynamics in the plane perpendicular to the x direction. The wave packet launched by tunnel ionization is confined laterally during tunneling. The Heisenberg uncertainty relationship implies transverse velocity distribution, which has a width [see, for example, Ivanov *et al.* (2005)]

$$v_{\perp,\mathrm{spr}} \approx E_L^{1/2}/(2I_p)^{1/4} \tag{18}$$

in each transverse direction.[35] This reduces the probability of recollision as

$$|\psi_c|^2 \propto w_i[v_{\perp,\mathrm{spr}}(t_r - t_0)]^{-2}, \tag{19}$$

where t_r is the moment of return (recollision). A further reduction factor of $1/(t_r - t_0)$ in the recollision probability at the moment t_r arises due to the longitudinal spreading of the wave packet.

C. Single-electron interference within the wave cycle of light

The interaction of atoms with strong fields provides textbook examples for one manifestation of quantum physics: the interference of a particle with itself. Electron wave packets released from the same valence electronic state (ground state) near subsequent wave crests and moving in the same direction will interfere with each other, resulting in the appearance of electrons at discrete energies spaced by the laser photon energy [above-threshold ionization (ATI)]; see Fig. 18(a) (Agostini *et al.*, 1979). Similarly, interference of the photon wavepackets emitted upon each recollision (twice per laser cycle) gives rise to periodic modulations of the emitted energetic photon spectrum, leading to HHG (McPherson *et al.*, 1987; Ferray *et al.*, 1988); see Fig. 18(b). The discrete ATI and HHG emission structure results from single-electron interference and the temporal coherence of the driving light wave.

The probability of tunnel ionization decreases exponentially with decreasing field strength. Consequently, atoms driven by a few-cycle wave eject only a few electron wave packets with significantly differing energy distributions due to significant differences between the field amplitudes in the subsequent half cycles of the pulse. This has two major implications: the emission of the

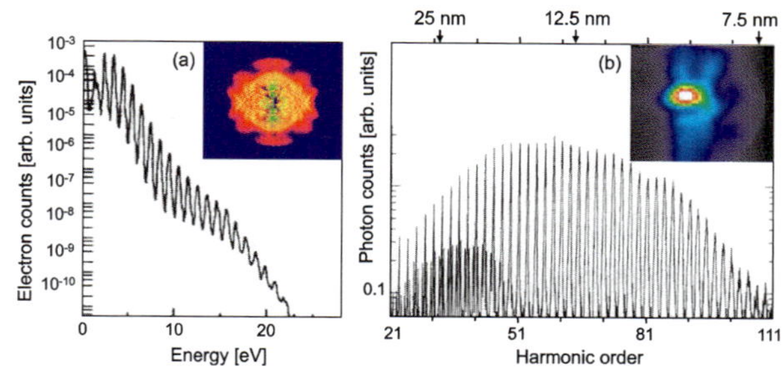

FIG. 18. (Color) Implications of single-electron interference in a strongly driven atom. (a) Energy distribution of above-threshold ionized (ATI) electrons produced by a multicycle femtosecond laser pulse. Courtesy of G. Paulus. Inset: Angular interference pattern of ATI electrons. Courtesy of M. Vrakking. (b) Spectral distribution of high order harmonic emission. Adapted from Macklin *et al.*, 1993. Inset: Spatial intensity profile of few-cycle driven harmonics at a photon energy of ~100 eV, implying a coherent near-diffraction-limited laser-like beam. Courtesy of M. Uiberacker.

most energetic electrons (i) can be confined to a single wave cycle at the peak of the pulse, and (ii) its temporal profile is most sensitive to the CE phase of the pulse. This opens the door to controlling electron dynamics in the sub-fs domain (lightwave electronics) and—by allowing measurement of the CE phase—control of light waveforms (light-wave engineering).

1. Above-threshold ionization and sub-fs electron pulses

An easily accessible physical quantity is the final velocity v_f of electrons ejected from strongly driven atoms. According to Eq. (14), it is given by the vector potential of the field $A_L(t)$ at the instant of release t_0. Figure 19(a) plots the laser-induced change of electron velocity, $\Delta v = -A_L(t_0)$, for the "sine-shaped" electric field waveform of Fig. 17. The dashed line sketches the ionization probability, i.e., the initial temporal profile of the released electron wave packets. As indicated in Fig. 17, the four central half cycles of $E_L(t)$ have sufficiently large amplitude to open a window—a "slit" in the time domain—for releasing the wave packet. The electron's final momentum is dictated by $A_L(t_0)$, which—for the sine-shaped electric field waveform—opens either one or two slits for high-energy electron emission toward the left and right detectors,[36] respectively. Figure 19(b) reveals that emission of the highest-energy ATI electrons indeed occurs in a twin or in a single sub-fs pulse, depending on the few-cycle waveform driving the interaction (Lindner *et al.*, 2005).

2. High-order harmonic emission and sub-fs photon pulses

In the previous section, we dealt with electrons whose motion toward the detector is undisturbed by the parent

[34] Lewenstein *et al.*, 1995; Salieres *et al.*, 2001; Corsi *et al.*, 2006; Kanai *et al.*, 2007 For an earlier reference, see Volkov, 1935.

[35] Spreading is reduced if part of the electron's ground-state wave function is driven into excited bound states; see Hu *et al.* (2001). This effect, which is ignored in the SFA, might play an important role in generating low-order harmonics, typically up to I_p.

[36] We use this terminology by assuming the laser fields to be polarized horizontally.

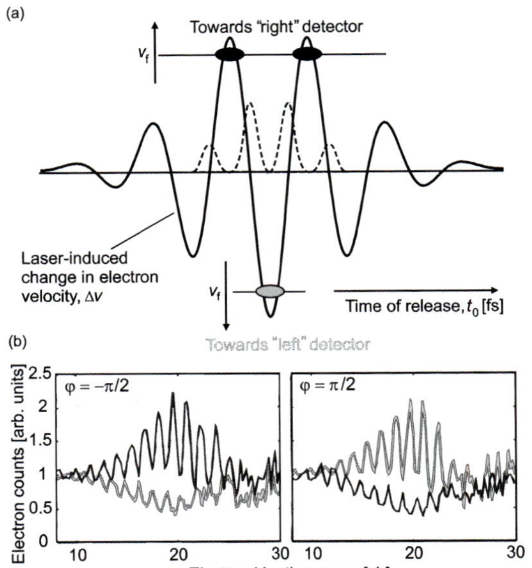

FIG. 19. Attosecond "double-slit" experiment (Lindner *et al.*, 2005). (a) A sine-shaped few-cycle pulse (not shown) ionizes atoms via optical-field ionization. Electrons are created with zero velocity at the instant of ionization, hence the change imposed by the laser field (solid line) is equal to their final velocity. "Slits" corresponding to a particular final velocity are shown by black and gray ellipses. The dashed line sketches the ionization rate. (b) Kinetic energy distributions of electrons arriving at the right and left detectors, depicted in black and gray, respectively. Adapted from Lindner *et al.*, 2005. The left panel shows spectra for the case shown in (a): the modulation of the spectrum on the right detector indicates interference between electrons released at two instants separated by the oscillation cycle, whereas the absence of pronounced modulation on the left detector reveals that electron emission in this direction is confined to a single ionization event, in accordance with the model shown in (a). By reversing the laser field (right panel), the situation is reversed. The measured spectra set an upper limit of 500 as for the duration of emitted electron wave packets. The experiment demonstrates attosecond control of electron emission from atoms using waveform-controlled few-cycle light.

atomic or molecular[37] ion. However, depending on the phase at the instant of ionization, ϕ_0, oscillations in the laser field may bring the electron back to its parent ion. In Eq. (13), the first term describes the drift velocity of the electron, which changes the sign near the maximum of the electric field, at $\phi_0=0$. Electrons released after the peak of the field, $\phi_0>0$, return to the core (see the shaded part of the electron emission probability in Fig. 20). The respective trajectories as a function of the instant of release t_0 or the phase ϕ_0 are given by Eq. (15). The return time t_r can be obtained by setting $x(t_r)=0$ in Eq. (15) for $t_r>t_0$. Figure 20 shows recolliding electron

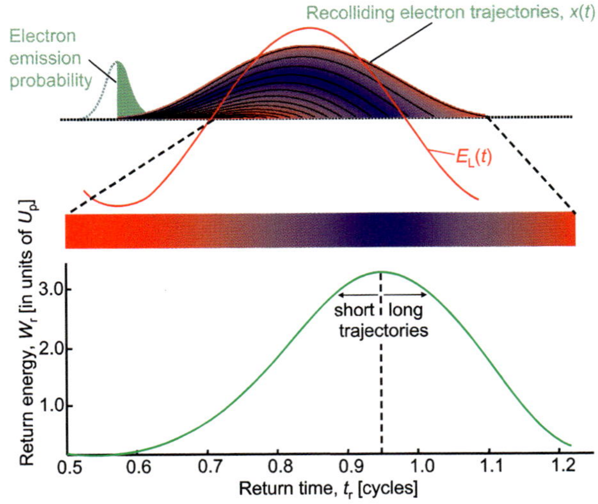

FIG. 20. (Color) Recolliding electron trajectories in atoms ionized with linearly polarized light. Top panel sketches electron trajectories—distance from the ion (vertical coordinate) vs time (horizontal coordinate). Depending on the moment of ionization, electrons return at different times with different energies, shown with changing color. Bottom panel shows energy of the returning electron vs return time $W_r(t_r)$ in units of U_p, see Eq. (20), for the selected laser cycle. This dependence indicates that electron wave packets returning along short and long trajectories carry positive and negative chirp, respectively. Courtesy of A. L'Huillier. (See also Varjú, Johnsson, *et al.*, 2005.)

trajectories[38] obtained with this procedure (upper panel) along with the energy of the electrons at t_r, in units of U_p, where (in atomic units)

$$U_p(t) = E_0^2 f^2(t)/4\omega_L^2 \tag{20}$$

is the cycle-averaged wiggle energy of the electron in the laser field given by Eq. (9), also known as the ponderomotive potential.

Figure 20 conveys several important messages. First, it shows that the moment of return is uniquely determined by the moment of ionization. Second, the maximum energy of the returning electron $W_{\max}$, reaches about $3.2U_p$, yielding for the maximum photon energy emitted during recollision (Kulander *et al.*, 1992; Corkum, 1993)[39]

$$\Omega_{\max} \approx 3.2U_p + I_p, \tag{21}$$

where I_p is the ionization potential. For few-cycle laser fields, valence electrons in rare gas atoms with high I_p, such as neon or helium, may survive in the ground state

[37]High-order harmonic generation in molecules has been reviewed by Bandrauk *et al.* (2007).

[38]For strong IR and visible fields the electron oscillation amplitude during its first excursion exceeds the Bohr radius by two orders of magnitude. Hence the initial offset of the electron's place of creation from the origin in Eq. (15) can be neglected on this scale, $x_0 \approx 0$.

[39]$\Omega_{\max}$ can be enhanced with chirped few-cycle pulses (Carrera and Chu, 2007).

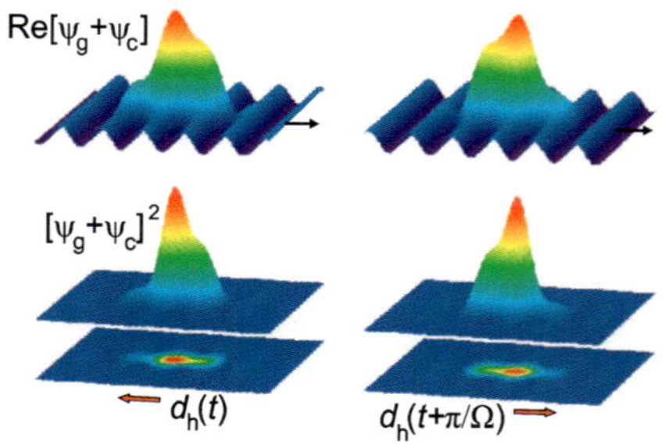

FIG. 21. (Color) Interference of recolliding wave packet with the ground state part of the same electron wave function, at two adjacent moments of time (left and right set of panels). Top and bottom panels show real part of the total wave function and the electron density, respectively. Interference leads to oscillations of electron density and hence of the induced dipole moment. Courtesy of D. Villeneuve.

up to intensities of several times 10^{15} W/cm^2. For NIR and IR light, $W_{\max}$ can reach hundreds and thousands of eV, respectively. Third, for each recollision energy, except for the maximum, there are two trajectories along which the electron can acquire the same energy by the time it returns to the core. These two sets of trajectories have been termed *short* and *long* trajectories. Last but not least, the electron wave packet returning to the core is heavily chirped. As a consequence, its duration is substantially longer (~1 fs) than the duration of the original wave packet released by optical field ionization [see, e.g., Niikura *et al.* (2005)].

This classical picture is reflected in the quantum analysis based on Eqs. (16) and (17).[40] As the returning electron reencounters its parent ion, the unbound (continuum) part of its wave function ψ_c, with its fast-oscillating phase, can interfere with the bound part of the electron wave function ψ_g. This interference creates fast-moving volumes of constructive and destructive interference, resulting in hyperfast oscillations of the electron density, as illustrated in Fig. 21. The one-active-electron dipole moment of our atom induced by the laser field, $d(t)=\langle\psi|d|\psi\rangle$, includes these fast oscillations via the term[41]

$$d_h(t) = \langle\psi_g|d|\psi_c\rangle + \text{c.c.}, \tag{22}$$

resulting in the harmonic emission intensity

$$I_h(t) \propto |\ddot{d}_h(t)|^2. \tag{23}$$

To assess the emission spectrum, we write the bound part of the wave function as $\psi_g=\phi_g(\mathbf{r})\exp(-iW_g t)$. After spatial integration in the expression of the matrix element given by Eq. (22), $d_h(t)$ will beat with the frequency $\Omega(t)$ given by the time derivative of the phase of ψ_c minus the time derivative of the phase of ψ_g. These derivatives are the energies of bound and returning wave packets $W_g=-I_p$, and, from Eq. (17), $W_r(t_r)=\mathbf{v}(t_r)^2/2$, respectively, so that $\Omega(t_r)=W_r(t_r)+I_p$. The beat frequency follows the curve $W_r(t_r)$ in Fig. 20, obtained from classical trajectories. The harmonic emission spectrum will contain all these frequencies up to the maximum frequency $\Omega_{\max}$ given by Eq. (21), determined by the maximum energy of the returning electron. Analysis of the spectrogram

$$H(\Omega,\tau) = \left|\int \ddot{d}_h(t)G(t-\tau)\exp(i\Omega t)dt\right|^2, \tag{24}$$

where $G(t-\tau)$ is a gating function (e.g., Gaussian), allows us to study when different frequencies are emitted (Yakovlev and Scrinzi, 2003).[42] Figure 22 shows the spectrogram of the harmonic emission from a macroscopic volume of gas exposed to a few-cycle NIR laser pulse. The time-frequency distribution of emission consists of branches resembling their classical counterpart, $W_r(t_r)$ in Fig. 20.

Figure 22(a) reveals a crucial correction to the simple classical approach. Quantum mechanically, the emission extends beyond the classical cutoff $\Omega_{\max}$ with exponentially decreasing spectral intensities for $\Omega>\Omega_{\max}$, as revealed in Fig. 22(b). In this range the temporal chirp of emission is almost negligible. This is of great practical importance to attosecond technology: filtering highest-energy emission around $\Omega_{\max}$ with a bandpass filter yields isolated [for a cosine-shaped pulse, Figs. 22(a) and 22(b)] or twin [for a sine-shaped pulse, Figs. 22(c) and 22(d)] emission bursts. These bursts are predicted to be near-Fourier-limited XUV pulses with sub-fs duration.

The disappearance of periodic modulation of the cutoff emission in the first HHG experiments with waveform-stabilized, few-cycle pulses [Fig. 23(a)] provided the first indication of the emergence of a single, isolated pulse of sub-fs duration from a few-cycle strong-field interaction (Baltuška, Udem, *et al.*, 2003). A shift of the CE phase of the few-cycle driver by $\pi/2$ introduced quasiperiodic modulation in the cutoff range with a maximum modulation depth, see Fig. 23(b), in accordance with the prediction of Fig. 22(d). These experiments have demonstrated the feasibility of controlling sub-fs coherent XUV emission with waveform-reproducible few-cycle light. Complete temporal characterization of the emitted radiation will be addressed in Sec. V.

---

[40] Lewenstein *et al.*, 1994; Kopold *et al.*, 2000; Salieres *et al.*, 2001; Becker *et al.*, 2002; Miloševic and Becker, 2002; Frolov *et al.*, 2007a, 2007b; Mineo *et al.*, 2007; Pérez-Hernández and Plaja, 2007.

[41] Equation (22) is also a matrix element describing the recombination of the returning electron back to its original ground state, a process observed recently via particle detection (Williams *et al.*, 2007).

[42] See also Kim *et al.*, 2001, 2006; Pukhov *et al.*, 2003; Chipperfield *et al.*, 2005; Carrera *et al.*, 2006.

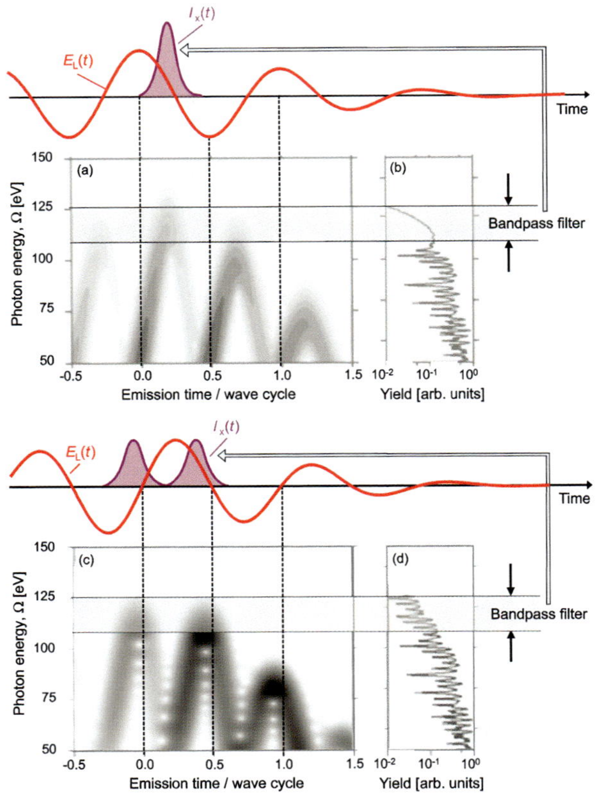

FIG. 22. (Color online) (a), (c) The spectrogram $H(\Omega,\tau)$ of the atomic high-harmonic emission from a gas of neon atoms exposed to a few-cycle NIR laser pulse (τ_L=5 fs, λ_L=800 nm); for further parameters used for the calculation, see Yakovlev and Scrinzi, 2003. (b), (d) Emission spectra. (a), (b) are calculated for a cosine-shaped driving pulse. From Yakovlev and Scrinzi, 2003. (c), (d) are for a sine-shaped driving pulse calculated for the same parameter set (Yakovlev, 2007). Propagation tends to eliminate the contribution of multiple returns. Filtering out high-energy emission yields a (a) single or (c) twin sub-fs XUV pulse. Courtesy of V. Yakovlev.

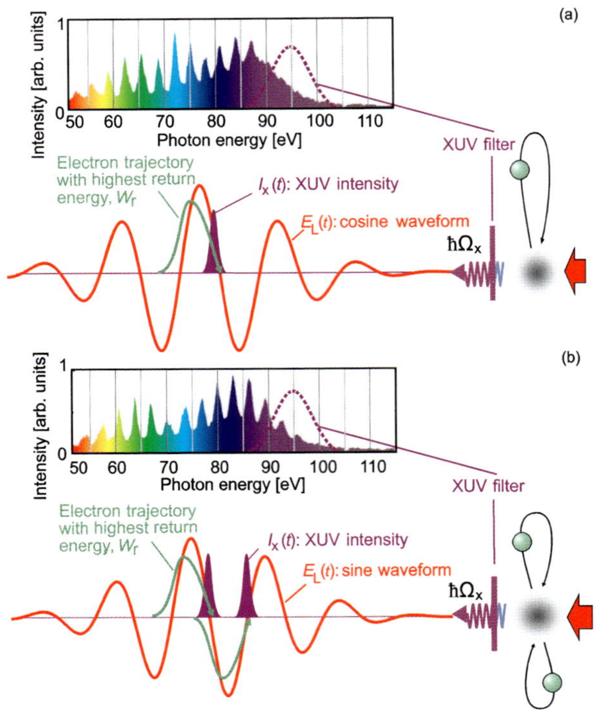

FIG. 23. (Color) High-order harmonic spectra obtained experimentally with a few-cycle NIR pulse (τ_L=5 fs, λ_L =750 nm) with (a) cosine-shaped and (b) sine-shaped waveforms (Baltuška, Udem, *et al.*, 2003). The cosine wave leads to a smooth spectrum in the cutoff range, indicating emission in a single burst of potentially sub-fs duration, whereas the appearance of modulation for the sine wave is in accordance with the prediction of Fig. 22(d) and suggests that emission is delivered in two bursts half a cycle apart.

D. Electron scattering

1. Elastic backscattering and high-energy electron emission

We now turn our attention to the elastic scattering of the recolliding electron (Fig. 15). The scattering breaks the phase of the electron's wiggling motion and abruptly changes the direction of the electron velocity. Equation (13) yields for the instant just before the recollision (denoted by t_r-0)

$$v(t_r-0)=A_L(t_r)-A_L(t_0). \tag{25}$$

As before, we assume that the laser field is linearly polarized and the electron is moving along the direction of the laser polarization for $t<t_r$. Upon elastic scattering, the electron changes the direction of its velocity while keeping its absolute value the same:

$$v_\parallel^2(t_r+0)+v_\perp^2(t_r+0)=[A_L(t_r)-A_L(t_0)]^2, \tag{26}$$

where $v_\parallel$ and $v_\perp$ stand for the momentum vector components parallel and perpendicular to the laser polarization. At the end of the laser pulse, these two components become

$$v_{\parallel f}=v_\parallel(t_r+0)-A_L(t_r), \quad v_{\perp f}=v_\perp(t_r+0). \tag{27}$$

As an example, take a closer look at backscattering. Then $v_\perp=0$ and the relationship between the momenta before and after the collision is

$$v_\parallel(t_r+0)=-v_\parallel(t_r-0)=-A_L(t_r)+A_L(t_0). \tag{28}$$

At the end of the pulse we then have

$$v_{\parallel f}=v_\parallel(t_r+0)-A_L(t_r)=-2A_L(t_r)+A_L(t_0). \tag{29}$$

This result is different from the final momentum $-A_L(t_0)$ for the case when no collision was present. The backscattered electron's final momentum is sensitive to the

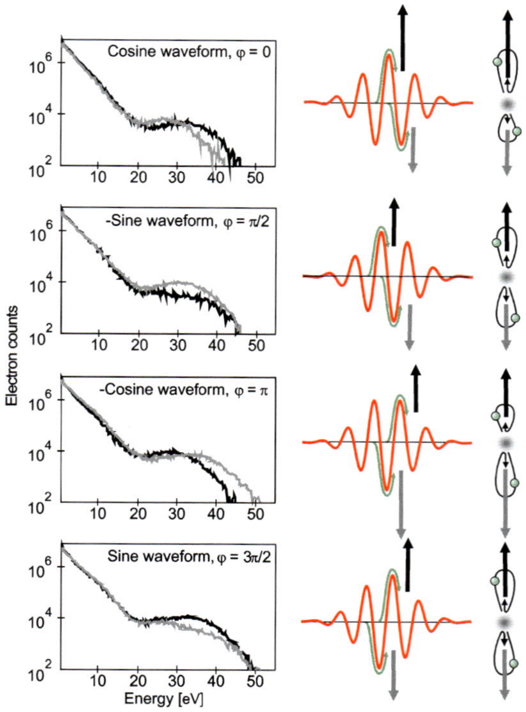

FIG. 24. (Color) Sensitivity of backscattered electrons to the CE phase. Left panels, energy distribution of ATI electrons emitted parallel to the electric field vector of the few-cycle ionizing field (τ_L=5 fs, λ_L=750 nm) towards the right and left detectors of Fig. 19, shown in black and gray colors, respectively. Adapted from Paulus *et al.*, 2003. The high energy shoulders originate from backscattering upon recollision. The highest-energy electrons are emitted for a "cosine-shaped" light waveform. Note that while the maximum electron energy is the same in both directions for a "sine-shaped" wave, the yield of electrons scattered to opposite directions (with energies above 20 eV) differ significantly (by a factor of more than 2) owing to the difference in the field strengths at the moment of ionization. As a result, the method also permits single-shot determination of the CE phase even for the symmetric case of a sine waveform, which was demonstrated experimentally (Wittmann *et al.*, 2008). The calibration of the stereo-ATI "phase meter" can be subjected to a self-consistency check by shifting the CE phase in a known way, by simply changing the path length in a thin glass plate; see Eq. (7).

values of the vector potential at the moment of ionization and the moment of recollision.[43]

In a few-cycle laser field the final energy and the number of the highest-energy backscattered electrons is a sensitive probe of the subcycle field evolution. Owing to this sensitivity and the intuitive insight provided by a simple "billiard-ball" model[44] into the underlying physical process, backscattered electrons constitute a reliable tool for determining the CE phase. Simultaneous detection of electrons ejected in both directions along the laser polarization offers robustness against laser amplitude noise. This "stereo-ATI" detection scheme was invented by Paulus (Paulus *et al.*, 2001) and subsequently used for the first unambiguous determination of the CE phase of few-cycle light (Paulus *et al.*, 2003; Miloševic *et al.*, 2006); see Fig. 24. Recently, it also permitted single-shot measurement of the CE phase (Wittmann *et al.*, 2008). The stereo-ATI technique is about to become a standard diagnostic tool in attosecond laboratories—an elegant example of how fundamental physics provides a practical tool for laser engineers.[45]

2. Inelastic scattering: Multiple electron emission and inner-shell excitation

Electron recollision does not have to be elastic. The recollision electron acts like an antenna (Kuchiev, 1987), which an atom extends into the laser field to absorb energy. The energy of the recolliding electron can (sometimes dramatically) exceed the binding energy of valence electrons. As a consequence, the recolliding electron may knock off a second electron from the electronic shell of the parent ion.

Efficient double ionization of noble gas atoms in infrared laser fields was first observed by L'Huillier *et al.* (1983). Recent experiments, in particular those using elliptically polarized light or correlated detection of both electrons, have established recollision as the origin of this process, now referred to as nonsequential double ionization (NSDI; Fittinghoff *et al.*, 1992, Walker *et al.*, 1993, 1994).[46] The elliptical field tends to deflect the returning electron wave packet from its target. The transverse velocity, induced by the transverse component of the field, is, in the limit of a slowly varying envelope, see Eq. (12),

$$v_\perp(t) = A_\perp(t) - A_\perp(t_0) \approx \frac{E_\perp}{\omega_L} f(t)\cos(\omega_L t + \varphi) - \frac{E_\perp}{\omega_L} f(t_0)\cos\phi_0, \qquad (30)$$

where $E_\perp = \varepsilon E_0$ and ε is the field ellipticity. When the resultant transverse drift exceeds $v_{\perp,\mathrm{spr}}$ given by Eq.

[43]If the recollision occurs near the zero of the electric field, i.e., when $|A_L(t_r)| \approx E_0/\omega_L$ see [Eq. (12)], the final energy after backscattering may reach values of up to $10U_p$. Detailed analysis can be found in Walker *et al.*, 1996 and Goreslavski and Popruzhenko, 1999. For earlier references, see, Corkum, 1993; Schafer *et al.*, 1993; Yang *et al.*, 1993.

[44]This is a reasonable approximation for energetic backscattered electrons, which spend little time near the core.

[45]Solid-state phase meters are also being developed: Apolonski *et al.*, 2004; Fortier *et al.*, 2004; Mücke *et al.*, 2004; Irvine *et al.*, 2006; Van Vlack and Hughes, 2007.

[46]Correlated measurement for the spectra of both electrons yields insight into the collision dynamics. See Becker and Faisal, 2005; furthermore, Moshammer *et al.*, 2000; Weber *et al.*, 2000a, 2000b; Rottke *et al.*, 2002; Weckenbrock *et al.*, 2003, 2004; Eremina *et al.*, 2003, 2004; Alnaser, Tong, *et al.*, 2004; Alnaser, Voss, *et al.*, 2004; de Jesus *et al.*, 2004; Rudenko *et al.*, 2004; Zeidler *et al.*, 2005; Zrost *et al.*, 2006. Recent work includes Figueira de Morisson Faria *et al.*, 2006; Ruiz *et al.*, 2006; Ho and Eberly, 2007.

(18), the returning electron misses the parent ion. Dietrich *et al.* (1994) demonstrated that increasing ellipticity gradually suppressed both HHG and NSDI in the same way, providing conclusive evidence for electron recollision being responsible for both phenomena. Recently, NSDI was also found to be controllable with the waveform of a few-cycle pulse.

Attosecond control over the recollision (Zeidler *et al.*, 2005) provides a unique opportunity for unraveling multielectron interactions and dynamics triggered by collisions (Rudenko *et al.*, 2004; Zrost *et al.*, 2006). Inelastic recollision in a molecule may induce structural dynamics in the molecular ion (see, e.g., Niikura *et al.*, 2002; Niikura, Légaré, *et al.*, 2003; Liu *et al.*, 2004; Kling *et al.*, 2006; Gräfe and Ivanov, 2007; Tong and Lin, 2007a, 2007b), for further discussion, see Sec. VII.

V. BASIC CONCEPTS FOR ATTOSECOND CONTROL AND METROLOGY

A. Conditions for experimental implementation

The atomic-scale electron dynamics discussed in the previous section unfold within the wave cycle. They are extraordinarily sensitive to the subcycle evolution of the field. Relatively small changes in the driving light waveform [see Fig. 10(b)] affect the energy distribution of high-energy photons (Fig. 23) and electrons (Fig. 24) originating from the strong-field interaction. This indicates two possible approaches to precision attosecond metrology.

For multicycle waves of laser light and attosecond pulse trains, one has to make sure that the subcycle electron-light interaction is repeated over several or many oscillation cycles under precisely the same conditions. Then, and only then, no information is lost due to the accumulation of attosecond signals originating from individual cycles during the multicycle interaction period. Multicycle attosecond metrology can draw either on periodically repeated recollisions or on sub-fs photon pulse trains emerging from them. In the former case, the multicycle light wave generates the recollision electron and triggers the process under study in a correlated fashion in one and the same interaction within each wave cycle. Then one of the two correlated processes can serve as a clock for the other. The concept, first demonstrated by Niikura *et al.* (2002) and Niikura, Legare, *et al.* (2003) for a diatomic molecule, relies on exploiting such correlations in the time domain.[47]

Ideally, this technique would require a light pulse with a constant amplitude (i.e., flat-top intensity profile) to make sure that both the processes under study as well as the reference process used as a clock unfold in the same way in each wave cycle.[48] Similar requirements hold when a train of sub-fs pulses, with adjacent bursts spaced by a half (or a full) cycle of the generating laser field, are used in conjunction with its multicycle driver. Ideally, a constant-amplitude light wave would be required to produce a PHz train of *identical* sub-fs pulses. Even with this in place, metrology would have to be restricted to processes terminating within the laser wave cycle to avoid multiple start and exposure while the process is under way. Realistic multicycle drivers have bell-shaped envelopes; consequently, the attosecond pulse trains produced by them have burst parameters varying significantly across the train. Complicated temporal profiles resulting even from simple processes triggered by these trains are rather difficult to retrieve, making high demand on the S/N ratio.[49]

These shortcomings can be overcome using few-cycle attosecond technology, which draws on a light wave consisting of easily distinguishable[50] oscillation cycles with well-controlled temporal evolution. This permits triggering of a microscopic process within any selected wave cycle with subcycle precision. Thus the process can be initiated with attosecond timing precision with respect to an entire light wave, rather than only to one of many possible cycles. An excellent example is a single sub-fs photon or electron pulse that can be generated with attosecond timing with respect to the generating few-cycle light wave. Subsequently it can be used either (i) for initiating motion with attosecond timing with respect to an extended light wave or (ii) for probing the motion that was triggered and steered by waveform-controlled light on subcycle as well as multicycle time scales.

In both scenarios, timing relative to the whole light waveform is critical for controlling and observing atomic-scale electronic motion. From the control perspective, it allows one to trigger the motion with attosecond accuracy with respect to the steering light force. From the measurement perspective, it enables a few-cycle light pulse to probe the triggered motion on both attosecond and, if needed, longer time scales.

B. Light waveform control

Energetic photon and electron emission from few-cycle-ionized atoms (Figs. 23 and 24, respectively) as

[47]Essentially the same concept underlies the core-hole-clock technique used for studying ultrafast (sub-fs to several femtoseconds) dynamics with the core hole decay serving as a clock. See Brühwiler *et al.*, 2002; Brena *et al.*, 2004; Föhlisch *et al.*, 2005; Nordlund *et al.*, 2007.

[48]It is challenging to produce a light wave with a rectangular temporal intensity profile (it would require the same bandwidth as that of a few-cycle pulse). Instead of using such a pulse, Niikura *et al.* (2002) were able to fulfill this condition by exploiting high-order nonlinearity involved in triggering the process: it confined the interaction to the central portion of the multicycle pulse, where the amplitude was nearly constant.

[49]Other than CRAB, see Sec. V.D.4, adiabatic phase expansion (Varjú *et al.*, 2005) and two-color probing (Huo, Zeng, Leng, *et al.*, 2005) have been proposed for this purpose.

[50]The cycles may be distinguished due to significant differences in either their period or their amplitude. Both options require a superposition of light frequencies covering about an octave or more.

well as terahertz emission from the emerging plasma (Kress, 2006; Gildenburg and Vvedenskii, 2007) carries direct information about the CE phase. With this parameter, the waveform is uniquely determined. Controlling of the shape of light waveforms is a first manifestation of attosecond control and metrology.

The stereo-ATI technique offers the highest control precision: it allows the CE phase to be determined with an accuracy of 100 mrad for a 6-fs, 750-nm laser pulse, implying a timing accuracy of the field oscillations cycles with respect to the pulse peak of approximately 40 as (Paulus *et al.*, 2005).[51] Forty attoseconds before or after the pulse peak the (cycle-averaged) intensity is reduced by merely $\approx 0.01\%$ with respect to its value on the pulse peak in a 6 fs pulse. This extraordinary sensitivity is a direct consequence of the highly nonlinear atomic response to the field. Recently, this sensitivity could be further improved by single-shot measurements (Wittmann *et al.*, 2008). Few-cycle-driven HHG opened another route to on-line monitoring of the CE phase (Cavalieri, Goulielmakis, *et al.*, 2007; Haworth *et al.*, 2007).[52]

Measurements often require signal accumulation over many laser shots. This prompts an important question regarding the robustness of the few-cycle waveforms generated. The degree of stabilization of the CE phase in the laser oscillator meanwhile reached the level of accuracy of its measurement (Fuji *et al.*, 2005). Delivery of few-cycle pulses with undistorted shape requires control of optical delay to within 1–2 fs over a spectral band of several hundred terahertz. Dispersion in optical components introduces delay differences of several or more *picoseconds* over this band. The required dispersion control with an accuracy of 10^{-3} or better over more than 100 THz of optical bandwidth can be afforded by chirped multilayer mirrors (Szipöcs *et al.*, 1994; Stingl *et al.*, 1995; Jung *et al.*, 1997), with their programmable counterparts: acousto-optic filters (Tournois, 1997; Kaplan and Tournois, 2004), with frequency-domain modulators (Zeek *et al.*, 1999; Yamane *et al.*, 2003; Yamashita *et al.*, 2006), or with their combinations (Baltuška *et al.*, 2002a, 2002b, 2002c). Preserving a selected waveform on target is even more challenging, given that a path length change by several microns through optical components shifts the CE phase considerably. Active compensation of CE phase drifts is therefore indispensable (Baltuška, Udem, *et al.*, 2003). The standard f–$2f$ interference technique, which was used for stabilizing the frequency comb of oscillators (Apolonski *et al.*, 2000; Holzwarth *et al.*, 2000; Jones *et al.*, 2000), is unable to accurately monitor extracavity CE phase drifts[53] (Dombi *et al.*, 2004; Paulus *et al.*, 2005). Precision high-field waveform synthesis relies on CE phase metrology based on strong-field interactions discussed above.

There is yet another effect that requires attention: the wave front of a continuous-wave laser beam delivered in the fundamental (Gaussian) mode of a stable laser resonator (Siegman, 1986) advances by a phase $\phi(z) = -\arctan(z/z_R)$ with respect to a plane wave carried at the same wavelength (Gouy, 1890) as the wave propagates through the focal region. Here $z_R = \pi w_0^2/\lambda_L$ denotes the Rayleigh distance and w_0 the beam radius in the focal plane. In a pulse, the Gouy effect shifts the CE phase as it propagates through the focal region. Within the Rayleigh range, the resultant shift was found to be a linear function of distance from the position ($z=0$) of the beam waist, $\Delta\varphi_{\text{Gouy}}(z) \approx (\pi/2)z/z_R$ (Fig. 25, Lindner *et al.*, 2004).[54] Within an interaction length of L, the CE phase gets shifted by $\Delta\varphi_{\text{Gouy}}(z=L)$ in addition to the dispersion-induced shift $\Delta\varphi(L)$, given by Eq. (7); hence constant waveform requires

$$|\Delta\varphi(L) + \Delta\varphi_{\text{Gouy}}(L)| \ll 1. \tag{31}$$

C. Attosecond metrology using nonlinear XUV optics

Conceptually, the most straightforward approach to characterizing a short pulse is to use the pulse to measure itself, e.g., using autocorrelation as discussed in Sec. II.B.1. In the case of an XUV pulse, this would require a suitable XUV nonlinearity (Kobayashi *et al.*, 2007). In the XUV or soft x-ray regime, multiphoton resonant absorption or ionization provides options for implementing nonlinear autocorrelation. For XUV photon energies less than the atomic ionization potential, measurement of the two-photon ionization yield as a function of delay between two replicas of the XUV pulse resulted in the second-order autocorrelation of this pulse (Kobayashi *et al.*, 1998; Sekikawa *et al.*, 2002). Recently this technique has been used to measure trains of sub-fs pulses via cycle-averaged (intensity) autocorrelation (Tzallas *et al.*, 2003, 2005; Nikolopoulos *et al.*, 2005) as well as the fringe-resolved (interferometric) autocorrelation (Nabekawa *et al.*, 2006). Detecting doubly charged (instead of singly charged) helium extended the photon energy range of the technique to ~79 eV (Nabekawa *et al.*, 2005).

Watanabe and co-workers recently succeeded in overcoming this limitation by realizing two-photon above-threshold ionization (ATI) and detecting photoelectrons instead of ions (Miyamoto *et al.*, 2004; Hasegawa *et al.*, 2005). They demonstrated the first autocorrelation and FROG measurement of a single, sub-fs pulse (Sekikawa *et al.*, 2004; Kosuge *et al.*, 2006). Unfortunately, the relevant two-photon cross sections become prohibitively low toward shorter wavelengths (Nikolopoulos and

---

[51] For a theoretical analysis, see Zhang *et al.*, 2007.

[52] Recently, Wu and Yang (2007) pointed out that bound-state dynamics at weak fields may be affected by the C-E phase and hence can possibly be used for its measurement.

[53] Even though it is currently widely used for this purpose.

[54] Note that $\Delta\varphi_{\text{Gouy}}(z)$ is different from $\phi(z)$ both in sign and slightly in magnitude, owing to the broad bandwidth and non-Gaussian beam profile of the few-cycle wave used (Lindner *et al.*, 2004).

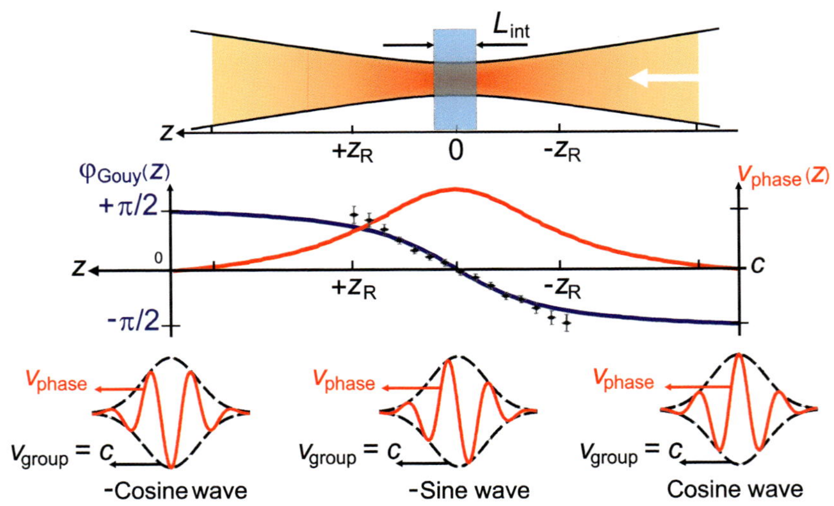

FIG. 25. (Color) Few-cycle pulse propagation through the beam focus. The pulse envelope propagates at the group velocity, which in vacuum equals the speed of light, $v_{group}=c$. The carrier wave suffers a phase advance due to the Gouy effect. As a consequence, it rushes on ahead of the envelope as the pulse propagates through the beam focus. Propagation from the far field to the focus and from the focus to the far field causes a Gouy phase shift of $\pi/2$ each, turning a "cosine"-shaped waveform into a "$-$sine"-shaped one at the beam waist and to a "$-$cosine"-shaped waveform in the far field behind focus. Courtesy of G. Paulus.

Lambropoulos, 2001), preventing straightforward extension of this metrology toward SXR frequencies.[55]

D. Light-field-controlled attosecond metrology

Two ways were found out of this corner. In one class of experiments a train of attosecond pulses was characterized by measuring the relative phase between the high-order harmonics comprising the train (Paul *et al.*, 2001; Aseyev *et al.*, 2003; Mairesse *et al.*, 2003; López-Martens *et al.*, 2005). The technique was dubbed RABBITT.[56] In the other class, a single sub-fs pulse was measured by attosecond streaking (Drescher *et al.*, 2001; Hentschel *et al.*, 2001; Kienberger *et al.*, 2004). These two apparently different approaches have common roots, as we show in this section.

From an experimental point of view, both techniques draw on cross correlation between the sub-fs pulse (or pulse train; henceforth, briefly, pulse) and a femtosecond laser pulse. The interaction medium is a gas of atoms from which the XUV pulse ejects photoelectrons (Fig. 26) in the presence of the strong laser field. Because the same laser field was previously used for generation of the sub-fs XUV pulse, the two fields are inherently synchronized. In the absence of resonances, the temporal profile of the XUV-induced photoemission rate follows the intensity profile $|a_X(t)|^2$ of the incident XUV pulse. The laser field modulates the final electron momentum and energy distribution. Measuring the modulated electron spectra as a function of the time delay between the XUV pulse and the laser pulse contains all information required for complete retrieval of both the laser field and the XUV pulse (Mairesse and Quéré, 2005; Quéré *et al.*, 2005). In contrast to autocorrelation, the technique

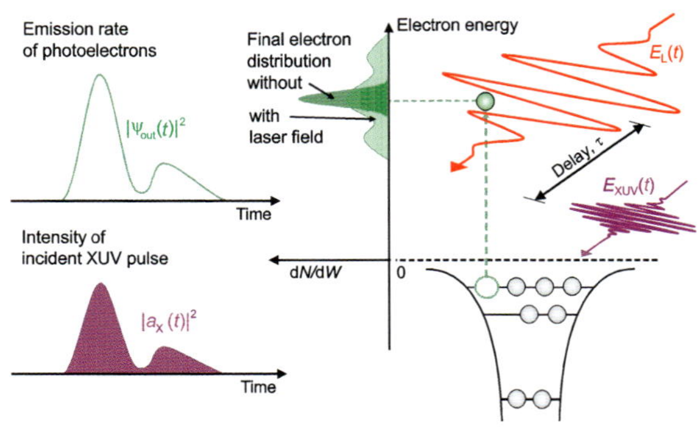

FIG. 26. (Color) Concept of light-field-controlled attosecond metrology. An ultrashort XUV pulse creates—in the presence of a strong laser field—a photoelectron, the temporal emission profile of which replicates that of the XUV pulse. The laser field modifies the final energy distribution of ejected photoelectrons. Measurement of this modification, as a function of delay between the XUV pulse and the laser field, offers the potential for sampling both processes (the XUV intensity and the laser field) with attosecond resolution.

[55]Novel techniques based on relativistic interactions addressed in Sec. VI.B may lead to femto- and attosecond x-ray pulses intense enough for nonlinear x-ray optics and x-ray-pump or x-ray-probe spectroscopy (Schweigert and Mukamel, 2007).

[56]The reconstruction of attosecond beating by interference of two-photon transitions. Together with its more advanced versions, developed by Salieres and co-workers and L'Huillier and co-workers, this technique has been used for characterizing and optimizing attosecond pulse trains in a number of experiments, e.g., Aseyev *et al.*, 2003; Dinu *et al.*, 2003; Mairesse *et al.*, 2003, 2004; López-Martens *et al.*, 2005; for recent reviews, see Agostini and DiMauro, 2004; Corkum and Krausz, 2007; Goulielmakis *et al.*, 2007.

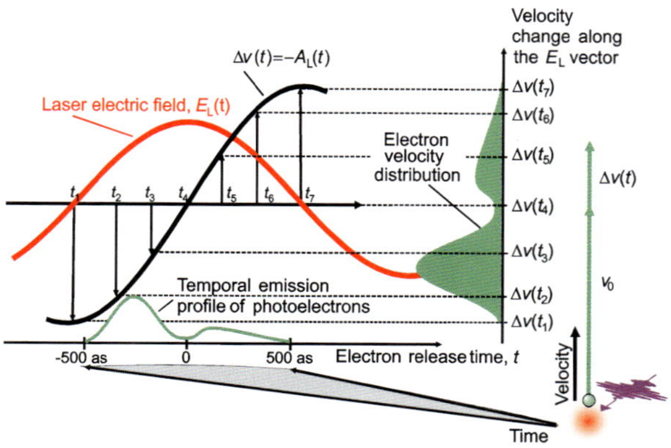

FIG. 27. (Color) Concept of optical-field-driven attosecond streaking. Photoelectrons released in a laser field parallel to the direction of electric field (red line) suffer a change of their initial velocities that is proportional to the vector potential of the field (black line) at the instant of release. This function is monotonic within a half wave cycle of the field, mapping the intensity profile of a sub-fs XUV pulse into a corresponding final velocity (or energy) distribution of photoelectrons.

can easily be scaled to shorter wavelengths, because the strong laser field obviates the need for an intense short-wavelength pulse.

A quantum analysis of XUV photoionization in the presence of the laser field reveals common grounds of this versatile attosecond metrology: the oscillating field of the laser acts as an attosecond phase gate, or phase modulator, on the electron wave packet released by the sub-fs pulse.[57] We show in this section how the generic concept of electron phase modulation, in specific circumstances, allows streaking, shearing, and spectral interferometry of the sub-fs electron wave packet and thereby leads (in a natural way) to the extension of streak imaging (Bradley *et al.*, 1971; Schelev *et al.*, 1971), FROG (Kane and Trebino, 1993; Trebino and Kane, 1993), SPIDER (Iaconis and Walmsley, 1998), and tomographic methods (Beck *et al.*, 1993; Walmsley and Wong, 1996) into the attosecond regime. The ultimate light-based ultrafast metrology, drawing on the oscillating light field as a gate, therefore naturally unifies apparently very different measurement concepts of time-resolved science.

1. Attosecond streak imaging

We time the strong laser field such that the XUV pulse fits within the time window of the central half cycle of the laser electric field, as shown in Fig. 27. The laser field is supposed to be weak enough not to ionize atoms, but strong enough to impart substantial momentum to the photoelectrons liberated by the XUV pulse (what is meant by "substantial" is discussed at the end of this section). We assume that the photoelectron is released with large initial kinetic energy,

$$W_0 = v_0^2/2 = \Omega_X - I_p \gg I_p, \tag{32}$$

where Ω_X is the carrier frequency of the incident XUV pulse

$$E_X(t) = a_X(t)e^{-i\Omega_X t} + \text{c.c.} \tag{33}$$

As a consequence, the electron quickly leaves the atom; therefore neglect of the Coulomb potential after the absorption of the energetic XUV photon is a reasonable approximation. The final velocity of an electron ejected at the moment t with initial velocity $\mathbf{v}_0$ is $\mathbf{v}_f = \mathbf{v}_0 - \mathbf{e}_x A_L(t)$ [see Eq. (11)], where the laser field is linearly polarized along the x direction, with $\mathbf{e}_x$ the corresponding unit vector. The change in the electron's velocity along the x direction is

$$\Delta v(t) = -A_L(t). \tag{34}$$

Within any half cycle between negative and positive maxima, $A_L(t)$ is a monotonic function of time (see Fig. 27). Consequently, the temporal distribution of the outgoing electron wave packet, which traces the temporal profile of the sub-fs XUV pulse, is mapped one to one onto a corresponding final velocity distribution of photoelectrons (Fig. 27). The resultant streak image provides direct time-domain information on the electron wave packet's emission time and hence the duration of the XUV pulse (Drescher *et al.*, 2001; Itatani *et al.*, 2002; Kitzler *et al.*, 2002).[58]

When contrasting this optical-field-driven "streak camera" with its microwave-driven predecessor (Fig. 3), the vastly increased streaking speed is most striking. However, there is a more profound difference. In contrast with its predecessor, here the streaking field is virtually jitter-free and acts directly at the location and instant of electron emission, preventing the initial velocity spread from impeding resolution as it does in conventional streak cameras. Moreover, the initial velocity (momentum) distribution can be measured together with the temporal emission profile, yielding information on the duration and chirp of the XUV pulse, as explained with the example of a linearly chirped XUV pulse in Fig. 28.

The optical-field-driven streak camera offers more degrees of freedom. One may observe electrons at an angle to the laser polarization. Assuming $|v_0| > |A_L|$, one obtains

$$\begin{aligned}\frac{v_f^2}{2} &= \frac{v_0^2}{2} + \frac{A_L^2(t)}{2}\cos 2\theta - A_L(t)\cos\theta\sqrt{v_0^2 - A_L^2(t)\sin^2\theta} \\ &\approx W_0 + 2U_p(t)\cos 2\theta\sin^2(\omega_L t + \varphi) \\ &\quad + \alpha\sqrt{8W_0U_p(t)}\cos\theta\sin(\omega_L t + \varphi),\end{aligned}$$

[57]Here the field acts upon a freed electron. Zheltikov (2006b) showed that it can also probe bound-state dynamics.

[58]For recent theoretical advances, see Kazansky and Kabachnik, 2006, 2007a; and Yudin *et al.*, 2007.

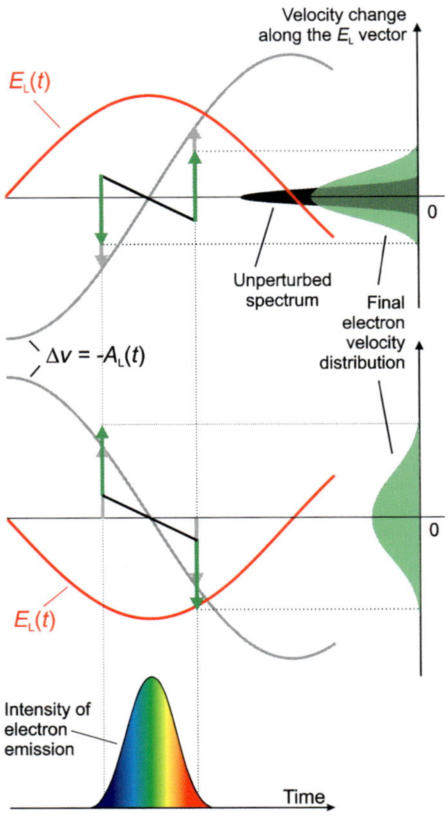

FIG. 28. (Color) Attosecond streak imaging of sub-fs pulses carrying lowest-order (linear) chirp. The linear chirp causes a linear sweep of initial photoelectron velocities. The laser field also sweeps the electron velocity, the sign and strength of which depends on the timing of the streaking field. During the positive half cycle of $E_L(t)$, see top panel, the two sweeps counteract each other, reducing the final width of the streaked spectrum as compared to the one expected for a Fourier-limited XUV pulse of identical bandwidth. In an adjacent half cycle $[E_L(t)<0]$ the sweeps add, enhancing the broadening of the "streak image," as shown in the lower panel. Thus comparison of streak images recorded with vector potentials of different slopes at the instant of release yields not only the duration but also the chirp of a sub-fs pulse.

$$\alpha = \sqrt{1-[2U_p(t)/W_0]\sin^2\theta\sin^2(\omega_L t+\varphi)} \tag{35}$$

(Drescher *et al.*, 2001; Itatani *et al.*, 2002; Quéré *et al.*, 2005). The observation angle θ is the angle the final velocity makes with the x direction and the approximate expression is valid in the limit of a slowly varying envelope; see Eq. (12). This expression allows us to select the optimum detection geometry for various applications.

Observation of electrons along the laser polarization direction is referred to as *parallel detection geometry*, $\theta\approx 0$; see Fig. 29(a). In the case of $W_0 \gg U_p$, this geometry offers the largest streaking owing to the third term in Eq. (35) and the advantage of direct correspondence of the streak image with the wave packet's temporal profile for a near-transform-limited pulse; see Fig. 27. These benefits come in combination with the possibility of a large collection angle. In fact, $\cos\theta$ does not change much within a detection cone as large as ±30°. The disadvantage of this geometry is that the laser-induced energy shift ΔW depends on the initial electron energy. This dependence becomes significant if the bandwidth of the electron wave packet is comparable to its mean energy. For this case, the *perpendicular detection geometry*, $\theta\approx\pi/2$, shown in Fig. 29(b) may be more advantageous, because the third term in Eq. (35) vanishes, and with it the dependence of streaking on the initial energy.

Attosecond streaking maps the temporal profile of a sub-fs pulse to a distribution of final electron energies. Hence the resolution limit Δt, indicating the duration of the shortest pulse that can be measured, is dictated by the condition that the laser-field-induced change in the electron energy should be comparable to the initial energy spread of the wave packet given by the ionizing XUV pulse. For the most favorable conditions, in the parallel detection geometry near the peak of the laser field, these considerations yield

$$\Delta t \sim 1/\sqrt{\omega_L \Delta W_{\max}}, \tag{36}$$

where $\Delta W_{\max}=v_0E_0/\omega_L$ is the maximum energy shift induced by the streaking field at t_1 and t_3 in Fig. 29(a) (Kienberger *et al.*, 2004). This is an order-of-magnitude estimate. As always, with good S/N this resolution limit can be improved (Smirnova *et al.*, 2005; Yakovlev *et al.*, 2005; Goulielmakis *et al.*, 2008).

2. Quantum analysis: The light field as an electron phase modulator

We now address the quantum treatment of photoionization in the presence of an intense laser field (Itatani *et al.*, 2002; Kitzler *et al.*, 2002). This will allow us to look at laser-affected photoelectron spectra from various perspectives, unifying all advanced measurement and reconstruction approaches of modern ultrafast metrology: streak imaging, frequency-resolved optical gating, spectral shearing interferometry, and tomographic techniques.

Consider the bound wave function of our atom $\psi_g=\phi_g(\mathbf{r})\exp(-iW_gt)=\phi_g(\mathbf{r})\exp(iI_pt)$. In the absence of the laser field, first-order perturbation theory in terms of the XUV field defined by Eq. (33) yields the amplitude of populating the state with final velocity v,

$$a(v) = -i\int_{-T}^{T}\exp\left(-i\frac{1}{2}\int_t^T v^2dt'\right)a_X(t) \times e^{-i\Omega t}d(v)\exp\left(i\int_{-T}^{t}I_pdt'\right)dt. \tag{37}$$

Here $d(v)$ is the transition matrix element between the ground state and the free state with momentum v (we detect only electrons ejected parallel to the laser polarization direction; thus the electron momentum can be described by a scalar variable). T is some distant moment in time, so that the XUV pulse begins past $t=-T$ and ends before $t=+T$. Performing the integrals in the

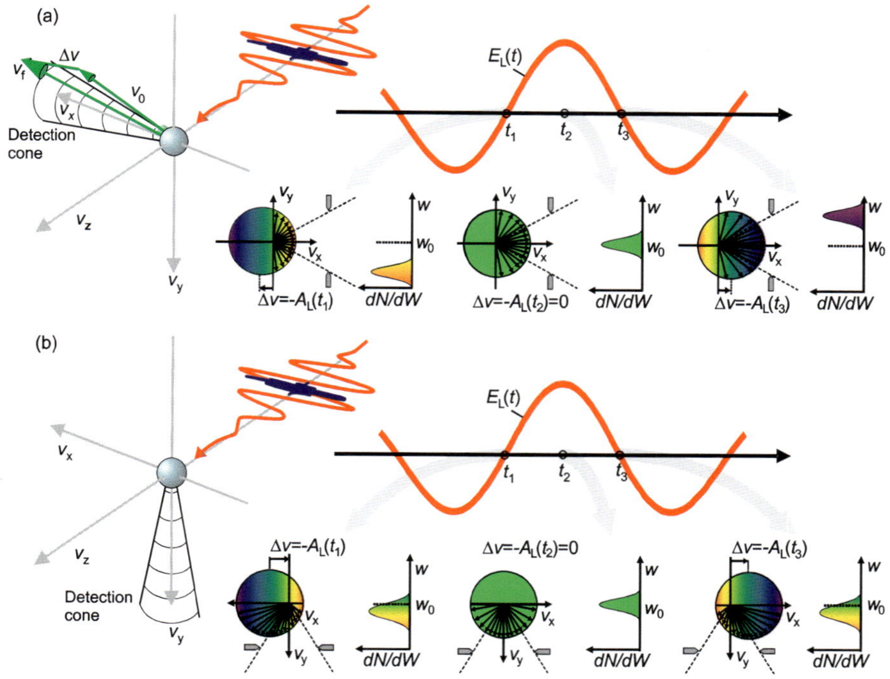

FIG. 29. (Color) Attosecond streaking: (a) parallel and (b) perpendicular detection geometries. Distribution of final electron velocities and corresponding energy distribution of electrons collected within the detection cone for three representative instants of electron release within the central half oscillations cycle of the few-cycle streaking light field. The laser-streaked electron spectra are illustrated for the hypothetical case of an XUV pulse duration much shorter than the half period of the streaking field.

exponents, using Eq. (32), dropping a constant phase factor, and letting $T\to\infty$, we obtain

$$a(v) = -i\int_{-\infty}^{\infty} \chi(v,t)e^{i(v^2/2)t}dt, \tag{38}$$

where we have introduced

$$\chi(v,t) = a_X(t)d(v)e^{-i(v_0^2/2)t}, \tag{39}$$

which plays the role of a quantum-mechanical distribution released by the XUV pulse (Yakovlev *et al.*, 2005);[59] here $v_0^2/2=\Omega-I_p$. In Eq. (38) we recognize the Fourier transform of $\chi(v,t)$. In the absence of resonances within the band of transition energies (determined by the spectral width of the XUV pulse), $d(v)$ is a smooth function with nearly constant phase [and $|d(v)|$ is known from experiment]. Hence the complex envelope of the XUV pulse $a_X(t)$ could be determined by a simple inverse Fourier transform of $a(v)$. Unfortunately, a simple photoelectron experiment measures $|a(v)|^2$ and not $a(v)$, with the phase of the transition amplitude remaining unknown.

To retrieve the unknown phase, we switch on the streaking field. In the framework of the SFA, the amplitude of populating the state with the final velocity v becomes

$$\begin{aligned} a(v) = -i\int_{-T}^{T} \exp\left(-i\frac{1}{2}\int_{t}^{T}[v+A_L(t')]^2dt'\right) \\ \times a_X(t)e^{-i\Omega_X t}d[v+A_L(t)] \\ \times\exp\left(i\int_{-T}^{t}I_pdt'\right)dt, \end{aligned} \tag{40}$$

where the first exponent in the integrand comes from the second term in Eq. (17). Assuming that the dipole matrix element is not affected much by the laser field, $d(v+A_L(t))\approx d(v)$, and with $T\to\infty$, we can rewrite Eq. (40) in the form

$$\begin{aligned} a_A(v) &= -i\int_{-\infty}^{\infty} \chi(v,t)\exp\left(-i\frac{1}{2}\int_{t}^{T}[v+A_L(t')]^2dt'\right)dt \\ &= -i\int_{-\infty}^{\infty} \chi(v,t)e^{i\Theta(t)}e^{i(v^2/2)t}dt, \end{aligned} \tag{41}$$

where

[59]A release distribution was also derived for the Auger decay (Smirnova *et al.*, 2003), forming the basis of attosecond Auger spectroscopy (Sec. VII.C).

$$\Theta(t) = -v\int_t^\infty A_L(t')dt' - \frac{1}{2}\int_t^\infty A_L^2(t')dt'$$
$$\approx \frac{vE_0}{\omega_L^2}f(t)\cos(\omega_L t + \varphi) + \frac{E_0^2 t}{4\omega_L^2}f^2(t)$$
$$+ \frac{E_0^2}{8\omega_L^3}f^2(t)\sin 2(\omega_L t + \varphi) \qquad (42)$$

with the approximation applicable if Eq. (12) is valid. According to Eq. (41), quantum mechanically the streaking laser field works as an attosecond electron phase modulator (Kosik *et al.*, 2005; Quéré *et al.*, 2005). As we show in the following sections, this fact allows us to benefit from well-established femtosecond pulse retrieval techniques in reconstructing both the XUV pulse and the streaking laser field from the measurement of a set of laser-streaked photoelectron spectra,

$$\sigma_A(v,t_0) = |a_A(v,t_0)|^2, \qquad (43)$$

recorded for different values of t_0, where t_0 is the moment of arrival of the sub-fs pulse.

We now relate this quantum treatment to the classical approach of the previous section. If the XUV pulse and hence the released electron wave packet are short compared with the laser oscillation half cycle, the phase $\Theta(t)$ can be Taylor expanded about the moment of release t_0. If we restrict expansion to the linear term $\Theta(t) \approx \Theta_0 + (\partial\Theta/\partial t)_{t_0}(t-t_0)$ in Eq. (41), the resultant linear phase modulation shifts the spectrum of the electron wave packet and hence its mean energy by $\Delta W = W_f - W_0 = v_f^2/2 - v_0^2/2 = -\partial\Theta/\partial t$. Calculating the first derivative of $\Theta(t)$ given by Eq. (42) using the approximation $\Delta W \ll W_0 \Rightarrow v \approx v_0$ in Eq. (41) leads us back to the classical expression for the laser-induced electron energy shift as given by Eq. (35). If the XUV pulse is longer than the laser wave cycle, the electron wave packet experiences several periodic modulations of its phase, leading to energy shifts in multiples of the laser photon energy ω_L. The final photoelectron spectrum is then composed of a series of sidebands separated by multiples of ω_L from the unperturbed spectrum.[60] This spectral structure is the result of quantum-mechanical interference and cannot be accounted for by our classical model. Only sub-fs confinement of the released electron wave packet, i.e., attosecond technology, enables classical physics to account for the main features of intense light-electron interaction (Drescher and Krausz, 2005).

3. Attosecond frequency-resolved optical gating

We rewrite Eq. (41) to explicitly include the adjustable delay τ between the phase-modulating laser pulse and the XUV pulse:

$$\sigma_A(v,\tau) = |a_A(v,\tau)|^2 = \left|\int_{-\infty}^{\infty}\chi(v,t)G(t-\tau)e^{i(v^2/2)t}dt\right|^2,$$
$$G(t-\tau) = e^{i\Theta(t-\tau)}. \qquad (44)$$

The analogy with Eq. (2) is evident. The streaked photoelectron spectrum is the gated Fourier transform of the electron release distribution $\chi(v,t)$. Here the gate is not an amplitude gate as in standard FROG techniques, but a phase gate. Due to its sub-fs gradients, it is well suited to resolving sub-fs transients. Hence $\sigma_A(v,\tau)$ can be viewed as an attosecond spectrogram and the iterative algorithms developed for complete retrieval of femtosecond pulses from FROG measurements[61] can be used to extract both $\chi(v,t)$ and the phase gate $\Theta(t)$. From them, the complex amplitude of the sub-fs XUV pulse $a_X(t)$, as well as the complete evolution of the streaking electric field $E_L(t)$, can be reconstructed. The method was dubbed CRAB-FROG (briefly, CRAB).[62] At least, in principle, it is applicable to retrieving arbitrarily complex and extended attosecond pulse structures (including trains of pulses) and femtosecond laser fields (including multicycle fields). CRAB generalizes RABBITT in that it is, in principle, capable of reconstructing not only the average characteristics of sub-fs pulses in the train, but also each individual pulse, although such a retrieval imposes increasing demands on S/N with increasing complexity of the spectrogram (see Fig. 30 for the case of a sequence of merely four XUV bursts). When implemented with a few-cycle NIR field and sub-optical-period XUV emission durations, CRAB constitutes an extremely powerful method of attosecond metrology. Due to a large redundancy of information in the spectrogram, it can, with good S/N, substantially improve the resolution given by Eq. (36).

4. Attosecond spectral shear interferometry

The attosecond FROG described in the previous section naturally leads us to the attosecond version of SPIDER. This becomes obvious by a closer glance at Figs. 30(c) and 30(d), showing streaking spectrograms of sub-fs XUV radiation composed of two identical bursts of 150 as duration spaced by the half period $T_L/2$ of the streaking field. The spectral fringes spaced by $2\omega_L \approx 3$ eV appear to be due to spectral interference between the two electron wave packets released by the two XUV replicas, which is caused by the difference in phase shift $\Delta\Theta = \Theta(t+T_L/2-\tau) - \Theta(t-\tau)$ imposed by the streaking field upon the wave packets. At selected delays τ, the two released wave packets coincide with adjacent zero

[60]They can be used for measuring the duration of femtosecond XUV pulses, see Schins *et al.*, 1994, 1996; Glover *et al.*, 1996; Bouhal *et al.*, 1997; Toma, *et al.*, 2000; Ge, 2006.

[61]Such as, e.g., the principal component generalized projections algorithm, PCGPA (Kane, 1999).

[62]Complete reconstruction of attosecond bursts via FROG (Mairesse and Quéré, 2005; Quéré *et al.*, 2005).

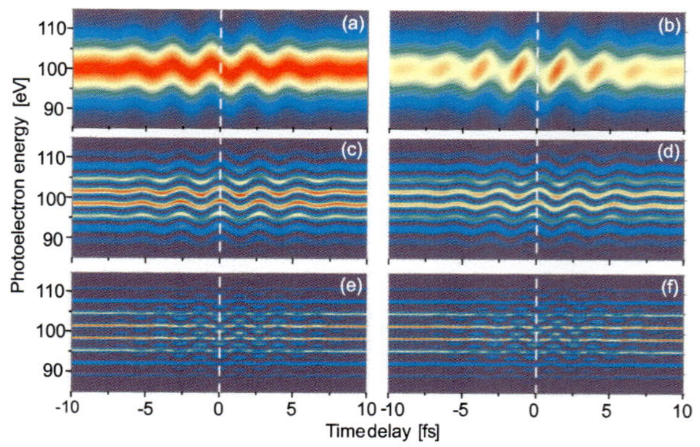

FIG. 30. (Color) Calculated attosecond streaking spectrograms. Top panels show spectrograms for (a) a single transform-limited 150-as pulse and (b) the same pulse dispersively stretched to a duration of 620 as. Middle panels show the same for two pulses separated by half the oscillation cycle of the streaking field. Bottom panels show corresponding traces for four pulses. The traces have been calculated for parallel detection geometry, see Fig. 29(a), and for a laser peak intensity of $I=5\times10^{10}$ W/cm^2 at a wavelength of $\lambda_L=800$ nm. Adapted from Quere *et al.*, 2005.

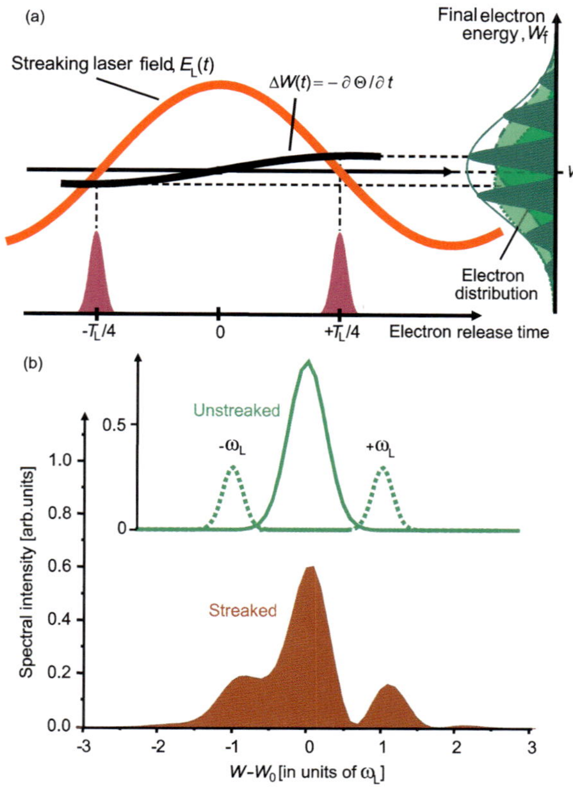

FIG. 31. (Color) SPIDER for attosecond XUV pulses. (a) SPIDER interferogram (in the form of a photoelectron energy distribution; see green area) for two delayed replicas of the XUV pulse, slightly sheared by a weak IR field (red line shows electric field of the shearing IR field). The amount of spectral shear between the two electron wave packets released by the two sub-fs XUV bursts can be adjusted by the strength of the streaking field. It is preferable that the spectral shear is small as compared to the width of the unperturbed electron energy spectrum, a demand precisely opposite to that required for the best possible resolution in streak imaging. (b) In the presence of a weak IR field, two spectral replicas of the same sub-fs XUV pulse corresponding to stimulated absorption or emission of a single photon (green dashed lines on the inset) may—together with the main (unperturbed) photoelectron spectrum—combine to SPIDER-like interferogram in the total spectrum.

transitions of the streaking field $E_L(t-\tau)$, when the vector potential passes its maximum and minimum. Recalling the relationship between the initial and final electron velocities $v_f=v_0-A_L(t_{\text{release}})$, we realize that one wave packet is shifted up in the final energy while the other is shifted down; see Fig. 31(a). Note that the streaking field strength is adjusted here to be much weaker than in the case of streak imaging (Fig. 27). Since the two replicas are also time delayed with respect to each other, SPIDER is, in principle, only a small step from here: one merely has to generate two XUV replicas from one and the same XUV pulse that we want to characterize (Cormier *et al.*, 2005). The modulated spectrum in Fig. 31(a) then becomes the perfect attosecond version of a SPIDER interferogram (Quéré *et al.*, 2003; Remetter *et al.*, 2006).

In attosecond metrology, there is a regime (Sekikawa *et al.*, 2003) where SPIDER- (Cormier *et al.*, 2005; Kosik *et al.*, 2005; Quéré *et al.*, 2005) and FROG-type characterization methods almost merge (Smirnova *et al.*, 2005; Yakovlev *et al.*, 2005). Let the original sub-fs pulse (without splitting it into two) interact with a weak streaking field. Without the streaking field, absorption of the XUV photon generates a photoelectron spectrum but the measurement is blind to the spectral phase of the photoelectron wave packet. With a perturbative streaking field present, the liberated electron can still absorb one photon from the IR field, creating an upshifted "replica" of the original photoelectron spectrum. Stimulated emission of a laser photon will create a downshifted replica. Interference of these replicas with the original spectrum transforms the spectral phase into amplitude modulation, just as in SPIDER, except that here one has to deal with three interfering spectral shapes and not two; see Fig. 31(b).

Despite complications in experimental implementation or numerical reconstruction, attosecond SPIDER may become the method of choice for pulses approaching the atomic unit of time (24 as), because SPIDER has no fundamental resolution limit. In practice, the time resolution will be limited by the precision of controlling the energy shear, which in turn is dictated by the quality of controlling the timing of the attosecond pulse with respect to the streaking field. Beyond attosecond pulse metrology, attosecond spectral shearing interferometry may also find intriguing applications in exploring correlated multielectron processes (Smirnova *et al.*, 2005); see Sec. VII.

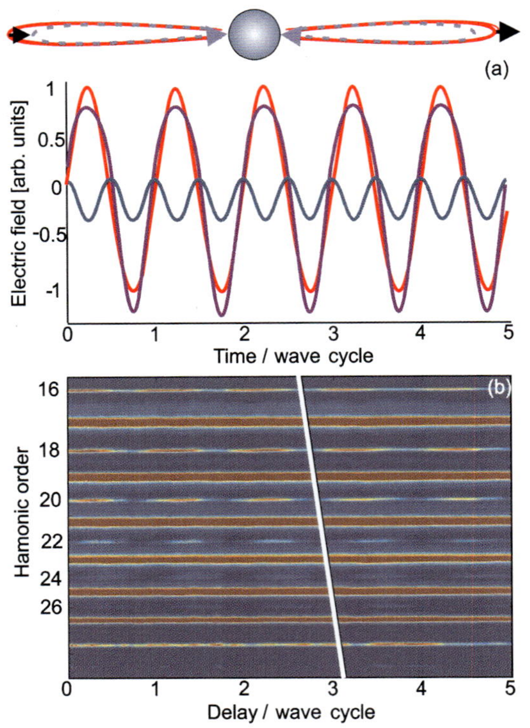

FIG. 32. (Color) Gating and control of electron recollision. (a) Adding weak second harmonic (blue solid line) to the fundamental field (red line) driving ionization and subsequent recollision breaks the symmetry between two electron trajectories starting at adjacent half cycles of the fundamental field (illustration on the top) due to a corresponding asymmetry in the total field (dashed line). This symmetry breaking yields even high-order harmonics. (b) Harmonic spectral intensity (in false-color representation) vs delay between the fundamental wave and the weak second harmonic modulating field. The white line indicates the variation of the delay which maximizes even harmonics of different order. From Dudovich, Smirnova, *et al.*, 2006.

E. Lightwave electronics at work: From measurement to control of sub-fs pulses

The quantum-mechanical description of a strong light field interacting with an electron ejected from an atom revealed that the light field modulates the phase of the outgoing electron's wave function. It is this sub-fs phase gate that light-field-controlled attosecond metrology is based upon. The electron can be released by either photoexcitation or collisional excitation. It must be timed, with sub-fs precision, to the streaking laser field, which modulates the electron's quantum phase. This requirement can be met only by optical synchronization. Therefore the "starter gun" must be derived from the streaking field. Photoemission can be linear, induced by an XUV photon pulse via single-photon absorption, or nonlinear, induced by a strong light field. Our observable can be a characteristic of the modulated electrons themselves, such as their final energy and momentum distributions (as considered in the previous section), or the energy and momentum distributions of the photons they produce upon recollision. The observables are measured as a function of the delay of the streaking field with respect to electron release. Using a waveform-controlled few-cycle streaking field, the complete history of the motion—including attosecond-scale dynamics and (possible) slower femtosecond-scale evolution—can be retrieved.

On the basis of different combinations of triggering of the electron release and observation of their modulated quantum phase, many variations of attosecond control and metrology are conceivable. Two combinations have been demonstrated so far. One of them provides a profound insight into the atomic process responsible for sub-fs XUV pulse generation. The other one permits complete temporal characterization of the sub-fs pulse at the location where it can be used for attosecond spectroscopy. We review them in this section.

1. Modulation of recollision electrons: Measuring the creation of a sub-fs pulse

Consider the electron liberated by a strong, linearly polarized laser field and its subsequent recollision with its parent ion (Fig. 15). Light-field-controlled attosecond metrology offers a simple recipe for gaining time-domain insight into recollision: (i) modulate the electron's phase or amplitude with a weak auxiliary light field and (ii) monitor the response of the products of recollision (e.g., photon emission) to this modulation. Measuring this response versus delay of the modulating field with respect to the electron's release provides in-depth time-domain information on the dynamics of recollision and concomitant sub-fs XUV emission (Dudovich, Levesque, *et al.*, 2006; Dudovich, Smirnova, *et al.*, 2006).

In the example sketched in Fig. 32(a), a weak light field oscillating at $2\omega_L$ is added to the fundamental light of frequency ω_L that drives ionization and recollision. This introduces a small additional phase shift Θ_2 to the quantum phase Θ_1 that the electron accumulates in the fundamental field from ionization at t_0 until recollision at t_r:

$$\Theta_{1+2}(t_r,\tau) \approx \Theta_1(t_r) + \Theta_2(t_r,\tau),$$

$$\Theta_2(t_r,\tau) \approx -\int_{t_r(t_0)}^{t_0} v(t',t_r)A_2(t'-\tau)dt', \qquad (45)$$

where $A_2(t-\tau)=A_2\sin[2\omega_L(t-\tau)]$ is the vector potential of the second field and $v(t,t_r)$ is the velocity of the electron returning to the core at t_r.[63] Although Θ_2 is much smaller than Θ_1, it has a large effect. Because $\Theta_2(t_r - \pi/\omega_L,\tau) = -\Theta_2(t_r,\tau)$, the weak field breaks the symme-

[63]The total phase is given by the second term in the classical action in Eq. (17), with the vector potential $A_L(t)=A_1(t)+A_2(t)$. Expanding the square in Eq. (17) and keeping terms linear in weak second harmonic field, we obtain Eq. (45).

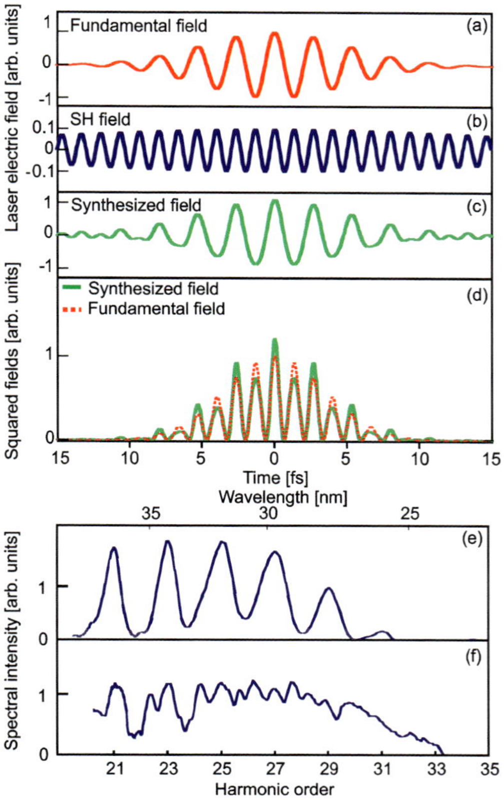

FIG. 33. (Color) A few-cycle field (c) and (d), synthesized, (a) from a fundamental wave, and (b) its second harmonic, is used to produce high-harmonic radiation with (f) a broad unmodulated continuum, for isolated sub-fs XUV pulse generation. For comparison, (e) shows the harmonic spectrum obtained with the fundamental driver field alone. From Oishi *et al.*, 2006.

try between the left and the right trajectories[64] of electrons returning to the core; see Fig. 32(a). As a result, even harmonics of ω_L appear in the harmonic spectrum, which can be measured as a function of the delay τ or equivalently of the relative phase $\phi=2\omega_L\tau$ of the modulating light with respect to the fundamental field releasing the electron wave packet; see Fig. 32(b). The measurement reveals the relative phase of the modulating field that maximizes the yield in any even harmonic: $\phi_{\max}(2N)$, see white line in Fig. 32(b). At the same time, for weak $2\omega_L$ field the relative phase $\phi_{\max}(t_r)$ maximizing Θ_2 in Eq. (45) and hence even harmonic emission at a given t_r can be reliably calculated using the SFA. Equating these (measured and calculated) quantities with each other, $\phi_{\max}(2N)=\phi_{\max}(t_r)$, maps emission frequency to emission time within the time interval of recollision, yielding the group delay versus frequency of the emitted sub-fs pulse (Dudovich, Smirnova, *et al.*, 2006).

[64]We use this terminology by assuming the laser fields to be polarized horizontally.

2. Modulation of recollision electrons: Controlling the creation of a sub-fs pulse

A simple increase of the modulating field amplitude turns the previous measurement into control. By varying ϕ, the emission can be switched completely between odd and even harmonics (Dudovich, Smirnova, *et al.*, 2006). Adjusting the modulating field so that even and odd harmonics are similar in strength implies that the sub-fs pulses repeat every cycle rather than the half cycle of the fundamental (Mauritsson *et al.*, 2006).[65]

If atoms are driven by a cosine-shaped few-cycle field [Fig. 33(a)], adding its second harmonic with appropriate phase [Fig. 33(b)] suppresses the half cycles neighboring the central one [Figs. 33(c) and 33(d)], confining highest-order harmonic emission to the recollision following the most intense, central half cycle. The result is a broad continuum near the cutoff, as shown in Fig. 33(f) (Oishi *et al.*, 2006).[66] Adding a weak sub-fs pulse or pulse train to the fundamental field provides yet another way of controlling HHG.[67]

The trajectory of recolliding electrons can also be manipulated by adding an orthogonally polarized moderate deflecting field to the fundamental field. By creating a pulse with temporally varying ellipticity of its polarization, linear polarization can be confined to the central wave cycle of the strong driving field, confining emission from a broad range of harmonics to a single recollision at the pulse peak.[68] The technique was dubbed polarization gating. Its recent implementation with waveform-controlled few-cycle light resulted in a continuum extending over some 25 eV at photon energies near 40 eV (Sola, Mével, *et al.*, 2006; Sola, Zair, *et al.*, 2006).

Ultrafast depletion of the ground state by ionization on the leading edge of a few-cycle laser pulse (Sekikawa *et al.*, 2004)[69] and propagation effects in wave guides (Christov *et al.*, 2000) as well as in gas jets (Gaarde *et al.*, 2006) can also select a single pulse.

[65]One attosecond burst per cycle may also emerge from HHG in asymmetric molecules (Lan, Lu, Cao, Li, and Wang, 2007).

[66]Related work includes Watanabe *et al.*, 1994; Kim, Kim, Kim, *et al.*, 2004; Taranukhin, 2004; Zamith *et al.*, 2004; Pfeifer *et al.*, 2006a, 2006b; Fleischer and Moiseyev, 2006; Cao, Lu, Lan, Hong, and Wang *et al.*, 2007; Lan, Lu, Cao, Li, and Wang, 2007; Zeng *et al.*, 2007.

[67]Bandrauk and Shon, 2002; Schafer *et al.*, 2004; Biegert *et al.*, 2006; Figueira de Morisson Faria *et al.*, 2006; Heinrich *et al.*, 2006; Chen, Li, Chi, and Yang, 2007; Figueira de Morisson Faria and Salieres, 2007.

[68]Budil *et al.*, 1993; Corkum *et al.*, 1994; Ivanov *et al.*, 1995; Altucci *et al.*, 1998; Platonenko and Strelkov, 1999; Kovacev *et al.*, 2003; Tcherbakoff *et al.*, 2003; Chang, 2004; López-Martens *et al.*, 2004; Strelkov *et al.*, 2004; Zair *et al.*, 2004; Huo, Zeng, Li, and Xu, 2005; Kim *et al.*, 2005; Shan *et al.*, 2005; Strelkov *et al.*, 2005; Strelkov, 2006; Altucci *et al.*, 2007.

[69]See also Bouhal *et al.*, 1997, 1998; Cao *et al.*, 2006; and Pfeifer *et al.*, 2007.

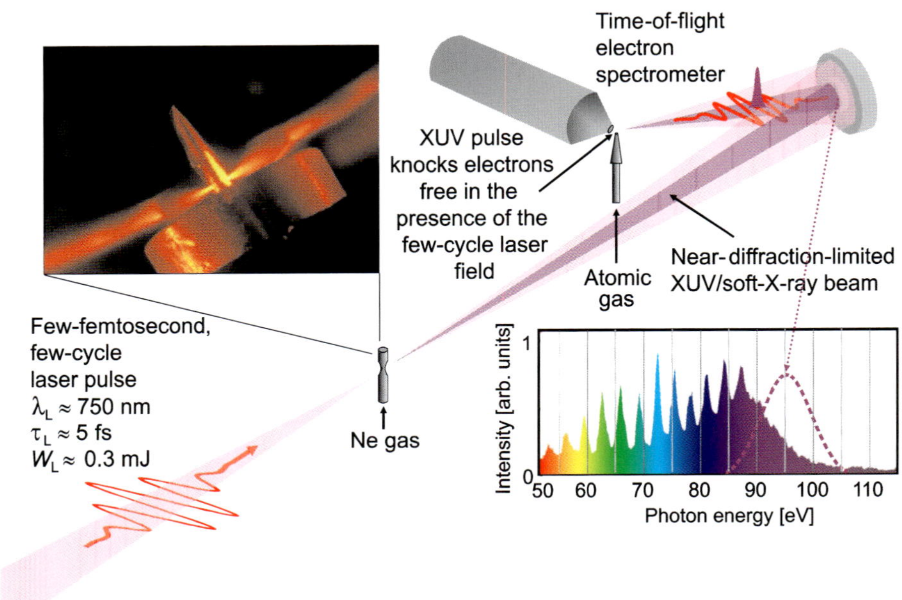

FIG. 34. (Color) Schematic of the first attosecond metrology beamline using waveform-controlled few-cycle light and isolated sub-fs XUV pulses (Vienna University of Technology).

3. Reproducible generation and measurement of single sub-fs XUV pulses

In this section we review the first experiments demonstrating the ability of controlled few-cycle light waves to control and measure subfemtosecond electronic transients. The laser pulse first generates the sub-fs pulse in a reproducible fashion (attosecond control) and subsequently samples it with sub-100-as resolution (attosecond measurement). Employing the same light wave for generation and measurement ensures sub-fs synchronism between the XUV pulse and the sampling field.

Figure 34 illustrates the apparatus developed for sub-fs pulse generation and measurement at Vienna University of Technology. Figure 35 provides information on the diagnostics required for making attosecond metrology work in a second-generation beamline. In

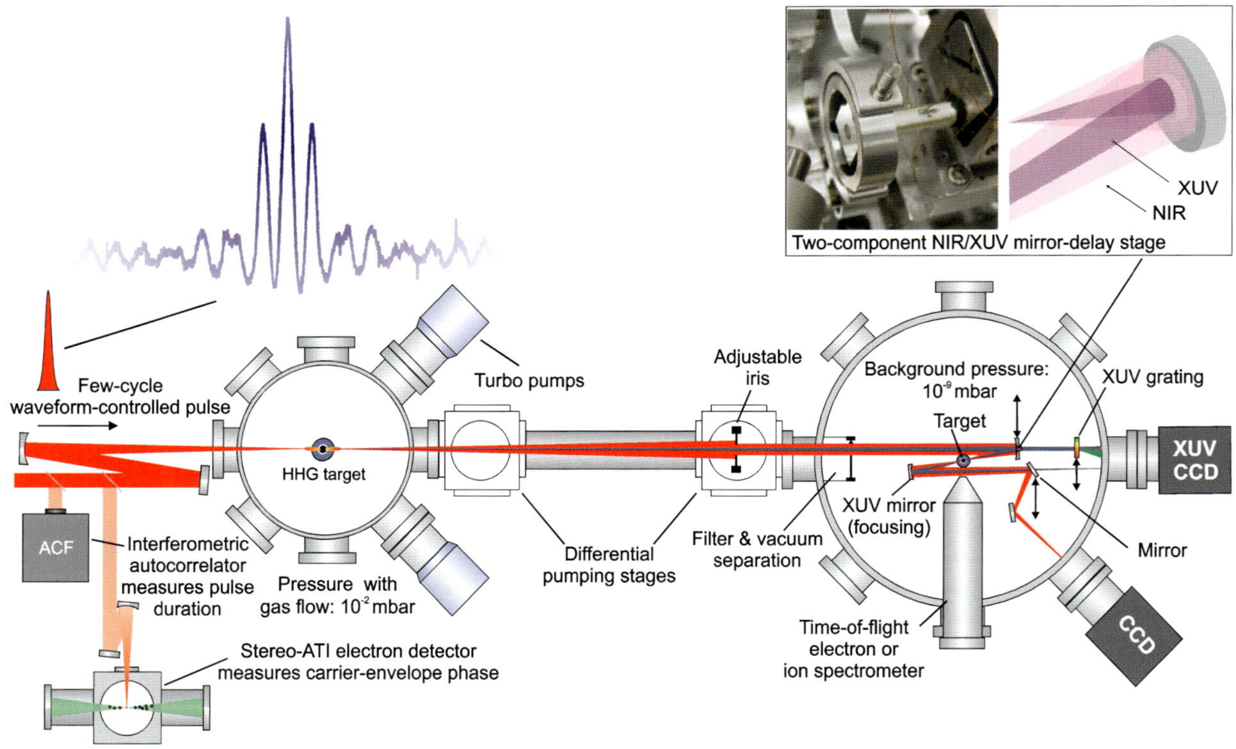

FIG. 35. (Color) Schematic of a the AS-1 attosecond metrology beamline at Max Planck Institute of Quantum Optics.

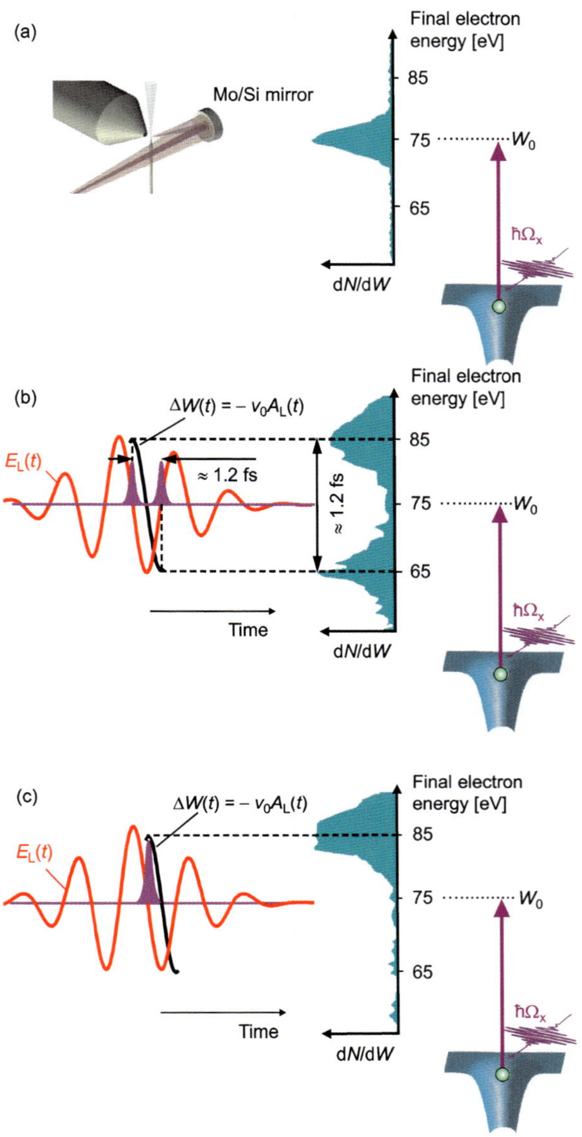

FIG. 36. (Color) Attosecond streak images of single and twin sub-fs XUV pulses. (a) In the absence of the streaking field $E_L(t)$, photoelectrons arrive at the detector with a kinetic energy equal to the XUV photon energy (~95 eV) minus their binding energy (~21.5 eV in neon). The width of their energy spread is determined by the bandwidth of the incident XUV pulse (~9 eV), which in turn is controlled by that of the Mo/Si mirror (dotted line in Fig. 34). (b) The streak image of twin pulses generated by a near-sinusoidal waveform allows calibration of the attosecond streak camera: 20-eV energy shift corresponds to a time interval of $T_L/2 \approx 1.2$ fs. (c) Streak image of a single sub-fs XUV pulse generated with a near-cosine waveform. No delay between the XUV pulse and its generating few-cycle wave has been introduced. The measurement yields an upper limit of 0.5 fs for the duration of the 95-eV XUV pulse and its absolute timing with respect to the generating wave: the sub-fs XUV pulse emerges near the zero transition of the laser field following the pulse peak. The laser-induced change in the electron's final energy, $\Delta W(t)$, is derived from the momentum change, $\Delta v(t) = -A_L(t)$, see Eq. (34) and Fig. 27, as $\Delta W(t) \approx v_0 \Delta v(t) = -v_0 A_L(t)$, by utilizing $\Delta v \ll v_0$.

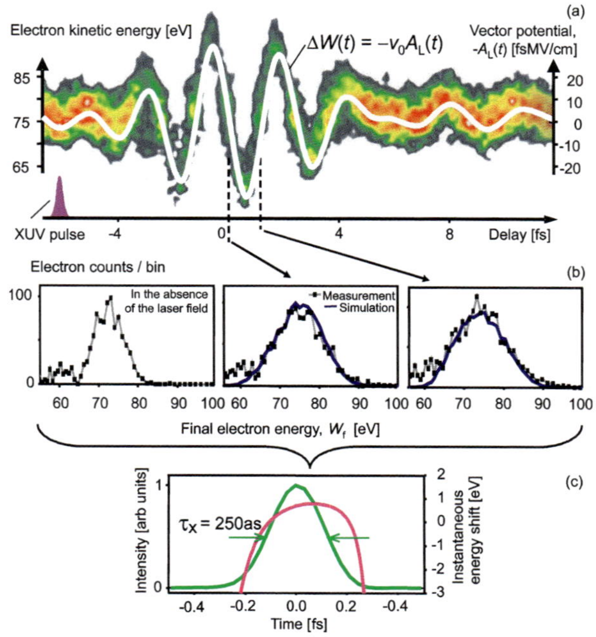

FIG. 37. (Color) Attosecond streaking spectrogram of an XUV pulse, recorded with a few-cycle streaking field. The streak images exhibit little broadening at any timing, indicating that the XUV pulse is near Fourier limited and is much shorter than $T_L/2$. As a consequence, one does not need to know its detailed structure to determine $E_L(t)$. The shift of the first moment ("center of mass") of the streaked spectra is proportional to $A_L(t)$. From the vector potential (white line) the electric field can be determined by differentiation. The result is plotted in Fig. 12. Since the XUV pulse appears to be near Fourier limited, its characteristics can be determined from a couple of streak images recorded at delays for which streaking is strongest, indicated with dashed lines in (a). As analyzed in Fig. 28, these streaked spectra (b) are very sensitive to lowest-order (linear) chirp. The absence of any measurable difference between the two streak images, see central and right panels of (b), indicate that the pulse is indeed close to its Fourier limit. Together with the unstreaked photoelectron spectrum shown on the left of (b), these images allow for complete reconstruction of the intensity profile (green line) and temporal chirp (violet line) of the sub-fs XUV pulse, as shown in (c).

these systems, few-cycle-driven high-order harmonics are generated in neon, with the peak intensity of the few-cycle driver adjusted to result in cutoff emission near 100 eV; see inset in Fig. 34. The XUV radiation, emitted in a collimated laserlike beam with a divergence much smaller than that of the focused laser beam, is passed through a metal filter blocking the laser light. The two collinear beams are reflected by a two-component Mo/Si multilayer mirror, which serves several purposes. It filters XUV photons in the ~90–100 eV energy band, allows an adjustable delay to be introduced between the laser field and the XUV pulse, and focuses both beams into a second target of atoms where the XUV pulse liberates electrons in the presence of the laser field. The photoelectrons' final en-

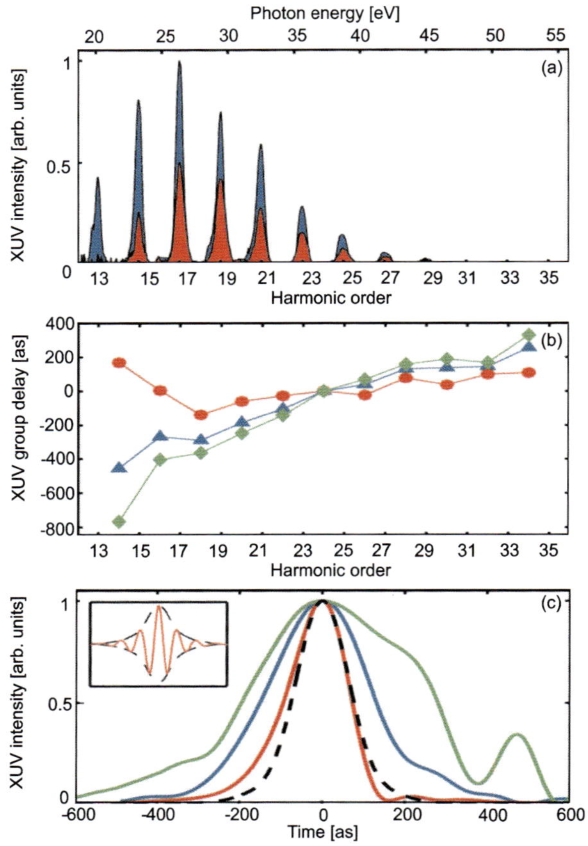

FIG. 38. (Color) Generation of trains of near-single-cycle 170-as pulses by trajectory and dispersion control (López-Martens *et al.*, 2005). (a) Emission from short trajectories, see Fig. 20, was selected by spatial filtering with an iris. (b) Measured group delay vs XUV frequency (harmonic order). The green curve was obtained without any dispersion added, indicating a positive frequency sweep, in accordance with the theoretical predictions for short trajectories; see Figs. 20 and 22. The blue and red curves were obtained with dispersion compensation via one and three 200-nm-thick aluminum filters placed in the XUV beam, respectively. (c) Corresponding temporal profiles of the transmitted XUV pulses. Courtesy of A. L'Huillier.

ergy distribution is analyzed in a time-of-flight spectrometer.

The first attosecond streaking experiment (Hentschel *et al.*, 2001) was performed using orthogonal detection geometry [Fig. 29(b)]. Although the CE phase of the few-cycle driver pulse was not stabilized, single sub-fs XUV pulses were observed in the experiment. Recently, Gaarde and co-workers have indeed found numerically that propagation effects and far-field spatial filtering can effectively isolate a single attosecond burst for a broad range of the CE phase values (Gaarde *et al.*, 2006; Gaarde and Schafer, 2006).

Figure 36 reviews the follow-up experiments, performed with phase-stabilized pulses in the parallel geometry in Fig. 29(a) (Kienberger *et al.*, 2004). The results reveal controlled emission of a single or double sub-fs XUV pulse[70] and corroborate predictions of the semiclassical theory summarized in Fig. 22. For an accurate measurement of the XUV pulse as well as the few-cycle laser wave, the XUV pulse has been delayed in steps of 150 as to scan its arrival time in the second target across the laser wave. Figure 37(a) shows the result of this measurement (Goulielmakis *et al.*, 2004), a particularly simple attosecond streaking spectrogram. As discussed in Sec. V.D.3, FROG-type algorithms allow complete reconstruction of the complex amplitude envelope $a_X(t)$ of the sub-fs XUV pulse [see Eq. (33)], and the few-cycle laser field $E_L(t)$.[71] Given the shortness and simplicity of the XUV pulse profile, from the measurements summarized in Fig. 37, reconstruction of $A_L(t)$ and $a_X(t)$ was straightforward and did not require iterative steps, yielding the laser field $E_L(t)$ shown in Fig. 12 and a near-transform-limited XUV pulse duration of 0.25 fs (full width at half maximum) as shown in Fig. 37(c) (Kienberger *et al.*, 2004).

4. Breaking the 100-as barrier

Chirp-free harmonic emission is restricted to a relatively narrow bandwidth of about 10–15 % of the cutoff energy. Below-cutoff harmonics result from electrons returning to the core along short and long trajectories (see Figs. 20 and 22). They produce emission at increasingly different moments, requiring trajectory selection[72] and dispersion control (Kim, Kim, Baik, *et al.*, 2004; López-Martens *et al.*, 2005); see Fig. 38. Along with few-cycle polarization gating, they permitted the production of isolated 36-eV pulses of 130 as in duration (Sansone *et al.*, 2006),[73] see the low-energy spectrogram in Fig. 39.

Using waveform-controlled sub-1.5-cycle NIR laser pulses and chirped XUV multilayers for spectral filtering and dispersion control (Wonisch *et al.*, 2004, 2006; Morlens *et al.*, 2005), isolated sub-100-as pulses at ~85 eV photon energy have also been demonstrated (Goulielmakis, Schultze, *et al.*, 2008); see the high-energy spectrogram in Fig. 39. These pulses are delivered with a flux of $\sim 10^{11}$ photons/s, allowing for time-resolved measurements of electron processes with a resolution approaching the atomic unit of time (~24 as).

[70]While the single pulse is needed for pump-probe studies, the double pulse is ideal for time-domain interferometry (Ishikawa, 2006; Remetter *et al.*, 2006).

[71]By detecting electrons ejected in opposite directions, the method can be generalized to measuring the field evolution in arbitrarily polarized laser pulses (Popruzhenko *et al.*, 2007; Shvetsov-Shilovski *et al.*, 2007).

[72]Antoine *et al.*, 1996, 1997; Lee *et al.*, 2001; López-Martens *et al.*, 2005; Merdji *et al.*, 2006; Cao, Lu, Lan, Wang, and Yang, 2007; Cao, Lu, Lao, Wang, and Li, 2007.

[73]There is an extensive literature on the optimization of single sub-fs pulse generation, see, e.g., Cao *et al.*, 2006; Chakraborty *et al.*, 2006; Lan *et al.*, 2006; Lan, Lu, Cao, Wang, and Hong, 2007; Ishikawa *et al.*, 2007; Popov *et al.*, 2007.

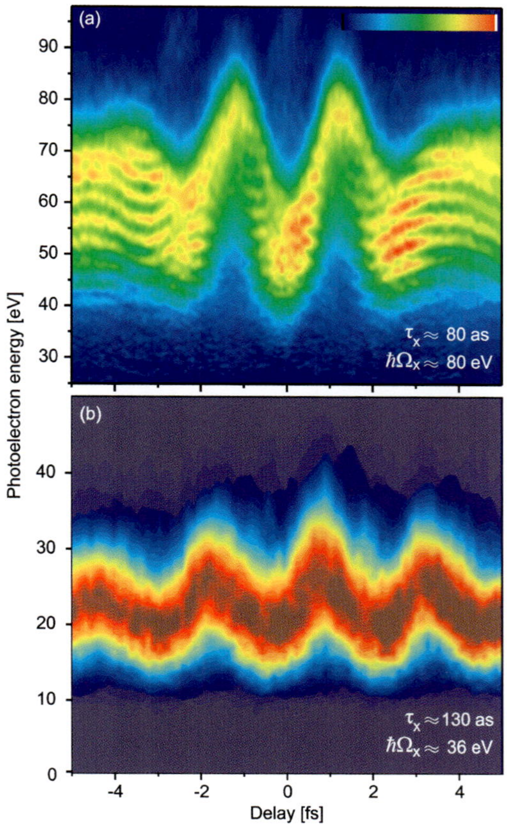

FIG. 39. (Color) Current frontiers of attosecond technology. (a) Attosecond streaking spectrograms of isolated XUV pulses generated and recorded with few-cycle, (a) ~3.5 fs and (b) ~5 fs ~750-nm pulses. Low-energy spectrogram: carrier photon energy ≈36 eV; target atoms: argon (Sansone *et al.*, 2006). High-energy spectrogram: carrier photon energy ~80 eV; target atoms: neon (Goulielmakis, Schultze, *et al.*, 2008). Measurements yielded XUV pulse durations of 78 and 130 as, respectively.

VI. SUB-fs ELECTRON AND PHOTON PULSES: CURRENT STATUS AND FUTURE PROSPECTS

High-order harmonic generation (HHG) from atoms and molecules has so far been the only source of sub-fs pulses suitable for spectroscopic use. The technique has played a central role in establishing attosecond technology, but the limited photon energy and photon flux currently available may stall its dynamic expansion. Promising routes to higher photon energy and yield include atomic HHG driven at longer wavelengths and exploitation of the interaction of relativistic electrons with intense laser light. They are discussed below.

A. Atomic electrons interacting with few-cycle fields: Atomic HHG

The physics of atomic HHG has been reviewed in Sec. IV.C.2. Here we address possible ways of scaling the technique to higher photon energies at experimentally useful flux levels. Such an advance will require us to overcome current limitations. The most fundamental is reabsorption of the harmonics by the generation medium itself. Absorption-limited HHG has been demonstrated up to photon energies of ~100 eV with conversion efficiencies ranging from 10^{-4} to 10^{-6} over the spectral range of 10–100 eV, respectively.[74] Absorption rapidly decreases for increasing harmonic photon energies above 100 eV. Here another limiting effect becomes dominant: dephasing between the driving wave and the harmonic wave due to free-electron-induced dispersion, limiting the coherent growth of the harmonic wave. However, unlike absorption, dephasing does not constitute an ultimate limitation. It can be overcome by periodic readjustment of the phase difference between the driving and the harmonic wave, a technique called quasi-phase-matching (QPM).[75]

If we can maintain phasing of the driving wave with the growing harmonic wave (by a suitable QPM scheme, see, e.g., Seres *et al.*, 2007) and prevent self-defocusing from occurring (by some guiding structure or beam shaping), dispersion-induced destruction of the driving pulse will become a major limitation. For ionizing fields, free electrons lead to two dispersion-related processes: temporal broadening and blueshift of the carrier frequency. Both effects lower the maximum ponderomotive potential [see Eq. (20)] and hence decrease Ω_{max} [see Eq. (21)] during propagation, limiting the growth of the near-cutoff harmonics. We denote the dispersion-induced pulse destruction length as L_{disp}.

How can plasma-induced dispersion be minimized? By lowering ionization. This, however, lowers the maximum harmonic photon energy. Use of a longer, i.e., infrared (IR) driver wavelength may offer a way out. Then, according to Eq. (21), any selected Ω_{max} achievable at λ_{NIR} can be produced at reduced intensity $|E_{IR}|^2=|E_{NIR}|^2(\lambda_{NIR}/\lambda_{IR})^2$ as compared to the intensity $|E_{NIR}|^2$ required at λ_{NIR}, and consequently at an exponentially reduced ionization rate w_i. This weakens the radiating atomic dipole, $d_h \propto \sqrt{w_i}$ [Eqs. (16) and (22)], but the enhanced coherent buildup may overcompensate this detrimental effect (Yakovlev *et al.*, 2007; Ivanov *et al.*, 2008).

Indeed, the harmonic photon yield F_h at Ω_{max} is proportional to $|N_h d_h|^2$, with N_h denoting the number of

[74]Constant *et al.*, 1999; Schnürer *et al.*, 1999; Hergott *et al.*, 2002; Takahashi *et al.*, 2002, 2004; Goulielmakis, Schultze, *et al.*, 2008.

[75]Phase-matched and quasi-phase-matched (QPM) HHG in the VUV-XUV-SXR spectral range has been experimentally demonstrated by Kapteyn, Murnane, and co-workers, see, e.g., Rundquist *et al.*, 1998; Durfee *et al.*, 1999; Gibson *et al.*, 2003, 2004; Paul *et al.*, 2003; Zhang *et al.*, 2004, 2005 (for a review, see, Paul *et al.*, 2006), and Zepf *et al.*, 2007. Rapid destruction of the driver pulse due to strong ionization prevented the potential of QPM HHG from being fully exploited.

dipoles radiating $\Omega_{\max}$ in phase. Ultimately, N_h is proportional to L_{disp}. The difference in refractive index between the fundamental and high-order harmonic waves induced by free electrons of density N_e scales as $\Delta n(\omega_L) \propto N_e \propto w_i$. As a consequence, $L_{\text{disp}} \propto 1/\Delta n \propto 1/N_e \propto 1/w_i$. The maximum achievable photon yield at frequency $\Omega_{\max}$ therefore scales as

$$F_h \propto |N_h d_h|^2 \propto 1/w_i(E_L). \tag{46}$$

This result predicts favorable scaling with wavelength, given that a moderate increase of λ_L allows dramatic reduction of $w_i(E_L)$ owing to the extremely nonlinear dependence of the tunnelling rate on the laser field strength E_L, which scales as $1/\lambda_L$. Equation (46) is valid as long as dispersion is dominated by free electrons. A more elaborate analysis of Yakovlev *et al.* (2007) corroborates the above conclusions.

IR-driven HHG was pioneered by L'Huillier and coworkers (Balcou *et al.*, 1992) and DiMauro and coworkers (Sheehy *et al.*, 1999), followed by studies of Bellini (2000) and Shan and Chang (2001). As pointed out by DiMauro, a long driver wavelength benefits HHG and strong-field interaction in many ways (Schultz *et al.*, 2007; Tate *et al.*, 2007) and may result in keV attosecond pulses (Seres *et al.*, 2005).

B. Relativistic electrons interacting with few-cycle light

1. Intense attosecond VUV to soft-x-ray pulses via surface HHG?

So far, we have been concerned with light-matter interactions at intensities where the electric field is strong enough to ionize atoms, but too weak to accelerate the freed electrons to speeds approaching the speed of light in vacuum, c. The electron's quiver energy in the light field becomes comparable to its rest energy when the normalized amplitude of the vector potential

$$a_0 = |e|A_0/m_e c \tag{47}$$

becomes comparable to or larger than unity, defining the regime of relativistic light-electron interactions. In terms of laser intensity I_L and laser wavelength λ_L, it can be expressed as $a_0^2 = I_L \lambda_L^2/(1.37\times 10^{18}\ \text{W}\ \mu\text{m}^2/\text{cm}^2)$. Present-day femtosecond laser technology allows a_0 to approach 100 (Mourou *et al.*, 2006). As ionization occurs at much lower intensities, bound-electron nonlinear optics is virtually "switched off" for $a_0 \gg 1$. That is, nonlinear response of matter associated with the motion of electrons near the atomic core is relevant only at the leading edge of ultraintense laser pulses, with only a tiny fraction of the laser pulse "feeling" this response. However, in this regime, free electrons respond in a nonlinear fashion. Indeed, the magnetic component of the Lorentz force $\mathbf{v}\times\mathbf{B}_L$ becomes comparable to that of the electric component $\mathbf{E}_L$. As the electron's velocity $\mathbf{v}$ is initially acquired from $\mathbf{E}_L$, modification of this velocity by $\mathbf{B}_L$ implies a nonlinear response of the electron to the field.

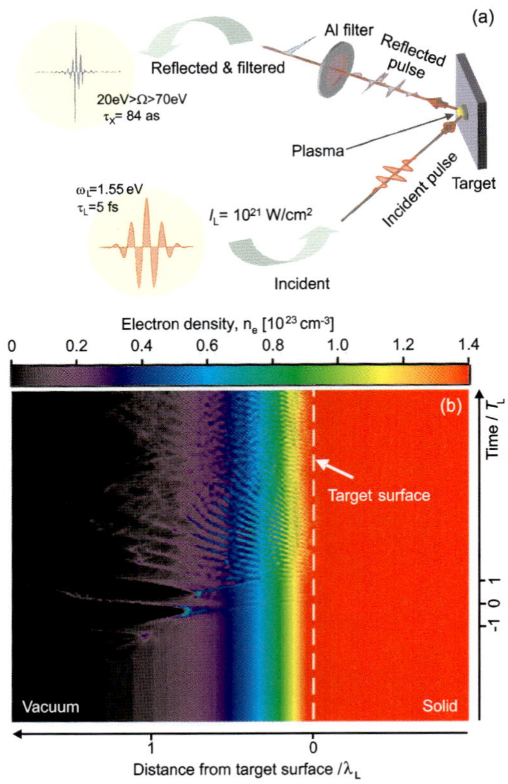

FIG. 40. (Color) High-order harmonic generation on a surface exposed to ultrashort-pulsed laser radiation at relativistic intensities. (a) Incident and reflected few-cycle laser wave, with the latter containing high-frequency components. Their isolation by a band-pass filter (in this example between 20 eV and 70 eV) is predicted to result in a single burst of attosecond duration. (b) Temporal evolution of the electron density of a preformed plasma with scale length $L_p = \lambda_L/4$ upon exposure to a p-polarized, cosine-shaped few-cycle NIR pulse ($\tau_L = 5$ fs, $\lambda_L = 750$ nm) incident at 45° at the surface with its peak at moment $t=0$ and a peak amplitude of the normalized vector potential of $a_0 = 3$. The relativistic motion of the plasma surface (excursions on the order of $\lambda_L/4$ over periods of $T_L/4$) is evident in the evolution of the electron density distribution (shown in false-color representation). Sharpening the oscillation peaks of the reflected pulse are conspicuous, indicative of a highly efficient conversion (~10%) of the incident IR light into UV or VUV light with sub-fs temporal structure. From Tsakiris *et al.*, 2006.

The main advantage of free-electron (or relativistic) nonlinear optics over its counterpart based on bound electrons is that the field strength can be increased virtually without limitation. Higher intensities tend to induce a more pronounced nonlinear response, offering the potential for more efficient conversion of (low-energy) laser photons into high-energy photons and electrons. Striking examples of this potential include the predicted conversion of more than 10% of the laser pulse energy into that of a relativistic electron bunch accelerated in a broken plasma wave ("bubble") in the

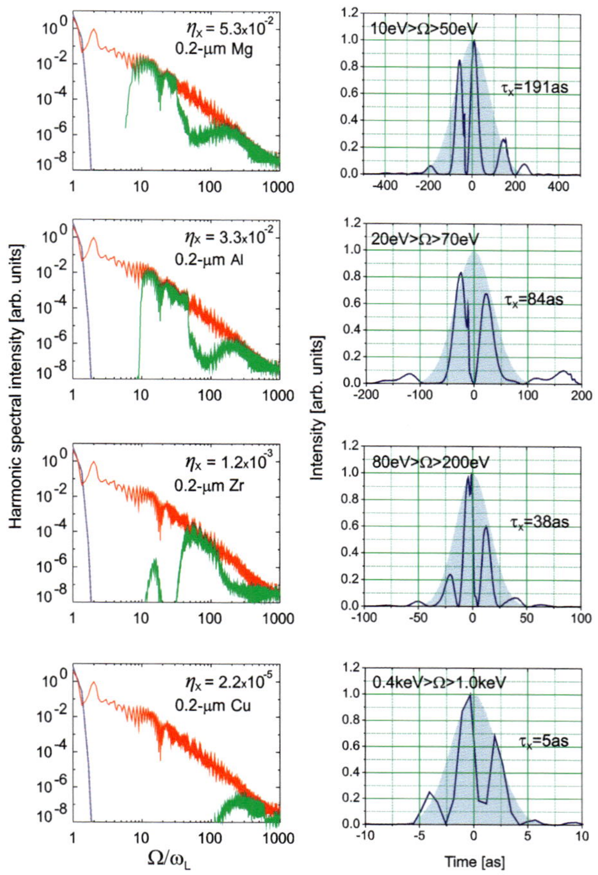

FIG. 41. (Color) Attosecond pulses from surface HHG, predicted by numerical simulations. Left panels, high-order harmonic spectra generated by a Gaussian-shaped few-cycle NIR laser pulse (τ_L=5 fs, λ_L=0.8 μm, a_0=20) impinging at a 45° angle of incidence on a planar target with a step-like density profile (red lines) and as transmitted through different spectral filters (green lines). Right panels, corresponding temporal profile of the reflected radiation transmitted through the filters. The simulations predict *isolated* attosecond pulses in different spectral ranges. From Tsakiris *et al.*, 2006.

forward direction[76] (Pukhov and Meyer-ter-Vehn, 2002) or into sub-fs and as VUV and XUV emission upon reflection from or transmission through a solid surface.[77]

In particular, the relativistic interaction of a few-cycle pulse with a solid surface (Fig. 40) has been predicted to result in attosecond pulses with unprecedented power

[76]Invented by Pukhov and Meyer-ter-Vehn (2002) in numerical studies and first demonstrated experimentally by Faure *et al.*, 2004; Geddes *et al.*, 2004; Mangles *et al.*, 2004. For a review, see Pukhov and Gordienko, 2006. Geissler and Meyer-ter-Vehn (2006) were the first to point out the benefits of few-cycle driver pulses.

[77]Reflection: Naumova, Sokolov, *et al.*, 2004; Pirozhkov *et al.*, 2006; Tsakiris *et al.*, 2006; Geissler *et al.*, 2007. Transmission: Mikhailova *et al.*, 2005. For a discussion of relativistic nonlinear optical phenomena, see Bulanov *et al.*, 1994; Lichters *et al.*, 1996; Gordienko *et al.*, 2004; Mourou *et al.*, 2006.

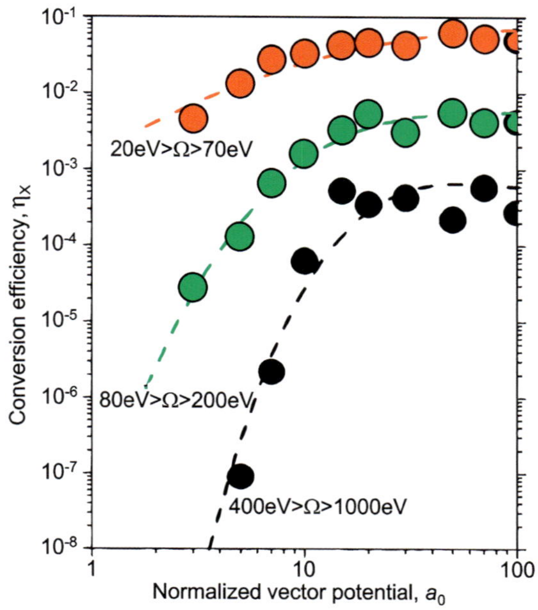

FIG. 42. (Color) Production efficiency of filtered attosecond pulses emerging from few-cycle-driven surface harmonic generation for three different spectral ranges (isolated by the filters specified in Fig. 41). as a function of the normalized vector potential. Simulation parameters other than the vector potential are the same as in Fig. 41. From Tsakiris *et al.*, 2006.

(Naumova, Nees, *et al.*, 2004; Gordienko *et al.*, 2005; Baeva *et al.*, 2006a, 2006b; Tsakiris *et al.*, 2006). The theoretically predicted slow roll-off of the harmonic intensity according to the power law $I_h \propto \omega^{-5/2}$ (Gordienko *et al.*, 2004), later corrected to $I_h \propto \omega^{-8/3}$ in Baeva *et al.* (2006a) was recently corroborated experimentally to keV photon energies (Dromey *et al.*, 2006, 2007).[78] Theoretical analysis also indicates that (i) reflected light from the oscillating plasma mirror suffers a Doppler upshift up to

$$\Omega_{\max} \approx 4\gamma_{\max}^2 \omega_L = 4(1 + a_0^2)\omega_L \tag{48}$$

and the emission of these photons is confined to small fractions of the laser period, of the order of

$$\tau_{\text{cutoff}} \approx T_L / 4\gamma_{\max}^2, \tag{49}$$

resulting in a periodic train of high-energy photon bursts. Recently Baeva *et al.* (2006a) have refined this simple model for the super-relativistic regime, $a_0 \gg 1$. Driving the interaction with a few-cycle laser field and suitable spectral filtering (Tsakiris *et al.*, 2006) or longer pulses with time-dependent ellipticity of their polarization (Baeva *et al.*, 2006b) may lead to single attosecond pulse generation;[79] see Fig. 41 (Tsakiris *et al.*, 2006; Geissler *et al.*, 2007). A train of sub-fs pulses was recently

[78]Different contributions to the process are discussed in Tarasevitsch *et al.*, 2007, and Thaury *et al.*, 2007.

[79]Spatial filtering can also isolate a single pulse in case of λ_L-sized interaction volume (Naumova, Nees, *et al.*, 2004).

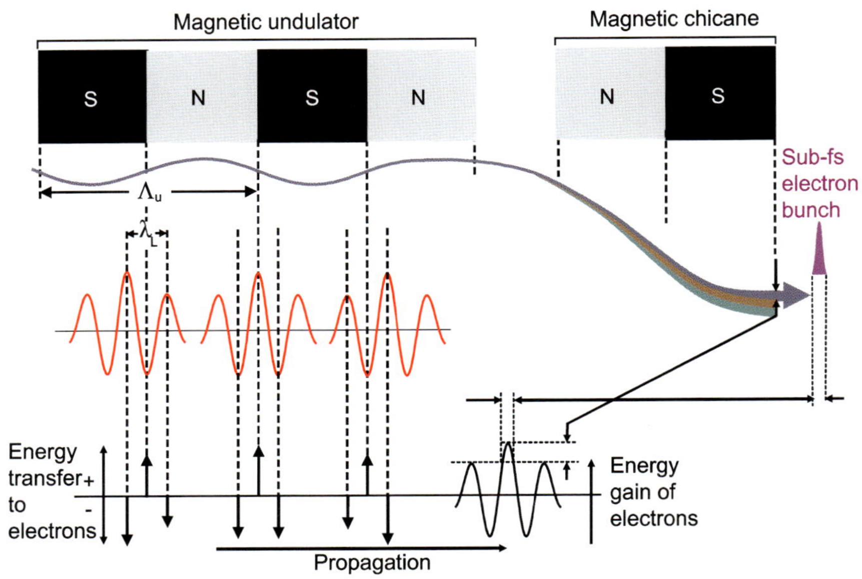

FIG. 43. (Color) Principle of sub-fs electron bunch slicing, by light-field-induced energy modulation of a relativistic electron beam. Energy transfer between the wiggling electron beam and the copropagating laser beam $E_L(t)$ polarized along the direction of the wiggling motion can take place owing to the transverse component $v_x(t)$ of the electron's velocity: $dW/dt = -v_x(t)E_L(t)$. The direction of energy transfer at the interface of bending magnets, where $v_x(t)$ is maximum, is depicted by arrows in the bottom panel. The undulator period Λ_u is chosen such that the electron beam is delayed by the laser wavelength λ_L with respect to the laser wave upon traveling one undulator period. As a consequence, those sub-laser-wavelength, sub-fs portions of the electron bunch that gain energy in one undulator period will also gain energy in the next one. In between, electrons will lose energy, repeatedly, from one undulator period to the next. This inverse free-electron laser (IFEL) effect results in an electron beam carrying an energy spectrum modulated following the oscillating electric field of the laser field at the output of the undulator. If IFEL is implemented with a cosine-shaped few-cycle laser field, maximum energy shift will be confined to a fraction of the half laser cycle at the pulse peak. This sub-fs portion of the electron bunch can be sliced out with a magnetic chicane and an aperture.

observed from the interaction driven by a multicycle laser pulse (Nomura *et al.*, 2009).

Figure 41 reveals several striking differences when compared with atomic HHG. First, isolated sub-fs pulse generation appears to be feasible for a wide range of frequencies including bands far below Ω_{max}. Second, increase of the filter bandwidth does not compromise the quality of single-pulse isolation up to bandwidths of more than an octave, implying isolated soft-x-ray pulses of a few attoseconds in duration. Last but not least, the efficiency of few-cycle-driven surface HHG is predicted to surpass that of atomic HHG by several orders of magnitude (Fig. 42). If these predictions can be verified, few-cycle-driven surface HHG will open up a new chapter in attosecond science.

2. Attosecond electron and hard-x-ray pulses?

Several theoretical studies suggest that the interaction of ultraintense few-cycle light with relativistic electrons may result in the production of isolated attosecond relativistic electron (Naumova, Sokolov, *et al.*, 2004; Ma *et al.*, 2006; Sakai *et al.*, 2006; Varin and Piché, 2006) and hard-x-ray (Zhang *et al.*, 2006) pulses. As an example, here we review a scheme proposed by Saldin *et al.* (2004) and Zholents and Fawley (2004), in which a few-cycle pulse acts as a sub-fs electron energy modulator, permitting sub-fs bunch slicing. The concept draws on the interaction of a strong, but nonrelativistic, few-cycle light field with an ultrashort (femtosecond) bunch of ultrarelativistic electrons ($W \gg 1$ MeV), in the static, periodic magnetic field of an undulator[80] and explained in Fig. 43. A several-millijoule energy, cosine-shaped, few-cycle near-infrared pulse (τ_L=5 fs at λ_L=800 nm) is capable—within merely two magnet periods—of introducing a sub-fs, central peak energy offset of some 40–50 MeV, which exceeds by a factor of 2 the energy shift induced in any other portion of the bunch. If the initial energy spread of the electron bunch is a small fraction of this light-induced energy shift, selecting the upshifted energy band at the output of a magnetic chicane (Fig. 43) results in a sub-fs ultrarelativistic electron bunch. Injection of this sub-fs electron bunch into a second undulator with a much shorter magnet period may allow generation of sub-fs hard-x-ray pulses up to peak powers at the 100-GW level (Saldin *et al.*, 2004) by free-electron lasing (Kroll *et al.*, 1981; Feldhaus *et al.*, 2005).

[80]The utility of this interaction, known as the inverse free-electron laser (Palmer, 1972; Courant *et al.*, 1985; Sears *et al.*, 2005), has been demonstrated for manipulating relativistic electron bunches, including microbunching (Liu *et al.*, 1998; Sung *et al.*, 2006), phase locking to a laser wave (Kimura *et al.*, 2001, 2004) as well as acceleration (Musumeci *et al.*, 2005).

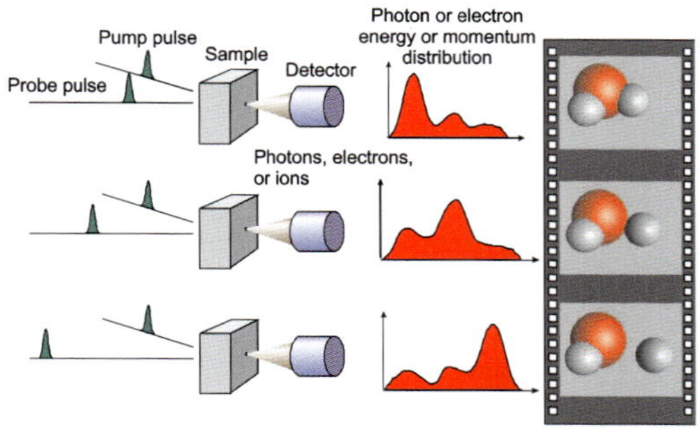

FIG. 44. (Color) Principle of pump-probe (time-resolved) spectroscopy.

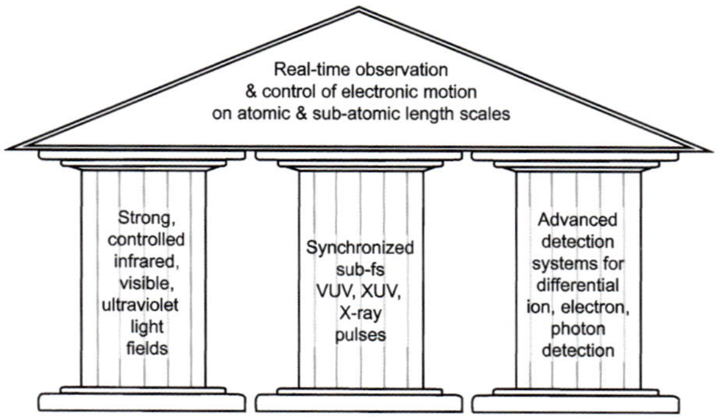

FIG. 45. Major technological pillars of attosecond science.

3. Next-generation accelerators driven by few-cycle light?

Byer and his co-workers recognized that the peak acceleration gradient of $\sim$50 MeV/m of conventional radio-frequency (rf) accelerators can potentially be increased to GeV/m by using laser (instead of rf) fields in dielectric (instead of metal) structures for acceleration.[81] Efficient operation of such a laser-driven vacuum accelerator relies on few-cycle driver pulses with a controlled CE phase.[82] The use of intense waveform-controlled few-cycle light in future linear electron accelerators will naturally lead to high-energy attosecond electron bunches and thereby extend the frontier of attosecond science into high-energy physics.

VII. REAL-TIME OBSERVATION AND CONTROL OF ATOMIC-SCALE ELECTRON AND NUCLEAR DYNAMICS

Real-time observation of microscopic motion requires the ability to trigger and subsequently probe the process under scrutiny.[83] The instants of triggering and probing must be defined with an accuracy sufficient to resolve details of the motion. This accuracy is dictated by the duration of the triggering and probing events, the accuracy of controlling the delay between them, and the S/N ratio, see Fig. 4. Dynamical information is provided by an observable varying as a function of the delay between triggering and probing events in a pump-probe measurement (Fig. 44). This quantity varies on the time scale at which the motion occurs, affording the observer *real-time access* to the process. If the observable also yields information on the location of the moving particles, a series of freeze-frame pump-probe images allows retrieval of the microscopic motion. We refer to this as four-dimensional (4D) imaging.

Microscopic dynamics outside the atomic core fall into two major categories: (i) motion of atoms as a whole leading to structural rearrangements in molecules and condensed matter and (ii) motion of electrons inside and between atoms. The natural time scales for these particles to travel atomic distances extend from several femtoseconds to several picoseconds and from several attoseconds to tens of femtoseconds, respectively (Figs. 1 and 2). Real-time observation and control of atomic motion relies on femtosecond laser techniques, using cycle-averaged quantities such as field amplitude and carrier frequency for initiating, probing, and controlling dynamics. Attosecond technology constitutes a radical shift: the oscillatory laser field along with a sub-fs XUV pulse takes over this job, potentially improving the resolution of probing and the precision of control by orders of magnitude.

The main technological pillars for attosecond science are shown in Fig. 45. Depending on the type of microscopic process to be studied and the questions to be asked, these tools are to be used in different combinations. Tables I–III provide an overview of the toolbox within attosecond technology, summarizing the different options of how to trigger, steer, probe, and image electronic and nuclear dynamics. Many of the tools listed are already available, other will hopefully emerge in the near future. In this section we review the status and address some prospects of attosecond spectroscopy and control. In doing so, we first consider the simplest case of isolated atoms (Sec. VII.A) and then move towards ever more complex systems, from simple diatomic molecules, through large biomolecules and supramolecular assemblies, to mesoscopic systems and condensed matter. We show how tools from our attosecond toolbox (Tables I–III), referred to using the code (I–III/A–E/1–2), are to be combined to address a wealth of questions in a wide range of different systems.

A. Electronic motion in atoms: Excitation, relaxation, and correlations

Electronic motion inside or around atoms can be triggered with sub-fs precision either by optical-field ionization or by a sub-fs VUV, XUV, or x-ray pulse. In this section we address schemes for attosecond spectroscopy based on these scenarios and review experiments.

---

[81] Huang *et al.*, 1996; Salamin and Keitel, 2000; Wang *et al.*, 2001, 2007; Plettner *et al.*, 2005.

[82] Plettner *et al.*, 2006.

[83] Kinematically complete collision experiments also provide insight into attosecond dynamics, but a delayed probe for real-time observation is not available in these experiments. See, e.g., Moshammer *et al.*, 1997; Chatzidimitriou-Dreismann *et al.*, 2003.

TABLE I. Toolbox of attosecond technology: triggering and steering the motion of electrons.

	Ultrashort-pulsed tools for	1 triggering	2 steering
		electronic and concomitant atomic motion with sub-femtosecond or attosecond precision	
A	Strong, multicycle, envelope-controlled IR/NIR light	periodic launching of positive-energy electron wave packets, periodic recollision, and recollision-induced dynamics, generation of a train of sub-fs XUV pulses via optical field ionization intensity requirement:[1] $\gamma < 1$ or $\gamma \approx 1$ timing precision:[2] ~ $T_L/20$	periodic subcycle control of strong-field atomic processes (ionization, recollision, recollision-induced dynamics); shaping sub-fs pulse trains by superimposing harmonic fields intensity requirement: $\gamma < 1$ or $\gamma \approx 1$ steering precision:[3] ~ $T_{min}/4$
B	Train of sub-fs VUV/XUV pulses	affecting and controlling periodic strong-field electron dynamics induced by a synchronous, multicycle laser field, e.g., by seeding desired electron trajectories via single-photon (possibly two-photon) bound-bound or bound-free transition timing precision:[4] ~ $\tau_X/5$	
C	Strong, few-cycle, waveform-controlled IR/NIR/VIS/UV light	initiation of electronic and possibly nuclear rearrangements in ionized systems with "absolute" timing precision, generation of a single sub-fs XUV/SXR pulse via optical field ionization requirement: definite waveform, $\gamma \ll 1$ or $\gamma \approx 1$ timing precision: ~ $T_L/20$	steering the motion of valence-band electrons in atomic and molecular systems with synthesized waveforms requirements: bandwidth > 1 octave and/or intensity:[5] $\Delta W_{Stark} \approx \Delta W_{valence}$ steering precision: ~ $T_{min}/4$
D	High-flux, sub-fs VUV/ XUV/ SXR pulse	triggering electronic motion in matter, especially localized, site-selective electronic excitation in molecular and condensed systems via single-photon bound-bound or bound-free transition flux requirement:[6] $F \approx 1/\sigma$ timing precision: ~ $\tau_X/5$	(possibly) coherent control of electronic excitation in atoms, molecules, and solids via two-photon transition induced by a sub-fs pulse with shaped amplitude and swept frequency intensity requirement[7]: $I\sigma^{(2)}\tau_X \approx 1$ steering precision: ~ $\tau_X/5$
E	Low-flux ($F << 1/\sigma$) sub-femtosecond XUV/X-ray pulse	time-resolved inner-shell spectroscopy via single-photon bound-bound or bound-free transition timing precision: ~ $\tau_X/5$ to be implemented in conjunction with (IIC1)	

[1] γ : Keldysh parameter, for definition, see, e.g., Brabec and Krausz, 2000.

[2] T_L : oscillation period of carrier wave.

[3] T_{min} : oscillation period of shortest wavelength spectral component of the wave.

[4] τ_X : pulse duration (FWHM) of XUV/SXR/X-ray pulse.

[5] $\Delta W_{Stark} \approx \Delta W_{valence}$: Stark shift of valence-band energy levels comparable to level spacing.

[6] $F \approx 1/\sigma$: Photon flux (number of photons/beam cross-sectional area) comparable to the inverse single-photon transition cross section. If, for example, $\sigma \approx 1$ Mbarn (10^{-18} cm^2) and the beam cross section ~ 1 μm^2, this condition requires ~ 10^{10} photons to be carried by the pulse.

[7] I: intensity (number of photons/beam cross-sectional area/time); $\sigma^{(2)}$: two-photon cross section of the transition one is attempting to control, which depends very sensitively on frequency and system. Some feeling for the required intensity can be gained from requiring that one-photon absorption is saturated.

1. Electronic excitation and relaxation dynamics: Attosecond streaking and tunneling spectroscopy

Core excitation of atoms causes a wealth of multielectron processes. Figure 46 presents a few examples. They unfold on a time scale ranging from tens of attoseconds to tens of femtoseconds. Direct time-domain access promises to yield deeper insight into these processes and to identify ways of affecting them. This, in turn, may

TABLE II. Toolbox of attosecond technology: probing and imaging the motion of electrons.

	Ultrashort-pulsed tools for	1 probing	2 imaging
	electronic and concomitant atomic motion with sub-femtosecond or attosecond precision		
A	Strong, multicycle, envelope-controlled IR/NIR light	probing periodic electron emission induced by a synchronized train of sub-fs XUV pulses (IB1) by measuring side-band intensity versus delay between IR pulse and sub-fs pulse train via single-photon free-free transition: RABBITT, CRAB intensity requirement:[8] $I_L \approx 10^{11}$ W/cm^2 resolution ~ $T_L/20$ electronic response restricted to shorter than the period of the sub-fs train	imaging dynamic change in molecular structure induced by optical-field ionization (IA1) via analyzing high-order harmonics and/or rescattered ATI electrons intensity requirement: $\gamma < 1$ or $\gamma \approx 1$ resolution: sub-$T_L/2$, sub-Angström delay of imaging with respect to the start of the dynamics restricted to less than T_L
B	Train of sub-fs VUV/XUV pulses	probing periodic strong-field electron dynamics induced (IA1) and possibly steered (IA2) by a synchronous, multicycle laser field via single-photon bound-free transition resolution ~ $\tau_X/5$	seeding desired electron trajectories for probing electronic and nuclear dynamics via single-photon bound-free transition resolution ~ $\tau_X/5$
C	Strong, few-cycle, waveform-controlled IR/NIR/VIS/UV light	probing multielectron excitation (shake-up) and relaxation (Auger, Coster-Kronig) processes following core excitation with a (low-flux) sub-fs XUV/X-ray pulse (see IE1) via one-photon or multiphoton free-free transition: attosecond streaking spectroscopy (AST) via multiphoton bound-free transition: attosecond tunneling spectroscopy (ATS) intensity requirement: $I_L \approx 10^{11}$ W/cm^2 or higher for free-free transitions[9], $\gamma \approx 1$ for bound-free transitions resolution ~ $T_L/20 - T_L/10$	imaging dynamic change in molecular structure induced by optical field ionization with a preceding cosine-shaped waveform-controlled few-cycle light (see IC1) or by a synchronized single high-flux, sub-fs VUV/XUV pulse (see ID1) via ATI and HHG resolution: sub-$T_L/2$, sub-Angström 4D imaging of electron and structural dynamics on surfaces via tunneling from a tip induced by a cosine waveform resolution ~ $T_L/20$, potentially sub-Å
D	Low-flux ($F << 1/\sigma$) sub-femtosecond XUV/SXR/hard-X-ray or electron pulse	probing (i) inner-atomic electron-electron interactions initiated and possibly steered via ID1-2 (ii) valence-band electron dynamics and simultaneous nuclear dynamics in molecular systems, initiated and possibly steered via (IC1-2) by observing dynamic changes in valence-band and inner-shell photoelectron spectra, respectively, via single-photon bound-free transition (photoionization) resolution ~ $T_L/5$	4D imaging of electron and structural dynamics atoms, molecules, solids initiated by either IC1 or ID1 with XUV/SXR pulse in combination with photoelectron emission microscopy: attosecond PEEM resolution ~ $\tau_{XUV}/5$, ~ 10 nm with hard-x-ray or electron pulse: by recording freeze-frame diffraction pattern: attosecond diffraction imaging resolution ~ $\tau_{XUV}/5$, ~ sub-Angström

[8] Critical parameter for free-free transitions is $vA_L \approx \omega_L$ where v is the electron momentum after ionization and A_L is the vector potential of the IR field. For $v \approx 1$ a.u. and 800 nm one obtains $I_L \approx 4\text{x}10^{11}$ W/cm^2.

[9] As in the previous footnote, the criticial parameter for free-free transitions is $vA_L \approx \omega_L$.

pave the way toward the development of compact, efficient XUV and x-ray lasers.

We assume that the atoms are hit by a sub-fs XUV or x-ray (henceforth, briefly, XUV) pulse. Depending on the XUV photon energy, atoms may be ionized or excited in several ways, as shown in Fig. 46. Core-level excitation typically triggers numerous intra-atomic and molecular processes including shakeup of one or more electrons and a number of subsequent relaxation processes, such as Auger decay, Coster-Kronig decay, or cascaded decay phenomena. Inner-valence ionization, on the other hand, usually does not open a channel for double ionization, but can—if the ion is in close proximity to other atoms (e.g., in a cluster or in an endohedral fullerene)—result in energy transfer to its neighbor via

TABLE III. Toolbox of attosecond technology: detecting observables in attosecond time-resolved experiments.

Detectors for		1 probing	2 imaging
		by observing and analyzing final products of a pump-probe excitation	
A	Ion detectors	study of electron excitation & relaxation processes, multielectron dynamics in highly excited (core-excited) atoms by charge-state detection observation of molecular fragments during chemical and biochemical reactions by mass spectrometry	measurement of 3D momentum distribution of ionic fragments by velocity map imaging (for emission possessing cylindrical symmetry)
B	Electron detectors	measuring the energy distribution of ejected electrons [both primary (photo) and secondary (Auger) electrons] by time-of-flight analysis /magnetic-bottle spectrometers or by hemispherical detection	measurement of 3D momentum distribution of ejected electrons by velocity map imaging (for emission possessing cylindrical symmetry) measurement of 2D momentum distribution of scattered electrons by position-sensitive (e.g., multichannel-plate detector)
C	Electron-ion detectors	coincidence detection of electrons and ions along with measurement of the electron energy distribution and charge state detection and ion mass spectrometry by some combination of IIIA1 and IIIB1	Coincidence measurement of momenta of all products (electrons and ionic fragments) of a reaction: reaction microscope by **COLTRIMS**: cold-target recoil-ion-momentum spectroscopy
D	Photon detectors	measurement of transmitted (absorption spectroscopy) or ejected (fluorescence spectroscopy) photon energy distribution by wavelength-dispersive spectrometry (WDS) drawing on grating or crystal spectrometers and CCD or channel plate detectors or by energy-dispersive spectrometry (EDS) drawing on single-photon counting semiconductor detectors	measurement of 2D momentum distribution of scattered photons: time-resolved x-ray diffraction with CCD or channel plate detectors

interatomic and molecular Coulombic decay.[84]

Two approaches to accessing sub-fs electron dynamics in the time domain have been demonstrated so far: attosecond streaking spectroscopy (AST) is an extension of streaking to positive-energy electrons containing atom-specific dynamic information such as Auger electrons, whereas in attosecond tunneling spectroscopy (ATS) electrons shaken up to the valence band are released from negative-energy (bound) states by tunnel ionization; see Fig. 46. Both approaches have in common that the sub-fs electric field transients of VIS and NIR light sample the process under study. As the probability of laser-field-induced transitions in these techniques may approach or even reach 100%, the sub-fs XUV pulse initiating the dynamics is allowed to have a low flux (type IE1-IIC1 spectroscopy; see Tables I and II), leading to a weak-pump–strong-probe scenario.

Energetic excitation of atoms or molecules usually leads to ejection of one or more electrons. Their temporal evolution often carries information about inner-atomic dynamic processes. For instance, the emission of Auger electrons mirrors the decay of a core hole. Therefore resolving the emission process in time by AST provides direct information on the temporal evolution of inner-shell dynamics. Figure 47 shows simulated streak images of Auger electron spectra versus delay between the sub-fs triggering XUV pulse and the few-cycle probing laser field (Drescher *et al.*, 2002). These spectrograms, if recorded with satisfactory S/N ratio, allow retrieval of the temporal evolution of the electron emission and hence of the core-hole decay, as discussed in Sec. V.D.3. Drawing on this concept, AST resulted in the first real-time observation of the decay of

[84]Cederbaum *et al.*, 1997; Marburger *et al.*, 2003; Averbukh *et al.*, 2004; Jahnke *et al.*, 2004; Averbukh and Cederbaum, 2006; Morishita *et al.*, 2006; Öhrwall *et al.*, 2004; Barth *et al.*, 2006; Kuleff and Cederbaum, 2007; Vaval and Cederbaum, 2007.

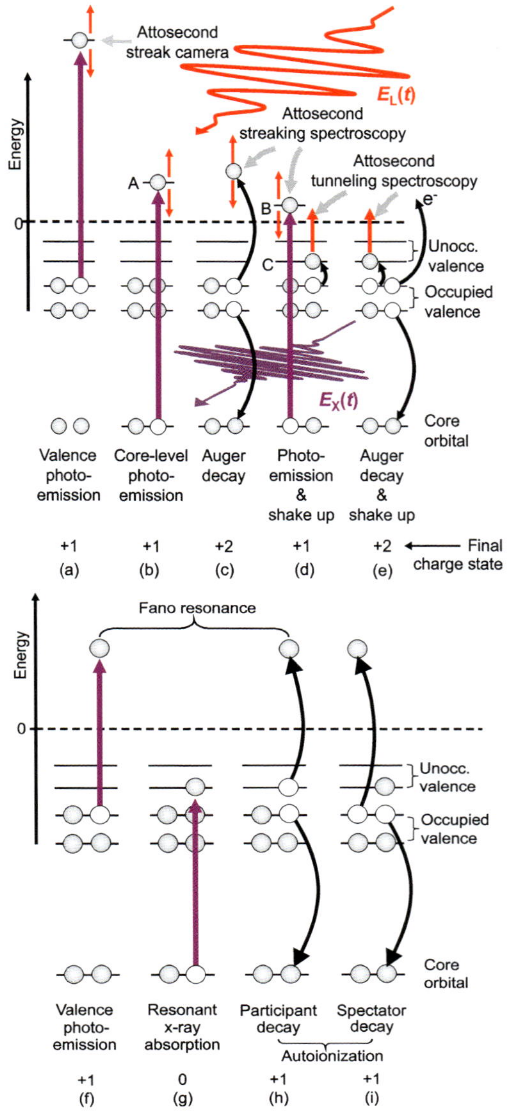

FIG. 46. (Color) Electronic excitation via XUV and x-ray photons followed by relaxation, which can be probed in real time via attosecond streaking spectroscopy and attosecond tunneling spectroscopy. The new spectroscopies draw on nonlinear free-free and bound-free electronic transitions, respectively, induced by waveform-controlled few-cycle light. (a), (f) Valence-electron emission, (b) core-electron emission, accompanied by (c) Auger decay or (d) valence-band excitation (shakeup), or both (e). (g) Resonant core-level excitation and subsequent (h) participant and (i) spectator decay. The basic idea for the graphical representation is borrowed from Brühwiler *et al.*, 2002.

an inner-shell vacancy (Drescher *et al.*, 2002).[85]

The lifetime of the inner-shell vacancy acquired in the above time-resolved measurement can also be inferred

[85]Short-pulse-induced Auger decay was realized before (Schins *et al.*, 1994, 1996), nevertheless with the decay remaining unresolved because of the long (sub-ps) duration of the x-ray excitation.

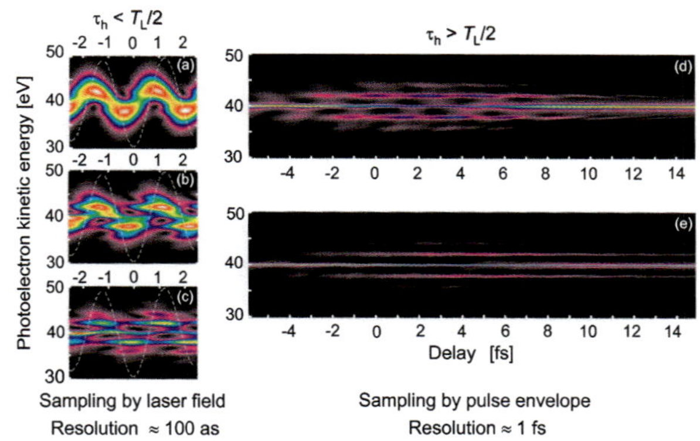

FIG. 47. (Color) Streaking spectrograms of Auger electron emission, simulated by V. Yakovlev and A. Scrinzi for different core-hole decay times, (a) $\tau_h=0.2$ fs, (b) $\tau_h=0.5$ fs, (c) $\tau_h=1$ fs, (d) $\tau_h=2$ fs, (e) $\tau_h=5$ fs, using a few-cycle NIR streaking laser field ($\tau_L=5$ fs, $\lambda_L=0.75$ μm) and a sub-fs ($\tau_X=0.5$ fs) XUV excitation pulse. Depending on whether the electron emission time τ_h is shorter or longer than the half wave cycle $T_L/2$, the spectrograms look different. For $\tau_h<T_L/2$, we see variations of the streaked spectra within the laser cycle (left panels). This subcycle variation in the streaked spectra tends to disappear as the emission time becomes longer than $T_L/2$. Instead, sidebands spaced by the laser photon energy appear for a delay range dictated by the duration of emission and the laser pulse. Variation of the sideband amplitude vs delay is the convolution of the emission process and the amplitude envelope of the laser field. If the laser pulse is short enough, the duration of electron emission can be determined by deconvolution. In the sub-fs domain, the process is sampled by the oscillating field, whereas in the femtosecond domain, the pulse envelope takes over the role of the "sampling function." It has to offer a resolution on the order of $T_L/2$, if all details of the motion, possibly ranging from the sub-fs to multi-fs regime, are to be uncovered. This is ensured only by a few-cycle streaking field. From Drescher *et al.*, 2003.

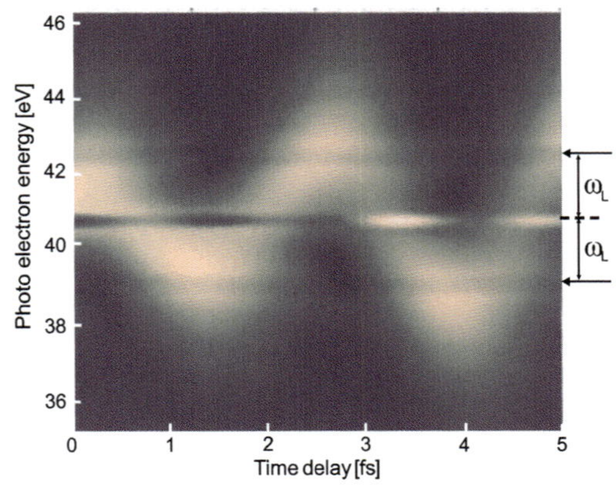

FIG. 48. (Color) Attosecond streaking spectrogram of two coalescent Lorentz resonances decaying over disparate time scales: $\tau_{h,1}=0.4$ fs and $\tau_{h,2}=40$ fs. Simultaneous launching of Auger electrons with different emission times manifests itself in the appearance of a broad oscillating feature displaying the short-lived emission and sidebands spaced by ω_L from the peak of the main lines (indicated by arrows) with their (long) temporal extension reflecting the duration of the long-lived emission. From Wickenhauser *et al.*, 2006.

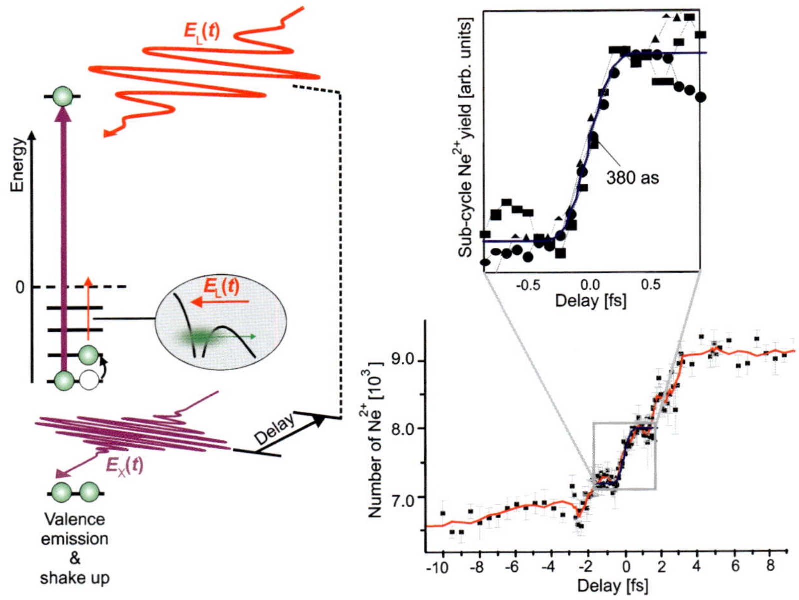

FIG. 49. (Color) Attosecond real-time observation of light-field-induced electron tunneling (Uiberacker *et al.*, 2007). A 250-as, 95-eV XUV pulse sets a valence electron free from neon atoms. The excess photon energy gives way to shakeup, promoting an electron to an excited state in the valence band. A time-delayed, waveform-controlled, few-cycle pulse (τ_L=5 fs, λ_L=0.75 μm) probes the transient population of the shakeup states by field ionization, resulting in doubly charged neon ions, Ne^{2+}. The squares and the line (five-point average) depict the number of Ne^{2+} ions measured as a function of delay between the XUV excitation and the probing NIR laser field. The steplike increase in Ne^{2+} yield with the steps approximately separated by half the laser period provides conclusive evidence for optical field ionization responsible for the depletion of excited valence states, in accordance with the theory of nonadiabatic tunneling by Yudin and Ivanov (2001a). The inset zooms in onto the central subcyle ionization step, plotting data (dots, squares, triangles) obtained from different measurements, which reproducibly indicate that shakeup and tunneling occur within less than 400 attoseconds. Kazansky and Kabachnik (2007c) proposed a modification of the above scheme for studying tunneling in the absence of shakeup.

from time-integral, spectrally resolved (briefly, frequency-domain) measurements (Jurvansuu *et al.*, 2001). However, retrieval of dynamics from frequency-domain spectroscopy becomes challenging when the final state of the dynamical evolution can be reached via several channels, resulting in complex spectral structures due to resonances between competing quantum transitions. A prominent example is the Fano resonance, which emerges whenever a positive-energy state can be accessed by more than one quantum pathways; see Figs. 46(f)–46(i). Fano resonances have been ubiquitous in time-integral studies.[86] AST was found to be capable of exploring details of complex temporal dynamics related to coherent excitation of several nearby Fano resonances, which are only partially accessible by frequency-domain spectroscopy (Wickenhauser *et al.*, 2005, 2006; Zhao and Lin, 2005), see Fig. 48, and to be sensitive to off-diagonal elements of the excitation density matrix, opening the way to monitor decoherence.

[86]Examples include photoabsorption (Beutler, 1935; Fano, 1935, 1961), electron, neutron, and Raman scattering (Adair *et al.*, 1949; Simpson and Fano, 1963; Cerdeira *et al.*, 1973), and show up in such disparate systems as atoms (Lambropoulos and Zoller, 1981; Rzazewski and Eberly, 1981), quantum-well structures and quantum dots (Göres *et al.*, 2000), and scanning tunneling microscopes (Madhaven *et al.*, 1998).

AST can probe not only relaxation but also excitation of the electronic shell. Indeed, detailed comparison of the temporal history of the electron emission processes populating states *A* and *B* in Figs. 46(b) and 46(d) may shed light on how part of the absorbed photon energy ends up promoting another electron into an excited state (shakeup) in Fig. 46(d).

AST probes positive-energy (freed) electrons; see Figs. 46(b)–46(d). Alternatively, intra-atomic or intramolecular processes can also be interrogated by probing intermediate states of negative-energy (bound) electrons. This is exactly what ATS does by probing the population of valence-band energy levels that are unoccupied before excitation; see Figs. 46(d) and 46(e). In this way, ATS is capable of tracing how valence states are being populated and subsequently depleted, following a core-level excitation, allowing real-time observation of both excitation and subsequent relaxation of the electronic shell of a core-excited atom (Uiberacket *et al.*, 2007), or, e.g., the dynamics of excited electronic states in a dissociating molecule (Gagnon *et al.*, 2007).

While AST relies on electron energy analysis (IIIB1), ATS draws on counting ions of different charge states (IIIA1). Again, as a first step, a core electron is ejected by a sub-fs XUV pulse [Fig. 46(b)]. The first question that arises is how the electronic shell of the ion gets excited, i.e., how does "shakeup" occur? Is it instanta-

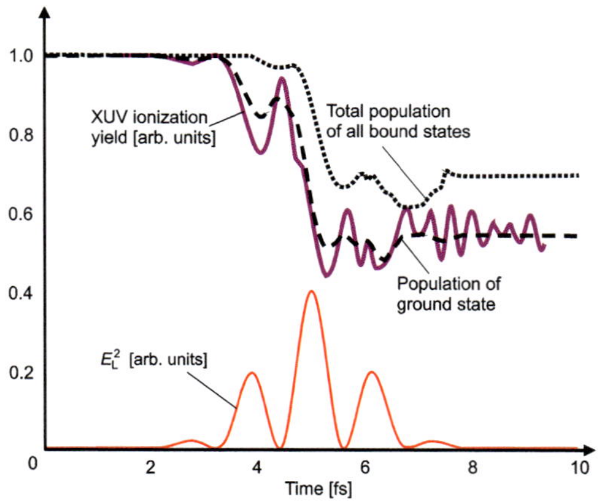

FIG. 50. (Color) Attosecond XUV probing of strong-field dynamics of a bound electron. Solid curve shows how the yield of ions created by XUV-induced ionization depends on the arrival time of the XUV pulse, relative to oscillations of the two-cycle, NIR driving field. As the bound electron wave packet $\psi_b(x,t)$ is stretched away from the core near the NIR field oscillation peaks, the high-momentum components decrease, because stretching the wave packet in coordinate space implies compression in momentum space. This reduces the bound-free transition matrix element and accounts for dips in the XUV yield near the maxima of the NIR field. Oscillations of the total yield at the end of the laser pulse suggest that nonadiabatic excitation during the laser pulse results in bound-state dynamics following the laser pulse. From Smirnova *et al.*, 2006.

neous (as it is currently understood) or is it possibly delayed? In principle, ATS may provide the answer by sampling the buildup of population in previously unoccupied valence states, such as state C in Fig. 46(d). But how can this population be probed with attosecond resolution using a femtosecond laser?

Theory[87] (Yudin and Ivanov, 2001b) predicts that not only in the conventional tunneling regime of $W_b \ll 2U_p$, but also in the intermediate regime of $W_b \approx U_p$ liberation of electrons with binding energies of $W_b \geq 10$ eV,[88] by intense several-cycle NIR or IR pulses occurs via nonadiabatic tunneling, with subcycle yield localized within small fractions of the half oscillation cycles near the wave crests. Proof-of-principle ATS experiments (Uiberacker *et al.*, 2007) corroborated this prediction (see Fig. 49). The observed 400-as rise time of the Ne^{2+} yield indicates that shakeup occurs too fast to be resolved with the current tools. Nevertheless, this sub-fs probe is sufficiently fast to sample a wide range of inner-shell relaxation processes. This potential was proven by tracing multielectron relaxation in xenon (Penent *et al.*, 2005) following a 95 eV as excitation (Uiberacker *et al.*, 2007). With recently improved tools, sub-1.5-cycle NIR waveforms (Schultze *et al.*, 2007) and sub-100-as XUV pulses (Goulielmakis *et al.*, 2008), details of shakeup and nonadiabatic tunneling may also come to light.

With the increase of the excitation photon energy towards the keV frontier and possibly beyond, AST and ATS will provide direct, time-domain access to a vast range of electronic dynamics, including those in magnetic materials. With their time resolution approaching the atomic unit, comparative ATS and AST studies may shed light on how energy absorbed by an atom is distributed among its electrons and how the electronic shell of the atom is re-arranged following the loss of one or more electrons (Breidbach and Cederbaum, 2005) and thereby on the fundamental question of electron correlations.

[87]For a review of strong-field ionization, see Popov, 2004. For earlier papers, see Keldysh, 1964; Perelomov *et al.*, 1966a, 1966b; Perelomov and Popov, 1967; Reiss, 1980a, 1980b, and Faisal, 1987.

[88]For pulses approaching the single-cycle limit, which may become available in the foreseeable future, this condition may be further relaxed toward a few electron-volts.

2. Electronic relaxation and rearrangement: Attosecond absorption spectroscopy

In the previous section we discussed how attosecond electron dynamics can be probed with strong, controlled light fields. Alternatively, sub-fs XUV pulses can also be used as probe pulses. Transient absorption spectroscopy (Loh *et al.*, 2007, 2008) with isolated attosecond XUV and x-ray pulses (IID1–IIID1; see Tables II and III) will also provide unprecedented insight into the relaxation dynamics and rearrangement of the electronic system of atoms or molecules following controlled excitation by a waveform-controlled few-cycle IR or VIS pulse (IC1-2; see Table I) or by an intense (possibly shaped) sub-fs XUV or x-ray pulse (ID1-2; see Table I). Isolated attosecond XUV pulses with a spectral width of several 10 eV (Schultze *et al.*, 2007; Goulielmakis *et al.*, 2008) provide ideal prerequisites for implementing attosecond absorption spectroscopy (AAS) by extending the experiments of Loh *et al.* (2007, 2008) into the time domain of the electronic response of matter (Breidbach and Cederbaum, 2005). This spectroscopy may give an answer to a long-standing question in atomic and molecular physics: How does the electronic system rearrange after the sudden removal or excitation of an electron?

3. Electron wave-packet motion: Attosecond photoelectron spectroscopy

Consider the simplest case when a single-electron bound wave packet is formed in an atom. Angle-resolved sub-fs XUV photoelectron spectra have been predicted to be sensitive to the instantaneous intra-atomic or intramolecular electron density near the core, enabling us to trace its dynamic changes with attosecond resolution (Yudin *et al.*, 2005, 2006). As an example, consider the electron wave-packet dynamics induced by a strong linearly polarized IR field during optical field ionization (Smirnova, Spanner, and Ivanov, 2006). The laser field depletes the ground state stepwise, every half cycle, via tunneling. While doing so, it also "shakes up" the electron, populating several excited states; see Fig. 50.

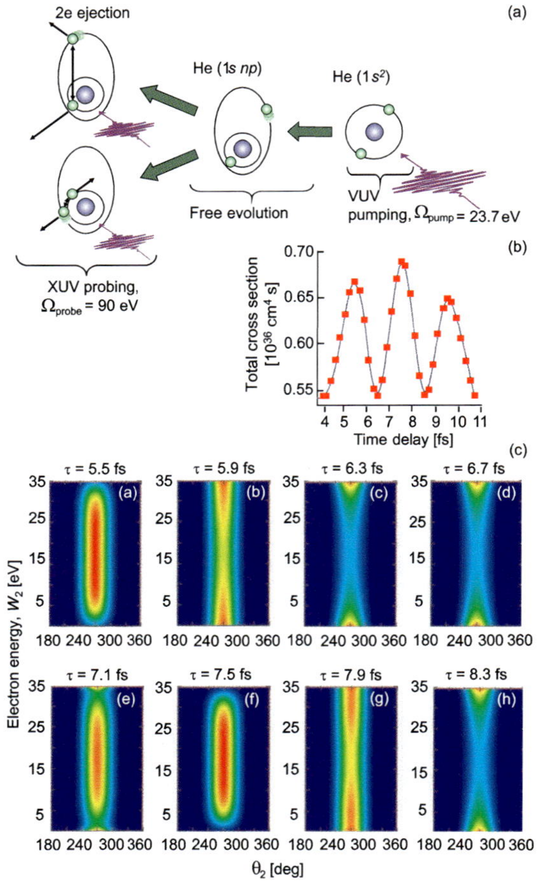

FIG. 51. (Color) Tracing intra-atomic electron motion and electron-electron interactions in helium (Hu and Collins, 2006). (a) Relevant interactions with the VUV pump ($\Omega_{\text{pump}} = 23.7$ eV, $\tau_{\text{pump}} = 1.5$ fs) and XUV probe pulse ($\Omega_{\text{probe}} = 90$ eV, $\tau_{\text{probe}} = 250$ as). (b) Calculated probability of XUV-induced double ionization vs pump-probe delay. (c) Contour plots showing the angular and energy distribution of the second electron for the first electron emitted along the XUV field polarization ($\theta_1 = 90$ deg) vs delay τ. The second electron is ejected with the largest probability in the opposite direction ($\theta_2 = 270$ deg). The total energy the two electrons carry is ≈ 35 eV. If ionization occurs when the excited electron wave packet is at its outer turning point (around $\tau = 6.5$ and 8.5 fs), the inner electron benefits from the vicinity of the core and picks up almost all of the incoming XUV photon energy. The outer electron is set free by the sudden change of the potential as the inner electron leaves the atom (shakeoff). 1 fs sooner or later, when the outer electron arrives at the core, the two electrons share the available excess energy, leaving the atom with comparable energy. Adapted from Hu and Collins, 2006.

This implies the formation and subsequent dynamics of an electronic wave packet bound to the core. The steplike decrease of the bound state population due to field ionization is, not surprisingly, also reflected in the XUV ionization yield (Fig. 50), which, however, shows much richer dynamics. Its origin becomes transparent when we realize that XUV absorption occurs near the core, from the high-momentum components of the wave function. Consequently, attosecond photoelectron spectroscopy (APS) uncovers the dynamics of these components in the bound electron wave packet $\psi_b(x,t)$.[89] Extended with spin-sensitive detection, spin-orbit wave-packet dynamics (Santra *et al.*, 2006) can also be studied in real time. Experimental implementation will draw on techniques IC1-IID1-IIIB1/2 for triggering, probing, and detection, respectively; see Tables I–III.

In the above example, a sub-fs XUV pulse was used to explore the motion of a bound-electron wave packet inside an atom. Smirnova, Patchkovskii, and Spanner (2007) proposed that—along with some control over the ionizing IR field—it can also be used for determining the spatiotemporal characteristics of the detached electron wave packet, as it recollides with its parent ion. In this case, the atom works like the tip of a STM microscope—XUV photons can only be absorbed by free electrons in its vicinity.

Attosecond atomic (molecular) electron dynamics can also be initiated via resonant single-photon excitation for tracing intra-atomic electron motion and electron-electron interaction in real time. Consider the prototypical case of a helium atom (Hu and Collins, 2006, 2007) exposed to a broadband VUV pulse which promotes one of the two electrons from the ground state (1*s*) into the superposition state *np* with $n = 2, 3, 4, \ldots$. In this way, an electron wave packet is launched into an atomic orbit; see Fig. 51(a). This electron wave packet oscillates between the core and its outer turning point, while the other electron resides close to the nucleus in its ground state. The strength of interaction between these two electrons is therefore varying in time. The simulations of Hu and Collins (2006) predict that the probability of double ionization induced by a time-delayed sub-fs XUV probe pulse significantly varies with the center of gravity position of the outer electron. If the attosecond probe arrives when the outer electron is closest to the inner one, i.e., correlation between the two electrons is strongest, e.g., at a delay of $\tau = 5.5$ fs in Fig. 51(b), double ionization is most probable. On the other hand, it is least likely when the electron reaches its outer turning point, e.g., at $\tau = 6.5$ fs in Fig. 51(b). The double ionization yield can be measured by collecting He^{2+} as a function of delay τ between the UV pump and the XUV probe pulse. Implementation therefore relies on techniques ID1-IID1-IIIA1 for pumping, probing, and detection, respectively; see Tables I–III. The variation of the pump-probe-induced double ionization yield versus τ, once measured, will display the motion of the excited electron wave packet in real time. Detection of ejected electrons in coincidence, discussed in the next section, will yield

[89]This dynamics is even richer in multielectron systems; see Eichmann *et al.*, 2000; Zon, 2000; Lezius *et al.*, 2001; Markevitch *et al.*, 2003; Bryan *et al.*, 2006. These studies can also be extended to situations when the electron wave packet is promoted into the continuum and may provide access to effects such as Rabi flopping from continuum states (Dimitrovski *et al.*, 2004).

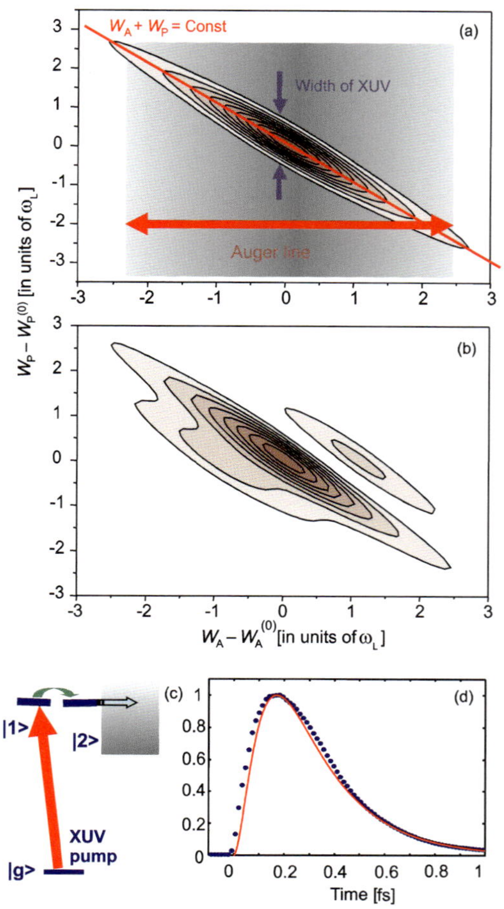

FIG. 52. (Color) Tracing sub-fs Auger decay triggered with a 1.2-fs XUV pulse and probed with a weak ($I_{peak} \approx 10^{10}$ W/cm^2) NIR ($\lambda_L = 800$ nm) few-cycle streaking field via concidence detection (Smirnova *et al.*, 2005). (a) Correlated spectrum of the liberated photoelectron and Auger electron. (b) Correlated electron spectrum recorded in the presence of a weak IR field, which creates sidebands of the original spectrum. (c) Model Auger decay. (d) Reconstruction of the decay dynamics: solid line is the exact decay, dots represent numerical reconstruction from a simulated spectrum containing 10^5 coincidence counts and recorded with a 160-meV energy resolution with 100-as jitter between the pump and probe pulses. The result indicates that a temporal resolution below the atomic unit of time (24 as) may be feasible using a ~1 fs XUV pump pulse. From Smirnova *et al.*, 2005.

further insight into interaction between the two electrons and the way they share the energy of the absorbed XUV photon.

4. Shedding light on electron-electron interactions: Attosecond coincidence spectroscopy

Measurement of the final momenta of all correlated particles produced in a pump-probe exposure *in coincidence*[90] offers unprecedented insight into correlated electron dynamics.[91] In what follows, three prototypical interactions will exemplify the concept: (i) simultaneous two-electron ejection induced by a single XUV photon, (ii) photoelectron and Auger electron emission following absorption of a single XUV photon, and (iii) XUV-pulse-triggered, strong-IR-field-controlled electron-electron collision.

Simultaneous two-electron photoemission. The intriguing questions of how the energy of the absorbed XUV photon is shared between the two ejected electrons and how this energy sharing depends on the distance between electrons in the intra-atomic dynamics discussed in the previous section can be answered by combining the attosecond pump-probe experiments (ID1-IID1; see Tables I and II) outlined in Fig. 51(a) with coincidence detection (IIIB2 or IIIC2; see Table III). The results of Hu and Collins (2006) summarized in Fig. 51(c) predict pronounced variations of the energy sharing between the two electrons as the moment of ionization is varied. Correlation between two free electrons released by an attosecond XUV pump pulse can also be resolved in time by the same method (Hu and Collins, 2007).

Auger decay. Inner shell XUV photoionization often results in the ejection of two electrons, one via photoemission [Fig. 46(b)], with energy W_P, the other via Auger decay [Fig. 46(c)], with energy W_A. When probing Auger decay via standard AST, the XUV pump pulse should be shorter than the process resolved. Smirnova *et al.* (2005) showed how coincidence detection allows one to abandon this stringent requirement. Consider a sub-fs Auger decay initiated by a substantially longer, few-fs XUV pulse. The total energy of the two electrons liberated in the correlated process is fixed by the relatively well-defined energy of the absorbed XUV photon: $W_P + W_A = \Omega_X - I_p^{++}$. The large uncertainty in W_A (caused by the ultrafast decay) translates into an equally large uncertainty in W_P; see Fig. 52(a). Correlated spectra of such shape appear whenever a two-particle energy conservation law is present.[92]

The correlated process can be completely characterized by attosecond streaking of the two-electron spectra in a *weak* IR laser field. The field creates energy-shifted "replicas" of the original spectrum via one-photon absorption and emission, with the spectral phase encoded in the interference of these replicas with the original spectrum [Fig. 52(b)]. Just as in SPIDER (Fig. 31), the time resolution is not limited by the XUV pulse duration, but rather by the jitter between the XUV and IR pulses and by noise.

---

[90]Cold-target recoil-ion-momentum spectroscopy (COLTRIMS) enables one to perform kinematically complete experiments. See, e.g., Cocke and Olson (1991), Ullrich *et al.* (1997), and Dörner *et al.* (2000).

[91]Nicolaides *et al.*, 2002; Mercouris *et al.*, 2004, 2007a, 2007b; Emmanouilidou and Rost, 2007; Morishita, Watanabe, and Lin, 2007.

[92]See, e.g., Chan *et al.*, 2002, 2003, 2004; and Fedorov *et al.*, 2004, 2005.

This approach can be used for any process resulting in the emission of two charged particles with fixed total energy, such as two-electron single-photon ionization. Processes of particular interest include shakeoff, Zeno and anti-Zeno stages of decay (Kofman and Kurizki, 2000), electron rearrangements in the core, and nonexponential decay due to a structured continuum.

Nonsequential double ionization. With coincidence detection, strong-field-controlled recollision provides time-domain access to the process of nonsequential double ionization. If the temporal evolution of the strong laser field is well controlled and known, the vector potential connects the moments of liberation of the two electrons after recollision with their final momenta. The SFA along with a few additional assumptions[93] offers a simple means of tagging the moments of liberation by the vector potential $\mathbf{A}_L(t)$, which induces changes in the final velocities of the ejected electrons that depends on their moments of creation. Coincidence measurement of the final electron velocities then allows retrieval of the history of their liberation. This approach provided insight[94] into NSDI and inneratomic electron thermalization following the recollision (Liu *et al.*, 2006).

If the recolliding electron is previously freed by the strong laser field, there is a direct connection between its energy W_r and the instant of recollision t_r, as shown by Fig. 20. Recent theoretical studies backed by experiments suggest that liberation of the electron with a sub-fs XUV pulse and its subsequent control with a polarization-shaped laser pulse will allow one to manipulate t_r and W_r of the recolliding electron independently.[95] Along with advances in theory[96] required for reliable retrieval of the moments at which the two electrons appear in the continuum after collision, this attosecond control may provide insight into the dynamics of electron-electron collision, as well as into more complex multielectron collision phenomena.

B. Electronic and nuclear motion in complex systems: From simple diatomics to biomolecules and clusters

The fundamental theory for molecules is that of the electronic structure. In the Born-Oppenheimer approximation electrons are considered to instantly adjust to the current position of the nuclei. The equilibrium molecular structure is determined by the minimum of the electron energy. Change in this energy with nuclear position creates the force setting the atoms in motion. The resultant rearrangement of atoms is responsible for a change of chemical composition and biological function. Consequently, the fundamental time scale of chemistry and biochemistry is defined by the atomic motion in molecules. It unfolds on a femtosecond scale (Fig. 1) and has been traced by femtosecond laser pulses in real time.[97]

Recent work suggests that attosecond-scale electronic dynamics in molecules may affect chemical changes.[98] Insight into electronic wave-packet motion in molecules (Breidbach and Cederbaum, 2003; Krause *et al.*, 2005; Remacle and Levine, 2006a) may also shed light on the microscopic mechanisms behind many chemical and biological phenomena such as electron delocalization in aromatic molecules (Poater *et al.*, 2005), biological energy conversion processes (Mirkin and Ratner, 1992; Miyashita *et al.*, 2005), DNA damage and repair (Yavin *et al.*, 2005), long-range electron transfer in biomolecules (Szent-Györgyi, 1941; Gray and Winkler, 2005), and its role in biological signal transduction. Transferring electrons efficiently inside and between molecules (Goldsmith *et al.*, 2005) is also a prerequisite for molecular electronics[99] and molecular photovoltaics. Molecule-substrate electron transfer, on the other hand, may be relevant to fighting radiation damage in biological imaging with atomic resolution (Fill *et al.*, 2008). Insight into the interplay between collective and collisional processes is of key importance to explore and possibly utilize collective phenomena in many-body systems; clusters offer an ideal test bed for addressing these phenomena (Fennel *et al.*, 2007).

Motivated by a number of questions, we review approaches to accessing electron dynamics in ever more complex atomic assemblies directly in the time domain.

1. Attosecond photoelectron spectroscopy

Just as in atoms, electronic motion in molecules can be triggered by creating a coherent superposition of two or more electronic states. Here we review how this motion can be tracked by APS, by measuring the angular momentum distribution of photoelectrons produced by

[93]Supported by extensive classical simulations of double ionization in strong laser field performed by Haan, Eberly, and co-workers: Haan *et al.*, 2002, 2006; Panfili *et al.*, 2002; Ho *et al.*, 2005.

[94]See, e.g., Weckenbrock *et al.*, 2004, and Zeidler *et al.*, 2005.

[95]Bandrauk and Shon, 2002; Schafer *et al.*, 2004; Johnsson *et al.*, 2005, 2006; Kitzler and Lezius, 2005; Biegert *et al.*, 2006; Figueira de Morisson Faria *et al.*, 2006; Kitzler *et al.*, 2006.

[96]See, e.g., Becker *et al.*, 1994, 2000, 2001; Brabec *et al.*, 1996; Becker *et al.*, 1997, 2002: Lohr *et al.*, 1997; Becker and Faisal, 2000, 2002; Maeda *et al.*, 2000; Goreslavskii *et al.*, 2001; Yudin and Ivanov, 2001b; Figueira de Morisson Faria *et al.*, 2002; Haan *et al.*, 2002; Panfili *et al.*, 2002; Popruzhenko *et al.*, 2002; Figueira de Morisson Faria, Liu, *et al.*, 2004; Figueira de Morisson Faria, Schomerus, *et al.*, 2004; Jaron-Becker *et al.*, 2004, 2006; Ho *et al.*, 2005; Prauzner-Bechcicki *et al.*, 2005; Haan *et al.*, 2006, 2007; Haan and Smith, 2007.

[97]For a comprehensive review of femtochemistry, see Zewail, 2000. The state of the art of this evolving field is well represented by Gessner *et al.*, 2006.

[98]The theoretical literature on coupled electronic and nuclear dynamics in molecules is expanding towards the sub-fs domain. Relevant work include Martinez *et al.*, 1996, 1997; Kawata *et al.*, 1999; Harumiya *et al.*, 2002; Kono *et al.*, 2004, 2006; Zhu *et al.*, 2004; Amano and Takatsuka, 2005; Awasthi *et al.*, 2005; Hu *et al.*, 2005; Nguyen and Bandrauk, 2006; Takahashi and Takatsuka, 2006; Takatsuka, 2006, 2007; Nest *et al.*, 2007. For a review, see Levine, 2005.

[99]For recent reviews, see, e.g., Jortner and Ratner, 1997; Feringa, 2001; Nitzan, 2001.

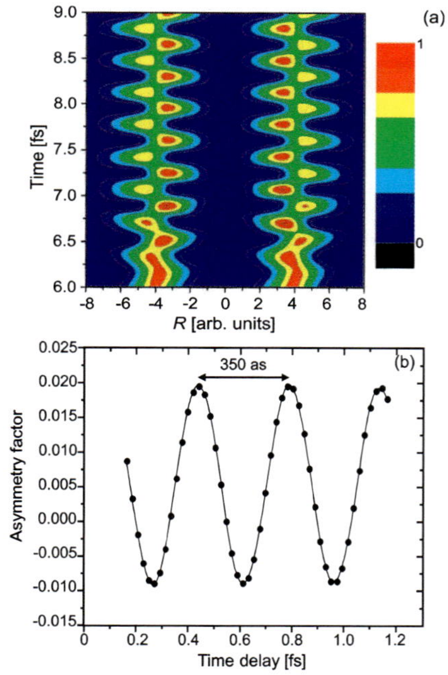

FIG. 53. (Color) Proposal for inducing attosecond electron wave-packet dynamics by a 0.8-fs, 115-nm VUV pump pulse in H_2^+ and probing it with a time-delayed 0.1-fs, 20-nm XUV pulse (Bandrauk *et al.*, 2004). Both pulses are polarized parallel to the molecular axis. (a) Contour plot of the electron probability distribution along the molecular axis for an internuclear distance of eight atomic units vs pump-probe delay. (b) Asymmetry factor $(P_- - P_+)/(P_- + P_+)$ vs delay, where P_+ and P_- represent the probability of observing the electron liberated by the XUV probe in the positive or negative direction (along the molecular axis), respectively. Adapted from Bandrauk *et al.*, 2004.

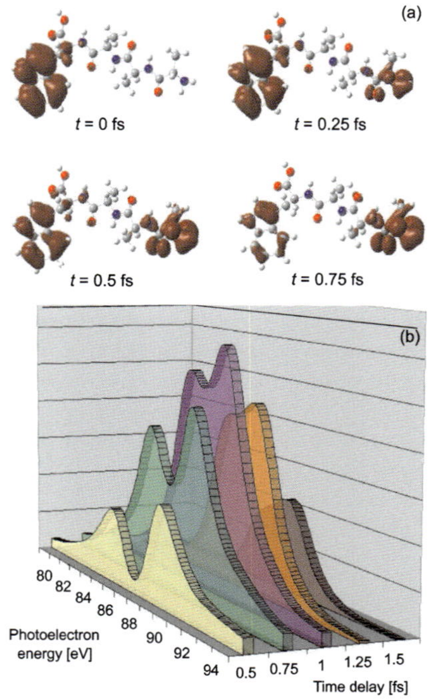

FIG. 54. (Color) Computed ultrafast positive charge (hole) migration in a tryptophane-terminated tetrapeptide (Remacle and Levine, 2006a, 2007). (a) The hole density shown in red indicates that the charge swings across the entire peptide from the aromatic amino acid on the left to the *N* end on the right within less than one femtosecond, following excitation of the electronic wave packet on an attosecond time scale. This hyperfast charge migration is proposed to be probed by measuring the kinetic energy distribution of photoelectrons released by a time-delayed sub-fs XUV pulse. (b) A series of such freeze-frame spectra calculated for a 250-as, 95-eV probe pulse at different pump-probe delays. From Remacle and Levine, 2007.

a time-delayed sub-fs XUV pulse (probing, IID1; detection, IIIB2; see Tables II and III).[100] Consider the simplest case: a coherent superposition of the two lowest-energy states of the H_2^+ molecular ion,

$$\psi(\mathbf{r},t) = c_1\phi_1(\mathbf{r})e^{-iW_1t} + c_2\phi_2(\mathbf{r})e^{-iW_2t}, \tag{50}$$

where indices 1 and 2 represent the $1\sigma_g$ (ground) electronic state and $2\sigma_u$ (the lowest excited) electronic state, respectively. In their simulations, Bandrauk *et al.* (2004) created this state with a sub-fs VUV pulse and probed the unfolding wave-packet dynamics with a 100 as XUV probe via photoionization. They found that the unfolding intramolecular electronic motion [Fig. 53(a)] can be traced by collecting photoelectrons in two opposite directions along the molecular axis aligned with the light fields [Fig. 53(b)].[101] [102]

Inspired by the experiments of Weinkauf *et al.* (1995, 1996, 1997), theoretical studies of electronic wave-packet dynamics have been extended for the charge (hole or electron) transport in large, biologically relevant systems.[103] Complexity arises here in several ways: the initially localized wave packet (created by photoexcitation) is composed of a large number of stationary electronic states and electron correlation must be considered during charge transport. Beyond its relevance to molecular electronics, photovoltaics, and biology, elec-

[100]Scrinzi *et al.*, 2001a, 2001b; Bandrauk *et al.*, 2004; Yudin *et al.*, 2005, 2006; Chelkowski *et al.*, 2006. In the femtosecond regime, the technique was demonstrated by Strasser *et al.*, 2007.

[101]The measurement also reveals the spectral phase of the attosecond probing pulse (Itakura, 2007).

[102]Studies were extended to asymmetric molecules (Remacle *et al.*, 2007).

[103]Martinez *et al.*, 1996, 1997; Remacle and Levine, 1997, 1999, 2006a; Cederbaum and Zobeley, 1999; Remacle *et al.*, 1999; Breidbach and Cederbaum, 2003, 2005; Hennig *et al.*, 2005; Kuleff *et al.*, 2005.

tron charge transfer in extended molecular systems may also play a key role in the fragmentation of large molecules observed by Weinkauf, Schlag, and co-workers (1995, 1996). The concept of "charge-directed reactivity" (Weinkauf *et al.*, 1997; Remacle *et al.*, 1998) may offer a new route to chemical reaction control.

APS, possibly combined with coincidence detection of the reaction products (IIIC2; see Table III), lends itself to unraveling the mysteries and implications of charge transfer in molecular systems. Viability of this approach depends on the sensitivity of this probing in *large systems*. Results of Levine and Remacle (2006) give us grounds to be optimistic. Figure 54(a) depicts hole migration in a tryptophane-terminated tetrapeptide following sub-fs photoionization (implemented either with a cosine-shaped few-cycle wave: IC1, or with an intense VUV pulse: ID1; see Table I). Figure 54(b) reveals that the sub-fs photoelectron probe is highly sensitive to the unfolding charge migration (Remacle and Levine, 2007), offering a means of monitoring it with attosecond resolution.[104] In our view, real-time observation (and possibly control) of electron transfer in large molecular systems, such as DNA (Kato *et al.*, 2004), is among the most exciting challenges attosecond technology faces today.

APS was proposed for observing ultrafast electron and structural motion in clusters (Georgescu *et al.*, 2007). Experiments of this kind may address fundamental questions about the way collisions destroy collective dynamics in many-body systems (Fennel *et al.*, 2007) and offer information that no conventional (time-integrated) measurement can provide due to dissipation (Georgescu *et al.*, 2007).

2. Attosecond high harmonic imaging

HHG from laser-atom or laser-molecule interaction carries information on the structure (Lein, Hay, *et al.*, 2002) and dynamics (Averbukh, 2004) of the bound electronic wave function interfering with the returning wave packet.[105]

Look at the component of the dipole moment responsible for high-frequency emission during recollision,

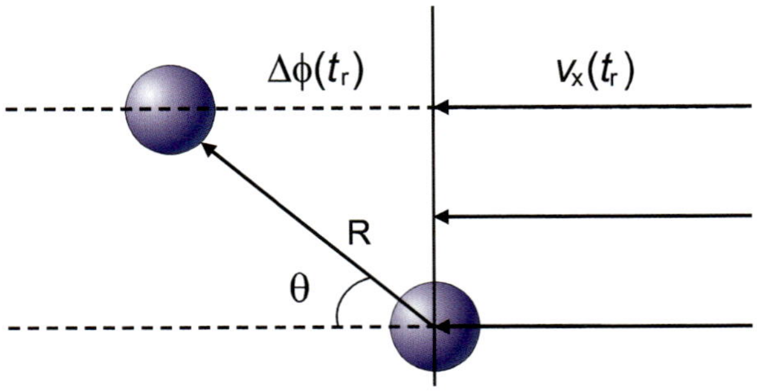

FIG. 55. (Color) Origin of interference in harmonic emission from a diatomic molecule. Returning electron with velocity v_x interferes with the ground state wave function—a superposition of atomic orbitals centered around the two nuclei spaced by R. The phase delay accumulated by the returning electron between the two nuclei is $\Delta\phi(t_r)=v_x(t_r)R\cos\theta$. From Lein, Hay, *et al.*, 2002.

$d_h(t)=\langle\psi_b|d|\psi_c\rangle$+c.c. As compared with Eq. (22), we replaced $|\psi_g\rangle$ with $|\psi_b\rangle$ to allow the bound portion of the wave function to be different from the ground state $|\psi_g\rangle$. In the framework of the SFA, stationary phase analysis of $d_h(t)$ (Ivanov *et al.*, 1996) implies that, in good approximation, at every moment t the transition matrix element is proportional to the matrix element between the bound wave packet and a *single* continuum state corresponding to the instantaneous velocity $v_x(t_r)$ of the classical electron recolliding with the parent ion at this moment in time. This velocity $v_x(t_r)$ of the recolliding electron determines the frequency of the emitted radiation:

$$\Omega(t_r) = v_x^2(t_r)/2 + I_p. \tag{51}$$

Now consider harmonic generation by a diatomic molecule (see Fig. 55), and approximate the electronic ground state as a linear superposition of two atomic orbitals $|\psi^{(A)}\rangle$ centered at each of the two nuclei:

$$\psi_b = \frac{1}{\sqrt{2}}[\psi^{(A)}(\mathbf{r}-\mathbf{R}/2) + \Lambda\psi^{(A)}(\mathbf{r}+\mathbf{R}/2)], \tag{52}$$

where $\Lambda=1$ and $\Lambda=-1$ correspond to even and odd superposition, respectively, and $\mathbf{R}$ is the vector connecting the two nuclei; see Fig. 55. The recombination matrix element becomes a sum of two identical contributions from each atomic orbital (each of the two nuclei). These two amplitudes interfere with each other much as in Young's double-slit experiment:

$$d_h(t_r) \propto d_h^{(A)}(t_r)\{1 + \Lambda\exp[i\Delta\phi(t_r)]\}, \tag{53}$$

where $d_h^{(A)}(t_r)$ is the contribution from a single atom and $\Delta\phi(t_r)=v_x(t_r)R\cos\theta$. For more complex molecules, the induced dipole will contain interference of several amplitudes with relative phases determined by the scattering centers' positions. The resultant interference in the

[104]The sub-fs photoelectron probe is also sensitive to structural changes (which did not occur in this example), due to dynamic shifts of core levels following rearrangement. In case of fragmentation, coincidence detection of photoelectrons and fragments may help in retrieving the complex electron-nuclear dynamics. Another option is to map out the kinetic energy release of the molecular ion fragments following a Coulomb explosion after the probe pulse (Stapelfeldt *et al.*, 1998; Lin *et al.*, 2006; Tong and Lin, 2006).

[105]Lein, Hay, *et al.*, 2002; Niikura *et al.*, 2002; Lein *et al.*, 2003; Averbukh, 2004; Itatani *et al.*, 2004; Kanai *et al.*, 2005; Lein, 2005; Niikura *et al.*, 2005; Vozzi *et al.*, 2005; Zimmerman *et al.*, 2005; Baker *et al.*, 2006; Chirila and Lein, 2006a, 2006b; Le, Morishita, and Lin, 2008, Morishita *et al.*, 2008; for a review, see Lein, 2007.

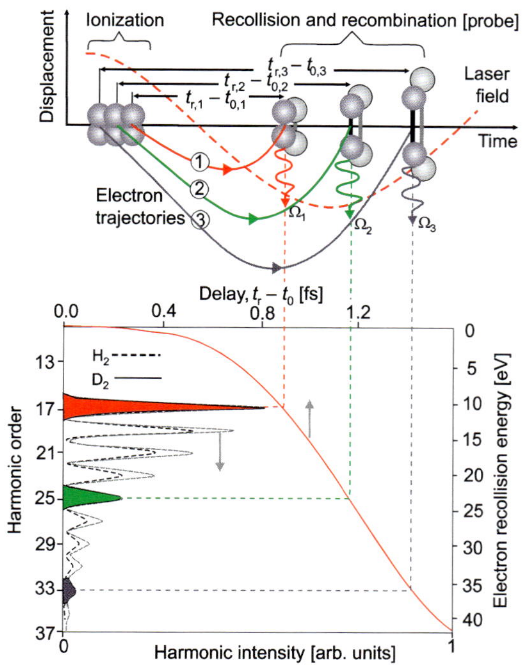

FIG. 56. (Color) Probing initial temporal evolution of nuclear coordinates in a molecule, following sudden excitation by ionization. The nuclear motion is mapped onto the spectral intensity distribution of high-order harmonics via the energy chirp of the recolliding electron responsible for the harmonic emission (Baker *et al.*, 2006). For the simplest case of a diatomic molecule, such as H_2 or D_2, the internuclear separation starts increasing at the moment of ionization. For increasing recollision instants $t_{r,1}$, $t_{r,2}$, and $t_{r,3}$ it increasingly deviates from the initial value (upper panel). The induced dipole responsible for the harmonic emission includes the overlap between vibrational wave functions in the neutral molecule and its ion. This overlap decreases with increasing internuclear separation in the ion, reducing the harmonic intensity. The monotonic dependence of the recollision energy W_r on the recollision time t_r for either short or long trajectories (Fig. 20) gives a one-to-one map between harmonic energy $\Omega(t_r)=W_r(t_r)+I_p$ and the recollision time t_r. The red line in the lower panel connects $\Omega(t_r)$ with t_r for short trajectories. From the decrease of the harmonic signal and knowledge of $W_r(t_r)$, the different internuclear separations at different moments of time following ionization can be determined. The technique is called PACER: probing attosecond dynamics by chirp encoded recollisions. Adapted from Baker *et al.*, 2007.

harmonic emission spectrum therefore reflects the structure of the molecule. This theoretical prediction (Lein, Hay, *et al.*, 2002) has stimulated further theoretical activity.[106] Experiments for diatomic and triatomic molecules (Itatani *et al.*, 2004; Kanai *et al.*, 2005; Vozzi *et al.*, 2005) have demonstrated clear minima in the harmonic spectra, even though their exact origin is currently unclear (the minima appear to shift with the intensity of the driving field). With molecular alignment (Torres *et al.*, 2007) and/or elliptic polarization of the driving field (Kitzler *et al.*, 2007), attosecond high harmonic imaging (AHI) hold promise for imaging molecular structures and multielectron dynamics. However, theory will have to meet a formidable challenge, especially for polyatomic molecules: quantitative description of rich laser-driven multielectron dynamics.[107]

AHI also provides access to nuclear motion (Lein, 2005; Baker *et al.*, 2006, 2008; Wagner *et al.*, 2006). For example, one can excite vibrations in a neutral molecule and probe them via a high-order harmonic generation signal, which shows modulations as a function of the vibrational motion (Wagner *et al.*, 2006). Alternatively, one can probe the dynamics in the ion. As soon as the electron is liberated from the molecule, a molecular ion is formed. The equilibrium configuration of the molecular ion is usually different from that of the neutral molecule. As a consequence, the nuclei are set in motion at the same instant t_0 as the electron is liberated. This correlation along with the chirped nature of the recolliding electron wave packet, see Fig. 20, permits one to map the nuclear motion onto the frequency spectrum of the harmonic emission. Drawing on the theoretical analysis of Lein (2005), Baker *et al.* (2006, 2008) put this concept into practice. This pioneering experiment is reviewed in Fig. 56. Implemented with a few-cycle (ideally single-cycle) IR laser pulse, AHI may allow one to capture both electronic and nuclear rearrangement in molecules within the first few femtoseconds following excitation.

3. Attosecond electron diffraction

Observation of short-lived transient structures calls for the combination of ultrafast probing techniques with those of conventional diffraction: following excitation by an ultrashort pump pulse the sample is exposed to a delayed electron or x-ray probe pulse. Time-resolved x-ray and electron diffraction as well as microscopy have recently attained sub-ps resolution with x-ray pulses emitted from laser-driven plasma sources[108] and with electron bunches (with energies ranging from ~30 to ~300 keV, corresponding to de Broglie wavelengths of ~0.06 to 0.02 Å) that are produced by femtosecond

[106]See, e.g., Zhou *et al.*, 2005a, 2005b; Chirila and Lein, 2006a, 2006b; Gordon *et al.*, 2006; Le *et al.*, 2006; Santra and Gordon, 2006; Patchkovskiim *et al.*, 2006; and, for a review, Lein, 2007.

[107]In atoms and diatomic molecules, correlated multielectron (mostly two-electron) dynamics has been extensively studied; see, e.g., Watson *et al.*, 1997; Tong and Chu, 1998; Chu and Chu, 2001; Lein, Engel, *et al.*, 2001; Lein, Gross, *et al.*, 2001, 2002; Lein, Kreibich, *et al.*, 2002; Becker and Faisal, 2005; Caillat *et al.*, 2005; Krause *et al.*, 2005; Klamroth, 2006; Gordon *et al.*, 2006; Jordan *et al.*, 2006; Patchkovskiim *et al.*, 2006; Rohringer *et al.*, 2006; Santra and Gordon, 2006; Gräfe and Ivanov, 2007; Volkova *et al.*, 2007; Walters *et al.*, 2007; Xie *et al.*, 2007.

[108]For reviews, see Rousse *et al.*, 2001; Bressler and Chergui, 2004; Pfeifer, Spielmann, and Gerber, 2006.

photoemission followed by acceleration via a dc field.[109]

Electron scattering cross sections exceed those of x-ray scattering typically by five to six orders of magnitude, but this comes at the expense of rapid pulse broadening due to Coulomb repulsion and/or velocity dispersion, which has so far limited resolution to a few hundred femtoseconds. In this section we discuss possible ways of improving the temporal resolution of ultrafast electron diffraction to the few-fs and sub-fs regime using the recolliding electron in an ionizing molecule or a microwave-accelerated, phase-focused single-electron pulse for recording diffraction images of transient molecular structure.

Electron diffraction upon recollision can be used to record and retrieve information on molecular structure (Lein, Marangos, *et al.*, 2002; Spanner *et al.*, 2004; Yurchenko *et al.*, 2004; Morishita *et al.*, 2008; Murray and Ivanov, 2008; Okunishi *et al.*, 2008; Ray *et al.*, 2008; Zhou *et al.*, 2008) and was recently demonstrated experimentally (Meckel *et al.*, 2008). When the electron returns to the parent ion, "elastic" scattering takes a diffraction image of the parent molecule.[110] In laser-induced electron diffraction, temporal confinement of the recolliding electron wave packet adds sub-fs temporal resolution to the spatial resolution dictated by the de Broglie wavelength of the returning electron. With NIR ($\lambda_L \approx 0.8\ \mu$m) and IR ($\lambda_L \approx 2\ \mu$m) light one may achieve recollision energies W_r in excess of 0.1 and 1 keV, respectively (see Fig. 20), implying a de Broglie wavelength of ≈ 1 Å and well below 1 Å, respectively. Few-cycle (ideally single-cycle) laser pulses will benefit the technique by maximizing W_r and suppressing multiple recollision and related undesirable interference effects (Spanner *et al.*, 2004; Yurchenko *et al.*, 2004; Hu and Collins, 2005).

Figure 57 shows results of the calculations of Spanner *et al.* (2004) for a diatomic molecule aligned perpendicularly to laser polarization. The diffraction image is clear when proper directions and energies corresponding to a fixed moment of recollision are selected; see Fig. 57(b). When implemented with a sub-two-cycle IR laser pulse, this technique offers subfemtosecond resolution combined with subangstrom resolution.

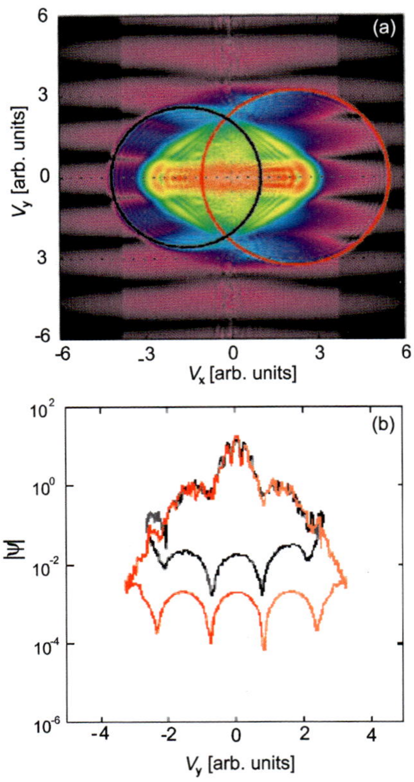

FIG. 57. (Color) Laser-induced recollision imaging of molecular structure by diffraction of strong-field-driven recollision electron (Spanner *et al.*, 2004). Driving field: $\tau_L = 5$ fs, $\lambda_L = 0.8\ \mu$m. (a) Transverse momentum distribution of the recolliding electron diffracted off its parent diatomic molecule. The molecule is aligned parallel to the y axis and the driving laser field is polarized parallel to the x axis. Different colors in (a) correspond to different orders of magnitude, from red to green to blue to violet to black. (b) Selected cuts of the spectrum, which correspond to the fixed (maximum) electron energy during recollision, show clear diffraction pattern. The pattern is particularly clear for electrons scattered at large angles (left part of the black circle, right part of the red circle in panel (a). Adapted from Spanner *et al.*, 2004.

Both AHI and laser-induced diffraction imaging rely on ionization. However, in many cases ionization is an invasive probe, inducing significant atomic rearrangements between the instants of release and recollision; see Fig. 56. The following proposal aims at overcoming these restrictions.

Microwave-accelerated sub-fs single-electron pulses. The electron bunches used by conventional ultrafast electron diffraction typically contain several thousand electrons. Repulsion (space charge) between them limits the pulse duration to several hundred femtoseconds on target. Zewail and co-workers were the first to propose the use of single-electron pulses produced at megahertz repetition rates with femtosecond laser oscillators (Lobastov *et al.*, 2005). In order to advance the state of the

[109]Since first demonstrations in the 10–100-ps range (Mourou and Williamson, 1982; Williamson *et al.*, 1984; Elsayed-Ali and Mourou, 1988) the technique evolved to sub-ps resolution owing to developments by Ahmed Zewail and Dwayne Miller (Williamson and Zewail, 1991; Williamson *et al.*, 1992, 1997; Ihee *et al.*, 2001; Siwick *et al.*, 2003; Dwyer *et al.*, 2007a, 2007b; Shorokov and Zewail, 2008). For reviews, see Srinivasan *et al.*, 2003; Zewail, 2006.

[110]Laser-induced electron diffraction is related to a well-established imaging technique, in which an electron freed by a picosecond synchrotron pulse takes an image of the core potential of the parent ion on its way out of the molecule in what is often referred to as a "half collision." Angle-resolved photoelectron spectra then reveal structural dynamics with picosecond resolution; see, e.g., Becker and Shirley, 1996; Gessner *et al.*, 2002; Kugeler *et al.*, 2004; Weber *et al.*, 2004; Rolles *et al.*, 2005.

art of ultrafast electron diffraction by this approach, two conditions must be met. First, sufficiently intense few-fs pump pulses at megahertz rate must be available. Recent advances in scaling the pulse energy from femtosecond Ti:sapphire oscillators afford promise of fulfilling this condition (Naumov *et al.*, 2005). Second, broadening of the electron pulse on its way from the photocathode to the target due to a spread of the initial electron velocity must be minimized. In the nonrelativistic limit this spread limits the duration of a dc-field-accelerated electron bunch on target to[111]

$$\tau_e = 2.34\sqrt{\Delta W_i}/E_{\mathrm{acc}}\ \mathrm{ps}, \tag{54}$$

where ΔW_i is the initial energy spread of electrons in eV and E_{acc} is the acceleration field in MV/m (Schelev *et al.*, 1971). Note that the temporal spread due to velocity dispersion is independent of the distance of the target from the photocathode. Vacuum breakdown limits E_{acc} to less than 10 MV/m (Kinoshita *et al.*, 1987), whereas ΔW_i typically ranges between 0.25 and 1 eV, preventing the delivery of electron pulses on target with a duration substantially less than 0.2 ps by applying a dc accelerating field.

Fill *et al.* (2006) proposed to accelerate single electrons released from a photocathode by a femtosecond laser pulse in a microwave cavity to overcome this limitation (Fig. 58). The microwave acceleration permits applying higher fields due to increased breakdown threshold and, more importantly, time dependence of the accelerating field, $E_{\mathrm{acc}}(t) = E_0 \sin(\omega t + \phi)$, results in different energy gain for particles with different initial energies or release moments. With proper choice of ϕ different initial energies and release instants can be simultaneously corrected (Monastyrskiy *et al.*, 2005).

Figure 58(b) depicts the simulated single-electron wave-packet duration on target along with the corresponding optimum phase as a function of target distance from the accelerating cavity for a 10 fs spread (full width at half maximum) of the instants of release (achievable with a 10-fs laser pulse). For a target distance of $L \approx 1$ cm, a 40-kV electron pulse can have a duration of less than 1 fs on target, demonstrating the power of microwave phase focusing. However, this pulse will be locked to the microwave field rather than to the fs laser pulse (Veisz *et al.*, 2007). Therefore the resolution of a laser-pump–electron-probe experiment will be limited by the jitter between the fs laser and the microwave field. They could already be synchronized to better than 10 fs (Ma *et al.*, 2001). Further progress along with techniques for measurement (Baum and Zewail, 2006; Reckenthaeler *et al.*, 2008) may allow single-electron diffraction to access transient electronic structures with atomic resolution in space and time.

[111]From an experimental perspective, Eq. (54) gives the arrival time jitter on target of single electrons with respect to the femtosecond pulse that released them. In quantum mechanics, it is the duration of the electron wave packet on target.

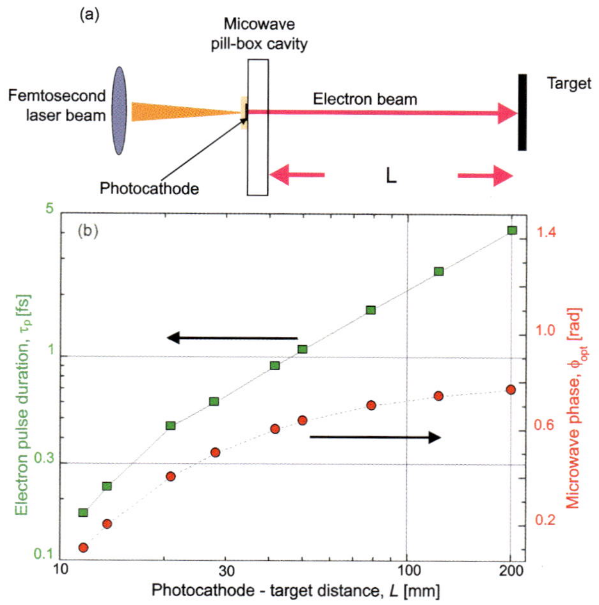

FIG. 58. (Color online) Proposal for a miniature microwave electron accelerator based on a pill-box microwave cavity. (a) The thin photocathode is illuminated from the back side. Accelerated electrons leave the cavity through a small hole on the right side. (b) Minimum electron wave-packet duration on target that can be achieved by optimizing phase focusing by the microwave field (squares); optimum microwave phase, ϕ_{opt}, at the instant of release of the electron (circles). From Fill *et al.*, 2006.

C. Electronic motion in solids and systems on surfaces

The broadest variety of electronic phenomena occurs in condensed matter. The spectrum of excitations localized in isolated particles is enriched with a number of delocalized transitions, charge-transfer processes in guest-host systems, a wealth of collective excitations such as excitons, plasmons or polaritons, charge screening on surfaces, and hot electron as well as electron-hole dynamics. These dynamics often unfold on a few-fs to sub-fs time scale. Capturing them in real time and controlling them on the electronic time scale with the electric field of light will allow researchers to explore and approach the limits of electron-based information technologies. Here we address the first steps towards attosecond resolution becoming commonplace in solid-state and surface science and related challenges.

1. Extension of attosecond spectroscopy to condensed matter

Time-resolved (two-photon) photoemission spectroscopy has been the most widely used approach to capturing transient electron states in solids and surfaces. In its first implementations an ultrashort VIS or UV pump pulse excites electrons to states above the Fermi level and a time-delayed VIS or UV probe pulse raises the electrons' energy to positive values, resulting in photoemission. The time evolution of the transient population

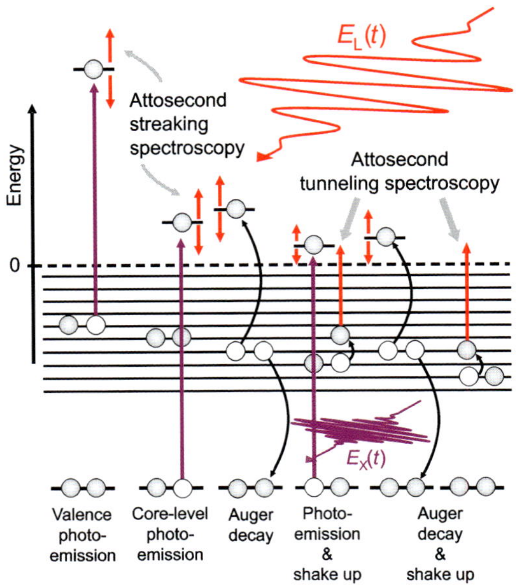

FIG. 59. (Color) Electronic excitation and relaxation processes in solids and possible ways of tracing these dynamics in real time.

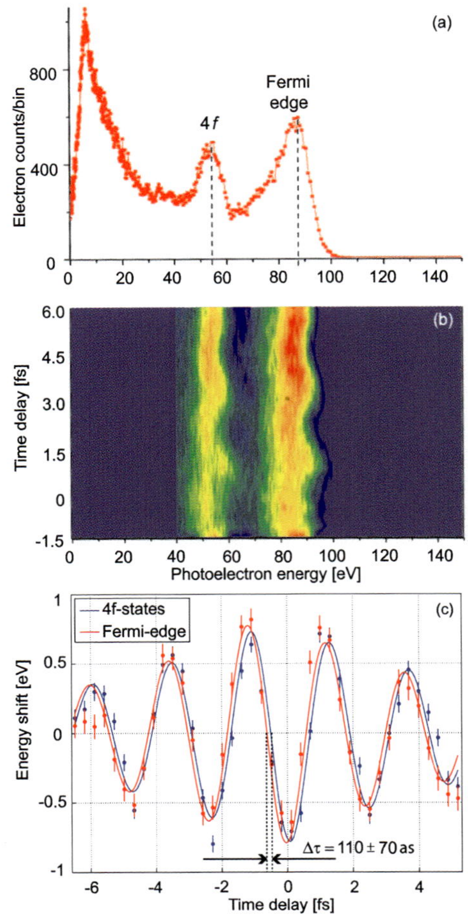

FIG. 60. (Color) Attosecond real-time observation of electronic charge transport in a solid (Cavalieri, Müller, *et al.*, 2007). Electrons were liberated by a single, 300-as, 95-eV XUV pulse in the presence of 5-fs, 750-nm, waveform-controlled pulses in a tungsten crystal. Both p-polarized pulses impinged on the sample at grazing incidence. (a), (b) The unperturbed XUV-induced photoelectron spectrum and the corresponding streaking spectrogram, respectively, of core-level electron emission (from the 4f level) and valence-band electron emission (from the Fermi edge) recorded with the few-cycle NIR field. (c) The center-of-mass energy shift of the recorded streaked spectra, revealing a ~110 as delay in emission of core electrons with respect to the emission of conduction band electrons. Courtesy of A. Cavalieri.

of excited states can then be retrieved by analyzing the photoelectrons' energy and momentum distribution versus pump-probe delay.[112]

At increasing excitation intensities, a pump-pulse-induced electron emission background (Riffe *et al.*, 1993) will emerge and extend to ever higher kinetic energies. This undesirable background tends to blur the photoelectron spectrum created by a low-photon-energy probe pulse. Pioneered by Haight and co-workers, extension of time-resolved photoemission to VUV probing photon energies elegantly circumvented this problem.[113] In principle, femtosecond VUV-XUV photoelectron spectroscopy can readily be extended to the attosecond domain by using a UV pump pulse with a duration of about 1 fs or less and a sub-fs XUV probe pulse (excitation, ID1; probing, IID1; detection, IIIB1-2; see Tables I–III, respectively). The former tool is not yet available but is expected to enrich the attosecond toolbox in the foreseeable future (Graf *et al.*, 2008). Alternatively, one may attempt to extend attosecond spectroscopies that have already been successfully demonstrated in the gas phase (Fig. 46) to solid matter; see Fig. 59. ATS may not be readily applicable to solids because the intense probe tends to generate large bound-excited-state populations before and during inducing photoemission (Quéré *et al.*, 2000). AST (triggering, IE1; probing, IIC1; detection, IIIB1-2; see Tables I–III, respectively) does not suffer from this problem.

Following the first observation of the laser-assisted photoelectric effect from a solid by Miaja-Avila *et al.* (2006), Cavalieri, Müller, *et al.* (2007) performed an attosecond spectroscopic experiment in a solid; see Fig. 60. The experiment has revealed (i) a sub-fs emission time both for the delocalized conduction-band electrons re-

[112]The technique was been first demonstrated in the picosecond time domain by Williams *et al.* (1982) and Yen *et al.* (1982) and with the improvement of the time resolution into the femtosecond (tens of fs) region by Schoenlein *et al.*, 1988; Bokor, 1989; Fann *et al.*, 1992; Schmuttenmaer *et al.*, 1994; Wolf *et al.*, 1996; Höfer *et al.*, 1997; Ogawa *et al.*, 1997. Phase-coherent detection has provided sub-optical-cycle resolution (Ogawa *et al.*, 1997), but retrieval of the dynamic processes from these interferometric time-resolved measurements is a nontrivial problem. For a review, see Petek and Ogawa, 1997.

[113]Haight and Silbermann, 1989; Haight, 1995; Bauer *et al.*, 2001; Siffalovic *et al.*, 2002. For a review, see Bauer, 2005.

leased from states near the Fermi edge as well as for the localized core electrons originating from the $4f$ state, and (ii) a delay in emission of the latter by approximately 100 as. Real-time access to atomic-scale charge transport dynamics in solids is now a reality.

2. Challenges

A thorough review of electron motions unfolding within a few femtoseconds or less in solids and on surfaces is beyond the scope of this paper. Here we choose a few examples to demonstrate the tremendous potential for gaining new, in-depth insight into fundamental processes in condensed matter with the help of attosecond tools and techniqes.

Charge screening in solids is one manifestation of many-body interactions (Mahan, 1993). It is predicted to establish within the inverse plasma frequency, on time scales ranging from hundreds of attoseconds in metals (Alducin *et al.*, 2004; Borisov *et al.*, 2004; Silkin *et al.*, 2007) to tens of femtoseconds in semiconductors (Huber *et al.*, 2001). One long-standing mystery in solid-state physics has been the high-energy tail of metallic x-ray spectra. Screening of the hole and electron created upon x-ray absorption has been assumed to be responsible for the observed anomalous behavior (see, e.g., Yue and Doniach, 1973; Canright, 1988), but lack of reliable data on the screening time has prevented validation of theoretical models so far. Attosecond probing of screening will also clarify the existence of unscreened bare Coulomb collisions on ultrashort time scales (Kwong and Bonitz, 2000) and that of transient excitonic states in metals before the onset of screening (Cao *et al.*, 1997; Schöne and Ekardt, 2000). The dynamics of surface excitations is yet another area offering many challenges (Echenique *et al.*, 2004).

Electron transfer across interfaces plays a central role in surface chemistry. Insight into charge transfer dynamics in macromolecular assemblies (Siffalovic *et al.*, 2004) and between adsorbate and bulk (conduction-band) states will contribute to the advancement of molecular electronics (Nitzan and Ratner, 2003; Zhu, 2004; Remacle and Levine, 2006b) and benefit the development of more efficient solar cells (O'Reagan and Graetzel, 1991). So far, the core-hole-clock technique has been the only approach providing access to charge transfer dynamics on a few-femtosecond or shorter time scale.[114] It is based on high-resolution spectroscopy of autionization following resonant core excitation. The lifetime of the core hole provides an internal reference for clocking charge transfer from the excited bound state, allowing one to determine electron transfer times with sub-fs resolution (Schnadt *et al.*, 2002; Föhlisch *et al.*, 2005), but require careful analysis of competing relaxation channels in molecular adsorbates (Kirchmann *et al.*, 2006). Direct sub-fs probing will therefore be particularly helpful in complex molecular architectures (Barth *et al.*, 2005).

Collective electronic motion in nanosystems. Nanoscopic materials have attracted interest because of their engineerable electrical and optical properties. Collective electron density fluctuations play a role in most of their potential applications. Of particular interest are those that can be excited at metal-vacuum or metal dielectric interfaces: surface plasmons (Raether, 1988). The prospect of nanoplasmonic devices (Barnes *et al.*, 2003; Van Duyne, 2004) and applications ranging from medical and pharmaceutical screening through environmental monitoring to subwavelength optics has stimulated efforts to gain an insight into the fundamental dynamics of plasmon excitations (Kubo *et al.*, 2005, 2007) and identify ways of controlling it. The natural time scale of collective electron dynamics in metallic nanosystems is less than 1 fs. A combination of photoelectron emission microscopy with AST will provide access to these dynamics with nanometer and sub-fs resolution (Stockman *et al.*, 2007).

D. Future prospects: Steering and imaging atomic-scale electronic motion

1. Steering electrons with the electric field of synthesized light

Here we address two questions related to extending attosecond control of electronic motion to complex systems ranging from simple (diatomic) molecules through complex (bio)molecules to nanostructures and assemblies. First, can synthesized (waveform-controlled) light fields be used for steering electron wave packets in molecules on the electronic time scale where nuclear motion is essentially frozen? Second, can such light field control on electronic to chemical time scales affect the outcome of structural re-arrangements or chemical reactions?

Barth and Manz (2006a, 2006b) suggested that the former question may be answered affirmatively. In their numerical experiment, a circularly polarized few-cycle UV laser pulse was applied to a Mg-porphyrin model system. This pulse induced a giant unidirectional ring current, with its strength and direction controlled by the instantaneous amplitude and the direction of rotation of the UV electric field; see Fig. 61. Beyond demonstrating the control of electronic motion in a molecule, this study also revealed a route to generating—for a period of a few femtoseconds—a unipolar (nonoscillating) magnetic field of unprecedented strength. Barth and Manz predicted that a UV pulse of moderate intensity ($\sim 10^{12}$ W/cm^2) would be able to generate a giant ring current on the order of 0.1 mA within a molecule. A magnetic field on the order of 10^4 T would be required to induce the same ring current (Steiner *et al.*, 2005).

We begin our discussion of the second question by reviewing a recent experiment on sub-fs control of electron localization during breakup of a D_2^+ molecule

[114]For reviews, see Wurth and Menzel, 2000; Brühwiler *et al.*, 2002; Föhlisch, 2006.

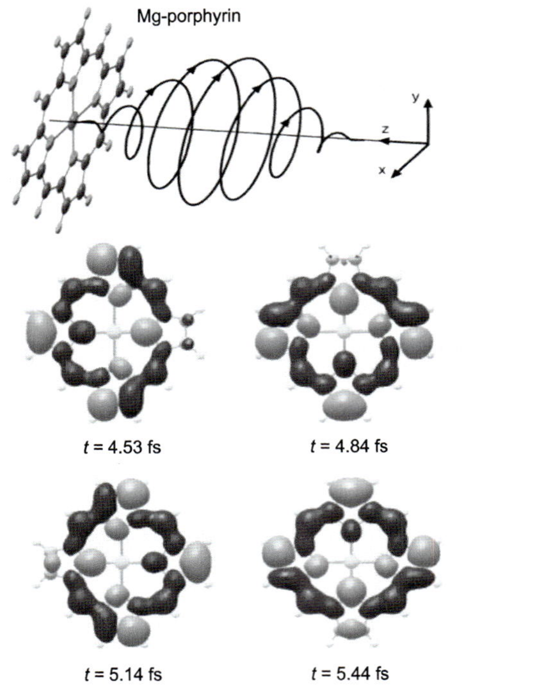

FIG. 61. Induction of unidirectional electronic ring currents in molecules. The upper panel shows an eight-cycle circularly polarized UV pulse propagating along the z axis to impinge upon a Mg-porphyrin molecule, with its plane aligned perpendicular to the z axis. The arrows indicate the time evolution of the field acting on the molecule. The lower panel shows several snapshots of instantaneous electron density distribution during the central part of the driver pulse, revealing a ring current driven by the rotating electric field. From Barth and Manz, 2006a, 2006b.

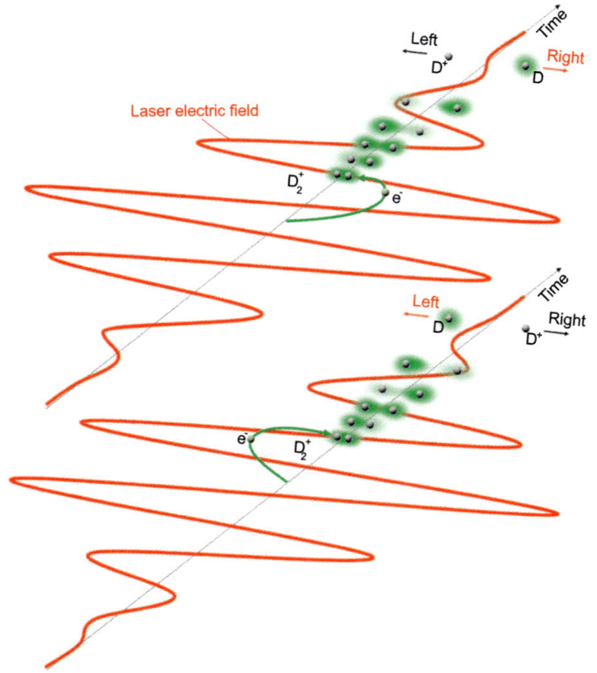

FIG. 62. (Color) Controlling electron localization during molecular dissociation with the electric field of light (Kling *et al.*, 2006). Laser-induced ionization and subsequent recollision of the freed electron triggers dissociation of the molecular ion (D_2^+). Controlling the electric field evolution in the ionizing few-cycle NIR laser field allows one to control the motion of the remaining electron (with its probability distribution shown in green) around the nuclei (spheres) while they move apart. A cosine-shaped waveform with its peak field pointing to the left or to the right forces the electron to stick—with a high probability—to the atom leaving the molecule to the right or left directions, respectively.

(Kling *et al.*, 2006).[115] Ionization of D_2 was followed by electron recollision with the molecular ion D_2^+, exciting the remaining bound electron into the dissociative $|\sigma_u\rangle$ state, still in the presence of the ionizing few-cycle laser pulse, which creates a coherent superposition of the ground $|\sigma_g\rangle$ and first excited $|\sigma_u\rangle$ state. This implies electron oscillation between the left and right nuclei as the nuclei move apart. Beyond some critical internuclear distance, the electron can no longer tunnel through the potential barrier emerging between the two nuclei. Which of the two nuclei will catch the electron is determined by the phase of the oscillatory electron motion relative to the dissociation. This phase, in turn, is controlled with attosecond precision by the carrier-envelope phase of the waveform;[116] see Fig. 62. The experiment reveals that (i) attosecond electron motion does affect chemical changes and (ii) attosecond shaping of the light field in a femtosecond pulse can be used to control it (Weitzel, 2007).

Nevertheless, an electron wave packet composed of many stationary states means population of many potential energy surfaces (PESs). Different PESs often lead to different products. That is, exciting a large number of electronic states will create multiple multidimensional rovibrational wave packets which will evolve along their potential energy surfaces towards different outcomes. However, this does not necessarily mean that attosecond waveform engineering and sub-fs electron wave packet formation and steering has no future in controlling molecular dynamics.

One may speculate on a new approach: creating an electron wave packet with a sub-fs UV pulse and subsequently steering it with the shaped waveform of a light field that is strong enough to induce substantial dynamic shifts of the potential surfaces. Under these circumstances stationary PESs lose their meaning and the light-field-steered electron wave-packet motion may possibly have a dominant influence on the outcome of reactions

[115]For theoretical research, see Haljan *et al.*, 1997; Bandrauk *et al.*, 2004; Roudnev *et al.*, 2004; Hu *et al.*, 2006; Roudnev and Esry, 2007; Tong and Lin, 2007a, 2007b.

[116]He *et al.* (2007) proposed that electron localization can be made more efficient by independent start using a two-pulse control scheme.

in complex systems.[117] This approach, if it were to work, would be new in that synthesized light fields (Fig. 14) would shape the density of bonding electrons in a molecule[118] in a desired way on the electronic, i.e., attosecond, time scale while maintaining this control over molecular, i.e., femtosecond, time scales, allowing the desired unimolecular reaction to be completed. Only experiments will tell us whether reality lives up to these visions.

An important question for the development of science and technology is whether electron-based information processing and storage can possibly be down-scaled to atomic dimensions and sped up to the atomic time scale, i.e., to optical frequencies. Can these ultimate limits be approached, or reached and implemented in practical devices by exploiting electric interactions (electronics), magnetic interactions (spintronics), or collective electron motion (plasmonics)? Can the electric field of infrared or visible light be used to control electric signals in future atomic-scale chips,[119] just as microwave fields do in current state-of-the-art nanoscale circuits, realizing the ultimate electron-based information technology: (solid-state) lightwave electronics? Answering these questions may be one of the central missions of attosecond science.

2. 4D imaging of electrons with subatomic resolution in space and time

In the experiments discussed in this section, tracing of electronic motion meant real-time observation of transitions of electrons between quantum states of different energies. No explicit information about changes of the probability distribution of the electrons' position has been acquired. Angle-resolved attosecond photoelectron spectra may provide access to dynamic changes in the electron distribution in simple systems, but this approach may encounter difficulties in more complex structures. Only microscopy and diffraction allows acquisition of direct information on the electron density distribution irrespective of the complexity of the system.

First, consider possibilities in microscopy. The availability of light pulses with characteristics controlled on a sub-fs scale offers several ways of furnishing well-established microscopic techniques with attosecond resolution. In photoemission electron microscopy (PEEM) the continuous-wave UV or XUV light source must simply be replaced with one emitting this radiation in the form of attosecond bursts (Stockman *et al.*, 2007). In scanning tunneling microscopy (STM), on the other hand, electron emission from the nanometer-sized tip may be confined to several hundred attoseconds by launching electrons with a cosine-shaped few-cycle light pulse via optical field ionization (Hommelhoff *et al.*, 2006). Last but not least, ultrafast electron microscopy (UEM, Shorokhov and Zewail, 2008) may be advanced into the few-fs and possibly sub-fs regime by microwave (Fill *et al.*, 2006; Veisz *et al.*, 2007) and optical (Plettner *et al.*, 2006) acceleration. In this way several-nanometer (PEEM) and angstrom (STM, UEM) resolution in space can be combined with attosecond resolution in time.

The extension of diffraction imaging from the three spatial dimensions to the fourth, temporal, dimension is conceptually even more straightforward. Again, the electron or x-ray beam used for mapping the electron density in molecules or crystals must be replaced with a short pulse. The high (multi-keV) particle energy reconciles the apparently conflicting requirements of narrow relative energy distribution ($\leq 1\,\%$) and ultrashort pulse duration. The ultrashort electron or x-ray pulse allows recording snapshots of the dynamic evolution of the electron density distribution following excitation by a short pump pulse. Dynamic changes in the electron distribution may occur due to (i) the motion of the nuclei; the electron cloud virtually instantly adjusts to this motion; or (ii) electronic excitation. The former process evolves on a multi-fs time scale and mirrors atomic rearrangement in molecules or solids. It has already been imaged with sub-ps time resolution.[120] Electronic rearrangements may unfold within attoseconds. An example is shown in Fig. 63. The above-discussed techniques for sub-fs x-ray (Sec. VI) and electron (Sec. VII.B.3) pulse generation will open the way to 4D imaging of electronic as well as nuclear motion with picometer resolution in space and attosecond resolution in time.

VIII. CONCLUSIONS AND OUTLOOK

The emerging experimental and theoretical tools of attoseconds physics provide, for the first time, real-time access to the motion of electrons on atomic and subatomic scales. Access means both capture and steering with a speed matched to the rapidity of the fastest electronic phenomena. Microscopic motion of electrons plays a key role in advancing the technology of compact x-ray sources, or pushing electronics and magnetic storage to ever smaller dimensions and ever higher speeds; in triggering chemical reactions; in biological signal transduction and the damage and repair mechanisms of DNA; and in the undesirable and desired radiation-

[117]The concept of relating site-selective reactivity to charge transfer, i.e., the existence of nonstationary electronic states, in large molecules (Weinkauf *et al.*, 1997), based on the assumption: "reactivity follows charge," is a first push in this direction.

[118]By superimposing many orbitals and adjusting this superposition during the pulse.

[119]Note that this light-field control of electric current is different from earlier demonstrations, where quantum interference governed by the cycle-averaged intensity (rather than the electric field) of light has controlled injection of electric current in semiconductors (see Costa *et al.*, 2007, and references therein).

[120]For reviews, see Pfeifer, Spielmann, *et al.*, 2006; Zewail, 2006.

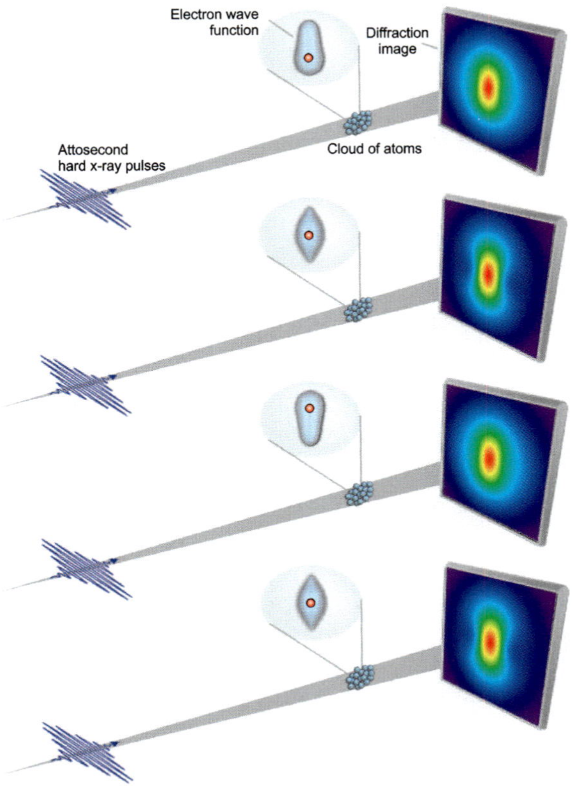

FIG. 63. (Color) Schematic of attosecond diffraction imaging of dynamic changes of atomic-scale electron distribution. As an example, hydrogen atoms excited into the $1S$-$2P$ coherent superposition state have been exposed to 100-as, 1-Å x-ray pulses in a numerical experiment (Yakovlev, 2007). From the recorded freeze-frame diffraction images shown on the screens, the instantaneous electron density distribution can be determined. From a series of such images electronic motion can be reconstructed with attosecond resolution in time and picometer resolution in space. Courtesy of V. Yakovlev.

induced damage to biological matter in cancer diagnostics and therapy, respectively.[121]

The emergence and expected future evolution of attosecond science demonstrate how different scientific disciplines and technologies, several of which have been recognized with the Nobel Prize, build upon each other to push the frontiers of science and technology. Lasers (Nobel Prize: Basov, Prokhorov, Townes) and perturbative nonlinear optics (Nobel Prize: N. Bloembergen) have permitted the generation and metrology of femtosecond laser pulses with well-controlled amplitude, opening the way to controlling and tracking chemical dynamics (Nobel Prize: A. H. Zewail). Chirped-pulse amplification and chirped multilayer mirrors have permitted routine generation of intense light pulses and pushing—with the generation of quasi-single-cycle laser light—the frontiers of ultrafast optics to its ultimate limit set by the oscillation cycle of the light field. Adding the frequency-comb technique (Nobel Prize: T. W. Hänsch) to these technologies has permitted control over the attosecond evolution of ultrastrong light fields, which along with photoelectron spectroscopy (Nobel Prize: K. M. Siegbahn) has opened the door to controlling and measuring the atomic-scale motion of electrons.

With the advent of lightwave electronics (Fig. 5), sub-fs tools have become available for direct attosecond real-time observation of the motion of electrons in atoms, molecules, and solids. Based on strong-field interactions, current sub-fs electron and photon pulses are delivered with particle energies limited to ~100 eV. Future expansion of attosecond science to relativistic light-electron interactions affords promise of a dramatic increase in energy and further decrease in duration of the probing particles. The capability of recording movies of any microscopic motion outside the atomic core will be the consequence of these advances.

[121]For questions that may serve as a compass for the possible future evolution of attosecond physics, see www.attoworld.de

GLOSSARY OF ABBREVIATIONS

AAS	Attosecond absorption spectroscopy
AHI	Attosecond harmonic imaging
APS	Attosecond photoelectron spectroscopy
AST	Attosecond streaking spectroscopy
ATI	Above-threshold ionization
ATS	Attosecond tunneling spectroscopy
CE	Carrier envelope (phase of ultrashort pulses)
COLTRIMS	Cold-target recoil-ion-momentum spectroscopy
CRAB	Complete reconstruction of attosecond bursts
FROG	Frequency-resolved optical gating
FWHM	Full width at half maximum (e.g., of a pulse intensity profile)
HHG	High-order harmonic (radiation)
IFEL	Inverse free electron laser
IR	Infrared (radiation)
NSDI	Nonsequential double ionization
NIR	Near-infrared (radiation)
PES	Potential energy surface
QPM	Quasi-phase-matching (of high-order harmonic generation)
RABBITT	Reconstruction of attosecond beating by interference of two-photon transitions
SAE	Single active electron (model)
SFA	Strong-field approximation
SPM	Self-phase modulation
S/N	Signal-to-noise (ratio)
SPIDER	Spectral phase interferometry for direct electric field reconstruction
UEM	Ultrafast electron microscopy
VIS	Visible (radiation)
VUV	Vacuum ultraviolet (radiation)
XUV	Extreme ultraviolet (radiation)

ACKNOWLEDGMENTS

We are grateful to Andre Bandrauk, Joachim Burgdörfer, Paul Corkum, Lorenz Cederbaum, Raphy Levine, and Albert Stolow for illuminating discussions and Adrian Cavalieri, Reinhard Kienberger, and Anne L'Huillier for careful reading of the manuscript and making numerous valuable comments. We thank Adrian Cavalieri, Eleftherios Goulielmakis, Ulrich Heinzmann, Raphy Levine, Anne L'Huillier, Katsumi Midorikawa, Gerhard Paulus, Valer Tosa, Matthias Uiberacker, and Dave Villeneuve for making artwork and/or results prior to publication available for this work. We gratefully acknowledge several calculations and simulations of Vlad Yakovlev for this review. Last but not least, we are indebted to Barbara Ferus for drawing and editing the artwork and to Bettina Schütz for her support in making the manuscript ready for submission. F.K. acknowledges support of the Munich Centre for Advanced Photonics (www.munich-photonics.de). M.I. acknowledges support of the Bessel prize of the A. v. Humboldt Foundation and the NSERC Special Research Opportunity fund.

REFERENCES

Abbamonte, P., K. D. Finkelstein, M. D. Collins, and S. M. Gruner, 2004, Phys. Rev. Lett. **92**, 237401.

Abraham, H., and T. Lemoine, 1899, Compt. Rend. **129**, 206.

Adair, R., C. Bockelman, and R. Peterson, 1949, Phys. Rev. **76**, 308.

Agostini, P., and L. F. DiMauro, 2004, Rep. Prog. Phys. **67**, 813.

Agostini, P., F. Fabre, G. Mainfray, G. Petite, and N. K. Rahman, 1979, Phys. Rev. Lett. **42**, 1127.

Akhmanov, S. A., V. A. Vysloukh, and A. S. Chirkin, 1992, *Optics of Femtosecond Laser Pulses* (AIP, New York).

Alducin, M., J. I. Juaristi, and P. M. Echenique, 2004, Surf. Sci. **559**, 233.

Alfano, R. R., and S. L. Shapiro, 1970, Phys. Rev. Lett. **24**, 584.

Alnaser, A. S., X. M. Tong, T. Osipov, S. Voss, C. M. Maharjan, P. Ranitovic, B. Ulrich, B. Shan, Z. Chang, C. D. Lin, and C. L. Cocke, 2004, Phys. Rev. Lett. **93**, 183202.

Alnaser, A. S., S. Voss, X.-M. Tong, C. M. Maharjan, P. Ranitovic, B. Ulrich, T. Osipov, B. Shan, Z. Chang, and C. L. Cocke, 2004, Phys. Rev. Lett. **93**, 113003.

Altucci, C., Ch. Delfin, L. Roos, M. B. Gaarde, A. L'Huillier, I. Mercer, T. Starczewski, and C.-G. Wahlström, 1998, Phys. Rev. A **58**, 3934.

Altucci, C., V. Tosa, and R. Velotta, 2007, Phys. Rev. A **75**, 061401(R).

Amano, M., and K. Takatsuka, 2005, J. Chem. Phys. **122**, 084113.

Antoine, Ph., A. L'Huillier, and M. Lewenstein, 1996, Phys. Rev. Lett. **77**, 1234.

Antoine, Ph., D. B. Milosevic, A. L'Huillier, M. B. Gaarde, P. Salieres, and M. Lewenstein, 1997, Phys. Rev. A **56**, 4960.

Apolonski, A., P. Dombi, G. G. Paulus, M. Kakehata, R. Holzwarth, T. Udem, Ch. Lemell, K. Torizuka, J. Burgdörfer, T. W. Hänsch, and F. Krausz, 2004, Phys. Rev. Lett. **92**, 073902.

Apolonski, A., A. Poppe, G. Tempea, C. Spielmann, T. Udem, R. Holzwarth, T. Hänsch, and F. Krausz, 2000, Phys. Rev. Lett. **85**, 740.

Aseyev, S. A. Y. Ni, L. J. Frasinski, H. G. Muller, and M. J. J. Vrakking, 2003, Phys. Rev. Lett. **91**, 223902.

Assion A., T. Baumert, M. Bergt, T. Brixner, B. Kiefer, V. Seyfried, M. Strehle, and G. Gerber, 1998, Science **282**, 919.

Assion A., T. Baumert, U. Weichmann, and G. Gerber, 2001, Phys. Rev. Lett. **86**, 5695.

Assion, A., M. Geisler, J. Helbing, and V. Seyfried, T. Baumert, 1996, Phys. Rev. A **54**, R4605.

Averbukh, V., 2004, Phys. Rev. A **69**, 043406.

Averbukh, V., and L. S. Cederbaum, 2006, Phys. Rev. Lett. **96**, 053401.

Averbukh, V., I. B. Müller, and L. S. Cederbaum, 2004, Phys. Rev. Lett. **93**, 263002.

Awasthi, M., Y. V. Vanne, and A. Saenz, 2005, J. Phys. B **38**, 3973.

Baeva, T., S. Gordienko, and A. Pukhov, 2006a, Phys. Rev. E **74**, 046404.

Baeva, T., S. Gordienko, and A. Pukhov, 2006b, Phys. Rev. E **74**, 065401(R).

Baker, S., J. S. Robinson, C. A. Haworth, C. C. Chirila, M. Lein, J. W. G. Tisch, and J. P. Marangos, 2007, J. Mod. Opt. **54**, 1011.

Baker, S., J. S. Robinson, C. A. Haworth, H. Teng, R. A. Smith, C. C. Chirila, M. Lein, J. W. G. Tisch, and J. P. Marangos, 2006, Science **312**, 424.

Baker, S., J. S. Robinson, M. Lein, C. C. Chirilă, R. Torres, H. C. Bandulet, D. Comtois, J. C. Kieffer, D. U. Villenenve, J. W. Tisch, and J. P. Marangos, 2008, Phys. Rev. Lett. **101**, 053901.

Balcou, P., C. Cornaggia, A. S. L. Gomes, L. A. Lompre, and A. L'Huillier, 1992, J. Phys. B **25**, 4467.

Baltuška, A., T. Fuji, and T. Kobayashi, 2002a, Opt. Lett. **27**, 306.

Baltuška, A., T. Fuji, and T. Kobayashi, 2002b, Opt. Lett. **27**, 1241.

Baltuška, A., T. Fuji, and T. Kobayashi, 2002c, Phys. Rev. Lett. **88**, 133901.

Baltuška, A., T. Udem, M. Uiberacker, M. Hentschel, E. Goulielmakis, C. Gohle, R. Holzwarth, V. S. Yakovlev, A. Scrinzi, T. W. Hänsch, and F. Krausz, 2003, Nature (London) **421**, 611.

Baltuška, A., M. Uiberacker, E. Goulielmakis, R. Kienberger, V. S. Yakovlev, T. Udem, T. W. Hänsch, and F. Krausz, 2003, IEEE J. Sel. Top. Quantum Electron. **9**, 972.

Bandrauk, A. D., E. Aubanel, and J. M. Gauthier, 1994, *Molecules in Laser Fields* (Dekker, New York).

Bandrauk, A. D., S. Barmaki, S. Chelkowski, and G. Lafmago Kamta, 2007, in *Progress in Ultrafast Laser Science Volume III*, edited by K. Yamanouchi (Kluwer, Amsterdam), p. 1.

Bandrauk, A. D., S. Chelkowski, and H. S. Nguyen, 2004, Int. J. Quantum Chem. **100**, 834.

Bandrauk, A. D., and N. H. Shon, 2002, Phys. Rev. A **66**, 031401.

Barnes, W. L., A. Dereux, and T. W. Ebbesen, 2003, Nature (London) **424**, 824.

Bartels, R., S. Backus, E. Zeek, L. Misoguti, G. Vdovin, I. P. Christov, M. M. Murnane, and H. C. Kapteyn, 2000, Nature (London) **406**, 164.

Bartels, R., H. C. Kapteyn, M. M. Murnane, I. P. Christov, and H. Rabitz, 2004, Phys. Rev. A **70**, 043404.

Barth, I., and J. Manz, 2006a, J. Am. Chem. Soc. **128**, 7043.

Barth, I., and J. Manz, 2006b, Angew. Chem., Int. Ed. **45**, 2962.

Barth, J. V., G. Costantini, and K. Kern, 2005, Nature (London) **437**, 671.

Barth, S., S. Marburger, S. Joshi, V. Ulrich, O. Kugeler, and U. Hergenhahn, 2006, Phys. Chem. Chem. Phys. **8**, 3218.
Bauer, M., 2005, J. Phys. D **38**, R253.
Bauer, M., C. Lei, K. Read, R. Tobey, J. Gland, M. M. Murnane, and H. Kapteyn, 2001, Phys. Rev. Lett. **87**, 025501.
Baum, P., and A. H. Zewail, 2006, Proc. Natl. Acad. Sci. U.S.A. **103**, 16105.
Baumert, T., R. Thalweiser, and G. Gerber, 1993, Chem. Phys. Lett. **209**, 29.
Beck, M., M. G. Raymer, I. A. Walmsley, and V. Wong, 1993, Opt. Lett. **18**, 2041.
Becker, A., and F. H. M. Faisal, 1994, Phys. Rev. A **50**, 3256.
Becker A., and F. H. M. Faisal, 2000, Phys. Rev. Lett. **84**, 3546.
Becker A., and F. H. M. Faisal, 2002, Phys. Rev. Lett. **89**, 193003.
Becker, A., and F. H. M. Faisal, 2005, J. Phys. B **38**, R1.
Becker, A., F. H. M. Faisal, Y. Liang, S. Augst, Y. Beaudoin, M. Chaker, and S. L. Chin, 2000, J. Phys. B **33**, L547.
Becker, A., L. Plaja, P. Moreno, M. Nurhuda, and F. H. M. Faisal, 2001, Phys. Rev. A **64**, 023408.
Becker, U., and D. Shirley, 1996, Eds., *Vuv and Soft X-Ray Photoionization* (Plenum, New York).
Becker, W., F. Grasbon, R. Kopold, D. B. Milosevic, G. G. Paulus, and H. Walther, 2002, Adv. At., Mol., Opt. Phys. **48**, 35.
Becker, W., A. Lohr, M. Kleber, and M. Lewenstein, 1997, Phys. Rev. A **56**, 645.
Bellini, M., 2000, Appl. Phys. B: Lasers Opt. **70**, 773.
Beutler, H., 1935, Z. Phys. **93**, 177.
Biegert, J., A. Heinrich, C. P. Hauri, W. Kornelis, P. Schlup, M. P. Anscombe, M. B. Gaarde, K. J. Schafer, and U. Keller, 2006, J. Mod. Opt. **53**, 87.
Binhammer, T., E. Rittweger, R. Ell, F. X. Kärntner, and U. Morgner, 2005, IEEE J. Quantum Electron. **41**, 1552.
Binhammer, T., E. Rittweger, U. Morgner, R. Ell, and F. X. Kärntner, 2006, Opt. Lett. **31**, 1552.
Blanchet, V., M. Z. Zgierski, T. Seideman, and A. Stolow, 1999, Nature (London) **401**, 52.
Bokor, J., 1989, Science **246**, 1130.
Borisov, A. D. Sánchez-Portal, R. Díez Muino, and P. M. Echenique, 2004, Chem. Phys. Lett. **387**, 95.
Bouhal, A., R. Evans, G. Grillon, A. Mysyrowicz, P. Breger, P. Agostini, R. C. Constantinescu, H. G. Muller, and D. von der Linde, 1997, J. Opt. Soc. Am. B **14**, 950.
Bouhal, A., P. Salières, P. Breger, P. Agostini, G. Hamoniaux, A. Mysyrowicz, A. Antonetti, R. Constantinescu, and H. G. Muller, 1998, Phys. Rev. A **58**, 389.
Boyd, R., 2003, *Nonlinear Optics*, 2nd ed. (Academic, New York).
Brabec, T., M. Yu. Ivanov, and P. B. Corkum, 1996, Phys. Rev. A **54**, R2551.
Brabec, T., and F. Krausz, 1997, Phys. Rev. Lett. **78**, 3282.
Brabec, T., and F. Krausz, 2000, Rev. Mod. Phys. **72**, 545.
Bradley, D. J., B. Liddy, and W. E. Sleat, 1971, Opt. Commun. **2**, 391.
Bradley, D. J., and G. H. C. New, 1974, Proc. IEEE **62**, 313.
Breidbach, J., and L. S. Cederbaum, 2003, J. Chem. Phys. **118**, 3983.
Breidbach, J., and L. S. Cederbaum, 2005, Phys. Rev. Lett. **94**, 033901.
Brena, B., D. Nordlund, M. Odelius, H. Ogasawara, A. Nilsson, and L. G. M. Pettersson, 2004, Phys. Rev. Lett. **93**, 148302.
Bressler, C., and M. Chergui, 2004, Chem. Rev. (Washington, D.C.) **104**, 1781.
Brixner, T., N. H. Damrauer, and G. Gerber, 2001, Adv. At., Mol., Opt. Phys. **46**, 1.
Brixner, T., N. H. Damrauer, G. Krampert, P. Niklaus, and G. Gerber, 2003, J. Mod. Opt. **50**, 539.
Brixner, T., N. H. Damrauer, P. Niklaus, and G. Gerber, 2001, Nature (London) **414**, 57.
Brixner, T., and G. Gerber, 2001, Opt. Lett. **26**, 557.
Brixner, T., G. Krampert, T. Pfeifer, R. Selle, G. Gerber, M. Wollenhaupt, O. Graefe, C. Horn, D. Liese, and T. Baumert, 2004, Phys. Rev. Lett. **92**, 208301.
Brühwiler, P. A., O. Karis, and N. Martensson, 2002, Rev. Mod. Phys. **74**, 703.
Brumer, P. W., and M. Shapiro, 2003, *Principles of the Quantum Control of Molecular Processes* (Wiley, New York).
Bryan, W. A., S. L. Stebbings, J. McKenna, E. M. L. English, M. Suresh, J. Wood, B. Srigengan, I. C. E. Turcu, J. M. Smith, E. J. Divall, C. J Hooker, A. J. Langley, J. L. Collier, I. D. Williams, and W. R. Newell, 2006, Nat. Phys. **2**, 379.
Budil, K. S., P. Salieres, M. D. Perry, and A. L'Huillier, 1993, Phys. Rev. A **48**, R3437.
Bulanov, S. V., N. M. Naumova, and F. Pegoraro, 1994, Phys. Plasmas **1**, 745.
Caillat, J., J. Zanghellini, M. Kitzler, O. Koch, W. Kreuzer, and A. Scrinzi, 2005, Phys. Rev. A **71**, 012712.
Canright, G. S., 1988, Phys. Rev. B **38**, 1647.
Cao, J., Y. Gao, R. J. D. Miller, and D. A. Mantell, 1997, Phys. Rev. B **56**, 1099.
Cao, W., P. Lu, P. Lan, W. Hong, and X. Wang, 2007, J. Phys. B **40**, 869.
Cao, W., P. Lu, P. Lan, X. Wang, and Y. Li, 2007, Phys. Rev. A **75**, 063423.
Cao, W., P. Lu, P. Lan, X. Wang, and G. Yang, 2006, Phys. Rev. A **74**, 063821.
Cao, W., P. Lu, P. Lan, X. Wang, and G. Yang, 2007, Opt. Express **15**, 530.
Carrera, J. J., and S. Chu, 2007, Phys. Rev. A **75**, 033807.
Carrera, J. J., X. M. Tong, and S. Chu, 2006, Phys. Rev. A **74**, 023404.
Cavalieri, A. L., E. Goulielmakis, B. Horvath, W. Helml, M. Schultze, M. Fiess, V. Pervak, L. Veisz, V. S. Yakovlev, M. Uiberacker, A. Apolonski, F. Krausz, and R. Kienberger, 2007, New J. Phys. **9**, 242.
Cavalieri, A. L., N. Müller, Th. Uphues, V. S. Yakovlev, A. Baltuška, B. Horvath, B. Schmidt, L. Blümel, R. Holzwarth, S. Hendel, M. Drescher, U. Kleineberg, P. M. Echenique, R. Kienberger, F. Krausz, and U. Heinzmann, 2007, Nature (London) **449**, 1029.
Cederbaum. L., and J. Zobeley, 1999, Chem. Phys. Lett. **307**, 205.
Cederbaum, L. S., J. Zobeley, and F. Tarantelli, 1997, Phys. Rev. Lett. **79**, 4778.
Cerdeira, F., T. Fjeldly, and M. Cardona, 1973, Phys. Rev. B **8**, 4734.
Cerullo, G., M. Nisoli, and S. De Silvestri, 1997, Appl. Phys. Lett. **71**, 3616.
Cerullo, G., M. Nisoli, S. Stagira, and S. De Silvestri, 1998, Opt. Lett. **23**, 1283.
Cerullo, G., M. Nisoli, S. Stagira, S. De Silvestri, G. Tempea, F. Krausz, and K. Ferencz, 1999, Opt. Lett. **24**, 1529.
Chakraborty, H. S., M. B. Gaarde, and A. Couairon, 2006, Opt. Lett. **31**, 3662.

Chan, K. W., C. K. Law, and J. H. Eberly, 2002, Phys. Rev. Lett. **88**, 100402.

Chan, K. W., C. K. Law, and J. H. Eberly, 2003, Phys. Rev. A **68**, 022110.

Chan, K. W., C. K. Law, M. V. Fedorov, and J. H. Eberly, 2004, J. Mod. Opt. **51**, 1779.

Chang, B. Y., H. Rabitz, and I. Sola, 2003, Phys. Rev. A **68**, 031402.

Chang, Z., 2004, Phys. Rev. A **70**, 043802.

Chatzidimitriou-Dreismann, C. A., M. Vos, C. Kleiner, and T. Abdul-Redah, 2003, Phys. Rev. Lett. **91**, 057403.

Chelkowski, S., A. D. Bandrauk, and P. B. Corkum, 2004, Phys. Rev. Lett. **93**, 083602.

Chelkowski, S., G. L. Yudin, and A. D. Bandrauk, 2006, J. Phys. B **39**, S409.

Chen, J. G., C. Li, F. P. Chi, and Y. J. Yang, 2007, Chin. Phys. Lett. **24**, 86.

Chen, X., X. Li, J. Liu, P. Wei, X. Ge, R. Li, and Z. Xu, 2007, Opt. Lett. **32**, 2402.

Chipperfield, L. E., L. N. Gaier, P. L. Knight, J. P. Marangos, and J. W. G. Tisch, 2005, J. Mod. Opt. **52**, 243.

Chirilă C. C., and M. Lein, 2006a, J. Mod. Opt. **53**, 113.

Chirilă C. C., and M. Lein, 2006b, Phys. Rev. A **73**, 023410.

Christov, I. P., H. C. Kapteyn, and M. M. Murnane, 2000, Opt. Express **7**, 362.

Christov, I. P., M. M. Murnane, and H. C. Kapteyn, 1997, Phys. Rev. Lett. **78**, 1251.

Chu, X., and S.-I. Chu, 2001, Phys. Rev. A **64**, 063404.

Cocke, C. L., and R. E. Olson, 1991, Phys. Rep. **205**, 153.

Constant, E., D. Garzella, P. Breger, E. Mevel, C. Dorrer, C. Le Blanc, F. Salin, and P. Agostini, 1999, Phys. Rev. Lett. **82**, 1668.

Constant, E., V. D. Taranukhin, A. Stolow, and P. B. Corkum, 1997, Phys. Rev. A **56**, 3870.

Corkum, P. B., 1993, Phys. Rev. Lett. **71**, 1994.

Corkum, P. B., N. Burnett, and F. Brunel, 1989, Phys. Rev. Lett. **62**, 1259.

Corkum, P. B., N. H. Burnett, and M. Yu. Ivanov, 1994, Opt. Lett. **19**, 1870.

Corkum, P. B., M. Yu. Ivanov, and J. Wright, 1997, Annu. Rev. Phys. Chem. **48**, 387

Corkum, P. B., and F. Krausz, 2007, Nat. Phys. **3**, 381.

Cormier, E., and P. Lambropoulos, 1998, Eur. Phys. J. D **2**, 15.

Cormier, E., I. Walmsley, E. M. Kosik, A. S. Wyatt, L. Corner, and L. F. DiMauro, 2005, Phys. Rev. Lett. **94**, 033905.

Corsi, C., A. Pirri, E. Sali, A. Tortora, and M. Bellini, 2006, Phys. Rev. Lett. **97**, 023901.

Costa, L., M. Betz, M. Spasenovic, A. D. Bristow, and H. M. Van Driel, 2007, Nat. Phys. **3**, 632.

Couairon, A., J. Biegert, C. P. Hauri, W. Kornelis, F. W. Helbing, U. Keller, and A. Mysyrowicz, 2006, J. Mod. Opt. **53**, 75.

Courant, E. D., C. Pellegrini, and W. Zakowicz, 1985, Phys. Rev. A **32**, 2813.

Cundiff, S. T., 2002, J. Phys. D **35**, R43.

Cundiff, S. T., and J. Ye, 2003, Rev. Mod. Phys. **75**, 325.

Cyr, D. R., and C. C. Hayden, 1996, J. Chem. Phys. **104**, 771.

Daniel, C., J. Full, L. González, C. Lupulescu, J. Manz, A. Merli, Š. Vajda, and L. Wöste, 2003, Science **299**, 536.

de Boeij, W. P., M. Pshenichnikov, and D. Wiersma, 1998, Annu. Rev. Phys. Chem. **49**, 99.

de Bohan, A., Ph. Antoine, D. B. Milosevic, and B. Piraux, 1998, Phys. Rev. Lett. **81**, 1837.

de Bohan, A., B. Piraux, L. Ponce, R. Taieb, V. Veniard, and A. Maquet, 2002, Phys. Rev. Lett. **89**, 113002.

De Jesus, V. L. B., B. Feuerstein, K. Zrost, D. Fischer, A. Rudenko, F. Afaneh, C. D. Schröter, R. Moshammer, and J. Ullrich, 2004, J. Phys. B **37**, L161.

Delone, N. B., and V. P. Krainov, 1985, *Atoms in Strong Light Fields* (Springer, Berlin).

DeMaria, A. J., W. H. Glenn, M. J. Brienza, and M. E. Mack, 1969, Proc. IEEE **57**, 2.

Diels, J. C., and W. Rudolph, 1996, *Ultrashort Laser Pulse Phenomena* (Academic, New York).

Dietrich, P., N. H. Burnett, M. Yu. Ivanov, and P. B. Corkum, 1994, Phys. Rev. A **50**, R3585.

Dietrich, P., F. Krausz, and P. Corkum, 2000, Opt. Lett. **25**, 16.

Dimitrovski, D., E. A. Solov'ev, and J. S. Briggs, 2004, Phys. Rev. Lett. **93**, 083003.

Dinu, L. C., H. G. Muller, S. Kazamias, G. Mullot, F. Augé, Ph. Balcou, P. M. Paul, M. Kovacev, P. Breger, and P. Agostini, 2003, Phys. Rev. Lett. **91**, 063901.

Dombi, P., A. Apolonski, Ch. Lemell, G. G. Paulus, M. Kakehata, R. Holzwarth, T. Udem, K. Torizuka, J. Burgdörfer, T. W. Hänsch, and F. Krausz, 2004, New J. Phys. **6**, 39.

Dörner, R., V. Mergel, O. Jagutzki, L. Spielberger, J. Ullrich, R. Moshammer, and H. Schmidt-Böcking, 2000, Phys. Rep. **330**, 95.

Drescher, M., M. Hentschel, R. Kienberger, G. Tempea, Ch. Spielmann, G. A. Reider, P. B. Corkum, and F. Krausz, 2001, Science **291**, 1923.

Drescher, M., M. Hentschel, R. Kienberger, M. Uiberacker, V. Yakovlev, A. Scrinzi, Th. Westerwalbesloh, U. Kleineberg, U. Heinzmann, and F. Krausz, 2002, Nature (London) **419**, 803.

Drescher, M., and F. Krausz, 2005, J. Phys. B **38**, S727.

Dromey, B., S. Kar, C. Bellei, D. C. Carroll, R. J. Clarke, J. S. Green, S. Kneip, K. Markey, S. R. Nagel, P. T. Simpson, L. Willengale, P. McKenna, D. Neely, Z. Najmudin, K. Krushelnick, P. A. Norreys, and M. Zepf, 2007, Phys. Rev. Lett. **99**, 085001.

Dromey, B., M. Zepf, A. Gopal, K. Lancaster, M. S. Wei, K. Krushelnick, M. Tatarakis, N. Vakakis, S. Moustaizis, R. Kodama, M. Tampo, C. Stoeckl, R. Clarke, H. Habara, D. Neely, S. Karsch, and P. Norreys, 2006, Nat. Phys. **2**, 456.

Dubietis, A., R. Butkus, and A. P. Piskarskas, 2006, IEEE J. Sel. Top. Quantum Electron. **12**, 163.

Dudovich, N., J. Levesque, O. Smirnova, D. Zeidler, D. Comtois, M. Yu. Ivanov, D. M. Villeneuve, and P. B. Corkum, 2006, Phys. Rev. Lett. **97**, 253903.

Dudovich, N., O. Smirnova, J. Levesque, M. Yu. Ivanov, Y. Mairesse, D. M. Villeneuve, and P. B. Corkum, 2006, Nat. Phys. **2**, 781.

Durfee, C. G., A. R. Rundquist, S. Backus, C. Herne, M. M. Murnane, and H. C. Kapteyn, 1999, Phys. Rev. Lett. **83**, 2187.

Dwyer, J. R., R. E. Jordan, C. T. Hebeisen, M. Harb, R. Ernstorfer, T. Dartigalongue, and R. J. D. Miller, 2007a, J. Mod. Opt. **54**, 905.

Dwyer, J. R., R. E. Jordan, C. T. Hebeisen, M. Harb, R. Ernstorfer, T. Dartigalongue, and R. J. D. Miller, 2007b, J. Mod. Opt. **54**, 923.

Echenique, P. M., R. Berndt, E. V. Chulkov, T. Fauster, A. Goldmann, and U. Höfer, 2004, Surf. Sci. Rep. **52**, 219.

Eden, J. G., 2004, Prog. Quantum Electron. **28**, 197.

Eichmann, U., M. Dörr, H. Maeda, W. Becher, and W. Sandner, 2000, Phys. Rev. Lett. **84**, 3550.

Eigen, M., 1954, Discuss. Faraday Soc. **17**, 194.

Elsayed-Ali, H. E., and G. Mourou, 1988, Appl. Phys. Lett. **52**,

103.
Emmanouilidou, A., and J. M. Rost, 2007, Phys. Rev. A **75**, 022712.
Eremina, E., X. Liu, H. Rottke, W. Sandner, A. Dreischuh, F. Lindner, F. Grasbon, G. G. Paulus, H. Walther, R. Moshammer, B. Feuerstein, and J. Ullrich, 2003, J. Phys. B **36**, 3269.
Eremina, E., X. Liu, H. Rottke, W. Sandner, M. G. Schätzel, A. Dreischuh, G. G. Paulus, H. Walther, R. Moshammer, and J. Ullrich, 2004. Phys. Rev. Lett. **92**, 173001.
Faisal, F. H. M., 1973, J. Phys. B **6**, L89.
Faisal, F. H. M., 1987, *Theory of Multiphoton Processes* (Plenum, New York).
Fann, W. S., R. Storz, H. W. K. Tom, and J. Bokor, 1992, Phys. Rev. Lett. **68**, 2834.
Fano, U., 1935, Nuovo Cimento **12**, 154.
Fano, U., 1961, Phys. Rev. **124**, 1866.
Farkas, Gy., and Cs. Tóth, 1992, Phys. Lett. A **168**, 447.
Faure, J., Y. Glinec, A. Pukhov, S. Kiselev, S. Gordienko, E. Lefebvre, J.-P. Rousseau, F. Burgy, and V. Malka, 2004, Nature (London) **431**, 541.
Fedorov M. V., M. A. Efremov, A. E. Kazakov, K. W. Chan, C. K. Law, and J. H. Eberly, 2004, Phys. Rev. A **69**, 052117.
Fedorov M. V., M. A. Efremov, A. E. Kazakov, K. W. Chan, C. K. Law, and J. H. Eberly, 2005, Phys. Rev. A **72**, 032110.
Feldhaus, J., J. Arthur, and J. B. Hastings, 2005, J. Phys. B **38**, S799.
Fennel, T., L. Ramunno, and T. Brabec, 2007, Phys. Rev. Lett. **99**, 233401.
Feringa, B. L., 2001, Ed., *Molecular Switches* (Wiley-VCH, Weinheim).
Ferray, M., A. L'Huillier, X. F. Li, L. A. Lompré, G. Mainfray, and C. Manus, 1988, J. Phys. B **21**, L31.
Figueira de Morisson Faria, C., X. Liu, W. Becker, and H. Schomerus, 2004, Phys. Rev. A **69**, 021402.
Figueira de Morisson Faria, C., and P. Salieres, 2007, Laser Phys. **17**, 390.
Figueira de Morisson Faria, C., P. Salieres, P. Villain, and M. Lewenstein, 2006, Phys. Rev. A **74**, 053416.
Figueira de Morisson Faria, C., H. Schomerus, and W. Becker, 2002, Phys. Rev. A **66**, 043413.
Figueira de Morisson Faria, C., H. Schomerus, X. Liu, and W. Becker, 2004, Phys. Rev. A **69**, 043405.
Fill, E. E., F. Krausz, and M. G. Raizen, 2008, New J. Phys. **10**, 093015.
Fill, E., L. Veisz, A. Apolonski, and F. Krausz, 2006, New J. Phys. **8**, 272.
Fittinghoff, D., P. R. Bolton, B. Chang, and K. C. Kulander, 1992, Phys. Rev. Lett. **69**, 2642.
Fleischer, A., and N. Moiseyev, 2006, Phys. Rev. A **74**, 053806.
Föhlisch, A., 2006, Appl. Phys. A: Mater. Sci. Process. **85**, 351.
Föhlisch, A., P. Feulner, F. Hennies, A. Fink, D. Menzel, D. Sanchez-Portal, P. M. Echenique, and W. Wurth, 2005, Nature (London) **436**, 373.
Fortier, T. M., P. A. Roos, D. J. Jones, S. T. Cundiff, R. D. R. Bhat, and J. E. Sipe, 2004, Phys. Rev. Lett. **92**, 147403.
Franco, I., and P. Brumer, 2006, Phys. Rev. Lett. **97**, 040402.
Frolov, M. V., A. V. Flegel, N. L. Manakov, and A. F. Starace, 2007a, Phys. Rev. A **75**, 063407.
Frolov, M. V., A. V. Flegel, N. L. Manakov, and A. F. Starace, 2007b, Phys. Rev. A **75**, 063408.
Fuji, T., N. Ishii, C. Y. Teisset, X. Gu, Th. Metzger, A. Baltuška, N. Forget, D. Kaplan, A. Galvanauskas, and F. Krausz, 2006, Opt. Lett. **31**, 1103.
Fuji, T., J. Rauschenberger, Ch. Gohle, A. Apolonski, T. Udem, V. S. Yakovlev, G. Tempea, T. W. Hänsch, and F. Krausz, 2005, New J. Phys. **7**, 116.
Gaarde, M. B., M. Murakami, and R. Kienberger, 2006, Phys. Rev. A **74**, 053401.
Gaarde, M. B., and K. J. Schafer, 2002, Phys. Rev. Lett. **89**, 213901.
Gaarde, M. B., and K. J. Schafer, 2006, Opt. Lett. **31**, 3188.
Gaarde, M. B., K. J. Schafer, K. C. Kulander, B. Sheehy, D. Kim, and L. F. DiMauro, 2000, Phys. Rev. Lett. **84**, 2822.
Gagnon, E., P. Ranitovic, X-M. Tong, C. L. Cocke, M. Murnane, H. Kapteyn, and A. Sandhu, 2007, Science **317**, 1374.
Gagnon, E., I. Thomann, A. Paul, A. Lytle, S. Backus, M. Murnane, H. Kapteyn, and A. Sandhu, 2006, Opt. Lett. **31**, 1866.
Gale, G. M., M. Cavallari, T. J. Driscoll, and F. Hache, 1995, Opt. Lett. **20**, 1562.
Ge, Y., 2006, Phys. Rev. A **74**, 015803.
Geddes, C. G. R., Cs. Toth, J. van Tilborg, E. Esarey, C. B. Schroeder, D. Bruhwiler, C. Nieter, J. Cary, and W. P. Leemans, 2004, Nature (London) **431**, 538.
Geissler, M., and J. Meyer-ter-Vehn, 2006, private commnication.
Geissler, M., S. Rykovanov, J. Schreiber, J. Meyer-ter-Vehn, and G. D. Tsakiris, 2007, New J. Phys. **9**, 218.
Georgescu, I., U. Saalmann, and J. M. Rost, 2007, Phys. Rev. Lett. **99**, 183002.
Gessner, O., Y. Hikosaka, B. Zimmermann, A. Hempelmann, R. R. Lucchese, J. H. D. Eland, P.-M. Guyon, and U. Becker, 2002, Phys. Rev. Lett. **88**, 193002.
Gessner, O., A. M. D. Lee, J. P. Shaffer, H. Reisler, S. V. Levchenko, A. I. Krylov, J. G. Underwood, H. Shi, A. L. L. East, D. M. Wardlaw, E. t. H. Chrysostom, C. C. Hayden, and A. Stolow, 2006, Science **311**, 219.
Gibson, E. A., A. Paul, N. Wagner, R. Tobey, S. Backus, I. P. Christov, A. Aquila, E. M. Gullikson, D. T. Attwood, M. M. Murnane, and H. C. Kapteyn, 2003, Science **302**, 95.
Gibson, E. A., A. Paul, N. Wagner, R. Tobey, D. Gaudiosi, S. Backus, I. P. Christov, M. M. Murnane, and H. C. Kapteyn, 2004, Phys. Rev. Lett. **92**, 033001.
Gildenburg, V. B., and N. V. Vvedenskii, 2007, Phys. Rev. Lett. **98**, 245002.
Glover, T. E., R. W. Schoenlein, A. H. Chin, and C. V. Shank, 1996, Phys. Rev. Lett. **76**, 2468.
Göres, J., D. Goldhaber-Gordon, S. Heemeyer, M. A. Kastner, Hadas Shtrikman, D. Mahalu, and U. Meirav, 2000, Phys. Rev. B **62**, 2188.
Goldsmith, R. H., L. E. Sinks, R. F. Kelley, L. J. Betzen, W. Liu, E. A. Weiss, M. A. Ratner, and M. R. Wasielewski, 2005, Proc. Natl. Acad. Sci. U.S.A. **102**, 3540.
Gordienko, S., A. Pukhov, O. Shorokhov, and T. Baeva, 2004, Phys. Rev. Lett. **93**, 115002.
Gordienko, S., A. Pukhov, O. Shorokhov, and T. Baeva, 2005, Phys. Rev. Lett. **94**, 103903.
Gordon, A., F. Kaertner, N. Rohringer, and R. Santra, 2006, Phys. Rev. Lett. **96**, 223902.
Goreslavskii S. P., and S. V. Popruzhenko, 1999, J. Phys. B **32**, L531.
Goreslavskii S. P., S. V. Popruzhenko, R. Kopold, and W. Becker, 2001, Phys. Rev. A **64**, 053402.
Goulielmakis, E., S. Koehler, B. Reiter, M. Schultze, A. J. Verhoef, E. E. Serebryannikov, A. M. Zheltikov, and F. Krausz, 2008, Opt. Lett. **33**, 1407.

Goulielmakis, E., E. M. Schultze, M. Hofstetter, V. S. Yakovlev, J. Gagnon, M. Uiberacker, A. L. Aquila, E. M. Gullikson, D. T Attwood, R. Kienberger, F. Krausz, and U. Kleineberg, 2008, Science **320**, 1614.
Goulielmakis, E., M. Uiberacker, R. Kienberger, A. Baltuška, V. Yakovlev, A. Scrinzi, Th. Westerwalbesloh, U. Kleineberg, U. Heinzmann, M. Drescher, and F. Krausz, 2004, Science **305**, 1267.
Goulielmakis, E., V. Yakovlev, A. L. Cavalieri, M. Uiberacker, V. Pervak, A. Apolonskiy, R. Kienberger, U. Kleineberg, and F. Krausz, 2007, Science **317**, 769.
Gouy, C. R., 1890, Acad. Sci., Paris, C. R. **110**, 1251.
Graf, U., M. Fiess, M. Schultze, R. Kienberger, F. Krausz, and E. Goulielmakis, 2008, Opt. Express **16**, 18956.
Gräfe, S., and M. Yu. Ivanov, 2007, Phys. Rev. Lett. **99**, 163603.
Gray, H. B., and J. R. Winkler, 2005, Proc. Natl. Acad. Sci. U.S.A. **102**, 3534.
Greenblatt, B. J., M. T. Zanni, and D. M. Neumark, 1997, Science **276**, 1675.
Greenhow, R. C., and A. J. Schmidt, 1974, in *Advances in Quantum Electronics*, edited by D. W. Goodwin (Academic, London), Vol. 2, p. 158.
Guandalini, A., P. Eckle, M. Anscombe, P. Schlup, J. Biegert, and U. Keller, 2006, J. Phys. B **39**, S257.
Haan, S. L., L. Breen, A. Karim, and J. H. Eberly, 2006, Phys. Rev. Lett. **97**, 103008.
Haan, S. L., L. Breen, A. Karim, and J. H. Eberly, 2007, Opt. Express **15**, 767.
Haan, S. L., and Z. S. Smith, 2007, Phys. Rev. A **76**, 053412.
Haan, S. L., P. S. Wheeler, R. Panfili, and J. H. Eberly, 2002, Phys. Rev. A **66**, 061402.
Haight, R., 1995, Surf. Sci. Rep. **21**, 275.
Haight, R., and J. A. Silbermann, 1989, Phys. Rev. Lett. **62**, 815.
Haljan, P., M. Yu. Ivanov, and P. B. Corkum, 1997, Laser Phys. **7**, 839.
Hänsch, T. W., 1990, Opt. Commun. **80**, 71.
Hänsch, T. W., 1997, private communication.
Harris, S. E., J. J. Macklin, and T. W. Hänsch, 1993, Opt. Commun. **100**, 487.
Harris, S. E., and A. V. Sokolov, 1997, Phys. Rev. A **55**, R4019.
Harris, S. E., and A. V. Sokolov, 1998, Phys. Rev. Lett. **81**, 2894.
Harumiya, K., H. Kono, Y. Fujimura, I. Kawata, and A. D. Bandrauk, 2002, Phys. Rev. A **66**, 043403.
Hasegawa, H., E. J. Takahashi, Y. Nabekawa, K. L. Ishikawa, and K. Midorikawa, 2005, Phys. Rev. A **71**, 023407.
Hauri, C. P., W. Kornelis, F. W. Helbing, A. Heinrich, A. Couairon, A. Mysyrowicz, J. Biegert, and U. Keller, 2004, Appl. Phys. B: Lasers Opt. **79**, 673.
Hauri, C. P., *et al.*, 2007, Opt. Lett. **32**, 868.
Haworth, C. A., L. E. Chipperfield, J. S. Robinson, P. L. Knight, J. P. Marangos, and J. W. G. Tisch, 2007, Nat. Phys. **3**, 52.
He, F., C. Ruiz, and A. Becker, 2007, Phys. Rev. Lett. **99**, 083002.
Heinrich, A., W. Kornelis, M. P. Anscombe, C. P. Hauri, P. Schlup, J. Biegert, and U. Keller, 2006, J. Phys. B **39**, S275.
Hennig, H., J. Breidbach, and L. S. Cederbaum, 2005, J. Chem. Phys. **122**, 134104.
Hentschel, M., R. Kienberger, Ch. Spielmann, G. A. Reider, N. Milosevic, T. Brabec, P. Corkum, U. Heinzmann, M. Drescher, and F. Krausz, 2001, Nature (London) **414**, 509.
Hergott, J.-F., M. Kovacev, H. Merdji, C. Hubert, Y. Mairesse, E. Jean, P. Breger, P. Agostini, B. Carré, and P. Salières, 2002, Phys. Rev. A **66**, 021801(R).
Ho, P. J., and J. H. Eberly, 2007, Opt. Express **15**, 1845.
Ho, P. J., R. Panfili, S. L. Heen, and J. H. Eberly, 2005, Phys. Rev. Lett. **94**, 093002.
Höfer, U., I. L. Shumay, Ch. Reuß, U. Thomann, W. Wallauer, and T. Fauster, 1997, Science **277**, 1480.
Holzwarth, R., T. Udem, T. W. Hänsch, J. C. Knight, W. J. Wadsworth, P. St. J. Russell, 2000, Phys. Rev. Lett. **85**, 2264.
Hommelhoff, P., Y. Sortais, A. Aghajani-Talesh, and M. A. Kasevich, 2006, Phys. Rev. Lett. **96**, 077401.
Hong, K. H., J. Lee, B. Hou, J. A. Nees, E. Power, and G. A. Mourou, 2006, Appl. Phys. Lett. **89**, 031113.
Hu, J., K.-L. Han, and G.-Z. He, 2005, Phys. Rev. Lett. **95**, 123001.
Hu, J., M.-S. Wang, K.-L. Han, and G.-Z. He, 2006, Phys. Rev. A **74**, 063417.
Hu, S. X., and L. A. Collins, 2005, Phys. Rev. Lett. **94**, 073004.
Hu, S. X., and L. A. Collins, 2006, Phys. Rev. Lett. **96**, 073004.
Hu, S. X., and L. A. Collins, 2007, J. Mod. Opt. **54**, 943.
Hu, S. X., D. B. Milošević, W. Becker, and W. Sandner, 2001, Phys. Rev. A **64**, 013410.
Huang, Y. C., D. Zheng, W. M. Tullock, and R. L. Byer, 1996, Appl. Phys. Lett. **68**, 753.
Huber, R., F. Tauser, A. Brodschelm, M. Bichler, G. Abstreiter, and A. Leitenstorfer, 2001, Nature (London) **414**, 286.
Huebener, R., 2005, *Electrons in Action: Roads to Modern Computers and Electronics* (Wiley-VCH Verlag, Weinheim).
Huo, Y., Z. Zeng, Y. Leng, R. Li, Z. Xu, C. Guo, Z. Sun, and Y. Rhee, 2005, Opt. Lett. **30**, 564.
Huo, Y., Z. Zeng, R. Li, and Z. Xu, 2005, Opt. Express **13**, 9897.
Iaconis, C., and I. Walmsley, 1998, Opt. Lett. **23**, 792.
Iaconis, C., and I. Walmsley, 1999, IEEE J. Quantum Electron. **35**, 501.
Ihee, H., V. A. Lobastov, U. M. Gomez, B. M. Goodson, R. Srinivasan, C.-Y. Ruan, and A. H. Zewail, 2001, Science **291**, 458.
Imran, T., Y. S. Lee, C. H. Nam, K. H. Hong, T. J. Yu, and J. H. Sung, 2007, Opt. Express **15**, 104.
Irvine, S. E., P. Dombi, Gy. Farkas, and A. Y. Elezzabi, 2006, Phys. Rev. Lett. **97**, 146801.
Ishikawa, K. L., 2006, Phys. Rev. A **74**, 023806.
Ishikawa, K. L., E. J. Takahashi, and K. Midorikawa, 2007, Phys. Rev. A **75**, 021801(R).
Itakura, R., 2007, Phys. Rev. A **76**, 033810.
Itatani, J., J. Levesque, D. Zeidler, H. Niikura, H. Pepin, J. C. Kieffer, P. B. Corkum, and D. M. Villeneuve, 2004, Nature (London) **432**, 867.
Itatani, J., F. Quéré, G. L. Yudin, M. Yu. Ivanov, F. Krausz, and P. Corkum, 2002, Phys. Rev. Lett. **88**, 173903.
Ivanov, M. Yu., T. Brabec, and N. Burnett, 1996, Phys. Rev. A **54**, 742.
Ivanov, M. Yu., P. B. Corkum, T. Zuo, and A. Bandrauk, 1995, Phys. Rev. Lett. **74**, 2933.
Ivanov, M. Yu., D. Matusek, and J. Wright, 1996, Phys. Rev. A **54**, 5159.
Ivanov, M. Yu., M. Spanner, and O. Smirnova, 2005, J. Mod. Opt. **52**, 165.
Ivanov, M. Yu., V. Yakovlev, and F. Krausz, 2008, in *Mid-Infrared Coherent Sources and Applications*, edited by M.

Ebrahim-Zade and I. Sorokina (Springer, Dordrecht, The Netherlands), p. 589.
Jahnke, T., A. Czasch, M. S. Schöffler, S. Schössler, A. Knapp, M. Käsz, J. Titze, C. Wimmer, K. Kreidi, R. E. Grisenti, A. Staudte, O. Jagutzki, U. Hergenhahn, H. Schmidt-Böcking, and R. Dörner, 2004, Phys. Rev. Lett. **93**, 163401.
Jaroń-Becker, A., A. Becker, and F. H. M. Faisal, 2004, Phys. Rev. A **69**, 023410.
Jaroń-Becker, A., A. Becker, and F. H. M. Faisal, 2006, Phys. Rev. Lett. **96**, 143006.
Joachain, C. J., M. Dörr, and N. Klystra, 2000, Adv. At., Mol., Opt. Phys. **42**, 225.
Johnsson, P., R. López-Martens, S. Kazamias, J. Mauritsson, C. Valentin, T. Remetter, K. Varjú, M. B. Gaarde, Y. Mairesse, H. Wabnitz, P. Salières, Ph. Balcou, K. J. Schafer, and A. L'Huillier, 2005, Phys. Rev. Lett. **95**, 013001.
Johnsson, P., K. Varjú, T. Remetter, E. Gustafsson, J. Mauritsson, R. López-Martens, S. Kazamias, C. Valentin, Ph. Balcou, M. B. Gaarde, K. J. Schafer, and A. L'Huillier, 2006, J. Mod. Opt. **53**, 233.
Jones, D. J., S. A. Diddams, J. K. Ranka, A. Stentz, R. S. Windeler, J. L. Hall, and S. T. Cundiff, 2000, Science **288**, 635.
Jordan, G., J. Caillat, C. Ede, and A. Scrinzi, 2006, J. Phys. B **39**, S341.
Jortner, J., and M. Ratner, 1997, Eds., *Molecular Electronics* (Blackwell, Oxford).
Judson, R. S., and H. Rabitz, 1992, Phys. Rev. Lett. **68**, 1500.
Jung, J. D., F. X. Kärnter, N. Matuschek, D. H. Sutter, F. Morier-Genoud, G. Zhang, U. Keller, V. Scheuer, M. Tilsch, and T. Tschudi, 1997, Opt. Lett. **22**, 1009.
Jurvansuu, M., A. Kivimäki, and S. Aksela, 2001, Phys. Rev. A **64**, 012502.
Kakehata, M., H. Takada, Y. Kobayashi, and K. Torizuka, 2004, Opt. Express **12**, 2070.
Kakehata, M., H. Takada, Y. Kobayashi, and K. Torizuka, 2006, J. Photochem. Photobiol., A **182**, 220.
Kakehata, M., H. Takada, Y. Kobayashi, K. Torizuka, Y. Fujihira, T. Homma, and H. Takahashi, 2001, Opt. Lett. **26**, 1436.
Kan, C., N. H. Burnett, C. E. Capjack, and R. Rankin, 1997, Phys. Rev. Lett. **79**, 2971.
Kane, D. J., 1999, IEEE J. Quantum Electron. **35**, 421.
Kane, D. J., and R. Trebino, 1993, Opt. Lett. **18**, 823.
Kanai, T., S. Minemoto, and H. Sakai, 2005, Nature (London) **435**, 470.
Kanai, T., E. J. Takahashi, Y. Nabekawa, and K. Midorikawa, 2007, Phys. Rev. Lett. **98**, 153904.
Kaplan, A. E., 1994, Phys. Rev. Lett. **73**, 1243.
Kaplan, A. E., and P. Shkolnikov, 1995, Phys. Rev. Lett. **75**, 2316.
Kaplan, D., and P. Tournois, 2004, *Ultrafast Optics IV*, edited by F. Krausz, G. Korn, P. B. Corkum, and I. A. Walmsley (Springer-Verlag, New York), p. 105.
Karsch, S., 2004, private communication.
Kato, H. S., M. Furukawa, M. Kawai., M. Taniguchi, T. Kawai, T. Hatsui, and N. Kosugi, 2004, Phys. Rev. Lett. **93**, 086403.
Kawata, I., H. Kono, and Y. Fujimura, 1999, J. Chem. Phys. **110**, 11152.
Kazansky, A. K., and N. M. Kabachnik, 2006, J. Phys. B **39**, 5173.
Kazansky, A. K., and N. M. Kabachnik, 2007a, J. Phys. B **40**, 2163.
Kazansky, A. K., and N. M. Kabachnik, 2007b, J. Phys. B **40**, 3413.
Kazansky, A. K., and N. M. Kabachnik, 2007c, J. Phys. B **40**, F299.
Keldysh, L. V., 1964, Zh. Eksp. Teor. Fiz. **47**, 1945 [Sov. Phys. JETP **20**, 1307 (1965)].
Keller, U., 2003, Nature (London) **424**, 831.
Kienberger, R., E. Goulielmakis, M. Uiberacker, A. Baltuška, V. Yakovlev, F. Bammer, A. Scrinzi, Th. Westerwalbesloh, U. Kleineberg, U. Heinzmann, M. Drescher, and F. Krausz, 2004, Nature (London) **427**, 817.
Kim, C. M., I. J. Kim, and C. H. Nam, 2005, Phys. Rev. A **72**, 033817.
Kim, H. T., V. Tosa, and C. H. Nam, 2006, J. Phys. B **39**, S265.
Kim, I. J., C. M. Kim, H. T. Kim, G. H. Lee, Y. S. Lee, J. Y. Park, D. J. Cho, and C. H. Nam, 2004, Phys. Rev. Lett. **94**, 243901.
Kim, J.-H., D. G. Lee, H. J. Shin, and C. H. Nam, 2001, Phys. Rev. A **63**, 063403.
Kim, T. K., C. M. Kim, M. G. Baik, G. Umesh, and C. H. Nam, 2004, Phys. Rev. A **69**, 051805(R).
Kimura, W., M. Babzien, I. Ben-Zvi, L. P. Campbell D. B. Cline, C. E. Dilley, J. C. Gallardo, S. C. Gottschalk, K. P. Kusche, R. H. Pantell, I. V. Pogorelsky, D. C. Quimby, J. Skaritka, L. C. Steinhauer, V. Yakimenko, and F. Zhou, 2004, Phys. Rev. Lett. **92**, 054801.
Kimura, W., A. van Steenbergen, M. Babzien, I. Ben-Zvi, L. P. Cambell, D. B. Cline, C. E. Dilley, J. C. Gallardo, S. C. Gottschalk, P. He, K. P. Kusche, Y. Liu, R. H. Pantell, I. V. Pogorelsky, C. C. Quimby, J. Skaritka, L. C. Steinhauer, and V. Yakimenko, 2001, Phys. Rev. Lett. **86**, 4041.
Kinoshita, K., M. Ito, and Y. Suzuki, 1987, Rev. Sci. Instrum. **58**, 932.
Kirchmann, P. S., P. A. Loukakos, U. Bovensiepen, M. Wolf, S. Vijayalakshmi, F. Hennies, A. Pietsch, M. Nagasono, A. Föhlisch, and W. Wurth, 2007, in *Ultrafast Phenomena XV, Proceedings of the 15th International Conference*, edited by P. Corkum *et al.*, Springer Series in Chemical Physics Vol. 88 (Springer, Berlin), p. 612.
Kitzler, M., and M. Lezius, 2005, Phys. Rev. Lett. **95**, 253001.
Kitzler, M., N. Milosevic, A. Scrinzi, F. Krausz, and T. Brabec, 2002, Phys. Rev. Lett. **88**, 173904.
Kitzler, M., K. O'Keeffe, and M. Lezius, 2006, J. Mod. Opt. **53**, 57.
Kitzler, M., X. Xie, A. Scrinzi, and A. Baltuška, 2007, Phys. Rev. A **76**, 011801(R).
Klamroth, T., 2006, J. Chem. Phys. **124**, 144310.
Kling, M. F., Ch. Siedschlag, A. J. Verhoef, J. I. Khan, M. Schultze, Th. Uphues, Y. Ni, M. Uiberacker, M. Drescher, F. Krausz, and M. J. J. Vrakking, 2006, Science **312**, 246.
Kobayashi, T., A. Shirakawa, and T. Fuji, 2001, IEEE J. Sel. Top. Quantum Electron. **7**, 525.
Kobayashi, Y., T. Sekikawa, Y. Nabekawa, and S. Watanabe, 1998, Opt. Lett. **23**, 64.
Kobayashi, Y., D. Yoshitomi, K. Iwata, H. Takada, and K. Torizuka, 2007, Opt. Express **15**, 9748.
Kofman, A. G., and G. Kurizki, 2000, Nature (London) **405**, 546.
Kono, H., Y. Sato, M. Kanno, K. Nakai, and T. Kato, 2006, Bull. Chem. Soc. Jpn. **79**, 196.
Kono, H., Y. Sato, N. Tanaka, T. Kato, K. Nakai, S. Koseki, and Y. Fujimura, 2004, Chem. Phys. **304**, 203.
Kopold, R., D. B. Milošević, and W. Becker, 2000, Phys. Rev. Lett. **84**, 3831.
Kosik, E. M., L. Corner, A. S. Wyatt, E. Cormier, I. A. Walms-

ley, and L. F. DiMauro, 2005, J. Mod. Opt. **52**, 361.
Kosuge, A., T. Sekikawa, X. Zhou, T. Kanai, S. Adachi, and S. Watanabe, 2006, Phys. Rev. Lett. **97**, 263901.
Kovacev, M., Y. Mairesse, E. Priori, H. Merdji, O. Tcherbakoff, P. Monchicourt, P. Breger, E. Mevel, E. Constant, P. Salieres, B. Carré, and P. Agostini, 2003, Eur. Phys. J. D **26**, 79.
Krause, J., K. Schafer, and K. Kulander, 1992, Phys. Rev. Lett. **68**, 3535.
Krause, P., T. Klamroth, and P. Saalfrank, 2005, J. Chem. Phys. **123**, 074105.
Krehl. P., and S. Engemann, 1995, Shock Waves **5**, 1.
Kress, M., T. Löffler, M. D. Thomson, R. Dörner, H. Gimpel, K. Zrost, T. Ergler, R. Moshammer, U. Morgner, J. Ullrich, and H. G. Roskos, 2006, Nat. Phys. **2**, 327.
Kroll, N. M., P. L. Morton, M. N. Rosenbluth, 1981, IEEE J. Quantum Electron. **QE-17**, 1436.
Kruer, W. L., 2003, *The Physics of Laser-Plasma Interactions* (Westview, Boulder).
Kubo, A., Y. S. Jung, H. K. Kim, and H. Petek, 2007, J. Phys. B **40**, S259.
Kubo, A., K. Onda, H. Petek, Z. Sun, Y. S. Jung, and H. K. Kim, 2005, Nano Lett. **5**, 1123.
Kuchiev, M. Yu, 1987, Pis'ma Zh. Eksp. Teor. Fiz. **45**, 404 [JETP Lett. **45**, 319 (1987)].
Kugeler, O., G. Prümper, R. Hentges, J. Viefhaus, D. Rolles, U. Becker, S. Marburger, and U. Hergenhahn, 2004, Phys. Rev. Lett. **93**, 033002.
Kulander, K. C., K. J. Schafer, and J. L. Krause, 1992, *Atoms in Intense Laser Fields*, edited by M. Gavrila (Academic, New York), p. 247.
Kulander, K. C., K. J. Schafer, and J. L. Krause, 1993, in *Super-Intense Laser-Atom Physics*, edited by B. Piraux, A. L'Huillier, and K. Rzazewski, Vol. 316 of NATO Advanced Study Institute Series B: Physics (Plenum, New York), p. 95.
Kuleff, A. I., J. Breidbach, and L. S. Cederbaum, 2005, J. Chem. Phys. **123**, 044111.
Kuleff, A. I., and L. S. Cederbaum, 2007, Phys. Rev. Lett. **98**, 083201.
Kwong, N.-H., and M. Bonitz, 2000, Phys. Rev. Lett. **84**, 1768.
Lambropoulos, P., 1985, Phys. Rev. Lett. **55**, 2141.
Lambropoulos, P., and P. Zoller, 1981, Phys. Rev. A **24**, 379.
Lan, P., P. Lu, W. Cao, Y. Li, and X. Wang, 2007, Phys. Rev. A **76**, 021801(R).
Lan, P., P. Lu, W. Cao, X. Wang, and W. Hong, 2007, Opt. Lett. **32**, 1186.
Lan, P., P. Lu, W. Cao, X. Wang, and G. Yang, 2006, Phys. Rev. A **74**, 063411.
Lappas, D. G., and A. L'Huillier, 1998, Phys. Rev. A **58**, 4140.
Le, A.-T., T. Morishita, and C. D. Lin, 2008, Phys. Rev. A **78**, 023814.
Le, Anh-Thu, X.-M. Tong, and C. D. Lin, 2006, Phys. Rev. A **73**, 041402.
Lee, D. G., H. J. Shin, Y. H. Cha, K. H. Hong, J.-H. Kim, and C. H. Nam, 2001, Phys. Rev. A **63**, 021801(R).
Lee, Y. S., J. H. Sung, C. H. Nam, T. J. Yu, and K. H. Hong, 2005, Opt. Express **13**, 2969.
Lehr, L., M. T. Zanni, C. Frischkorn, R. Weinkauf, and D. M. Neumark, 1999, Science **284**, 635.
Lein, M., 2005, Phys. Rev. Lett. **94**, 053004.
Lein, M., 2007, J. Phys. B **40**, R135.
Lein, M., P. P. Corso, J. P. Marangos, and P. L. Knight, 2003, Phys. Rev. A **67**, 023819.
Lein, M., V. Engel, and E. K. U. Gross, 2001, Opt. Express **8**, 411.
Lein, M., E. K. U. Gross, and V. Engel, 2001, Phys. Rev. A **64**, 023406.
Lein, M., E. K. U. Gross, and V. Engel, 2002, Laser Phys. **12**, 487.
Lein, M., N. Hay, R. Velotta, J. P. Marangos, and P. L. Knight, 2002, Phys. Rev. Lett. **88**, 183903.
Lein, M., T. Kreibich, E. K. U. Gross, and V. Engel, 2002, Phys. Rev. A **65**, 033403.
Lein, M., J. P. Marangos, and P. L. Knight, 2002, Phys. Rev. A **66**, 051404(R).
Levine, R. D., 2005, *Molecular Reaction Dynamics* (Cambridge University Press, Cambridge, England).
Levine, R. D., and F. Remacle, 2006, private communication.
Levis, R. J., G. Menkir, and H. Rabitz, 2001, Science **292**, 709.
Lewenstein, M., Ph. Balcou, M. Yu. Ivanov, A. L'Huillier, and P. B. Corkum, 1994, Phys. Rev. A **49**, 2117.
Lewenstein, M., P. Salieres, and A. L'Huillier, 1995, Phys. Rev. A **52**, 4747.
Lezius, M., V. Blanchet, D. M. Rayner, D. M. Villeneuve, A. Stolow, and M. Yu. Ivanov, 2001, Phys. Rev. Lett. **86**, 51.
L'Huillier, A., and P. Balcou, 1993, Phys. Rev. Lett. **70**, 774.
L'Huillier, A., L. A. Lompre, G. Mainfray, and C. Manus, 1983, Phys. Rev. A **27**, 2503.
Li, C., E. Moon, H. Mashiko, C. M. Nakamura, P. Ranitovic, C. M. Maharjan, C. L. Cocke, and Z. Chang, 2006, Opt. Express **14**, 11468.
Li, C., E. Moon, H. Wang, H. Mashiko, C. M. Nakamura, J. Tackett, and Z. Chang, 2007, Opt. Lett. **32**, 796.
Lichters, R., J. Meyer-ter-Vehn, and A. Pukhov, 1996, Phys. Plasmas **3**, 3425.
Lin, C. D., X. M. Tong, and T. Morishita, 2006, J. Phys. B **39**, S419.
Lindner, F., G. G. Paulus, H. Walther, A. Baltuška, E. Goulielmakis, M. Lezius, and F. Krausz, 2004, Phys. Rev. Lett. **92**, 113001.
Lindner, F., M. G. Schätzel, H. Walther, A. Baltuška, E. Goulielmakis, F. Krausz, D. B. Milosevic, D. Bauer, W. Becker, and G. G. Paulus, 2005, Phys. Rev. Lett. **95**, 040401.
Liu, X., C. Figueira de Morisson Faria, W. Becker, and P. B. Corkum, 2006, J. Phys. B **39**, L305.
Liu, X., H. Rottke, E. Eremina, W. Sandner, E. Goulielmakis, K. O. Keeffe, M. Lezius, F. Krausz, F. Lindner, M. G. Schätzel, G. G. Paulus, and H. Walther, 2004, Phys. Rev. Lett. **93**, 263001.
Liu, Y., X. J. Wang, D. B. Cline, M. Babzien, J. M. Fang, J. Gallardo, K. Kusche, I. Pogorelsky, J. Skaritka, and A. van Steenbergen, 1998, Phys. Rev. Lett. **80**, 4418.
Lobastov, V. A., R. Srinivasan, and A. H. Zewail, 2005, Proc. Natl. Acad. Sci. U.S.A. **102**, 7069.
Loh, Z.-H., C. H. Greene, and S. R. Leone, 2008, Chem. Phys. **350**, 7.
Loh, Z.-H., M. Khalil, R. E. Correa, R. Santra, C. Buth, and S. R. Leone, 2007, Phys. Rev. Lett. **98**, 143601.
Lohr, A., M. Kleber, R. Kopold, and W. Becker, 1997, Phys. Rev. A **55**, R4003.
López-Martens, R., J. Mauritsson, P. Johnsson, and A. L'Huillier, 2004, Phys. Rev. A **69**, 053811.
López-Martens, R., K. Varjú, P. Johnsson, J. Mauritsson, Y. Mairesse, P. Salieres, M. B. Gaarde, K. J. Schafer, A. Persson, S. Svanberg, C.-G. Wahlström, and A. L'Huillier, 2005, Phys. Rev. Lett. **94**, 033001.
Ma, L.-S., R. K. Shelton, H. C. Kapteyn, M. M. Murnane, and

J. Ye, 2001, Phys. Rev. A **64**, 021802.
Ma, Y.-Y., Z.-M. Sheng, Y.-T. Li, W.-W. Chang, X.-H. Yuan, M. Chen, H.-C. Wu, J. Zheng, and J. Zhang, 2006, Phys. Plasmas **13**, 110702.
Macklin, J., J. D. Kmetec, and C. L. Gordon III, 1993, Phys. Rev. Lett. **70**, 766.
Madhaven, V., W. Chen, T. Jamneala, M. F. Crommie, and N. S. Wingreen, 1998, Science **280**, 567.
Maeda, H., M. Dammasch, U. Eichmann, W. Sandner, A. Becker, and F. H. M. Faisal, 2000, Phys. Rev. A **62**, 035402.
Mahan, G. D., 1993, *Many-Particle Physics*, 2nd ed. (Plenum, New York).
Mairesse, Y., A. De Bohan, L. J. Frasinski, H. Merdji, L. C. Dinu, P. Monchicourt, P. Breger, M. Kovacev, T. Auguste, B. Carré, H. G. Muller, P. Agostini, and P. Salieres, 2004, Phys. Rev. Lett. **93**, 163901.
Mairesse, Y., A. De Bohan, L. J. Frasinski, H. Merdji, L. C. Dinu, P. Monchicourt, P. Brégger, M. Kovacev, R. Taieb, B. Carré, H. G. Muller, P. Agostini, and P. Salieres, 2003, Science **302**, 1540.
Mairesse, Y., and F. Quéré, 2005, Phys. Rev. A **71**, 011401(R).
Mangles, S. P. D., C. D. Murphy, Z. Najmudin, A. G. R. Thomas, J. L. Collier, A. E. Dangor, E. J. Divall, P. S. Foster, J. G. Gallacher, C. J. Hooker, D. A. Jaroszynski, A. J. Langley, W. B. Mori, P. A. Norreys, F. S. Tsung, R. Viskup, B. R. Walton, and K. Krushelnick, 2004, Nature (London) **431**, 535.
Manzoni, C., C. Vozzi, E. Benedetti, G. Sansone, S. Stagira, O. Svelto, S. De Silvestri, M. Nisoli, and G. Cerullo, 2006, Opt. Lett. **31**, 963.
Marburger, S., O. Kugeler, U. Hergenhahn, and T. Möller, 2003, Phys. Rev. Lett. **90**, 203401.
Markevitch, A. N., S. M. Smith, D. A. Romanov, H. B. Schlegel, M. Yu. Ivanov, and R. J. Levis, 2003, Phys. Rev. A **68**, 011402(R).
Martinez, T. J., M. Ben-Nun, and R. D. Levine, 1996, J. Phys. Chem. **100**, 7884.
Martinez, T. J., M. Ben-Nun, and R. D. Levine, 1997, J. Phys. Chem. **101**, 6389.
Mashiko, H., C. M. Nakamura, C. Li, E. Moon, H. Wang, J. Tackett, and Z. Chang, 2007, Appl. Phys. Lett. **90**, 161114.
Matsubara, E., K. Yamane, T. Sekikawa, and M. Yamashita, 2007, J. Opt. Soc. Am. B **24**, 985.
Matuschek, N., F. X. Kärtner, and U. Keller, 1998, IEEE J. Sel. Top. Quantum Electron. **4**, 197.
Mauritsson, J., P. Johnsson, E. Gustafsson, A. L'Huillier, K. J. Schafer, and M. B. Gaarde, 2006, Phys. Rev. Lett. **97**, 013001.
McPherson, A, G. Gibson, H. Jara, U. Johann, T. S. Luk, I. A. McIntyre, K. Boyer, C. K. Rhodes, 1987, J. Opt. Soc. Am. B **4**, 595.
Meckel, M., D. Comtois, D. Zeidler, A. Staudte, D. Pavicic, H. C. Bandulet, H. Pépin, J. C. Kieffer, R. Dörner, D. M. Villeneuve, and P. B. Corkum, 2008, Science **320**, 1478.
Mercouris, T., Y. Komninos, and C. A. Nicolaides, 2004, Phys. Rev. A **69**, 032502.
Mercouris, T., Y. Komninos, and C. A. Nicolaides, 2007a, Phys. Rev. A **75**, 013407.
Mercouris, T., Y. Komninos, and C. A. Nicolaides, 2007b, Phys. Rev. A **76**, 033417.
Merdji, H., M. Kovacev, W. Boutu, P. Salieres, F. Vernay, and B. Carré, 2006, Phys. Rev. A **74**, 043804.
Miaja-Avila, L., C. Lei, M. Aeschlimann, J. L. Gland, M. M. Murnane, H. C. Kapteyn, and G. Saathoff, 2006, Phys. Rev. Lett. **97**, 113604.
Mikhailova, Yu. M., V. T. Platonenko, and S. G. Rykovanov, 2005, JETP Lett. **81**, 571.
Miloševic D., and W. Becker, 2002, Phys. Rev. A **66**, 063417.
Miloševic D., G. G. Paulus, D. Bauer, and W. Becker, 2006, J. Phys. B **39**, R203.
Mineo, H., K. Nagaya, M. Hayashi, and S. H. Lin, 2007, J. Phys. B **40**, 2435.
Mirkin, C. A., and M. A. Ratner, 1992, Annu. Rev. Phys. Chem. **43**, 719.
Miyamoto, N., M. Kamei, D. Yoshitomi, T. Kanai, T. Sekikawa, T. Nakajima, and S. Watanabe, 2004, Phys. Rev. Lett. **93**, 083903.
Miyashita, O., M. Y. Okamura, and J. N. Onuchic, 2005, Proc. Natl. Acad. Sci. U.S.A. **102**, 3558.
Monastyrskiy, M., S. Andreev, D. Greenfield, G. Bryukhnevich, V. Tarasov, and M. Yu Schelev, 2005, in *High-Speed Photography and Photonics*, edited by B. J. Thompson (SPIE, Bellingham, WA), Vol. 5580 p. 324.
Morishita, T., A.-T. Le, Z. Chen, and C. D. Lin, 2008, Phys. Rev. Lett. **100**, 013903.
Morishita, T., S. Watanabe, and C. D. Lin, 2007, Phys. Rev. Lett. **98**, 083003.
Morishita, Y., X.-J. Liu, N. Saito, T. Lischke, M. Kato, G. Prümper, M. Oura, H. Yamaoka, Y. Tamenori, I. H. Suzuki, and K. Ueda, 2006, Phys. Rev. Lett. **96**, 243402.
Morlens, A.-S., P. Balcou, P. Zeitoun, C. Valentin, V. Laude, and S. Kazamias, 2005, Opt. Lett. **30**, 1554.
Moshammer, R., B. Feuerstein, W. Schmitt, A. Dorn, C. D. Schröter, J. Ullrich, H. Rottke, C. Trump, M. Wittmann, G. Korn, K. Hoffmann, and W. Sandner, 2000, Phys. Rev. Lett. **84**, 447.
Moshammer, R., W. Schmitt, J. Ullrich, H. Kollmus, A. Cassimi, R. Dörner, O. Jagutzki, R. Mann, R. E. Olson, H. T. Prinz, H. Schmidt-Böcking, and L. Spielberger, 1997, Phys. Rev. Lett. **79**, 3621.
Mourou, G., T. Tajima, and S. V. Bulanov, 2006, Rev. Mod. Phys. **78**, 309.
Mourou, G., and S. Williamson, 1982, Appl. Phys. Lett. **41**, 44.
Mücke, O. D., T. Tritschler, M. Wegener, U. Morgner, F. X. Kärtner, G. Khitrova, H. M. Gibbs, 2004, Opt. Lett. **29**, 2160.
Muller, H. G., and F. C. Kooiman, 1998, Phys. Rev. Lett. **81**, 1207.
Murray, R., and M. Ivanov, 2008, J. Mod. Opt. **55**, 2513.
Musumeci, P., S. Ya. Tochitsky, S. Boucher, C. E. Clayton, A. Doyuran, R. J. England, C. Joshi, C. Pellegrini, J. E. Ralph, J. B. Rosenzweig, C. Sung, S. Tolmachev, G. Travish, A. A. Varfolomeev, A. A. Varfolomeev, Jr., T. Yarovoi, and R. B. Yoderl, 2005, Phys. Rev. Lett. **94**, 154801.
Nabekawa, Y., H. Hasegawa, E. J. Takahashi, and K. Midorikawa, 2005, Phys. Rev. Lett. **94**, 043001.
Nabekawa, Y., T. Shimizu, T. Okino, K. Furusawa, H. Hasegawa, K. Yamanouchi, and K. Midorikawa, 2006, Phys. Rev. Lett. **97**, 153904.
Nakajima, T., and S. Watanabe, 2006, Opt. Lett. **31**, 1920.
Naumov, S., A. Fernandez, R. Graf, P. Dombi, F. Krausz, and A. Apolonskiy, 2005, New J. Phys. **7**, 216.
Naumova, N. M., J. A. Nees, I. V. Sokolov, B. Hou, and G. Mourou, 2004, Phys. Rev. Lett. **92**, 063902.
Naumova, N., I. Sokolov, J. Nees, A. Maksimchuk, V. Yanovsky, and G. Mourou, 2004, Phys. Rev. Lett. **93**, 195003.
Nest, M., R. Padmanaban, and P. Saalfrank, 2007, J. Chem. Phys. **126**, 214106.
Neumark, D. M., 2001, Annu. Rev. Phys. Chem. **52**, 255.

Neutze, R., R. Wouts, D. van der Spoel, E. Weckert, and J. Hajdu, 2000, Nature (London) **406**, 752.
Nicolaides, C. A., T. Mercouris, and Y. Komninos, 2002, J. Phys. B **35**, L271.
Niikura, H., P. B. Corkum, and D. M. Villeneuve, 2003, Phys. Rev. Lett. **90**, 203601.
Niikura, H., F. Légaré, R. Hasbani, A. D. Bandrauk, M. Y. Ivanov, D. M. Villeneuve, and P. B. Corkum, 2002, Nature (London) **417**, 917.
Niikura, H., F. Légaré, R. Hasbani, M. Y. Ivanov, D. M. Villeneuve, and P. B. Corkum, 2003, Nature (London) **421**, 826.
Niikura, H., D. M. Villeneuve, and P. B. Corkum, 2005, Phys. Rev. Lett. **94**, 083003.
Nikolopoulos, L. A. A., E. P. Benis, P. Tzallas, D. Charalambidis, K. Witte, and G. D. Tsakiris, 2005, Phys. Rev. Lett. **94**, 113905.
Nikolopoulos, L. A. A., and P. Lambropoulos, 2001, J. Phys. B **34**, 545.
Nishioka, H., W. Odajima, K. Ueda, and H. Takuma, 1995, Opt. Lett. **20**, 2505.
Nisoli, M., S. De Silvestri, and O. Svelto, 1996, Appl. Phys. Lett. **68**, 2793.
Nisoli, M., S. De Silvestri, O. Svelto, R. Szipöcs, K. Ferencz, C. Spielmann, S. Sartania, and F. Krausz, 1997, Opt. Lett. **22**, 522.
Nitzan, A., 2001, Annu. Rev. Phys. Chem. **52**, 681.
Nitzan, A., and M. A. Ratner, 2003, Science **300**, 1384.
Nguyen, N. A., and A. D. Bandrauk, 2006, Phys. Rev. A **73**, 032708.
Nomura, Y., R. Hörlein, P. Tzallas, B. Dromey, S. Rykovanov, Zs. Major, J. Osterhoff, S. Karsch, L. Veisz, M. Zepf, D. Charalambidis, F. Krausz, and G. D. Tsakiris, 2009, Nat. Phys. (to be published).
Nordlund, D., H. Ogasawara, H. Bluhm, O. Takahashi, M. Odelius, M. Nagasono, L. G. M. Pettersson, and A. Nilsson, 2007, Phys. Rev. Lett. **99**, 217406.
Norrish, R. G. W., and G. Porter, 1949, Nature (London) **164**, 658.
Ogawa, S., H. Nagano, H. Petek, and A. Heberle, 1997, Phys. Rev. Lett. **78**, 1339.
Öhrwall, G., M. Tchaplyguine, M. Lundwall, R. Feifel, H. Bergersen, T. Rander, A. Lindblad, J. Schulz, S. Peredkov, S. Barth, S. Marburger, U. Hergenhahn, S. Svensson, and O. Björneholm, 2004, Phys. Rev. Lett. **93**, 173401.
Oishi, Y., M. Kaku, A. Suda, F. Kannari, and K. Midorikawa, 2006, Opt. Express **14**, 7230.
Oishi, Y., A. Suda, F. Kannari, and K. Midorikawa, 2005, Rev. Sci. Instrum. **76**, 093114.
O'Keeffe, K., P. Jöckl, H. Drexel, V. Grill, F. Krausz, and M. Lezius, 2004, Appl. Phys. B: Lasers Opt. **78**, 583.
Okunishi, M., T. Morishita, G. Prümper, K. Shimada, C. D. Lin, S. Watanabe, and K. Ueda, 2008, Phys. Rev. Lett. **100**, 143001.
O'Reagan, B., and M. Grätzel, 1991, Nature (London) **353**, 737.
Palmer, R., 1972, J. Appl. Phys. **43**, 3014.
Panfili, R., S. L. Haan, and J. H. Eberly, 2002, Phys. Rev. Lett. **89**, 113001.
Patchkovskii, S., Z. X. Zhao, T. Brabec, and D. Villeneuve, 2006, Phys. Rev. Lett. **97**, 123003.
Paul, A., R. Bartels, I. P. Christov, H. C. Kapteyn, M. M. Murnane, and S. Backus, 2003, Nature (London) **421**, 51.
Paul, A., E. A. Gibson, X. Zhang, A. Lytle, T. Popmintchev, X. Zhou, M. M. Murnane, I. P. Christov, and H. C. Kapteyn, 2006, IEEE J. Quantum Electron. **42**, 14.
Paul, P. M., E. S. Toma, P. Berger, G. Mullot, F. Auge, Ph. Balcou, H. G. Muller, and P. Agostini, 2001, Science **292**, 1689.
Paulus, G. G., F. Grasbon, H. Walther, P. Villoresi, M. Nisoli, S. Stagira, E. Priori, and S. De Silvestri, 2001, Nature (London) **414**, 182.
Paulus, G. G., F. Lindner, H. Walther, A. Baltuška, E. Goulielmakis, M. Lezius, and F. Krausz, 2003, Phys. Rev. Lett. **91**, 253004.
Paulus, G. G., F. Lindner, H. Walther, A. Baltuška, and F. Krausz, 2005, J. Mod. Opt. **52**, 221.
Penent, F., J. Paladoux, J. P. Lablanquie, L. Andric, R. Feitel, and J. H. D. Eland, 2005, Phys. Rev. Lett. **95**, 083002.
Perelomov, A. M., and V. S. Popov, 1967, Zh. Eksp. Teor. Fiz. **52**, 514 [Sov. Phys. JETP **25**, 336 (1967)].
Perelomov, A. M., V. S. Popov, and M. V. Terent'ev, 1966a, Zh. Eksp. Teor. Fiz. **50**, 1393 [Sov. Phys. JETP **23**, 924 (1966)].
Perelomov, A. M., V. S. Popov, and M. V. Terent'ev, 1966b, Zh. Eksp. Teor. Fiz. **51**, 309 [Sov. Phys. JETP **24**, 207 (1967)].
Pérez-Hernández, J. A., and L. Plaja, 2007, Phys. Rev. A **76**, 023829.
Petek, H., and S. Ogawa, 1997, Prog. Surf. Sci. **56**, 239.
Pfeifer, T., L. Gallmann, M. J. Abel, D. M. Neumark, and S. R. Leone, 2006a, Opt. Lett. **31**, 975.
Pfeifer, T., L. Gallmann, M. J. Abel, D. M. Neumark, and S. R. Leone, 2006b, Phys. Rev. Lett. **97**, 163901.
Pfeifer, T., A. Jullien, M. J. Abel, P. M. Nagel, L. Gallmann, D. M. Neumark, and S. R. Leone, 2007, Opt. Express **15**, 17120.
Pfeifer, T., C. Spielmann, and G. Gerber, 2006, Rep. Prog. Phys. **69**, 443.
Pirozhkov, A. S., S. V. Bulanov, T. Zh. Esirkepov, M. Mori, A. Sagisaka, and H. Daido, 2006, Phys. Plasmas **13**, 013107.
Platonenko, V. T., and V. Strelkov, 1999, J. Opt. Soc. Am. B **16**, 435.
Plettner, T., R. L. Byer, E. Colby, B. Cowan, C. M. S. Sears, J. E. Spencer, and R. H. Siemann, 2005, Phys. Rev. Lett. **95**, 134801.
Plettner, T., P. P. Lu, and R. L. Byer, 2006, Phys. Rev. ST Accel. Beams **9**, 111301.
Poater, J., M. Duran, M. Solá, and B. Silvi, 2005, Chem. Rev. (Washington, D.C.) **105**, 3911.
Popov, A. M., O. V. Tikhonova, and E. A. Volkova, 2007, Laser Phys. **17**, 103.
Popov, V. S., 2004, Phys. Usp. **47**, 855.
Popruzhenko, S. V., Ph. A. Korneev, S. P. Goreslavski, and W. Becker, 2002, Phys. Rev. Lett. **89**, 023001.
Popruzhenko, S. V., N. I. Shvetsov-Shilovski, S. P. Goreslavski, W. Becker, and G. G. Paulus, 2007, Opt. Lett. **32**, 1372.
Pozar, D. M., 2004, *Microwave Engineering*, 3rd ed. (Wiley, New York).
Prauzner-Bechcicki, J. S., K. Sacha, B. Eckhardt, and J. Zakrzewski, 2005, Phys. Rev. A **71**, 033407.
Protopapas, M., D. G. Lappas, C. H. Keitel, and P. L. Knight, 1996, Phys. Rev. A **53**, R2933.
Pukhov, A., and S. Gordienko, 2006, Philos. Trans. R. Soc. London, Ser. A **364**, 623.
Pukhov, A., S. Gordienko, and T. Baeva, 2003, Phys. Rev. Lett. **91**, 173002.
Pukhov, A., and J. Meyer-ter-Vehn, 2002, Appl. Phys. B **74**, 355.
Quéré, F., S. Guizard, Ph. Martin, G. Petite, H. Merdji, B.

Carré, J.-F. Hergott, and L. Le Déroff, 2000, Phys. Rev. B **61**, 9883.
Quéré, F., J. Itatani, G. L. Yudin, and P. B. Corkum, 2003, Phys. Rev. Lett. **90**, 073902.
Quéré, F., Y. Mairesse, and J. Itatani, 2005, J. Mod. Opt. **52**, 339.
Raether, H., 1988, Ed., *Surface Plasmons* (Springer, Berlin).
Ray, D., B. Ulrich, I. Bocharova, C. Maharjan, P. Ranitovic, B. Gramkow, M. Magrakvelidze, S. De, I. V. Litvinyuk, A.-T. Le, T. Morishita, C. D. Lin, G. G. Paulus, and C. L. Cocke, 2008, Phys. Rev. Lett. **100**, 143002.
Reckenthaeler, P., M. Centurion, V. S. Yakovlev, M. Lezius, F. Krausz, and E. E. Fill, 2008, Phys. Rev. A **77**, 042902.
Reichert, J., R. Holzwarth, T. Udem, and T. W. Hänsch, 1999, Opt. Commun. **172**, 59.
Reid, K. L., 2003, Annu. Rev. Phys. Chem. **54**, 397.
Reiss, H. R., 1980a, Phys. Rev. A **22**, 1786.
Reiss, H. R., 1980b, J. Opt. Soc. Am. **7**, 574.
Reiss, H. R., 1992, Prog. Quantum Electron. **16**, 1.
Remacle, F., R. Kienberger, F. Krausz, and R. D. Levine, 2007, Chem. Phys. **338**, 342.
Remacle, F., and R. D. Levine, 1997, J. Chem. Phys. **107**, 3382.
Remacle, F., and R. D. Levine, 1999, J. Chem. Phys. **110**, 5089.
Remacle, F., and R. D. Levine, 2006a, Proc. Natl. Acad. Sci. U.S.A. **103**, 6793.
Remacle, F., and R. D. Levine, 2006b, Faraday Discuss. **131**, 45.
Remacle, F., and R. D. Levine, 2007, Z. Phys. Chem. (Munich) **221**, 647.
Remacle, F., R. D. Levine, and M. A. Ratner, 1998, Chem. Phys. Lett. **285**, 25.
Remacle, F., R. D. Levine, E. W. Schlag, and R. Weinkauf, 1999, J. Phys. Chem. A **103**, 10149.
Remetter, T., P. Johnsson, J. Mauritsson, K. Varjú, Y. Ni, F. Lépine, E. Gustafsson, M. Kling, J. Kahn, R. López-Martens, K. J. Schafer, M. J. J. Vrakking, and A. L'Huillier, 2006, Nat. Phys. **2**, 323.
Renault, A., D. Z. Kandula, S. Witte, A. L. Wolf, R. Th. Zinkstok, W. Hogervorst, and K. S. E. Eikema, 2007, Opt. Lett. **32**, 2363.
Rice, S., and M. Zhao, 2000, *Optical Control of Molecular Dynamics* (Wiley, New York).
Riedle, E., M. Beutter, S. Lochbrunner, J. Piel, S. Schenkl, S. Sporlein, and W. Zinth, 2000, Appl. Phys. B: Lasers Opt. **71**, 457.
Riffe, D. M., X. Y. Wang, M. C. Downer, D. L. Fisher, T. Tajima, J. L. Erskine, and R. M. More, 1993, J. Opt. Soc. Am. B **10**, 1424.
Rohringer, N., A. Gordon, and R. Santra, 2006, Phys. Rev. A **74**, 043420.
Rolles, D., M. Braune, S. Cvejanović, O. Geßner, R. Hentges, S. Korica, B. Langer, T. Lischke, G. Prümper, A. Reinköster, J. Viefhaus, B. Zimmermann, V. McKoy, and U. Becker, 2005, Nature (London) **437**, 711.
Rottke, H., C. Trump, M. Wittmann, G. Korn, W. Sandner, R. Moshammer, A. Dorn, C. D. Schröter, D. Fischer, J. P. Crespo Lopez-Urrutia, P. Neumayer, J. Diepenwisch, C. Höhr, B. Feuerstein, and J. Ullrich, 2002, Phys. Rev. Lett. **89**, 013001.
Roudnev, V., and B. D. Esry, 2007, Phys. Rev. A **76**, 023403.
Roudnev, V., B. D. Esry, and I. Ben-Itzhak, 2004, Phys. Rev. Lett. **93**, 163601.
Rousse, A., C. Rischel, and J.-C. Gauthier, 2001, Rev. Mod. Phys. **73**, 17.
Rudenko, A., K. Zrost, B. Feuerstein, V. L. B. de Jesus, C. D. Schröter, R. Moshammer, and J. Ullrich, 2004, Phys. Rev. Lett. **93**, 253001.
Ruiz, C., L. Plaja, L. Roso, and A. Becker, 2006, Phys. Rev. Lett. **96**, 053001.
Rundquist, A., C. G. Durfee, III, S. Backus, C. Herne, Z. Chang, M. M. Murnane, and H. C. Kapteyn, 1998, Science **280**, 1412.
Rzazewski, K., and J. H. Eberly, 1981, Phys. Rev. Lett. **47**, 408.
Sakai, K., S. Miyazaki, S. Kawata, S. Hasumi, and T. Kikuchi, 2006, Laser Part. Beams **24**, 321.
Salamin, Y. I., and C. H. Keitel, 2000, Appl. Phys. Lett. **77**, 1082.
Saldin, E. L., E. A. Schneidmiller, and M. V. Yurkov, 2004, Opt. Commun. **239**, 161.
Salieres, P., B. Carre, L. Le Deroff, F. Grasbon, G. G. Paulus, H. Walther, R. Kopold, W. Becker, D. B. Milosevic, A. Sanpera, and M. Lewenstein, 2001, Science **292**, 902.
Salieres, P., A. L'Huillier, P. Antoine, and M. Lewenstein, 1999, Adv. At., Mol., Opt. Phys. **41**, 83.
Sansone, G., E. Benedetti, F. Calegari, C. Vozzi, L. Avaldi, R. Flammini, L. Poletto, P. Villoresi, C. Altucci, R. Velotta, S. Stagira, S. De Silvestri, and M. Nisoli, 2006, Science **314**, 443.
Santra, R., and R. W. Dunford, and L. Young, 2006, Phys. Rev. A **74**, 043403.
Santra, R., and A. Gordon, 2006, Phys. Rev. Lett. **96**, 073906.
Sartania, S., Z. Cheng, M. Lenzer, G. Tempea, C. Spielmann, and F. Krausz, 1997, Opt. Lett. **22**, 1562.
Schafer, K. J., M. B. Gaarde, A. Heinrich, J. Biegert, and U. Keller, 2004, Phys. Rev. Lett. **92**, 023003.
Schafer, K. J., and K. C. Kulander, 1990, Phys. Rev. A **42**, 5794.
Schafer, K. J., and K. C. Kulander, 1997, Phys. Rev. Lett. **78**, 638.
Schafer, K. J., B. Yang, L. F. DiMauro, and K. C. Kulander, 1993, Phys. Rev. Lett. **70**, 1599.
Schelev, M. Ya., M. C. Richardson, and A. J. Alcock, 1971, Appl. Phys. Lett. **18**, 354.
Schins, J. M., P. Breger, P. Agostini, R. C. Constantinescu, H. G. Muller, A. Bouhal, G. Grillon, A. Antonetti, and A. Mysyrowicz, 1996, J. Opt. Soc. Am. B **13**, 197.
Schins, J. M., P. Breger, P. Agostini, R. C. Constantinescu, H. G. Muller, G. Grillon, A. Antonetti, and A. Mysyrowicz, 1994, Phys. Rev. Lett. **73**, 2180.
Schmuttenmaer, C. A., M. Aeschlimann, H. E. Elsayed-Ali, R. J. D. Miller, D. A. Mantel, J. Cao, and Y. Gao, 1994, Phys. Rev. B **50**, 8957.
Schnadt, J., P. A. Brühwiler, L. Patthey, J. N. O'Shea, S. Södergren, M. Odellus, R. Ahuja, O. Karis, M. Bässler, P. Persson, H. Siegbahn, S. Lunell, and N. Martensson, 2002, Nature (London) **418**, 620.
Schneider, T., P. L. Chocian, and J.-M. Rost, 2002, Phys. Rev. Lett. **89**, 073002.
Schnürer, M., Z. Cheng, M. Hentschel, G. Tempea, P. Kálmán, T. Brabec, and F. Krausz, 1999, Phys. Rev. Lett. **83**, 722.
Schoenlein, R. W., J. G. Fujimoto, G. L. Eesley, and T. W. Capehart, 1988, Phys. Rev. Lett. **61**, 2596.
Schöne, W.-D., and W. Ekardt, 2000, Phys. Rev. B **62**, 13464.
Schultz, K. D., C. I. Blaga, R. Chirla, P. Colosimo, J. Cryan, A. M. March, C. Roedig, E. Sistrunk, J. Tate, J. Wheeler, P. Agostini, and L. F. DiMauro, 2007, J. Mod. Opt. **54**, 1075.
Schultze, M., E. Goulielmakis, M. Uiberacker, M. Hofstetter, J. Kim, D. Kim, F. Krausz, and U. Kleineberg, 2007, New J. Phys. **9**, 243.

Schweigert, I. V., and S. Mukamel, 2007, Phys. Rev. A **76**, 012504.
Scrinzi, A., M. Geissler, and T. Brabec, 2001a, Phys. Rev. Lett. **86**, 412.
Scrinzi, A., M. Geissler, and T. Brabec, 2001b, Laser Phys. **11**, 169.
Scrinzi, A., M. Yu. Ivanov, R. Kienberger, and D. M. Villeneuve, 2005, J. Phys. B **39**, R1.
Sears, C. M. S., E. R. Colby, B. M. Cowan, R. H. Siemann, J. E. Spencer, R. L. Byer, and T. Plettner, 2005, Phys. Rev. Lett. **95**, 194801.
Seel, M., and W. Domcke, 1991, J. Chem. Phys. **95**, 7806.
Sekikawa, T., T. Kanai, and S. Watanabe, 2003, Phys. Rev. Lett. **91**, 103902.
Sekikawa, T., T. Katsura, S. Miura, and S. Watanabe, 2002, Phys. Rev. Lett. **88**, 193902.
Sekikawa, T., A. Kosuge, T. Kanai, and S. Watanabe, 2004, Nature (London) **432**, 605.
Seres, J., E. Seres, A. J. Verhoef, G. Tempea, C. Streli, P. Wobrauschek, V. Yakovlev, A. Scrinzi, C. Spielmann, and F. Krausz, 2005, Nature (London) **433**, 596.
Seres, J., V. S. Yakovlev, E. Seres, Ch. Streli, P. Wobrauschek, C. Spielmann, and F. Krausz, 2007, Nat. Phys. **3**, 878.
Shan, B., and Z. Chang, 2001, Phys. Rev. A **65**, 011804(R).
Shan, B., S. Ghimire, and Z. Chang, 2005, J. Mod. Opt. **52**, 277.
Shapiro, M., and Y. Zeiri, 1986, J. Chem. Phys. **85**, 6449.
Sheehy, B., J. D. D. Martin, L. F. DiMauro, P. Agostini, K. J. Schafer, M. B. Gaarde, and K. C. Kulander, 1999, Phys. Rev. Lett. **83**, 5270.
Shirakawa, A., and T. Kobayashi, 1998, Appl. Phys. Lett. **72**, 147.
Shirakawa, A., I. Sakane, and T. Kobayashi, 1998, Opt. Lett. **23**, 1292.
Shirakawa, A., I. Sakane, M. Takasaka, and T. Kobayashi, 1999, Appl. Phys. Lett. **74**, 2268.
Shorokov, D., and A. H. Zewail, 2008, Phys. Chem. Chem. Phys. **10**, 2879.
Shverdin, M. Y., D. R. Walker, D. D. Yavuz, G. Y. Yin, and S. E. Harris, 2005, Phys. Rev. Lett. **94**, 033904.
Shvetsov-Shilovski, N. I., S. P. Goreslavski, S. V. Popruzhenko, W. Becker, and G. G. Paulus, 2007, Laser Phys. Lett. **4**, 726.
Siegman, A., 1986, *Lasers* (University Science Books, Mill Valley).
Siffalovic, P., M. Drescher, and U. Heinzmann, 2002, Europhys. Lett. **60**, 924.
Siffalovic, P., M. Michelswirth, P. Bartz, B. Decker, C. Agena, C. Schäfer, S. Molter, R. Ros, M. Bach, M. Neumann, D. Andelmetti, J. Mattay, U. Heinzmann, and M. Drescher, 2004, J. Biotechnol. **112**, 139.
Silkin, V. M., M. Quijada, R. Díez Muino, E. V. Chulkov, and P. M. Echenique, 2007, Surf. Sci. **601**, 4546.
Simpson, J., and U. Fano, 1963, Phys. Rev. Lett. **11**, 158.
Siwick, B. J., J. R. Dwyer, R. E. Jordan, and R. J. D. Miller, 2003, Science **302**, 1382.
Smirnova, O., S. Patchkovskii, and M. Spanner, 2007, Phys. Rev. Lett. **98**, 123001.
Smirnova, O., M. Spanner, and M. Ivanov, 2006, J. Phys. B **39**, S323.
Smirnova, O., V. S. Yakovlev, and M. Ivanov, 2005, Phys. Rev. Lett. **94**, 213001.
Smirnova, O., V. S. Yakovlev, and A. Scrinzi, 2003, Phys. Rev. Lett. **91**, 253001.
Smith, P. W., M. A. Duguay, and E. P. Ippen, 1974, Prog. Quantum Electron. **3**, 107.
Sokolov, A., M. Y. Shverdin, D. R. Walker, D. D. Yavuz, A. M. Burzo, G. Y. Yin, and S. E. Harris, 2005, J. Mod. Opt. **52**, 285.
Sokolov, A., D. R. Walker, D. D. Yavuz, G. Y. Yin, and S. E. Harris, 2001, Phys. Rev. Lett. **87**, 033402.
Sola, I., 2004, Phys. Rev. A **69**, 033401.
Sola, I. J., E. Mével, L. Elouga, E. Constant, V. Strelkov, L. Poletto, P. Villoresi, E. Benedetti, J-P. Caumes, S. Stagira, C. Vozzi, G. Sansone, and M. Nisoli, 2006, Nat. Phys. **2**, 319.
Sola, I. J., A. Zair, R. López-Martens, P. Johnsson, K. Varjú, E. Cormier, J. Mauritsson, A. L'Huillier, V. Strelkov, E. Mével, and E. Constant, 2006, Phys. Rev. A **74**, 013810.
Spanner, M., O. Smirnova, and M. Yu. Ivanov, 2004, J. Phys. B **37**, L243.
Spielmann, C., C. Kan, N. H. Burnett, T. Brabec, M. Geissler, A. Scrinzi, M. Schnürer, and F. Krausz, 1998, IEEE J. Sel. Top. Quantum Electron. **4**, 249.
Srinivasan, R., V. A. Lobastov, C.-Y. Ruan, and A. H. Zewail, 2003, Helv. Chim. Acta **86**, 1761.
Stapelfeldt, H., E. Constant, H. Sakai, and P. B. Corkum, 1998, Phys. Rev. A **58**, 426.
Steiner, E., A. Soncini, and P. W. Fowler, 2005, Org. Biomol. Chem. **3**, 4053.
Steinmeyer, G., D. H. Sutter, L. Gallmann, N. Matuschek, and U. Keller, 1999, Science **286**, 1507.
Stingl, A., M. Lenzner, C. Spielmann, and F. Krausz, 1995, Opt. Lett. **20**, 602.
Stockman, M., M. Kling, U. Kleineberg, and F. Krausz, 2007, Nat. Photonics **1**, 539.
Stolow, A., 2003a, Annu. Rev. Phys. Chem. **54**, 89.
Stolow, A., 2003b, Int. Rev. Phys. Chem. **22**, 377.
Stolow, A., A. E. Bragg, and D. M. Neumark, 2004, Chem. Rev. (Washington, D.C.) **104**, 1719.
Strasser, D., F. Goulay, and S. R. Leone, 2007, J. Chem. Phys. **127**, 184305.
Strelkov, V. V., 2006, Phys. Rev. A **74**, 013405.
Strelkov, V., A. Zair, O. Tcherbakoff, R. López-Martens, E. Cormier, E. Mével, and E. Constant, 2004, Appl. Phys. B: Lasers Opt. **78**, 879.
Strelkov, V., A. Zair, O. Tcherbakoff, R. López-Martens, E. Cormier, E. Mevel, and E. Constant, 2005, J. Phys. B **38**, L161.
Strickland, A. D., and G. Mourou, 1985, Opt. Commun. **56**, 219.
Suda, A., M. Hatayama, K. Nagasaka, and K. Midorikawa, 2005, Appl. Phys. Lett. **86**, 111116.
Sung, C., S. Ya. Tochitsky, S. Reiche, J. B. Rosenzweig, C. Pellegrini, and C. Joshi, 2006, Phys. Rev. ST Accel. Beams **9**, 120703.
Sussman, B. J., M. Yu. Ivanov, and A. Stolow, 2005, Phys. Rev. A **71**, 051401.
Sussman, B. J., D. Townsend, M. Yu. Ivanov, and A. Stolow, 2006, Science **314**, 278.
Suzuki, T., 2006, Annu. Rev. Phys. Chem. **57**, 555.
Szent-Györgyi, A., 1941, Science **93**, 609.
Szipöcs, R., K. Ferencz, C. Spielmann, and F. Krausz, 1994, Opt. Lett. **19**, 201.
Takahashi, E., Y. Nabekawa, and K. Midorikawa, 2004, Appl. Phys. Lett. **84**, 4.
Takahashi, E., Y. Nabekawa, T. Otsuka, M. Obara, and K. Midorikawa, 2002, Phys. Rev. A **66**, 021802(R).
Takahashi, S., and K. Takatsuka, 2006, J. Chem. Phys. **124**, 144101.

Takatsuka, K., 2006, J. Chem. Phys. **124**, 064111.

Takatsuka, K., 2007, J. Phys. Chem. A **111**, 10196.

Takeuchi, S., and T. Kobayashi, 1994, J. Appl. Phys. **75**, 2757.

Taranukhin, V. D., 2004, J. Opt. Soc. Am. B **21**, 419.

Tarasevitsch, A., K. Lobov, C. Wünsche, and D. von der Linde, 2007, Phys. Rev. Lett. **98**, 103902.

Tate, J., T. Auguste, H. G. Muller, P. Salieres, P. Agostini, and L. F. DiMauro, 2007, Phys. Rev. Lett. **98**, 013901.

Tavella, F., A. Marcinkevicius, and F. Krausz, 2006, Opt. Express **14**, 12822.

Tavella, F., Y. Nomura, L. Veisz, V. Pervak, A. Marcinkevicius, and F. Krausz, 2007, Opt. Lett. **32**, 2227.

Tcherbakoff, O., E. Mevel, D. Descamps, J. Plumridge, and E. Constant, 2003, Phys. Rev. A **68**, 043804.

Telle, H. R., G. Steinmeyer, A. E. Dunlop, J. Stenger, D. H. Sutter, and U. Keller, 1999, Appl. Phys. B: Lasers Opt. **69**, 327.

Tempea, G., M. Geissler, and T. Brabec, 1999, J. Opt. Soc. Am. B **16**, 669.

Tempea, G., F. Krausz, Ch. Spielmann, and K. Ferencz, 1998, IEEE J. Sel. Top. Quantum Electron. **4**, 193.

Thaury, C., F. Quéré, J.-P. Geindre, A. Levy, T. Ceccotti, P. Monot, M. Bougeard, F. Réau, P. d'Oliveira, P. Audebert, R. Marjoribanks, and Ph. Martin, 2007, Nat. Phys. **3**, 424.

Toma, E. S., H. G. Muller, P. M. Paul, P. Breger, M. Cheret, P. Agostini, C. Le Blanc, G. Mullot, and G. Cheriaux, 2000, Phys. Rev. A **62**, 061801(R).

Tong, X. M., and S-I. Chu, 1998, Phys. Rev. A **57**, 452.

Tong, X. M., and C. D. Lin, 2006, Phys. Rev. A **73**, 042716.

Tong, X. M., and C. D. Lin, 2007a, Phys. Rev. Lett. **98**, 123002.

Tong, X. M., and C. D. Lin, 2007b, J. Phys. B **40**, 641.

Torres, R., N. Kajumba, J. G. Underwood, J. S. Robinson, S. Baker, J. W. G. Tisch, R. de Nalda, W. A. Bryan, R. Velotta, C. Altucci, I. C. E. Turcu, and J. P. Marangos, 2007, Phys. Rev. Lett. **98**, 203007.

Tournois, P., 1997, Opt. Commun. **140**, 245.

Trebino, R., 2000, *Frequency-Resolved Optical Gating: The Measurement of Ultrashort Laser Pulses* (Kluwer Academic, Norwell, MA).

Trebino, R., K. W. DeLong, D. N. Fittinghoff, J. N. Sweetser, M. A. Krumbügel, B. A. Richman, and D. J. Kane, 1997, Rev. Sci. Instrum. **68**, 3277.

Trebino, R., and D. J. Kane, 1993, J. Opt. Soc. Am. A **10**, 1101.

Tsakiris, G. D., K. Eidmann, J. Meyer-ter-Vehn, and F. Krausz, 2006, New J. Phys. **8**, 19.

Tzallas, P., D. Charalambidis, N. A. Papadogiannis, K. Witte, and G. Tsakiris, 2003, Nature (London) **426**, 267.

Tzallas, P., D. Charalambidis, N. A. Papadogiannis, K. Witte, and G. D. Tsakiris, 2005, J. Mod. Opt. **52**, 321.

Udem, T., R. Holzwarth, and T. W. Hänsch, 2002, Nature (London) **416**, 233.

Uiberacker, M., T. Uphues, M. Schultze, A. J. Verhoef, V. Yakovlev, M. F. Kling, J. Rauschenberger, N. M. Kabachnik, H. Schröder, M. Lezius, K. L. Kompa, H.-G. Muller, M. J. J. Vrakking, S. Hendel, U. Kleineberg, U. Heinzmann, M. Drescher, and F. Krausz, 2007, Nature (London) **446**, 627.

Ullrich, J., R. Moshammer, R. Doerner, O. Jagutzki, V. Mergel, H. Schmidt-Böcking, and L. Spielberger, 1997, J. Phys. B **30**, 2917.

Van Duyne, R. P., 2004, Science **306**, 985.

Van Vlack, C., and S. Hughes, 2007, Phys. Rev. Lett. **98**, 167404.

Varin, C., and M. Piché, 2006, Phys. Rev. E **74**, 045602(R).

Varjú, K., P. Johnsson, R. López-Martens, T. Remetter, E. Gustafsson, J. Mauritsson, M. B. Gaarde, K. J. Schafer, Ch. Erny, I. Sola, A. Zaïr, E. Constant, E. Cormier, E. Mével, and A. L'Huillier, 2005, Laser Phys. **15**, 888.

Varjú, K., Y. Mairesse, P. Agostini, P. Breger, B. Carré, L. J. Frasinski, E. Gustafsson, P. Johnsson, J. Mauritsson, H. Merdji, P. Monchicourt, A. L'Huillier, and P. Salieres, 2005, Phys. Rev. Lett. **95**, 243901.

Vaval, N., and L. S. Cederbaum, 2007, J. Chem. Phys. **126**, 164110.

Veisz, L., G. Kurkin, K. Chernov, V. Tarnetsky, A. Apolonski, F. Krausz, and E. Fill, 2007, New J. Phys. **9**, 451.

Volkov, D. M., 1935, Z. Phys. **94**, 250.

Volkova, E. A., A. M. Popov, and O. V. Tikhonova, 2007, Opt. Spectrosc. **102**, 159.

Vogt, G., G. Krampert, P. Niklaus, P. Nuernberger, and G. Gerber, 2005, Phys. Rev. Lett. **94**, 068305.

Vozzi, C., F. Calegari, E. Benedetti, J.-P. Caumes, G. Sansone, S. Stagira, M. Nisoli, R. Torres, E. Heesel, N. Kajumba, J. P. Marangos, C. Altucci, and R. Velotta, 2005, Phys. Rev. Lett. **95**, 153902.

Wagner, N., A. Wuest, I. Christov, T. Popmintchev, X. Zhou, M. Murnane, and H. Kapteyn, 2006, Proc. Natl. Acad. Sci. U.S.A. **103**, 13279.

Walker, B., E. Mevel, B. Yang, P. Breger, J. P. Chambaret, A. Antonetti, L. F. DiMauro, and P. Agostini, 1993, Phys. Rev. A **48**, R894.

Walker, B., B. Sheehy, L. F. DiMauro, P. Agostini, K. J. Schafer, and K. C. Kulander, 1994, Phys. Rev. Lett. **73**, 1227.

Walker, B., B. Sheehy, K. C. Kulander, and L. F. DiMauro, 1996, Phys. Rev. Lett. **77**, 5031.

Walmsley, I. A., and V. Wong, 1996, J. Opt. Soc. Am. B **13**, 2453.

Walters, Z. B., S. Tonzani, and C. Greene, 2007, J. Phys. B **40**, F277.

Wang, P. X., Y. K. Ho, X. Q. Yuan, Q. Kong, N. Cao, A. M. Sessler, E. Esarey, and Y. Nishida, 2001, Appl. Phys. Lett. **78**, 2253.

Wang, W., P. X. Wang, Y. K. Ho, Q. Kong, Y. Gu, and S. J. Wang, 2007, Rev. Sci. Instrum. **78**, 093103.

Watanabe, S., K. Kondo, Y. Nabekawa, A. Sagisaka, and Y. Kobayashi, 1994, Phys. Rev. Lett. **73**, 2692.

Watson, J. B., A. Sanpera, D. G. Lappas, P. L. Knight, and K. Burnett, 1997, Phys. Rev. Lett. **78**, 1884.

Weber, T., A. O. Czasch, O. Jagutzki, A. K. Müller, V. Mergel, A. Kheifets, E. Rotenberg, G. Meigs, M. H. Prior, S. Daveau, A. Landers, C. L. Cocke, T. Osipov, R. Díez Muiño, H. Schmidt-Böcking, and R. Dörner, 2004, Nature (London) **431**, 437.

Weber, T., M. Weckenbrock, A. Staudte, L. Spielberger, O. Jagutzki, V. Mergel, F. Afaneh, G. Urbasch, M. Vollmer, H. Giessen, and R. Dörner, 2000a, Phys. Rev. Lett. **84**, 443.

Weber, T., M. Weckenbrock, A. Staudte, L. Spielberger, O. Jagutzki, V. Mergel, F. Afaneh, G. Urbasch, M. Vollmer, H. Giessen, and R. Dörner, 2000b, J. Phys. B **33**, L127.

Weckenbrock, M., A. Becker, A. Staudte, S. Kammer, M. Smolarski, V. R. Bhardwaj, D. M. Rayner, D. M. Villeneuve, P. B. Corkum, and R. Dörner, 2003, Phys. Rev. Lett. **91**, 123004.

Weckenbrock, M., D. Zeidler, A. Staudte, Th. Weber, M. Schoeffler, M. Meckel, S. Kammer, M. Smolarski, O. Jagutzki, V. R. Bhardwaj, D. M. Rayner, D. M. Villeneuve, P. B. Corkum, and R. Dörner, 2004, Phys. Rev. Lett. **92**, 213002.

Weiner, A. M., 1995, Prog. Quantum Electron. **19**, 161.

Weiner, A. M., J. P. Heritage, and E. M. Kirschner, 1988, J. Opt. Soc. Am. B **5**, 1563.

Weinkauf, R., P. Schanen, A. Metsala, E. W. Schlag, M. Bürgle, and H. Kessler, 1996, J. Chem. Phys. **100**, 18567.

Weinkauf, R., P. Schanen, D. Yang, S. Soukara, and E. W. Schlag, 1995, J. Chem. Phys. **99**, 11255.

Weinkauf, R., E. W. Schlag, T. J. Martinez, and R. D. Levine, 1997, J. Chem. Phys. **99**, 11255.

Weitzel, K.-M., 2007, ChemPhysChem **8**, 213.

Wernet, Ph., D. Nordlund, U. Bergmann, M. Cavalleri, M. Odelius, H. Ogasawara, L. A. Näslund, T. K. Hirsch, L. Ojamäe, P. Glatzel, L. G. M. Pettersson, and A. Nilsson, 2004, Science **304**, 995.

Wickenhauser, M., J. Burgdörfer, F. Krausz, and M. Drescher, 2005, Phys. Rev. Lett. **94**, 023002.

Wickenhauser, M., J. Burgdörfer, F. Krausz, and M. Drescher, 2006, J. Mod. Opt. **53**, 247.

Wilhelm, T., J. Piel, and E. Riedle, 1997, Opt. Lett. **22**, 1494.

Wille, K., 2000, *The Physics of Particle Accelerators* (Oxford University Press, Oxford).

Williams, I. D., J. McKenna, J. Wood, M. Suresh, W. A. Bryan, S. L. Stebbings, E. M. L. English, C. R. Calvert, B. Srigengan, E. J. Divall, C. J. Hooker, A. J. Langley, and W. R. Newell, 2007, Phys. Rev. Lett. **99**, 173002.

Williams, R. T., R. R. Royt, J. C. Rife, J. P. Long, and M. N. Kabler, 1982, J. Vac. Sci. Technol. **21**, 509.

Williamson, J. C., J. Cao, H. Ihee, H. Frey, and A. H. Zewail, 1997, Nature (London) **386**, 159.

Williamson, J. C., M. Dantus, S. B. Kim, and A. H. Zewail, 1992, Chem. Phys. Lett. **196**, 529.

Williamson, J. C., and A. H. Zewail, 1991, Proc. Natl. Acad. Sci. U.S.A. **88**, 5021.

Williamson, S., G. Mourou, and J. C. M. Li, 1984, Phys. Rev. Lett. **52**, 2364.

Witte, S., R. Th. Zinkstok, A. L. Wolf, W. Hogervorst, W. Ubachs, and K. S. E. Eikema, 2006, Opt. Express **14**, 8168.

Wittmann, T., B. Horvath, W. Helml, M. G. Schätzel, X. Gu, A. L. Cavalieri, G. G. Paulus, and R. Kienberger, 2008, unpublished.

Wohlleben, W., T. Buckup, J. L. Herek, R. J. Cogdell, and M. Motzkus, 2003, Biophys. J. **85**, 442.

Wohlleben, W., T. Buckup, J. L. Herek, H. Hashimoto, R. J. Cogdell, and M. Motzkus, 2004, in *Femtochemistry and Femtobiology: Ultrafast Events in Molecular Science*, edited by J. L. Hynes and M. Martin (Elsevier, Burlington MA), p. 91.

Wohlleben, W., T. Buckup, J. L. Herek, and M. Motzkus, 2005, ChemPhysChem **6**, 850.

Wolf, M., E. Knoesel, and T. Hertel, 1996, Phys. Rev. B **54**, R5295.

Wonisch, A., U. Neuhäusler, N. M. Kabachnik, T. Uphues, M. Uiberacker, V. Yakovlev, F. Krausz, M. Drescher, U. Kleineberg, and U. Heinzmann, 2006, Appl. Opt. **45**, 4147.

Wonisch, A., Th. Westerwalbesloh, W. Hachmann, N. Kabachnik, U. Kleineberg, and U. Heinzmann, 2004, Thin Solid Films **464-465**, 473.

Wu, K., X. Yang, and H. Zeng, 2007, Appl. Phys. B: Lasers Opt. **88**, 189.

Wu, Y., and X. Yang, 2007, Phys. Rev. A **76**, 013832.

Wurth, W., and D. Menzel, 2000, Chem. Phys. **251**, 141.

Xie, X., G. Jordan, M. Wickenhauser, and A. Scrinzi, 2007, J. Mod. Opt. **54**, 999.

Xu, J., and X.-C. Zhang, 2006, *THz Science and Technology* (Peking University Press, Beijing, China).

Xu, L., C. Spielmann, A. Poppe, T. Brabec, F. Krausz, and T. W. Hänsch, 1996, Opt. Lett. **21**, 2008.

Yakovlev, V., 2007, private communication.

Yakovlev, V. S., F. Bammer, and A. Scrinzi, 2005, J. Mod. Opt. **52**, 395.

Yakovlev, V. S., M. Ivanov, and F. Krausz, 2007, Opt. Express **15**, 15351.

Yakovlev, V. S., and A. Scrinzi, 2003, Phys. Rev. Lett. **91**, 153901.

Yamane, K., Z. Zhang, K. Oka, R. Morita, M. Yamashita, and A. Suguro, 2003, Opt. Lett. **28**, 2258.

Yamashita, M., K. Yamane, and R. Morita, 2006, IEEE J. Sel. Top. Quantum Electron. **12**, 213.

Yang, B., K. J. Schafer, B. Walker, K. C. Kulander, P. Agostini, and L. F. DiMauro, 1993, Phys. Rev. Lett. **71**, 3770.

Yavin, E., A. K. Boal, E. D. A. Stemp, E. M. Boon, A. L. Livingston, V. L. O'Shea, S. S. David, and J. K. Barton, 2005, Proc. Natl. Acad. Sci. U.S.A. **102**, 3546.

Ye, J., and S. T. Cundiff, 2005, in *Femtosecond Optical Frequency Comb Technology*, edited by J. Ye and S. T. Cundiff (Springer Science + Business Media, New York), p. 12.

Yen, R., J. M. Liu, N. Bloembergen, T. K. Yee, J. G. Fujimoto, and M. M. Salour, 1982, Appl. Phys. Lett. **40**, 185.

Yu, T. J., K.-H. Hong, H.-G. Choi, J. H. Sung, I. W. Choi, D.-K. Ko, J. Lee, J. Kim, D. E. Kim, C. H. Nam, 2007, Opt. Express **15**, 8203.

Yudin, G. L., A. D. Bandrauk, and P. Corkum, 2006, Phys. Rev. Lett. **96**, 063002.

Yudin, G. L., S. Chelkowski, J. Itatani, A. D. Bandrauk, and P. Corkum, 2005, Phys. Rev. A **72**, 051401(R).

Yudin, G. L., and M. Ivanov, 2001a, Phys. Rev. A **64**, 013409.

Yudin, G. L., and M. Y. Ivanov, 2001b, Phys. Rev. A **63**, 033404.

Yudin, G. L., S. Patchkovskii, P. B. Corkum, and A. D. Bandrauk, 2007, J. Phys. B **40**, F93.

Yue, J. T., and Doniach, 1973, Phys. Rev. B **8**, 4578.

Yurchenko, S. N., S. Patchkovskii, I. Litvinyuk, P. Corkum, and G. Yudin, 2004, Phys. Rev. Lett. **93**, 223003.

Zair, A., O. Tcherbakoff, E. Mével, E. Constant, R. López-Martens, J. Mauritsson, P. Johnsson, and A. L'Huillier, 2004, Appl. Phys. B: Lasers Opt. **78**, 869.

Zamith, S., Y. Ni, A. Gürtler, L. D. Noordam, H. G. Muller, and M. J. J. Vrakking, 2004, Opt. Lett. **29**, 2303.

Zavelani-Rossi, M., G. Cerullo, S. De Silvestri, L. Gallmann, N. Matuschek, G. Steinmeyer, U. Keller, G. Angelow, V. Scheuer, and T. Tschudi, 2001, Opt. Lett. **26**, 1155.

Zeek, E., K. Maginnis, S. Backus, U. Russek, M. Murnane, G. Mourou, and H. Kapteyn, 1999, Opt. Lett. **12**, 1540.

Zeidler, D., A. Staudte, A. B. Bardon, D. M. Villeneuve, R. Doerner, and P. B. Corkum, 2005, Phys. Rev. Lett. **95**, 203003.

Zeng, Z., Y. Cheng, X. Song, R. Li, and Z. Xu, 2007, Phys. Rev. Lett. **98**, 203901.

Zepf, M., B. Dromey, M. Landreman, P. Foster, and S. M. Hooker, 2007, Phys. Rev. Lett. **99**, 143901.

Zewail, A., 2000, J. Phys. Chem. A **104**, 5660.

Zewail, A., 2006, Annu. Rev. Phys. Chem. **57**, 65.

Zewail, A. H., 1994, *Femtochemistry Volume I and II* (World Scientific, Singapore).

Zhang, J., L. Bai, S. Gong, Z. Xu, and D.-S. Guo, 2007, Opt. Express **15**, 7261.

Zhang, P., Y. R. Song, and Z. G. Zhang, 2006, Acta Phys. Sin.

55, 6208.
Zhang, X., A. R. Libertun, A. Paul, E. Gagnon, S. Backus, I. P. Christov, M. M. Murnane, and H. C. Kapteyn, R. A. Bartels, Y. Liu, and D. T. Attwood, 2004, Opt. Lett. **29**, 1357.
Zhang, X., A. Lytle, T. Popmintchev, A. Paul, N. Wagner, M. M. Murnane, H. C. Kapteyn, and I. P. Christov, 2005, Opt. Lett. **30**, 1971.
Zhao, Z. X., and C. D. Lin, 2005, Phys. Rev. A **71**, 060702(R).
Zhou, X., Z. Chen, T. Morishita, A.-T. Le, and C. D. Lin, 2008, Phys. Rev. A **77**, 053410.
Zhou, XiaoXin, X. M. Tong, Z. X. Zhao, and C. D. Lin, 2005a, Phys. Rev. A **72**, 033412.
Zhou, XiaoXin, X. M. Tong, Z. X. Zhao, and C. D. Lin, 2005b, Phys. Rev. A **71**, 061801.
Zheltikov, A. M., 2006a, Phys. Usp. **49**, 605.
Zheltikov, A. M., 2006b, Phys. Rev. A **74**, 053403.
Zholents, A. A., and W. M. Fawley, 2004, Phys. Rev. Lett. **92**, 224801.
Zhu, X.-Y., 2004, Surf. Sci. Rep. **56**, 1.
Zhu, C., S. Nangia, A. W. Jasper, and D. G. Truhlar, 2004, J. Chem. Phys. **121**, 7658.
Zimmermann, B., M. Lein, and J. M. Rost, 2005, Phys. Rev. A **71**, 033401.
Zon, B. A., 2000, J. Exp. Theor. Phys. **91**, 899.
Zrost, K., A. Rudenko, T. Ergler, B. Feuerstein, V. L. B. de Jesus, C. D. Schröter, R. Moshammer, and J. Ullrich, 2006, J. Phys. B **39**, S371.

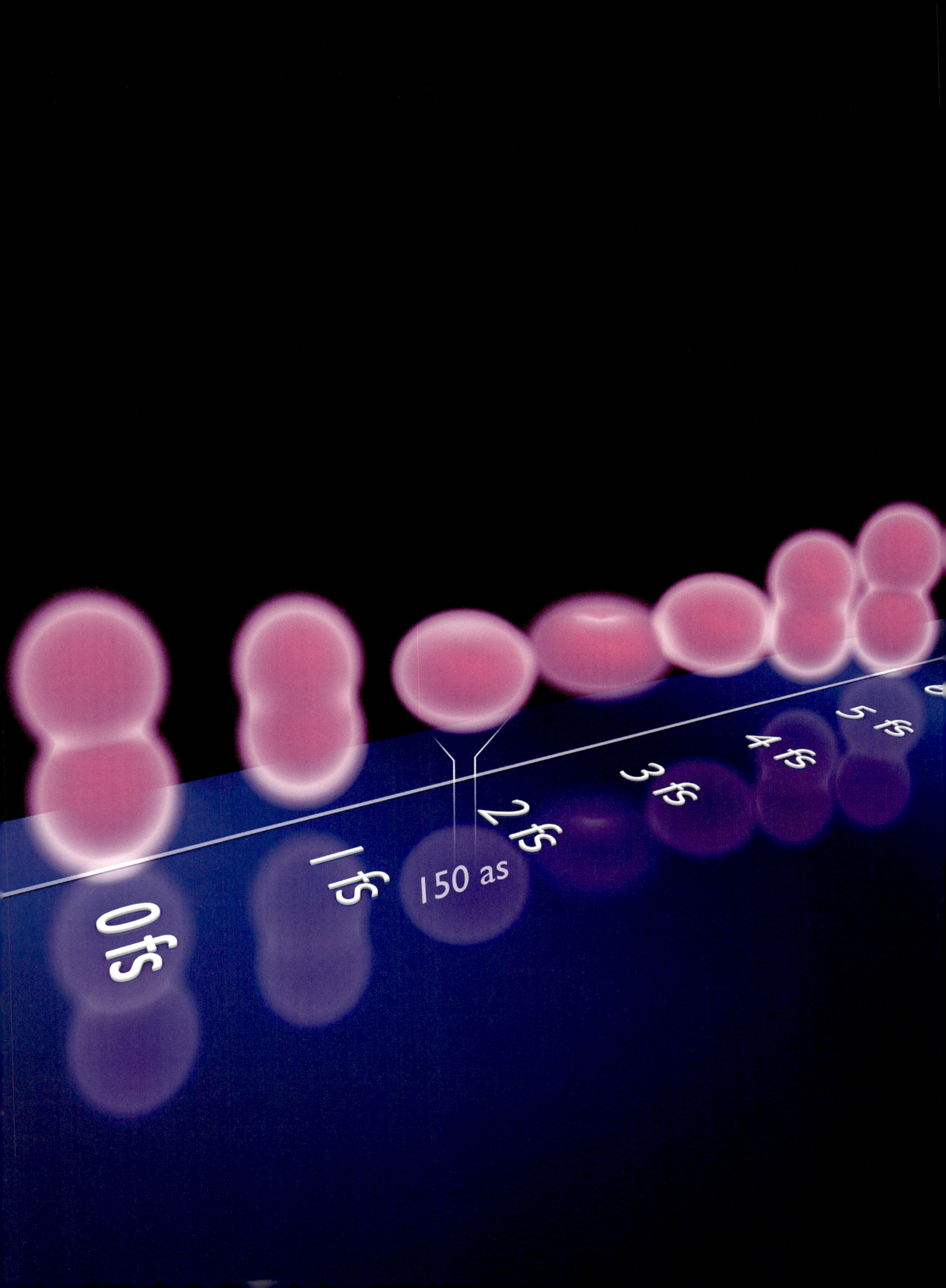
0 fs
1 fs
150 as
2 fs
3 fs
4 fs
5 fs

Tracking Electronic Motions at the Atomic Scale

8

Opposite: Ensemble-averaged probability distributions of the valence hole wavepacket created in the 4p subshell of krypton atoms following strong-field (tunneling) ionization by a few-cycle NIR laser pulse.The probability distributions have been reconstructed from attosecond XUV transient absorption spectra recorded with 150-attosecond XUV pulses [42]. © Christian Hackenberger

Direct time-domain investigations of electronic motions (via pump-probe spectroscopy) rely on the well-reproducible, well-characterized attosecond tools described in Chapter 7. These tools, few-cycle visible/near-infrared (NIR) strong laser fields with controlled waveform and isolated attosecond XUV pulses, come in perfect temporal synchronism. As a result, they lend themselves from pump-probe studies. For attosecond resolution, both the "starter gun" (pump) and the interrogation (probe) must be performed with attosecond "sharpness".

The attosecond temporal gradients for the pumping and the interrogating process can be provided by the temporal intensity profile of the XUV pulse and the controlled sub-femtosecond evolution of the NIR laser field. Hence, electron dynamics can be captured by triggering an event and probing it with the intensity profile of the XUV pulse and the oscillating electric field of the NIR pulse or vice versa. The attosecond "snapshots" recorded with such pump-probe exposures come in the form of physical observables such as the spectral/angular distributions of photoelectrons and/or of the photons transmitted through the sample under scrutiny (Fig. 8-1).

This section reviews the different approaches for attosecond real-time observation of the most fundamental quantum dynamics, ranging from electron tunneling [40] and electron-electron interactions through the photo effect [41] to valence electron wavepacket dynamics in atoms [42], and atomic-scale transport phenomena in crystalline solids [43-44].

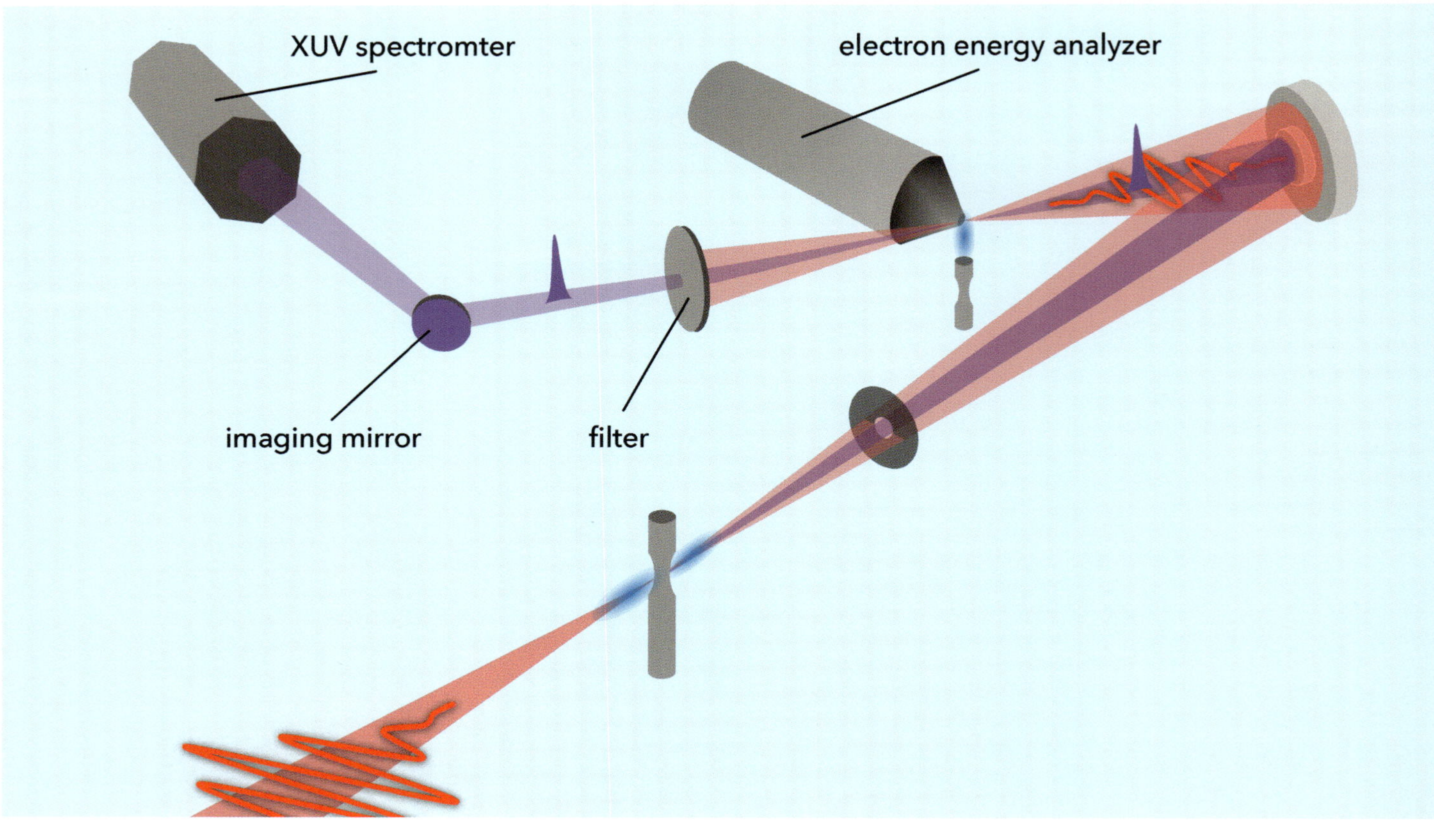

Figure 8-1 Real-time access to electron phenomena by attosecond probe or streaking spectroscopy. The few-cycle laser electric field (red) generates coherent high-harmonic XUV light, from which the inner part of the two-component mirror on the right spectrally filters out an isolated attosecond pulse (blue). Translating this inner mirror with respect to the external one reflecting (and focusing) the NIR laser light introduces a delay between the XUV pulse and the NIR laser field, which can be adjusted with attosecond accuracy. Using the strong laser field as a pump, the apparatus permits the implementation of attosecond photoelectron and/or transient absorption spectroscopy by recording the energy/momentum distribution of the ejected photoelectron spectra and/or the transmitted XUV photoelectron spectra, respectively, as a function of the delay of the attosecond XUV probe with respect to the NIR laser pump. Alternatively, electron processes can also be triggered with the attosecond XUV pulse and probed, with attosecond resolution, by streaking the emitted photoelectron spectra (attosecond streaking spectroscopy). The setup allows the implementation of both types of spectroscopies and formed the basis of the first attosecond beamlines in Vienna and Munich.

Electron tunneling has been a textbook example of a microscopic quantum effect lacking any analogous phenomenon in the realm of classical physics describing macroscopic phenomena. An external electric field comparable in strength to the atomic Coulomb field, binding outer electrons to the nucleus, suppresses the binding field to form a barrier. If the field is strong enough, the barrier can become narrow enough to allow the bound electron to "tunnel" through it and get free. Ultrashort-pulse lasers have been providing the conditions for tunneling ionization to occur for more than half a century. However, because the barrier exists only for a fraction of the laser half cycle, i.e. for a fraction of a femtosecond, it could never be observed in real time. Attosecond technology permitted Matthias Uiberacker, Thorsten Uphues and their co-workers this feat [40].

Electron-electron interactions that influence, among other things, the electronic properties of solids had been elusive due to the lack of direct experimental probes and the inaccessible computational power required to perform ab-initio calculations accounting for these interactions. In the absence of these tools, physicists had to draw on simplified models, the validity of which had to be tested by measurements accessing the phenomenon under scrutiny as directly as possible. The photoeffect in a multi-electron atom is one of the simplest probes of electron-electron interactions. Independent theoretical work from Cleanthes Nicolaides, Joachim Burgdörfer, Vladislav Yakovlev and their co-workers has predicted photoemission to be delayed by a few attoseconds with respect to the absorption of an energetic (XUV or X-ray) photon. This delay was directly related to electron-electron interactions occurring within the first few attoseconds following the instant of absorption [41]. Martin Schultze and his co-workers used attosecond streaking to measure the relative timing between electrons liberated from two different quantum levels, for the first time [41]. Attosecond chronoscopy hence provides a powerful probe for intra-atomic electron correlations and a way of testing theoretical models designed to account for them.

Valence electron motion is the key to understanding the breaking and formation of chemical bonds at the most fundamental level, in terms of the electrons constituting the glue for those bonds. One of the important remaining questions of molecular science is how the initial electronic motions, occurring at a time scale where the nuclei are still frozen, i.e. within the first few femtoseconds, affect the fate of the molecule at the end of all the structural dynamics (including possible bond breakage and formation) that may follow the excitation and are concluded typically only on a multi-femtosecond to picosecond time scale. Answering this question calls for real-time access to valence electron dynamics. This access was first demonstrated by Eleftherios Goulielmakis and his co-workers in collaboration with Zi-Heng Loh and Stephen Leone and supported theoretically by Nina Rohringer, Robin Santra, and their colleagues, providing insight into how valence hole wavepacket motion unfolds in the first femtoseconds following the strong-field ionization of a multi-electron atom (Chapter 8 title image).

Atomic-scale electron transport in solids is at the heart of modern electronics. Ever faster transport in ever smaller nano-scale circuits at the expense of ever smaller heat dissipation is the key to advancing electronic chips to ever higher information processing speeds, eventually to frequencies approaching those of infrared and visible light. In order for this to become reality, sub-femtosecond atomic-scale charge carrier transport in solid-state structures needs to be understood and controlled. Attosecond metrology combined with ultrahigh-vacuum technology has, for the first time, rendered real-time observation of electron transport across atomic layers of crystalline solids feasible in experiments conducted by Adrian Cavalieri, Stefan Neppl and Reinhard Kienberger in collaboratoration with Ulrich Heinzmann, Peter Feulner, and their colleagues [43-44].

[40] M. Uiberacker, T. Uphues, M. Schultze, A. Verhoef, V. Yakovlev, M. Kling, J. Rauschenberger, N. M. Kabachnik, H. Schröder, M. Lezius, K. L. Kompa, H. G. Müller, M. Vrakking, S. Hendel, U. Kleineberg, U. Heinzmann, M. Drescher, F. Krausz. Attosecond real-time observation of electron tunnelling in atoms. *Nature* **446**, 627 (2007).
(Page: 353)

[41] M. Schultze, M. Fieß, N. Karpowicz, J. Gagnon, M. Korbman, M. Hofstetter, S. Neppl, A. Cavalieri, Y. Komninos, T. Mercouris, C. Nicolaides, R. Pazourek, S. Nagele, J. Feist, J. Burgdörfer, A. Azzeer, R. Ernstorfer, R. Kienberger, U. Kleineberg, E. Goulielmakis, F. Krausz, V. Yakovlev. Delay in photoemission. *Science* **328**, 1658 (2010).
(Page: 359)

[42] E. Goulielmakis, Z. Loh, A. Wirth, R. Santra, N. Rohringer, V. Yakovlev, S. Zherebtsov, T. Pfeifer, A. Azzeer, M. Kling, S. Leone, F. Krausz. Real-time observation of valence electron motion. *Nature* **466**, 739 (2010).
(Page: 364)

[43] A. Cavalieri, N. Müller, T. Uphues, V. Yakovlev, A. Baltuška, B. Horváth, B. Schmidt, L. Blümel, R. Holzwarth, S. Hendel, M. Drescher, U. Kleineberg, P. M. Echenique, R. Kienberger, F. Krausz, U. Heinzmann. Attosecond spectroscopy in condensed matter. *Nature* **449**, 1029 (2007).
(Page: 370)

[44] S. Neppl, R. Ernstorfer, A. Cavalieri, C. Lemell, G. Wachter, E. Magerl, E. Bothschafter, M. Jobst, M. Hofstetter, U. Kleineberg, J. Barth, D. Menzel, J. Burgdörfer, P. Feulner, F. Krausz, R. Kienberger. Direct observation of electron propagation and dielectric screening on the atomic length scale. *Nature* **517**, 342 (2015).
(Page: 374)

Vol 446|5 April 2007|doi:10.1038/nature05648 nature

ARTICLES

Attosecond real-time observation of electron tunnelling in atoms

M. Uiberacker[1,2], Th. Uphues[3], M. Schultze[2], A. J. Verhoef[2,4], V. Yakovlev[1], M. F. Kling[5], J. Rauschenberger[1,2], N. M. Kabachnik[3,6], H. Schröder[2], M. Lezius[2], K. L. Kompa[2], H.-G. Muller[5], M. J. J. Vrakking[5], S. Hendel[3], U. Kleineberg[1], U. Heinzmann[3], M. Drescher[7] & F. Krausz[1,2,4]

Atoms exposed to intense light lose one or more electrons and become ions. In strong fields, the process is predicted to occur via tunnelling through the binding potential that is suppressed by the light field near the peaks of its oscillations. Here we report the real-time observation of this most elementary step in strong-field interactions: light-induced electron tunnelling. The process is found to deplete atomic bound states in sharp steps lasting several hundred attoseconds. This suggests a new technique, attosecond tunnelling, for probing short-lived, transient states of atoms or molecules with high temporal resolution. The utility of attosecond tunnelling is demonstrated by capturing multi-electron excitation (shake-up) and relaxation (cascaded Auger decay) processes with subfemtosecond resolution.

With the invention of the laser in 1960 and subsequent advances in ultrashort light pulse generation, light fields with a strength of several volts per angstrom have become routinely available. They rival the fields acting on valence electrons in atomic systems, allowing their release from atoms, molecules and solids. These advances sparked a revolution in studying the interaction of electrons with light. The primary step in strong-field interactions is the liberation of electrons from their atomic bound state. The revolutionary theory of Keldysh[1] and subsequent work[2–6] suggested that a valence electron may escape by tunnelling through its atomic binding potential suppressed by the light field (Fig. 1a). If the dimensionless parameter:

$$\gamma = \frac{\omega_L \sqrt{2mW_b}}{|e|E_0} \qquad (1)$$

is less than one, under the assumption of $\hbar\omega_L \ll W_b$ ionization is predicted to be confined to short intervals lasting a fraction of the half oscillation cycle of the light field (Fig. 1b). Here E_0 and ω_L stand for the amplitude and angular frequency of the oscillations of the laser electric field $E_L(t) = E_0\varepsilon(t)\cos(\omega_L t + \varphi)$, with $\varepsilon(t)$ being the amplitude envelope function, and e, m and W_b the charge, mass and binding energy of the electron. Recent studies[6] suggest that tunnelling remains the dominant ionization mechanism even for γ substantially exceeding one, that is, under conditions when the potential barrier formed by the atomic binding potential and the ionizing light field varies during tunnelling (non-adiabatic regime).

In this work we report what we believe is the first real-time observation of light-induced electron tunnelling. The observation of ionization occurring in subfemtosecond steps spaced by the half laser cycle up to values of the Keldysh parameter as high as three is in good agreement with analytic and numerical calculations. Our approach provides experimental access to all forms of optical field ionization—both adiabatic and nonadiabatic tunnelling, as well as barrier-suppression ionization—and allows us to test models of these processes for the first time.

Once the process of field ionization is fully understood, the technique of attosecond tunnelling will provide direct time-domain insight into a wide range of multi-electron dynamics and electron–electron interactions, ultimately with a resolution approaching the atomic unit of time (~24 as). We demonstrate this potential by probing shake-up and Auger cascade processes with subfemtosecond resolution.

Attosecond probing of electron dynamics

Figure 2 illustrates different options for attosecond sampling of electronic motion in atoms or molecules. A subfemtosecond extreme ultraviolet (XUV) pulse triggers the motion by exciting a valence or core electron (Fig. 2a, b). The unfolding excitation and relaxation processes (Fig. 2) could, in principle, be probed by a delayed replica of the pulse. However, the low flux of currently available subfemtosecond XUV pulses and the low two-photon transition probabilities in the XUV and X-ray regimes have thwarted this straightforward extension of conventional pump–probe techniques into the XUV–X-ray spectral range.

A few-cycle wave of visible or near-infrared (NIR) light with controlled waveform[7] in combination with a highly nonlinear process may replace the subfemtosecond pulse either in probing or starting electron dynamics. This was first demonstrated by attosecond streaking[8,9]: the strong-field interaction of a few-cycle light wave with free electrons released by a subfemtosecond XUV excitation pulse results in broadening and shifting of their final momentum distribution. Recording the streaked spectra of the emitted photo- and Auger electrons versus delay between the XUV pump and the few-cycle laser probe allowed us to retrieve the XUV pulse[10,11] and the laser field[12] as well as the inner-atomic relaxation dynamics[13] with subfemtosecond resolution.

Here, we demonstrate that nonlinear interaction of the same light wave with bound electrons ionizes in subfemtosecond steps and hence offers a means of probing intra-atomic and intra-molecular electron dynamics—including when no free electrons are released—by means of attosecond tunnelling. This approach relies on the fact that energetic photo-excitation as well as subsequent rearrangement

[1]Department für Physik, Ludwig-Maximilians-Universität, Am Coulombwall 1, [2]Max-Planck-Institut für Quantenoptik, Hans-Kopfermann-Strasse 1, D-85748 Garching, Germany. [3]Fakultät für Physik, Universität Bielefeld, Universitätsstrasse 25, D-33615 Bielefeld, Germany. [4]Technische Universität Wien, Gusshausstrasse 27, A-1040 Vienna, Austria. [5]FOM-Instituut voor Atoom- en Molecuulfysica (AMOLF), Kruislaan 407, 1098 SJ, Amsterdam, The Netherlands. [6]Institute of Nuclear Physics, Moscow State University, Moscow 119992, Russia. [7]Institut für Experimentalphysik, Universität Hamburg, Luruper Chaussee 149, D-22671 Hamburg, Germany.

ARTICLES NATURE|Vol 446|5 April 2007

via Auger decay is accompanied by transitions to unoccupied orbitals via 'shake-up' (Fig. 2d–f)[14–16]. Here, we use the term 'shake-up' in a broad sense, standing for all possible processes populating excited ionic states (including instantaneous and non-instantaneous ones), henceforth referred to as shake-up states (represented by levels 1, 2 and 3 in the example of Fig. 2). These populations can be probed via optical field ionization by a strong, few-cycle NIR pulse of variable delay Δt, with respect to the subfemtosecond XUV excitation (Fig. 1c) by measuring the number of ions resulting from the XUV pump–NIR probe exposure as a function of Δt (Fig. 1d).

Shake-up usually populates several quantum states in the valence band, from which electrons can be freed by the NIR probe. Hence, the ion yield will constitute an integral signal, with contributions from all shake-up states up to a certain binding energy from which ionization is feasible for the intensity chosen. Population dynamics of individual states of significantly differing binding energy can be retrieved by pump–probe scans repeated at different NIR probe intensities and/or from the temporal separation of the depletion of the states in the same delay scan, as explained in Fig. 1d and demonstrated in Fig. 4. The ion yield constitutes an integral signal in a temporal sense, too. The shake-up states are exposed to the ionizing NIR field from the moment they have been populated until the end of the NIR pulse. The beginning of this time interval within the NIR pulse is adjusted with the delay Δt between the XUV pump and the NIR probe. As the delay is scanned from large negative values (NIR probe first) to large positive values (XUV pump first) the measured ion yield (Fig. 1d) starts increasing at $\Delta t < 0$ owing to ionization on the trailing edge of the NIR pulse and continues to increase with increasing Δt because the XUV pump shifts towards the peak of the ionizing NIR probe and the shake-up states are exposed to ever higher NIR probe intensities.

In the absence of Auger decay, shake-up excitation results from photoionization only (Fig. 2d). The time-dependent ionization dynamics sketched in Fig. 1d can then be traced by measuring the yield of doubly charged ions as a function of the delay between the XUV pump and the sampling NIR light field. The temporal ionization gradients sketched in Fig. 1d incorporate the finite duration of the XUV excitation, a possible delayed response of shake-up and the tunnelling dynamics. With a sufficiently rapid ($\ll$1 fs) excitation, these measurements can thus provide direct insight into the temporal evolution of shake-up (in the presence of a strong optical field) and light-induced tunnelling. In what follows, first we prove this concept by exposing neon atoms to our subfemtosecond XUV pump and few-cycle NIR probe pulses (Fig. 3) and measuring the yield of the product of the pump–probe exposure, Ne^{2+}, versus Δt (Fig. 4). Then we extend the approach to probing electrons shaken up in xenon atoms during Auger decay. Our primary observables in this case are higher charged states, Xe^{N+} (for $N > 2$). Measuring their yield versus Δt displays the course of an Auger cascade.

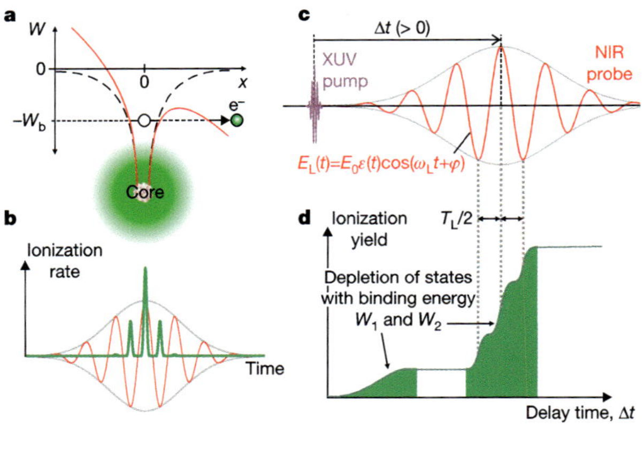

Figure 1 | Strong-field ionization and pump-probe setting for its real-time observation. a, Exposing an atom to a strong NIR laser field will result in a modified potential (solid curve) composed of the Coulomb potential (dashed curve) and the time-dependent effective potential of the laser pulse. The laser is polarized along the x direction and W_b is the binding energy of the electron. At sufficiently high laser field strengths the atomic binding potential is suppressed to a small barrier in the x or $-x$ direction for the maxima and minima of the laser electric field, respectively, allowing optical tunnelling to become the dominant ionization mechanism. **b,** The highly nonlinear dependence of the tunnelling rate on the width of the potential barrier confines ionization to time intervals of very short duration near the field oscillation maxima. In the electric field of a few-cycle NIR laser pulse (thin line) ionization is predicted to be restricted to several subfemtosecond intervals (thick line). **c,** Concept for tracing optical-field ionization: a subfemtosecond XUV pulse generates ions in excited (shake-up) states, from which a time-delayed NIR few-cycle probe pulse liberates electrons to produce doubly charged ions. A delay $\Delta t = 0$ is defined as the coincidence of the peaks of the envelopes of the NIR and XUV pulse. $\Delta t > 0$ implies that the peak of the XUV pulse envelope precedes that of the NIR pulse. **d,** Yield of doubly charged ions versus delay Δt between the XUV pump and the NIR probe, as predicted qualitatively for the case of electrons being prepared in two shake-up states of significantly differing binding energy (W_1 and $W_2 \gg W_1$) for liberation via strong-field ionization (see text for discussion). The state of low binding energy is depleted at low intensities, where multi-photon ionization dominates and hence sub-cycle structure in the ionization dynamics is absent. By contrast, the state of high binding energy is emptied by tunnelling, resulting in pronounced sub-half-cycle steps in the ionization profile.

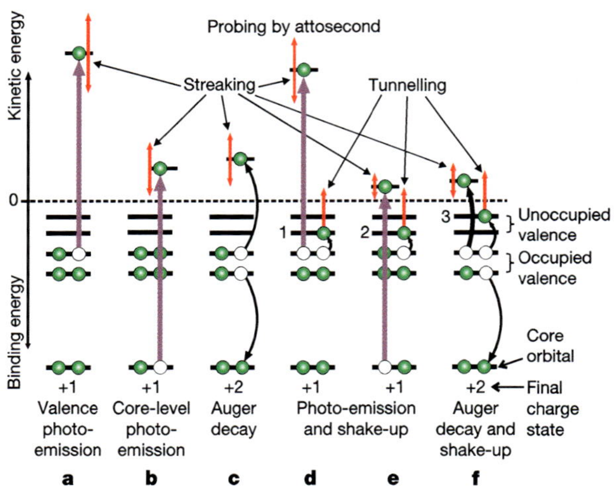

Figure 2 | Probing electron dynamics in atoms, molecules or solids with attosecond sampling techniques. A subfemtosecond XUV pulse triggers the motion by inducing valence (process **a**) or core photoelectron emission (process **b**). The temporal evolution of photo- and Auger electron emission (process **c**) can be probed via attosecond streaking to retrieve the XUV pump pulse or the sampling NIR field and trace inner-shell relaxation dynamics. XUV photoexcitation as well as subsequent Auger decay processes are usually accompanied by shake-up of another electron to a previously unoccupied level (processes **d**, **e** and **f**). In this case the liberated electrons will be ejected at a reduced kinetic energy compared to the cases without shake-up processes **a**, **b** and **c**. The difference in energy is used for shaking up bound electrons (represented by curved black arrows to levels 1, 2 and 3). For sufficiently strong probing laser fields, the shake-up electrons can be liberated by tunnelling ionization. The temporal evolution of the tunnelling current will provide information about the inner-atomic electron dynamics that populates and/or depopulates the interrogated shake-up states and the duration of the process that have populated the levels on atto- and femtosecond timescales. Note that the final charge state given in the figure is increased by attosecond tunnelling, while it remains unchanged in the case of attosecond streaking. The observable for streaking is the momentum distribution of liberated electrons, whereas in tunnelling it is the number of ions in different charge states.

NATURE|Vol 446|5 April 2007 ARTICLES

Experimental set-up

For a detailed description of the attosecond pump–probe apparatus, see the Supplementary Information. Briefly, the XUV pump originates from high-harmonic generation in a neon gas jet exposed to 300-μJ, 750-nm waveform-controlled laser pulses with a duration of $\tau_L \approx (5.5 \pm 0.5)$ fs, at a repetition rate of 3 kHz. The collinear, linearly polarized XUV and NIR light beams are passed through several filters and reflected by a concentric double-mirror arrangement (Mo/Si multilayer: ~ 9 eV bandwidth at photon energy of 91 eV). The mirror introduces a variable delay between the XUV and the NIR fields, isolates a single or twin XUV pulse (depending on the carrier-envelope phase of the laser pulse) of $\tau_X \approx 250$ as duration[11] by filtering the cut-off part of the harmonic emission spectrum and focuses both beams into a jet of atoms under scrutiny. The absolute delay between the XUV and the NIR signals is determined with an accuracy of better than ±0.5 fs; for details see Supplementary Information. Ions created in the common focus of the two beams are detected by a time-of-flight ion spectrometer (reflectron) that combines high mass resolution ($\Delta m/m \approx 10^{-3}$) with the capability of analysing particles within a micrometre-scale detection volume[17–19]. The target and background pressures were ~10^{-2} and ~10^{-8} mbar, respectively.

Shake-up and tunnelling

To probe shake-up and light-induced electron tunneling, we ionized neon atoms with our subfemtosecond XUV pulse. Figure 3 shows the level structure and transitions relevant to our experiments. The core shell was not accessed by our XUV photons, and hence Auger decay is absent[20]. The threshold energies for single and double ionization from the outer shell of Ne are 21.56 and 62.53 eV, respectively[21]. The XUV photons produced Ne^{1+} and Ne^{2+} ions with a ratio of

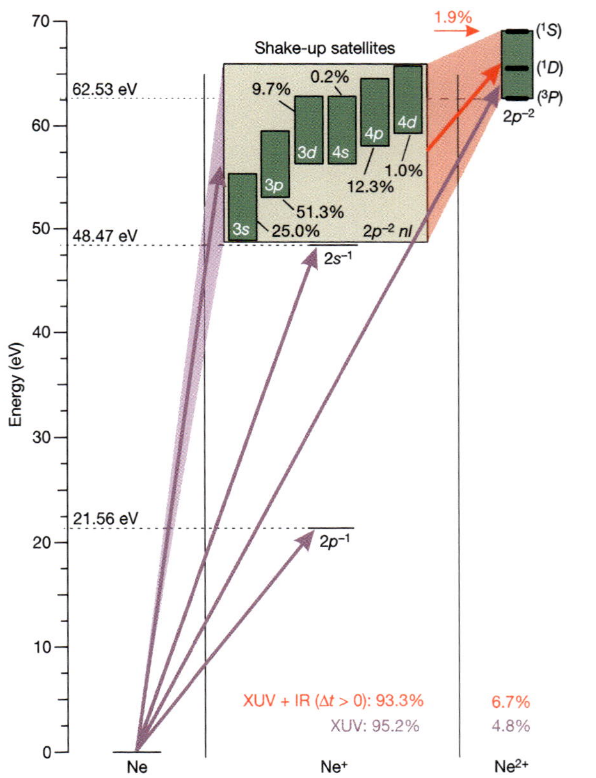

Figure 3 | Energy levels and transitions in Ne^{1+} and Ne^{2+} ions relevant to this study. The plotted levels represent energies required to ionize and possibly excite a neutral atom from its ground state. Absorption of photons of about 91-eV energy can produce singly or doubly charged ions with probabilities of 95.2% and 4.8%, respectively. Owing to shake-up a small fraction of the singly-charged ions is produced in $2p^{-2}nl$ configurations, where the probabilities of different channels are known from electron spectroscopy (see Supplementary Information). Each of these configurations, represented by a green box, consists of $2p^{-2}(^3P)nl$, $2p^{-2}(^1D)nl$, and $2p^{-2}(^1S)nl$ states (the atomic terms in parentheses describe electrons not involved in the interaction). Our few-cycle laser field following the XUV pulse ($\Delta t > 0$) can remove electrons from these shake-up states, thus increasing the probability of double ionization to 6.7%. Depending on the initial $2p^{-2}nl$ state, the doubly ionized ion is left in one of the ^3P, ^1D or ^1S states. On its own, the laser pulse produces only singly charged ions (in configuration $2p^{-1}$) above the detection limit.

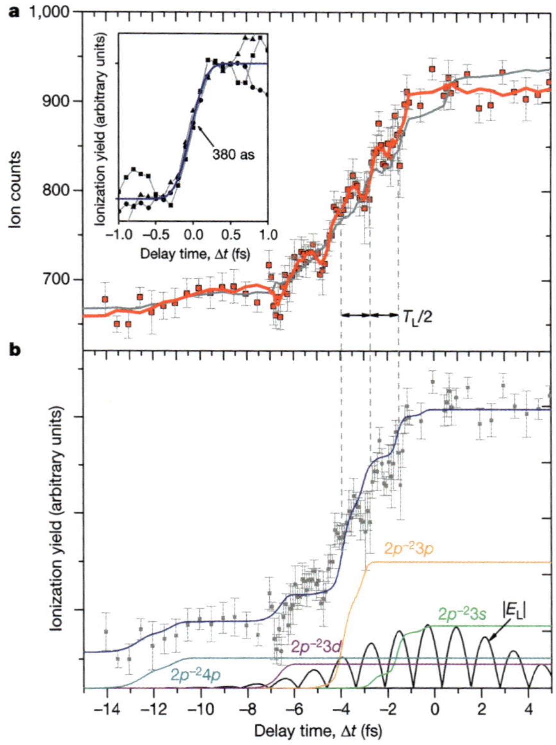

Figure 4 | Ne^{2+} ion yield versus delay between the subfemtoscond XUV pump and the few-cycle NIR probe: experiment and modelling. The peak intensity of the NIR probe was $(7 \pm 1) \times 10^{13}$ W cm^{-2}. **a**, The experimental data were acquired from six delay scans repeated under the same experimental conditions. The signal was accumulated for 3 s at each delay setting. The squares and the error bars show the average and the standard error (s.e.m.) of the results of the six measurements. The thick red line shows the average of five adjacent data points; the thin grey line shows the same as the thick line but recorded with NIR probe pulses of random carrier-envelope phase. In the inset, squares, triangles, and circles depict an ionization step extracted from three different measurements normalized to give the same change in the ionization yield. The solid line shows an error-function fit to the data yielding a rise time of 380 as (full-width at half-maximum, FWHM, of the gaussian function derived from the error function). The zero of delay is set arbitrarily to the centre of the tunnelling process. **b**, Simulation of the pump–probe experiment based on the nonadiabatic theory of tunnel ionization[6]. The thin coloured lines show the calculated fractional ionization yields contributed by electrons liberated from different shake-up states. The thick blue line depicts the overall ionization rate obtained by totalling the fractional rates (and by adding the result to the background measured at large negative delays). The simulations were carried out for a gaussian 250-as XUV pulse and a gaussian 5.5-fs laser pulse with a peak intensity of 7×10^{13} W cm^{-2}. The black solid curve represents the absolute value of the laser field. For discussion of the results, see text.

(19.7 ± 0.5):1 (with ~2,500 Ne^{1+} ions created per second), in good agreement with the results of synchrotron measurements[22]. A few per cent of the Ne^{1+} ions were promoted into $2p^{-2}nl$ (principal quantum number *n*: 3 or 4; quantum orbit *l*: *s*, *p* or *d*) configurations[23] (Fig. 3). These satellite states can only decay radiatively on a picosecond timescale.

Double ionization by the NIR field was not observed at the intensity level chosen. The XUV-generated Ne^{2+} yield was therefore not affected by the NIR probe for $\Delta t \ll -\tau_L$ but was significantly enhanced by the laser field for Δt approaching zero and becoming positive. The NIR-induced Ne^{2+} yield enhancement amounted to $(40 \pm 4)\%$ of the XUV-produced Ne^{2+} yield at a NIR peak intensity of $(7 \pm 1) \times 10^{13}\,W\,cm^{-2}$. The absence of this enhancement for $\Delta t \ll -\tau_L$ clearly indicates that the laser sets electrons free from states excited by the XUV pulse as sketched in Fig. 2d. The laser-induced change in the Ne^{2+} yield amounts to some 2% of the XUV-produced Ne^{1+} ions. This implies that a substantial fraction of the population of the $2p^{-2}nl$ shake-up satellites must have been depleted by field ionization.

Figure 4a shows the number of Ne^{2+} ions detected as a function of delay Δt between the XUV pump and NIR probe. Figure 4b compares the prediction of the Yudin–Ivanov theory[6] (lines) with the experimental data (squares). In our modelling the shake-up states were populated instantly during XUV photoionization (for more details, see Supplementary Information). The calculations are in reasonable agreement with our measurements and reveal how the different shake-up states are depleted sequentially by laser-field ionization. The signal starts increasing at large negative delays owing to depletion of the $2p^{-2}4p$ state (relative population ~12%) and the $2p^{-2}3d$ state (~10%) at NIR intensity levels reached some 10 and 6 fs after the peak of the NIR probe ($\Delta t \sim -10$ and -6 fs), respectively. A more dramatic increase in the Ne^{2+} yield is observed as the delay approaches zero, as a consequence of the depletion of the most highly populated $2p^{-2}3p$ (~50%) and $2p^{-2}3s$ (~25%) states. In spite of their relatively high binding energy (~10 and 13 eV, respectively), these states are also depleted before zero delay, that is, by the field oscillation cycles comprised in the trailing edge of the pulse, leaving no room for increasing the Ne^{2+} yield with increasing Δt beyond 0. This main contribution to the Ne^{2+} yield emerges within approximately one and a half wave cycles of the NIR field, $\sim(3/2)T_L = 3\pi/\omega_L$, in several sharp steps that are spaced by $\sim T_L/2$; this clearly shows that field-induced tunnelling is the main cause of the observed increase in the Ne^{2+} yield. This conclusion is also supported by the disappearance of the steps in a pump–probe scan performed with a randomly varying carrier-envelope phase of the NIR probe pulses (grey line in Fig. 4a).

The main shake-up population (residing in the $2p^{-2}3p$ and $2p^{-2}3s$ states) is depleted at NIR intensities corresponding to a Keldysh parameter γ of the order of three. Hence, our experiment verifies not only the existence of light-field-induced tunnelling, as predicted by Keldysh some four decades ago[1], but also confirms the dominant role of this ionization mechanism up to γ values substantially exceeding 1, as predicted recently by Yudin and Ivanov[6]. This conclusion is also backed by numerical solutions of the time-dependent Schrödinger equation (see Supplementary Information).

The steepness of the ionization steps and the dips preceding them in the measured data are not well reproduced by our model, which neglects the influence of electron–electron interactions and that of the strong NIR field on the XUV-induced transitions populating the shake-up states. Recent work[24] and our TDSE simulations (see Supplementary Information) indicate that the influence of the strong laser field on the XUV excitation process may (at least partially) be responsible for this discrepancy.

We feel that these experiments afford profound insight into fundamental electronic processes such as tunnelling and shake-up by contrasting theoretical models with time-domain data. To exploit this potential both (1) accurate models of shake-up in the presence of a strong laser field need to be developed and (2) the temporal resolution needs to be improved by using shorter XUV pulses[25] and improving the signal-to-noise ratio as well as the accuracy of determining the zero of delay. These advances will allow determination of the attosecond temporal evolution of the light-field-induced tunnelling current and they will provide deep insight into the nature of the electron–electron interactions responsible for shake-up.

Once models for shake-up and tunnelling have been tested and verified with attosecond precision, the technique of attosecond tunnelling will provide direct time-domain insight into a wide range of multi-electron dynamics inside atoms and molecules by probing the transient population of excited valence states while these dynamics are unfolding. With improved signal-to-noise and XUV pulse duration, the temporal resolution may potentially approach the atomic unit of time (~24 as). At present, the observed rise time of the Ne^{2+} yield of less than 400 as (which sets a corresponding upper limit on the time it takes the excited electronic states to become populated during XUV ionization and on tunneling; see inset in Fig. 4a) dictates the temporal resolution of our pump–probe approach. In the next section we demonstrate its applicability to probing intra-atomic multi-electron dynamics in real time.

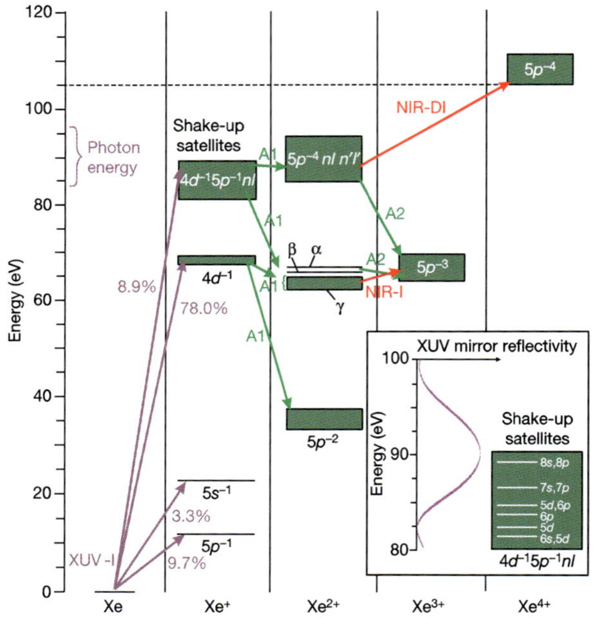

Figure 5 | Energy levels and transitions in xenon ions relevant to the current study. The relative population of states in Xe^{1+} is given for an XUV excitation energy of 90 eV from ref. 30. The XUV light preferably ionizes from the $4d^{-1}$ shell, creating an inner-shell vacancy. A subsequent Auger process (green arrows labelled A1) will follow with 99% probability[26] and is predominantly decaying to Xe^{2+} in configuration $5p^{-2}$. Some of the populated states, α and β (presumably $5s^{-1}5p^{-2}\,7p$ states[28,31]) can further decay to Xe^{3+} via a second Auger process (green arrow labelled A2), leading to triply charged ions. The A1 process also populates states below the threshold for Xe^{3+}. These are denoted by γ and represent $5s^{-1}5p^{-2}\,6p$ as well as $5p^{-3}nl$ configurations. The laser pulse can ionize these states (red arrow labelled NIR-I). Furthermore, a series of $4d^{-1}$ shake-up satellites in Xe^{1+} is also populated—with a small probability—by the XUV pulse. The inset shows the possible configurations[29] together with the XUV mirror reflectivity determining the XUV excitation spectrum. The satellites mainly decay to the α, β, γ states, with a small fraction of the Xe^{2+} population ending up in $5p^{-4}nln'l'$ configurations. These states are short-lived and decay via the A2 process. Before this occurs, the $nln'l'$ electrons can be liberated by the laser field to yield Xe^{4+} in the $5p^{-4}$ state (red arrow labelled NIR-DI).

NATURE|Vol 446|5 April 2007 ARTICLES

Multi-electron relaxation

As a first application of the technique of attosecond tunnelling, we captured Auger cascade xenon atoms following excitation by a sub-femtosecond XUV pulse. Figure 5 sketches the relevant energy levels and transitions. Energy-resolved synchrotron measurements have revealed that (1) the 91-eV XUV pulse will preferably liberate electrons from the $4d$ orbital[22], (2) the vacancy decays by subsequent single (A1 in Fig. 5) and cascaded (A1 and A2 in Fig. 5) Auger processes, leading to Xe^{2+} and Xe^{3+}, respectively[26], and (3) the lifetimes of the $4d_{3/2}$ and $4d_{5/2}$ holes are 6.3 ± 0.2 and 5.9 ± 0.2 fs, respectively[27]. These time-integral measurements have hitherto been able to set only a lower limit of 23 fs for the time constant of A2 (ref. 27).

To trace this dynamics in real time, we simultaneously recorded the number of Xe ions emerging in different charged states as a function of Δt. At the laser intensity of $(7 \pm 1) \times 10^{13}$ W cm^{-2} used in this experiment, the XUV-induced Xe^{1+} yield is buried in laser-generated background, preventing delay-dependent effects from coming to light in the Xe^{1+} signal. This is not the case for higher charged states. With increasing delay, rapid exponential increase was observed in the Xe^{3+} signal near $\Delta t = 0$ (Fig. 6b), concurrent with a significant decrease of the Xe^{2+} yield. The background in the Xe^{3+} signal arises from the A1–A2 Auger cascade discussed above. From Fig. 5 we infer that the laser-induced increase in the Xe^{3+} yield is due to ionization from the γ-states in Xe^{2+}, which cannot spontaneously decay into Xe^{3+}. This NIR-probe-induced transition (denoted by NIR-I in Fig. 5) yields Xe^{3+} in configuration $5p^{-3}$. Because these are populated by the first Auger process, A1, the exponential increase in the Xe^{3+} signal is the convolution of the A1 decay and the NIR-induced ionization process. No decrease of the enhanced Xe^{3+} yield was observed up to our maximum delay of 300 fs, indicating that the lifetime of the γ states was longer than 1 ps.

Charge-states higher than Xe^{3+} cannot be created with the XUV pulse alone, because the XUV photon energy is below the threshold for Xe^{4+} production (~105 eV). However, with the probe switched on, we did observe Xe^{4+} ions for $\Delta t > -\tau_L$ (Fig. 6a). The Xe^{4+} signal first grows within a few femtoseconds, followed by a longer decay. The Xe^{4+} ions are created by NIR-induced double ionization (denoted by NIR-DI in Fig. 5) from the intermediate doubly excited $5p^{-4}nln'l'$ (Fig. 5) and/or singly excited $5s^{-2}5p^{-1}nl$ (not shown in Fig. 5) states of Xe^{2+}. These states are populated by A1 from the satellites $4d^{-1}5p^{-1}nl$ of the $4d^{-1}$ state upon emission of low-energy electrons[28,29] and emptied by A2 to states of larger binding energy in Xe^{3+}, which cannot be reached by the NIR probe.

From these results we conclude that the exponential rise and decay of the Xe^{4+} signal in Fig. 6a reveal the evolution of the A1 and A2 Auger decays, respectively. The laser-induced double ionization may be either sequential or non-sequential (with the second step induced by recollision of the first electron). In either case, in the first step a binding energy similar to that of the most highly populated shake-up states in the neon experiment must be overcome by tunnel ionization. Hence, the probing (ionization) process may be assumed to be the same as measured in the neon experiment, see Fig. 4a. With this sampling function, fitting the result of a simple rate-equation analysis including the transitions A1 and A2 to the Xe^{4+} data shown in Fig. 6 yields decay times of $\tau_{A1} = 6.0 \pm 0.7$ fs and $\tau_{A2} = 30.8 \pm 1.4$ fs for the A1 and A2 Auger processes, respectively. Both time constants are in accordance with the results of energy-resolved measurements[27].

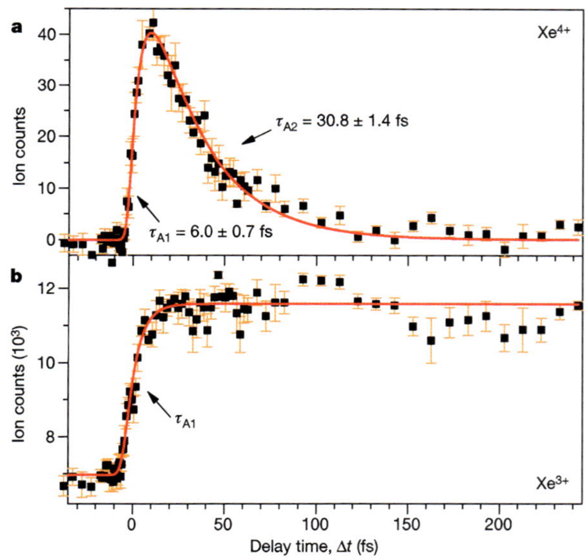

Figure 6 | Xe^{4+} and Xe^{3+} ion yields versus delay between the subfemtosecond XUV pump and the few-cycle NIR probe. The data have been compiled from the results of five delay scans repeated under the same experimental conditions. The signals were accumulated for 20 s at each delay setting. The squares and the error bars show the average and the s.e.m. of the results of these measurements. **a**, Assuming a sampling function identical to the ionization profile measured in the neon experiment (see solid line in Fig. 4a), a double exponential fit (see Supplementary Information) to the measured Xe^{4+} yield versus Δt (solid line) yields the Auger decay times $\tau_{A1} = 6.0 \pm 0.7$ fs and $\tau_{A2} = 30.8 \pm 1.4$ fs. **b**, With the temporal evolution of the A1 process acquired from the Xe^{4+} data and using the fact that the rise of the Xe^{3+} yield with τ comprises both A1 and the laser-induced ionization process, we can obtain the temporal evolution of the laser-induced transition from that of the Xe^{3+} yield. The exponential rise from about 7,000 to 11,500 counts is consistent with a laser-induced ionization time of 5.8 ± 2.5 fs (FWHM). This is comparable to τ_L, indicating that low-order, one-photon and/or two-photon transitions may promote electrons from the γ states of Xe^{2+} to the $5p^{-3}$ states of Xe^{3+} (see Fig. 5). Note that the fluctuations in the Xe^{3+} signal can be largely accounted for by XUV intensity variations, which can be efficiently eliminated by normalization to, for example, the Xe^{2+} ion yield (see Supplementary Information).

Conclusions and outlook

We have reported the observation of light-induced electron tunnelling from atoms in real time. Electrons are found to escape from their atomic binding potential within several subfemtosecond time intervals near the oscillation peaks of the ionizing few-cycle near-infrared laser field. Our results are in good agreement with the predictions of the theory Keldysh put forward four decades ago. The observed sub-femtosecond ionization steps provide a powerful means of probing the transient population of short-lived valence electronic states in excited atoms or molecules, offering direct, time-domain access to a wide range of multi-electron dynamics unfolding on an attosecond to femtosecond timescale. Proof-of-principle attosecond tunnelling experiments in neon and xenon demonstrate this potential. Simultaneous implementation of attosecond tunnelling and attosecond streaking spectroscopy along with scaling of the techniques to higher photon energies and shorter X-ray pulse durations will provide unprecedented insight into the transient electronic states of matter.

Received 2 November 2006; accepted 26 January 2007.

1. Keldysh, L. V. Ionization in the field of a strong electromagnetic wave. *Sov. Phys. JETP* **20**, 1307–1314 (1965).
2. Faisal, F. H. M. Multiple absorption of laser photons by atoms. *J. Phys. B* **6**, L89–L92 (1973).
3. Reiss, H. R. Effect of an intense electromagnetic field on a weakly bound system. *Phys. Rev. A* **22**, 1786–1813 (1980).
4. Brabec, T. & Krausz, F. Intense few-cycle laser fields: frontiers of nonlinear optics. *Rev. Mod. Phys.* **72**, 545–591 (2000).
5. Scrinzi, A., Geissler, M. & Brabec, T. Ionization above the coulomb barrier. *Phys. Rev. Lett.* **83**, 706–709 (1999).
6. Yudin, G. L. & Ivanov, M. Yu. Nonadiabatic tunnel ionization: looking inside a laser cycle. *Phys. Rev. A* **64**, 013409 (2001).

7. Baltuska, A. *et al.* Attosecond control of electronic processes by intense light fields. *Nature* **421**, 611–615 (2003).
8. Itatani, J. *et al.* Attosecond streak camera. *Phys. Rev. Lett.* **88**, 173903 (2002).
9. Kitzler, M., Milosevic, N., Scrinzi, A., Krausz, F. & Brabec, T. Quantum theory of attosecond XUV pulse measurement by laser-dressed photoionization. *Phys. Rev. Lett.* **88**, 173904 (2002).
10. Hentschel, M. *et al.* Attosecond metrology. *Nature* **414**, 509–513 (2001).
11. Kienberger, R. *et al.* Atomic transient recorder. *Nature* **427**, 817–821 (2004).
12. Goulielmakis, E. *et al.* Direct measurement of light waves. *Science* **305**, 1267–1269 (2004).
13. Drescher, M. *et al.* Time-resolved atomic inner-shell spectroscopy. *Nature* **419**, 783–787 (2002).
14. Svensson, S., Eriksson, B., Martensson, N., Wendin, G. & Gelius, U. Electron shake-up and correlation satellites and continuum shake-off distributions in x-ray photoelectron spectra of the rare gas atoms. *J. Electron Spectrosc. Related Phenomena* **47**, 327–384 (1988).
15. Aksela, H., Aksela, S. & Kabachnik, N. Resonant and nonresonant Auger recombination. In *VUV and Soft X-Ray Photoionization* (eds Becker, U. & Shirley, D. A.) 401–440 (Plenum, New York, 1996).
16. Istomin, A. Y., Manakov, N. L. & Starace, A. F. Perturbative analysis of the triply differential cross section and circular dichroism in photo-double-ionization of He. *Phys. Rev. A* **69**, 032713 (2004).
17. Schröder, H., Wagner, M., Kaesdorf, S. & Kompa, K. L. Surface-analysis by laser ionization. *Ber. Bunsenges. Phys. Chem.* **97**, 1688–1692 (1993).
18. Wagner, M. & Schröder, H. A novel 4 grid ion reflector for saturation of laser multiphoton ionization yields in a time-of-flight mass-spectrometer. *Int. J. Mass Spectrom.* **128**, 31–45 (1993).
19. Witzel, B., Schröder, H., Kaesdorf, S. & Kompa, K. L. Exact determination of spatially resolved ion concentrations in focused laser beams. *Int. J. Mass Spectrom.* **172**, 229–238 (1998).
20. Larkins, F. P. Charge state dependence of x-ray and Auger electron emission spectra for rare-gas atoms—II. The neon atom. *J. Phys. B* **4**, 14–19 (1971).
21. National Institute of Standards and Technology *Physical Reference Data* ⟨http://physics.nist.gov/PhysRefData/⟩ (1994).
22. Holland, D. M. P., Codling, K., West, J. B. & Marr, G. V. Multiple photoionization in the rare gases from threshold to 280 eV. *J. Phys. B* **12**, 2465–2484 (1979).
23. Becker, U. & Shirley, D. A. Partial Cross Sections and Angular Distributions. In *VUV and Soft X-Ray Photoionization* (eds Becker, U. & Shirley, D. A.) 135–173 (Plenum, New York, 1996).
24. Smirnova, O., Spanner, M. & Ivanov, M. Y. Coulomb and polarization effects in laser-assisted XUV ionization. *J. Phys. B* **39**, 323–339 (2006).
25. Sansone, G. *et al.* Isolated single-cycle attosecond pulses. *Science* **314**, 443–446 (2006).
26. Kämmerling, B., Krässig, B. & Schmidt, V. Direct measurement for the decay probabilities of $4d_j$ hole states in xenon by means of photoelectron-ion coincidences. *J. Phys. B* **25**, 3621–3629 (1992).
27. Penent, F., Palaudoux, J., Lablanquie, P. & Andric, L. Multielectron spectroscopy: the xenon 4d hole double Auger decay. *Phys. Rev. Lett.* **95**, 083002 (2005).
28. Lablanquie, P. *et al.* Photoemission of threshold electrons in the vicinity of the xenon 4d hole: dynamics of Auger decay. *J. Phys. B* **35**, 3265–3295 (2002).
29. Hayaishi, T. *et al.* Manifestation of Kr 3d and Xe 4d conjugate shake-up satellites in threshold-electron spectra. *Phys. Rev. A.* **44**, R2771–R2774 (1991).
30. Becker, U. *et al.* Subshell photoionization of Xe between 40 and 1000 eV. *Phys. Rev. A* **39**, 3902–3911 (1989).
31. Viefhaus, J. *et al.* Auger cascades versus direct double Auger: relaxation processes following photoionization of the Kr 3d and Xe 4d, 3d inner shells. *J. Phys. B* **38**, 3885–3903 (2005).

Supplementary Information is linked to the online version of the paper at www.nature.com/nature.

Acknowledgements We thank A. F. Starace for discussions. We are grateful for financial support from the Volkswagenstiftung (Germany), the Marie Curie Research Training Network XTRA, Laserlab Europe, and a Marie Curie Intra-European Fellowship. F.K. acknowledges support from the FWF (Austria). The research of M.F.K. and M.J.J.V. is part of the research programme of the Stichting voor Fundamenteel Onderzoek der Materie, which is financially supported by the Nederlandse Organisatie voor Wetenschappelijk Onderzoek. This research was supported by the cluster of excellence Munich Centre for Advanced Photonics (www.munich-photonics.de).

Author Contributions M.U., Th.U., M.S. and A.J.V. contributed equally to this work.

Author Information Reprints and permissions information is available at www.nature.com/reprints. The authors declare no competing financial interests. Correspondence and requests for materials should be addressed to M.U. (matthias.uiberacker@mpq.mpg.de) or F.K. (ferenc.krausz@mpq.mpg.de).

RESEARCH ARTICLES

Delay in Photoemission

M. Schultze,[1,2*] M. Fieß,[2] N. Karpowicz,[2] J. Gagnon,[1,2] M. Korbman,[2] M. Hofstetter,[1] S. Neppl,[3] A. L. Cavalieri,[2] Y. Komninos,[4] Th. Mercouris,[4] C. A. Nicolaides,[4] R. Pazourek,[5] S. Nagele,[5] J. Feist,[5,6] J. Burgdörfer,[5] A. M. Azzeer,[7] R. Ernstorfer,[3] R. Kienberger,[2,3] U. Kleineberg,[2] E. Goulielmakis,[2] F. Krausz,[1,2] V. S. Yakovlev[1,2*]

Photoemission from atoms is assumed to occur instantly in response to incident radiation and provides the basis for setting the zero of time in clocking atomic-scale electron motion. We used attosecond metrology to reveal a delay of 21 ± 5 attoseconds in the emission of electrons liberated from the 2*p* orbitals of neon atoms with respect to those released from the 2*s* orbital by the same 100–electron volt light pulse. Small differences in the timing of photoemission from different quantum states provide a probe for modeling many-electron dynamics. Theoretical models refined with the help of attosecond timing metrology may provide insight into electron correlations and allow the setting of the zero of time in atomic-scale chronoscopy with a precision of a few attoseconds.

The emission of an electron from an atom upon the absorption of an energetic photon (photoemission, or the photoelectric effect) is one of the most elementary quantum-mechanical phenomena (*1–3*). In the absence of resonances, the formation of the outgoing wave packet is usually presumed to instantly follow temporal variations of the incident light field. A delayed response, the possibility of which was recognized in the 1950s (*4*, *5*), would have several far-reaching consequences. If a single electron is set free when an atom absorbs a photon, it is, strictly speaking, not a single-electron process. Rather, it is the result of the correlated motion of all the electrons; hence, electron correlations generally affect the properties of the emitted photoelectron (*6*, *7*). As a consequence, comprehensive attosecond-scale temporal characterization of photoemission promises new insights into intra-atomic electron correlations and can bolster theoretical models of dynamic changes of atomic structure (*8*) and of correlated many-electron dynamics.

Furthermore, a delay in photoemission implies a change in the timing of ejection of the electron pulse with respect to the arrival of the photon pulse. This would compromise the accuracy of setting the zero of time for clocking microscopic processes on the atomic time scale. Capturing the outgoing electron with the controlled electric field of a laser pulse (*9*) via attosecond streaking (*10*, *11*) provides the only means of timing the arrival of a photon pulse with a precision comparable to the atomic unit of time. This timing is of fundamental importance because attosecond light pulses (*12–15*) presently constitute the most precise trigger for timing electronic processes. Hence the zero of time in atomic-scale chronoscopy, the time-resolved observation of electronic and concomitant processes, is to coincide with the arrival time of this photon pulse. Any unknown delay Δt in the emission of the photoelectron wave packet will result in an apparent shift of the origin of the time arrow (Fig. 1A). This introduces an unknown systematic error in clocking the microscopic motion under scrutiny, preventing us from revealing its details. Reliable absolute timing information would be essential, for example, in understanding the multielectron response of metals and semiconductors to sudden excitations (*16*).

Our measurements reveal a small delay time between the formation of electron wave packets originating from different atomic orbitals. The temporal properties of both wave packets were recorded by means of attosecond streaking (*10*), and the measured spectrograms were found to be shifted with respect to each other along the pump-probe delay axis (henceforth referred to as temporally shifted), as if one of the wave packets were emitted later than the other. Comprehensive quantum simulations show that a delay of one wave packet with respect to the other indeed results in a temporal shift between spectrograms, even though the delay between wave packets and the shift between spectrograms may not be exactly equal to each other. The delay considered here is of a different physical origin than that previously observed in photoemission from a solid (*17*), where a temporal shift was attributed to transport effects that occur before an electron reaches the surface (*18*, *19*).

Attosecond streaking measurements on Ne. Ne was chosen for our study for two reasons. First, its small polarizability precludes the possibility of an artificial temporal shift of the spectrogram due to the Stark effect (*20*). Second, the relatively high photoionization cross-section (*21*) benefits the signal-to-noise ratio (*S*/*N*) of attosecond streaking, which is of crucial importance for achieving the required timing accuracy.

Even the most advanced measurements in photoelectron spectroscopy can only provide energy and angle-resolved probability distributions (*22*, *23*). Probing the photoionization cross-section corresponds to measuring the absolute values of complex-valued matrix elements that quantum mechanics assigns to transitions from a bound atomic state to states where a photoelectron propagates toward the detector. The quantum phase of the matrix element, the energy derivative of which defines the emission timing (*5*), has largely remained experimentally inaccessible (*24*). Our work demonstrates that the energy derivative of the quantum phase, which is closely related to the Wigner-Smith time delay (*4*, *25*), can now be accessed by attosecond streaking.

Pushing attosecond timing metrology into the sub–10-as regime has been enabled by waveform-controlled near–single-cycle pulses (duration ~3.3 fs) of near-infrared (NIR, carrier wavelength ~750 nm) light (*26*). These pulses permit the generation of isolated sub–100-as extreme ultraviolet (XUV) pulses via high-order harmonic generation (*12*, *14*). More importantly for the present study, they increase the XUV photon yield by two orders of magnitude, resulting in an attosecond photon flux (at ~100 eV) in excess of 10^{11} photons/s (*14*). The resulting more than 10-fold improvement in *S*/*N* has advanced the timing accuracy of attosecond streaking to a fraction of the atomic unit of time.

High-order harmonic generation in Ne atoms driven by sub–4-fs NIR pulses yields robust XUV continua in the energy range of 100 to 140 eV, which are stable against substantial changes in the carrier-envelope phase (*14*). This bandwidth supports isolated pulses below 100 as in duration. Simultaneous temporal characterization of photoemission pathways with distinguishable final ionic states calls for recording streaking spectrograms that are well separated in energy and sets a limit on the XUV bandwidth. To fulfill this condition for the photoelectrons arising from the 2*p* and 2*s* subshells of Ne atoms, the spectrum of attosecond pulses is confined with a specially designed lanthanum/molybdenum multilayer XUV mirror and a 150-nm-thick metallic foil. In combination, they act as a bandpass filter centered at 106 eV, with a full width at half maximum of 14 eV, supporting a Fourier-limited XUV pulse duration well below 200 as.

Photoelectrons ejected from Ne atoms by 106-eV attosecond pulses and streaked by the electric field of near–single-cycle NIR pulses were recorded with a time-of-flight spectrometer versus the delay between the XUV and NIR pulses (Fig. 2A). The electron spectrometer

[1]Department für Physik, Ludwig-Maximilians-Universität, Am Coulombwall 1, D-85748 Garching, Germany. [2]Max-Planck-Institut für Quantenoptik, Hans-Kopfermann-Straße 1, D-85748 Garching, Germany. [3]Physik Department, Technische Universität München, James-Franck-Straße, D-85748 Garching, Germany. [4]Theoretical and Physical Chemistry Institute, National Hellenic Research Foundation, 48 Vassileos Constantinou Avenue, Athens 11635, Greece. [5]Institute for Theoretical Physics, Vienna University of Technology, Wiedner Hauptstraße 8-10, 1040 Vienna, Austria. [6]Institute for Theoretical Atomic, Molecular and Optical Physics (ITAMP), Harvard-Smithsonian Center for Astrophysics, Cambridge, MA 02138, USA. [7]Physics and Astronomy Department, King Saud University, Riyadh 11451, Saudi Arabia.

*To whom correspondence should be addressed. E-mail: Martin.Schultze@mpq.mpg.de (M.S.), Vladislav.Yakovlev@physik.uni-muenchen.de (V.S.Y.)

was equipped with an electrostatic lens, enhancing electron counts in the spectral range between 40 and 70 eV. Figure 2B shows the streaking spectrogram reconstructed by the use of an algorithm based on frequency-resolved optical gating (FROG) (*27–29*). The measured and calculated spectrograms show excellent agreement. Quantitative comparison of streaked spectral lineouts in Fig. 2D demonstrates the high fidelity with which the characteristics of the emitted electron wave packets can be retrieved (*30*) (Fig. 2C). The evaluated group delay versus frequency reveals that the emission of the electron wave packet from the *2p* state lags behind that originating from the *2s* state by approximately 20 as. The linear slope of the group delay representing an energy sweep of the emitted electrons potentially provides more insight into photoemission dynamics. We reproducibly observed a difference between the energy sweeps of the *2s* and *2p* wave packets on the order of several thousand attoseconds squared. Simulations show, however, that the retrieved energy sweep, in contrast to the average group delay, is sensitive to the electrostatic lens. Hence, the energy sweeps cannot be reliably inferred from the current measurements.

In addition to the FROG approach, we have developed a simplified procedure in order to extract the delay from a large set of measured attosecond streaking spectrograms [for details, see the caption of Fig. 3 and (*29*)]. Figure 3 summarizes the relative delay extracted from 40 measurements via this method. The result of this analysis agrees within the experimental error with the one obtained with the FROG technique (*29*). We found a clear correlation between the uncertainty of the evaluated delay and the fractional energy carried by a satellite XUV pulse accompanying the main attosecond pulse. Therefore, we selected a set of delay values evaluated from streaking spectrograms acquired with attosecond XUV pulses exhibiting less than 3% satellite pulse content (red diamonds, Fig. 3). These data result in a mean relative delay of 21 as between *2p* and *2s* photoelectrons, with a standard deviation of 5 as. In order to exclude any influence of shakeup states on the evaluated delay, the measurements were repeated using an XUV multilayer mirror with a bandwidth of 4.5 eV centered at ~121 eV (supporting Fourier-limited pulses of ~400 as) in another attosecond beamline. The narrower spectrum of these XUV pulses allowed us to spectrally separate the direct *2s* emission line from most of its much weaker high-order shakeup satellites. The results are also plotted in Fig. 3 (green circles) and found, within our error bars, to be in agreement with the ones achieved with sub–200-as pulses.

The measurements summarized in Fig. 3 were performed by applying streaking laser electric fields of different strength, varying by a factor of 6 in amplitude; that is, by a factor of 36 in intensity. In spite of this large dynamic range, the observed delay shows no dependence on the amplitude of the streaking field. The error

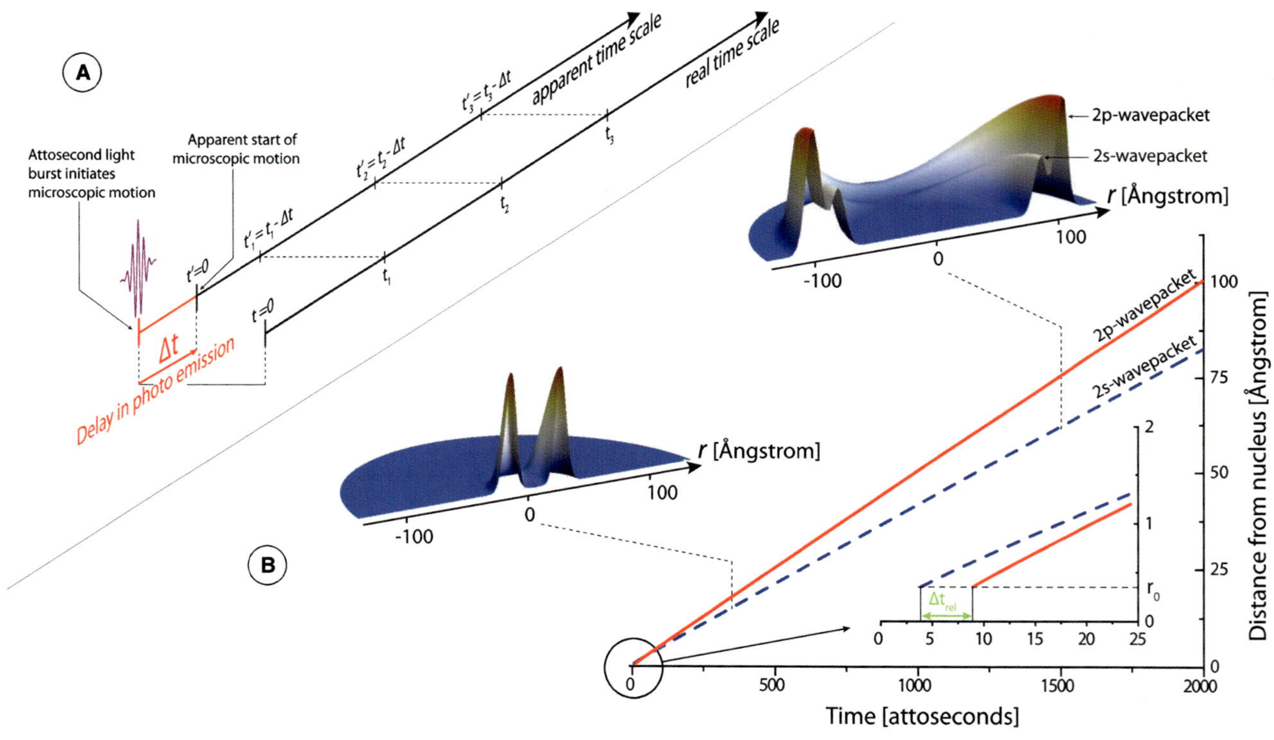

Fig. 1. Attosecond delay in photoemission and its consequences. (**A**) The "real" time scale begins at the maximum of the XUV pulse, whereas the "apparent" time scale in the measurement starts with the release of an electron wave packet and is temporally shifted by a possible delay Δt in photoemission. A delay between the arrival time of the attosecond XUV excitation pulse and the instant of emission would falsify the conclusions reached from measurements with an atomic chronoscope, which is triggered by the emission of an electron wave packet. A microscopic event that occurred at $t = t_{event}$ is indicated by this chronoscope to have apparently happened at $t' = t_{event} - \Delta t$, thereby tainting a comparison between theory and experiment with an unknown systematic error of Δt. (**B**) The surface plots show the spatial distribution of the photoelectron density around the atomic core at $t_1 = 300$ as and $t_2 = 1500$ as after the maximum of the XUV pulse, evaluated by solving the time-dependent Schrödinger equation with the aid of the state-specific expansion approach. As time progresses, the wave packets released from the *2s* and *2p* subshells become spatially separated because of their different velocities. Far from the nucleus, where the overlap with ionic orbitals is negligible, their motion can be described semi-classically. Therefore, knowing the average position and velocity of a wave packet that propagates toward the detector, we can illustrate a possible delay in its emission by tracing a classical electron trajectory back to the ion. The red solid and blue dashed lines show the classical trajectories of the *2p* and *2s* photoelectrons, respectively. The lines terminate at a distance $r_0 \approx 0.3$ Å, which is equal to the radius of the valence shell. At this distance, the trajectories behave as if they started with a relative delay $\Delta t_{rel} = 5$ as, which is in reasonable agreement with the value obtained by a more rigorous theoretical analysis.

RESEARCH ARTICLES

bars are reduced at higher intensities, primarily because of the increased spectral shift.

Theoretical discussion. First, we show that the measured delay of ~20 as cannot be explained by a delayed onset of streaking, which was the dominant effect in (*17*). The streaking NIR field may be significantly screened by bound electrons at small distances from the nucleus. After the absorption of an XUV photon, it takes the positive-energy electron a finite time to leave this screened volume, and this time interval may be different for electrons originating from different orbitals. However, for an atom, this difference cannot exceed a few attoseconds. The characteristic scales can be extracted from the classical trajectories shown in Fig. 1B. If we assume that the 2*s* and 2*p* electrons are set in motion at the same moment, their classical trajectories would acquire a relative delay of 20 as after traveling over 5 Å, whereas significant screening from the streaking field is limited to a distance of less than 1 Å from the nucleus. Furthermore, if screening played a dominant role, the faster 2*p* electrons would be exposed to the streaking field earlier than the slower 2*s* ones, whereas measurements and quantum simulations show that the slower electron is emitted first.

Now we turn our attention to the quantum-mechanical description. First of all, we need a definition for the photoemission delay. Consider a photoelectron wave function $|\psi(t)\rangle$ created by an XUV pulse centered at $t = 0$. The motion of the wave packet after photoionization is conveniently described in a basis of continuum states $|\varepsilon\rangle$, each of which has a well-defined energy ε and describes a wave that propagates in the direction of the detector. In this basis, each probability amplitude has a simple time dependence: $\langle\varepsilon|\psi(t)\rangle = \chi(\varepsilon)e^{-i\frac{\varepsilon}{\hbar}t}$, where the complex-valued function $\chi(\varepsilon)$ fully describes the properties of the wave packet. In this representation, a delay Δt in photoemission, shown as a shift of the electron's trajectory in Fig. 1B, adds $\frac{\varepsilon}{\hbar}\Delta t$ to the phase of $\chi(\varepsilon)$. It is therefore meaningful to define the group delay of the outgoing electron wave packet, in accordance with earlier work (*4*, *5*, *25*), as $\alpha(\varepsilon) = \hbar\frac{d}{d\varepsilon}\arg[\chi(\varepsilon)]$. Analyzing our simulations, we average $\alpha(\varepsilon)$over the bandwidth of the XUV pulse (*29*) and denote the result as $\bar{\alpha}$.

As the first and most important task, we validate the experimental methodology. Intuitively, one expects that a delay in the formation of a wave packet causes a corresponding temporal shift of the streaking spectrogram. This holds true

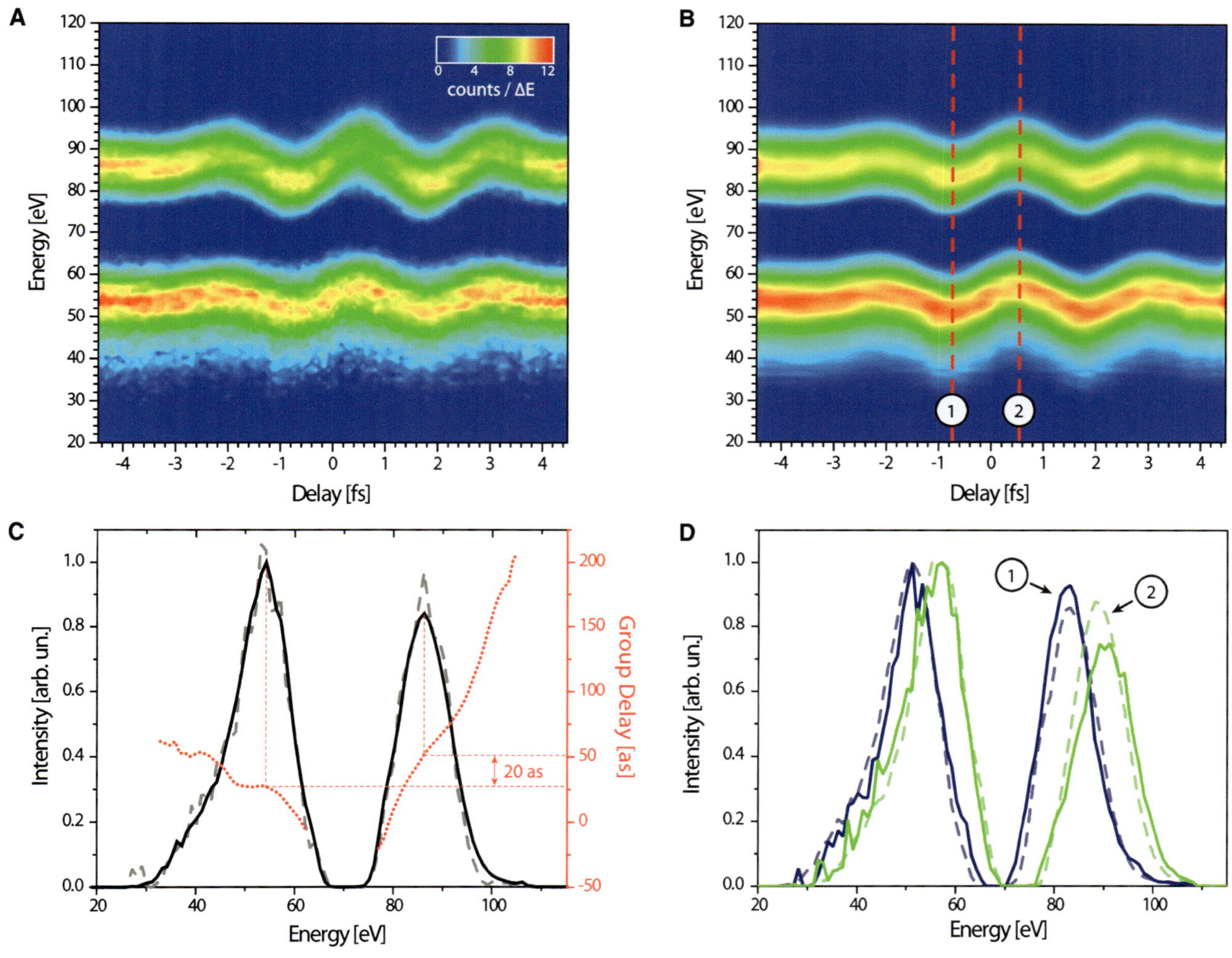

Fig. 2. Attosecond streaking spectrograms (**A** and **B**), evaluated photoelectron wave packets (**C**), and streaked spectra (**D**). The spectrograms in (A) are composed of a series of photoelectron energy spectra recorded by releasing 2*s* and 2*p* electrons from Ne with an attosecond XUV pulse in the presence of a strong NIR few-cycle laser field, as a function of the delay between the XUV and NIR fields. The spectrogram is processed with a FROG algorithm tailored for streaking measurements (*30*). (B) shows the spectrogram reconstructed by this algorithm. The retrieved 2*s* and 2*p* spectra, together with the respective group delays, are plotted in (C) (black solid line and red dotted line, respectively). The reconstructed energy spectra are in excellent agreement with the measured ones (gray dashed line). The average difference between the group delays corresponds to a 20-as retardation of the 2*p* emission with respect to the 2*s* emission. (D) compares reconstructed and measured streaked spectra at two delays, which exhibit the largest positive and negative shifts of the electron energy distribution.

within the Coulomb-Volkov approximation (CVA) (*29*). To quantify the accuracy of the CVA, the single-electron time-dependent Schrödinger equation in three spatial dimensions was numerically solved with an effective potential that modeled Ne (*29*). The analysis of wave packets yielded a spectrally averaged relative group delay of $\overline{\alpha}_{2p} - \overline{\alpha}_{2s} = 4.5$ as. Although the CVA yields the same value of the temporal shift between spectrograms, the numerical solution of the Schrödinger equation results in simulated spectrograms that are shifted with respect to each other by 6.8 as. The origin of this discrepancy lies in the fact that the photoelectron interacts with both the streaking field and the ion, resulting in a quantum motion that is not exactly described by known analytical approaches. Thus, for the current experimental parameters, the small deviations between the electron's exact motion and that modeled via the CVA give rise to a 2-as discrepancy in the relative delay.

Accepting this small discrepancy, many-electron models were applied to investigate the effects of electron correlation. As a first attempt, the multiconfigurational Hartree-Fock method was used to evaluate transition matrix elements from the ground state of Ne to states where the electron wave asymptotically propagated along the direction of the streaking NIR electric field. These matrix elements predict a relative delay of $\overline{\alpha}_{2p} - \overline{\alpha}_{2s} = 4.0$ as. The major drawback of this model is that it does not account for interchannel coupling (*6*). This deficiency was overcome by modeling the interaction with the XUV pulse using the state-specific expansion approach (*31*, *32*). This model accounts for electron correlations before and after photoionization and predicts a relative group delay of $\overline{\alpha}_{2p} - \overline{\alpha}_{2s} = 6.4$ as. Our modeling successfully predicts that the emission of 2*s* electrons precedes that of 2*p* electrons, but the computed relative delay is ~15 as (3 SD) smaller than the measured value.

So far, the theoretical discussion has focused on the relative delay between two photoemission channels, which can be acquired experimentally. Precise determination of the zero of time for allowing us to track the history of microscopic phenomena accurately (Fig. 1A) calls for precise knowledge of the delay between the XUV pulse and an outgoing electron wave packet (henceforth, absolute delay). This can only be inferred from theory. For multielectron systems, such as Ne, physical description of the discrepancies revealed by this work proved to be a challenge. The sensitive experimental test to which time-dependent many-electron models can now be subjected will benefit their development.

Meanwhile, it is possible to obtain reliable absolute emission times for He, with which truly ab initio simulations (*33*) can be carried out with the help of supercomputers. Such simulations were performed for the He ($1s^2$) ground state, and for direct ionization with a 100-eV photon, a 5-as temporal shift of the spectrogram was found. Such modeling will allow precise timing calibration of attosecond measurements, once sufficiently powerful attosecond sources will allow the recording of spectrograms for He with sufficiently good statistics in spite of its small photoionization cross-section.

Conclusions and outlook. Establishing the zero of time in atomic chronoscopy is currently tainted with an error of up to several tens of attoseconds. Because attosecond streaking can measure only relative delays between different photoemission channels, the knowledge of absolute delays relies on the predictions of thoroughly tested time-dependent multielectron models. Presently, only two-electron ab initio simulations provide this degree of reliability, but the low photoionization cross-section of He limits (because of low *S/N*) the timing accuracy. For more complex systems, phase-sensitive measurements of the photoelectron wave packets via attosecond streaking will put many-electron models of atomic photoionization to comprehensive, highly sensitive tests, which is a prerequisite for gradually improving them and gaining trust in their predictions. These developments will improve our understanding of subatomic electron correlations and will make the absolute timing precision of atomic chronoscopy approach the 1-as frontier.

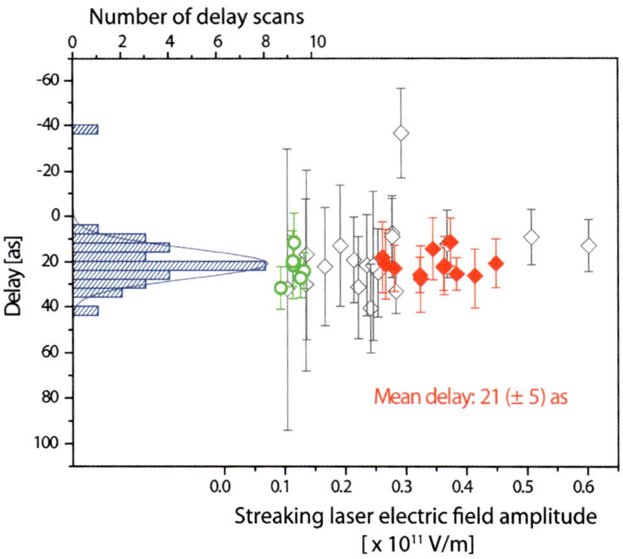

Fig. 3. The relative delay between photoemission from the 2*p* and 2*s* subshells of Ne atoms, induced by sub–200-as, near–100-eV XUV pulses. The depicted delays are extracted from measured attosecond streaking spectrograms by fitting a spectrogram, within the strong-field approximation, with parameterized NIR and XUV fields. Our optimization procedure matches the first derivatives along the time delay dimension of the measured and reconstructed spectrograms, thereby eliminating the influence of unstreaked background electrons [for details on the fitting algorithm, see (*29*)]. From the analysis of a set of spectrograms, the measured delays and associated retrieval uncertainties are plotted against the amplitude of the vector potential applied in the attosecond streak camera. Spectrograms measured in the presence of a satellite attosecond pulse were found to exhibit a less accurate retrieval of the delay value. When a subset of data (red diamonds) that represents scans with less than 3% satellite pulse content was evaluated, a mean delay value of 21 as with a standard deviation of ~5 as was found. The green circles represent the result of analyzing spectrograms recorded with an XUV pulse with narrower bandwidth in order to exclude the potential influence of shakeup states contributing to the electron kinetic energy spectrum.

References and Notes

1. H. Hertz, *Annalen Physik Chem.* **267**, 983 (1887).
2. W. Hallwachs, *Annalen Physik Chem.* **269**, 301 (1888).
3. A. Einstein, *Annalen Physik* **17**, 132 (1905).
4. E. P. Wigner, *Phys. Rev.* **98**, 145 (1955).
5. C. A. A. de Carvalho, H. M. Nussenzveig, *Phys. Rep.* **364**, 83 (2002).
6. A. F. Starace, in *Handbuch der Physik*, W. Mehlhorn, Ed. (Springer, Berlin, 1982), vol. 31.
7. S. T. Manson, *Radiat. Phys. Chem.* **75**, 2119 (2006).
8. M. Y. Ivanov, J. P. Marangos, *J. Mod. Opt.* **54**, 899 (2007).
9. A. Baltuška *et al.*, *Nature* **421**, 611 (2003).
10. R. Kienberger *et al.*, *Nature* **427**, 817 (2004).
11. M. Nisoli, G. Sansone, *Prog. Quantum Electron.* **33**, 17 (2009).
12. G. Sansone *et al.*, *Science* **314**, 443 (2006).
13. M. Schultze *et al.*, *N. J. Phys.* **9**, 243 (2007).
14. E. Goulielmakis *et al.*, *Science* **320**, 1614 (2008).
15. M. Hentschel *et al.*, *Nature* **414**, 509 (2001).
16. A. Borisov, D. Sánchez-Portal, R. Díez Muiño, P. M. Echenique, *Chem. Phys. Lett.* **387**, 95 (2004).
17. A. L. Cavalieri *et al.*, *Nature* **449**, 1029 (2007).
18. A. K. Kazansky, P. M. Echenique, *Phys. Rev. Lett.* **102**, 177401 (2009).
19. C. Lemell, B. Solleder, K. Tőkési, J. Burgdörfer, *Phys. Rev. A* **79**, 062901 (2009).
20. J. C. Baggesen, L. B. Madsen, *Phys. Rev. Lett.* **104**, 043602; and erratum, 209903 (2010).
21. U. Becker, D. A. Shirley, in *VUV and Soft X-Ray Photoionization*, U. Becker, D. A. Shirley, Eds. (Plenum, New York, 1997), chap. 5.
22. A. Rudenko *et al.*, *Phys. Rev. Lett.* **101**, 073003 (2008).
23. J. Mauritsson *et al.*, *Phys. Rev. Lett.* **100**, 073003 (2008).

24. The relative phases of interfering partial waves are accessible to measurement because they determine angular photoelectron distributions; still, this information is not sufficient to retrieve either the energy dependence of a phase in a particular photoionization channel or the relative phase between channels that do not interfere.
25. F. T. Smith, *Phys. Rev.* **118**, 349 (1960).
26. A. L. Cavalieri *et al.*, *N. J. Phys.* **9**, 242 (2007).
27. Y. Mairesse, F. Quéré, *Phys. Rev. A* **71**, 011401 (2005).
28. J. Gagnon, E. Goulielmakis, V. S. Yakovlev, *Appl. Phys. B* **92**, 25 (2008).
29. Information on technology and methods is available as supporting material on *Science* Online.
30. J. Gagnon, V. S. Yakovlev, *Opt. Express* **17**, 17678 (2009).
31. C. A. Nicolaides, Th. Mercouris, Y. Komninos, *J. Phys. At. Mol. Opt. Phys.* **35**, L271 (2002).
32. Th. Mercouris, Y. Komninos, C. A. Nicolaides, *Phys. Rev. A* **76**, 033417 (2007).
33. J. S. Parker, B. J. S. Doherty, K. J. Meharg, K. T. Taylor, *J. Phys. At. Mol. Opt. Phys.* **36**, L393 (2003).
34. Supported by the Max Planck Society and the Deutsche Forschungsgemeinschaft Cluster of Excellence: Munich Centre for Advanced Photonics (www.munich-photonics.de) and the King Saud University–Max-Planck-Institut für Quantenoptik collaboration. Supported in part by NSF through TeraGrid resources provided by the National Institute for Computational Sciences and Texas Advanced Computing Center under grant TG-PHY090031. R.P., S.N., and J.B. acknowledge support from the Austrian Science Fund (FWF) under grant FWF-SFB016. J.F. acknowledges support by NSF through a grant to ITAMP. E.G. acknowledges a Marie-Curie Reintegration grant (MERG-CT-2007-208643). R.K. acknowledges support from the Sofia Kovalevskaja award of the Alexander von Humboldt Foundation and an ERC starting grant. We thank A. Maquet and R. Taieb for fruitful discussions.

Supporting Online Material
www.sciencemag.org/cgi/content/full/328/5986/1658/DC1
SOM Text
Fig. S1
References

10 March 2010; accepted 10 May 2010
10.1126/science.1189401

Reconstituting Organ-Level Lung Functions on a Chip

Dongeun Huh,[1,2] Benjamin D. Matthews,[2,3] Akiko Mammoto,[2] Martín Montoya-Zavala,[1,2] Hong Yuan Hsin,[2] Donald E. Ingber[1,2,4]*

Here, we describe a biomimetic microsystem that reconstitutes the critical functional alveolar-capillary interface of the human lung. This bioinspired microdevice reproduces complex integrated organ-level responses to bacteria and inflammatory cytokines introduced into the alveolar space. In nanotoxicology studies, this lung mimic revealed that cyclic mechanical strain accentuates toxic and inflammatory responses of the lung to silica nanoparticles. Mechanical strain also enhances epithelial and endothelial uptake of nanoparticulates and stimulates their transport into the underlying microvascular channel. Similar effects of physiological breathing on nanoparticle absorption are observed in whole mouse lung. Mechanically active "organ-on-a-chip" microdevices that reconstitute tissue-tissue interfaces critical to organ function may therefore expand the capabilities of cell culture models and provide low-cost alternatives to animal and clinical studies for drug screening and toxicology applications.

One of the causes of the high cost of pharmaceuticals and the major obstacles to rapidly identifying new environmental toxins is the lack of experimental model systems that can replace costly and time-consuming animal studies. Although considerable advances have been made in the development of cell culture models as surrogates of tissues and organs for these types of studies (*1*), cultured cells commonly fail to maintain differentiation and expression of tissue-specific functions. Improved tissue organization can be promoted by growing cells in three-dimensional extracellular matrix (ECM) gels (*2*); however, these methods still fail to reconstitute structural and mechanical features of whole living organs that are central to their function. In particular, existing model systems do not recreate the active tissue-tissue interface between the microvascular endothelium and neighboring parenchymal tissues where critical transport of fluids, nutrients, immune cells, and other regulatory factors occur, nor do they permit application of dynamic mechanical forces (e.g., breathing movements in lung, shear in blood vessels, peristalsis in gut, tension in skin, and the like) that are critical for the development and function of living organs (*3*).

Microscale engineering technologies first developed to create microchips, such as microfabrication and microfluidics, enable unprecedented capabilities to control the cellular microenvironment with high spatiotemporal precision and to present cells with mechanical and biochemical signals in a more physiologically relevant context (*4*–*7*). This approach has made it possible to microfabricate models of blood vessels (*8*, *9*), muscles (*10*), bones (*11*), airways (*12*), liver (*13*–*16*), brain (*17*, *18*), gut (*19*), and kidney (*20*, *21*). However, it has not yet been possible to engineer integrated microsystems that replicate the complex physiological functionality of living organs by incorporating multiple tissues, including active vascular conduits, and placing them in a dynamic and mechanically relevant organ-specific microenvironment.

To provide the proof of principle for a biomimetic microsystems approach, we developed a multifunctional microdevice that reproduces key structural, functional, and mechanical properties of the human alveolar-capillary interface, which is the fundamental functional unit of the living lung. This was accomplished by microfabricating a microfluidic system containing two closely apposed microchannels separated by a thin (10 μm), porous, flexible membrane made of poly(dimethylsiloxane) (PDMS). The intervening membrane was coated with ECM (fibronectin or collagen), and human alveolar epithelial cells and human pulmonary microvascular endothelial cells were cultured on opposite sides of the membrane (Fig. 1A). Once the cells were grown to confluence, air was introduced into the epithelial compartment to create an air-liquid interface and more precisely mimic the lining of the alveolar air space. The compartmentalized channel configuration of the microdevice makes it possible to manipulate fluid flow, as well as delivery of cells and nutrients, to the epithelium and endothelium independently.

During normal inspiration, intrapleural pressure decreases, causing the alveoli to expand; this pulls air into the lungs, resulting in stretching of the alveolar epithelium and the closely apposed endothelium in adjacent capillaries (Fig. 1B). We mimicked this subatmospheric, pressure-driven stretching by incorporating two larger, lateral microchambers into the device design. When vacuum is applied to these chambers, it produces elastic deformation of the thin wall that separates the cell-containing microchannels from the side chambers; this causes stretching of the attached PDMS membrane and the adherent tissue layers (Fig. 1A, right versus left). When the vacuum is released, elastic recoil of PDMS causes the membrane and adherent cells to relax to their original size. This design replicates dynamic mechanical distortion of the alveolar-capillary interface caused by breathing movements.

These hollow microchannels were fabricated by soft lithography in conjunction with a new method that uses chemical etching of PDMS (*22*) to form the vacuum chambers. Fabrication begins with alignment and permanent bonding of a 10-μm-thick porous PDMS membrane (containing 10-μm-wide pentagonal pores) and two PDMS layers containing recessed microchannels (Fig. 1C). A PDMS etching solution composed of tetrabutyl-

[1]Wyss Institute for Biologically Inspired Engineering at Harvard University, Boston, MA 02115, USA. [2]Vascular Biology Program, Departments of Pathology and Surgery, Children's Hospital Boston, and Harvard Medical School, Boston, MA 02115, USA. [3]Department of Medicine, Children's Hospital Boston, Boston, MA 02115, USA. [4]School of Engineering and Applied Sciences, Harvard University, Cambridge, MA 02138, USA.

*To whom correspondence should be addressed. E-mail: don.ingber@wyss.harvard.edu

Vol 466 | 5 August 2010 | doi:10.1038/nature09212 nature

LETTERS

Real-time observation of valence electron motion

Eleftherios Goulielmakis[1]*, Zhi-Heng Loh[2,3]*, Adrian Wirth[1], Robin Santra[4,5], Nina Rohringer[6], Vladislav S. Yakovlev[1,7], Sergey Zherebtsov[1], Thomas Pfeifer[2,3]†, Abdallah M. Azzeer[8], Matthias F. Kling[1], Stephen R. Leone[2,3] & Ferenc Krausz[1,7]

The superposition of quantum states drives motion on the atomic and subatomic scales, with the energy spacing of the states dictating the speed of the motion. In the case of electrons residing in the outer (valence) shells of atoms and molecules which are separated by electronvolt energies, this means that valence electron motion occurs on a subfemtosecond to few-femtosecond timescale (1 fs = 10^{-15} s). In the absence of complete measurements, the motion can be characterized in terms of a complex quantity, the density matrix. Here we report an attosecond pump–probe measurement of the density matrix of valence electrons in atomic krypton ions[1]. We generate the ions with a controlled few-cycle laser field[2] and then probe them through the spectrally resolved absorption of an attosecond extreme-ultraviolet pulse[3], which allows us to observe in real time the subfemtosecond motion of valence electrons over a multifemtosecond time span. We are able to completely characterize the quantum mechanical electron motion and determine its degree of coherence in the specimen of the ensemble. Although the present study uses a simple, prototypical open system, attosecond transient absorption spectroscopy should be applicable to molecules and solid-state materials to reveal the elementary electron motions that control physical, chemical and biological properties and processes.

The millielectronvolt-scale spacing of vibrational energy levels implies that changes in molecular structure occur on a multifemtosecond timescale and can be accessed by femtosecond pump–probe spectroscopy[4]. Electronic phenomena in the valence band are one hundred to one thousand times faster and have remained elusive so far. Electronic coherence is the key to accessing them in real time. It has been studied in kinematically complete experiments[5], from which the motion can be inferred but not directly observed. Apart from in the simplest systems, experimental techniques are unable to probe all degrees of freedom. In these cases, we have to deal with an open system, which can only be characterized in terms of ensemble-averaged quantities (observables) predicted by the system's density matrix. Under these circumstances, the synchrony of wave-packet dynamics in the specimens of the ensemble (that is, coherence) is indispensable and only time-resolved measurements can provide direct access to the observables of the motion. Combination of the powerful concepts of correlated measurement and high-harmonic spectroscopy[6–9] has recently uncovered signatures of electronic coherence and resultant dynamics in an ensemble of ionizing molecules within a temporal window of ~1 fs following ionization[10]. The degree and the persistence of coherence have not been measured and the method is limited to the scrutiny of systems with large (≥10 eV) ionization potentials and of processes under strong-field influence. Isolated attosecond extreme-ultraviolet (EUV) pulses[3,11–13] lend themselves as a probe to overcome these limitations.

In this work, we introduce attosecond probe spectroscopy to study—as one of the simplest open quantum systems—krypton atoms ionized by a strong field. Strong-field ionization has been studied extensively[14], without answering the question of whether it is able to create long-lived coherences. Insufficient temporal confinement of ionization, recollision and/or electron correlations under strong-field influence may affect electronic coherence in the emerging ions. Our study reveals that strong-field ionization by a waveform-controlled[2], near-single-cycle[13] laser pulse is capable of launching a broadband (~0.7-eV splitting) valence electron wave packet with a high degree of coherence ($g \approx 0.6$) that persists without notable decay for much longer than 10 fs.

Attosecond probe spectroscopy can rely on the momentum or energy distribution of the liberated photoelectrons[15] or transmitted photons as observables. Here we choose the latter option, transient absorption spectroscopy[16–19], which can be extended to the scrutiny of condensed-matter phenomena whereas photoelectron spectroscopy is restricted to surfaces and gas-phase samples. Figure 1a illustrates our attosecond probing concept applied to the multiple ionization of krypton. Conventional spectroscopy[20] shows that the ions are created in manifolds of states depicted by green boxes and denoted by nl^{-i}, indicating that, relative to the ground-state configuration of the atom, i electrons are missing from the nl subshell. An attosecond pulse carried at a photon energy of ~80 eV promotes the krypton ions created in their $4p^{-i}$ manifolds to the $3d^{-1}4p^{-(i-1)}$ core-hole states. The transitions lead to dips in the spectrum of the broadband EUV radiation transmitted through the ionized atomic ensemble (Fig. 1b), revealing populations and coherences on an attosecond–femtosecond timescale.

In our experiments, we ionized krypton atoms with sub-4-fs, waveform-controlled NIR laser pulses carried at a wavelength of ~750 nm (ref. 13; wave period, $T_L \approx 2.5$ fs) and probed the emergence of ions and electronic dynamics within their $4p^{-i}$ valence subshells with isolated sub-150-as EUV pulses carried at ~80 eV (Fig. 1c; for details, see Methods Summary). The target of the NIR pump/EUV probe exposure is a quasistatic cell filled with krypton atoms at densities on the order of ~10^{18} cm^{-3}. The EUV beam size at the focus is a small fraction of that of the NIR beam and, hence, ionization and accompanying electron dynamics are probed near the optical axis, where the radial dependence of the laser intensity can be neglected.

EUV absorption spectra are recorded as functions of the pump–probe delay (negative delay means that the EUV probe precedes the NIR excitation) by a spectrometer downstream of the target. No

[1]Max-Planck-Institut für Quantenoptik, Hans-Kopfermann-Strasse 1, D-85748 Garching, Germany. [2]Departments of Chemistry and Physics, University of California, Berkeley, California 94720, USA. [3]Chemical Sciences Division, Lawrence Berkeley National Laboratory, Berkeley, California 94720, USA. [4]Argonne National Laboratory, Argonne, Illinois 60439, USA. [5]Department of Physics, University of Chicago, Chicago, Illinois 60637, USA. [6]Lawrence Livermore National Laboratory, Livermore, California 94551, USA. [7]Department für Physik, Ludwig-Maximilians-Universität München, Am Coulombwall 1, D-85748 Garching, Germany. [8]Physics and Astronomy Department, King Saud University, Riyadh, 11451, Kingdom of Saudi Arabia. †Present address: Max-Planck-Institut für Kernphysik, Saupfercheckweg 1, D-69117 Heidelberg, Germany.
*These authors contributed equally to this work.

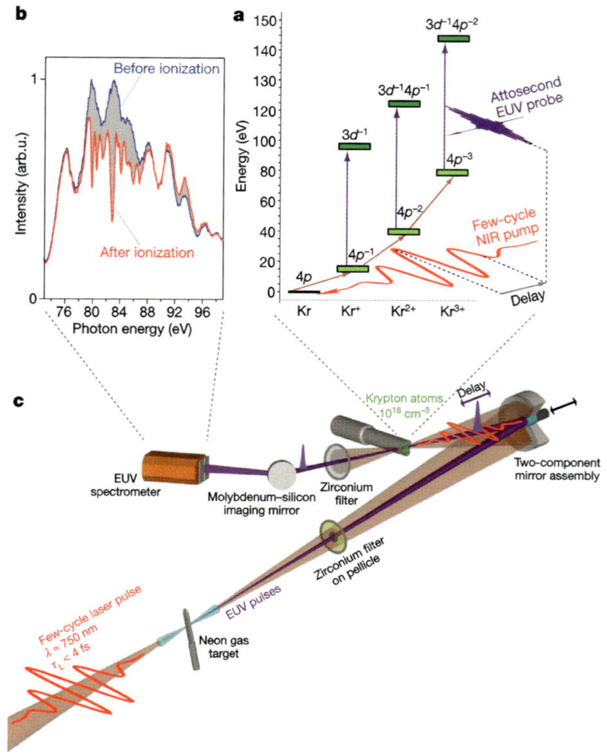

Figure 1 | Probing intra-atomic electron motion by attosecond absorption spectroscopy. a, The strong electric field of a near-infrared (NIR) laser pulse (in red) with a duration of $\tau_L < 4$ fs liberates electrons from the $4p$ valence subshell of krypton atoms to generate singly charged $4p^{-1}$, doubly charged $4p^{-2}$ or triply charged $4p^{-3}$ ions in the $4p^{-1}$, $4p^{-2}$ and $4p^{-3}$ manifolds of quantum states, respectively, by means of optical field ionization (indicated by red arrows). A subfemtosecond EUV pulse (in violet) with a carrier photon energy of ~80 eV is passed through the ions and promotes them to core-hole excited-state manifolds $3d^{-1}$, $3d^{-1}4p^{-1}$ and $3d^{-1}4p^{-2}$ (as indicated by the violet arrows). Transient EUV absorption spectra are acquired by recording the attosecond EUV pulse spectrum transmitted through the ionized gas target as a function of pump–probe delay with an EUV spectrometer. **b, c,** Spectral intensity distribution of the relevant part of the broadband attosecond probe pulse (**b**) and schematic of the experimental set-up (**c**). The pulse is transmitted through an ensemble of krypton atoms (~10^{18} cm^{-3}; interaction length, ~1 mm) before their ionization (blue curve) and after their ionization (red curve) by the laser pulse with a peak intensity of ~7×10^{14} W cm^{-2}, as well as by several optical elements shown in **c** and discussed in detail in Methods. arb.u., arbitrary units.

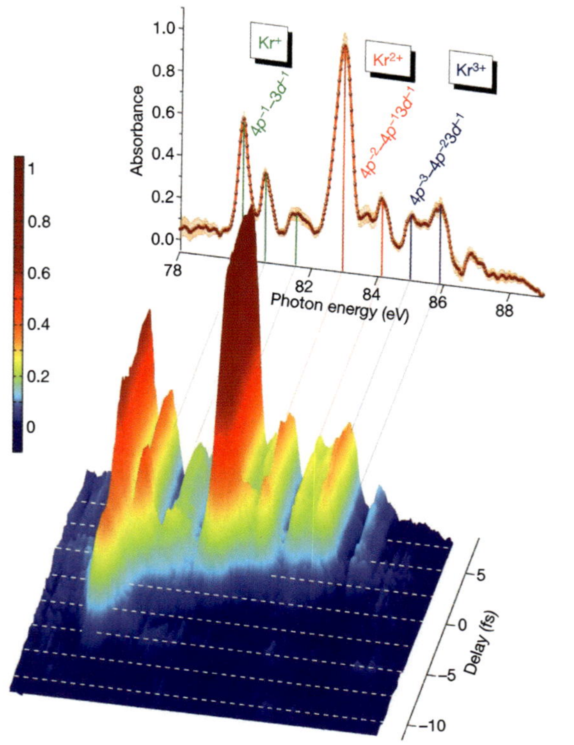

Figure 2 | Transient absorption spectra of krypton ions. Absorbance is defined as $A(E, \tau) = \ln(I_{trans}(E, \tau)/I_0(E))$, where $I_0(E)$ is the spectral density recorded at a negative delay of −10 fs, that is, the attosecond probe precedes the ionizing laser pulse by 10 fs in the atomic sample, and $I_{trans}(E, \tau)$ is the spectral density recorded at a pump–probe delay τ. The delay is varied in steps of 200 as. Error bars indicate the standard error of the mean values acquired from several spectra recorded at the same delay. The EUV probe pulse shows the formation of charge states up to Kr^{3+} as indicated in the spectrum shown in the background, which is recorded at $\tau \approx 10$ fs. Disregarding a forerunner in the main Kr^+ line, the origin of which is unclear, the Kr^{2+} lines appear with a significant, well-resolved delay of about one-half the laser period ($T_L/2 \approx 1.25$ fs) after the Kr^+ lines, and the Kr^{3+} lines appear with approximately the same delay after the Kr^{2+} lines. The decrease in the neutral krypton population in the atomic sample manifests itself as a reduction of the absorption in the range 91–93 eV, where neutral krypton atoms absorb resonantly. Relative occupations between the ionic states Kr^+, Kr^{2+} and Kr^{3+} are estimated as $N_{Kr^+} : N_{Kr^{2+}} : N_{Kr^{3+}} = 1:0.875:0.25$.

high-order harmonic radiation emerging from the krypton target was detected in the spectral range of interest (>60 eV). In Fig. 2, we plot a series of transient EUV absorption spectra for an on-axis peak laser intensity of ~7×10^{14} W cm^{-2}, which is sufficiently high to produce singly, doubly and triply charged ions, as inferred from the appearance of absorption lines associated with electronic transitions in these ions. Our spectrometer has a resolution of ~0.45 eV and hence is unable to resolve the ~88-meV width[21] of the observed lines resulting from the ~8-fs core-hole decay of the $3d^{-1}4p^{-(i-1)}$ states[22].

The appearance of different charge states delayed by approximately half the laser cycle with respect to each other is consistent with our understanding that the primary process in strong-field ionization is electron release near the field oscillation peaks by means of tunnelling or above-barrier escape. Attosecond transient absorption spectroscopy offers unprecedented insight into the dynamics of multiple ionization but theory must be further developed before this potential can be exploited. This is because the narrow natural linewidths of the absorption lines in the transmitted EUV light imply that they result from a coherent polarization response of the atomic ensemble that extends over several femtoseconds after the EUV pulse hits the sample. During this time, the strong laser field is still present and may affect the atomic polarization response. Modelling of this interaction is beyond the scope of this work; here we focus on dynamics occurring in Kr^+ ions after the ionizing field has vanished, where this complication does not arise.

It is known from photoelectron[23,24] and photoabsorption[18,25] spectroscopy that strong-field ionization populates not only the ground state but also excited electronic states of the emerging ions. Exposure of krypton atoms to a strong, low-frequency laser field is expected to generate Kr^+ ions predominantly in their $4p^{-1}_{j=3/2}$ ground-state manifold and the $4p^{-1}_{j=1/2}$ excited-state manifold, comprising four ($m_{j=3/2} = -3/2, -1/2, 1/2, 3/2$) and two ($m_{j=1/2} = -1/2, 1/2$) states, respectively ($j$ denotes the total angular momentum and m_j denotes its projection on the z axis). The fine-structure (or spin–orbit) energy splitting between the two manifolds is $\Delta E_{SO} = 0.67$ eV (ref. 26). However, the critical question is whether they can be populated coherently during strong-field ionization so as to allow the creation of subsequent wave-packet motion in the valence shell[27].

NATURE|Vol 466|5 August 2010 LETTERS

We have numerically modelled the interaction of krypton atoms with a strong laser field, as described in ref. 1. The populations of the six relevant quantum states of Kr^+ and the coherence between these states are respectively given by the diagonal elements $\rho_{j,j'=j}^{(m_j)}$ and the corresponding off-diagonal elements of the reduced density matrix of the atomic ensemble (Supplementary Information, section I). Solving the time-dependent Schrödinger equation for a few-cycle ionizing NIR field, linearly polarized along the quantization axis, yields the evolution of the density matrix elements as depicted in Fig. 3. By the end of the laser pulse ($t \gtrsim 3$ fs), ~30% of the atoms are ionized, with a hole emerging in the ($4p_j$, m_j) or ($4p_j$, $-m_j$) orbital with a relative population of $\rho_{3/2,3/2}^{(1/2)} + \rho_{3/2,3/2}^{(-1/2)} = 2\rho_{3/2,3/2}^{(1/2)} = 0.69$, $\rho_{1/2,1/2}^{(1/2)} + \rho_{1/2,1/2}^{(-1/2)} = 2\rho_{1/2,1/2}^{(1/2)} = 0.26$ and $\rho_{3/2,3/2}^{(3/2)} + \rho_{3/2,3/2}^{(-3/2)} = 2\rho_{3/2,3/2}^{(3/2)} = 0.05$ (Fig. 3a). The hole populations increase in subfemtosecond steps near the oscillation peaks of the ionizing laser field depicted by the dashed line. Only states with the same value of m_j can form a coherent superposition state characterized by a non-zero value of the corresponding off-diagonal element of the density matrix. In our case, this condition is fulfilled by the ($4p_{3/2}^{-1}$, $m_j = \pm 1/2$) and ($4p_{1/2}^{-1}$, $m_j = \pm 1/2$) states[1]. The degree of coherence between these states can be characterized by the parameter

$$g(t) = \frac{\left|\rho_{3/2,1/2}^{(1/2)}\right|}{\sqrt{\rho_{3/2,3/2}^{(1/2)}\rho_{1/2,1/2}^{(1/2)}}} \tag{1}$$

Perfect coherence corresponds to $g = 1$, whereas incoherent superpositions yield $g = 0$. In Fig. 3b, we plot the half-cycle-averaged $g(t)$ for pulse durations of 3.8 fs and 7.6 fs (full-width at half-maximum of the intensity envelope).

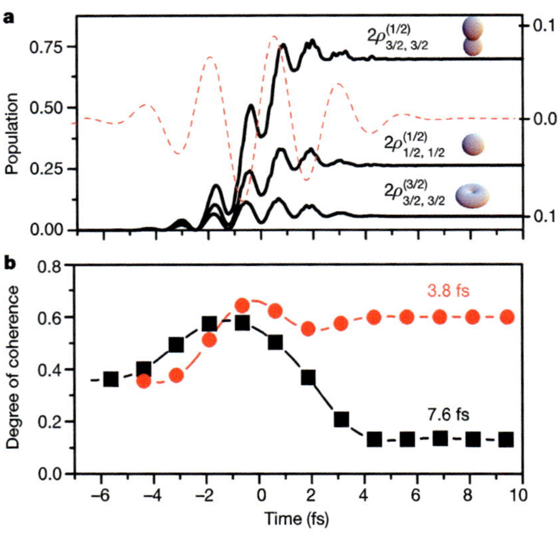

Figure 3 | Build-up of electronic coherence in Kr^+ produced by optical field ionization (theory). **a**, Temporal evolution of the diagonal elements of the reduced density matrix of Kr^+ in the presence of a 750-nm, 3.8-fs laser pulse with a peak intensity of $\sim 3 \times 10^{14}$ W cm^{-2} and a sinusoidal waveform. The notation of the matrix elements is explained in the text. The populations have been normalized such that the trace of the reduced density matrix at $\tau = 10$ fs equals one. The dashed line shows the electric field of the laser pulse (in atomic units (a.u.)) used in the simulations. The hole density distributions of the corresponding orbitals of the $4p$ subshell are also depicted. **b**, The degree of coherence, $g(t)$ (see equation (1)), averaged over time intervals of the laser half-cycle (~1.25 fs), is shown with red dots. Squares show the half-cycle-averaged degree of coherence for an ionizing laser pulse with a duration of 7.6 fs at the same peak intensity. The degree of coherence is defined such that, after the NIR pump pulse, it equals one for a perfectly coherent hole wave packet. The detailed behaviour of the degree of coherence during ion formation is not yet understood.

The evolution of coherence shown in Fig. 3b can be understood intuitively. At every intense wave crest, a fraction of the atoms is ionized and a hole spin–orbit wave packet is launched in the $4p$ valence subshell by populating the $4p^{-1}$ manifold. Spin–orbit wave packets launched in ions that are generated at a given wave crest have a fairly well-defined phase with respect to the wave crest and evolve with a half-period of 3.1 fs. As a consequence, their phase has changed by 0.4π by the time the next wave crest arrives (after one half of the laser period, $T_L/2 = 1.25$ fs), such that ions produced at this wave crest are phase-shifted with respect to the ions generated previously. In the 3.8-fs pulse, ionization is confined to two to three wave crests, that is, to within a single laser period. Therefore, the final population of the $4p^{-1}$ manifold builds up within less than the 3.1-fs half-period, ensuring that the ions produced at the individual wave crests are substantially in phase. This gives rise to a final ion ensemble with a high degree of coherence ($g \approx 0.6$). By contrast, in the 7.6-fs pulse, the build-up of $4p^{-1}$ population takes longer, leading to a wider distribution of relative phases among the ions created and, hence, to a strongly reduced degree of coherence of the ion ensemble ($g \approx 0.13$). Our simulations reveal that, in the absence of a tunable excitation period, the creation of a persisting electronic coherence critically relies on the confinement of excitation to a time interval that is comparable to the characteristic timescale of the resultant wave-packet motion. These simulations also reveal that field ionization by multicycle pulses[10] generally creates incoherent ion ensembles.

When the coherent broadband EUV probe promotes the system from the components of its superposition state into a common final state, quantum interference leads to a temporal modulation of the transition probability: its depth is indicative of the degree of coherence of the probed superposition state, and its variation with pump–probe delay reveals the temporal evolution of the system. In our case, strong-field ionization is predicted to create a hole superposition in the $4p_{j=3/2}^{-1}$ ground-state manifold and the $4p_{j=1/2}^{-1}$ excited-state manifold of Kr^+ ions (Fig. 4a). A broadband pulse carried at ~80 eV promotes the ions from these states—by means of dipole-allowed transitions—into the $3d_{3/2}^{-1}$ manifold by creating a core-level vacancy. A transition from the $4p_{3/2}^{-1}$ states to the $3d_{5/2}^{-1}$ states is also possible, but the $4p_{1/2}^{-1} \rightarrow 3d_{5/2}^{-1}$ transition is forbidden by electric dipole selection rules. Figure 4b shows the simulated absorption cross-section of Kr^+ for the above transitions (see Supplementary Information, section I, for details) as a function of pump–probe delay, for the simulation parameters given in Fig. 3.

The high degree of coherence between the $4p_{1/2}^{-1}$ and $4p_{3/2}^{-1}$ states (Fig. 3b) is predicted to translate into modulation of the absorption cross-sections as functions of pump–probe delay for the $4p_{3/2}^{-1} \rightarrow 3d_{3/2}^{-1}$ and $4p_{1/2}^{-1} \rightarrow 3d_{3/2}^{-1}$ transitions (Fig. 4b), owing to a temporal variation of the relative phase, ϕ, between the $4p_{1/2}^{-1}$ and $4p_{3/2}^{-1}$ states

$$\begin{aligned} \phi(t) &= \frac{E_{|4p,j=1/2\rangle} - E_{|4p,j=3/2\rangle}}{\hbar} t + \phi_0 \\ &\equiv \frac{\Delta E_{SO}}{\hbar} t + \phi_0 \equiv \frac{2\pi}{T_{SO}} t + \phi_0 \end{aligned} \tag{2}$$

where $E_{|4p,j=1/2\rangle}$ is the energy of the $4p_{1/2}^{-1}$ state, $E_{|4p,j=3/2\rangle}$ is the energy of the $4p_{3/2}^{-1}$ state and $\hbar$ is Planck's constant divided by 2π. Here $\Delta E_{SO} = 0.67$ eV implies a wave-packet oscillation period of $T_{SO} = 6.2$ fs. The modulation is most pronounced in the weaker $4p_{3/2}^{-1} \rightarrow 3d_{3/2}^{-1}$ transition and is absent from the $4p_{3/2}^{-1} \rightarrow 3d_{5/2}^{-1}$ transition, which is insensitive to the quantum superposition. In equation (2), $\phi(t)$ describes the relative phase between the two states after the NIR pulse. For a given NIR waveform, ϕ_0 is a nontrivial, well-defined quantity, which will become experimentally measurable by combining attosecond absorption spectroscopy with attosecond streaking to determine the NIR waveform and its timing with respect to the attosecond probe pulse with attosecond accuracy. Access to the phase during the laser pulse has been possible for short-lived coherences[10]. The experimental data (Fig. 4c) clearly show the predicted consequences of the

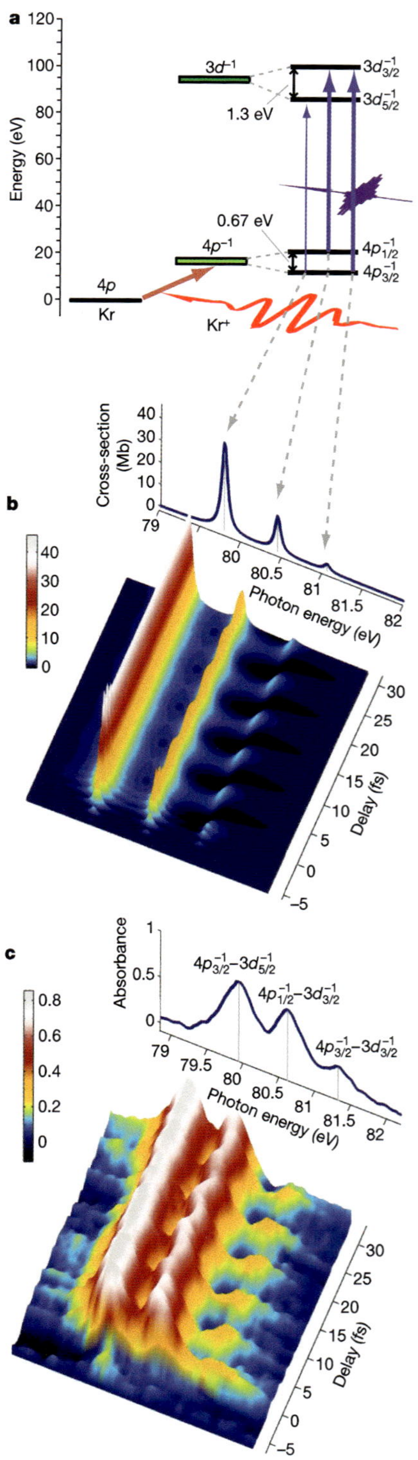

Figure 4 | Attosecond absorption spectroscopy reveals intra-atomic electron wave-packet motion in Kr^+. a, Energy-level diagram showing the spin–orbit splitting of the 4*p* and 3*d* subshells in Kr^+. A sub-4-fs NIR laser pulse (red wave) liberates an electron from the 4*p* subshell and leaves the ensemble of ions in a coherent superposition of $4p^{-1}_{3/2}$ and $4p^{-1}_{1/2}$ states. Single-photon EUV absorption promotes the ions from these states to the core-excited $3d^{-1}_{3/2}$ state. **b,** Simulated transient EUV absorption spectra reveal characteristic modulations present in the $4p^{-1}_{3/2}\rightarrow 3d^{-1}_{3/2}$ and $4p^{-1}_{1/2}\rightarrow 3d^{-1}_{3/2}$ transitions as functions of pump–probe delay. The modulation depth is highly sensitive to the degree of coherence, $g(t)$. Mb, megabarn. **c,** False-colour plot of an attosecond absorption spectrogram comprising 40 transient absorption spectra recorded at delays increased in steps of 1 fs with a sub-4-fs, ~750-nm laser pump and a sub-150-as, ~80-eV EUV probe. The reference spectrum was recorded at −6 fs. The absorption spectrum plotted in the background is taken at a delay of 30 fs. The linewidths are determined by the ~0.45-eV resolution of our EUV spectrometer. The lower modulation depth of the $4p^{-1}_{1/2}\rightarrow 3d^{-1}_{3/2}$ transition relative to the calculations shown in **b** is a result of the spectral resolution being limited in comparison with the ~88-meV natural linewidth of the studied transitions. The zero of the delay scale is set to coincide with the instant when the main $4p^{-1}_{3/2}\rightarrow 3d^{-1}_{5/2}$ absorption line reaches 95% of its quasistationary value.

wave-packet motion described by equation (2), including the deep amplitude modulation of the $4p^{-1}_{3/2}\rightarrow 3d^{-1}_{3/2}$ absorption line as well as the energy modulation of the lines, with the latter resulting from absorptive and dispersive terms in the transient absorption cross-section[16] (see equation (1) in Supplementary Information).

Components of the reduced density matrix of the system are determined by adjusting them along with T_{SO} such that the model's prediction agrees best with spectra recorded over the photon energy range 78.5–82 eV and the pump-probe delay range 3–33 fs (Supplementary Information, section III). This procedure has yielded (for $t\geq 3$ fs) $\rho^{(1/2)}_{3/2,3/2}+\rho^{(-1/2)}_{3/2,3/2}=0.42\pm0.10$, $\rho^{(1/2)}_{1/2,1/2}+\rho^{(-1/2)}_{1/2,1/2}=0.35\pm0.03$, $\rho^{(3/2)}_{3/2,3/2}+\rho^{(-3/2)}_{3/2,3/2}=0.23\pm0.08$, $g=0.63\pm0.17$ and $T_{SO}=6.3\pm0.1$ fs. This value for T_{SO} is in excellent agreement with the one derived from ΔE_{SO} (ref. 26). The degree of coherence agrees well with the predictions in Fig. 3, verifying that few-cycle ionization is capable of creating robust electronic coherence. It survives over a timescale of multiple tens of femtoseconds, allowing electron wave packets to affect structural dynamics once launched in molecular orbitals.

Fig. 5a displays the mean optical density of the $4p^{-1}_{3/2}\rightarrow 3d^{-1}_{3/2}$ absorption line as a function of pump–probe delay over the energy range 81.20–81.45 eV (black dots) in comparison with the prediction of our model with the parameters given above (red line). To reconstruct the quantum mechanical motion of the electron wave packet, the temporal evolution of the quantum phase $\phi(t)$ defined by equation (2) must also be retrieved. Attosecond absorption spectroscopy allows this retrieval over a multifemtosecond time interval with a resolution matched to the electronic timescale (Fig. 5b). The uncertainty in the obtained values of the quantum phase (of $\sim\pi/5$) translates—using equation (2)—into a temporal resolution of ~0.6 fs. In Fig. 5b, we plot the evolution of the quantum phase versus the time elapsed since the main absorption line of Kr^+ reached 95% of its stationary value. In this frame of reference, the electron (or hole) density distribution can be reconstructed at any instant after the ionizing laser pulse with subfemtosecond accuracy. We plot the result of this reconstruction at a few selected instants separated by 1 fs (Supplementary Information, section II). Measuring ϕ_0 along with the amplitude of the off-diagonal matrix element as a function of the laser waveform, which can be determined using attosecond streaking[11], will provide insight into the dynamics of optical field ionization and the concomitant formation of intra-atomic electron wave packets. Through comparisons with model predictions, this measured phase will provide a new, sensitive test of the modelling of strong-field ionization of multi-electron atoms, a process that is far from being well understood.

Attosecond transient absorption spectroscopy extends real-time insight into microscopic motion, from nuclear wave-packet[4] and Rydberg wave-packet[28] dynamics to electrons in the valence shell. Its unique features, that is, gentle probing owing to the absence of strong fields, the capacity to study processes within condensed matter, and wide applicability to materials with arbitrary ionization potential, render this approach ideal for attosecond real-time observation of electronic and concomitant processes in atoms, molecules and solids[29,30].

NATURE|Vol 466|5 August 2010 LETTERS

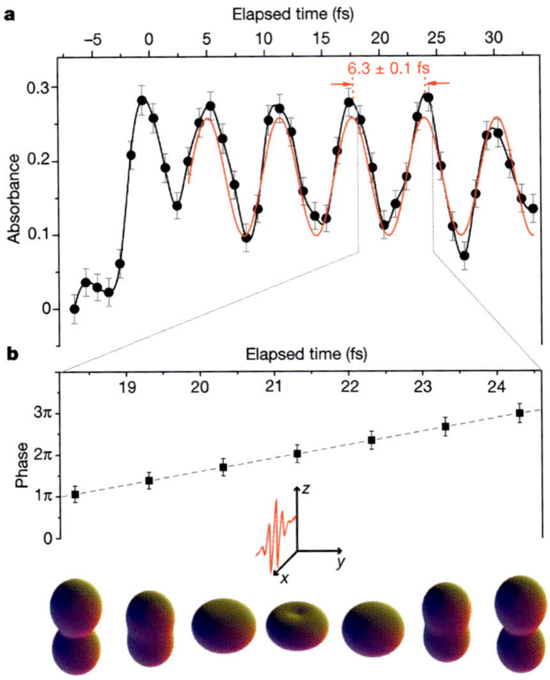

Figure 5 | Reconstruction of valence-shell electron wave-packet motion. **a**, Absorbance (dots) averaged over the photon energy range 81.20–81.45 eV corresponding to the $4p_{3/2}^{-1} \to 3d_{5/2}^{-1}$ transition (see Fig. 4c), as a function of time elapsed since zero as defined in the text. The full line shows the result of our modelling for the values of the fit parameters given in the text. The modulation occurs with a period of 6.3 ±0.1 fs. Error bars depict the standard error of the values extracted from several data sets recorded under identical experimental conditions. **b**, Quantum phase of the 4*p* superposition state $\phi(t)$ (see equation (2)), as retrieved from the measured attosecond absorption spectrogram shown in Fig. 4c. Uncertainty in the values, resulting from our measurement and modelling, indicates accuracy of reconstruction of the superposition of ~π/5. The lower diagram shows ensemble-averaged hole density distributions in the 4*p* subshell of Kr^+ reconstructed from the measured $\phi(t)$ and the measured components of the density matrix, at instants separated by 1 fs, within an interval of 17–25 fs following ionization.

METHODS SUMMARY

We sent sub-4-fs, 0.3-mJ NIR laser pulses at a wavelength of ~750 nm into a neon-filled tube to generate EUV pulses by means of high-harmonic generation. The collinear NIR and EUV beams were then passed through a filter assembly and focused in a quasistatic gas cell containing the krypton gas at a pressure of about 80 mbar. The EUV pulse hitting the target was delayed with respect to the NIR laser pulse and had a duration of less than 150 as. A broadband molybdenum–silicon multilayer mirror imaged, in one transverse dimension, the attosecond EUV beam transmitted through the krypton gas target to the entrance slit of an EUV spectrometer used for measuring the spectral intensity distribution of the transmitted beam.

Full Methods and any associated references are available in the online version of the paper at www.nature.com/nature.

Received 25 September 2009; accepted 24 May 2010.

1. Rohringer, N. & Santra, R. Multichannel coherence in strong-field ionization. *Phys. Rev. A* **79**, 053402 (2009).
2. Baltuska, A. *et al.* Attosecond control of electronic processes by intense light fields. *Nature* **421**, 611–615 (2003).
3. Hentschel, M. *et al.* Attosecond metrology. *Nature* **414**, 509–513 (2001).
4. Zewail, A. H. Femtochemistry: atomic-scale dynamics of the chemical bond. *J. Phys. Chem. A* **104**, 5660–5694 (2000).
5. Schöffler, M. S. *et al.* Ultrafast probing of core hole localization in N_2. *Science* **320**, 920–923 (2008).
6. Niikura, H. *et al.* Sub-laser-cycle electron pulses for probing molecular dynamics. *Nature* **417**, 917–922 (2002).
7. Niikura, H. *et al.* Probing molecular dynamics with attosecond resolution using correlated wave packet pairs. *Nature* **421**, 826–829 (2002).
8. Niikura, H., Villeneuve, D. M. & Corkum, P. B. Mapping attosecond electron wave packet motion. *Phys. Rev. Lett.* **94**, 083003 (2005).
9. Baker, S. *et al.* Probing proton dynamics in molecules on an attosecond time scale. *Science* **312**, 424–427 (2006).
10. Smirnova, O. *et al.* High harmonic interferometry of multi-electron dynamics in molecules. *Nature* **460**, 972–977 (2009).
11. Kienberger, R. *et al.* Atomic transient recorder. *Nature* **427**, 817–821 (2004).
12. Sansone, G. *et al.* Isolated single-cycle attosecond pulses. *Science* **314**, 443–446 (2006).
13. Goulielmakis, E. *et al.* Single-cycle nonlinear optics. *Science* **320**, 1614–1617 (2008).
14. Brabec, T. *Strong Field Laser Physics* (Springer, 2008).
15. Yudin, G. L. *et al.* Attosecond photoionization of coherently coupled electronic states. *Phys. Rev. A* **72**, 051401 (2005).
16. Pollard, W. T., Lee, S.-Y. & Mathies, R. A. Wave packet theory of dynamic absorption spectra in femtosecond pump-probe experiments. *J. Chem. Phys.* **92**, 4012–4029 (1990).
17. Mathies, R. A. *et al.* Direct observation of the femtosecond excited-state cis-trans isomerization in bacteriorhodopsin. *Science* **240**, 777–779 (1988).
18. Loh, Z.-H. *et al.* Quantum state-resolved probing of strong-field-ionized xenon atoms using femtosecond high-order harmonic transient absorption spectroscopy. *Phys. Rev. Lett.* **98**, 143601 (2007).
19. Loh, Z.-H. & Leone, S. R. Ultrafast strong-field dissociative ionization dynamics of CH_2Br_2 probed by femtosecond soft X-ray transient absorption spectroscopy. *J. Chem. Phys.* **128**, 204302 (2008).
20. Southworth, S. H. *et al.* *K*-edge X-ray-absorption spectroscopy of laser-generated Kr^+ and Kr^{2+}. *Phys. Rev. A* **76**, 043421 (2007).
21. Jurvansuu, M., Kivimäki, A. & Aksela, S. Inherent lifetime widths of Ar $2p^{-1}$, Kr $3d^{-1}$, Xe $3d^{-1}$, and Xe $4d^{-1}$ states. *Phys. Rev. A* **64**, 012502 (2001).
22. Drescher, M. *et al.* Time-resolved atomic inner-shell spectroscopy. *Nature* **419**, 803–807 (2002).
23. Rottke, H., Ludwig, J. & Sandner, W. 'Short' pulse MPI of xenon: the $^2P_{1/2}$ ionization channel. *J. Phys. B* **29**, 1479 (1996).
24. Gubbini, E. *et al.* Core relaxation in atomic ultrastrong laser field ionization. *Phys. Rev. Lett.* **94**, 053602 (2005).
25. Young, L. *et al.* X-ray microprobe of orbital alignment in strong-field ionized atoms. *Phys. Rev. Lett.* **97**, 083601 (2006).
26. Saloman, E. B. Energy levels and observed spectral lines of krypton, Kr I through Kr XXXVI. *J. Phys. Chem. Ref. Data* **36**, 215–386 (2007).
27. Santra, R., Dunford, R. W. & Young, L. Spin-orbit effect on strong-field ionization of krypton. *Phys. Rev. A* **74**, 043403 (2006).
28. Jones, R. R. & Noordam, L. D. Electronic wavepackets. *Adv. At. Mol. Opt. Phys.* **38**, 1–38 (1998).
29. Bucksbaum, P. H. The future of attosecond spectroscopy. *Science* **317**, 766–769 (2007).
30. Krausz, F. & Ivanov, M. Y. Attosecond physics. *Rev. Mod. Phys.* **81**, 163–234 (2009).

Supplementary Information is linked to the online version of the paper at www.nature.com/nature.

Acknowledgements We thank U. Kleineberg, M. Hofstetter and M. Fiess for invaluable contributions. This work was supported by the Max Planck Society, the Nobel Program of King Saud University and the DFG Cluster of Excellence: Munich Centre for Advanced Photonics (http://www.munich-photonics.de). E.G. acknowledges a Marie-Curie Reintegration grant (MERG-CT-2007-208643). A.W., S.Z. and M.F.K. acknowledge support by the Emmy Noether programme of the DFG. Z.-H.L., T.P. and S.R.L. acknowledge support from the Air Force Office of Scientific Research (FA9550-04-1-0242), the National Science Foundation (CHE-0742662 and EEC-0310717) and the Director, Office of Science, Office of Basic Energy Sciences, US Department of Energy (DE-AC02-05-CH11231). T.P. acknowledges support from the MPRG program of the MPG. R.S. is supported by the Office of Basic Energy Sciences, Office of Science, US Department of Energy (DE-AC02-06CH11357). Part of this work was performed under the auspices of the US Department of Energy by Lawrence Livermore National Laboratory (DE-AC52-07NA27344). S.R.L. gratefully acknowledges appointment as a Miller Research Professor in the Miller Institute for Basic Research in Science.

Author Contributions E.G., Z.-H.L. and A.W. conceived and designed the experiments; E.G., A.W. and Z.-H.L. performed the measurements; A.W., Z.-H.L., E.G., T.P., S.Z., A.M.A., M.F.K., S.R.L and F.K. evaluated, analysed and interpreted the experimental data; and R.S., N.R. and V.S.Y. performed the theoretical modelling. All authors discussed the results and contributed to the final manuscript.

Author Information Reprints and permissions information is available at www.nature.com/reprints. The authors declare no competing financial interests. Readers are welcome to comment on the online version of this article at www.nature.com/nature. Correspondence and requests for materials should be addressed to E.G. (elgo@mpq.mpg.de), S.R.L. (srl@berkeley.edu) or F.K. (krausz@lmu.de).

doi:10.1038/nature09212 nature

METHODS

Carrier-envelope-phase-controlled, sub-4-fs, 0.3-mJ NIR laser pulses carried at a wavelength of ~750 nm and delivered at a repetition rate of 3 kHz are focused into a neon-filled tube to generate EUV pulses by means of high-harmonic generation (Fig. 1c). On their way towards a two-component, concentric mirror module, the collinear NIR and EUV beams pass through a filter assembly consisting of a 150-nm-thick zirconium foil and an ultrathin pellicle. The small-divergence EUV beam is transmitted through the zirconium filter covering a circular spot ~3 mm in diameter, whereas the NIR beam is efficiently blocked by this filter. The outer part of the (more divergent) NIR beam is transmitted by the pellicle carrying the zirconium foil.

The EUV beam transmitted through the circular zirconium filter has a diameter of ~3 mm when hitting the internal part of the double mirror assembly, which also has a diameter of ~3 mm. This inner mirror is covered with a molybdenum–silicon multilayer with a reflectance of ~2.5% over a 28-eV (full-width at half-maximum) band centred at ~80 eV. In these experiments, spectral filtering and the intensity of the driving field yield pulses with ~15-eV bandwidth. The NIR beam is reflected by the outer part of the mirror assembly, which is coated with silver. Both mirrors are deposited on a super-polished substrate with a radius of curvature of 25 cm. Both beams are focused in a quasistatic gas cell formed by a nickel tube, containing the krypton gas at a pressure of about 80 mbar. The effective interaction length in the krypton gas target is ~1 mm. The EUV pulses hitting the target have a duration of less than 150 as as verified by attosecond streak-camera measurements. It is delayed with respect to the laser pulse by the focusing molybdenum–silicon mirror mounted on a piezo-controlled translation stage. A motorized aperture installed downstream of the source adjusts the on-axis NIR intensity on target (not shown) between zero and $\sim 7 \times 10^{14}\,\mathrm{W\,cm^{-2}}$. A second zirconium foil, installed behind the krypton gas cell, prevents the NIR light from entering the EUV spectrometer. A broadband molybdenum–silicon multilayer mirror images, in one transverse dimension, the transmitted EUV beam from the krypton gas target to the entrance slit of an EUV spectrometer used for measuring the spectral intensity distribution of the attosecond EUV beam transmitted through the ionized target.

Vol 449 | 25 October 2007 | doi:10.1038/nature06229 nature

LETTERS

Attosecond spectroscopy in condensed matter

A. L. Cavalieri[1], N. Müller[2], Th. Uphues[1,2], V. S. Yakovlev[3], A. Baltuška[1,4], B. Horvath[1], B. Schmidt[5], L. Blümel[5], R. Holzwarth[5], S. Hendel[2], M. Drescher[6], U. Kleineberg[3], P. M. Echenique[7], R. Kienberger[1], F. Krausz[1,3] & U. Heinzmann[2]

Comprehensive knowledge of the dynamic behaviour of electrons in condensed-matter systems is pertinent to the development of many modern technologies, such as semiconductor and molecular electronics, optoelectronics, information processing and photovoltaics. Yet it remains challenging to probe electronic processes, many of which take place in the attosecond (1 as = 10^{-18} s) regime. In contrast, atomic motion occurs on the femtosecond (1 fs = 10^{-15} s) timescale and has been mapped in solids in real time[1,2] using femtosecond X-ray sources[3]. Here we extend the attosecond techniques[4,5] previously used to study isolated atoms in the gas phase to observe electron motion in condensed-matter systems and on surfaces in real time. We demonstrate our ability to obtain direct time-domain access to charge dynamics with attosecond resolution by probing photoelectron emission from single-crystal tungsten. Our data reveal a delay of approximately 100 attoseconds between the emission of photoelectrons that originate from localized core states of the metal, and those that are freed from delocalized conduction-band states. These results illustrate that attosecond metrology constitutes a powerful tool for exploring not only gas-phase systems, but also fundamental electronic processes occurring on the attosecond timescale in condensed-matter systems and on surfaces.

Photoemission spectroscopy is based on the photoelectric effect, first explained by Einstein more than 100 years ago[6]. According to Einstein's law, photoelectrons ejected from a metal surface by light will have a kinetic energy that depends on the incident photon energy and the electron's original bound state energy. Photoelectron spectra will thus provide information about the electronic structure of the metal if well-characterized light sources are used[7]. Indeed, experiments using a broad range of photon energies have now been performed to determine the steady-state electronic properties of many bulk materials, thin films, and surfaces. The photoemission process itself involves three steps: excitation, transport, and ultimately escape of the photoelectron through the surface[8]. Here, in a proof-of-principle experiment, we combine this spectroscopy with attosecond temporal resolution to obtain time-domain insight into the electron transport stage of the photoemission process. The measurements represent (to our knowledge) the first direct attosecond time-resolved observation of electron transport in a condensed-matter system, and we expect that they will trigger other experimental research into the dynamics of processes that have attracted interest in solid-state and surface science. Such processes include charge transfer[9,10], charge screening[11], image charge creation and decay[12], electron–electron scattering[13], and collective electronic motion[14].

Time-resolved photoemission spectroscopy was originally implemented in the picosecond (1 ps = 10^{-12} s) to femtosecond regime, using first visible[15–17] and then extreme ultraviolet (XUV)[18,19] radiation. These experiments utilize one light pulse to trigger the dynamics, followed by a second light pulse to induce photoemission and thereby probe the transient state. Experiments using the laser-assisted photoelectric effect have been carried out[20–22]; but the XUV photoemission lasted over several wave cycles of the coincident near-infrared (NIR) light, limiting the time resolution to >10 fs. To overcome this limitation, we use single sub-femtosecond XUV pulses[4,5] for pumping, and coincident NIR waveform-controlled few-cycle laser pulses[23] as a probe[5]. The XUV pulse triggers the photoemission process, with only those photoelectron wave packets initiated from the uppermost atomic layers escaping without inelastic collision.

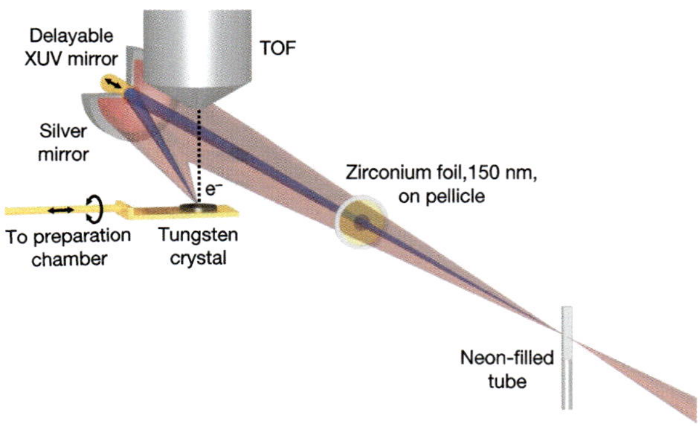

Figure 1 | Experimental set-up. Waveform-controlled, ~5-fs, 750-nm, 400-μJ laser pulses are focused with a mirror of 500-mm focal length into a ~2-mm-diameter tube filled with neon to generate XUV radiation by high-harmonic generation. The collinear XUV and NIR beams co-propagate towards a cored two-part mirror in the measurement chamber that is maintained under ultrahigh vacuum in a 1-m-length differential vacuum pumping stage, until the beams are separated by a pellicle/zirconium foil assembly. The zirconium foil transmits the XUV but blocks the NIR. The XUV radiation, indicated by the blue beam, is incident on a 6 eV (full-width at half-maximum FWHM) broad multilayer band-pass reflector centred at ~91 eV, which is mounted on a piezo-electric delay stage. With a proper XUV spectrum, the multilayer mirror reflects and focuses ~300-as (FWHM) XUV pulses. The NIR pulse, indicated by the violet beam, is reflected by a stationary (silver) outer annular mirror confocal with the inner mirror (f = 12.5 cm). Both pulses are focused onto the (110) surface of a tungsten single crystal that is mounted on a manipulator to control the angle of incidence. The manipulator is also used to retract the crystal into a preparatory chamber for cleaning. Resultant XUV-induced photoemission, which is detected by the time-of-flight spectrometer (TOF), is streaked by the coincident NIR laser pulse.

[1]Max-Planck-Institut für Quantenoptik, Hans-Kopfermann-Str. 1, D-85748 Garching, Germany. [2]Fakultät für Physik, Universität Bielefeld, D-33615 Bielefeld, Germany. [3]Department für Physik, Ludwig-Maximilians-Universität, Am Coulombwall 1, D-85748 Garching, Germany. [4]Institut für Photonik, Technische Universität Wien, Gußhausstr. 27, A-1040 Wien, Austria. [5]Menlo Systems GmbH, Am Klopferspitz 19, D-82152 Martinsried, Germany. [6]Institut für Experimentalphysik, Universität Hamburg, Luruper Chaussee 149, D-22761 Hamburg, Germany. [7]Dpto. Fisica de Materiales UPV/EHU, Centro Mixto CSIC-UPV/EHU and Donostia International Physics Center (DPIC), Paseo Manual de Lardizabal 4, 20018 San Sebastian, Spain.

Photoexcited electron wave packets propagate through the material in upper conduction bands, ultimately leaving the surface with an average kinetic energy determined by the XUV photon energy, the initial binding energy, and the material work function. An attosecond transient recorder (ATR), previously developed and used in gas-phase experiments[5], is used to observe the emitted photoelectron wave packet. In this scheme, the photoelectron momentum is further influenced by the electric field of a coincident few-cycle NIR laser pulse, giving rise to a 'streaked' final momentum distribution[24,25].

An ATR spectrogram is compiled by measuring a series of streaked photoelectron spectra with a time-of-flight detector, recorded as a function of time delay between the XUV pump and NIR streaking field. Important characteristics of the emitted electron wave packets, including their duration and frequency sweep, or 'chirp', can be determined from the spectra[24,25]. If measured for two or more different types of electrons, the complete ATR spectrograms can also yield relative timing information about the arrival of the wave packets on the surface, because the streaking effect is negligible until the electrons emerge from the surface (see Methods). The resolution of the ATR depends on the duration of the XUV excitation, the gradient of the streaking NIR field, and the signal-to-noise ratio in the photoelectron spectra.

In comparison to experiments performed on isolated atoms, ATR measurements in condensed-matter systems are more complicated because the photoelectron wave packets can be released from energy bands containing many distinct states rather than from a single, isolated energy level. Unoccupied conduction-band states just above the Fermi energy (defined by the highest occupied energy level in the absence of thermal excitation) might become populated by single-photon absorption of the leading edge of the NIR probe field prior to XUV photoemission, which is an unwanted complication. In contrast to conduction-band states, the localized 4*f* core states of tungsten are deeply bound and fully populated. Therefore, these states are unsusceptible to this potential influence of the streaking field, and constitute an ideal test case for proof of the extension of attosecond metrology to solids.

As a further challenge, above-threshold ionization (ATI) by the streaking field can, in condensed matter, generate energetic photoelectrons, obscuring detection of XUV-induced photoelectrons. ATI is favoured by the low work function of metals (as compared to the relatively high ionization potential of isolated atoms), which limits the intensity of the applied streaking field to levels far below those that can be used in gas-phase experiments.

As indicated by the diagram of the experimental set-up in Fig. 1, streaked photoemission spectra from a tungsten (110) crystal surface were recorded by collecting electrons within a narrow cone aligned perpendicularly to the surface. The relative delay between the XUV pulse and the NIR waveform-controlled streaking field was varied in 300-as steps in a sequence chosen to minimize systematic error, with

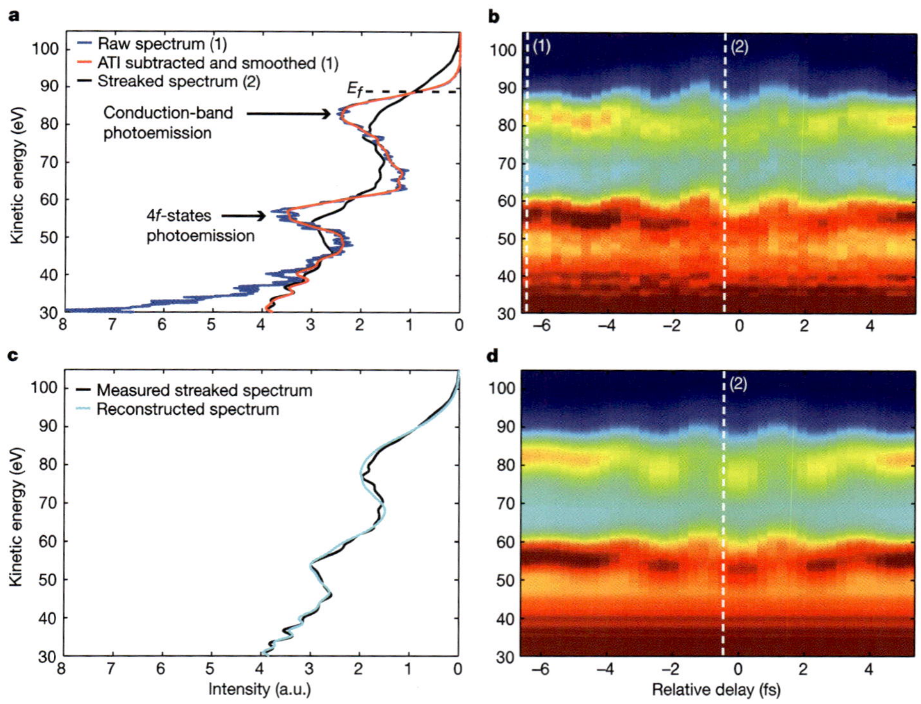

Figure 2 | Attosecond time-resolved photoemission spectra. **a**, Photoelectron spectra collected at two different relative delays. The spectrum recorded at the delay indicated by the white dashed line labelled '(1)' in **b** is far from the zero of delay as defined by the overlap of the maximums of the NIR and XUV pulse envelopes. This spectrum shows pronounced peaks corresponding to the 4*f*-state and conduction-band photoemission. The blue line shows the raw spectrum as recorded by the time-of-flight detector, and the red line shows the corresponding spectrum after subtraction of NIR-induced ATI background and numerical smoothing. The 4*f* photoemission is peaked near 56 eV. The conduction-band photoemission is peaked near 83 eV, owing to a high density of *d*-band conducting states just below the Fermi energy. E_f denotes the kinetic energy of a photoelectron excited from the Fermi energy level. The other spectrum in **a** (displayed only after ATI subtraction and smoothing) was recorded near zero delay, as indicated by the white dashed line labelled '(2)'. At this delay, the XUV pulse peak coincides with the NIR field maximum on our target. Consequently, the NIR vector potential crosses zero, giving rise to the strongest streaking and a clear broadening of the photoemission peaks is observed. **b**, The full ATI subtracted spectrogram for both the 4*f*-states and conduction-band photoemission. The streaking waveform (vector potential) is evident in both spectrograms, proving the extension of attosecond metrology to condensed-matter systems. We can expect that the energy modulation of the conduction-band peak should be a little larger than that of the 4*f* peak owing to their initial photoelectron kinetic energy. The amplitude of the spectral shift has a square-root dependence on initial kinetic energy (KE)[24], yielding an expected ratio in spectral shift of $\sqrt{\mathrm{KE}_{\mathrm{cond}}}\Big/\sqrt{\mathrm{KE}_{4f}}\approx 1.2$, which is consistent with our measurements. **c**, The reconstructed spectrum corresponding to that measured and displayed in **a**. **d**, The full simulated spectrogram.

spectra integrated for 60 s at each delay. (For a brief description of the experiment, see Methods Summary; full details regarding set-up, measurements and data analysis are provided as Supplementary Information.) Characteristic spectra obtained with our system are shown in Fig. 2a, indicating that emission from the conduction band occurs at a kinetic energy of ~83 eV while emission from the localized $4f$ states occurs at ~56 eV. At kinetic energies significantly below the $4f$ peak, the measured spectrum is due to NIR-induced ATI photoelectrons and XUV-generated photoelectrons that have undergone inelastic scattering.

The two distinct background components were distinguished by recording an additional photoelectron spectrum without the NIR streaking field. The ATI component was subsequently subtracted from the measured data (see Supplementary Information). This subtraction is illustrated for a fixed delay in Fig. 2a, and was performed at each of the delay steps, resulting in the full spectrogram presented in Fig. 2b. Here, a positive relative delay corresponds to the XUV pulse arriving earlier with respect to the streaking field at the surface. Both the $4f$ and conduction-band photoemission exhibit a pronounced periodic upshift and downshift in energy as a function of relative delay and, as in previous gas-phase experiments, the spectrogram reveals the waveform (vector potential) of the streaking field[5,26,27]. Our ability to resolve the field oscillation indicates that the photoemission from the $4f$ core states and from the conduction band is subfemtosecond in duration, and proves that attosecond metrology has been successfully extended to condensed-matter systems.

Further examination reveals that the $4f$ spectrogram is shifted along the delay coordinate with respect to the conduction-band spectrogram. This effect is readily apparent upon inspection of the smoothed spectrograms that are obtained by interpolation of the measured data and shown in Fig. 3a. We quantify the temporal shift in the measured data by evaluating, for each delay step, the centre-of-mass (COM) of the spectral regions spanning the $4f$ and conduction-band peaks that cover the energy intervals 47–66 eV and 66–110 eV, respectively. Characterizing the periodic motion of the peaks through their COM requires no assumptions or fitting parameters, yet yields timing information that is invariant to fluctuations in the instantaneous laser parameters. The approach is also relatively insensitive to inelastic scattered background photoelectrons, which could not be subtracted from our measurements. As a result, the COM accurately describes the streaking-induced time-dependence of the energy shift of the $4f$ and conduction-band peaks, as shown in Fig. 3b.

By comparing the COM trajectories of the $4f$ and conduction band at the seven zero-crossings of the vector potential, we obtain seven independent measurements of their relative timing. This yields a temporal shift of $\Delta\tau = 110 \pm 70$ as between the ATR spectrograms of the conduction-band and $4f$ photoelectrons. (The error estimate results from a straightforward extrapolation of the error in calculating the COM; see Supplementary Information.) This shift or delay was observed in different independent measurements made at different locations on the tungsten sample, with the results corroborating the above value of $\Delta\tau$ to within the measurement error. We note that the rather large error associated with our $\Delta\tau$ value could be most effectively reduced in future measurements by using higher XUV photon energies and fluxes.

The shift between the two spectrograms indicates that, on average, photoelectrons originating from the localized $4f$ states emerge from the tungsten surface approximately 100 as later than those originating from the delocalized conduction band—even though the photoemission process for both types of electrons is initiated simultaneously by the same XUV pulse. The delay effect thus occurs during transport of the excited photoelectrons to the surface, illustrating that our technique provides a means to directly observe features of electron wave packet propagation towards the surface with attosecond precision.

By adapting a quantum mechanical model used in previous gas-phase streaking experiments[28], we are able to reconstruct the measured spectra and spectrograms. The modelling of the streaking experiment requires some assumptions, leaving several parameters (such as duration of the electron wave packets, their chirp, and their emission time) for optimization. Figure 2c and d shows the reconstructions that best agree with experiment. These were obtained for wave packets with a duration of ~300 as (full-width at half-maximum, FWHM) and assuming a delay of ~100 as between the emission times of the electron wave packets, which supports the conclusions drawn from the COM analysis.

Our measurements also indicate that electron wave packets launched from both the localized $4f$ and delocalized conduction-band states are nearly undistorted on propagation to the surface. To explain the observed delay, we consider the group velocities for the two different photoelectron wave packets travelling in the solid. The crucial point is that after absorption of an XUV photon, the electron is excited into an upper conduction band region that depends on the electron's

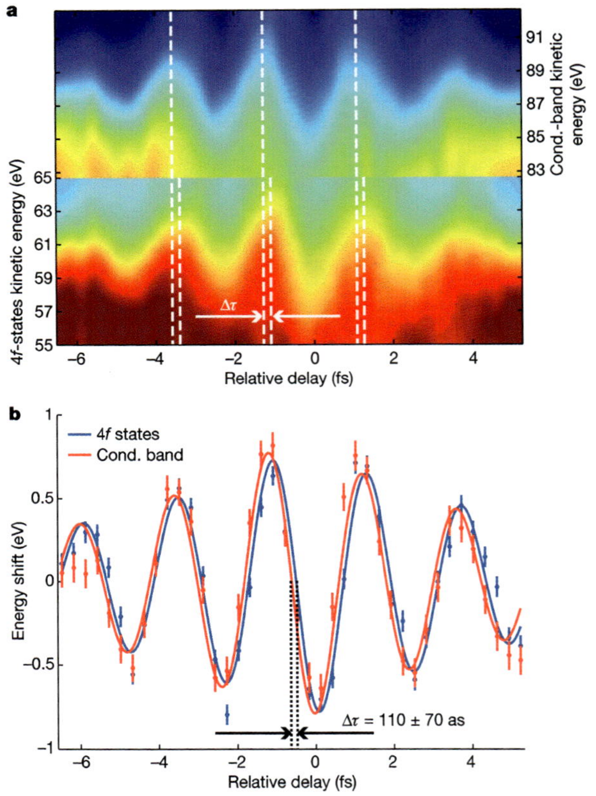

Figure 3 | Evidence of delayed photoemission. a, The $4f$ and conduction-band spectrograms, following cubic-spline interpolation of the measured data (but without background subtraction). The spectral region between ~65 eV and ~83 eV has been omitted to more easily compare the edges of the $4f$ and conduction-band peaks. A small shift in the relative delay is evident, as indicated by the white dashed lines through the fringes, and can be seen at each fringe maximum and minimum. Quantification of the shift of the $4f$ with respect to the conduction-band spectrogram is made by COM analysis, and is summarized in **b**. The energy intervals, within which the COMs were calculated, are 47–66 eV for the $4f$ photoemission peak and 66–110 eV for the conduction-band photoemission peak. Vertical error bars (±1 s.d.) are calculated from noise in the measured spectra (see Supplementary Information for details). For ease of visual comparison, the COM energy shift of the $4f$ spectral region was scaled by a factor of 2.5, to offset the stabilizing effect of the background plateau underneath the $4f$ peak (see Supplementary Information), in order to illuminate the ~100 as delay in emission. Rescaling these COM data points along the energy axis cannot influence the measured delay. The COM data points were fitted with a damped sinusoid, which corresponds to the NIR streaking field, to guide the eye.

initial binding energy and the XUV photon energy. If the relationship between momentum and energy of such an electron were that of a free electron, the ratio of velocities of the conduction-band and $4f$ electrons travelling with energies of 85 eV and 58 eV, respectively, would be a factor of ~1.2 (energy is defined with respect to the Fermi energy inside the material). However, elastic interactions with atoms in the crystal lattice, which give rise to electronic band structure, modify the momentum–energy relationship. This implies that in tungsten for XUV photon energies of ~91 eV, the mean velocity of the conduction-band photoelectrons is approximately twice that of the $4f$ photoelectrons (see Supplementary Information). We also note that due to their longer inelastic mean free paths[29], the slow $4f$ photoelectrons originate, on average, from ~1 Å deeper in the tungsten crystal than the fast conduction-band photoelectrons. On the basis of these considerations, we estimated the absolute delay between the initial excitation of a photoelectron and its escape through the surface to be ~60 as and ~150 as for the conduction and $4f$ photoelectrons, respectively. These values suggest a relative delay between photoelectron emissions from the surface of ~90 as, which is in good agreement with the observed delay in the emission of the $4f$ and conduction-band photoelectrons.

In summary, our observations demonstrate the successful extension of attosecond metrology to condensed-matter systems. Although the current experimental apparatus provides access to only the relative group delay in electron wave packet propagation, future measurements of absolute emission delays may be feasible using the same methods, for example by direct comparison with gas-phase ATR data. At this point, attosecond photoemission spectroscopy presents a clear path toward ultimately uncovering the intermediate processes leading to ejection of a photoelectron. This information might shed new light on data previously obtained with conventional time-integral photoemission spectroscopy and allow for accurate description of charge dynamics on the electronic timescale in both condensed matter and on surfaces.

METHODS SUMMARY

A 1-kHz-repetition-rate, waveform-controlled, few-cycle, ~5-fs, 400-μJ, 750-nm Ti:sapphire laser system is the front-end of our apparatus and is used in combination with proper spectral filtering to efficiently generate isolated attosecond pulses of XUV radiation by high-harmonic generation. As shown in Fig. 1, the XUV and NIR pulses co-propagate towards a two-part focusing mirror at near normal incidence. The inner component is a Mo/Si multilayer mirror and reflects the XUV radiation over a bandwidth of ~6 eV (FWHM) centred at ~91 eV, supporting 300-as transform-limited pulses[30]. The XUV mirror is mounted on a translation stage, providing a precise delay between the XUV pump and the NIR streaking pulse. The temporal resolution that can be achieved in pump-probe experiments using these pulses and this apparatus is expected to be a small fraction of the pulse half-width because XUV pulses generated with similar spectra and multilayer optics have previously been fully characterized and observed to be gaussian[24].

The tungsten surface must be sufficiently free of contamination to minimize photoelectron scattering on emission. Therefore, the measurement chamber is maintained under ultrahigh vacuum conditions with typical background pressures of $< 10^{-9}$ mbar, which suppresses the accumulation of contaminates on the crystal surface to a level permitting a full ATR spectrogram to be recorded without interruption.

In our application of the ATR, detection of electrons occurs in the direction normal to the tungsten crystal surface, and the NIR streaking field is incident on the tungsten (110) crystal surface near Brewster's angle (~75°). For this angle of incidence, owing to the refractive index of tungsten, photoelectron wave packets are not efficiently accelerated in the direction of observation until they emerge from the surface. Even though the streaking field penetrates the tungsten crystal, inside the material the electric field component along the surface normal is weaker by a factor of approximately 16, allowing us to neglect streaking effects until the electron wave packets emerge from the surface. The absence of effective streaking until the photoelectron emerges from the surface is generally the case for solids, allowing us to time processes occurring within the material.

Additional, detailed description of the experimental apparatus, measurement technique, and data analysis is provided in Supplementary Information.

Received 20 June; accepted 3 September 2007.

1. Reis, D. A. & Lindenberg, A. M. in *Light Scattering in Solids IX* (eds Cardona, M. & Merlin, R.) 371–422 (Topics in Applied Physics 108, Springer, Berlin, 2007).
2. Fritz, D. M. *et al.* Ultrafast bond softening in bismuth: Mapping a solid's interatomic potential with X-rays. *Science* **315**, 633–636 (2007).
3. Pfeifer, T., Spielmann, C. & Gerber, G. Femto-second X-ray science. *Rep. Prog. Phys.* **69**, 443–505 (2006).
4. Hentschel, M. *et al.* Attosecond metrology. *Nature* **414**, 509–513 (2001).
5. Kienberger, R. *et al.* Atomic transient recorder. *Nature* **427**, 817–821 (2004).
6. Einstein, A. Über einen die Erzeugung und Verwandlung des Lichts betreffenden heuristischen Gesichtspunkt. *Ann. Phys.* **17**, 132–148 (1905).
7. Siegbahn, K. Electron-spectroscopy — outlook. *J. Electron Spectrosc. Relat. Phenom.* **5**, 3–97 (1974).
8. Berglund, C. N. & Spicer, W. E. Photoemission studies of copper and silver: theory. *Phys. Rev.* **136**, A1030–A1044 (1964).
9. Brühwiler, P. A., Karis, O. & Martensson, N. Charge-transfer dynamics studied using resonant core spectroscopies. *Rev. Mod. Phys.* **74**, 703–740 (2002).
10. Föhlisch, A. *et al.* Direct observation of electron dynamics in the attosecond domain. *Nature* **436**, 373–376 (2005).
11. Borisov, A., Sánchez-Portal, D., Díez-Muino, R. & Echenique, P. M. Dimensionality effects in time-dependent screening. *Chem. Phys. Lett.* **387**, 132–137 (2004).
12. Huber, R. *et al.* How many-particle interactions develop after ultrafast excitation of an electron-hole plasma. *Nature* **414**, 286–289 (2001).
13. Haight, R. Electron dynamics at surfaces. *Surf. Sci. Rep.* **21**, 277–325 (1995).
14. Cavalleri, A. *et al.* Tracking the motion of charges in a terahertz light field by femtosecond X-ray diffraction. *Nature* **442**, 664–666 (2006).
15. Yen, R. *et al.* Picosecond laser interaction with metallic zirconium. *Appl. Phys. Lett.* **40**, 185–187 (1982).
16. Höfer, U. *et al.* Time-resolved coherent photoelectron spectroscopy of quantized electronic states on metal surfaces. *Science* **277**, 1480–1482 (1997).
17. Petek, H. & Ogawa, S. Femtosecond time-resolved two-photon photoemission studies of electron dynamics in metals. *Prog. Surf. Sci.* **56**, 239–310 (1997).
18. Haight, R. & Peale, D. R. Tunable photoemission with harmonics of subpicosecond lasers. *Rev. Sci. Instrum.* **65**, 1853–1857 (1994).
19. Siffalovic, P. *et al.* Laser-based apparatus for extended ultraviolet femtosecond time-resolved photoemission spectroscopy. *Rev. Sci. Instrum.* **72**, 30–35 (2001).
20. Schins, J. M. *et al.* Observation of laser-assisted Auger decay in argon. *Phys. Rev. Lett.* **73**, 2180–2183 (1994).
21. Glover, T. E., Schoenlein, R. W., Chin, A. H. & Shank, C. V. Observation of laser assisted photoelectric effect and femtosecond high order harmonic radiation. *Phys. Rev. Lett.* **73**, 2180–2183 (1994).
22. Miaja-Avila, L. *et al.* Laser-assisted photoelectric effect from surfaces. *Phys. Rev. Lett.* **97**, 113604 (2006).
23. Baltuska, A. *et al.* Attosecond control of electronic processes by intense light fields. *Nature* **422**, 611–615 (2003).
24. Quéré, F., Mairesse, Y. & Itatani, J. Temporal characterization of attosecond XUV fields. *J. Mod. Opt.* **52**, 339–360 (2005).
25. Yakovlev, V., Bammer, F. & Scrinzi, A. Attosecond streaking measurements. *J. Mod. Opt.* **52**, 395–410 (2005).
26. Goulielmakis, E. *et al.* Direct measurement of light waves. *Science* **305**, 1267–1269 (2004).
27. Sansone, G. *et al.* Isolated single-cycle attosecond pulses. *Science* **314**, 443–446 (2006).
28. Kitzler, M., Milosevic, N., Scrinzi, A., Krausz, F. & Brabec, T. Quantum theory of attosecond XUV pulse measurement by laser dressed photoionization. *Phys. Rev. Lett.* **88**, 173904 (2002).
29. Tanuma, S., Powell, C. J. & Penn, D. R. Calculations of electron inelastic mean free paths. 2. Data for 27 elements over the 50–2000-eV range. *Surf. Interface Anal.* **17**, 911–926 (1991).
30. Wonisch, A. *et al.* Design, fabrication, and analysis of chirped multilayer mirrors for reflection of extreme-ultraviolet attosecond pulses. *Appl. Opt.* **45**, 4147–4156 (2006).

Supplementary Information is linked to the online version of the paper at www.nature.com/nature.

Acknowledgements We thank W. Hachmann for expeditious preparation of the XUV multilayer optical substrate. We acknowledge partial financial support by the Deutsche Forschungsgemeinschaft through the DFG Cluster of Excellence Munich Centre for Advanced Photonics, and through the SFB 613, and by the Volkswagen Stiftung Germany, and by the EURYI scheme award. P.M.E. acknowledges support from the Basque and Spanish Governments. R.K. acknowledges a fellowship from the Austrian Academy of Sciences and additional support from the Sofja Kovalevskaja Award of the Alexander von Humboldt Foundation. The apparatus to generate attosecond pulses was constructed at Technische Universität Wien, thanks to the support of the FWF.

Author Information Reprints and permissions information is available at www.nature.com/reprints. Correspondence and requests for materials should be addressed to A.L.C. (adrian.cavalieri@mpq.mpg.de) or F.K. (krausz@lmu.de) or U.H. (uheinzm@physik.uni-bielefeld.de).

LETTER

doi:10.1038/nature14094

Direct observation of electron propagation and dielectric screening on the atomic length scale

S. Neppl[1,2]†, R. Ernstorfer[3], A. L. Cavalieri[4,5,6], C. Lemell[7], G. Wachter[7], E. Magerl[2], E. M. Bothschafter[8], M. Jobst[1,2], M. Hofstetter[2,8], U. Kleineberg[2,8], J. V. Barth[1], D. Menzel[1,3], J. Burgdörfer[7,9], P. Feulner[1], F. Krausz[2,8] & R. Kienberger[1,2]

The propagation and transport of electrons in crystals is a fundamental process pertaining to the functioning of most electronic devices. Microscopic theories describe this phenomenon as being based on the motion of Bloch wave packets[1]. These wave packets are superpositions of individual Bloch states with the group velocity determined by the dispersion of the electronic band structure near the central wavevector in momentum space[1]. This concept has been verified experimentally in artificial superlattices by the observation of Bloch oscillations[2]—periodic oscillations of electrons in real and momentum space. Here we present a direct observation of electron wave packet motion in a real-space and real-time experiment, on length and time scales shorter than the Bloch oscillation amplitude and period. We show that attosecond metrology[3] (1 as = 10^{-18} seconds) now enables quantitative insight into weakly disturbed electron wave packet propagation on the atomic length scale without being hampered by scattering effects, which inevitably occur over macroscopic propagation length scales. We use sub-femtosecond (less than 10^{-15} seconds) extreme-ultraviolet light pulses[3] to launch photoelectron wave packets inside a tungsten crystal that is covered by magnesium films of varied, well-defined thicknesses of a few ångströms[4]. Probing the moment of arrival of the wave packets at the surface with attosecond precision reveals free-electron-like, ballistic propagation behaviour inside the magnesium adlayer—constituting the semi-classical limit of Bloch wave packet motion. Real-time access to electron transport through atomic layers and interfaces promises unprecedented insight into phenomena that may enable the scaling of electronic and photonic circuits to atomic dimensions. In addition, this experiment allows us to determine the penetration depth of electrical fields at optical frequencies at solid interfaces on the atomic scale.

A detailed microscopic understanding and control of electronic and optical properties of solids depends on our ability to access the dynamics of electrons on atomic time and length scales. Tracking the propagation of electrons in real time requires the ability to pinpoint their position at a rate comparable to the time on which interactions with other electrons and the crystal lattice may affect their trajectories inside the material. In a classical picture, the upper bound for the necessary temporal resolution is therefore the time it takes the electrons to travel the several-ångström distance between neighbouring atoms. This implies transit times well below one femtosecond even for kinetic energies as low as $E_{\rm kin} \approx 1$ eV. Previous time-resolving studies on electron propagation in condensed matter employed laser-based spectroscopic techniques to reveal ballistic currents and drift motion of charge carriers[5–8]. Restricted to the pico- and femtosecond timescales, they are able to probe carrier dynamics averaged over several hundreds of nanometres. In contrast, the experimental approach demonstrated here offers quantitative real-time access to electron transport on the inter-atomic length scale.

Figure 1 illustrates the basic principle of the experiment. Attosecond extreme-ultraviolet (XUV) light pulses generate Bloch electron wave packets with final-state energies substantially above the vacuum level. Electron wave packets with a sufficiently large momentum component along the surface-normal direction z contribute to the photoelectron current reaching the time-of-flight detector. Conventional photoelectron spectroscopy is restricted to measuring the energy and momentum distributions of photoelectrons. Here we additionally capture the temporal profile of the photoemission process by having the ejected electrons interact with the controlled few-cycle electric field of a visible/near-infrared (henceforth referred to as 'NIR') laser pulse covering the 500–1,000 nm spectral range and synchronized, with attosecond precision, to the XUV pulse.

As this field modifies ('streaks') the momentum of a photoelectron in proportion to the laser vector potential $A_{\rm L}(z, t)$ at the instant t the electron enters the NIR field[3,9,10], the temporal profile of the electron wave packet leaving the sample is mapped onto its final momentum distribution. Full streaking spectrograms obtained by recording these laser-modulated electron energy distributions as a function of delay Δt between the XUV and the NIR pulses are therefore highly sensitive to the spatio-temporal characteristics of both the photoelectron wave packet and the streaking laser field on the ångström length scale and the attosecond timescale[9–11].

Direct (time-domain) access to these electronic and optical wave packets promises a unique insight into the photoelectric effect, including underlying electron propagation and phenomena as fundamental as dielectric screening of light fields at solid surfaces. Here we show that this can be achieved by combining state-of-the-art attosecond timing metrology (chronoscopy)[3,9–11] with sample engineering on the ångström level[4,12].

When excited from Bloch states inside the crystal to positive-energy states[13], photoelectrons are not immediately exposed to the streaking field because the electric field amplitude is substantially reduced (or screened) at the topmost layer by the response of the metal electrons. Therefore, the time delay associated with the propagation of the respective Bloch wave packets towards the surface (included in both the quantum-mechanical one-step[13] and the semi-classical three-step[14,15] description of photoemission) is encoded in the streaking spectrogram[9–11]. Differences in the propagation times of electrons ejected from different initial states manifest themselves as a temporal offset between the respective streaking traces[9–11] (see Fig. 1a). Previous studies on single crystals revealed a considerable time delay between the emission of core-level and conduction band (CB) photoelectrons from the transition metal tungsten W(110)[9], whereas such a delay was found to be absent

[1]Physik-Department, Technische Universität München, 85747 Garching, Germany. [2]Max-Planck-Institut für Quantenoptik, Hans-Kopfermann-Straße 1, 85748 Garching, Germany. [3]Fritz-Haber-Institut der Max-Planck-Gesellschaft, Faradayweg 4–6, 14195 Berlin, Germany. [4]Max Planck Institute for the Structure and Dynamics of Matter, Luruper Chaussee 149, 22761 Hamburg, Germany. [5]Fakultät für Mathematik, Informatik und Naturwissenschaften, University of Hamburg, Luruper Chaussee 149, 22761 Hamburg, Germany. [6]Center for Free-Electron Laser Science (CFEL), Luruper Chaussee 149, 22761 Hamburg, Germany. [7]Institute for Theoretical Physics, Vienna University of Technology, Wiedner Hauptstrasse 8-10/E136, A-1040 Vienna, Austria. [8]Fakultät für Physik, Ludwig-Maximilians-Universität München, Am Coulombwall 1, D-85748 Garching, Germany. [9]Institute of Nuclear Research of the Hungarian Academy of Sciences (ATOMKI), 4001 Debrecen, Hungary. †Present address: Lawrence Berkeley National Laboratory, Chemical Sciences Division, Berkeley, California 94720, USA.

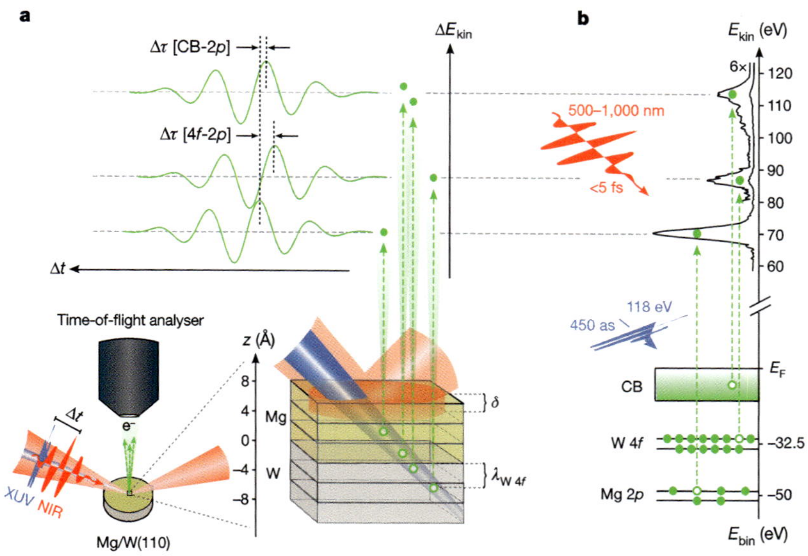

Figure 1 | Spatio-temporal dynamics in attosecond photoemission from Mg/W(110). **a**, Principle of the experiment: photoelectrons (green dots) are launched inside a tungsten W(110) crystal and a magnesium (Mg) overlayer a few ångströms thick by an XUV pulse of ~450 as, and are detected in ultrahigh vacuum with a time-of-flight analyser. At the surface, the arrival times of electrons released from different initial states are probed by streaking their associated electron energy distributions with a $2 \times 10^{11}\,\mathrm{W\,cm^{-2}}$ strong electric field delivered by a sub-5 fs broadband linearly polarized visible/near-infrared laser pulse. Relative time delays $\Delta\tau$ developing during the propagation of the photoelectrons to the metal–vacuum interface are detected as temporal shifts between their streaked energy distributions. The time shifts $\Delta\tau$ are sensitive to the atomic-scale electron transport characteristics (quantified by the inelastic mean free path λ; indicated only for the W 4*f* electrons), the Mg overlayer thickness and the screening behaviour of the laser field at the solid–vacuum interface. **b**, Schematic energy-level diagram for the probed electronic transitions. The central XUV photon energy of ~118 eV allows the simultaneous excitation of Mg 2*p*, W 4*f* and the joint CB states (binding energy E_{bin} as indicated). A background-corrected photoelectron spectrum of $n = 4$ adlayers of Mg on W(110) in the absence of the laser field is shown as the black solid line. For better visibility, the strength of the CB and W 4*f* signals are magnified by a factor of six.

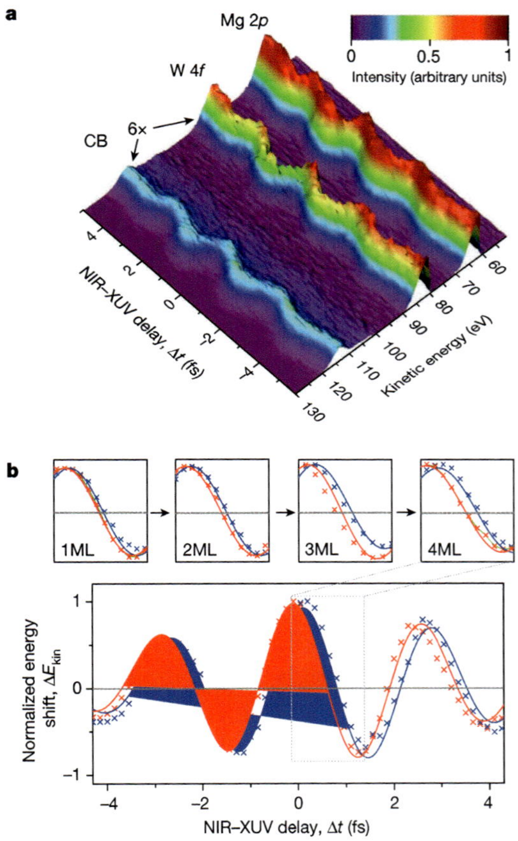

Figure 2 | Attosecond time-resolved photoemission from Mg/W(110). **a**, Representative streaking spectrogram for $n = 4$ Mg monolayers (ML). All photoelectron spectra are corrected for the inelastic electron background signal. The strength of the CB and W 4*f* signals is magnified by a factor of 6 for better visibility. **b**, Exemplary timing analysis of the Mg 2*p* and W 4*f* core-level electrons: the first moments calculated from their respective kinetic energy distributions are shown as red crosses (Mg 2*p*) and blue crosses (W 4*f*) as functions of NIR–XUV delay Δt. A global fit of the resultant streaking traces to a parameterized waveform for the NIR vector potential (solid lines) reveals a relative time shift $\Delta\tau[4f\text{-}2p]$, which can be identified with the time delay occurring during the release of the electrons from the metal surface. Insets illustrate the evolution of $\Delta\tau[4f\text{-}2p]$ for $0 < n \leq 4$. Regions exhibiting the largest gradient of the streaking field (corresponding to the highest temporal resolution) are highlighted. An analogous evaluation of $\Delta\tau[\mathrm{CB}\text{-}2p]$ is presented in the Supplementary Information.

in the photoemission from the free-electron metal magnesium Mg(0001)[10]. Theoretical models have addressed different contributions such as may arise from the band structure of the material[9,16,17], the spatial characteristics of the initial-state wavefunctions[18–21], and elastic and inelastic scattering effects[9,22]. These models also differ from each other in the way that the screening of the laser field at the surface is taken into account when calculating the photoemission time delays[19,22,23]. To isolate the atomic-scale electron propagation process from this multitude of disparate effects, we investigate hybrid metallic samples consisting of a controllable number *n* of Mg adlayers on a W(110) crystal[12,24] (see Fig. 1a and Supplementary Information for details) and contrast the measured time shifts with electron transport calculations.

In our experiments, XUV pulses with a duration of about 450 as and a photon energy of $\hbar\omega_{XUV} = 118$ eV simultaneously generate photoelectrons from core states of the substrate (W 4*f*) and adlayer (Mg 2*p*), as well as from the energetically overlapping CB states of both materials (Fig. 1b). A representative streaking spectrogram for $n = 4$ Mg adlayers on W(110) is shown in Fig. 2a. Despite the ~80% attenuation of the W substrate photoemission due to inelastic scattering in the Mg overlayer, the streaked W 4f and CB photoemission lines are clearly discernible. They are also sufficiently separated from each other and from the Mg 2p line over the entire range of XUV–NIR delays, Δt, to guarantee an accurate quantitative analysis of their relative emission dynamics[9–11]. In what follows, we reference the emission times of the W 4f and CB electrons to the Mg 2p emission from the Mg overlayer and denote the resultant relative delays as $\Delta\tau[4f-2p]$ and $\Delta\tau[\mathrm{CB}-2p]$, respectively.

We begin with an analysis of $\Delta\tau[4f-2p]$ because the involved photoelectrons originate from atomic-like states that are entirely localized in the W(110) substrate or the Mg overlayer. This allows unambiguous interrelation of the measured time shifts and the well-defined propagation distances in the Mg adlayer systems, which may not perfectly apply to CB electrons owing to the delocalized character of the CB initial-state wavefunctions[19–21]. Relative time shifts $\Delta\tau[4f-2p]$ extracted with a robust quantum-mechanical fitting scheme[9–11] (see Supplementary Information) from streaking spectrograms at different Mg coverages are summarized in Fig. 3a (blue diamonds). They reveal a distinct monotonic increase of $\Delta\tau[4f-2p]$ with the number n of Mg adlayers, reaching $\Delta\tau[4f-2p] = 215 \pm 20$ as for a Mg film 10.4 Å ($n = 4$) thick. We verified this trend using a simpler analysis that compares the first moments of the streaked W 4f and Mg 2p energy distributions as functions of Δt (Fig. 2b; see Supplementary Information).

The simplest description of the electron propagation is to consider ballistic motion of the centroid of a Bloch wave packet as a free point-like electron in one dimension[25]. For the Mg layers, $\Delta\tau[4f-2p]$ is then dominated by the average propagation time τ_{4f} of the 4f-derived wave packets with central wavevectors $k = k_0$ travelling at a group velocity $v_{4f} = \frac{\mathrm{d}E(k)}{\hbar \mathrm{d}k}\big|_{k=k_0}$. Owing to the free-electron-like band-structure of Mg[10,26], we can assume that $v_{4f} \approx \sqrt{2E_{\mathrm{kin}}/m_e}$, where m_e is the free electron mass. The kinetic energy of the photoelectrons inside the Mg layer amounts to $E_{\mathrm{kin}} \approx 93$ eV (the photon energy is $\hbar\omega_{\mathrm{XUV}} = 118$ eV, the binding energy of the 4f electrons is $E_b \approx 32.5$ eV, the Fermi energy of bulk Mg is $E_{\mathrm{Fermi}} \approx 7$ eV), leading to an average group velocity of $v_{4f} \approx 0.057$ Å as^{-1}. We therefore expect τ_{4f}—which is the upper limit for $\Delta\tau[4f-2p]$—to increase almost linearly with the number of Mg adlayers n, according to $\tau_{4f} \approx n \times d/v_{4f} \approx n \times 45$ as, where $d = 2.6$ Å is the interlayer spacing of the epitaxial Mg films[12]. A linear fit to the experimental data of Fig. 3a yields a delay of ~42 as per adlayer. The good agreement between experiment and model prediction provides conclusive evidence for the atomic-scale ballistic propagation of the 4f electrons being the microscopic origin of the observed time shifts in the spectrograms and also corroborates free-electron-like transport in Mg.

This interpretation is substantiated by electron transport simulations[22] of the ballistic motion of the W 4f, Mg 2p and CB electrons in the Mg/W(110) systems. In general, time delays obtained from such transport calculations are sensitive to (1) the average group velocities of the electrons at the relevant energies, (2) their energy-dependent inelastic mean free path $\lambda(E_{\mathrm{kin}})$ in the traversed materials and (3) the spatio-temporal profile of the streaking field near the surface[22]. The average group velocities can be deduced from electronic structure calculations[9,17] and all relevant values of $\lambda(E_{\mathrm{kin}})$ are known from synchrotron experiments[10]

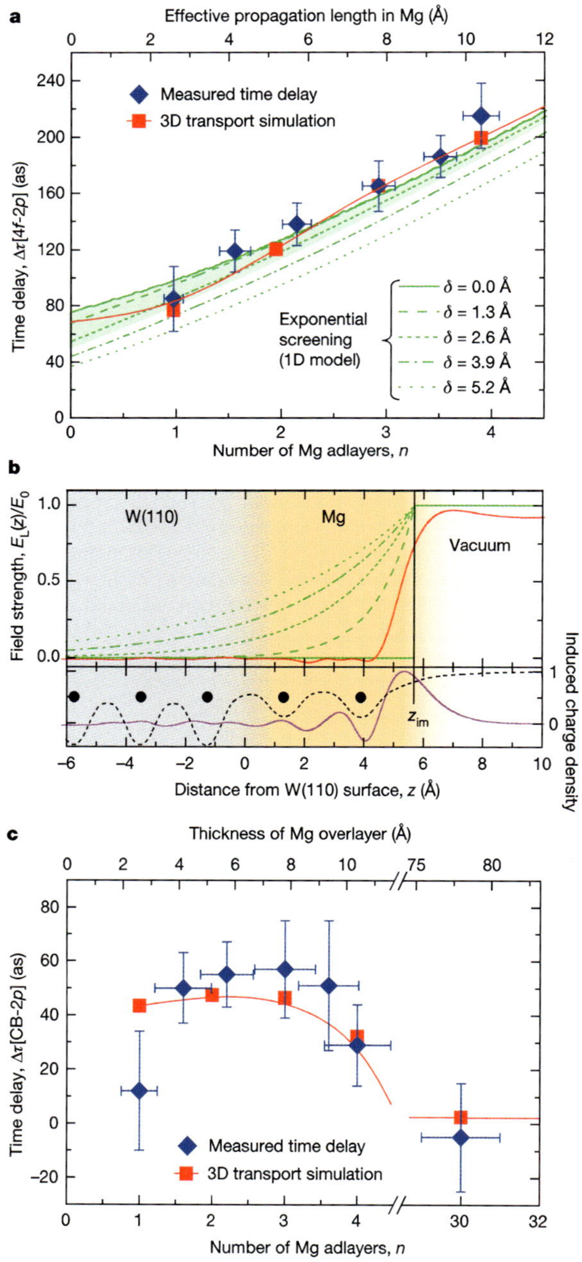

Figure 3 | Atomic-scale photoelectron transport and screening of the incident light field. a, Time delays $\Delta\tau[4f-2p]$ between the release of W 4f and Mg 2p electrons extracted from a large set of streaking spectrograms with different numbers of Mg adlayers are shown as blue diamonds. Error bars denote full standard deviations and are obtained by averaging measurements performed under similar experimental conditions. Fractional adlayers correspond to dispersed two-dimensional islands (on top of a completed Mg layer) that coalesce upon further Mg deposition. Green lines are time delays predicted by our one-dimensional (1D) simulation of the photoelectron release dynamics for different screening lengths δ assuming an exponentially decaying normal component of the incident NIR streaking field at the metal–vacuum interface according to $E_{\mathrm{L}}(z, t) = E_0 e^{-(z-z_{\mathrm{im}})/\delta}$. The light-green-shaded area highlights the screening scenarios compatible with the experiment. Red squares indicate time delays derived from a full three-dimensional electron transport model (the red line is a guide to the eye). **b**, The upper panel illustrates the different screening scenarios for $E_{\mathrm{L}}(z, t)$ considered in **a** for the example of $n = 2$ Mg adlayers (orange shading) on W(110) (grey shading). The red line is the spatial variation of $E_{\mathrm{L}}(z, t)$ at the interface predicted by TDDFT. The lower panel is a snapshot of the NIR-induced charge density at the metal–vacuum interface at the maximum of the laser pulse derived by TDDFT (magenta line). The position of the dynamic image plane z_{im} is indicated as a vertical black solid line. The lattice potential (averaged parallel to the crystal surface) employed in the DFT calculations is shown as a dotted black line. The positions of the Mg and W atoms at the interface along the surface normal are indicated as black dots. The gradual transitions from the W electronic structure (grey) to the Mg electronic structure (orange) and to the vacuum (white) are indicated in the background colours. **c**, Comparison of time shifts $\Delta\tau[\mathrm{CB}-2p]$ measured between the emission of CB and Mg 2p electrons with time delays predicted by the three-dimensional (3D) electron transport model.

or theory[27]. As a consequence, our experiments open up the opportunity to explore the spatial variation of the laser field's $E_L(z)$ component normal to the surface (z axis) on the ångström length scale.

Screening at metallic surfaces becomes effective near the so-called image plane z_{im}, located about a half-layer spacing outside the centre of the topmost atomic layer[28,29]. As our method probes the optical near-field at the metal–vacuum interface, the commonly used Fresnel equations based on macroscopic properties of target components with perfectly sharp interfaces cannot be applied. Instead, a phenomenological exponential decay $E_L(z,t) = E_0 e^{-(z-z_{im})/\delta}$ of the surface-normal component of the field inside the material appears to be a reasonable assumption[18,23]. We therefore modelled the impact of different screening lengths δ, that is, the length scale on which the stepwise prediction of the Fresnel formula does not apply, on the time delays $\Delta\tau[4f - 2p]$ using one-dimensional electron transport simulations (see Supplementary Information). The time delays $\Delta\tau[4f - 2p]$ predicted by this simple model as a function of n and δ are plotted as green lines in Fig. 3a. Apparently, only the range $0 \le \delta \le 3$ Å is compatible with the experiment and the associated error bars, indicating a screening within one atomic layer of Mg.

To scrutinize the origin of this rapid interfacial screening, we used time-dependent density functional theory (TDDFT)[30] to calculate $E_L(z, t)$ for the Mg/W(110) systems. The surface-normal component of the incident laser field induces a polarization charge layer at the metal surface that shields the interior of the solid against the external electric field. The centroid of the induced screening charge density defines the exact position z_{im} of the image plane, which marks the microscopic onset of the local screening process[30] (see Fig. 3b). Both the positions $z_{im,n}$ and the width δ_n of the induced screening charge are found to be almost independent of the number n of Mg layers. The key finding is that the laser field is already fully screened at the plane defined by the centre of the atoms of the topmost layer for all Mg/W(110) systems, in agreement with the conclusion drawn from the comparison of our experimental data with the phenomenological modelling.

Finally, we incorporated the abrupt screening of the streaking field at the surface revealed by TDDFT in a full three-dimensional streaking simulation[22] of the electron propagation in Mg/W(110) (see Supplementary Information for a detailed description). Similar to the above-mentioned one-dimensional model, wave packet propagation is simulated by the transport of an ensemble of point-like charges taking stochastic inelastic and elastic scattering events into account. The time delays $\Delta\tau[4f - 2p]$ predicted by these calculations (red squares in Fig. 3a) are in good agreement with the experiment.

Compared to the core-level photoemission time delay $\Delta\tau[\mathrm{W}\ 4f - 2p]$, the temporal shift of the conduction band emission $\Delta\tau[\mathrm{CB} - 2p]$ (Fig. 3c) is distinctly smaller and exhibits a strikingly different dependence on the number of Mg adlayers. A detailed analysis of $\Delta\tau[\mathrm{CB} - 2p]$ within electron transport models is complicated by different (spectrally unresolved) contributions of W(110)- and Mg-derived states to the joint CB feature at $E_{kin} \approx 115$ eV. However, by weighting the excitation probabilities from these different initial states according to their atomic photoexcitation cross-sections (see Supplementary Information), we achieved good overall agreement with the experimental results, and correctly reproduced the vanishing $\Delta\tau[\mathrm{CB} - 2p]$ time delay for bulk Mg[10]. This suggests that our approximate treatment of the Bloch wave packet propagation and the dielectric screening response remains valid also for more delocalized initial electronic states, in contrast to recent predictions[23,24].

We emphasize that $\Delta\tau[\mathrm{CB} - 2p]$ for $n = 1$ is overestimated in our transport model and lies outside the experimental error margin. A deviation from the present model appears likely for the Mg/W(110) monolayer system, since strong mixing of band states at the interface may lead to a deviation of the initial-state band structure and excitation cross-sections from their bulk characteristics[24]. A detailed discussion of this phenomenon is beyond the scope of the present study, but indicates the potential of attosecond photoelectron spectroscopy to probe interfacial hybridization between electronic states directly in the time domain. Therefore this technique may be applied to phenomena not describable by a simple combination of separate electronic structure models for bulk and surface layers.

This work extends the realm of attosecond spectroscopy to the direct observation of atomic-scale propagation and damping of electronic and optical wave packets at solid surfaces. The resultant insight into attosecond temporal and—simultaneously—ångström spatial dimensions opens the door for understanding and exploring electron transport phenomena on the atomic scale and the dielectric response of solids at optical frequencies. Applied to overlayer materials with non-free-electron-like positive-energy states, such studies will shed light on whether stationary band structure can be used to predict atomic-scale electron propagation on ultrashort timescales. Extrapolation of coverage-dependent streaking spectroscopy to the sub-monolayer regime will provide access to absolute photoemission times and possible intrinsic (atomic) retardation effects in the photoemission process. Beyond addressing these fundamental questions, attosecond electron transport chronoscopy may prove instrumental in advancing electronic and photonic circuits towards atomic dimensions.

Received 15 September; accepted 14 November 2014.

1. Bloch, F. Über die Quantenmechanik der Elektronen in Kristallgittern. *Z. Phys.* **52,** 555–600 (1929).
2. Leo, K., Bolivar, P. H., Brüggemann, F., Schwedler, R. & Köhler, K. Observation of Bloch oscillations in a semiconductor superlattice. *Solid State Commun.* **84,** 943–946 (1992).
3. Hentschel, M. *et al.* Attosecond metrology. *Nature* **414,** 509–513 (2001).
4. Schiller, F., Heber, M., Servedio, V. D. P. & Laubschat, C. Electronic structure of Mg: from monolayers to bulk. *Phys. Rev. B* **70,** 125106 (2004).
5. Gremillet, L. *et al.* Time-resolved observation of ultrahigh intensity laser-produced electron jets propagating through transparent solid targets. *Phys. Rev. Lett.* **83,** 5015–5018 (1999).
6. Sha, W., Norris, T. B., Schaff, W. J. & Meyer, K. E. Time-resolved ballistic acceleration of electrons in a GaAs quantum-well structure. *Phys. Rev. Lett.* **67,** 2553–2556 (1991).
7. Sha, W., Rhee, J.-k., Member, S., Norris, T. B. & Schaff, W. J. Transient carrier and field dynamics in quantum- well parallel transport: from the ballistic to the quasi-equilibrium regime. *IEEE J. Quantum Electron.* **28,** 2445–2455 (1992).
8. Shaner, E. & Lyon, S. Picosecond time-resolved two-dimensional ballistic electron transport. *Phys. Rev. Lett.* **93,** 037402 (2004).
9. Cavalieri, A. L. *et al.* Attosecond spectroscopy in condensed matter. *Nature* **449,** 1029–1032 (2007).
10. Neppl, S. *et al.* Attosecond time-resolved photoemission from core and valence states of magnesium. *Phys. Rev. Lett.* **109,** 22–26 (2012).
11. Schultze, M. *et al.* Delay in photoemission. *Science* **328,** 1658–1662 (2010).
12. Aballe, L., Barinov, A., Locatelli, A., Mentes, T. O. & Kiskinova, M. Initial stages of heteroepitaxial Mg growth on W(110): early condensation, anisotropic strain, and self-organized patterns. *Phys. Rev. B* **75,** 115411 (2007).
13. Mahan, G. D. Theory of photoemission in simple metals. *Phys. Rev. B* **2,** 4334–4350 (1970).
14. Berglund, C. N. & Spicer, W. E. Photoemission studies of copper and silver: theory. *Phys. Rev.* **136,** A1030–A1044 (1964).
15. Feibelman, P. J. & Eastman, D. E. Photoemission spectroscopy—correspondence between quantum theory and experimental phenomenology. *Phys. Rev. B* **10,** 4932–4947 (1974).
16. Borisov, G., Sánchez-Portal, D., Kazansky, K. & Echenique, P. M. Resonant and nonresonant processes in attosecond streaking from metals. *Phys. Rev. B* **87,** 121110 (2013).
17. Krasovskii, E. E. Attosecond spectroscopy of solids: streaking phase shift due to lattice scattering. *Phys. Rev. B* **84,** 195106 (2011).
18. Liao, Q. & Thumm, U. Attosecond time-resolved photoelectron dispersion and photoemission time delays. *Phys. Rev. Lett.* **112,** 023602 (2014).
19. Kazansky, A. K. & Echenique, P. M. One-electron model for the electronic response of metal surfaces to subfemtosecond photoexcitation. *Phys. Rev. Lett.* **102,** 177401 (2009).
20. Zhang, C. H. & Thumm, U. Attosecond photoelectron spectroscopy of metal surfaces. *Phys. Rev. Lett.* **102,** 123601 (2009).
21. Zhang, C. H. & Thumm, U. Effect of wave-function localization on the time delay in photoemission from surfaces. *Phys. Rev. A* **84,** 033401 (2011).
22. Lemell, C., Solleder, B., Tőkési, K. & Burgdörfer, J. Simulation of attosecond streaking of electrons emitted from a tungsten surface. *Phys. Rev. A* **79,** 62901 (2009).
23. Zhang, C. H. & Thumm, U. Probing dielectric-response effects with attosecond time-resolved streaked photoelectron spectroscopy of metal surfaces. *Phys. Rev. A* **84,** 1–7 (2011).
24. Vinogradov, N., Marchenko, D., Shikin, A., Adamchuk, V. & Rader, O. Size effects in ultrathin Mg/W(110) films: quantum electronic states. *Phys. Solid State* **51,** 179–188 (2009).
25. Kroemer, H. On the group velocity of Bloch waves. *Proc. IEEE* **63,** 988 (1975).

RESEARCH LETTER

26. Bartynski, R. A., Gaylord, R. H., Gustafsson, T. & Plummer, E. W. Angle-resolved photoemission study of the surface and bulk electronic structure of Mg(0001) and Mg(112-bar0). *Phys. Rev. B* **33,** 3644–3665 (1986).
27. Tanuma, S., Powell, C. J. & Penn, D. R. Calculations of electron inelastic mean free paths. IX. Data for 41 elemental solids over the 50 eV to 30 keV range. *Surf. Interf. Anal.* **43,** 689–713 (2011).
28. Lang, N. D. & Kohn, W. Theory of metal surfaces: induced surface charge and image potential. *Phys. Rev. B* **7,** 3541–3550 (1973).
29. Liebsch, A. Electronic screening at metal surfaces and the connection with physical phenomena. *Phys. Scr.* **35,** 354 (1987).
30. Wachter, G. *et al.* Electron rescattering at metal nanotips induced by ultrashort laser pulses. *Phys. Rev. B* **86,** 035402 (2012).

Supplementary Information is available in the online version of the paper.

Acknowledgements This research was supported by the Munich-Centre for Advanced Photonics. C.L., G.W. and J.B. acknowledge support by the FWF special research programs SFB-041 (ViCoM) and SFB-049 (NextLite) and project P21141-N16. G.W. is supported by the International Max Planck Research School for Advanced Photon Science (IMPRS-APS). R.K. acknowledges an ERC Starting Grant. Calculations have been performed on the Vienna Scientific Cluster. S.N. and P.F. thank the Helmholtz Zentrum Berlin for support. We thank P. Echenique, E. E. Krasovskii, A. Kazansky and A. D. Sanchez-Portal for discussions.

Author Contributions S.N. conceived the material system for this study and performed preparatory experiments. S.N., A.L.C., P.F., E.M., R.E. and R.K. designed and developed the experiment. S.N., R.E. and A.L.C. performed the measurements (with the assistance of E.M., M.J. and E.M.B.). S.N. and R.E. analysed the data. C.L. and S.N. performed the ballistic electron simulations. G.W. and C.L. performed the TDDFT calculations. M.H. and U.K. developed and prepared the XUV multilayer optics. S.N., R.E., C.L., J.B. and R.K. wrote the manuscript with input from the other authors. R.K. and F.K. initiated the project and R.K., F.K. and P.F. supervised the project. All authors discussed the results and conclusions drawn from them.

Author Information Reprints and permissions information is available at www.nature.com/reprints. The authors declare no competing financial interests. Readers are welcome to comment on the online version of the paper. Correspondence and requests for materials should be addressed to S.N. (sneppl@lbl.gov) and R.K. (reinhard.kienberger@tum.de).

A three-channel optical waveform synthesizer using ultrabroad-band dicroic chirped mirrors for subdividing the super-octave continuum delivered by the white beam into sub-octave (red, yellow, blue) spectral channels and guiding the beams through the device [50]. Extended with a fourth (violet) channel, the apparatus produced the first sub-optical-cycle radiation in the visible, ultraviolet and infrared spectral domain [51]. © Thorsten Naeser

9 Steering Electronic Motions at the Attosecond Scale

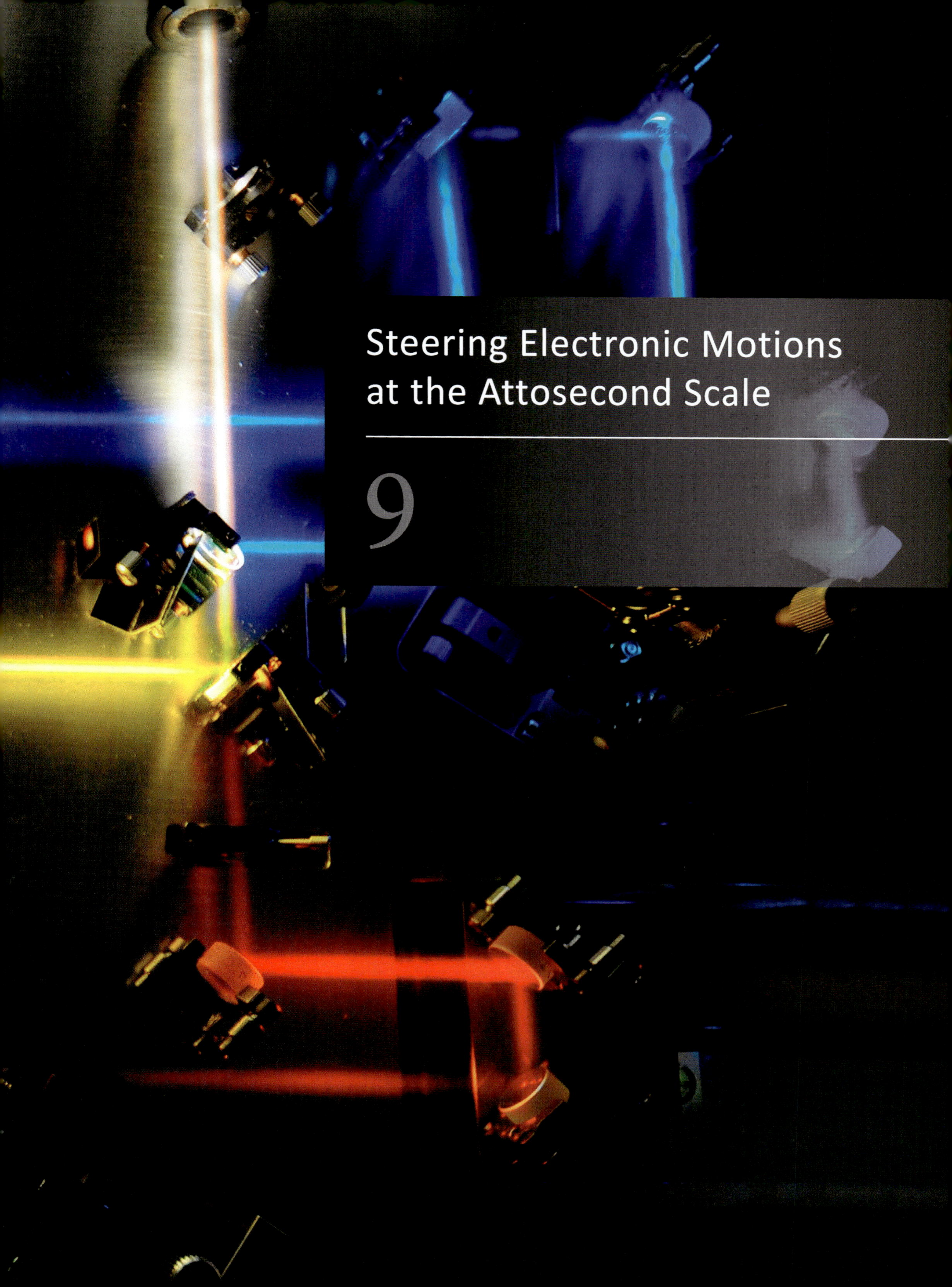

By exposing atomic systems to strong controlled laser fields [30,36], one can exert a force with a controlled and known temporal evolution on electrons. Its strength can rival those of intra-atomic and intra-molecular forces and is varying on the time scale characteristic of electronic motions occurring inside atoms and molecules.

A first spectacular manifestation of this control has been the generation of isolated attosecond XUV pulses coming in a single burst or as twin pulses depending on the carrier-envelope phase (CEP) of the few-cycle laser field driving the (high-order harmonic) generation process. In fact, a cosine waveform (Fig. 1-1, page 32) of the few-cycle laser field with one distinct strongest half cycle at the center of the pulse was found to give rise to a single attosecond burst emerging from the high-harmonic generation process after spectral filtering. By contrast, a sine waveform (Fig. 1-1, page 32) yields twin attosecond pulses in the same setting. The experiments demonstrating these striking implications of a slight reshaping of the waveform are reviewed in [45]. This review also addresses the question of how attosecond physics sheds light on wave-particle duality.

The same differences in the evolution of strong few-cycle fields have been shown to lead to isolated emission of photoelectron wavepackets in the form of single vs. twin pulses. Control of the sub-femtosecond temporal structure of electron emission is closely related to that of attosecond pulse generation, given that the latter process follows the former in the very same field-atom interaction. A group led by Gerhard Paulus and Herbert Walther took the lead to demonstrate how strong few-cycle light with a controlled waveform is able to release electrons via strong-field-induced ionization in one or two time slots of sub-cycle, i.e. sub-femtosecond duration [46].

In the experiments previously reviewed, attosecond control was demonstrated in the simplest possible experimental setting – in the interaction of few-cycle light fields with isolated atoms. However, the attosecond light force also offers the potential for exercising control over atomic-scale electron motions in molecular systems. This prompts the question, to what extent the possible outcome of these motions, such as changes in the structure and/or atomic constitution of molecules, may be affected by the attosecond temporal evolution of the field driving those motions.

Matthias Kling, Marc Vrakking and their collaborators have performed the first proof-of-principle experiment (Fig. 9-1) to address this question in the simplest of all molecules, in H_2 [47]. The study reveals that full control in photo-chemistry may have to start in the sub-femtosecond domain, with the launching of electron wavepackets and subsequent control of their motion within the molecule during the initial time window

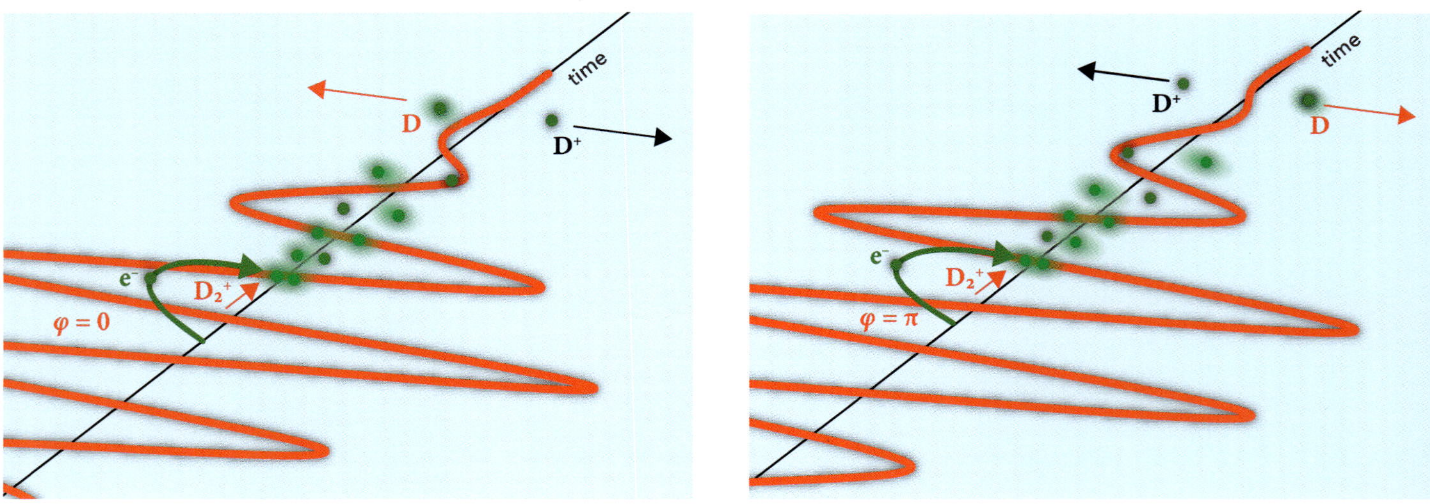

Figure 9-1 Attosecond optical field control of molecular dissociation [47]. Optical field ionization by a few-cycle laser pulse leaves the simplest molecule, H_2, or its isotope, D_2, behind with only one electron "connecting" the nuclei. This electron, the probability distribution of which is depicted in green, was promoted to an excited state during the interaction, in which it was unable to bind the two nuclei to each other to form a stable molecule. As a consequence, the molecule breaks apart. During dissociation, the electron is driven back and forth between the nuclei by the electric field of the few-cycle light wave until it sticks to one of them. To which one, was experimentally found to be controllable with the light waveform. A cosine waveform and its inverted counterpart maximized the probability of the electron teaming up with the nucleus ejected to the left (left panel) and right (right panel), respectively.

where the nuclei are still frozen. This proof-of-principle study also suggests the feasibility of directing charge in molecules by the electric force of light on a sub-fs timescale.

A multi-lateral collaboration initiated and conducted by Matthias Kling extended sub-femtosecond photochemistry to an organic molecule [48], demonstrating that the outcome of a photo-chemical process can also be affected in a more complex system by the dynamics occurring before the nuclei actually start moving. To what extent the fate of the molecule can be biased or even predetermined by these initial electronic motions is among the principal questions attosecond science will have to address in the years to come.

Sub-cycle shaping of the strong laser-induced force was miniscule in the experiments reviewed in [45-48], as revealed by Figure 3b (page xviii). The substantial effect of this small change on attosecond XUV pulse generation and on the outcome of photochemical processes results from a highly nonlinear response to the driving field. Yet, versatile atomic-scale electron steering depends on much more freedom in tailoring the driving force-field's sub-femtosecond temporal evolution than offered merely by a shift of the carrier-envelope phase. Sub-cycle temporal shaping requires a bandwidth of the force-field that substantially exceeds an octave and ways of precisely controlling the group-delay dispersion over this bandwidth. The former condition can be readily fulfilled via optical-Kerr-effect-induced self-phase modulation (and some other nonlinear interactions) driven by intense few-cycle laser pulses. The latter requirement constitutes a formidable challenge, namely control of the group delay with sub-femtosecond precision over the entire spectral range of frequencies carried in the laser pulse.

There is only one technology that offers a remedy: chirped multilayers. Their permanent advancement ever since their invention [8] permitted the advancement of femtosecond technology to its limit by the generation of near-single-cycle pulses of visible-infrared light (Chapter 2). Vladimir Pervak (technological development) and Michael Trubetskov (computational modeling) have been at the forefront of the world-wide activities that have been driving this progress. Thanks to their and other researchers' persistent efforts, chirped multilayers are now pushing the frontiers of attosecond technology more than two decades after their birth by permitting synthesis of super-octave waveforms of visible-infrared light. The progress article [49] reviews how they do so. Advances in the design, manufacturing, and characterization of broadband multilayer structures over two decades have culminated in dichroic chirped multilayers permitting spectral separation, recombination as well as group-delay control. These optical elements can now subdivide a multi-octave spectral band into several sub-octave channels that each offer independent, individual control of the group delay with sub-femtosecond precision.

By the end of the first decade of attosecond physics, these novel broadband optical tools permitted the development of a multi-channel light waveform synthesizer [50]. This feat, achieved by Adrian Wirth, Mohammed Hassan, Eleftherios Goulielmakis and their collaborators has brought about the technical capability of versatile shaping of strong force fields within the optical wave cycle, i.e. on an attosecond time scale. An immediate implication has been the synthesis of powerful optical (ultraviolet-to-infrared) transients with sub-femtosecond duration, which opens the door to attosecond-pump/attosecond-probe spectroscopy [51].

Pioneering research led by Peter Baum has brought to light the fact that the controlled force of optical fields can also be utilized to temporally compress accelerated free electron wavepackets. Catherine Kealhofer and her co-workers carried out the first experiment proving this concept. By using THz control fields, the team was able to compress free multi-keV electron packets to tens of femtoseconds [52]. Scaling the concept towards hundreds of THz with synthesized visible-infrared fields [50] will give rise to multi-keV, ångström-wavelength electron pulses with sub-femtosecond duration, which are synchronized to the controlling optical field with sub-femtosecond precision. Achieving this milestone will open the door to capturing electron distributions in atomic systems with ångström resolution in space and attosecond resolution in time by means of *attosecond electron diffraction.*

The strong electric force of light engineerable on atomic-to-molecular time scales that is now becoming available, opens up fascinating prospects in manipulating, controlling and capturing processes of the microcosm.

[45] M. Drescher, F. Krausz. Attosecond physics: facing the wave-particle duality. *Journal of Physics B* **38**, 727 (2005).
(Page: 386)

[46] F. Lindner, M. G. Schätzel, H. Walther, A. Baltuška, E. Goulielmakis, F. Krausz, D. B. Milosevic, D. Bauer, W. Becker, G. G. Paulus. Attosecond double-slit experiment. *Physical Review Letters* **95**, 040401 (2005).
(Page: 400)

[47] M. Kling, C. Siedschlag, A. Verhoef, J. I. Khan, M. Schultze, T. Uphues, Y. F. Ni, M. Uiberacker, M. Drescher, F. Krausz, M. Vrakking. Control of electron localization in molecular dissociation. *Science* **312**, 246 (2006).
(Page: 404)

[48] A. Alnaser, M. Kübel, R. Siemering, B. Bergues, N. Kling, K. Betsch, Y. Deng, J. Schmidt, Z. Alahmed, A. Azzeer, J. Ullrich, I. Ben-Itzhak, R. Moshammer, U. Kleineberg, F. Krausz, R. de Vivie-Riedle, M. Kling. Subfemtosecond steering of hydrocarbon deprotonation through superposition of vibration modes. *Nature Communications* **5**, 3800 (2014).
(Page: 407)

[49] O. Razskazovskaya, F. Krausz, V. Pervak. Multilayer coatings for femto- and attosecond technology. *Optica* **4**, 129 (2017).
(Page: 413)

[50] A. Wirth, M. Hassan, I. Grguras, J. Gagnon, A. Moulet, T. Luu, S. Pabst, R. Santra, Z. Alahmed, A. Azzeer, V. Yakovlev, V. Pervak, F. Krausz, E. Goulielmakis. Synthesized light transients. *Science* **334**, 195 (2011).
(Page: 423)

[51] M. Hassan, T. Luu, A. Moulet, O. Razskazovskaya, P. Zhokhov, M. Garg, N. Karpowicz, A. Zheltikov, V. Pervak, F. Krausz, E. Goulielmakis. Optical attosecond pulses and tracking the nonlinear response of bound electrons. *Nature* **530**, 66 (2016).
(Page: 429)

[52] C. Kealhofer, W. Schneider, D. Ehberger, A. Ryabov, F. Krausz, P. Baum. All-optical control and metrology of electron pulses. *Science* **352**, 429 (2016).
(Page: 445)

INSTITUTE OF PHYSICS PUBLISHING JOURNAL OF PHYSICS B: ATOMIC, MOLECULAR AND OPTICAL PHYSICS

J. Phys. B: At. Mol. Opt. Phys. **38** (2005) S727–S740 doi:10.1088/0953-4075/38/9/019

Attosecond physics: facing the wave–particle duality

Markus Drescher[1,2] **and Ferenc Krausz**[3,4]

[1] Institut für Experimentalphysik, Universität Hamburg, Luruper Chaussee 149, 22761 Hamburg, Germany
[2] Fakultät für Physik, Universität Bielefeld, Universitätsstr. 25, 33615 Bielefeld, Germany
[3] Max-Planck-Institut für Quantenoptik, Hans-Kopfermann-Strasse 1, D-85748 Garching, Germany
[4] Department für Physik, Ludwig-Maximilians-Universität München, Am Coulombwall 1, D-85478 Garching, Germany

Received 17 December 2004
Published 25 April 2005
Online at stacks.iop.org/JPhysB/38/S727

Abstract

Recent progress in generation and control of intense optical fields has given rise to isolated soft x-ray pulses with durations significantly below 1 fs. These constitute a tool of unprecedented temporal definition for attosecond physics—the study and manipulation of electronic motion on a time scale approaching the atomic unit of time. A key mechanism in such experiments is the well-defined momentum transfer between a quasi-free electron, released from an atom following irradiation by a short x-ray pulse, and a precisely controlled strong visible light field. The electrons' final kinetic energy thus sensitively depends on the timing of electron release with respect to the field oscillations and reveals the ejected electrons' confinement in time with sub-cycle, i.e. attosecond, resolution. Experiments resulting in electron emission of different durations can be interpreted in terms of a particle-like or wave-like electron, depending on whether the emission duration is considerably shorter or longer than the wave period of the probing light.

1. Introduction

The interpretation [1] of disturbing experimental findings [2] in the photoeffect in conjunction with the subsequent fundamental work in quantum physics at the turn of the 20th century forced us to accept the dual nature of light and matter as having both particle and wave properties. As confirmed in many experiments, matter reveals a particle or wave character depending on the question asked, i.e. the type of experiment conducted. In this paper, we discuss how observation of the interplay between light and matter *in the time domain* allows one—in the same experiment—to interrogate the particle and the wave nature of an electron simultaneously. The length of the temporal interval over which electrons are set free to interact with the light field (in the terminology of quantum mechanics: the duration of the electron wave packet) in relation to the oscillation period of the light wave decides which property

0953-4075/05/090727+14$30.00

dominates. By varying the temporal width of the probing light 'gate' or the duration of the particle wave packet, the transition between classical and quantum-mechanical behaviour can be studied.

A comprehensive formal consideration of both wave properties (e.g. oscillation and interference) and particle properties (e.g. momentum and spatial confinement) is provided by the wave packet concept [3] as a suitable superposition of extended waves within a certain frequency interval. Via Fourier transform, the interval of contributing wave frequencies determines the temporal and longitudinal spatial extents of the wave packet. Well-defined freely propagating light wave packets can be obtained from modern pulsed laser systems [4] with a time-bandwidth product close to its minimum possible value (Fourier-limited pulse). The photoeffect serves as a mediator for converting a light wave packet into a matter (= electron) wave packet in a highly coherent fashion. In our approach, an electron wave packet is launched from atoms ionized by an XUV pulse to interact with the strong field of a pulse of visible laser light. The relative time scales of the electron wave packet and the probing light field appear to determine the observed character of the electron.

If the electron wave packet's duration τ_e is considerably shorter than the light oscillation period T_L, $\tau_e \ll T_L$, particle-like behaviour is observed; if $\tau_e \gg T_L$, the wave nature will dominate. From an experimental point of view, the latter condition is much easier to meet because it relies on rather long—fs to ps—ionizing pulses which can be delivered from a number of sources [5]. Observation of particle-like behaviour, however, is much more demanding; probing with a light field of period $T_L \approx 2.5$ fs (red light) calls for XUV pulses with a duration $\tau_x < 1$ fs. It is only the recent progress in generation of attosecond (1 as $= 10^{-18}$ s) XUV pulses that allows us to unveil the particle character of electrons in such experiments and study the gradual transition in their behaviour to propagation in the form of waves.

In section 2, we introduce the photonic tools necessary for our technique: (i) phase-controlled few-cycle intense laser pulses and (ii) isolated sub-femtosecond XUV pulses. In the following sections, the electron's particle or wave character will manifest itself in experimental examples for the cases $\tau_e \ll T_L$ and $\tau_e \gg T_L$, respectively. The former case provides an illustrative picture for introducing the basic concept of probing the electron emission with a rapidly varying electromagnetic field. The interaction of sub-femtosecond electron emission with a fully controlled (few-cycle) optical wave permits the time-domain characterization of sub-fs XUV pulses as well as of the electric field of visible light in a classical framework, constituting the first manifestations of attosecond metrology. The same type of experiment reveals the wave-like nature of the electron if its emission lasts longer than T_L. In our experiments, an electron wave packet with this property results from secondary electron emission due to atomic Auger decay following ionization. A suitable choice of experimental parameters allows one to generate situations where $\tau_e \sim T_L$. For such intermediate cases, the electron exhibits particle and wave properties simultaneously.

2. Basic concepts and tools of attosecond metrology

Access to ultrafast phenomena relies on the measurement of brief time intervals, which, in turn, requires a physical quantity that varies fast in a controlled and reproducible manner. In femtosecond metrology, the amplitude envelope $E_a(t)$ of the electric field

$$E_L(t) = E_a(t)\cos(\omega_L t + \varphi) \tag{1}$$

of a femtosecond laser pulse (of carrier frequency ω_L) takes over the role of this quantity. In the shortest pulses, the rise time of the field amplitude is as brief as a few femtoseconds [6] and permits metrology with a resolution approaching 1 fs. Pump–probe techniques furnished with this time resolution opened the way to observing for the first time the breakage and creation of chemical bonds by tracing the motion of atoms within molecules in real time [7]. Extension of this capability to electronic motion deep in the interior of atoms and molecules requires new tools, because the characteristic time scale for this class of microscopic dynamics is defined by the classical Bohr orbit time of the electron in the ground state of hydrogen, which amounts to some 150 as.

Metrology on this time scale requires controlled variation of a physical quantity within 1 fs. Given the fact that the field oscillation period of visible light is longer than 1 fs, this definitely cannot be achieved with $E_a(t)$. However, the hyperfast variation of the oscillating electric field of the laser pulse does meet the speed criterion for attosecond metrology. In fact, the electric field of 750 nm radiation (emitted by titanium:sapphire-laser-based femtosecond systems) changes from zero to its maximum within some 625 as, providing a sufficiently rapid slope for attosecond measurements.

In principle, time-domain measurements referenced to the laser electric field do not strictly rely on a short duration of the laser pulse. Triggering the microscopic motion to be studied and performing the attosecond measurement repeatedly cycle-by-cycle by exploiting correlation [8] may obviate the need for a very short laser pulse. Because the measurement is accumulated over several or many laser cycles in this case, it has to be ensured that (i) the sub-laser-cycle dynamics evolves the same way and (ii) is referenced to the same clock from cycle-to-cycle during the measurement period. In general, it is hard to meet these requirements with high precision, which means that the implementation of this concept to a wide range of problems is not straightforward.

The above difficulties can be avoided in a natural manner by triggering the sub-cycle dynamics only once per laser pulse. This requires a cycle that can be distinguished from all the others in the pulse. Isolation of a single cycle is straightforward in a pulse comprising merely a few oscillation cycles. Intense few-cycle pulses can now be routinely produced [9]. The amplitude of the central wave cycle in such a pulse is significantly higher than that of any other cycles, allowing discrimination. The central wave cycle of a few-cycle pulse therefore lends itself to *both* getting an attosecond process going *and* providing the reference clock for measuring it. Both of these operations call for well-controlled, well-reproducible field evolution.

Reproducible field evolution from one laser pulse to the next requires that the timing of the oscillations with respect to the pulse peak can be reproduced. In the carrier–envelope representation of the light field as given by (1), this timing is mathematically denoted by φ, which has been referred to as the carrier–envelope phase. $\varphi = 0$ in the representation of (1) implies a waveform that has one single most intense half oscillation cycle at the centre of the pulse. We call this a *cosine waveform*. $\varphi = -\pi/2$, on the other hand, results in a waveform with two equally strong half cycles. In our terminology, this is a *sine waveform*. These waveforms (figure 1) constitute two specific examples out of an infinite number of possibilities for the same amplitude envelope $E_a(t)$ and carrier frequency ω_L. Few-cycle pulses with a controlled waveform, i.e. controlled value of φ, open the way to attosecond control and metrology in a natural way.

Unfortunately, conventional ultrashort-pulse lasers do not deliver pulses with a constant waveform. Rather, the carrier–envelope phase is subject to a pulse-to-pulse shift as a consequence of the difference between the group and phase delays in the laser cavity. This pulse-to-pulse phase shift had been measured as early as 1996 [10], but stabilization of the shift [11–13] became possible only years later, drawing on a proposal by Hänsch [14–16]. These carrier–envelope-phase-stabilized femtosecond oscillators [11, 13] revolutionized frequency

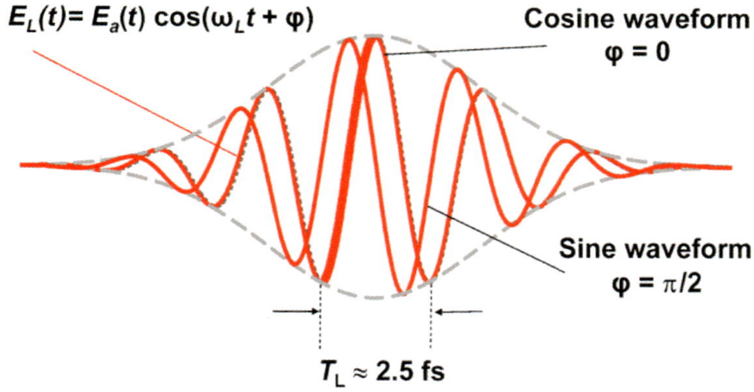

Figure 1. Few-cycle light waves of identical amplitude envelope and carrier frequency carried at a wavelength of 750 nm. Different values of the carrier–envelope phase result in different, in our example cosine- and sine-shaped, waveforms. If the waveform is reproducible (from laser pulse to laser pulse), the rapidly varying field within, for example, the half cycle highlighted can be used as a reference clock for time-domain measurements on a sub-femtosecond time scale.

metrology owing to their precisely controlled frequency comb, but had virtually no impact in the time domain because the sensitivity of light–matter interactions to the waveform relies on field strengths higher than that available from oscillators. Collaboration between one of the authors' groups with that of Hänsch recently led to the generation of waveform-reproducible intense few-cycle pulses [17].

The first source of phase-stabilized high-power few-cycle laser pulses delivers 5 fs, 0.1 TW pulses at a wavelength of 750 nm and a repetition rate of 1 kHz (figure 2). Whilst the control loops in figure 2 permit stabilization of φ as well as changing it by a known amount, the absolute value of φ is still unknown and remains to be determined from strong-field experiments. These waveform-controlled few-cycle pulses have opened the door to control the motion of electrons in and around atoms within the wave cycle, i.e. on a sub-femtosecond time scale. Sub-wave-cycle electron dynamics naturally unfold in the process of optical field ionization. This phenomenon is predicted to set in at laser field strengths yielding a cycle-averaged wiggling energy of a free electron (ponderomotive potential) comparable to or larger than the binding energy of valence electrons, which is typically fulfilled at intensity levels beyond 10^{14} W cm^{-2} in the near-infrared spectral range.

A linearly polarized intense light wave detaches electrons from the irradiated atoms whenever the electric field exceeds the threshold for tunnel ionization (near the oscillation peaks). The liberated electrons are first removed from the atomic core, but the laser field changes its direction and pulls the electrons back to their original location less than a period later, resulting in a recollision of the electron wave packets with their parent ions. This recollision gives rise, with some probability, to recombination of the electron into its original bound state and to simultaneous emission of an energetic (extreme ultraviolet or soft-x-ray) photon. The phenomenon has been modelled by Lewenstein *et al* [18] on the basis of the above-outlined intuitive picture put forward by Corkum [19] and independently by Schafer *et al* [20]. If the driving laser pulse consists of many wave cycles, this process is repeated quasi-periodically each half cycle. Hence, theory predicts the light emission in the form of a quasi-periodic train of sub-fs bursts with a spectrum consisting of high-order harmonics of the incident laser field. This time structure was recently confirmed experimentally [21, 22].

In the case of a few-cycle driver, only a few recollisions occur. The electron wave packet recolliding near the zero transition of the laser field following the pulse peak is

Attosecond physics: facing the wave–particle duality

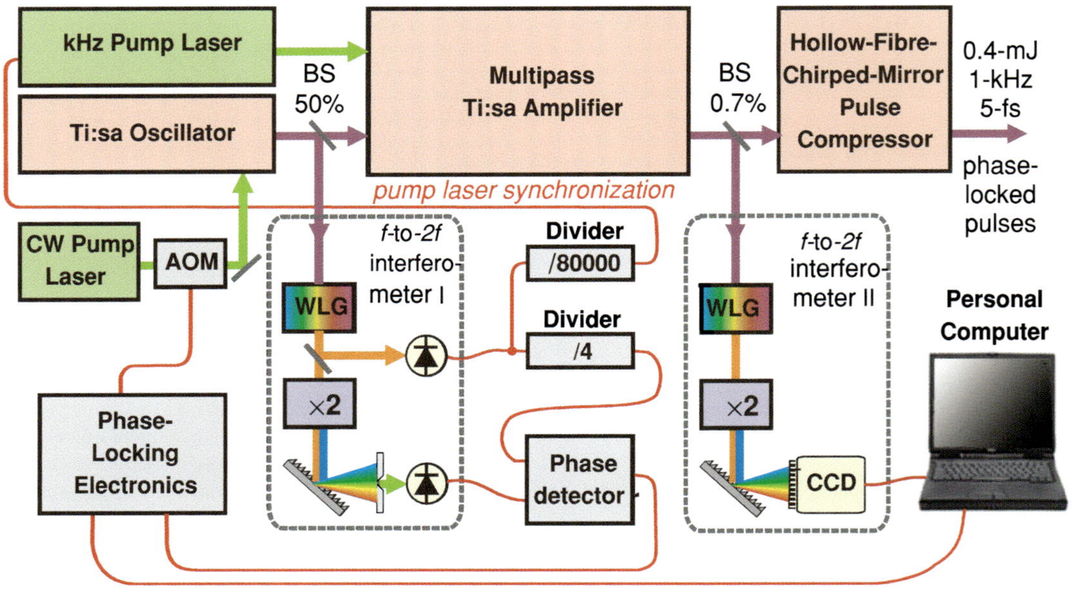

Figure 2. Overview of a carrier–envelope-phase-stabilized, i.e. waveform-controlled, few-cycle laser source. The phase of the pulses emitted from the Ti:sapphire oscillator is maintained by a frequency-domain (f-to-$2f$) technique originally developed for frequency-domain metrology, through a feedback loop controlling the pump power with an acousto-optical modulator (AOM). The energy of the phase-stabilized pulses is boosted by approximately six orders of magnitude in a multipass amplifier (Femtopower Compact Pro, Femtolasers GmbH) and compressed to a duration of 5 fs in a gas-filled-hollow-fibre/chirped-mirror compressor. Random phase drifts in the amplifier are monitored with Interferometer II and pre-compensated in the oscillator to yield waveform-controlled few-cycle pulses with excellent long-term stability.

significantly more energetic than the others, because it is pulled back by the most intense half cycle to the vicinity of the core. This implies the emergence of a burst at higher photon energies than other recollision events. For a cosine waveform, we expect one single highest energy XUV burst to be emitted (figure 3(a)), whereas a sine waveform—comprising two half cycles of highest amplitude—will create a couple of pulses at the highest photon energies (figure 3(b)).

These considerations suggest that waveform-controlled few-cycle light can excite electronic motion in a controlled manner within the oscillation cycle of visible light, i.e. on a sub-femtosecond time scale. The resultant sub-femtosecond extreme ultraviolet emission, in turn, can be used to test the ability of controlled light field oscillations to perform measurements on an attosecond time scale. To this end, the XUV radiation together with the few-cycle laser pulse used for generating it are focused into a second atomic gas jet (figure 4). The laser field is strong enough to change significantly the momentum of the electrons knocked off by XUV photons. The momentum change—measured by time-of-flight spectrometry—is directed parallel to the electric field vector and depends on the instant of release of the electron wave packet within the laser cycle.

The sensitive dependence of the momentum change of the freed electrons on the phase of the field at the instant of their release *and* precise control over the hyperfast light field oscillations permit probing of the ejected wave packet with sub-femtosecond resolution. Depending on the emission duration of the electron wave packet as compared with the wave period of the probing laser light, the electron reveals particle-like or wave-like behaviour in its interaction with the light field, as is discussed in the following sections.

M Drescher and F Krausz

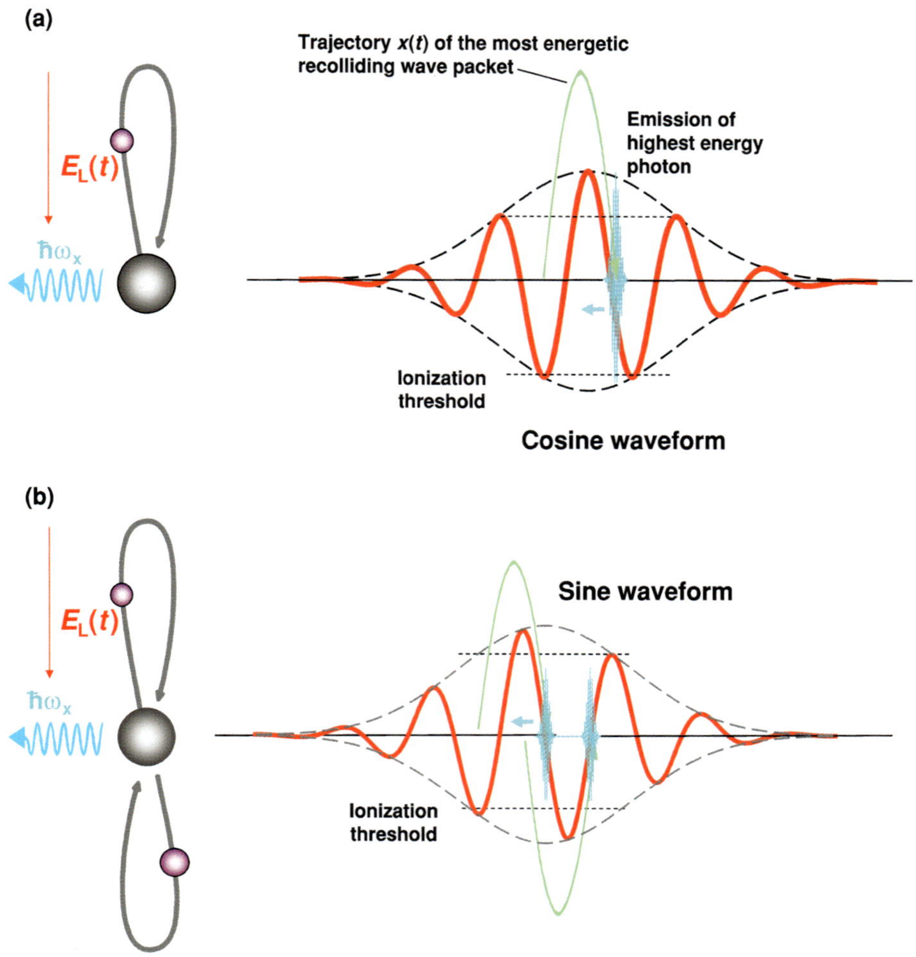

Figure 3. Schematic illustration of XUV emission from atoms ionized by a linearly polarized few-cycle laser pulse. The most-weakly bound atomic electron is detached with some probability from the atom as the laser electric field exceeds the optical-field-ionization threshold (black dashed line). The electron wave packet is first removed and then pushed back to the vicinity of the core. The most energetic part of the electron wave packet recollides at the zero transition of the driving electric field, resulting in emission of the most energetic photons in one or two bursts of sub-femtosecond duration for (a) a cosine- and (b) a sine-shaped driver wave, respectively.

3. Particle-like behaviour: sub-femtosecond pulse characterization

Let us consider photoionization of an atom in the presence of a visible laser field too weak to induce directly or even to influence the electron emission process. As the ionization can be assumed to occur instantaneously on the time scale under consideration, excitation with an ultrashort XUV pulse of duration $\tau_x \ll T_L$ forms an electron wave packet with an envelope mimicking the photon wave packet, $\tau_e \sim \tau_x$. Owing to this temporal confinement, we could consider this escaping photoelectron as a particle with a defined initial momentum p_i. Once liberated, however, the electron experiences the laser field and is accelerated in the direction of the light polarization, thus gaining or losing momentum, $\Delta p = p_f - p_i$. In momentum space, this transfer can be regarded as an alternating linear displacement of the momentum sphere [23] as depicted in figure 5, where isotropic electron emission is assumed. From this classical scheme, it becomes clear that the momentum transfer Δp depends on the angle between p_i and the laser polarization plane, but most importantly on the phase and amplitude of the laser field

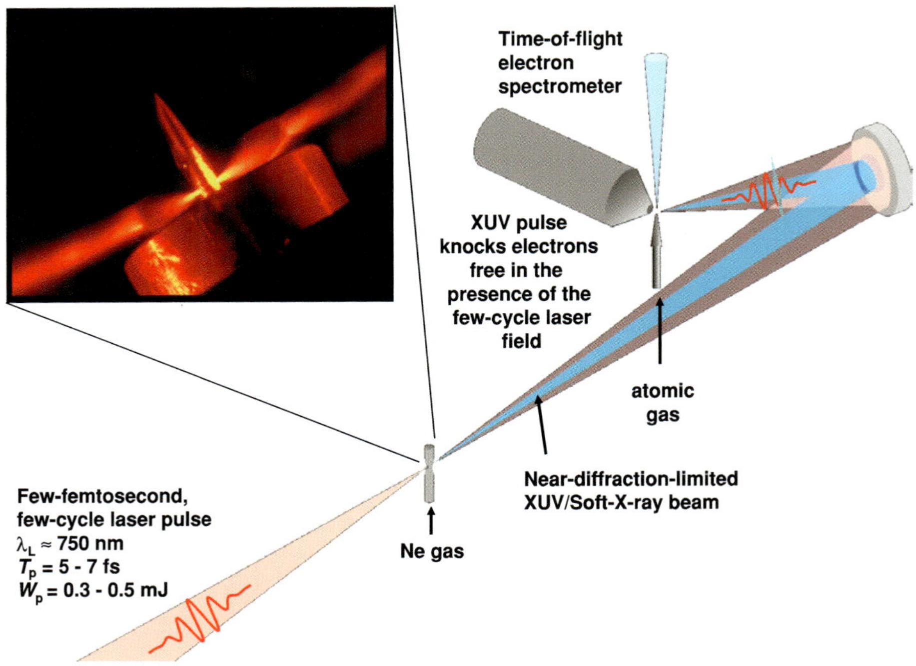

Figure 4. Schematic of the experimental set-up for generation of sub-femtosecond XUV pulses and their use for impulsive atomic excitation in the presence of a waveform-controlled few-cycle light wave. The extreme ultraviolet radiation emerging from a tube filled with neon gas as a result of the interaction depicted in figure 3 is emitted in a highly collimated laser-like beam propagating collinearly with the generating laser beam in an evacuated beamline. The two beams are reflected by two different parts of a composite molybdenum–silicon multilayer mirror. The two concentric parts can be translated with respect to each other, allowing one beam to be delayed with attosecond precision with respect to the other. Both beams are focused into a second atomic gas target which is excited by the XUV pulses in the presence of the laser field. The freed electrons are detected and analysed with a time-of-flight spectrometer.

at the instance of electron injection, i.e. the ionization event. Δp is parallel with the electric field vector in a linearly polarized light field with its magnitude given by

$$\Delta p = \mathrm{e} \int_t^\infty E_\mathrm{L}(t')\,\mathrm{d}t' = \mathrm{e}A_\mathrm{L}(t), \tag{2}$$

where $A_\mathrm{L}(t)$ is the vector potential. Δp vanishes when the ionization takes place in the field maximum and is strongest for the zero field transitions a quarter of a light period earlier or later. This corresponds to a 90° phase shift between the light oscillation and the observed momentum transfer, independent of the exact form of the pulse envelope $E_\mathrm{a}(t)$.

The validity of this classical picture of particle-like electrons interacting with a synchronized light field is put to the test in an experiment depicted in figure 6, utilizing sub-fs XUV pulses well synchronized with a phase-stabilized few-cycle laser field as introduced in the previous section. Free atoms are ionized with the (pump) XUV pulse and the escaping photoelectrons are probed with a delayed light field. Varying the delay Δt between the XUV and laser pulses, one should expect an alternating bipolar shift of the final kinetic electron energy W_f measured in the spectrometer according to

$$W_\mathrm{f}(\Delta t) = W_\mathrm{i} + \sqrt{8W_\mathrm{i}kE_\mathrm{a}^2(\Delta t)}\,\sin(\omega_\mathrm{L}\Delta t), \tag{3}$$

with $k = \mathrm{e}^2/4m_\mathrm{e}\omega_\mathrm{L}^2$.

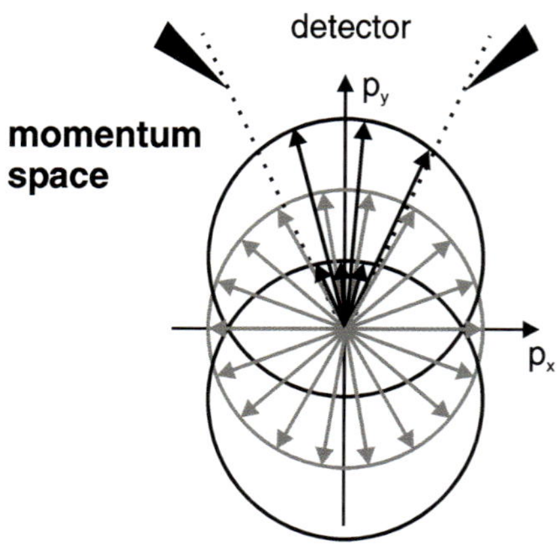

Figure 5. Schematic picture of the momentum transfer imposed on a photoemitted electron by a strong light field, polarized vertically in this figure. An initially isotropic distribution (grey) of photoelectrons with equal momentum (length of arrows) is modulated by the addition of a constant momentum equivalent to the vector potential of the light field at the instant of electron emission. Accordingly, the momentum transfer directly follows the oscillation of the vector potential. The most pronounced variation is observed in the light polarization direction.

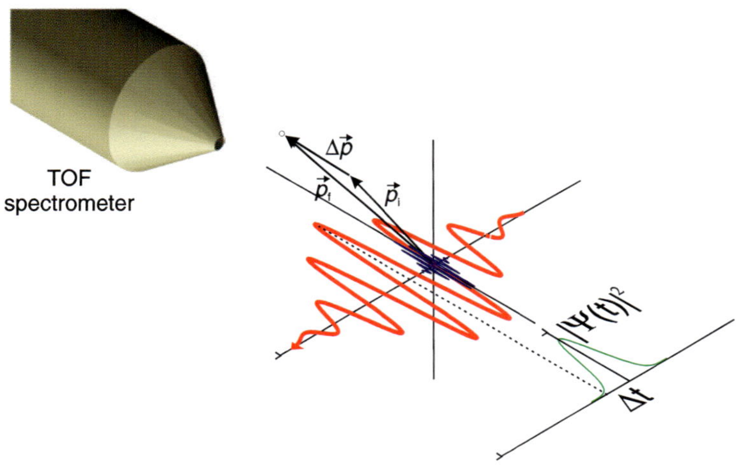

Figure 6. Observation of electron wave packets shorter than the light oscillation cycle. XUV and laser pulses travel collinearly and with fixed relative delay through an atomic gas. The photoeffect transforms the XUV photon wave packet into an electron wave packet of the same temporal shape. Upon emission, the electron immediately experiences the laser field and gains or loses kinetic energy, depending on the temporal delay. This kinetic energy shift is most pronounced in the direction of the laser field polarization, where a time-of-flight electron spectrometer is situated.

Figure 7 presents the result for a corresponding experiment on neon atoms excited with an 8 eV wide XUV pulse centred at a photon energy of 92 eV [24]. The particle-like behaviour is clearly evident in the oscillatory pattern of the Ne 2p photoelectrons. The extreme sensitivity of the electron energy on the light phase provides the footing for temporal measurements with a resolution of a fraction of the laser period. Analysis of figure 7 results in a XUV pulse duration of $250 + 30/-5$ as and proves that our XUV source delivers isolated pulses rather than pulse trains. The shortness of these pulses has provided a δ-like gate for a first-time sampling of the temporal evolution of a visible light field [25].

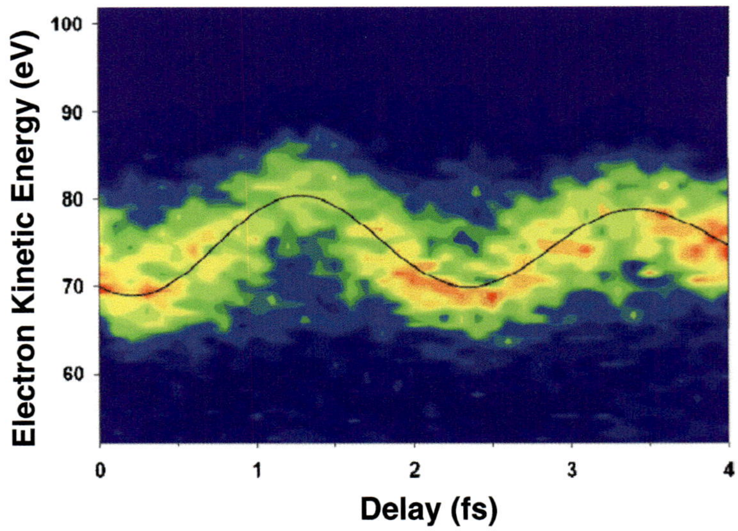

Figure 7. Experimental result from a measurement as depicted in figure 6. Excitation of neon atoms in an 8 eV wide band centred around a photon energy of 93 eV yields Ne-2p photoelectrons at a kinetic energy of approximately 72 ± 4 eV. Momentum transfer between the photoelectron and the laser field leads to the observed oscillatory energy shift directly mapping the vector potential of the light field. This is a signature for particle-like behaviour of the electron. Analysis of this distribution results in a duration of 250 +30/−5 as for the electron wave packet, i.e. the XUV pulse, and so the condition $\tau_e < T_L$ is safely met.

4. Wave-like behaviour: sampling secondary electrons

Relaxation processes in atoms give rise to delayed electron emission in the form of electron wave packets with a shape and temporal extension characteristic of a specific decay (figure 8). Sampling the temporal profile of such wave packets thus reveals details of the underlying electronic dynamics with attosecond precision. For many atomic shells, the typical decay time is of the order of several femtoseconds, thus extending over a few laser field cycles. For such cases, the outgoing electron wave packet will interact with the light field at a multitude of different relative phases as depicted in figure 9. Coherent superposition of all contributions leads to the formation of sidebands [26] in the electron kinetic energy spectra—a signature for wave-like behaviour.

One of the most comprehensively studied (in the energy domain) systems of atomic relaxation is nonresonant MNN Auger decay in krypton following excitation of the 3d shell above the ionization threshold at 93.8 eV [27]. Time-resolved measurement of the characteristic exponential decay time τ_h of the core-hole decay relies on sampling of the Auger electron wave packet formed after excitation of krypton atoms with an XUV pulse in a 3 eV wide band centred at 97 eV, corresponding to a (Fourier-limited) duration of 0.9 fs. A typical electron spectrum acquired in the time-of-flight spectrometer is shown in the upper right corner of figure 10. Combining such spectra taken at different delays Δt between the XUV and the laser pulse provides a map of the temporal evolution of the electron emission, displayed as a surface plot in figure 10. According to the above discussion for electron emission with $\tau_e \gg T_L$, interaction with the laser field leads to the emergence of sidebands separated by multiples of $h\nu_L \sim 1.6$ eV from each Auger line. For the case of the $M_{4,5}N_1N_{2,3}$ Auger lines investigated here, this spacing approximately coincides with the typical separation between the Auger lines [28], thereby obscuring most of the sidebands. Only for the least energetic $(^1P_1)$ line is the first sideband well isolated and taken into consideration. Since the

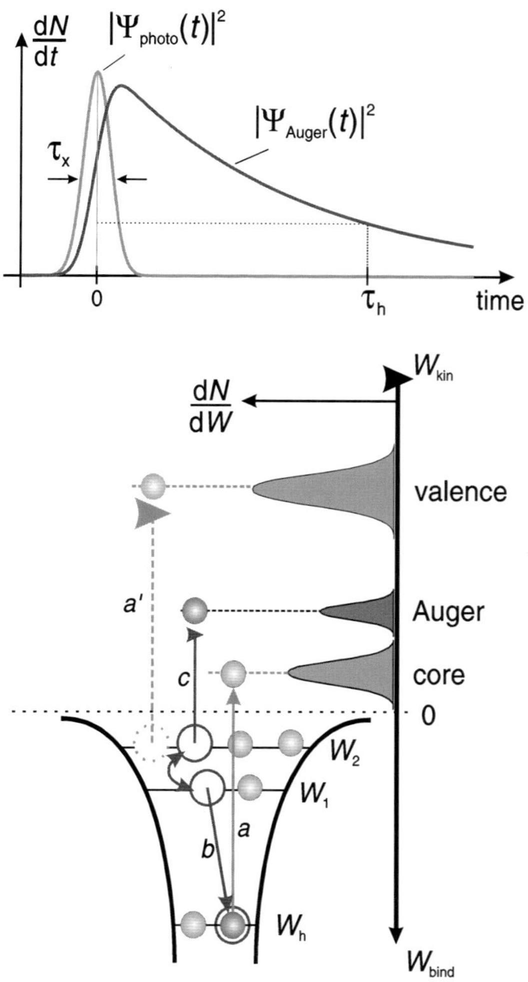

Figure 8. Lower panel: scheme of the Auger decay. Sufficiently high photon energies can create an inner-shell hole upon emission of a photoelectron (a). The vacancy is spontaneously filled with a less bound electron (b). The binding energy of the latter is transferred to a third (Auger) electron leaving the ion with defined kinetic energy (c). Upper panel: since the Auger decay is a spontaneous relaxation process, the secondary Auger wave packet $\Psi_{Auger}(t)$ should follow an exponential decay law. It will therefore, in general, be more extended in time than the primary photoelectron wave packet $\Psi_{Photo}(t)$.

sideband area represents a convolution of the Auger emission rate and the laser pulse envelope $E_a(t)$, the latter must be precisely known to extract the desired value of τ_h. In accordance with the technique for light field mapping introduced in the previous section, this calibrating information is simultaneously embedded in the acquired electron spectra of the 4p photo line. With this information as input, a fit procedure with τ_h as the only varied parameter is capable of extracting the temporal evolution of the Auger emission rate. This analysis yields $\tau_h = 7.9(+1.0/-0.9)$ fs for the lifetime of $M(3d_{5/2})$ vacancies in krypton [29]. Assuming monoexponential decay, this lifetime corresponds to a natural linewidth of $\Gamma = \hbar/\tau_h = 84 \pm 10$ meV, well in accordance with the energy-domain result of 88 ± 4 meV from [30].

As evident from figure 10, a wave-like character is clearly perceptible already for $\tau_e \sim 3T_L$. Typical time constants of atomic decay processes, however, cover a range from 0.1 fs to 10 fs [31]. In many cases, one therefore faces situations where $\tau_e \sim T_L$ suggests behaviour intermediate between the particle and wave limits. In a quantum mechanical calculation, the

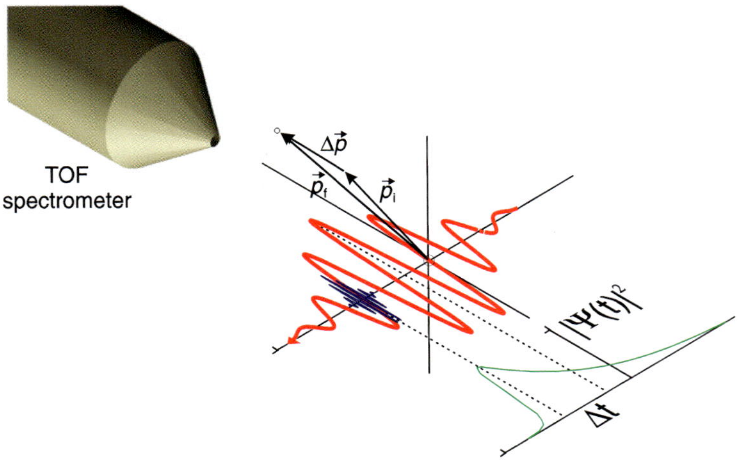

Figure 9. Experimental arrangement for observation of electron wave packets longer than the light oscillation cycle. XUV and laser pulses travel collinearly and with fixed relative delay through an atomic gas. In contrast to the experiment depicted in figure 6, the electron wave packet is longer than the exciting XUV pulse due to an intrinsic atomic process, e.g. Auger decay. Different parts of the wave packet experience different phases of the laser field; the resulting coherent superposition leads to an interference pattern in the electron spectra.

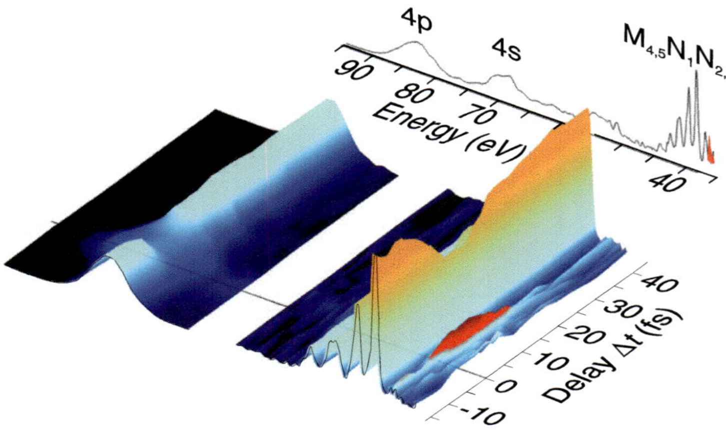

Figure 10. Experimental result from measurement of the time evolution of MNN Auger decay in Kr using the set-up depicted in figure 9. The delay-dependent plot is composed of spectra as shown in the upper right corner. The temporally extended coherent interaction between Auger electrons and the laser field leads to the transient formation of spectral sidebands of the Auger lines (red), from which the inner-shell relaxation dynamics can be retrieved. To this end, simultaneously acquired broadening of the 4p photoelectron line provides important information about the temporal shape of the laser field envelope.

steady transition of time-resolved spectra from the particle to the wave case has been followed [29, 32]. Figure 11 presents quantum mechanical simulations of Scrinzi and Yakovlev [29] of electron spectra assuming different lifetimes of a model Auger decay with a field-free kinetic energy of 40 eV. At $\tau_e = 5$ fs ($\tau_e = 2T_L$) the sideband pattern discussed above clearly dominates. For $\tau_e = 200$ as ($\tau_e = T_L/12.5$), on the other hand, the well-expressed oscillation closely resembling the photoelectron study in figure 7 reveals the particle character of secondary electrons emitted almost instantaneously after excitation. Already for $\tau_e = 500$ as ($\tau_e = T_L/5$), contributions from the coherent interaction with different visible field phases

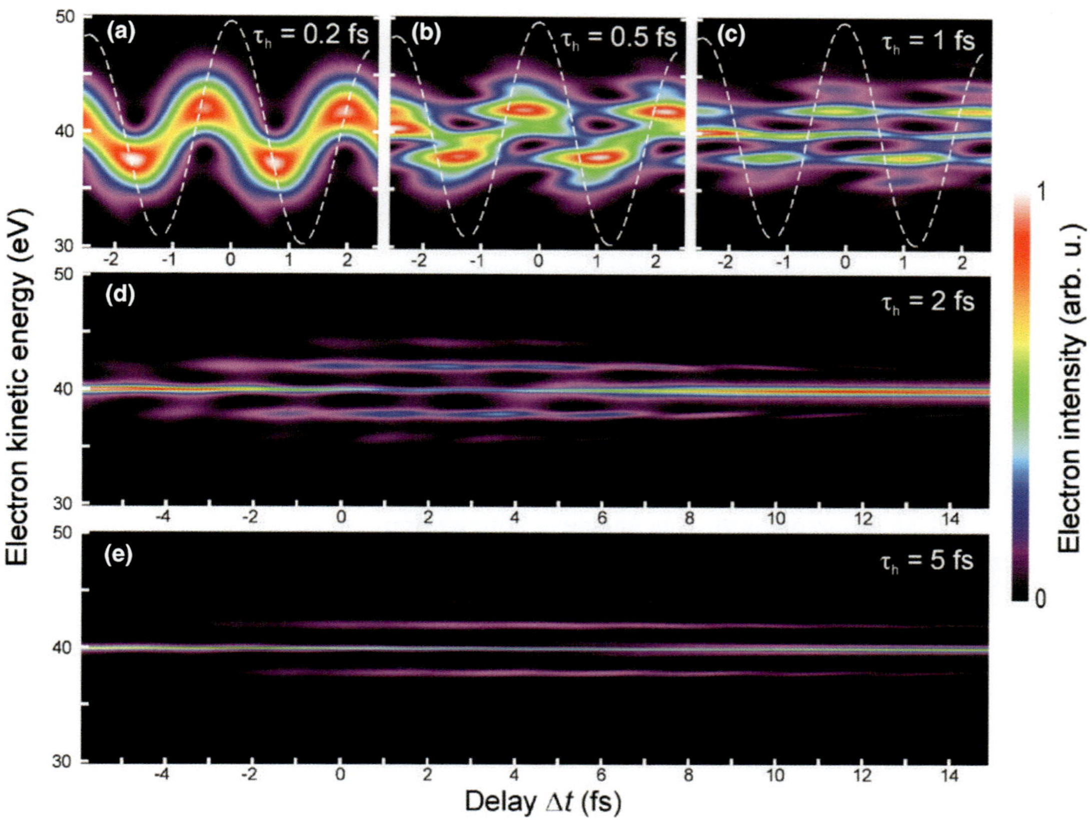

Figure 11. Simulation of time-resolved electron spectra for hypothetical Auger decay at various time constants for the atomic relaxation. An initial kinetic energy centred around 40 eV is assumed. The figure visualizes the gradual transition from a particle-like character of the electrons for very short electron wave packets, revealed by an oscillatory dependence of the spectra on the temporal delay, to a clear signature of wave-like behaviour in the form of spectral sidebands.

become clearly visible. A complex pattern in the range of $T_L/5 < \tau_e < T_L$ fully reveals the dual nature of matter with contributions from the particle (oscillation) and the wave (sidebands) behaviour becoming simultaneously observable.

5. Conclusion

Electric and/or magnetic fields are often used to affect the motion of electrons and probe their characteristics. Attosecond metrology now allows one to uncover wave- or particle-like behaviour of electrons emanating from the photoeffect induced by ultrashort XUV excitation. If the emission time of the electron is long in relation to the wave cycle of probing light ($\tau_e \gg T_L$), sidebands spaced by the laser photon energy from the main line appear in the energy spectrum of electrons emitted in the presence of a strong light field. This phenomenon results from interference between electron wave packets released at instants differing by the laser period (or its integer multiple) and is a clear manifestation of the wave nature of electrons. *The electrons behave like a coherent collective matter wave.* In the limit of $\tau_e \ll T_L$, this temporal interference is completely suppressed and the electrons' motion can be modelled and their momentum and energy can be precisely predicted in the framework of classical physics. *The electrons behave like a shower of particles with a point-like charge.* These two limits can, in principle, also be realized by varying the frequency of the light wave used for probing the electrons released by the photoeffect or Auger effect. It may thus be concluded that, depending on the oscillation period of the probing field relative to the emission time, the electrons behave as a coherent matter wave or as point-like particles in their interaction with the electromagnetic field.

From a practical point of view, the intermediate regime $\tau_e < T_L$ or $\tau_e \sim T_L$ appears to be most important for the emerging field of attosecond science. In this regime, the electrons' energy and momentum distributions are modified gently and in a well-defined manner by the rapidly varying field. From the modified momentum and/or energy distributions, it is possible to determine the temporal structure of the electron emission. Probing the emission of primary (photo) and secondary (Auger) electrons created by impulsive (attosecond) excitation in this way yields direct time-domain information of the evolution of the underlying atomic excitation and relaxation processes, respectively.

The resolution of optical-field sampling of electron emission is dictated by the initial kinetic energy W_i of the ejected atomic electrons, by T_L and (as always) by the signal-to-noise (S/N) ratio. The briefest time interval within which two different atomic events can be recognized as different by this technique is ultimately limited by quantum-mechanical uncertainty, which dictates that any short time structure comes with a broad energy spectrum. To be able to resolve the spectral 'images' of two events separated in time by δt, they must be shifted with respect to each other in the final energy spectrum of the electrons by as much as at least their own spectral width $\delta W \approx \hbar/\delta t$. This requirement makes the resolving power

$$\delta t = \frac{T_L}{2\pi}\sqrt{\frac{\hbar\omega_L}{\Delta W_{max}}}, \qquad (4)$$

where ΔW_{max} stands for the energy shift suffered by the electron ejected at the peak of $A_L(t)$. In recent measurements ΔW_{max} approached 30 eV, yielding (for $T_L = 2.5$ fs) $\delta t \approx 100$ as [24]. Attosecond metrology based on interaction of electrons with a strongly controlled wave of intense few-cycle light is able to distinguish two ultrafast atomic events following each other within 100 as, constituting the shortest interval of time directly measurable to date.

Recent experimental advances afford promise of generation of single attosecond bursts with durations well below 100 as and photon energies approaching 1 keV [33]. At these excitation energies ΔW_{max} can be enhanced by at least an order of magnitude, yielding $\delta t \approx 30$ as. Extension of the presented experiments to exploring simultaneously the sub-femtosecond temporal variation of emission intensity and momentum distribution of primary (photo) *and* secondary (Auger) electron emission will provide unprecedented insight into the complex excitation and relaxation dynamics of the electronic shell of atoms and molecules with a resolution that will approach the atomic unit of time (24 as).

Acknowledgments

For their invaluable contributions to the reported work, we thank A Baltuska, F Bammer, P Corkum, E Goulielmakis, U Heinzmann, M Hentschel, R Kienberger, U Kleineberg, N Milosevic, G A Reider, M Uiberacker, A Scrinzi, C Spielmann, G Tempea, V Yakovlev and Th Westerwalbesloh. The research has been sponsored by the FWF (Austria, grants no F016, Z63, P15382), the DFG (Germany, grants no SPP1053, HE1049, KL1077) and the VW-Foundation (Germany, grant no I/79 227).

References

[1] Einstein A 1905 *Ann. Phys., Lpz.* **17** 132

[2] Lenard P 1902 *Ann. Phys., Lpz.* **8** 149

[3] Bransden B H and Joachain C J 1983 *Physics of Atoms and Molecules* (London: Longman) p 55
[4] Diels J C and Rudolph W 1996 *Ultrashort Laser Pulse Phenomena: Fundamentals, Techniques, and Applications on a Femtosecond Time Scale* (New York: Academic)
[5] Attwood D 1999 *Soft X-Rays and Extreme Ultraviolet Radiation: Principles and Applications* (Cambridge: Cambridge University Press)
[6] Steinmeyer G D, Sutter H, Gallmann L, Matuschek N and Keller U 1999 *Science* **286** 1507
[7] Zewail A 2000 *J. Chem. Phys. A* **104** 5660
[8] Niikura H *et al* 2002 *Nature* **417** 917
[9] Sartania S *et al* 1997 *Opt. Lett.* **22** 1562
[10] Xu L, Spielmann Ch, Poppe A, Brabec T, Krausz F and Hänsch T W 1996 *Opt. Lett.* **21** 2008
[11] Jones D J 2000 *Science* **288** 635
[12] Apolonski A *et al* 2000 *Phys. Rev. Lett.* **85** 740
[13] Holzwarth R *et al* 2000 *Phys. Rev. Lett.* **85** 2264
[14] Hänsch T W 1997 private communication
[15] Telle H R *et al* 1999 *Appl. Phys.* B **69** 327
[16] Reichert J, Holzwarth R, Udem Th and Hänsch T W 1999 *Opt. Commun.* **172** 59
[17] Baltuska A *et al* 2003 *Nature* **421** 611
[18] Lewenstein M, Balcou P, Ivanov M Y, 'Huillier A L and Corkum P B 1994 *Phys. Rev.* A **49** 2117
[19] Corkum P B 1993 *Phys. Rev. Lett.* **71** 1994
[20] Schafer K J, DiMauro L F and Kulander K C 1993 *Phys. Rev. Lett.* **70** 1599
[21] Paul P M *et al* 2001 *Science* **292** 1689
[22] Tzallas P, Charalambidis D, Papadogiannis N A, Witte K and Tsakiris G D 2003 *Nature* **426** 267
[23] Drescher M, Hentschel M, Kienberger R, Tempea G, Spielmann C, Reider G, Corkum P and Krausz F 2001 *Science* **291** 1923
[24] Kienberger R *et al* 2004 *Nature* **427** 817
[25] Goulielmakis E *et al* 2004 *Science* **305** 1267
[26] Glover T E, Schoenlein R W, Chin A H and Shank C V 1996 *Phys. Rev. Lett.* **76** 2468
[27] Aksela H, Aksela S and Pulkkinen H 1984 *Phys. Rev.* A **30** 2456
[28] Schmidtke B, Khalil T, Drescher M, Müller N, Kabachnik N M and Heinzmann U 2001 *J. Phys. B: At. Mol. Opt. Phys.* **34** 4293
[29] Drescher M *et al* 2002 *Nature* **419** 803
[30] Jurvansuu M, Kivimäki A and Aksela S 2001 *Phys. Rev.* A **64** 012502
[31] Martensson N and Nyholm R 1981 *Phys. Rev.* B **24** 7121
[32] Smirnova O, Yakovlev V and Scrinzi A 2003 *Phys. Rev. Lett.* **91** 253001
[33] Seres J 2005 *Nature* **433** 596

PRL 95, 040401 (2005) PHYSICAL REVIEW LETTERS week ending 22 JULY 2005

Attosecond Double-Slit Experiment

F. Lindner,[1] M. G. Schätzel,[1] H. Walther,[1,2] A. Baltuška,[1] E. Goulielmakis,[1] F. Krausz,[1,2,3] D. B. Milošević,[4] D. Bauer,[5] W. Becker,[6] and G. G. Paulus[1,2,7]

[1]*Max-Planck-Institut für Quantenoptik, 85748 Garching, Germany*
[2]*Ludwig-Maximilians-Universität München, 85748 Garching, Germany*
[3]*Institut für Photonik, Technische Universität Wien, Gusshausstr. 27, A-1040 Wien, Austria*
[4]*Faculty of Science, University of Sarajevo, Zmaja od Bosne 35, 71000 Sarajevo, Bosnia and Herzegovina*
[5]*Max-Planck-Institut für Kernphysik, Saupfercheckweg 1, 69117 Heidelberg, Germany*
[6]*Max-Born-Institut, Max-Born-Str. 2a, 12489 Berlin, Germany*
[7]*Department of Physics, Texas A&M University, College Station, Texas 77843-4242, USA*
(Received 22 March 2005; published 19 July 2005)

A new scheme for a double-slit experiment in the time domain is presented. Phase-stabilized few-cycle laser pulses open one to two windows (slits) of attosecond duration for photoionization. Fringes in the angle-resolved energy spectrum of varying visibility depending on the degree of which-way information are measured. A situation in which one and the same electron encounters a single and a double slit at the same time is observed. The investigation of the fringes makes possible interferometry on the attosecond time scale. From the number of visible fringes, for example, one derives that the slits are extended over about 500 as.

DOI: 10.1103/PhysRevLett.95.040401 PACS numbers: 03.65.Ta, 07.60.Ly, 32.80.Rm, 42.50.Xa

The conceptually most important interference experiment is the double-slit scheme, which has played a pivotal role in the development of optics and quantum mechanics. In optics its history goes back to Young's double-slit experiment. Its scope was greatly expanded by Zernike's work and continues to deliver new insights into coherence to the present day [1]. One of the key postulates of quantum theory is interference of matter waves, experimentally confirmed by electron diffraction [2,3]. More than 30 years later, Jönsson was the first to perform a double-slit experiment with electrons [4]. Of particular importance for interpreting quantum mechanics have been experiments with a single particle at any given time in the apparatus [5,6]. More recent work has illuminated the fundamental importance of complementarity in which-way experiments [7] and of quantum information in quantum-eraser schemes [8].

In this Letter a novel realization of the double-slit experiment is described. It is distinguished from conventional schemes by a combination of characteristics: (i) The double slit is realized not in position-momentum but in time-energy domain. (ii) The role of the slits is played by windows in time of attosecond duration. (iii) These "slits" can be opened or closed by changing the temporal evolution of the field of a few-cycle laser pulse. (iv) At any given time there is only a single electron in the double-slit arrangement. (v) The presence and absence of interference are observed for the same electron at the same time.

Interference experiments in the time-energy domain are not entirely new. Interfering electron wave packets were created by femtosecond laser pulses [9]. Accordingly, the windows in time (or temporal slits) during which these wave packets are launched were comparable to the pulse duration. In the present experiment, in contrast, the slits are open during a small fraction of an optical cycle, which gives the attosecond width. A number of experiments, in particular, in intense-laser atom physics but also for Rydberg atoms and microwaves [10], can and have been interpreted in this spirit (for a review see, for example, [11]). In this Letter, however, the optical cycles are precisely tailored by controlling the phase of few-cycle laser pulses (also known as absolute or carrier-envelope phase). This provides an unprecedented degree of control for the double-slit arrangement. Not only are the principles of quantum mechanics beautifully demonstrated, it is also likely that applications exploiting interferometric techniques for measuring attosecond dynamics will emerge.

In the present experiment, argon atoms are ionized by intense (1×10^{14} W/cm^2) few-cycle 760 nm laser pulses. Photoelectrons emitted in opposite directions ("left" and "right") are detected by two opposing electron detectors placed symmetric to the laser focus. The laser field is horizontally polarized, i.e., parallel to an axis defined by the electron detectors. Interference of temporally separated wave packets leads to a fringe pattern in the energy domain because time and energy are conjugated variables. Therefore, the electron kinetic energy needs to be measured, in our case via the time-of-flight method. The carrier-envelope phase of the field and thus its temporal evolution can be controlled by delaying the envelope of the pulse with respect to the carrier. This is accomplished by shifting a glass wedge into or out of the beam. The phase of the field is measured as described in [12].

Photoionization of atoms with an ionization threshold much greater than the photon energy is a highly nonlinear process. For intense fields, the first step can be described by optical field ionization. This immediately explains the generation of one attosecond window (or slit) in time per half-cycle close to its extremum; see Fig. 1. By using phase-controlled few-cycle laser pulses [13], it is possible to manipulate the temporal evolution of the field, thus gradually opening or closing the slits, and controlling which-way information. Depending on the field, one or two half-cycles (or anything in between) contribute to the electron amplitude for a given direction and electron energy. This corresponds to a varying degree of which-way information and, accordingly, to a varying contrast of the interference fringes. The temporal slits leading to electrons of given final momentum are spaced by approximately the optical period. This results in a fringe spacing close to the photon energy.

Figure 2 displays measured electron spectra. In Fig. 2(a) the spectra recorded at the left and the right detectors are shown for $\pm$ cos-like and $\pm$ sin-like pulses as defined in Fig. 1. A problem in presenting such spectra is that they quickly roll off with increasing electron energy. This roll-off was eliminated by dividing the spectra by the average of all spectra over the pulse's phase. Clear interference fringes with varying visibility are observed as expected from the discussion above. The highest visibility is observed for $-$ sin-like pulses in the positive (right) direction. For the same pulses, the visibility is very low in the opposite direction. Changing the phase by π interchanges the role of left and right as expected. The most straightforward explanation—which will be detailed by a simple model below—is to assume that, for $-$ sin-like pulses, there are two slits and no which-way information for the positive direction and just one slit and (almost) complete which-way information in the negative direction. The fact that the interference pattern does not entirely disappear is caused by the pulse duration, which is still slightly too long to create a perfect single slit.

Under the conditions of this experiment, each argon atom emits at most one electron [14], whose various options of how to reach a given final state lead to interference. For sin-like pulses, these options correspond to a double slit in time in one direction and to a single slit in the other, and are created for each atom separately by the few-cycle laser pulse. Therefore, even though there is more than one argon atom in the laser focus, the experiment operates under single-electron conditions. On the scale of the electron's deBroglie wavelength, other atoms are far away and, moreover, randomly distributed. This is in contrast to the double slit in space where the beam has to be sufficiently dilute to ensure a one-electron measurement.

The fringe pattern exhibits an envelope. From Fig. 2 a width of this envelope of about 4 fringes is inferred. Just as for a double-slit experiment, the width of this envelope can

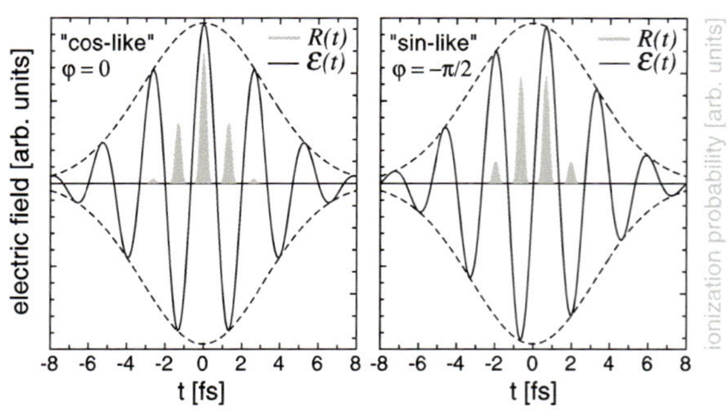

FIG. 1. Temporal variation of the electric field $\mathcal{E}(t) = \mathcal{E}_0(t) \times \cos(\omega t + \varphi)$ of few-cycle laser pulses with phase $\varphi = 0$ ("cos-like") and $\varphi = 3\pi/2$ ("sin-like"). In addition, the field ionization probability $R(t)$, calculated at the experimental parameters, is indicated. Note that an electron ionized at $t = t_0$ will not necessarily be detected in the opposite direction of the field $\mathcal{E}$ at time t_0 due to deflection in the oscillating field.

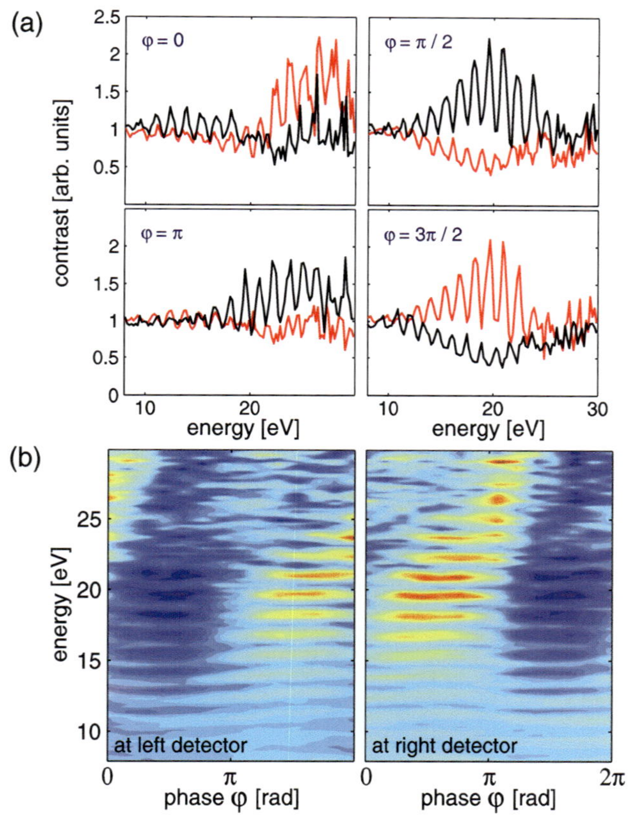

FIG. 2 (color). Photoelectron spectra of argon measured with 6 fs laser pulses for intensity 1×10^{14} W/cm^2 as a function of the phase. Panel (a) displays the spectra for $\pm$ sin- and $\pm$ cos-like laser fields. The red curves are spectra recorded with the left detector (negative direction), while the black curves relate to the positive direction. For $\varphi = \pi/2$ the fringes exhibit maximum visibility for electron emission to the right, while in the opposite direction minimum fringe visibility is observed. In addition, the fringe positions are shifted. Panel (b) displays the entire measurement where the fringe visibility is coded in false colors. The fringe positions vary as the phase φ of the pulse is changed. This causes the wavelike bending of the stripes in these figures. Both panels, in principle, show the same information because a phase shift of π mirrors the pulse field in space and thus reverses the role of positive and negative direction. However, the data shown were recorded simultaneously but independently as the phase φ was varied between 0 and 2π.

PRL 95, 040401 (2005) PHYSICAL REVIEW LETTERS week ending 22 JULY 2005

be associated with the width of the slits. It will turn out, however, that what is seen here is not the width of the slit. Rather, each slit can be resolved into a pair of slits whose separation is inversely proportional to the width of the envelope.

Disregarding the changing visibility, the peaks observed in the spectra resemble the well-known above-threshold ionization (ATI) peak pattern and they are certainly related to them. However, the relationship is nontrivial: Besides the visibility of the fringes, their positions also change as the phase of the field is varied. Details of the fringe shifts can be seen in Fig. 2(b). For conventional ATI, one would try to explain this in terms of the ponderomotive potential U_P. This does not work here, because the concept of the ponderomotive potential, which is defined as the cycle-averaged kinetic energy of an electron quivering in an oscillating electric field, is questionable in the few-cycle regime.

In contrast, an interpretation based on the double-slit analogy is obvious. In a spatial double slit, the fringe pattern would shift if a phase shifter (for light, simply a glass plate) were placed in front of one of the slits. For nontrivial particle trajectories one needs to consider the action S along the particles' paths and use the fact that the particles' phases are given by $S/\hbar$.

In order to exclude other scenarios, we compare the experimental data with results obtained by numerically solving the time-dependent Schrödinger equation (TDSE) in three spatial dimensions. The calculation was done at the experimental conditions for an effective argon potential in the single active electron approximation [15]. The calculated spectra were divided by the phase-averaged spectrum, using the same procedure that was applied to the experimental data underlying Fig. 2. The numerical TDSE result for the right-going electrons is shown in Fig. 3, to be compared with the experimental result in the right panel of Fig. 2(b). Virtually all details found in the measurement can also be found in the calculation. This confirms that single-electron dynamics are sufficient to explain the fringes.

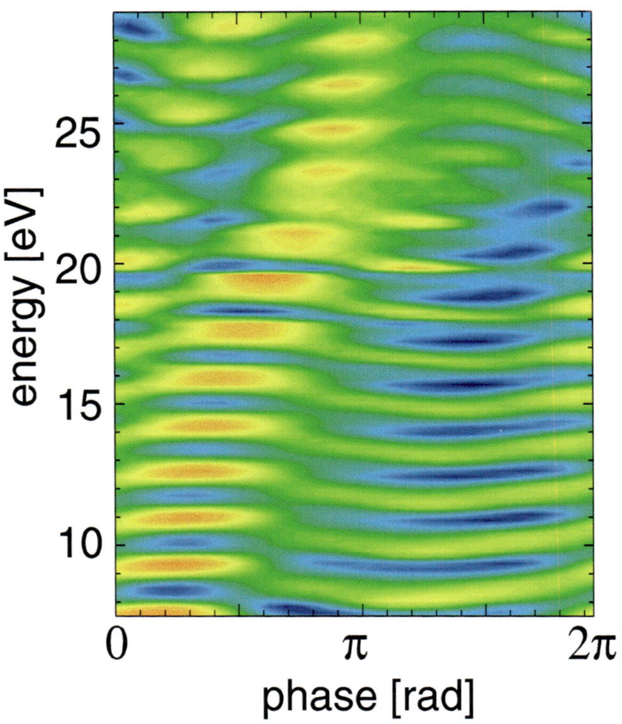

FIG. 3 (color). The result of a numerical solution of the time-dependent Schrödinger equation, as described in the text. This figure should be compared with the right panel of Fig. 2(b).

For an interpretation we resort to a classical model, the so-called simple-man's model [16], which—together with various extensions and modifications—has proven to be extremely helpful for understanding strong-field laser-atom interaction; for a review see, for example, [11]. Alternatively, Keldysh-type models, which can be interpreted as an approximation of Feynman's path integral [17], could be used. Respective results can be found in the literature: Ref. [18] predicts effects analogous to those described in this Letter for circular polarization. References [13,19,20] explain related classical effects for electromagnetic extreme ultraviolet radiation produced by high-harmonic generation. For the present problem, the classical and the quantum model lead to qualitatively the same results. The various assumptions of the model outlined in the following are fulfilled only approximately. Indeed, systematic deviations between the experiment and the model are the consequence.

The classical model assumes that an electron is launched into the continuum at some time t_0. Another crucial assumption of the model is that the electron's velocity is zero at $t = t_0$. This means that $p - eA(t_0) = 0$, where p is the momentum of the electron at the detector, $A(t)$ the vector potential of the field, and $e = -|e|$ the electron's charge. It is largely this relationship that explains the double-slit behavior of few-cycle photoionization.

The strength of the classical model is the intuitive insight it provides. In the following, hardly more than the number and position of the solutions of $p - eA(t_0) = 0$ for given p will be used in order to explain the double-slit behavior. The respective solutions $t_0(p)$ in a Keldysh-type model are complex, thus allowing access to classically forbidden electron energies. However, the symmetry of these solutions stays the same and so do the results qualitatively.

In Fig. 4 the vector potential $A(t)$ is drawn for a $-$sin-like pulse. The solutions of $p - eA(t_0) = 0$ and thus all trajectories of momentum p that could interfere can be found by intersecting $A(t)$ with a horizontal line at p/e. It is now important to recall that a fringe pattern of maximal visibility requires *equally strong* slits, i.e., minimal which-way information. For a few-cycle pulse whose envelope is maximal at $t = 0$, the "strength" of a slit decreases very quickly with increasing $|t_0|$ and is essentially zero for $|t_0| > 2\pi/\omega$ because of the highly nonlinear dependence of photoionization on the field strength. As the

PRL 95, 040401 (2005) PHYSICAL REVIEW LETTERS week ending 22 JULY 2005

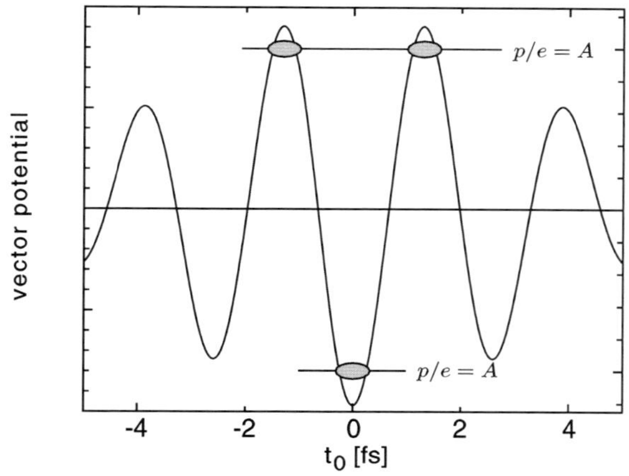

FIG. 4. Vector potential of a − sin-like few-cycle pulse. The temporal slits are given by the condition $p - eA(t_0) = 0$. For a − sin-like pulse, this leads to a double slit in the negative (since $e = -|e|$) direction and a single slit in the opposite direction. Each slit can be resolved into a pair of slits.

maximum of the pulse envelope was chosen to be at $t = 0$, the condition of equally strong slits is identical to requiring that the solutions of $p - eA(t_0) = 0$ be symmetric with respect to $t = 0$. This is the case for − sin-like pulses with electrons emitted in the negative direction and for + sin-like pulses with electrons emitted in the positive direction. For both cases, the respective opposite direction can be considered to act like a single slit as long as the pulse is short enough.

Figure 4 also shows that each slit is, on closer inspection, a pair of slits and that the temporal separation of these subslits depends on the electron energy [21]. The experimental data actually provide a measurement of the time difference of the two subslits, which is approximately 500 as. This is a first simple example for using interferometry on the attosecond time scale in order to investigate electronic dynamics. In addition, Fig. 2(a) shows that the relative phase of the subslits is different for sin-like and cos-like pulses, resulting in a shift of the fringe envelope.

It should be noted that the simple-man's model does not reproduce the dependence of the fringe visibility on electron energy as observed experimentally and in the solution of the TDSE. Therefore, the direction for which interference is predicted by the simple model may be wrong, depending on the energy. Using several theoretical models (3D TDSE, 1D TDSE, Keldysh-type, and classical), we were able to show that this is not a fundamental problem of the classical theory. Rather, it is an effect of the atomic binding potential, which obviously deflects the outgoing photoelectrons. The solution of the *one-dimensional* TDSE (which cannot deflect) with a soft-core potential, for example, agrees qualitatively very well with the classical and a Keldysh-type model. In particular, it does not show a pronounced energy dependence of the fringe visibility, and it predicts the interferences in the same direction as the simple models.

In conclusion, we have realized an intriguing implementation of the double slit in the time domain. The observation of interference and its absence at the same time for the same electron is a beautiful demonstration of the principles of quantum mechanics. It should also be noted that attosecond slits were used and that the interferograms reflect the attosecond dynamics of electronic transitions. Further experimental and theoretical progress should make it possible to use interferometric techniques for attosecond science.

We gratefully acknowledge fruitful discussions with J. Burgdörfer. This work has been supported by the Austrian Science Fund (Grants No. F016, No. Z63, and No. P15382), the German Research Foundation (Grants No. PA730/2 and No. BA2190/1), the Welch Foundation (Grant No. A-1562), and VolkswagenStiftung.

[1] L. Mandel and E. Wolf, *Optical Coherence and Quantum Optics* (Cambridge, Cambridge, England, 1995).
[2] C. J. Davisson, and L. H. Germer, Nature (London) **119**, 558 (1927).
[3] G. P. Thomson and A. Reid, Nature (London) **119**, 890 (1927).
[4] C. Jönsson, Z. Phys. **161**, 454 (1961).
[5] P. G. Merli, G. F. Missiroli, and G. Pozzi, Am. J. Phys. **44**, 306 (1976).
[6] A. Tonomura, J. Endo, T. Matsuda, T. Kawasaki, and H. Ezawa, Am. J. Phys. **57**, 117 (1989).
[7] M. O. Scully, B.-G. Englert, and H. Walther, Nature (London) **351**, 111 (1991).
[8] Y.-H. Kim, R. Yu, S. P. Kulik, Y. Shih, and M. O. Scully, Phys. Rev. Lett. **84**, 1 (2000).
[9] M. Wollenhaupt *et al.*, Phys. Rev. Lett. **89**, 173001 (2002).
[10] T. F. Gallagher, Phys. Rev. Lett. **61**, 2304 (1988).
[11] W. Becker, F. Grasbon, R. Kopold, D. B. Milošević, G. G. Paulus, and H. Walther, Adv. At. Mol. Opt. Phys. **48**, 35 (2002).
[12] G. G. Paulus *et al.*, Phys. Rev. Lett. **91**, 253004 (2003).
[13] A. Baltuška *et al.*, Nature (London) **421**, 611 (2003).
[14] The probability for double ionization is known to be several orders of magnitude smaller than for single ionization under the present conditions.
[15] D. Bauer, Phys. Rev. Lett. **94**, 113001 (2005).
[16] H. B. v. Linden, v. d. Heuvell, and H. G. Muller, in *Multiphoton Processes, Studies in Modern Optics*, Vol. 8 (Cambridge University Press, Cambridge, England, 1988), p. 25.
[17] P. Salières *et al.*, Science **292**, 902 (2001).
[18] D. B. Milošević, G. G. Paulus, and W. Becker, Phys. Rev. Lett. **89**, 153001 (2002).
[19] M. Nisoli *et al.*, Phys. Rev. Lett. **91**, 213905 (2003).
[20] G. Sansone *et al.*, Phys. Rev. Lett. **92**, 113904 (2004).
[21] The time separation has to be calculated from the Keldysh-type theory. Note that the classical model of Fig. 4 does not apply for momenta larger than $|e \max A(t)|$.

REPORTS

22. A simple calculation indicates that the magnitudes of V_s^+ and V_s^- are on the order of a few tens to hundreds of volts. In practice, if we consider the polarization and dielectric screening in the calculation, the local potential is much smaller than the numbers given by the equations here. An accurate calculation of the potential distribution as a result of the ionic charges introduced by the PZ effect and the surface charges caused by boundaries must be solved numerically and self-consistently. In our analysis, a correct magnitude and sign of the potential is sufficient for illustrating the physical model.
23. S. Hasegawa, S. Nishida, T. Yamashita, H. Asahi, *J. Ceramic Proc. Res.* **6**, 245 (2005).
24. R. F. Pierret, *Semiconductor Device Fundamentals* (Addison-Wesley, Reading, MA, 1996), chapter 14.
25. W. I. Park, G. C. Yi, J. W. Kim, S. M. Park, *Appl. Phys. Lett.* **82**, 4358 (2003).
26. Y. Huang *et al.*, *Science* **294**, 1313 (2001).
27. A. Bachtold, P. Hadley, T. Nakanishi, C. Dekker, *Science* **294**, 1317 (2001); published online 4 October 2001 (10.1126/science.1065824).
28. J. Chen *et al.*, *Science* **310**, 1171 (2005).
29. U.S. patent pending.
30. Supported by NSF grant DMR 9733160, the NASA Vehicle Systems Program and Department of Defense Research and Engineering, and the Defense Advanced Research Projects Agency. We thank X. Wang, W. L. Hughes, J. Zhou, and J. Liu for their help.

Supporting Online Material
www.sciencemag.org/cgi/content/full/312/5771/242/DC1
SOM Text
Figs. S1 to S6
Table S1
Movies S1 and S2

19 December 2005; accepted 10 March 2006
10.1126/science.1124005

Control of Electron Localization in Molecular Dissociation

M. F. Kling,[1] Ch. Siedschlag,[1] A. J. Verhoef,[2] J. I. Khan,[1] M. Schultze,[2] Th. Uphues,[3] Y. Ni,[1] M. Uiberacker,[4] M. Drescher,[3,5] F. Krausz,[2,4] M. J. J. Vrakking[1]

We demonstrated how the subcycle evolution of the electric field of light can be used to control the motion of bound electrons. Results are presented for the dissociative ionization of deuterium molecules ($D_2 \rightarrow D^+ + D$), where asymmetric ejection of the ionic fragment reveals that light-driven intramolecular electronic motion before dissociation localizes the electron on one of the two D^+ ions in a controlled way. The results extend subfemtosecond electron control to molecules and provide evidence of its usefulness in controlling reaction dynamics.

Few-cycle laser light with a controlled evolution of the electric field $E(t) = a(t) \times \cos(\omega t + \varphi)$, with amplitude $a(t)$, frequency ω, and carrier envelope phase φ (*1*), has recently allowed the steering of the motion of electrons in and around atoms on a subfemtosecond time scale. Manifestations of this control include the reproducible generation and measurement of single subfemtosecond pulses (*2*, *3*) and controlled electron emission from atoms (*4*, *5*). Here we address the question of whether this control can be extended to electron wave packets in molecules and, if so, can light-field–driven electronic motion affect reaction dynamics?

Many of the processes in terms of which strong-field molecular interactions are presently interpreted (such as bond softening and enhanced ionization) were discovered in experimental and theoretical work on H_2 and its isotopes HD and D_2 [see (*6*) and references therein]. The role of phase control in the dissociation of hydrogen has recently been addressed in a few theoretical studies (*7*–*9*). We present experiments on the dissociation of D_2^+ into $D^+ + D$ by intense few-cycle laser pulses with controlled field evolution and report a pronounced dependence of the direction of the D^+ ejection (and hence of the localization of the electron in the system) on the waveform driving the reaction. Quantum-classical computations reveal that light-field control of molecular electron dynamics is responsible for the observed phenomenon.

The dynamics of molecules in intense laser fields typically includes ionization and dissociation. The dissociation of D_2 in intense laser fields is known to involve several pathways whose relative importance depends on intensity and pulse duration (*6*). The formation of fragment ions occurs via a two-step mechanism (Fig. 1A) in which initially the molecule is ionized by the laser field (Fig. 1A, red arrow) and a vibrational wave packet is launched in the $1s\sigma_g^+$ state. Breakup of the D_2^+ ion is triggered by excitation to a repulsive state or after double ionization.

In the single-ionization pathways, excitation of bound D_2^+ (such as to the $2p\sigma_u^+$ state in Fig. 1A) by recollision of the first electron [recollision excitation (RCE), green line] or directly by the laser field [sequential excitation (SE), blue line] leads to dissociation and the formation of a D^+ ion and a D atom. For example, in recent molecular clock studies, vibrational motion in D_2^+ was time-resolved by exploiting RCE (*10*, *11*). Additional dissociation mechanisms can be understood by considering that molecular potentials are modified by strong laser fields. Bond softening (BS, purple line) (*12*) occurs when energy gaps open up at avoided crossings between adiabatic field-dressed potential energy curves.

In double-ionization pathways, the formation of D_2^+ is followed by a second ionization

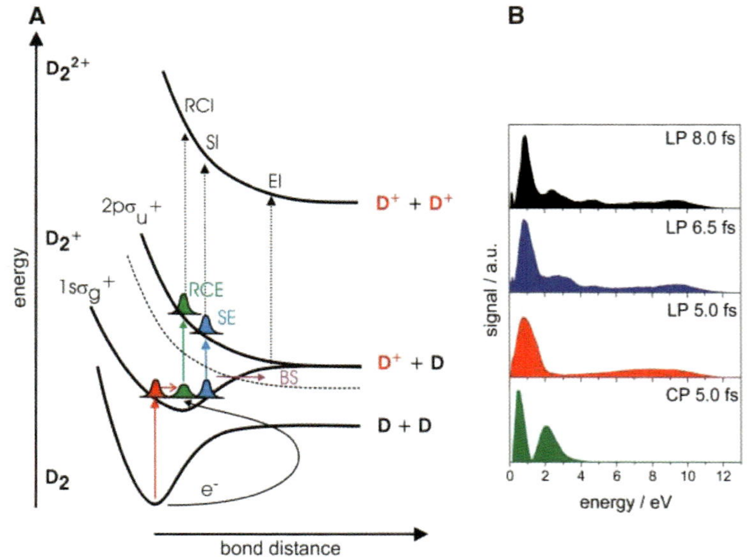

Fig. 1. (A) Pathways for the production of D^+ ions from D_2 by dissociation of the molecular ion (through BS, SE, or RCE) or by Coulomb explosion (through RCI, SI, or EI). BS occurs when the avoided crossing between diabatic potentials that are dressed by the laser field gives rise to dissociation from vibrational levels that were originally bound (*12*). **(B)** D^+ kinetic energy spectra for dissociation of D_2 by 5- to 8-fs linearly polarized (LP) and 5-fs circularly polarized (CP) laser pulses without phase stabilization, at $I = 1.2 \pm 0.2 \times 10^{14}$ W cm^{-2} and $I = 2.4 \pm 0.2 \times 10^{14}$ W cm^{-2}, respectively.

[1]FOM Instituut voor Atoom en Molecuul Fysica (AMOLF), Kruislaan 407, 1098 SJ Amsterdam, Netherlands. [2]Max-Planck-Institut für Quantenoptik, Hans-Kopfermann-Strasse 1, D-85748 Garching, Germany. [3]Fakultät für Physik, Universität Bielefeld, Universitätsstrasse 25, D-33615 Bielefeld, Germany. [4]Department für Physik, Ludwig-Maximilians-Universität München, Am Coulombwall 1, D-85748 Garching, Germany. [5]Institut für Experimentalphysik, Universität Hamburg, Luruper Chaussee 149, D-22761 Hamburg, Germany.

step via recollision [recollision ionization (RCI)] or via the laser field [sequential ionization (SI)]. Enhanced ionization (EI) takes place when the vibrational wave packet reaches a transition region where electron localization occurs and where (anti)-bonding molecular orbitals turn into atomic orbitals. Double ionization (via RCI, SI, or EI) leads to the breakup of the molecule by Coulomb repulsion between the two D^+ ions. Two momentum-matched D^+ ions are emitted symmetrically along the molecular axis, irrespective of the shape of the driving fields. To address the feasibility of reaction control by means of steering electronic motion, we therefore have to investigate the BS, SE, and RCE pathways and keep double ionization at a minimum.

We have studied the dissociation of D_2^+ in a few-cycle laser field by measuring the emission of D^+ as a probe of the location (or absence) of the electron. D^+ fragment kinetic energy and angular distributions were retrieved using the velocity-map imaging technique (*13*). For a description of the experimental apparatus, see the supporting online material (SOM). In general, all the pathways illustrated in Fig. 1 contribute to the D^+ kinetic energy spectra. Their relative importance sensitively depends on the intensity and duration of the laser (*6*). Figure 1B shows D^+ kinetic energy spectra recorded for linearly and circularly polarized laser pulses with durations of 5 to 8 fs at peak intensities (*I*) of $I = 1.2 \pm 0.2 \times 10^{14}$ W cm^{-2} and $I = 2.4 \pm 0.2 \times 10^{14}$ W cm^{-2}, respectively.

Previous studies (*14–16*) suggest that D^+ ions with energies below 3 eV originate from BS (at ~ 0.9 eV) and EI (at ~ 2.5 eV). Above 3 eV, contributions may arise from SE/SI and RCE/RCI. In our experiment, the disappearance of the D^+ signal above 3 eV upon changing the polarization from linear to circular (while maintaining the peak electric field of the laser) indicates that recollision is responsible for the creation of the high-energy fragments (Fig. 1B). Decreasing the pulse duration from 8 to 5 fs suppresses spectral components assigned to double ionization. Light-field control of molecular dissociation was therefore pursued with phase-stabilized 5-fs laser pulses.

Figure 2 shows a cut through the D^+ three-dimensional momentum distribution in Cartesian coordinates (p_x, p_y) at $p_z = 0$, for a 5-fs, 1×10^{14} W cm^{-2} laser field without phase stabilization. The laser propagated along the *x* axis ($\theta = 90°/270°$) and was polarized along the *y* axis ($\theta = 0°/180°$). In agreement with previous studies (*17*), the BS (0 to 2 eV) and the weak EI channel (2 to 3 eV) that appear in the center of the image show relatively narrow angular distributions. A nearly isotropic distribution is measured for higher energies (3 to 10 eV) and is a typical signature of recollision-induced fragmentation (*15*). No difference in the up versus down emission of D^+ ions (along the laser polarization axis) is observed without phase stabilization.

In an attempt to control and probe the final location of the electron in the D_2^+ dissociation, D^+ images were recorded while changing the carrier envelope phase φ. The angle-integrated asymmetry in the D^+ emission was evaluated as a function of energy $W = p^2/2m$ and φ via

$$A(W,\varphi) = \frac{P_{up}(W,\varphi) - P_{down}(W,\varphi)}{P_{up}(W,\varphi) + P_{down}(W,\varphi)}$$

with

$$P_{up}(W,\varphi) = \int_{330}^{360} d\theta \int_{0}^{360} d\phi\, P(W,\theta,\phi,\varphi)\sin\theta + \int_{0}^{30} d\theta \int_{0}^{360} d\phi\, P(W,\theta,\phi,\varphi)\sin\theta$$

and

$$P_{down}(W,\varphi) = \int_{150}^{210} d\theta \int_{0}^{360} d\phi\, P(W,\theta,\phi,\varphi)\sin\theta$$

with θ and ϕ being the polar and azimuthal angles, respectively.

We chose to analyze the ion emission within a restricted angular range because (as will become clear) our ability to control electron motion in D_2^+ requires that the laser couple the two lowest-lying electronic states. For molecules aligned orthogonally to the laser polarization axis, this coupling is absent.

Figure 3A depicts the angle-integrated D^+ energy spectrum obtained with 5-fs laser pulses with a peak intensity of $I = 1 \times 10^{14}$ W cm^{-2} and a randomly varying carrier envelope phase. Figure 3B reveals how phase locking results in a nonzero asymmetry $A(W,\varphi)$ as a function of the kinetic energy *W* of the D^+ fragments (*y* axis) and the laser phase φ (*x* axis). Regions where the asymmetry oscillates as a function of the phase represent conditions where the direction of the D^+ emission, and hence the localization of the electron in the dissociation process, is effectively controlled by the subcycle evolution of the laser field driving the dissociation. The extent of the control is further illustrated in Fig. 3C, which displays a series of curves where $A(W,\varphi)$ is integrated over selected energy intervals. The highest degree of asymmetry with a modulation depth of ~50% is observed between 3 and 8 eV. A very small phase dependence is seen between 0 and 3 eV. Asymmetric D^+ ejection is observed predominantly at kinetic energies that are virtually absent in a circularly polarized field, suggesting that electron-ion recollision is a vital element in the mechanism responsible for the observed phase control.

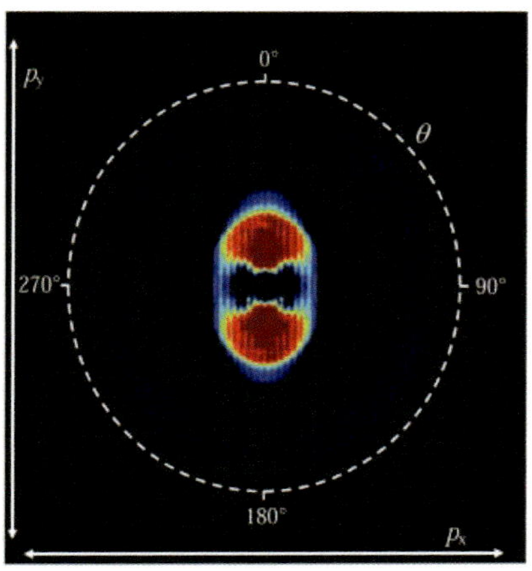

Fig. 2. Two-dimensional D^+ momentum image for D_2 dissociation in a 5-fs, 1×10^{14} W cm^{-2} laser field without phase stabilization.

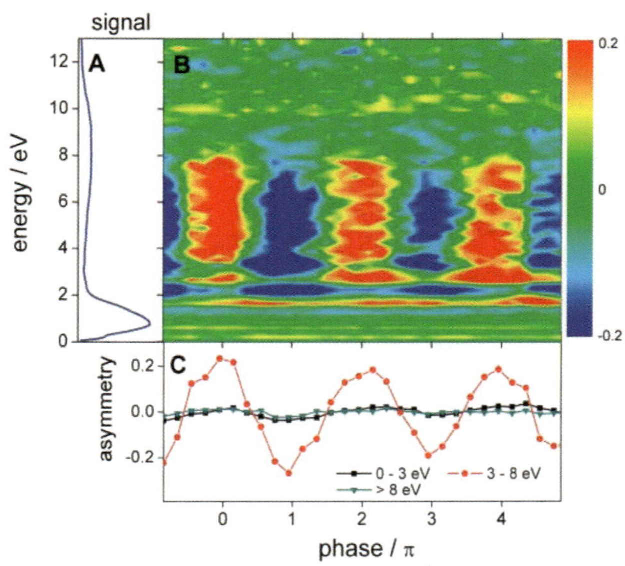

Fig. 3. **(A)** D^+ kinetic energy spectrum for D_2 dissociation with 5-fs, 1×10^{14} W cm^{-2} laser pulses without phase stabilization. **(B)** Map of asymmetry parameter $A(W,\varphi)$ as a function of the D^+ kinetic energy and carrier envelope phase φ (measured over a range of 6π with a step size of $\Delta\varphi = 0.1\pi$). **(C)** Integrated asymmetry over several energy ranges versus carrier envelope phase φ.

REPORTS

To gain qualitative insight into the mechanism responsible for the observed phase-controlled asymmetry in the dissociation of D_2^+, we modeled the laser-driven motion of the two nuclei and the bound electron by numerically solving the time-dependent Schrödinger equation (see the SOM for details). As in previous studies (*18*), the process is modeled in terms of the $1s\sigma_g^+$ and $2p\sigma_u^+$ electronic states, and the molecule is assumed to be aligned along the laser polarization axis.

In our modeling, the D_2^+ molecular ion is formed in a single ionization event that occurs at the maximum of the laser electric field. This ionization produces a vibrational wave packet in the $1s\sigma_g^+$ ground electronic state that mimics the ($v = 0$) vibrational wave function of the D_2 ground state. Population transfer from the $1s\sigma_g^+$ ground electronic state to the $2p\sigma_u^+$ excited electronic state is introduced at a delay of 1.7 fs after ionization [corresponding to the first recollision time (*10*)]. Later recollision events are efficiently suppressed with short laser pulses (*15*). Because of the strongly repulsive nature of the $2p\sigma_u^+$ state, the excited D_2^+ molecule dissociates and the momentum-matched D and D^+ fragments acquire a large kinetic energy (E_k up to 10 eV). During the dissociation, the laser field transfers part of the $2p\sigma_u^+$ population to the $1s\sigma_g^+$ state, producing a dissociative wave packet with a large excess kinetic energy. The emerging coherent superposition of the two electronic states results in a time-dependent localization of the electron density on the upper or lower nucleus due to the gerade and ungerade nature of the two states (*7, 9, 19*).

In Fig. 4, the temporal evolution of the laser field and the occupations of the $1s\sigma_g^+$ and $2p\sigma_u^+$ electronic states are depicted in (A) and (B), respectively. A time-dependent electron localization parameter quantifying the localization on the upper/lower nucleus (for definition, see the SOM) is displayed in (C). By the time the molecule has dissociated (with the internuclear distance reaching ~15 atomic units), the electron density is found to localize predominantly on the lower D atom. A simple shift of the carrier envelope phase φ by π reverses the field and turns the direction of emission of the ionic/atomic fragment opposite, in agreement with our experimental observation. To the best of our knowledge, this is the first demonstration of direct light-field control of a chemical reaction via the steering of electronic motion.

Let us consider the electron localization dynamics after the recollision in more detail. Initially, the $1s\sigma_g^+$ state is hardly populated by the laser-induced coupling, and the energy gap $\Delta W(t) = \hbar\omega(t)$ between the binding $1s\sigma_g^+$ and repulsive $2p\sigma_u^+$ states is much larger than the laser photon energy. Hence, the electronic wave function can respond nearly instantaneously to changes in the laser field, and the electron localization parameter (Fig. 4C) oscillates with a small amplitude at the frequency of the light pulse. As the eigenfrequency $\omega(t)$ of the two-level system decreases because of the increasing bond length, the laser field starts populating the lower state substantially, strongly increasing the extent of electron localization. In our calculations, the electron wave packet keeps evolving with the instantaneous eigenfrequency $\omega(t)$ as the laser field strength approaches zero (for $t > 7$ fs). The oscillation of the electron localization ceases when the interatomic barrier that builds up between the two D^+ ions can no longer be overcome by the electron. In the experiment, because of a nonzero background of the few-cycle pulses (<10% in intensity), a laser field may still be present when the molecule breaks up and contribute to the asymmetry (*20*).

The field-controlled electron dynamics demonstrated in this work does not rely on electron recollision. In fact, control of electron dynamics in molecules can also be achieved by preparing a coherent electronic superposition state by photoexcitation (*7, 19, 21, 22*). In contrast with the current experiment, the photoexcitation will permit the creation of an electron wave packet at the beginning of the reaction, thereby offering a greater flexibility for its subsequent light-field control. This approach will also be applicable in neutral molecules.

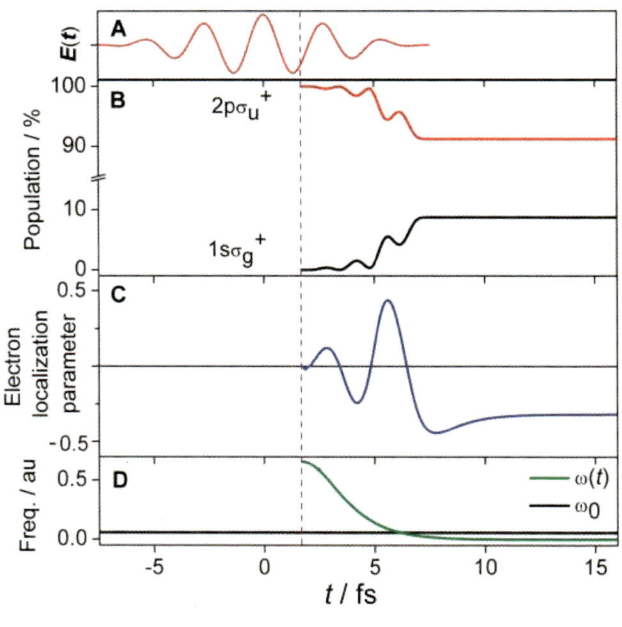

Fig. 4. (A) Electric field of the 5-fs, $1{\cdot}10^{14}$ W cm^{-2} pulse used in the calculations, with carrier envelope phase $\varphi = 0$. **(B)** Time-dependent populations of the $1s\sigma_g^+$ and $2p\sigma_u^+$ states of D_2^+ after excitation through recollision. **(C)** Temporal evolution of the electron localization parameter starting from the time of recollision. **(D)** Time dependence of the eigenfrequency $\omega(t)$ of the two-level system versus the laser frequency ω_0 (frequencies are given in atomic units). The vertical dashed purple line marks the time of recollision in all panels.

References and Notes

1. A. Baltuška *et al.*, *Nature* **421**, 611 (2003).
2. R. Kienberger *et al.*, *Nature* **427**, 817 (2004).
3. E. Goulielmakis *et al.*, *Science* **305**, 1267 (2004).
4. G. G. Paulus *et al.*, *Phys. Rev. Lett.* **91**, 253004 (2003).
5. F. Lindner *et al.*, *Phys. Rev. Lett.* **95**, 040401 (2005).
6. J. H. Posthumus, *Rep. Prog. Phys.* **67**, 623 (2004).
7. A. D. Bandrauk, S. Chelkowski, H. S. Nguyen, *Int. J. Quant. Chem.* **100**, 834 (2004).
8. V. Roudnev, B. D. Esry, I. Ben-Itzhak, *Phys. Rev. Lett.* **93**, 163601 (2004).
9. H. Niikura, D. M. Villeneuve, P. B. Corkum, *Phys. Rev. A* **73**, 021402 (2006).
10. H. Niikura *et al.*, *Nature* **417**, 917 (2002).
11. H. Niikura *et al.*, *Nature* **421**, 826 (2003).
12. P. H. Bucksbaum, A. Zavriyev, H. G. Muller, D. W. Schumacher, *Phys. Rev. Lett.* **64**, 1883 (1990).
13. A. T. J. B. Eppink, D. H. Parker, *Rev. Sci. Instr.* **68**, 3477 (1997).
14. A. S. Alnaser *et al.*, *Phys. Rev. Lett.* **91**, 163002 (2003).
15. A. S. Alnaser *et al.*, *Phys. Rev. Lett.* **93**, 183202 (2004).
16. H. Sakai *et al.*, *Phys. Rev. A* **67**, 063404 (2003).
17. K. Sändig, H. Figger, T. W. Hänsch, *Phys. Rev. Lett.* **85**, 4876 (2000).
18. I. Kawata, H. Kono, Y. Fujimura, *J. Chem. Phys.* **110**, 11152 (1999).
19. G. L. Yudin, S. Chelkowski, J. Itatani, A. D. Bandrauk, P. B. Corkum, *Phys. Rev. A* **72**, 051401 (2005).
20. M. Y. Ivanov, P. B. Corkum, P. Dietrich, *Laser Phys.* **3**, 375 (1993).
21. H. Niikura, D. M. Villeneuve, P. B. Corkum, *Phys. Rev. Lett.* **94**, 083003 (2005).
22. P. Krause, T. Klamroth, P. Saalfrank, *J. Chem. Phys.* **123**, 074105 (2005).

23. We thank J. Rauschenberger for experimental support and acknowledge valuable discussions with B. Esry, P. Corkum, and H. Niikura. We are grateful for financial support from the Marie Curie Research Training Network XTRA (grant no. MRTN-CT-2003-505138) and a Marie Curie Intra-European Fellowship (no. MEIF-CT-2003-500947) (M.F.K.). The research of M.F.K., Ch.S., J.I.K., Y.N., and M.J.J.V. is part of the research program of the Stichting voor Fundamenteel Onderzoek der Materie (FOM), which is financially supported by the Nederlandse Organisatie voor Wetenschappelijk Onderzoek (NWO).

Supporting Online Material

www.sciencemag.org/cgi/content/full/312/5771/246/DC1
SOM Text
Figs. S1 and S2
References

15 February 2006; accepted 15 March 2006
10.1126/science.1126259

nature COMMUNICATIONS

ARTICLE

Received 21 Dec 2013 | Accepted 3 Apr 2014 | Published 8 May 2014 DOI: 10.1038/ncomms4800

Subfemtosecond steering of hydrocarbon deprotonation through superposition of vibrational modes

A.S. Alnaser[1,2,3], M. Kübel[4], R. Siemering[5], B. Bergues[2], Nora G. Kling[6], K.J. Betsch[2], Y. Deng[2], J. Schmidt[4], Z.A. Alahmed[3], A.M. Azzeer[3], J. Ullrich[7,8], I. Ben-Itzhak[6], R. Moshammer[7], U. Kleineberg[4], F. Krausz[2,4], R. de Vivie-Riedle[5] & M.F. Kling[2,4,6]

Subfemtosecond control of the breaking and making of chemical bonds in polyatomic molecules is poised to open new pathways for the laser-driven synthesis of chemical products. The break-up of the C-H bond in hydrocarbons is an ubiquitous process during laser-induced dissociation. While the yield of the deprotonation of hydrocarbons has been successfully manipulated in recent studies, full control of the reaction would also require a directional control (that is, which C-H bond is broken). Here, we demonstrate steering of deprotonation from symmetric acetylene molecules on subfemtosecond timescales before the break-up of the molecular dication. On the basis of quantum mechanical calculations, the experimental results are interpreted in terms of a novel subfemtosecond control mechanism involving non-resonant excitation and superposition of vibrational degrees of freedom. This mechanism permits control over the directionality of chemical reactions via vibrational excitation on timescales defined by the subcycle evolution of the laser waveform.

[1] Physics Department, American University of Sharjah, POB26666 Sharjah, UAE. [2] Max-Planck-Institut für Quantenoptik, D-85748 Garching, Germany. [3] Department of Physics & Astronomy, King-Saud University, Riyadh 11451, Saudi Arabia. [4] Department für Physik, Ludwig-Maximilians-Universität München, D-85748 Garching, Germany. [5] Department für Chemie und Biochemie, Ludwig-Maximilians-Universität München, 81377 München, Germany. [6] J. R. Macdonald Laboratory, Physics Department, Kansas-State University, Manhattan, Kansas 66506, USA. [7] Max-Planck-Institut für Kernphysik, D-69117 Heidelberg, Germany. [8] Physikalisch-Technische Bundesanstalt, D-38116 Braunschweig, Germany. Correspondence and requests for materials should be addressed to A.S.A. (email: aalnaser@aus.edu) or to M.F.K. (email: matthias.kling@lmu.de).

The nuclear kinematics that are responsible for making and breaking chemical bonds occur on femtosecond timescales[1]. The steering of chemical reactions has been demonstrated for a variety of systems in the gas and condensed phase using modulated light fields[2–6], with the amplitude, phase and polarization used as a control parameter. Breaking of the hydrogen bonds is one of the most important dynamic reactions in the chemistry related to biology, combustion and catalysis, motivating considerable efforts to monitor and ultimately control the process. It has been studied in great detail, where efficient strong-field-induced proton ejection was reported[7–14]. Dynamic charge localization[11] and concerted fragmentation[12] scenarios were suggested to explain the origin of the high kinetic energies of the ejected protons. These studies corroborated the need for ultrashort light pulses, consisting of a few cycles, for steering the deprotonation process. As an additional challenge, the aforementioned control schemes do not provide means for a bond-specific manipulation of the deprotonation in symmetric hydrocarbons.

Few-cycle pulses with well-defined waveforms[15] offer a new quality of control on direct manipulation of the wavefunction of a molecular system[16]. Few-cycle waveforms with adjustable carrier-envelope-phase (CEP) have permitted control of charge localization in the ionization and dissociation of diatomic molecules on a subfemtosecond timescale[17–21]. The approach has also been successful in altering the yields of ionization and fragmentation in hydrocarbons[13]. Despite its relevance to stereochemistry, that is, the rearrangements of atoms in a molecule, full control of the deprotonation of hydrocarbons, including the directional steering of the emission of the protons, has, however, not been reported.

Here, we demonstrate the subfemtosecond control of the proton-ejection direction from acetylene molecules with the waveform of near-single-cycle laser pulses. On the basis of quantum dynamical calculations, we show that the demonstrated directional control does not originate from electronic motion effects considered so far[16]. Rather, the mechanism is based on the manipulation of the phases of individual components of the multi-mode vibrational wave packet by the laser field in the neutral molecule and cation. While control of vibrational superpositions and their phase after resonant excitation has been performed in the context of molecular vibrational qubits[22,23], the superposition between vibrational modes is formed through a non-resonant process in our study and is controlled on an attosecond timescale. The process described here constitutes a novel mechanism for steering the reactions of complex molecules.

Results

Laser-induced deprotonation of the acetylene dication. The electric field waveform of (transform-limited) few-cycle light pulses can be described as $E(t) = E_0(t)\cos(\omega t + \phi)$, where $E_0(t)$ is the amplitude envelope, ω the carrier angular frequency and ϕ is the CEP. The CEP offers a way to tailor the electric field waveform on subfemtosecond timescales and has proven to be a powerful parameter for steering electron motion[16].

In our present experimental studies on the CEP control of the deprotonation of acetylene, we employ CEP-tagged reaction microscopy (REMI)[24] (for details see the Methods section). The interaction of acetylene molecules with intense few-cycle pulses results in the dissociative ionization and isomerization of the molecule yielding a variety of reaction products[7]. One of the reactions that we concentrate on here is the deprotonation of acetylene after two consecutive ionization steps, see Fig. 1a, where a 1D cut of the potential energy surfaces (PES) is shown. After

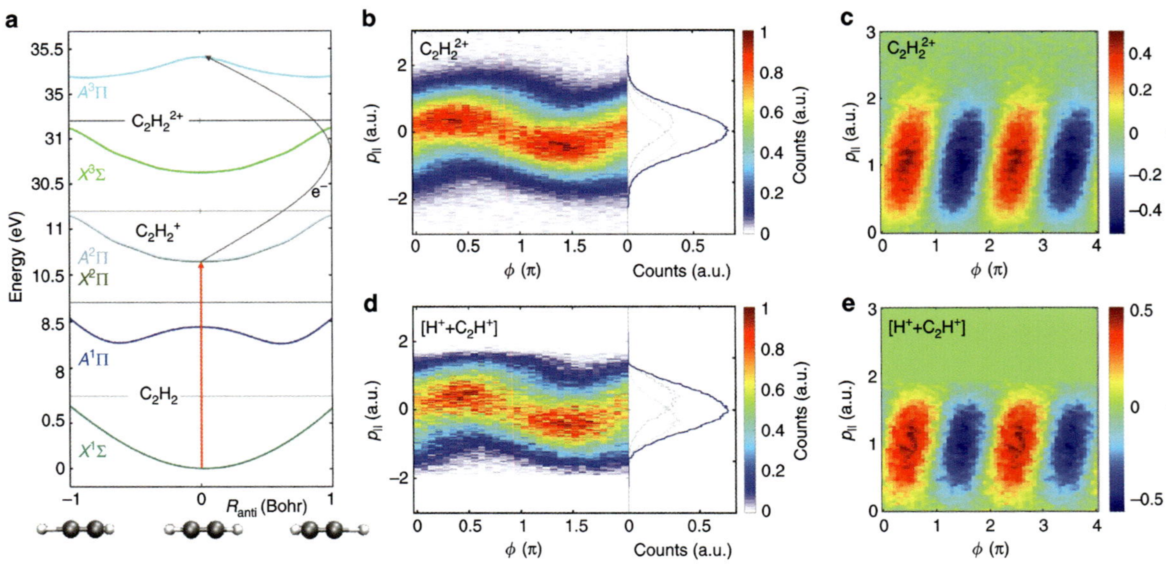

Figure 1 | Generation of the acetylene dication and the deprotonation fragments. (**a**) 1-D cut of potential energy curves of the considered states in the ionization and deprotonation: ground ($X^1\Sigma_g^+$) and first excited state ($A^1\Pi_u$) of the neutral molecule, the cation ground ($X^2\Pi_u$) and degenerate first excited state ($A^2\Pi_u$), and the dication ground ($X^3\Sigma_g^-$) and first excited state ($A^3\Pi_u$) along the anti-symmetric stretching coordinate R_{anti} at $R_{sym} = 0$. The red vertical line indicates the tunnelling ionization and the grey curved line marks the second ionization step by electron recollision. (**b**) Density plot of the acetylene dication longitudinal recoil momentum as a function of the experimental CEP, ϕ, for 4 fs pulses at an intensity of 2×10^{14} W cm^{-2}. CEP-integrated momenta are shown in the right panel. The dashed (dotted) lines shown correspond to CEP's with maximum positive (negative) asymmetry in the ion emission, respectively. (**c**) Density plot for the asymmetry parameter, $A(p_{||},\phi)$, for $C_2H_2^{2+}$ ions. (**d**) Density plot of the vector momentum sum of the recoil of the two (H^+,C_2H^+) deprotonation fragments. (**e**) Density plot of the asymmetry parameter for the momentum sum of the deprotonation fragments (H^+,C_2H^+).

population of the first excited state of $C_2H_2{}^{2+}$, the dication breaks up into H^+ and C_2H^+ fragments, which are detected with the REMI in coincidence.

To confirm the population scenario of the dication, which involves ionization by the laser field followed by subsequent recollisional ionization and excitation (Fig. 1a), high-resolution spectra of the longitudinal (along the laser polarization axis) momentum $p_{||}$ for the doubly charged $C_2H_2{}^{2+}$ recoil ions as a function of the CEP for 4-fs laser pulses at a peak intensity of $2\times10^{14}\,\mathrm{W\,cm^{-2}}$ were recorded and are shown in Fig. 1b. Clear modulation of the dication's momentum with the CEP is visible.

To enhance the visibility of CEP-effects, it is useful to define an asymmetry parameter, $A(p_{||},\phi)$, as

$$A(p_{||},\phi)=\frac{N_{\mathrm{pos}}(p_{||},\phi)-N_{\mathrm{neg}}(p_{||},\phi)}{N_{\mathrm{pos}}(p_{||},\phi)+N_{\mathrm{neg}}(p_{||},\phi)} \quad (1)$$

with $N_{\mathrm{pos}}(p_{||},\ \phi)$ and $N_{\mathrm{neg}}(p_{||},\ \phi)$ representing the yields for positive and negative momentum components $p_{||}$, respectively. The 2-dimensional map in Fig. 1c for the dication shows high-amplitude asymmetry. Corroborated by previous studies on atoms[25] and diatomic molecules[26], the dependence of the recoil-ion momentum on the CEP indicates that the doubly charged ion, at this laser intensity, is formed via an electron recollision process. Here, the free electron created in the first tunnelling event is guided by the laser field, and, for certain CEPs, the electron will return with sufficient energy to ionize the $C_2H_2^+$ ion. Because of momentum conservation during ionization, the momentum sum of the two ionized electrons is equal and opposite to the measured recoil momentum of the $C_2H_2{}^{2+}$ ion (neglecting the small momentum contribution of the absorbed photons).

The data in Fig. 1b,c indicate that—for the near-single cycle pulses used in the experiment—the electron recollision and the corresponding double ionization and excitation of $C_2H_2{}^{2+}$ are controlled by the light waveform. This is analogous to previous findings on the non-sequential double ionization of argon with 4-fs pulses[25], where it was shown that electron recollision can be limited to a single event by using sufficiently short, CEP-controlled laser pulses.

Exploiting the applied coincident multi-particle detection scheme, the deprotonation channel has been identified by the coincidence of the H^+ and C_2H^+ fragments following the dissociation of the dication. We can investigate the dependence of their yield on the CEP. Figure 1d,e shows the CEP-dependent momentum sum, p_{sum}, of the coincident H^+ and C_2H^+ fragments parallel to the laser field and the asymmetry that is derived from these data.

A clear CEP dependence is visible in Fig. 1d,e exhibiting the same oscillatory behaviour as seen for the dication. The similarity of the two plots indicates that the recollision process is responsible for the recoil momentum of both the dication and the centre of mass motion of the reaction products after deprotonation.

Directional laser-waveform control of the deprotonation. As a signature of the deprotonation we investigate H^+ fragments from coincident creation of H^+ and C_2H^+ upon the break-up of the dication. Figure 2a,b shows the momentum and asymmetry for H^+ ions. The directional control of the deprotonation reaction becomes visible when analysing the asymmetry of the H^+ emission, which is shown in Fig. 2b. Figure 2c shows the asymmetry parameter $A(\phi)$ that has been integrated over all momenta of the H^+ fragments. The observed preferential CEP-dependent ejection of protons to the left and right of the laser polarization axis illustrates the subfemtosecond steering of the hydrogen ejection.

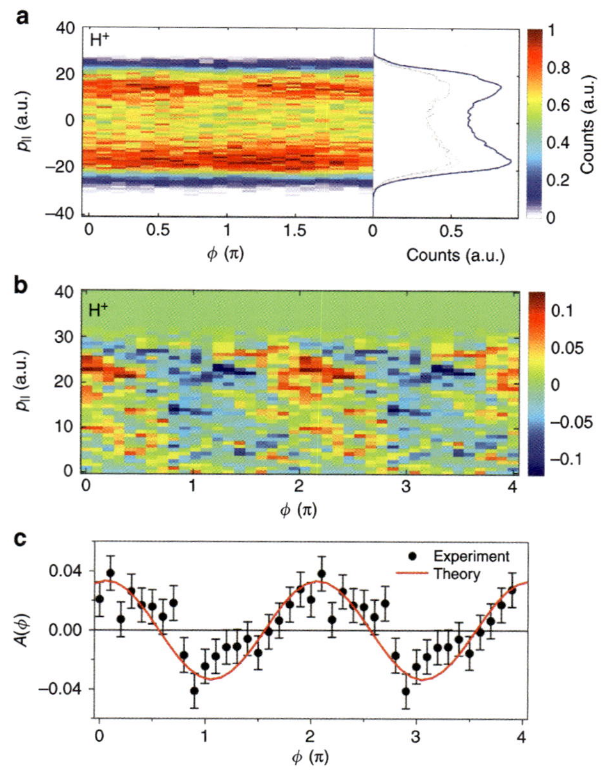

Figure 2 | Subfemtosecond directional control of deprotonation. (**a**) Density plot of the momentum of H^+ fragments as a function of CEP. CEP-integrated momenta are shown in the right panel. The dashed (dotted) lines shown correspond to CEP with maximum positive (negative) asymmetry in the ion emission, respectively. (**b**) Density plot for the asymmetry parameter, $A(p_{||},\phi)$, for the H^+ fragments. (**c**) The asymmetry parameter, $A(\phi)$, integrated over all momenta for H^+ ions. The error bars were derived via error propagation using the s.d. in the ion yields with positive and negative momenta. The clear oscillation in the asymmetry for the directional emission of H^+ fragments as a function of CEP under the given laser conditions demonstrates the CEP-control of the hydrogen emission direction. Also shown is the calculated asymmetry (solid red line) obtained from our quantum dynamical calculations (see text for details), which has been shifted in absolute CEP for best overlap with the experimental data since the absolute CEP is not known from the experiment.

Discussion

To account for the asymmetric deprotonation, the well-known Tannor-Rice control scheme exploiting excited state dynamics[27] might appear to be a first candidate. However, for the current case, where the involved electronic state potentials are symmetric with respect to the deprotonation direction, an additional asymmetry has to be introduced by the laser field. We have performed quantum dynamical calculations to model the relevant molecular dynamics. The details of our calculations are given in the Methods section. The deprotonation of the dication is described in the basis of the symmetric $|0m>$ and anti-symmetric C-H stretching $|n0>$ modes with m, n being the number of vibrational quanta, and $|0m>$ and $|n0>$ implicitly including the respective time evolution factor $e^{-i\frac{E_{m(n)}}{\hbar}t}$. Around the minimum of PES, these modes correspond to the normal modes and their respective overtones. A linear combination of these modes, $1/\sqrt{2}(|0m\rangle \pm |n0\rangle)$, will induce a wave packet

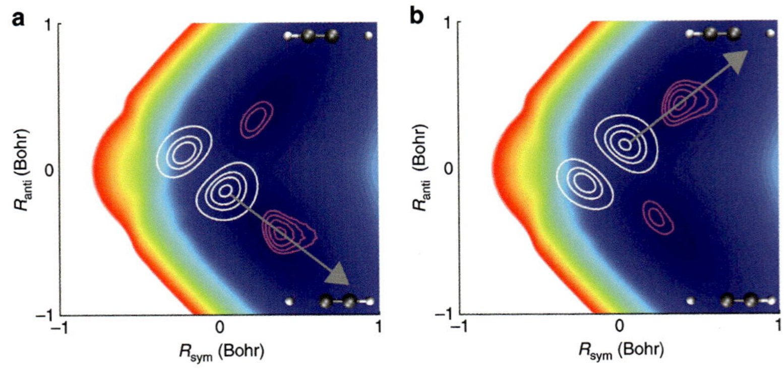

Figure 3 | The relevant first excited state of the dication for deprotonation. The state ($A^3\Pi_u$) is shown as a 2D-surface plotted against the symmetric and anti-symmetric stretching coordinates. The arrows in **a** and **b** indicate the two directions, into which protons can be emitted. To illustrate the control process, we plotted two wave packets representing the ideal superposition of $1/\sqrt{2}(|01\rangle - |10\rangle)$ (**a**) and $1/\sqrt{2}(|01\rangle + |10\rangle)$ (**b**). The initial wave packets are shown by white topographic lines, a snapshot of the same wave packet after 2.4 fs in magenta. The negative superposition (**a**) yields dominant cleavage of the left C-H bond, while the positive superposition (**b**) yields dominant cleavage of the right C-H bond.

motion along either one of the C-H stretching coordinates. When such a wave packet experiences a repulsive potential either the deprotonation of the left or the right H-atom is initiated. Two-dimensional (2D) PES along the normal mode coordinates (R_{sym}, R_{anti}) are calculated for the neutral, cationic, dicationic ground and excited states. The first excited state, the $A^3\Pi_u$ state, of the dication turns out to be the important state for the reaction (see Fig. 1a and Methods section) and its 2D surface is shown in Fig. 3.

From the stretching modes, only the anti-symmetric mode $|10>$ is IR-active, and the incident light field is not resonant with the vibrational transitions. To prepare the required superposition, it is sufficient to address and control only the anti-symmetric $|10>$ mode, via IR transitions. For the given light field, this happens in the neutral and cation. The symmetric mode contributions in the vibrational wave packet occur solely due to the ionization steps. The time-dependent population of the anti-symmetric mode $|10>$ follows the electric field evolution of the few-cycle pulse (see Methods section). The CEP of the pulse imprints a certain phase (φ_{CEP}) on the addressed anti-symmetric mode $|10>(|10>e^{-i\varphi_{CEP}t})$. When the few-cycle pulse is able to modify the phase of the anti-symmetric stretching mode, the sign in the superposition of the symmetric and anti-symmetric modes can be varied with the CEP and exhibits a modulation with a periodicity of 2π.

To mimic the effect of the second ionization including the recolliding electron, the cationic wave packet is projected onto the dication. During the second ionization step, the intensity of the remaining light field is too low to significantly enhance the control of the asymmetric stretching mode $|10>$ components. The main role of the recolliding electron is to provide the energy that is sufficient to reach the excited triplet state $A^3\Pi_u$. The energy of the recolliding electron can be evaluated for the given laser parameters to about 33 eV, corresponding to 3.17 U_p[28] with the ponderomotive potential $U_p = I/(4\omega^2)$ (in atomic units) that depends on the laser intensity I and the angular frequency ω. An electron energy of 25 eV is enough to reach the $A^3\Pi_u$ state of the dication. Because of the significant differences in the potential shape of the dicationic $A^3\Pi_u$ state and the cationic ground state (Fig. 1a), higher vibrational quanta ($|n0>$ and $|0m>$) up to the dissociation limit become now the dominant components of the vibrational wave packet and dissociation of the C-H bond can occur[13]. The preparation of the CEP-dependent features in the vibrational wave packet takes place in the neutral and cation, and is transferred to the dication in the second ionization. We observe that the propagation of the vibrational wave packet in the dication to either one of the dissociative channels ($H^+ + CCH^+$ and $HCC^+ + H^+$, see Fig. 3) can be manipulated by the CEP. Evaluation of the fraction of the wave packet correlating with the individual dissociation channels provides the theoretical asymmetry plotted in Fig. 2c. The theoretical curve shows excellent agreement with the experimental data, which lends support to the described novel subfemtosecond control mechanism.

The presented experimental and theoretical data support a new and very general coherent-control scheme, where the direction of proton ejection in a symmetric hydrocarbon is steered through manipulating the phases of the individual components of the vibrational wave packet. The CEP control-induced mechanism is substantially different from all previous studies, where electronic wave packets dominated the outcomes of the molecular dynamics[16], in the context of which the enhancement of the asymmetry through selection of vibrational states has been reported[29,30]. While in the current proof-of-principle experiment, the degree of control was rather limited, it can be enhanced by using near-single cycle waveforms that are resonant with the optically active mode. With the advent and current development towards the generation of ultrashort, phase-stable light pulses in the mid-infrared[31], substantial improvement of control is within reach. We expect the control scheme to be widely applicable to a variety of chemical processes, where the reactions of not only asymmetric but also symmetric molecules can be manipulated.

Methods

Experimental set-up. Linearly polarized laser pulses with a central wavelength of 750 nm, an energy of 150 μJ and a full width at half maximum of the temporal intensity envelope of 4 fs are generated at a repetition rate of 10 kHz in an amplified laser system. The laser intensity was kept stable on the level of 10%. The CEP of each laser pulse was recorded by sending a fraction of the laser beam (20 μJ) into a single-shot stereographic above-threshold ionization (Stereo-ATI) phase meter[32]. A detailed description of the CEP measurement and CEP tagging techniques can be found in ref. 33. In brief, the time-of-flight spectra of electrons emitted along both directions of the laser polarization axis are recorded on either side of the apparatus. For each laser shot, the CEP is inferred from the asymmetry between the left and the right TOF spectra. Note that the CEP is measured up to a constant offset in the phase meter.

In a recent study, we showed that slow drifts of the laser intensity can markedly affect the results of phase-tagged ionization yield measurements with long acquisition time[34]. To correct for any effects of small drifts of the laser parameters on the CEP measurement, we have applied the procedure outlined in ref. 34, which had been tested to give reliable phase-tagged ion yields for the non-sequential double ionization of argon. The uncertainty in determining the CEP was on the order of 300 mrad during the entire measurement.

A fraction (up to 15 μJ) of the remaining part of the laser beam is focused with a spherical mirror ($f = 25$ cm) into a well-localized (1 mm width) and cold supersonic gas jet of acetylene molecules in the centre of a REMI. Ions and electrons created in the laser focus (40 μm diameter) can be extracted by an electric field towards time- and position-sensitive MCP detectors at the left and right side of the instrument, respectively. In the current experiments, only ion signals were recorded. The ions' impact positions were determined utilizing a two-dimensional delay line anode mounted behind the MCP. Pulses from the ends of the x and y delay lines and a time signal from the microchannel plate were registered by a multi-hit time-to-digital converter and recorded in event mode. From these data, off-line analysis provided the 3D vector momentum of each ionic fragment at the instant of double ionization. The deprotonation fragments are identified and isolated from other fragments by applying tight momentum sum conditions using the coincidence high-resolution PIPICO (Photo-Ion Photo-Ion) spectra[7].

Statistical evaluation of the asymmetry data. To assess the statistical significance of the observed asymmetry, the data were subjected to a statistical test. Asymmetry curves were generated from the data for 1,000 different random re-orderings of the simultaneously recorded CEP values, that is, the CEP measured for an event was randomly associated with a different event before calculating the asymmetry. Any re-ordering should destroy the correlation of the data with the CEP, and those data can therefore be regarded as providing an upper limit for the noise of the measurement. The amplitude of the asymmetry oscillation was extracted from each test curve by a sinusoidal fit of the resulting asymmetry curve and plotted against the R^2 value in Fig. 4, which is defined as:

$$R^2 = 1 - \frac{\sum_i [A(\phi_i) - f(\phi_i)]^2}{\sum_i [A(\phi_i) - \bar{A}]^2} \tag{2}$$

where $\bar{A}$ denotes the mean value of the asymmetry over all bins and $f(\phi_i)$ the fitted asymmetry for bin i. The point corresponding to the true experimental data (non-randomized CEP ordering) is displayed as a blue point indicated by an arrow. It can be seen that this point shows a significantly larger goodness of the fit compared with the randomized test data. Therefore, despite its small amplitude, we conclude that the observed directional emission of protons is statistically significant.

Quantum chemistry methods. PES for ground and excited states of the neutral, cation and dication states of acetylene were calculated on the CASSCF[10,10], [9,10] and [8,10] level of theory, respectively, using the program package MOLPRO[35] with the 6-311++G** basis set as well as the dipole moments and transition dipole moments. Two-dimensional PES were calculated along the symmetric and anti-symmetric stretching coordinate with a step size of 0.05 a.u.. The PES were interpolated to 256 × 256 points for the evaluation of the vibrational eigenstates and the propagation of the vibrational wave packets. In the neutral molecule, only the IR-transitions play a role since the first excited state is separated by 7.8 eV (vertically) or 6.2 eV (adiabatically) and does not influence the ground state wave packet.

Wave packet propagation. The Chebychev propagator was used to solve the time-dependent Schrödinger equation. The time-dependent Schrödinger equation for the neutral molecule under the influence of the light field reads:

$$(H_n + \mu_{nn}\varepsilon(t))\Psi_n(t) = \widetilde{\Psi}_n(t) \tag{3}$$

with H_n the Hamiltonian of $X^1\Sigma_g^+$, μ_{nn} the associated dipole moment and $\varepsilon(t)$ the near-single cycle laser field.

The light field $\varepsilon(t)$ was included in the dipole approximation:

$$\varepsilon(t) = E_0 \exp\left[-2\left(\frac{t - t_1}{\mathrm{FWHM}/\sqrt{2\ln 2}}\right)^2\right]\cos(\omega_0(t - t_1)) \tag{4}$$

and was characterized by a full width at half maximum of 4 fs with a cycle-averaged intensity of 2×10^{14} W cm^{-2}. The light field was discretized with a time step of 0.024 fs. The part of the laser interaction experienced by the neutral molecule has a duration of 4.8 fs (200 time steps) in correspondence to the time when the intensity maximum is reached and the system is ionized. During that period, the initial superposition is formed. All ionization steps are approximated by setting the vibrational wave packet onto the corresponding ionic state.

In the first ionization step, the vibrational wave packet of the neutral molecule is transferred to the lowest cation states. The first two electronic states ($X^2\Pi_u$ and $A^2\Pi_u$) in the cation are nearly degenerate and we assume an equal distribution over both states in the ionization of the neutral molecule. A second type of interaction with the laser field via electronic transitions back and forth between the states is possible and included for completeness in the total Hamiltonian given below. However, the transition dipole moment between those states is almost non-existent along the deprotonation modes and its effect is negligible. The time-dependent Schrödinger equation for the cation reads:

$$\begin{pmatrix} H_1 + \mu_{11}\varepsilon(t) & -\mu_{12}\varepsilon(t) \\ -\mu_{21}\varepsilon(t) & H_2 + \mu_{22}\varepsilon(t) \end{pmatrix}\begin{pmatrix} \Psi_1(t) \\ \Psi_2(t) \end{pmatrix} = \begin{pmatrix} \widetilde{\Psi}_1(t) \\ \widetilde{\Psi}_2(t) \end{pmatrix} \tag{5}$$

with H_1, H_2 the Hamiltonians of $X^2\Pi_u$, $A^2\Pi_u$ of the cation, μ_{11}, μ_{22} the associated dipole moments, and μ_{12} the almost zero transition dipole moment.

The dissociation of the wave packet in the excited triplet state $A^3\Pi_u$ of the dication was propagated for 480 fs with time steps of 0.24 fs. The part of the wave packet leaving the PES (Fig. 3) on the top/bottom corresponds to the right/left deprotonation. Its amount is quantified by summing up all parts during the propagation

Mechanism of the CEP control. The neutral molecule experiences the part of the light field from time t_0 to t_1 (see Fig. 5a). The population of, for example, the

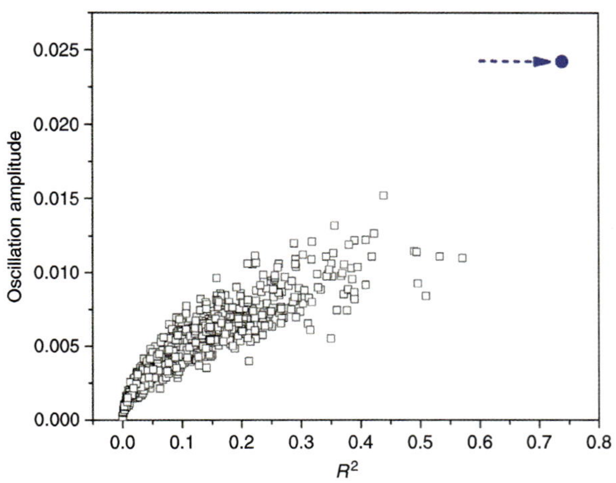

Figure 4 | Statistical evaluation of the asymmetry. In the measured data, the CEPs were randomized and the resulting asymmetry curve was fitted by a sinusoidal to extract the amplitude. This was repeated 1,000 times and the fitted amplitudes and R^2 values (see text) for each are plotted. The true experimental data are displayed as a blue point indicated by an arrow, demonstrating the statistical significance of the observed CEP-dependence.

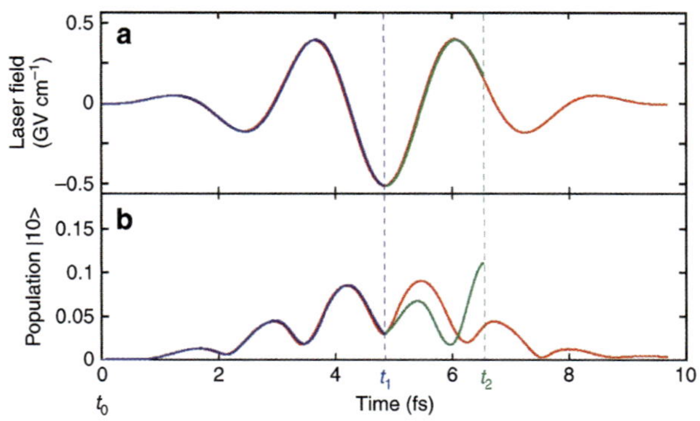

Figure 5 | Laser-field control of asymmetric hydrogen bond breaking. **(a)** Electric field waveform of the near-single-cycle laser pulse used in the calculations (red line). The parts of the field that act on the neutral and cation are displayed as blue and green lines, respectively. **(b)** Population of the IR-active mode |10>. The population induced in the neutral molecule is shown as blue line. At the peak of the laser field (t_1), the |10> population is transferred to the cation and its evolution is shown as the green line. Recollisional excitation, 1.7 fs (t_2) later, transfers the population to the $A^3\Pi_u$ state of the dication, from where dissociation can occur. Now the eigenfunctions constituting the vibrational wave packet are continuum wavefunctions. Without ionization, interaction with the entire field would induce only a transient population shown as red line, which vanishes with the end of the laser pulse.

anti-symmetric mode |10> (Fig. 5b) follows the electric field of the few-cycle pulse. The CEP of the few-cycle pulse imprints a certain phase (φ_{CEP}) on the eigenfunction of | 10> (| 10> $e^{-i\varphi_{CEP}t}$), which leads through superposition with the |01> mode (populated in the ionization) to the directional deprotonation. Ionization is assumed to happen at maximal field strength (see blue line at t_1 in Fig. 5b). From the evolution of the |10> population, it becomes obvious that the ionization step is crucial to populate the |10> mode as the interaction with the complete non-resonant light field would induce neither population (shown in red) nor phase information transfer. The first ionization transfers the vibrational wave packet to the cation. The major part of the wave packet is still the |00> component with no directional preference. However, because of the slight difference in the shape of the neutral and cationic potential energy curves now also the IR inactive mode |01> is slightly populated with a system dependent, but uncontrollable phase $e^{-i\frac{E_{01}}{\hbar}t}$. The population of the IR-active mode |10> is further increased by the interactions with the remaining light field (Fig. 5b, green line) and the imprinted phase in the superposition, defining the direction of the wave packet oscillation, is strengthened. The cation experiences the part of the light field from time t_1 to t_2 (Fig. 5a). Contributions from higher vibrational quantum numbers may occur, but play a minor role. The prepared wave packet retains the memory of direction, when transferred to the dication in the final step.

References

1. Zewail, A. H. Laser femtochemistry. *Science* **242,** 1645–1653 (1988).
2. Brixner, T. & Gerber, G. Quantum control of gas-phase and liquid-phase femtochemistry. *ChemPhysChem.* **4,** 418–438 (2003).
3. Brif, C., Chakrabarti, R. & Rabitz, H. Control of quantum phenomena: past, present and future. *New J. Phys.* **12,** 075008 (2010).
4. Judson, R. S. & Rabitz, H. Teaching lasers to control molecules. *Phys. Rev. Lett.* **68,** 1500–1503 (1992).
5. Assion, A. *et al.* Control of chemical reactions by feedback-optimized phase-shaped femtosecond laser pulses. *Science* **282,** 919–922 (1998).
6. Daniel, C. *et al.* Deciphering the reaction dynamics underlying optimal control laser fields. *Science* **299,** 536–539 (2003).
7. Alnaser, A. S. *et al.* Momentum-imaging investigations of the dissociation of D_2^+ and the isomerization of acetylene to vinylidene by intense short laser pulses. *J. Phys. B* **39,** S485–S492 (2006).
8. Hoshina, K., Furukawa, Y., Okino, T. & Yamanouchi, K. Efficient ejection of H_3^+ from hydrocarbon molecules induced by ultrashort intense laser fields. *J. Chem. Phys.* **129,** 104302 (2008).
9. Itakura, R. *et al.* Two-body Coulomb explosion and hydrogen migration in methanol induced by intense 7 and 21fs laser pulses. *J. Chem. Phys.* **127,** 104306 (2007).
10. Xu, H., Okino, T., Nakai, K. & Yamanouchi, K. Ultrafast hydrogen migration in hydrocarbon molecules driven by intense laser fields. in *Progress in Ultrafast Intense Laser Science VII* 35–52 (Springer, 2011).
11. Markevitch, A. N., Romanov, D. A., Smith, S. M. & Levis, R. J. Coulomb explosion of large polyatomic molecules assisted by nonadiabatic charge localization. *Phys. Rev. Lett.* **92,** 063001 (2004).
12. Roither, S. *et al.* High energy proton ejection from hydrocarbon molecules driven by highly efficient field ionization. *Phys. Rev. Lett.* **106,** 163001 (2011).
13. Xie, X. *et al.* Attosecond-recollision-controlled selective fragmentation of polyatomic molecules. *Phys. Rev. Lett.* **109,** 243001 (2012).
14. Okino, T., Watanabe, A., Xu, H. & Yamanouchi, K. Ultrafast hydrogen scrambling in methylacetylene and methyl-d_3-acetylene ions induced by intense laser fields. *Phys. Chem. Chem. Phys.* **14,** 10640–10646 (2012).
15. Baltuška, A. *et al.* Attosecond control of electronic processes by intense light fields. *Nature* **421,** 611–615 (2003).
16. Kling, M. F., von den Hoff, P., Znakovskaya, I. & de Vivie-Riedle, R. (Sub-) femtosecond control of molecular reactions via tailoring the electric field of light. *Phys. Chem. Chem. Phys.* **15,** 9448–9467 (2013).
17. Kremer, M. *et al.* Electron localization in molecular fragmentation of H_2 by carrier-envelope phase stabilized laser pulses. *Phys. Rev. Lett.* **103,** 213003 (2009).
18. Betsch, K. J. *et al.* Controlled directional ion emission from several fragmentation channels of CO driven by a few-cycle laser field. *Phys. Rev. A* **86,** 063403 (2012).
19. Xu, H. *et al.* Carrier-envelope-phase-dependent dissociation of hydrogen. *New J. Phys.* **15,** 023034 (2013).
20. Sansone, G. *et al.* Electron localization following attosecond molecular photoionization. *Nature* **465,** 763–766 (2010).
21. Siu, W. *et al.* Attosecond control of dissociative ionization of O_2 molecules. *Phys. Rev. A* **84,** 063412 (2011).
22. Tesch, C. M. & de Vivie-Riedle, R. Quantum computation with vibrationally excited molecules. *Phys. Rev. Lett.* **89,** 157901 (2002).
23. de Vivie-Riedle, R. & Troppmann, U. Femtosecond lasers for quantum information technology. *Chem. Rev.* **107,** 5082–5100 (2007).
24. Johnson, N. G. *et al.* Single-shot carrier-envelope-phase-tagged ion-momentum imaging of nonsequential double ionization of argon in intense 4-fs laser fields. *Phys. Rev. A* **83,** 013412 (2011).
25. Bergues, B. *et al.* Attosecond tracing of correlated electron-emission in non-sequential double ionization. *Nat. Commun.* **3,** 813 (2012).
26. Kübel, M. *et al.* Non-sequential double ionization of N_2 in a near-single-cycle laser pulse. *Phys. Rev. A* **88,** 023418 (2013).
27. Tannor, D. J., Kosloff, R. & Rice, S. A. Coherent pulse sequence induced control of selectivity of reactions: exact quantum mechanical calculations. *J. Chem. Phys.* **85,** 5805 (1986).
28. Lein, M. Molecular imaging using recollision electrons. *J. Phys. B* **40,** R135–R173 (2007).
29. Anis, F. & Esry, B. D. Enhancing the intense field control of molecular fragmentation. *Phys. Rev. Lett.* **109,** 133001 (2012).
30. Lan, P., Takahashi, E. J., Liu, K., Fu, Y. & Midorikawa, K. Carrier envelope phase dependence of electron localization in the multicycle regime. *New J. Phys.* **15,** 063023 (2013).
31. Cerullo, G., Baltuška, A., Mücke, O. D. & Vozzi, C. Few-optical-cycle light pulses with passive carrier-envelope phase stabilization. *Laser & Photon. Rev.* **5,** 323–351 (2011).
32. Wittmann, T. *et al.* Single-shot carrier-envelope phase measurement of few-cycle laser pulses. *Nat. Phys.* **5,** 357–362 (2009).
33. Rathje, T. *et al.* Review of attosecond resolved measurement and control via carrier-envelope phase tagging with above-threshold ionization. *J. Phys. B* **45,** 074003 (2012).
34. Kübel, M. *et al.* Carrier-envelope-phase tagging in measurements with long acquisition times. *New J. Phys.* **14,** 093027 (2012).
35. Werner, H. J. *et al.* *Molpro, Version 2006.1, a Package of Ab-Initio Programs* (2006).

Acknowledgements

We are grateful for support by the DFG via the grant Kl-1439/3 and the Cluster of Excellence: Munich Center for Advanced Photonics (MAP). A.S.A. acknowledges support from the American University of Sharjah. N.G.K., I.B. and M.F.K. were partially supported by the Chemical Sciences, Geosciences, and Biosciences Division, Office of Basic Energy Sciences, Office of Science, U.S. Department of Energy under DE-FG02-86ER13491 and the NSF under CHE-0822646. We are also grateful for support from the King-Saud University in the framework of the MPQ-KSU-LMU collaboration and the visiting professorship program.

Author contributions

A.S.A. and M.F.K. conceived the experiment. A.S.A., M.K., N.G.K., K.J.B. and Y.D. carried out the measurements; the data analysis and interpretation were done by A.S.A., M.K., B.B, N.G.K., K.J.B., R.S., R.d.V.-R. and M.F.K.; R.S. and R.d.V.-R. performed the calculations; part of the REMI was provided by and used in cooperation with R.M. and J.U.; the 10-kHz laser system was provided by and used in cooperation with J.S. and U.K.; Z.A.A., A.M.A., I.B. and F.K. contributed to the discussion of the results; the manuscript was prepared by A.S.A., M.K., R.S., B.B., F.K., R.d.V.-R. and M.F.K, and was discussed among all authors.

Additional information

Competing financial interests: The authors declare no competing financial interests.

Reprints and permission information is available online at http://npg.nature.com/reprintsandpermissions/

How to cite this article: Alnaser, A. S. *et al.* Subfemtosecond steering of hydrocarbon deprotonation through superposition of vibrational modes. *Nat. Commun.* 5:3800 doi: 10.1038/ncomms4800 (2014).

Review Article

Vol. 4, No. 1 / January 2017 / Optica

optica

Multilayer coatings for femto- and attosecond technology

O. Razskazovskaya,[1] F. Krausz,[1,2] and V. Pervak[2,3,*]

[1]Max-Planck Institute of Quantum Optics, Hans-Kopfermann-Str. 1, D-85748 Garching, Germany
[2]Ludwig-Maximilian Universität München, Am Coulombwall 1, D-85748 Garching, Germany
[3]Ultrafast Innovations GmbH, Am Coulombwall 1, D-85748 Garching, Germany
*Corresponding author: vladimir.pervak@physik.uni-muenchen.de

Received 5 October 2016; revised 20 December 2016; accepted 20 December 2016 (Doc. ID 278214); published 12 January 2017

Chirped multilayer mirrors have permitted mode-locked lasers to routinely generate pulses in the sub-10 fs regime. Continuous progress in the design, manufacturing, and characterization of multilayer structures has led to ever more precise group-delay dispersion control over ever broader spectral ranges. The resultant few-cycle laser fields have opened the door to the generation and measurement of isolated attosecond pulses and led to the birth of attosecond metrology. Precision multilayer dispersion control over increasing bandwidth has gradually pushed the frontier of femtosecond technology to what has been thought to be its ultimate limit: the wave cycle of visible light, allowing routine generation of sub-100 attosecond pulses. Next-generation attosecond technology will be based on synthesized multi-octave waveforms; they are expected to advance the field in several ways: first, by permitting control of electronic motions with a force variable on the atomic time scale; second, by providing sub-femtosecond optical transients for attosecond nonlinear pump-probe spectroscopy; third, by producing attosecond pulses at Angstrom wavelengths, opening the door to four-dimensional imaging with atomic resolution in space and time. In this article, we address the enabling technology: chirped multilayers for spectral separation and recombination as well as precise dispersion control of multi-octave optical radiation spanning from the ultraviolet to the mid-infrared range. This cutting-edge optical technology provides the force engineerable on atomic-to-molecular time scales and brings about the next revolution in ultrafast science.

OCIS codes: (310.4165) Multilayer design; (310.6845) Thin film devices and applications; (320.5520) Pulse compression; (320.1590) Chirping.

https://doi.org/10.1364/OPTICA.4.000129

1. INTRODUCTION

Observing and controlling fast-evolving processes relies on light flashes with a duration substantially shorter than the time evolution of the process of interest. The quest for ever-shorter pulses started shortly after the first laser was demonstrated [1] and rapidly led to sub-100 fs pulses by passively mode-locked dye lasers [2–6].

After a reign of two decades, femtosecond dye lasers were replaced by Kerr-lens mode-locked Ti:sapphire lasers, in which self-phase modulation and group-delay dispersion (GDD) control by a pair of prisms allowed the pulse duration to approach [7,8] and surpass the 10 fs regime for the first time [9]. Uncompensated higher-order dispersion was found to be the limiting factor in further pulse shortening [9,10], while the delicacy of the prism arrangement challenged the reproducible operation. Replacing the prisms with aperiodic multilayer resonator mirrors with tailored GDD [11], which became known as chirped mirrors, allowed these limitations to be overcome. With their use, Stingl *et al.* [12,13] demonstrated the generation of highly stable sub-10 fs optical pulses from a Kerr-lens mode-locked Ti:sapphire laser. Over the next decade, the capability of chirped mirrors to reflect light over a broad range of wavelengths with controlled phase, combined with simplicity and high efficiency, had been instrumental in advancing Ti:sapphire oscillators to the few-femtosecond, few-cycle regime [14–19].

Discovery of nonlinear pulse compression [20–24] slowly pushed the achievable spectral bandwidths to an optical octave. Precise and versatile techniques for manipulation of the temporal profile of ultrashort pulses were in demand. Programmable pulse shapers [25] got into the spotlight. They include liquid crystal arrays [26,27], deformable mirrors [28], and acousto-optic programmable dispersive filters (AOPDFs) [29,30]. Programmable pulse shapers provide unprecedented flexibility for amplitude and phase shaping of the spectra, along with the possibility of on-demand tuning. However, they come at the expense of non-negligible disadvantages. Liquid crystal arrays and deformable mirrors need to be placed in the Fourier plane of a 4-f system, thus often becoming a cumbersome arrangement. In addition, their efficiency is limited by the efficiency of involved diffraction gratings. Deformable mirrors are also constrained by the

Review Article Vol. 4, No. 1 / January 2017 / Optica

maximum amplitude of the deformation. AOPDF has rather low efficiency while operating on the broadband spectra. As a result, despite their lack of flexibility, dispersive multilayer mirrors retained their leading positions in the field of dispersion control over broadband spectra.

By the turn of the millennium, nonlinear pulse compression of amplified pulses along with the progress of dispersive mirror technology [31] extended the frontiers of ultrashort pulse generation to what has been believed to be an ultimate border—one cycle of the optical carrier wave. Advanced chirped-mirror-based ultrafast systems nowadays routinely deliver powerful near-single-cycle pulses [32,33], which in turn enable the generation of trains of and isolated attosecond extreme ultraviolet pulses [34–36].

To this end and more generally for the control of electronic phenomena with the electric force of light, precise control and shaping of the electric field evolution (henceforth "optical field") is of crucial importance [37,38]. Shifting the phase of the carrier wave with respect to the pulse envelope, referred to as the carrier-envelope phase (CEP), constitutes the simplest means of modifying the temporal evolution of the optical field. This modification preserves the sinusoidal field evolution and introduces a sub-cycle change in the amplitude evolution for pulse duration comparable to the carrier wave cycle.

More versatile control of atomic-scale electronic motion requires sculpting of the field evolution on an attosecond time scale, i.e., *within the wave cycle of light*. This calls for super-octave bandwidth and control schemes more sophisticated than simple adjustment of the CEP. Sub-cycle tailoring of optical fields can be implemented by changing the relative time delays between several coherently recombined CEP-stable wavepackets carried at different wavelengths. This generic approach constitutes the basis of optical waveform synthesis (OWS). The first successful implementations of the OWS concept were based on stimulated Raman scattering [39], the synchronization of independent ultrafast sources [40], or fiber amplifiers jointly seeded with a spectral continuum [41].

More recently, multi-octave, phase-stable supercontinua produced by self-phase modulation of sub-millijoule femtosecond pulses in gas-filled hollow fibers have been spectrally subdivided and, after precision GDD and CEP control as well as delay of the separated wavepackets, coherently recombined. The approach has yielded optical field transients with non-sinusoidal waveforms [42], including some with durations shorter than 1 fs [43]. The output energy of this "passive" non-amplifying (henceforth "passive") scheme is limited to a fraction of the energy of the seed continuum, typically in the multi-microjoule to sub-millijoule regime. This passive scheme can be turned into an active one by incorporating optical parametric amplifiers (OPAs) in the individual spectral channels. In this case, the synthesizer can then be seeded at lower energy, which is then boosted by many orders of magnitude in broadband OPA stages [44–48]. Such an OPA-based, "active" optical waveform synthesizer (henceforth "active") holds promise for delivering multi-octave transients with multi-millijoule energy and multi-TW peak power.

Both the passive and active schemes are similar in that they rely on chirped mirrors. Following the demands of the OWS, chirped mirror technology had to be extended from the visible (VIS) and near-infrared (NIR) range around the spectrum of Ti:sapphire toward the ultra violet (UV) and the infrared (IR), where chirped mirrors have been hardly available [49]. In this work, we discuss the specific demands on the optics required for OWS, challenges associated with the different spectral domains spreading over several octaves, and design approaches that address these challenges, and we demonstrate the most recent progress in the field of dispersive optics for OWS.

We start by addressing the problem of dividing multi-octave radiation into different spectral channels while preserving the spectral phase and amplitude distribution of the light wave, and then move on to discuss the broadband dispersive multilayer mirrors required for GDD control over the individual spectral ranges. We give an overview of multilayer optics for the VIS and NIR range of Ti:sapphire lasers and then address the latest advances that extend precision GDD control into the UV and toward the mid IR.

2. MULTI-OCTAVE GROUP DELAY CONTROL

Independent of the specific scheme, OWS relies on seed pulses with ultra-broad spectral coverage, spanning preferably a couple of octaves or more. The spectrum is separated into several channels, of which each is treated individually. In the case of the passive scheme, each channel is compressed to near its Fourier-transform-limit (FTL) in a dispersive delay line, before being recombined to synthesize the transients. In active schemes, each wavepacket remains temporally stretched while being amplified in one or more OPA stages. In [50], Chia and co-workers suggested keeping the individual channels slightly chirped until after the final recombination in order to decrease the peak intensity and thus reduce the B-integral. The ultra-broadband dispersive mirror pair was meant to compensate the dispersion of the combined waveform afterward. The laser-induced damage threshold of the final mirror pair still remains a concern.

The channels of any OWS apparatus are initially separated with customized dichroic beam splitters (DBS). There are several major requirements to these key components: (1) a high contrast between reflected and transmitted light; (2) a "cross-talk" region between channels that is used for channel-to-channel synchronization; and (3) smooth, well controlled (preferably flat) phase of *both* reflected *and* transmitted spectra in order to ensure compressibility of the individual channels. Fulfillment of all three requirements simultaneously is a formidable task [50]. Conventional DBSs are based on quarter-wave stacks of alternating high- and low-refractive-index dielectric layers. Their high-reflectance (HR) band is determined by the contrast between the refractive indices of the coating materials and is typically much narrower than the desired spectral span of the individual synthesizer channel. While exhibiting negligible GDD within the central region of their HR band, they are plagued by strong group delay (GD)/GDD ripples at the edges of the zone, which is intolerable in applications for OWS. In order to extend the HR zone to the desired bandwidth and suppress dispersion ripples in the transition zones, the layer thicknesses are numerically optimized. The resultant slight detuning from a quarter-wave structure also creates the required "cross-talk" zones.

Beyond the just-mentioned criteria, the DBSs of the OWS have to fulfill general requirements for mirrors used in demanding applications, such as compensation for the tensile stress caused by the multilayer coating itself. In the case of an interferometric setup such as an OWS, this is particularly important in order to match the wavefronts of the individual channels and ensure their optimal superposition across the entire beam. This critical

Review Article Vol. 4, No. 1 / January 2017 / Optica

requirement is met by applying a compensatory coating on the back surface of the mirror substrate.

The designed reflectance and GD for the class of DBSs used for separating the UV, VIS-UV, and VIS channels in OWS in [43] and VIS and NIR from IR in an OPA-based active OWS under development are presented in Fig. 1. The DBSs characterized by the curves in the Fig. 1 well fulfill the above-mentioned major requirements demanded by OWS applications: (1) there is a high contrast between high-reflectance and high-transmittance zones, yet (2) there are relatively narrow transition zones between the channels and (3) the GD is smooth for both the high-reflectance and high-transmittance zones, even including the transition regions. These DBSs can be used for splitting as well as for recombining the beams.

3. VISIBLE DISPERSIVE OPTICS

Broadband dispersive multilayer mirrors in the VIS and NIR have matured over the decades. In their first applications, dispersive mirrors aimed to introduce constant negative GDD over a selected spectral range [51,52]. Several design approaches have been applied to improve mirror performance [53,54].

In the pursuit of near-single-cycle light, the required working bandwidth rapidly approached an optical octave [21,55]. Under

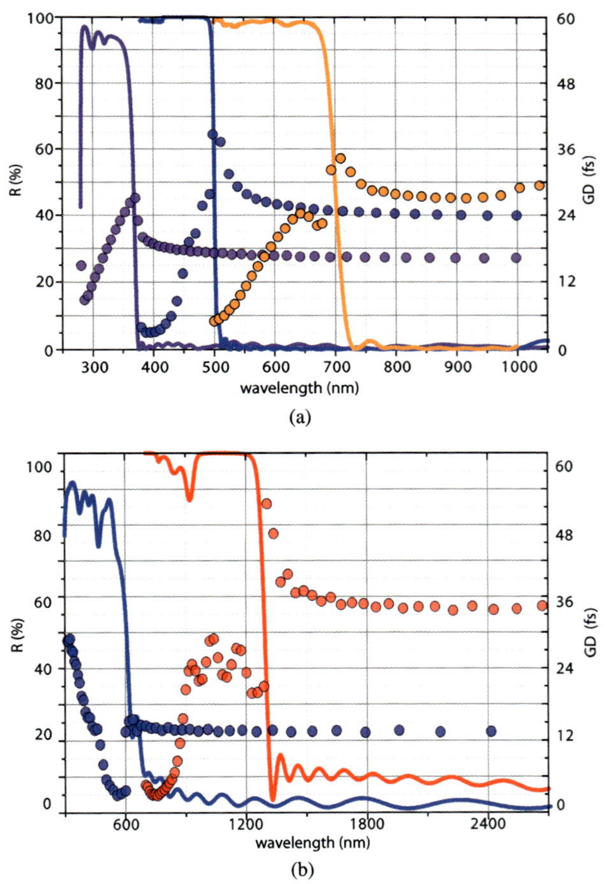

Fig. 1. Dichroic beam splitters for optical waveform synthesis. (a) Class of dichroic beam splitters for separation of ultraviolet, visible, and near-infrared bands in the passive synthesizer in [43]. (b) Class of dichroic beam splitters for separation of visible and near-infrared bands in the passive and active synthesizers under development. Solid curves, designed reflectance; dotted curves, designed GD.

these conditions, control of GDD to the lowest order became insufficient for achieving transform-limited pulses and more sophisticated designs were needed. Double-chirped mirrors [56–58], complementary mirror pairs [59], and double-angle mirrors [60] enabled significant extension of the bandwidth and simultaneous control over the higher-order dispersion. Evolution of the design tools and computational power has also been instrumental in the progress of the thin-film multilayer coatings for ultrafast optics [61–63].

As a result, the state-of-the-art chirped mirrors nowadays are able to operate over super-octave spectra and compensate the phase to an arbitrary dispersion order [50,64]. However, the accurate initial definition of the dispersion target remains a major challenge for the development of broadband dispersive multilayers.

A. Phase Compensation Versus Material Compensation

The main goal of the dispersion management in the case of pulse compression is to compensate for spectral phase variations; therefore, the precise control of it is of utmost importance. A major prerequisite is accurate knowledge of the spectral phase to be compensated for. Several techniques have been developed for measuring this quantity, allowing (together with the spectral amplitude) complete characterization of the ultra-short laser pulses [65–67]. Unfortunately, most of the techniques are not without shortcomings. These include the presence of ambiguities, insufficient temporal and/or spectral resolution and/or range, inherent inability to measure the complete pulse intensity and/or phase, inability to measure complex pulses, and artifacts due to multi-shot averaging over pulses with fluctuating characteristics [68]. As a result, the phase retrievals may be inconsistent and/or inaccurate, and thus compromise the accuracy of the target for dispersive multilayer development (henceforth “dispersion target”).

Whenever the chirp carried by the laser pulse to be compressed is dominated by material dispersion, it may be more reliable to aim at canceling this material dispersion, which is well known. Therefore, in the material compensation approach, the dispersion target is chosen to compensate for the dispersion accumulated by the initially compressed pulse after propagation through known materials (including air). Since the optical constants of the commonly used optical materials and gases are well known, the resulting target can be calculated with a high accuracy, with the only remaining source of uncertainty being nonlinear effects.

Figure 2 presents two group delay versus wavelength dispersion targets for mirrors to be used in the VIS spectral range. The material-based dispersion target (blue curve) in Fig. 2 was designed to compensate for dispersion introduced by 2 mm of fused silica and 1 m of air with four bounces on double-angle dispersive mirrors. The phase-based target (red curve) was derived from a measured spectral phase. While the material-based target is clearly dominated by the second- and third-order dispersion, the phase-based target is noticeably influenced by the higher orders. These higher orders are likely inherited from the paternal nonlinear process (for example, broadening in a hollow-core fiber) or accumulated upon propagation due to induced nonlinearities.

In rare cases, when the accurate measurement of the spectral phase is hindered and compensation for material dispersion only does not yield the desired quality of compression, a hybrid approach may be helpful. In this case, the initial target is approximated by the known material dispersion. With the first set of

Review Article Vol. 4, No. 1 / January 2017 / Optica

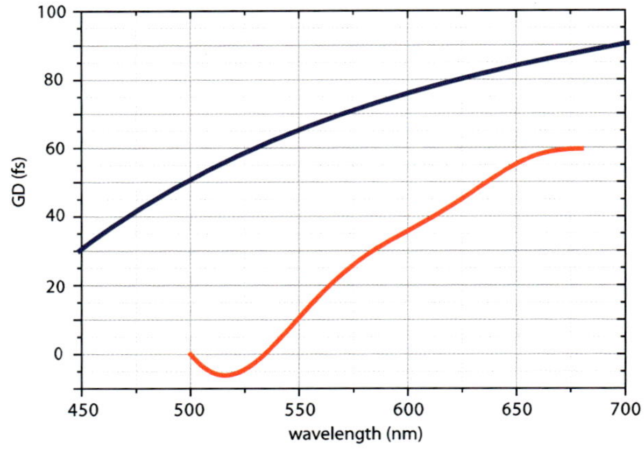

Fig. 2. Dispersion targets developed based on the phase compensation (red curve) and material compensation (dark blue curve) approach. The vertical offset is introduced for illustration purposes.

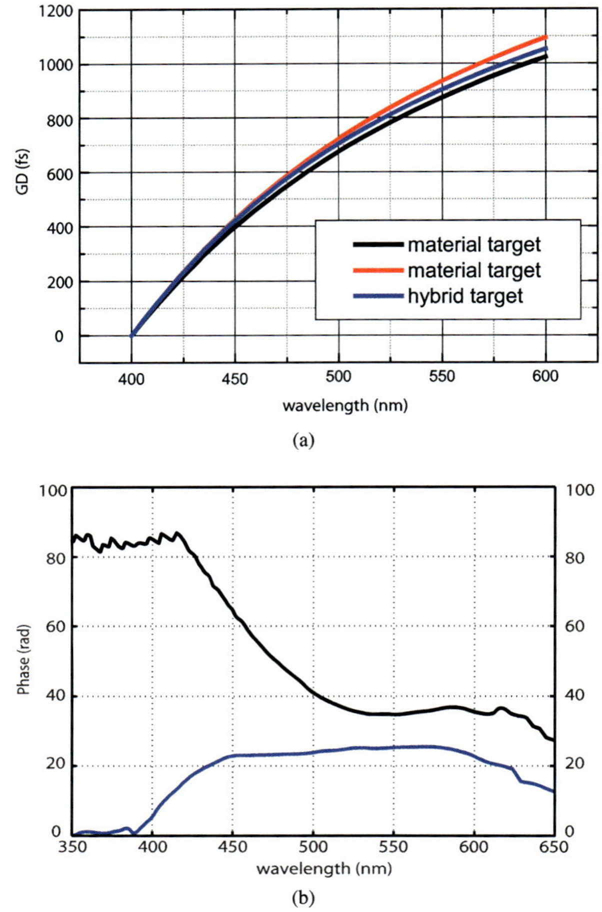

Fig. 3. Hybrid dispersion target approach for the design of the visible and near-infrared dispersive mirrors. (a) Hybrid dispersion target for the design of visible and near-infrared dispersive mirrors. (b) Pulse compression with hybrid target. Black curve shows spectral phase of the 15 fs pulse achieved with material compensation only. Blue curve shows spectral phase of the 6.1 fs pulse (FTL = 6 fs) achieved with hybrid compensation.

dispersive optics produced in this way, the residual phase is either carefully measured or compensated for by a programmable dispersive filter [29,30]. The new target is corrected for this residual phase. A dispersion target developed following the hybrid approach is presented in Fig. 3. The black curve in Fig. 3(a) is the initial dispersion target calculated to compensate for a particular amount of material dispersion in the optical system. After the first optics were produced and implemented, it was found that the blue edge of the spectrum (wavelength range 400 to 450 nm) required additional dispersion [red curve in Fig. 3(a)]. The new hybrid dispersion target was then designed to introduce the missing residual dispersion identified in the just-mentioned procedure (blue curve). The reconstructed spectral phases of the ultrashort pulses achieved with material-only and hybrid approaches are presented in Fig. 3(b). The spectral phase of the 6.1 fs pulse achieved via hybrid compression (blue curve) is noticeably flatter in the working spectral range than the spectral phase of the 15 fs pulse achieved with material compensation only, thus obviously yielding the better compression. The hybrid approach is powerful at the expense of increased time and material consumption.

The production of multiple layers that precisely reproduce the dispersion target over the entire bandwidth is challenging. Dispersive multilayer coatings are very sensitive to deposition errors, as even slight changes in the layer thicknesses affect the spectral phase, so high-precision deposition techniques are required to ensure reliable production. Magnetron and ion beam sputtering are well-established processes for the production of dielectric multilayer optics for the VIS and NIR spectral ranges that offer higher accuracy and reproducibility than electron beam evaporation. Consequently, the coating materials are often chosen to be compatible with either of the methods.

The working spectral range of a thin-film coating is defined by the transparency regions of the coating materials. A pair of layer materials with large contrast between their refractive indices is highly beneficial for broadband control of optical radiation. Currently, Ta_2O_5/SiO_2 and Nb_2O_5/SiO_2 are the two most commonly used material pairs for production of the dielectric thin-film coatings for the VIS and NIR spectral ranges. Their low absorptance and high refractive index contrast (2.12/1.47 for Ta_2O_5/SiO_2 and 2.28/1.47 for Nb_2O_5/SiO_2 at 800 nm [69]) permit the production of multilayer structures with high reflectance and low loss in the VIS spectral range.

The multilayer mirrors following the dispersion targets in Fig. 2 were produced by magnetron sputtering deposition process. The agreement between designed and measured characteristics demonstrates the reliability of the technology (Fig. 4). Upon implementation, both designs yielded well-compressed, nearly transform-limited pulses: 6.5 fs (6.1 fs FTL) [Fig. 4(a)] in the case of phase-based target and 2.9 fs (2.7 fs FTL) [Fig. 4(b)] in the case of the material-based target, therefore validating the suitability of both approaches.

Overall, significant developments in the field of dispersive coatings stimulated by the demands of the ultrafast optics allow for a versatile and reliable dispersion management in the VIS and NIR spectral ranges.

4. ULTRAVIOLET DISPERSIVE OPTICS

For almost two decades after the first demonstration of dispersive multilayer mirrors [11], their evolution responded to the demands of Ti:sapphire-based systems, with little effort being

Review Article Vol. 4, No. 1 / January 2017 / Optica

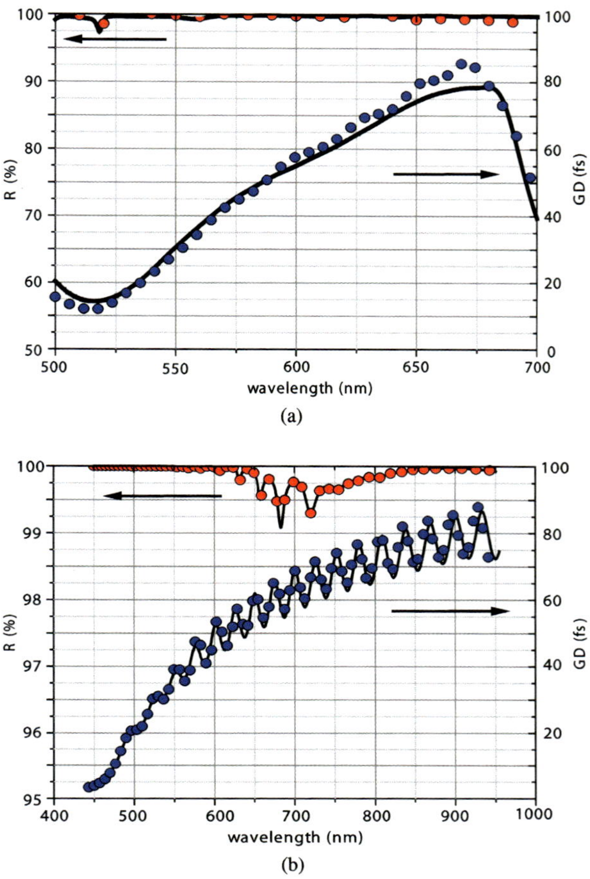

Fig. 4. Dispersive multilayers in visible spectral range. (a) Dispersive multilayer developed upon phase compensation dispersion target. (b) Dispersive multilayer developed upon material compensation dispersion target. Black solid curves, designed reflectance/GD; dotted red curve, measured reflectance; dotted blue curve, measured GD.

invested into the extension of the technology into the UV spectral range [70]. Generation of the synthesizer's spectrum via broadening of the femtosecond Ti:sapphire pulses in a gas-filled hollow-core fiber [42,43] gives rise to a significant extension of the spectral range into the UV. Consequently, it is possible—and for many applications desirable—to extend the synthesizer's spectral coverage to the UV spectral range.

UV is a notoriously difficult working range in terms of dispersion control for broadband radiation, because the refractive index and absorption rapidly increase as the frequency of the incident radiation approaches atomic resonances in the UV, resulting in strong dispersion and often non-negligible absorption over even relatively short propagation distances.

Manufacturing of precision dispersive multilayers in the UV is significantly more challenging. The physical thickness of the individual layers scales with the central wavelength, and thus it becomes smaller for shorter wavelengths, while the absolute deposition error remains approximately unchanged. Consequently, the relative error in layer deposition tends to be higher in the UV than in the VIS-IR range. As a result, electron-beam evaporation of the HfO_2/SiO_2 material pair, the well-established standard technology for the production of high-reflectivity mirrors and filters in the UV, is unsuitable for the manufacturing of dispersive optics for UV. The accuracy of this deposition technique entails intolerable errors for chirped multilayers in the UV. Fortunately, the deposition of this material pair has recently been adapted for magnetron sputtering [71] and ion-beam sputtering, enabling the deposition of layers with higher accuracy. This progress has been instrumental in extending broadband dispersive multilayers into the UV range. Another major challenge has turned out to be the optical properties of the thin-film materials themselves.

A. Induced Nonlinear Effects

While progress in ultra-short pulse generation, assisted by dispersive multilayers, has been beneficial for advances in the field of nonlinear optics, modeling of thin-film multilayer coatings has so far excluded nonlinear effects. Their reflectance, transmittance, and dispersion were evaluated on the basis of the linear optical properties of the layer materials. Our recent work has demonstrated that this approach may be inappropriate while working in the UV.

The dispersive multilayer mirrors employed in strong field experiments are typically exposed to intensities on the order of 10^{10} W/cm^2 to 10^{12} W/cm^2. This regime is dominated by the bound electronic nonlinearities related to the second- and third-order electric susceptibility. The appearance of these nonlinearities in bulk media is known and taken into account upon designing of the apparatus. However, as the average thickness of the thin-film multilayer for the visible spectral range is about 7 μm and nonlinear refractive indices of the most commonly used coating materials are rather small in this range, it was assumed and accepted that thin-film coatings are not susceptible to nonlinearities until intensities reaching nearly the laser-induced damage threshold.

In our recent work, we uncovered the fact that dispersive multilayer structures might be affected by a nonlinear polarization response for incident intensities significantly lower than the damage threshold due to strong local field enhancement. In the UV range, when nonlinear refraction and two-photon absorption rapidly increases [72], this enhancement becomes sufficient to trigger the appearance of the third-order nonlinearities that are otherwise not observed [73]. However, the effect is not only bound to the UV [74]. Therefore, the optics of the future high-energy active OWS systems might be affected and these effects should be taken into account.

B. Establishing Control Over Induced Nonlinearities

To prevent small nonlinearities from adversely affecting the dispersive properties of chirped multilayers, it is desirable to take the nonlinear absorptance and the nonlinear refractive index into account. In the absence of reliable experimental data on nonlinear refractive indices and two-photon absorption coefficients of commonly used dielectric coating materials over the broad spectral ranges of interest, accurate incorporation of these parameters into the design routine is unfortunately not possible. At the present, only approximate modeling based on estimated values of nonlinear coefficients is feasible.

In our recent work [73], we have found a way to estimate the two-photon absorption coefficient for a spectral range of interest from the experimental data and to incorporate it into the design procedure. We showed that undesirable nonlinear absorptance can be efficiently suppressed by substituting a significant amount of the high-index-layer material with low-index-layer material. As the two-photon absorption coefficients are inversely proportional to the electronic band gap of the materials [72], the low-index materials have much weaker induced absorptance and therefore their properties are much less intensity dependent.

We have recently implemented this approach in the design of a UV dispersive multilayer for a passive OWS [43], using an "effective" spectrally averaged two-photon absorption coefficient estimated from the experiment [75]. Figure 5 illustrates the key points of the approach. It shows designed reflectance of the optimized and non-optimized UV dispersive multilayer. The gray curve shows the reflectance of the non-optimized design without any of the nonlinear absorptance taken into account, the blue curve shows the reflectance of the non-optimized design affected by the nonlinear absorptance, and the red curve shows the reflectance of the optimized design with nonlinear absorptance taken into account. There is a clear improvement of averaged reflectance from the non-optimized to optimized design. The improvement is reached by modification of the initial, non-optimized design depicted in the Fig. 5(b), with a significant amount of high-index material being substituted by low-index material. The resultant multilayer stack of the optimized design is presented in Fig. 5(c).

The optimization for minimum two-photon-induced absorptance has to preserve the dispersion properties of the mirror [Fig. 5(a)]. This was achieved by suitable adjustment of the low-index replacement layers to ensure the required optical delay. Consequently, the replacement layers need to be thicker, yielding an increased physical thickness of the overall coating. This, in turn, increases the risk for excessive tensile stress.

Modification of the stack alternates the distribution of the electric field in the multilayer coating (Fig. 6). The enhancement is reduced and the peaks of the electric field are re-situated in low-index layers. As the distribution of the electric filed is also known to strongly influence the damage thresholds of the thin-film coatings [76,77], the modified distribution yields a potentially higher damage threshold.

Extending dispersive mirror technology further into the UV calls for exploration of new materials. One of the prospective alternatives is the Al_2O_3/SiO_2 pair. However, the much lower refractive index contrast (1.84/1.50 versus 2.25/1.50 for HfO_2/SiO_2 at 250 nm [69]) will require more layers and overall thicker coatings to be deposited in order to achieve a similar introduced GD. At the same time, the high precision of the deposition process needs to be maintained. The fulfillment of the both criteria is possible by transferring the technology to ion-beam sputtering deposition plants.

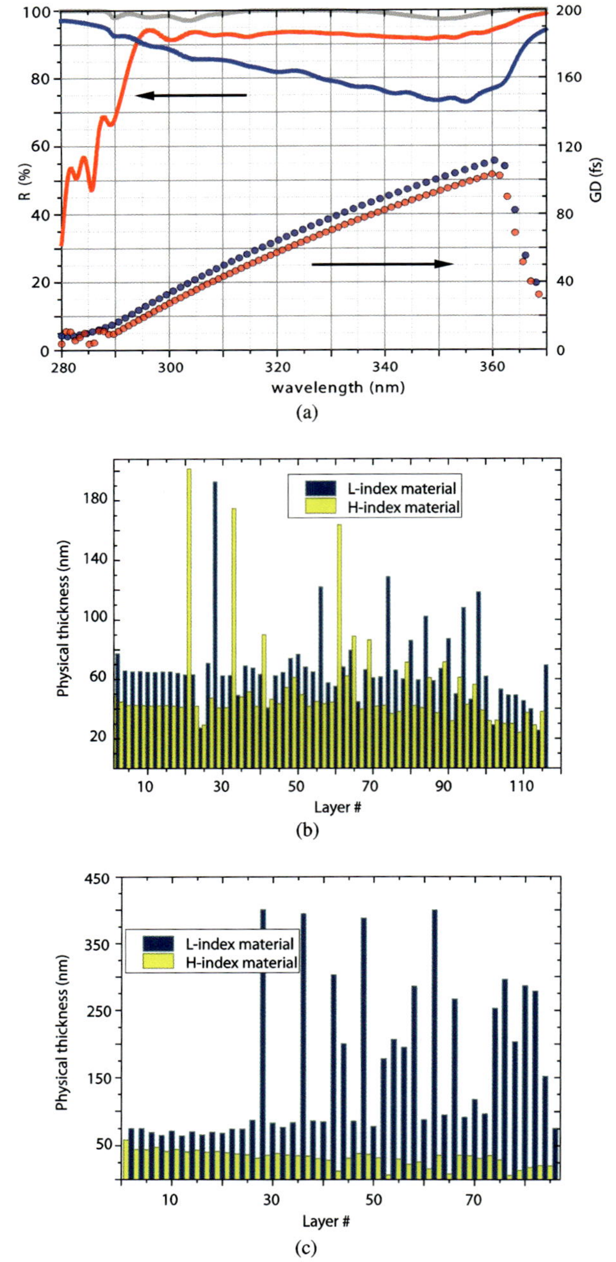

Fig. 5. Gaining control over induced nonlinear effects. (a) Solid curves, designed reflectance of the non-optimized multilayer mirror without (gray) and with (blue) nonlinear absorptance taken into account, and of the optimized design with nonlinear absorptance taken into account (red). Working intensity is $0.25 * 10^{10}$ W/cm^2. Dotted curves, designed GD of the optimized (red) and non-optimized (blue) mirrors. Vertical offset of the GD curves is for illustration purposes. (b) Layer sequence of the non-optimized mirror. (c) Layer sequence of the optimized mirror. Figures (b), (c), and partially (a) reproduced from [75].

5. INFRARED DISPERSIVE OPTICS

There are a number of reasons for extending the capability of optical waveform synthesis into the IR spectral range; they include the exploration of strong-field phenomena in solids [78] and the generation of isolated attosecond x-ray pulses via extending high-harmonic generation into keV photon energies [79–81]. These prospects sparked interest in the development of ultrafast sources with spectral coverage extending into the mid IR [82–84], including schemes for OWS [46,48].

Just as in the UV, the development of dispersive optics for IR lags behind, but for a different reason: the lack of scientific interest. Starting from the wavelength of 1.5 μm, most of the commonly used optical materials change from positive to negative group-velocity dispersion (GVD), while Si and Ge maintain positive GVD through the mid-IR range. Consequently, it becomes possible to perform compression in bulk by carefully balancing the materials with opposite GVD in the beam path [83,85,86]. Unfortunately, the approach works well only for dispersion compensation to lowest-order (GDD), with high-order

Review Article Vol. 4, No. 1 / January 2017 / Optica

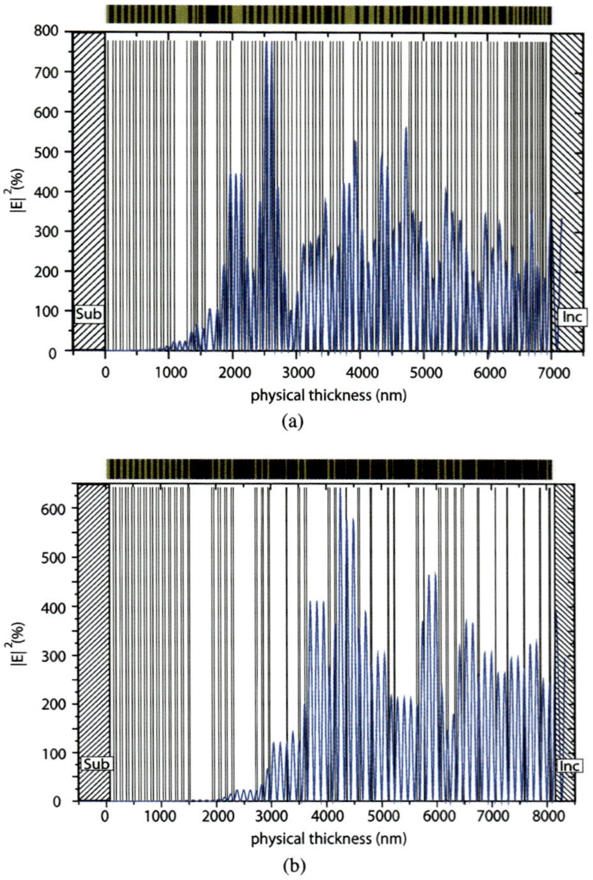

Fig. 6. Gaining control over induced nonlinear effects: distribution of the electric field inside the multilayer structures. (a) Electric field distribution inside the non-optimized dispersive multilayer stack. (b) Electric field distribution inside the optimized dispersive multilayer stack. Dark yellow bands, high-index material; brown bands, low-index material.

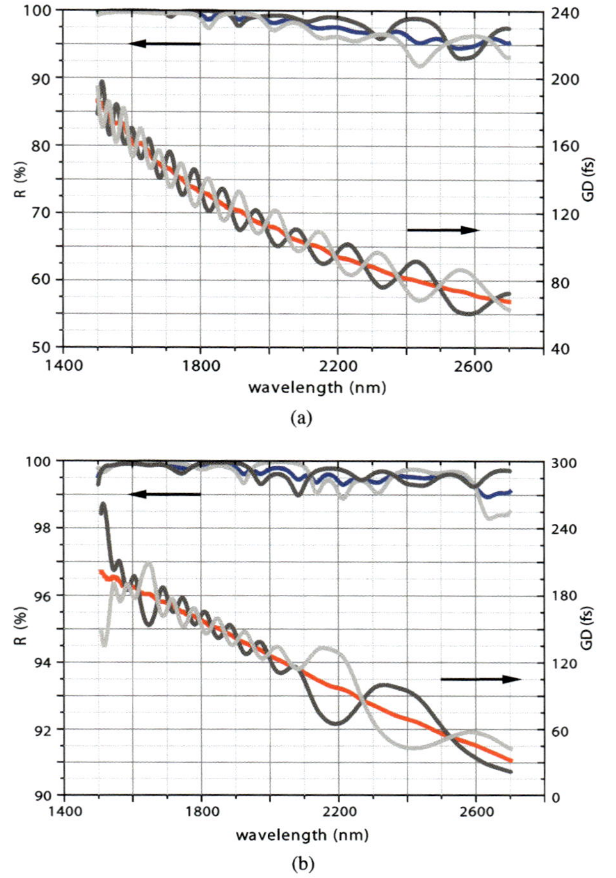

Fig. 7. Dispersive multilayers in the infrared spectral range designed to compensate the material dispersion of 0.25 mm Si to the lowest order. (a) Dispersion target and design imitating 0.25 mm Si. (b) Dispersion target and design introducing second-order identical to 0.25 mm Si, yet with negative third-order dispersion. Gray, dark gray curves, the designed GD curves of an individual mirror of the mirror pair; red curve, resulting designed averaged GD; gray, dark gray curves, reflectance of the individual mirror of the mirror pair; blue curve, resulting averaged reflectance.

contributions to dispersion, which are essential for few-cycle pulses, left uncompensated [85]. In addition, with its band gap of 1.11 eV (at 300 K), Si is susceptible to two-photon absorption, restricting the maximum incident intensity. Implementation of dispersive multilayer mirrors offers an elegant solution for circumventing these shortcomings.

Two alternative chirped mirror designs covering the IR range spanning from 1.6 to 2.7 μm are presented in Fig. 7. The design illustrated in Fig. 7(a) imitates dispersion introduced by 0.25 mm of Si; the design in Fig. 7(b) follows an alternative target. While it introduces the GDD equivalent of 0.25 mm Si at the central wavelength, the sign of the third-order dispersion is changed to negative.

As the thickness of the individual layer of the multilayer design scales with the central wavelength of the incident spectrum, multilayer structures in the IR tend to be substantially thicker than their counterparts designed for VIS or UV wavelengths. For example, the physical thickness of the coatings in Fig. 7 approaches 20 μm. The risk of excessive tensile stress, which might affect the mechanical properties of the deposited structure and make it peel off from the substrate, increases. The overall physical thickness must therefore be traded off against bandwidth, dispersion, and reflectance of the dispersive mirror, all of which tend to increase with increasing number of layers and hence structure thickness. For instance, the design in Fig. 7(b) was allowed to be slightly thicker and contain more layers than the design in Fig. 7(a), resulting in higher reflectance and improved pulse-compression efficiency.

Fortunately, the limitation to the maximum physical thickness is not a fundamental one. The mechanical stress may be reduced by tuning the parameters of the deposition process at the expense of changes in the optical properties of the layers. These changes must be well understood and controlled before the approach can be used for the production of thicker structures.

Unlike in the UV, there are several layer materials excellently suited for multilayer development in the IR range. The already-mentioned material pair exploited in VIS/NIR, Nb_2O_5/SiO_2, is transparent until 8 μm. For the spectral region beyond 2 μm, Si is an excellent high-index material, while SiO_2 and Al_2O_3 are used as low-index counterparts and are transparent until 7 μm [87]. The large electronic band gaps of Nb_2O_5 and SiO_2—4 and 7.5 eV, respectively—allow for the avoidance of nonlinear effects in the infrared range, permitting development and production of

Review Article Vol. 4, No. 1 / January 2017 / Optica

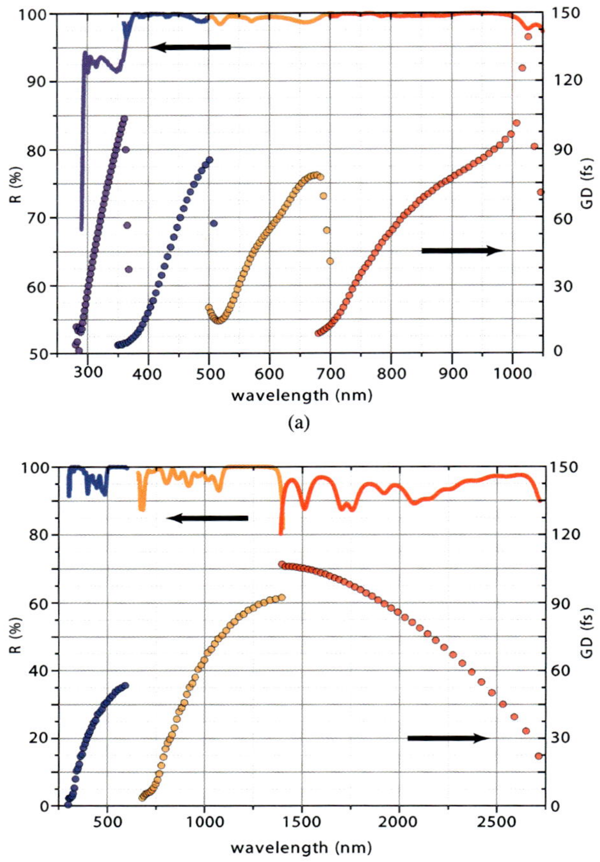

Fig. 8. Dispersive multilayers spanning over three optical octaves developed and produced for OWS. (a) Chirped mirrors developed and produced for the passive OWS in [43,91]. (b) Chirped mirrors under development and production for the passive and active infrared OWSs. Solid curves, designed reflectance; dotted curves, designed GD.

a broad class of efficient, precise, and flexible multilayer designs essential for broadband pulse manipulation.

6. CONCLUSIONS

Over the last two decades, dispersive multilayer optics have become standard components of ultrashort pulse laser systems and have defined the frontiers of ultrafast technology. In cooperation with well-controlled optical nonlinearities they permitted, for the first time, the generation of near-bandwidth-limited, near-single-cycle optical pulses. These in turn opened the door (by yet another nonlinear process: high-order harmonic generation) to the production of isolated attosecond extreme ultraviolet (XUV) pulses, the dispersion control of which also benefits from aperiodic multilayer structures [88–90].

Next-generation attosecond metrology calls for versatile control of the electromagnetic force of light within a half cycle of its oscillation. This calls for multi-octave superposition of optical fields with adjustable phase and amplitude—OWS in brief. Again, aperiodic dielectric multilayers constitute an enabling technology for this latest frontier of ultrafast science.

Figure 8(a) summarizes the dispersive mirrors controlling few-cycle pulses in the four channels of a two-octave-spanning UV-VIS-NIR synthesizer, which has recently been demonstrated to be capable of generating a wide range of sub-cycle to few-cycle light transients, including sub-femtosecond optical pulses for the first time [43]. Figure 8(b), in turn, depicts dispersive mirrors developed for a prototypical three-channel device covering the entire near IR and reaching out into the mid IR, which is currently under construction.

State-of-the-art multilayer deposition technology had to rise to several challenges to be able to produce dispersive mirrors with the required precision in previously inaccessible wavelength ranges. Extension of the dispersive multilayer technology into the UV spectral range required both improved precision in depositing the nanometer-thick layers and new design paradigms to avoid undesirable nonlinear effects coming into play. On the other hand, pushing the opposite spectral frontier of dispersive multilayer optics to ever-longer wavelengths calls for new parameter optimization of coating algorithms for providing an optimal balance between the best possible mechanical and optical properties of ever-broader multilayer structures.

At present, precision dispersive multilayer designs span the working range of three octaves, allowing for precise and efficient dispersion control of light waveforms composed of signals all the way from ultraviolet (0.29 μm) to mid-infrared (2.7 μm) wavelengths. They define the cutting edge of ultrafast technology and pave the way toward new research directions based on dramatic advances in our capability to control atomic-scale electronic motions. In spite of all these advances, the technology is far from approaching its ultimate limits. Ongoing persistent technological efforts, driven by a range of scientific applications, are likely to expand the frontiers of multilayer-based optical dispersion control by several more octaves in the years to come.

Funding. Army Research Office (ARO) (W911NF-15-1-0360); Max-Planck-Gesellschaft (MPG); Munich-Centre for Advanced Photonics (MAP); Deutsche Forschungsgemeinschaft (DFG).

Acknowledgment. The authors thank Dr. M. Trubetskov for his continuous and invaluable contribution to the field; Drs. M. Th. Hassan, T.T. Luu, and E. Goulielmakis for fruitful discussions that yielded development of the dispersive designs in Fig. 1(a) and Fig. 8(a); Dr. A. Sommer for discussions that resulted in the designs of Fig. 3; and Dr. N. Karpowicz for the ideas of Fig. 7. We are also grateful to Dr. M. Weidman for his assistance with the styling of the text.

REFERENCES

1. T. H. Maiman, "Stimulated optical radiation in ruby," Nature **187**, 493–494 (1960).
2. A. J. DeMaria, "Self mode-locking of lasers with saturable absorbers," Appl. Phys. Lett. **8**, 174–176 (1966).
3. E. Ippen, "Passive mode locking of the cw dye laser," Appl. Phys. Lett. **21**, 348–350 (1972).
4. R. L. Fork, "Generation of optical pulses shorter than 0.1 psec by colliding pulse mode locking," Appl. Phys. Lett. **38**, 671–673 (1981).
5. J. A. Valdmanis, R. L. Fork, and J. P. Gordon, "Generation of optical pulses as short as 27 femtoseconds directly from a laser balancing self-phase modulation, group-velocity dispersion, saturable absorption, and saturable gain," Opt. Lett. **10**, 131–133 (1985).
6. D. E. Spence, P. N. Kean, and W. Sibbett, "60-fsec pulse generation from a self-mode-locked Ti:sapphire laser," Opt. Lett. **16**, 42–44 (1991).

7. C.-P. Huang, H. Nathel, M. T. Asaki, S. Backus, M. M. Murnane, and H. C. Kapteyn, "17-fs pulses from a self-mode-locked Ti:sapphire laser," Opt. Lett. **17**, 1289–1291 (1992).
8. M. T. Asaki, C.-P. Huang, D. Garvey, J. Zhou, H. C. Kapteyn, and M. M. Murnane, "Generation of 11-fs pulses from a self-mode-locked Ti:sapphire laser," Opt. Lett. **18**, 977–979 (1993).
9. J. Zhou, I. P. Christov, G. Taft, C.-P. Huang, M. M. Murnane, and H. C. Kapteyn, "Pulse evolution in a broad-bandwidth Ti:sapphire laser," Opt. Lett. **19**, 1149–1151 (1994).
10. E. Treacy, "Optical pulse compression with difraction gratings," IEEE J. Quantum Electron. **5**, 454–458 (1969).
11. R. Szipöcs, C. Spielmann, F. Krausz, and K. Ferencz, "Chirped multilayer coatings for broadband dispersion control in femtosecond lasers," Opt. Lett. **19**, 201–203 (1994).
12. A. Stingl, R. Szipöcs, C. Spielmann, and F. Krausz, "Generation of 11-fs pulses from a Ti:sapphire laser without the use of prisms," Opt. Lett. **19**, 204–206 (1994).
13. A. Stingl, R. Szipöcs, M. Lenzner, C. Spielmann, and F. Krausz, "Sub-10-fs mirror-dispersion-controlled Ti:sapphire laser," Opt. Lett. **20**, 602–604 (1995).
14. C. Spielmann, P. F. Curley, T. Brabec, and F. Krausz, "Ultrabroadband femtosecond lasers," IEEE J. Quantum Electron. **30**, 1100–1114 (1994).
15. R. Fluck, I. D. Jung, G. Zhang, F. X. Kärtner, and U. Keller, "Broadband saturable absorber for 10-fs pulse generation," Opt. Lett. **21**, 743–745 (1996).
16. I. D. Jung, F. X. Kärtner, N. Matuschek, D. H. Sutter, F. Morier-Genoud, G. Zhang, U. Keller, V. Scheuer, M. Tilsch, and T. Tschudi, "Self-starting 6.5-fs pulses from a Ti:sapphire laser," Opt. Lett. **22**, 1009–1011 (1997).
17. U. Morgner, F. X. Kärtner, S.-H. Cho, Y. Chen, H. A. Haus, J. G. Fujimoto, E. P. Ippen, V. Scheuer, G. Angelow, and T. Tschudi, "Sub-two-cycle pulses from a Kerr-lens mode-locked Ti:sapphire laser," Opt. Lett. **24**, 411–413 (1999).
18. D. H. Sutter, G. Steinmeyer, L. Gallmann, N. Matuschek, F. Morier-Genoud, U. Keller, V. Scheuer, G. Angelow, and T. Tschudi, "Semiconductor saturable-absorber mirror-assisted Kerr-lens mode-locked Ti:sapphire laser producing pulses in the two-cycle regime," Opt. Lett. **24**, 631–633 (1999).
19. R. Ell, U. Morgner, F. X. Käärtner, J. G. Fujimoto, E. P. Ippen, V. Scheuer, G. Angelow, T. Tschudi, M. J. Lederer, A. Boiko, and B. Luther-Davies, "Generation of 5-fs pulses and octave-spanning spectra directly from a Ti:sapphire laser," Opt. Lett. **26**, 373–375 (2001).
20. M. Nisoli, S. De Silvestri, G. Valiulis, and A. Varanavicius, "Fivefold femtosecond pulse compression by sum frequency generation," Appl. Phys. Lett. **68**, 3540–3542 (1996).
21. M. Nisoli, S. De Silvestri, and O. Svelto, "Generation of high energy 10 fs pulses by a new pulse compression technique," Appl. Phys. Lett. **68**, 2793–2795 (1996).
22. F. Théberge, N. Aközbek, W. Liu, A. Becker, and S. L. Chin, "Tunable ultrashort laser pulses generated through filamentation in gases," Phys. Rev. Lett. **97**, 023904 (2006).
23. T. Fuji and T. Suzuki, "Generation of sub-two-cycle mid-infrared pulses by four-wave mixing through filamentation in air," Opt. Lett. **32**, 3330–3332 (2007).
24. S. A. Trushin, K. Kosma, W. Fub, and W. E. Schmid, "Sub-10-fs supercontinuum radiation generated by filamentation of few-cycle 800 nm pulses in argon," Opt. Lett. **32**, 2432–2434 (2007).
25. A. M. Weiner, "Femtosecond pulse shaping using spatial light modulators," Rev. Sci. Instrum. **71**, 1929–1960 (2000).
26. A. M. Weiner, D. E. Leaird, J. S. Patel, and J. R. Wullert, "Programmable femtosecond pulse shaping by use of a multielement liquid-crystal phase modulator," Opt. Lett. **15**, 326–328 (1990).
27. E. Matsubara, K. Yamane, T. Sekikawa, and M. Yamashita, "Generation of 2.6 fs optical pulses using induced-phase modulation in a gas-filled hollow fiber," J. Opt. Soc. Am. B **24**, 985–989 (2007).
28. E. Zeek, K. Maginnis, S. Backus, U. Russek, M. Murnane, G. Mourou, H. Kapteyn, and G. Vdovin, "Pulse compression by use of deformable mirrors," Opt. Lett. **24**, 493–495 (1999).
29. P. Tournois, "Acousto-optic programmable dispersive filter for adaptive compensation of group delay time dispersion in laser systems," Opt. Commun. **140**, 245–249 (1997).
30. F. Verluise, V. Laude, Z. Cheng, C. Spielmann, and P. Tournois, "Amplitude and phase control of ultrashort pulses by use of an acousto-optic programmable dispersive filter: pulse compression and shaping," Opt. Lett. **25**, 575–577 (2000).
31. V. Pervak, "Recent development and new ideas in the field of dispersive multilayer optics," Appl. Opt. **50**, C55–C61 (2011).
32. A. Verhoef, J. Seres, K. Schmid, Y. Nomura, G. Tempea, L. Veisz, and F. Krausz, "Compression of the pulses of a Ti:sapphire laser system to 5 femtoseconds at 0.2 terawatt level," Appl. Phys. B **82**, 513–517 (2006)
33. W. Schweinberger, A. Sommer, E. Bothschafter, J. Li, F. Krausz, R. Kienberger, and M. Schultze, "Waveform-controlled near-single-cycle milli-joule laser pulses generate sub-10 nm extreme ultraviolet continua," Opt. Lett. **37**, 3573–3575 (2012).
34. P. Paul, E. Toma, P. Breger, G. Mullot, F. Auge, P. Balcou, H. Muller, and P. Agostini, "Observation of a train of attosecond pulses from high harmonic generation," Science **292**, 1689–1692 (2001).
35. M. Hentschel, R. Kienberger, C. Spielmann, G. A. Reider, N. Milosevic, T. Brabec, P. Corkum, U. Heinzmann, M. Drescher, and F. Krausz, "Attosecond metrology," Nature **414**, 509–513 (2001).
36. G. Sansone, E. Benedetti, F. Calegari, C. Vozzi, L. Avaldi, R. Flammini, L. Poletto, P. Villoresi, C. Altucci, R. Velotta, S. Stagira, S. De Silvestri, and M. Nisoli, "Isolated single-cycle attosecond pulses," Science **314**, 443–446 (2006).
37. A. Baltuska, T. Udem, M. Uiberacker, M. Hentschel, E. Goulielmakis, C. Gohle, R. Holzwarth, V. S. Yakovlev, A. Scrinzi, T. Haensch, and F. Krausz, "Attosecond control of electronic processes by intense light fields," Nature **421**, 611–615 (2003).
38. E. Goulielmakis, M. Schultze, M. Hofstetter, V. S. Yakovlev, J. Gagnon, M. Uiberacker, A. L. Aquila, E. M. Gullikson, D. T. Attwood, R. Kienberger, F. Krausz, and U. Kleineberg, "Single-cycle nonlinear optics," Science **320**, 1614–1617 (2008).
39. S. E. Harris and A. V. Sokolov, "Subfemtosecond pulse generation by molecular modulation," Phys. Rev. Lett. **81**, 2894–2897 (1998).
40. R. K. Shelton, L.-S. Ma, H. C. Kapteyn, M. M. Murnane, J. L. Hall, and J. Ye, "Phase-coherent optical pulse synthesis from separate femtosecond lasers," Science **293**, 1286–1289 (2001).
41. G. Krauss, S. Lohss, T. Hanke, A. Sell, S. Eggert, R. Huber, and A. Leitenstorfer, "Synthesis of a single cycle of light with compact erbium-doped fibre technology," Nat. Photonics **4**, 33–36 (2010).
42. A. Wirth, M. T. Hassan, I. Grguras, J. Gagnon, A. Moulet, T. T. Luu, S. Pabst, R. Santra, Z. A. Alahmed, A. M. Azzeer, V. S. Yakovlev, V. Pervak, F. Krausz, and E. Goulielmakis, "Synthesized light transients," Science **334**, 195–200 (2011).
43. M. T. Hassan, T. T. Luu, A. Moulet, O. Raskazovskaya, P. Zhokhov, M. Garg, N. Karpowicz, A. M. Zheltikov, V. Pervak, F. Krausz, and E. Goulielmakis, "Optical attosecond pulses and tracking the nonlinear response of bound electrons," Nature **530**, 66–70 (2016).
44. S.-W. Huang, G. Cirmi, J. Moses, K.-H. Hong, S. Bhardwaj, J. R. Birge, L.-J. Chen, E. Li, B. J. Eggleton, G. Cerullo, and F. X. Kärtner, "High-energy pulse synthesis with sub-cycle waveform control for strong-field physics," Nat. Photonics **5**, 475–479 (2011).
45. S.-W. Huang, G. Cirmi, J. Moses, K.-H. Hong, S. Bhardwaj, J. R. Birge, L.-J. Chen, I. V. Kabakova, E. Li, B. J. Eggleton, G. Cerullo, and F. X. Kärtner, "Optical waveform synthesizer and its application to high-harmonic generation," J. Phys. B **45**, 074009 (2012).
46. H. Fattahi, H. G. Barros, M. Gorjan, T. Nubbemeyer, B. Alsaif, C. Y. Teisset, M. Schultze, S. Prinz, M. Haefner, M. Ueffing, A. Alismail, L. Vámos, A. Schwarz, O. Pronin, J. Brons, X. T. Geng, G. Arisholm, M. Ciappina, V. S. Yakovlev, D.-E. Kim, A. M. Azzeer, N. Karpowicz, D. Sutter, Z. Major, T. Metzger, and F. Krausz, "Third-generation femtosecond technology," Optica **1**, 45 (2014).
47. C. Manzoni, O. D. Mücke, G. Cirmi, S. Fang, J. Moses, S.-W. Huang, K.-H. Hong, G. Cerullo, and F. X. Kärtner, "Coherent pulse synthesis: towards sub-cycle optical waveforms," Laser Photon. Rev. **9**, 129–171 (2015).
48. O. D. Mucke, S. Fang, G. Cirmi, G. M. Rossi, S.-H. Chia, H. Ye, Y. Yang, R. Mainz, C. Manzoni, P. Farinello, G. Cerullo, and F. X. Kartner, "Toward waveform nonlinear optics using multimillijoule sub-cycle waveform synthesizers," IEEE J. Sel. Top. Quantum Electron. **21**, 1–12 (2015).
49. V. Pervak, O. Razskazovskaya, I. B. Angelov, K. L. Vodopyanov, and M. Trubetskov, "Dispersive mirror technology for ultrafast lasers in the range 220-4500 nm," Adv. Opt. Technol. **3**, 55–63 (2014).

50. S.-H. Chia, G. Cirmi, S. Fang, G. M. Rossi, O. D. Mücke, and F. X. Kärtner, "Two-octave-spanning dispersion-controlled precision optics for sub-optical-cycle waveform synthesizers," Optica **1**, 315–333 (2014).
51. G. Tempea, F. Krausz, C. Spielmann, and K. Ferencz, "Dispersion control over 150 THz with chirped dielectric mirrors," IEEE J. Sel. Top. Quantum Electron. **4**, 193–196 (1998).
52. R. Szipöcs, "Dispersive properties of dielectric laser mirrors and their use in femtosecond pulse lasers," Ph.D. dissertation (University of Szeged, 2000).
53. N. Matuschek, L. Gallmann, D. Sutter, G. Steinmeyer, and U. Keller, "Back-side-coated chirped mirrors with ultra-smooth broadband dispersion characteristics," Appl. Phys. B **71**, 509–522 (2000).
54. G. Tempea, V. Yakovlev, B. Bacovic, F. Krausz, and K. Ferencz, "Tilted-front-interface chirped mirrors," J. Opt. Soc. Am. B **18**, 1747–1750 (2001).
55. M. Nisoli, S. De Silvestri, O. Svelto, R. Szipöcs, K. Ferencz, C. Spielmann, S. Sartania, and F. Krausz, "Compression of high-energy laser pulses below 5 fs," Opt. Lett. **22**, 522–524 (1997).
56. F. X. Kärtner, N. Matuschek, T. Schibli, U. Keller, H. A. Haus, C. Heine, R. Morf, V. Scheuer, M. Tilsch, and T. Tschudi, "Design and fabrication of double-chirped mirrors," Opt. Lett. **22**, 831–833 (1997).
57. N. Matuschek, F. X. Kartner, and U. Keller, "Analytical design of double-chirped mirrors with custom-tailored dispersion characteristics," IEEE J. Quantum Electron. **35**, 129–137 (1999).
58. N. Matuschek, F. X. Kärtner, and U. Keller, "Theory of double-chirped mirrors," IEEE J. Sel. Top. Quantum Electron. **4**, 197–208 (1998).
59. F. X. Kärtner, U. Morgner, R. Ell, T. Schibli, J. G. Fujimoto, E. P. Ippen, V. Scheuer, G. Angelow, and T. Tschudi, "Ultrabroadband double-chirped mirror pairs for generation of octave spectra," J. Opt. Soc. Am. B **18**, 882–885 (2001).
60. V. Pervak, I. Ahmad, M. K. Trubetskov, A. V. Tikhonravov, and F. Krausz, "Double-angle multilayer mirrors with smooth dispersion characteristics," Opt. Express **17**, 7943–7951 (2009).
61. A. V. Tikhonravov, M. K. Trubetskov, and G. W. DeBell, "Application of the needle optimization technique to the design of optical coatings," Appl. Opt. **35**, 5493–5508 (1996).
62. A. V. Tikhonravov, M. K. Trubetskov, and G. W. DeBell, "Optical coating design approaches based on the needle optimization technique," Appl. Opt. **46**, 704–710 (2007).
63. A. V. Tikhonravov and M. K. Trubetskov, "Modern design tools and a new paradigm in optical coating design," Appl. Opt. **51**, 7319–7332 (2012).
64. V. Pervak, A. Tikhonravov, M. Trubetskov, S. Naumov, F. Krausz, and A. Apolonski, "1.5-octave chirped mirror for pulse compression down to sub-3 fs," Appl. Phys. B **87**, 5–12 (2007).
65. R. Trebino, K. W. DeLong, D. N. Fittinghoff, J. N. Sweetser, M. A. Krumbügel, B. A. Richman, and D. J. Kane, "Measuring ultrashort laser pulses in the time-frequency domain using frequency-resolved optical gating," Rev. Sci. Instrum. **68**, 3277–3295 (1997).
66. C. Iaconis and I. Walmsley, "Spectral phase interferometry for direct electric-field reconstruction of ultrashort optical pulses," Opt. Lett. **23**, 792–794 (1998).
67. S. Keiber, S. Sederberg, A. Schwarz, M. Trubetskov, V. Pervak, F. Krausz, and N. Karpowicz, "Electro-optic sampling of near-infrared waveforms," Nat. Photonics **10**, 159–162 (2016).
68. M. Rhodes, G. Steinmeyer, and R. Trebino, "Standards for ultrashort-laser-pulse-measurement techniques and their consideration for self-referenced spectral interferometry," Appl. Opt. **53**, D1–D11 (2014).
69. M. N. Polyanskiy, "Refractive index database," http://refractiveindex.info.
70. C. A. Rivera, S. E. Bradforth, and G. Tempea, "Gires-Tournois interferometer type negative dispersion mirrors for deep ultraviolet pulse compression," Opt. Express **18**, 18615–18624 (2010).
71. V. Pervak, F. Krausz, and A. Apolonski, "Hafnium oxide thin films deposited by reactive middle-frequency dual-magnetron sputtering," Thin Solid Films **515**, 7984–7989 (2007).
72. R. DeSalvo, A. A. Said, D. J. Hagan, E. W. Van Stryland, and M. Sheik-Bahae, "Infrared to ultraviolet measurements of two-photon absorption and *n* 2 in wide bandgap solids," IEEE J. Quantum Electron. **32**, 1324–1333 (1996).
73. O. Razskazovskaya, T. T. Luu, M. Trubetskov, E. Goulielmakis, and V. Pervak, "Nonlinear absorbance in dielectric multilayers," Optica **2**, 803–811 (2015).
74. E. Fedulova, M. Trubetskov, T. Amotchkina, K. Fritsch, P. Baum, O. Pronin, and V. Pervak, "Kerr effect in multilayer dielectric coatings," Opt. Express **24**, 21802–21817 (2016).
75. O. Razskazovskaya, M. T. Hassan, T. Luu, E. Goulielmakis, and V. Pervak, "Efficient broadband highly dispersive HfO_2/SiO_2 multilayer mirror for pulse compression in near ultraviolet," Opt. Express **24**, 13628–13633 (2016).
76. P. G. Antal and R. Szipöcs, "Relationships among group delay, energy storage, and loss in dispersive dielectric mirrors," Chin. Opt. Lett. **10**, 053101 (2012).
77. I. B. Angelov, M. von Pechmann, M. K. Trubetskov, F. Krausz, and V. Pervak, "Optical breakdown of multilayer thin-films induced by ultrashort pulses at MHz repetition rates," Opt. Express **21**, 31453–31461 (2013).
78. F. Krausz and M. I. Stockman, "Attosecond metrology: from electron capture to future signal processing," Nat. Photonics **8**, 205–213 (2014).
79. A. Gordon and F. X. Kärtner, "Scaling of keV HHG photon yield with drive wavelength," Opt. Express **13**, 2941–2947 (2005).
80. T. Popmintchev, M.-C. Chen, P. Arpin, M. M. Murnane, and H. C. Kapteyn, "The attosecond nonlinear optics of bright coherent X-ray generation," Nat. Photonics **4**, 822–832 (2010).
81. T. Popmintchev, M.-C. Chen, D. Popmintchev, P. Arpin, S. Brown, S. Alisauskas, G. Andriukaitis, T. Balciunas, O. D. Mucke, A. Pugzlys, A. Baltuska, B. Shim, S. E. Schrauth, A. Gaeta, C. Hernandez-Garcia, L. Plaja, A. Becker, A. Jaron-Becker, M. M. Murnane, and H. C. Kapteyn, "Bright coherent ultrahigh harmonics in the keV X-ray regime from mid-infrared femtosecond lasers," Science **336**, 1287–1291 (2012).
82. C. Vozzi, G. Cirmi, C. Manzoni, E. Benedetti, F. Calegari, G. Sansone, S. De Silvestri, M. Nisoli, and G. Cerullo, "High-energy, few-optical-cycle pulses at 1.5 m with passive carrier-envelope phase stabilization," Opt. Express **14**, 10109–10116 (2006).
83. X. Gu, G. Marcus, Y. Deng, T. Metzger, C. Teisset, N. Ishii, T. Fuji, A. Baltuska, R. Butkus, V. Pervak, H. Ishizuki, T. Taira, T. Kobayashi, R. Kienberger, and F. Krausz, "Generation of carrier-envelope-phase-stable 2-cycle 740-J pulses at 2.1-m carrier wavelength," Opt. Express **17**, 62–69 (2009).
84. G. Andriukaitis, T. Balčinas, S. Ališauskas, A. Pugzlys, A. Baltuska, T. Popmintchev, M.-C. Chen, M. M. Murnane, and H. C. Kapteyn, "90 GW peak power few-cycle mid-infrared pulses from an optical parametric amplifier," Opt. Lett. **36**, 2755–2757 (2011).
85. N. Demirdöven, M. Khalil, O. Golonzka, and A. Tokmakoff, "Dispersion compensation with optical materials for compression of intense sub-100-fs mid-infrared pulses," Opt. Lett. **27**, 433–435 (2002).
86. J. A. Gruetzmacher and N. F. Scherer, "Few-cycle mid-infrared pulse generation, characterization, and coherent propagation in optically dense media," Rev. Sci. Instrum. **73**, 2227–2236 (2002).
87. M. Friz and F. Waibel, "Coating materials," in *Optical Interference Coatings* (Springer, 2003), pp. 105–130.
88. C. Bourassin-Bouchet, S. De Rossi, J. Wang, E. Meltchakov, A. Giglia, N. Mahne, S. Nannarone, and F. Delmotte, "Shaping of single-cycle sub-50-attosecond pulses with multilayer mirrors," New J. Phys. **14**, 023040 (2012).
89. M. Hofstetter, M. Schultze, M. Fiess, B. Dennhardt, A. Guggenmos, J. Gagnon, V. S. Yakovlev, E. Goulielmakis, R. Kienberger, E. M. Gullikson, F. Krausz, and U. Kleineberg, "Attosecond dispersion control by extreme ultraviolet multilayer mirrors," Opt. Express **19**, 1767–1776 (2011).
90. A. Guggenmos, M. Hofstetter, R. Rauhut, B. Nickel, E. Gullikson, and U. Kleineberg, "Aperiodic multilayer mirrors for attosecond soft x-ray pulses," Proc. SPIE **8502**, 850204 (2012).
91. M. T. Hassan, A. Wirth, I. Grgura, A. Moulet, T. T. Luu, J. Gagnon, V. Pervak, and E. Goulielmakis, "Invited article: attosecond photonics: synthesis and control of light transients," Rev. Sci. Instrum. **83**, 111301 (2012).

RESEARCH ARTICLE

Synthesized Light Transients

A. Wirth,[1] M. Th. Hassan,[1,2] I. Grguraš,[1] J. Gagnon,[1] A. Moulet,[1] T. T. Luu,[1] S. Pabst,[3,4] R. Santra,[3,4] Z. A. Alahmed,[2] A. M. Azzeer,[2] V. S. Yakovlev,[1,5] V. Pervak,[5] F. Krausz,[1,5] E. Goulielmakis[1]*

Manipulation of electron dynamics calls for electromagnetic forces that can be confined to and controlled over sub-femtosecond time intervals. Tailored transients of light fields can provide these forces. We report on the generation of subcycle field transients spanning the infrared, visible, and ultraviolet frequency regimes with a 1.5-octave three-channel optical field synthesizer and their attosecond sampling. To demonstrate applicability, we field-ionized krypton atoms within a single wave crest and launched a valence-shell electron wavepacket with a well-defined initial phase. Half-cycle field excitation and attosecond probing revealed fine details of atomic-scale electron motion, such as the instantaneous rate of tunneling, the initial charge distribution of a valence-shell wavepacket, the attosecond dynamic shift (instantaneous ac Stark shift) of its energy levels, and its few-femtosecond coherent oscillations.

The generation and measurement (briefly, synthesis) of electric field transients permitted the characterization of electric circuits with sub-picosecond temporal resolution (*1*) and constitutes a base technology for advancing high-speed electronics and electron-based information technologies. Electronic processes on the atomic scale typically evolve on a few-femtosecond to sub-femtosecond time scale. Time-domain access to these dynamics requires the extension of electric field control to optical frequencies.

As a first step to this end, measurement (*2*, *3*) and control (*4*–*9*) of the phase of field oscillations relative to their envelope [carrier-envelope phase (CEP)] yielded reproducible few-cycle light waveforms (*10*). Attosecond metrology (*11*) was further advanced by the reproducible generation and measurement of isolated attosecond pulses

[1]Max-Planck-Institut für Quantenoptik (MPQ), Hans-Kopfermann-Strasse 1, D-85748 Garching, Germany. [2]Department of Physics and Astronomy, King Saud University, Riyadh 11451, Kingdom of Saudi Arabia. [3]Center for Free-Electron Laser Science, Deutsches Elektronen Synchrotron, Notkestrasse 85, 22607 Hamburg, Germany. [4]Department of Physics, University of Hamburg, Jungiusstrasse 9, 20355 Hamburg, Germany. [5]Department für Physik, Ludwig-Maximilians-Universität (LMU), Am Coulombwall 1, D-85748 Garching, Germany.

*To whom correspondence should be addressed. E-mail: elgo@mpq.mpg.de

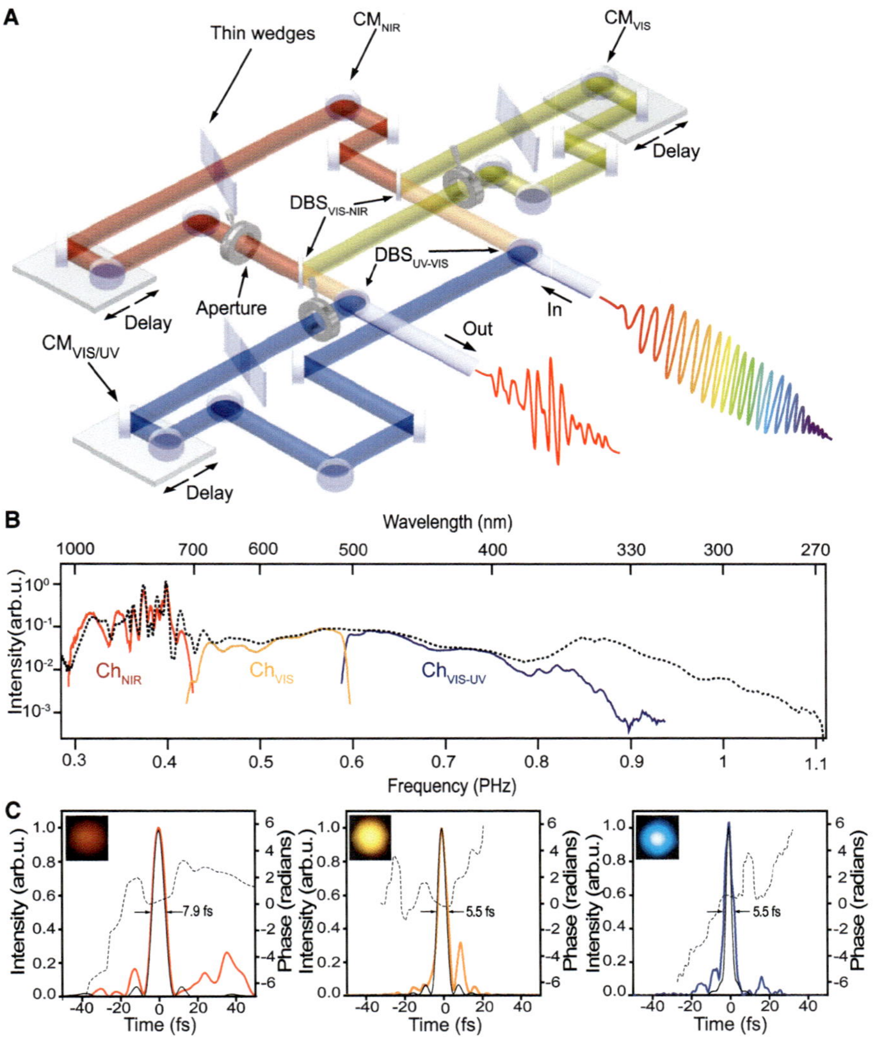

Fig. 1. Apparatus for infrared-visible-ultraviolet field synthesis. **(A)** Schematic representation of a prototypical three-channel light field synthesizer. **(B)** Spectrum of the coherent radiation at the exit of the hollow-core fiber (dashed line). Spectra exiting the individual channels (not to scale) are shown in red for Ch_{NIR} (700 to 1100 nm), yellow for Ch_{VIS} (500 to 700 nm), and blue for Ch_{VIS-UV} (350 to 500 nm). **(C)** Temporal intensity (solid lines) and phase profiles (dashed curves) of the respective pulses. The thin black lines depict the intensity profiles of the corresponding bandwidth-limited pulses, with durations of $\tau_{Ch(NIR)}$ = 6.8 fs, $\tau_{Ch(VIS)}$ = 5 fs, and $\tau_{Ch(VIS-UV)}$ = 4.5 fs. Insets show photos of the respective beam profiles taken at the exit of the apparatus.

RESEARCH ARTICLE

(*12–14*). Control (*15*) and real-time observation (*16–21*) of electronic processes would greatly benefit from sub-femtosecond sculpting and confinement of strong light fields. We are able to demonstrate this capability along with some of its consequences.

Tailoring light fields on the electronic time scale requires the coherent superposition and manipulation of frequencies over more than an octave in the visible and flanking spectral ranges. So far, this demand could only be met through the technique of molecular modulation (*22–27*). This approach recently allowed the subcycle shaping of optical fields via the superposition of quasi-monochromatic waves in the infrared-visible range (*28*). Although these periodic waveforms are highly relevant to advancing modern electronics, time-domain access to electronic phenomena calls for the temporal confinement of the sculpted waveform to a single cycle or just a few oscillation cycles. We refer to such super-octave optical waveforms as light transients. Recent experiments have paved the way toward the synthesis of light transients (*29–33*), but they have not yet achieved the goal of subcycle field shaping and measurement.

Here, we report on the shaping, confinement, and attosecond sampling of the fields of intense light transients within their carrier wave cycle (~2.4 fs) over the frequency band of 0.3 to 0.9 PHz. A variety of on-demand waveforms with controlled subcycle field evolution, yielding sub-femtosecond rise times or subcycle confinement of instantaneous intensity, demonstrate the power of PHz field synthesis. As an application of enhanced atomic-scale electron control, we field-ionized atoms within a single wave crest and triggered valence electron motion on a sub-femtosecond scale. By providing a sub-femtosecond optical field trigger and a robust attosecond probe, subcycle light transients establish sub-femtosecond pump-probe spectroscopy.

1.5-octave optical field synthesizer. We produced coherent supercontinua by propagating ~0.8 mJ, ~25-fs pulses carried at a wavelength of λ_0 ~ 780 nm in a hollow-core fiber filled with neon gas. Spectral broadening was enhanced with respect to previous experiments (*34*) by raising the gas pressure to ~3.5 bar, resulting in a nearly uniform (to within 20 dB) energy distribution over a bandwidth of >0.6 PHz (330 to 1100 nm) (Fig. 1B). Manipulation of individual spectral components requires their spatial separation and subsequent recombination. The conventional approach, based on prisms and liquid crystal modulators (*31*), is hardly scalable for super-octave–spanning operation; therefore, we instead implemented chirped multilayer mirror technology, proposed in (*34*), which offers scalability to several octaves in the visible and nearby spectral regions. Our prototypical three-channel device (Fig. 1A) subdivides the aforementioned ~0.6-PHz spectral range into three bands of nearly equal width—Ch_{NIR}, 700 to 1100 nm; Ch_{VIS}, 500 to 700 nm; and $Ch_{VIS\text{-}UV}$, 350 to 500 nm—with the help of dichroic beamsplitters $DBS_{VIS\text{-}NIR}$ and $DBS_{UV\text{-}VIS}$ (fig. S1). Dispersive chirped mirrors $CM_{VIS/UV}$, CM_{VIS}, and CM_{NIR} compensate for the chirp carried by the pulse as well as that introduced by the thin fused silica wedge pairs incorporated in each channel and the beamsplitters near the edges of the spectral bands. As a result, the pulses in the individual channels are compressed close to their bandwidth-limited durations (Fig. 1C).

The chirp, the CEP, and the delay of the pulses formed in Ch_{NIR}, Ch_{VIS}, and $Ch_{VIS\text{-}UV}$ can be precisely controlled by wedges and nanometer-precision delay stages, respectively. The adjustment of the beam size in each channel—via an iris—allows control of the pulse's energy [supporting online material (SOM) text, section 1]. These control knobs offer both subcycle shaping of the generated fields and compression close to their bandwidth limit. Owing to the high efficiency of the chirped multilayer optics used (fig. S2), the device transmits some ~83% of the incident continuum beam (Fig. 1B, dotted line), resulting in a pulse energy of ~0.3 mJ/pulse at the exit of the apparatus (Ch_{NIR} ~ 250 μJ, Ch_{VIS} ~ 35 μJ, and $Ch_{VIS\text{-}UV}$ ~ 15 μJ). The setup is assembled on a monolithic aluminum base plate, with active thermal and interferometric path-length stabilization (fig. S3).

By analogy with femtosecond electro-optic sampling of THz transients (*1*), we can use attosecond streaking (*12*, *35*) for sampling the electric field of PHz transients. To this end, the PHz transients exiting our three-channel synthesizer were gently focused (to a peak intensity of ~10^{14} W/cm^2) into a neon gas jet, where they generated broadband extreme ultraviolet (XUV) radiation (*14*) emitted in a near-diffraction-limited beam collinear with the driving radiation (SOM text, section 2). Bandpass filtering (width of ~13 eV centered at ~85 eV) near the cut-off energy (~90 eV), implemented with multilayer

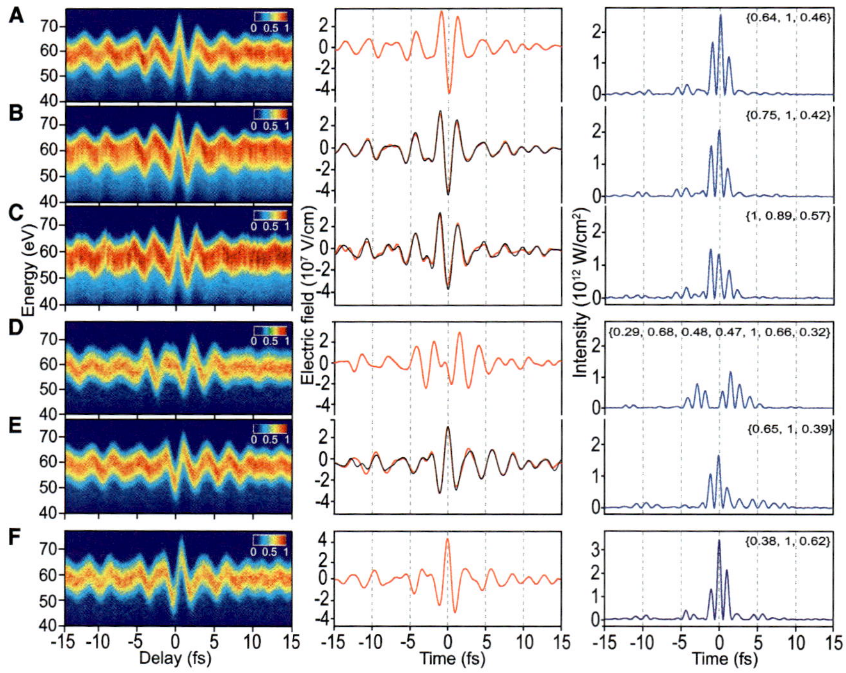

Fig. 2. Synthesis of petahertz light field transients. (**A** to **F**) Attosecond streaking spectrograms composed of photoelectron spectra normalized to their integral (left) and the respective retrieved electric fields (middle) and instantaneous intensity (right). Relative intensities for the most intense field crests—normalized to the maximum—are given in brackets. From (A) to (C), the delay of $Ch_{VIS\text{-}UV}$ is varied in steps of 200 as (~π/4). Dashed black lines in (B) and (C) show the field transients calculated from the reference waveform of (A). (D) Relative delays and CEPs of the individual channels are adjusted so as to create twin transients with a field minimum in between them. (E) Ch_{NIR} is delayed by 1.45 fs (~π), resulting in a high-frequency leading transient followed by a low-frequency tail. The dashed line in (E) shows the field transients calculated from the reference waveform of (D). Transients in (A) to (C), (E), and (F) carry less than one cycle within the FWHM of their temporal intensity profile. (F) τ_{FWHM} ~ 2.1 fs, incorporating only ~0.88 field cycles at the carrier wavelength of λ_0 ~ 710 nm.

optics and thin metal foils, isolated a single attosecond pulse (*12*, *14*). Both pulses were then focused into a second neon gas jet placed near the entrance of a time-of-flight electron spectrometer (TOF) for measuring the XUV-induced, laser-field–streaked photoelectron spectra versus delay (an attosecond streaking spectrogram) (*35*).

Attosecond streaking spectrograms of subcycle waveforms synthesized from near-bandwidth–limited fields exiting Ch_{NIR}, Ch_{VIS}, and $Ch_{VIS\text{-}UV}$ are shown in Fig. 2, left. The spectrograms are composed of a series of laser-field–streaked XUV photoelectron spectra recorded as a function of delay between the XUV pulse and the subcycle field. A delay step of 0.2 fs was used, which safely allows sampling up to the highest frequency components (~0.9 PHz) in the waveform. Remarkably, each spectrogram reveals an isolated <200-as XUV pulse. This is a direct consequence of the substantial subcycle variation of field amplitude (Fig. 2, right), which provides an efficient temporal gate—via ionization confinement and/or energy filtering—for isolating a single attosecond burst in the XUV radiation emitted by the ionizing atoms. The feasibility of generating a robust isolated attosecond probe for a wide range of waveforms not only is required for the sampling of the transients but is most important also for the interrogation of processes triggered and/or controlled by the optical field transients.

Any of the retrieved electric field waveforms (Fig. 2, middle) permits full characterization of the properties of the apparatus and subsequently on-demand synthesis of prescribed fields. To this end, we used the waveform in Fig. 2A as a reference and retrieved the values of the control parameters of the system such as the field amplitudes, phase delays, and the CEPs of the three channels through the numerical band-pass filtering of the measured output waveform within the spectral ranges defined by Ch_{NIR}, Ch_{VIS}, and $Ch_{VIS\text{-}UV}$. The field transients shown in Fig. 2, B and C, were then synthesized by delaying $Ch_{VIS\text{-}UV}$ in steps of $\pi/4$ (~200 as) with respect to the reference waveform. A more complex, nonsinusoidal transient, which is generated by delaying Ch_{NIR} so that the fields from the three channels cancel each other at the center of the waveform, is shown in Fig. 2D. The transient shown in Fig. 2E is generated via delay of Ch_{NIR} by $\sim\pi$ with respect to the waveform shown in Fig. 2D, resulting in a single intense field crest pointing in the opposite direction to the peak field in Fig. 2A. The full red and the dashed black lines in Fig. 2, middle, depict, respectively, waveforms measured or calculated from the constituent Ch_{NIR}, Ch_{VIS}, and $Ch_{VIS\text{-}UV}$ fields retrieved from the reference waveform (Fig. 2A), with control parameters changed by known amounts with respect to those of the reference waveform. The agreement between prediction and measurement demonstrates controlled sub-femtosecond shaping, complete characterization, and reproducibility (synthesis) of petahertz field transients.

The instantaneous intensity shown in Fig. 2, right, reveals substantial variations of the strength of consecutive wave crests upon these transformations. The reference waveform (Fig. 2A) exhibits half cycles with relative intensities of (0.64, 1, 0.46) underpinning the subcycle character of the transient. Delaying $Ch_{VIS\text{-}UV}$ gradually transforms the field into the highly asymmetric transient of Fig. 2C, with the temporal extension left almost unchanged. This transient carries its most intense field crest right at its leading edge, followed by half cycles of decreasing intensity (1, 0.89, 0.57), resulting in a sub-femtosecond rise time of its instantaneous intensity. On the other hand, a transient with its two most intense field crests separated by ~4.5 fs and a half-cycle virtually annihilated in between is revealed in Fig. 2D. The transient with the largest degree of temporal energy confinement is shown in Fig. 2F, with field crest intensities of (0.38, 1, 0.62) and with ~35% of its energy carried in a single wave crest.

Field ionization and its real-time sampling. We used transients with a central field crest ~1.7 times more intense than the adjacent half cycles (Fig. 3B) to ionize Kr atoms enclosed in a quasi-static gas cell (length $l \approx 0.74$ mm) at a density of $\approx 5.6 \times 10^{18}$ cm^{-3}. The gas cell was positioned at the laser focus, replacing the neon gas jet previously used for recording the streaking spectrograms shown in Fig. 3A (the experimental setup is shown in fig. S4). This procedure reveals the absolute timing of any process initiated or affected by the field transients with attosecond precision (SOM text, section 8).

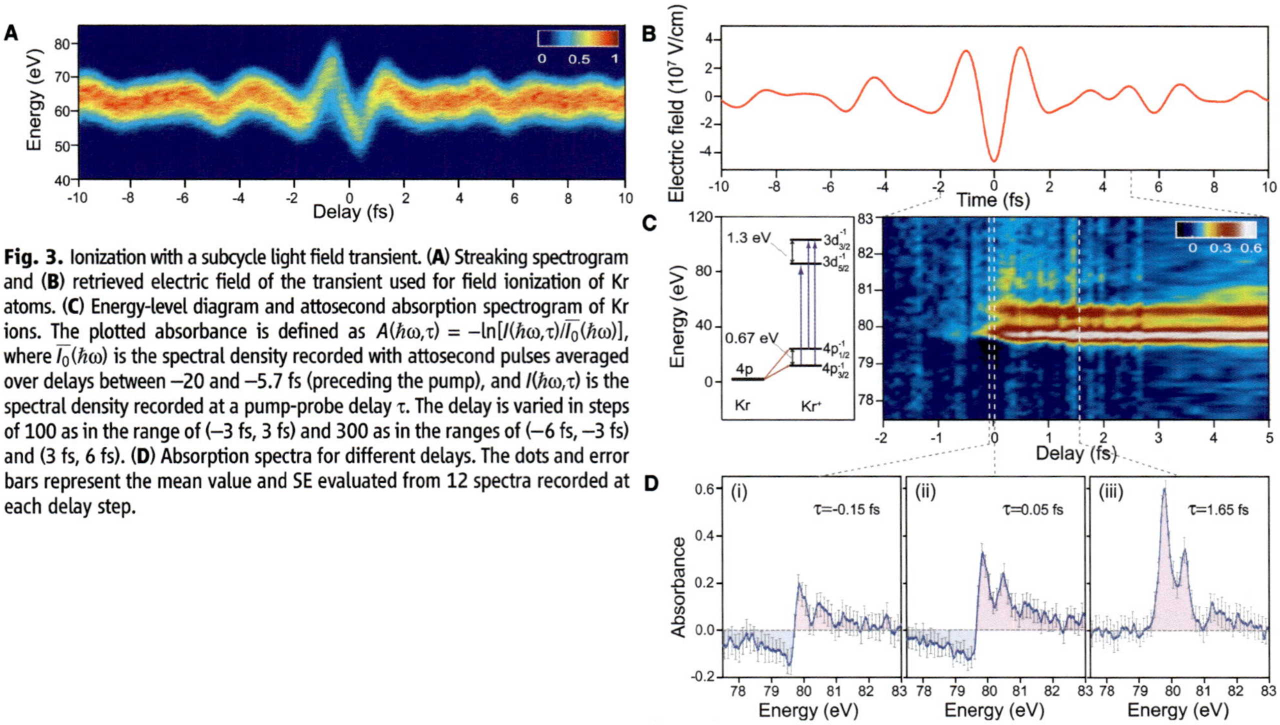

Fig. 3. Ionization with a subcycle light field transient. **(A)** Streaking spectrogram and **(B)** retrieved electric field of the transient used for field ionization of Kr atoms. **(C)** Energy-level diagram and attosecond absorption spectrogram of Kr ions. The plotted absorbance is defined as $A(\hbar\omega,\tau) = -\ln[I(\hbar\omega,\tau)/\overline{I_0}(\hbar\omega)]$, where $\overline{I_0}(\hbar\omega)$ is the spectral density recorded with attosecond pulses averaged over delays between −20 and −5.7 fs (preceding the pump), and $I(\hbar\omega,\tau)$ is the spectral density recorded at a pump-probe delay τ. The delay is varied in steps of 100 as in the range of (−3 fs, 3 fs) and 300 as in the ranges of (−6 fs, −3 fs) and (3 fs, 6 fs). **(D)** Absorption spectra for different delays. The dots and error bars represent the mean value and SE evaluated from 12 spectra recorded at each delay step.

RESEARCH ARTICLE

We increased the intensity with an iris to $\approx 4.8 \times 10^{14}$ W/cm^2 and ionized ~16% of Kr atoms in the gas cell. We then probed strong-field ionization with a time-delayed attosecond pulse by measuring transient absorption spectra (*19*). Krypton ions created in their $4p^{-1}_{j=3/2}$ ground-state manifold and in the $4p^{-1}_{j=1/2}$ excited-state manifold, comprising four ($m_{j=3/2} = -3/2, -1/2, 1/2, 3/2$) and two ($m_{j=1/2} = -1/2, 1/2$) states, respectively, are promoted to the $3d^{-1}$ core-hole excited states by absorption of XUV photons from the attosecond pulse (Fig. 3C, left) (*36*). The resulting spectra transmitted through the gas cell were recorded as a function of the delay between the attosecond probe (pulse duration of ~200 as, centered at ~85 eV) and the ionizing field transient (pump), yielding the absorption spectrogram shown in Fig. 3C, right. The relevant transitions, $4p^{-1}_{3/2} \rightarrow 3d^{-1}_{5/2}$, $4p^{-1}_{1/2} \rightarrow 3d^{-1}_{3/2}$, and $4p^{-1}_{3/2} \rightarrow 3d^{-1}_{3/2}$, are indicated with arrows in the Fig. 3C level diagram. The absorption lines recorded at the leading edge of the ionizing field transient (Fig. 3D, i and ii, for example) have a characteristic profile revealing negative absorbance at photon energies below the resonances and positive absorbance above them (*37*). These transient line shapes gradually evolve to quasi-steady-state profiles toward the trailing edge of the ionizing field (Fig. 3D, iii).

State-selective sub-femtosecond tracing of field ionization. In order to relate the attosecond transient absorption spectra with ion population dynamics, we have performed numerical simulations by treating the generated Kr ions within a simplified, three-level model using the density matrix formalism (SOM text, section 3). The simulations reproduce well the experimental data (fig. S5) and demonstrate—in agreement with previous studies (*38*, *39*)—that the distortions of the transient absorption spectra are due to the synthesized pump field acting on the XUV-initiated, time-dependent ionic polarization response. Moreover, our simulations (based on an adiabatic tunnel-ionization calculation) suggest that—under the conditions of our experiments—the emerging absorption lines coincide with the population dynamics of the relevant ionic states, permitting retrieval of their transient evolution from the peak absorbances with good (~10%) accuracy (fig. S9). Figure 4A shows the sub-femtosecond evolution of the effective transient population in the ground state manifold $4p^{-1}_{j=3/2}$ (black dots), defined as $\rho^{\mathrm{eff}}_{3/2,3/2}(t) = \rho^{(\pm 1/2)}_{3/2,3/2}(t) + a\rho^{(\pm 3/2)}_{3/2,3/2}(t)$, where $a = 2/3$ reflects the higher transition cross-section for the ($m_j = \pm 1/2$) transitions between $4p^{-1}_{j=3/2}$ and $3d^{-1}_{5/2}$ states; the $4p^{-1}_{j=1/2}$ manifold population $\rho^{(\pm 1/2)}_{1/2,1/2}(t)$ is represented by black diamonds. These populations are retrieved from the $4p^{-1}_{3/2} \rightarrow 3d^{-1}_{5/2}$ and $4p^{-1}_{1/2} \rightarrow 3d^{-1}_{3/2}$ absorbances versus pump-probe delay, respectively. Because attosecond streaking is performed in the same apparatus, this evolution can be directly timed and contrasted with the evolution of the ionizing field [$|E_L(t)|^2$, shown by the dashed line].

The buildup of the retrieved ionic populations exhibits steps that are in synchrony with the field crests of the transient. This becomes even more evident from the population rates obtained by taking the time-derivative of the ionic populations in Fig. 4A. The ionization rate, which is estimated as $\frac{d}{dt}\rho^{\mathrm{eff}}_{3/2,3/2}(t)$ and shown in Fig. 4B (dots and red line), exhibits three main features at the crests of the ionizing field. The main ionization burst is responsible for approximately 80% of the ion population and has a full width at half maximum (FWHM) of <0.7 fs, indicating a sub-femtosecond confinement of field ionization. It is this confinement to a single field crest that allows quantitative evaluation of the time-dependent rate of optical field ionization and state-selective population dynamics. For the populations depicted, we evaluate a peak production rate for $\rho^{\mathrm{eff}}_{3/2,3/2}(t)$ of $\Gamma_{\mathrm{peak}} = (0.12 \pm 0.01)$ fs^{-1} and for $\rho^{(\pm 1/2)}_{1/2,1/2}(t)$ of $\Gamma_{\mathrm{peak}} = (0.059 \pm 0.009)$ fs^{-1}, which is in excellent agreement with results obtained by numerically integrating the Schrödinger equation of a single-active electron model in three dimensions (SOM text, section 6), yielding $\Gamma_{\mathrm{peak}}(\rho^{\mathrm{eff}}_{3/2,3/2}) = 0.13$ fs^{-1} and $\Gamma_{\mathrm{peak}}(\rho^{(\pm 1/2)}_{1/2,1/2}) = 0.059$ fs^{-1}, which also well reproduces details of the temporal evolution

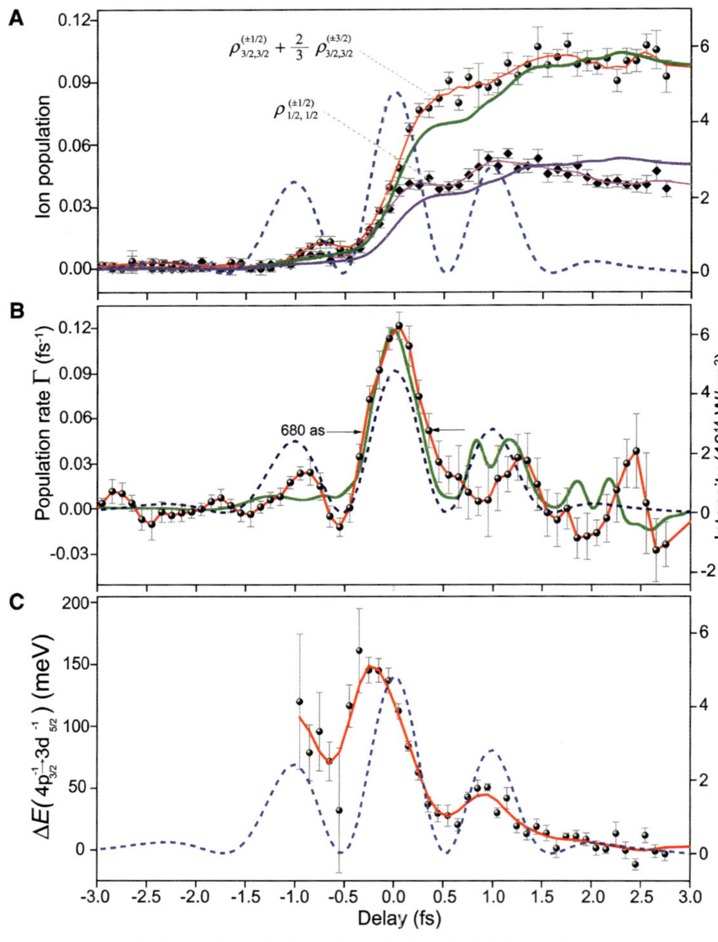

Fig. 4. Attosecond ionization and Stark effect dynamics in Kr$^+$. **(A)** Population dynamics in the ground-state $4p^{-1}_{j=3/2}$ manifold (dots), $\rho^{\mathrm{eff}}_{3/2,3/2}(t)$, and in the excited-state $4p^{-1}_{j=1/2}$ manifold (diamonds), $\rho^{(\pm 1/2)}_{1/2,1/2}(t)$, retrieved from the absorption spectrogram of Fig. 3C and contrasted with the instantaneous intensity (dashed line) as well as with the prediction of numerical simulations convolved with the XUV probe pulse duration (green and purple lines). **(B)** Ionization (population) rate $d\rho^{\mathrm{eff}}_{3/2,3/2}(t)/dt$ evaluated from the data in (A) (dots and red line) in comparison with the theoretical prediction (green line). **(C)** Shift of the central energy of the $4p^{-1}_{3/2} \rightarrow 3d^{-1}_{5/2}$ transition evaluated by fitting a Lorentzian profile to the attosecond transient absorption spectra shown in Fig. 3C. Red curves in (A) to (C) and magenta curve in (A) are guides to the eye obtained by three adjacent point fast Fourier transform–smoothing.

RESEARCH ARTICLE

of the ionic populations (Fig. 4A, green and purple lines, and B, green line).

Observation of the instantaneous optical Stark shift. The shift ΔE of quantum energy levels of atoms, molecules, or solids induced by an optical field $E_L(t)$—the ac Stark shift (*40*)—plays a central role in fundamental dynamical processes. So far, only cycle-averaged Stark effects have been accessible to experiments (*41*). If the laser frequency ω_L is much smaller than atomic resonance frequencies, the Stark shift of a nondegenerate atomic level is expected to instantly follow variations of the laser field: $\Delta E(t) = -1/2\mu E_L^2(t)$, where $\mu = \mu(\omega_L)$ is the atomic polarizability (*42*). When an XUV pulse creates a coherent superposition of two states with a difference in their respective polarizabilities $\Delta\mu(\omega_L)$, the induced polarization oscillations experience a phase shift, which is approximately given by

$$\Delta\varphi_{\text{dipole}}(t) \approx -\frac{\Delta\mu}{2\hbar}\int_{t_0}^{t} E_L^2(t')dt' \qquad (1)$$

where t_0 denotes the moment of arrival of the attosecond XUV pulse (with $\tau_{\text{XUV}} \ll 2\pi/\omega_L$). The instantaneous Stark shift detunes the energy at which the atom most efficiently absorbs photons from the XUV probe pulse—in our case, by the $3d \rightarrow 4p$ transition in Kr^+ ions—and modulates the frequency of the respective coherent dipole emission. Because the decay of the emission lasts several field cycles ($\tau = \tau_{3d} \sim 7.5$ fs), the Stark effect does not merely shift but also reshapes the transient absorption lines shown in Fig. 3D, i and ii. Thanks to the nearly instantaneous triggering of the polarization oscillations and their subsequent rapid decay, signatures of the instantaneous ac Stark shift come to light in our transient absorption spectra (Fig. 4C), which is in agreement with our simulations (figs. S8 and S9).

Valence wavepacket with well-defined quantum phase. Attosecond probing of few-cycle–driven field ionization of Kr atoms has revealed the emergence of a valence electron wavepacket in the 4*p* subshell of the Kr^+ ensemble—as a consequence of liberation of electrons from the $4p_{j=3/2}$ as well as the $4p_{j=1/2}$ manifolds—separated in energy by spin-orbit coupling in the Kr atoms (*19*). By repeating this attosecond absorption spectroscopic experiment with our subcycle transient shown in Fig. 3B, and supplementing it with attosecond streaking, we can now launch a valence wavepacket within a sub-femtosecond interval and with sub-femtosecond absolute timing accuracy.

From the recorded absorption spectrogram (Fig. 5A), we retrieve (SOM text, section 4) the fractional populations of the six ionic states, in the form of the diagonal matrix elements $\rho_{j,j'=j}^{(m_j)}$ of the reduced density matrix—$\rho_{3/2,3/2}^{(3/2)} + \rho_{3/2,3/2}^{(-3/2)} = 0.315 \pm 0.024$, $\rho_{3/2,3/2}^{(1/2)} + \rho_{3/2,3/2}^{(-1/2)} = 0.400 \pm 0.024$, $\rho_{1/2,1/2}^{(1/2)} + \rho_{1/2,1/2}^{(-1/2)} = 0.285 \pm 0.004$—and a degree of coherence of $g = 0.85 \pm 0.06$, which exceeds that measured in the previous experiment (*19*) by ~1.4 times and is unparalleled for long-lived (>1 fs) coherences in the valence shell. Our simulations [based on a state-of-the-art three-dimensional time-dependent configuration-interaction singles approach (*43*) that includes correlation dynamics between the field-generated hole and the photoelectron and has been extended to include spin-orbit interaction (*44*)] well predict the measured coherence as well as fractional populations in the $4p_{j=1/2}^{-1}$ and $4p_{j=3/2}^{-1}$ manifolds (SOM text, section 7).

The nearly perfect coherence is, once again, a direct consequence of the sub-femtosecond width of the ionization gate. This confinement, along with sub-femtosecond absolute timing information from streaking, has far-reaching consequences. The former allows launching of the wavepacket with a well-defined initial phase, whereas the latter permits reliable determination of this initial quantum phase. The retrieved phase $\phi(t)$ [equation 2 in (*19*)] of the quantum superposition is shown in Fig. 5, B and C, along with representative snapshots of the generated ensemble-averaged hole density distributions, as evaluated from our data in Fig. 5A. Linear extrapolation of $\phi(t)$ to "time zero," the birth of the hole at the peak of the ionizing field transient (Fig. 5B, blue line), yields $\phi(t_0) = (0.99 \pm 0.04)\pi$, which is in very good agreement with the prediction of our configuration-interaction–based simulations: $\phi(t_0) = 1.06\pi$. This initial quantum phase implies an elongated initial hole-density distribution aligned with the ionizing field vector, which is commensurate with our intuitive expectation.

Outlook. Subcycle engineering of optical field transients opens new prospects for steering the atomic-scale motion of electrons (*15*) with the electric force of light and for driving complex valence-shell dynamics in molecules (*45*). As a simple manifestation of enhanced control over valence shell dynamics, they allow sub-femtosecond temporal confinement of ionization and precise associated triggering of a wealth of subsequent electronic phenomena. They also provide an isolated attosecond photon probe for interrogating the unfolding electronic and—in molecules—nuclear motions by means of attosecond absorption and/or photoelectron spectroscopy, as well as an isolated electron probe for tracing these dynamics via electron diffraction (*46*) or high-harmonic interferometry (*47*). This constitutes a substantial extension of the repertoire of attosecond science, which was previously restricted to either triggering or probing electronic processes within a sub-femtosecond time window. The feasibility

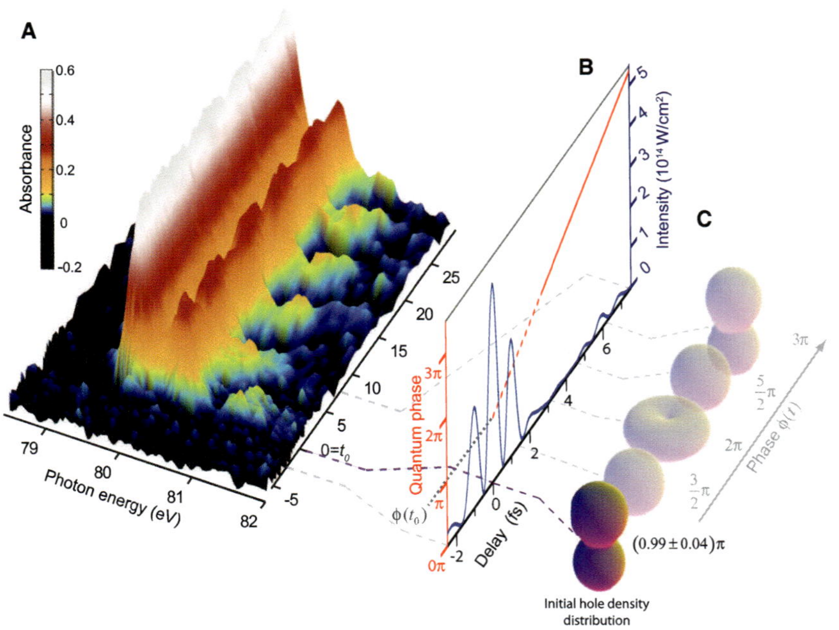

Fig. 5. Initial quantum phase and density distribution of a valence electron wavepacket. **(A)** Attosecond XUV transient absorption spectrogram of Kr atoms field-ionized by a subcycle field transient shown by the blue line in **(B)**. Linear extrapolation of the retrieved quantum phase $\phi(t)$ [shown by the red line in (B)] to time zero as determined through attosecond streaking, allows access to the initial quantum phase $\phi(t_0) = (0.99 \pm 0.04)\pi$ of the valence electron wavepacket. **(C)** Ensemble-averaged initial hole density distribution in the valence shell at the instant of ionization and its subsequent evolution, as evaluated from (A).

of sub-femtosecond pump-probe interrogation of strong-field phenomena opens exciting research prospects. Real-time insight into multiple ionization in multi- to single-cycle laser fields, or into strong-field–induced electron correlations in atoms, molecules, or solids (*48*, *49*), are but a few examples.

References and Notes

1. J. F. Whitaker *et al.*, *Microelectron. Eng.* **12**, 369 (1990).
2. L. Xu *et al.*, *Opt. Lett.* **21**, 2008 (1996).
3. T. Brabec, F. Krausz, *Rev. Mod. Phys.* **72**, 545 (2000).
4. J. Reichert *et al.*, *Opt. Commun.* **172**, 59 (1999).
5. D. J. Jones *et al.*, *Science* **288**, 635 (2000).
6. S. A. Diddams *et al.*, *Phys. Rev. Lett.* **84**, 5102 (2000).
7. A. Apolonski *et al.*, *Phys. Rev. Lett.* **85**, 740 (2000).
8. T. Udem, R. Holzwarth, T. W. Hänsch, *Nature* **416**, 233 (2002).
9. S. T. Cundiff, J. Ye, *Rev. Mod. Phys.* **75**, 325 (2003).
10. A. Baltuška *et al.*, *Nature* **421**, 611 (2003).
11. M. Hentschel *et al.*, *Nature* **414**, 509 (2001).
12. R. Kienberger *et al.*, *Nature* **427**, 817 (2004).
13. G. Sansone *et al.*, *Science* **314**, 443 (2006).
14. E. Goulielmakis *et al.*, *Science* **320**, 1614 (2008).
15. M. F. Kling *et al.*, *Science* **312**, 246 (2006).
16. G. Sansone *et al.*, *Nature* **465**, 763 (2010).
17. M. Schultze *et al.*, *Science* **328**, 1658 (2010).
18. J. Mauritsson *et al.*, *Phys. Rev. Lett.* **105**, 053001 2010).
19. E. Goulielmakis *et al.*, *Nature* **466**, 739 (2010).
20. S. Zherebtsov *et al.*, *Nat. Phys.*, published online 24 April 2011 (10.1038/nphys1983).
21. P. Eckle *et al.*, *Science* **322**, 1525 (2008).
22. S. E. Harris, A. V. Sokolov, *Phys. Rev. Lett.* **81**, 2894 (1998).
23. A. V. Sokolov, D. R. Walker, D. D. Yavuz, G. Y. Yin, S. E. Harris, *Phys. Rev. Lett.* **85**, 562 (2000).
24. J. Q. Liang, M. Katsuragawa, F. L. Kien, K. Hakuta, *Phys. Rev. Lett.* **85**, 2474 (2000).
25. A. V. Sokolov, D. R. Walker, D. D. Yavuz, G. Y. Yin, S. E. Harris, *Phys. Rev. Lett.* **87**, 033402 (2001).
26. T. Suzuki, M. Hirai, M. Katsuragawa, *Phys. Rev. Lett.* **101**, 243602 (2008).
27. Z.-M. Hsieh *et al.*, *Phys. Rev. Lett.* **102**, 213902 (2009).
28. H.-S. Chan *et al.*, *Science* **331**, 1165 (2011).
29. R. K. Shelton *et al.*, *Science* **293**, 1286 (2001).
30. M. Yamashita, K. Yamane, R. Morita, *IEEE J. Sel. Top. Quantum Electron.* **12**, 213 (2006).
31. S. Rausch, T. Binhammer, A. Harth, F. X. Kärtner, U. Morgner, *Opt. Express* **16**, 17410 (2008).
32. G. Krauss *et al.*, *Nat. Photonics* **4**, 33 (2010).
33. K. Okamura, T. Kobayashi, *Opt. Lett.* **36**, 226 (2011).
34. E. Goulielmakis *et al.*, *Science* **317**, 769 (2007).
35. E. Goulielmakis *et al.*, *Science* **305**, 1267 (2004).
36. This indicates that relative to the ground state configuration, a hole (or electron vacancy) is created in the nl_j^{-1} nl subshell, where j denotes the total angular momentum and m_j its projection on the z axis, the latter being aligned with the laser polarization.
37. B. R. Mollow, *Phys. Rev. A* **5**, 1522 (1972).
38. T. Unold, K. Mueller, C. Lienau, T. Elsaesser, A. D. Wieck, *Phys. Rev. Lett.* **92**, 157401 (2004).
39. C. H. B. Cruz, J. P. Gordon, P. C. Becker, R. L. Fork, C. V. Shank, *Int. J. Quant. Elec.* **24**, 261 (1988).
40. S. H. Autler, C. H. Townes, *Phys. Rev.* **100**, 703 (1955).
41. B. J. Sussman, *Am. J. Phys.* **79**, 477 (2011).
42. N. B. Delone, V. P. Krainov, *Phys. Uspekhi* **169**, 753 (1999).
43. L. Greenman *et al.*, *Phys. Rev. A* **82**, 023406 (2010).
44. S. Pabst, L. Greenman, P. J. Ho, D. A. Mazziotti, R. Santra, *Phys. Rev. Lett.* **106**, 053003 (2011).
45. F. Remacle, M. Nest, R. D. Levine, *Phys. Rev. Lett.* **99**, 183902 (2007).
46. M. Meckel *et al.*, *Science* **320**, 1478 (2008).
47. O. Smirnova *et al.*, *Nature* **460**, 972 (2009).
48. M. Gertsvolf, M. Spanner, D. M. Rayner, P. B. Corkum, *J. Phys. B* **43**, 131002 (2010).
49. M. Durach, A. Rusina, M. F. Kling, M. I. Stockman, *Phys. Rev. Lett.* **105**, 086803 (2010).

Acknowledgments: We acknowledge the development of dedicated attosecond soft x-ray optics by M. Hofstetter and U. Kleineberg (LMU, MPQ). This work was supported by the Max Planck Society, the European Research Council grant (Attoelectronics-258501), the Deutsche Forschungsgemeinschaft Cluster of Excellence: Munich Centre for Advanced Photonics (www.munich-photonics.de), the King Saud University–MPQ collaboration, and the European Research Training Network ATTOFEL.

Supporting Online Material
www.sciencemag.org/cgi/content/full/science.1210268/DC1
SOM Text
Figs. S1 to S10
References (*50–59*)

24 June 2011; accepted 30 August 2011
Published online 1 September 2011;
10.1126/science.1210268

REPORTS

Observation of Correlated Particle-Hole Pairs and String Order in Low-Dimensional Mott Insulators

M. Endres,[1*] M. Cheneau,[1] T. Fukuhara,[1] C. Weitenberg,[1] P. Schauß,[1] C. Gross,[1] L. Mazza,[1] M. C. Bañuls,[1] L. Pollet,[2] I. Bloch,[1,3] S. Kuhr[1,4]

Quantum phases of matter are characterized by the underlying correlations of the many-body system. Although this is typically captured by a local order parameter, it has been shown that a broad class of many-body systems possesses a hidden nonlocal order. In the case of bosonic Mott insulators, the ground state properties are governed by quantum fluctuations in the form of correlated particle-hole pairs that lead to the emergence of a nonlocal string order in one dimension. By using high-resolution imaging of low-dimensional quantum gases in an optical lattice, we directly detect these pairs with single-site and single-particle sensitivity and observe string order in the one-dimensional case.

The realization of strongly correlated quantum many-body systems using ultracold atoms has enabled the direct observation and control of fundamental quantum effects (*1–3*). A prominent example is the transition from a superfluid (SF) to a Mott insulator (MI), occurring when interactions between bosonic particles on a lattice dominate over their kinetic energy (*4–8*). At zero temperature and in the limit where the ratio of kinetic energy over interaction energy vanishes, particle fluctuations are completely suppressed and the lattice sites are occupied by an integer number of particles. However, at a finite tunnel coupling but still in the Mott insulating regime, quantum fluctuations create correlated particle-hole pairs on top of this fixed-density background, which can be understood as virtual excitations. These particle-hole pairs fundamentally determine the properties of the MI, such as its residual phase coherence (*9*), and lie at the heart of superexchange-mediated spin interactions that form the basis of quantum magnetism in multicomponent quantum gas mixtures (*10–12*).

In a one-dimensional system, the appearance of correlated particle-hole pairs at the transition point from a SF to a MI is intimately connected to the emergence of a hidden string-order parameter $\mathcal{O}_P$ (*13*, *14*):

$$\mathcal{O}_P^2 = \lim_{l\to\infty} \mathcal{O}_P^2(l) = \lim_{l\to\infty} \left\langle \prod_{k\leq j\leq k+l} e^{i\pi\delta\hat{n}_j} \right\rangle \quad (1)$$

Here, $\delta\hat{n}_j = \hat{n}_j - \bar{n}$ denotes the deviation in occupation of the jth lattice site from the average background density, and k is an arbitrary position along the chain. In the simplest case of a MI with unity filling ($\bar{n} = 1$), relevant to our experiments, each factor in the product of operators in Eq. 1 yields −1 instead of +1 when a single-particle fluctuation from the unit back-

[1]Max-Planck-Institut für Quantenoptik, 85748 Garching, Germany. [2]Theoretische Physik, Eidgenössische Technische Hochschule (ETH) Zurich, 8093 Zurich, Switzerland. [3]Ludwig-Maximilians-Universität, 80799 Munich, Germany. [4]University of Strathclyde, Scottish Universities Physics Alliance, Glasgow G4 0NG, UK.

*To whom correspondence should be addressed. E-mail: manuel.endres@mpq.mpg.de

LETTER

doi:10.1038/nature16528

Optical attosecond pulses and tracking the nonlinear response of bound electrons

M. Th. Hassan[1]*, T. T. Luu[1]*, A. Moulet[1], O. Raskazovskaya[2], P. Zhokhov[3,4], M. Garg[1], N. Karpowicz[1], A. M. Zheltikov[3,4], V. Pervak[2], F. Krausz[1,2] & E. Goulielmakis[1]

The time it takes a bound electron to respond to the electromagnetic force of light sets a fundamental speed limit on the dynamic control of matter and electromagnetic signal processing. Time-integrated measurements of the nonlinear refractive index[1] of matter indicate that the nonlinear response of bound electrons to optical fields is not instantaneous; however, a complete spectral characterization of the nonlinear susceptibility tensors[2]—which is essential to deduce the temporal response of a medium to arbitrary driving forces using spectral measurements—has not yet been achieved. With the establishment of attosecond chronoscopy[3–5], the impulsive response of positive-energy electrons to electromagnetic fields has been explored through ionization of atoms[6] and solids[7] by an extreme-ultraviolet attosecond pulse[8] or by strong near-infrared fields[9–11]. However, none of the attosecond studies carried out so far have provided direct access to the nonlinear response of bound electrons. Here we demonstrate that intense optical attosecond pulses synthesized in the visible and nearby spectral ranges allow sub-femtosecond control and metrology of bound-electron dynamics. Vacuum ultraviolet spectra emanating from krypton atoms, exposed to intense waveform-controlled optical attosecond pulses, reveal a finite nonlinear response time of bound electrons of up to 115 attoseconds, which is sensitive to and controllable by the super-octave optical field. Our study could enable new spectroscopies of bound electrons in atomic, molecular or lattice potentials of solids[12], as well as light-based electronics operating on sub-femtosecond timescales and at petahertz rates[13–15].

Intramolecular atomic motion dynamically distorts the interatomic electron cloud and thereby gives rise to a transient electronic polarizability[16] and a finite nonlinear response of matter to light fields. Time-resolved spectroscopies can now probe this response with femtosecond lasers[17]. A finite nonlinear response of matter to external fields can also emerge in the absence of nuclear degrees of freedom, purely as a result of electron dynamics caused by coherent excitation of electronic states and concomitant spatiotemporal evolution of the electron cloud[18,19]. However, quantum mechanics implies that these dynamics occur on an attosecond (as) timescale of $\Delta t \approx h/\Delta E \approx 400$–800 as (in which ΔE is the energy spacing between the relevant electronic levels, typically about 5–10 eV, and h is Planck's constant). Consequently, it has long been conjectured that the build-up of the nonlinear polarization of bound electrons under non-resonant driving by light fields is instantaneous[20]. For electron dynamics to influence substantially the nonlinear response of a medium to an optical field and to enable its real-time tracing, both induction of electronic coherence and nonlinear probing of concomitant polarization dynamics must occur within time intervals shorter or comparable to Δt. In the visible and nearby spectral ranges, where bound electrons are typically probed and the required field strengths for nonlinear studies can be most readily produced, this condition implies a light pulse with its energy substantially being confined to a half wave cycle. We refer to such a pulse as an optical attosecond pulse.

To illustrate how such a pulse can enable the study of the dynamics of the bound-electron response, we theoretically investigate the interaction of an intense ($4 \times 10^{13}\,\mathrm{W\,cm^{-2}}$) half-cycle optical field (Fig. 1a, red dashed line), identical to that used in our experiments described below, with a system lacking nuclear degrees of freedom—a krypton (Kr) atom. We calculate the induced nonlinear dipole moment $\rho(t)$ by numerical integration of the three-dimensional time-dependent Schrödinger equation (TDSE; Fig. 1a, blue line; $\rho_{\mathrm{TDSE}}(t)$) and by using an adiabatic model describing Kr (Fig. 1a, black line; $\rho_{\mathrm{ad}}(t)$); see Methods. The instantaneous response that is inherent to the adiabatic model is manifested by the accurate synchronization between the optical field $E(t)$ (Fig. 1a, red dashed line) and the nonlinear dipole moment $\rho_{\mathrm{ad}}(t)$ (Fig. 1a, black line) induced in the system, and is further corroborated by the synchronous emission of the $\rho_{\mathrm{ad}}(\omega)$ spectrum (where ω is frequency) obtained from time–frequency analysis of the dipole (Fig. 1b, left panel). In contrast, when we account for the detailed electronic structure of Kr when using the TDSE to simulate the interaction with the ultrafast optical field, coherent oscillations of the electron cloud (Fig. 1a, grey dashed line) markedly affect the nonlinear dipole $\rho_{\mathrm{TDSE}}(t)$. The leading-edge and maximum of $\rho_{\mathrm{TDSE}}(t)$ are noticeably retarded, relative to those of the instantaneous response $\rho_{\mathrm{ad}}(t)$ and the driving electric field $E(t)$. The time–frequency analysis of $\rho_{\mathrm{TDSE}}(t)$ (Fig. 1c, left panel) reveals an attosecond-scale, non-uniform retardation (group delay dispersion) of the nonlinear emission (Fig. 1c, left panel, white dashed line) that is evident across the entire dipole spectrum and extends to the resonant area (10–15 eV) of Kr. Here, the nonlinear response is dominated by long-lasting coherences that typically evolve over substantially longer timescales (Fig. 1a, grey dashed line). In contrast, the nonlinear emission at lower energies (<9 eV) is virtually confined within the time interval of the interaction of the optical field with Kr. Hence, this low-energy spectral range offers an adequate basis for exploring the dynamics of the nonlinear response of the electron cloud to the optical driver.

A comparison between the Fourier-filtered (<9 eV) nonlinear dipoles $\rho_{\mathrm{ad}}(t)$ and $\rho_{\mathrm{TDSE}}(t)$ predicted by the adiabatic model (Fig. 1d, black line) and the TDSE model (Fig. 1d, blue line), respectively, reveals a finite relative delay of $\tau \approx 48$ as (Fig. 1a inset). This delay represents a substantial fraction of the time span of the induced dipole $\rho_{\mathrm{TDSE}}(t)$, which is about 250 as. The observed retardation is sensitive to, and is monotonically increasing with, applied field strength (compare the blue, orange and green lines in Fig. 1d); with an optical-field intensity of $8 \times 10^{13}\,\mathrm{W\,cm^{-2}}$, it reaches a value of $\tau \approx 122$ as. Figure 1e shows a similar sensitivity of the dynamics of the dipole when the field intensity is adjusted by varying the global phase ϕ_{G} of the driving field (justification for why the term carrier-envelope phase (CEP) is inappropriate for multi-octave waveforms is given in Methods).

[1]Max-Planck-Institut für Quantenoptik, Hans-Kopfermann-Straße 1, D-85748 Garching, Germany. [2]Department für Physik, Ludwig-Maximilians-Universität, Am Coulombwall 1, D-85748 Garching, Germany. [3]Department of Physics and Astronomy, Texas A&M University, College Station, Texas 77843, USA. [4]Physics Department, International Laser Center, M.V. Lomonosov Moscow State University, 119992 Moscow, Russia.
*These authors contributed equally to this work.

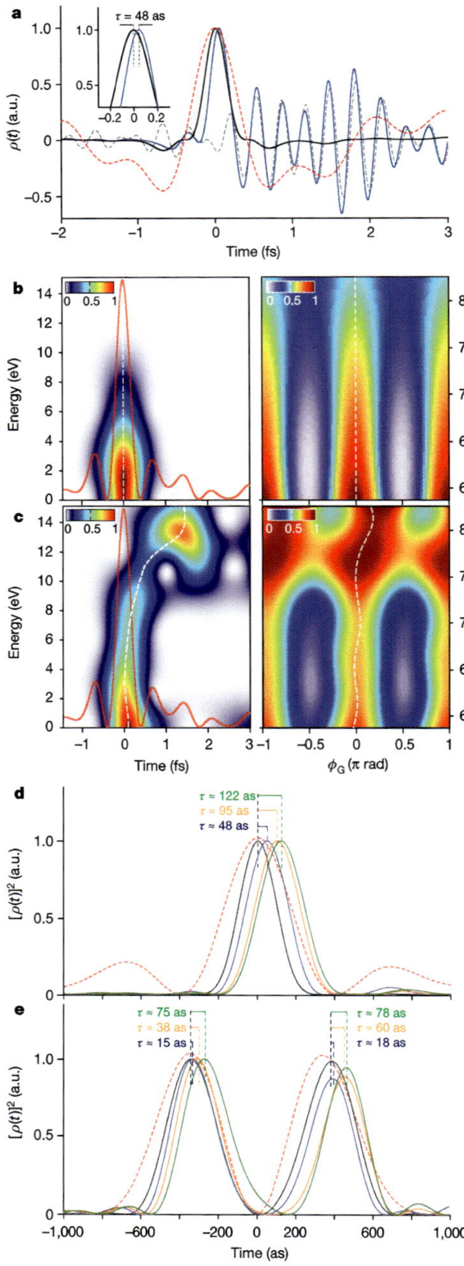

Figure 1 | Nonlinear response of bound electrons of Kr to an optical attosecond pulse. a, Nonlinear dipole moment $\rho(t)$ of Kr atoms subjected to an intense half-cycle optical field (dashed red line) calculated using the adiabatic ($\rho_{ad}(t)$, black line) and the TDSE ($\rho_{TDSE}(t)$, blue line) model. The dashed grey line represents the bound-electron response in the energy range (10–15 eV) that is calculated using the TDSE model. The inset shows a close-up of the peak at time $t = 0$, highlighting the delay τ between the predictions of the two models. **b, c,** Time–frequency analysis of the nonlinear dipoles $\rho_{ad}(t)$ (**b**) and $\rho_{TDSE}(t)$ (**c**) (left panels); simulated nonlinear dipole spectra (5.7–8.2 eV) as a function of the global phase ϕ_G of the optical attosecond pulse predicted by the adiabatic (**b**) and TDSE (**c**) Kr models (right panels). The red line represents the instantaneous intensity of the driving field. The dashed white lines represent the centre of mass along the x axis of each plot. The colour scale represents spectral intensity in arbitrary units (a.u.). **d, e,** Low-pass-filtered (0–8 eV) normalized nonlinear dipoles for three peak intensities of the driving field: 4×10^{13} W cm^{-2} (blue lines), 6×10^{13} W cm^{-2} (orange lines) and 8×10^{13} W cm^{-2} (green lines) for global phases $\phi_G \approx 0$ (**d**) and $\phi_G \approx \pi/2$ rad (**e**).

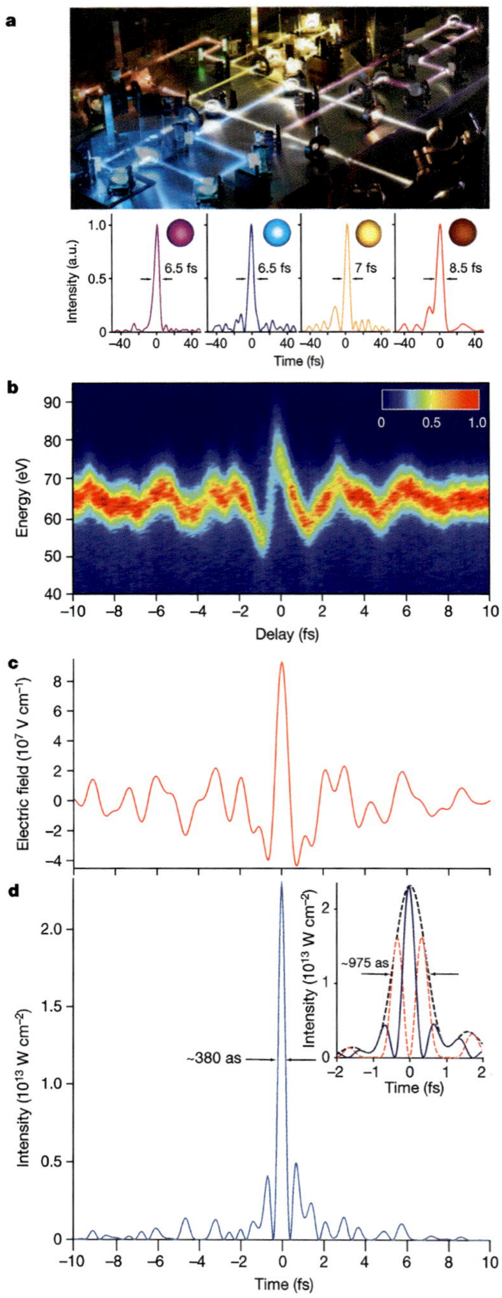

Figure 2 | Synthesis of an optical attosecond pulse. a, Photograph of an attosecond light-field synthesizer (top). The beams (artificially visualized) of ultra-wide-band pulses (about 1.1–4.6 eV) are divided by dichroic beam-splitters into four, almost equally wide, spectral bands. The pulses in each band are compressed so that they have durations of a few femtoseconds (the temporal intensity profiles are shown in the bottom panels), before they are spatiotemporally superimposed to yield a single beam/pulse at the exit of the apparatus. The insets in the bottom panels show representative beam profiles in the far field. **b,** Attosecond-streaking spectrogram of an optical attosecond pulse. The colour bar represents the yield of photoelectron counts in arbitrary units. **c, d,** Evaluated electric field (**c**) and instantaneous intensity profile (**d**; blue line) of the pulse. The intensity profile has a FWHM duration of approximately 380 as. The inset of **d** shows a close-up of this instantaneous intensity profile (blue line, $\phi_G \approx 0$) along with that of the same pulse, but with $\phi_G \approx \pi/2$ rad (dashed red line), and their common intensity envelope (dashed black line), which has a FWHM duration of approximately 975 as.

Phase-gating[21] methods have successfully been used for tracing ionized-electron dynamics. To determine whether these methods can be extended to trace bound-electron dynamics with optical attosecond pulses, we simulate nonlinear dipole spectra as a function of the global phase ϕ_G of the optical attosecond pulse using the adiabatic model ($\rho_{ad}(\omega, \phi_G)$, Fig. 1b, right panel) and the TDSE model ($\rho_{TDSE}(\omega, \phi_G)$, Fig. 1c, right panel). The instantaneity of the nonlinear response underlying the adiabatic model (Fig. 1b, left panel) results in uniform spectral modulations of the emitted radiation with ϕ_G (Fig. 1b, right panel, shown for the range 5.7–8.2 eV to ease comparison with experiments). Indeed, the noticeable deviation between the instantaneous and TDSE time-domain nonlinear dipoles (Fig. 1d, e), and the dependence of this deviation on ϕ_G, give rise to easily discernible features in the corresponding spectrogram (Fig. 1c, left panel). These features are the manifestation of the delayed nonlinear response and include the non-synchronous amplitude modulation of the emitted frequencies with global phase (highlighted by the white dashed line) and the non-monotonic variation of the amplitude of the spectral emission along the energy axis.

We describe the nonlinear response of bound electrons in Kr with a simple heuristic model that expresses the nonlinear dipole $\rho(t, \phi_G)$ as a sum of instantaneous and delayed nonlinearities:

$$\rho(t, \phi_G) = aE^3(t, \phi_G) + bE^5(t, \phi_G) + cE^5(t - \mathrm{d}t, \phi_G) \quad (1)$$

where a, b and c are coefficients, dt represents a delay and $E(t, \phi_G)$ is the electric field of the optical attosecond pulse for a global phase ϕ_G (see Methods). Provided $E(t, \phi_G)$ is known, equation (1) allows retrieval of the complete nonlinear dipole $\rho(t, \phi_G)$ from the numerical reconstruction of the corresponding ϕ_G spectrogram (Fig. 1c, right panel). This procedure underlines the suitability of the model for retrieving the nonlinear bound-electron response, and is instrumental for unveiling such dynamics from the experiments presented below.

To experimentally study the sub-femtosecond nonlinear response of bound electrons to light fields, we synthesized optical attosecond pulses. Two major advances of our synthesis method with respect to previously used methods[22,23] have enabled the confinement and manipulation of optical pulses to <1 fs. First, the development of a second-generation light-field synthesizer (Fig. 2a), which extends optical synthesis to the deep ultraviolet range and manipulates pulses with spectra extending over more than two optical octaves over the visible and neighbouring ranges (about 1.1–4.6 eV), was crucial. Second, and equally important, was the realization of a broadband and virtually dispersion-free spectral equalization of the synthesized pulses, to give pulses with twice the initial bandwidth (see Methods section 'Optical attosecond pulse synthesis'). These advances allow synthesis of optical attosecond pulses, which are highly energetic (about 50 μJ) and several

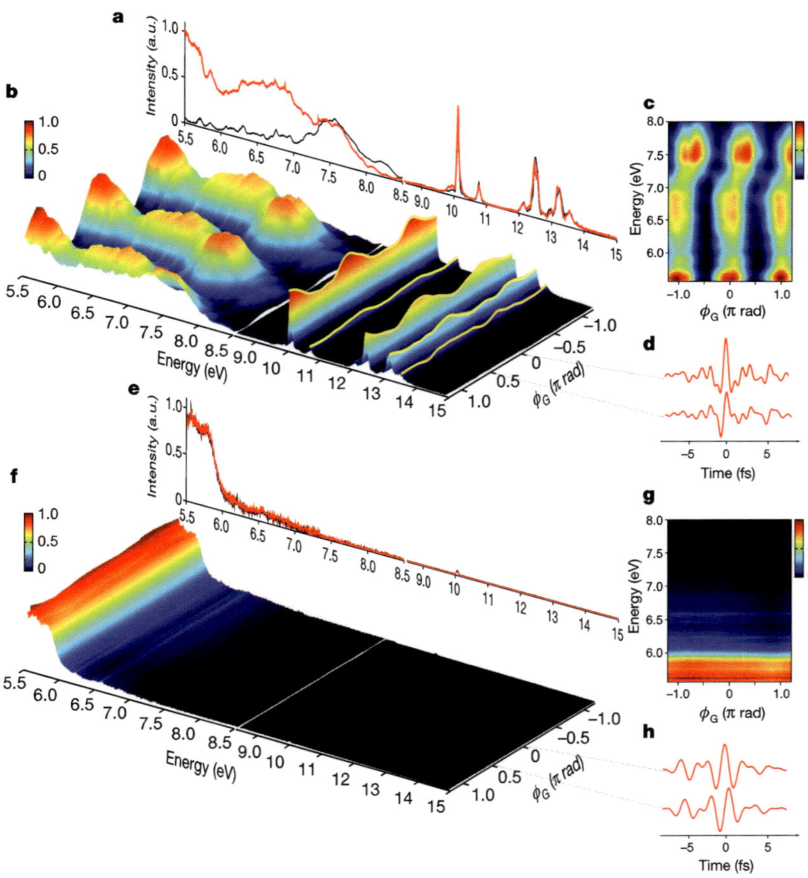

Figure 3 | Attosecond control of bound electrons in Kr. a, e, VUV spectra generated in neutral Kr atoms driven by an optical attosecond pulse (**a**) and a single-cycle pulse (**e**) of the same peak intensity (about 5×10^{13} W cm^{-2}), at global phase settings of $\phi_G \approx 0$ (red line) and $\phi_G \approx \pi/2$ rad (black line). **b, f,** Spectrograms composed of 25 VUV spectra (about 5.5–15 eV) recorded as a function of ϕ_G for an optical attosecond pulse (**b**) and a single-cycle pulse (**f**). **c, g,** Sections (about 5.5–8 eV) of the global-phase spectrograms in **b** and **f**, respectively. **d, h,** Representative driving waveforms at $\phi_G \approx 0$ (**d**) and $\phi_G \approx \pi/2$ rad (**h**). The colour scales in **b, c, f** and **g** indicate intensity (in arbitrary units).

orders of magnitude more intense (about 50 GW) than those obtainable with current extreme-ultraviolet (EUV) technology[24].

Figure 2b shows a representative streaking spectrogram of an optical attosecond pulse. The field waveform (Fig. 2c) retrieved from the data reveals a nearly isolated, half-cycle peak (centroid wavelength of approximately 530 nm), which has an intensity (Fig. 2d) approximately five times that of the adjacent peaks and carries more than about 50% of the energy of the waveform. The instantaneous intensity profile of the dominant half-cycle in Fig. 2d has a full-width at half-maximum (FWHM) duration of approximately 380 as, whereas an evaluation based on the conventional definition of the intensity envelope (Fig. 2d inset, dashed black line) yields a FWHM of approximately 975 as.

In our experiments (see Extended Data Fig. 1b), we probe the nonlinear polarization response of Kr atoms to optical attosecond pulses by recording the generated vacuum-ultraviolet (VUV) spectra. As shown in Fig. 3a, the spectra are quasi-continuous in the 5.5–10 eV range and exhibit discrete peaks coinciding with Kr excited-state energies in the 10–15 eV range. Variation of the global phase ϕ_G of the pulses yields a pronounced broadband modulation of the VUV emission (Fig. 3b), verifying our TDSE model predictions (Fig. 1c, right panel). In contrast, a pulse synthesized to consist of approximately a single optical cycle of commensurable intensity fails to generate broadband nonlinear excitation or its ϕ_G control (Fig. 3e–g). These results clearly demonstrate that in the absence of the extreme nonlinearities that arise under conditions of ionization, optical attosecond pulses are the key to manipulating bound electrons on a sub-femtosecond timescale.

The non-uniform modulation of the amplitude of the emitted dipole spectra (Fig. 3c) with ϕ_G and the uneven distribution of spectral amplitudes along the energy axis differ from the predictions of the adiabatic model for the same driving waveform (Fig. 1b, right panel), but resemble key features of the spectrograms constructed using TDSE simulations (Fig. 1c, right panel). These features offer experimental evidence of the non-instantaneous character of the response of the bound electrons of Kr in our experiments.

To apply the methodology presented above to retrieve the dynamics of the bound-electronic response, we supplemented our measurements with attosecond streaking of the driving field $E(\tau, \phi_G)$. By studying the dependence of the VUV emission as a function of the intensity of the optical driver, we verified that bound-electronic nonlinearities dominate the response (see Methods). Figure 4 shows the measured (left panels) and the corresponding reconstructed (right panels) spectrograms recorded at gradually increasing optical-field intensities: about $5 \times 10^{13}\,\mathrm{W\,cm^{-2}}$ (Fig. 4a); about $7 \times 10^{13}\,\mathrm{W\,cm^{-2}}$ (Fig. 4b); and about $9 \times 10^{13}\,\mathrm{W\,cm^{-2}}$ (Fig. 4c). The retrieved nonlinear dipoles (Fig. 4d) exhibit delays with respect to the instantaneous dipole (black line) and driving field $E(\tau, 0)$ (red dashed line) of $\tau \approx 45 \pm 4$ as (Fig. 4a), $\tau \approx 70 \pm 5$ as (Fig. 4b) and $\tau \approx 115 \pm 7$ as (Fig. 4c). The data in Fig. 4d verify the attosecond control of the bound-electronic response induced by the field, and emphasize the non-trivial character of the induced dynamics, which is manifested by a non-uniform delay (chirp). The latter is highlighted by the delay values for consecutive half-field-cycles shown in Fig. 4e. These findings, which are in excellent agreement with the predictions of the TDSE simulations performed for identical excitation fields, provide conclusive evidence for the feasibility of tracing and control of the nonlinear response of bound electrons on a sub-femtosecond timescale with high fidelity.

The findings of this study are compatible with an intuitive physical picture of the delayed nonlinear response, which we can now extend from the femtosecond molecular scale[16] to attosecond electron dynamics. The optical-field-induced superposition of the ground state and excited states—which are typically more polarizable because the electron resides further from the atomic core—gives rise to an increase in the total nonlinear polarizability of the electron cloud during its interaction with the field. But this superposition requires a finite time to evolve, resulting in a delay in the nonlinear response of the system.

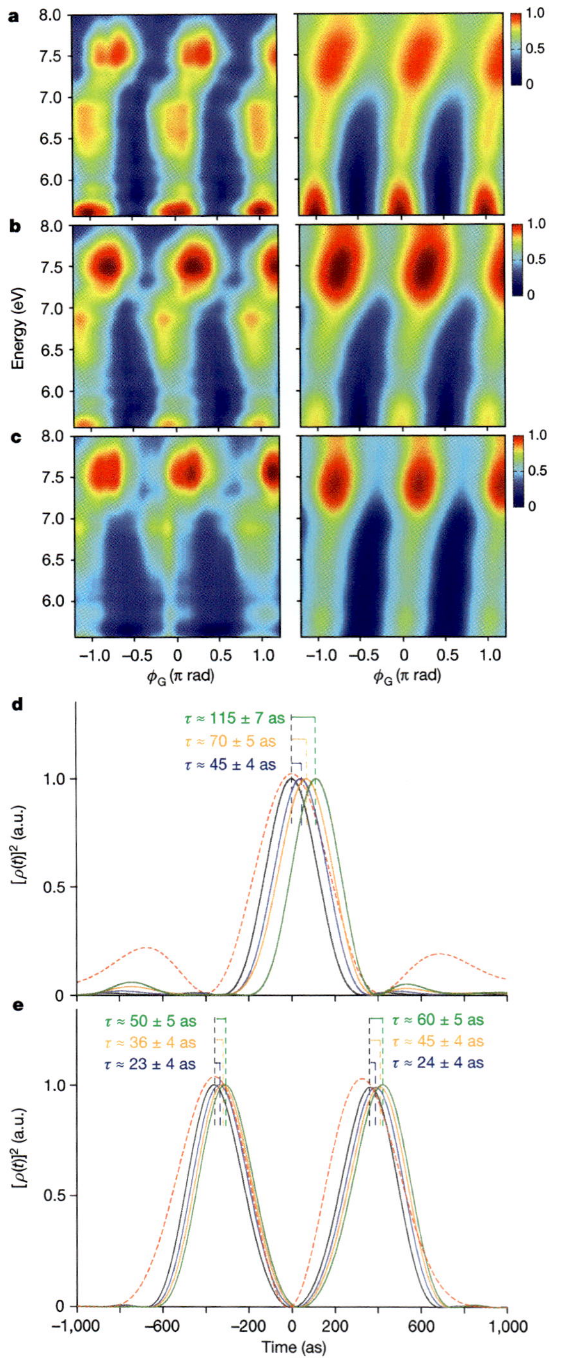

Figure 4 | Sub-femtosecond, delayed nonlinear response of bound electrons in Kr. a–c, Global-phase spectrograms of the optical attosecond pulse recorded at peak-intensity settings of about $5 \times 10^{13}\,\mathrm{W\,cm^{-2}}$ (**a**), $7 \times 10^{13}\,\mathrm{W\,cm^{-2}}$ (**b**) and $9 \times 10^{13}\,\mathrm{W\,cm^{-2}}$ (**c**) are shown in the left panels. Corresponding reconstructed spectrograms based on equation (1) are shown in the right panels. The colour bar represents spectral intensity in arbitrary units. **d, e**, Low-pass-filtered (0–8 eV) nonlinear dipoles obtained using the intensity settings in **a–c** (dark blue, orange, green lines, respectively) are shown along with the instantaneous response (black line) simulated for global-phase settings of the optical attosecond pulse of $\phi_G \approx 0$ (**d**) and $\phi_G \approx \pi/2$ rad (**e**). Standard errors of the mean for the delays (τ) indicated are evaluated from the reconstruction of three data sets recorded under identical experimental conditions. The dashed red lines are the normalized instantaneous intensities of the input electric fields.

Because both the amplitude of this superposition and the Stark shift[22] of the induced electronic dipoles scale with field strength, the delay in the nonlinear response will also increase, in agreement with our theoretical and experimental studies.

It should be possible to extend the approach demonstrated here to the condensed phase. This will allow the electronic response in the bulk of complex systems subjected to strong fields to the probed[14,25,26], and will enable the possibility to trace electronic dephasing in such systems. By providing a probe, which is as short as the lifetime and dephasing times of core-hole electronic states in solids generated under attosecond X-ray excitation[15], our tools could provide new ways to explore the nonlinear properties of X-ray excited solids. These novel spectroscopic capabilities may turn out to be instrumental—among others—to exploring the ultimate frontiers of electromagnetic signal processing in optical and X-ray regimes.

Online Content Methods, along with any additional Extended Data display items and Source Data, are available in the online version of the paper; references unique to these sections appear only in the online paper.

Received 10 June; accepted 3 December 2015.

1. Sheik-Bahae, M., Hutchings, D. C., Hagan, D. J. & Van Stryland, E. W. Dispersion of bound electron nonlinear refraction in solids. *IEEE J. Quantum Electron.* **27,** 1296–1309 (1991).
2. Boyd, R. W. *Nonlinear Optics* Ch. 1 (Academic Press, 1992).
3. Hentschel, M. *et al.* Attosecond metrology. *Nature* **414,** 509–513 (2001).
4. Niikura, H. *et al.* Sub-laser-cycle electron pulses for probing molecular dynamics. *Nature* **417,** 917–922 (2002).
5. Corkum, P. B. & Krausz, F. Attosecond science. *Nature Phys.* **3,** 381–387 (2007).
6. Schultze, M. *et al.* Delay in photoemission. *Science* **328,** 1658–1662 (2010).
7. Cavalieri, A. L. *et al.* Attosecond spectroscopy in condensed matter. *Nature* **449,** 1029–1032 (2007).
8. Goulielmakis, E. *et al.* Single-cycle nonlinear optics. *Science* **320,** 1614–1617 (2008).
9. Uiberacker, M. *et al.* Attosecond real-time observation of electron tunnelling in atoms. *Nature* **446,** 627–632 (2007).
10. Eckle, P. *et al.* Attosecond ionization and tunneling delay time measurements in helium. *Science* **322,** 1525–1529 (2008).
11. Shafir, D. *et al.* Resolving the time when an electron exits a tunnelling barrier. *Nature* **485,** 343–346 (2012).
12. Ivanov, M. & Smirnova, O. Opportunities for sub-laser-cycle spectroscopy in condensed phase. *Chem. Phys.* **414,** 3–9 (2013).
13. Krüger, M., Schenk, M. & Hommelhoff, P. Attosecond control of electrons emitted from a nanoscale metal tip. *Nature* **475,** 78–81 (2011).
14. Schiffrin, A. *et al.* Optical-field-induced current in dielectrics. *Nature* **493,** 70–74 (2013).
15. Schultze, M. *et al.* Controlling dielectrics with the electric field of light. *Nature* **493,** 75–78 (2013).
16. Weiner, A. M. *Ultrafast Optics* Ch. 6 (John Wiley & Sons, 2009).
17. Zewail, A. H. Femtochemistry: atomic-scale dynamics of the chemical bond. *J. Phys. Chem. A* **104,** 5660–5694 (2000).
18. Goulielmakis, E. *et al.* Real-time observation of valence electron motion. *Nature* **466,** 739–743 (2010).
19. Smirnova, O. *et al.* High harmonic interferometry of multi-electron dynamics in molecules. *Nature* **460,** 972–977 (2009).
20. Zewail, A. H. *Femtochemistry: Ultrafast Dynamics of the Chemical Bond* Vol. 1 (World Scientific, 1994).
21. Dudovich, N. *et al.* Measuring and controlling the birth of attosecond XUV pulses. *Nature Phys.* **2,** 781–786 (2006).
22. Wirth, A. *et al.* Synthesized light transients. *Science* **334,** 195–200 (2011).
23. Hassan, M. Th. *et al.* Invited Article: attosecond photonics: synthesis and control of light transients. *Rev. Sci. Instrum.* **83,** 111301 (2012).
24. Sansone, G., Poletto, L. & Nisoli, M. High-energy attosecond light sources. *Nature Photon.* **5,** 655–663 (2011).
25. Vampa, G. *et al.* Linking high harmonics from gases and solids. *Nature* **522,** 462–464 (2015).
26. Hohenleutner, M. *et al.* Real-time observation of interfering crystal electrons in high-harmonic generation. *Nature* **523,** 572–575 (2015).

Acknowledgements This work was supported by European Research Council grant (Attoelectronics-258501), the Deutsche Forschungsgemeinschaft Cluster of Excellence: Munich Centre for Advanced Photonics (http://www.munich-photonics.de), the Max Planck Society and the European Research Training Networks ATTOFEL and MEDEA, the Russian Foundation for Basic Research (projects numbers 13-02-01465, 13-02-92115 and 13-04-40335) and the Welch Foundation (grant number A-1801).

Author Contributions M.Th.H., T.T.L., A.M. and M.G. conducted the experiments; O.R. and V.P. developed the dispersive optics for the experiments; E.G. planned the experiments and supervised the project; M.Th.H., T.T.L., P.Z., A.M. and N.K. conducted the simulations; and M.Th.H., T.T.L., A.M., A.M.Z., F.K. and E.G. interpreted the data and contributed to the preparation of the manuscript.

Author Information Reprints and permissions information is available at www.nature.com/reprints. The authors declare no competing financial interests. Readers are welcome to comment on the online version of the paper. Correspondence and requests for materials should be addressed to E.G. (elgo@mpq.mpg.de).

METHODS

Optical attosecond pulse synthesis. Light waveforms whose spectra extend over more than two optical octaves in the visible and adjacent spectral range (about 1.1–4.6 eV) (Extended Data Fig. 1a) are generated by the nonlinear broadening of laser pulses (approximately 22 fs, 1 mJ, 790 nm) through a hollow-core fibre (HCF) filled with Ne gas (about 2.3 bar). The energy of the generated supercontinuum at the exit of the HCF is about 550 μJ. The spectra of these pulses are divided by dichroic beam-splitters into four almost equally wide spectral bands centred in the near infrared (NIR; about 1.1–1.75 eV), visible (vis; about 1.75–2.5 eV), visible-ultraviolet (vis-UV; about 2.5–3.5 eV) and deep ultraviolet (DUV; about 3.5–4.6 eV). The pulses in these bands are individually compressed by dispersive mirrors to durations of a few femtoseconds before they are spatially and temporally superimposed to yield a single beam/pulse at the exit of the apparatus. The pulses are temporally characterized by a transient-grating frequency-resolved optical gating (TG-FROG) apparatus. The durations T of the pulses in different channels of the synthesizer were measured to be $T_{\rm NIR} \approx 8.5$ fs, $T_{\rm vis} \approx 7$ fs, $T_{\rm vis\text{-}UV} \approx 6.5$ fs and $T_{\rm DUV} \approx 6.5$ fs (see inset to Fig. 2a). The synthesizer apparatus transmits about 82% of the energy of the incoming supercontinuum. As a result, the pulse energy at the exit of the apparatus is about 320 μJ, and is distributed among the four channels as $\mathrm{Ch}_{\rm NIR} \approx 255$ μJ, $\mathrm{Ch}_{\rm vis} \approx 45$ μJ, $\mathrm{Ch}_{\rm vis\text{-}UV} \approx 15$ μJ and $\mathrm{Ch}_{\rm DUV} \approx 4$ μJ, where Ch_i denotes the energy of channel i.

For the synthesis of optical attosecond pulses, both precise control of the relative delay between the constituent pulses in the synthesizer as well as an intensity control over the spectral channels is required. To this end, we followed a new approach, which effectively enables the passive spectral intensity control and facilitates the attosecond streaking characterization of the generated optical waveforms in the same set-up. The EUV attosecond probe is generated first (Extended Data Fig. 1b) by focusing the light transients from the synthesizer into a quasi-static Ne gas cell. The EUV radiation, which emerges collinearly to the driver waveform, is transmitted through a thin, round Zr foil, while the optical pulse, which is transmitted around the geometrical margins of this foil, forms an annular beam. A double-mirror module consisting of a concave, multilayer, coated inner mirror and a metal–dielectric–metal (MDM)-coated concave annular sector (outer mirror) (Extended Data Fig. 2) of the same focal length ($f = 12.5$ cm), focuses the light transients and the EUV attosecond probe into a second Ne gas nozzle placed near the entrance of a time-of-flight (TOF) spectrometer (Extended Data Fig. 1b). One of the essential characteristics of the MDM is that the imposed spectral control results in negligible phase distortions over the whole spectral range of the supercontinuum pulse. This was experimentally verified by FROG measurements of the pulses in the constituent channels upon reflection off the MDM unit. At the same time, because the EUV probe is generated before the spectral intensity control of the optical transient, it overcomes a fundamental limitation of a half-cycle field: the efficient generation of intense EUV-probe pulses. This is because a half-cycle field does not involve at least two intense field crests, which are required for ionization and subsequent acceleration of electrons[27] leading to high harmonic generation.

Single-cycle optical pulses are generated by physically suppressing a part (>3 eV) of the high-frequency spectrum of the optical attosecond pulse.

Generation of VUV spectra. We focus optical attosecond pulses into a quasi-static cell, which replaces the Ne gas jet used for performing attosecond-streaking characterization of their fields, filled with Kr atoms at a moderate pressure (about 80 mbar) (Extended Data Fig. 1b). We probe the nonlinear polarization of the system by recording VUV spectra that emerge collinearly with the driving optical field using a spectrometer placed downstream from the cell. The spectra are sampled at energies higher than about 5 eV, that is, beyond the constituent spectrum of the optical attosecond pulses, and extend no higher than the ionization threshold of neutral Kr ($I_p \approx 14$ eV).

Ultimate limit of carrier-envelope phase decomposition. The concept of the carrier-envelope phase $\phi_{\rm CE}$ (CEP) is most readily understood in the time domain[28]. The carrier-envelope phase is the interval $\Delta t_{\rm peak}$ between the maximum of the envelope and the maximum of the instantaneous field, translated in phase at the centroid frequency $\omega_{\rm L}$ of the spectrum: $\phi_{\rm CE} = \omega_{\rm L} \times \Delta t_{\rm peak}$. Any electric field can be decomposed, according to Hilbert's transform as

$$E(t) = A(t) cos(w_{\rm L}t + \phi_{\rm G})$$

in which $A(t)$ is the envelope (the modulus of the analytical field) and $\phi_{\rm G}$ is the global (or absolute) phase.

The maximum of the field is determined by solving

$$\dot{E}(t) = 0 \Leftrightarrow \frac{\dot{A}(t)}{\omega_{\rm L} A(t)} = \tan(\omega_{\rm L}t + \phi_{\rm G}) \qquad (2)$$

On the basis of this equation and the estimation $\frac{\dot{A}(t)}{\omega_{\rm L} A(t)} \approx \frac{\Delta\omega}{\omega_{\rm L}}$, we distinguish two different regimes:

(1) For pulses with durations longer than one cycle, $\Delta\omega / \omega_{\rm L} = 1$ and the solution of equation (2) yields $\phi_{\rm G} = \omega_{\rm L} \times \Delta t_{\rm peak} = \phi_{\rm CE}$, which confirms the equivalence between the global ($\phi_{\rm G}$) and carrier-envelope ($\phi_{\rm CE}$) phases.

(2) For sub-cycle pulses, $\Delta\omega/\omega_{\rm L} > 1$ and so the solution of equation (2) is not trivial because, in a fraction of a cycle, the envelope varies substantially. Consequently, $\omega_{\rm L} \times \Delta t_{\rm peak} < \phi_{\rm G}$ and the global and carrier-envelope phases are no longer equivalent.

Extended Data Fig. 3a presents the case of a half-cycle pulse—an optical attosecond pulse. The sinusoidal waveform (red) reaches its maximum at 0.2 periods instead of 0.25, which is expected for a wave with a phase of π/2 rad. Extended Data Fig. 3b shows the CEP determined using the method of the field maximum depending on the pulse duration, for three settings of the global phase. This comparison reveals a considerable difference between $\phi_{\rm CE}$ and $\phi_{\rm G}$ in the short pulse regime. In view of the above justification, it is clear that the CEP is an accurate description of the global phase only for pulses longer than one cycle.

Time-dependent Schrödinger equation and adiabatic models. To theoretically study the nonlinear dipole dynamics in Kr atoms exposed to intense optical attosecond pulses, we used two models. In the first, we solve the three-dimensional time-dependent Schrödinger equation (TDSE) within the single-active-electron approximation. To this end, we used a central potential for Kr, which was calculated using optimized effective-potential methods[29]. In the second, and in order to describe instantaneous response, we assumed an adiabatic model based on a two-level system[12,30], in which the dipole moment, in the quasi-static approximation, can be expressed as:

$$\rho(t) = \frac{2d^2 E(t)}{\sqrt{4d^2[E(t)]^2 + \omega_0^2}}$$

where d is chosen to match the nonlinear polarizability of Kr (refs 31, 32), ω_0 is the excitation energy $\hbar\omega_0 \approx 10$ eV and $E(t)$ denotes the electric field of the optical attosecond pulse. To access the nonlinear component of the induced electronic dipole moment at a given intensity of the driving field in both models, we perform a second calculation at a much lower (about six orders of magnitude) intensity. As a next step, we subtract the virtually linear dipole calculated at the lower intensity from the original one, after multiplying it by the corresponding ratio between the two intensities. The calculated global-phase spectrograms (spectral emission as a function of the global phase) using the adiabatic and TDSE models are shown in Extended Data Fig. 4a, b. In accordance with the discussion in the main text, the adiabatic model (Extended Data Fig. 4a) predicts uniform modulations of the spectral amplitude of the emitted spectral components as a function of the global phase $\phi_{\rm G}$. In contrast, the spectrogram calculated using the TDSE model embodies the signatures of the delayed electronic response (Extended Data Fig. 4b) in the form of asynchronous amplitude modulations between different frequencies/energies of the emitted dipole. These features are present in the entire emitted spectrum, not only close to the resonant area (10–14 eV). We show that these effects can be used to extract the dynamics of the nonlinear response by reconstructing the global-phase ($\phi_{\rm G}$) spectrograms recorded in our experiments.

Extended Data Fig. 4c shows a global-phase spectrogram simulated for a single-cycle pulse using the TDSE Kr model. In this regime of single-cycle pulses, the global-phase spectrogram does not show discernible variation over $\phi_{\rm G}$, which is experimentally verified (see Fig. 3e–g).

Ionization-free strong field polarization of bound electrons. An essential innovation introduced by using optical attosecond pulses is their unique capability to drive nonlinear dynamics in quantum systems without inducing a substantial degree of ionization or excitation, that is, without greatly altering the original system. In experiments where the polarization of the system is to be probed, both excessive ionization and excitations markedly modify the system, resulting in a considerable degree of 'contamination' in the emitted signal from the new atomic entities; such contamination is challenging to resolve both experimentally and theoretically.

This capability is unique to isolated sub-cycle pulse structures and is not observed for trains of sub-cycle field modulation[33] because the nanosecond-long exposure of atoms to such fields yields a substantial degree of ionization, which is actually the means to trace the waveform in these experiments[34]. The ionization and excitation probability of Kr atoms calculated by the solution of the TDSE and the waveform used in our experiments for a range of intensities (summarized in Extended Data Table 1) verify this conjecture. Experimentally, we also verified this fact by using the previously established and highly sensitive technique of attosecond transient absorption spectroscopy[18].

RESEARCH LETTER

The interaction of optical attosecond pulses with matter, even at excessive intensities such as those used in our experiments and simulations, can be mostly considered as a scattering process. The system and the pulse virtually do not exchange energy; rather, the pulse probes the system via nonlinear scattering and the emission of coherent radiation.

Nonlinearities in the interaction of Kr with intense optical attosecond pulses. Identifying the dominant nonlinearities in the interaction between the optical attosecond pulses and the Kr atoms, and understanding the dependence of these nonlinearities on the variation of the global phase (ϕ_G) of the optical attosecond pulse—the key control point of the interaction in experiments with such waveforms—is essential for the development of intuitive models that can describe the nonlinear dynamics. As we show below, it allows the development of a robust methodology that permits the reconstruction of the dynamics of the response from global-phase spectrograms.

To this end, we theoretically (using TDSE simulations in Kr) studied the intensity dependence of the yield of the spectral emission (averaged over the range 0–8 eV) under optical attosecond pulses for the range of intensity settings used in our experiments. The corresponding nonlinearities can be evaluated from the slope of the linear fit of the data in the log–log diagram of Extended Data Fig. 5. A linear fitting over the entire range of intensities studied (2×10^{13}–$10\times10^{13}\,\mathrm{W\,cm^{-2}}$) reveals a slope of about 4, and suggests the dominance and coexistence of the two most essential nonlinearities in centrosymmetric systems: the third- (E^3) and the fifth- (E^5) order nonlinearities—broadly know as bound-electronic nonlinearities.

Simple model of the field-driven bound-electron nonlinear response of Kr. The study presented in the previous paragraph highlights the dominance of bound-electron nonlinearities in the response and the low or negligible sensitivity of the nonlinearity to the global phase of our pulses.

To account for the non-instantaneous response revealed in our TDSE simulations, and inspired by previous approaches in ultrafast spectroscopy[35], in which the nonlinear response is decomposed into instantaneous and delayed components, we describe the nonlinear dipole moment as a sum over instantaneous third- and fifth-order nonlinearities, as well as the fifth-order delayed nonlinearity according to equation (1):

$$\rho(t,\phi_G) = aE^3(t,\phi_G) + bE^5(t,\phi_G) + cE^5(t-\mathrm{d}t,\phi_G)$$

Here, a, b and c are coefficients, $\mathrm{d}t$ represents a delay of the fifth-order response and $E(t,\phi_G)$ is the electric field of the optical attosecond pulse for a global phase ϕ_G.

One would generally expect delayed terms to be considered for all nonlinearities involved (including the third-order); the energy diagram of Extended Data Fig. 6 offers an intuitive explanation of our choice to limit the delayed terms to only fifth-order nonlinearities. Indeed, delayed response in the range 0–9 eV will primarily involve virtual transitions, which can coherently couple at least two electronic states of the system (ground and excited states or combinations of excited states). The diagram demonstrates this assuming virtual transition compatible with the energy spectrum of our optical attosecond pulse (1.1–4.6 eV). Such transitions can only occur within the fifth-order response (higher-order Kerr effects). The diagram also highlights—through the multiphoton picture—that at this extreme limit of pulse duration and corresponding bandwidth, non-resonant and resonant response are virtually inseparable. As a result, the coherent dynamics induced between two or more states of the system will be manifested at each nonlinearly emitted spectral component in the process.

Extended Data Fig. 7a–c shows representative, synthetic, global-phase spectrograms for three values of the parameter $\mathrm{d}t$ ($\mathrm{d}t=0$ (instantaneous response), $\mathrm{d}t=20$ as and $\mathrm{d}t=30$ as), generated by equation (1) in the spectral range of our experiments (see Fig. 4). The synthetic spectrograms highlight the capability of the model to capture key features of the experimental spectrograms (for example, Figs 1 or 4), such as the profoundly asynchronous modulation of the emission in the range 7–8 eV and the weakening of the amplitude in the range 6–6.5 eV—both are the result of dynamic nonlinear interference between delayed and instantaneous terms in equation (1).

To further verify the validity of our model, we used the TDSE simulations in Kr as a basis to explore how a dipole described by the above equations could represent the (below resonances) nonlinear response of the system.

Extended Data Fig. 7d shows the TDSE simulated dipoles (black line) and their fitting with equation (1). We investigated both the capability of the model to fit only a fraction (spectrally filtered from 5.5–8 eV; Extended Data Fig. 7d) of the dipole as well the entire dipole (0–8 eV) below resonances (Extended Data Fig. 7e). A single set of the parameters a, b, c and $\mathrm{d}t$ in equation (1), and a given peak intensity of the pulse, can precisely reproduce the nonlinear dipole (black lines) for all settings of the global phase. Representative examples for two (extreme) settings of the global phase $\phi_G\approx0$ (left panels of Extended Data Fig. 7d and e) and $\phi_G\approx\pi/2$ rad (right panels of Extended Data Fig. 7d and e) are shown. Furthermore, the parameters a, b, c and $\mathrm{d}t$ extracted from fitting part of the nonlinear spectrum (5.5–8 eV) (Extended Data Fig. 7d) are identical to those required to fit the entire range (0–8 eV) (Extended Data Fig. 7e). As a conclusion, even a limited fraction of the nonlinear spectrum contains information about the interfering terms in equation (1) The model works well for a wide range of peak intensities (not shown); therefore, we conclude that it is adequate for reconstructing the response.

Nonlinear dipole reconstructions. The most critical test of the capability of our model to reconstruct experimental data and to trace the time-domain nonlinear dipole dynamics, is to perform a numerical experiment. To do so, we create a theoretical spectrogram using the TDSE model (where the nonlinear dipole is known *a priori*) and attempt to reconstruct this nonlinear dipole using (i) equation (1), (ii) the spectrogram constructed by the TDSE simulation, and (iii) the *a priori* measured driver field waveform.

For both experimental and synthetic data, we used a quickly converging nonlinear algorithm to perform the reconstructions. It is based on a commercial, highly optimized numerical routine that uses the 'trust-region' method[36], which is usually used for constrained problems. The root-mean-square deviation (r.m.s.d.) was determined, and defined as the main parameter to optimize our reconstruction via its minimization. The r.m.s.d. is calculated as:

$$\mathrm{r.m.s.d.} = \sqrt{\frac{\sum_{i=1}^{n_{\phi_G}}\int_{\omega_1}^{\omega_2}(\hat{X}_i - X_i)^2\,\mathrm{d}\omega}{n_{\phi_G}\times n_\omega}}$$

Where $\hat{X}$ and X are the computed and original (measured) spectra, respectively, at certain global-phase settings, $\omega_1=5.5$ eV, $\omega_2=8$ eV, n_{ϕ_G} is the number of global-phase settings of the measured VUV spectrogram, and n_ω is the number of frequencies involved (from ω_1 to ω_2).

The results of this study are summarized in (Extended Data Fig. 8a, b), in which the original and the reconstructed spectrograms are shown. The reconstruction parameters in equation (1) are $a=0.180$, $b=0.096$, $c=-0.177$ and $\mathrm{d}t=67$ as. Extended Data Figure 8c, d compares the reconstructed (red line) and the *a priori* known nonlinear dipole (black line) for two different settings of the global phase. Blue lines show the instantaneous response for the same waveforms for comparison.

Intensity dependent nonlinear delay. We study the delayed nonlinear bound-electronic response over a wide range of intensities ($(2–8)\times10^{13}\,\mathrm{W\,cm^{-2}}$). The evaluated delays between the instantaneous and the TDSE-simulated dipoles as function of the driver field intensity are shown in Extended Data Fig. 9a; the delays between the instantaneous dipoles and those reconstructed from our measured spectrograms (see Fig. 4) are shown in Extended Data Fig. 9b.

27. Corkum, P. B. Plasma perspective on strong field multiphoton ionization. *Phys. Rev. Lett.* **71,** 1994–1997 (1993).
28. Krausz, F. & Ivanov, M. Attosecond physics. *Rev. Mod. Phys.* **81,** 163–234 (2009).
29. Muller, H. G. An efficient propagation scheme for the time-dependent Schrödinger equation in the velocity gauge. *Laser Phys.* **9,** 138–148 (1999).
30. Gauthey, F. I., Garraway, B. M. & Knight, P. L. High harmonic generation and periodic level crossings. *Phys. Rev. A* **56,** 3093–3096 (1997).
31. Lundeen, T., Hou, S. Y. & Nibler, J. W. Nonresonant third order susceptibilities for various gases. *J. Chem. Phys.* **79,** 6301–6305 (1983).
32. Shelton, D. P. Nonlinear-optical susceptibilities of gases measured at 1064 and 1319 nm. *Phys. Rev. A* **42,** 2578–2592 (1990).
33. Harris, S. E. & Sokolov, A. V. Subfemtosecond pulse generation by molecular modulation. *Phys. Rev. Lett.* **81,** 2894–2897 (1998).
34. Baker, S., Walmsley, I. A., Tisch, J. W. G. & Marangos, J. P. Femtosecond to attosecond light pulses from a molecular modulator. *Nature Photon.* **5,** 664–671 (2011).
35. Blow, K. J. & Wood, D. Theoretical description of transient stimulated Raman scattering in optical fibers. *IEEE J. Quantum Electron.* **25,** 2665–2673 (1989).
36. Coleman, T. F. & Li, Y. Y. An interior trust region approach for nonlinear minimization subject to bounds. *SIAM J. Optim.* **6,** 418–445 (1996).

LETTER RESEARCH

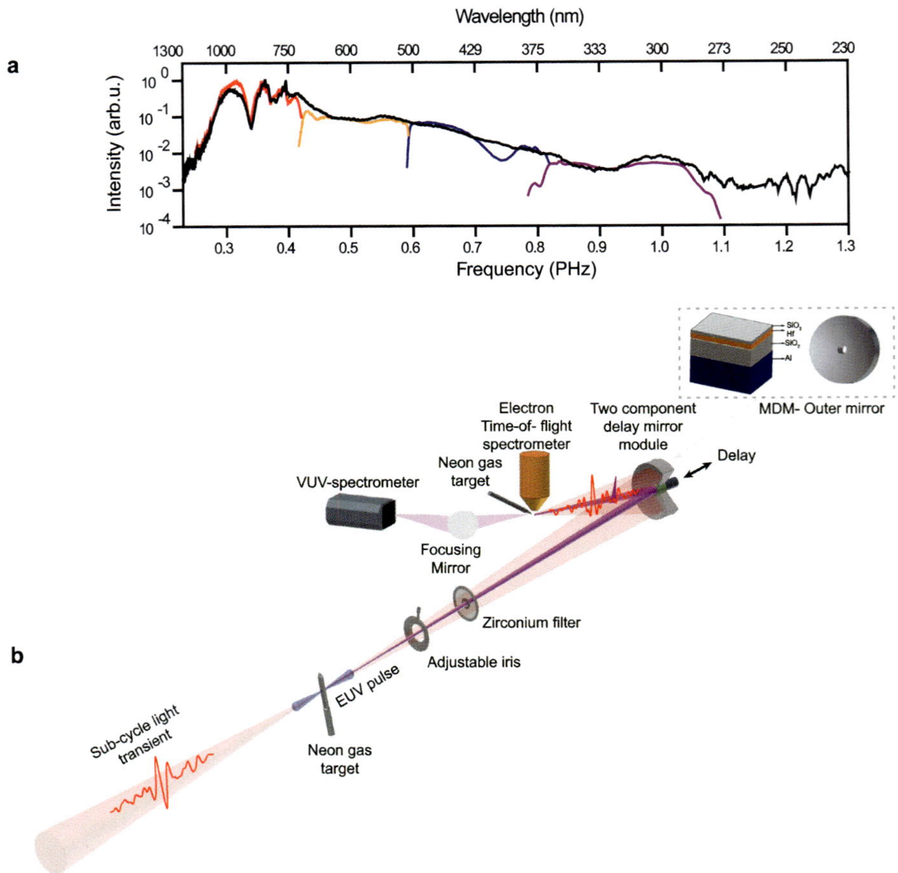

Extended Data Figure 1 | Supercontinuum spectra and experimental set-up. a, The supercontinuum spectra spans more than two octaves (1.1–4.6 eV), and is shown by the black line; the spectra of the individual channels Ch_{NIR}, Ch_{vis}, $Ch_{vis\text{-}UV}$ and Ch_{DUV} are shown by the red, orange, blue and violet lines, respectively. **b**, Schematic illustration of the experiment set-up.

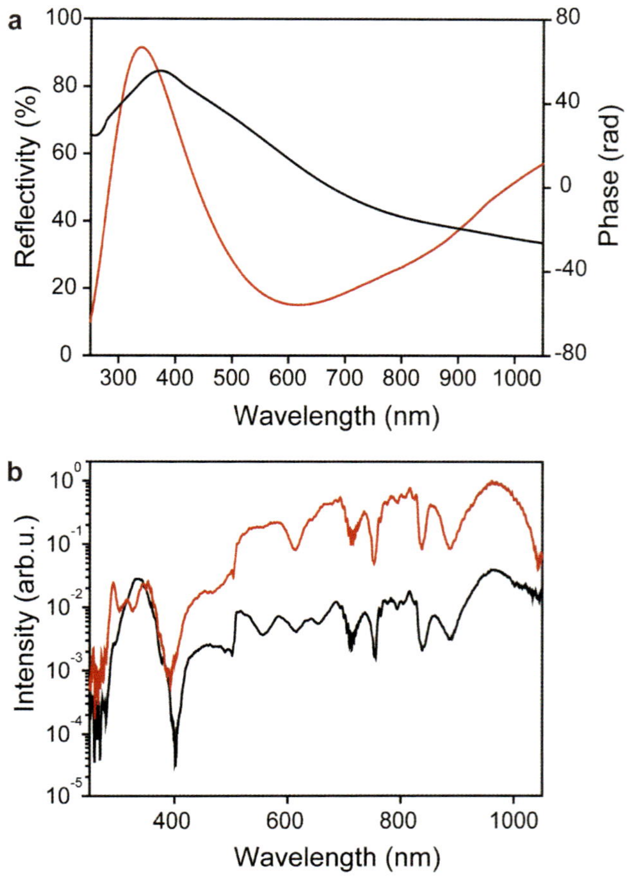

Extended Data Figure 2 | Spectral equalization by a metal–dielectric–metal (MDM) coating. a, The designed reflectivity (red line) and spectral phase (black) of the MDM coating applied on the outer mirror of the double mirror module shown in Extended Data Fig. 1. **b,** Multi-octave spectrum at the exit of the synthesizer (red line) and upon reflection off the MDM mirror (black line).

LETTER RESEARCH

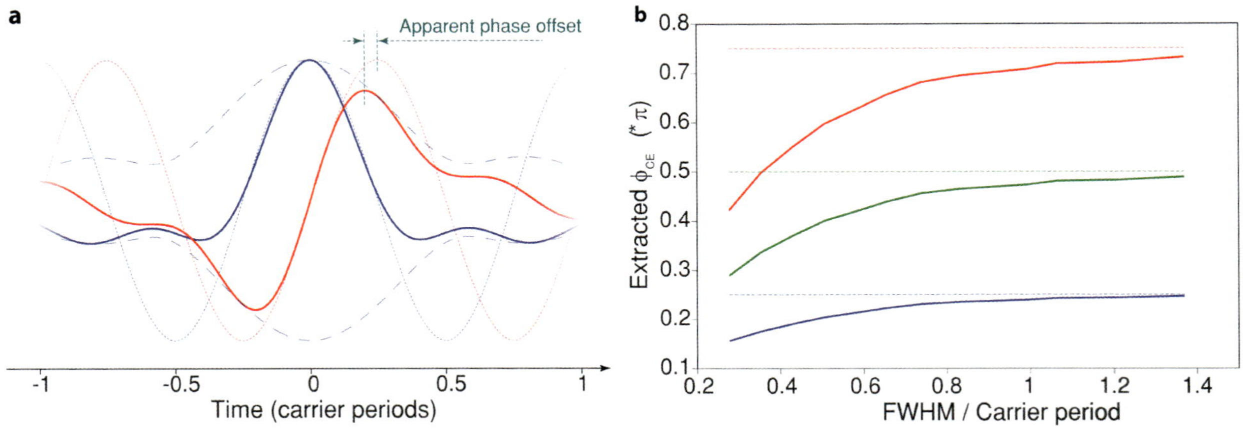

Extended Data Figure 3 | Global phase and CEP difference in a sub-cycle regime. a, Half-cycle, cosine-like (blue line) and sine-like (red line) waveforms, compared with the carrier wave (dotted blue and red lines) and envelope (dashed blue lines). **b,** Limits of CEP decomposition in the time domain. Retrieved CEP (solid lines) versus global-phase offset (dashed lines) as a function of the ratio between the FWHM of the intensity envelope and the period of the carrier wave for $\phi_G = \pi/4$ rad (blue line), $\phi_G = \pi/2$ rad (green line) and $\phi_G = 3\pi/4$ rad (red).

RESEARCH LETTER

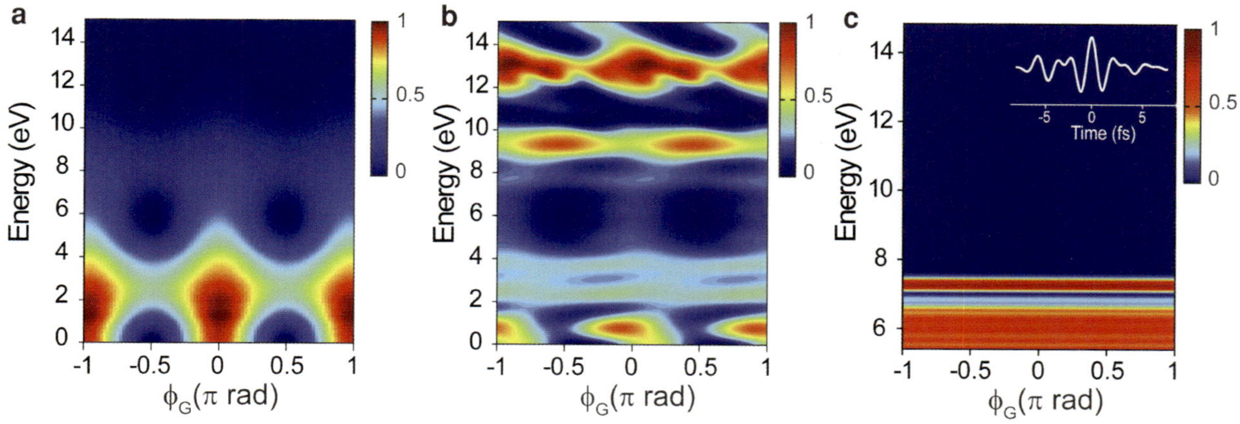

Extended Data Figure 4 | Simulated global-phase spectrograms. **a**, **b**, Calculated global-phase (ϕ_G) spectrograms using the adiabatic model (**a**) and the TDSE model (**b**) for experimentally sampled waveforms (optical attosecond pulse) and an intensity of about 4×10^{13} W cm^{-2}. **c**, Simulated global-phase spectrograms for a sub-cycle pulse (about 2 fs; shown in inset). The polarization response is insensitive to global-phase variation of the waveform; see also experiments in Fig. 3. The colour scale represents the spectral intensity in arbitrary units.

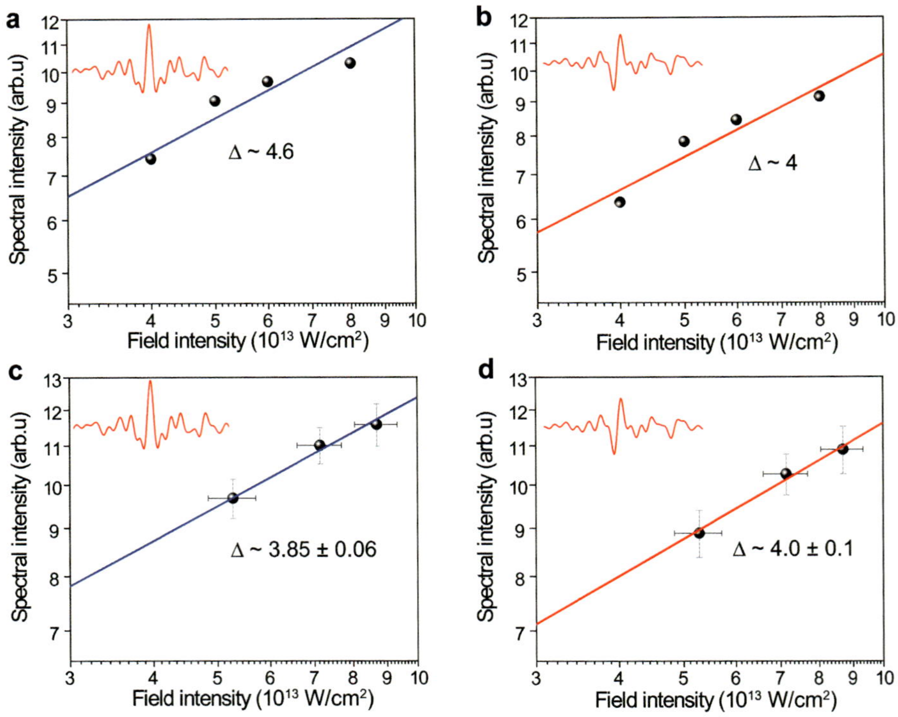

Extended Data Figure 5 | Nonlinearity study of the interaction of an optical attosecond pulse with a Kr atom. a–d, Data points show the spectrally integrated yield of the nonlinear polarization (0–8 eV) as a function of the intensity of the optical attosecond pulse for two global phases ϕ_G of the driving waveform, $\phi_G \approx 0$ (**a, c**) and $\phi_G \approx \pi/2$ rad (**b, d**), evaluated from TDSE simulations (**a, b**) and our experiments in Fig. 4 (**c**, d). The error bars represent the standard error of the mean of the input field intensity and the calculated integrated spectral yield of the nonlinear polarization. Δ is the slope of the linear fitting of the log–log plots; the errors shown indicate the standard error of mean, with $n = 3$. Insets show

RESEARCH LETTER

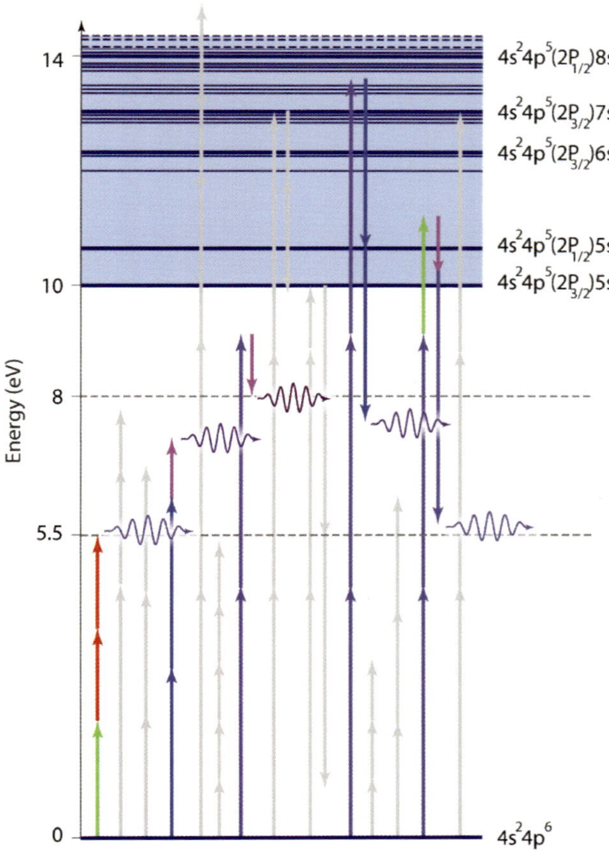

Extended Data Figure 6 | Representation of the possible multiphoton transitions on a Kr atom energy level diagram assuming third- and fifth-order nonlinear processes.

LETTER RESEARCH

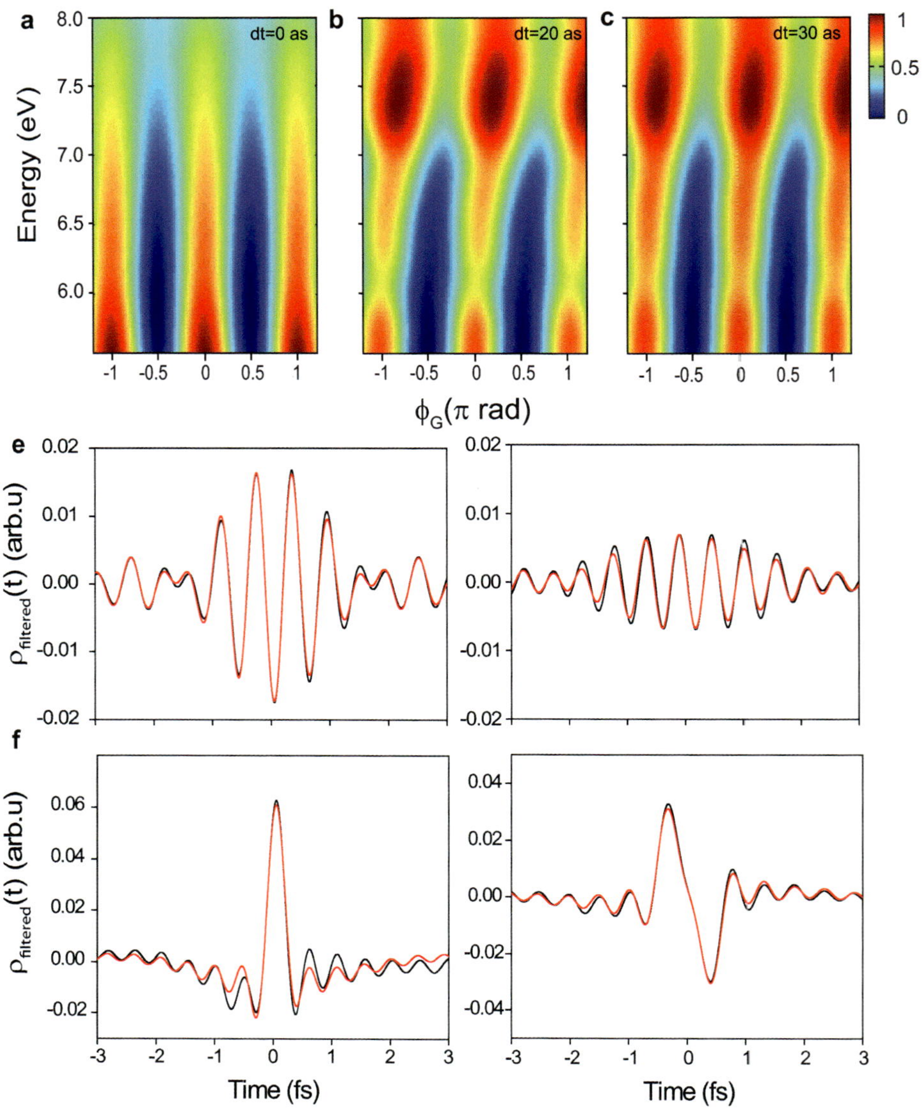

Extended Data Figure 7 | Synthetic global-phase spectrograms and nonlinear dipole fittings. a–c, Synthetic global-phase (ϕ_G) spectrograms determined using the simple model (equation (1) with the same set of parameters (a, b, c) and delays $dt = 0$ as (**a**), $dt = 20$ as (**b**) and $dt = 30$ as (**c**). The colour scale represents spectral intensity in arbitrary units. **d**, **e**, Fits (red lines) of the spectrally filtered (**d**, 5.5–8 eV; **e**, 0–8 eV) nonlinear dipoles ($\rho(t)$, black lines) with $\phi_G \approx 0$ (left column) and $\phi_G \approx \pi/2$ rad (right column) using equation (1) and 6×10^{13} W cm^{-2}.

RESEARCH LETTER

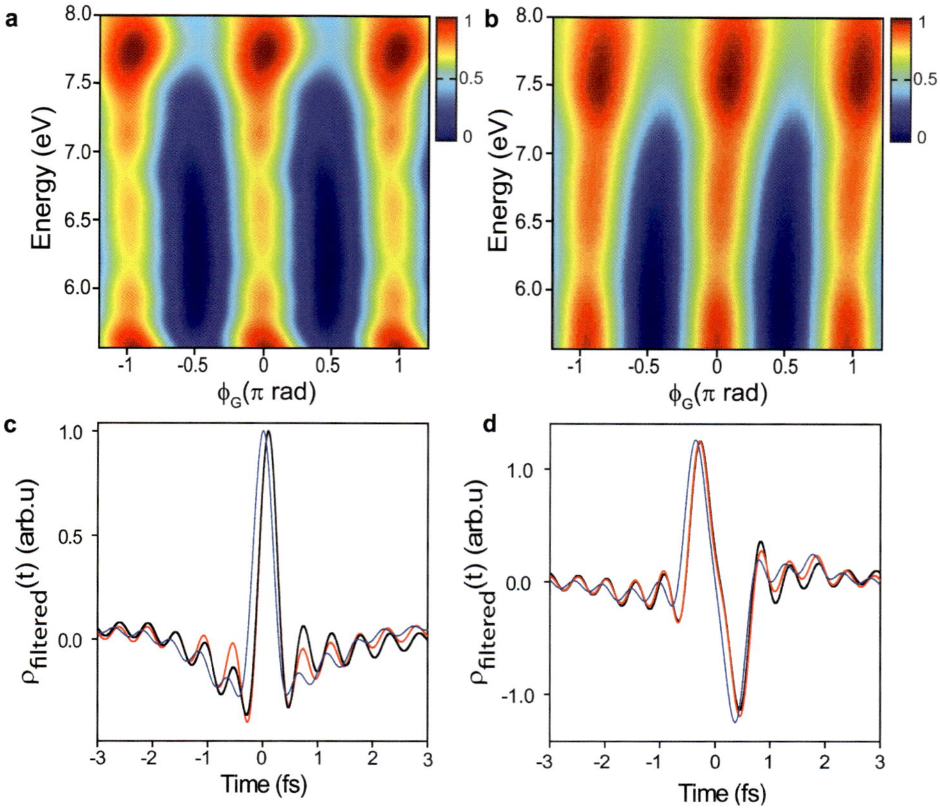

Extended Data Figure 8 | Reconstruction of the global-phase spectrogram calculated by the TDSE model. a, Calculated spectrogram at different global-phase settings ($\phi_G = -1.2\pi - 1.2\pi$ rad) of the optical attosecond pulse using the TDSE model. **b**, Reconstructed spectrogram using the model of equation (1). The colour bar represents spectral intensity in arbitrary units. **c**, **d**, Calculated nonlinear dipoles using the TDSE model at $\phi_G \approx 0$ (**c**) and $\phi_G \approx \pi/2$ rad (**d**) are plotted in black; the reconstructed nonlinear dipoles and the adiabatic dipoles are shown in red and blue, respectively.

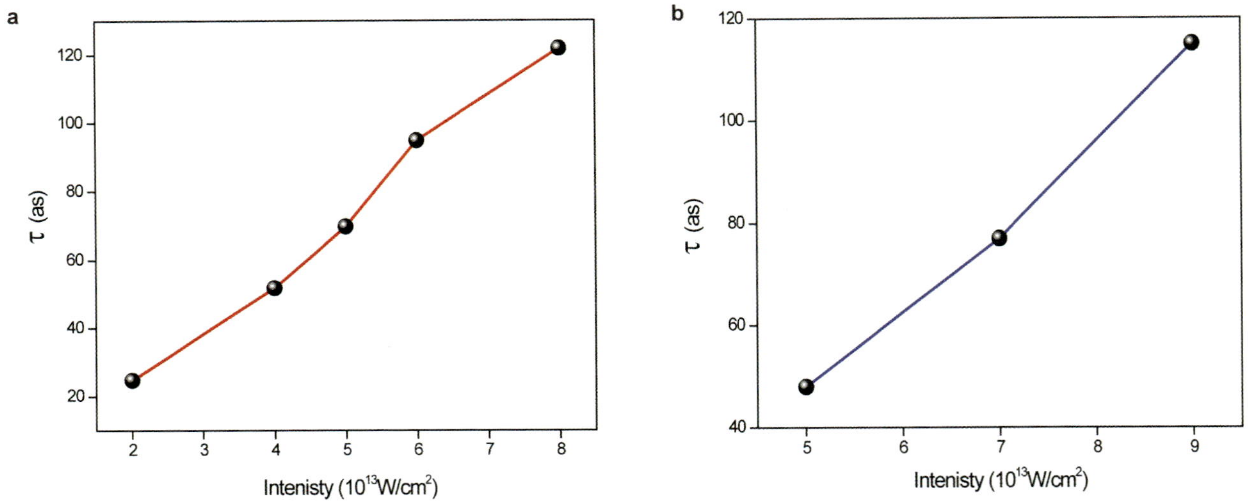

Extended Data Figure 9 | Intensity dependence of the nonlinear dipole delay. a, b, The delay between the adiabatic dipole and the calculated dipole using the TDSE model as a function of the peak intensity according to TDSE simulations (**a**) and evaluated from the experimental data (**b**; see Fig. 4).

Extended Data Table 1 | Ionization and excitation probabilities calculated using TDSE simulations of Kr for a range of peak intensities of the optical attosecond pulse

Peak intensity (10^{13} W/cm^2)	1	2	4	6	8	10
Ionization probability (%)	0.0018	0.06	0.18	0.84	1.38	2.4
Excitation probability (10^{-4} %)	0.004	0.05	0.825	1.88	3.63	3.8

RESEARCH ARTICLE

ULTRAFAST OPTICS

All-optical control and metrology of electron pulses

C. Kealhofer,[1,2] W. Schneider,[1,2] D. Ehberger,[1,2] A. Ryabov,[1,2] F. Krausz,[1,2*] P. Baum[1,2*]

Short electron pulses are central to time-resolved atomic-scale diffraction and electron microscopy, streak cameras, and free-electron lasers. We demonstrate phase-space control and characterization of 5-picometer electron pulses using few-cycle terahertz radiation, extending concepts of microwave electron pulse compression and streaking to terahertz frequencies. Optical-field control of electron pulses provides synchronism to laser pulses and offers a temporal resolution that is ultimately limited by the rise-time of the optical fields applied. We used few-cycle waveforms carried at 0.3 terahertz to compress electron pulses by a factor of 12 with a timing stability of <4 femtoseconds (root mean square) and measure them by means of field-induced beam deflection (streaking). Scaling the concept toward multiterahertz control fields holds promise for approaching the electronic time scale in time-resolved electron diffraction and microscopy.

Electron beams controlled with temporally varying electric fields enabled the first oscilloscopes and electronic computers; today, they are essential to free-electron lasers (*1, 2*), ultrafast streak cameras (*3–5*), and femtosecond imaging and diffraction techniques (*6–8*). Microwave technology at gigahertz frequencies has been the workhorse for ultrafast electron pulse control thus far, with applications ranging from particle acceleration to ultrafast pulse compression and high-resolution streak cameras. Unfortunately, microwave components suffer from appreciable phase drifts, and laser-microwave synchronization below 10 fs becomes technologically challenging (*9–11*). Laser field–driven dielectric accelerator structures operating near petahertz frequencies are being developed for next-generation particle accelerators (*12, 13*), but the short wavelength (~1 μm) and oscillation period (~1 fs) place extreme requirements on the input beam emittance and pulse duration (*12, 13*). In between, there is the regime of terahertz frequencies. Femtosecond laser–generated plasma waves exhibit terahertz longitudinal fields, which can accelerate electrons efficiently to the relativistic and ultrarelativistic energy domain (*14*). However, the excessive energy, divergence, and insufficient timing control of the resultant several- to multi-megaelectron volt electron pulses impair their utility for most of the above applications and particularly for ultrafast electron diffraction and microscopy.

Pulsed terahertz fields can be generated in a much more controlled way from the coherent nonlinear polarization of matter, such as by optical rectification (*15*). The pulses can reach field strengths on the order of 10^{10} V/m (*16*), and their half-cycle durations ideally match practical electron bunch dimensions (femtosecond-picosecond in time and micrometer-millimeter in diameter). In addition, terahertz control fields derived from a single ultrafast laser via nonlinear optics result in near-perfect (potentially subfemtosecond) temporal synchronization, offering the potential to substantially surpass the performance of microwave-laser synchronization and obviating the need for locking electronics. Therefore, terahertz and infrared radiation generated from laser-driven coherent nonlinear polarization is ideally suited for controlling electron pulses, as indicated by simulations (*17–21*) and recent experiments on controlling nanoscale photoemission (*22*) and electron acceleration in a dielectric waveguide (*23*).

[1]Ludwig-Maximilians-Universität München, Am Coulombwall 1, 85748 Garching, Germany. [2]Max-Planck-Institute of Quantum Optics, Hans-Kopfermann-Straße 1, 85748 Garching, Germany. *Corresponding author. Email: ferenc.krausz@lmu.de (F.K.); peter.baum@lmu.de (P.B.)

Concept and implementation

The experimental implementation of our terahertz-field–controlled electron beamline is depicted in Fig. 1 and consists of two functional units, one for pulse compression and one for temporal characterization by streaking (*24*). Near-1-ps pulses from a ytterbium:yttrium aluminum garnet (Yb:YAG)–based regenerative amplifier (*25*) drive two optical rectification stages for generating few-cycle terahertz pulses. With the same laser, electron pulses with on average approximately one electron per pulse are generated by two-photon photoemission from a gold thin-film photocathode and electrostatically accelerated to 70 keV. The optical rectification stages produce near-single-cycle pulses at a carrier frequency of 0.3 THz with pulse energies up to 40 nJ, implying field strengths of up to ~10^6 V/m under gentle focusing conditions. These are used to control the electron pulse's momentum, energy, duration, or timing.

We used butterfly-shaped metal resonators (Fig. 1, inset) to mediate the interaction between the electrons and the terahertz fields. The terahertz electric field is enhanced in the plane of the resonators and confined to subwavelength dimensions so that energy and momentum conservation in the electron-photon interaction can be satisfied (*26*).

Electrons passing through the resonator structures experience a net change in momentum that corresponds to the integral of the Lorentz force along their trajectories. This change varies sinusoidally as a function of arrival time of the electron in the terahertz field. The first terahertz control stage uses a tilted resonator to provide force components longitudinal to the electron beam for temporal compression. The second terahertz control stage, oriented normal to the beam, provides a delay-dependent deflection for temporal characterization.

The root-mean-square (RMS) electron beam radius is 3 μm at the compression resonator and 11 μm at the streaking resonator, at least 8 times smaller than the resonator mode dimensions. Because the electron pulse from the source is shorter than 1 ps and hence shorter than the half cycle of the terahertz field, the latter exerts a uniform but time-dependent force on the electron wavepacket in all dimensions. This force imparts acceleration and/or compression at the first stage and time-dependent deflection (streaking) at the second stage. As a result, the pulsed electron beam is entirely under the control of a single intense-pulse laser system and its optical fields.

Electron pulse compression

Even without any space charge, electron pulses unavoidably have a finite duration after acceleration because of dispersion (*27*). Electron pulse compression is therefore indispensable for reaching subphonon or electronic time scales in diffraction (*28*). Electron pulse compression requires deceleration and acceleration of the electron when arriving before or after the mean electron arrival time, respectively. For pulse compression with terahertz fields, the first butterfly microstructure is oriented at an angle of 45° with respect to the electron beam. Terahertz pulses are incident at an angle of 45° at the metal surface and hence normal to the electron beam (Fig. 1), so that the terahertz-excited resonance provides time-dependent force components parallel to the electron beam. In analogy to microwave compression, the time-dependent fields permit compression of the electron pulse substantially below its initial duration (*28*) and hence down to attosecond duration in the absence of space charge (*27*). In the limit of one- or few-electron pulses, the pulse duration is understood as the ensemble distribution of electron arrival times with respect to the laser reference.

The effective strength of the compression stage is quantified by g_E, the energy imparted to the electrons in the forward direction with changing delay time. In the experiment, g_E is proportional to the incoming terahertz peak field strength divided by the cycle period. After the interaction, the electron pulse becomes shorter as it propagates, reaching a minimum duration at a distance

$$f_c = m_e(\gamma\beta c)^3/g_E \quad (1)$$

where c is the speed of light, m_e is the mass of the electron, β is the ratio of the electron speed to the speed of light, and $\gamma = 1/\sqrt{(1-\beta^2)}$. Localized terahertz fields enhanced to merely 10^6 V/m are sufficient to yield a g_E of 50 eV/ps and hence produce a convenient temporal focus at a distance of tens of centimeters from the compression stage, depending on central energy (24).

Temporal characterization

To measure the temporal profile of the electron pulse, the second stage is configured to provide a terahertz-driven delay-dependent deflection, which we dub "streaking," in analogy to microwave-driven and laser field–driven devices. We first tested the temporal characterization with the uncompressed pulse. The time-dependent deflection of the beam yields a deflectogram (Fig. 2B), the beam profile (in the plane of deflection) plotted versus delay of the electron pulse with respect to the terahertz field. At the zero crossings of the field, the deflection is a steep function of time, and the beam width increases because of the finite-duration electron pulse. Raw images of the electron beam were recorded on the phosphor-covered camera with the terahertz deflecting field switched on and off and timed so as to yield maximum streaking of the beam (Fig. 2A). At the maxima and minima, minimal spreading occurs. The electron pulse duration and the time-dependent optical forces are determined by fitting the data of Fig. 2B assuming an instantaneous momentum transfer transverse to the electron beam induced by the terahertz field as a function of the electron arrival time.

The deflection is proportional to the integral of the Lorentz force along the electron's trajectory and closely follows the electric field's temporal profile inside the slit because of near-field confinement, as a consequence of the electron traversing the field-enhanced region within a tiny fraction of the field oscillation period (22, 29). This is in sharp contrast with attosecond streaking, which operates in the opposite limit (transit time much longer than the streaking field duration) so that the streaking spectrogram mimics the vector potential of the streaking field (30). A spline interpolation through four support points per picosecond is used for modeling time-dependent deflection, and a Gaussian temporal profile is assumed for the electron pulse. The calculated deflectogram with the fitted time-dependent deflection is shown in Fig. 2C. The electron pulse duration is 930 fs at full width at half maximum (FWHM), which is consistent with a propagation-induced temporal spread resulting from a 0.6-eV initial energy spread of the electron pulse.

The deflectogram displays sustained deflection oscillations far beyond the duration of the near-single-cycle incident terahertz driving pulse. The best fit yields a resonance frequency of 0.29 ± 0.01 terahertz of the resonator with a decay rate of 4.5 ± 0.4 ps resulting from radiative and resistive damping.

The resultant maximum streaking speed as a function of incident peak terahertz field strength in this range is shown in Fig. 2D. The highest value measured exceeds 8 μrad/fs, corresponding to a streaking speed of 4.4 μm/fs at the camera in our experimental geometry. The distance over which all momentum change accumulates is determined by the terahertz near-field decay length (~100 μm) and the foil thickness (~30 μm). The resulting interaction length is some 100 times smaller than in conventional streak cameras (3, 4), removing a substantial limitation of their temporal resolution.

In the terahertz field–driven streak camera, temporal resolution is dictated by the beam quality and signal-to-noise ratio. At an incident terahertz field strength of 1.3×10^6 V/m, the RMS beam width on the camera of 1.6 pixels (25 μm) corresponds to 6-fs resolution (RMS). With ~1000 detected electrons, the beam quality and signal-to-noise ratio allow detection of displacements or spot width changes of 0.1 pixels (1.56 μm). A displacement of 0.1 pixels corresponds to 0.4 fs accuracy of arrival time measurements, whereas a 0.1-pixel increase of the spot size from the unstreaked value corresponds to a 2-fs (RMS) or ~4.5-fs (FWHM) pulse duration. These resolutions are achieved with a signal accumulation

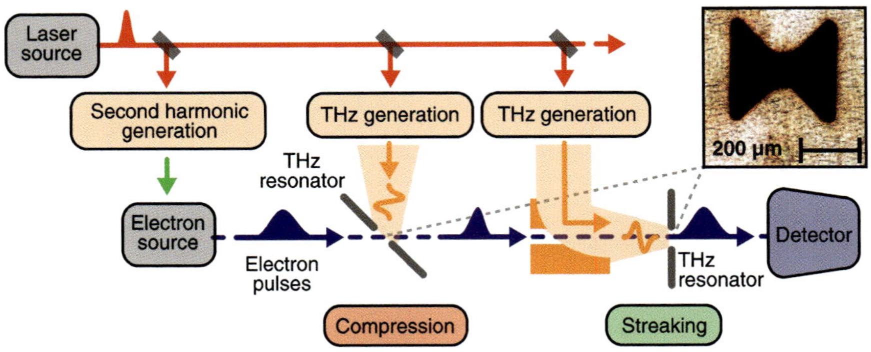

Fig. 1. Experimental setup. A 1-ps laser pulse from a Yb:YAG regenerative amplifier is frequency-doubled and generates an electron pulse from a thin-film gold photocathode. The laser also drives two optical-rectification stages, each generating single-cycle terahertz pulses with energy of up to 40 nJ. (Inset) Terahertz resonator structures are laser-machined in a 30-μm-thick aluminum foil. A first element, used for compression, is oriented at 45° to the electron beam, providing time-dependent longitudinal forces on the electrons. The second terahertz resonator, used for streaking, is oriented normal to the beam, resulting in time-dependent transverse deflection.

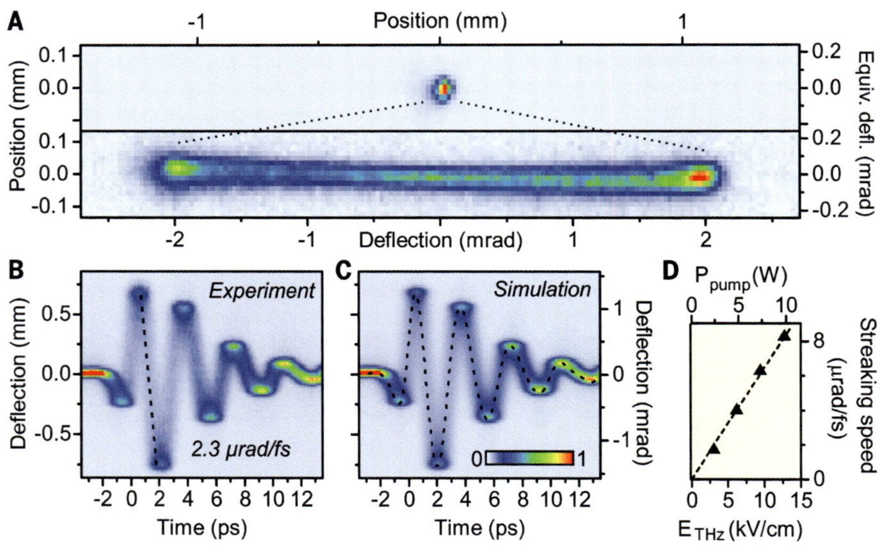

Fig. 2. All-optical terahertz streak camera. (**A**) Images of the electron beam on the camera with and without the terahertz field. The FWHM of the streaked beam is ~50 times larger than that of the unstreaked spot. (**B**) Time-dependent deflection (deflectogram) measured by varying the delay between the electron pulse and the streaking terahertz. Images of the beam are integrated along the unstreaked direction so as to determine a one-dimensional profile at each time delay. (**C**) Simulated deflectogram, result of a fit simultaneously characterizing the electron pulse duration and the streaking field (dotted line). The fitted electron pulse duration is 930 fs. (**D**) Streaking speed versus input terahertz field strength. The linear relationship supports a direct, field-driven interaction.

time of 6 s and improve with longer integration. These values represent a near-two-order-of-magnitude improvement over the state of the art of microwave streak cameras (*3, 4*), obtained already at terahertz fields of ~10^6 V/m as opposed to 10^{10} V/m available (*16*). As a consequence, our terahertz streaking concept may allow the direct measurement of subfemtosecond electron pulses (*31, 32*) and their timing drifts.

Compression and measurement

Streaking deflectograms are shown in Fig. 3, A and B, with and without terahertz field compression, respectively. In the latter, the terahertz field strength of the compression stage has been adjusted to produce a temporal focus at the streaking stage. A pronounced sharpening of the trace can be seen in the vertical (streaking) direction, indicating a substantial reduction in electron pulse duration.

The systematic evolution of measured electron temporal profile for a varying compression strength g_E is shown in Fig. 3C (*33*). The deflection as a function of time is reconstructed and used to perform the nonlinear transformation between the spatial profile of the streaked pulse and the underlying temporal profile of the terahertz field–manipulated incident pulse (*24*). The dotted lines depict the results of three-dimensional electron trajectory simulations (*33*).

The electron pulse (Fig. 3C, blue) first shortens to a minimum duration (Fig. 3C, fifth trace from bottom) and afterward lengthens again, with a double-peaked shape that is characteristic of overcompression by a sinusoidal field (*31*). The experimental and theoretical pulse durations versus terahertz-field strength and average power of the driving laser are shown in Fig. 3D. The shortest pulse (Fig. 3D, inset) has a FWHM duration of 75 fs, which is in excellent agreement with the simulation and a factor of 12 shorter than the original, 930-fs pulses and shorter than the half period of many fundamental phonon modes and molecular vibrations. Comparison of the shortest measured pulse shape with a Gaussian profile (dotted line) reveals a deviation; this is a consequence of the terahertz field's residual curvature over the time scale of the incoming, uncompressed electron pulse. Optimized electron sources with sub-100-fs duration (*34*) will reduce this effect; the particle-tracing simulations (*24*) show that 3-fs (FWHM) compressed pulses can be generated with a smaller source size and the 100-fs incoming pulses achievable if the photoemission energy is matched to the work function (*7, 27*). This would be an order of magnitude shorter than the 28-fs (FWHM) pulses generated in the single-electron regime so far (*28*) and allow the study of light-driven electronic motion via subcycle diffraction (*35, 36*).

Electron-laser timing metrology

Femtosecond pump-probe crystallography with electrons (*7*) or with x-ray free-electron lasers (*1, 2*) suffers from laser-electron timing jitter at the sample location, which is typically caused by imperfect laser-microwave synchronization (*10, 11*). In contrast to the statistical electronic processes in the photodiodes used for microwave synchronization (*9*), laser-generated terahertz fields are, within the attosecond response time of the underlying nonlinear polarization, perfectly locked in time to the intensity profile of the pump pulses. Timing jitter/drift between the laser pump pulse and the electron probe pulse can only originate from extrinsic effects, such as fluctuations in path lengths or laser pulse energy.

Experimentally, we studied drifts of the electron pulse with respect to the terahertz field by setting the near-field–enhanced streak camera to a constant delay and recording the beam deflection over time (Fig. 4A). First, we scrutinized a possible role of amplitude-to-timing conversion in the nonlinear optical rectification process. The change in electron–terahertz field timing when varying the laser pump pulse intensity before the terahertz generation crystal is shown in Fig. 4B. We observed a systematic change in timing, but the slope at the operation conditions (7 W) is only 1.0 fs per 1% change in laser power, which is negligible for our laser system, whose intensity drifts and shot-to-shot energy fluctuations are below 2% (*25*).

In a second experiment, we measured drifts of the photoemission electrons with respect to the streaking terahertz field by turning the terahertz control field off. For integration times longer than 1 s, we obtained 4.6 fs (RMS) over 15 min (Fig. 4C). Last, we repeated this measurement with the terahertz-driven pulse compression activated; with integration times of 6 s, and after ~2 hours of laboratory thermalization, we measured 3.7 fs RMS over 3 hours (Fig. 4E). This vastly outperforms the long-term stability of state-of-the-art laser-microwave synchronization with feedback loops (*7*) and also compares favorably with the record value (5 fs RMS) achieved with direct microwave extraction from a laser pulse train and data post-processing (*9*). We expect that the measured, residual 3.7-fs drift could be further decreased by minimizing the (currently meter-scale) optical path lengths and/or stabilizing them interferometrically. Synchronization in the all-optical scheme is entirely single-pulse–based, and therefore, few-femtosecond stability can be maintained at any—and especially at very low—laser repetition rate. A clock or master oscillator is not required.

Direct terahertz-electron interaction at a foil

Enhancement of the terahertz field by a microstructure resonator has proven highly beneficial for electron pulse control and metrology, but some experiments might require more direct control,

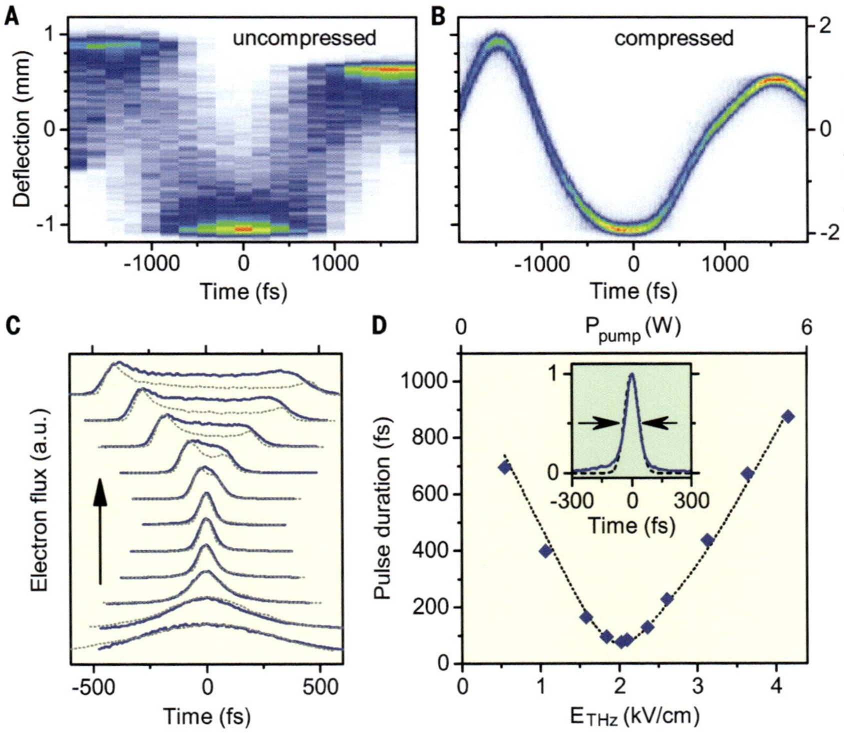

Fig. 3. Terahertz-driven electron pulse compression. (**A** and **B**) Comparison of deflectograms for (A) uncompressed (930 fs) and (B) compressed (75 fs) pulses at time steps of 200 and 10 fs, respectively. The deflectograms were measured sequentially and have a temporal resolution of ~9 fs (RMS). (**C**) Measured temporal profiles of the electron pulses (blue traces) as the compression strength is increased (black arrow), taking into account for long pulses the curvature of the time-dependent deflection. Simulated temporal profiles (10^4 particle trajectories) (*24*) are superimposed (dotted black lines). Differences between the measured and simulated pulses are due in part to the curvature of the streaking field, which reduces the streaking time-resolution at times far from zero. (**D**) Electron pulse duration (FWHM) versus incident terahertz field strength and optical pump power. The measured values are plotted as blue diamonds, and the simulation is shown as a dotted black line. (Inset) The shortest measured pulse profile [fifth trace from the bottom of (C)].

avoiding the localization and limited spectral response inextricably linked to resonators. For a direct electron-terahertz interaction, we realized a resonator-free interaction geometry based on a symmetry-breaking surface (*29*, *37*). In the depicted concept (Fig. 5A), a thin metal foil (70-nm aluminum) acts as a mirror for terahertz radiation, and electrons transmitted through it experience an abrupt extinction of the electromagnetic field, leading to net deflection. The combination of angles is chosen so that extended beam profiles experience a homogeneous time dependence owing to lateral phase matching (*29*).

Measured and calculated streaking deflectograms of an uncompressed electron pulse are depicted in Fig. 5, C and D, at an energy of 90 keV, exhibiting excellent agreement. Although the peak deflection is smaller than for the resonator-based setting, the deflectogram clearly exhibits single-cycle behavior, preserving the incident terahertz pulse's ultrabroadband spectrum without alteration (*25*). The peak-to-peak deflection as a function of the incoming terahertz field strength is shown in Fig. 5B, revealing the expected linear dependence. Electrons delivered in a nearly collimated beam are directly (and spatially uniformly) deflected directly with the field cycles, advancing the century-old cathode-ray-tube oscilloscope to the terahertz domain. Owing to the 1-fs-scale timing accuracy of the sampling electron pulse, the cut-off frequency of this terahertz cathode-ray oscilloscope is dictated by the inverse electron pulse duration, which is >10 THz with the demonstrated compressed pulses, >100 THz with the few-femtosecond electron pulses predicted by the simulations, and >1 PHz with the isolated attosecond electron pulses feasible with multistage compression.

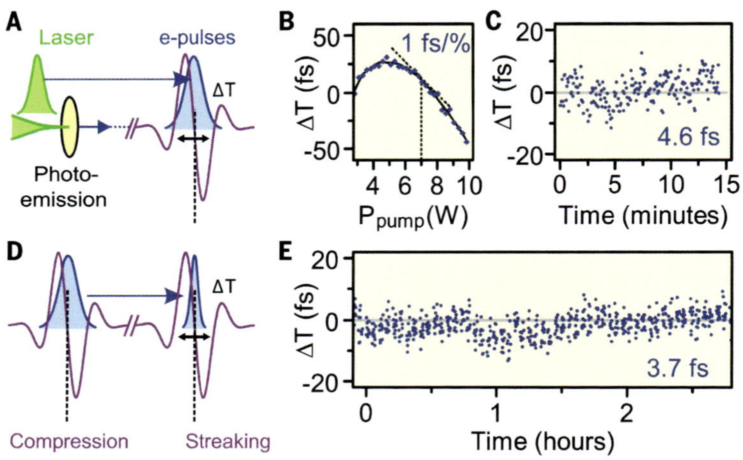

Fig. 4. Passive few-femtosecond synchronization. (**A**) Concept for measuring arrival-time drifts between the uncompressed electron pulses and the optical streaking field. (**B**) Systematic coupling of laser fluctuations to timing drifts. Measuring shifts in the terahertz zero-crossing time with increasing laser pump power reveals a slope of less than 1 fs per 1% change in laser power in a wide range of operation conditions. (**C**) Measured timing drift between photoemitted electron pulses and the terahertz field cycles at 10-s intervals. The integration time for each data point is 1 s. (**D**) Concept for measuring the arrival-time drift between terahertz-compressed electron pulses and the streaking field cycles. (**E**) Result with 6-s integration times reveals passive longer-term synchronization. The jitter values in (C) and (E) are RMS deviations; statistical errors are 3.5 and 2.7 fs, respectively.

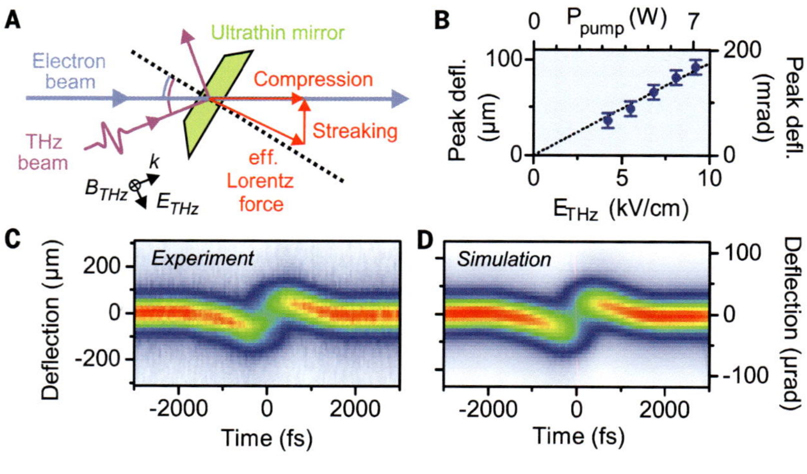

Fig. 5. Resonance-free single-cycle streaking. (**A**) Resonator-free concept for mediating the electron-terahertz interaction by using an ultrathin terahertz mirror. The angles of the terahertz and the electron beams with respect to the foil's surface normal (dotted line) are chosen for lateral phase matching across arbitrarily large electron and terahertz beam profiles. (**B**) Measurement of peak deflection versus incident terahertz field strength. (**C**) Experimental deflectogram revealing single-cycle behavior. (**D**) Simulated deflectogram with an electron pulse duration of 800 fs.

Conclusions and outlook

Our demonstrated generic and scalable all-optical methodology for the control and characterization of ultrafast electron pulses used laser-generated subcycle terahertz transients for 12-fold electron pulse compression, followed by temporal profile characterization with 10-fs resolution. Electron-transparent foils mediating the electron-field interaction permit time-energy phase-space manipulation of collimated electron beams of any size and, conversely, the temporal characterization of optical field transients up to frequencies only limited by the electron pulse duration, which—as opposed to the photon pulse in electro-optical sampling—may be shortened to <1 fs (*27*). Alternatively, resonant structures dramatically lower the terahertz power by a factor >1000 for effective compression and characterization of ultrashort electron pulses. Whereas the former opens the prospect of a petahertz-bandwidth cathode-ray oscilloscope as an alternative to attosecond photon-pulse metrology (*30*), the latter may lead to unprecedented spatiotemporal resolutions in pump-probe electron diffraction and imaging.

The moderate field strengths applied in our experiment offer the potential for increasing the interaction strengths by more than two orders of magnitude, limited by the high values of electrical breakdown at terahertz frequencies. This will result in correspondingly reduced propagation distances between subsequent terahertz control stages, which in turn should improve the passive few-femtosecond timing stability into the subfemtosecond regime. Simultaneous compression of 100-fs-scale input electron pulses to near-1-fs duration will permit spatiotemporal imaging of the fastest structural and infrared field-driven electron dynamics in molecular systems (*36*) as well as condensed matter (*35*) through single-electron diffraction (*38*) or microscopy (*6*) and allow measuring microscopic/atomic-scale electric field waveforms up to frequencies of visible light.

The demonstrated all-optical control may also be helpful for manipulating ultrabright multi-electron bunches. Compact electron guns have demonstrated sub-200-fs electron pulses with sufficient charge for single-shot structural dynamics studies (*7*). Our concept is directly applicable to them as well. Alternatively, single-shot ultrafast electron diffraction may be advanced into the few-femtosecond regime and possibly beyond by terahertz-streaking 100-fs-scale probe pulses after passage through the sample (*4*). The unprecedented temporal resolution of the terahertz-driven streak camera could also be used to characterize electron microbunching in free-electron lasers, supplementing terahertz-based diagnostics for the x-ray output (*39*, *40*). The demonstrated concept is scalable to higher terahertz frequencies and multiple stages, offering the potential for cascaded compression into the subfemtosecond regime or direct injection into a single optical cycle of a laser-field accelerator (*12*). This may, in the long run, lead to isolated attosecond electron pulses for recording dynamic changes of electron distribution in complex systems, including biological molecules and solid-state nanostructures.

REFERENCES AND NOTES

1. W. Ackermann *et al.*, *Nat. Photonics* **1**, 336–342 (2007).
2. P. Emma *et al.*, *Nat. Photonics* **4**, 641–647 (2010).
3. G. H. Kassier *et al.*, *Rev. Sci. Instrum.* **81**, 105103 (2010).
4. C. M. Scoby, R. K. Li, E. Threlkeld, H. To, P. Musumeci, *Appl. Phys. Lett.* **102**, 023506 (2013).
5. C. Kealhofer *et al.*, *Opt. Lett.* **40**, 260–263 (2015).
6. D. J. Flannigan, A. H. Zewail, *Acc. Chem. Res.* **45**, 1828–1839 (2012).
7. G. Sciaini, R. J. D. Miller, *Rep. Prog. Phys.* **74**, 096101 (2011).
8. J. Tenboer *et al.*, *Science* **346**, 1242–1246 (2014).
9. M. Walbran, A. Gliserin, K. Jung, J. Kim, P. Baum, *Phys. Rev. Appl.* **4**, 044013 (2015).
10. S. Schulz *et al.*, *Nat. Commun.* **6**, 5938 (2015).
11. G. J. H. Brussaard *et al.*, *Appl. Phys. Lett.* **103**, 141105 (2013).
12. R. J. England *et al.*, *Rev. Mod. Phys.* **86**, 1337–1389 (2014).
13. E. Hemsing, G. Stupakov, D. Xiang, A. Zholents, *Rev. Mod. Phys.* **86**, 897–941 (2014).
14. E. Esarey, C. B. Schroeder, W. P. Leemans, *Rev. Mod. Phys.* **81**, 1229–1285 (2009).
15. M. C. Hoffmann, J. A. Fülöp, *J. Phys. D Appl. Phys.* **44**, 083001 (2011).
16. M. Shalaby, C. P. Hauri, *Nat. Commun.* **6**, 5976 (2015).
17. S. D. Vartak, N. M. Lawandy, *Opt. Commun.* **120**, 184–188 (1995).
18. R. B. Yoder, J. B. Rosenzweig, *Phys. Rev. ST Accel. Beams* **8**, 111301 (2005).
19. L. J. Wong, A. Fallahi, F. X. Kärtner, *Opt. Express* **21**, 9792–9806 (2013).
20. J. Fabiańska, G. Kassier, T. Feurer, *Sci. Rep.* **4**, 5645 (2014).
21. S. R. Greig, A. Y. Elezzabi, *Appl. Phys. Lett.* **105**, 241115 (2014).
22. L. Wimmer *et al.*, *Nat. Phys.* **10**, 432–436 (2014).
23. E. A. Nanni *et al.*, *Nat. Commun.* **6**, 8486 (2015).
24. Materials and methods are available as supplementary materials on *Science* Online.
25. W. Schneider *et al.*, *Opt. Lett.* **39**, 6604–6607 (2014).
26. F. J. García de Abajo, M. Kociak, *New J. Phys.* **10**, 073035 (2008).
27. E. Fill, L. Veisz, A. Apolonski, F. Krausz, *New J. Phys.* **8**, 272 (2006).
28. A. Gliserin, M. Walbran, F. Krausz, P. Baum, *Nat. Commun.* **6**, 8723 (2015).
29. F. O. Kirchner, A. Gliserin, F. Krausz, P. Baum, *Nat. Photonics* **8**, 52–57 (2014).
30. R. Kienberger *et al.*, *Nature* **427**, 817–821 (2004).
31. P. Baum, A. H. Zewail, *Proc. Natl. Acad. Sci. U.S.A.* **104**, 18409–18414 (2007).
32. A. Feist *et al.*, *Nature* **521**, 200–203 (2015).
33. General Particle Tracer code (www.pulsar.nl/gpt) .
34. J. Hoffrogge *et al.*, *J. Appl. Phys.* **115**, 094506 (2014).
35. V. S. Yakovlev, M. I. Stockman, F. Krausz, P. Baum, *Sci. Rep.* **5**, 14581 (2015).
36. H. C. Shao, A. F. Starace, *Phys. Rev. Lett.* **105**, 263201 (2010).
37. T. Plettner *et al.*, *Phys. Rev. Lett.* **95**, 134801 (2005).
38. S. Lahme, C. Kealhofer, F. Krausz, P. Baum, *Struct. Dyn.* **1**, 034303 (2014).
39. U. Frühling *et al.*, *Nat. Photonics* **3**, 523 (2009).
40. I. Grguraš *et al.*, *Nat. Photonics* **6**, 852 (2012).

ACKNOWLEDGMENTS

This work was supported by the European Research Council and the Munich-Centre for Advanced Photonics. We thank D. Frischke for preparing ultrathin aluminum foils. The authors declare no competing financial interests.

SUPPLEMENTARY MATERIALS

www.sciencemag.org/content/352/6284/429/suppl/DC1
Materials and Methods
Fig. S1
References (*41*–*43*)

4 December 2015; accepted 2 March 2016
10.1126/science.aae0003

REPORTS

ORGANIC CHEMISTRY

Catalytic asymmetric addition of Grignard reagents to alkenyl-substituted aromatic *N*-heterocycles

Ravindra P. Jumde, Francesco Lanza, Marieke J. Veenstra, Syuzanna R. Harutyunyan*

Catalytic asymmetric conjugate addition reactions represent a powerful strategy to access chiral molecules in contemporary organic synthesis. However, their applicability to conjugated alkenyl-*N*-heteroaromatic compounds, of particular interest in medicinal chemistry, has lagged behind applications to other substrates. We report a highly enantioselective and chemoselective catalytic transformation of a wide range of β-substituted conjugated alkenyl-*N*-heteroaromatics to their corresponding chiral alkylated products. This operationally simple methodology can introduce linear, branched, and cyclic alkyl chains, as well as a phenyl group, at the β-carbon position. The key to this success was enhancement of the reactivity of alkenyl-heteroaromatic substrates via Lewis acid activation, in combination with the use of readily available and highly reactive Grignard reagents and a copper catalyst coordinated by a chiral chelating diphosphine ligand.

The majority (88%) of all known active pharmaceutical ingredients (APIs) contain functionalized heterocyclic aromatic rings with a preponderance of *N*-containing aromatic heterocycles (*1*). Furthermore, approximately half of all APIs are chiral (*1*). Because the two enantiomers of a chiral drug can exhibit markedly different bioactivity, any new chiral API must be produced as a single enantiomer. Catalytic asymmetric carbon-carbon (C-C) bond formation represents the most straightforward and atom-efficient strategy for the construction of organic chiral molecules (*2*–*4*). Organometallic reagents are used in a substantial fraction of the C-C bond–forming reactions used to construct API molecules (*5*–*7*). The conjugate addition of organometallic reagents to electron-deficient substrates (Michael acceptors) has proven to be a powerful method for creating new C-C bonds in a catalytic asymmetric manner for more than 20 years (*7*–*12*). In this context, the catalytic asymmetric addition of organometallics to conjugated alkenyl-heteroaromatic compounds represents an attractive strategy to access valuable chiral heterocyclic aromatic compounds in enantiopure form. Addition of carbon nucleophiles to conjugated vinyl-substituted heteroaromatic compounds, leading mainly to achiral molecules, is well known (*13*, *14*). In contrast, there are only a handful of reports of nucleophilic additions to β-substituted analogs, in particular when organometallics are considered.

An early attempt at such a catalytic asymmetric reaction, reported in 1998, was the nickel-catalyzed addition of Grignard reagents to substituted 4-(1-alkenyl)pyridines (*15*). Although the reaction did not appear to be ligand-accelerated and provided only 0 to 15% enantiomeric excess (ee), these results suggested that catalytic asymmetric versions of such transformations were feasible. It was not until 2010 that a highly enantioselective, Rh-catalyzed addition of an organometallic reagent to alkenyl-substituted heteroaromatic compounds—namely arylation with an arylboronic acid—was realized, furnishing a wide range of chiral products with high yields and enantioselectivities (*16*–*19*).

The limited number of reports on transition metal–catalyzed asymmetric conjugate addition of nucleophiles to β-substituted alkenyl-heteroaromatic compounds, restricted to precious metal–catalyzed arylations, is in striking contrast to the plethora of methodologies available for the catalytic asymmetric alkylations, arylations, alkynylations, and allylations of common Michael acceptors, activated by carbonyl, nitrile, sulfonyl, and nitro groups (*8*–*12*). The paucity of methodologies for the nucleophilic addition to β-substituted alkenyl-heteroaromatic compounds is rooted in the intrinsically lower reactivity of these compounds, due to the relatively weak activation from the heteroaromatic moiety (*13*, *14*). Furthermore, the numerous examples of nonasymmetric additions to vinyl-substituted heteroaromatic compounds indicate that the presence of a β-substituent decreases the reactivity further, presumably because of the steric hindrance it introduces.

We decided to explore the addition of Grignard reagents to β-substituted conjugated alkenyl-heteroaromatic compounds. Inexpensive and readily available Grignard reagents are some of the most commonly used organometallics in synthetic chemistry (*20*), especially in copper-catalyzed asymmetric conjugate addition to a variety of Michael acceptors (*9*–*12*). We reasoned

Stratingh Institute for Chemistry, 9747 AG Groningen, Netherlands.
*Corresponding author. Email: s.harutyunyan@rug.nl

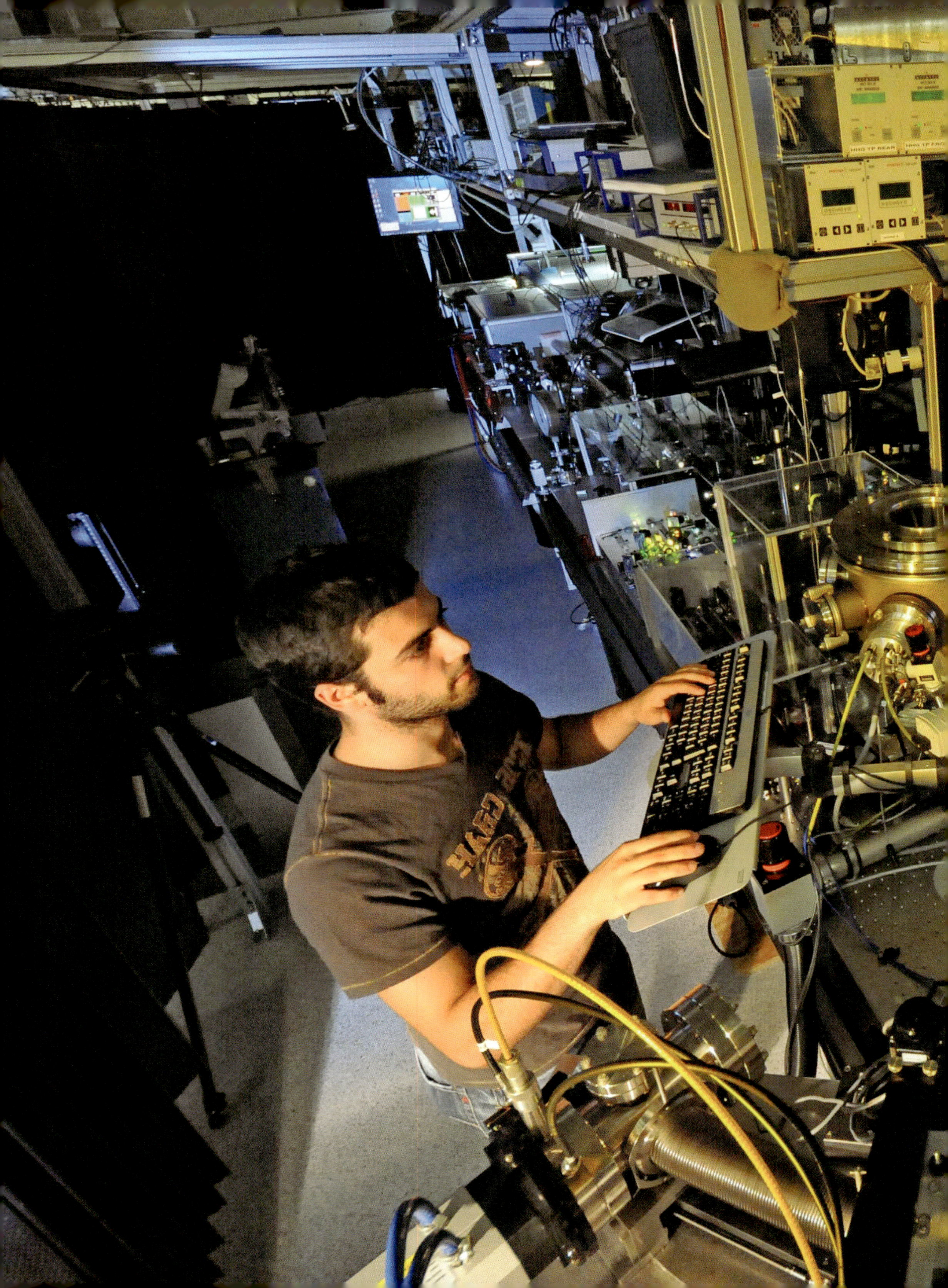

10 The Technology Platform: Vacuum Beamlines

The AS-1 attosecond beamline at the Max Planck Institute of Quantum Optics. © Thorsten Naeser

Attosecond XUV pulses have been one of the key tools of first-generation attosecond technology, which emerged at the turn of the third millennium. As they are strongly attenuated in atmospheric air, their generation and application need to be performed in vacuum chambers. The locations of generation and application are usually separated by several meters. The need for this separation stems from the high intensity of the laser pulse required for the generation of the attosecond XUV light. This intensity exceeds the threshold for damage of solid-state optical components and hence can only be lowered by beam expansion due to free propagation to levels tolerated by optical components used for steering and focusing the few-cycle laser pulse on the experimental target under scrutiny. The system, usually consisting of separate vacuum chambers for the generation of the attosecond pulses and their subsequent application and connected by a vacuum tube, is referred to as an *attosecond beamline*.

The papers in this chapter describe three attosecond beamlines at different levels of sophistication along with some of their key elements that were developed at the Max Planck Institute of Quantum Optics (MPQ) during the first decade of the new millennium. These efforts were undertaken by a number of co-workers under the leadership of Ralph Ernstorfer, Peter Feulner, Eleftherios Goulielmakis, Reinhard

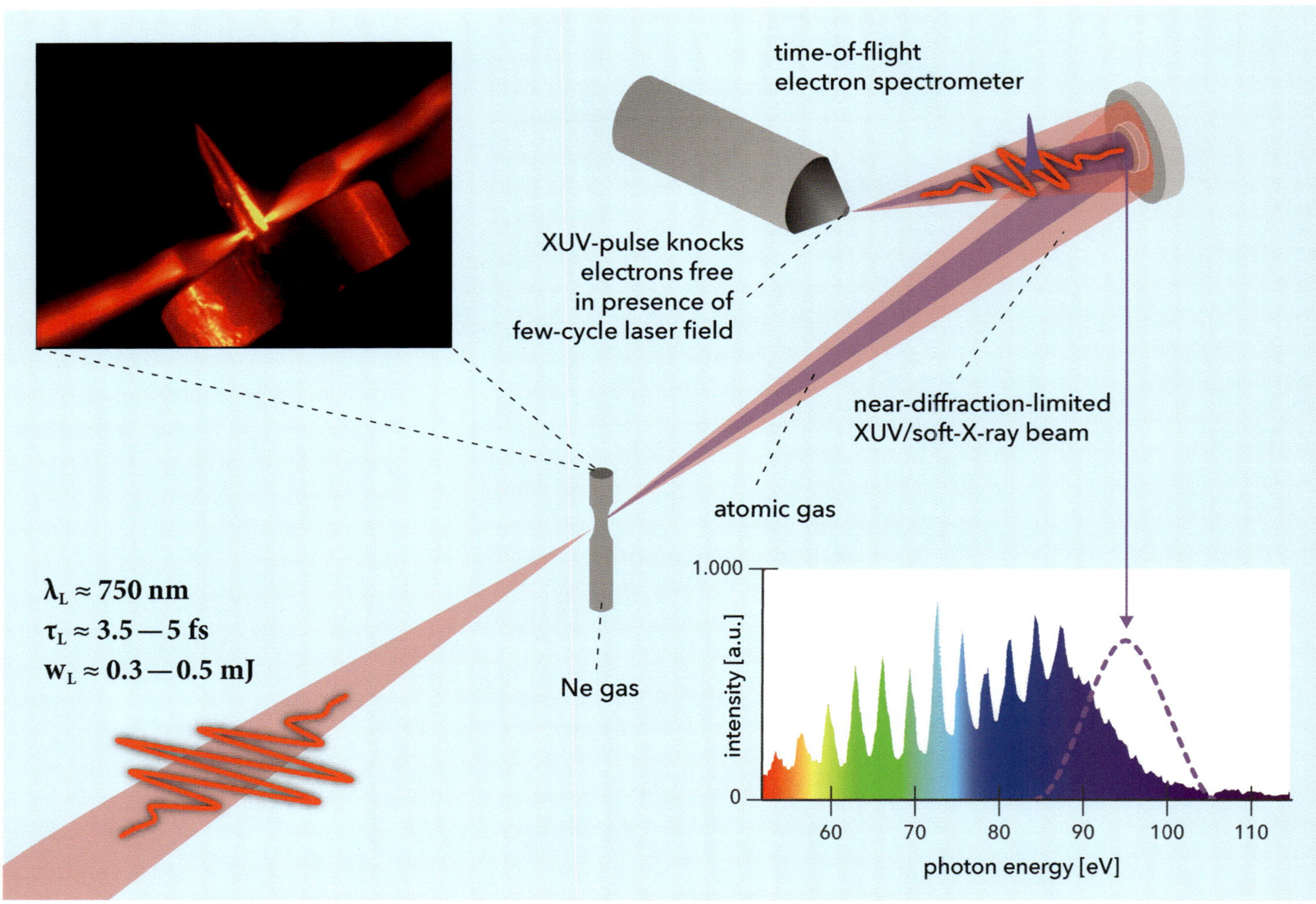

Figure 10-1 Vacuum beamline for the generation and measurement of isolated attosecond pulses, following the first-generation setup realized at the Vienna University of Technology at the turn of the new millenium. High-order harmonic radiation has been generated by exposing neon atoms to waveform-controlled few-cycle NIR laser pulses. The emerging XUV light is emitted in a near-diffraction-limited laser-like beam with a divergence much smaller than that of the pump laser beam. The two collinear beams are reflected by a two-component mirror, which filters XUV photons near 100 eV, delays the XUV pulse with respect to the laser field and focuses the two beams into a second target of neon atoms for pulse characterization via attosecond streaking or of some other substance, in which electron dynamics is to be scrutinized via attosecond streaking or photoelectron spectroscopy.

Kienberger, and Martin Schultze.

Reprint [53] is devoted to the first-generation beamline AS-1 delivering the XUV and NIR beams collinearly on a target placed in a high-vacuum environment.

Reprint [54] demonstrates how attosecond XUV and VUV pulses can be produced collinearly in such a geometry to broaden the range of possibilities for attosecond spectroscopy.

Reprint [55] describes the AS-2 beamline, in which the XUV beam is separated from the NIR beam after its generation, to offer flexibility for frequency conversion of the NIR beam. Moreover, this geometry has allowed the first demonstration of *attosecond polarization spectroscopy*, in which the nonlinear polarization response of matter induced by the strong laser field can be directly read from its temporal distortion via subsequent attosecond streaking. This novel technique is reviewed in the next chapter.

Reprint [56] addresses AS-3, the world's first ultra-high-vacuum attosecond beamline which allows attosecond streaking and photoelectron spectroscopy to be performed in solids and on surfaces.

Last but not least, in reprint [57] Ulf Kleineberg and his co-workers describe the key elements they contributed to all of these attosecond beamlines, namely XUV multilayer optics, which they used for steering and focusing the broadband attosecond pulses.

[53] M. Schultze, A. Wirth, I. Grguras, M. Uiberacker, T. Uphues, A. Verhoef, J. Gagnon, M. Hofstetter, U. Kleineberg, E. Goulielmakis, F. Krausz. State-of-the-art attosecond metrology. *Journal of Electron Spectroscopy and Related Phenomena* **184**, 68 (2011).
(Page: 454)

[54] E. Bothschafter, A. Schiffrin, V. Yakovlev, A. Azzeer, F. Krausz, R. Ernstorfer, R. Kienberger. Collinear generation of ultrashort UV and XUV pulses. *Optics Express* **18**, 9173 (2010).
(Page: 464)

[55] M. Fieß, M. Schultze, E. Goulielmakis, B. Dennhardt, J. Gagnon, M. Hofstetter, R. Kienberger, F. Krausz. Versatile apparatus for attosecond metrology and spectroscopy. *Review of Scientific Instruments* **81**, 093103 (2010).
(Page: 472)

[56] E. Magerl, S. Neppl, A. Cavalieri, E. Bothschafter, M. Stanislawski, T. Uphues, M. Hofstetter, U. Kleineberg, J. Barth, D. Menzel, F. Krausz, R. Ernstorfer, R. Kienberger, P. Feulner. A flexible apparatus for attosecond photoelectron spectroscopy of solids and surfaces. *Review of Scientific Instruments* **82**, 063104 (2011).
(Page: 480)

[57] M. Hofstetter, M. Schultze, M. Fieß, B. Dennhardt, A. Guggenmos, J. Gagnon, V. Yakovlev, E. Goulielmakis, R. Kienberger, E. M. Gullikson, F. Krausz, U. Kleineberg. Attosecond dispersion control by extreme ultraviolet multilayer mirrors.
Optics Express **19**, 1767 (2011).
(Page: 486)

Journal of Electron Spectroscopy and Related Phenomena 184 (2011) 68–77

Contents lists available at ScienceDirect

Journal of Electron Spectroscopy and Related Phenomena

journal homepage: www.elsevier.com/locate/elspec

State-of-the-art attosecond metrology

M. Schultze[a,b,*], A. Wirth[a], I. Grguras[a], M. Uiberacker[a], T. Uphues[a], A.J. Verhoef[a], J. Gagnon[a], M. Hofstetter[b], U. Kleineberg[b], E. Goulielmakis[a], F. Krausz[a,b]

[a] *Max-Planck-Institut für Quantenoptik, Hans-Kopfermann-Str. 1, D-85748 Garching, Germany*
[b] *Department für Physik, Ludwig-Maximilians-Universität, Am Coulombwall 1, D-85748 Garching, Germany*

ARTICLE INFO

Article history:
Available online 22 January 2011

Keywords:
Attosecond spectroscopy
Streaking
Tunnel ionization
High-harmonic-generation
Ultrashort laser
Transient absorption

ABSTRACT

Tracking and controlling electron dynamics in the interior of atoms, molecules as well as in solids is at the forefront of modern ultrafast science [1–5]. Time-resolved studies of these dynamics require attosecond temporal resolution that is provided by an ensemble of techniques consolidated under the term "attosecond metrology" [6,7]. This work reports the development and commissioning of what we refer to as next-generation attosecond beamline technology: the AS-1 attosecond beamline at the Max-Planck Institute of Quantum Optics. It consists of a phase-stabilized few-cycle laser system, for the generation of XUV radiation, and modules tailored for the spectral filtering and isolation of attosecond pulses as well as for their temporal characterization. The setup produces the shortest attosecond pulses demonstrated to date and combines them with advanced spectroscopic instrumentation (electron-, ion- and XUV-spectrometers). These pulses serve as temporally confined trigger events (attosecond streaking and tunneling spectroscopy) or probe pulses (attosecond absorption and photoelectron spectroscopy) enabling attosecond chronoscopy to be applied to a broad range of systems belonging to the microcosm.

1. Introduction

Experiments with attosecond resolution, particularly those utilizing the generation of high harmonic radiation, are necessarily very sensitive to the overall waveform of the driving laser field. The measurement, the precise and reproducible control of the carrier envelope phase, or more generally the waveform of the laser pulses combined with high temporal stability is the key aspect of the laser sources employed in the experiments. With decreasing pulse durations, experimental results become significantly more sensitive to variations of the carrier envelope phase. Besides the reproducibility of the characteristics of the laser pulses, the mechanical stability of the experimental system has to be a main focus of attention to perform attosecond pump–probe experiments in an interferometric setup. Investigations with a temporal resolution of 100 as require mechanical stability better than a few-nanometers over several hours.

2. Generation and measurement of few-cycle laser pulses

To generate the few-cycle NIR pulses required by experiments providing contemporary attosecond resolution we employ a commercially available chirped-pulse-amplification system (Femtopower Compact Pro, Femtolasers Produktions GmbH) [8,9] that we adapted according to the needs of attosecond experiments. The amplifier system (Fig. 1) is seeded by an ultra broadband Ti:sapphire oscillator, emitting pulses of about 6 fs with an output power of 300 mW at 78 MHz [10]. Before seeding the amplifier (seed pulse energy ~1 nJ) the oscillator pulses are dispersively stretched to 15 ps in bulk material. After the fourth pass of the amplification, a Pockels cell is installed in the beam path, reducing the repetition rate to 3 kHz. A total of 9 passes through a Ti:Sa crystal pumped with ~20 W/3 kHz at 532 nm result in a pulse energy of 1.2 mJ. Subsequent recompression takes place in a hybrid prism/positive dispersive mirror-compressor. The prism compressor is modified to overcompensate the positive chirp the amplified pulses carry; the pulses therefore maintain their negative chirp throughout the final prism and are not subject to self-phase modulation (SPM) which would narrow and modulate the spectrum and thus hinder the final temporal compression of the pulses. The full compression takes place by reflection off positive-dispersion chirped mirrors where SPM does not occur, thereby, sub-25 fs pulses (full-width at half maximum – FWHM) extending over a spectral width of 65 nm and energy of 900 μJ [11] are achieved.

The amplified pulses are focused (f = 1.8 m) into a 1 m long hollow-core fiber (FiberTech GmbH) with an inner diameter of 250 μm, that is filled with neon at a pressure of ~2 bar. Nonlinear propagation of these pulses in the hollow-core waveguide results in (see Fig. 1) the broadening of their spectrum, primar-

* Corresponding author at: Max-Planck-Institut für Quantenoptik, Hans-Kopfermann-Str. 1, D-85748 Garching, Germany.
E-mail address: martin.schultze@mpq.mpg.de (M. Schultze).

0368-2048/$ – see front matter © 2011 Elsevier B.V. All rights reserved.
doi:10.1016/j.elspec.2011.01.003

M. Schultze et al. / Journal of Electron Spectroscopy and Related Phenomena 184 (2011) 68–77

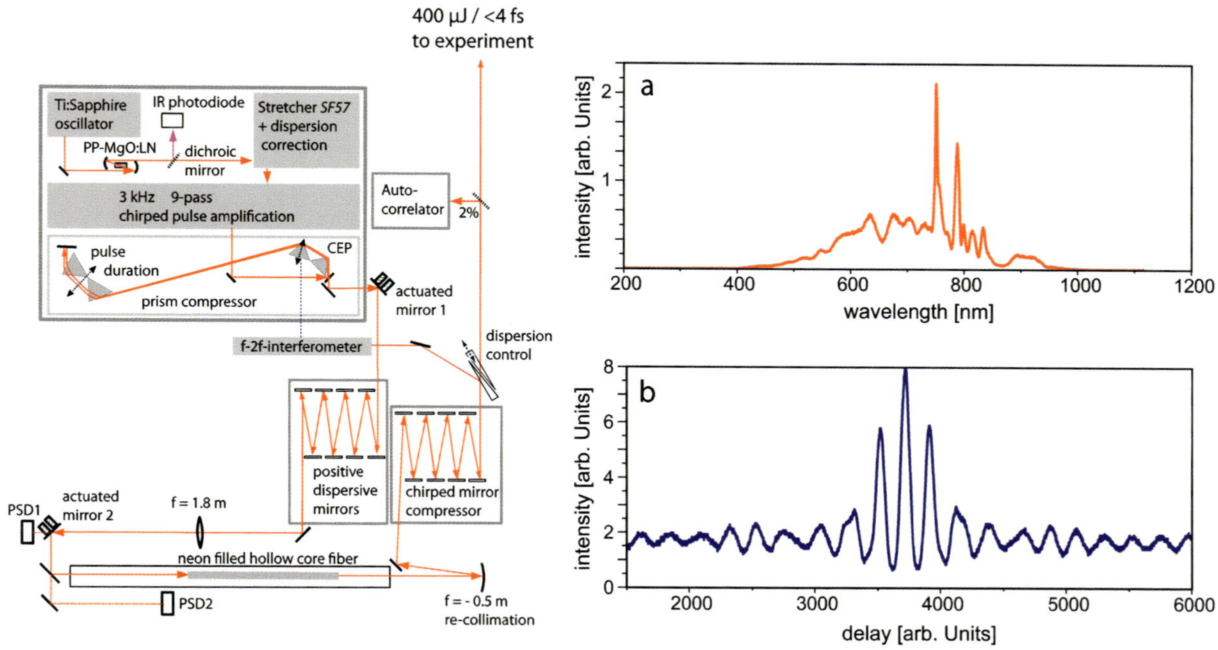

Fig. 1. Schematic laser setup. The left panel shows the Ti:Sa oscillator/9-pass amplifier system with the subsequent pulse-broadening and compression stages including a gas filled hollow core fiber and chirped mirrors for the dispersion management. On the right a typical spectrum achieved after spectral broadening the Ti:Sa output in the fiber is shown (panel a) and the corresponding second order autocorrelation measured after recompression of the pulses in a chirped mirror compressor. The calculated pulse duration is below 4 fs and the pulses carry energy of about 400 μJ at the repetition rate of 3 kHz.

ily due to self-phase-modulation but also because of processes like self-steepening and ionization-induced blue shift [12,13]. The broadened pulses are recompressed to a duration of <4-fs following 8 reflections off low-loss chirped mirrors optimized for the wavelength range of 500–1000 nm [14]. The transmission through the fiber is above 50%. The additional losses the chirped mirrors introduce result in a total transmission of ~45% and pulse energies of 400 μJ after recompression. For the long-term stability for attosecond experiments, a commercial beam stabilization system [15] is used to track the beam pointing with two position sensitive detectors (PSD in Fig. 1), and compensates for beam displacements at the exit of the laser with two piezo-actuated mirrors.

3. Controlling the carrier envelope phase (CEP)

For laser pulses that last only a few oscillations of the light field, the actual evolution of the electric field inside the pulse envelope, rather than just the cycle-averaged intensity, becomes relevant to the experiment. Such ultrashort pulses are usually described in terms of the offset between the maximum of the intensity envelope and the peak of the electric field waveform that oscillates at the central carrier frequency. This temporal offset divided by the period of the carrier wave, is dubbed carrier envelope phase (CEP). For broadband laser pulses, the propagation in a dispersive medium produces a difference in phase and group velocity resulting in a slippage of the carrier wave under the envelope. Two consecutive pulses may suffer a phase shift such that a certain waveform has a recurrence period of $2\pi/\Delta\varphi$, corresponding to a offset frequency f_{offset} of the oscillator's frequency comb [16,17].

In the case of the ultra broadband output of femtosecond oscillators, the so-called f-to-zero technique can be employed to track the CEP [18]. This scheme relies on focusing the broadband pulses of the oscillator into a periodically poled nonlinear medium (here periodically poled lithium niobate – PPLN) to generate the difference frequency of its extreme spectral components. Besides that, the bandwidth of the pulses is slightly increased via self-phase modulation. The optical f-to-zero beat signal carrying the CEP information is band-pass filtered by a suitable infrared low pass filter and detected on a photodiode. Locking electronics (Menlo Systems) are used to detect the beat frequency and to generate an error signal proportional to the deviation of f_{offset} from the desired value. The error signal is used in a closed loop as feedback signal to drive an acousto-optic modulator that is installed into the beam path of the pump laser (Verdi Coherent). Modulating the pump laser intensity influences the intra cavity nonlinear phase shift providing a feedback loop for the stabilization of f_{offset}. Fig. 1 shows the location of the crystal (PP-MgO:LN) and the IR sensitive photodiode inside the laser system.

In the presented setup, the intra-cavity dispersion is adjusted such that every fourth pulse has, in principle, the same CE phase and therefore an identical waveform. The stabilization is used to compensate for phase drifts that occur due to instabilities introduced by thermal changes or drifts in the pump laser parameters. Details on the stability performance of these systems can be found in [8]. The amplification process of the pulses to mJ level comes along with additional sources of noise. The amplification in 9 passes through a Ti:Sa crystal includes extensive propagation through air (~15 m) and numerous optical components. Earlier schemes used for the stabilization of the CEP phase of the amplified pulses have been based on an f-to-$2f$ interferometer that was seeded by a small fraction of energy (<1%) of the compressed pulses at the exit of the amplifier [19]. The relatively narrow output spectrum of the laser amplifier does not allow for a direct f-to-$2f$ measurement but needs the spectrum to be broadened in a filament by means of white light generation. This filament is usually realized in a sapphire crystal and is very sensitive to fluctuations in the intensity; this dependency translates intensity fluctuations into artificial phase noise in the detection.

Here, we present an upgrade of the scheme for the detection and compensation of the carrier-envelope phase of few-cycle pulses. The detection module is installed at the exit of a chirped mirror compressor after the hollow core fiber. The octave-spanning bandwidth of the spectrum negates the need for additional spectral broadening, and furthermore, the phase noise accumulated inside

M. Schultze et al. / Journal of Electron Spectroscopy and Related Phenomena 184 (2011) 68–77

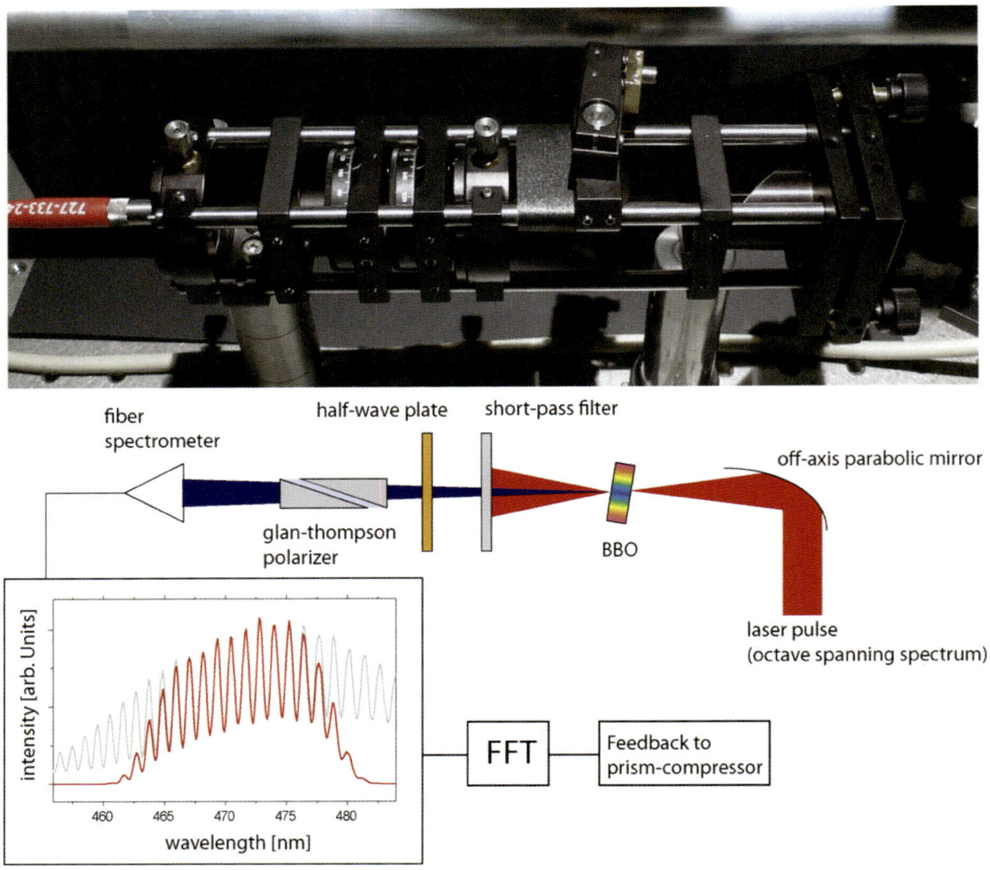

Fig. 2. Compact *f*-to-2*f* setup. Around 4% of the amplified and broadened light after the fiber and the chirped mirror compressor are focused by a parabolic mirror into a BBO crystal. The second harmonic light is filtered by a short-pass filter and its polarization is turned to be parallel to the blue part of the fundamental spectrum. A Glan–Thompson polarizer is used to balance the relative intensity and the interference pattern carrying the CE-phase information is observed by a fiber spectrometer. The observed spectral modulation is shown in the inset (grey spectrum). Proper filtering (red spectrum) and subsequent fast-Fourier transformation (FFT) yield a peak at the modulation frequency. The corresponding FFT phase of this frequency peak depends on the carrier-envelope phase. By affecting the insertion depth of one of the prisms of the prism compressor into the beam (Fig. 1) the electronics keep the peak position constant and thus stabilize the CE-phase of the laser pulses. (For interpretation of the references to color in this figure legend, the reader is referred to the web version of this article.)

the hollow-fiber is measured and compensated as well. Experimentally, the carrier envelope phase φ_{CE} is observed as an interference pattern in the spectral region where the fundamental and the frequency doubled light overlap; its implementation is shown in Fig. 2. The period and phase of this interference pattern is retrieved by Fourier analysis of this part of the spectrum and by its evolution, a feedback signal is processed (coded in a LabView program). As implied in Fig. 1, the feedback signal is converted into a displacement of one of the prisms inside the prism compressor [20]. Displacing the prism perpendicular to the beam using a piezo translation stage leaves the beam direction almost unchanged, but at the same time it adjusts the amount of material dispersion the pulses accumulate. A standard proportional controller algorithm takes care of adjusting the prism position such that the CE phase changes are counteracted. The achieved phase stability is typically better than 100 mrad RMS which has been confirmed by single shot measurements [21].

4. Temporal and spatial pulse characterisation

For a continuous monitoring of the pulse characteristics, around 4% of the pulse energy is split off behind the chirped mirror compressor and is sent to a second harmonic interferometric autocorrelator. Fig. 1 shows a representative autocorrelation trace that upon evaluation yields a pulse duration of 3.9 fs [22].

Whereas the interferometric second-order autocorrelation comprises a fast diagnostic apparatus and allows fine tuning of the dispersion of the pulses on a daily basis, more elaborate access into the properties of the pulses can be gained via frequency resolved pulse characterization techniques.

To this end, an all-reflective transient grating frequency-resolved optical gating (FROG) device has been incorporated into the setup that is based on the formation of a transient density grating when two time delayed replicas of the pulse impinge on a thin nonlinear medium [23]. The nonlinear medium is a 50 μm thick UV grade fused silica substrate, and introduces limited dispersion to the pulses. The device is therefore extremely broadband and virtually dispersion-free in the spectral range of the laser pulses. Fig. 3 shows a spectrogram recorded with the transient grating apparatus as well as its reconstruction based on commercial software (Femtosoft Technologies) [24]. The FROG error for a 512 × 512 grid was 0.0075, indicating the excellent quality of the reconstruction.

The evaluated pulse length amounts to ~4.2 fs, and the retrieval ascertained a positive chirp of 30 fs^2 at 800 nm of the pulses whereas the FL of the spectrum was 3.8 fs. Shorter waveforms that can be composed by these spectra have been sampled via the streaking technique [25]. The information acquired about the spectral phase facilitates the design of particularly adapted chirped mirrors in future upgrades of the beamline.

M. Schultze et al. / Journal of Electron Spectroscopy and Related Phenomena 184 (2011) 68–77

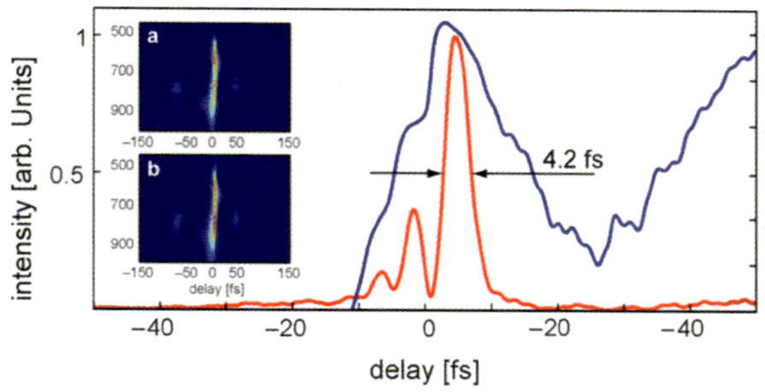

Fig. 3. Time domain picture resulting from a frequency-resolved cross correlation (FROG) measurement of the amplified and compressed laser pulses. The analysis of the trace resulting from the all reflective transient-grating FROG (see text) yields a half-width duration of 4.2 fs. The insets show the measured (a) and the retrieved (b) FROG trace. Across the pulse the spectral phase is adequately flat, indicating the compression of the pulses to a duration close to the Fourier limit (3.7 fs) that is achievable by controlling the dispersion by a pair of thin glass wedges.

In experiments that require high intensities an excellent beam quality is essential. The beam quality of the compressed laser pulses was characterized by means of the M^2 parameter. Using a CCD-camera in combination with a microscope objective, the focal spot size can be accurately determined and the beam diameter as a function of the distance to the focus is obtained using the knife-edge method. The measurement resulted in a very small beam propagation factor of $M^2 = 1.8 \pm 0.1$ with a Bessel-like beam profile caused by the hollow core fiber acting as spatial filter that considerably improves the beam characteristic. These beam parameters come along with a virtually stationary propagation direction since the fiber exit is fixed in space and facilitate a smooth focussing ideal for the generation of high-order harmonics.

5. Pump–probe apparatus

Since a duration of 100 as corresponds to only 30 nm of spatial displacement, special attention has to be turned to the implementation of vibration damping and isolation of the most sensitive units from external mechanical noise sources.

To provide the required stability of the system, the optical breadboards inside the vacuum chambers are mounted on separate pedestals that are directly clamped on the optical table that carries the experiment. These posts are connected to the bottom of the chamber and the walls only by edge welded bellows. The transmission of unavoidable vibrations of the chamber walls, due to the mounted turbo pumps, to optical components located on the breadboards is therefore suppressed. Furthermore, transmission of mechanical vibrations emerging from the roughing pumps to the system is dumped via long hydroformed flexible bellows that are installed into bins filled with silica sand.

6. High-order-harmonic source

The generation of the high harmonic radiation takes place by focusing the few cycle driver pulses onto a thin nickel tube (2 mm inner diameter) filled with neon and kept at a constant backing pressure. The physics of harmonic generation at the atomic and macroscopic levels has been extensively studied [26,27]. The laser drills two holes across the tube, allowing it to enter and exit the tube while ensuring that the hole size is the smallest that transmits the main laser mode. By choosing the tube diameter, the interaction length can be tuned in order maximize the XUV photon flux. For the sub-4 fs driver pulses with 300 μJ pulse energy, a propagation length of 2–3 mm at a pressure of ~250 mbar was identified to maximize the flux of the XUV in the spectral range (70–130 eV).

The efficiency of conversion of the fundamental into XUV photons can theoretically reach the level of 10^{-7} to 10^{-8} [28].

To have a useful amount of photons in the experimental target reabsorption has to be minimized. A significant amount of generated photons is reabsorbed if the partial pressure of the generation gas exceeds 10^{-3} mbar [29]. Two hybrid turbo pumps (Alcatel ATH31+) remove the expanding gas jet close to the interaction volume maintaining a pressure below 10^{-2} mbar in the HHG chamber during operation. For the laser parameters presented here, focus intensities around $1–8 \times 10^{14}$ W/cm^2 at a focus diameter of ~150 μm (achieved by focusing with spherical mirrors of focal length between 50 and 70 cm) yield the maximum photon flux in an isolated pulse in the energy range between 70 and 130 eV. For high harmonic emission in this energy range neon exhibits the optimum combination between ionization potential and absorption cross section [28]. The XUV generation can be described with a three-step process: electron liberation through tunnel ionization, acceleration of the electron in the laser electric field and subsequent recombination of electron and ion. This three-step model [26,30] relies on tunnel ionization as a first step and therefore predominantly occurs in the most intense on-axis part of the laser focus [31]. Due to the short wavelength of the generated light, together with its high degree of coherence, the XUV radiation is emitted with a small divergence angle as compared to the driving laser pulse. This allows for the spatial separation of the two beams.

After 10 cm propagation in the vacuum chamber that hosts the generation target, the XUV beam and the co-propagating residual IR pulses enter the first differential pumping stage. The reabsorption up to this point is estimated to be below 2% and negligible at any other location of the experimental setup since the pressure of the residual gas drops to less than 10^{-3} mbar.

Between the two differential pumping stages (see figure) an ionization based detector for the XUV ("HH-Meter") is implemented. Its design is inspired by the ionization chambers that are broadly used in synchrotron facilities [32]. A conductive wire along the tube with a small applied voltage collects the charged fragments that are liberated by ionization of the residual gas and an amplifier suitable for nA currents provide the desired measure of the XUV flux. The signal of the HH-Meter is proportional to the number of electrons and ions created by absorption of XUV photons across the entire emitted spectrum down to energies equal or higher to the ionization potential of neon. Therefore the stability of this signal is only partially related to that of the intensity of the attosecond pulse which is extracted from the cut-off of the emerging spectra. However, as an independent measure of the source stability and flux it is a useful tool that allows one to monitor the XUV signal and identify ideal and stable conditions for performing sensitive experiments. Typically the XUV flux stability recorded with the HH detector is better than 2.5% (standard deviation/mean value).

In the second differential pumping stage, the pressure is further reduced to $\sim 5 \times 10^{-7}$ mbar and a motorized aperture allows for controlling the intensity of the few-cycle pulse that enters the experimental chamber between 0 and ~50 μJ.

7. Experimental chamber

At the entrance of the experimental chamber (see Fig. 4), the two collinear beams pass a variable filter setup that permits separation of the beams into the central XUV part and the annular IR part. A laser grade pellicle (15 μm thickness) is prepared to carry in its center a ⌀3.5 mm free-standing metallic foil to block the remaining NIR light on the beam axis. At the same time, this foil acts as a high-pass filter for the high-harmonic radiation and is chosen to have transmittance characteristics matching the spectral bandwidth required for the experiments. The XUV transmission of thin

M. Schultze et al. / Journal of Electron Spectroscopy and Related Phenomena 184 (2011) 68–77

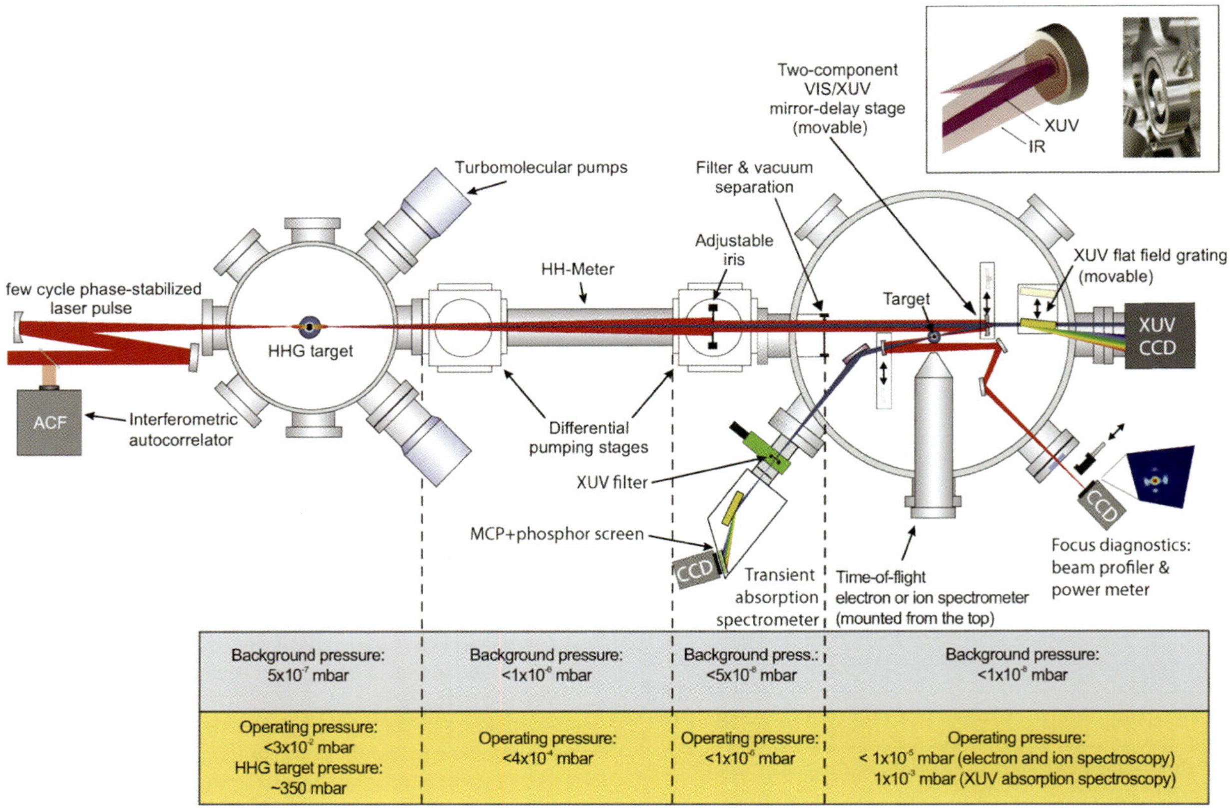

Fig. 4. Experimental setup: few-cycle phase-stabilized NIR laser pulses are focused into a noble gas target (Ne) to produce high order harmonic spectra from which attosecond XUV pulses can be isolated. The remaining NIR pulse and the XUV light propagate through differential pumping stages to prevent the gas used in the generation process from entering the experimental chamber. The "HH-Meter" tracks the ionization of the background gas and therefore provides a direct measure of the XUV flux. The NIR intensity is adjusted by an aperture. A thin metal foil mounted at the center of a laser grade pellicle acts as high pass filter to extract the continuum part of the generated XUV spectrum. The fundamental is transmitted through the pellicle in form of an annular beam around this filter. The two concentric beams are then focused by a double mirror assembly (inset) consisting of a XUV multilayer mirror in the inner part and a silver-coated perforated mirror of the same focal length. XUV and NIR pulses are spatially and temporally overlapped in the Fourier plane while their delay can be adjusted over a range of ±150 fs by translating the inner mirror along the beam axis using a closed loop piezo stage. Pump–probe experiment can be performed on a variety of gaseous samples and besides the time-of-flight spectra of the emitted photoelectrons (streaking measurements), mass spectra (ionization studies) and the transmitted photon spectrum can be recorded (transient absorption spectroscopy).

metal foils typically used in attosecond experiments is given in Fig. 5 [33]. After that filtering, the two beams impinge on a two-component delay mirror assembly shown in the inset of Fig. 4. The inner mirror is coated by a multilayer structure designed for high reflectivity in the XUV range. The outer part of the mirror assembly is coated with silver to reflect the laser pulses virtually dispersion-free. Both concave mirrors have a radius of curvature (ROC) of −250 mm. The inner mirror [34] is mounted on a closed

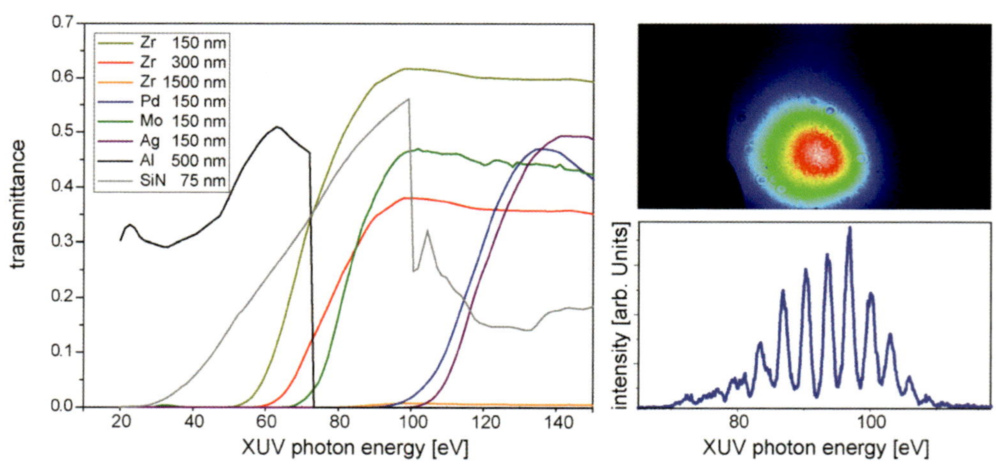

Fig. 5. Filter transmittance vs. photon energy (left panel) [33]. The black and the grey curve depict transmittance edges of aluminum and silicon-nitride that are suited for a calibration of both the electron- as well as the photon spectrometer. The colored curves show the transmission properties for different metals (material and thickness specified in the legend of the graph) that are used in the experiment for wavelength selection of the radiation. High-harmonic beam profile (right panel) and spectrum. The upper image shows the beam profile of the HHG beam on the XUV sensitive camera. The lower panel shows the HHG spectrum as observed in the experimental chamber after transmission of the incident XUV light through a 1.5 μm zirconium foil. To demonstrate the spectrometer resolution, the laser bandwidth was restricted to achieve multicycle pulses that generate modulated high-harmonic spectra with the highest possible fringe contrast.

M. Schultze et al. / Journal of Electron Spectroscopy and Related Phenomena 184 (2011) 68–77

loop controlled piezo stage that allows translation of the mirror along the propagation axis with a resolution better than 2 nm setting the ultimate limit for the smallest possible delay step to the equivalent of 14 as. The double mirror assembly is motorized to move perpendicularly along the incoming beam, and features two independently adjustable angles to precisely position the focus of the two beams in the acceptance volume of the detection device.

For the experiments, it is crucial to perfectly overlap the two pulses reflected off the inner and the outer mirror not only along the beam, i.e. in time but also to accurately superpose them in space. To this end, an imagining system is installed that is based on a concave lens (f = 15 cm, $M \sim 20$) and a CCD camera that is placed outside the experimental chamber. Due to its shorter wavelength, the focus of the XUV pulse is safely smaller than the laser focus ($\sim$25 μm FWHM).

The double mirror stack focuses the two beams under a narrow angle (<5°) such that the focus is located a bit to the side of the incoming beams, the astigmatism and the waveform distortion that result from this aberration are small enough not to bias the experiments. The focal region is located within the acceptance volume of an electron time-of-flight and/or a reflectron mass spectrometer detecting ionic charge states. Given the low absorption cross sections in this energy range and the limited number of photons, the gas pressure in the focal volume should be as high as possible without deteriorating the vacuum conditions over an extended region. To this end, a carbon-coated and grounded glass nozzle (to minimize charging effects) with an inner diameter of $\sim$30 μm effuses a gas as a cone perpendicular to the laser beams. This approach limits the interaction of the attosecond and laser pulses to within a fraction of the focal volume and minimizes averaging effects that could emerge due to a longitudinal phase delay (Gouy phase shift) or non-uniform intensity distributions. Another class of experiments relies on the XUV spectrum transmitted through an ensemble of atoms. For these attosecond transient absorption experiments, a quasi-static gas cell which is similar to the high-harmonic generation target discussed earlier can be installed into the focus. All the in-vacuum motorization is done using vacuum suited translation stages [35].

8. Spatial and spectral characterization of the XUV source

To optimize the XUV source, the direct observation of the direction and spectral distribution of the beam is essential. For these purposes, an imaging spectrometer based on a grazing-incidence flat field grating [36,37] and a back-illuminated CCD camera are located at the exit of the experimental chamber. After the double-mirror assembly is moved out of the beam path, the XUV light can be monitored (Fig. 4). The XUV camera observes either the direct beam, or, after inserting the grating, the spectral extent of the generated XUV light (Fig. 5). The imaging spectrometer has an overall efficiency (grating $\sim$85%, CCD-chip $\sim$40%) of 34%. Its minimal acquisition time to achieve a statistically significant spectrum is limited by the shortest exposure time of the camera, which is $\sim$50 ms. The spectrometer therefore constitutes one of the fastest CE-phase monitoring devices by requiring the accumulation of approximately 150 pulses to provide information about changes in the CE-phase. The spectrometer's resolution is estimated to be around 0.5 eV, therefore the typical spacing of twice the fundamental photon energy (3.2 eV for 750 nm photons) in the harmonic spectra can be clearly resolved as shown in Fig. 5. In order to avoid saturating the CCD chip, the laser light is rejected and the XUV light attenuated by a 1.5 μm zirconium foil at the entrance of the spectrometer.

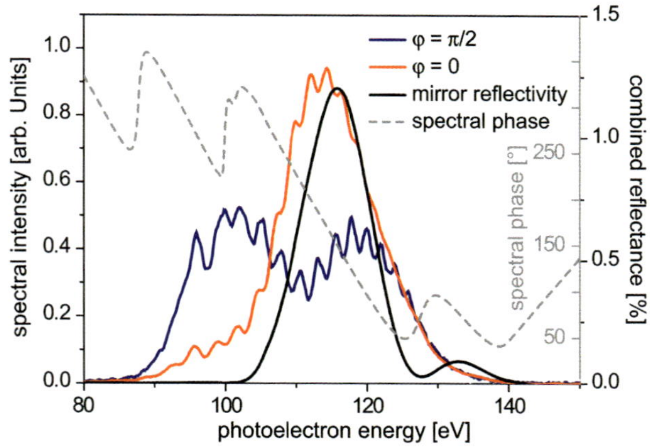

Fig. 6. XUV spectra emanating from the interaction of few-cycle phase stabilized laser pulses with Ne atoms after being transmitted through 1.5 μm of a free standing Zr foil. The spectra are integrated over $\sim$300 shots of the laser. Red/blue curve: spectra of the XUV pulses emitted by few-cycle driver pulses with a CE phase different by $\pi/2$. The high degree of modulation in the spectrum associated to the waveform labeled as $\pi/2$ constitutes the spectral manifestation of a substantial fraction of intensity contained in a satellite pulse in the time domain. To isolate attosecond pulses from the generated spectrum, multilayer mirrors with band-pass characteristics are used. The combined reflectance/transmittance characteristics of a multilayer mirror and the palladium foil are indicated by the black line. The grey dashed curve shows the calculated spectral phase response of the multilayer mirror. (For interpretation of the references to color in this figure legend, the reader is referred to the web version of this article.)

9. Spectral filtering of the XUV light

To make use of the extreme temporal resolution that attosecond pulses offer, an experiment needs to be capable of spectrally selecting and focusing the generated radiation. Given the low conversion efficiency, the focusing should not drastically compromise the energy contained in the attosecond pulse and the temporal characteristics should be preserved. Furthermore, for isolating single attosecond pulses it is furthermore necessary that the XUV optics filter out the modulated part of the spectrum that give rise to a train of attosecond pulses [38].

Elaborate XUV multilayer optics can meet these ambitious demands. These mirrors consist of a series of layers of alternating materials (in the energy range discussed here e.g. molybdenum/silicon or molybdenum/lanthanum) coated on a finely polished concave substrate. By adjusting the relative layer thicknesses, the sequence of materials and the number of layers, the central energy, the spectral response as well as the reflectance of the mirrors can be adjusted to meet the requirements of attosecond experiments. In addition, a proper adjustment of the mirror's spectral dispersion properties of the mirror allows to achieve pulses with durations close to the Fourier limit of the reflected spectra or to set arbitrary spectral phase values ("chirp") according to the needs of a certain experiment [39]. The combined transmittance/reflectance characteristics of the metallic filter and the mirror facilitate the selection of a spectral band in which, with carefully chosen parameters of the generating laser field, the emission corresponds to only one single half cycle of the laser pulse and thus forms an isolated pulse.

The sensitivity of this method is displayed in Fig. 6 showing XUV spectra recorded for different settings of the CE phase of the driver laser pulses matched to the reflectance band of a multilayer mirror supporting 160 as pulses. Recent advances in the laser source development towards pulse durations containing less than two cycles of the laser field [11] allow for the generation of ever broader unmodulated XUV spectra. The combination of state-of-the-art short-pulse lasers with ultra-broadband multilayer mirrors

M. Schultze et al. / Journal of Electron Spectroscopy and Related Phenomena 184 (2011) 68–77

led to the realization of the first sources that routinely generate XUV pulses of less than 100 as duration [25].

10. XUV pulse intensity

Core-hole excitations, followed by dynamics evolving on an attosecond timescale generally possess low absorption cross sections in the XUV. For example, in the case of direct photoionization of helium 1s electrons at a 90 eV photon energy, the partial cross section amounts to only 5×10^{-1} Mb, while for the first shake-up satellite ($n = 2$) it is only 4×10^{-2} Mb [40].

Experimentally, these low cross sections translate into low count rates in the cross correlation experiments described later in the manuscript. Modifications in the generation scheme of the high harmonic light therefore have to be reviewed in consideration of the generated flux within the useful energy band. The number of photons that form the attosecond pulse, however, is not directly measurable and hence is often obtained through rough estimations. The XUV-sensitive camera used to observe photon spectra shown in Fig. 4 can record images of the direct XUV beam. After proper background subtraction and considering the known quantum efficiency of the camera and the integration time, an absolute number of photons after the filters in front of the camera can be extracted. The spectral shape of the attosecond pulse is accessible using the electron spectrometer, since neither the ionization cross section of neon nor the detection efficiency for the electrons vary substantially throughout the bandwidth of interest.

Attosecond streaking together with a FROG analysis of the spectrogram (an example is shown in Fig. 7) can certify the absence of satellite pulses implying that all the energy is contained within a single attosecond pulse [25,41]. The knowledge of the filter transmittance (Fig. 5) and the mirror reflectance (Fig. 6), which can be independently measured at synchrotron sources, allow for the identification of the integral of the photon spectrum after the mirror. The number of photons generated fades dramatically towards higher energies. Due to the fast intensity drop of the harmonic radiation in the cut-off part of the spectrum, only a vigorous suppression of the low-energetic photons allows for the generation of attosecond pulses truly isolated in time. Choosing the right filter material corresponding to a given mirror reflectivity is therefore a crucial point in the reproducible generation of isolated attosecond pulses. The typical XUV flux attainable in isolated pulses amounts to $\sim 5 \times 10^8$ photons per second. A single pulse therefore contains 10^5 photons corresponding to pulse energies of a few pJ [41].

Attosecond time-resolved experiments can be carried out based on three different detection schemes that can be incorporated in the presented apparatus. An electron time-of-flight spectrometer was used for the timing studies of photoionization, while for time resolved experiments on tunnel ionization, the ions were detected by a reflectron type time-of-flight ion spectrometer. Quantum level population dynamics of valence shell electrons are tracked by means of NIR pump–XUV probe absorption spectroscopy.

11. Attosecond streaking metrology and spectroscopy

In the case of electron time of flight spectrometry, the kinetic energy of the electrons under scrutiny is determined by measuring their arrival time on a detector placed in a known distance from the interaction volume. Since the range of kinetic energies of interest is set by the photon energy, causing the excitation to be lowered by the ionization potential of the atom, the most energetic photons detected have a kinetic energy of about 100 eV. The resolution of ~1% of the electron kinetic energy is satisfactory for the experiments carried out, even though it sets a limit to the observable structures in the recorded spectra, which might give additional insight to the processes involved.

In the TOF spectrometer [42], the electrons pass a magnetic shielded drift tube, after which they undergo acceleration. This post acceleration ensures a uniform response of the microchannel plates (MCP), irrespective of the electrons kinetic energy. At the entrance of the flight tube, an electrostatic lens increases the acceptance angle of electrons within a certain range of kinetic energies. The signal measured at the MCP is digitized using a multiscaler card [43] with a minimal time discrimination of 250 ps. The red line in panel c of Fig. 7 shows a typical electron time-of-flight spectrum recorded when neon is ionized with radiation reflected by the multilayer mirror described in Fig. 6. The electron TOF scheme typically yields one count per laser pulse within the experimental bandwidth, using neon as a target gas. This rate is almost independent of the bandwidth response of the multilayer mirror because the mirror's reflectance is inversely related to its bandwidth. The spectrogram shown in Fig. 7 comprises 100 individual spectra and is taken within 30 min, at a delay step size of 100 as. The two spectral contributions can be identified as electrons released from the 2p and 2s subshells of the neutral neon atom. If the attosecond XUV pulse releases electrons in the presence of the laser electric field, the final kinetic energy of the electron is modified.

The net momentum change imparted to the electrons by the laser field depends on the time delay between the two pulses. A set of electron TOF spectra are recorded over a range of delays between them. The resulting spectrogram is shown in Fig. 7 and it is immediately evident that this so called "streaking spectroscopy" [2,6,7,44] samples the temporal evolution of the laser electric field (more precisely its associated vector potential). A more involved analysis along the lines of frequency-resolved optical gating (FROG) [45–47] additionally allows for reconstructing the temporal structure of the attosecond pulse, including its temporal and spectral phase. Streaking experiments enable the temporal characterization of the attosecond burst in the high harmonic process when performed on electrons liberated from only one specific atomic state [48,49]. The growing ability to shape the XUV emission according to the needs of the experiment extends the technique to a wider spectral range and thus to the study of multiple photoemission lines. For a spectrogram as shown in Fig. 7, it is found that the emission of electrons from the 2s shell of neon precedes the liberation of those from the 2p shell [50] by about 20 as. These results highlight the unprecedented temporal resolution offered by attosecond streaking spectroscopy, and how (in combination with the powerful FROG technique) it gives access to subtle timing features in the photoemission occurring in gases and solids [3] that hitherto remained experimentally inaccessible.

12. Attosecond ionization spectroscopy

In combination with a mass spectrometer, the presented pump–probe system constitutes a sampling system capable of tracing the relaxation dynamics of core excited ions with attosecond resolution [51,52]. The ion population transfer following the attosecond XUV excitation is tracked with a reflectron type TOF ion spectrometer [53]. Its working principle is based on time-of-flight mass spectrometry where ions are accelerated in an electric field to the same kinetic energy with the velocity of the ion depending on the mass-to-charge ratio. After the ionization event, the ratio between different charge states can be determined as shown in Fig. 8. We have chosen a reflectron spectrometer because the ion mirror setup grants two important facilities. On the one hand it is possible to set its dispersion characteristics for the drifting ions, to compensate for an energy spread of the generated ion packet due to e.g. delayed ion formation and space charge effects and thus

M. Schultze et al. / Journal of Electron Spectroscopy and Related Phenomena 184 (2011) 68–77

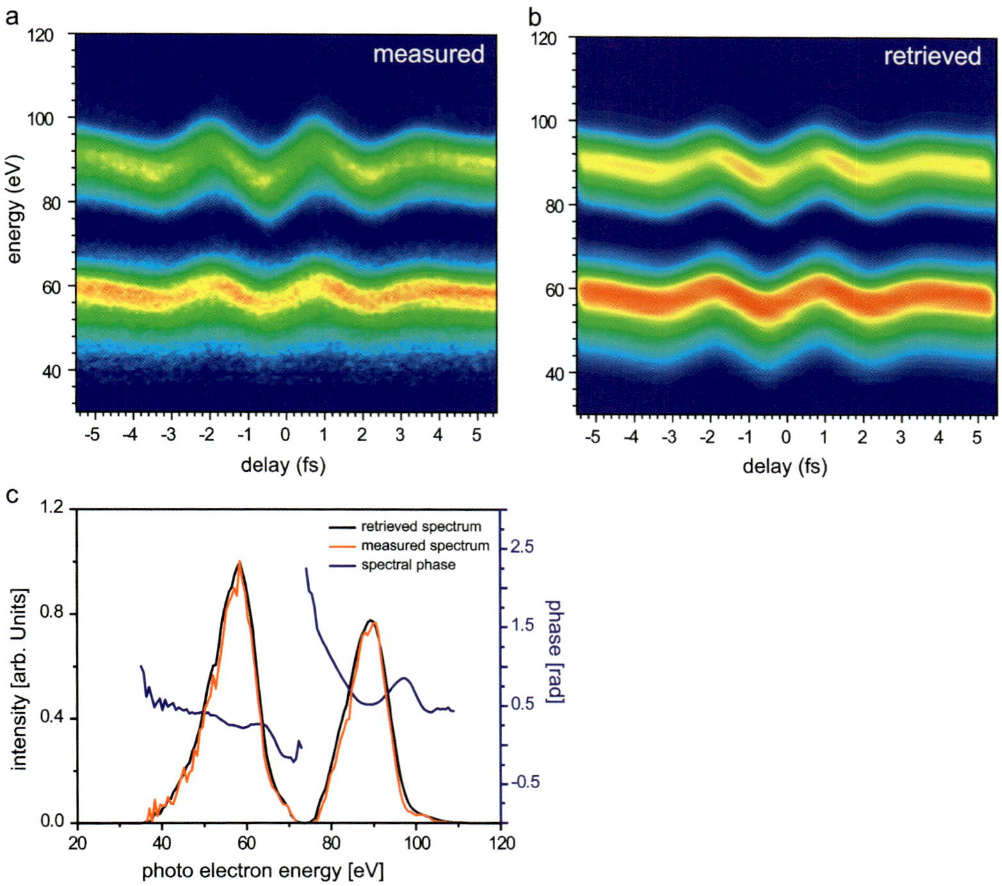

Fig. 7. Measured (a) streaking spectrogram for the ionization of neon from the 2p and the 2s shell after absorption of a 200 as XUV pulse. The streaking spectrogram is processed with a frequency-resolved optical gating (FROG) code to identify the temporal structure of the XUV pulse. The similarity of the measured (red) and the retrieved (black line) photoelectron spectrum shown in panel c highlights the accuracy of the FROG analysis that additionally yields the spectral phase (blue line) of the electron wave packet released after XUV absorption. This spectral phase comprises information about the relative timing of the two emissions and the phase of the emission itself. (For interpretation of the references to color in this figure legend, the reader is referred to the web version of this article.)

improve the attainable resolution. On the other hand, the ion mirror can act as an energy filter that permits the detection of ions exclusively created in a confined volume inside the laser focus providing access to truly intensity resolved ionization yields [54].

The photon energy of the XUV pulses used for that experiment was centered at 93 eV and exhibited a bandwidth of 9 eV, corresponding to pulses of ~250 as in duration. The data are accumulated using an amplifier in combination with a multiscaler card. The overall resolution of the spectrometer and the detection electronics allows a $\Delta m/m$ ratio of better than 10^{-3}.

An illustrative application of attosecond ionization spectroscopy is to explore the inner-shell relaxation times of xenon atoms as presented in Fig. 8. For the 91 eV XUV photons applied in this experiment, the cross-section for single ionization of xenon is much smaller than for double and even triple ionization accompanied by secondary processes such as Auger decay or autoionization. The resulting temporal evolution of the number of atoms in the Xe^{4+} state reveals the characteristic time structure of two subsequent Auger decays [51]. In this experiment, one attosecond pulse released about 5 ions distributed over the different charge states.

13. Attosecond transient absorption spectroscopy

As depicted in Fig. 4, a second XUV spectrometer enables the observation of the spectral content of attosecond pulses after their propagation through an ensemble of atoms within a confined interaction volume. By retracting a movable mirror which was directing the beam into the imaging system and onto the CCD, both beams impinge on a concave (ROC = −340 mm) multilayer broadband XUV optic whose reflectivity was optimized for an angle of incidence of 67°. This concave optic creates a line focus in the plane of the entrance slit of the spectrometer. A suitable thin metal filter (here zirconium of 150 nm thickness) in front of the entrance slit blocks the fundamental NIR radiation. Spectral dispersion of the XUV light takes place in a commercially available Rowland-circle based spectrometer (MCPherson, Model 248/310, no exit slit), utilizing a concave grating with 2400 grooves per millimeter. Its detection is realized by means of a microchannel plate in combination with a phosphor screen and fiber-coupled CCD camera. This detection scheme inherits a heavily reduced sensitivity to the fundamental NIR light than compared to XUV sensitive CCD cameras, which is a prerequisite for obtaining a high signal-to-noise ratio of the XUV spectra. In this class of experiments, the NIR pulse serves as the pump and the attosecond XUV pulse acts as the probe pulse which can read out characteristic information of the target atoms, which in turn translates to resonant and non-resonant absorption features in its spectrum, as shown in Fig. 9 (upper panel). Here, a sub-4 fs NIR pulse strong-field ionizes at 5.5×10^{14} W/cm^2 neutral krypton (Kr) atoms in the experimental interaction volume comprising a gas pressure of ~80 mbar. The remaining neutral atoms and strong-field-created ions are then probed via a delayed ~150 as XUV pulse centered at ~80 eV, as indicated in the inset of Fig. 9

M. Schultze et al. / Journal of Electron Spectroscopy and Related Phenomena 184 (2011) 68–77

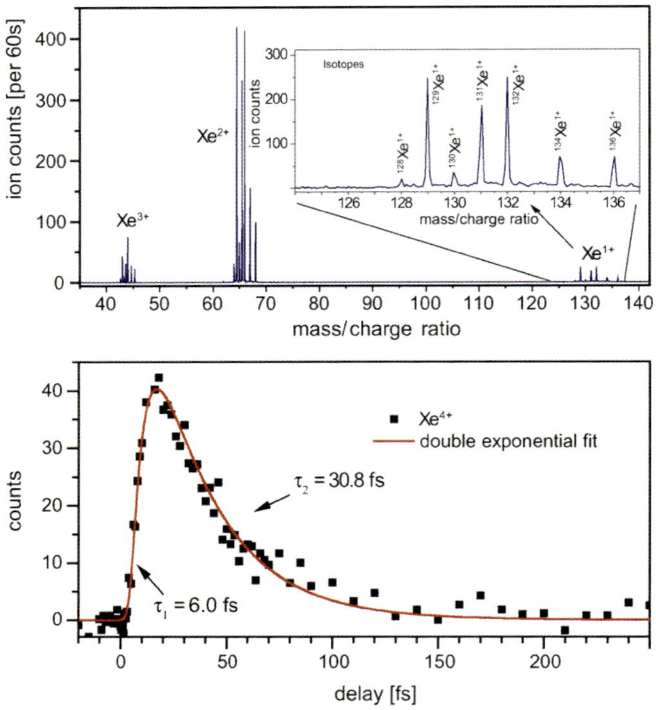

Fig. 8. Upper panel: mass-spectrum recorded in xenon after attosecond XUV excitation with photon energy around 100 eV. Three charge states of xenon are displayed, the relative intensity shows a distinguished correspondence with the interaction cross sections known from synchrotron studies. The inset shows a close up of the generated Xe^{+} ions. Seven isotopes can be discriminated; there are two more (mass 124 and 126, respectively) with natural abundances below 0.1%. The resolution of the spectrometer is specified as $\Delta m/m = 10^{-3}$. In the presence of a strong laser pulse, a part of the Xe^{3+} population can be transferred to Xe^{4+}. The temporal evolution of this population reveals the characteristic time structure of two Auger processes.

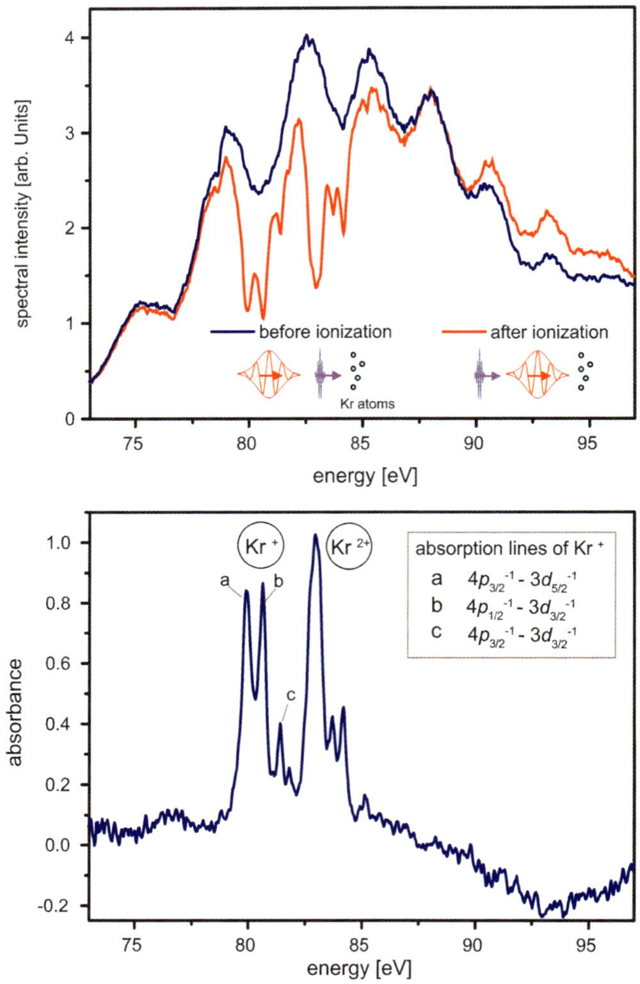

Fig. 9. Exemplary XUV attosecond transient absorption spectra of an attosecond probe pulse centered at ~80 eV and propagated through a krypton (Kr) gas target (upper panel). Blue (orange) curve shows the recorded XUV spectrum before (after) ionization of the gas target by a sub-4 fs NIR pump pulse at 5.5×10^{14} W/cm^2. System-specific absorption signatures in the XUV spectrum arise from the absorption properties of the probed ensemble of neutral and strong-field generated charged states of Kr as displayed in the lower panel. Besides absorption lines of Kr^{+} and Kr^{2+} also a region of negative absorbance between 89 eV and 98 eV arises from the NIR-reduced number density of neutral Kr absorbers. Identified absorption lines of Kr^{+} are further specified in terms of transitions in the inset of the lower panel where the superscript -1 denotes one missing electron in the corresponding term. The total angular momentum of the ion is given by the subscript. As shown above, the broad XUV spectrum of the attosecond probe pulse allows for simultaneous tracking of absorption features of different charge states as a function of NIR pump–XUV probe delay. (For interpretation of the references to color in this figure legend, the reader is referred to the web version of this article.)

(upper panel). The exposure time of each of the recorded spectra amounted to 30 s. In order to achieve the reported visibility of characteristic absorption features in the spectrum of the XUV pulse upon its propagation through the gas target, a quasi static gas target, similar to the one used for high-order harmonic generation, had been used instead of a diffusive gas nozzle which is used for attosecond streaking experiments. Due to the inner diameter of the quasi-static gas cell, confinement of the interaction volume between NIR pump and attosecond probe pulses amounts to 1 mm along the beam axis. Fig. 9 (lower panel) displays the extracted absorbance of the XUV pulse based on the two XUV spectra shown in the graph above. Several resonant absorption lines are clearly discernible from non-resonant absorption areas, whereas we can identify the distinct absorption lines as core-to-valence shell transitions triggered by the XUV pulse in Kr^{+} and Kr^{2+}. The spectral resolution of ~350 meV at 80 eV was limited by the almost fully opened entrance slit of the spectrometer.

Extending this experiment to other atomic or molecular [1,4,5] systems by customizing, within certain limitations, the central energy of the attosecond pulse to the spectral needs of the system of interest, the characteristic information of quantum level populations and, if applicable, subsequent coherent dynamics within the irradiated system can be extracted by tracing the characteristic absorption features on an attosecond time scale. Since the determination of quantum level populations after the strong-field interaction are also accessible by conventional time integrated spectroscopy or transient absorption studies utilizing trains of attosecond pulses [55], this attosecond transient absorption spectroscopic approach is able to address sub-cycle ionization dynamics as well as to time-resolve coherent superpositions in the valence shell of atoms. Thus, it can completely and quantitatively retrieve strong-field created electronic coherences as reported in [56].

14. Conclusion and outlook

We have presented a complete experimental setup for attosecond metrology and spectroscopy based on a laser system that provides intense quasi-monocycle laser pulses and a beamline geared for experiments that require attosecond resolution. The beamline combines the merits of attosecond sources with conventional spectroscopic approaches that can enable complete insight into electronic processes in atoms, molecules and solids.

M. Schultze et al. / Journal of Electron Spectroscopy and Related Phenomena 184 (2011) 68–77

Acknowledgements

This work is supported by the Max Planck Society and the Deutsche Forschungsgemeinschaft Cluster of Excellence: Munich Centre for Advanced Photonics (www.munich-photonics.de) and the KSU-MPQ collaboration.

References

[1] B.H. Muskatel, F. Remacle, R.D. Levine, Phys. Scr. 80 (2009) 048101.
[2] M. Drescher, M. Hentschel, R. Kienberger, M. Uiberacker, V. Yakovlev, A. Scrinzi, Th. Westerwalbesloh, U. Kleineberg, U. Heinzmann, F. Krausz, Nature 419 (2002) 803.
[3] A.L. Cavalieri, et al., Nature 449 (2007) 1029.
[4] P. Eckle, et al., Science 322 (2008) 1525.
[5] T. Remetter, P. Johnsson, J. Mauritsson, K. Varju, Y. Ni, F. Lepine, E. Gustafsson, M. Kling, J. Khan, R. Lopez-Martens, K.J. Schafer, M.J.J. Vrakking, A. L'Huillier, Nat. Phys. 2 (2006) 323.
[6] M. Hentschel, R. Kienberger, C. Spielmann, G.A. Reider, N. Milosevic, et al., Nature 414 (2001) 509–513.
[7] R. Kienberger, E. Goulielmakis, M. Uiberacker, A. Baltuška, V.S. Yakovlev, et al., Nature 427 (2004) 817–821.
[8] T. Fuji, J. Rauschenberger, A. Apolonski, V.S. Yakovlev, G. Tempea, et al., Opt. Lett. 30 (2005) 332–334.
[9] J. Rauschenberger, T. Fuji, M. Hentschel, A.-J. Verhoef, T. Udem, et al., Laser Phys. Lett. 3 (2006) 37–42.
[10] T. Fuji, J. Rauschenberger, C. Gohle, A. Apolonski, T. Udem, et al., New J. Phys. 7 (2005) 116.
[11] A.L. Cavalieri, E. Goulielmakis, B. Horvath, W. Helml, M. Schultze, M. Fieß, V. Pervak, L. Veisz, V.S. Yakovlev, M. Uiberacker, A. Apolonski, F. Krausz, R. Kienberger, New J. Phys. 9 (2009) 242.
[12] M. Nisoli, S. De Silvestri, O. Svelto, R. Szipöcs, K. Ferencz, Ch. Spielmann, S. Sartania, F. Krausz, Opt. Lett. 22 (1997) 522–524.
[13] A.B. Fedotov, E.E. Serebryannikov, A.M. Zheltikov, Phys. Rev. A 76 (2007) 053811.
[14] V. Pervak, A.V. Tikhonravov, M.K. Trubetskov, S. Naumov, F. Krausz, et al., Appl. Phys. B 87 (2007) 5–12.
[15] TEM Messtechnik GmbH, http://www.tem-messtechnik.de/.
[16] R. Holzwarth, Th. Udem, T.W. Hänsch, J.C. Knight, W.J. Wadsworth, P.St.J. Russell, Phys. Rev. Lett. 85 (2000) 2264.
[17] D.J. Jones, S.A. Diddams, J.K. Ranka, A. Stentz, R.S. Windeler, J.L. Hall, S.T. Cundiff, Science 288 (2000) 635.
[18] C. Gohle, J. Rauschenberger, T. Fuji, T. Udem, A. Apolonski, et al., Opt. Lett. 30 (2005) 2487–2489.
[19] A. Baltuska, et al., IEEE J. Sel. Top. Quantum Electron. 9 (4 (July–August)) (2003) 972–989.
[20] A.J. Verhoef, Method and device for carrier envelope phase stabilization, Patent No. PCT/EP2006/11561.
[21] T. Wittmann, B. Horváth, W. Helml, M.G. Schätzel, X. Gu, A. Cavalieri, G.G. Paulus, R. Kienberger, Nat. Phys. 5 (2009) 357.
[22] C. Rullière (Ed.), Femtosecond Laser Pulses: Principles and Experiments, second ed., Springer, New York, USA, 2005.
[23] D. Kane, R. Trebino, IEEE J. Quantum Electron. 29 (1993) 571–579.
[24] U. Graf, M. Fieß, M. Schultze, R. Kienberger, F. Krausz, E. Goulielmakis, Opt. Expr. 16 (23) (2008) 18956–18963.
[25] E. Goulielmakis, et al., Science 320 (2008) 1614.
[26] P. Corkum, Phys. Rev. Lett. 71 (1993) 1994–1997.
[27] T. Brabec, F. Krausz, Rev. Mod. Phys. 72 (545) (2000).
[28] M. Schnürer, Z. Cheng, M. Hentschel, F. Krausz, T. Wilhein, D. Hambach, G. Schmahl, M. Drescher, Y. Lim, U. Heinzmann, Appl. Phys. B 70 (2000) S227–S232.
[29] M. Schnürer, Z. Cheng, M. Hentschel, F. Krausz, T. Wilhein, D. Hambach, G. Schmahl, M. Drescher, Y. Lim, U. Heinzmann, Appl. Phys. B 70 (7) (2000) 227.
[30] M. Lewenstein, P. Balcou, M.Y. Ivanov, A. L'Huillier, P. Corkum, Phys. Rev. A 49 (1994) 2117–2132.
[31] L. Poletto, G. Tondello, P. Villoresi, Rev. Sci. Instrum. 72 (2001) 2868.
[32] H. Geiger, W. Müller, Naturwissenschaften 16 (31) (1928) 617.
[33] The Center for X-Ray Optics at the Lawrence Berkeley National Laboratory, http://www-cxro.lbl.gov/.
[34] A. Wonisch, U. Neuhäusler, N.M. Kabachnik, T. Uphues, M. Uiberacker, V. Yakovlev, F. Krausz, M. Drescher, U. Kleineberg, U. Heinzmann, Appl. Opt. 45 (17) (2005) 4147–4156.
[35] Micos GmbH, http://www.micos-online.com/.
[36] T. Harada, H. Sakuma, K. Takahashi, T. Watanabe, H. Hara, T. Kita, Appl. Opt. 37 (1998) 6803–6810.
[37] N. Nakano, H. Kuroda, T. Kita, T. Harada, Appl. Opt. 23 (1984) 2386.
[38] D. Charalambidis, P. Tzallas, E.P. Benis, E. Skantzakis, G. Maravelias, L.A.A. Nikolopoulos, A.P. Conde, G.D. Tsakiris, New J. Phys. 10 (2008) 025018.
[39] M. Hofstetter, M. Schultze, M. Fieß, et al., Optics Express 19 (3) (2011) 1767–1776.
[40] U. Becker, D.A. Shirley, VUV and Soft-X-ray Photoionization, Plenum Press, New York, 1996.
[41] M. Schultze, E. Goulielmakis, M. Uiberacker, M. Hofstetter, J. Kim, D. Kim, F. Krausz, U. Kleineberg, New J. Phys. 9 (243) (2007).
[42] Time-of-flight (TOF) Spectrometers manufactured by Stefan Kaesdorf, Geräte für Forschung und Industrie, http://www.kaesdorf.brd.de/.
[43] FAST ComTec Communication Technology GmbH, http://www.fastcomtec.com/.
[44] J. Itatani, F. Quéré, G.L. Yudin, M.Yu. Ivanov, F. Krausz, P.B. Corkum, Phys. Rev. Lett. 88 (2002) 173903.
[45] Y. Mairesse, F. Quéré, Phys. Rev. A 71 (2005) 0011401(R).
[46] J. Gagnon, E. Goulielmakis, V.S. Yakovlev, Appl. Phys. B 92 (2008) 25–32.
[47] J. Gagnon, V.S. Yakovlev, Opt. Expr. 17 (2009) 17678–17693.
[48] G. Sansone, E. Benedetti, F. Calegari, C. Vozzi, L. Avaldi, et al., Science 314 (2006) 443–446.
[49] E. Goulielmakis, M. Uiberacker, R. Kienberger, A. Baltuška, V.S. Yakovlev, et al., Science 305 (2004) 1267–1269.
[50] M. Schultze, et al., Science 328 (2010) 1658–1662.
[51] M. Uiberacker, et al., Nature 446 (2007) 627–632.
[52] Th. Uphues, M. Schultze, M.F. Kling, M. Uiberacker, S. Hendel, U. Heinzmann, N.M. Kabachnik, M. Drescher, New J. Phys. 10 (2008) 2.
[53] M. Wagner, H. Schröder, J. Mass Spectrom. 128 (1–2) (1993) 31–45.
[54] M. Schultze, B. Bergues, H. Schröder, F. Krausz, K.L. Kompa, New J. Phys. in press.
[55] Z.-H. Loh, et al., Phys. Rev. Lett. 98 (2007) 143601.
[56] E. Goulielmakis, et al., Nature 466 (2010) 739–743.

Collinear generation of ultrashort UV and XUV pulses

E. M. Bothschafter[1,2], A. Schiffrin[1], V.S. Yakovlev[1,3], A.M. Azzeer[4], F. Krausz[1,3], R. Ernstorfer[2,*] and R. Kienberger[1,2]

[1] *Max-Planck-Institut für Quantenoptik, Hans-Kopfermann-Str. 1, D-85748 Garching, Germany.*
[2] *Physik-Department E11, Technische Universität München, James Franck Str., D-85748 Garching, Germany.*
[3] *Department für Physik, Ludwig-Maximilians-Universität, Am Coulombwall 1, D-85748 Garching, Germany.*
[4] *Department of Physics and Astronomy, King Saud University, P.O. Box 2455, Riyadh 11451, Kingdom of Saudi Arabia.*
* *ralph.ernstorfer@mpq.mpg.de*

Abstract: We demonstrate the collinear generation of few-femtosecond ultraviolet and attosecond extreme ultraviolet pulses via a combination of third-harmonic and high harmonic generation in noble gases. The ultrashort coherent light bursts are produced by focusing a sub-1.5-cycle near-infrared/visible laser pulse in two subsequent quasi-static noble gas targets. This approach provides an inherently synchronized pair of UV and XUV pulses, where the UV radiation has a photon energy of ~5 eV and a pulse energy of up to 1 μJ and the XUV radiation contains up to $3.5 \cdot 10^6$ XUV photons per shot with a photon energy exceeding 100 eV. This source represents a novel tool for future UV pump/XUV probe experiments with unprecedented time-resolution.

OCIS codes: (190.7110) Ultrafast nonlinear optics; (190.7220) Upconversion, (190.4180) Multiharmonic generation, (260.7190) Ultraviolet, (260.7200) Ultraviolet, extreme

References and links

1. M. Uiberacker, T. Uphues, M. Schultze, A. J. Verhoef, V. Yakovlev, M. F. Kling, J. Rauschenberger, N. M. Kabachnik, H. Schröder, M. Lezius, K. L. Kompa, H. G. Muller, M. J. J. Vrakking, S. Hendel, U. Kleineberg, U. Heinzmann, M. Drescher, and F. Krausz, "Attosecond real-time observation of electron tunnelling in atoms," Nature **446**(7136), 627–632 (2007).
2. P. Eckle, M. Smolarski, P. Schlup, J. Biegert, A. Staudte, M. Schoffler, H. G. Muller, R. Dorner, and U. Keller, "Attosecond angular streaking," Nat. Phys. **4**(7), 565–570 (2008).
3. A. L. Cavalieri, N. Müller, T. Uphues, V. S. Yakovlev, A. Baltuska, B. Horvath, B. Schmidt, L. Blümel, R. Holzwarth, S. Hendel, M. Drescher, U. Kleineberg, P. M. Echenique, R. Kienberger, F. Krausz, and U. Heinzmann, "Attosecond spectroscopy in condensed matter," Nature **449**(7165), 1029–1032 (2007).
4. R. Kienberger, E. Goulielmakis, M. Uiberacker, A. Baltuska, V. Yakovlev, F. Bammer, A. Scrinzi, T. Westerwalbesloh, U. Kleineberg, U. Heinzmann, M. Drescher, and F. Krausz, "Atomic transient recorder," Nature **427**(6977), 817–821 (2004).
5. M. Hentschel, R. Kienberger, C. Spielmann, G. A. Reider, N. Milosevic, T. Brabec, P. Corkum, U. Heinzmann, M. Drescher, and F. Krausz, "Attosecond metrology," Nature **414**(6863), 509–513 (2001).
6. J. Itatani, F. Quéré, G. L. Yudin, M. Y. Ivanov, F. Krausz, and P. B. Corkum, "Attosecond streak camera," Phys. Rev. Lett. **88**(17), 173903 (2002).
7. P. S. Kirchmann, P. A. Loukakos, U. Bovensiepen, and M. Wolf, "Ultrafast electron dynamics studied with time-resolved two-photon photoemission: intra- and interband scattering in C6F6/Cu(111)," N. J. Phys. **7**, 113 (2005).
8. M. Bauer, "Femtosecond ultraviolet photoelectron spectroscopy of ultra-fast surface processes," J. Phys. D Appl. Phys. **38**(16), R253–R267 (2005).
9. G. Saathoff, L. Miaja-Avila, M. Aeschlimann, M. M. Murnane, and H. C. Kapteyn, "Laser-assisted photoemission from surfaces," Phys. Rev. A **77**(2), 022903 (2008).
10. P. Baum, S. Lochbrunner, and E. Riedle, "Tunable sub-10-fs ultraviolet pulses generated by achromatic frequency doubling," Opt. Lett. **29**(14), 1686–1688 (2004).
11. N. Aközbek, A. Iwasaki, A. Becker, M. Scalora, S. L. Chin, and C. M. Bowden, "Third-harmonic generation and self-channeling in air using high-power femtosecond laser pulses," Phys. Rev. Lett. **89**(14), 143901 (2002).
12. C. G. Durfee 3rd, S. Backus, H. C. Kapteyn, and M. M. Murnane, "Intense 8-fs pulse generation in the deep ultraviolet," Opt. Lett. **24**(10), 697–699 (1999).
13. T. Fuji, T. Horio, and T. Suzuki, "Generation of 12 fs deep-ultraviolet pulses by four-wave mixing through filamentation in neon gas," Opt. Lett. **32**(17), 2481–2483 (2007).

#124076 - $15.00 USD Received 9 Feb 2010; revised 29 Mar 2010; accepted 5 Apr 2010; published 16 Apr 2010

14. K. Kosma, S. A. Trushin, W. E. Schmid, and W. Fuß, "Vacuum ultraviolet pulses of 11 fs from fifth-harmonic generation of a Ti:sapphire laser," Opt. Lett. **33**(7), 723–725 (2008).
15. U. Graf, M. Fiess, M. Schultze, R. Kienberger, F. Krausz, and E. Goulielmakis, "Intense few-cycle light pulses in the deep ultraviolet," Opt. Express **16**(23), 18956–18963 (2008).
16. J. J. Macklin, J. D. Kmetec, and C. L. Gordon 3rd, "High-order harmonic generation using intense femtosecond pulses," Phys. Rev. Lett. **70**(6), 766–769 (1993).
17. A. L'Huillier, and P. Balcou, "High-Order Harmonic Generation in Rare Gases with a 1-Ps 1053-Nm Laser," Phys. Rev. Lett. **70**(6), 774–777 (1993).
18. P. B. Corkum, "Plasma Perspective on Strong Field Multiphoton Ionization," Phys. Rev. Lett. **71**(13), 1994–1997 (1993).
19. F. Brunel, "Harmonic-Generation Due to Plasma Effects in a Gas Undergoing Multiphoton Ionization in the High-Intensity Limit," J. Opt. Soc. Am. B **7**(4), 521–526 (1990).
20. A. L. Cavalieri, E. Goulielmakis, B. Horvath, W. Helml, M. Schultze, M. Fiess, V. Pervak, L. Veisz, V. S. Yakovlev, M. Uiberacker, A. Apolonski, F. Krausz, and R. Kienberger, "Intense 1.5-cycle near infrared laser waveforms and their use for the generation of ultra-broadband soft-x-ray harmonic continua," N. J. Phys. **9**(7), 242 (2007).
21. E. Goulielmakis, M. Schultze, M. Hofstetter, V. S. Yakovlev, J. Gagnon, M. Uiberacker, A. L. Aquila, E. M. Gullikson, D. T. Attwood, R. Kienberger, F. Krausz, and U. Kleineberg, "Single-cycle nonlinear optics," Science **320**(5883), 1614–1617 (2008).
22. M. Nisoli, E. Priori, G. Sansone, S. Stagira, G. Cerullo, S. De Silvestri, C. Altucci, R. Bruzzese, C. de Lisio, P. Villoresi, L. Poletto, M. Pascolini, and G. Tondello, "High-brightness high-order harmonic generation by truncated bessel beams in the sub-10-fs regime," Phys. Rev. Lett. **88**(3), 033902 (2002).
23. P. Salieres, T. Ditmire, K. S. Budil, M. D. Perry, and A. L'Huillier, "Spatial Profiles of High-Order Harmonics Generated by a Femtosecond Crlisaf Laser," J. Phys. At. Mol. Opt. Phys. **27**(9), L217–L222 (1994).
24. T. Bartel, P. Gaal, K. Reimann, M. Woerner, and T. Elsaesser, "Generation of single-cycle THz transients with high electric-field amplitudes," Opt. Lett. **30**(20), 2805–2807 (2005).
25. Y. Chen, M. Yamaguchi, M. Wang, and X.-C. Zhang, "Terahertz pulse generation from noble gases," Appl. Phys. Lett. **91**(25), 251116 (2007).
26. M. Kreß, T. Löffler, M. D. Thomson, R. Dörner, H. Gimpel, K. Zrost, T. Ergler, R. Moshammer, U. Morgner, J. Ullrich, and H. G. Roskos, "Determination of the carrier-envelope phase of few-cycle laser pulses with terahertz-emission spectroscopy," Nat. Phys. **2**(5), 327–331 (2006).
27. IRD, "International Radiation Detectors," *http://www.ird-inc.com/axuvwdd/axuv-100_zr_c.htm.*
28. D. L. Windt, "IMD - Software for modeling the optical properties of multilayer films," Comput. Phys. **12**(4), 360–370 (1998).
29. F. J. Himpsel, P. Heimann, T. C. Chiang, and D. E. Eastman, "Geometry-Dependent Si(2p) Surface Core-Level Excitations for Si(111) and Si(100) Surfaces," Phys. Rev. Lett. **45**(13), 1112–1115 (1980).
30. A. Bideaumehu, Y. Guern, R. Abjean, and A. Johanningilles, "Measurement of refractive-indexes of neon, Argon, krypton and xenon in the 253.7-140.4 nm wavelength range - Dispersion-relations and estimated oscillator-strengths of the resonance lines," J. Quant. Spectrosc. Radiat. Transf. **25**(5), 395–402 (1981).
31. V. Tosa, H. T. Kim, I. J. Kim, and C. H. Nam, "High-order harmonic generation by chirped and self-guided femtosecond laser pulses. I. Spatial and spectral analysis," Phys. Rev. A **71**(6), 063807 (2005).
32. M. Mlejnek, E. M. Wright, and J. V. Moloney, "Femtosecond pulse propagation in argon: A pressure dependence study," Phys. Rev. E Stat. Phys. Plasmas Fluids Relat. Interdiscip. Topics **58**(4), 4903–4910 (1998).

1. Introduction

Within the last decade attosecond metrology has been established as a tool for studying some of the fastest processes in nature, for example, field-induced tunneling of electrons in atoms [1, 2] and electron transport in solids on sub-nm length scales [3]. One spectroscopic technique for accessing this time domain is the attosecond transient recorder (ATR) [4]. This method is based on the availability of isolated attosecond (1 as $=10^{-18}$ s) pulses in the extreme ultraviolet (XUV) that are synchronized to a carrier envelope phase-stabilized few-cycle near-infrared (NIR) laser pulse. The momentum of photoelectrons emitted by the attosecond XUV pulse is altered depending on the strength of the vector potential at the time when the electron is released [5, 6]. Based on this concept, the ATR is a technique for studying the time of photoemission of electrons originating from ground states. In order to investigate the dynamics of excited electronic wave-packets, for instance, resolving the formation and relaxation of photo-excited electrons in valence states of atoms, molecules or solids on the attosecond to few-femtosecond timescale, it is highly desirable to generate an intense femtosecond light pulse in the ultraviolet (UV) spectral range in addition to an attosecond XUV pulse. Employing the conventional pump-probe scheme of time-resolved two-photon photoemission spectroscopy (TR-2PPE) [7–9], ultraviolet light-induced transient electron populations are photoemitted by the subsequent synchronized XUV pulse at well-defined time

#124076 - $15.00 USD Received 9 Feb 2010; revised 29 Mar 2010; accepted 5 Apr 2010; published 16 Apr 2010

delays. With the availability of isolated attosecond XUV pulses, this UV-XUV TR-2PPE scheme offers the intriguing possibility to study excited state dynamics with a time-resolution shorter than the optical period of the pump pulse

Our work aims to provide the prerequisites, i.e., highly synchronized UV and XUV pulses with the required pulse duration for this type of experiments. Since no suitable broadband laser material is available in the ultraviolet domain, the generation of ultrashort UV laser pulses relies on nonlinear frequency mixing and frequency up-conversion of femtosecond laser pulses in solids [10] and gases [11–15]. When multi-cycle laser pulses are used for generation [10–14], pulse durations below 10 fs can only be reached by compressing the generated UV pulses and precisely controlling and tuning the dispersion in the ultraviolet after the generation process. Recently, it has been demonstrated that third-harmonic generation from Fourier-limited few-cycle NIR laser pulses in noble gases directly provides intense sub-4 fs pulses in the deep ultraviolet spectral range [15]. While this process relies on the induced polarization due to the third-order electric susceptibility $\chi^{(3)}$ of the gas and therefore is favored in high-density media, higher-order harmonics can be generated with a very similar technique in low-density gases in a different regime [16, 17]. High harmonic generation can be described in a semi-classical three-step model [18, 19] involving tunnel ionization of the least bound electron of a noble gas atom [19], acceleration of the electron in the electric field followed by recombination with the atomic core, upon which the electron emits its excess energy in a short, high energetic burst of light. This process is triggered and completed every half-cycle of the driving laser field, so that by using NIR light pulses approaching the single-oscillation limit [20] the photons with the highest energy are emitted during a single half cycle. Filtering these so-called high-harmonic cut-off photons yields isolated XUV pulses with pulse durations below 100 as [21].

In the following we present a novel approach for generating ultrashort UV and attosecond XUV pulses that are inherently synchronized using a collinear configuration of two quasi-static gas cells in a single laser focus.

2. Experimental setup

The vacuum chamber employed in this work is shown schematically in Fig. 1. Two static gas cells (nickel tube, 2-3 mm diameter, 0.1 mm wall thickness) are placed in the focus (f=600 mm) of a few-cycle NIR laser beam. Both targets can be independently supplied with arbitrary gases and pressures exceeding 10 bar. In practice, the first gas target is optimized for third-harmonic generation and therefore it is supplied with several bar of either argon (Ar) or neon (Ne). The second gas target is typically supplied with several hundreds of mbar of Ne with the exact pressure being optimized for high harmonic generation. The holes in the targets through which the laser beam and the generated harmonics propagate are drilled by the focused fundamental laser beam and have a diameter of ~100 μm.

The gas load emerging from the gas targets is pumped differentially by using a second vacuum chamber nested inside the main vacuum setup with two small apertures (Ø 0.7 mm) for the laser beams and a total width of 50 mm in direction of propagation of the laser beam. This measure minimizes absorption and group velocity dispersion of the generated UV and XUV pulses after their generation. The two gas targets are placed inside of this small chamber and therefore the propagation distances of the fundamental laser beam and of the generated harmonic radiation in the region with a background pressure of 10^{-2}–1 mbar are approximately 50 mm and 25 mm, respectively. The gas target pressures given in the experimental section apply to the backing pressure of the gas targets. The fact that the background pressure is at least two orders of magnitude smaller than the backing pressure assures that the interaction region of the laser is mostly confined to the width of the gas targets.

Both targets can be positioned independently inside this small chamber in all three dimensions for alignment. The positions of the gas targets in direction of the laser beam (x_1, x_2 in Fig. 1) with respect to the laser focus and to each other are optimized for collinear generation. The optimal distance between x_1 and x_2 is typically a few millimeters and changes

#124076 - $15.00 USD Received 9 Feb 2010; revised 29 Mar 2010; accepted 5 Apr 2010; published 16 Apr 2010

in dependence of the gas pressure applied to the UV target as this affects the fundamental laser field as discussed below in section 3.3.

The parameters of the NIR laser pulses are similar to those in [20]. The nearly Fourier-limited sub-4 fs 400 μJ pulses have a 450 nm broad spectrum centered at 720 nm. The focusing geometry corresponds to the one used for the generation of isolated attosecond pulses. The peak intensity in the focus can be tuned with an iris in front of the focusing mirror and its maximum is $7 \cdot 10^{14}$ W/cm^2.

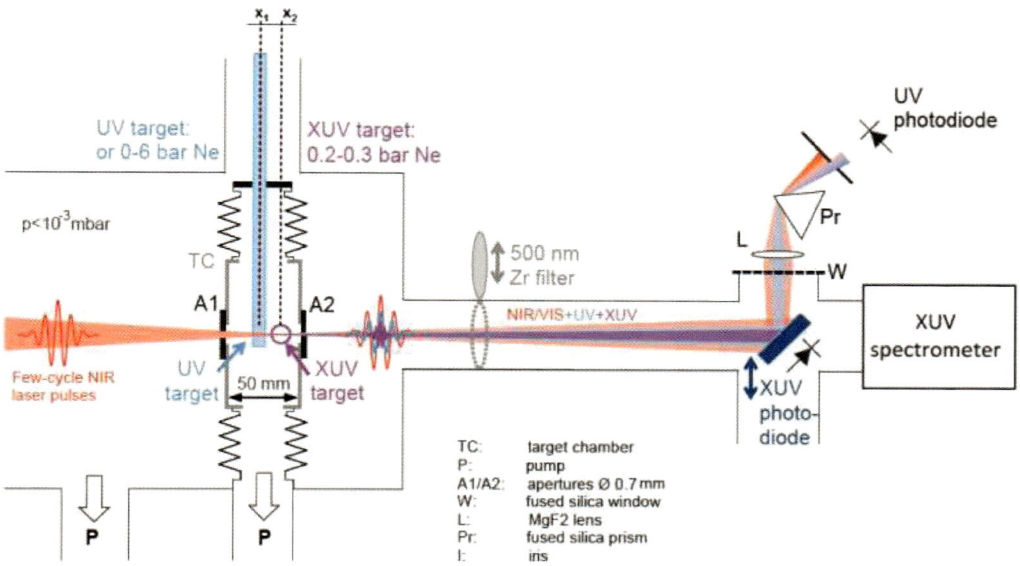

Fig. 1. Collinear generation of UV and XUV pulses in two subsequent gas targets. The targets, formed by 2 to 3 mm thick nickel tubes, can be positioned independently (x1, x2). They are situated inside a main vacuum setup in a small, differentially pumped vacuum chamber with two apertures (Ø 0.7 mm) for the laser. The few-cycle 720 nm NIR pulses (sub-4 fs, 400 μJ) are focused into the XUV gas target (purple) which is supplied with 0.2 - 0.3 bar Ne. The UV target (blue) is placed in front of the XUV target and supplied with Ar or Ne at variable pressure. The number of XUV photons per pulse is measured with an XUV diode placed under 45° in the beam path of the co-propagating NIR, UV and XUV beams. The diode is coated with a filter layer that partially reflects the NIR and the UV through a fused silica window outside the vacuum system, where the UV can be characterized in spectrum and pulse energy after separation from the IR with a prism. For the measurement of the XUV an additional 500 nm zirconium filter is inserted in front of the diode. The diode can be retracted for spectral characterization of the XUV with a grazing incidence grating (600 l/mm) spectrometer.

Behind the gas targets, the NIR, the UV and the XUV laser pulses propagate collinearly with different divergence depending on the respective wavelength [22, 23]. According to recent reports of THz generation from gases driven by femtosecond laser pulses [24–26], we also expect high field amplitude THz emission from the gas targets. This would add yet another source of ultrashort pulses in a different spectral range and will be subject of further investigations.

The number of XUV photons per pulse is measured with an XUV diode (International Radiation Detectors, AXUV100) placed under an angle of 45° in the co-propagating NIR, UV and XUV beams. The diode is coated with a directly deposited filter consisting of 200 nm zirconium (Zr), 10 nm amorphous carbon (a-C), 100 nm molybdenum (Mo) and 5 nm boron carbide (B_4C). In addition to that, a 500 nm Zr filter was inserted perpendicularly to the propagation direction to block the fundamental laser completely.

The quantum efficiency of the diode with a 200 nm Zr and 50 nm a-C is specified by the manufacturer [27] and shown as dashed curve in Fig. 2. The transmission of the various filter layers in the setup under the respective angle of incidence is calculated with IMD [28]. A correction for the reduced transmission through the custom filter layers of the specified

#124076 - $15.00 USD Received 9 Feb 2010; revised 29 Mar 2010; accepted 5 Apr 2010; published 16 Apr 2010

quantum efficiency gives the effective quantum efficiency of the diode, shown as solid line in Fig. 2. The average quantum efficiency of 0.45 in the spectral window between 6.5 nm and 13.5 nm was used to calculate the absolute value of XUV photons per pulse from the amplified diode signal.

The change in transmission of the filter layer between normal and 45° incidence was also evaluated experimentally and shows a significantly stronger attenuation under 45° incidence than the calculations predict. Therefore our results represent a conservative measure of the number of XUV photons per pulse in the transmission window of the filter layers.

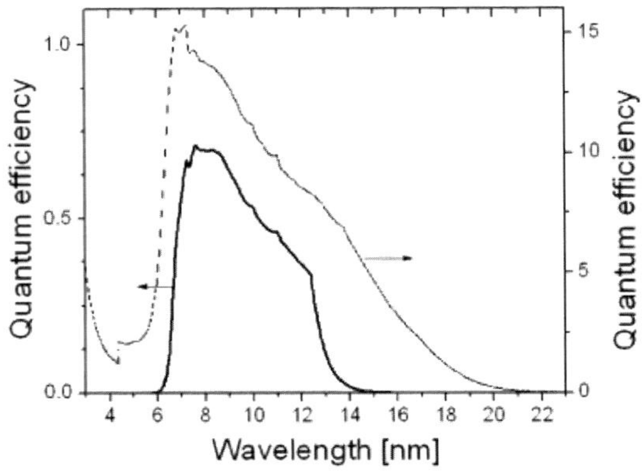

Fig. 2. XUV diode effective quantum efficiency (solid line) calculated from the quantum efficiency specified for the diode with a 200 nm Zr/ 50 nm C coating by manufacturer (dashed line) with corrections for custom filter layer.

The XUV spectra are measured with a grazing incidence grating (600 lines/mm) spectrometer (McPherson, 248/100G). The energy axes of the spectra are calibrated using the L-absorption edge of silicon at ~100 eV [29] by comparing spectra transmitted through a 200 nm thick silicon filter with the simulated transmission curve [28].

The filter layer on the XUV diode partially reflects the NIR and the UV outside the vacuum system through a fused silica window, where the UV is separated from the NIR with a prism and an aperture after recollimation with a lens. The UV pulse energy is measured with a low energy photodiode head (Ophir Optronics Ldt., PD10) with reflection losses on the diode, the window, the prism and the lens being taken into account. For the spectral characterization of the UV pulses with a VIS/UV spectrometer (Maya Pro 2000, 220 - 1100 nm, Ocean Optics), the NIR radiation is suppressed by two reflections at a Si wafer surface under Brewster's angle for 720 nm. As described in [15], this method suppresses the NIR up to two orders of magnitude of the NIR laser. The UV spectra have been corrected for the spectral modification induced by two reflections on silicon.

3. Experimental results

We characterized the influence of the UV gas target on the XUV generation in order to evaluate the suitability of collinear generation for ultrafast UV pump/XUV probe experiments.

3.1 Measurement of the UV pulse energy and number of XUV photons per pulse

The UV pulse energy was measured in dependence of the gas pressure in the UV target for Ar and Ne; the results are plotted as full squares in Figs. 3(a) and 3(b), respectively. The conversion efficiency of the NIR to the third harmonic can be tuned with the UV target pressure and we can achieve UV pulses with energies up to 1 μJ, which is in good agreement with previous results [15]. The number of photons per XUV pulse, collinearly generated in the second target filled with 270 mbar Ne is shown in Fig. 3(a) and (b) as circles. Without any gas being applied in the UV target, the high harmonic generation provides up to $3.5 \cdot 10^6$ photons per pulse corresponding to a pulse energy exceeding 50 pJ. The Ne pressure in the XUV target was optimized for XUV generation and kept constant at 270 mbar. For Ne being

#124076 - $15.00 USD Received 9 Feb 2010; revised 29 Mar 2010; accepted 5 Apr 2010; published 16 Apr 2010

used in the UV target, the XUV signal (Fig. 3(b), circles) decreases almost linearly when the pressure in the UV target is increased to 1.5 bar. For further increasing pressure, the slope of the curve gets smaller. When Ar is used in the UV target, a quantitatively comparable but much steeper decrease with increasing pressure is observed. This is followed by an almost constant XUV signal for pressures above 600 mbar.

We find that the UV pulse energy can be tuned over a several hundred nJ range and still 10^6 photons per pulse with photon energies exceeding 100 eV can be achieved in co-propagating XUV pulses that are synchronized with the UV pulses.

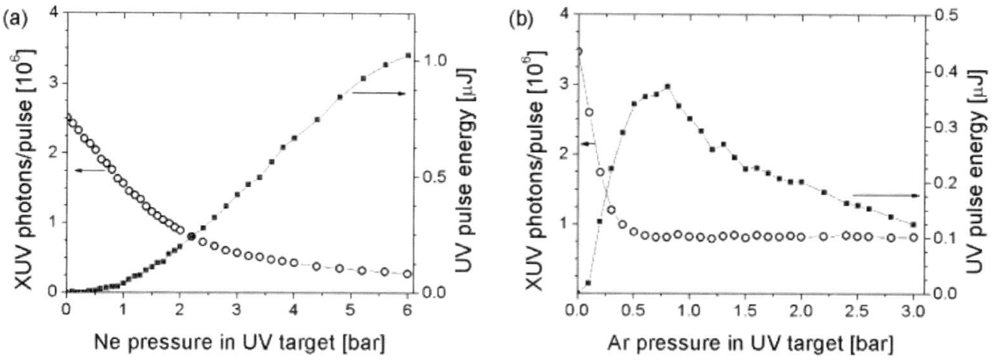

Fig. 3. XUV photon number per pulse (circles) and UV pulse energy (full squares) as functions of the gas pressure in the UV target for Ne (a) and Ar (b). The Ne pressure in the XUV target was kept constant at 270 mbar.

3.2 Spectral characterization

The spectral characterization of the UV and XUV pulses under identical conditions during collinear generation is shown in Fig. 4 and Fig. 5. The spectrum of the UV light generated in Ne (Fig. 4(a)) shows no dependence on the UV target pressure. The spectra from the bottom to the top correspond to the UV pulse energies of 30%, 65% and 100% of the maximal achievable value in Ne. The spectrum of Ar-generated UV pulses narrows with increasing pressure (Fig. 4(b)). This can be attributed to a loss in phase matching bandwidth that also affects the conversion efficiency at higher Ar pressures.

The normalized spectra of the third harmonic (UV) have been corrected for two reflections on Si wafers under Brewster's angle for 720 nm. The spectra are centered at approximately 5 eV and range from 4.5 eV to more than 5.5 eV. On the high energy side, the spectral characterization is restricted by the sensitivity range of the spectrometer.

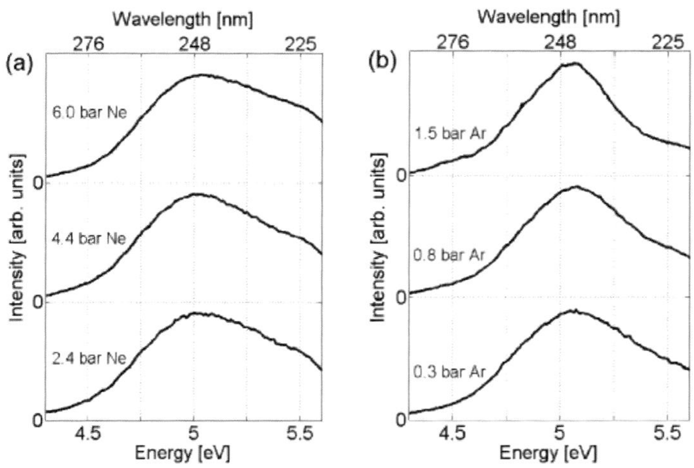

Fig. 4. UV spectra after both gas targets. The XUV radiation was generated in 270 mbar Ne and the UV was generated in Ne (a) and Ar (b) at the respective pressures given in the figure.

#124076 - $15.00 USD Received 9 Feb 2010; revised 29 Mar 2010; accepted 5 Apr 2010; published 16 Apr 2010

In order to estimate the temporal spread of the UV pulses due to propagation through the XUV gas target, we calculated the group velocity dispersion of Ne at 1 bar according to the Sellmeier formula taken from [30] to be $0.06 \cdot fs^2/mm$ at 245 nm. As the XUV target is only 3 mm thick and the pressure in the target is around 300 mbar, we expect that the UV pulse duration is comparable to the results achieved previously with this technique [15]. For instance, the full-width half-maximum bandwidth of the spectrum generated in 6 bar Ne supports pulse durations down to 2.3 fs for a Fourier-limited pulse.

The normalized XUV spectra show in both cases that the cut-off is shifted towards lower photon energies as the pressure in the UV target is increased. This red-shift of the high harmonic cut-off occurs both when Ar or Ne is being used in the UV target.

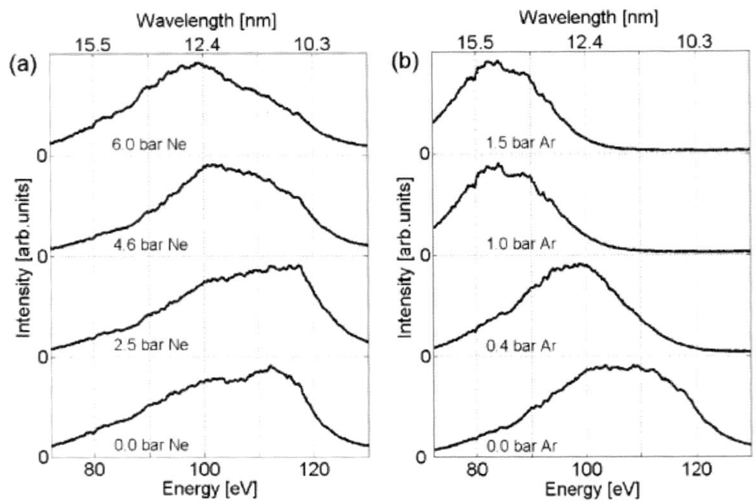

Fig. 5. XUV spectra generated in 230 mbar Ne. In the first gas jet, UV was generated simultaneously in Ne (a) or Ar (b) at the respective pressures given in the figures. (a) The detector strongly attenuates spectra above 115 eV, so that the cut-off of the spectrum lies outside of the detector's range for the spectra at 0 bar and 2.5 bar. With rising pressure in the UV target, the cut-off shifts toward lower energies with rising Ne pressure (bottom to top) in the UV target. (b) When Ar is being used for UV generation, the XUV cut-off also shifts toward lower energies when the pressure rises from 0 to 1 bar but remains constant for higher pressures.

3.3 Interpretation of the results

Qualitatively, the red shift of the XUV cut-off and the decrease of the XUV signal in dependence of the pressure and the gas in the UV target can be explained by changes of the fundamental laser pulse properties such as pulse shape, instantaneous frequency and peak intensity. These are mainly caused by intensity-dependent ionization of the gas in the UV target. Assuming a radial intensity profile with its peak at the center, the negative free-electron dispersion in the resulting plasma is initially non-uniform in the plane of the pulse front. The free-electron density is higher on the axis of the laser propagation and decreases radially. Therefore, the ionized gas acts like a defocusing lens and changes the radius of curvature of the converging laser beam [31, 32] which eventually results in a flat-top-like intensity profile. Once this is reached, the ionization probability becomes uniform over the central part of the beam and the laser pulse propagates in a self-guided fashion.

Due to the difference in ionization potential between Ar and Ne, the ionization probability at low pressures is higher for Ar than it is for Ne at the same laser intensity. The XUV signal measured in dependence of the pressure in the UV target gives an indication that the laser pulse propagating through the UV target reaches the stable flat-top-like profile already at pressures around 0.5 bar in Ar whereas in Ne this is only approached at pressures above 5 bar. The red-shift of the XUV cut-off is also in good agreement with this interpretation, as highest achievable photon energy depends linearly on the laser intensity [18].

#124076 - $15.00 USD Received 9 Feb 2010; revised 29 Mar 2010; accepted 5 Apr 2010; published 16 Apr 2010

In addition to the defocusing of the NIR beam, we also expect a contribution to the red-shift of the XUV cut-off energy resulting from a dynamic blue-shift of the instantaneous frequency of the NIR pulse [5] during propagation through the ionized gas in the UV target.

4. Conclusions

We have successfully demonstrated the collinear generation of UV and XUV pulses by means of third-harmonic generation and high-order harmonic generation in a noble gas with a quasi single-cycle NIR laser pulse. While simultaneously providing XUV pulses with up to $3.5\ 10^6$ photons per pulse and a cut-off photon energy of more than 100 eV, our approach allows obtaining UV pulses with a pulse energy tunable up to ~1 μJ generated in either Ne or Ar by varying the gas target pressure. The shape of the UV spectra, centered at ~5 eV, was shown to be preserved as the gas pressure was varied. As we are able to minimize the dispersion of the generated harmonic radiation due to background noble gas pressure, the collinearly generated UV and XUV pulses are expected to have a temporal structure comparable to the ones produced independently and reported in previous studies [15,21], with respective durations in the sub-4 fs and attosecond timescales. The presented method of collinear generation provides an inherent synchronization between the UV and the XUV pulses. Such findings represent a novel real-time optical tool providing a few-fs time resolution with which the dynamics of excited electronic states in atomic, molecular and condensed matter systems exhibiting absorption resonances in the UV or XUV can be addressed.

Acknowledgements

The authors thank Adrian Cavalieri for valuable experimental support and Ulrich Graf and Michael Hofstetter for helpful discussions and assistance. A.S. acknowledges funding from the Swiss National Science Foundation. R.K. acknowledges financial support from the Sofja Kovalevskaja Award of the Alexander von Humboldt Foundation and an ERC Starting Grant. The work was supported by the DFG Cluster of Excellence: Munich Centre for Advanced Photonics (www.munich-photonics.de).

#124076 - $15.00 USD Received 9 Feb 2010; revised 29 Mar 2010; accepted 5 Apr 2010; published 16 Apr 2010

REVIEW OF SCIENTIFIC INSTRUMENTS **81**, 093103 (2010)

Versatile apparatus for attosecond metrology and spectroscopy

M. Fieß,[1,a)] M. Schultze,[1,2] E. Goulielmakis,[1] B. Dennhardt,[3] J. Gagnon,[1] M. Hofstetter,[1] R. Kienberger,[1,3] and F. Krausz[1,2]
[1]*Max-Planck-Institut für Quantenoptik, Hans-Kopfermann-Straße 1, 85748 Garching, Germany*
[2]*Fakultät für Physik, Ludwig-Maximilians-Universität München, Am Coulombwall 1, 85748 Garching, Germany*
[3]*Physik Department E11, Technische Universität München, James-Franck-Straße, 85748 Garching, Germany*

(Received 10 May 2010; accepted 15 July 2010; published online 14 September 2010)

We present the AS-2 Attosecond Beamline at the Joint Laboratory for Attosecond Physics of the Max-Planck-Institut für Quantenoptik and Ludwig-Maximilians-Universität for time resolved pump/probe experiments with attosecond resolution. High harmonic generation and subsequent filtering of the generated extreme ultraviolet (XUV) continuum by means of metal filters and XUV multilayer mirrors serve for the generation of isolated attosecond laser pulses. After high harmonic generation, the remaining fundamental laser pulse is spatially separated from the attosecond XUV pulse, to what is to our knowledge for the first time, by means of a perforated mirror in a Mach–Zehnder interferometer. Active stabilization of this interferometer guarantees the necessary temporal resolution for tracking attosecond dynamics in real time. As a proof-of-principle, photoelectron streaking experiments are performed and experimental techniques for their realization are summarized. Finally we highlight the potential of the presented beamline system for future experiments in comparison with previously demonstrated attosecond beamlines. © *2010 American Institute of Physics.* [doi:10.1063/1.3475689]

I. INTRODUCTION

To date, attosecond laser pulses are the shortest controllable events. They offer access to the time evolution of the fastest phenomena we are currently able to investigate. These attosecond pulses are generated by a frequency conversion of ultrashort, few-cycle laser pulses in the visible/near infrared range. Events evolving on attosecond time scales are electron dynamics in atoms,[1] solids,[2] and molecules.[3] Up to now, pump-probe experiments exploring the attosecond time scale have been performed with isolated extreme ultraviolet (XUV) attosecond pulses as short as 80 asec (Ref. 4) as pump and field gradients of few-cycle carrier-envelope-phase (CEP) stabilized femtosecond laser pulses as probe. The limitation on near infrared CEP-stabilized femtosecond laser pulses with an electric field gradient on an attosecond time scale as a precisely synchronized attosecond reference besides the attosecond XUV pulse has the following two reasons: first, the currently achievable attosecond-pulse energies prevent the utilization of 2 attosecond pulses as pump and probe pulses since the two photon absorption probability scales with the fifth to sixth power of the wavelength,[5] which means that it is extremely low for the XUV regime. Second, the attosecond pulse generation is based on high-order harmonic generation (HHG) of the few-cycle infrared femtosecond laser pulse. Up to now, the separation of the generated high harmonics from the remaining fundamental infrared light is problematic due to the lack of beam splitting optics conserving the short pulse duration and high pulse energy of both the XUV attosecond pulse and the infrared femtosecond pulse. In principle, the use of doughnut-shaped beam profiles after placing a circular block in the fundamental beam allows for the spatial separation of the generated high harmonics from the fundamental light, however, at the expense of a substantially reduced fundamental intensity.[6] For that reason, the present attosecond experimental systems rely on the collinear propagation of both the attosecond and the femtosecond pulse. This collinear setup has been realized in first-generation attosecond beamlines,[7,8] which have limited potential for further manipulating the fundamental laser pulse in order to extend the developed techniques toward the use of near UV pulses in combination with XUV pulses. As a consequence, real-time attosecond pump-probe experiments of electron dynamics with waveforms other than the fundamental near-infrared (NIR)-waveform were not possible. Accordingly, experiments in more complex systems such as small molecules consisting of only a few atoms were prevented due to their typically very low absorption cross section in the visible and NIR part of the electromagnetic spectrum.

So far, interferometer setups for attosecond pump-probe experiments are based on splitting the fundamental beam before HHG.[9,10] This comes to the expense of a substantial reduction of the XUV photon flux and photon energy since only a fraction of the fundamental pulse energy is used for HHG.

For the first time, the AS-2 Attosecond Beamline at the Joint Laboratory for Attosecond Physics of the Max-Planck-Institut für Quantenoptik and Ludwig-Maximilians-

[a)]Author to whom correspondence should be addressed. Electronic mail: markus.fiess@mpq.mpg.de.

0034-6748/2010/81(9)/093103/8/$30.00

Fieß *et al.* Rev. Sci. Instrum. **81**, 093103 (2010)

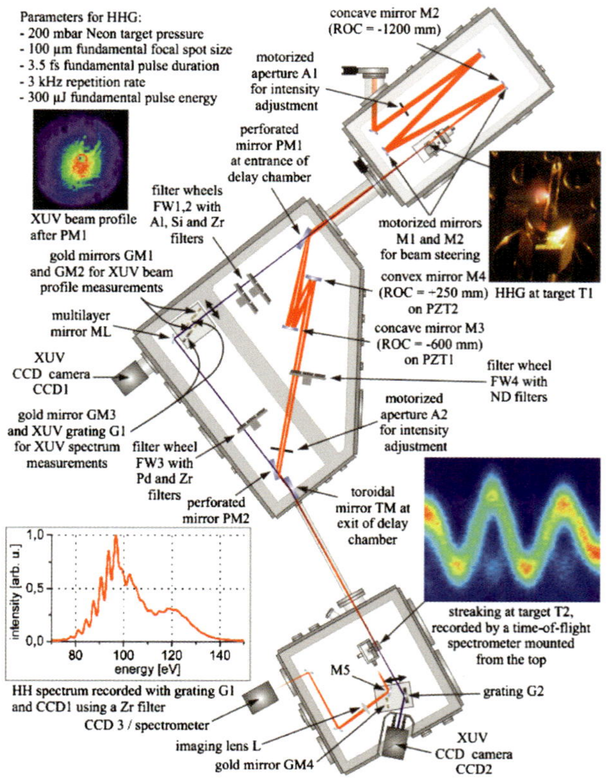

FIG. 1. (Color online) Experimental setup consisting of three vacuum chambers: chamber for HHG (top right), delay chamber (middle), and experiment chamber (bottom right).

Universität presented in this article allows for the spatial separation of the XUV attosecond laser pulse from the NIR femtosecond laser pulse after HHG. This feature opens the door for a conversion of the fundamental pulse to novel waveforms triggering a broad range of dynamics in different systems while using the full fundamental power for the generation of intense XUV pulses.

II. EXPERIMENTAL SETUP

Figure 1 shows the experimental setup: The laser system used for attosecond pulse generation is based on a commercial core (Femtolasers Produktions GmbH) adapted to the needs of the experiment. It generates CEP-stabilized laser pulses with 3.5 fs pulse duration, 300 μJ pulse energy at 3 kHz repetition rate, a spectrum spanning from 450 to 1050 nm with a central wavelength of 760 nm, and a vertical polarization.[11] Vertical polarization is more appropriate for designing XUV mirrors with high reflectance at non-normal incidence. These laser pulses, propagating in a vacuum system, are focused on a gas target T1 by a concave mirror M2, radius of curvature (ROC)=−1200 mm. The target consists of a nickel tube (3 mm diameter, 300 μm wall thickness, Good Fellow) filled with neon at a backing pressure of 200 mbar. In the 100 μm diameter focus, high harmonics with photon energies up to 140 eV are generated. Most of the optical components are motorized and can be adjusted under vacuum conditions: the flat mirror M1 and the concave mirror M2 in the HHG chamber are motorized for steering the direction of the beam. The gas target T1 can be moved in all three directions in space in order to position the gas target relative to the laser focus for optimum high harmonic generation. The movement along the beam axis allows to position the target into the part of the confocal parameter where preferably short-trajectory harmonics are generated.[12]

After HHG, both the generated high harmonic beam and the remaining fundamental light propagate collinearly into the second vacuum chamber, the delay chamber. Here, the fundamental light and the high harmonics are separated into the two paths of a Mach–Zehnder interferometer. One leg of the interferometer provides the possibility of adjusting the delay between the fundamental driving pulse and the generated attosecond pulse. The much smaller divergence of the high harmonic beam compared to the fundamental beam allows for the spatial separation of the two beams by means of a perforated mirror PM1, with a hole diameter of 1.0 mm through which the HH beam can pass. The distance between the target T1 and PM1 amounts to 900 mm. The fundamental light transmitted through the hole of PM1 is blocked by thin metal filters selected according to their transmittance (e.g., 500 nm Zr for the energy range of 70–140 eV, Lebow). These metal foils act as high pass filters for the high harmonic radiation. In combination with the reflectance characteristics of the multilayer mirror ML, a part of the HH spectrum is reflected. For few cycle laser pulses at a certain CEP the cutoff part of the continuum is smooth.[13] Filtering this region of the spectrum corresponds to the isolation of one single attosecond pulse out of an attosecond pulse train in the time domain.[14] HHG is optimized in terms of beam profile using an XUV sensitive camera [XUV camera PIXIS XO, Roper Scientific, charge-coupled device (CCD)1 in Fig. 1] and in terms of spectral distribution with a home-made grating spectrometer (grating 001–0266, Hitachi[15]). In order to preserve the interferometer stability, the multilayer mirror ML is not moved for beam profile measurements or spectral measurements of the generated HH. Instead, mirrors coated with 80% Au and 20% Pd (Georg Albert physical vapor deposition) on superpolished silicon substrates (Gooch & Housego) are moved in with a translation stage. At large incidence angles (>80°, measured from the normal axis resulting in a reflectance >60%) the HH radiation is reflected on two gold mirrors, bypassing the multilayer mirror, and is detected for beam profile measurements on a camera chip of an XUV CCD camera (CCD1 in Fig. 1). For spectral measurements the high harmonics are reflected off a third gold mirror, which can be inserted, and subsequently dispersed by a grating. The dispersed beam again circumvents the multilayer mirror and spectra are measured with the same XUV camera CCD1. For propagation of the HH to the experiment, all gold mirrors are moved out of the XUV beam.

The mirrors M3 and M4 in the other path of the Mach–Zehnder interferometer combine three functions:

(1) A telescope that consists of M3 (ROC=−600 mm) and M4 (ROC=+250 mm) for reducing the beam diameter.
(2) The delay between the fundamental pulse and the attosecond pulse in the other path of the interferometer is set by moving mirror M3, which is mounted on a

nanometer-resolution piezotranslation stage (PZT).

(3) Instabilities of the Mach–Zehnder interferometer are compensated by movements of mirror M4, which is mounted on a piezotranslation stage as well. M4 is the active part of a stabilization system of the interferometer, which is described in Sec. III.

The paths of both the fundamental beam and the high harmonics beam are recombined by a second perforated mirror PM2 with a hole diameter of 2.0 mm, which is approximately the XUV beam diameter at this point of the diverging XUV beam. The beams are subsequently focused on the experimental target T2 by a toroidal mirror TM (Société Européenne de Systèmes Optiques). The optical path length of the legs of the interferometer is 1700 mm. The distance between the second perforated mirror PM2 and the toroidal mirror is 100 mm. The toroidal mirror is coated with nickel, the angle of incidence is 86° measured from the normal axis, the effective focal length is 720 mm, and the accepted fundamental beam diameter is 6 mm. The divergence of the fundamental beam after the telescope formed by mirrors M3 and M4 is set such that the toroidal mirror TM focuses both the fundamental and the HH beam at the position of the experimental target T2. The distance between TM and T2 is 982 mm. The focal spot size of the fundamental focus generated by the toroidal mirror in the experiment chamber is 150 μm, and the maximum intensity is 5×10^{13} W/cm^2. The focal spot size of the XUV beam, measured with the knife-edge method, is 40 μm and therefore by a factor of 3 to 4 smaller than the focal spot size of the intense fundamental beam.

The experimental target T2 consists of a gas nozzle with motorized movement along all three dimensions. In pump-probe experiments, photoelectron spectra, ion mass spectra, and transient HH spectra can be measured for different delays between the fundamental pulse and the XUV pulse. Photoelectron spectra are recorded by means of an electron time-of-flight spectrometer (Stefan Käsdorf, Geräte für Forschung und Industrie) with an energy resolution of about 1% of the total kinetic photoelectron energy. The ion mass spectra are measured with a reflectron spectrometer (Stefan Käsdorf, Geräte für Forschung und Industrie).[16] Transient HH spectra can be measured with a second XUV-spectrometer after dispersion of the HH beam on grating G2 and detection of the spectra on CCD2 (grating 541 00 220, Jobin Yvon; XUV camera PIXIS XO, Roper Scientific).

Typical gas loads for HHG result in a pressure of 5×10^{-3} mbar in the HHG chamber. Differential pumping between the three vacuum chambers allows for a pressure of 5×10^{-4} mbar in the delay chamber and 8×10^{-6} mbar in the experiment chamber. The microchannel plate (MCP) detector of the electron time-of-flight spectrometer, which is mounted to the experiment chamber, is differentially pumped by a separate turbomolecular pump. The maximum gas load of the experimental target T2 is set such that the pressure of 3×10^{-6} mbar for neon as target sample is not exceeded in order to ensure safe operation of the MCP detector.

Summarizing all these features, four distinct advantages of the second generation attosecond beamline compared to its predecessors can be stated:

(1) The spatial separation of the fundamental light from the generated XUV light allows for a tailoring of the pump laser pulse according to the requirements of the experimental target, e.g., by nonlinear frequency conversion. In the previous attosecond beamlines[7,8] a manipulation of the remaining fundamental light after HHG was not possible due to the collinear propagation of fundamental light and XUV light after HHG.

(2) The angle of incidence on the flat multilayer mirror ML is 45°, which allows for mirror designs with significant higher reflectivity and larger spectral bandwidth for *s*-polarized XUV light as compared to near-normal incidence mirror designs as used in the previous attosecond beamlines.

(3) Focusing both the fundamental and the HH beam by means of a toroidal mirror at grazing incidence offers simultaneous access to transient absorption measurements and photoelectron streaking experiments at large XUV photon flux.

(4) A much longer Rayleigh range of both the fundamental and the XUV beam in the focus of the experimental target simplifies the alignment for streaking experiments considerably.

III. CALIBRATION OF THE XUV GRATING SPECTROMETER

The XUV spectrometers (G1 and G2) are calibrated using the absorption edges of aluminum and silicon foils. The XUV radiation is horizontally dispersed by the grating and the different wavelength components are recorded on the CCD-chip. Full vertical binning of the CCD chip results in raw spectra with the number of counts versus the pixel column number n. In the narrow wavelength range of interest it is a reasonable assumption that the pixel number n is proportional to the wavelength λ. The pixel number of the silicon absorption edge is n_1; the pixel number of the aluminum absorption edge is n_2. With this follows the corresponding wavelength λ for each pixel n

$$\lambda(n) = \frac{\lambda(n_1) - \lambda(n_2)}{n_1 - n_2}(n - n_2) + \lambda(n_2). \quad (1)$$

The conversion from wavelengths λ in nanometer to photon energies E in eV is calculated by $E(\text{eV}) = 1239/\lambda(\text{nm})$. Typical raw spectra recorded after transmission through thin aluminum, silicon, and zirconium foils and the corresponding calibrated XUV spectra are shown in Fig. 2.

IV. ALIGNMENT OF THE TEMPORAL AND SPATIAL OVERLAP

Finding the temporal overlap of the pump and the probe pulses is straightforward in a collinear setup like in the first generation attosecond beamline.[7,8] The metal filters, which block the fundamental light that propagates along the XUV beam path, are removed and two fundamental laser pulses

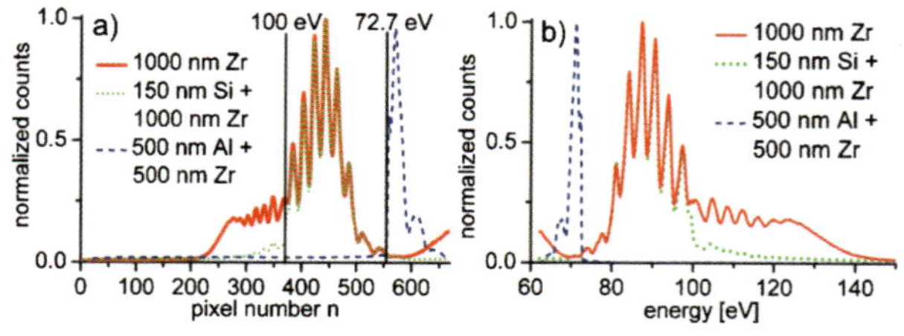

FIG. 2. (Color online) (a) Raw HH spectra and (b) calibrated HH spectra measured after transmission through thin metal foils made of Zr (solid line), Si (dotted line), and Al (dashed line). The spectrum below 70 eV in case of a 1000 nm zirconium filter shows the second diffraction order of the grating.

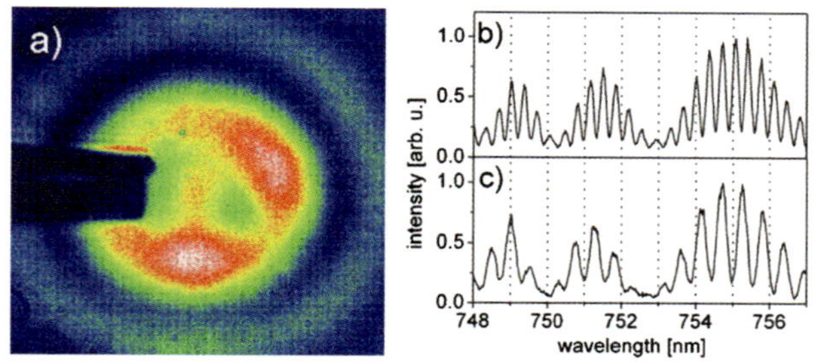

FIG. 3. (Color online) (a) Spatial interference with the shadow of the gas nozzle, (b) and (c) spectral interference of two fundamental laser pulses for (b) τ=5.4 ps and for (c) τ=3.5 ps.

are interfered in the focus of the experimental target. Each of the two laser pulses propagates along one of the two paths of the interferometer. A lens, which images the focus at the position of the experimental target through a glass window on a CCD-camera outside the vacuum chamber, is sufficient for the detection of the interference pattern between two fundamental laser pulses (CCD3 in Fig. 1). In case of a large interferometer setup, however, this method causes severe difficulties since the coherence length of the fundamental laser pulses amounts to only a few microns and determines the delay range within which this spatial interference is visible. Finding this micrometer-size overlap region in a Mach–Zehnder interferometer is nontrivial. Using a high-resolution spectrometer, one can observe the interference in the spectral domain over a much wider delay range. The same imaging system in combination with a spectrometer (HR4000, 0.05 nm resolution, Ocean Optics) instead of a CCD camera observes spectral fringes as follows:

Fourier transformation of the temporal electric field of two laser pulses

$$E_1(t) = \frac{1}{2}\left[\frac{1}{2\pi}\int_{\omega=-\infty}^{+\infty} d\omega\sqrt{I(\omega)}\exp(-i\omega t) + \text{c.c.}\right], \quad (2)$$

$$E_2(t) = \frac{1}{2}\left\{\frac{1}{2\pi}\int_{\omega=-\infty}^{+\infty} d\omega\sqrt{I(\omega)}\exp[-i\omega(t+\tau)] + \text{c.c.}\right\} \quad (3)$$

with identical spectrum $I(\omega)$, flat phase but delayed by τ, gives the modulated total spectrum

$$I_{\text{tot}}(\omega) = \left|\int_{t=-\infty}^{+\infty} dt[E_1(t)+E_2(t)]\exp(i\omega t)\right|^2$$
$$= 2I(\omega)[1+\cos(\omega\tau)]. \quad (4)$$

The modulation of $I_{\text{tot}}(\omega)$ is determined by the delay τ between the two laser pulses. Since the number of fringes per frequency interval increases with τ, this method allows the definition of an increasing or decreasing delay if the delay is changed, e.g., by moving PZT1 in Fig. 1. In case of a central wavelength of 750 nm and a spectral resolution of 0.05 nm, delays of up to 38 ps corresponding to 11 mm difference of optical path length can be resolved.

A prerequisite for the observation of modulated spectra as shown in Figs. 3(b) and 3(c) is that the intensities of the two interfering laser pulses are of the same magnitude. Reducing the intensity of the more intense laser pulse by inserting a neutral density filter into the beam, e.g., with filter wheel FW4 in Fig. 1, is not a solution, since this filter would introduce additional group delay. Instead, the intensities are balanced by adjusting aperture A2 in Fig. 1. Closing aperture A2, however, slightly steers the beam path. Therefore, the spatial overlap of both laser beams at the position of target T2 is adjusted by moving in a neutral density filter with filter wheel FW4 and observing the image of the two foci on CCD camera CCD3 [Fig. 3(a)].

V. ACTIVE STABILIZATION OF THE MACH–ZEHNDER INTERFEROMETER

In the past, attosecond pump-probe experiments have been realized in collinear[7,8] and in Mach–Zehnder[9,10] interferometer setups. Pump-probe experiments with a temporal resolution on an attosecond timescale in an interferometer setup require nanometer-scale relative stability of the optical beam path lengths of the two paths, since temporal delays between the pump and the probe pulse are set by controlling the difference of the optical path lengths of both pulses. This can be ensured by an active feedback stabilization of the interferometer,[10,17] where the error signal is acquired in a reference interferometer. This reference interferometer is formed by half inch optical components mounted on top of those forming the XUV-IR interferometer. By attaching the components on the mounts that support the XUV-IR interferometer, we ensure that the behavior in terms of drift of the reference beam paths exactly mimic the drift motion of the beam paths of the fundamental and XUV laser pulses, which are used in the actual experiment. The thermal stability of the whole interferometer setup benefits from the use of stainless steel for most of the components due to its small thermal expansion coefficient in comparison with, e.g., aluminum.

In the following, the two beam splitters mounted on the two perforated mirrors are called BS1 and BS2. The half inch mirrors of the upper reference interferometer, which are the little brothers of ML, M3, and M4 in Fig. 1, are called stabilization mirrors SML, SM3, and SM4, respectively. A continuous-wave (cw) laser beam of a low noise, vacuum compatible diode laser (ULN diode laser, Coherent) is split into two beams by BS1, which is mounted on the mirror mount of PM1 in Fig. 1. Both beams are recombined by a second half inch beam splitter BS2, which is mounted on the mirror mount of PM2. The beam, which is transmitted by BS1 propagates above the beam path of the high harmonics

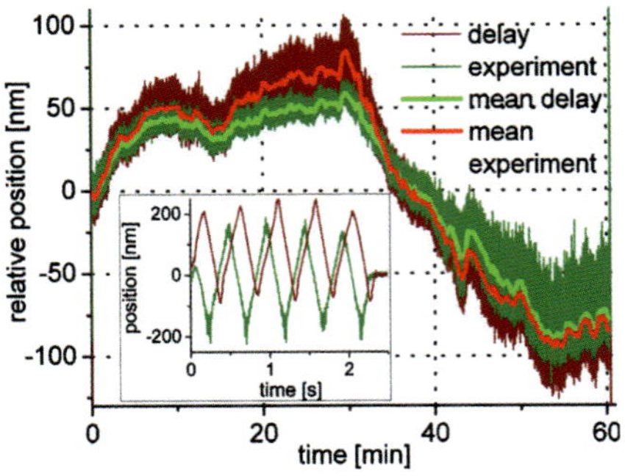

FIG. 4. (Color) Relative change of the optical path length of the reference interferometer (delay, red) and the lower interferometer, which has to be stabilized (experiment, green). The thick red and green lines show the mean value per second of both interferometers. The inset shows the calibration by means of a linear movement of the piezotranslation stage PZT2 before the active stabilization is started (see text).

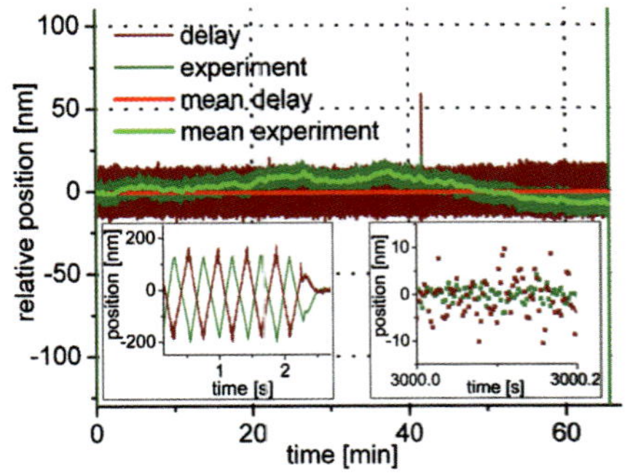

FIG. 5. (Color) Variation of the difference of the optical path length of the reference interferometer (delay, red) and the lower interferometer (experiment, green), in case of active stabilization. The thick red and green lines show the mean value per second of both interferometers. As in Fig. 4 the left inset shows the calibration by means of a linear movement of the piezostage PZT2 before the active stabilization is started. The right inset illustrates the statistical noise of 80 data points.

beam and is reflected by a half inch mirror SML attached on the mirror mount of ML in Fig. 1. Since M3 in Fig. 1 is used for controlling the delay between the fundamental laser pulse and the XUV laser pulse, M3 is not directly coupled with the reference interferometer. Instead, the mirror mount of M3 is mounted on a piezotranslation stage PZT1. PZT1 and the stainless steel post, on which the mount of the half inch mirror SM3 of the reference interferometer is installed, are assembled on the same base plate. M4 and SM4 are mounted on a second piezotranslation stage PZT2 (NanoX200CAP, Piezosystem Jena), which is used for the active stabilization of both the lower interferometer and the upper reference interferometer.

After BS2, the intensity of the interference between both cw beams of the reference interferometer is measured by means of a small active surface photodiode (silicon pin detector ET-2000, EOT). It is operated in photoconductive mode with an active area of only 80×80 μm^2 in order to avoid averaging over several interference fringes. A change of the intensity measured by the photodiode corresponds to a change of the difference of the optical path lengths of the two paths of the Mach–Zehnder interferometer. Therefore, the photodiode voltage was stabilized by means of a PID-algorithm written in LABVIEW and controls the voltage, which sets the position of piezotranslation stage PZT2. The PID-algorithm performs 400 corrections per second of the position of PZT2. The limiting factor for the speed of the active stabilization is the resonance frequency of PZT2 at approximately 500 Hz in case of a mass load of only 55 g for the mirrors M4, SM4, and their mounts. Therefore, only slow drifts due to thermal expansion of the breadboard, mirror mounts, etc. can be compensated but not fast vibrations introduced by, e.g., turbomolecular pumps. Such fast vibrations have been efficiently suppressed by using dampers between the turbomolecular pumps and the walls of the vacuum chambers, by a decoupling of the walls of the chamber and the optical table inside the delay chamber on which the interferometer is mounted, and by using turbomolecular pumps with magnetic bearings.

In order to test the quality of the coupling between both interferometers, a second cw laser was set up in the HHG chamber and propagated through the lower interferometer from the focus for HHG in the HHG chamber all the way up to the focus in the experiment chamber. The focus in the experiment chamber was imaged outside on a second photodiode identical to the one used for the reference interferometer in the delay chamber. Figure 4 shows the relative change of the optical path lengths in both the upper reference interferometer and the lower interferometer, which has to be actively stabilized. This measurement proves, first, that both interferometers are coupled with an accuracy of less than 15 nm and, second, that the observed long-term drift motion of more than 100 nm is not negligible and has to be compensated by means of an active stabilization.

Before the active stabilization is started and after the stabilization is stopped, PZT2 is moved linearly over a distance of several microns in order to generate some fringes, which are detected by the photodiode of the reference interferometer in the delay chamber. These fringes are used for the calibration of the changes of the path length from the measured photodiode voltages. The linear change of the photodiode voltage at the points of inflection of the sinusoidal curve is extrapolated to the minima and maxima and the photodiode voltage and converted to a path length change assuming a distance of $\lambda/2$ between adjacent extrema of the sinusoidal voltage curve. After this calibration, the sinusoidal voltage curve changes to a zigzag curve as stipulated by the linear movement of PZT2 (see inset of Figs. 4 and 5).

Figure 5 shows the same measurement as displayed in Fig. 4 in case of active stabilization of the Mach–Zehnder interferometer. It proves the capability of the presented stabilization scheme to suppress long term drift motions of the interferometer paths to within 50 as on a time scale of 1 h. The residual drift motion of the IR-XUV interferometer, which is shown in Fig. 5, might be due to imperfect coupling of both interferometers. For time intervals of 1 s, which is usually the minimum integration time for the data acquisition of, e.g., photoelectron spectra, the standard deviation has been calculated in case of the lower interferometer. The mean value of all standard deviations per 1 s interval is only 1.8 nm which corresponds to 6.0 as and defines the short term stability of the Mach–Zehnder interferometer. Different

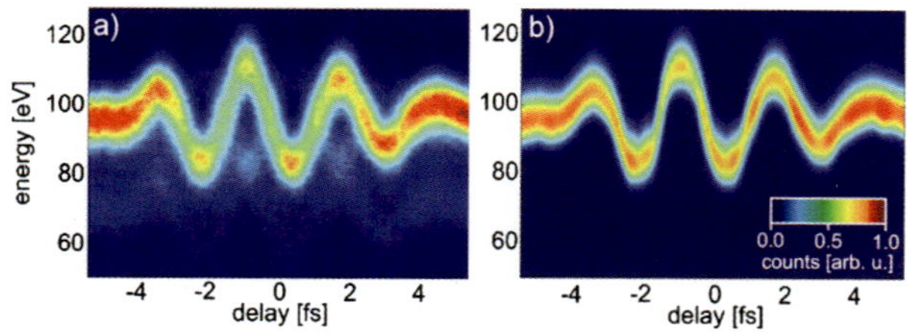

FIG. 6. (Color online) (a) Streaking spectrogram of $2p$ photoelectrons of neon measured with the active stabilization of the interferometer setup. The photoline at lower energies, which is hardly visible, originates from electrons released from the $2s$ orbital. It is suppressed due to the setting of the electrostatic lens of the electron time-of-flight spectrometer, which was optimized for highest $2p$ photoelectron count rate. (b) Retrieved streaking spectrogram after background subtraction and a FROG retrieval, which reproduces the streaking curve of the $2p$ photoelectrons.

electromagnetic shielding of the photodiodes and cables explains the different statistical noise for both interferometers.

VI. STREAKING MEASUREMENTS

The kinetic energy distribution of photoelectrons, which are released by XUV photons of an attosecond pulse, can be modified by the electric field of a synchronized and intense NIR laser pulse. A measurement of the kinetic energy of the photoelectrons for different delays between the XUV pulse and the NIR pulse is called "streaking", and the measured streaking curve can be approximately described by the vector potential of the NIR electric field.[18,19] A streaking measurement of the NIR electric field was performed to prove the functionality of the presented setup. Photoelectron spectra for different delays between the XUV and NIR pulse were recorded by focusing both the NIR and XUV beams in neon gas with simultaneous active stabilization of the Mach–Zehnder interferometer in the delay chamber. A typical streaking scan resulting from such a measurement is shown in Fig. 6(a).

The measured streaking curve shown in Fig. 6(a) proves the necessary short-term stability of the Mach–Zehnder interferometer in the delay chamber. The active stabilization of the interferometer setup guarantees the absence of such long-term drift motions as shown in Fig. 4 within the accuracy of the active stabilization. A long term drift motion would manifest itself as an apparent chirp of the NIR electric field in the streaking spectrogram. The obvious absence of a change in the periodicity of the sampled vector potential witnesses the reliability of the active stabilization.

In the presented data, the integration time per delay step was 10 s, the total acquisition time amounted to 15 min. The spectrogram was recorded using a XUV multilayer mirror supporting a bandwidth of 13 eV with a peak reflectance of 10% at 113 eV and a flat spectral phase. A 150 nm thick palladium foil was used as a high-pass filtering material. In the experiment, an electron time-of-flight spectrometer records the photoelectron flight time versus delay between the attosecond XUV pulse and the femtosecond NIR pulse. The maximum number of photoelectron counts per flight time channel of the electron time-of-flight spectrometer amounted to 814. Since one spectrum represents the sum over 30000 laser shots, multiple events resulting in detection nonlinearities can be safely excluded.

The conversion of the measured quantity of photoelectron counts $n_t(t)$ per flight time t to photoelectron counts $n_E(E)$ per kinetic energy E has to comply with the conservation of the total number of photoelectrons

$$\int_0^\infty n_t(t)dt = \int_\infty^0 n_E(E)dE \Leftrightarrow n_E\left[\frac{m}{2}\left(\frac{s}{t}\right)^2\right] = n_t(t)\frac{t^3}{ms^2}. \tag{5}$$

Here, m is the electron mass, t is the flight time, and s is the flight distance. This correction of the number of photoelectron counts was applied to the raw data of the spectrogram, as shown in Fig. 6(a).

For the measured streaking spectrogram of Fig. 6(a), the energy range of the electrostatic lens, within which the photoelectron count rate is increased by several orders of magnitude, was optimized for the kinetic energy of photoelectrons emitted from the $2p$ shell of the neon atoms. Therefore, the streaking curve generated by photoelectrons, which are emitted from the $2s$ shell of the neon atoms, is hardly visible at lower energies around 70 eV. Scattered photoelectrons produce a delay-independent background at photoelectron energies below the $2p$ streaking curve. Photoelectrons generated via above-threshold-ionization do not appear in the presented energy range for the laser intensities employed in the experiment. A small XUV satellite pulse containing ~12.5% of the total XUV power causes a weak $2p$ streaking curve, which is time-shifted by half a laser period with respect to the intense $2p$ streaking curve generated by the main XUV pulse. Due to this half-period shift, the minima of the $2p$ streaking curve caused by the XUV satellite pulse coincide with the maxima of the $2s$ streaking curve and are further accentuated by the background of scattered photoelectrons. Therefore, the minima and maxima of the weak $2p$ satellite streaking curve do not show the same number of photoelectron counts. The presence of a satellite pulse must manifest itself as a modulation of the XUV photon spectrum that leads to the according intensity modulation of the photoelectron spectra.[4] The spectral resolution of the electron time-of-flight spectrometer in the energy range of the presented data is not sufficient for resolving these spectral fringes. However, FROG requires streaking to be fully coherent. Therefore, the spectral fringes must be measured for the satellite pulse to be characterized. For this reason, the FROG algorithm has ignored the satellite pulse, which does not appear in the retrieved spectrogram, shown in Fig. 6(b).

The measured streaking spectrogram displayed in Fig. 6(a) clearly demonstrates that the attosecond pulses isolated from the XUV continuum carry no visible chirp, which would require an absolute GDD of about 5000 asec^2 to be resolved in the presented energy range. The FROG analysis, presented in Fig. 6(b), was performed after background subtraction of the measured streaking spectrogram, and reproduces the higher photoelectron count rates around each crest of the vector potential. In case of an unchirped XUV pulse, the peaking count rates for those delay steps is caused by the

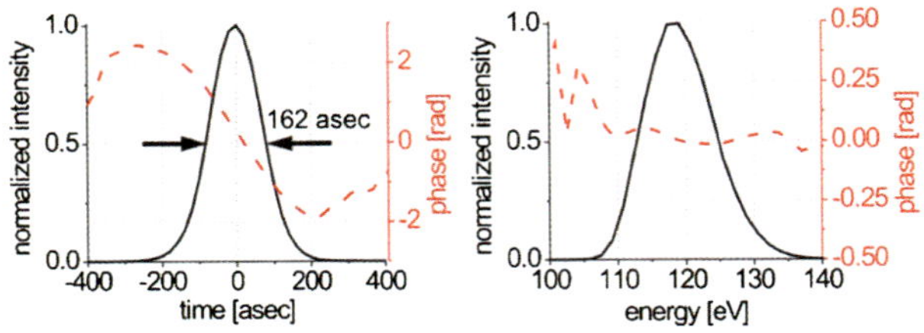

FIG. 7. (Color online) (a) Retrieved temporal intensity (solid line) and phase (dashed line) of the XUV pulse. (b) Retrieved spectrum (solid line) and spectral phase (dashed line) of the XUV pulse.

fact that the XUV ionization takes place when the vector potential is approximately constant within the XUV pulse duration. Since the change of the photoelectron momentum is proportional to the vector potential, this results in a minimal spectral broadening of the photoelectron spectra around the crests of the vector potential. The maximum count rate within the photoelectron bandwidth is therefore modulated along the delay axis even though the stability of the XUV source ensures an approximately constant ($\pm 3\%$) number of photoelectrons per delay step.

Figures 7(a) and 7(b) show the retrieved NIR electric field in the time domain and spectral domain, respectively. The analysis yields a flat spectral phase resulting in an average GDD of 240 asec^2 and a retrieved XUV pulse duration of 162 asec [full width at half maximum (FWHM)] close to the Fourier limit of 155 asec (FWHM). The FROG error $P=0.1694$ defines a measure of the difference between the background subtracted measured spectrogram with elements $A_{i,j}$, and the retrieved spectrogram with elements $B_{i,j}$, and indicates the accuracy of the FROG retrieval independent from the used array size. We define it as

$$P=\sqrt{\frac{2N}{N_1+N_2}},$$
$$N=\sum_{i,j}(A_{i,j}-B_{i,j})^2, \quad N_1=\sum_{i,j}A_{i,j}^2, \quad N_2=\sum_{i,j}B_{i,j}^2. \tag{6}$$

The high reflectance of the XUV multilayer mirrors at the incidence angle of 45° results in a fivefold XUV photon flux, being quantified by a fivefold photoelectron count rate compared to previously demonstrated attosecond sources based on near normal incidence reflectors. This not only allows for quick scans to characterize both the XUV and the laser pulse but will also facilitate experiments that combine attosecond temporal resolution with samples that presently remained inaccessible due to their low absorption cross section, such as helium or secondary atomic emission lines. The resolution of attosecond cross-correlation experiments critically depends on the amplitude of the laser electric field used to streak the released photoelectrons.[20] In the presented setup, the long focal length used to focus the two beams results in a large volume with only small variations of the laser intensity and negligible Gouy-phase shift within the interaction region of the gas target and the laser beams. By virtue of this geometry, unprecedented ponderomotive modifications of the photoelectron kinetic energy are possible (as large as five times the spectral extend of the XUV pulse) and result in a very accurate pulse characterization.

VII. SUMMARY

In conclusion, we could show that our new second generation beamline allows the spatial separation of the remaining fundamental laser pulse from the generated HH after the HHG target. This enables a further manipulation of the fundamental laser pulse after HHG, for instance by nonlinear frequency conversion, and the preparation of a laser pulse suitable for an efficient interaction with various atomic or molecular target systems. The complication of the remaining instability of the interferometer setup could be solved by means of an active stabilization, which ensures the necessary long-term stability for pump-probe experiments with attosecond resolution. State-of-the-art streaking measurements with active stabilization of the interferometer demonstrate the feasibility of attosecond pump-probe experiments with this novel setup. Compared to previous attosecond beamlines both flux and spectral bandwidth of the XUV pulses could be increased by a 45° angle of incidence of the XUV beam on the reflecting multilayer optic. Future XUV sources will benefit from an extended reflectance ranging above 200 eV. Furthermore, this setup offers easy access to measurements of the transmitted XUV spectrum as a function of delay between the two pulses bringing attosecond transient absorption measurements into range.

ACKNOWLEDGMENTS

We thank B. Reiter, S. Köhler, M. Uiberacker, H. P. Schönauer, M. Fischer, A. Horn, and B. Horvath for support in setting up the beamline. This work was supported by the Munich Center for Advanced Photonics (MAP). R. Kienberger acknowledges funding from the Sofja Kovalevskaja Award of the Alexander von Humboldt foundation and an ERC Starting Grant. E.G. acknowledges a MC Reintegration Grant (Grant No. MERG-CT-2007-208643). M. Fieß is grateful to Studienstiftung des deutschen Volkes.

[1] M. Drescher, M. Hentschel, R. Kienberger, M. Uiberacker, V. Yakovlev, A. Scrinzi, Th. Westerwalbesloh, U. Kleineberg, U. Heinzmann, and F. Krausz, Nature (London) **419**, 803 (2002).
[2] A. L. Cavalieri, N. Müller, Th. Uphues, V. S. Yakovlev, A. Baltuska, B. Horvath, B. Schmidt, L. Blümel, R. Holzwarth, S. Hendel, M. Drescher, U. Kleineberg, P. M. Echenique, R. Kienberger, F. Krausz, and U. Heinzmann, Nature (London) **449**, 1029 (2007).
[3] B. H. Muskatel, F. Remacle, and R. D. Levine, Phys. Scr. **80**, 048101 (2009).
[4] E. Goulielmakis, M. Schultze, M. Hofstetter, V. S. Yakovlev, J. Gagnon, M. Uiberacker, A. L. Aquila, E. M. Gullikson, D. T. Attwood, R. Kienberger, F. Krausz, and U. Kleineberg, Science **320**, 1614 (2008).
[5] M. Schnürer, Z. Cheng, M. Hentschel, G. Tempea, P. Kálmán, T. Brabec, and F. Krausz, Phys. Rev. Lett. **83**, 722 (1999).
[6] J. Peatross, J. L. Chaloupka, and D. D. Meyerhofer, Opt. Lett. **19**, 942 (1994).
[7] M. Drescher, M. Hentschel, R. Kienberger, G. Tempea, C. Spielmann, G. A. Reider, P. B. Corkum, and F. Krausz, Science **291**, 1923 (2001).
[8] T. Pfeifer, M. J. Abel, P. M. Nagel, W. Boutu, M. J. Bell, Y. Liu, D. M. Neumark, and S. R. Leone, Opt. Lett. **34**, 1819 (2009).
[9] G. Sansone, E. Benedetti, F. Calegari, C. Vozzi, L. Avaldi, R. Flammini, L. Poletto, P. Villoresi, C. Altucci, R. Velotta, S. Stagira, S. De Silvestri, and M. Nisoli, Science **314**, 443 (2006).
[10] M. Chini, H. Mashiko, H. Wang, S. Chen, C. Yun, S. Scott, S. Gilbertson,

and Z. Chang, Opt. Express **17**, 21459 (2009).
[11] A. L. Cavalieri, E. Goulielmakis, B. Horvath, W. Helml, M. Schultze, M. Fieß, V. Pervak, L. Veisz, V. Yakovlev, M. Uiberacker, A. Apolonskiy, F. Krausz, and R. Kienberger, New J. Phys. **9**, 242 (2007).
[12] P. Antoine, A. L'Huillier, and M. Lewenstein, Phys. Rev. Lett. **77**, 1234 (1996).
[13] I. P. Christov, M. M. Murnane, and H. C. Kapteyn, Phys. Rev. Lett. **78**, 1251 (1997).
[14] M. Hentschel, F. Krausz, R. Kienberger, C. Spielmann, G. A. Reider, N. Milosevic, T. Brabec, P. Corkum, U. Heinzmann, and M. Drescher, Nature (London) **414**, 509 (2001).
[15] T. Kita, T. Harada, N. Nakano, and H. Kuroda, Appl. Opt. **22**, 512 (1983).
[16] M. Wagner and H. Schröder, Int. J. Mass Spectrom. Ion Process. **128**, 31 (1993).
[17] F. M. Böttcher, B. Manschwetus, H. Rottke, N. Zhavoronkov, Z. Ansari, and W. Sandner, Appl. Phys. B: Lasers Opt. **91**, 287 (2008).
[18] J. Itatani, F. Quéré, G. L. Yudin, M. Yu. Ivanov, F. Krausz, and P. B. Corkum, Phys. Rev. Lett. **88**, 173903 (2002).
[19] R. Kienberger, E. Goulielmakis, M. Uiberacker, A. Baltuska, V. Yakovlev, F. Bammer, A. Scrinzi, Th. Westerwalbesloh, U. Kleineberg, U. Heinzmann, M. Drescher, and F. Krausz, Nature (London) **427**, 817 (2004).
[20] J. Gagnon, E. Goulielmakis, and V. S. Yakovlev, Appl. Phys. B: Lasers Opt. **92**, 25 (2008).

REVIEW OF SCIENTIFIC INSTRUMENTS **82**, 063104 (2011)

A flexible apparatus for attosecond photoelectron spectroscopy of solids and surfaces

E. Magerl,[1] S. Neppl,[2] A. L. Cavalieri,[1,3] E. M. Bothschafter,[1,4] M. Stanislawski,[1] Th. Uphues,[1] M. Hofstetter,[1,5] U. Kleineberg,[1,5] J. V. Barth,[2] D. Menzel,[2] F. Krausz,[1,5] R. Ernstorfer,[1,4] R. Kienberger,[1,4] and P. Feulner[2]

[1]*Max-Planck-Institut für Quantenoptik, Hans-Kopfermann-Straße 1, 85748 Garching, Germany*
[2]*Physik Department E20, Technische Universität München, James-Franck-Straße, 85748 Garching, Germany*
[3]*Max-Planck Research Department for Structural Dynamics, Universität Hamburg, Notkestrasse 85, 22607 Hamburg, Germany*
[4]*Physik Department E11, Technische Universität München, James-Franck-Straße, 85748 Garching, Germany*
[5]*Ludwig-Maximilians-Universität München, Fakultät für Physik, Am Coulombwall 1, 85748 Garching, Germany*

(Received 13 January 2011; accepted 12 May 2011; published online 15 June 2011)

We describe an apparatus for attosecond photoelectron spectroscopy of solids and surfaces, which combines the generation of isolated attosecond extreme-ultraviolet (XUV) laser pulses by high harmonic generation in gases with time-resolved photoelectron detection and surface science techniques in an ultrahigh vacuum environment. This versatile setup provides isolated attosecond pulses with photon energies of up to 140 eV and few-cycle near infrared pulses for studying ultrafast electron dynamics in a large variety of surfaces and interfaces. The samples can be prepared and characterized on an atomic scale in a dedicated flexible surface science end station. The extensive possibilities offered by this apparatus are demonstrated by applying attosecond XUV pulses with a central photon energy of ~125 eV in an attosecond streaking experiment of a xenon multilayer grown on a Re(0001) substrate. © *2011 American Institute of Physics*. [doi:10.1063/1.3596564]

I. INTRODUCTION

Over the last decades, time-resolved photoelectron spectroscopy utilizing laser pulses reaching from the infrared to the ultraviolet (UV) spectral range has evolved into a tool to routinely study chemical and electronic processes at surfaces on the femtosecond timescale.[1–3] However, many fundamental phenomena including dynamical electronic screening,[4,5] electron transfer from chemisorbates,[6,7] and atomic-scale motion of electrons in or near surfaces[8] proceed on an attosecond timescale, which cannot be resolved even if interferometric two-photon photoelectron spectroscopy with well-characterized femtosecond laser pulses are employed.[9–11]

Alternatively, the attosecond transient recorder (ATR) is a newly established technique that provides direct time-resolved access to these processes. The ATR uses carrier-envelope phase (CEP)-stabilized few-cycle laser pulses in the near infrared (NIR) to reproducibly modulate or "streak" the kinetic energy of photoelectrons generated by a synchronized isolated sub-fs XUV pulse.[12,13] This technique has been used extensively to study attosecond electron dynamics in atoms in the gas phase,[13–16] and recently the method has been generalized to include condensed matter systems.[8]

We note that photon energies in the extreme-ultraviolet (XUV) range allow excitation of innershell levels, whereas time-resolved photoelectron spectroscopy in the UV range is restricted to investigations of valence states. Valence orbitals depend on chemistry; in most cases, they are delocalized over several or even many atoms. Innershell levels, however, are elemental fingerprints; their binding energies are only moderately affected by the environment; apart from highly symmetric molecules, delocalization can be neglected. In contrast to time-resolved photoelectron spectroscopy in the UV range, the XUV approach allows a site-selective, atom-specific look. Several methods are under development trying to combine high time resolution with high photon energies.[17–19] Currently, our ATR detection scheme applied to solid surfaces enables photon energies up to 140 eV and time-resolution in the attosecond regime.

Here, we describe a flexible apparatus that enables systematic studies of attosecond phenomena in well-ordered solid state model systems by merging XUV pulse production through high-order harmonic generation (HHG) with standard surface science preparation and detection schemes in ultrahigh vacuum (UHV). In comparison to an apparatus for gas phase streaking experiments,[15,20] an apparatus designed for condensed matter experiments must satisfy additional requirements. One challenge is the combination of HHG, in a relatively high-density noble gas target,[21] with surface sensitive photoelectron spectroscopy.[22] In order to exclude the influence of surface contamination during typical data acquisition periods, including measurements at cryogenic temperatures, a background pressure of $\leq 10^{-10}$ mbar in the measurement chamber area is needed. Since the pressure inside the HHG target exceeds 100 mbar, and because even noble gases, with the exception of helium, would adsorb on the sample at cryogenic conditions, a gradient of many orders of magnitude in pressure must be achieved by differential pumping. Further design goals include: tunable XUV photon energies, independent control of the angle of incidence of the laser pulses and the detection angle of the photoelectrons, compatibility with state-of-the-art surface preparation methods in UHV and the possibility for gas phase measurement for calibration and reference purposes.

0034-6748/2011/82(6)/063104/6/$30.00

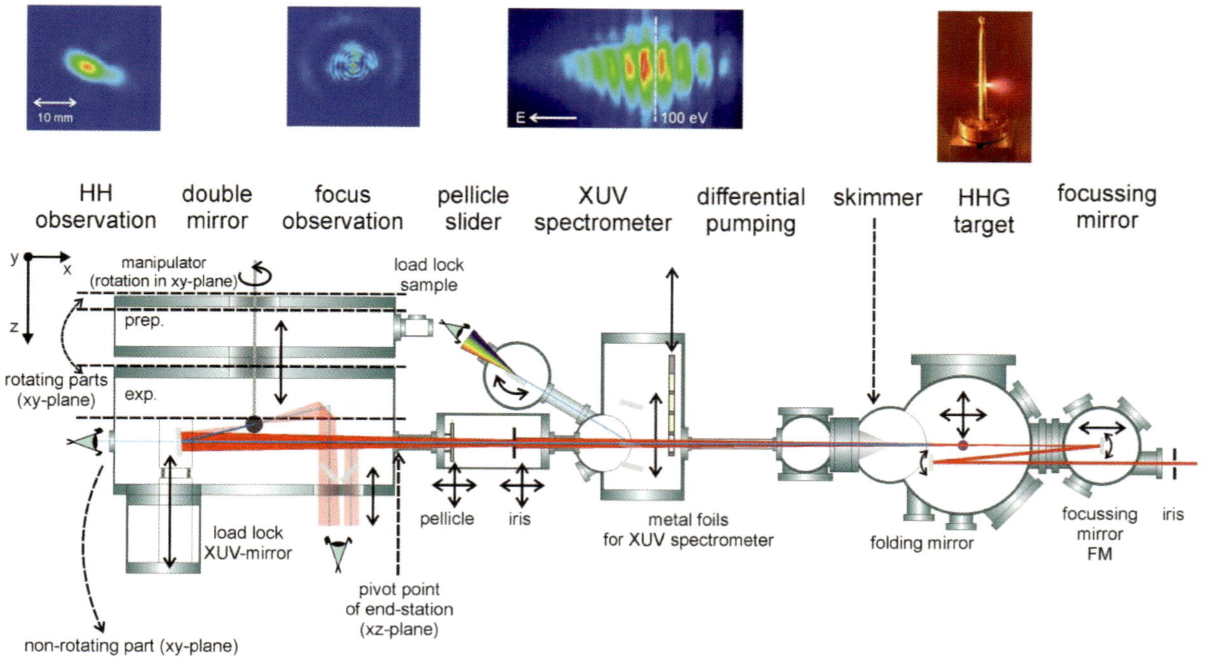

FIG. 1. (Color online) Schematic top-view of the AS-3 beamline and end-station. Details of the various degrees of freedom of sample, preparation tools, detectors, and the end-station are given in Fig. 3. The back plate of the preparation chamber is rotatable so that the manipulator with the sample can be moved to different ports for target preparation and characterization before the sample is inserted into the measurement chamber.

Different approaches for time- and angle-resolved photoelectron spectroscopy, employing femtosecond high harmonic radiation, have been described previously.[23–25] In contrast, the apparatus described here utilizes attosecond XUV pulses and corresponding attosecond time resolution, increased XUV photon energy range, and enhanced flexibility regarding angle of incidence and detection angle.

In order to meet these requirements, a new attosecond beamline and measurement end station at the Max Planck Institute of Quantum Optics, referred to as AS-3, has been built. To the best of our knowledge, this is currently the only operational beamline providing attosecond time resolution for solid state experiments in UHV. A schematic overview of this apparatus is shown in Fig. 1 and a more detailed description of the beamline and the end station is given below.

II. BEAMLINE

A 3 kHz titanium-sapphire laser system in conjunction with hollow-core fiber spectral broadening provides CEP-stable, 400 μJ, octave-spanning pulses centered at 750 nm. These spectrally broadened pulses are subsequently compressed to less than 4 fs in a negative dispersion chirped mirror pulse compressor and are ideal for use in HHG.[10] Following compression, the laser propagates in an evacuated beam tube ($p < 10^{-3}$ mbar) and enters the AS-3 beamline through an adjustable iris aperture (Fig. 1). The laser pulses are focused onto the HHG target by a spherical mirror (FM) in a z-folded geometry. The focal length of the optic can be varied from 300 to 800 mm, resulting in peak intensities from $\sim 2 \times 10^{14}$ W/cm^2 to $\sim 2 \times 10^{15}$ W/cm^2, respectively. High-harmonic radiation is generated in a small volume of noble gas, typically Ne, with a pressure of ~100 mbar. The gas is confined by a nickel tube with an inner diameter of ~2 mm. The tube is sealed on one end and supplied with a constant gas pressure from the other, as described before.[26] The target is mounted on two translation stages for precise alignment of the HHG target along the longitudinal and one transverse axis of the laser beam. Two apertures in the nickel tube along the beam axis are created by laser ablation. The inner-diameter of the nickel tube is adjusted for optimal phase matching conditions for HHG and maximum XUV flux.[27,28] The laser intensity and the gas pressure of the HHG gas target are also adjusted for optimizing the generation of cut-off harmonics in the range from 90 to 140 eV (see Fig. 2) – in combination

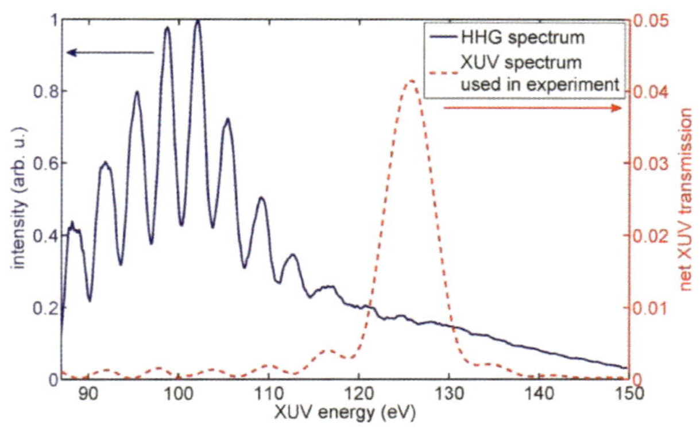

FIG. 2. (Color online) Measured high harmonic spectrum (corrected for reflectivity of the grating and metal filter transmission of 950 nm of Zr) (solid curve). Calculated net XUV transmission used in the reported xenon experiment (based on XUV mirror reflectivity and transmission of a 200 nm Zr metal filter) (dotted curve).

with the CEP-stabilization of the NIR laser a prerequisite for the generation of isolated attosecond pulses in this spectral region.[29–31] Resultant XUV radiation can be monitored with a home-built spectrometer for the energy region between 80 eV and 140 eV. Two grazing incidence gold-palladium mirrors are mounted on a translation stage with direction of motion perpendicular to the beam path. One mirror reflects the direct beam for alignment purposes; the other reflects the XUV beam onto a grating, which disperses the radiation horizontally. A gold-coated flat-field grating (Hitachi No. 001–0437 (Refs. 32 and 33)) suitable for photon energies between 60 and 250 eV is used for dispersion. The detector unit consists of a double microchannel plate assembly in chevron geometry with cesium iodide coating for an enhanced XUV sensitivity and a phosphor screen on the backside for optical detection (Beam Imaging Solutions). Two stages of spectral and spatial filters can be moved in front of the spectrometer. The first stage consists of a 450 nm thick zirconium foil used for suppression of the fundamental driving laser. A second stage consists of a sliding filter plate, offering four insets for additional filters: free-standing foils of 150 nm thick silicon, 200 nm thick aluminum, 200 nm thick palladium or zirconium, and a 500 μm wide slit to improve the resolution of the spectrometer. Figure 2 shows the spectrum of the broadband XUV radiation resulting from HHG in Ne using a CEP-stabilized, sub-4 fs NIR driving laser pulse. The peak intensity in the HHG target gas was $\sim 5.7 \times 10^{14}$ W/cm^2.

The gas load emerging from the HHG target divides the beamline in a high vacuum part (HV, 10^{-2}–10^{-9} mbar) near the gas target and an ultrahigh vacuum section ($<10^{-9}$ mbar) near the end-station. Most of the HHG target gas load is pumped by a 600 l turbo pump directly attached to the HHG chamber. Behind the HHG chamber, a differential pumping stage is realized by a skimmer with an opening of 1.5 mm and a 200 l turbo pump limiting the downstream gas flow. From here a narrow beam pipe in combination with a 300 l turbo pump is sufficient to keep the pressure below 5×10^{-9} mbar. The final transition to a pressure of $<10^{-10}$ mbar is accomplished by an additional differential pumping stage that is pumped by a 400 l turbo pump. This last differential stage does not contain sensitive components and can be baked. In addition, a two component pellicle/Zr foil filter, which is used to separate the XUV and NIR pulses as described below, can be placed in front of the last differential pumping stage and acts as an additional barrier to residual gas emerging from the beamline.

Attosecond streaking experiments require isolated attosecond pulses, filtered from broadband high harmonic radiation,[12] to be additionally isolated from the driving fundamental NIR pulses. Once separated, the relative timing between the isolated XUV pulse and the NIR pulse must also be precisely adjusted. A two-component pellicle/Zr filter, in combination with a precision-controlled double-mirror assembly, is used for this purpose.[26] The two-component filter consists of a nitrocellulose pellicle (thickness: 2, 5, or 15 μm) whose inner part has been replaced by a 200 nm thick zirconium or palladium metal filter with a diameter of 5 mm. When the co-propagating XUV and NIR beams are incident on this filter, the central metal component suppresses the NIR by approximately seven orders of magnitude, effectively blocking it,[34,35] while transmitting a significant fraction of the XUV radiation (the transmission is 52% at 100 eV for 200 nm of zirconium and 37% at 135 eV for 200 nm of palladium[36]). On the other hand, the outer nitrocellulose component allows for nearly full transmission of the NIR (~92%), while completely blocking the XUV radiation. Furthermore, due to the fact that the XUV radiation has a much smaller divergence than the NIR radiation, almost all of the XUV radiation, at the relevant photon energies, is incident on the metal, transparent portion of the filter, allowing for a very efficient method for isolation.

The two-component filter is mounted in a sealed translation sleeve for vacuum separation, and can be moved in the plane perpendicular to the beam path for proper alignment and therefore, maximum transmission of the XUV beam. Directly in front of the pellicle, a motorized iris on a three-axis manipulator is used to adjust the intensity of the co-propagating NIR pulse incident on the sample. Behind the pellicle, the annular NIR pulse and the central XUV beam enter the end station (see Sec. III) where they are reflected by a coaxial assembly of two focusing spherical mirrors consisting of a central XUV-reflecting part, typically 5 mm diameter, and an NIR-reflecting outer mirror.[26] The multilayer surface structure of the inner mirror is designed to reflect XUV radiation with a given bandwidth and central photon energy, which defines the pulse duration of the isolated attosecond pulse.[37,38] In conjunction with the metal foil in the two-component filter, it reflects a part of the cut-off region of the spectrum with a flat phase, to avoid temporal dispersion, which would broaden the attosecond XUV pulse.[38] The outer mirror is typically coated with 13 nm of boron carbide acting as a partial reflector for the NIR pulses with a reflectivity of ~20%. The pointing of the outer mirror is motorized, enabling spatial overlap of the focused beams. Temporal delay between the XUV and the NIR pulses is adjusted by translating the inner mirror along the incident beam with a piezo-driven translation stage, which has a minimum step size of 0.2 nm corresponding to ~1.3 attoseconds of delay.

The entire double mirror assembly is mounted on a stack of 4 piezo-driven stages (Nanomotion), which allow positioning of the mirrors relative to the laser beams as well as aligning the reflected beams onto the solid sample. The complete mirror stack can be retracted from the main experimental chamber into a load-lock in order to make exchanging the mirrors possible without breaking vacuum in the main chamber.

III. END STATION

The end-station consists of two cylindrical stainless steel vessels separated by a gate valve. One of the chambers is used for sample preparation and characterization and the other for photoelectron spectroscopy (measurement chamber). The longitudinal axes of the chambers are parallel. The measurement chamber is designed in an in-axis geometry, i.e., the sample, the laser focus, and the foci of all detectors in working position coincide with the chamber axis (see below). Off-axis mounting is used for the preparation chamber such that the foci of the preparation/diagnostic instruments lie on a circle of 250 mm radius about the chamber axis for maximum solid-angle range of each station. The sample is brought to the individual prep-stations by moving the manipulator along this circle by means of a large differentially pumped rotary feedthrough. The preparation chamber accommodates standard equipment for surface preparation and controlled layer growth: an ion sputter gun, a low-energy electron diffraction system, LEED (SPECS), a Knudsen effusion cell, a modified Bayard-Alpert pressure gauge for temperature programmed desorption (TPD) studies, and a gas dosing system composed of a multi-capillary array doser attached to a gas manifold via a flow-controller.

The samples are mounted on a holder clamped onto the head of a home-built XYZ-manipulator. A second two-stage differentially pumped feedthrough at the manipulator head enables 360° rotation around the Z axis. The sample temperature is controlled with a thermocouple. The reference junction of this thermocouple is attached to the sample holder, the temperature of which is measured independently by a second K-type thermocouple. This design does not require feedthroughs from thermocouple materials at the vacuum interface and allows application of any thermocouple combination as dictated by the requirements of the experiment. An available temperature range of 15–2500 K is achieved by thermal contact to a liquid He flow-cryostat integrated in the manipulator, and by radiative or electron-beam heating of the sample. Arbitrary temperature profiles can be generated by software with a proportional integral derivative (PID) -controlled feedback loop. Fast sample exchange without degradation of the UHV conditions is provided by a load lock system.

The measurement chamber is composed of a detector ring that is rotatable around its longitudinal Z axis (enabled by two differentially pumped rotary feedthroughs) and a fixed section (see Fig. 1). The rotating part is equipped with an electron time-of-flight spectrometer (Stefan Kaesdorf, Geraete fuer Forschung und Industrie), a hemispherical electron energy analyzer HEA (SPECS), and a Mg/Al-twin anode x-ray source (PSP Vacuum Technology) for standard CW photoelectron spectroscopy for sample characterization. These components are all equipped with translation and tilt mechanisms enabling optimal alignment with the sample surface (see Fig. 3).

This flexible configuration, combined with the sample rotation allows independent choice of both XUV/NIR angle of incidence and of the electron detection angle (see Fig. 3). Considering the linear polarization of both the NIR and the XUV pulses, this arrangement can be used to distinguish polarization from angular effects in the photoemission process.

The NIR and XUV beams enter the end station through the fixed part of the measurement chamber and are reflected by the double mirror assembly onto the sample. For optimization and alignment purposes, the fixed part also houses an MCP/fluorescent screen system (see above) to observe the XUV beam. In normal operation, this detector is obscured by the double mirror and the chamber wall; therefore, the entire end-station can be rotated horizontally around the entry port using air pads in order to send the beam to this detector. Furthermore, to enable gas phase measurements, a gas nozzle supported by an XYZ-manipulator mounted on the fixed part can be brought to the center of the chamber to replace the solid sample.

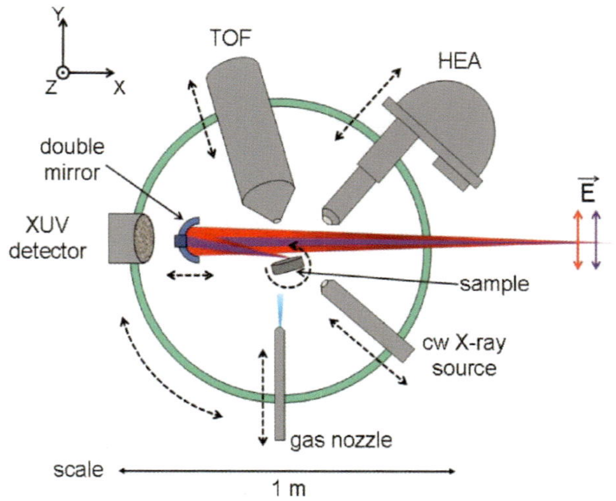

FIG. 3. (Color online) Cross-sectional view of the rotary part of the measurement chamber. The various degrees of freedom for positioning are indicated by dashed arrows.

Both chambers and the rotary feedthroughs are individually pumped by a system of turbomolecular and scroll pumps. In addition, the chambers are also pumped by liquid nitrogen cooled titanium sublimation pumps. After baking the UHV system and with the use of l-N_2-cooled titanium sublimation pumps, a base pressure below 10^{-10} mbar can be routinely achieved and maintained during the measurements.

IV. STREAKING OF CONDENSED XENON

A xenon multilayer serves as a benchmark system to demonstrate the performance of the new apparatus. Xenon is condensed onto a clean Re(0001) surface cooled to 20 K with a final coverage of 200 layers. The xenon coverage has been calibrated by comparing the areas of multilayer and monolayer desorption maxima in TPD.[39] The atomically clean and well ordered Re(0001) surface was produced according to established procedures consisting of Ne^+ sputtering followed by several oxidation and annealing cycles and final flashes to 2400 K.[40]

A MoSi-multilayer mirror acting as a 6.2 eV broad bandpass filter centered at 125 eV filters the XUV continuum produced by HHG and focuses it onto the sample. Photoelectron spectra from 200 atomic layers of xenon taken at different relative delays are shown as a red and a blue curve in Fig. 4(a). Photoemission from the Re substrate is suppressed by the Xe adsorbate and two emission lines at kinetic energies of ~115 eV and ~60 eV with respect to the vacuum level, originating from the Xe valence band, which is derived from Xe $5p$ electrons, and Xe $4d_{5/2,\,3/2}$ core levels, appear. Emission from the Xe $5s$ state cannot be resolved due to its low cross section. The weak features in the region between the two visible Xe-states are due to satellite reflections of the multilayer mirror (cf. Fig. 2). The peak widths of the Xe states reflect the broad energy range of the exciting attosecond XUV pulse and the energetic structure of the Xe

Magerl *et al.* Rev. Sci. Instrum. **82**, 063104 (2011)

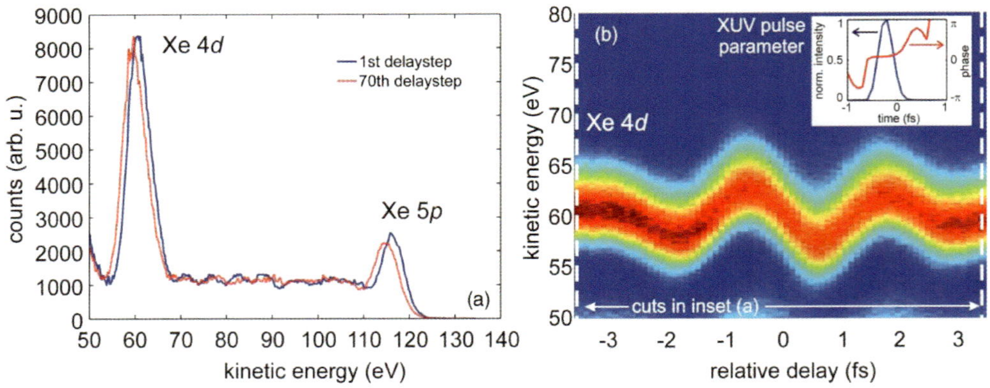

FIG. 4. (Color online) (a) Photoelectron spectra of a multilayer of xenon condensed on a Re(0001) substrate taken at the start and the end of a streaking scan with an integration time of 0.5 min (35 min total duration) illustrating the long-term stability of the xenon multilayer. (b) False-color plot showing the full streaking scan of the xenon 4*d* state. The retrieved attosecond XUV pulse is shown in the small inset. The kinetic energy is referenced to the vacuum level, the spectra, and the spectrogram have been corrected for NIR-induced space charge effects.

condensate. A convolution of a high-resolution photoelectron spectrum obtained by synchrotron radiation with the photon energy dependent reflectivity of the multilayer mirror yields peak widths of 6.6 eV and 6.7 eV for the 5*p* and 4*d* peak, respectively (not shown). These values get slightly modified in the spectrogram in Fig. 4 due to the influence of the NIR streaking field and a non-zero chirp of the XUV pulse.[13]

All displayed photoelectron spectra were corrected for NIR-induced space charge effects. Space charge in the context here refers to energetically low photoelectrons, excited non-resonantly via multiple photo ionization by the NIR pulse, which accelerate the XUV photoelectrons towards the detector. In our case, this space charge leads to an overall shift of the kinetic energy of the XUV photoelectrons, which we correct for by adding an offset value derived from spectra obtained without the NIR field. This method is justified because the NIR focus on our sample is much larger than the XUV focus, i.e., the space charge effect is laterally homogeneous. A possible linear change of the amount of space charge during the acquisition of the spectrogram can be accounted for by fitting the effect with a linear function of $a + bt$. As bt turned out to be in the meV range, this term was not included in the correction.

The large bandwidth of the XUV pulse also sets limits to the minimal energy separation of two different electronic states, which are investigated with the ATR. In order to be able to extract information on their relative timing,[8,41] both peaks have to be spectrally resolved, limiting the separation of the two states to at least the XUV pulse width. In reality, the XUV pulse will display a non-zero chirp which will cause a further broadening of the peaks when the NIR is introduced.[13] Multiplets or a broad conduction band give rise to additional broadening. A general expression for this additional broadening cannot be given as it depends both on the XUV and NIR pulse parameters and on the investigated electronic structure of the sample.

Due to the UHV conditions, the signal strength of the photolines remained unchanged for more than 2 h – even at a substrate temperature of 20 K. The xenon multilayers proved to be stable under simultaneous illumination of both the XUV and the NIR driver/streaking pulses enabling the acquisition of high quality streaking spectrograms[13] as shown in Fig. 4(b). Due to the higher XUV photon energy, in comparison to previously reported streaking experiments,[8] electronic states with a high binding energy like the Xe 4*d* line are much less affected by the NIR-background photoelectrons, resulting from above-threshold ionization photoemission and inelastically scattered electrons.

The estimated NIR streaking intensity, based on the modulation depth of the photoelectron kinetic energies,[12] is $\sim 4 \times 10^{11}$ W/cm^2. The high signal-to-noise ratio achieved in this measurement allows for accurate retrieval of important characteristics of the attosecond XUV pulses. The result of a frequency-resolved optical gating (FROG) routine applied to a xenon multilayer spectrogram is shown in the inset of Fig. 4(b).[42] This analysis reveals an XUV pulse duration of 346 as with a central photon energy of 125 eV and a group delay dispersion of -9.2×10^3 as^2. The figure of merit of the algorithm is 0.23, meaning the root-mean-square deviation between the retrieved and the measured spectrograms.[42] To date, these are the highest energy XUV photons used in any attosecond streaking experiment.

V. SUMMARY

We describe a versatile apparatus dedicated to attosecond time-resolved photoelectron spectroscopy of solids and surfaces. The system is capable of generating sub-fs attosecond XUV pulses with energies ranging from 90 to 140 eV and provides UHV conditions in a flexible end-station, where a wide range of standard surface science tools is available for preparation and characterization of well-defined surfaces and more sophisticated model systems. By performing an attosecond streaking experiment on a xenon multilayer, we demonstrate the long-term stability of our system, which allows us to record high quality streaking spectrograms even from a comparably fragile solid-state sample. Prolonged acquisition times, offered by the UHV environment, now enables streaking measurements to be performed on solid surfaces with equivalent quality to those made in simpler gas phase experiments. The flexibility of the end-station offers possibilities for both angle- and polarization-resolved attosecond electron spectroscopic experiments. Combined with the large variety of possible sample preparation schemes, this new apparatus constitutes the experimental basis for systematic studies of attosecond phenomena in solids and on surfaces. In the future, this apparatus can be further supplemented by a scheme to collinearly generate few-fs UV and attosecond XUV pulses[43] allowing UV-XUV two-photon photoelectron spectroscopy in addition to attosecond streaking.

ACKNOWLEDGMENTS

The authors thank K. Kölbl, K. Eberle, and H. P. Schönauer for their valuable contributions to this work. Supported by the DFG Cluster of Excellence: Munich Centre for Advanced Photonics (www.munich-photonics.de). R.K. acknowledges funding from the Sofja Kovalevskaja Award of the Alexander von Humboldt Foundation and an ERC Starting Grant.

[1]C. Frischkorn and M. Wolf, Chem. Rev. **106**(10), 4207 (2006).
[2]H. Petek and S. Ogawa, Prog. Surf. Sci. **56**(4), 239 (1997).
[3]M. J. Weinelt, J. Phys.: Condens. Matter **14**(43), R1099 (2002).
[4]A. Borisov, D. Sanchez-Portal, R. D. Muino, and P. M. Echenique, Chem. Phys. Lett. **387**(1–3), 95 (2004).
[5]R. D. Muino, D. Sanchez-Portal, V. M. Silkin, E. V. Chulkov, and P. M. Echenique, Proc. Natl. Acad. Sci. **108**, 971 (2011).
[6]C. Keller, M. Stichler, G. Comelli, F. Esch, S. Lizzit, W. Wurth, and D. Menzel, Phys. Rev. Lett. **80**(8), 1774 (1998).
[7]A. Föhlisch, P. Feulner, F. Hennies, A. Fink, D. Menzel, D. Sanchez-Portal, P. M. Echenique, and W. Wurth, Nature (London) **436**(7049), 373 (2005).
[8]A. L. Cavalieri, N. Müller, T. Uphues, V. S. Yakovlev, A. Baltuska, B. Horvath, B. Schmidt, L. Bluemel, R. Holzwarth, S. Hendel, M. Drescher, U. Kleineberg, P. M. Echenique, R. Kienberger, F. Krausz, and U. Heinzmann, Nature (London) **449**(7165), 1029 (2007).
[9]S. Ogawa, H. Nagano, H. Petek, and A. P. Heberle, Phys. Rev. Lett. **78**(7), 1339 (1997).
[10]A. L. Cavalieri, E. Goulielmakis, B. Horvath, W. Helml, M. Schultze, M. Fiess, V. Pervak, L. Veisz, V. S. Yakovlev, M. Uiberacker, A. Apolonski, F. Krausz, and R. Kienberger, New J. Phys. **9**, 242 (2007).
[11]U. Graf, M. Fiess, M. Schultze, R. Kienberger, F. Krausz, and E. Goulielmakis, Opt. Express **16**(23), 18956 (2008).
[12]J. Itatani, F. Quere, G. L. Yudin, M. Y. Ivanov, F. Krausz, and P. B. Corkum, Phys. Rev. Lett. **88**(17), 173903 (2002).
[13]R. Kienberger, E. Goulielmakis, M. Uiberacker, A. Baltuska, V. Yakovlev, F. Bammer, A. Scrinzi, T. Westerwalbesloh, U. Kleineberg, U. Heinzmann, M. Drescher, and F. Krausz, Nature (London) **427**(6977), 817 (2004).
[14]M. Uiberacker, T. Uphues, M. Schultze, A. J. Verhoef, V. Yakovlev, M. F. Kling, J. Rauschenberger, N. M. Kabachnik, H. Schröder, M. Lezius, K. L. Kompa, H. G. Muller, M. J. J. Vrakking, S. Hendel, U. Kleineberg, U. Heinzmann, M. Drescher, and F. Krausz, Nature (London) **446**(7136), 627 (2007).
[15]M. Drescher, M. Hentschel, R. Kienberger, M. Uiberacker, V. S. Yakovlev, A. Scrinzi, T. Westerwalbesloh, U. Kleineberg, U. Heinzmann, and F. Krausz, Nature (London) **419**(6909), 803 (2002).
[16]E. Goulielmakis, M. Uiberacker, R. Kienberger, A. Baltuska, V. S. Yakovlev, A. Scrinzi, T. Westerwalbesloh, U. Kleineberg, U. Heinzmann, M. Drescher, and F. Krausz, Science **305**(5688), 1267 (2004).
[17]A. Pietzsch, A. Fohlisch, M. Beye, M. Deppe, F. Hennies, M. Nagasono, E. Suljoti, W. Wurth, C. Gahl, K. Dobrich, and A. Melnikov, New J. Phys. **10**, 033004 (2008).
[18]M. Stockman and M. F. Kling, Nat. Photonics **1**, 539 (2007).
[19]A. Mikkelsen, J. Schwenke, T. Fordell, G. Luo, K. Klunder, E. Hilner, N. Anttu, A. A. Zakharov, E. Lundgren, J. Mauritsson, J. N. Andersen, H. Q. Xu, and A. L'Huillier, Rev. Sci. Instrum. **80**, 123703 (2009).
[20]M. Fiess, M. Schultze, E. Goulielmakis, B. Dennhardt, J. Gagnon, M. Hofstetter, R. Kienberger, and F. Krausz, Rev. Sci. Instrum. **81**, 093103 (2010).
[21]A. L'Huillier and Ph. Balcou, Phys. Rev. Lett. **70**(6), 774 (1993).
[22]S. Hüfner, *Photoelectron Spectroscopy* (Springer-Verlag, Berlin, 1996).
[23]S. Mathias, L. Miaja-Avila, M. M. Murnane, H. C. Kapteyn, M. Aeschlimann, and M. Bauer, Rev. Sci. Instrum. **78**(8), 083105 (2007).
[24]G. L. Dakovski, Y. Li, T. Durakiewicz, and G. Rodriguez, Rev. Sci. Instrum. **81**, 073108 (2010).
[25]B. H. Christensen, M. K. Raarup, and P. Balling, Nucl. Instrum. Methods Phys. Res. A **615**, 114 (2010).
[26]M. Drescher, M. Hentschel, R. Kienberger, G. Tempea, C. Spielmann, G. A. Reider, P. B. Corkum, and F. Krausz, Science **291**, 1923 (2001).
[27]E. Constant, D. Garzella, P. Breger, E. Mevel, C. Dorrer, C. Le Blanc, F. Salin, and P. Agostini, Phys. Rev. Lett. **82**(8), 1668 (1999).
[28]G. Tempea, M. Geissler, M. Schnürer, and T. Brabec, Phys. Rev. Lett. **84**(19), 4329 (2000).
[29]A. Baltuska, T. Udem, M. Uiberacker, M. Hentschel, E. Goulielmakis, C. Gohle, R. Holzwarth, V. S. Yakoviev, A. Scrinzi, T. W. Hänsch, and F. Krausz, Nature (London) **421**(6923), 611 (2003).
[30]P. B. Corkum, Phys. Rev. Lett. **71**(13), 1994 (1993).
[31]M. Lewenstein, P. Balcou, M. Y. Ivanov, A. Lhuillier, and P. B. Corkum, Phys. Rev. A **49**(3), 2117 (1994).
[32]N. Nakano, H. Kuroda, T. Kita, and T. Harada, Appl. Opt. **23**(14), 2386 (1984).
[33]T. Kita, T. Harada, N. Nakano, and H. Kuroda, Appl. Opt. **22**(4), 512 (1983).
[34]T. A. Johnson, R. Soufli, E. M. Gullikson, and M. Clift, Proc. SPIE **5538**, 119 (2004).
[35]E. Palik, *Handbook of Optical Constants* (Academic, New York, 1991).
[36]B. L. Henke, E. M. Gullikson, and J. C. Davis, At. Data Nucl. Data Tables **54**(2), 181 (1993).
[37]A. Wonisch, U. Neuhausler, N. M. Kabachnik, T. Uphues, M. Uiberacker, V. S. Yakovlev, F. Krausz, M. Drescher, U. Kleineberg, and U. Heinzmann, Appl. Opt. **45**(17), 4147 (2006).
[38]M. Hofstetter, M. Schultze, M. Fiess, A. Guggenmos, J. Gagnon, E. Magerl, E. Bothschafter, R. Ernstorfer, R. Kienberger, E. M. Gullikson, F. Krausz, and U. Kleineberg, in *Prooceedings of Postdeadline Paper of the Ultrafast Phenomena XVII*, paper PDP8, 2010.
[39]R. Opila and R. Gomer, Surf. Sci. **112**, 1 (1981).
[40]M. Stindtmann, M. Farle, T. S. Rahman, L. Benabid, and K. Baberschke, Surf. Sci. **381**(1), 12 (1997).
[41]M. Schultze, M. Fiess, N. Karpowicz, J. Gagnon, M. Korbman, M. Hofstetter, S. Neppl, A. L. Cavalieri, Y. Komninos, T. Mercouris, C. A. Nicolaides, R. Pazourek, S. Nagele, J. Feist, J. Burgdoerfer, A. M. Azzeer, R. Ernstorfer, R. Kienberger, U. Kleineberg, E. Goulielmakis, F. Krausz, and V. S. Yakovlev, Science **328**(5986), 1658 (2010).
[42]J. Gagnon, E. Goulielmakis, and V. S. Yakovlev, Appl. Phys. B **92**(1), 25 (2008).
[43]E. M. Bothschafter, A. Schiffrin, V. S. Yakovlev, A. M. Azzeer, F. Krausz, R. Ernstorfer, and R. Kienberger, Opt. Express **18**, 9173 (2010).

Attosecond dispersion control by extreme ultraviolet multilayer mirrors

Michael Hofstetter,[1,2,*] Martin Schultze,[1,2] Markus Fieß,[2] Benjamin Dennhardt,[2,3] Alexander Guggenmos,[1,2] Justin Gagnon,[1,2] Vladislav S. Yakovlev,[1,2] Eleftherios Goulielmakis,[2] Reinhard Kienberger,[2,3] Eric M. Gullikson,[4] Ferenc Krausz,[1,2] and Ulf Kleineberg[1,2]

[1]*Ludwig-Maximilians-Universität München, Fakultät für Physik, Am Coulombwall 1, D-85748 Garching, Germany*
[2]*Max Planck Institut für Quantenoptik, Hans-Kopfermann-Str. 1, D-85748 Garching, Germany*
[3]*Technische Universität München, Physik Department E11, James-Franck-Straße, D-85748 Garching, Germany*
[4]*Center for X-Ray Optics, Lawrence Berkeley National Lab 2-400, Berkeley, California 94720 USA*
* *michael.hofstetter@mpq.mpg.de*

Abstract: We report the first experimental demonstration of a-periodic multilayer mirrors controlling the frequency sweep (chirp) of isolated attosecond *XUV* pulses. The concept was proven with about 200-attosecond pulses in the photon energy range of 100-130 *eV* measured via photoelectron streaking in neon. The demonstrated attosecond dispersion control is engineerable in a wide range of *XUV* photon energies and bandwidths. The resultant tailor-made attosecond pulses with highly enhanced photon flux are expected to significantly advance attosecond metrology and spectroscopy and broaden their range of applications.

OCIS codes: (320.0320) Ultrafast optics; (320.1590) Chirping; (320.5540) Pulse shaping; (340.7480) X-rays, soft x-rays, extreme ultraviolet (EUV); (310.4165) Multilayer design.

References and links

1. M. Drescher, M. Hentschel, R. Kienberger, G. Tempea, C. Spielmann, G. A. Reider, P. B. Corkum, and F. Krausz, "X-ray pulses approaching the attosecond frontier," Science **291**(5510), 1923–1927 (2001).
2. A. L. Cavalieri, N. Müller, T. Uphues, V. S. Yakovlev, A. Baltuška, B. Horvath, B. Schmidt, L. Blümel, R. Holzwarth, S. Hendel, M. Drescher, U. Kleineberg, P. M. Echenique, R. Kienberger, F. Krausz, and U. Heinzmann, "Attosecond spectroscopy in condensed matter," Nature **449**(7165), 1029–1032 (2007).
3. M. Uiberacker, Th. Uphues, M. Schultze, A. J. Verhoef, V. Yakovlev, M. F. Kling, J. Rauschenberger, N. M. Kabachnik, H. Schröder, M. Lezius, K. L. Kompa, H.-G. Muller, M. J. J. Vrakking, S. Hendel, U. Kleineberg, U. Heinzmann, M. Drescher, and F. Krausz, "Attosecond real-time observation of electron tunnelling in atoms," Nature **446**(7136), 627–632 (2007).
4. M. Schultze, M. Fieß, N. Karpowicz, J. Gagnon, M. Korbman, M. Hofstetter, S. Neppl, A. L. Cavalieri, Y. Komninos, Th. Mercouris, C. A. Nicolaides, R. Pazourek, S. Nagele, J. Feist, J. Burgdörfer, A. M. Azzeer, R. Ernstorfer, R. Kienberger, U. Kleineberg, E. Goulielmakis, F. Krausz, and V. S. Yakovlev, "Delay in photoemission," Science **328**(5986), 1658–1662 (2010).
5. E. Goulielmakis, Z. H. Loh, A. Wirth, R. Santra, N. Rohringer, V. S. Yakovlev, S. Zherebtsov, T. Pfeifer, A. M. Azzeer, M. F. Kling, S. R. Leone, and F. Krausz, "Real-time observation of valence electron motion," Nature **466**(7307), 739–743 (2010).
6. A. Wonisch, U. Neuhäusler, N. M. Kabachnik, T. Uphues, M. Uiberacker, V. Yakovlev, F. Krausz, M. Drescher, U. Kleineberg, and U. Heinzmann, "Design, fabrication, and analysis of chirped multilayer mirrors for reflection of extreme-ultraviolet attosecond pulses," Appl. Opt. **45**(17), 4147–4156 (2006).
7. A.-S. Morlens, P. Balcou, P. Zeitoun, C. Valentin, V. Laude, and S. Kazamias, "Compression of attosecond harmonic pulses by extreme-ultraviolet chirped mirrors," Opt. Lett. **30**(12), 1554–1556 (2005).
8. M. Suman, G. Monaco, M.-G. Pelizzo, D. L. Windt, and P. Nicolosi, "Realization and characterization of an XUV multilayer coating for attosecond pulses," Opt. Express **17**(10), 7922–7932 (2009).
9. M. Hentschel, R. Kienberger, Ch. Spielmann, G. A. Reider, N. Milosevic, T. Brabec, P. Corkum, U. Heinzmann, M. Drescher, and F. Krausz, "Attosecond metrology," Nature **414**(6863), 509–513 (2001).
10. I. P. Christov, M. M. Murnane, and H. C. Kapteyn, "High-Harmonic Generation of Attosecond Pulses in the "Single-Cycle" Regime," Phys. Rev. Lett. **78**(7), 1251–1254 (1997).
11. R. Kienberger, E. Goulielmakis, M. Uiberacker, A. Baltuška, V. Yakovlev, F. Bammer, A. Scrinzi, Th. Westerwalbesloh, U. Kleineberg, U. Heinzmann, M. Drescher, and F. Krausz, "Atomic transient recorder," Nature **427**(6977), 817–821 (2004).
12. F. Krausz, and M. Ivanov, "Attosecond physics," Rev. Mod. Phys. **81**(1), 163–234 (2009).

#138629 - $15.00 USD Received 23 Nov 2010; revised 31 Dec 2010; accepted 3 Jan 2011; published 14 Jan 2011

13. E. Goulielmakis, M. Uiberacker, R. Kienberger, A. Baltuška, V. Yakovlev, A. Scrinzi, Th. Westerwalbesloh, U. Kleineberg, U. Heinzmann, M. Drescher, and F. Krausz, "Direct measurement of light waves," Science **305**(5688), 1267–1269 (2004).
14. V. Pervak, A. V. Tikhonravov, M. K. Trubetskov, S. Naumov, F. Krausz, and A. Apolonski, "1.5-octave chirped mirror for pulse compression down to sub-3 fs," Appl. Phys. B **87**(1), 5–12 (2007).
15. E. Gustafsson, T. Ruchon, M. Swoboda, T. Remetter, E. Pourtal, R. López-Martens, Ph. Balcou, and A. L'Huillier, "Broadband attosecond pulse shaping," Opt. Lett. **32**(11), 1353–1355 (2007).
16. R. López-Martens, K. Varjú, P. Johnsson, J. Mauritsson, Y. Mairesse, P. Salières, M. B. Gaarde, K. J. Schafer, A. Persson, S. Svanberg, C.-G. Wahlström, and A. L'huillier, "Amplitude and phase control of attosecond light pulses," Phys. Rev. Lett. **94**(3), 033001 (2005).
17. G. Sansone, E. Benedetti, F. Calegari, C. Vozzi, L. Avaldi, R. Flammini, L. Poletto, P. Villoresi, C. Altucci, R. Velotta, S. Stagira, S. De Silvestri, and M. Nisoli, "Isolated single-cycle attosecond pulses," Science **314**(5798), 443–446 (2006).
18. E. Goulielmakis, M. Schultze, M. Hofstetter, V. S. Yakovlev, J. Gagnon, M. Uiberacker, A. L. Aquila, E. M. Gullikson, D. T. Attwood, R. Kienberger, F. Krausz, and U. Kleineberg, "Single-cycle nonlinear optics," Science **320**(5883), 1614–1617 (2008).
19. M. Drescher, M. Hentschel, R. Kienberger, M. Uiberacker, V. Yakovlev, A. Scrinzi, Th. Westerwalbesloh, U. Kleineberg, U. Heinzmann, and F. Krausz, "Time-resolved atomic inner-shell spectroscopy," Nature **419**(6909), 803–807 (2002).
20. Y. Ménesguen, S. de Rossi, E. Meltchakov, and F. Delmotte, "Aperiodic multilayer mirrors for efficient broadband reflection in the extreme ultraviolet," Appl. Phys., A Mater. Sci. Process. **98**(2), 305–309 (2010).
21. M. Hofstetter, M. Schultze, M. Fieß, A. Guggenmos, J. Gagnon, E. Magerl, E. Botschafter, R. Ernstorfer, R. Kienberger, E. M. Gullikson, F. Krausz, and U. Kleineberg, "First attosecond pulse control by multilayer mirrors above 100 eV photon energy," in *Ultrafast Phenomena XVII*, 11.17, M. Chergui, D. Jonas, E. Riedle, B. Schönlein, A. Taylor (Oxford University Press, New York, 2010).
22. A. V. Tikhonravov, M. K. Trubetskov, and G. W. DeBell, "Optical coating design approaches based on the needle optimization technique," Appl. Opt. **46**(5), 704–710 (2007).
23. B. L. Henke, E. M. Gullikson, and J. C. Davis, "X-ray interactions: photoabsorption, scattering, transmission, and reflection at E=50-30000 eV, Z=1-92," At. Data Nucl. Data Tables **54**(2), 181–342 (1993).
24. E. M. Gullikson, J. H. Underwood, P. Batson, and V. Nikitin, "A soft x-ray/EUV reflectometer based on a laser produced plasma source," J. XRay Sci. Technol. **3**(4), 283–299 (1992).
25. A. Aquila, F. Salmassi, and E. Gullikson, "Metrologies for the phase characterization of attosecond extreme ultraviolet optics," Opt. Lett. **33**(5), 455–457 (2008).
26. M. Kitzler, N. Milosevic, A. Scrinzi, F. Krausz, and T. Brabec, "Quantum theory of attosecond XUV pulse measurement by laser dressed photoionization," Phys. Rev. Lett. **88**(17), 173904 (2002).
27. R. Kienberger, M. Hentschel, M. Uiberacker, Ch. Spielmann, M. Kitzler, A. Scrinzi, M. Wieland, Th. Westerwalbesloh, U. Kleineberg, U. Heinzmann, M. Drescher, and F. Krausz, "Steering attosecond electron wave packets with light," Science **297**(5584), 1144–1148 (2002).
28. M. Schultze, E. Goulielmakis, M. Uiberacker, M. Hofstetter, J. Kim, D. Kim, F. Krausz, and U. Kleineberg, "Powerful 170-attosecond XUV pulses generated with few-cycle laser pulses and broadband multilayer optics," N. J. Phys. **9**(7), 243 (2007).
29. M. Fieß, M. Schultze, E. Goulielmakis, B. Dennhardt, J. Gagnon, M. Hofstetter, R. Kienberger, and F. Krausz, "Versatile apparatus for attosecond metrology and spectroscopy," Rev. Sci. Instrum. **81**, 1 (2010).
30. R. Trebino, K. W. DeLong, D. N. Fittinghoff, J. N. Sweetser, M. A. Krumbügel, B. A. Richman, and D. J. Kane, "Measuring ultrashort laser pulses in the time-frequency domain using frequency-resolved optical gating," Rev. Sci. Instrum. **68**(9), 3277 (1997).
31. A download link of the program can be found at http://www.attoworld.de/Home/ourResearch/Downloads/index.html
32. Y. Mairesse, A. de Bohan, L. J. Frasinski, H. Merdji, L. C. Dinu, P. Monchicourt, P. Breger, M. Kovačev, R. Taïeb, B. Carré, H. G. Muller, P. Agostini, and P. Salières, "Attosecond synchronization of high-harmonic soft x-rays," Science **302**(5650), 1540–1543 (2003).
33. J. Gagnon, E. Goulielmakis, and V. S. Yakovlev, "The accurate FROG characterization of attosecond pulses from streaking measurements," Appl. Phys. B **92**(1), 25–32 (2008).
34. J. Gagnon, and V. S. Yakovlev, "The robustness of attosecond streaking measurements," Opt. Express **17**(20), 17678–17693 (2009).
35. L. Y. Peng, F. Tan, Q. Gong, E. A. Pronin, and A. F. Starace, "Few-cycle attosecond pulse chirp effects on asymmetries in ionized electron momentum distributions," Phys. Rev. A **80**(1), 013407 (2009).
36. V. S. Yakovlev, J. Gagnon, N. Karpowicz, and F. Krausz, "Attosecond streaking enables the measurement of quantum phase," Phys. Rev. Lett. **105**(7), 073001 (2010).
37. Y. Li, J. Lewellen, Z. Huang, V. Sajaev, and S. V. Milton, "Time-resolved measurement of a self-amplified free-electron laser," NIM A **507**(1-2), 413–416 (2003).

1. Introduction

Attosecond (1 *as* = 10^{-18} *s*) light pulses at short wavelength from the extreme ultraviolet (*XUV*) to the soft *X*-ray regime have recently gained fundamental importance in studying electron motions in atoms, molecules or at surfaces [1–5]. Here, fundamental ultrafast

#138629 - $15.00 USD Received 23 Nov 2010; revised 31 Dec 2010; accepted 3 Jan 2011; published 14 Jan 2011

electron dynamics in excitation, transport and emission processes on time scales of a few tens of attoseconds can be observed. These ultrashort time-scales of electron motion are inevitably connected to atomic or mesoscopic spatial scales and nano-scaled microscopic resolution can be provided by the short wavelength nature of the soft *X*-ray pulses itself. The performance and applicability of attosecond spectroscopy critically depends on the flux, photon energy, bandwidth and dispersion control of attosecond pulses. Multilayer-coated *XUV* mirrors [6–8] have served attosecond metrology from its very beginning [9], by focusing and isolating single attosecond pulses from coherent high-order harmonic radiation [10–13]. Chirped multilayer mirrors in the visible/infrared range, are inevitable components in generating few cycle laser pulses of a few femtosecond pulse-length only. Those are the necessary driving pulses for attosecond pulse generation [14]. In contrast to the development of chirped dielectric mirrors in the visible, the realization of chirped *XUV* multilayer mirrors requires new approaches in computational design, manufacturing and characterization, which have not been available and combined until now. So far, *XUV* dispersion control over multi-electron volt bandwidths could only be accomplished for photon energies below 100 *eV* [15–19]. Further, control of the spectral distribution and chirp of an attosecond extreme ultraviolet pulse have so far been realized by means of ultrathin metal filters (typical thickness ~100 *nm*) [15–17] and *XUV* multilayer mirrors [18,19] with chirp control mainly relying on the metal filter components. Their applicability is limited due to strong *XUV* attenuation and the availability of only few materials, thus only specific spectral ranges, where the real part of the complex index of refraction provides sufficient (usually only negative) dispersion.

Specially designed non-periodic binary or ternary multilayer-coated *XUV* mirrors allow for a large degree of freedom in influencing the *XUV* pulse characteristics by reflection of those mirrors [20,21]. So far, multilayer mirrors in attosecond physics applications were used as band-pass reflectors preserving the spectral phase of the attosecond pulse that is defined in the generation process and by the dispersion characteristics of all *XUV* optical elements passed by the pulse. In this work we demonstrate, how specially designed non-periodic multilayer *XUV* mirrors furnish attosecond technology with the ability to control the spectral phase in addition to spectral intensity of attosecond pulses, offering a great degree of freedom in influencing the *XUV* pulse characteristics, such as the pulse shape, duration, frequency sweep and central wavelength.

2. The Experiment

2.1 Concept

Within this paper, the group delay (*GD*) τ is defined as the *negative* derivative of the spectral phase $\phi(\omega)$ with respect to angular frequency ω, $\tau = -d\phi/d\omega$. Its derivative is the group delay dispersion $GDD = d\tau/d\omega$, known to be the lowest order phase term that temporally broadens the pulse.

Figure 1 displays a computational example for group-delay dispersion (*GDD*) management by non-periodic multilayer coatings optimized to introduce positive (left panel) or negative (right panel) *GDD* to an initially unchirped *XUV* pulse (higher order dispersion control like third order dispersion *TOD* by non-periodic multilayer is also feasible, although not demonstrated here). The square of the electric field amplitude of the standing wave resulting from the superposition of the incident and reflected radiation depicted by the false-colour plots reveals a penetration depth varying nearly linearly with photon energy in the centre of the high-reflectivity (*HR*) range of the multilayer Bragg peak ($\approx$ 100-130 *eV*). The penetration depth increasing and decreasing with photon energy imposes a positive and negative chirp, respectively, on an originally bandwidth-limited attosecond pulse reflected by the mirror. Thus, introducing positive *GDD* results in a positive chirp of an initially unchirped pulse and vice versa.

#138629 - $15.00 USD Received 23 Nov 2010; revised 31 Dec 2010; accepted 3 Jan 2011; published 14 Jan 2011

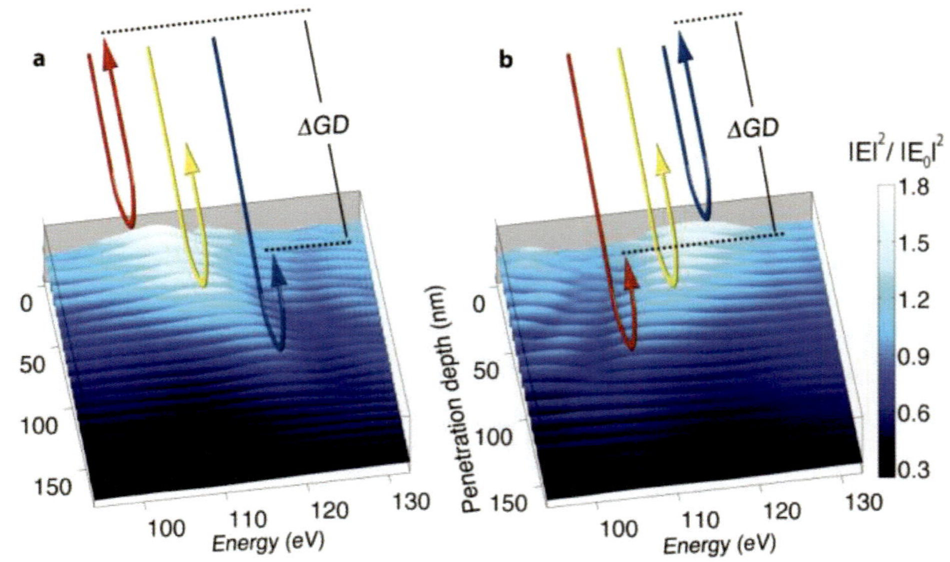

Fig. 1. Operational principle of *XUV* chirped multilayers: Calculated square of the electric field amplitude of the standing wave formed in the multilayer structure upon reflection in the 94–133 *eV* photon energy range, normalized to that of the incident wave. Following the energy dependent maxima of the standing field, it is obvious that the penetration depth is wavelength dependent. Thus the mirrors introduce positive (a) and negative (b) chirp, resulting in an increasing and decreasing group delay for increasing photon energy respectively as the energy-dependent penetration depth signifies as depicted by the coloured arrows. The multilayer structure assumed here resembles the experimentally realized a-periodic multilayer mirrors as displayed in Fig. 2 (a/d) and (c/f).

2.2 Design

To prove the validity of the concept for precision chirp control of attosecond *XUV* pulses, we have developed a set of three multilayer mirrors exhibiting distinctly different *GDD* but similar reflectivity characteristics. Optimized for an incident angle of 45° they exhibit high reflectivity within the range of 100-130 *eV* with comparable full widths at half maximum energy bandwidth: ΔE = 11-13 *eV FWHM* and peaking at different central energies within the range of 107-122 *eV*.

The multilayer designs have been calculated and optimized by a Fresnel equation thin film code coupled to a needle optimization algorithm [22]. The mirrors have been designed such, that the *GDD* is almost maximum and mostly linear within the final attosecond pulse spectrum. Small shifts of the mirror spectrum due to the thin metal filter or the shape of the high harmonic cut-off spectrum have been estimated and taken into account in the design. The calculated designs have been optimized for additional suppressed reflectivity contributions in the near vicinity of the main Bragg peak and have been tested for stability against small layer thickness deviations. *X*-ray optical constants in the *XUV* are based on the tabulated data of Henke and Gullikson [23]. The final multilayer coating designs are displayed in the lower panels (*d-f*) of Fig. 2. While the designs of the positively- and the negatively-chirped mirrors contain around 30 layers Mo/B_4C the almost unchirped mirror is a 17 layer $Mo/Si/B_4C$ stack. The mirrors have been deposited by means of Dual Ion Beam Deposition (*DIBD*) on flat high-polished glass substrates. Interface losses and compound formation have been included in the calculation and could be compensated within the coating procedure.

#138629 - $15.00 USD Received 23 Nov 2010; revised 31 Dec 2010; accepted 3 Jan 2011; published 14 Jan 2011

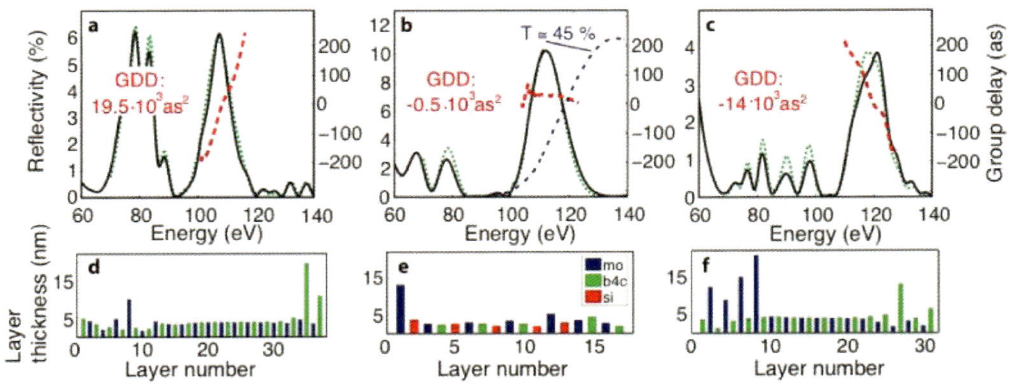

Fig. 2. Characteristics of *XUV* chirped multilayer coatings designed and manufactured for the experiments: Set of *XUV* multilayers with positive (a), near zero (b) and negative (c) group delay dispersion. The diagrams compare the measured (solid black) with the designed (dotted green) reflectivity curves. The calculated group delays (derivative of the spectral phase with respect to angular frequency) are also displayed (dashed red curve). The transmittivity of a 150-nm palladium foil is additionally qualitatively plotted in panel (b), depicted by the blue dash-dotted line. The lower panels (d), (e) and (f) show the corresponding mirror-designs of the appropriate mirrors above. Each colour represents one material as explained by the legend in 2(e).

Figure 2(a)–2(c) shows the calculated reflectivity and group delay of the three multilayer mirrors versus photon energy, compared with the reflectivity measured by *X*-ray reflectometry using synchrotron radiation [24]. The measured and computed *XUV* reflectivities are in excellent agreement, revealing maximum peak reflectivities between 4 and 10 percent. Notable deviations in the peak shape appear only in Fig. 2(c) near 120 *eV*, possibly due to minor uncertainties in the deposition layer thicknesses. In our proof-of-concept experiments, radiation at photon energies below 100 *eV* is suppressed by a 150 *nm* thick Palladium (*Pd*) filter (its transmittivity is shown in Fig. 2(b) by the blue dash-dotted line). The throughput of this high-pass filter combination could be enhanced by fine-tuning the filter thickness and the mirrors' high-reflectivity band. Since isolated attosecond pulses can be extracted by spectral filtering from the cut-off part of the generated *HH* spectrum [9], the high energy extend of the spectrum has to coincide with the high reflectivity range of the mirror. Reflectivity above this energy window thus does not affect the attosecond pulse generation. The three mirrors are designed to introduce substantial positive *GDD*, negligible *GDD*, and substantial negative *GDD*, with computed values of $19.5\text{x}10^3\ as^2$, $-0.5\text{x}10^3\ as^2$, and $-14\text{x}10^3\ as^2$, respectively.

2.3 Streaking spectrograms and analyses

In contrast to the reflectivity, there is no standard frequency-domain technique available for measuring *GDD* in the extreme ultraviolet spectral range. Aquila *et. al.* have characterized the *GDD* of a multilayer mirror by measuring the photoelectron yield from its surface [25]. Here we have resorted to attosecond metrology for measuring the attosecond pulse dispersion directly and assessing the mirrors' capability of controlling the chirp of isolated attosecond *XUV* pulses [12,15,26].

Unlike in the well known collinear setup [27,28], where a concentric double mirror is used to introduce a delay between the inner (lower divergent) *XUV* beam and the outer (higher divergent) laser beam, in our setup the laser and *XUV* beam are completely spatially separated in two paths of an interferometric setup before recollimation [29]. Thus focusing of the *XUV* radiation (by means of a grazing incidence parabola) and time delay introduction between the *XUV* and the laser pulse (by means of a delay stage in the optical beam path) is decoupled from the plane *XUV* multilayer mirror. This technique allows for an easy and rapid exchange and alignment of the 45 *deg XUV* mirrors, which is inevitable for future *XUV* phase sensitive attosecond experiments.

#138629 - $15.00 USD Received 23 Nov 2010; revised 31 Dec 2010; accepted 3 Jan 2011; published 14 Jan 2011

We have implemented attosecond streaking by liberating photoelectrons from the *2p* sub-shell of an ensemble of neon atoms with sub-300-attosecond *XUV* pulses filtered by the combination of a *Pd* foil and one of the band-pass multilayer mirrors described above. The energy distribution of the ejected electrons has been streaked by the controlled linearly polarized electric field of near-single-cycle near-infrared (*NIR*) laser pulses [11,18,26]. The streaked spectra for electrons collected in a narrow cone aligned with the laser polarization were recorded as a function of the delay between the ionizing *XUV* pulse and the streaking *NIR* field.

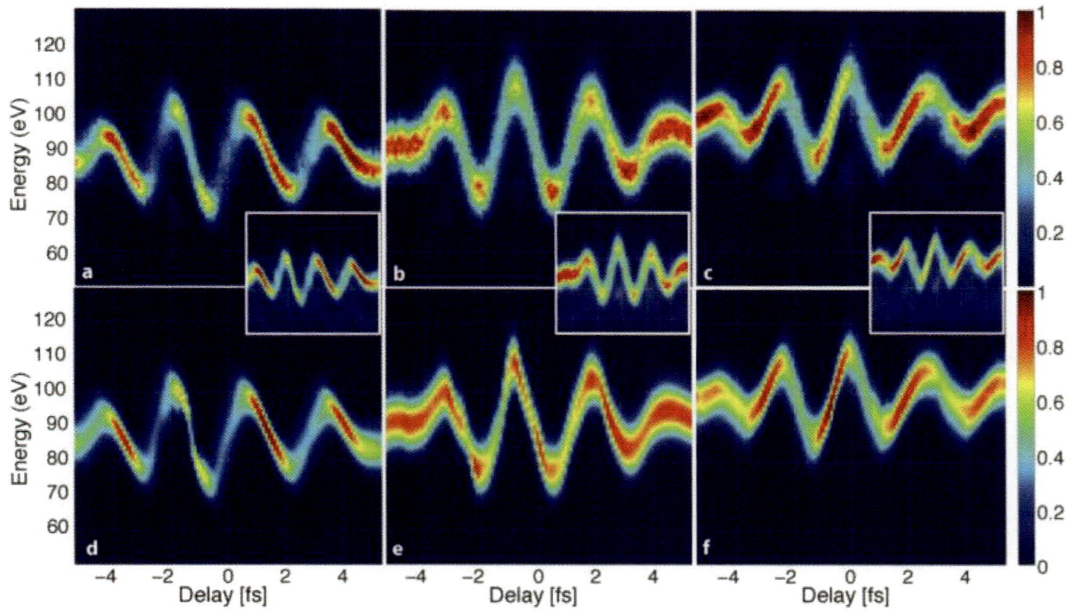

Fig. 3. Measured and retrieved attosecond electron streaking spectrograms, upper and lower row, respectively. Panels (a)-(c) and (d)-(f) display spectrograms recorded with attosecond pulses reflected off the mirrors described in Fig. 2(a)–2(c), respectively. The energy shift in the recorded spectra is proportional to the vector potential $A_L(t)$ of the streaking laser field at the instant of photoemission, hence the dependence of this energy shift on the delay between the attosecond *XUV* pulse and the *NIR* laser field reflects the temporal evolution of the *NIR* laser field's vector potential, $A_L(t)$. Narrowing and broadening in the streaked electron spectra at the zero crossings is displayed by enhanced and diminished electron count rates, respectively. The small insets show the energy calibrated raw data before subtraction of the background.

To access the temporal intensity profile of the synthesized attosecond pulses and their frequency sweep we performed a frequency-resolved optical gating (*FROG*) analysis [30,31] of the acquired streaking spectrograms. This method gives not only access to a full characterization of the laser vector potential, it allows us as well to characterize the spectral intensity distribution and the group delay variation [32,33] of the final attosecond *XUV* pulse reflected from the mirrors after passing the filter, allowing for a direct comparison of its *GDD* with the design values of the mirrors' *GDD*. The resultant streaking spectrograms recorded with the three mirrors of Fig. 2(a)–2(c) are displayed in Fig. 3(a)–3(c) along with their corresponding *FROG* retrievals in Fig. 3(d)–3(f), respectively. The streaked energy spectra of electrons ejected from the *2p* sub-shell are accompanied by a low- energy background originating from various processes, including inelastic scattering, above-threshold ionization via the *NIR* field, contribution from a small satellite pulse and shakeup channels. The *2p* spectrogram also slightly overlaps with that corresponding to *2s* electrons. We subtracted this undesirable background in order to isolate the *2p* spectrogram for a *FROG* analysis. The insets in Fig. 3 show the raw data before subtraction of that background. All features of the spectrograms remain obviously untouched. Thus the choice of the background function does not affect the qualitative result but has some influence on the details of the quantitative

#138629 - $15.00 USD Received 23 Nov 2010; revised 31 Dec 2010; accepted 3 Jan 2011; published 14 Jan 2011

analyses. Using the *FROG CRAB* algorithm *Attogram* described in refs [33,34], we retrieved the vector potential of the streaking *NIR* laser field together with the spectral phase and intensity distribution of the attosecond *XUV* pulse. Figure 3(d)–3(f) depict the corresponding streaking spectrograms calculated from the retrieved *NIR* waveform and *XUV* pulse. The calculated spectrograms appear to be in excellent agreement with the measured ones.

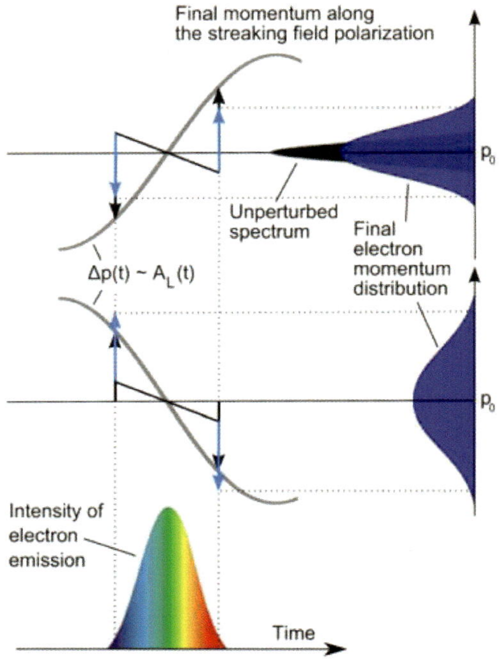

Fig. 4. Qualitative analysis of attosecond streak images of a linearly negatively-chirped attosecond pulse. For explanation, see text.

The spectrograms clearly reveal signatures of the chirp imposed to the *XUV* pulse after being reflected by the chirped *XUV* multilayer mirrors. In order to recognize these signatures, we analyse, with the help of Fig. 4, qualitatively the implications of lowest-order (linear) frequency sweep of the attosecond *XUV* pulse on its streaking spectrogram recorded with a linearly polarized few-cycle field. The linear chirp causes a linear sweep of initial photoelectron momenta depicted by the slanted black straight lines in Fig. 4. The laser field transfers a momentum Δp to the *XUV*-released electron that is proportional to the vector potential $A_L(t)$ of the streaking field at the instant of release. As a consequence, $A_L(t)$ also sweeps the electron momentum near its zero crossings. This laser-induced sweep is positive (negative) at the positive (negative) slope of $\Delta p(t)$, respectively. Depending on its sign, it increases or decreases the initial sweep of photoelectron momentum imposed by the chirp of the ionizing attosecond *XUV* pulse. This, in turn, gives rise to a broadening or narrowing of the corresponding final momentum (energy) distribution (streak image), respectively, which can be approximated at moderate streaking intensities as $\Delta E \sim 2ep(t_0)A_L(t_0)$, where $p(t_0)$ represents the electron momentum and $A_L(t_0)$ the vector field amplitude of the streaking laser field at time t_0 of its birth. Hence, a negatively-chirped pulse (taken in the example considered in Fig. 4) results in a streaking spectrogram exhibiting narrower/broader streaked spectra at the positive/negative slope of $\Delta p(t)$, exactly as revealed by Fig. 3(c) and 3(f). Conversely, a positively-chirped pulse is predicted to produce a streaking spectrogram with broader/narrower streaked spectra at the positive/negative slope of $\Delta p(t)$, as displayed by Fig. 3(a) and 3(d).

We can now turn our attention to a quantitative evaluation of the chirp carried by the attosecond *XUV* pulse in the three experiments performed with the three different mirrors. Figure 5 shows the retrieved intensity and *GD* of the attosecond pulses after reflection off the

#138629 - $15.00 USD Received 23 Nov 2010; revised 31 Dec 2010; accepted 3 Jan 2011; published 14 Jan 2011

XUV multilayer and filtering by the transmission filter as evaluated from the photo-electron spectrograms in Fig. 3. The solid black lines in Fig. 5(a)–5(c) show a direct measurement of the *XUV* pulse spectrum at the target after passing the filter and the appropriate mirror. Comparison of the retrieved *XUV* spectra (green dotted lines) with those directly measured photon spectra shows remarkable agreement of the bandwidth, spectral profile and central energy position. This agreement between retrieved and measured spectra highlights the precision with which *XUV* mirrors are designed and manufactured and demonstrates the power of attosecond streaking measurements coupled with the *FROG CRAB* retrieval procedure.

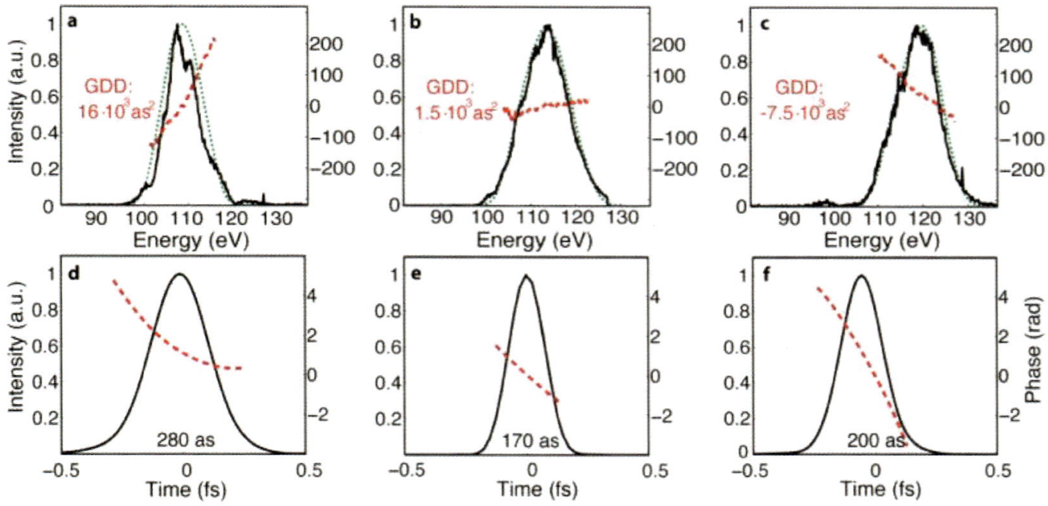

Fig. 5. Retrieved spectral and temporal characteristics of the attosecond *XUV* pulse, upper row and lower row, respectively. Panels (a), (b) and (c) display the intensity spectrum (green dotted line) and group delay (red dashed line) of the attosecond *XUV* pulse reflected off the mirrors described in Fig. 2(a), 2(b), and 2(c), respectively, as retrieved from the measured streaking spectrograms shown in Fig. 3(a), 3(b), and 3(c), respectively. The black full lines depict the corresponding *XUV* spectra measured directly with an *XUV* spectrometer. The evaluated effective group-delay dispersions weighted by the final spectral *XUV* intensity are also shown. Panels (d), (e), (f) show the temporal intensity profile (black line) and temporal phase (red dashed line) retrieved from the streaking spectrograms of Fig. 3(a), 3(b), and 3(c), respectively. The *XUV* pulse duration (full width at intensity half maximum) has been evaluated and is displayed at the bottom of each panel.

Table 1 summarizes the evaluated pulse durations and the *GDD* carried by the reflected pulses, in comparison with the designed *GDD* of the mirrors. Small deviations from the average *GDD* influence the final pulse more if its spectral intensity is high at that energy. Thus, all *GDD* values in Table 1 have been weighted with the spectral profiles $I(\omega)$ of the final pulses for a more meaningful comparison: $GDD_{\text{average}}=\int I(\omega)GDD_\omega d\omega/\int I(\omega)d\omega$. Comparing the retrieved averaged *GDD* values with the design goals shows an excellent quantitative agreement for two of the mirrors, namely the positive *GDD* (to within 20%), and the near zero *GDD* mirror (to within $\approx 2\text{x}10^3$ as^2, causing negligible pulse broadening in the 100-*as* domain), while the quantitative agreement is compromised by the same layer thickness errors in the negative *GDD* mirror giving rise to the deviations in Fig. 2(c). Note that even thickness deviations on an atomic scale result in significant changes of the mirrors spectral phase response though the reflectance suffers only minor modifications.

Table 1. Comparison of *XUV* mirror designed and *FROG* retrieved pulse parameters*

	Positive *GDD* mirror	Zero *GDD* mirror	Negative *GDD* mirror
GDD design goal	$D_{(+)}$ = **19.5x10^3** *as*^{2}	$D_{(0)}$ = **-0.5x10^3** *as*^{2}	$D_{(-)}$ = **-14x10^3** *as*^{2}
FROG retrieved *GDD*	$D_{(+)}$ = **16x10^3** *as*^{2}	$D_{(0)}$ = **1.5x10^3** *as*^{2}	$D_{(-)}$ = **-7.5x10^3** *as*^{2}
FROG retrieved pulse duration (Fourier limit)	$\tau_{(+)}$ = **280** *as* (200 *as*)	$\tau_{(0)}$ = **170** *as* (165 *as*)	$\tau_{(-)}$ = **200** *as* (175 *as*)

*The average *GDD* values and the appropriate standard deviations are calculated by weighting the calculated and the retrieved discrete *GDD* values with the intensity of the retrieved *XUV* spectrum (see text for a more detailed description). For the positive *GDD* mirror we measure 16x10^3 ± 4x10^3 *as*^{2} versus designed 19.5x10^3 ± 4x10^3 *as*^{2}, for the zero *GDD* mirror we find 1.5x10^3 ± 2x10^3 *as*^{2} versus a design value of –0.5 x10^3 ± 3x10^3 *as*^{2} and for the negative *GDD* mirror a *GDD* of –14x10^3 ± 7.5 x10^3 *as*^{2} was designed and –7.5x10^3 ± 2x10^3 *as*^{2} has been retrieved. The lower line shows the retrieved pulse-lengths in comparison to their Fourier limits. It is obvious that the two chirped mirrors broaden the pulse significantly due to the introduced chirp, while there is hardly an effect on the pulse duration by the near zero *GDD* mirror.

Uncertainties in the *XUV* optical constants of the multilayer materials *Mo*, B_4C and *Si* in the 100-130 *eV* energy range may also contribute to the residual discrepancies between design and measurement, with dispersion of the *Pd* filter appearing to be less than 1.5x10^3 *as*^{2} in the 100-130 *eV* photon energy range. The appearance of a single attosecond pulse in the fringe-less spectrograms (Fig. 3) ensures that our mirror reflectivity peak really coincides with the cut-off spectrum. The near-vanishing chirp of the attosecond pulse filtered from the cut-off range of the high-harmonic spectrum and reflected by the near-zero-*GDD* mirror (b) at the same time validates theoretical considerations that predict the attochirp to be negligible in the cut-off range [32,12]. Finally the good agreement between the retrieved *GDD* values of the produced attosecond pulses and the designed mirror dispersion confirms that the chirp of the attosecond pulse is mainly affected by the mirror, while the intrinsic chirp of the high harmonic generation and the additional filter play subordinate roles.

3. Conclusions

In summary, we have demonstrated control of the chirp and temporal profile of isolated attosecond *XUV* pulses with non-periodic multilayer mirrors in an energy range above 100 *eV*. This technique can be extended to third and higher order phase term treatment as well and is applicable to any spectral range were high reflecting multilayer mirrors are producible, it is important to mention that this range by far exceeds the photon energy range accessible with contemporary high-harmonic-sources. Precision control of the characteristics of attosecond pulses adds an important technical capability to the toolbox of attosecond technology, affording promise for pushing its frontiers to higher temporal resolutions down to the atomic unit of time (24 *as*) and extending coherent control from structural dynamics towards electronic motion in the valence band. High harmonic radiation contains, due to its generation process, intrinsically an essential amount of chirp in a spectral range out of the cut-off region. Precise dispersion control of attosecond pulses could allow us to push the limits to even shorter Fourier-limited pulses, taking spectral bandwidth exceeding the cut-off range and thus entering this temporal regime. Moreover, control over electronic motion in photo-ionization by chirped few-cycle attosecond pulses was recently predicted in a theoretical work by Peng et al. [35]. Furthermore important insight into fundamental electronic structures as the quantum phase can be gained from chirp analyses in electronic states via *XUV* streaking [36]. Absolute control of the chirp of an attosecond pulse becomes even more interesting when we reach the regime of nonlinear processes in the *XUV* range, where for example coherent control by chirped pulses could be achieved. Future seeding of free electron lasers (*FEL*) in the *XUV* by coherent High Harmonic radiation allows for controlling the spectral and temporal *FEL* pulse output with sufficiently high photon numbers [37] and thus requires precise control over the spectral seed chirp as can be provided by the multilayer mirrors described in this paper.

#138629 - $15.00 USD Received 23 Nov 2010; revised 31 Dec 2010; accepted 3 Jan 2011; published 14 Jan 2011

Acknowledgements

This work was supported by the DFG Excellence Cluster "Munich Centre for Advanced Photonics" (MAP). R.K. acknowledges funding from the Sofja Kovalevskaja Award of the Alexander von Humboldt Foundation and an ERC Starting Grant.

#138629 - $15.00 USD Received 23 Nov 2010; revised 31 Dec 2010; accepted 3 Jan 2011; published 14 Jan 2011

The electron distribution of atoms (green clouds) is distorted by the electric field of light impinging on the sample (red wave on the right). The temporal evolution of the resultant polarization (green curve) can be retrieved from measuring the waveforms of the input and transmitted laser pulse (red wave on the left) via a sampling device that has petahertz bandwidth. Such a device is now available in the form of a simple solid-state circuit [58], providing access to the attosecond polarization of matter induced by light. © Dennis Luck

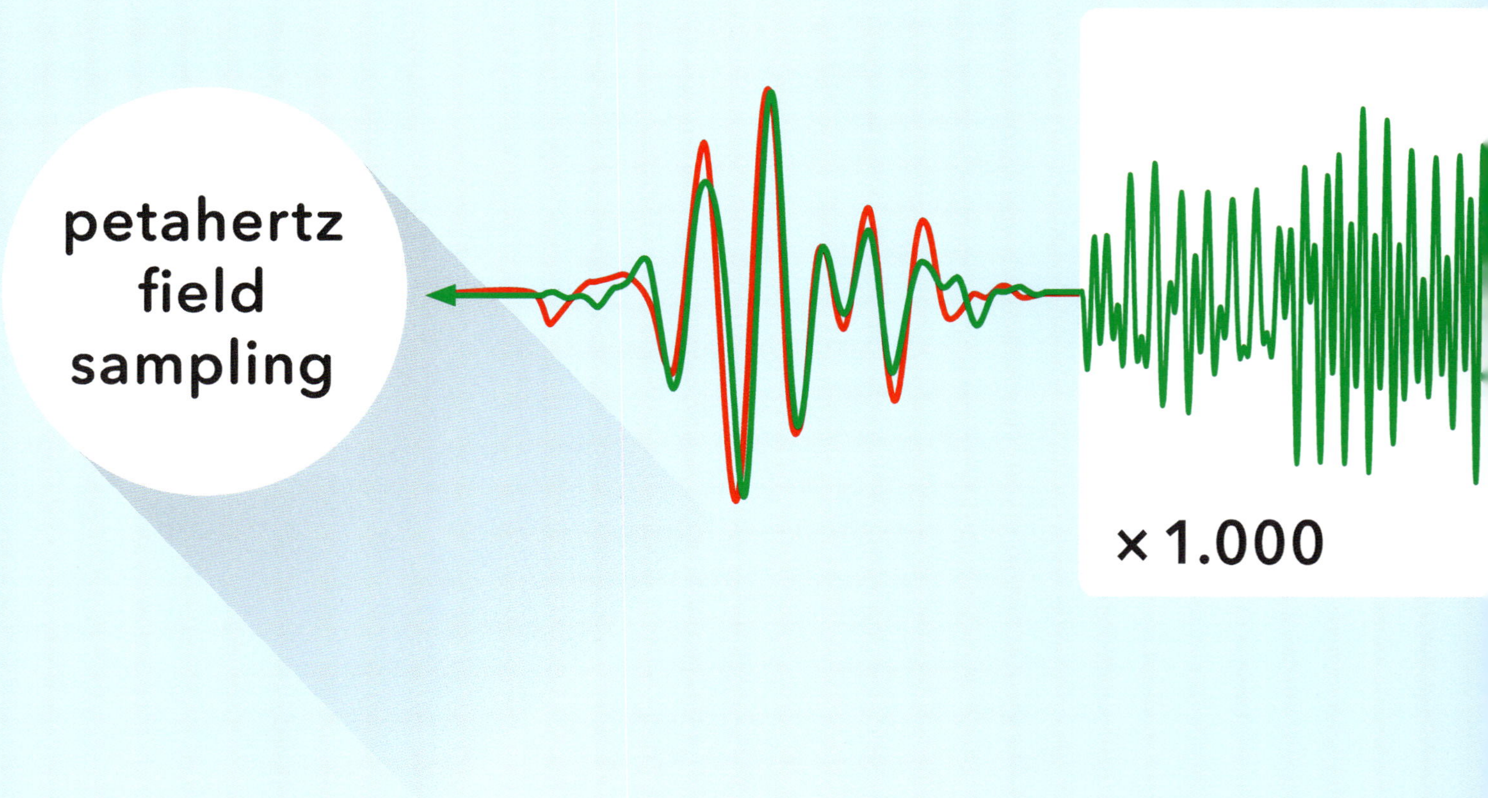

Towards the Next Generation: AM 2.0

11

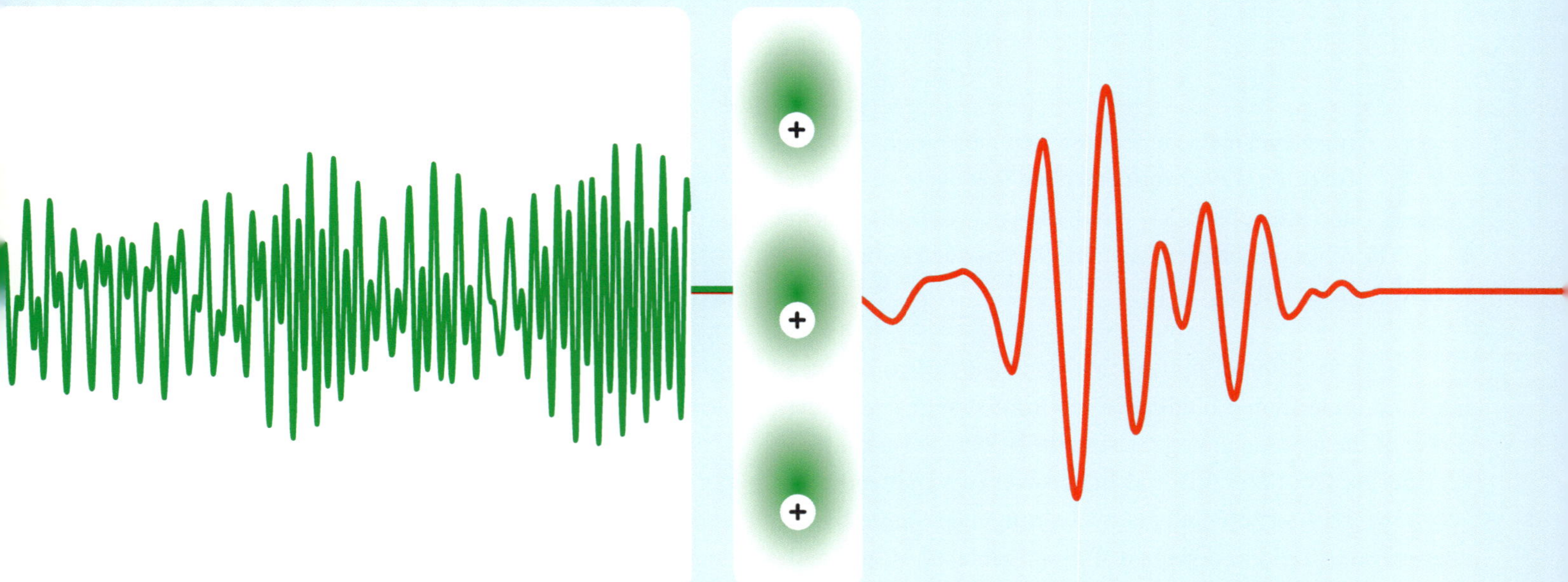

Figure 11-1 The MPQ AS-1 attosecond beamline, operated for the metrology of few-cycle visible/infrared laser light and synchronized attosecond XUV pulses as well as for their applications for tracking electron dynamics. Martin Schultze (left), Matthias Uiberacker (middle) and Eleftherios Goulielmakis (right) prepare the apparatus for an attosecond pump-probe experiment.

Attosecond metrology of the first generation, AM 1.0, based on waveform-controlled few-cycle laser light and attosecond XUV pulses synchronized to them relies on complex vacuum systems such as the AS-1 beamline at MPQ (Fig. 11-1).

The observation of optical-field-induced electron tunneling from their atomic binding in real time demonstrates the power of the new technology. Using the MPQ AS-1 beamline, Eleftherios Goulielmakis and his co-workers induced this very process in krypton atoms by a strong near-single-cycle, near-infrared field and probed it by an isolated attosecond pulse. It was sent through the atoms during their exposure to the ionizing field to measure the change in absorptivity caused by electron release. Probing this change as a function of delay with respect to the ionizing laser field, provides direct insight into the emergence of ions, or equivalently, electron release via tunneling in real time. The result for a near-single-cycle laser field is shown in Figure 11-2a. It confines tunneling ionization, in this case of krypton atoms, to a fraction of the central half cycle, to less than one femtosecond, in accordance with the predicted highly nonlinear dependence of the tunneling rate on the strength of the driving field [50].

This process is the primary step in creating an attosecond XUV pulse, one of the key elements of AM 1.0, as discussed previously. Generation of attosecond XUV light relies on a complex laser system and a vacuum beamline for their creation and applications. This complexity has severely impeded wide proliferation of the technology as well as its prospects for future real-world applications. The very same process, light-induced electron tunneling, has permitted changing this unsatisfactory state of matters.

By using few-cycle infrared fields, valence-to-conduction-band electron tunneling or multiphoton-induced transition (Fig. 11-2b) can be induced below the optical breakdown threshold, turning an insulator into a conducting state on the time scale of 1 fs or shorter without damage, as revealed by recent experiments[11]. The resultant ultrafast conductivity can be utilized for optical field sampling by creating the carriers in an insulating gap between two electrodes

[11] S. Sederberg *et al.* Attosecond optoelectronic field measurement in solids. Submitted for publication.

(Fig. 11-2c). The strong injection field is polarized parallel to the electrodes to prevent it from driving the created charge carriers, electrons and holes, towards the electrodes. It is the weak laser field to be sampled that is to be polarized across the electrodes so that it can separate the suddenly created charge to an extent that is proportional to the vector potential of the field at the instant of carrier injection. The separated charge then drives a current through an external circuit, which was first reported in [58]. In addition to optical-field-induced conductivity, the same strong field also modifies the optical

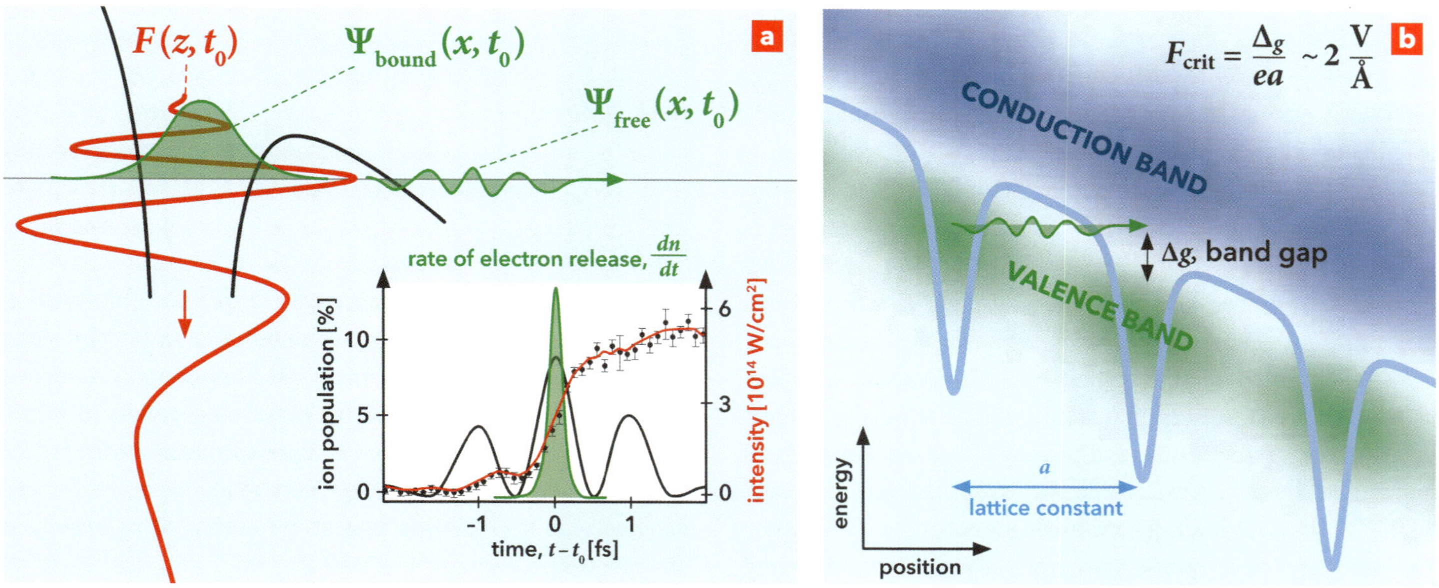

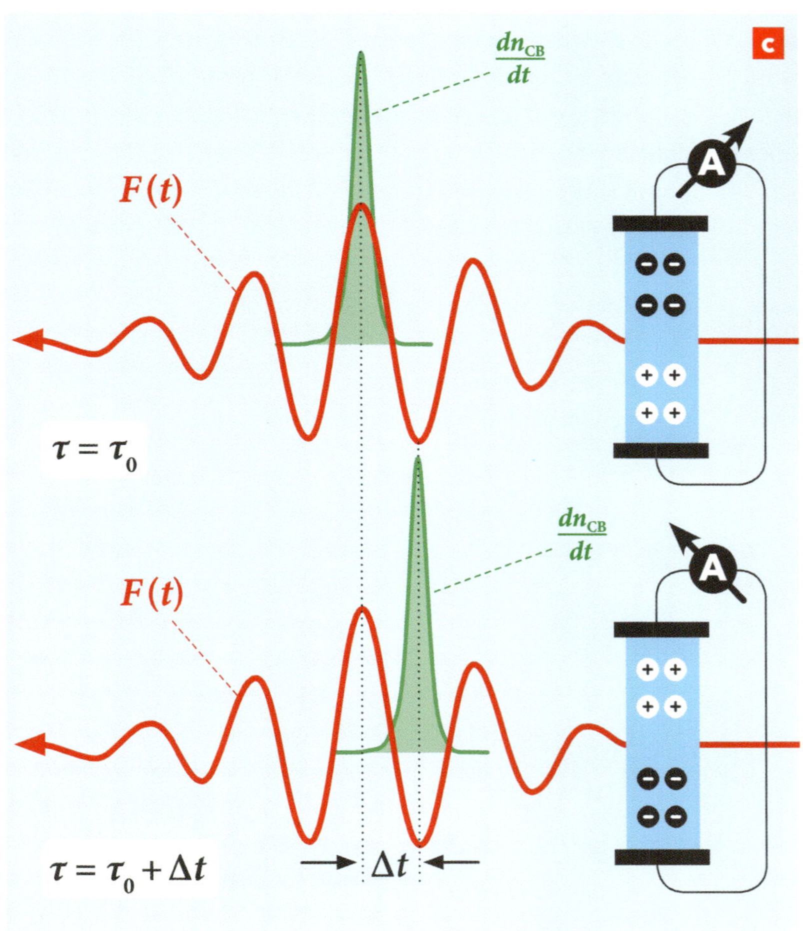

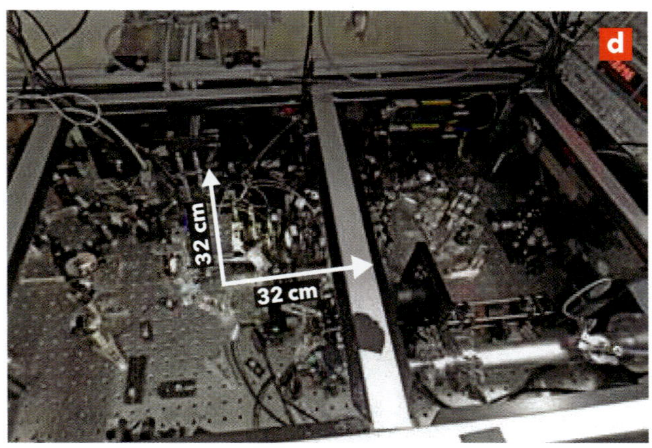

Figure 11-2 Optical-field-induced tunneling in (a) atoms and (b) wide-gap solids. Induced by a near-single-cycle laser field, process (c) injects charge carriers, electrons and holes, in the conduction band and the valence band, respectively, within a single half cycle of the ionizing field. This sub-femtosecond carrier injection opens the door for sampling the electric field of light pulses up to petahertz frequencies, creating the basis for next-generation attosecond metrology, AM2.0, relying on all-solid-state instrumentation (d).

properties, such as the absorption coefficient and refractive index, on an attosecond time scale [59]. The papers reprinted as [58] and [59] were published in the first issue of *Nature* in 2013 and have opened the door for extending signal metrology and processing, respectively, up to petahertz frequencies.

Several years later, after utilizing optical-field-induced conductivity for measuring the light phase [60] and the scaling of electro-optic sampling close to its ultimate limit [61], solid-state petahertz-bandwidth signal metrology has become a reality. Shawn Sederberg, Sabine Keiber, Dmitry Zimin and Nicholas Karpowicz measured this current as a function of delay of the carrier injection with respect to the laser field and observed oscillations following those of the driving laser field up to frequencies exceeding 1 PHz. The team has also observed a linear response of this petahertz sampler up to critical fields, over an intensity dynamic range of some three orders of magnitude and proved it to be sensitive over more than four octaves, all the way from the mid infrared to the deep ultraviolet. Theoretical modelling of Vladislav Yakovlev and benchmarking against attosecond streaking by Martin Schultze and his co-workers revealed that the current flowing through the external circuit is proportional to the vector potential of the driving field at the instant of carrier injection. As a consequence, the laser electric field to be sampled can be simply derived from the current vs. delay function via temporal derivation.

Petahertz-bandwidth field measurements constitutes a radically new way of attosecond metrology, which deserves a new name, AM2.0, obviating the need for a complex vacuum apparatus. Attosecond metrology 2.0 relies on compact all-solid-state instrumentation (Fig. 11-2d) for probing electron processes in most forms of matter (in particular inside solids the platform of modern electronics), which is inaccessible to XUV probe light.

Furnished with the capability of measuring the electric field waveform of light pulses, we can now directly access the optical-field-induced polarization of matter, including its magnitude, evolution and timing with respect to the driving field by simply comparing the electric field of few-cycle laser pulses before and after traversing a thin sample under scrutiny (Fig. 11-3). The polarization induced in the sample can be directly derived from the difference of the two measured fields.

We have dubbed the approach *attosecond polarization spectroscopy*, which was first demonstrated in [62] by using attosecond streaking from optical field sampling. Attosecond polarization spectroscopy may be regarded as a generalization of pump-probe spectroscopy in that it provides both the amplitude and the phase of the dynamic response of matter and does so with a resolution that offers access to both nuclear and electronic dynamics.

As a first application, the team working with Nicholas Karpowicz demonstrated precision probing of the linear polarization of matter by propagating a complex super-octave optical waveform through fused silica (Fig. 11-4). The spectral phase shift, evaluated from the difference between the input and output waveforms, is in excellent agreement with the prediction from the Sellmeier equation, revealing high-fidelity sampling of complex petahertz waveforms.

Measurement of the material polarization along with its attosecond timing with respect to the driving optical field

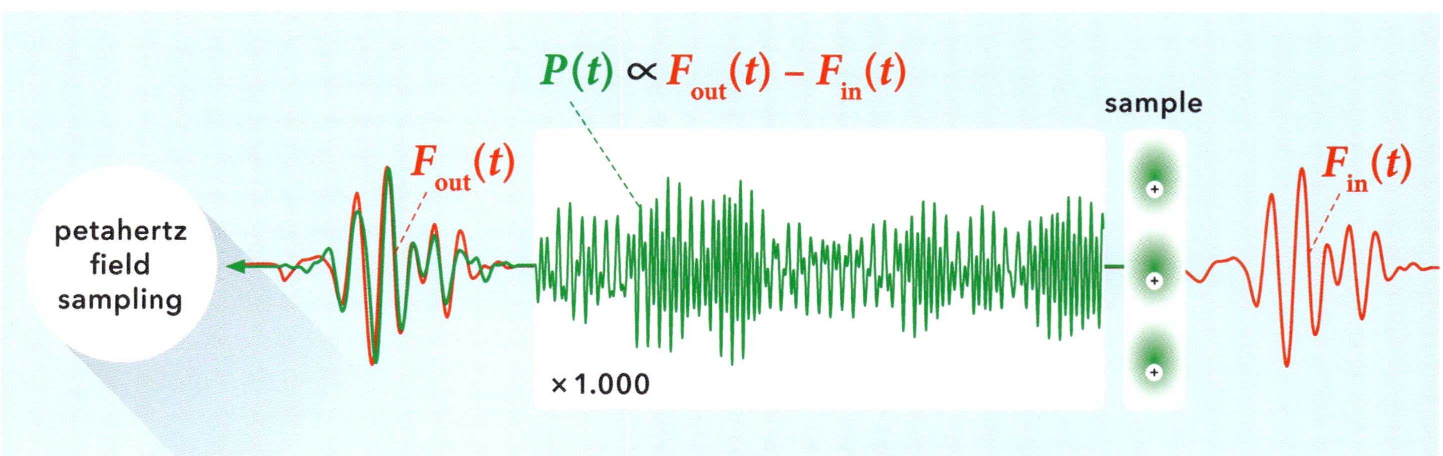

Figure 11-3 Schematic illustration of attosecond polarization sampling.

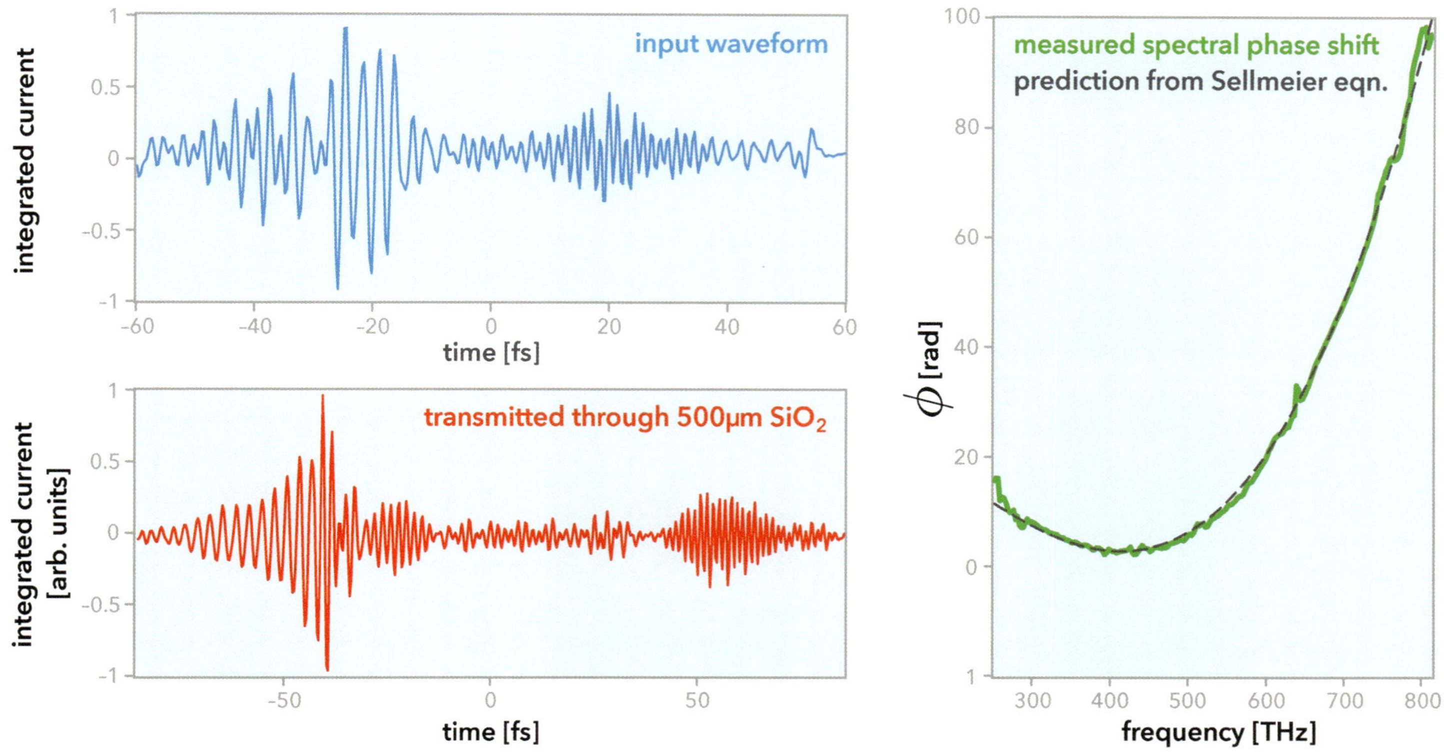

Figure 11-4 Attosecond time-resolved measurement of frequency-dependent phase shifts of optical radiation.

provides access to the instantaneous energy flow between light and matter via the simple formula

$$dW/dt = E_L(t)\,\frac{d}{dt}P(t)$$

multiplying the driving electric field with the electric current density given by the derivative of the polarization. The electric field waveforms of our near-single-cycle NIR laser pulse transmitted at low and high intensity, respectively, through a 120 nm thin sample of fused silica as recorded by our PHz sampling circuit are compared below (Fig. 11-5a).

The difference between the two normalized waveforms yields the nonlinear polarization induced by the strong field (Fig. 11-5b), which together with its driving field, permits the calculation of the attosecond energy transfer from the incident light field to the silica sample (Fig. 11-5c).

Combining this novel sampling technique with few-cycle laser technology forms the basis of attosecond metrology of the next generation (AM2.0). It is based entirely on solid-state technology and obviates the need for a complex vacuum apparatus as was previously indispensible for attosecond measurements. AM2.0 provides controlled and fully characterized terahertz-to-petahertz optical transients for interrogating the dynamic response of matter including both electronic and nuclear motions. Direct access to light-matter energy-transfer dynamics (Fig. 11-5c) will be instrumental in the development of future THz-to-PHz-rate processors and can now be performed with a simple solid-state instrumentation. Yet another future real-world application with potentially far-reaching importance is addressed in the next chapter.

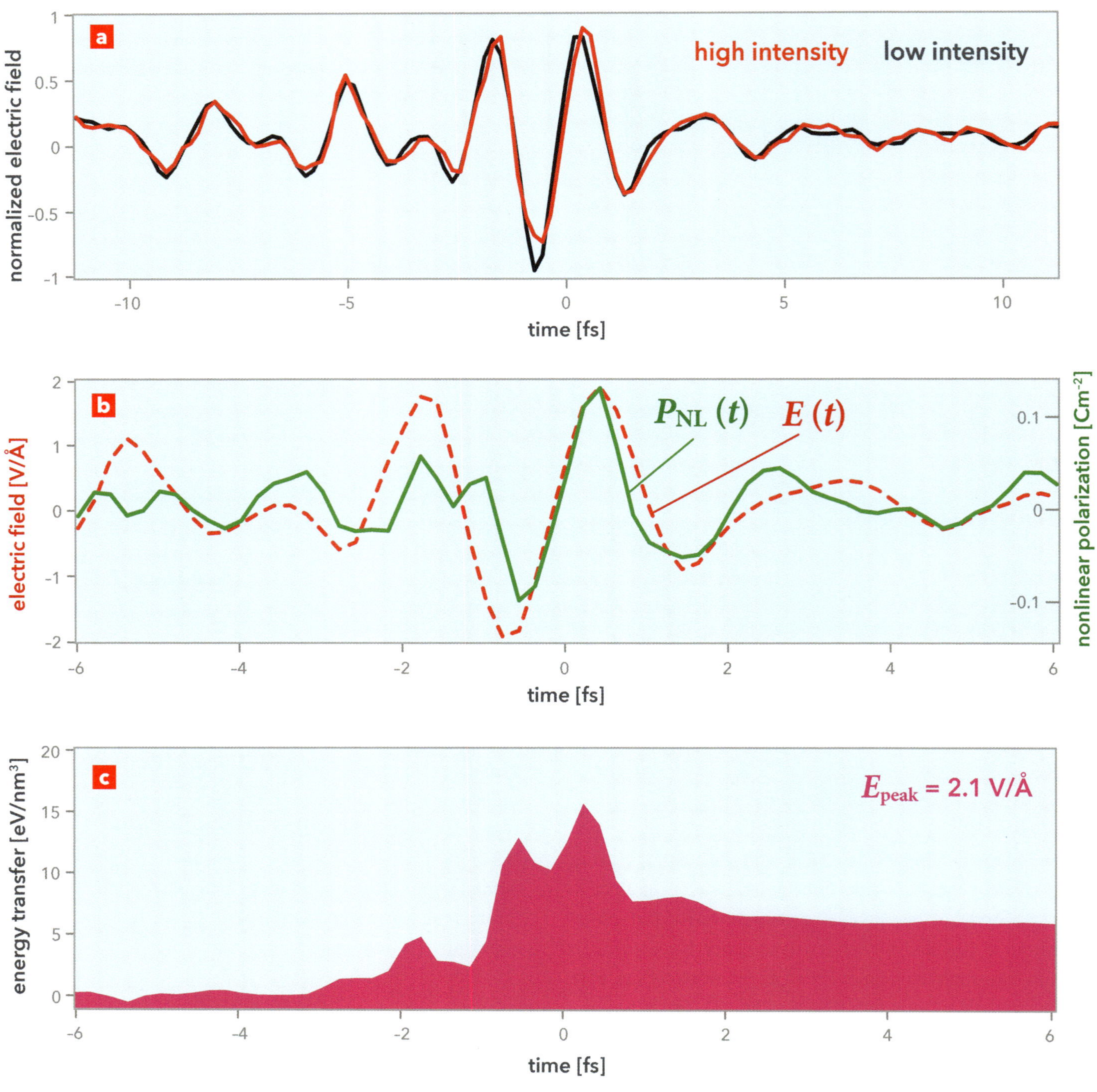

Figure 11-5 Attosecond real-time measurement of the nonlinear polarization induced in silica by a few-cycle visible/near-infrared laser pulse. (a) Temporal evolution of electric field transmitted through a 120-nm-thick fused silica sample at low (black line) and high (red line) intensity, with the latter corresponding to a peak electric field strength of 2.1 volt/ångström. (b) Temporal evolution of the nonlinear polarization induced in the sample by the intense wave depicted in panel (a). (c) Temporal evolution of the energy density transferred from the incident laser field to the silica sample.

[58] A. Schiffrin, T. Paasch-Colberg, N. Karpowicz, V. Apalkov, D. Gerster, S. Mühlbrandt, M. Korbman, J. Reichert, M. Schultze, S. Holzner, J. Barth, R. Kienberger, R. Ernstorfer, V. Yakovlev, M. Stockman, F. Krausz. Optical-field-induced current in dielectrics. *Nature* **493**, 70 (2013).
(Page: 504)

[59] M. Schultze, E. Bothschafter, A. Sommer, S. Holzner, W. Schweinberger, M. Fieß, M. Hofstetter, R. Kienberger, V. Apalkov, V. Yakovlev, M. Stockman, F. Krausz. Controlling dielectrics with the electric field of light. *Nature* **493**, 75 (2013).
(Page: 509)

[60] T. Paasch-Colberg, A. Schiffrin, N. Karpowicz, S. Kruchinin, Özge Sağlam, S. Keiber, O. Razskazovskaya, S. Mühlbrandt, A. Alnaser, M. Kübel, V. Apalkov, D. Gerster, J. Reichert, T. Wittmann, J. Barth, M. Stockman, R. Ernstorfer, V. Yakovlev, R. Kienberger, F. Krausz. Solid-state light-phase detector. *Nature Photonics* **8**, 214 (2014).
(Page: 513)

[61] S. Keiber, S. Sederberg, A. Schwarz, M. Trubetskov, V. Pervak, F. Krausz, N. Karpowicz. Electro-optic sampling of near-infrared waveforms. *Nature Photonics* **10**, 159 (2016).
(Page: 518)

[62] A. Sommer, E. Bothschafter, S.A. Sato, C. Jakubeit, T. Latka, O. Razskazovskaya, H. Fattahi, M. Jobst, W. Schweinberger, V. Shirvanyan, V. Yakovlev, R. Kienberger, K. Yabana, N. Karpowicz, M. Schultze, F. Krausz. Attosecond nonlinear polarization and light-matter energy transfer in solids. *Nature* **534**, 86 (2016).
(Page: 523)

LETTER

doi:10.1038/nature11567

Optical-field-induced current in dielectrics

Agustin Schiffrin[1]†, Tim Paasch-Colberg[1], Nicholas Karpowicz[1], Vadym Apalkov[2], Daniel Gerster[3], Sascha Mühlbrandt[1,3], Michael Korbman[1], Joachim Reichert[3], Martin Schultze[1,4], Simon Holzner[1], Johannes V. Barth[3], Reinhard Kienberger[1,3], Ralph Ernstorfer[1,3,5], Vladislav S. Yakovlev[1,4], Mark I. Stockman[2] & Ferenc Krausz[1,4]

The time it takes to switch on and off electric current determines the rate at which signals can be processed and sampled in modern information technology[1–4]. Field-effect transistors[1–3,5,6] are able to control currents at frequencies of the order of or higher than 100 gigahertz, but electric interconnects may hamper progress towards reaching the terahertz (10^{12} hertz) range. All-optical injection of currents through interfering photoexcitation pathways[7–10] or photoconductive switching of terahertz transients[11–16] has made it possible to control electric current on a subpicosecond timescale in semiconductors. Insulators have been deemed unsuitable for both methods, because of the need for either ultraviolet light or strong fields, which induce slow damage or ultrafast breakdown[17–20], respectively. Here we report the feasibility of electric signal manipulation in a dielectric. A few-cycle optical waveform reversibly increases—free from breakdown—the a.c. conductivity of amorphous silicon dioxide (fused silica) by more than 18 orders of magnitude within 1 femtosecond, allowing electric currents to be driven, directed and switched by the instantaneous light field. Our work opens the way to extending electronic signal processing and high-speed metrology into the petahertz (10^{15} hertz) domain.

Three basic types of solid—metals, semiconductors and dielectrics—fundamentally differ in their reaction to an applied electric field. Metals respond to a small field with a current linearly proportional to it. This implies a strong screening that prevents high fields and charge density gradients from forming inside metals, which makes it fundamentally difficult to control their responses. By contrast, semiconductors, in which there is a relatively small but non-zero energy gap, Δ_g, between the valence and conduction bands, allow electric fields to penetrate and charge gradients to build up. This forms the basis of contemporary digital electronics[1–3,21]. Dielectrics, with their large bandgap and conduction and valence band widths, offer the fastest response. However, because of their reaction to applied fields (that is, extremely low conductivity at low fields and breakdown at high fields[17–20]), they have been thought to be unsuitable for electronic signal processing in field-effect devices. Here we demonstrate that a strong, few-cycle optical field is capable of transforming the dielectric into a state of highly increased polarizability, allowing optical currents to flow and resulting in macroscopic charge separation that is detectable in an external circuit.

In our experiments, we exposed a fused silica[22] sample ($\Delta_g \approx 9$ eV and conduction band width $\Delta_c \approx 10$ eV) to a strong, waveform-controlled, few-cycle field $F_i(t)$ with a photon energy of $\hbar\omega_L \approx 1.7$ eV (where $\hbar$ denotes Planck's constant divided by 2π and ω_L is the carrier angular frequency of the optical field), which transforms the dielectric into an optically conducting state, that is, injects carriers. We therefore refer to this field as the injection field. By carriers, we mean electrons in highly polarizable states at optical frequencies (rather than ones capable of conducting d.c. current, requiring delocalized electrons). Two unbiased gold electrodes collect the charge separated by the optical-field-induced current (Fig. 1a, Methods Summary and Supplementary Information). First, the applied external field $F_i(t) = F_0 f(t) \cos(\omega_L t + \varphi_{CE})$, with a controlled carrier-envelope phase (CEP) φ_{CE}, a sub-4-fs envelope $f(t)$ with $f(0) = 1$, and $F_0 \approx 1.7$ V Å^{-1}, was polarized perpendicularly to the metal–dielectric interface, along coordinate x, to drive the generated carriers towards the electrodes (Fig. 1a). Fig. 2 plots Q_P, the charge transferred through the ammeter shown in Fig. 1 during exposure to a single laser pulse, as a function of the change in φ_{CE} (Fig. 2a) and as a function of F_0 (Fig. 2b). The CEP of the laser pulse is shifted by $\Delta\varphi_{CE}$ on changing the

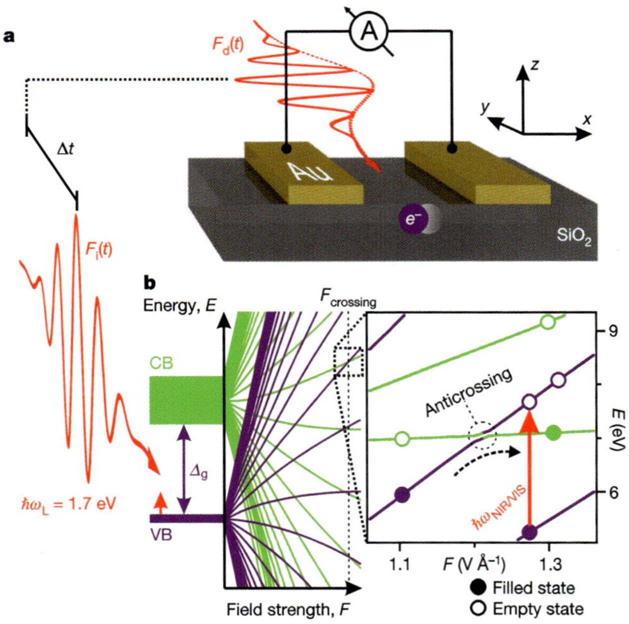

Figure 1 | Optical-field-induced conductivity and current control in a dielectric. a, Schematic of the metal–dielectric nanojunction. **b**, Schematic illustration of the adiabatic energy levels of the electronic states in the valence band (VB, purple) and conduction band (CB, green) of the dielectric under the influence of a static or slowly varying strong electric field. The eigenenergies fan out as the strength of the electric field increases, resulting in avoided crossings, that is, anticrossings (inset). At low fields, the valence band states are occupied, and the conduction band states are empty. As the field strength increases, the valence band and conduction band levels cross, but the respective Wannier–Stark states are localized at distant sites, and the anticrossings are passed diabatically (that is, the conduction band states remain unpopulated) until the field approaches or exceeds ~1 V Å^{-1}. At these field strengths, electrons may be promoted into the conduction band via Zener tunnelling, leaving the electron in the lower-energy state after the passage of the anticrossing (adiabatic transition, depicted by dashed arrow). The resultant unoccupied valence band states mediate strong single-photon resonances at visible/near-infrared angular frequencies $\omega_{NIR/VIS}$ within the valence band (red arrow). The emergence of these resonances results in a strong transient polarizability.

[1]Max-Planck-Institut für Quantenoptik, Hans-Kopfermann-Strasse 1, D-85748 Garching, Germany. [2]Department of Physics and Astronomy, Georgia State University, Atlanta, Georgia 30340, USA. [3]Physik-Department, Technische Universität München, James-Franck-Strasse, D-85748 Garching, Germany. [4]Fakultät für Physik, Ludwig-Maximilians-Universität, Am Coulombwall 1, D-85748 Garching, Germany. [5]Fritz-Haber-Institut der Max-Planck-Gesellschaft, Faradayweg 4–6, 14195 Berlin, Germany. †Present addresses: Department of Physics and Astronomy, University of British Columbia, Vancouver, British Columbia, V6T 1Z1 Canada; Quantum Matter Institute, University of British Columbia, Vancouver, British Columbia, V6T 1Z4 Canada.

LETTER RESEARCH

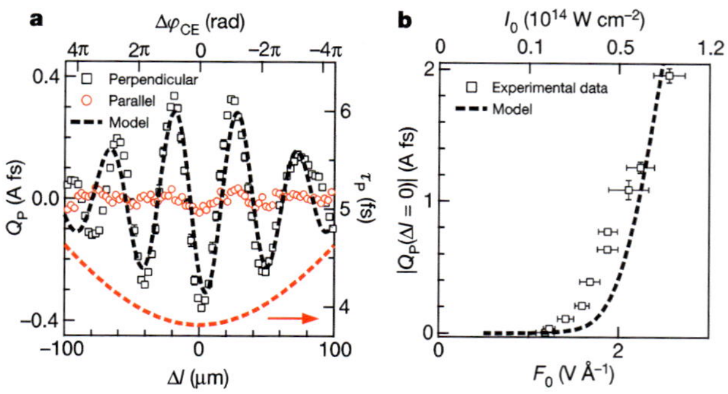

Figure 2 | Carrier-envelope-phase control and intensity dependence of optical-field-generated electric current in SiO_2. **a**, Plot of the φ_{CE}-dependent component of $Q_P(\Delta l)$ against the change, Δl, in the propagation length through a fused silica wedge with respect to the propagation length yielding the minimum pulse duration ($\Delta l = 0$), for polarizations perpendicular (along x; squares) and parallel (in the y–z plane; circles) to the metal–dielectric interface ($F_0 \approx 1.7\,\mathrm{V\,Å^{-1}}$). The data represent an average of several consecutively acquired values at a given value of Δl. Error bars show the standard deviation, and most are smaller than the size of the symbols representing the mean values of the data. The black dashed line shows the prediction of our quantum mechanical model. For the parallel polarization, the signal is suppressed by more than an order of magnitude. The residual signal is attributed to microscopic imperfections of the macroscopically plane metal–dielectric interface, which result in locations with a non-zero perpendicular component of the field. The red dashed line (right axis) depicts the change in pulse duration τ_P (full width at half maximum intensity) as a function of Δl, taking into account the group velocity dispersion of the visible/near-infrared pulse in the fused silica wedge. **b**, Plot of the maximum amplitude, $|Q_P(\Delta l = 0)|$, of the transferred charge against the peak amplitude, F_0, of the applied external field polarized along x: measurement (squares) and theoretical prediction (dashed line). For a given F_0 value, $|Q_P(\Delta l = 0)|$ is determined by fitting the most pronounced oscillation of $Q_P(\Delta l)$ (**a**, squares) with a sine function. The vertical error bars account for the standard deviation of such fit. Different values of F_0 correspond to different beam sizes. Data points have been normalized accordingly. F_0 is determined by monitoring the pulse energy and the laser beam waist at the focus, and the horizontal error bars quantify random fluctuations in these parameters. I_0 is the corresponding peak intensity of the optical pulse. Data from our model have been multiplied by the effective cross-section, A_{eff}, of the metal–dielectric interfaces confining the active volume of the dielectric (main text).

propagation length by Δl in a pair of thin, fused silica wedges[23] (Methods Summary and Supplementary Fig. 1).

We find that Q_P varies quasi-periodically with Δl. Its minimum at $\Delta l = 0$ turns into a maximum at $\Delta l = \pm\Delta l_{deph}$, with $\Delta l_{deph} = 23\,\mu m$ shifting[23] the CEP by $\Delta\varphi_{CE} = \pi$, which results in an approximate inversion of the optical waveform: $F_i(t, \Delta l_{deph}) \approx -F_i(t, 0)$. This reversal of Q_P and its order-of-magnitude reduction when $F_i(t)$ is polarized parallel to the metal–dielectric interface (that is, in the y–z plane) provide the first significant indication that this current is induced directly by the instantaneous light field. The maximum value of Q_P scales nearly exponentially with the field amplitude and reaches ~10^4 electrons before breakdown (Fig. 2b). This nonlinear dependence is responsible for the decay of $Q_P(\Delta l)$ with increasing Δl, as a consequence of dispersive pulse broadening.

To decouple the injection and driving processes, we irradiated the hybrid junction with two synchronized, orthogonally polarized fields. An injection field with a peak amplitude of $F_0^{(i)} \approx 2\,\mathrm{V\,Å^{-1}}$ was polarized parallel to the electrode–insulator interface (that is, in the y–z plane) to prevent it from driving current through the circuit. This driving was accomplished by a weaker version of the same field, the drive field $F_d(t)$, which had an amplitude of $F_0^{(d)} \approx 0.2\,\mathrm{V\,Å^{-1}}$ (Fig. 1a). The delay, Δt, between the peaks of $F_i(t)$ and $F_d(t)$ determines the timing of carrier injection with respect to the drive field and thereby controls the momentum that $F_d(t)$ transfers to the charge carriers. Fig. 3a shows Q_P versus Δt for the CEPs of the injection ($\varphi_{CE}^{(i)}$) and drive ($\varphi_{CE}^{(d)}$) pulses corresponding to one of the zero-crossings of Q_P near $\Delta l = 0$ in Fig. 2a. Such values of $\varphi_{CE}^{(i)}$ and $\varphi_{CE}^{(d)}$ are chosen to avoid a residual (background) current originating in any of the fields independently (Fig. 2a).

The transferred charge periodically changes its sign as a function of Δt, revealing two major current reversal cycles equal to the field oscillation cycle of $F_d(t)$ (Fig. 3a). The magnitude of the signal rapidly decays outside this few-femtosecond delay interval. Fig. 3b plots the result of the same measurement performed with an 'inverted' drive field, in which the field oscillations are reversed by shifting the CEP by $\Delta\varphi_{CE}^{(d)} = \pi$. For any value of Δt, the transferred charge is reversed with respect to its value at the same delay in Fig. 3a. This is exactly the behaviour we intuitively expect for a current driven directly by $F_d(t)$. The observed $Q_P(\Delta t)$ reflecting the oscillating behaviour of $F_d(t)$ provides compelling evidence that the light field governs the current that emerges from the dielectric medium and is measured in the external circuit. The transferred charge was also found to depend sensitively on the CEP of the injection field $\varphi_{CE}^{(i)}$ (Supplementary Fig. 7). This also corroborates the evidence that carrier injection is induced directly by the instantaneous field $F_i(t)$ rather than by effects governed by the cycle-averaged intensity. Experiments on free-standing electrodes kept under helium flux did not yield any φ_{CE}-dependent electronic signal under identical irradiation conditions (Supplementary Fig. 10), and so we identify the insulator as the primary source of current in our experiments.

Fig. 3c plots the electric field waveform of the drive pulse retrieved from attosecond streaking[24]. Such streaking experiments were performed in a parallel ultrahigh vacuum set-up independently from the light-field-induced current measurements and under identical laser conditions. The good correspondence between the solid lines in Fig. 3a and Fig. 3c suggests that the injected carriers sample the optical field with good fidelity. Carrier injection must therefore be substantially confined to a half-period, that is, to a time window of ~1 fs. This confinement is also indicative of a strong nonlinearity, in accordance with Fig. 2b.

To describe our observations theoretically, we calculate the current density, $j_x(t)$, generated in the bulk insulator along coordinate x by solving the time-dependent Schrödinger equation for a dielectric film with a thickness in the range 50–500 nm exposed to a strong optical field. We neglect the Coulomb interaction between the electrons, which is justified by the extremely short (~1 fs) timescale on which the processes occur. The time dependence of the applied electric fields is taken from attosecond-streaking experiments (Fig. 3c). We compute $j_x(t)$ as the temporal derivative of the field-induced polarization, multiplying it by the effective cross-section, A_{eff}, of the metal–dielectric interfaces confining the active volume of the dielectric. More details are given in the Supplementary Information.

Our computations predict the dependence of

$$Q_P = A_{eff} \int_{-\infty}^{\infty} j_x(t)\,dt$$

on the CEP (Fig. 2a, black dashed line), the peak electric field (Fig. 2b), and the delay between the injection and drive fields for different $F_d(t)$ (Fig. 3) and $F_i(t)$ (Supplementary Fig. 7) waveforms. All the observed dependencies have been excellently reproduced using parameters—defining the band structure and three relevant dipole matrix elements—that we chose once and did not subsequently adjust. The parameter A_{eff} was adjusted to fit the magnitude of the collected charge, and $A_{eff} \approx 5 \times 10^{-12}\,m^2$ gave the best agreement with the measurements. This value matches the experimentally estimated metal–dielectric cross-section within an order of magnitude. Hence, our microscopic model predicts our observations both qualitatively and, within an order of magnitude, quantitatively. This quantitative agreement also implies that a significant fraction of the charge separated by the optical currents in the dielectric is transferred to the metal leads, in

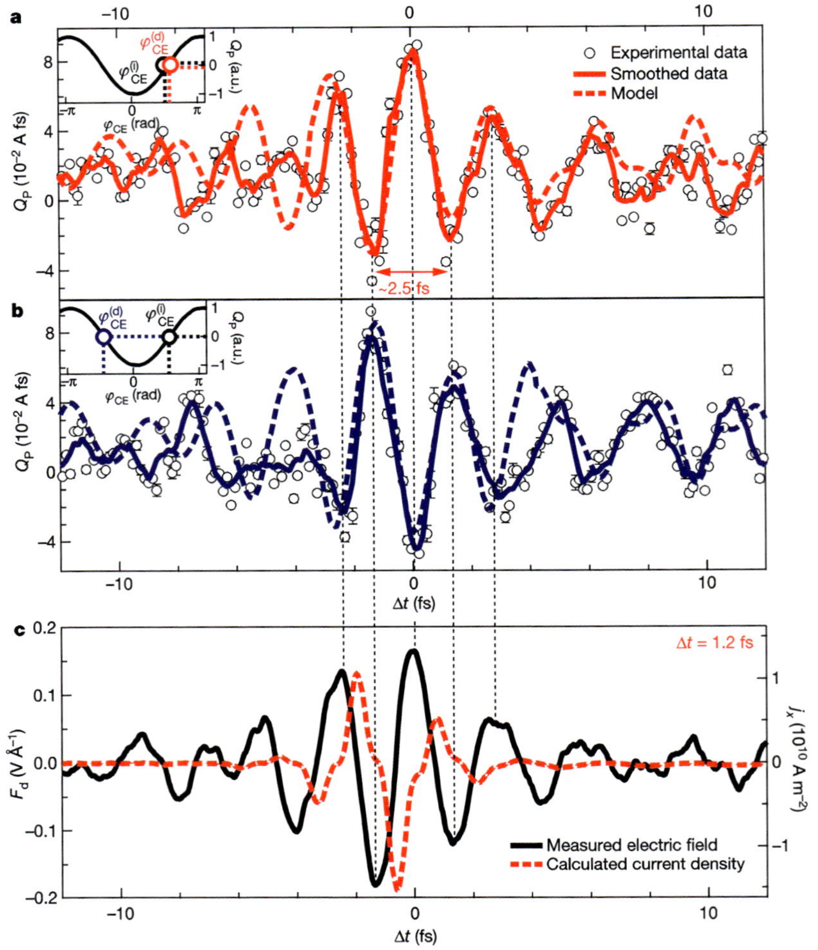

Figure 3 | Subfemtosecond control of electric current with the electric field of light. a, b, Transferred charge, Q_P, versus delay, Δt, between the injection ($F_i(t)$) and drive ($F_d(t)$) fields, with parameters as described in the text. For $\Delta t < 0$, $F_d(t)$ precedes $F_i(t)$. The CE phases of the injection and drive fields, $\varphi_{CE}^{(i)}$ and $\varphi_{CE}^{(d)}$ respectively, are set such that Q_P induced independently by any of the two fields vanishes (insets). The experimental data shown were recorded under identical conditions except for $\varphi_{CE}^{(d)}$, which is shifted by π between the two measurements (compare the insets). The experimental data (circles) represent an average of several consecutively acquired values at a given Δt value. The error bars are the respective standard deviations. The solid lines correspond to the smoothed data obtained by averaging over adjacent points. The dashed lines show the prediction of our microscopic quantum mechanical model. Data from our model have been multiplied by the effective cross-section, A_{eff}, of the metal–dielectric interfaces confining the active volume of the dielectric (main text). The slightly asymmetric polarity of the experimental signal could be due to the non-zero residual current for each individual pulse. The dependence of the current on Δt demonstrates that the charge-carrier generation takes place predominantly in the silica medium and that electrons of the gold electrodes, which are not sensitive to the component of the field parallel to the metal–insulator interface, that is, $F_i(t)$, do not contribute to the optically injected current. **c,** Real-time optical electric field of the visible/near-infrared pulses retrieved from attosecond-streaking measurements[24] (solid black line), which were performed under identical laser conditions. The magnitude of the field is normalized to the strength of the drive pulse, $F_d(t)$, used in the photocurrent experiment. The red dashed curve shows the time-dependent current density, $j_x(t)$, as calculated from our quantum mechanical model for $\Delta t = 1.2$ fs. The vertical black dashed lines provide a guide to the eye for comparing the temporal evolution of the measured electronic signal with that of the drive electric field.

line with the predictions of a theoretical study of optically induced currents in a single-molecule nanojunction[10] (Supplementary Information, section 3.4).

The calculated $j_x(t)$ in Fig. 3c and Supplementary Fig. 3 reveals ultrafast turn-on and turn-off behaviour, allowed by the extremely wideband energy spectrum of the dielectric. This predicted temporal behaviour enables us to evaluate the peak current density from our measurements. With a maximum value of $Q_P(\Delta t)$ on the order of 0.1 A fs (Fig. 3a, b) and A_{eff} as specified above, our experimental data yield a transferred charge density of the order of 10^{10} A fs m^{-2}. Given that the duration of the current density pulse is of the order of 1 fs (see $j_x(t)$ in Figs 3c and Supplementary Fig. 3), we estimate a peak current density perpendicular to the electrodes of $J_x \approx 10^{10}$ A m^{-2}, driven by a peak field amplitude of $F_0^{(d)} \approx 0.2 \times 10^{10}$ V m^{-1}. We may now introduce an effective electrical conductivity at optical frequencies as $\sigma_{eff}(\omega_L) = J_x / F_0^{(d)} \approx 5\ \Omega^{-1}\ \mathrm{m}^{-1}$, which exceeds the static d.c. conductivity of amorphous silica, $\sigma_{a\text{-}SiO_2} < 10^{-18}\ \Omega^{-1}\ \mathrm{m}^{-1}$, by more than 18 orders of magnitude.

Our microscopic theory correctly predicts the field-induced increase in polarizability that is responsible for the observed current, but does not provide intuitive insight into its origin. We suggest, with reference to a detailed discussion in the Supplementary Information, that this field-enhanced polarizability can be qualitatively interpreted in terms of the dynamic formation of Wannier–Stark states[25–28] localized at a certain site, l, of the lattice along the direction of the applied field, F, with an energy of $E_l = E_n - eaFl$ (e is the electron charge, a is the lattice constant and E_n is the band offset) in the adiabatic limit (Fig. 1b). For high fields, $F > 1\,\mathrm{V\,Å^{-1}}$, the eigenenergies of the Wannier–Stark states from the valence band and conduction band that reside at neighbouring lattice sites $|l_v\rangle$ and $|l_c\rangle$ become equal; see the avoided crossing and anticrossing of the respective purple and green plots of energy against field the inset of Fig. 1b. For increasing field strength, adiabatic passage of such an anticrossing via Zener-type tunnelling[17] may occur with a significant probability (Fig. 1b inset, dashed arrow), emptying a valence band state. The resultant unoccupied valence band states give rise to strong visible/near-infrared single-photon resonances within the valence band (Fig. 1b inset, red arrow). The emergence of these resonances directly implies a strong transient polarizability[29,30].

This field-induced transformation is predicted to be reversible and highly nonlinear (as suggested by the resultant $j_x(t)$ in Fig. 3c and Supplementary Fig. 3). The consequence is a sub-cycle increase in the polarizability of the system, which leads to an asymmetric charge displacement along the field vector: the averaging of the current to zero that is inherent in linear processes is eliminated. Consequently, net charge accumulations of opposite sign—dependent on the waveform (Fig. 2) or on the delay between injection and driving fields (Fig. 3)—form at the opposite facets of the dielectric and transfer to the electrodes.

Our experiments reveal an ultrafast 'turning on' of the measured current. At the same time, they do not provide direct evidence for the similarly fast 'turning off' predicted by our theory. A proof of the ultrafast turn-off behaviour of the underlying field-induced nonlinear polarization and conduction band population was provided by recent time-resolved absorption and reflection experiments[31]. This suggests that conductivity and, consequently, current can be switched on and off in a dielectric using optical fields on a timescale of less than, or of the order of, 1 fs, and that this can be done without incurring much dissipation. This operation cycle is similar to that occurring in a field-effect transistor, but is fundamentally more energy efficient: when switching, the field-effect transistor dissipates by electron–hole recombination all energy stored in it, whereas the dielectric in our experiment returns almost all stored energy to the injection field. These possibilities may have ramifications for overcoming the speed limits of semiconductor electronics. In the shorter term, our work holds promise for the development of a solid-state device for direct sampling of electric transients with bandwidths extending to optical frequencies.

METHODS SUMMARY

Source of strong ultrashort optical waveforms. A customized titanium–sapphire chirped-pulse amplifier produces linearly polarized, visible/near-infrared laser pulses with controllable φ_{CE}, a pulse energy of ~400 μJ and a repetition rate of 3 kHz. The ultrabroad spectrum of the laser spans the 450–1,100-nm spectral range, supporting a sub-4-fs pulse duration (full width at half maximum of the intensity temporal profile), which corresponds to a ~1.5-cycle pulse at the carrier wavelength of ~750 nm. The laser pulses can be focused onto the investigated solid-state system such that the cycle-averaged peak intensity reaches $10^{14}\,\mathrm{W\,cm^{-2}}$ (see Supplementary Fig. 2 for properties of the laser).

Photoactive metal–dielectric nanocircuit. The photoactive circuit on which the experiments have been performed consists of a metal–dielectric–metal nanojunction. This structure is fabricated by cleaving amorphous SiO_2 (Crystec GmbH) and coating the adjacent surfaces of the atomically sharp cleaved silica edge with ~50 nm of gold evaporated at grazing incidence. We obtain regular, straight, metal electrodes isolated from each other by dielectric nanoscale trenches with widths of the order of ~50 nm. No voltage bias is applied to the electrodes during the experiment. The junction is coated with a supplementary sputtered silica nanofilm such that the metal electrodes are embedded in a homogeneous silica matrix. See Supplementary Information for more details on the sample structure and experimental set-up. Identical measurements were made on samples composed of cleaved monocrystalline $SiO_2(001)$ (Crystec GmhH) and on flat, ~500-nm gold–SiO_2–gold nanogaps. Both yielded results very similar to those reported in this work, but it was possible to distinguish possible effects arising from the lack of long-range order in the amorphous insulator and from local plasmonic resonances at the metal–dielectric interface. These points are discussed in Supplementary Information.

Detection scheme for the optically induced φ_{CE}-dependent electric signal. In this work, we present experimental data on the φ_{CE}-dependent component of the optically generated electronic signals in the circuit. We isolate the electric current induced by the electric field of the optical waveform from the φ_{CE}-independent contributions (for example electric currents caused by photo-ionization of gold due to imbalanced irradiation of the electrodes). The observable of interest is filtered by synchronizing the pulse train from the chirped-pulse amplifier in such a way that two consecutive ultrashort light bursts have a φ_{CE}-change of π. This results in a modulation of φ_{CE} at half the repetition rate of the pulse train. The optically generated electronic signal in the nanocircuit is measured using a current–voltage converter with a variable high gain and a bandwidth supporting the laser pulse train. The φ_{CE}-dependent component of the current is extracted with a lock-in amplifier locked at the φ_{CE} modulation frequency. The presented data correspond to experiments performed in ambient atmosphere at ~20 °C. We carried out identical measurements in vacuum, yielding the same results.

Received 20 December 2011; accepted 31 August 2012.
Published online 5 December 2012.

1. Kahng, D. Electric field controlled semiconductor device. US patent 3,102 230 (1963).
2. Taur, Y. & Ning, T. H. *Fundamentals of modern VLSI devices* (Cambridge Univ. Press, 1998).
3. Liou, J. J. & Schwierz, F. *Modern Microwave Transistors: Theory, Design and Performance* (Wiley-Interscience, 2003).
4. Caulfield, H. J. & Dolev, S. Why future supercomputing requires optics. *Nature Photon.* **4,** 261–263 (2010).
5. Schwierz, F. & Liou, J. J. RF transistors: recent developments and roadmap toward terahertz applications. *Solid-State Electron.* **51,** 1079–1091 (2007).
6. Deal, W. R. *et al.* Low noise amplification at 0.67 THz using 30 nm InP HEMTs. *IEEE Microw. Wirel. Compon. Lett.* **21,** 368–370 (2011).
7. Kurizki, G., Shapiro, M. & Brumer, P. Phase-coherent control of photocurrent directionality in semiconductors. *Phys. Rev. B* **39,** 3435–3437 (1989).
8. Atanasov, R., Hache, A., Hughes, J. L. P., van Driel, H. M. & Sipe, J. E. Coherent control of photocurrent generation in bulk semiconductors. *Phys. Rev. Lett.* **76,** 1703–1706 (1996).
9. Prechtel, L. *et al.* Time-resolved picosecond photocurrents in contacted carbon nanotubes. *Nano Lett.* **11,** 269–272 (2011).
10. Franco, I., Shapiro, M. & Brumer, P. Robust ultrafast currents in molecular wires through stark shifts. *Phys. Rev. Lett.* **99,** 126802 (2007).
11. Valley, G. C. Photonic analog-to-digital converters. *Opt. Express* **15,** 1955–1982 (2007).
12. Nagatsuma, T. Photonic measurement technologies for high-speed electronics. *Meas. Sci. Technol.* **13,** 1655–1663 (2002).
13. Auston, D. H. Picosecond optoelectronic switching and gating in silicon. *Appl. Phys. Lett.* **26,** 101–103 (1975).
14. Auston, D. H. Ultrafast optoelectronics. *Top. Appl. Phys.* **60,** 183–233 (1988).
15. Shimosato, H., Ashida, M., Itoh, T., Saito, S. & Sakai, K. in *Ultrafast Optics V* (eds Watanabe, S. & Midorikawa, K.) 317–323 (Springer Ser. Opt. 132, Springer, 2007).
16. Katzenellenbogen, N. & Grischkowsky, D. Efficient generation of 380 fs pulses of THz radiation by ultrafast laser-pulse excitation of a biased metal-semiconductor interface. *Appl. Phys. Lett.* **58,** 222–224 (1991).
17. Zener, C. A theory of the electrical breakdown of solid dielectrics. *Proc. R. Soc. Lond. A* **145,** 523–529 (1934).
18. Rethfeld, B. Free-electron generation in laser-irradiated dielectrics. *Phys. Rev. B* **73,** 035101 (2006).
19. Jones, S. C., Braunlich, P., Casper, R. T., Shen, X. A. & Kelly, P. Recent progress on laser-induced modifications and intrinsic bulk damage of wide-gap optical-materials. *Opt. Eng.* **28,** 1039–1068 (1989).
20. Lenzner, M. *et al.* Femtosecond optical breakdown in dielectrics. *Phys. Rev. Lett.* **80,** 4076–4079 (1998).
21. Schwierz, F., Wong, H. & Liou, J. J. *Nanometer CMOS* (Pan Stanford, 2010).
22. Koslowski, T., Kob, W. & Vollmayr, K. Numerical study of the electronic structure of amorphous silica. *Phys. Rev. B* **56,** 9469–9476 (1997).
23. Xu, L. *et al.* Route to phase control of ultrashort light pulses. *Opt. Lett.* **21,** 2008–2010 (1996).
24. Kienberger, R. *et al.* Atomic transient recorder. *Nature* **427,** 817–821 (2004).
25. Wannier, G. H. Wave functions and effective Hamiltonian for Bloch electrons in an electric field. *Phys. Rev.* **117,** 432–439 (1960).
26. Bleuse, J., Bastard, G. & Voisin, P. Electric-field-induced localization and oscillatory electro-optical properties of semiconductor superlattices. *Phys. Rev. Lett.* **60,** 220–223 (1988).
27. Mendez, E. E., Agulló-Rueda, F. & Hong, J. M. Stark localization in GaAs-GaAlAs superlattices under an electric field. *Phys. Rev. Lett.* **60,** 2426–2429 (1988).
28. Bar-Joseph, I. *et al.* Room-temperature electroabsorption and switching in a GaAs/AlGaAs superlattice. *Appl. Phys. Lett.* **55,** 340–342 (1989).
29. Durach, M., Rusina, A., Kling, M. F. & Stockman, M. I. Metallization of nanofilms in strong adiabatic electric fields. *Phys. Rev. Lett.* **105,** 086803 (2010).
30. Durach, M., Rusina, A., Kling, M. F. & Stockman, M. I. Predicted ultrafast dynamic metallization of dielectric nanofilms by strong single-cycle optical fields. *Phys. Rev. Lett.* **107,** 086602 (2011).
31. Schultze, M. *et al.* Controlling dielectrics with the electric field of light. *Nature* doi:10.1038/nature11720 (this issue).

RESEARCH LETTER

Supplementary Information is available in the online version of the paper.

Acknowledgements We thank P. Altpeter and Y. Deng for technical support and discussions, and we thank the Munich-Centre for Advanced Photonics for financial support. A.S. acknowledges the Alexander von Humboldt Foundation and the Swiss National Science Foundation. N.K. acknowledges the Alexander von Humboldt Foundation. The work of M.I.S. and V.A. was supported by the Chemical Sciences, Biosciences and Geosciences Division (grant no. DEFG02-01ER15213) and by the Materials Sciences and Engineering Division (grant no. DE-FG02-11ER46789) of the Office of the Basic Energy Sciences, Office of Science, US Department of Energy. R.K. acknowledges an ERC Starting Grant.

Author Contributions A.S., R.K., R.E. and F.K. designed and supervised the experiments. A.S., T.P.-C., D.G., S.M., J.R. and J.V.B. participated in sample design and fabrication. A.S., T.P.-C., N.K., D.G., S.M., M.S. and S.H. performed the measurements. A.S., N.K., V.A., M.K., V.S.Y. and M.I.S. took part in the theoretical modelling. A.S., T.P.-C., N.K., R.K., R.E., V.S.Y. and F.K. analysed and interpreted the experimental data. All authors discussed the results and contributed to the final manuscript.

Author Information Reprints and permissions information is available at www.nature.com/reprints. The authors declare no competing financial interests. Readers are welcome to comment on the online version of the paper. Correspondence and requests for materials should be addressed to A.S. (aschiffr@phas.ubc.ca), M.I.S. (mstockman@gsu.edu) or F.K. (ferenc.krausz@mpq.mpg.de).

LETTER

doi:10.1038/nature11720

Controlling dielectrics with the electric field of light

Martin Schultze[1,2], Elisabeth M. Bothschafter[1,3], Annkatrin Sommer[1], Simon Holzner[1], Wolfgang Schweinberger[1], Markus Fiess[1], Michael Hofstetter[2], Reinhard Kienberger[1,3], Vadym Apalkov[4], Vladislav S. Yakovlev[2], Mark I. Stockman[4] & Ferenc Krausz[1,2]

The control of the electric and optical properties of semiconductors with microwave fields forms the basis of modern electronics, information processing and optical communications. The extension of such control to optical frequencies calls for wideband materials such as dielectrics, which require strong electric fields to alter their physical properties[1–5]. Few-cycle laser pulses permit damage-free exposure of dielectrics to electric fields of several volts per ångström[6] and significant modifications in their electronic system[6–13]. Fields of such strength and temporal confinement can turn a dielectric from an insulating state to a conducting state within the optical period[14]. However, to extend electric signal control and processing to light frequencies depends on the feasibility of reversing these effects approximately as fast as they can be induced. Here we study the underlying electron processes with sub-femtosecond solid-state spectroscopy, which reveals the feasibility of manipulating the electronic structure and electric polarizability of a dielectric reversibly with the electric field of light. We irradiate a dielectric (fused silica) with a waveform-controlled near-infrared few-cycle light field of several volts per angström and probe changes in extreme-ultraviolet absorptivity and near-infrared reflectivity on a timescale of approximately a hundred attoseconds to a few femtoseconds. The field-induced changes follow, in a highly nonlinear fashion, the turn-on and turn-off behaviour of the driving field, in agreement with the predictions of a quantum mechanical model. The ultrafast reversibility of the effects implies that the physical properties of a dielectric can be controlled with the electric field of light, offering the potential for petahertz-bandwidth signal manipulation.

A dielectric subjected to a weak optical field reacts to its change instantly (adiabatically) as long as the laser frequency $\omega_L \ll \Delta_{gap}/\hbar$, where Δ_{gap} is the gap between the valence band and conduction band; for silica $\Delta_{gap} \approx 9$ eV (numerical values below are given for this material). When the strength of the electric field F approaches the critical field strength, inducing a change in electron potential energy by Δ_{gap} over the lattice period $a \approx 5$ Å then

$$F_{crit} = \frac{\Delta_{gap}}{|e|a} \approx 2\text{V}\text{Å}^{-1} \quad (1)$$

(where e is the electron charge), and Zener-type transitions[1] inject electrons into the entire conduction band within its spectral width $\Delta_c \approx 10$ eV. This by itself, irrespective of ω_L, leads to ultrafast dynamics of the induced broadband polarization on a timescale of $\hbar/\Delta_c \approx 0.1$ fs. Hence, real-time access to strong-field-induced dynamics in dielectrics calls for sub-femtosecond temporal resolution. In our work, this was provided by the envelope of sub-100-attosecond extreme-ultraviolet (XUV) pulses and the controlled instantaneous field of few-femtosecond near-infrared (NIR) laser pulses.

In a first set of experiments, the waveform-controlled linearly polarized field, $F_L(t)$, of NIR laser pulses of less than 4 fs (carrier wavelength $\lambda_L = 780$ nm and wave cycle $T_L = 2\pi/\omega_L = 2.6$ fs), along with isolated 72-as XUV pulses carried at $\hbar\omega_{XUV} \approx 105$ eV and polarized parallel to the laser field, impinged collinearly on a free-standing 125-nm SiO_2 film (see Fig. 1a and Methods). The XUV pulse probed the processes induced by $F_L(t)$ by promoting electrons from the L-shell of silicon into the conduction band states. Transient changes in the conduction band are reflected in the XUV spectra recorded as a function of delay of the XUV probe with respect to the NIR-field excitation; see Fig. 1c. We note that the XUV pulse induces polarization over the entire band $\Delta_c/\hbar$, which largely dephases within $\tau_{dephasing} \sim \hbar/\Delta_c \sim 100$ as after passage of the XUV pulse through the sample. The XUV light radiated from the sample coherently with the XUV probe field therefore decayed within about 100 as after the XUV pulse. This permitted—in contrast with previous experiments in atomic gases[15]—us to record attosecond transient absorption spectra that provide information about the evolving state of the quantum system at the instant of probing and are not appreciably affected by the ongoing strong-field excitation afterwards.

We supplemented attosecond probing with simultaneous streaking[16] of the excitation field $F_L(t)$ (Fig. 1b) in a neon gas jet in front of the SiO_2 sample. Figure 1a displays the laser field and the XUV pulse envelope retrieved from this measurement. Simultaneous implementation of these two attosecond techniques permitted assignment of the transient absorption spectra to well-defined moments within the excitation field, $F_L(t)$, providing absolute timing of the processes under scrutiny.

For a detailed analysis, we evaluate the absorbance A, given by $A(\hbar\omega_{XUV}) = \alpha(\hbar\omega_{XUV})d$, where $\alpha(\hbar\omega_{XUV})$ is the absorption coefficient at the photon energy $\hbar\omega_{XUV}$, and d is the thickness of the sample. We first discuss XUV absorption near the edge of the conduction band. The solid line in Fig. 2b shows the transient change in A induced by the NIR field plotted in Fig. 2a; its measured peak electric field strength is $F_0 = 2.5 \pm 0.5$ V Å^{-1}. It reveals a large (>10%) increase of the sample's XUV transmittance over sub-femtosecond intervals. The oscillations in absorbance at a frequency of $2\omega_L$ are accompanied by similar oscillations in the position of the conduction-band edge (Fig. 2c). The measurement resolves condensed-matter processes driven by the instantaneous field of visible light. The changes emerge within a single half-cycle of the driving field and terminate with similar abruptness several femtoseconds later; see also the inset of Fig. 2b. The XUV absorption bleaching extends, with similar temporal behaviour, over the entire conduction band; see Supplementary Fig. 1.

To analyse the observed field-induced modifications, we adapted—with exactly the same parameters—the quantum-mechanical model developed for describing electric currents in dielectrics induced by strong, few-cycle NIR fields under physical conditions identical to the experiments of this work[14]. Briefly, it is based on the time-dependent Schrödinger equation in the presence of the measured field $F_L(t)$, incorporating field screening but neglecting electron collisions and excitonic effects. For a detailed description, see Supplementary Information. The model accurately predicts the magnitude and the sub-femtosecond temporal structure of the observed modulation in strength (Fig. 2b) and position (Fig. 2c) of the 109-eV absorption line, as well as the bandwidth and the strength of the XUV absorption bleaching (Supplementary Fig. 1). Agreement in the phase of these modulations with respect to the oscillating laser field (Fig. 2b, c)

[1]Max-Planck-Institut für Quantenoptik, Hans-Kopfermann-Strasse 1, D-85748 Garching, Germany. [2]Fakultät für Physik, Ludwig-Maximilians-Universität, Geschwister-Scholl-Platz 1, D-80539 München, Germany. [3]Physik-Department, Technische Universität München, James-Franck-Strasse, D-85748 Garching, Germany. [4]Department of Physics, Georgia State University, Atlanta, Georgia 30340, USA.

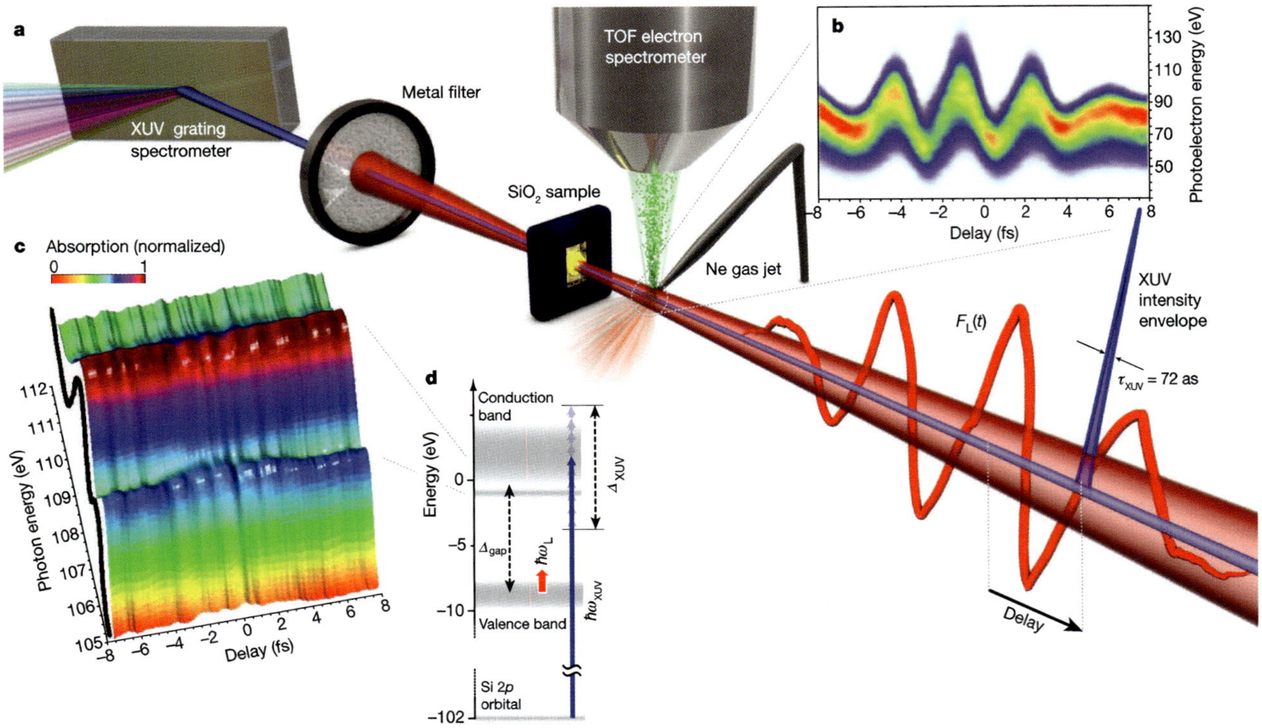

Figure 1 | Simultaneous attosecond absorption and streaking spectroscopy. **a**, Schematic illustration of the experimental set-up. TOF, time of flight. **b**, Attosecond streaking spectrogram of the 72-attosecond (full-width at half maximum) isolated XUV pulse and a near-single-cycle NIR pulse used in the experiments. The spectrogram was recorded by measuring the energy spectrum of photoelectrons released by the XUV pulse into the NIR field as a function of NIR–XUV delay. **c**, Attosecond transient absorption spectrogram of a 125-nm-thick SiO_2 membrane in the range 105–112 eV as a function of delay between the sub-100-attosecond XUV pulse and the NIR laser pulse with a delay step of $\Delta t_{delay} = 100$ as (the black line shows the synchrotron-XANES absorption data[24] for comparison). **d**, Schematic energy level diagram of SiO_2. The blue arrow represents the XUV absorption from Si L-states to the SiO_2 conduction band. The 9-eV bandgap between valence and conduction band exceeds the NIR photon energy by a factor of 6, as indicated by the red arrow.

supports our understanding that it is the instantaneous field that drives the processes.

In a second set of experiments, we scrutinized the influence of the strong field on the polarizability in the NIR range by studying the transmitted and reflected NIR light (for details, see Methods and Supplementary Information). Within our experimental accuracy of about ±1.2% (dictated by the laser pulse-to-pulse stability), the transmittance of our NIR beam did not vary up to the intensity threshold for breakdown. In contrast, the power of the reflected beam, detected in a z-scan geometry, increased by nearly an order of magnitude for field strengths approaching the breakdown threshold as shown in Fig. 3a. To ensure the highest contrast for the subsequent pump–probe study (see below), this measurement was implemented with

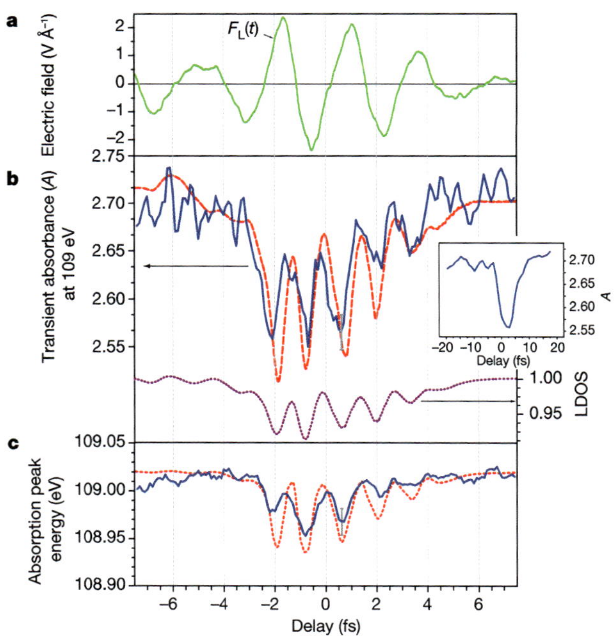

Figure 2 | Attosecond time-resolved strong-field-induced effects in SiO_2. Solid lines are experimental results; dashed lines are predictions of theoretical modelling. **a**, Electric field of the few-cycle NIR laser pulse impinging on the SiO_2 sample, $F_L(t)$, as extracted from attosecond streaking (see Fig. 1b). **b**, Transient change of the absorbance $A(\hbar\omega_{XUV}) = \alpha(\hbar\omega_{XUV})d$ integrated over a 1-eV bandwidth at $\hbar\omega = 109$ eV, as a function of the delay (the delay step is $\Delta t_{delay} = 100$ as) between the 72-as XUV probe and the NIR laser pulse (blue solid line), along with the prediction of our quantum mechanical model (red dashed line). The inset shows the evolution of A in a more extended delay range, recorded with larger delay steps ($\Delta t_{delay} = 0.5$ fs). The error bar in **b** represents the standard error of the average over 15 spectral lineouts within the energy range 108.5–109.5 eV; it exhibits little variation over the delay scan. The dashed violet line is the calculated local density of states (LDOS) at the position of a Si atom (integrated over the energy range accessed by the XUV pulse, for more details, see Supplementary Information) versus delay of the XUV probe. **c**, Energy of the absorption peak at 109 eV subject to an optical-field-induced (dynamic Stark) shift (the blue solid line shows the measurement, the red dashed line shows the calculation). The error bar in **c** represents the 95% confidence interval of the peak position of a least-squares fit of the area of a Gaussian function to the area under the absorption line; it exhibits little variation over the delay scan.

LETTER RESEARCH

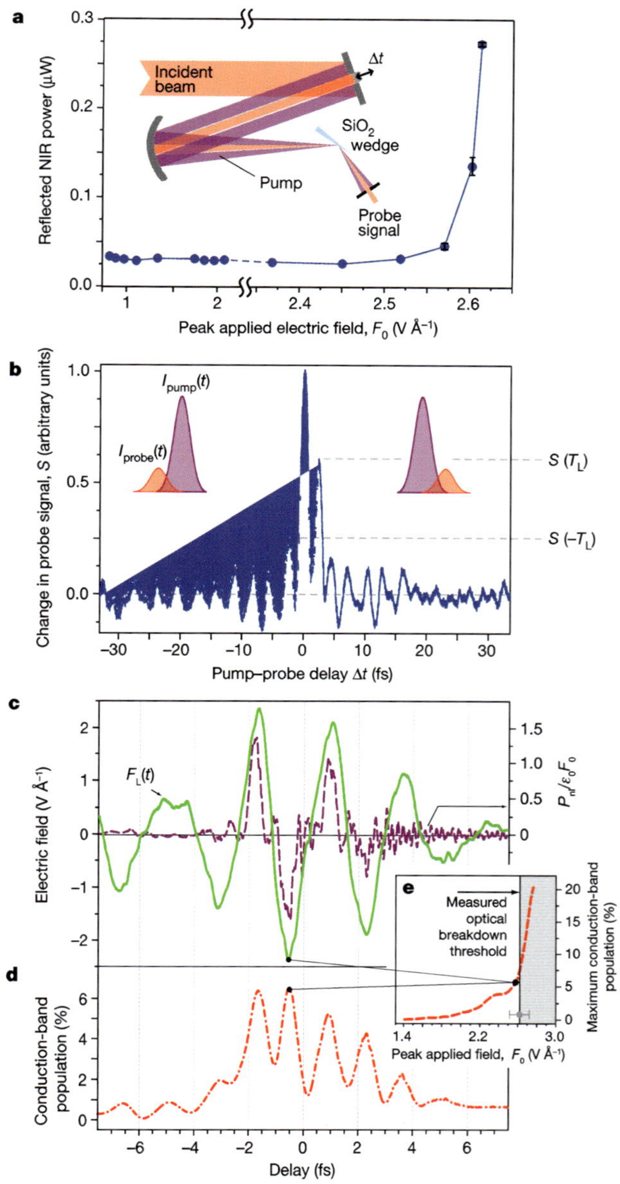

Figure 3 | Wave-cycle-resolved NIR femtosecond probing of strong-field-induced nonlinear reflectivity of SiO_2. a, Reflected power of *p*-polarized NIR laser pulses of less than 4 fs incident at Brewster's angle on the thin-wedged sample of fused silica as a function of peak electric field (penetrating into the sample). The process is fully reversible for several thousand laser shots before irreversible damage occurs owing to self-focusing inside the sample rather than on the surface. The data points are averaged over 3,000 laser pulses (error bars represent the standard deviation). Inset, schematic illustration of the pump–probe transient reflectivity set-up (for description, see Methods). **b,** Reflected power on axis of the probe pulse of less than 4 fs (represented by the red beam; intensity envelope I_{probe}) as a function of the delay with respect to the strong pump pulse of less than 4 fs (illustrated by the violet beam; intensity envelope I_{pump}). The curve is the result of an average of ten scans; for representative individual scans and details, see Supplementary Fig. 2 and the related description in the Supplementary Information. The reflected probe signal at delays of $\Delta t = \pm T_L$ is denoted by $S(\pm T_L)$. **c,** Electric field of the few-cycle NIR laser pulse impinging on the SiO_2 sample, $F_L(t)$, as extracted from attosecond streaking (see green solid line in Fig. 1b), and the resulting nonlinear polarization $P_{nl}(t)$, normalized to $\varepsilon_0 F_0$, predicted by our model (purple dashed line). **d,** Computed transient evolution of the population of (unperturbed) conduction band states timed to the laser field. **e,** Peak transient conduction band population versus peak applied field strength. The grey shading represents the experimental value of the optical breakdown threshold, as determined from the reflectance measurement yielding the data shown in **a.**

light impinging at Brewster's angle and polarized in the plane of incidence (*p*-polarized). The dependence on the field strength was extremely nonlinear, closely resembling that of the field-induced current[14]. Our model is not yet applicable to an oblique angle of incidence but did correctly predict the field-strength-dependence of the current[14]. The similarity in the intensity scaling of the optical current and reflectivity points to their common physical origin: the field-enhanced polarizability.

Is this effect mainly due to a conduction-band population surviving the few-cycle field or to reversible effects being not only turned on but also turned off by the field? Although the XUV bleaching already suggests that the latter is more likely, we sought additional evidence using time-resolved optical reflectometry. To this end, we split the incident pulse into a strong pump and a weak probe beam, with the sum of the two field maxima being slightly below the breakdown threshold, and measured the reflected probe intensity as a function of the delay Δt; see the sketch of the apparatus in Fig. 3a and the result in Fig. 3b.

Comparing the reflected probe signal S at delays $\Delta t = \pm T_L$, we found the signal to be significantly greater in the immediate aftermath of the strong pulse, $S(T_L)/S(-T_L) = 2.1 \pm 0.2$ (with no measurable dependence on the carrier-envelope phase to within the measurement accuracy limited by intensity fluctuations). However, comparing signals on opposite fringe peaks at $\Delta t \approx \pm 2T_L$, $\pm 3T_L$,..., showed no appreciable asymmetry—$S(nT_L) \approx S(-nT_L)$, $n > 1$—within our experimental precision. If a significant fraction of the induced nonlinear reflectivity (and, consequently, polarization) survived the pulse peak over several cycles, this would imply $S(nT_L)$ significantly larger than $S(-nT_L)$, owing to a much larger fraction of the probe pulse experiencing the increased reflectivity. Our observation thus provides compelling evidence, in a model-independent manner, for the substantial reversibility of the strong-field-induced increase in polarizability on the timescale of the optical wave cycle.

Figure 3c contrasts the applied electric field $F_L(t)$ (solid line) with the calculated field-induced nonlinear polarization $P_{nl}(t)$ normalized to $\varepsilon_0 F_0$ (dashed line). The strongly sharpened half cycles of $P_{nl}(t)$ point to an extremely nonlinear scaling with driving field strength, consistent with the data in Fig. 3a. The nonlinear response is largely confined to the central laser cycle and disappears completely before the end of the pulse. This effect is accompanied by a field-induced transient population and subsequent depopulation of the conduction band, confined, again, to several femtoseconds (see Fig. 3d). For field strengths beyond $F_0 \approx 2.5\,\text{V Å}^{-1}$, Fig. 3e shows a near-exponential growth of conduction band population, causing breakdown. In fact, the maximum field we could safely apply to our samples before breakdown was $F_0 \approx 2.5\,\text{V Å}^{-1}$. Thus, our model also correctly predicts the threshold for optical breakdown (Fig. 3e). Both the nonlinear polarization and the conduction band population induced by the strong field return to near-zero immediately after the laser pulse for $F_0 \leq 2.5\,\text{V Å}^{-1}$. This is fully consistent with the abrupt decay of field-induced transient NIR reflectivity and XUV absorption bleaching: see Figs 3b and 2b, respectively. These results imply that the sample exposed to fields as high as $F_0 \leq 2.5\,\text{V Å}^{-1}$ resumes its original (field-free) state immediately after exposure. This points towards the reversible, Hamiltonian nature of the laser-induced dynamics up to the critical field strength.

Our quantum mechanical model identifies light-field-induced reversible changes of the local density of states (Fig. 2b), the conduction-band population, and the polarizability as the mechanisms responsible for transient XUV absorption bleaching, NIR reflectivity enhancement, optical breakdown and optical-field-induced currents[14]. The model accounts for all these phenomena with a set of parameters adjusted in ref. 14, which we did not modify in this work. The ability

RESEARCH LETTER

to predict such a large aggregate of disparate physical effects properly without parameter adjustment provides convincing evidence of the theory's validity.

Considerations presented in ref. 14 provide a hint towards a possible intuitive picture to explain the above effects in terms of the well-known phenomenon of Wannier–Stark localization[17–21]. The field-induced strong transient localization of states is expected to occur in all bands, allowing XUV-induced transitions predominantly between states localized at the same site of the lattice. The energy levels of such colocalized Wannier–Stark states of different bands are shifted identically by the field, leaving the XUV transition energies (between the L-band and conduction band) nearly unchanged. Hence, XUV probing is insensitive to these energy shifts, apart from the small dynamic shifts observed in Fig. 2c, but uncovers dynamic changes in the density of states. The Wannier–Stark ladder formation delivers an intuitive account for the computed decrease of the local density of states (Fig. 2b), which appears to be mainly responsible for the decrease of the XUV absorption. The field-enhanced polarizability of the system can also be related to the strong Stark shifts in the conduction band and valence band: they pull interband transition frequencies into resonance with the visible–NIR spectrum of the driving field[14]. The resultant enhancement in polarizability appears to be the common origin of the optical-field-induced reflectivity observed here and the currents observed in ref. 14. In spite of the consistency of this argument, the intuitive Wannier–Stark picture remains hypothetical until directly verified by experiment, for example, by probing valence band-conduction band transitions with sub-femtosecond resolution.

In conclusion, the synergistic use of attosecond streaking and absorption spectroscopy and wave-cycle-resolved optical reflectivity measurements provide real-time insight into strong-field phenomena in dielectrics on sub-femtosecond to few-femtosecond timescales. The experiments reveal, in agreement with quantum-mechanical modelling, that field-induced changes of all the physical quantities studied are reversible up to the critical field strength, that is, they can be turned on and off on the timescale of the optical period. Controlling the properties of dielectrics with the instantaneous electric field of light points the way towards petahertz signal sampling and processing technologies.

METHODS SUMMARY

XUV transmittivity. Time-resolved XUV absorption spectroscopy was performed with broadband, isolated attosecond XUV pulses generated via high harmonic generation of visible–infrared laser pulses of less than 4 fs (FemtoPower Compact Pro, Femtolasers) in neon. The collinear laser and XUV beam were separated into two arms of a Mach–Zehnder-type interferometer. The relative timing between the two arms could be adjusted with a mirror moved by a piezo translator, and a variable aperture controls the infrared intensity on target. For details, see the Supplementary Information and ref. 22. The XUV and NIR beams were focused with a toroidal mirror on free-standing chemical-vapour-deposited SiO_2 samples (either 125 nm or 250 nm thick). The XUV beam transmitted through the sample was spectrally dispersed by a flat-field grating and projected on a XUV-sensitized camera[23]. Our measurements were performed near (to within <10%) optical breakdown. NIR-field-induced breakdown manifested itself as macroscopic mechanical damage of the sample, which became immediately visible in the magnified image of the focal region, which was permanently monitored.

NIR transmittivity and reflectivity. Field-induced changes in NIR transmittivity through 125-nm films and the reflectivity of thin (~0.2 mm) wedged plates were studied with laser pulses of less than 4 fs gently focused and the intensity on the sample was varied by translating it longitudinally along the optical axis (*z*-scan). The pump–probe measurement is sketched schematically in Fig. 3a. The incident NIR beam was split into two collinear beams by a two-component plane mirror consisting of a 5-mm-diameter inner section and a concentric external section. The inner and outer sections reflected the weak probe and the strong pump beam with a reflectivity of 10% and 97.5%, respectively. The energy of the pump pulse was controlled by an aperture. The probe pulse was delayed by translating the inner section with a piezo stage and focused along with the pump pulse onto the sample. The probe beam was isolated with an iris in front of the detector measuring the reflected probe pulse energy as a function of pump–probe delay. A more detailed description of the Methods is given in the Supplementary Information.

Received 20 December 2011; accepted 24 October 2012.
Published online 5 December 2012.

1. Zener, C. A theory of the electrical breakdown of solid dielectrics. *Proc. R. Soc. Lond. A* **145,** 523–529 (1934).
2. Wannier, G. Wave functions and effective Hamiltonian for Bloch electrons in an electric field. *Phys. Rev.* **117,** 432–439 (1960).
3. Franz, W. Einfluß eines elektrischen Feldes auf eine optische Absorptionskante. *Z. Naturforsch. A* **13,** 484 (1958).
4. Keldysh, L. V. Behavior of non-metallic crystals in strong electric fields. *Sov. J. Exp. Theor. Phys.* **6,** 763 (1958).
5. Mizumoto, Y., Kayanuma, Y., Srivastava, A., Kono, J. & Chin, A. H. Dressed-band theory for semiconductors in a high-intensity infrared laser field. *Phys. Rev. B* **74,** 045216 (2006).
6. Lenzner, M. *et al.* Femtosecond optical breakdown in dielectrics. *Phys. Rev. Lett.* **80,** 4076–4079 (1998).
7. Gertsvolf, M., Spanner, M., Rayner, D. M. & Corkum, P. B. Demonstration of attosecond ionization dynamics inside transparent solids. *J. Phys. At. Mol. Opt. Phys.* **43,** 131002 (2010).
8. Mitrofanov, A. *et al.* Optical detection of attosecond ionization induced by a few-cycle laser field in a transparent dielectric material. *Phys. Rev. Lett.* **106,** 147401 (2011).
9. Shih, T., Winkler, M. T., Voss, T. & Mazur, E. Dielectric function dynamics during femtosecond laser excitation of bulk ZnO. *Appl. Phys. A* **96,** 363–367 (2009).
10. Ghimire, S. *et al.* Redshift in the optical absorption of ZnO single crystals in the presence of an intense midinfrared laser field. *Phys. Rev. Lett.* **107,** 167407 (2011).
11. Ghimire, S. *et al.* Observation of high-order harmonic generation in a bulk crystal. *Nature Phys.* **7,** 138 (2011).
12. Durach, M., Rusina, A., Kling, M. & Stockman, M. Metallization of nanofilms in strong adiabatic electric fields. *Phys. Rev. Lett.* **105,** 086803 (2010).
13. Durach, M., Rusina, A., Kling, M. & Stockman, M. Predicted ultrafast dynamic metallization of dielectric nanofilms by strong single-cycle optical fields. *Phys. Rev. Lett.* **107,** 086602 (2011).
14. Schiffrin, A. *et al.* Optical-field-induced current in dielectrics. *Nature* doi:10.1038/nature11567 (this issue).
15. Goulielmakis, E. *et al.* Real-time observation of valence electron motion. *Nature* **466,** 739–743 (2010).
16. Kienberger, R. *et al.* Atomic transient recorder. *Nature* **427,** 817–821 (2004).
17. Bloch, F. Über die Quantenmechanik der Elektronen in Kristallgittern. *Z. Phys.* **52,** 555–600 (1929).
18. Wannier, G. H. *Elements of Solid State Theory. Elements* 173–177 (Cambridge Univ. Press, 1959).
19. Bleuse, J., Bastard, G. & Voisin, P. Electric-field-induced localization and oscillatory electro-optical properties of semiconductor superlattices. *Phys. Rev. Lett.* **60,** 220–223 (1988).
20. Mendez, E., Agulló-Rueda, F. & Hong, J. Stark localization in GaAs-GaAlAs superlattices under an electric field. *Phys. Rev. Lett.* **60,** 2426–2429 (1988).
21. Mendez, E. E. & Bastard, G. Wannier-Stark ladders and Bloch oscillations in superlattices. *Phys. Today* **46,** 34, http://dx.doi.org/10.1063/1.881353 (1993).
22. Fiess, M. *et al.* Versatile apparatus for attosecond metrology and spectroscopy. *Rev. Sci. Instrum.* **81,** 093103 (2010).
23. Schultze, M. *et al.* State-of-the-art attosecond metrology. *J. Electron Spectrosc. Relat. Phenom.* **184,** 68–77 (2011).
24. Li, D. *et al.* X-ray absorption spectroscopy of silicon dioxide (SiO_2) polymorphs; the structural characterization of opal. *Am. Mineral.* **79,** 622–632 (1994).

Supplementary Information is available in the online version of the paper.

Acknowledgements This work was supported by the Max Planck Society and the Deutsche Forschungsgemeinschaft Cluster of Excellence: Munich Centre for Advanced Photonics (www.munich-photonics.de).The work of M.I.S. and V.A. was supported by grant number DEFG02-01ER15213 from the Chemical Sciences, Biosciences and Geosciences Division and by grant number DE-FG02-11ER46789 from the Materials Sciences and Engineering Division of the Office of the Basic Energy Sciences, Office of Science, US Department of Energy. We thank K. Yabana, R. Ernstorfer and N. Karpowicz for discussions.

Author Contributions M.S., R.K., M.I.S. and F.K. conceived and supervised the study. M.S., E.M.B., A.S., S.H., W.S., M.F. and M.H. prepared and performed the experiment. V.A. and M.I.S. accomplished the theoretical modelling. M.S., E.M.B., A.S., V.S.Y. and F.K. analysed and interpreted the experimental data. All authors discussed the results and contributed to the final manuscript.

Author Information Reprints and permissions information is available at www.nature.com/reprints. The authors declare no competing financial interests. Readers are welcome to comment on the online version of the paper. Correspondence and requests for materials should be addressed to M.S. (martin.schultze@mpq.mpg.de), M.I.S. (mstockman@gsu.edu) and F.K. (krausz@lmu.de).

LETTERS
PUBLISHED ONLINE: 12 JANUARY 2014 | DOI: 10.1038/NPHOTON.2013.348
nature photonics

Solid-state light-phase detector

Tim Paasch-Colberg[1]*, Agustin Schiffrin[1†], Nicholas Karpowicz[1], Stanislav Kruchinin[1], Özge Sağlam[2], Sabine Keiber[1], Olga Razskazovskaya[1], Sascha Mühlbrandt[1†], Ali Alnaser[1,3,4], Matthias Kübel[1], Vadym Apalkov[5], Daniel Gerster[2], Joachim Reichert[2], Tibor Wittmann[1,6], Johannes V. Barth[2], Mark I. Stockman[5], Ralph Ernstorfer[7], Vladislav S. Yakovlev[1,6], Reinhard Kienberger[1,2] and Ferenc Krausz[1,6]*

Attosecond science relies on the use of intense, waveform-controlled, few-cycle laser pulses[1] to control extreme nonlinear optical processes taking place within a fraction of an optical period. A number of techniques are available for retrieving the amplitude envelope and chirp of such few-cycle laser pulses. However, their full characterization requires detection of the absolute offset between the rapidly oscillating carrier wave and the pulse envelope, the carrier-envelope phase (CEP). So far, this has only been feasible with photoelectron spectroscopy, relying on complex vacuum set-ups[2-4]. Here, we present a technique that enables the detection of the CEP of few-cycle laser pulses under ambient conditions. This is based on the CEP-dependence of directly measurable electric currents generated by the electric field of light in a metal-dielectric-metal nanojunction. The device holds promise for routine measurement and monitoring of the CEP in attosecond laboratories.

The electric field of a laser pulse can be described as $F(t) = \frac{1}{2}\tilde{F}(t)e^{-i(\omega_L t+\varphi_{CE})} + \text{c.c.}$, where ω_L is the carrier frequency and the complex amplitude $\tilde{F}(t) = F_0 f(t)e^{-i\varphi(t)}$ comprises the normalized real-field envelope, the peak amplitude F_0 and the time-dependent phase $\varphi(t)$. The CEP φ_{CE} describes the offset between the carrier wave oscillating at ω_L and the maximum of the real-field envelope $f(t)$. If the pulse duration, defined as the full-width at half-maximum of $f^2(t)$ (the square of the envelope) becomes comparable with the carrier period $2\pi/\omega_L$ of the waveform, φ_{CE} affects the evolution of nonlinear electron phenomena driven by $F(t)$, including high-order harmonic generation[5,6], ionization of atoms and molecules[7-9], interaction with plasmas[10], as well as photoemission from metals[11,12] and nanoparticles[13]. Attosecond control over these processes calls for control and full characterization of the applied electric field $F(t)$.

Attosecond streaking[14] permits complete measurement of the vector potential of a laser pulse, from which $F(t)$ is easily obtained as its temporal derivative. Alternatively, the combination of conventional pulse characterization techniques[15-17] acquiring the normalized complex amplitude $\tilde{F}(t)/F_0 = f(t)e^{-i\varphi(t)}$ and a measurement of the CEP, φ_{CE}, may provide access to the full waveform. Although relative changes of the CEP can be detected by spectral characterization of mixed fields using simple and inexpensive set-ups[18-21], tracking down its absolute value is currently based on nonlinear photoemission from atoms or solids with stereo above-threshold ionization (stereo-ATI)[2-4] and attosecond streaking[14,17], both of which require rather complex, space-consuming vacuum apparatus.

Here, we report the first method to enable absolute CEP detection with a solid-state detector that is applicable in ambient conditions. Recently, we have shown that the strong electric field of an intense, linearly polarized, visible/near-infrared (vis/NIR), few-cycle laser pulse can rapidly increase the (a.c.) conductivity of a solid insulator, allowing electric currents to be induced and switched with the field of visible light[22]. In these experiments, we exposed amorphous silicon dioxide (bandgap $E_g \approx 9$ eV) to a strong, controlled electric field $F(t)$ of a few-cycle pulse with a carrier photon energy of $\hbar\omega_L \approx 1.7$ eV. The resultant transient nonlinear increase in electric polarization allowed the few-cycle field to drive a current through two unbiased gold electrodes connected to the sample. Because the effect is controlled directly by $F(t)$, it is sensitive to the CEP[22].

This sensitivity can be utilized for the measurement of φ_{CE} if, and only if, the following two conditions are fulfilled: (1) the current versus φ_{CE} dependence is calibrated and (2) this calibration is robust against variations in the pulse intensity. This work addresses and fulfils these conditions by applying a series of careful experiments and numerical simulations. We used sub-4 fs (~1.5 cycle) vis/NIR laser pulses carried at $\lambda_L \approx 0.75$ μm to study the φ_{CE} dependence of optical-field-induced electric currents in a metal–dielectric–metal junction and compared it with measurements simultaneously performed with a stereo-ATI phasemeter[5] (Fig. 1). The junction was exposed to a strong optical field at normal incidence, with perpendicular polarization of $F(t)$ (Fig. 1a,b). The peak electric field amplitude was set at $F_0 \approx 1.24$ V Å^{-1} by focusing pulses of ~4 μJ energy. This is only ~1% of the total pulse energy of the utilized few-cycle laser system[23]. All experiments reported herein were performed in vacuum to meet the requirements of the stereo-ATI measurements. However, CEP detection via optical field-induced currents in a solid dielectric can be achieved under ambient conditions, as demonstrated in ref. 22.

The stabilized CEP of the laser pulses was modulated to allow every second pulse in the train to have the same φ_{CE} (that is, φ_{CE} was altered by π for consecutive pulses). This variation of φ_{CE} is beneficial for the isolation of CEP-dependent electric currents using a lock-in amplifier for removal of CEP-independent

[1]Max-Planck-Institut für Quantenoptik, Hans-Kopfermann-Strasse 1, D-85748 Garching, Germany, [2]Physik-Department, Technische Universität München, James-Franck-Strasse, D-85748 Garching, Germany, [3]Physics Department, POB2666 American University of Sharjah-Sharjah, United Arab Emirates, [4]Faculty of Physics and Astronomy, King Saud University, Riyadh, Saudi-Arabia, [5]Department of Physics and Astronomy, Georgia State University, Atlanta, Georgia 30340, USA, [6]Fakultät für Physik, Ludwig-Maximilians-Universität, Am Coulombwall 1, D-85748 Garching, Germany, [7]Fritz-Haber-Institut der Max-Planck-Gesellschaft, Faradayweg 4-6, D-14195 Berlin, Germany; †Present addresses: Quantum Matter Institute, University of British Columbia, Vancouver, British Columbia, V6T 1Z4 Canada, and Department of Physics and Astronomy, University of British Columbia, Vancouver, British Columbia, V6T 1Z1 Canada (A.S.), Institute of Microstructure Technology (IMT), Karlsruhe Institute of Technology (KIT), D-76021 Karlsruhe, Germany (S.M.).
*e-mail: tim.paasch-colberg@mpq.mpg.de; ferenc.krausz@mpq.mpg.de

NATURE PHOTONICS DOI: 10.1038/NPHOTON.2013.348 LETTERS

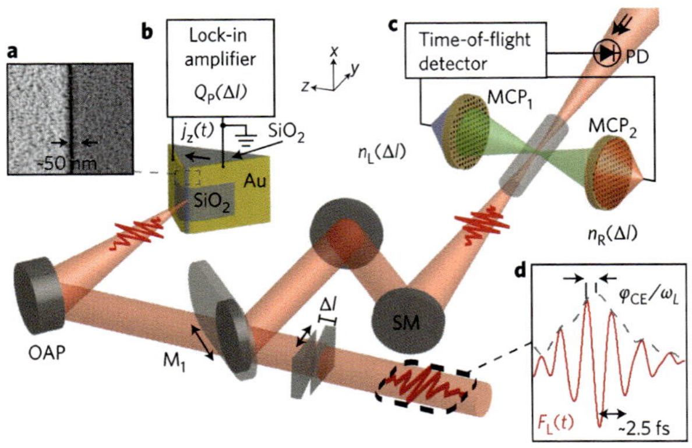

Figure 1 | Measurement of the absolute CEP of light pulses by optical-field-induced electric currents in a solid-state device. A metal-dielectric-metal junction in vacuum is exposed to a few-cycle vis/NIR laser pulse, focused with an off-axis parabolic mirror (OAP), with its polarization oriented perpendicular to the junction (that is, in the *z*-direction). The CEP is varied by tuning the propagation length Δl of the pulses inside a pair of fused silica wedges. The light field $F_L(t)$ mediates the increased conductivity of the dielectric, polarizing the junction and creating a current density $j_z(t)$ between the electrodes. **a**, SEM image of an example junction. **b**, The CEP-dependent fraction Q_P of the total charge is detected with a lock-in amplifier. Q_P is calibrated with respect to the absolute CEP of the laser pulse via parallel stereo-ATI measurements. **c**, The laser field is focused with a spherical mirror (SM) into a gas-filled cell where it ionizes Xe atoms and accelerates the photoelectrons towards a pair of MCPs. The CEP-dependent time-of-flight distributions $n_{L,R}(t_{TOF}, \Delta l)$ of the photoelectrons are determined using a photodiode (PD) for zero time reference in the left (L) and right (R) directions for various values of Δl. **d**, $F_L(t)$ was determined in a previous experiment via attosecond streaking[14].

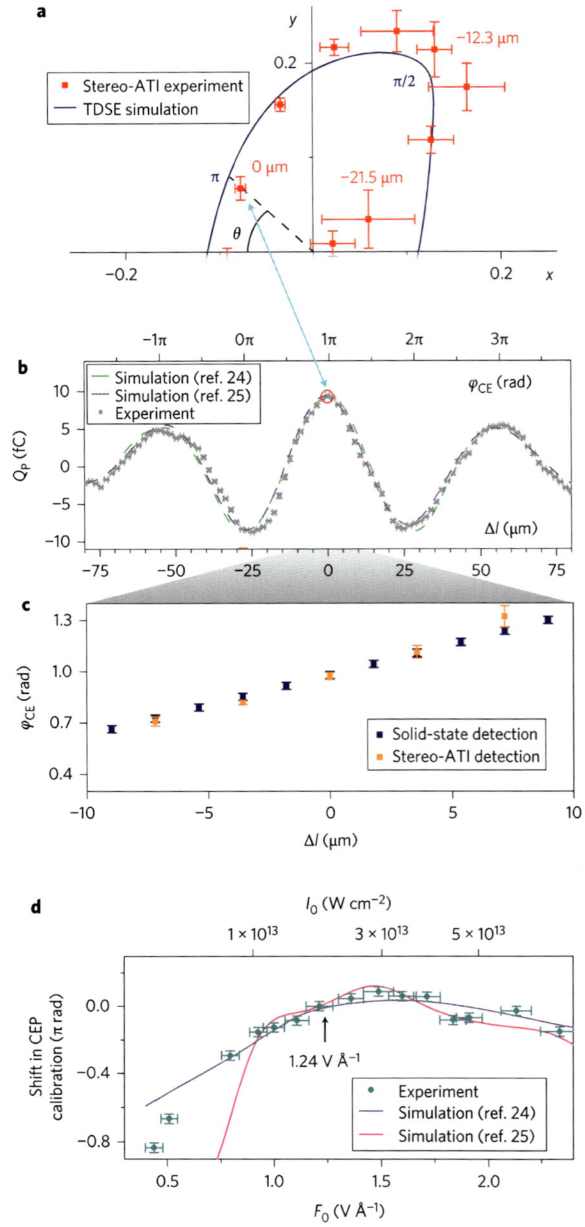

Figure 2 | Calibration of the CEP-dependent current. a, Parametric plot of the asymmetry parameters *X* and *Y* of the stereo-ATI measurements (red). Error bars indicate standard deviations of both measurements. The results of the simulation based on solving the TDSE are shown in blue. Radial angle θ is used to calibrate the experimental stereo-ATI data. **b**, Q_P (grey) shown as a function of the propagation length Δl of the laser pulse in the dispersive fused-silica wedge (lower horizontal axis). Data points represent an average of 1,500 consecutive measurements at a given Δl and $F_0 = 1.24\ \text{V Å}^{-1}$. Error bars in **b** and **c** indicate standard deviation. Using stereo-ATI measurements from **a** at $\Delta l = 0$ μm, Q_P is calibrated with respect to the absolute CEP (φ_{CE}, upper horizontal axis). Results of quantum-mechanical simulations (refs 24,25) are shown in green and purple. **c**, The method to detect φ_{CE} with a solid-state device (blue) is compared with detection using a stereo-ATI apparatus (orange) at different values of Δl. **d**, Relative shift of $Q_P(\varphi_{CE})$ for different values of laser field amplitude F_0 (lower axis) and of cycle-averaged peak intensity I_0 (upper), compared with the simulation results.

background signals (see Methods). In the following, 'φ_{CE}' always denotes the CEP of odd-numbered pulses. However, recent experiments with an improved geometry of the solid-state device have led to a reduction of the φ_{CE}-independent charge per pulse by about two orders of magnitude, enabling a direct readout of the signal without the requirement of a lock-in amplifier.

The CEP was adjusted continuously by changing the propagation length Δl of the pulses through a pair of fused-silica wedges[21]. The value of Δl for which the CEP is changed by $\Delta\varphi_{CE} = 2\pi$ is $\Delta l_{2\pi(\text{Sellmeier})} \cong 51$ μm according to the Sellmeier equation. The value $\Delta l = 0$ μm is defined as the propagation length where the pulse duration is minimum; that is, an increase of $|\Delta l|$ results in an increase of the pulse duration due to (small) changes in group-delay dispersion.

The CEP-dependent fraction Q_P of the total charge (that is, the induced current integrated over an interval longer than the laser pulse) flowing through the external circuit, per laser shot, was measured with a lock-in detection apparatus (see Methods) as a function of $\Delta\varphi_{CE}$. Figure 2b plots the resultant Q_P as a function of Δl. Note that the values of Q_P exceed (by more than an order of magnitude) those acquired in previous measurements (see Fig. 2a in ref. 22) due to improvements in sample quality. $Q_P(\Delta l)$ shows a clean oscillatory behaviour with its maximum at $\Delta l = 0$ μm and a period of $\Delta l_{2\pi(\text{exp})} \cong 56$ μm, which is close to $\Delta l_{2\pi(\text{Sellmeier})}$. This discrepancy is attributed to the non-optimal alignment of the dispersive wedge with respect to the beam propagation direction. Hence, Q_P switches polarity for $\Delta\varphi_{CE} = \pi$, as intuitively expected. This establishes the link between magnitude and direction of the current and the electric field $F(t)$ of the laser. The observed decay of the amplitude of $Q_P(\Delta l)$ with the increase in $|\Delta l|$ is a consequence of dispersive pulse broadening inside the glass wedges. However, in our experiments, $Q_P(\Delta l)$

LETTERS NATURE PHOTONICS DOI: 10.1038/NPHOTON.2013.348

was still detectable above the noise level for values of $|\Delta l| > 400$ μm, corresponding to a pulse duration of more than 9 fs (FWHM of the time-dependent cycle-averaged intensity).

Subsequently, Q_P was calibrated with respect to the absolute CEP of the laser pulse via stereo-ATI measurements performed with identical pulses[4]. After the measurement of $Q_P(\Delta l)$ with the solid-state device, a mirror was inserted into the beam path, deflecting the pulses into a stereo-ATI apparatus located, together with the solid-state detector, in the same vacuum chamber (Fig. 1). The CEP of the incident laser pulse was detected by analysing the kinetic energy distribution of the electrons that are photoemitted from Xe atoms (see Methods). Uncertainty due to a Gouy phase shift in both foci can be neglected, as in both experiments the sample was placed exactly in the region of the highest laser intensity.

We set 17 different propagation lengths Δl, ranging from −21.5 μm to +27.5 μm. For each, 500 single-shot stereo-ATI measurements were performed. Because consecutive laser pulses had a CEP shift of π, which is only required for the accurate detection of $Q_P(\Delta l)$, only spectra from odd-numbered pulses were considered for the stereo-ATI measurements. As shown in ref. 4, φ_{CE} can then be reconstructed by calculating two asymmetry parameters (X, Y) by integrating the averaged time-of-flight spectra $n_{L,R}(t_{TOF})$ of the electrons photoemitted from Xe atoms by the intense few-cycle vis/NIR pulses in two different regions. The parametric plot of (X, Y) in Fig. 2a was obtained by calculating, for each value of Δl, $X = (P_L^{(X)} - P_R^{(X)})/(P_L^{(X)} + P_R^{(X)})$ and $Y = (P_L^{(Y)} - P_R^{(Y)})/(P_L^{(Y)} + P_R^{(Y)})$, where $P_{L,R}^{(X,Y)} = \int_{t_1^{(X,Y)}}^{t_2^{(X,Y)}} n_{L,R}(t_{TOF}) dt_{TOF}$. The photoelectron spectra $n_{L,R}(t_{TOF})$ were measured with the left (L) and right (R) microchannel plates (MCPs) of the set-up in Fig. 1. The values for the integration were $t_1^X = 27$ ns, $t_2^X = t_1^Y = 31$ ns and $t_2^Y = 45$ ns, respectively. These intervals were chosen and optimized for high phase asymmetry.

A theoretical parametric plot (X_{TDSE}, Y_{TDSE}) (blue in Fig. 2a) was calculated by numerically solving the three-dimensional, time-dependent Schrödinger equation (TDSE) for valence electrons in isolated Xe atoms exposed to a strong few-cycle optical electric field $F(t)$ with well-defined absolute CEP (see Methods for details). By associating an experimental pair (X, Y) with a theoretical pair (X_{TDSE}, Y_{TDSE}) via a radial angle $\theta = \arctan(Y_{TDSE}/X_{TDSE}) = \arctan(Y/X)$ in Fig. 2a and, because each theoretical pair (X_{TDSE}, Y_{TDSE}) corresponds to an absolute CEP, it is possible to associate a well-defined φ_{CE} to an experimental pair of parameters (X, Y).

To calibrate $Q_P(\Delta l)$ with respect to the absolute CEP of the incident laser pulses, the experimental pair (X, Y) at the maximum value $Q_P(\Delta l = 0)$ of the CEP-dependent transferred charge was used. The upper horizontal scale in Fig. 2b depicts the obtained calibration. $Q_P(\varphi_{CE})$ oscillates with a period of 2π, showing a relatively small error and high contrast with respect to the background noise, and has its maximum at $\varphi_{CE}^{max} = (0.98 \pm 0.02)\pi$. An important self-consistency check can now be performed by comparing the CEP retrieved from the stereo-ATI and the (meanwhile calibrated) solid-state measurement at various values of Δl. Figure 2c shows this comparison and reveals a very good agreement between the two methods. The error of the reported φ_{CE} detection method is on the same order as the error of the established stereo-ATI technique. It therefore lends itself to the determination of the absolute φ_{CE} of single few-cycle laser pulses, as long as there is an unambiguous relationship between measured current and pulse φ_{CE}.

We compared the results of the solid-state-based phase retrieval with the predictions of two quantum-mechanical models. The first model, which was used previously[24] to describe the ultrafast increase in conductivity of SiO_2 nanojunctions, is based on the nearest-neighbour tight-binding approximation. The second model, presented in detail in ref. 25, describes quantum dynamics in a one-dimensional pseudopotential (see Methods for details). In both models, the electric field in the medium is evaluated self-consistently with the induced polarization. As shown in Fig. 2b, excellent agreement is achieved between the measured data and the predictions of both theoretical models.

Reliable determination of φ_{CE} at different laser intensities via the detection of Q_P calls for low sensitivity of the measurement result with respect to intensity variations. To assess this sensitivity, we performed a series of measurements of Q_P as a function of φ_{CE} (as in Fig. 2b) for different laser field strengths F_0. With varying F_0, we observe a displacement of the Q_P versus φ_{CE} curve with respect to φ_{CE}; that is, when F_0 varies, the value of φ_{CE} for which Q_P is maximum changes. Figure 2d shows this shift in the CEP scale with respect to the calibrated scale obtained at $F_0 \approx 1.24$ V Å^{-1}, together with the prediction of the theoretical models. For field strengths in excess of 1 V Å^{-1}, good agreement is found between theory and experiment.

Both theory and experiment reveal significant shifts below 1 V Å^{-1}, rendering this intensity regime unsuitable for CEP measurements by the solid-state detector. In stark contrast, for field strengths above 1 V Å^{-1}, shifts of the Q_P versus φ_{CE} curves with varying F_0 remain small (Fig. 2b). Only the amplitude of $Q_P(\varphi_{CE})$ changes significantly and nonlinearly (cf. Fig. 2b in ref. 22). The phase calibration is robust with an accuracy on the order of ±310 mrad in the range from $F_0 \approx 1$ to 2.5 V Å^{-1}, corresponding to $I_0 \approx 1 \times 10^{13}$ to 7×10^{13} W cm^{-2} (see upper axis in Fig. 2d for an intensity scale) or, in other words, to ±75% around 4×10^{13} W cm^{-2}. This is a remarkable result, because conventional techniques for CEP detection and stabilization show a much bigger sensitivity to fluctuations of the pulse energy[26,27], where an energy fluctuation of 1% induced a φ_{CE} shift of 160 mrad.

However, the measurement of $Q_P(\varphi_{CE})$ from a single metal–dielectric–metal junction, as described above, allows the determination of the absolute φ_{CE} of a single laser pulse with a remaining phase ambiguity only (cf. Fig. 2b). In fact, within the φ_{CE} interval of 0–2π, two different values of φ_{CE} can give the same $Q_P(\varphi_{CE})$, for example, $Q_P(0.7\pi) = Q_P(1.2\pi)$. For stabilized φ_{CE}, this can be resolved by subsequent measurements of $Q_P(\varphi_{CE})$ with slightly changed values of φ_{CE} around its nominal value (cf. Fig. 3b). In the absence of CEP stabilization, this ambiguity must be removed within one laser shot. This could be resolved by the use of two identical junctions, exposed simultaneously to pulses of equal field strength from the same source, for which the relative CEPs differ by $\pi/2$ (Fig. 3c). Analogously to the photoelectron signals of the two independent MCPs in the stereo-ATI set-up, the pair of measured electronic signals $Q_P^{(1)}(\varphi_{CE})$ and $Q_P^{(2)}(\varphi_{CE})$ from both junctions is unambiguously related to the absolute CEP of the initial laser pulse, φ_{CE}. For example, in Fig. 3a, $Q_P^{(1)}(0.7\pi) = Q_P^{(1)}(1.2\pi)$, but $Q_P^{(2)}(0.7\pi) \neq Q_P^{(2)}(1.2\pi)$, allowing for disambiguation and single-shot CEP detection via $\varphi_{CE} \simeq \varphi_{CE}^{max} + \arg(Q_P^{(1)} + iQ_P^{(2)})$. This method of using the signals from both junctions also minimizes any potential phase errors arising from amplitude changes between consecutive laser pulses.

In summary, we have demonstrated the feasibility of measuring the absolute CEP of few-cycle laser fields with a solid-state device. Its ability to function in ambient conditions at relatively low laser energy, its stability against fluctuations of the laser energy, and also the provision of a direct electronic readout make this approach uniquely powerful. Furthermore, the lifetime of such a device is not limited by laser-induced damage. Thus, this light-phase detector holds promise as a valuable addition to the toolbox of attosecond metrology and spectroscopy, not only for absolute CEP detection, but also for CEP stabilization.

NATURE PHOTONICS DOI: 10.1038/NPHOTON.2013.348 LETTERS

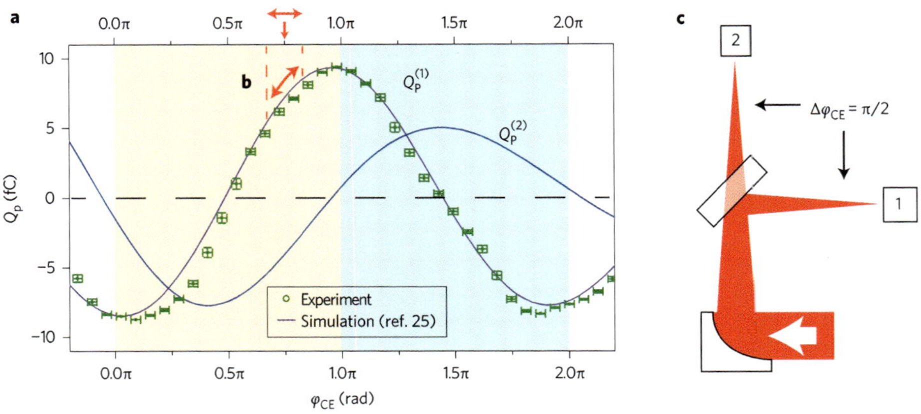

Figure 3 | Direct determination of absolute CEP via directly measurable photocurrents. a, The CEP-dependent component Q_P as a function of the absolute φ_{CE} of the incident laser pulse in the interval 0–2π rad. **b,** The φ_{CE} of repetitive identical pulses can be detected with a single junction by slight variations of φ_{CE} around its nominal value (here, $\varphi_{CE} = 0.75\pi$ is taken as an example). **c,** For the φ_{CE} detection of single laser pulses with very high signal-to-noise ratio, a device consisting of two Au-SiO_2-Au junctions can be used. The pulse irradiating the second junction has a CEP shift of $\pi/2$ after transmission through a dispersive beamsplitter, leading to an identical shift of the current from this junction (the corresponding simulation result is shown in blue). The sign (positive or negative) of the collected charge from the second junction $Q_P^{(2)}$ can be used to determine if φ_{CE} is in the interval 0–π or π–2π, and the measured charge value from the first junction, $Q_P^{(1)}$, can then be used to read out the actual value of φ_{CE}. Thus, the real-time electric field of the incident laser pulse can be reconstructed.

Methods

Source of ultrashort laser pulses. Linearly polarized laser pulses in the vis/NIR, with ~400 μJ pulse energy and controllable CEP, were generated at a repetition rate of 3 kHz with a customized Ti:sapphire chirped-pulse amplifier system[23]. The ultra-broadband spectrum of the laser pulses ranged from 500 to 1,050 nm, yielding a sub-4 fs pulse duration (full-width at half-maximum of the temporal intensity profile). This corresponds to an ~1.5 cycle pulse at a carrier wavelength of ~0.75 μm. The laser pulses could either be focused onto the solid-state system or into the gas cell of the stereo-ATI device, reaching a cycle averaged peak intensity of up to 1×10^{14} W cm^{-2}.

Photoactive metal–dielectric circuit. The solid-state device on which the experiments on ultrafast current in the solid state were performed comprised a nanoscale metal–dielectric–metal junction. These structures were fabricated by cleaving amorphous SiO_2 (Crystec GmbH) and coating the adjacent surfaces of the atomically sharp cleaved silica edge with ~50 nm of Au evaporated at grazing incidence. Using this procedure, a pair of regular straight Au electrodes, isolated from each other by a SiO_2 nanotrench (width on the order of ~50 nm), was obtained. This junction was coated with an additional sputtered SiO_2 nanofilm (~200 nm thick) to embed the Au electrodes in a homogeneous silica matrix. During the experiment, no bias voltage was applied to the electrodes.

Detection scheme for optically induced CEP-dependent electrical signals. The experimental data on optically induced ultrafast currents inside a solid-state device presented in this work are the CEP-dependent component of the optically generated electric signal in the nanojunction. This CEP-dependent component was filtered from CEP-independent contributions of the electric signal (for example, electric currents caused across the junction by CEP-independent photoionization of Au due to imbalanced irradiation of the pair of electrodes), which are up to two orders of magnitude bigger, by applying a lock-in technique. For this purpose the pulse train of the chirped-pulse amplifier system was synchronized such that two consecutive laser pulses had a CEP change of π, resulting in a modulation of the CEP at half the repetition rate of the laser pulse train. The optically induced electric signal in the solid-state device was amplified using a high-gain I/V converter with a bandwidth that supports the laser pulse train. With a lock-in amplifier referenced to the modulation frequency of the CEP modulation, the CEP-dependent component Q_P of the electric signal was extracted. The signal-to-noise ratio of the lock-in detection of Q_P was about 350.

Stereo-ATI device for detecting absolute CEP. The laser pulses were focused into a gas cell filled with Xe at a pressure of $\sim 3.5 \times 10^3$ mbar, in which the strong laser field leads to above-threshold ionization of the Xe atoms and subsequent acceleration of the released electrons in the laser field. A CEP-dependent signature was imprinted on the kinetic energy distribution of the electrons accelerated in the opposite direction along the laser polarization. Those energy distributions were recorded by detecting the electrons' flight time through a 155-mm-long vacuum tube (pressure, $\sim 2 \times 10^{-6}$ mbar) on both sides of the gas cell by using a MCP at the end of each vacuum tube and a photodiode behind the gas cell to detect the time of the laser pulse incidence. Thus, the absolute CEP of the incoming laser pulses could be reconstructed once the absolute dependence of the photoelectron spectrum on the CEP was determined. This was done by numerically solving the TDSE in the single active electron approximation[28,29] using a model atomic potential known to accurately reproduce the energy structure and rescattering behaviour in strong-field photoionization[30], and averaging the resulting photoelectron spectra over intensity. This map of photoelectron energy versus CEP was then related to the experimental data, as described in the main text.

Details of the quantum-mechanical simulation. Two models were used. The first is the same as that described in ref. 24. The second, described in ref. 25, used an effective one-dimensional periodic potential

$$U(z) = \sum_{n=-\infty}^{+\infty} c_1 \{\tanh^2[c_2(z + na)] - 1\}$$

with parameters that approximate the properties of α-quartz along the [0001] direction (lattice constant $a = 5.4$ Å, $c_1 = 54.86$ eV and $c_2 = 1.553$ Å^{-1}). The solution of the stationary problem yields a bandgap of $E_g \approx 9$ eV and an effective electron mass in the lowest conduction band of $m_c \approx 0.4\, m_0$. Parameters c_1 and c_2 were adjusted to obtain the best agreement between simulations and experimental data. Evaluating the transferred charge Q_P, we averaged over the intensity distribution in the laser spot, $Q_P \propto \int_0^\infty q_P\,[F_0(x)]\,dx$. Here, $q_P[F_0]$ is the charge density for a given peak amplitude and we assumed a Gaussian beam profile with a centre at $x = 0$ and a beam waist of 50 μm. Our simulations indicate that the observed relative phase shift in the dependence of $Q_P(\varphi_{CE})$ on the laser intensity (Fig. 2d) is related to the interplay between two competing forces acting on electrons: the field of the laser pulse and the screening field induced by the macroscopic polarization of the dielectric.

Received 17 July 2013; accepted 20 November 2013; published online 12 January 2014

References

1. Krausz, F. & Ivanov, M. Attosecond physics. *Rev. Mod. Phys.* **81,** 163–234 (2009).
2. Paulus, G. G. *et al.* Absolute-phase phenomena in photoionization with few-cycle laser pulses. *Nature* **414,** 182–184 (2001).
3. Paulus, G. G. *et al.* Measurement of the phase of few-cycle laser pulses. *Phys. Rev. Lett.* **91,** 253004 (2003).
4. Wittmann, T. *et al.* Single-shot carrier-envelope phase measurement of few-cycle laser pulses. *Nature Phys.* **5,** 357–362 (2009).
5. Chipperfield, L. E., Robinson, J. S., Knight, P. L., Marangos, J. P. & Tisch, J. W. G. The generation and utilisation of half-cycle cut-offs in high harmonic spectra. *Laser Photon. Rev.* **4,** 697–719 (2010).
6. Goulielmakis, E. *et al.* Single-cycle nonlinear optics. *Science* **320,** 1614–1617 (2008).

LETTERS NATURE PHOTONICS DOI: 10.1038/NPHOTON.2013.348

7. Bergues, B. *et al.* Attosecond tracing of correlated electron-emission in non-sequential double ionization. *Nature Commun.* **3,** 813 (2012).
8. Dietrich, P., Krausz, F. & Corkum, P. B. Determining the absolute carrier phase of a few-cycle laser pulse. *Opt. Lett.* **25,** 16–18 (2000).
9. Kling, M. F. *et al.* Control of electron localization in molecular dissociation. *Science* **312,** 246–248 (2006).
10. Borot, A. *et al.* Attosecond control of collective electron motion in plasmas. *Nature Phys.* **8,** 416–421 (2012).
11. Apolonski, A. *et al.* Observation of light-phase-sensitive photoemission from a metal. *Phys. Rev. Lett.* **92,** 073902 (2004).
12. Krüger, M., Schenk, M. & Hommelhoff, P. Attosecond control of electrons emitted from a nanoscale metal tip. *Nature* **475,** 78–81 (2011).
13. Zherebtsov, S. *et al.* Controlled near-field enhanced electron acceleration from dielectric nanospheres with intense few-cycle laser fields. *Nature Phys.* **7,** 656–662 (2011).
14. Kienberger, R. *et al.* Atomic transient recorder. *Nature* **427,** 817–821 (2004).
15. Trebino, R. *et al.* Measuring ultrashort laser pulses in the time-frequency domain using frequency-resolved optical gating. *Rev. Sci. Instrum.* **68,** 3277–3295 (1997).
16. Iaconis, C. & Walmsley, I. A. Spectral phase interferometry for direct electric-field reconstruction of ultrashort optical pulses. *Opt. Lett.* **23,** 792–794 (1998).
17. Witting, T. *et al.* Sub-4-fs laser pulse characterization by spatially resolved spectral shearing interferometry and attosecond streaking. *J. Phys. B* **45,** 074014 (2012).
18. Holzwarth, R. *et al.* Optical frequency synthesizer for precision spectroscopy. *Phys. Rev. Lett.* **85,** 2264–2267 (2000).
19. Reichert, J., Holzwarth, R., Udem, T. & Hänsch, T. W. Measuring the frequency of light with mode-locked lasers. *Opt. Commun.* **172,** 59–68 (1999).
20. Telle, H. R. *et al.* Carrier-envelope offset phase control: a novel concept for absolute optical frequency measurement and ultrashort pulse generation. *Appl. Phys. B* **69,** 327–332 (1999).
21. Xu, L. *et al.* Route to phase control of ultrashort light pulses. *Opt. Lett.* **21,** 2008–2010 (1996).
22. Schiffrin, A. *et al.* Optical-field-induced current in dielectrics. *Nature* **493,** 70–74 (2013).
23. Cavalieri, A. L. *et al.* Intense 1.5-cycle near infrared laser waveforms and their use for the generation of ultra-broadband soft-X-ray harmonic continua. *New J. Phys.* **9,** 242 (2007).
24. Apalkov, V. & Stockman, M. I. Theory of dielectric nanofilms in strong ultrafast optical fields. *Phys. Rev. B* **86,** 165118 (2012).
25. Kruchinin, S. Y., Korbman, M. & Yakovlev, V. S. Theory of strong-field injection and control of photocurrent in dielectrics and wide band gap semiconductors. *Phys. Rev. B* **87,** 115201 (2013).
26. Kübel, M. *et al.* Carrier-envelope-phase tagging in measurements with long acquisition times. *New J. Phys.* **14,** 093027 (2012).
27. Li, C. *et al.* Determining the phase-energy coupling coefficient in carrier-envelope phase measurements. *Opt. Lett.* **32,** 796–798 (2007).
28. Karpowicz, N. & Zhang, X.-C. Coherent terahertz echo of tunnel ionization in gases. *Phys. Rev. Lett.* **102,** 093001 (2009).
29. Muller, H. G. An efficient propagation scheme for the time-dependent Schrödinger equation in the velocity gauge. *Laser Phys.* **9,** 138–148 (1999).
30. Bian, X. B. *et al.* Subcycle interference dynamics of time-resolved photoelectron holography with midinfrared laser pulses. *Phys. Rev. A* **84,** 043420 (2011).

Acknowledgements

The authors thank Y. Deng for technical support and fruitful discussions as well as the Munich-Centre for Advanced Photonics for financial support. A.S. acknowledges the Alexander von Humboldt Foundation and the Swiss National Science Foundation. N.K. acknowledges the Alexander von Humboldt Foundation. Ö.S. acknowledges a Marie Curie International Incoming Fellowship (project NANOULOP, no. 302157). R.K. acknowledges an ERC starting grant.

Author contributions

T.P.-C., A.S., R.K., R.E. and F.K. conceived and supervised the experiments. A.S., T.P.-C., D.G., O.R., S.M., J.R. and J.V.B. participated in sample design and fabrication. T.P.-C., A.S., N.K., A.A., Sa.K. and Ö.S. performed the measurements. St.K., V.A., M.I.S. and V.S.Y. provided the theoretical description and numerical modelling of the solid-state device for phase detection. T.P.-C. and N.K. performed numerical simulations and analysis of the stereo-ATI measurements. T.W. provided the stereo-ATI phasemeter. T.P.-C., A.S., N.K., St.K., M.K., T.W., V.S.Y., R.K., R.E. and F.K. analysed and interpreted the experimental data. All authors discussed the results and contributed to the final manuscript.

Additional information

Reprints and permissions information is available online at www.nature.com/reprints. Correspondence and requests for materials should be addressed to T.P.C. and F.K.

Competing financial interests

The authors declare no competing financial interests.

nature photonics LETTERS
PUBLISHED ONLINE: 18 JANUARY 2016 | DOI: 10.1038/NPHOTON.2015.269

Electro-optic sampling of near-infrared waveforms

Sabine Keiber[1,2], Shawn Sederberg[1], Alexander Schwarz[1], Michael Trubetskov[1,3], Volodymyr Pervak[1,2], Ferenc Krausz[1,2] and Nicholas Karpowicz[1*]

Access to the complete electric field evolution of a laser pulse is essential for attosecond science in general[1], and for the scrutiny and control of electron phenomena in solid-state physics specifically[2–6]. Time-resolved field measurements are routine in the terahertz spectral range, using electro-optic sampling (EOS)[7–9], photoconductive switches[10,11] and field-induced second harmonic generation[12,13]. EOS in particular features outstanding sensitivity and ease of use, making it the basis of time-resolved spectroscopic measurements[14] for studying charge carrier dynamics[15–20] and active optical devices[21]. In this Letter, we show that careful optical filtering allows the bandwidth of this technique to be extended to wavelengths as short as 1.2 μm (230 THz) with half-cycle durations 2.3 times shorter than the sampling pulse. In a proof-of-principle application, we measure the influence of optical parametric amplification (OPA) on the electric field dynamics of a few-cycle near-infrared (NIR) pulse.

As the sampling pulse must usually be shorter than a half-cycle of the field oscillation to be observed, EOS, in which the electric field to be measured induces a polarization change in the sampling pulse, has been constrained to frequencies of 135 THz (2.2 μm) and below[9]. The direct recording of higher-frequency field oscillations has only recently been achieved using the attosecond streaking[22,23] and petahertz optical oscilloscope[24] techniques. These methods exploit the extreme nonlinearities underlying high harmonic generation in gases to induce strong temporal confinement, extending the temporal resolution of few-femtosecond laser pulses to the attosecond timescale. They rely on the generation and detection of extreme ultraviolet radiation, and thus require dedicated vacuum systems, reducing their flexibility and ease of integration with other experiments. An alternative approach is to combine techniques for the characterization of spectral amplitude and phase with carrier envelope phase (CEP) sensitive measurements to extract the complete field[25]. However, simple and direct access to NIR fields is unavailable.

The experimental setup used to demonstrate that EOS can directly provide unambiguous and complete field characterization of a NIR waveform is shown in Fig. 1, and described in detail in Methods. A broadband, phase-stable infrared pulse is generated through difference-frequency generation[26] (DFG), amplified in an optical parametric chirped pulse amplifier (OPCPA) and recombined with a 5-fs sampling pulse with adjustable time delay, after which they are co-focused in a beta barium borate (BBO) crystal. The infrared electric field is sampled via the polarization rotation it induces in the sampling pulse, producing the waveform shown in Fig. 1b. The accuracy of the waveform acquisition is verified by comparing the spectrum obtained by taking the squared modulus of the Fourier transformation of the sampled electric field with one measured independently by a calibrated InGaAs-based grating spectrometer (Ocean Optics NIRQuest 512). The excellent agreement between the spectra measured in the two methods suggests that the electric field is recorded without significant distortions, due to the association between a flat spectral amplitude and flat spectral phase in the response of the process (see Methods for details).

At first glance, this agreement is surprising, given that our 5-fs pulse duration would be expected to introduce a roll-off at approximately 100 THz, with rapidly decreasing sensitivity at higher frequencies as the polarization is rotated in opposite directions by adjacent half-cycles within the pulse, leading to a strongly diminished signal. This somewhat unexpected performance originates in the fact that, as depicted in the inset of Fig. 1a, the mechanism of the polarization rotation in high-frequency (broadband) EOS differs from the standard picture of the quasi-static Pockels effect, where the sampling pulse should be significantly shorter than a half-period of the oscillation to be observed. When sampling a high-frequency waveform, a perpendicularly polarized nonlinear signal is generated that—compared with the sampling pulse—is shifted to higher or lower frequencies in the case of phase-matching of the sum frequency generation (SFG) or DFG, respectively. To produce a polarization rotation, this nonlinearly generated signal should be degenerate with another reference wavelength present in the spectrum of the sampling pulse. We detect SFG in our study, as DFG would have considerable spectral overlap with the pulse we wish to sample.

The detection process requires two steps: first, a sum frequency between the infrared and sampling waves must be generated, and the new wave must then interfere with a local oscillator within the pre-existing sampling spectrum. Accordingly, the frequency response $S(\Omega)$ of EOS is given by

$$S(\Omega)= \int_{\Omega}^{\infty} d\omega_{LO} R(\omega_{LO})|E(\omega_{LO})|\,|E(\omega_{LO}-\Omega)|e^{i[\phi(\omega_{LO})-\phi(\omega_{LO}-\Omega)]} \quad (1)$$

integrating over local oscillator frequencies ω_{LO}, where $|E(\omega)|e^{i\varphi(\omega)}$ represents the complex spectral amplitudes of the sampling pulse, Ω is the frequency of the wave to be sampled and $R(\omega)$ is the response of the detection system.

The shortest wavelengths in the sampling spectrum function as local oscillators for the largest number of detected frequencies. Therefore, emphasizing these components of the pulse favours high-frequency detection. The spectrum of our sampling pulse is shown in Fig. 2—the wavelength region from 600–500 nm has been depleted by DFG, but nonetheless contains non-zero energy. The sampling pulse is transmitted through a short-pass filter with 600 nm cut-off wavelength before the polarization is detected, which firstly improves the signal-to-noise ratio[27] but also, as we will show, dramatically increases the detection bandwidth.

Applying equation (1) to the measured complex spectrum of the sampling pulse, it is found that when $R(\omega) = 1$ the response is substantially reduced above 100 THz, shown in Fig. 2b. However, if

[1]Max-Planck-Institut für Quantenoptik, Hans-Kopfermann-Strasse 1, D-85748 Garching, Germany. [2]Department für Physik, Ludwig-Maximilians-Universität, Am Coulombwall 1, D-85748 Garching, Germany. [3]Research Computing Center, Moscow State University, Leninskie Gory, 119992 Moscow, Russia.
*e-mail: nicholas.karpowicz@mpq.mpg.de

LETTERS NATURE PHOTONICS DOI: 10.1038/NPHOTON.2015.269

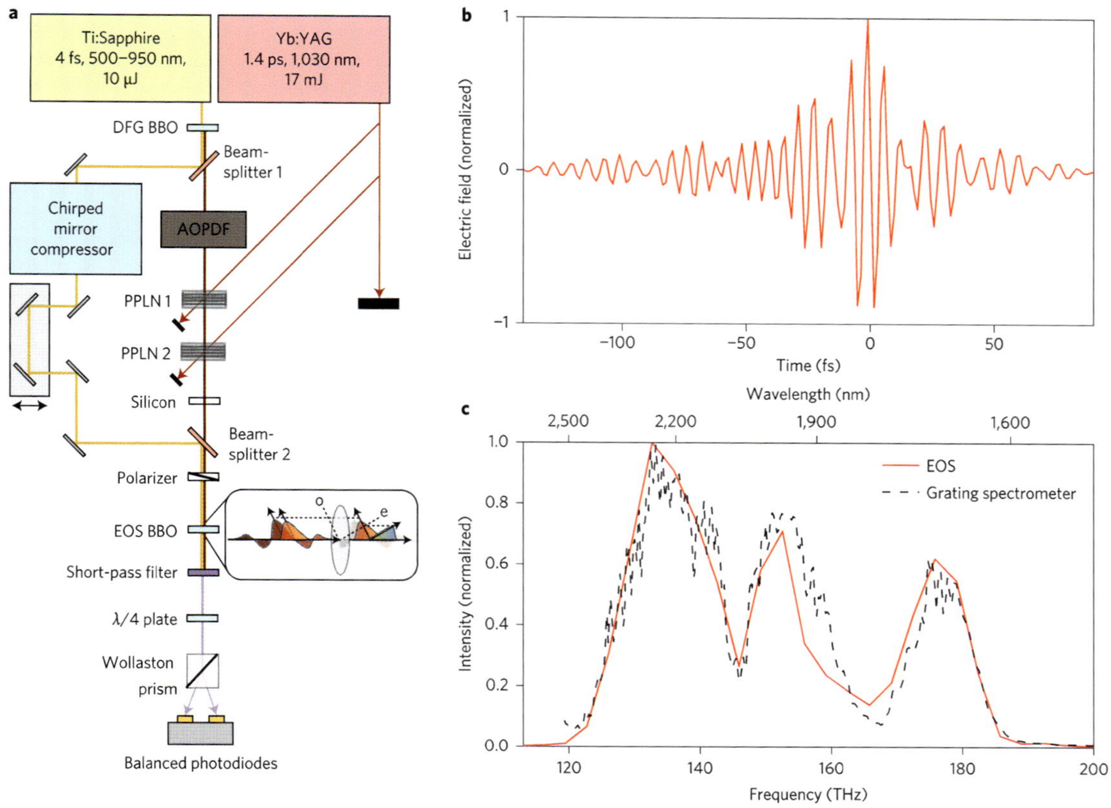

Figure 1 | System layout and representative results. a, The OPCPA and detection scheme, utilizing two optically synchronized laser systems operating at 3 kHz, seeded by the same Ti:Sapphire oscillator. PPLN: periodically poled lithium niobate. Inset, Illustration of the type-I EOS process employed, where the infrared and visible pulses enter the crystal on the ordinary (o) axis and produce a sum-frequency pulse on the extraordinary (e) axis, which has spectral and temporal overlap with the residual visible light on the o axis. **b**, Recorded electric field versus time delay measured via the polarization rotation detected by the Wollaston prism and balanced photodiodes. **c**, Comparison of the Fourier transform of the waveform in (**b**) and the spectrum measured by a calibrated InGaAs-based grating spectrometer.

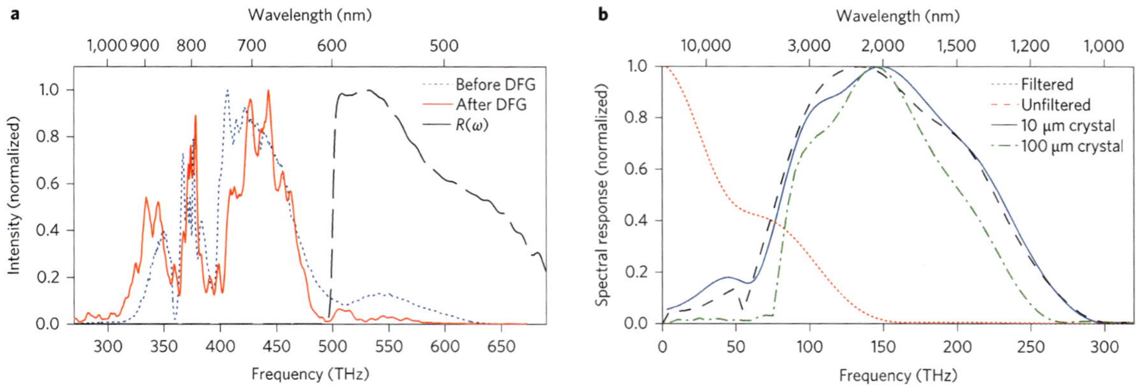

Figure 2 | Sampling pulse spectrum and calculated EOS response functions. a, The sampling pulse undergoes a significant red-shift during the DFG process, creating a depleted region in the spectral range where the detection system is active, with measured response function $R(\omega)$. **b**, The effect of spectral filtering on the spectral response function of EOS using the measured sampling pulse. The 'filtered' curve corresponds to the system employed, including the short-pass filter depicted in (**a**), while the 'unfiltered' curve assumes spectrally uniform detection. Calculations including nonlinear propagation and dispersion in the crystal show similar responses for both 10 and 100 μm crystal thicknesses when the same spectral filter is employed.

the response is recalculated with spectral weights determined by the transmission of the short-pass filter and the response of the silicon photodiodes in the balanced detector, there is a dramatic shift towards higher frequencies, with a peak at 150 THz and FWHM bandwidth over 150 THz. Numerically propagating the infrared and sampling pulses with the nonlinear wave equation through the birefringent crystal shows that this broad bandwidth phase-matches well in BBO, with the spectral width only slightly

NATURE PHOTONICS DOI: 10.1038/NPHOTON.2015.269 LETTERS

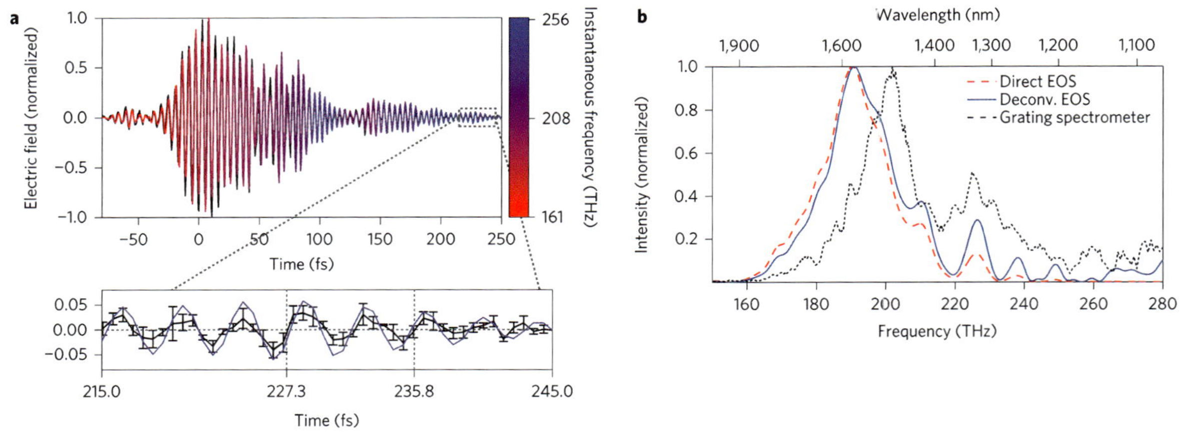

Figure 3 | Sampling of high-frequency NIR light. a, Waveform measured with AOPDF removed and DFG phase-matching set to short wavelengths, chirped in silicon wafer. The colour of the line portrays the instantaneous frequency of the wave. The black line shows the raw data without filtered deconvolution. Inset, Zoomed-in view of high-frequency oscillations at the trailing edge of pulse, with zero crossings two periods apart marked, exhibiting 4.25 fs period corresponding to 1.27 μm, or 235 THz. Error bars indicate the standard error of the mean of four consecutive measurements. **b**, Spectrum measured with EOS compared with that measured with InGaAs-based grating spectrometer, both with and without correction for calculated spectral response.

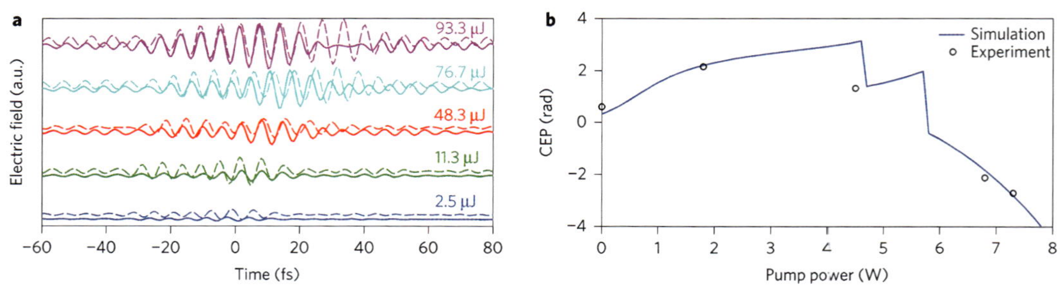

Figure 4 | OPCPA pulse amplification dynamics measured with EOS. a, Waveforms corresponding to (from top to bottom) pump powers of 7.3, 6.8, 4.5, 1.8 and 0 W in second OPCPA stage, corresponding to amplified pulse energies of 93.3, 76.6, 48.3, 11.3 and 2.5 μJ, respectively. Dashed lines are simulated waveforms. Waveforms are offset for clarity. **b**, CEP change versus pump power: black circles are measured values, blue line is simulated relation. Jumps in the simulated curve are due to pulse envelope reshaping causing a change in which oscillation peak has the greatest amplitude.

reduced in a 100-μm-thick crystal, as plotted in Fig. 2b. Therefore, this scheme allows for unattenuated detection of the whole OPA bandwidth and significantly extends the cut-off frequency. We note that this is only the case when the signal contributions of all local oscillators add coherently, requiring that the spectral components in the pulse involved in the detection process, the local oscillators themselves and the lower-frequency mixing inputs, have minimal group delay dispersion.

We tested the spectral limits of EOS with 5-fs sampling pulses by tuning the phase-matching of the DFG BBO crystal to generate a phase-stable NIR continuum favouring shorter wavelengths, and bypassing the OPA due to its limited transmission bandwidth. A positive chirp was applied to the pulse such that its spectral components were clearly separated in time, with faster oscillations appearing at the trailing edge. Owing to the large dynamic range provided by the filtered and balanced detection scheme (on the order of 10^5 in the case of the amplified pulse), this unamplified and temporally stretched field with energies on the tens of nanojoules scale is nonetheless easily detectable, as shown in Fig. 3. Characterization of the sampling pulse spectral phase by cross-correlation frequency-resolved optical gating (XFROG, see Methods), $\phi(\omega)$, enabled filtered deconvolution from the EOS response function. Cycles with 4.25 fs period are clearly visible in both the deconvolved and raw measurement, as marked in the zoomed-in area, corresponding to a half-cycle duration 2.3 times shorter than the sampling pulse duration. Detected wavelengths reach 1.2 μm, visible on a linear scale after Fourier transformation directly performed on the raw data, exhibiting good agreement with the spectrum measured with the grating spectrometer after deconvolution.

The ability to directly record NIR waveforms without the involvement of XUV radiation and complex vacuum apparatus presents numerous opportunities in both fundamental science and applied optical research. To demonstrate the ability of the technique to observe technologically relevant and complicated phenomena, we apply it to the OPA process itself, as pulses are amplified by nearly two orders of magnitude, from 2.5 to 93 μJ. In this scenario, the large dynamic range and absolute field information enable the study of the influence of the amplification process on the CEP. In Fig. 4, the significant changes in the amplified waveform indicate the complexity of the cascaded nonlinear optical process, especially as it saturates (at a pump power of 7.3 W in our case). A temporal shift and reshaping of the waveform are seen as it is amplified. This is due to a combination of the OPA phase[28] and the thermo-optic effect inside of the lithium niobate crystal. Most importantly for applications where CEP stability is paramount, the CEP changes significantly with pump power, over a 5-radian span as confirmed by simulations of nonlinear pulse propagation inside the crystal and subsequent linear propagation and focusing of the pulse (described in further detail in the Methods). This strong coupling between phase and pump power in a single OPA stage emphasizes the necessity of extreme pump laser stability for phase-dependent experiments relying on multi-stage OPA systems. The non-monotonic CEP dynamics present a possible means of CEP management, by setting the phase shift of multiple amplification stages such that positive and negative slopes of phase versus power nullify the net phase shift from fluctuations in pump power.

In summary, EOS is a powerful technique in the NIR, giving access to the temporal evolution of electric fields covering the entire telecommunications band with a single, coherent measurement in free-space, without retrieval algorithm or complex data processing. Access to the exact waveform, including spectral amplitude, spectral phase and absolute CEP, of few-cycle mid-to-near-infrared lasers will be an essential tool for the next generation of attosecond and strong-field physics measurements making use of the rising availability of reliable sources in this spectral range by allowing for direct comparison of the applied electric field in experiment and theory.

Methods

Methods and any associated references are available in the online version of the paper.

Received 15 August 2015; accepted 9 December 2015; published online 18 January 2016

References

1. Krausz, F. & Ivanov, M. Attosecond physics. *Rev. Mod. Phys.* **81,** 163–234 (2009).
2. Hohenleutner, M. *et al.* Real-time observation of interfering crystal electrons in high-harmonic generation. *Nature* **572,** 572–575 (2015).
3. Schultze, M. *et al.* Controlling dielectrics with the electric field of light. *Nature* **493,** 75–78 (2013).
4. Kuehn, W. *et al.* Coherent ballistic motion of electrons in a periodic potential. *Phys. Rev. Lett.* **104,** 146602 (2010).
5. Eisele, M. *et al.* Ultrafast multi-terahertz nano-spectroscopy with sub-cycle temporal resolution. *Nature Photon.* **8,** 841–845 (2014).
6. Schubert, O. *et al.* Sub-cycle control of terahertz high-harmonic generation by dynamical Bloch oscillations. *Nature Photon.* **8,** 119–123 (2014).
7. Wu, Q. & Zhang, X.-C. Free-space electro-optic sampling of terahertz beams. *Appl. Phys. Lett.* **67,** 3523–3525 (1995).
8. Leitenstorfer, A., Hunsche, S., Shah, J., Nuss, M. C. & Knox, W. H. Detectors and sources for ultrabroadband electro-optic sampling: experiment and theory. *Appl. Phys. Lett.* **74,** 1516–1518 (1999).
9. Sell, A., Scheu, R., Leitenstorfer, A. & Huber, R. Field-resolved detection of phase-locked infrared transients from a compact Er:fiber system tunable between 55 and 107 THz. *Appl. Phys. Lett.* **93,** 251107 (2008).
10. Auston, D. H., Cheung, K. P. & Smith, P. R. Picosecond photoconducting Hertzian dipoles. *Appl. Phys. Lett.* **45,** 284–286 (1984).
11. Ashida, M. Ultra-broadband terahertz wave detection using photoconductive antenna. *Japan. J. Appl. Phys.* **47,** 8221–8225 (2008).
12. Dai, J., Xie, X. & Zhang, X.-C. Detection of broadband terahertz waves with a laser-induced plasma in gases. *Phys. Rev. Lett.* **97,** 103903 (2006).
13. Karpowicz, N. *et al.* Coherent heterodyne time-domain spectrometry covering the entire 'terahertz gap'. *Appl. Phys. Lett.* **92,** 011131 (2008).
14. Pashkin, A. *et al.* Femtosecond response of quasiparticles and phonons in superconducting $YBa_2Cu_3O_7$ studied by wideband terahertz spectroscopy. *Phys. Rev. Lett.* **105,** 067001 (2010).
15. Waschke, C. *et al.* Coherent submillimeter-wave emission from Bloch oscillations in a semiconductor superlattice. *Phys. Rev. Lett.* **70,** 3319–3322 (1993).
16. Gaal, P. *et al.* Internal motions of a quasiparticle governing its ultrafast nonlinear response. *Nature* **450,** 1210–1213 (2007).
17. Kübler, C. *et al.* Coherent structural dynamics and electronic correlations during an ultrafast insulator-to-metal phase transition in VO_2. *Phys. Rev. Lett.* **99,** 116401 (2007).
18. Somma, C., Reimann, K., Flytzanis, C., Elsaesser, T. & Woerner, M. High-field terahertz bulk photovoltaic effect in lithium niobate. *Phys. Rev. Lett.* **112,** 146602 (2014).
19. Pashkin, A., Sell, A., Kampfrath, T. & Huber, R. Electric and magnetic terahertz nonlinearities resolved on the sub-cycle scale. *New J. Phys.* **15,** 065003 (2013).
20. Huber, R. *et al.* How many-particle interactions develop after ultrafast excitation of an electron-hole plasma. *Nature* **414,** 286–289 (2001).
21. Kröll, J. *et al.* Phase-resolved measurements of stimulated emission in a laser. *Nature* **449,** 698–701 (2007).
22. Itatani, J. *et al.* Attosecond streak camera. *Phys. Rev. Lett.* **88,** 173903 (2002).
23. Goulielmakis, E. *et al.* Direct measurement of light waves. *Science* **305,** 1267–1269 (2004).
24. Kim, K. T. el al. Petahertz optical oscilloscope. *Nature Photon.* **7,** 958–962 (2013).
25. Nomura, T., Shirai, H. & Fuji, T. Frequency-resolved optical gating capable of carrier-envelope phase determination. *Nature Commun.* **4,** 2820 (2013).
26. Fattahi, H., Schwarz, A., Keiber, S. & Karpowicz, N. Efficient, octave-spanning difference-frequency generation using few-cycle pulses in simple collinear geometry. *Opt. Lett.* **38,** 4216–4218 (2013).
27. Porer, M., Ménard, J.-M. & Huber, R. Shot noise reduced terahertz detection via spectrally postfiltered electro-optic sampling. *Opt. Lett.* **39,** 2435–2438 (2014).
28. Demmler, S. *et al.* Control of nonlinear spectral phase induced by ultra-broadband optical parametric amplification. *Opt. Lett.* **37,** 3933–3935 (2012).

Acknowledgements

We acknowledge support from LASERLAB-EUROPE (grant agreement no. 284464, the European Commission's Seventh Framework Programme) and the Munich-Centre for Advanced Photonics. S.S. acknowledges financial support from the Banting Postdoctoral Fellowship program.

Author contributions

The measurement was performed by S.K., S.S. and N.K. The OPCPA was prepared by A.S., S.K., F.K. and N.K. The specialized multilayer optics were designed and fabricated by M.T. and V.P. The simulations were performed by and the experimental concept was conceived by N.K. All authors reviewed and contributed to the final manuscript.

Additional information

Supplementary information is available in the online version of the paper. Reprints and permissions information is available online at www.nature.com/reprints. Correspondence and requests for materials should be addressed to N.K.

Competing financial interests

The authors declare no competing financial interests.

NATURE PHOTONICS DOI: 10.1038/NPHOTON.2015.269 LETTERS

Methods

Optical system. The sampled infrared waveforms are derived from a two-stage OPCPA with a central wavelength of 2.1 μm, operating at 3 kHz repetition rate[29]. The seed for the OPCPA is generated by focusing a 4-fs white-light pulse into a 500 μm thick type-II BBO crystal (Castech) for intrapulse DFG. The unconverted white light is separated with a dichroic beamsplitter, recompressed with a chirped mirror compressor, and reused as the sampling pulse in the EOS setup. The final white light pulse duration is measured to be 5 fs using XFROG, with measured spectrogram and waveform shown in Supplementary Fig. 1.

The spectral phase of the passively phase-stable infrared seed pulses is shaped in an acousto-optic programmable dispersive filter (AOPDF) (DAZZLER, Fastlite) and amplified via two OPCPA stages in 5% MgO-doped periodically poled lithium niobate (PPLN) crystals with 2 and 1 mm thicknesses for the first and second stage, respectively. The stages are pumped by a thin-disk Yb:YAG regenerative amplifier[30] with 0.4 and 3 mJ pulse energy, respectively. 4 mm of silicon are used to compress the 100 μJ infrared pulses below two optical cycles (13 fs).

The white light sampling and infrared pulses are recombined collinearly with a dichroic beamsplitter, and focused into a type-I BBO crystal (Innowit) for electro-optic sampling. Typical focus sizes are 80 and 180 μm ($1/e^2$) for sampling and sampled pulse, respectively. The polarization rotation imprinted on the sampling pulse is recorded by a balanced photodiode and read out by a lock-in amplifier as a function of delay between the two pulses. Modulation of the CEP with the AOPDF at the lock-in reference frequency allows for heterodyne detection by introducing a shift of the frequency comb of the infrared light. The waveform measurements were performed with a 50-μm-thick BBO in Fig. 1, 15-μm-thick BBO in Fig. 2, and 100-μm-thick BBO in Fig. 3.

Measurement time and dynamic range. A typical acquisition with 150 and 300 laser shots averaged per data point takes about ten minutes. The dynamic range is on the order of $>10^5$ in the electric field.

Measurement of unamplified, broadband continuum. In the case of the measurement of the short-wavelength pulse (Fig. 3), the AOPDF and PPLN amplification stages were removed because of their limited transmission bandwidth and the frequency modulation the AOPDF supplied was replaced with an amplitude modulation via an optical chopper. The slight discrepancy between the spectrum retrieved from EOS and the one measured with the grating spectrometer might stem from the slightly different spectral distributions in the focus that are probed by the EOS and the collimated beam. In the amplified infrared pulse this deviation is less pronounced due to mode filtering by the OPA stages.

Thermal effects in OPCPA crystals. We confirmed a temperature increase of 20 K in the second-stage crystal with a thermal camera as the pump power was increased to 7.3 W leading to the thermo-optic effect contributing to a temporal shift in the amplified waveforms in Fig. 4a.

Simulation of amplification and detection. The simulations are based on the nonlinear wave equation in the slowly-varying amplitude approximation[31] with no assumptions regarding the shape of the pulse envelope, expressed in the frequency domain as

$$\frac{\partial E_\omega(z)}{\partial z} = -ik_\omega E_\omega(z) - \frac{i\omega}{2n(\omega)\varepsilon_o c} P_\omega^{(NL)}(z)$$

where E_ω is the complex amplitude of an individual spectral component, assumed to be a plane wave with wave-vector k_ω propagating in the z direction in a given polarization axis, and $P_\omega^{(NL)}$ is the nonlinear polarization of the crystal on that axis, which is calculated in the time domain via inverse Fourier transformation, and then transformed back to obtain its individual frequency components. The refractive index $n(\omega)$ is frequency dependent, using the dispersion of BBO[32] and $LiNbO_3$[33] in the calculations of EOS and OPCPA, respectively. The propagation scheme is a one-dimensional exponential midpoint algorithm, and considers the birefringence of the crystal, tensor nature of the second order nonlinearity, Kerr effect and thermo-optic effect in the case of $LiNbO_3$. In the OPCPA simulations, a Gaussian spatial profile was assumed for the seed and pump (with 1D simulations performed for each point along the radial coordinate), and the resulting beam propagated in free space with the assumption of radial symmetry and sampled in the centre of the focus of a 200 mm focal length parabolic mirror. The CEP is calculated as the temporal phase at the temporal position of the maximum modulus of the complex analytic waveform.

Phase response of EOS. The response function $S(\Omega)$ is complex-valued. In Supplementary Fig. 2, both the phase and modulus are plotted, considering phase-matching, for 10 and 100 μm crystal thickness. The spectral phase response shows little variation within the FWHM of the spectral amplitude response, in both cases.

References

29. Deng, Y. *et al.* Carrier-envelope-phase-stable, 1.2 mJ, 1.5 cycle laser pulses at 2.1 μm. *Opt. Lett.* **37,** 4973–4975 (2012).
30. Metzger, T. *et al.* High-repetition-rate picosecond pump laser based on a Yb:YAG disk amplifier for optical parametric amplification. *Opt. Lett.* **34,** 2123–2125 (2009).
31. Brabec, T. & Krausz, F. Nonlinear optical pulse propagation in the single-cycle regime. *Phys. Rev. Lett.* **78,** 3282–3285 (1997).
32. Zhang, D., Kong, Y. & Zhang, J. Optical parametric properties of 532-nm-pumped beta-barium-borate near the infrared absorption edge. *Opt. Commun.* **184,** 485–491 (2000).
33. Gayer, O., Sacks, Z., Galun, E. & Arie, A. Temperature and wavelength dependent refractive index equations for MgO-doped congruent and stoichiometric $LiNbO_3$. *Appl. Phys. B* **91,** 343–348 (2008).

LETTER

doi:10.1038/nature17650

Attosecond nonlinear polarization and light–matter energy transfer in solids

A. Sommer[1]*, E. M. Bothschafter[1,2]*†, S. A. Sato[3], C. Jakubeit[1], T. Latka[1], O. Razskazovskaya[1], H. Fattahi[1], M. Jobst[1], W. Schweinberger[1,2], V. Shirvanyan[1], V. S. Yakovlev[1,4], R. Kienberger[5], K. Yabana[3,6], N. Karpowicz[1], M. Schultze[1,2] & F. Krausz[1,2]

Electric-field-induced charge separation (polarization) is the most fundamental manifestation of the interaction of light with matter and a phenomenon of great technological relevance. Nonlinear optical polarization[1,2] produces coherent radiation in spectral ranges inaccessible by lasers and constitutes the key to ultimate-speed signal manipulation. Terahertz techniques[3–8] have provided experimental access to this important observable up to frequencies of several terahertz[9–13]. Here we demonstrate that attosecond metrology[14] extends the resolution to petahertz frequencies of visible light. Attosecond polarization spectroscopy allows measurement of the response of the electronic system of silica to strong (more than one volt per ångström) few-cycle optical (about 750 nanometres) fields. Our proof-of-concept study provides time-resolved insight into the attosecond nonlinear polarization and the light–matter energy transfer dynamics behind the optical Kerr effect and multi-photon absorption. Timing the nonlinear polarization relative to the driving laser electric field with sub-30-attosecond accuracy yields direct quantitative access to both the reversible and irreversible energy exchange between visible–infrared light and electrons. Quantitative determination of dissipation within a signal manipulation cycle of only a few femtoseconds duration (by measurement and *ab initio* calculation) reveals the feasibility of dielectric optical switching at clock rates above 100 terahertz. The observed sub-femtosecond rise of energy transfer from the field to the material (for a peak electric field strength exceeding 2.5 volts per ångström) in turn indicates the viability of petahertz-bandwidth metrology with a solid-state device.

Matter responds to electromagnetic radiation by a displacement of its electrons with respect to the nuclei, turning its atomic constituents into dipole antennas. The overall strength of these dipoles per unit volume is characterized by the polarization vector, $\boldsymbol{P}$. Its dependence on the incident electric field, $\boldsymbol{E}(t)$, describes the macroscopic material response. Its nonlinear component, $\boldsymbol{P}_{NL}$, constitutes the basis for manipulating the electronic and optical properties with the electric field of light[1,2]. The energy transferred from the electromagnetic field to the medium per unit volume can be expressed as:

$$W(t)=\int_{-\infty}^{t}\boldsymbol{E}(t')\cdot\frac{\mathrm{d}}{\mathrm{d}t'}\boldsymbol{P}_{NL}(t')\mathrm{d}t' \qquad (1)$$

Here we assume that the contribution of linear polarization to $W(t)$ is negligible. This is a prerequisite for ultrahigh-rate signal manipulation, which relies on low dissipation. In fact, it is this dissipation that has limited the clock rate in contemporary integrated digital electronics to several gigahertz[15] for more than a decade[16,17].

A substantial increase of the electronic processing speed requires a new paradigm that is capable of greatly reducing the dissipation per switching cycle. Recent experiments indicated a possible way of advancing contemporary microwave electronics to the frequency of visible light by manipulating the electronic and optical properties of wide-bandgap materials with strong visible light fields at photon energies much smaller than the bandgap of the material[18,19]. However, the crucial question of how the energy density deposited irreversibly per switching cycle, $W_{irreversible}$, relates to the reversible energy exchange per unit volume, $W_{reversible}$, could not be answered. Pushing the frontiers of information processing to optical frequencies requires minimizing $W_{irreversible}$ while keeping $W_{reversible}$ high enough for reliable signal processing. Insight into field–matter energy exchange at optical frequencies requires access to $W(t)$ on a sub-femtosecond scale.

To this end, we propagated a strong, linearly polarized field $E(t)$ and its strongly attenuated replica $E_{ref}(t)=\beta E(t)$ through a thin sample of a transparent wide-bandgap material, in our case fused silica, of thickness ℓ. The attenuation factor β is sufficiently small to prevent any observable nonlinear material response to $E_{ref}(t)$. Both transmitted waveforms are recorded in a measurement sequence as outlined in Fig. 1, once attenuated after and once attenuated before the sample by the same attenuation factor β. We show in Supplementary Information section 1 how a difference between these transmitted waves, $\Delta E(t)=E(\ell,t)-\beta^{-1}E_{ref}(\ell,t)$, directly yields the nonlinear polarization $P_{NL}(t)$ induced by the strong field $E(t)$; see Supplementary equation (13).

In our experiments, we focus few-cycle near-infrared waveforms carried at a wavelength of $\lambda=750$ nm into thin fused silica samples ($\ell=10\,\mu$m); for details of the experimental setup and procedures see Supplementary Information section 2. The focus of the transmitted waveform (that is, the interaction region) is imaged into an attosecond streak camera[20,21]. Here the temporal evolution of the transmitted electric fields $E(z=\ell,t)$ and $E_{ref}(z=\ell,t)$ (henceforth referred to as $E(t)$ and $E_{ref}(t)$) is sampled with sub-250-as extreme-ultraviolet pulses. The peak intensities I_{peak} of the strong and attenuated waves have been set to $(1.3\pm0.1)\times10^{14}$ W cm^{-2} and $(6.7\pm0.3)\times10^{12}$ W cm^{-2}, respectively. Figure 2a compares the transmitted fields and reveals the evolution of the nonlinear phase shift, $\Delta\varphi_{NL}(t)$, induced by the strong field. $\Delta\varphi_{NL}(t)$ increases towards the pulse peak, tapers off on its tail and finally vanishes; see insets to Fig. 2a. For $E_{peak}\approx2.6\pm0.1$ V Å^{-1}, the induced phase shift at the field maximum amounts to $\Delta\varphi_{max}=0.7\pm0.1$ rad, which translates into a change of the refractive index by $\Delta n\approx(0.9\pm0.1)\times10^{-2}$.

The field-induced phase shift evaluated at the pulse centre, $\Delta\varphi_{peak}$, is depicted in Fig. 2b and exhibits a linear scaling with the applied peak intensity. In contrast to previous research[22–24], our time-resolved study reveals the absence of saturation of the optical Kerr effect up to $E_{peak}\approx2.7$ V Å^{-1}, close to the threshold for dielectric breakdown for few-cycle laser pulses. The Kerr nonlinearity therefore appears to be

[1]Max-Planck-Institut für Quantenoptik, Hans-Kopfermann-Strasse 1, 85748 Garching, Germany. [2]Fakultät für Physik, Ludwig-Maximilians-Universität, Am Coulombwall 1, 85748 Garching, Germany. [3]Graduate School of Pure and Applied Sciences, University of Tsukuba, Tsukuba 305-8571, Japan. [4]Center for Nano-Optics and Department of Physics and Astronomy, Georgia State University, Atlanta, Georgia 30303, USA. [5]Physik-Department, Technische Universität München, James-Franck-Strasse 1, 85748 Garching, Germany. [6]Center for Computational Sciences, University of Tsukuba, Tsukuba 305-8577, Japan. †Present address: Paul Scherrer Institut, 5232 Villingen, Switzerland.
*These authors contributed equally to this work.

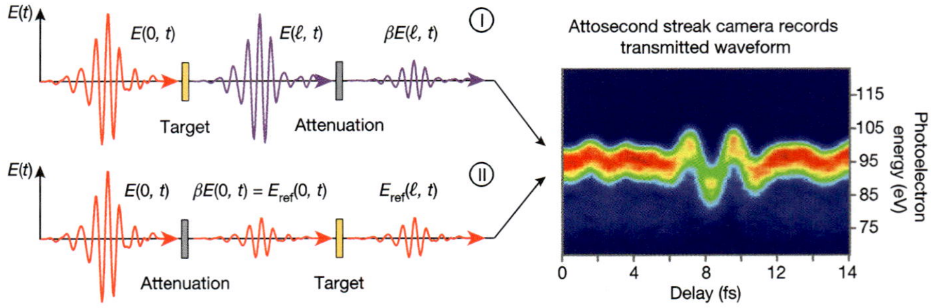

Figure 1 | Attosecond spectroscopy of the nonlinear polarization. To induce a nonlinear material response $P_{NL}(t)$, the incident strong field $E(0, t)$ is transmitted through the sample, and subsequently its amplitude is decreased by the attenuation factor β before sampling the transmitted electric field waveform in a streak camera setup (I). The nonlinear polarization response is 'deactivated' by attenuating the incident field before the sample and transmitting the weak reference field $E_{ref}(0, t) = \beta E(0, t)$ through the medium under scrutiny (II). The difference between the output waveforms, $\Delta E(t) = E(\ell, t) - \beta^{-1} E_{ref}(\ell, t)$, directly yields the nonlinear polarization of the medium, $P_{NL}(t)$, see Supplementary Information section 1. The false-colour plot shows a typical attosecond streaking spectrogram of the transmitted waveform used in the experiments.

potentially suitable for petahertz-scale signal manipulation and metrology beyond critical fields[19] $E_{crit} \approx \Delta_g/(ea)$ (where Δ_g denotes the bandgap, $e = |e|$ is the elementary charge, and a is the lattice period; for silica, $E_{crit} \approx 2\,\text{V}\,\text{Å}^{-1}$), provided that dissipation originating from carriers promoted into the conduction band during the nonlinear interaction

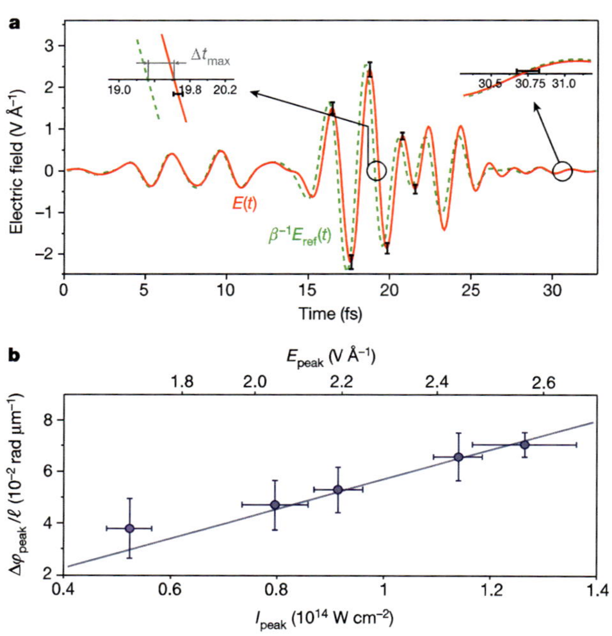

Figure 2 | Sub-femtosecond-resolved optical Kerr effect in silica. **a**, After passage through a 10-μm-thick fused silica sample, the electric field $E(t)$ of the few-cycle near-infrared pulse with a peak intensity of $1.3 \times 10^{14}\,\text{W cm}^{-2}$, approximately 10% below the threshold for optical damage, is modified as a result of the nonlinear light–matter interaction, as revealed by its comparison to a low-intensity ($I_{peak} = 7 \times 10^{12}\,\text{W cm}^{-2}$) reference waveform $E_{ref}(t)$ (for $\beta = 0.27$). This comparison yields a transient positive phase shift induced by the strong field, as anticipated from the dynamic increase of the refractive index owing to the optical Kerr effect. The two insets show close-ups of the comparison near the centre and at the end of the pulse, revealing the full reversibility of the effect. $E(t)$ and $E_{ref}(t)$ are obtained from averaging a set of three recordings performed under identical conditions on individual samples. **b**, The phase shift $\Delta\varphi_{peak}$ evaluated at the peak of the field envelope for different peak intensities I_{peak} of $E(t)$ is found to exhibit a linear dependence on the field intensity. Each data point represents the mean value of three individual recordings under identical conditions; the error bars indicate the standard deviation.

remains low. Although no lasting negative phase shift indicative of residual conduction band population is observable at the trailing edge of the waveform (where the Kerr effect vanishes), an accurate determination of the resultant $W_{irreversible}$ and the related $W_{reversible}$ requires evaluation of $P_{NL}(t)$.

The difference $\Delta E(t) = E(t) - \beta^{-1} E_{ref}(t)$ yields $P_{NL}(t)$ via Supplementary equation (13) (for details, see Supplementary Information section 1). Figure 3 depicts $P_{NL}(z = \ell/2, t)$ along with $E(z = \ell/2, t)$, both numerically propagated to the middle of the sample where their relative timing can be most precisely determined (see Supplementary Information section 1) for $E_{peak} = 2.6 \pm 0.1\,\text{V}\,\text{Å}^{-1}$. $P_{NL}(t)$ oscillates almost perfectly in phase with $E(t)$, indicating a dominant role of bound electrons. This is in strong contrast to the response of free electrons appearing in the ionization of neon atoms in the gas phase[25], exhibiting a 90° phase shift with respect to the driving field (Supplementary Information section 3). A closer inspection reveals that $P_{NL}(t)$ lags slightly behind $E(t)$ on the front edge and the peak of the pulse, indicating—according to equation (1)—energy transfer from the field to the electronic system of fused silica, both of which become of opposite sign on the trailing edge of the pulse.

The response time of the polarizing electronic system, $\tau_{response}$, can be evaluated from the central zero-crossing of the fields (Fig. 3, upper left panel) as $\tau_{response} \approx 80$ as for $E_{peak} = 2.6 \pm 0.1\,\text{V}\,\text{Å}^{-1}$. This is smaller than estimates from the Bohr orbit time[1] and from $\chi^{(3)}$ measurements in the range of 0.1–1 fs and decreases further with decreasing intensity, to well below 40 as for $E_{peak} < 2.2\,\text{V}\,\text{Å}^{-1}$, as displayed in Fig. 3b. This can be understood by connecting $\tau_{response}$ to the nonlinear (field-induced) absorption coefficient, α_{NL}. For $\tau_{response}$ much smaller than the laser period, equation (1) yields a simple linear relationship, $\alpha_{NL} \propto \tau_{response}$. For multi-photon absorption, α_{NL} scales highly nonlinearly with the intensity and, according to this relationship, so does $\tau_{response}$. Supplementary Information section 4 presents detailed modelling of the intensity scaling of $\tau_{response}$ as well as a derivation of $\alpha_{NL}(\tau_{response})$.

Rendering $P_{NL}(t)$ an experimental observable, attosecond polarization spectroscopy allows to explore the intricate dynamic exchange of energy during nonlinear light–matter interactions. Inserting the measured values of $P_{NL}(t)$ and $E(t)$ into equation (1) provides direct experimental access to the work $W(t)$ done on the electrons by the laser field per unit volume, that is, the energy density transferred from the field to the electronic system. Figure 4a plots the measured $W(t)$ for several different peak intensities and the results of time-dependent density functional theory (TD-DFT) modelling (see inset to Fig. 4, ref. 26 and Supplementary Information section 5). We find very good qualitative agreement between theory and experiment regarding all observables analysed, including the behaviour of the maximum phase shift, the change in refractive index and the evaluated amount of dissipated energy. Quantitative agreement is achieved only when the

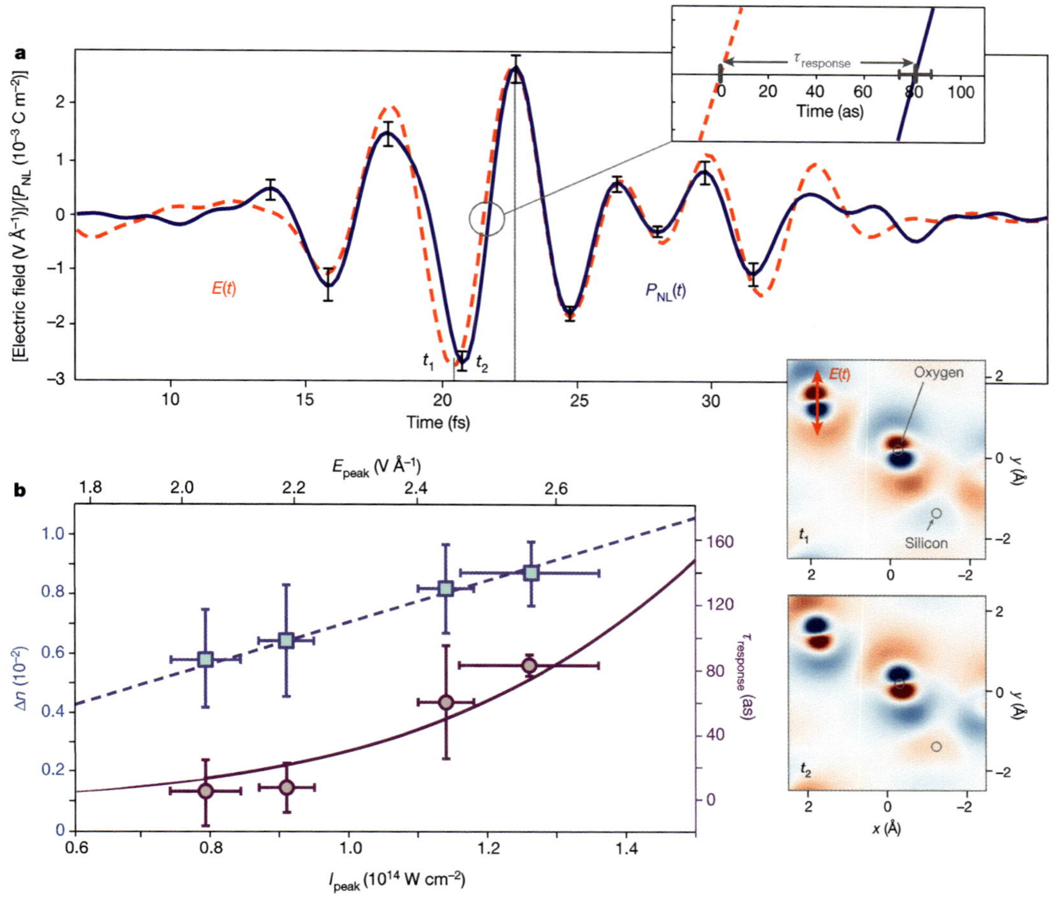

Figure 3 | The nonlinear optical polarization response of silica at critical field strengths. a, The strong electric field numerically back-propagated to the centre of the fused silica sample ($z = \ell/2 = 5\,\mu$m) is contrasted with the nonlinear polarization $P_{NL}(t)$ evaluated from $E(t)$ and $E_{ref}(t)$ (Fig. 2a) at the same position (see Supplementary Information section 1). The response time of the nonlinear polarization near optical breakdown is found to be about 100 attoseconds at the pulse peak (close-up in the top inset). The other two insets display the computed spatial rearrangement of the electron density distribution for two extrema of the electric field at instants t_1 and t_2 in false-colour representation (red indicates an increase and blue a decrease relative to the unperturbed state). Electrons located in the vicinity of the oxygen atoms appear to dominate the polarization response, whereas the electron cloud around the silicon centres remains largely unaffected. **b,** The response time of the nonlinear polarization is evaluated near the pulse peak as a function of the peak intensity of the applied field (circles) and compared to the results of a perturbation theory calculation (solid line; for details see Supplementary Information section 4). The field-induced change in refractive index Δn is evaluated from $P_{NL}(t)$ at the pulse peak as a function of the applied peak intensity I_{peak}. The nonlinear index n_2 determined from a linear regression (dashed line) is approximately one-third of the values acquired from time-integrated measurements using multi-cycle pulses[29]. All data points and error bars represent the average and the standard deviation, respectively, of the evaluation of three individual data sets recorded under identical conditions.

theoretically employed peak electric field is adjusted to values approximately 20% larger. This discrepancy can be attributed to inaccuracies of the exchange-correlation potential used in the TD-DFT calculations.

In all cases, the energy transferred from the field to the material increases up to the pulse peak and slightly beyond. This is because the field, while growing, needs to do ever more work to remove the electrons ever farther from their field-free location. The field amplitude decreasing after the pulse peak allows the displaced electrons to return gradually to their equilibrium position and radiate a part of the absorbed energy back into the driving laser field. This results in a negative slope for $W(t)$. The positive and negative slope are connected to the phase lag and phase advance of $P_{NL}(t)$ with respect to its driving field $E(t)$ before and after the pulse peak, respectively; these are clearly discernible in Fig. 3.

The energy density $W_{irreversible} = W(t \to \infty)$ irreversibly deposited in the system defines the charge carrier density promoted from the valence band into the conduction band according to $N_{carrier} \approx W_{irreversible}/\Delta_g$ (assuming population of the lowest-energy states of the conduction band). For $E_{peak} = 2.6\,\mathrm{V\,Å^{-1}}$ a residual relative carrier concentration of $N_{carrier}/N_{VB} = 2.6 \times 10^{-4}$ is found, where $N_{VB} = 1.4 \times 10^{23}\,\mathrm{cm^{-3}}$ is the density of the valence-band electrons. This small residual carrier concentration is pivotal for future ultrafast signal processing and can hardly be determined with similar sensitivity by any other experimental method. By analogy, we can define the reversibly exchanged energy density as the difference between the maximum transferred energy, W_{max}, and $W_{irreversible}$. $W_{reversible} = W_{max} - W_{irreversible}$ can be interpreted in terms of a virtual conduction-band population with a number of virtual carriers of $N_{virtual} \approx W_{reversible}/\Delta_g$. $N_{virtual}$ is the result of a projection of the laser-dressed and fully occupied valence-band states onto conduction-band states and hence it fully returns the energy density associated with it to the field upon its disappearance[27]. In contrast, the real population, $N_{carrier}$, survives the field and—upon its subsequent decay—causes dissipation.

Although $N_{virtual}$ seems rather elusive, attosecond metrology presents a very direct manifestation of the underlying reversible field–matter energy exchange. The initial energy flow dW/dt into the electronic system first extracts energy from the field on the leading edge of the

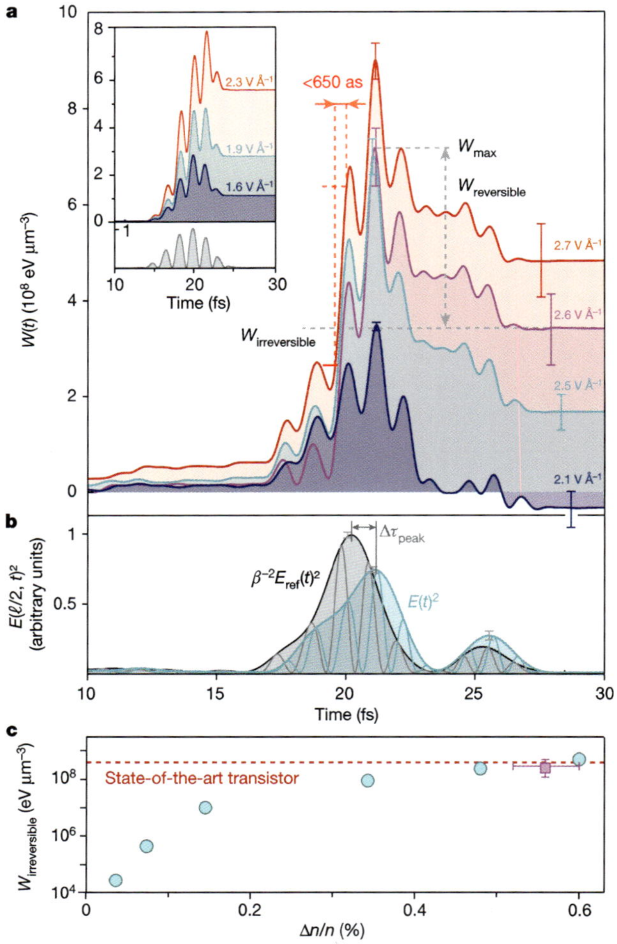

Figure 4 | Energy exchange between strong optical fields and electrons in real time. **a**, The amount of energy the few-cycle near-infrared laser field transfers into a unit volume of silica is obtained from the measured $E(t)$ and $P_{NL}(t)$ via equation (1). $W(t)$ shows signatures of a substantial transient virtual conduction-band population (which is proportional to $W_{max} - W_{irreversible}$) oscillating in synchrony with the driving electric field. In the steepest of these oscillations, energy is transferred into the material within less than 650 as at $E_{peak} = 2.7\,\mathrm{V\,Å^{-1}}$. The amount of energy irreversibly dissipated in the sample $W_{irreversible}$ depends critically on the maximum applied field strength E_{peak}. Shown are the results of recordings for three different field amplitudes with $E_{peak} = 2.5\,\mathrm{V\,Å^{-1}}$, $2.6\,\mathrm{V\,Å^{-1}}$ and $2.7\,\mathrm{V\,Å^{-1}}$, as indicated, and a measurement closest to the average of five recordings with E_{peak} set equal to $2.1\,\mathrm{V\,Å^{-1}}$ (the uncertainty in the stated values of E_{peak} is $\pm 0.1\,\mathrm{V\,Å^{-1}}$). At this field strength, $W_{irreversible}$ becomes immeasurably small (with its error exceeding its nominal value). In the inset, $W(t)$ is computed from the nonlinear polarization at $z = \ell/2 = 5\,\mu\mathrm{m}$ obtained by the TD-DFT calculations outlined in Supplementary Information section 5 for a set of three different values of the peak electric field[26], as indicated. The spectrum of the computed nonlinear polarization shows the emergence of odd harmonics of the fundamental radiation (high harmonic generation). The results shown here are computed from the low-pass filtered nonlinear polarization to mimic the frequency transfer characteristics of the optical setup employed for the experiments. **b**, The squared electric field evolution of the reference wave and its envelope (black line) in comparison to the squared field and its envelope of the wave transmitted at a peak field strength of $2.5\,\mathrm{V\,Å^{-1}}$, showing clear indications of energy redistribution and consequent reshaping of the pulse envelope caused by the nonlinear polarization (see text). **c**, The dissipated energy density, equal to the irreversibly transferred energy density shown in **a**, as a function of the relative refractive index change, as extracted from the results of the TD-DFT simulation (circles) and experimental data taken at $E_{peak} = 2.6 \pm 0.1\,\mathrm{V\,Å^{-1}}$ (square). The excellent agreement between theory and experiment verifies the simulation results, permitting reliable prediction of the relevant quantities for much lower field strengths. The dashed line marks the dissipated energy density of a state-of-the-art MOSFET; for details see Supplementary Information Fig. 16. All error bars indicate the standard deviation of the evaluation of three sets of recordings performed under identical conditions.

pulse. This is returned by a reversed energy flow on the trailing edge, resulting in a temporal shift of the pulse peak, $\Delta\tau_{peak} \approx 1$ fs, as shown in Fig. 4b. The phenomenon is widely known as self-steepening or optical shock wave formation[1]. Our study reveals that this phenomenon is an inherent consequence of the reversible field–matter energy transfer accompanying the field-induced change in the phase of the pulse. Hence, a field-induced change in the group index, Δn_g, is inextricably linked to that of the refractive index, Δn, implying a group delay and a phase shift, respectively.

Signal manipulation relies on the change of refractive index Δn (and Δn_g), which is characterized by $W_{reversible}$; in contrast, dissipation is detrimental to signal manipulation and is determined by $W_{irreversible}$. Hence, the scaling of $W_{irreversible}$ and $W_{reversible}$ (or, equivalently, Δn) with the applied field strength is of key importance for future signal-processing applications. We evaluated the dissipated energy per unit volume and per several-femtosecond optical switching/modulation cycle versus Δn from our *ab initio* TD-DFT calculations, which we verified against measurement at the highest field strength, near optical breakdown (see Fig. 4c and the discussion in Supplementary Information section 6). At measurable levels of Δn, $W_{irreversible}$ in a silica optical switch can be some four orders of magnitude smaller than the heat dissipation of a state-of-the-art metal oxide semiconductor field-effect transistor (MOSFET) operating at up to 10 GHz in integrated circuits. This very much reduced dissipation per switching cycle should therefore allow the operation of a dielectric switch/modulator at 100 THz or beyond.

An equally important discovery is the sub-femtosecond rise time of the transferred energy at $E_{peak} \geq 2.5\,\mathrm{V\,Å^{-1}}$ within each optical cycle. With slightly shorter pulses than those used in these experiments[28], more than 90% of W_{max} will be transferred within a single sub-femtosecond rise depicted in Fig. 4. The resultant buildup of carriers in the conduction band within less than 1 fs will permit sampling of electric-field waveforms beyond the petahertz frontier in the simple setting demonstrated recently[18].

Our proof-of-principle study on silica shows that careful choice of the peak electric field strength at $E < E_{crit}$ may open a route towards 100-THz-rate signal processing. The observed sub-femtosecond gradient in nonlinear energy transfer and the related change in electronic/optical properties at $E > E_{crit}$ may pave the way towards sampling optical fields (from the infrared to the ultraviolet) with a compact, cost-effective solid-state device. A petahertz solid-state oscilloscope should enable signal processing and metrology at visible light frequencies.

Traditional pump–probe spectroscopy makes use of the cycle-averaged amplitude envelope to resolve dynamics. In contrast, attosecond polarization spectroscopy uses the oscillating field as a probe, providing direct access to the full (linear and nonlinear) oscillating polarization and hence to the (reversible and irreversible) energy exchange between visible light and matter, as well as a delay in the system response. Hence, attosecond polarization spectroscopy is a generalization of pump–probe spectroscopy, yielding complete information about the dynamic electronic response of matter to strong visible light fields with attosecond resolution and, thanks to the intense attosecond field gradients, with a signal-to-noise ratio orders of magnitude better than that of any other attosecond technique demonstrated so far. Implemented with a probe waveform of sufficiently broad spectral coverage, the approach allows, in principle, complete retrieval of the nonlinear polarization and hence of the entire response of the electronic system to strong-field excitation.

Received 16 August 2015; accepted 7 March 2016.
Published online 23 May 2016.

1. Boyd, R. W. *Nonlinear Optics* (Academic Press, Elsevier, 2008).
2. Wegener, M. *Extreme Nonlinear Optics: an Introduction* (Springer, 2005).
3. Valdmanis, J. A., Mourou, G. & Gabel, C. W. Picosecond electro-optic sampling system. *Appl. Phys. Lett.* **41,** 211–212 (1982).
4. Wu, Q. & Zhang, X. C. Free-space electro-optic sampling of terahertz beams. *Appl. Phys. Lett.* **67,** 3523 (1995).
5. Sell, A., Leitenstorfer, A. & Huber, R. Phase-locked generation and field-resolved detection of widely tunable terahertz pulses with amplitudes exceeding 100 MV/cm. *Opt. Lett.* **33,** 2767–2769 (2008).
6. Hebling, J., Lo Yeh, K., Hoffmann, M. C. & Nelson, K. A. High-power THz generation, THz nonlinear optics, and THz nonlinear spectroscopy. *IEEE J. Sel. Top. Quantum Electron.* **14,** 345–353 (2008).
7. Huber, R. *et al.* Switching ultrastrong light–matter coupling on a subcycle scale. *J. Appl. Phys.* **109,** 102418 (2011).
8. Leitenstorfer, A., Nelson, K. A., Reimann, K. & Tanaka, K. Focus on nonlinear terahertz studies. *New J. Phys.* **16,** 045016 (2014).
9. Kuehn, W. *et al.* Terahertz-induced interband tunneling of electrons in GaAs. *Phys. Rev. B* **82,** 075204 (2010).
10. Junginger, F. *et al.* Nonperturbative interband response of a bulk InSb semiconductor driven off resonantly by terahertz electromagnetic few-cycle pulses. *Phys. Rev. Lett.* **109,** 147403 (2012).
11. Somma, C., Reimann, K., Flytzanis, C., Elsaesser, T. & Woerner, M. High-field terahertz bulk photovoltaic effect in lithium niobate. *Phys. Rev. Lett.* **112,** 146602 (2014).
12. Ulbricht, R., Hendry, E., Shan, J., Heinz, T. F. & Bonn, M. Carrier dynamics in semiconductors studied with time-resolved terahertz spectroscopy. *Rev. Mod. Phys.* **83,** 543–586 (2011).
13. Kampfrath, T., Tanaka, K. & Nelson, K. A. Resonant and nonresonant control over matter and light by intense terahertz transients. *Nature Photon.* **7,** 680–690 (2013).
14. Krausz, F. & Ivanov, M. Attosecond physics. *Rev. Mod. Phys.* **81,** 163–234 (2009).
15. Taur, Y. & Ning, T. H. *Fundamentals of Modern VLSI Devices* (Cambridge Univ. Press, 2009).
16. Markov, I. L. Limits on fundamental limits to computation. *Nature* **512,** 147–154 (2014).
17. Ionescu, A. M. & Riel, H. Tunnel field-effect transistors as energy-efficient electronic switches. *Nature* **479,** 329–337 (2011).
18. Schiffrin, A. *et al.* Optical-field-induced current in dielectrics. *Nature* **493,** 70–74 (2012).
19. Schultze, M. *et al.* Controlling dielectrics with the electric field of light. *Nature* **493,** 75–78 (2012).
20. Itatani, J. *et al.* Attosecond streak camera. *Phys. Rev. Lett.* **88,** 173903 (2002).
21. Kienberger, R. *et al.* Atomic transient recorder. *Nature* **427,** 817–821 (2004).
22. Pati, A. P., Wahyutama, I. S. & Pfeiffer, A. N. Subcycle-resolved probe retardation in strong-field pumped dielectrics. *Nature Commun.* **6,** 7746 (2015).
23. Loriot, V., Hertz, E., Faucher, O. & Lavorel, B. Measurement of high order Kerr refractive index of major air components. *Opt. Express* **17,** 13429–13434 (2009); erratum **18,** 3011–3012 (2010).
24. Brée, C., Demircan, A. & Steinmeyer, G. Saturation of the all-optical Kerr effect. *Phys. Rev. Lett.* **106,** 183902 (2011).
25. Geissler, M. *et al.* Light propagation in field-ionizing media: extreme nonlinear optics. *Phys. Rev. Lett.* **83,** 2930–2933 (1999).
26. Yabana, K., Sugiyama, T., Shinohara, Y., Otobe, T. & Bertsch, G. Time-dependent density functional theory for strong electromagnetic fields in crystalline solids. *Phys. Rev. B* **85,** 045134 (2012).
27. Yablonovitch, E., Heritage, J. P., Aspnes, D. E. & Yafet, Y. Virtual photoconductivity. *Phys. Rev. Lett.* **63,** 976–979 (1989).
28. Cavaleri, A. L. *et al.* Intense 1.5-cycle near infrared laser waveforms and their use for the generation of ultra-broadband soft-x-ray harmonic continua. *New J. Phys.* **9,** 242 (2007).
29. Milam, D. Review and assessment of measured values of the nonlinear refractive-index coefficient of fused silica. *Appl. Opt.* **37,** 546–550 (1998).

Supplementary Information is available in the online version of the paper.

Acknowledgements We acknowledge discussions with M. Stockman and V. Apalkov. This work was supported by the Max Planck Society and the Deutsche Forschungsgemeinschaft Cluster of Excellence: Munich Centre for Advanced Photonics (http://www.munich-photonics.de). M.S. was supported by a Marie Curie International Outgoing Fellowship (FP7-PEOPLE-2011-IOF). E.M.B. acknowledges funding from the European Community's Seventh Framework Programme (FP7/2007-2013) under grant agreement number 290605 (PSI-FELLOW/COFUND) and from the Swiss National Science Foundation through NCCR MUST. This research is based upon work supported by the US Air Force Office of Scientific Research under award number FA9550-16-1-0073 and used computational resources of the K computer provided by the RIKEN Advanced Institute for Computational Science through the HPCI System Research project (Project ID: hp140103).

Author Contributions F.K. and M.S. initiated, conceived and supervised the study. A.S. and E.M.B. developed the experimental method. A.S., E.M.B. and C.J. (in close cooperation with T.L., O.R., M.J., W.S. and V.S.) prepared and performed the experiment. S.A.S., H.F., K.Y. and N.K. accomplished the theoretical modelling. A.S., E.M.B., V.S.Y., R.K., N.K., M.S. and F.K. analysed and interpreted the experimental data. All authors discussed the results and contributed to the final manuscript.

Author Information Reprints and permissions information is available at www.nature.com/reprints. The authors declare no competing financial interests. Readers are welcome to comment on the online version of the paper. Correspondence and requests for materials should be addressed to M.S. (martin.schultze@mpq.mpg.de) or F.K. (ferenc.krausz@mpq.mpg.de).

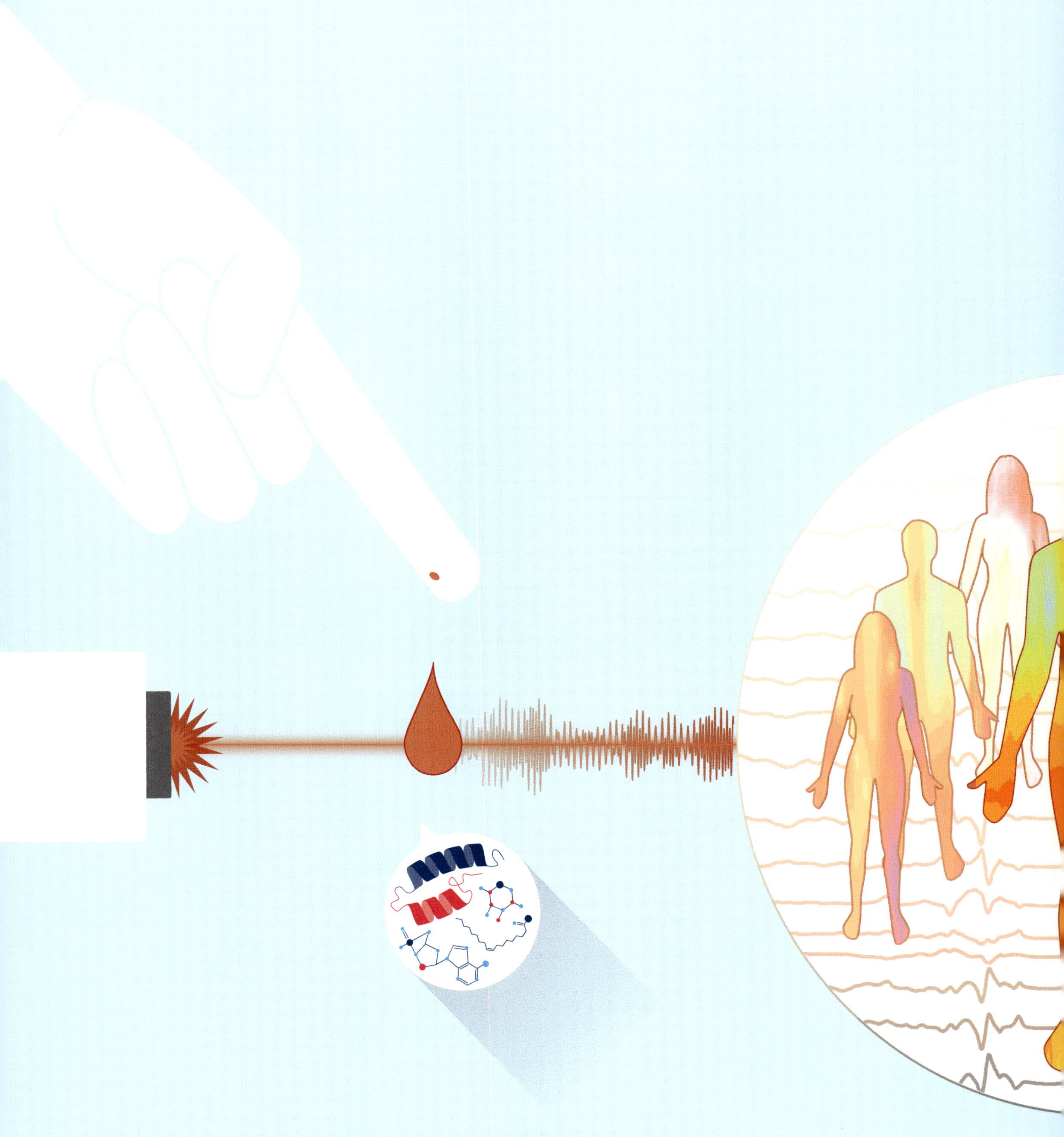

Attosecond metrology opens up a new avenue for detecting the physiological state of living organisms by sensing signals emitted by the molecules of biofluids following their sudden excitation. Sensitive, high-fidelity measurement of the attosecond-femtosecond temporal evolution of these signals permits the detection of miniscule changes in molecular composition. © Dennis Luck

From Electron Capture to Disease Detection

CO-AUTHORED BY MIHAELA ŽIGMAN

12

At the turn of the millennium, attosecond metrology rendered light field oscillations and atomic-scale electronic motions observable in real time. Its next generation, AM2.0, uses simple solid-state instrumentation to probe microscopic electronic currents. These currents govern the emission of light and the energy flow in light-matter interactions. Resolving them temporally in complex systems would provide basic insight into fundamental dynamic phenomena unfolding in atomic dimensions. This capability may also open new avenues towards detecting changes in the composition of complex (biological) matter, with the potential for real-world applications.

As one example of a possibly far-reaching impact, attosecond metrology may offer a new approach to probing molecular constituents of living organisms and thereby sensing miniscule changes in the molecular composition of biofluids, such as blood plasma/serum. Robust correlation of these changes with abnormal health states would open the door for disease detection. Here, we present a fresh perspective on how we envision attosecond metrology 2.0 to help disentangle the complexity of living matter, specifically physiological processes of a human organism and its regulation. This could ultimately serve biomedical objectives to detect any disease that leaves traces in the blood.

The abundance and structure of organic molecules to facilitate proper functioning of an organism, a system, is only allowed to vary within a narrow range. Alterations in the molecular constitution of a systemic body fluid, such as

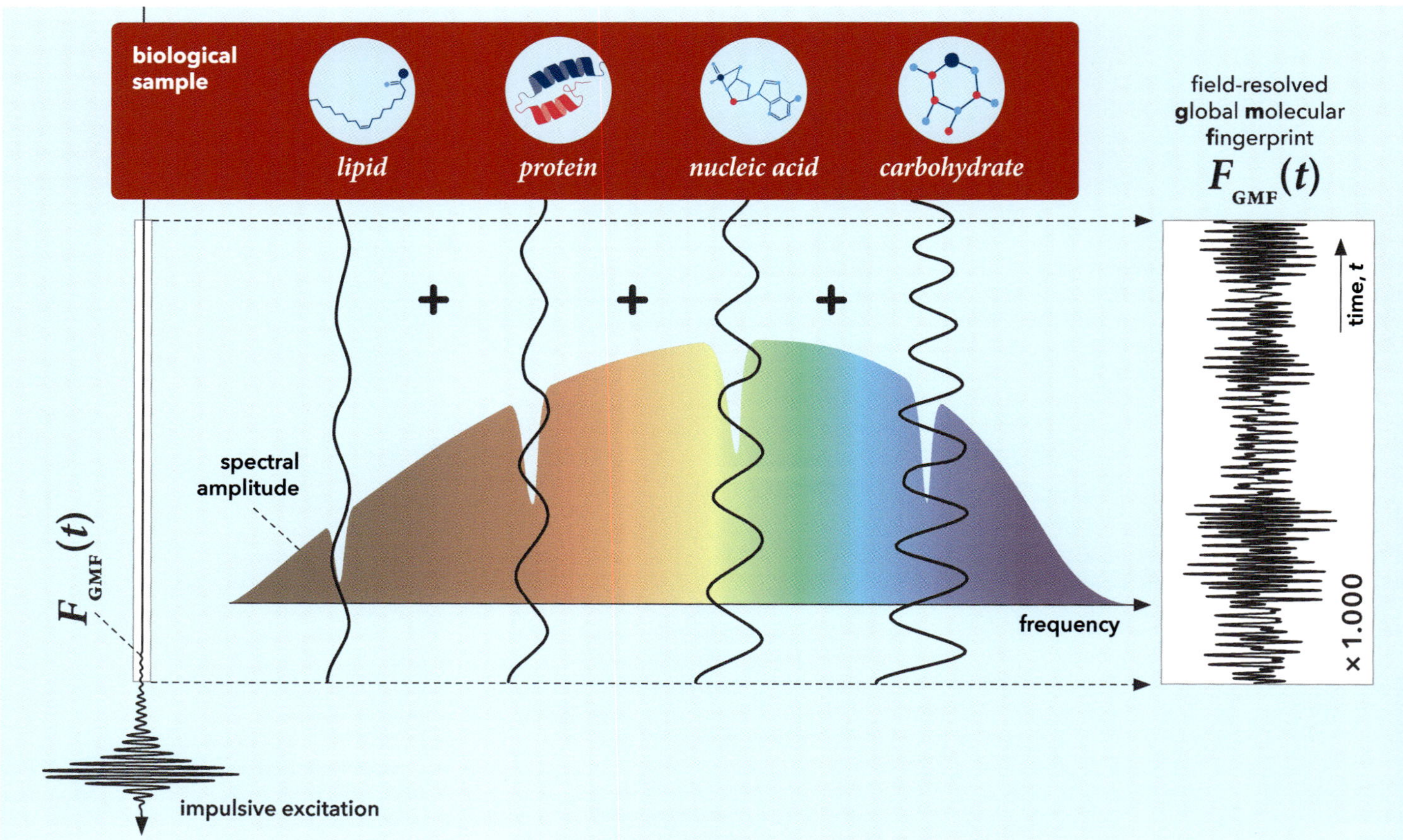

Figure 12-1 Schematic illustration of the concept of field-resolved infrared molecular fingerprinting based on time-resolved sampling of the electric field emitted by impulsively-excited molecular vibrations. With 1-fs-scale ultraviolet and deep-ultraviolet excitation and PHz field sampling, field-resolved spectroscopic fingerprinting can be extended to signals originating from coherent electronic oscillations. The suddenly-excited nuclear (and in the future also: electronic) oscillatory motions radiate a coherent electric field in the wake of the sudden (femtosecond or attosecond) excitation. This signal forms a "global molecular fingerprint" of the sample, incorporating vibrational (and in the future: electronic) signals from all molecules of the biological sample. Attosecond metrology 2.0 now allows sampling this signal up to petahertz frequencies.

blood serum or plasma, can be indicative of the emergence of a disease and can serve as the basis for molecular pathology or as a window to the innermost physiological state of a person.

Blood is composed of tens of thousands of different molecules, the concentration of which depends on the physiological state of the organism. Substantial relative changes in their abundance, particularly when occurring in a correlated fashion, may be able to reliably indicate the onset of a disease, such as cancer. In spite of a dynamic growth of tools and techniques for molecular cancer detection, the overriding still open question is how to best design an approach that would be sensitive and specific enough to allow blood to be used as a medium for (i) robust detection of cancerous changes early on during tumor growth and (ii) non-invasive liquid biopsies to guide clinical decision making. A new spectroscopic approach based on AM2.0 offers a promising route to highly sensitive broadband detection across all molecules present in the biological sample, which may turn out to be the key to reliable molecular diagnosis.

Each molecule exhibits a unique set of electronic and structural states of motion: charge migrations and vibrations, respectively. The microscopic currents induced by these motions radiate an electric field carrying all the important parameters – frequency, amplitude, and phase – of these periodic motions. These are characteristic of the molecule's atomic composition and structure, thus constituting a "molecular fingerprint". Albeit molecular complexity of a systemic biofluid such as blood serum/plasma, spectroscopic interrogation has the unique advantage of directly providing system-wide molecular fingerprints that we coined "global molecular fingerprints" (GMF).

GMFs directly integrate, without molecular identification, information from different molecular families into a single measurement, thus providing potential for advancing molecular fingerprinting and other "omic" approaches of analyzing highly complex biomedical samples. The molecular signal carrying the complete GMF including all the parameters of all electronic and vibrational motions could not be generated thus far. Signals carrying partial information (e.g. about vibrational frequencies and amplitudes) only recently became directly accessible. Time-domain sampling of the electric field carrying the GMF, $F_{\mathrm{GMF}}(t)$, from terahertz to petahertz frequencies now provides direct access to the relevant molecular signal.

In conventional frequency-domain spectroscopies, such as Fourier-transform infrared (FTIR) absorption spectroscopy[12], the electric field resulting from molecular vibrations destructively interferes with the resonant input field that activates these motions, "burning a hole" in the spectrum of the transmitted radiation (Fig. 12-1). The position and depth of this hole reveals the eigenfrequency of one (or more) vibrational mode(s) and the concentration of the corresponding molecular specimen respectively. As a consequence, the relative intensity noise of the illumination source limits the smallest detectable molecular concentrations as well as their smallest detectable changes in concentration. The moderate sensitivity resulting from this limitation has so far prevented molecular spectroscopy from offering a reliable approach to molecular fingerprinting that is useful for clinical applications.

Consider an extremely short (few-cycle) ultra-broadband (ideally super-octave) infrared laser pulse. Such a laser pulse can impulsively excite molecular vibrations, which induce intra-molecular currents that radiate fields coherently from the molecules vibrating in synchrony due to an excitation by a coherent laser field. The resultant $F_{\mathrm{GMF}}(t)$ is emitted over several picoseconds before decaying due to a loss of synchrony of the vibrating molecules. Given a much shorter, few-10-fs excitation, the molecular signal carrying the GMF can be measured directly, not plagued by any background from the excitation. AM2.0 offers direct time-resolved measurement of these fields, up to PHz frequencies to include – ultimately – the electronic response, allowing background-free detection of the molecular signal, plotted along the vertical time axis in the right inset of Fig. 12-1. Field-resolved molecular fingerprinting therefore offers an extensively improved detection limit, which has recently been demonstrated in proof-of-principle experiments.

The first directly measured molecular signals carrying a (partial) vibrational GMF of a biofluid, in this case of human blood sera are shown in Figure 12-2. In spite of the excitation bandwidth (several THz) being limited to merely a few per cent of the entire spectral range of molecular vibrations, the GMF signals show noticeable differences even between samples of sera from different healthy individuals. This not only documents biological variability between individuals, it holds promise for robust differentiation of abnormal physiologies as well.

[12] M. J. Baker *et al.* Using Fourier transform IR spectroscopy to analyze biological materials. *Nat. Protoc.* **9**, 1771 (2014).

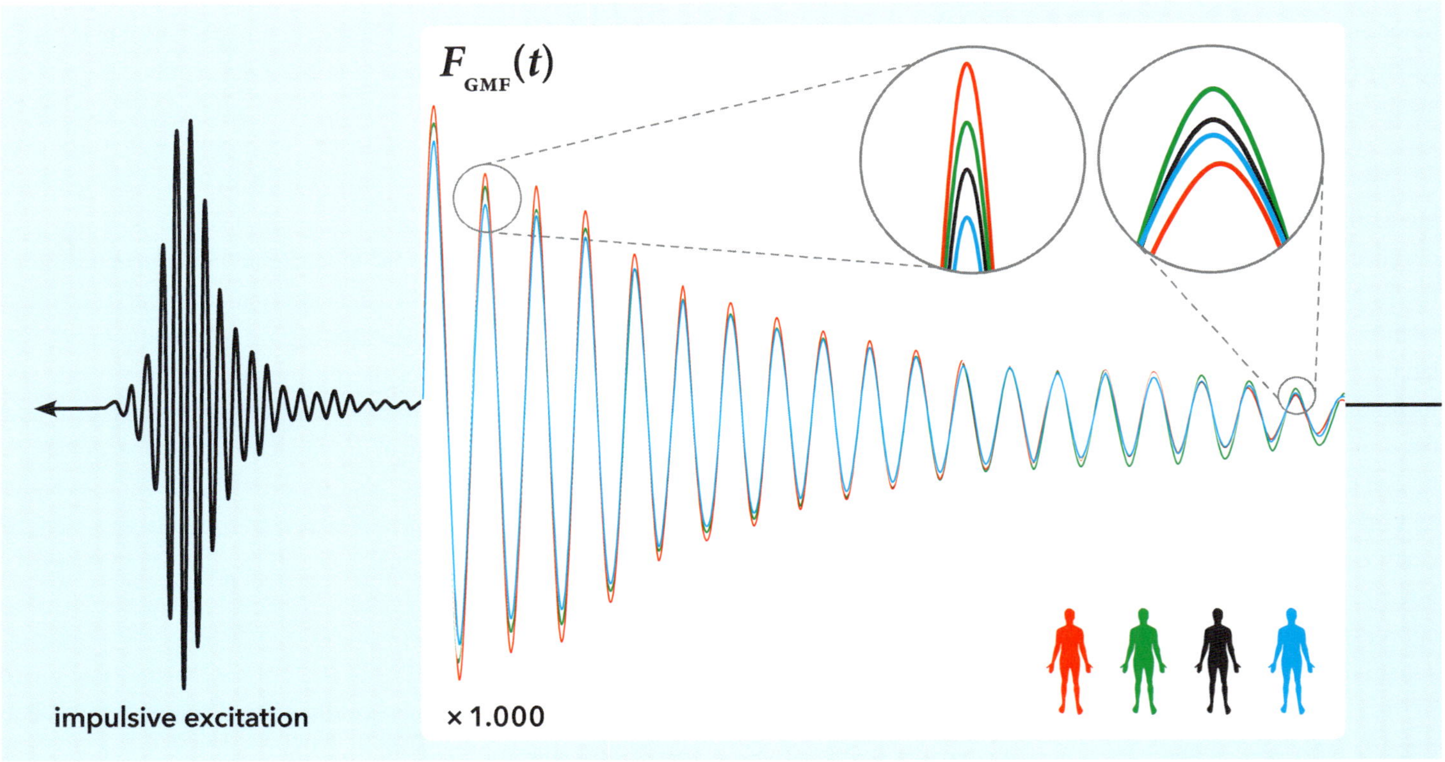

Figure 12-2 First infrared molecular fingerprints of human blood sera recorded by field-resolved spectroscopy of impulsively-excited molecular vibrations. Although the samples have been drawn from healthy individuals, they do differ. In spite of the narrow spectral coverage of this first-generation device, the molecular signals reveal significant differences between the case and control samples (see zoomed window in Fig. 12-3), allowing differentiation of lung cancer from benign pathology of lung already in this first pilot study.

In order to test this capability, the novel method of field-resolved infrared molecular fingerprinting was applied to a pilot case-control clinical study to compare the fingerprints of lung cancer (in this case non-small-cell lung carcinoma) in human blood serum with that of benign conditions (in this case lung hamartoma) in an attempt to assess the method's potential for detecting lung cancer and differentiating it from other abnormalities of the organ hosting the malignancy. To this end, the blood sera of 62 patients suffering from benign hamartoma were analyzed along with that of 48 patients with malignant lung tumors. The first time-resolved vibrational molecular fingerprints acquired in this proof-of-principle clinical study are shown in Figure 12-3. The first prototypical field-resolved infrared spectrometer employed covers only about 5% of the molecular fingerprint region within a spectral band centered at 10 μm wavelength.

Analyzing the signals within a very brief, sub-picosecond-duration temporal window following the impulsive femtosecond excitation reveals significant differences between the case-control GMF signals, allowing differentiation of lung cancer from a benign pathology of lung with an accuracy of 78% in this first pilot study. It is remarkable that the fingerprints of malignant tumors appear to be more heterogeneous than those of the benign reference pathology. The instrumential noise was evaluated in measurements of the same control sera over and over again (quality-controls) in between the measurements of blinded randomized case and control samples. There is clear evidence that the difference between the cases and controls is larger than experimental noise.

The fact that we are able to measure blood sera and plasma with this new technique is just the beginning. The molecular information accessed can be increased by several orders of magnitude via increasing the bandwidth of excitation and probing and decomposing the highly complex biological samples into fractions containing only specific groups of molecules and recording the fingerprints of these fractions in separate measurements.

Pumping energy into the internal degrees of freedom of a molecule within less than 1 fs (by field-ionizing it with a 1-fs-scale optical transient) will coherently excite both electronic and vibrational wavepacket motions over an overall bandwidth exceeding 1 PHz[13]. The resultant PHz-to-THz emission will be sampled with the novel technology of AM2.0 described previously. The molecular system's complete (vibrational-electronic, vibronic) dynamic response will, in combination with (chemical and/or chromatographic) fractionation of the sample, have the capacity to deliver unprecendented information about its molecular composition. Field-resolved *vibronic* molecular fingerprinting of biofluids (along with fractionation) hold the promise for an unparalleled degree of specificity for robust identification of abnormal physiologies.

Spectroscopic molecular fingerprinting is still in its infancy but carries a vast potential. The field-resolved approach based on AM2.0 has outperformed state-of-the-art FTIR by more than an order-of-magnitude in sensitivity already in first proof-of-principle experiments, performed by Marinus Huber, Ioachim Pupeza and their co-workers[14]. Superior sensitivity will be combined with ever improving specificity by exciting molecular vibrations impulsively over an ever increasing spectral range, eventually with a multi-octave infrared waveform covering the entire spectral range of vibrational frequencies from several THz to about 150 THz. Ultimately, sub-femtosecond excitation and sampling will add the coherent electronic response to the vibrational fingerprint. Background-free field-resolved detection of *vibronic molecular*

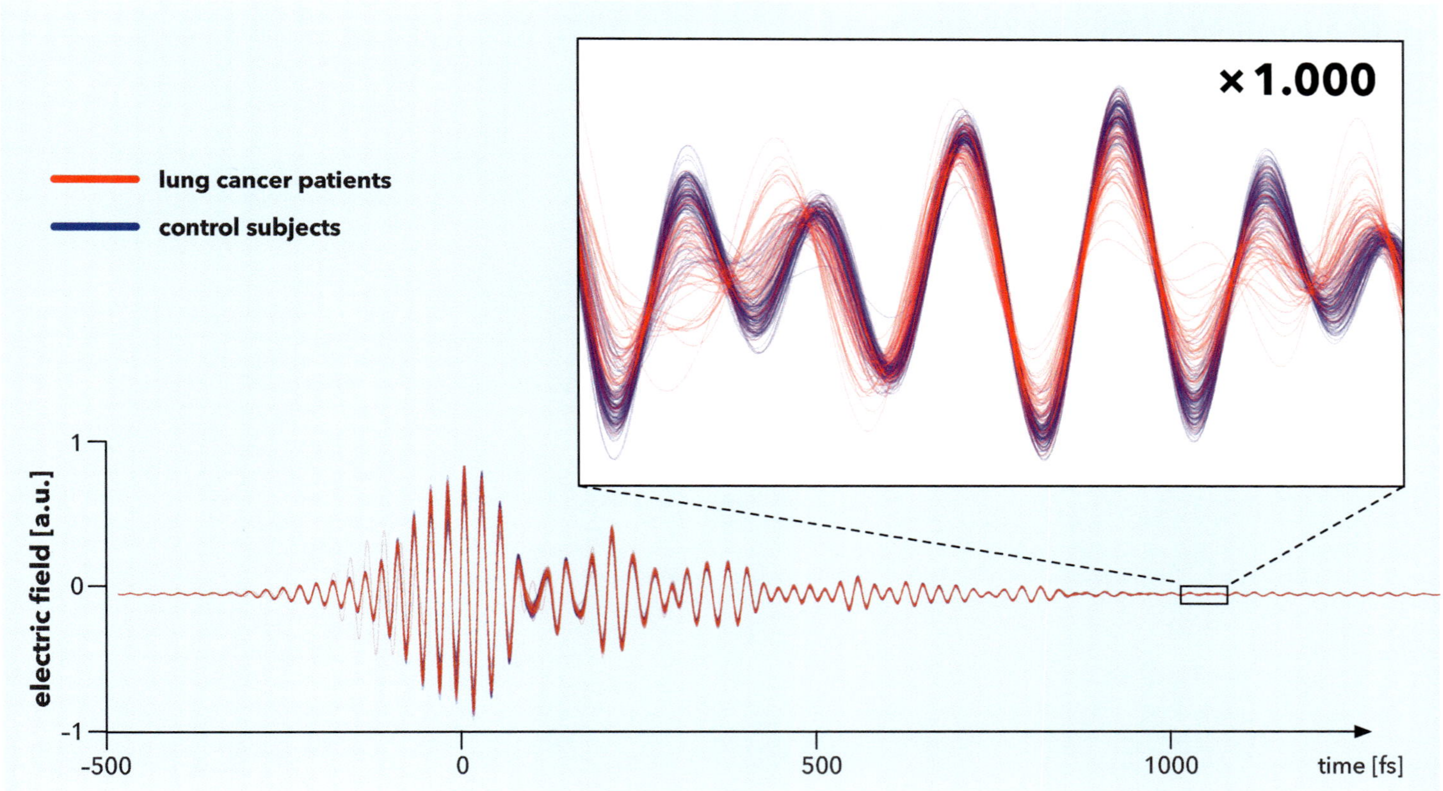

Figure 12-3 First case-control clinical study based on field-resolved molecular fingerprinting of blood sera. The figure summarizes preliminary molecular fingerprints acquired at the beginning of this pilot study performed in collaboration between the Laboratory of Attosecond Physics of the MPQ and LMU with Prof. J. Behr and his co-workers from the Clinic of Pulmonology at the LMU. The red curves represent the field-resolved molecular fingerprints of the blood serum samples of patients suffering from non-small-cell lung carcinoma (NSCLC) whereas the violet curves depict the fingerprints of subjects diagnosed with benign lung hamartoma. The lung cancer infrared fingerprints scatter more and are significantly different from those of the reference subjects.

[13] A. I. Kuleff and L. S. Cederbaum. Radiation generated by the ultrafast migration of a positive charge following the ionization of a molecular system. *Phys. Rev. Lett.* **106**, 053001 (2011).

[14] I. Pupeza, M. Huber *et al.* Field-resolved infrared spectroscopy of biofluids. Submitted for publication.

fingerprints is expected to provide a valuable addition to the existing toolbox for the detection of diseases in humans, with cancer being only the first example. With this valuable tool, one could envisage contributions to early detection, screening, health and disease monitoring as well as assessing the susceptibility of individuals to any disorders leaving their trace in biofluids.

We envision the future penetration of attosecond metrology into life sciences and medicine in an iterative manner, in which progress in physical technologies will need to be integrated with other molecular, possibly "omic" approaches and modern molecular medicine. This will further necessitate commensurate advances in machine learning and neural networks and their applications to obtain the ability to not only screen for physiological differences but also contribute to fundamental insights in how living systems function mechanistically.

The journey from recording the first field-resolved GMF of human blood sera (Fig. 12-2 – Fig. 12-3) to proving the ability of the approach to reliably detect the emergence of cancer and monitor its progression or decay and finally to a true understanding of how GMFs reflect human physiologies will be long. However, if successful, the return could be enormous and possibly even very different from what we envision today. The future of advancing metrology to an ever higher speed and sensitivity is bright.